Khanna's Multichoice Questions & Answers in

METALLURGICAL ENGINEERING

Third Edition

Om Prakash Gupta
B. Tech. (Chemical Engg.)
Ex DGM, HRD
Bokaro Steel Plant
Steel Authority of India Limited (SAIL)

KHANNA BOOK PUBLISHING CO. (P) LTD.
Publisher of Engineering and Computer Books
4C/4344, Ansari Road, Darya Ganj, New Delhi-110002
Phone : 011-23244447-48 **Mobile:** +91-9910909320
E-mail : contact@khannabooks.com
Website : www.khannabooks.com

Dear Readers,

To prevent the piracy, this book is secured with HIGH SECURITY HOLOGRAM on the front title cover. In case you don't find the hologram on the front cover title, please write us to at contact@khannabooks.com or whatsapp us at +91-99109 09320 and avail special gift voucher for yourself.

Specimen of Hologram on front Cover title:

Moreover, there is a SPECIAL DISCOUNT COUPON for you with EVERY HOLOGRAM.

How to avail this SPECIAL DISCOUNT:

Step 1: Scratch the hologram
Step 2: Under the scratch area, your "coupon code" is available
Step 3: Logon to www.khannabooks.com
Step 4: Use your "coupon code" in the shopping cart and get your copy at a special discount
Step 5: Enjoy your reading!

Price: ₹ 449.00

ISBN: 978-93-86173-11-9

Khanna's Multi-Choice Questions & Answers in Metallurgical Engineering
by O.P. Gupta

First Edition: 2017
Second Edition: 2021
Third Edition: 2023

To view complete list of books, Please scan the QR Code:

Published by:
Khanna Book Publishing Co. (P) Ltd.
CIN: U22110DL1998PTC095547

Visit us at: **www.khannabooks.com**
Write us at: **contact@khannabooks.com**

Printed in India by: Vaibhav Products, Delhi.

PREFACE TO THE SECOND EDITION

This book is meant for diploma & degree students of metallurgical engineering for their academic programs as well as for various competitive examinations . This book has been structured in four sections. First section contains multiple choice type questions of various subjects of metallurgical engineering .Second section contains chapter wise questions of GATE (Graduate Aptitude Test in Engineering) from 1991 to 2017. Third section contains SHORT QUESTIONS & ANSWERS IN METALLURGICAL ENGINEERING . Fourth section contains APPENDICES containing Glossary of terms related to Metallurgical Engineering . GATE Questions & Answers from 2018 to 2021 have been given at the end of the book.

This book has been designed to serve as "Hand Book of Metallurgical Engineering " which will be useful for various competitive examinations for recruitment in various public sector & Private Sector companies as well as for GATE Examination.

Question have been arranged subject wise and answers are given at the bottom of the page.

All the constructive & useful suggestion for further improvement of the book will be gladly accepted and incorporated in the future editions.

May, 2021

Om Prakash Gupta

PREFACE TO THE FIRST EDITION

This book is meant for diploma & degree students of metallurgical engineering for their academic programs as well as for various competitive examinations for securing jobs. This book has been structured in three sections. First section contains multiple choice type questions of various subjects of metallurgical engineering. Second section contains chapterwise questions of GATE (Graduate Aptitude Test in Engineering) from 1991 to 2016. Third section contains SHORT QUESTIONS & ANSWERS IN METALLURGICAL ENGINEERING. Fourth section contains APPENDICES containing Glossary of terms related to Metallurgical Engineering and Q&A of GATE-2017.

This book has been designed to serve as "Hand Book of Metallurgical Engineering" which will be useful for various competitive examinations for recruitment in various public sector & Private Sector companies as well as for GATE Examination.

Question have been arranged subjectwise and answers are given at the bottom of the page.

All the constructive & useful suggestion for further improvement of the book will be gladly accepted and incorporated in the future editions.

April, 2017 **Om Prakash Gupta**

CONTENTS

Section A
MCQ

Section

A

1

General Metallurgy

1. Metal hardness determines its resistance to
   (a) scratching, wear & penetration.
   (b) machinability.
   (c) cutting ability
   (d) all (a), (b) & (c).
2. Brinell & Vickers hardness values are almost identical upto a hardness of
   (a) 60 (b) 130
   (c) 235 (d) 300.
3. In which of the hardness tests, the height of first rebound of a hammer being dropped freely is taken as the index of hardness ?
   (a) Rockwell & Brinell tests
   (b) Vickers test
   (c) Shore test
   (d) None of these.
4. In general, metals at low temperatures (subzero), have more .... strength.
   (a) tensile (b) yield
   (c) both (a) & (b) (d) neither (a) nor (b).
5. Which of the following defects is not produced in steel ingots & casting due to solidification shrinkage ?
   (a) Piping (b) Gas holes
   (c) Both (a) & (b) (d) Neither (a) nor (b).
6. Which is not a requisite of moulding sand ?
   (a) High temperature resistance,
   (b) Satisfactory bond strength.
   (c) Permeability to gases.
   (d) None of these.
7. Which is not a deoxidiser of steel ?
   (a) Nickel (b) Ferro-manganese
   (c) Ferro-silicon (d) Aluminium.
8. Pressure of oxygen blown in L.D. converter is about .... Kgf/cm^2.
   (a) 1-5 (b) 15-20
   (c) 45-50 (d) 90-100.
9. Cementite is
   (a) a compound of iron & carbon (6.67%).
   (b) hard (Brinell hardness 600+) & brittle.
   (c) chemically Fe_3C
   (d) all (a), (b) & (c)
10. 0.87 percent carbon steel contains .... percent pearlite.
    (a) 20 (b) 50
    (c) 80 (d) 100.
11. Presence of free cementite in plain carbon steel
    (a) increases its hardness.
    (b) reduces its strength.
    (c) both (a) & (b).
    (d) neither (a) nor (b).

Answers

1. (d)	2. (d)	3. (c)	4. (c)	5. (d)	6. (d)
7. (a)	8. (b)	9. (d)	10. (d)	11. (c)	

12. Which allotropic form of iron has a face centred cubic (fcc) structure?
(*a*) α-iron (*b*) β-iron
(*c*) γ-iron (*d*) δ-iron.

13. Which Indian steel plant has the largest installed steel production capacity ?
(*a*) Rourkela steel plant.
(*b*) Vizag steel plant
(*c*) Bhilai steel plant
(*d*) Durgapur steel plant.

14. Razors are made of ....steel.
(*a*) mild (*b*) low carbon
(*c*) medium carbon (*d*) high carbon.

15. In high carbon steels, hardness is obtained at the expense of
(*a*) ductility
(*b*) toughness
(*c*) both (*a*) & (*b*)
(*d*) neither (*a*) nor (*b*).

16. Annealing can not
(*a*) soften the steel and improve machinability.
(*b*) relieve internal stresses induced by rolling, forging or uneven cooling.
(*c*) remove coarseness of grain.
(*d*) none of these.

17. Specific energy consumption in Indian steel plants is about ....GCal/ton crude steel.
(*a*) 6.5 -9.0 (*b*) 4-5
(*c*) 12 - 13 (*d*) 9 - 11.
(1 G Cal = 109 Calories)

18. Hadfield's. manganese steel
(*a*) contains 12-14% Mn and 1% C.
(*b*) has great resistance to wear.
(*c*) is used for rock drills & stone crushers.
(*d*) all (*a*), (*b*) & (*c*).

19. High speed steel is softened by
(*a*) tempering at 400°C followed by quenching.
(*b*) normalizing at 1250°C.
(*c*) annealing at 850°C (for 4 hours) followed by slow cooling.
(*d*) all (*a*), (*b*) & (*c*).

20. Which is the chief alloying element in stainless steel which inhibits corrosion ?
(*a*) Chromium (*b*) Silicon
(*c*) Nickel (*d*) Carbon.

21. Melting temperature of cast iron is........°C.
(*a*) 900 - 950 (*b*) 1140 - 1200
(*c*) 1350- 1400 (*d*) 1400 - 1500.

22. Which property of the material is the most important in its rolling or forging ?
(*a*) Ductility (*b*) Malleability
(*c*) Hardness (*d*) Creep.

23. Wrought iron is never shaped by
(*a*) hammering (*b*) pressing
(*c*) casting (*d*) forging

24. Which one contains the least percentage of carbon ?
(*a*) Pig iron (*b*) Wrought iron
(*c*) High speed steel (*d*) Cast iron.

25. Which is not a copper based alloy?
(*a*) Gun metal (*b*) Brass
(*c*) Bronze (*d*) Monel.

26. Which is not a tin based alloy ?
(*a*) Soft solder (*b*) Babbit metal
(*c*) Nichrome (*d*) Monel.

27. Annealing can not
(a) relieve internal stress.
(*b*) improve wear resistance.
(*c*) improve machinability.
(*d*) refine grain structure.

Answers

12. (c)	13. (c)	14. (d)	15. (c)	16. (d)	17. (a)
18. (d)	19. (c)	20. (a)	21. (b)	22. (b)	23. (c)
24. (b)	25. (d)	26. (c)	27. (b)		

28. Case hardening of low carbon steel can not be done by
    (a) cyaniding
    (b) nitriding
    (c) electroplating
    (d) induction hardening.

29. Ferrite is a solid solution of carbide in
    (a) pearlite (b) cementite
    (c) gamma iron (d) alpha iron.

30. Which of the following processes is most economical for achieving reasonable strength without developing internal stresses ?
    (a) Hardening.
    (b) Hardening followed by tempering.
    (c) Normalising.
    (d) Annealing.

31. Which alloying element does not impart hardness to steel ?
    (a) Silicon (b) copper
    (c) Nickel (d) Aluminium

32. Hardness (BHN) of ferrite is about
    (a) 10 (b) 50
    (c) 100 (d) 150.

33. Manganese content in carbon steel is limited to a maximum of ....percent.
    (a) 1 (b) 1.65
    (c) 2.2 (d) 2.6.

34. Gamma iron occurs in the temperature range of ....°C.
    (a) 0 – 770 (b) 770 – 910
    (c) 910 – 1400 (d) 1400 –1639

35. Which is the most suitable process for improving mechanical properties of steel casting ?
    (a) Hardening
    (b) Phase annealing
    (c) Tempering
    (d) Recrystallisation annealing.

36. Eutectoid steel has a structure of
    (a) pearlite (b) martensite
    (c) sorbite (d) none of these.

37. Annealing temperature used for relieving internal stresses in welding structure is about............°C.
    (a) 300-350 (b) 500-600
    (c) 800-900 (d) 1000-1100

38. Which of the following salts is never used in cyaniding process ?
    (a) Sodium cyanide.
    (b) Sodium chloride.
    (c) Sodium hydroxide.
    (d) Sodium carbonate.

39. Which is not a deoxidiser used for producing 'killed steel' ?
    (a) Copper.
    (b) Ferro-silicon.
    (c) Ferro-manganese.
    (d) None of these.

40. Presence of sulphur in pig iron makes it
    (a) soft (b) ductile
    (c) brittle (d) hard

41. Carbon content in commercially available cold rolled steel is about ....percent.
    (a) 0.1 (b) 0.35
    (c) 0.5 (d) 0.8.

42. Hardness (BHN) of cementite is about
    (a) 300 (b) 600
    (c) 1000 (d) 1400.

43. Hypo eutectoid steel has a structure of
    (a) pearlite & ferrite.
    (b) pearlite & cementite.
    (c) cementite & ferrite.
    (d) none of these.

44. Which alloying element does not improve the machinability of steel?
    (a) Lead (b) Sulphur
    (c) Silicon (d) Phosphorous.

Answers

28. (c)	29. (d)	30. (c)	31. (b)	32. (b)	33. (b)
34. (c)	35. (b)	36. (a)	37. (b)	38. (c)	39. (a)
40. (d)	41. (a)	42. (d)	43. (a)	44. (c)	

45. Depth of hard case obtained by cyaniding process of surface hardening (using sodium cyanide bath at 850°C for 3 hours) is about............mm.
(a) 0.4 (b) 1
(c) 1.5 (d) 2.2

46. Creep is predominant even at room temperature in
(a) gun metal bearings
(b) white metal bearings
(c) copper
(d) aluminium.

47. Surface hardness (VPN) of the order of............ is obtained by nitriding operation.
(a) 100 (b) 500
(c) 1000 (d) 2000.

48. Which of the following elements of steel reduces its brittleness by combining with sulphur ?
(a) Manganese (b) Silicon
(c) Magnesium (d) Vanadium

49. Austenite is not
(a) soft (b) ductile
(c) malleable (d) magnetic

50. Presence of manganese in steel increases its
(a) fluidity (b) tensile strength
(c) ductility (d) malleability

51. Wear resistanwce property of alloy steel is improved by the addition of
(a) nickel (b) chromium
(c) vanadium (d) molybdenum

52. Minimum percentage of chromium or nickel in austenitic stainless steel is
(a) 4 (b) 8
(c) 18 (d)24

53. Eutectoid steel has............percent carbon.
(a) 0.025 (b) 0.1
(c) 0.8 (d) 1.5

54. Gamma iron has............cubic crystal structure.
(a) body centred (b) face centred
(c) simple (d) none of these.

55. Pig iron is a product of the
(a) cupola
(b) open hearth furnace
(c) electric arc furnace
(d) none of these.

56. Blow holes in steel casting is not minimised by the addition of
a) magnesium (b) silicon
(c) aluminium (d) titanium.

57. Which is the most practised method of steel making ?
(a) L.D. process
(b) Open hearth furnace.
(c) Duplex process.
(d) Bessemer process

58. Carbon percentage in the steel used for boiler tube manufacture is about
(a) 0.5 (b) 0.15
(c) 0.65 (d) 0.35

59. Ductility of low carbon steel falls very rapidly at...........°C.
(a) - 40 (b) - 80
(c) - 130 (d) 0

60. Austenite can be present even at sub-zero temperatures in an alloy steel due to the presence of high percentage of
(a) C (b) Ni
(c) Cr (d) Co

Answers

45. (a)	46. (b)	47. (c)	48. (a)	49. (d)	50. (b)
51. (a)	52. (b)	53. (c)	54. (b)	55. (d)	56. (a)
57. (a)	58. (b)	59. (c)	60. (c)		

61. Structure obtained after austempering of steel (done to improve mechanical properties) is
    (*a*) martensite
    (*b*) bainite
    (*c*) lameller carbide
    (*d*) none of these.

62. Carbon content in carbon steel used for making heavy duty leaf and coil spring is about ....percent.
    (*a*) 0.1 (*b*) 0.5
    (*c*) 1 (*d*) 1.8.

63. The optimum hardening temperature of hyper eutectoid steel is 30 to 50°C .... critical temperature.
    (*a*) above upper
    (*b*) above lower
    (*c*) below lower
    (*d*) none of these.

64. Steel castings subjected to normalising (to improve mechanical properties) is cooled in
    (*a*) water (*b*) oil
    (*c*) air (*d*) none of these

65. The grain structure obtained after isothermal hardening is
    (*a*) pearlite (*b*) martensite
    (*c*) acicular troostite (*d*) sorbite.

66. Residual austenite in steel (which reduces the lifting power of steel magnet) can be removed (by transforming it into martensite) by cooling it to ....°C.
    (*a*) 0 (*b*) -50 to - 100
    (*c*) - 200 to 250 (*d*) 8

67. With decrease of temperature to -150°C from 0°C for most of the metals, the ultimate tensile & yield strength
    (*a*) decreases
    (*b*) increases
    (*c*) remains same
    (*d*) may increase or decrease.

68. Admirality brass (used for steam condenser tubes) comprises of
    (*a*) 70% Cu + 30% Zn
    (*b*) 30% Cu + 70% Zn
    (*c*) 70% Cu + 30% Sn
    (*d*) 30% Sn + 70% Zn.

69. Bath temperature of steel in L.D. process is measured by a/an
    (*a*) resistance thermometer.
    (*b*) optical pyrometer.
    (*c*) immersion thermocouple couple.
    (*d*) bimetallic thermometer.

70. A solder
    (*a*) wets the base metal and is drawn into the joints by capillary action.
    (*b*) flows freely over the surfaces i.e. spreads and makes contact with the joint opening.
    (*c*) solidifies as a sound, firmly adhering film and has an adequate mechanical strength.
    (*d*) all (*a*), (*b*) & (*c*).

71. Strain hardening of metal/alloy reduces its
    (*a*) strength
    (*b*) hardness
    (*c*) ductility & plasticity
    (*d*) none of these.

72. Hot working of metals
    (*a*) introduces residual stress in it.
    (*b*) is carried out above its recrystallisation temperature.
    (*c*) does not produce strain hardening.
    (*d*) is done to obtain very thin sheets.

73. Hardening treatment to steel imparts it
    (*a*) no internal strain.
    (*b*) no hardness.
    (*c*) minimum ductility or maximum brittleness.
    (*d*) none of these.

Answers

61. (b)	62. (c)	63. (a)	64. (c)	65. (c)	66. (b)
67. (b)	68. (a)	69. (c)	70. (d)	71. (c)	72. (b)
73. (c)					

74. With increase of carbon content, the hardening capacity of steels
(*a*) increases
(*b*) decreases
(*c*) remains same
(*d*) may increase or decrease.

75. Phosphorous in steel
(*a*) occurs as Fe_3P, which induces shortness.
(*b*) is less dangerous in low carbon than in high carbon-steel.
(*c*) upto 0.1% is sometimes deliberately added to prevent sticking during rolling of packs of sheets and cutting steels.
(*d*) all (*a*), (*b*) & (*c*).

76. Sulphur in steel
(*a*) occurs as FeS or MnS and induces hot shortness.
(*b*) generally should be < 0.05%.
(*c*) is sometimes deliberately added when used for screw stock and other free cutting steels.
(*d*) all (*a*), (*b*) & (*c*).

77. Alloying elements addition in steels does not improve their
(*a*) magnetic properties.
(*b*) depth of hardening.
(*c*) strength at elevated ttemperature.
(*d*) none of these.

78. Exposure to very high concentration of ....does not cause pitting corrosion in the immersed steel.
(*a*) H_2S0_4 (*b*) HNO_3
(*c*) O_2 (*d*) FeO

79. The grain structure obtained by tempering of steel (done to improve mechanical properties) is
(*a*) martensite
(*b*) bainite
(*c*) troostite
(*d*) carbide.

80. During cooling of steel (C=0.8%) from 1000°C, the pearlite formation occurs at... ....°C.
(*a*) 423 (*b*) 723
(*c*) 870 (*d*) 925

81. Presence of manganese in steel is desirable upto a certain limit, as it increases its
(*a*) ductility (*b*) fluidity
(*c*) tensile strength (*d*) malleability

82. Minimum percentage of carbon in cast iron is
(*a*) 0.05 (*b*) 0.5
(*c*) 1 (*d*) 2

83. The annealing temperature for relieving the internal stresses in welded structures to prevent distortion is about...........°C.
(*a*) 350-450 (*b*) 500-600
(*c*) 750 – 850 (*d*) 1000 - 1100

84. The property which makes wrought iron unsuitable for pipe manufacture is its
(*a*) hardness
(*b*) weldability
(*c*) thread cutting ability
(*d*) corrosion resistance

85. Which of the following alloying elements can not impart high strength at elevated temperature to steel ?
(*a*) Silicon (*b*) Manganese
(*c*) Nickel (*d*) Magnesium

86. Maximum percentage of carbon in ferrite is
(*a*) 0.025 (*b*) 0.25
(*c*) 0.05 (*d*) 0.5

87. Maximum percentage of carbon in austenite is
(*a*) 0.025 (*b*) 0.25
(*c*) 0.05 (*d*) 0.5

Answers

74. (a)	75. (d)	76. (d)	77. (d)	78. (b)	79. (a)
80. (b)	81. (c)	82. (d)	83. (b)	84. (a)	85. (d)
86. (a)	87. (c)				

88. Bath temperature maintained during isothermal annealing of chromium steel by the transformation of supercooled austenite is about...........°C.
(*a*) 350-450 (*b*) 600-750
(*c*) 800 – 950 (*d*) 1000 - 1150.

89. Addition of...........to steel improves its corrosion resistance property.
(*a*) cobalt (*b*) chromium
(*c*) copper (*d*) none of these.

90. Which of the following is used as 'etchant' in metallography of carbon steel ?
(*a*) 2% HNO_3 in water.
(*b*) 2% HNO_3 in C_2H_5OH.
(*c*) 2% HCl in water.
(*d*) none of these.

91. Normalising operation can not improve the ....of the metal.
(*a*) ductility
(*b*) mechanical properties
(*c*) yield strength
(*d*) all 'a', 'b' & 'c'.

92. Presence of...........in steel does not improve its machinability.
(*a*) sulphur (*b*) phosphorous
(*c*) silicon (*d*) lead

93. In an alloy steel, austenite can be present even at subzero temperatures because of the presence of high percentage of
(*a*) carbon (*b*) aluminium
(*c*) chromium (*d*) cobalt.

94. Pick out the correct statement.
(*a*) Copper is used as a deoxidiser to produce 'killed steel'.
(*b*) Fatigue is the cause of failure of a material under repeated applied fluctuating stresses.
(*c*) Normalising improves the ductility of a metal.
(*d*) Case carburising is the most widely used method of case hardening of steel parts with very high carbon content.

95. Pick out the wrong statement.
(*a*) Ferrite is a solid solution of carbon in the solvent of α-iron.
(*b*) Wear resistance property of alloy steel is improved by nickel.
(*c*) Addition of silicon does not impart hardness to steel.
(*d*) Admirability brass is used for making steam condenser tubes.

96. .... is essentially added to steel in order to inhibit grain growth in austenite during its heat treatment.
(*a*) Aluminium (*b*) Copper
(*c*) Tin (*d*) none of these.

97. Which of the following is a criteria for determination of hardness of steel ?
(*a*) Percentage of its alloying elements.
(*b*) Percentage of its carbon content.
(*c*) The distribution & shape of carbides in iron.
(*d*) None of these.

98. On severe deformation in a particular direction (as in rolling), a metal becomes
(*a*) anisotropic (*b*) isotropic
(*c*) homogeneous (*d*) ductile

99. For eutectoid steel, the optimum hardening temperature is
(*a*) T_1 + 30 to 50°C (*b*) T_2 + 30 to 50°C
(*c*) (T_1 - 30)°C (*d*) (T_2 + 30)°C.
where, T_1 = lower critical temperature, °C
T_2 = upper critical temperature, °C

100. The minimum percentage of nickel or chromium in austenitic stainless steel is
(*a*) 3 (*b*) 6
(*c*) 8 (*d*) 12

Answers

88. (b)	89. (b)	90. (b)	91. (a)	92. (c)	93. (c)
94. (b)	95. (c)	96. (a)	97. (c)	98. (a)	99. (b)
100. (c)					

101. Constantan is an alloy of nickel and
   (*a*) chromium (*b*) iron
   (*c*) copper (*d*) tin

102. Pick out the wrong statement.
   (*a*) Babbit metal is used as bearing lining.
   (*b*) 'Elinvar' (having nickel as the main alloying element) is used for making hair springs for watches and other precision instruments.
   (*c*) 'Perminvar' (an alloy of iron, copper & nickel) is widely used in electrical industry due to its exceptional magnetic properties of constant permeability.
   (*d*) Ferrite is a solid solution of carbide in gamma iron.

103. Pick out the correct statement.
   (*a*) Austenite is a solid solution of carbon in gamma iron.
   (*b*) Bell metal is not used for making bells.
   (*c*) Constantan is not used for making thermocouples.
   (*d*) Tempering of steel castings improves its mechanical properties.

104. In a case carburising process requiring hardening to be done for a thickness of 2 mm ; for a soaking temperature of 930°C, the holding time of heating is about...........hours.
   (*a*) 5-11 (*b*) 9-16
   (*c*) 19-24 (*d*) 26 -35.

105. Percentage of copper in 'Perminvar' is about
   (*a*) 10 (*b*) 25
   (*c*) 40 (*d*) 60

106. Babbit metal is an alloy of tin, copper.
   (*a*)and zinc.
   (*b*)antimony and bismuth.
   (*c*) and antimony.
   (*d*) nickel and chromium.

107. Ratio of copper to zinc in utensil grade brass is about
   (*a*) 1 : 1 (*b*) 2 : 1
   (*c*) 3 : 1 (*d*) 4 : 1

108. Statue grade bronze is composed of
   (*a*) 10% Cu + 90% Sn.
   (*b*) 10% Sn + 90% Cu.
   (*c*) 50% Cu + 50.% Sn.
   (*d*) 50% Cu + 50% Zn.

109. In flame hardening process, the amount of oxygen consumed is proportional to
   (*a*) t (*b*) 1/t
   (*c*) $\sqrt{t}$ (*d*) $1\sqrt{t}$
   where, t - thickness of the hardened layer.

110. In spite of the fact that nitriding process is expensive & complex, it is still used for surface hardening of steel machine parts due to the fact that
   (*a*) quenching is not needed.
   (*b*) high production rate can be achieved.
   (*c*) the hardened part is not distorted.
   (*d*) only 800°C furnace temperature is needed.

111. Rubber exhibits
   (*a*) plastic properties.
   (*b*) linear elastic stress-strain relationship.
   (*c*) non linear plastic stress-strain relationship.
   (*d*) none of these.

112. Percentage of nickel in 'Elinvar' is about
   (*a*) 16 (*b*) 32
   (*c*) 48 (*d*) 60

113. 'Alnico' used for making permanent magnets is an alloy comprising of
   (*a*) 50% Fe + 20% Al + 20% Ni + 10% Co
   (*b*) 50% Fe + 25% Ni + 25% Cr
   (*c*) 50% Fe + 25% Al + 25% Cr
   (*d*) none of these.

Answers

101. (c)	102. (d)	103. (a)	104. (c)	105. (b)	106. (b)
107. (b)	108. (b)	109. (c)	110. (a)	111. (c)	112. (b)
113. (a)					

114. Solder is an alloy comprising of tin and
(*a*) lead (*b*) antimony
(*c*) copper (*d*) nickel

115. Bell metal is an alloy comprising of copper and
(*a*) lead (*b*) tin
(*c*) zinc (*d*) nickel

116. The ductility of copper and aluminium .... when its temperature falls from 0 to - 150°C.
(*a*) increases
(*b*) decreases
(*c*) remains unchanged
(*d*) may increase or decrease ; depends on other factors.

117. The grain structure obtained after isothermal hardening is
(*a*) sorbite
(*b*) pearlite
(*c*) acicular troostite
(*d*) martensite

118. ....of steel castings results in improved mechanical properties.
(*a*) Phase annealing
(*b*) Hardening
(*c*) Recrystallisation annealing
(*d*) Tempering

119. Out of the following heat treatment processes, which one is the most economical for the attainment of reasonably good strength without developing internal stresses ?
(*a*) Hardening
(*b*) Hardening followed by tempering
(*c*) Normalising
(*d*) Annealing.

120. What will be the result of working of low carbon steel at a high strain rate ?
(*a*) Increase in softness.
(*b*) Increase in ultimate tensile strength (UTS).
(*c*) Decrease in toughness.
(*d*) None of these.

121. Percentages of vanadium, chromium & tungsten are respectively .... in high speed steel.
(*a*) 1, 4 & 18 (*b*) 4, 8 & 18
(*c*) 4, 14 & 18 (*d*) 4, 8 & 14.

122. Cemented carbide tool is an alloy comprising of tungsten, cobalt and
(*a*) molybdenum (*b*) carbon
(*c*) vanadium (*d*) chromium

123. Which of the following is characterised as a 'noble metal'?
(*a*) Lead (*b*) Silver
(*c*) Barium (*d*) Polonium

124. Amount of energy that a material can absorb before its fracture is a measure of its
(*a*) resilience (*b*) toughness
(*c*) malleability (*d*) ductility

125. High speed steel is commonly designated as
(*a*) 18 : 4 : 1 (*b*) 18 : 8 : 1
(*b*) 18 : 4 : 4 (*d*) 18: 8 : 4

126. Pick out the wrong statement.
(*a*) Silicon content in silicon transformer steel is about 4-5%.
(*b*) Rail steel is a medium carbon steel.
(*c*) Nichrome wire used for electrical heating elements comes under the category of nimonic alloys.
(*d*) The hardening capacity of steel decreases with increasing carbon content.

Answers

114. (a)	115. (b)	116. (c)	117. (c)	118. (a)	119. (c)
120. (b)	121. (a)	122. (b)	123. (b)	124. (b)	125. (a)
126. (d)					

127. High resistance electrical heating elements/ wires are not made of
   (a) Kanthal wire *(b)* nichrome
   (c) nimonic alloys *(d)* marageing steel

128. An alloy used as thermocouple material comprises of 40% nickel and 60% copper. It is called
   (a) German silver *(b)* chromel
   (c) constantan *(d)* Kanthal

129. Which of the following is not an ore dressing operation?
   (a) Classification *(b)* Smelting
   (c) Roasting *(d)* Both *(b)* & *(c)*

130. Which of the following is characterised as 'rare earth' metal?
   (a) Cessium *(b)* Molybdenum
   (c) Nickel *(d)* Aluminium

131. The capacity of a material to absorb/store energy is called its
   (a) brittleness *(b)* resilience
   (c) malleability *(d.)* none of these

132. Which of the following metals can be extracted from its sulphide ore using both the hydrometallurgical as well as pyrometallurgical method of metal extraction?
   (a) Iron *(b)* Copper
   (c) Aluminium *(d)* Lead

133. Which of the following is not a deoxidiser of steel?
   (a) Manganese *(b)* Silicon
   (c) Aluminium *(d)* None of these

134. A .... operation is termed as liquation.
   (a) roasting *(b)* smelting
   (c) refinning *(d)* classification

135. Which of the following is never used as an alloying element in steel?
   (a) Molybdenum *(b)* Tungsten
   (c) Lead *(d)* Silicon

136. Austenite of steel is transformed to......... after quenching operation.
   (a) martensite *(b)* bainite
   (c) pearlite *(d)* none of these

137. Potable/drinking water storage tanks and pipelines are never made of .... because of the reason that human body can not discharge it out in any form, if corroded/eroded particles of it is entrained with the drinking water and consumed by the human beings.
   (a) polythene *(b)* polyvinyl chloride
   (c) lead *(d)* cast iron

138. In iron-carbon equilibrium diagram, the carbon content of cast iron is assumed between .... per cent.
   (a) 1.5 and 4 *(b)* 2.0 and 4
   (c) 2 and 4.3. *(d)* 1.5 and 4.3

139. A coarse crystalline structure of steel in induces
   (a) strength *(b)* brittleness
   (c) ductility *(d)* none of these

140. .... is the hardest, strongest and least ductile of all the constituents of steel.
   (a) Austenite *(b)* Pearlite
   (c) Martensite *(d)* Bainite

141. The amount of pearlite in hyper eutectoid steel is generally .... per cent.
   (a) 25 *(b)* < 50
   (c) 65 *(d)* > 90

142. Cold working of steels increases its
   (a) yield point (b) tensile strength
   (c) brittleness (d) none of these

Answers

127. (d)	128. (c)	129. (d)	130. (a)	131. (b)	132. (b)
133. (b)	134. (c)	135. (c)	136. (a)	137. (c)	138. (c)
139. (b)	140. (c)	141. (d)	142. (c)		

143. Spheroidise annealing of medium/high carbon tool steel raises its
(*a*) tensile strength
(*b*) hardness
(*c*) elongation percentage
(*d*) none of these

144. In 'normalising' of the object as compared to 'annealing',
(*a*) time required is less.
(*b*) cooling is done in water.
(*c*) grain refinement is not achieved.
(*d*) internal stresses are not relieved.

145. Which of the following heat treatment processes are designed for increasing the hardness of steel with minimum loss of ductility ?
(*a*) Isothermal annealing
(*b*) Tempering
(*c*) Normalising
(*d*) Spheroidise annealing

146. During hardening of steel objects, sudden quenching in a single medium
(*a*) gives rise to internal stresses.
(*b*) promotes its tendency of cracking & warping.
(*c*) promotes transformation of austenite into martensite.
(*d*) all 'a', 'b' & 'c'.

147. With increase in tempering temperature of the hardened steel object, its .... reduces.
(*a*) toughness (*b*) hardness
(*c*) rigidity (*d*) both (*b*) & (*c*)

148. One of the microconstituents of cast iron has a Brinell hardness number of about 80. It may be
(*a*) pearlite
(*b*) austenite
(*c*) ferrite
(*d*) none of these

149. Pure iron can dissolve about .... percent carbon at a temperature little above its melting point.
(*a*) 3.5 (*b*) 4
(*c*) 4.5 (*d*) 5

150. Pick out the wrong statement.
(*a*) When cast iron is in the molten state, most of the carbon is in solution as Fe_3C.
(*b*) Slow cooling of molten cast iron favours the precipitation of graphite due to the decomposition of Fe_3C into Fe and graphite.
(*c*) Rapid cooling of molten cast iron allows the carbon to remain as Fe_3C.
(*d*) During cooling of molten cast iron, presence of high silicon resists the decomposition of Fe_3C while that of sulphur accelerates its decomposition.

151. Bell less top (BLT) charging system is the most modern method of raw material charging in a
(*a*) blast furnace
(*b*) L.D. converter
(*c*) coke oven
(*d*) sintering machine

152. Heterogeneous mixture obtained by smelting of arsenical ores resulting in the formation of aresnides of two more metals is called 'speiss'. Speiss is
(*a*) malleable (*b*) ductile
(*c*) brittle (*d*) slag

153. One of the micro constituents of cast iron has a Brinell hardness of about 200. It could be
(*a*) ferrite (*b*) cementite
(*c*) pearlite (*d*) either (*b*) or (*c*)

Answers

143. (c)	144. (a)	145. (b)	146. (d)	147. (d)	148. (c)
149. (d)	150. (d)	151. (a)	152. (c)	153. (d)	

154. Shot blasting process is meant for the
   (*a*) surface cleaning of castings.
   (*b*) 'salamander' tapping from blast furnace.
   (*c*) defect location in steel ingots.
   (*d*) dismantling of furnace refractory lining.

155. Electrical transformer steel uses .... as an alloying element.
   (*a*) silicon (b) lead
   (*c*) nickel (*d*) molybdenum

156. .... has high damping capacity.
   (*a*) Cast iron (*b*) Steel
   (*c*) Brass (*d*) Bronze

157. NIMONICS' are .... based alloys.
   (*a*) chromium (*b*) nickel
   (*c*) lead (*d*) vanadium

158. 'Active metals' like sodium, potassium, calcium, magnesium, aluminium etc. are extracted from their ores by
   (*a*) carbo-thermic reduction,
   (*b*) flash roasting.
   (*c*) electrolytic reduction.
   (*d*) froth floatation.

159. Which of the following metals is not found in 'native' form (free state) on earth?
   (*a*) Gold (*b*) Platinum
   (*c*) Magnesium (*d*) None of these

160. Pyrolusite is the oxide ore of
   (*a*) manganese (*b*) zinc
   (*c*) calcium (*d*) mercury

161. Anglesite is the sulphate ore of
   (*a*) calcium (*b*) cooper
   (*c*) magnesium (*d*) lead

162. Rocks formed in the crust of the earth by chemical changes such as hydrolysis, precipitation and oxidation & reduction are called .... rocks.
   (*a*) igneous (*b*) volcanic
   (*c*) metamorphic (*d*) sedimentary

163. When sulphide ores are roasted, the gas produced is
   (*a*) $C0_2$ (*b*) H_2S
   (*c*) $S0_2$ (*d*) $S0_3$

164. The co-ordination number of an atom in a crystal is indicated by the number of
   (*a*) its valance electrons.
   (*b*) its nucleons.
   (*c*) nearest atoms that it possesses,
   (*d*) possible slip planes in the crystal.

165. Which of the following alloys does not contain copper?
   (*a*) Bronze (*b*) Bell metal
   (*c*) Brass (*d*) Solder

166. Which of the following metals is not present in German Silver?
   (*a*) Copper (*b*) Zinc
   (*c*) Nickel (*d*) Silver

167. Landing gears of aircrafts are made of
   (*a*) high speed steel
   (*b*) babbit metals
   (*c*) precipitation hardened marageing steel
   (*d*) none of these

168. Which of the following methods can not be used for control of engineering properties of materials?
   (*a*) Nuclear radiation
   (*b*) Grain size control
   (*c*) Alloying
   (*d*) None of these

169. Alloying of a metal can not improve upon its
   (*a*) strength
   (*b*) toughness
   (*c*) corrosion resistance
   (*d*) none of these

Answers

154. (a)	155. (a)	156. (a)	157. (b)	158. (c)	159. (c)
160. (a)	161. (d)	162. (d)	163. (c)	164. (c)	165. (d)
166. (d)	167. (c)	168. (d)	169. (d)		

170. Stress causing a deformation of .... percent is called the yield strength of the material.
(a) 0.2 (b) 0.5
(c) 1 (d) 1.2

171. Pearlite contains about 0.8% carbon. Its hardness is about .... on BHN scale.
(a) 100 (b) 170
(c) 250 (d) 350

172. The limit of .... percent carbon is recognised as the dividing line between steel and cast iron.
(a) 1.2 (b) 1.5
(c) 1.8 (d) 2

173. Which of the following is characterised as a 'light metal'?
(a) Sodium (b) Zinc
(c) Gold (d) Copper

174. Which of the following metals is extracted from its ore by electrometallurgical process?
(a) Copper (b) Aluminium
(c) Zinc (d) Nickel

175. Corrosion of metals can not be obviated by its
(a) aluminising
(b) chromising/galvanising
(c) alloying
(d) tempering

176. Which of the following does not come under the category of heat treatment process?
(a) Pack carburising
(b) Nitriding and cyaniding
(c) Galvanising
(d) Spheroidising

177. Which is not a solid state semi-conductor?
(a) Tellurium (b) Cobalt
(c) Selenium (d) Germanium

178. Pick out the wrong statement.
(a) Styrene butadiene rubber (SBR) is a polymerisation product having very low modulus of elasticity.
(b) Phenol formaldehyde (also called bakelite) comes under the category of elastomer.
(c) Duralumin is a thermocouple making material.
(d) Corrosion rate of materials generally increases at higher temperature.

179. Percentage of silver in German silver is
(a) 0 (b) 20
(c) 40 (d) 60

180. Pick out the correct statement.
(a) Carbon content in high corbon steel may be upto 2%.
(b) Lead is the main alloying element which imparts heat resistant properties to steel.
(c) High carbon steel is used for making heavy duty springs as for locomotive wagons and heavy vechicles.
(d) none of these

181. Which of the following is characterised as a heavy metal?
(a) Nickel (b) Platinum
(c) Magnesium (d) Lithium

182. A popular high speed steel is designated as 18:4:1. 18, 4 and 1 respectively represent the percentages of .... in steel.
(a) tungsten, chromium & vanadium
(b) nickel, chromium & molybdenum
(c) nickel, chromium & vanadium
(d) nickel, chromium & tungsten

183. Phosphor bronze is used for making
(a) cutting tools (b) utensils
(c) thermocouples (d) bearings

Answers

170. (a)	171. (b)	172. (d)	173. (a)	174. (b)	175. (d)
176. (c)	177. (b)	178. (c)	179. (a)	180. (c)	181. (a)
182. (a)	183. (d))				

184. Which of the following is not a cutting tool material?
(*a*) Hadfield manganese steel
(*b*) Tungsten carbide
(*c*) Stellite alloy
(*d*) Both 'b'&'c'

185. An alloy of copper and aluminium is called
(*a*) hindalium (*b*) duralumin
(*c*) German silver (*d*) monel metal

186. Percentage of copper in German silver is about
(*a*) 0 (*b*) 20
(*c*) 50 (*d*) 80

187. Mirageing steels which have high yield & tensile strength are
(*a*) nickel, chromium & vanadium.
(*b*) nickel, cobalt & molybdenum.
(*c*) vanadium, molybdenum & chromium.
(*d*) none of these.

188. An alloy of aluminium and .... is called Hindalium.
(*a*) magnesium (*b*) silver
(*c*) manganese (*d*) nickel

189. Which of the following is characterised as a 'refractory metal'?
(*a*) Zirconium (*b*) Thorium
(*c*) Strontium (*d*) Tin

190. Coating of .... is done in galvanising process.
(*a*) tin (*b*) lead
(*c*) zinc (*d*) aluminium

191. .... flux is used for the extraction of metal from its self fluxing ores.
(*a*) Neutral (*b*) Acid
(*c*) Basic (*d*) No

192. .... of a metal is termed as 'dross'.
(*a*) Oxide (*b*) Carbonate
(*c*) Sulphide (*d*) Nitrate

193. Bearings are lined using .... metal.
(*a*) gun (*b*) babbit
(*c*) silver (*d*) lead

194. .... in a crystal is a vacancy defect.
(*a*) Stacking fault (*b*) Twin defect
(*c*) Dislocation (*d*) Point defect .

195. Which of the following heat treatment processes is applied to steel castings?
(*a*) Quenching
(*b*) Full annealing
(*c*) Normalising
(*d*) Sub-critical annealing

196. An age hardening alloy is exemplified by
(*a*) babbit metal (*b*) duralumin
(*c*) brass (*d*) bronze

197. .... is not a constituent of iron and steel.
(*a*) Pearlite (fe) Cementite
(*c*) Silumin (*d*) Bainite

198. An iron ore mine is not located at
(*a*) Purnapani (Orissa)
(*b*) Barsua (Orissa)
(*c*) Bailadila (Chattisgarh)
(*d*) Gua (Jharkhand)

199. Desulphurisation in blast furnace is promoted by high
(*a*) hearth temperature
(*b*) slag volume
(*c*) slag basicity
(*d*) all (*a*), (*b*) and (*c*)

200. High slag rate (about 300-400 Kg/ton pig iron) in Indian blast furnace is mainly due to the
(*a*) high ash content in coke.
(*b*) high percentage (about 70%) of sinter in the B.F. charge burden,
(*c*) use of quartz for viscosity reduction of slag.
(*d*) use of low hot blast temperature.

Answers

184. (a)	185. (b)	186. (c)	187. (b)	188. (a)	189. (a)
190. (c)	191. (d)	192. (a)	193. (b)	194. (d)	195. (b)
196. (b)	197. (c)	198. (a)	199. (d)	200. (a)	

201. Pouring of molten liquid steel in the ingot moulds is termed as the
 (*a*) casting (*b*) teeming
 (*c*) lancing (*d*) none of these

202. Hot top is used in
 (*a*) pig iron ladles (*b*) steel ladles
 (*c*) ingot moulds (*d*) torpedo ladles

203. Which of the following is not a non-ferrous alloy?
 (*a*) Meehanite (*b*) Muntz metal
 (*c*) Magnalium (*d*) Gun metal

204. .... ores are subjected to flash smelting.
 (*a*) Aluminium (*b*) Copper
 (*c*) Zinc (*d*) both 'b' & 'c'

205. An alloy of nickel and copper is called
 (*a*) gun metal (*b*) German silver
 (*c*) monel (*d*) muntz metal

206. Which of the following is an electro-refin-ning process?
 (*a*) Aluminising (*b*) Galvanising
 (*c*) Chromising (*d*) None of these

207. Silicon content in 'Silicon transformer steel' is about .... percent.
 (*a*) 1-1.5 (*b*) 4.5-5
 (*c*) 14.5-15 (*d*) 18.5-19

208. Silicon present in transformer steel
 (*a*) increases its electrical resistance.
 (*b*) decreases its hysterisis loss.
 (*c*) increases its hysterisis loss.
 (*d*) decreases its magnetic permeability.

209. 'Inconel' is used for insulating blanket in jet engines, and it has excellent resistance for high temperature corrosion, oxidation & creep properties. Its main alloying elements are
 (*a*) tungsten, vanadium & molybdenum.
 (*b*) chromium, vanadium & tungsten.
 (*c*) nickel, chromium & iron.
 (*d*) vanadium, molybdenum & nickel.

210. One of the most commonly used stainless steel is designated as 18:8 steel. 18 and 8 respectively represent the percentage of .... in the stainless steel.
 (*a*) nickle & chromium
 (*b*) chromium & tungsten
 (*c*) chromium & nickel
 (*d*) chromium & vanadium

211. Babbit metal used for making bearing is mainly .... based alloy.
 (*a*) lead (*b*) tin
 (*c*) Copper (*d*) antimony

212. 'Hastealloy' has excellent high temperature properties besides having high corrosion oxidation and creep properties. Its main alloing elements are
 (*a*) nickle, molybdenum & iron.
 (*b*) nickle, chromium & tungsten.
 (*c*) molybednum, chromium & tungsten.
 (*d*) iron, chromium & vanadium

213. Which among the following materials of construction has the highest resistance to atmospheric air corrosion?
 (*a*) Cast iron (*b*) Low carbon steel
 (*c*) Copper (*d*) Aluminium

214. Leaching solvent used in Baeyer's process for bauxite purification is
 (*a*) $NaOH$ (*b*) NH_4OH
 (*c*) $NaCl$ (*d*) dilute H_2SO_4

215. Extraction of .... employs thermal reduction process.
 (*a*) aluminium (*b*) copper
 (*c*) magnesium (*d*) none of these

Answers

201. (b)	202. (c)	203. (b)	204. (b)	205. (c)	206. (d)
207. (b)	208. (a)	209. (c)	210. (c)	211. (b)	212. (a)
213. (c)	214. (a)	215. (d)			

216. .... can not be extracted by carbon reduction process.
(*a*) Aluminium (*b*) Lead
(*c*) Zinc (*d*) none of these

217. The main purpose of smelting an ore is to
(*a*) oxidise it.
(*b*) separate out the volatile impurities.
(*c*) reduce it.
(*d*) obtain an alloy of the desired metal.

218. Carnallite is an ore of
(*a*) magnesium (*b*) zinc
(*c*) copper (*d*) aluminium

219. Which of the following is an ore of zinc ?
(*a*) Diaspore (*b*) Calamine
(*c*) Malachite (*d*) None of these

220. Heating of pyrite ores in presence of air to remove sulphur is termed as its
(*a*) smelting (*b*) roasting
(*c*) calcination (*d*) fluxing

221. Leaching of ore with dilute cyanide solution is done for the extraction of
(*a*) zinc (*b*) titanium
(*c*) silver (*d*) copper

222. Which of the following is not a characteristic of alkali metals ?
(*a*) High thermal conductivity.
(*b*) High melting point.
(*c*) High electrical conductivity.
(*d*) None of these.

223. The metal that can be extracted from sea water is
(*a*) beryllium (*b*) calcium
(*c*) magnesium (*d*) lithium

224. .... does not contain aluminium.
(*a*) Feldspar (*b*) Cryolite
(*c*) Fluorspar (*d*) Bauxite

225. Pick out the correct statement.
(*a*) All the ores are mineral, but all the minerals are not ores.
(*b*) Rocky impurities present in a mineral are called mattee.
(*c*) Electrolytic reduction method is used in the extraction of transition metals.
(*d*) In froth floatation process of concentration of sulphide ores, the ore particles float because they are insoluble.

226. Aluminium metal is refined by the .... process.
(*a*) Baeyer's (*b*) Hoop's
(*c*) Hall's (*d*) Solvay's

227. Blister copper is a/an .... of copper.
(*a*) pure form (*b*) ore
(*c*) alloy (*b*) impure form

228. With increase in temperature, the electrical conductivity of semiconductors
(*a*) increases (*b*) decreases
(*c*) remains constant (*d*) either 'a' or 'b

229. Superconductors are those substance which
(*a*) conduct electricity at low temperatures.
(*b*) offer no resistance to current flow.
(*c*) conduct electricity at high temperatures.
(*d*) offer high resistance to current flow.

230. Closenes of packing is maximum is case of the .... crystal lattice
(*a*) face centred (*b*) simple cubic
(*c*) body centred (*d*) none of these

231. .... iron is not an allotropic form of iron.
(*a*) Alpha (*b*) Beta
(*c*) Gamma (*d*) Delta

Answers

216. (a)	217. (d)	218. (a)	219. (c)	220. (a)	221. (b)
222. (b)	223. (c)	224. (c)	225. (a)	226. (b)	227. (d)
228. (a)	229. (b)	230. (a)	231. (b)		

232. In order to be called 'steel', an alloy should have iron percentage >
(a) 40 (b) 50
(c) 70 (d) 80

233. Carbon content in drills, dies & commonly employed taps should be about .... percent.
(a) 0.35 (b) 0.5
(c) 0.8 (d) > 1

234. Silicon steel is most widely used in
(a) making leaf springs
(b) electrical industry
(c) making cutting tools
(d) none of these

235. .... process is used for coating steel with a thin layer of phosphate.
(a) Parkerising (b) Sheradising
(c) Phosphorising (d) Anodising

236. Cast irons are generally specified by their
(a) iron perecentage (b) carbon percentage
(c) hardness (d) tensile strength

237. The range of Mho's scale of hardness is from
(a) 1 to 8 (b) 1 to 10
(c) 1 to 15 (d) 1 – 20

238. Production of articles having soft ductile interior and a very hard surface is done by
(a) case hardening
(b) tempering and hardening
(c) annealing and hardening
(d) hardening

239. Nitriding is a/an .... process.
(a) case hardening (b) spheroidising
(c) annealing (d) normalising

240. Conversion of metallic powder into articles of different forms is done by a technique called
(a) carbiding
(b) precipitation
(c) powder metallurgy
(d) high pressure casting

241. The process of heating the cold pressed metal powder in powder metallurgy technique is termed as the
(a) precipitation (b) deposition
(c) sintering (d) incineration

242. Pick the odd man out from the following metallurgical processes.
(a) Cyaniding (b) Nitriding
(c) Galvanising (d) Pack carburising

243. In amorphous materials, there is no definite atomic structure and atoms exit in a random pattern just as in a liquid. which of the following is an amorphous material?
(a) Brass (b) Glass
(c) Lead (d) Zinc

244. Steel pipes are generally manufactured by the .... process.
(a) electroforming (b) forging
(c) extrusion (d) cold working

245. The phenomenon of weld decay is found in
(a) cast iron (b) stainless steel
(c) brass (d) bronze

246. Galvanising is generally done on
(a) none-ferrous metals
(b) low carbon steel
(c) stainless steel
(d) cast iron

247. The alloying element whose percentage is maximum in high speed steel is
(a) chromium (b) tungsten
(c) molybdenum (d) vanadium

Answers

232. (d)	233. (d)	234. (b)	235. (a)	236. (d)	237. (b)
238. (a)	239. (a)	240. (c)	241. (c)	242. (c)	243. (b)
244. (c)	245. (b)	246. (b)	247. (b)		

248. Cemented carbide tools are not suitable for cutting
    (*a*) aluminium (*b*) brass
    (*c*)steel (*d*) cast iron
249. Tensile strength of steel can be safely increased by adding carbon upto .... percent.
    (*a*) 0.35 (*b*) 0.66
    (*c*) 0.83 (*d*) 1.2
250. Wrought iron does not have the
    (*a*) ability to hold protective coating
    (*b*) uniform strength in all directions
    (*c*) ductility property
    (*d*) high resistance to crushing & corrosion
251. The primary factor which determines the hardness of steel is the
    (*a*)carbon percentage in it
    (*b*) percentage of alloying elements in it
    (*c*) type of heat treatment employed
    (*d*) shape and distribution of carbides in iron
252. As the impurities are oxidised, the melting point of iron
    (*a*) decreases
    (*b*) increases
    (*c*) remains unchanged
    (*d*) unpredictable
253. Which of the following products are produced by powder metallurgy techniques?
    (*a*) Cemented carbides
    (*b*) Bearings and porous metallic parts
    (*c*) Tungsten and molybdenum
    (*d*) All (*a*), (*b*) & (*c*)
254. Machinability of a metal does not depend upon its
    (*a*) tensile strength (*b*) toughness
    (*c*) hardness (*d*) none of these
255. The imperfection in the crystal structure of metal is called
    (*a*) slip (*b*) impurity
    (*c*) clevage (*d*) dislocation
256. Iron has got .... isotopes.
    (*a*) two (*b*) three
    (*c*) four (*d*) no
257. .... wire is expected to be the strongest in tension.
    (a) Stainless steel (*b*) Piano
    (*c*) Mild steel (*d*) Brass
258. Line imperfection in a crystal is called the
    (*a*) Schottky defect
    (*b*) edge dislocation
    (*c*) Frenkel defect
    (*d*) none of these
259. The material for wire drawing should have high
    (*a*) stiffness (*b*) toughness
    (*c*) ductility (*d*) hardness
260. .... is used for making immitation jewellery.
    (*a*) Gun metal
    (*b*) Aluminium bronze
    (*c*) Silicon bronze
    (*d*) Phosphorous bronze
261. .... is a mesomorphous material.
    (*a*) Mica (*b*) Glass
    (*c*) Rubber (1i) Silver
262. The highest percentage of carbon that an iron-carbon alloy can have is .... per cent.
    (*a*) 2.4 (*b*) 3.6
    (*c*) 4.8 (*d*) 6.6
263. Cold worked steel parts are normally subjected to
    (*a*) annealing (*b*) shot peening
    (*c*) tempering (*d*) sheradising

Answers

248. (c)	249. (b)	250. (b)	251. (d)	252. (b)	253. (d)
254. (b)	255. (d)	256. (c)	257. (b)	258. (b)	259. (c)
260. (b)	261. (a)	262. (d)	263. (a)		

264. The ratio of the atoms to the lattice parameters for fcc unit cell is
(a) 1: 2
(b) 2 : 1
(c) 1:2√2
(d) 1 : 4

265. Silicon is doped with .... in a p-type semi-conductor.
(a) phosphorous
(b) aluminium
(c) arsenic
(d) none of these

266. .... cubic lattice structure has the maximum co-ordination number.
(a) Face centred
(b) Body centred
(c) Simple
(d) none of these

267. Silicon is doped with .... in a n-type semi-conductor.
(a) phosphorous
(b) carbon
(c) boron
(d) none of these

268. .... cubic lattice structure has the greatest atomic packing factor.
(a) Face centred
(b) Simple
(c) Body centred
(d) none of these

269. A crystal lattice is a/an
(a) random arrangement of molecules but an orderly arrangement of atoms.
(b) random arrangement of atoms in a crystal.
(c) orderly arrangement of atoms inside a crystal.
(d) piece of crystal.

270. Match the typical product yield percentage in an integrated steel plant in India.
List I
(a) Yield of gross coke from dry coal.
(b) Yield of blast furnace grade (i.e. 25 - 80 mm size) coke from dry coal carbonised.
(c) Yield of continuous cast slab from steel.
(a) Yield of tar from dry coal carbonised.
List II
I. 67
III. 96
II. 3
IV. 87

271. Match the temperature (°C) maintained in various processes/furnaces in an integrated steel plant.
List I
(a) Zinc bath used for galvanising steel.
(b) Cold rolled steel batch annealing furnace.
(c) Soaking pit for heating steel ingot.
(d) Sulphuric acid pickling bath for steel coil.
List II
I. 100
II. 1300
III. 700
IV. 450

Answers

264. (c) 265. (b) 266. (a) 267. (a) 268. (a) 269. (c)
270 (a)-IV (b)-I (c) -II (d) -III 271. (a)-IV (b)-III (c) -II (d) -I

272. Match the temperature (°C) encountered in various metallurgical processes.

List I

(*a*) Tempering oil bath temperature.
(*b*) L.D. steel tapping temperature.
(*c*) Glass tank furnace temperature.
(*d*) Oxy-acetylene flame temperature.

List II

I. 1650 II. 400
III. 2400 IV. 1400

273. Match the relevant terminologies in both the lists.

List I

(*a*) Occupies top position in electromotive series of metals.
(*b*) Metals whose oxides decompose by heat alone.
(*c*) Activators and depressors.
(*d*) Lowermost in the electromotive series of metals.

List II

I. Ag, Pt, Au
II. Froth floatation
III Ag and Hg
IV Alkali and alkaline earth metals

274. Match the mine locations where metallic ores are found.

List I

(*a*) Copper ore mine
(*b*) Iron ore mine
(*c*) Uranium ore mine
(*d*) Gold ore mine

List II

I. Jadugoda (Jharkhand)
II Khetri (Rajasthan)
III Kiriburu (Jharkhand)
IV Kolar (Karnataka)

275. Match the main alloying elements present in various alloys.

List I

(*a*) Transformer steel (*b*) Inconel
(*c*) Gun metal (*a*) Pewter type metal

List II

I. Nickel II. Silicon
III. Tin IV. Copper

276. Match the effects of various alloying elements added in steel.

List I

(*a*) Silicon (b) Nickel
(*c*) Vanadium (*d*) Tungsten

List II

I. Increases toughness of steel.
II. Confers red hardness to steel.
III. Increases hardenability of steel.
IV. Alloying element for electrical and magnetic steel.

277. Match the specific gravity of various alloying elements of steel.

List I	List II
(*a*) Molybdenum	I. 19.35
(*b*) Carbon	II. 10.2
(*c*) Tungsten	III. '7.2
(*d*) Manganese/chromium	IV. 2.3

278. Match the melting point (°C) of various alloying elements of steel.

List I	List II
(*a*) Nickel	I. 3410
(*b*) Vanadium	II. 1453
(*c*) Molybdenum	III. 1890
(*d*) Tungsten	IV. 2617

279. Match the temperature (°C) during heating of cast iron with its various allotropic forms existing.

List I	List II
(*a*) α-iron	I. 1400-1539
(*b*) β-iron	II. 910-1400
(*c*) γ- iron	III. 768-910
(*d*) δ-iron	IV. < 768

Answers

272. (a)-II (b)-I (c) -IV (d) -III
273. (a)-IV (b)-III (c) -II (d) -I
274. (a)-II (b)-III (c) -I (d) -IV
275. (a)-II (b)-I (c) -IV (d) -III
276. (a)-IV (b)-I (c) -III (d) -II
277. (a)-II (b)-IV (c) -I (d) -III
278. (a)-II (b)-III (c) -IV (d) -I
279. (a)-IV (b)-III (c) -II (d) -I

280. Match the effect of various alloying elements added in steel.

List I

(*a*) Manganese (*b*) Aluminium

(*c*) Chromium (*d*) Cobalt

List II

I. Adds to red hardness by hardening ferrite.

II. Counteracts brittleness from sulphur.

III. Acts as deoxidiser.

IV. Adds to depth hardenability with improved resistance to wear & abrasion.

281. Match the effects of various alloying elements added in cast iron.

List I

(*a*) chromium (0.15-1.0%).

(*b*) Vanadium (0.15-0.50%).

(*c*) Manganese (0.3-1.25%).

(*d*) Molybdenum (0.3-1.0%).

List II

I. Increases hardenability and freedom from cracking & distortion.

II. Decreases machinability.

III Improves tensile strength, transverse strength, hardness, & resistance to wear & heat.

IV Increases fluidity & density in castings.

282. Which of the following is not found on iron-carbon equilibrium diagram ?

(*a*) Curie point (*b*) Peritectic point

(*c*) Eutectic point (*d*) Eutectoid point

283. The temperature at which the magnetic property of iron disappears (i.e., it becomes non-magnetic) and its electrical conductivity & specific heat also changes is called the Curie point, which is .... °C.

(*a*) 768 (*b*) 908

(*c*) 1400 (*d*) 1539

284. Which of the following micro-constituents of steel is not a ferrite-cementite mixture having lamellar structure ?

(*a*) Pearlite (*b*) Sorbite

(*c*) Troostite (*d*) Ledeburite

285. Time-temperature-transformation (TTT) diagram is not the same as the

(*a*) Ellingham diagram.

(*b*) Isothermal transformation diagram.

(*c*) Bain's curve.

(*d*) C-curve or S-curve.

286. Cast iron compared to steel is better in

(*a*) ductility. (*b*) strength.

(*c*) malleability. (*d*) fluidity & castability.

287. Phosphorous exists in iron & steel as

(*a*) P (*b*) P_2O_5

(*c*) FeP (*d*) P_2O_3

288. Cryolite is

(*a*) a double flouride of sodium & aluminium (AlF_3. 3NaF).

(*b*) used as a flux in melting practices.

(*c*) used as an electrolyte in aluminium reduction cell.

(*d*) all a, b & c.

289. Carbon content in blast furnace flue dust is about .... percent.

(*a*) 1-5 (*b*) 20-25

(*c*) 45 - 50 (*d*) 75 - 80

290. Carbon is present in the uncombined (graphitic) form in case of

(*a*) cast iron (*b*) steel

(*c*) ferroalloy (*d*) none of these

291. Carbon is present in the combined form (carbide) in case of

(*a*) pig iron (b) steel

(*c*) ferroalloys (*d*) all (*a*), (*b*), & (*c*)

292. High speed cutting tool steels contain about .... percent of tungsten.

(*a*) 6 – 8 (*b*) 18 – 20

(*c*) 30 – 35 (*d*) 50 – 55

Answers

280. (a)-II (b)-III (c) -IV (d) -I 281. (a)-II (b)-III (c) -IV (d) -I 282. (a) 283. (a)
284. (d) 285. (a) 286. (d) 287. (c) 288. (d) 289. (b)
290. (a) 291 (d) 292. (b)

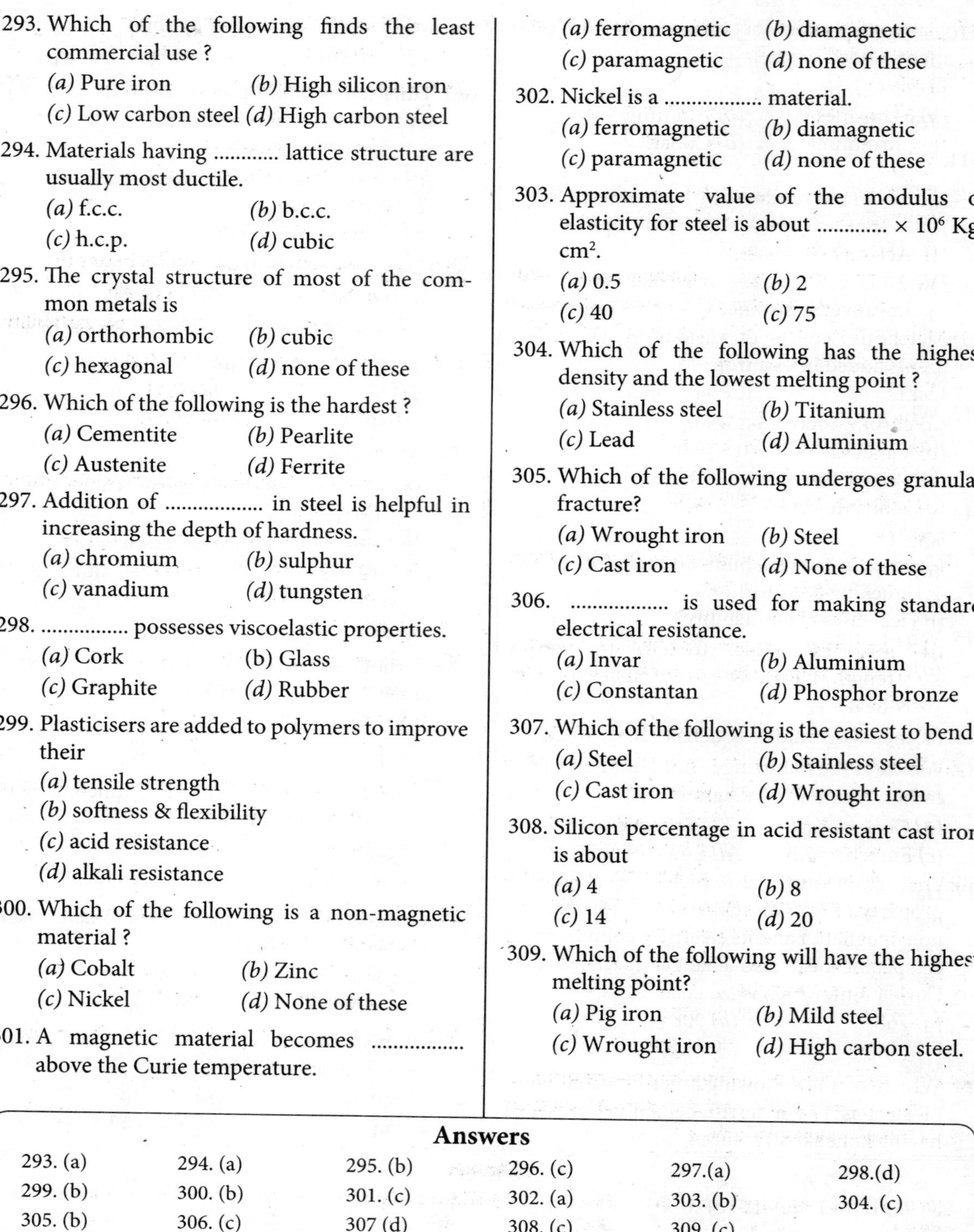

293. Which of the following finds the least commercial use ?
(*a*) Pure iron (*b*) High silicon iron
(*c*) Low carbon steel (*d*) High carbon steel

294. Materials having .... lattice structure are usually most ductile.
(*a*) f.c.c. (*b*) b.c.c.
(*c*) h.c.p. (*d*) cubic

295. The crystal structure of most of the common metals is
(*a*) orthorhombic (*b*) cubic
(*c*) hexagonal (*d*) none of these

296. Which of the following is the hardest ?
(*a*) Cementite (*b*) Pearlite
(*c*) Austenite (*d*) Ferrite

297. Addition of .... in steel is helpful in increasing the depth of hardness.
(*a*) chromium (*b*) sulphur
(*c*) vanadium (*d*) tungsten

298. .... possesses viscoelastic properties.
(*a*) Cork (b) Glass
(*c*) Graphite (*d*) Rubber

299. Plasticisers are added to polymers to improve their
(*a*) tensile strength
(*b*) softness & flexibility
(*c*) acid resistance
(*d*) alkali resistance

300. Which of the following is a non-magnetic material ?
(*a*) Cobalt (*b*) Zinc
(*c*) Nickel (*d*) None of these

301. A magnetic material becomes .... above the Curie temperature.
(*a*) ferromagnetic (*b*) diamagnetic
(*c*) paramagnetic (*d*) none of these

302. Nickel is a .... material.
(*a*) ferromagnetic (*b*) diamagnetic
(*c*) paramagnetic (*d*) none of these

303. Approximate value of the modulus of elasticity for steel is about .... $\times 10^6$ Kg/cm^2.
(*a*) 0.5 (*b*) 2
(*c*) 40 (*c*) 75

304. Which of the following has the highest density and the lowest melting point ?
(*a*) Stainless steel (*b*) Titanium
(*c*) Lead (*d*) Aluminium

305. Which of the following undergoes granular fracture?
(*a*) Wrought iron (*b*) Steel
(*c*) Cast iron (*d*) None of these

306. .... is used for making standard electrical resistance.
(*a*) Invar (*b*) Aluminium
(*c*) Constantan (*d*) Phosphor bronze

307. Which of the following is the easiest to bend?
(*a*) Steel (*b*) Stainless steel
(*c*) Cast iron (*d*) Wrought iron

308. Silicon percentage in acid resistant cast iron is about
(*a*) 4 (*b*) 8
(*c*) 14 (*d*) 20

309. Which of the following will have the highest melting point?
(*a*) Pig iron (*b*) Mild steel
(*c*) Wrought iron (*d*) High carbon steel.

Answers

293. (a)	294. (a)	295. (b)	296. (c)	297.(a)	298.(d)
299. (b)	300. (b)	301. (c)	302. (a)	303. (b)	304. (c)
305. (b)	306. (c)	307 (d)	308. (c)	309. (c)	

310. .... welding process is normally used in the fabrication of car bodies.
(a) Resistance (b) Thermit
(c) Arc (d) Brazing/soldering

311. Which of the following has the highest compressive strength?
(a) Wrought iron (b) Cast iron
(c) Mild steel (d) High carbon steel

312. Brinell Hardness number (BHN) for talc is approximately in the range of
(a) 1-5 (b) 20-30
(c) 100 - 150 (d) 200 - 250

313. Which of the following is the softest material ?
(a) Quartz (b) Calcite
(c) Corrundum (d) Flourite

314. Vicker's hardness number for diamond is about
(a) 4000 (b) 8000
(c) 12000 (d) 16000

315. Which of the following is the hardest material ?
(a) Topaz (b) Quartz
(c) Corrundum (d) Flourite

316. Slow plastic deformation of metals under a constant stress is termed as the .... failure.
(a) fatigue (b) endurance
(c) creep (c) none of these

317. Hammers and railway rails are normally made of
(a) mild steel
(b) dead mild steel
(c) medium carbon steel
(d) high carbon steel

318. Mild steel is used for making
(a) fish plates (b) die block
(c) channels (d) drop forging

319. Dead mild steel, which contains 0.10 to 0.15 percent carbon, is used for making
(a) shafts (b) flanges
(c) gears (d) shear blades

320. With increase in temperature, the electrical conductivity of a .... decreases.
(a) semi-conductor (b) metal or alloy
(c) dielectric (d) none of these

321. X-rays do not exhibit the property of
(a) reflection (b) scattering
(c) diffraction (d) dispersion

322. Cupola produces .... iron.
(a) pig (b) cast
(c) wrought (d) carbon free

323. Puddling process is used for converting pig iron into
(a) cast iron (b) wrought iron
(c) mild steel (d) semi-steel

324. The softest and the least strong constituent of steel is
(b) cementite (b) pearlite
(c) ferrite (d) bainite

325. Maximum permissible sulphur content in steel is .... percent.
(a) 0.01 (b) 0.055
(c) 0.50 (d) 0.80

326. Which of the following is a light alloy?
(a) Monel metal (b) Dow metal
(c) German silver (d) Babbit metal

327. .... of austenite decreases the hardenability in steel.
(a) Fine grains
(b) Coarse grains
(c) Homogeneity
(d) Dissolved elements (except cobalt)

Answers

310. (a)	311. (d)	312. (b)	313.(b)	314.(b)	315. (c)
316. (c)	317. (d)	318. (c)	319. (b)	320. (b)	321. (d)
322. (b)	323. (b)	324. (c)	325. (b)	326. (b)	327. (a)

328. Hot working of lead is carried out at
   (*a*) 75°C (*b*) 373°K
   (*c*) 150°C (*d*) room temperature

329. Drossing is a .... operation.
   (*a*) smelting (*b*) deslagging
   (*c*) roasting (*d*) dressing

330. Fire refining process is employed in case of
   (*a*) tin (*b*) copper
   (*c*) zinc (*d*) aluminium

331. The malleability of a material is the property by virtue of which, it can be rolled or hammered into thin sheets. Which of the following materials has the maximum malleability ?
   (*a*) Lead (*b*) Copper
   (*c*) Aluminium (*d*) Wrought iron

332. The iron ore mostly used for the production of pig iron in our country (containing 65 - 66% Fe) is
   (*a*) magnetite (*b*) hematite
   (*c*) limonite (*d*) siderite

333. Cast iron is
   (*a*) used for making shock resisting parts.
   (*b*) manufactured in cupola and is brittle.
   (*c*) having compressive strength more than its tensile strength.
   (*a*) all (*a*), (*b*), and (*c*).

334. Carbon content in plain carbon steel is .... percent.
   (*a*) 0.10 – 0.15 (*b*) 0.35 – 0.45
   (*c*) 0.65 – 0.80 (*d*) 0.85 – 1.2

335. .... is added in low carbon steel to raise its yield point.
   (*a*) Sulphur (*b*) Phosphorous
   (*c*) Silicon (*d*) Manganese

336. A steel alloy containing 36% nickel is called ...., which has a zero co-efficient of expansion.
   (*a*) austenitic stainless steel
   (*b*) heat resisting steel
   (*c*) invar
   (*d*) high speed steel

337. .... can replace tungsten in high speed steel.
   (*a*) Chromium (*b*) Vanadium
   (*c*) Cobalt (*d*) Molybdenum

338. Shock resisting steels should possess high
   (*a*) hardness
   (*b*) toughness
   (*c*) tensile strength
   (*d*) wear resistance

339. Which of the following alloying elements reduces the formation of iron sulphide in steel ?
   (*a*) Cobalt (*b*) Nickel
   (*c*) Manganese (*d*) Vanadium

340. A material in which the atoms are arranged regularly in some directions but not in others is termed as 'mesomorphous material'; an example of which is
   (*a*) lead
   (*b*) glass
   (*c*) mica
   (*d*) silver

341. Examination of crystal structure of a material is normally done by
   (*a*) metallurgical microscope.
   (*b*) X-ray technique.
   (*c*) spectroscope techniques.
   (*d*) optical microscope.

Answers

328. (d)	329. (d)	330. (b)	331. (a)	332. (b)	333.(d)
334.(b)	335. (b)	336. (c)	337. (d)	338. (b)	339. (c)
340. (c)	341. (b)				

342. Pick out the wrong statement.
(*a*) A coarse grained steel has low toughness and a greater tendency to distort during heat treatment.
(*b*) The lower critical temperature is the same for all steels, which is 723°C.
(*c*) A fine grained steel is more ductile and has a less tendency to distort during heat treatment.
(*d*) The upper critical temperature of steels does not vary with their carbon contents.

343. Which of the following processes improves the machinability of steel but decreases its hardness & tensile strength ?
(*a*) Spheroidising (*b*) Normalising
(*c*) Full annealing (*d*) None of these

344. When the steel is subjected to normalising, its .... decreases.
(*a*) yield point
(*b*) ductility
(*c*) ultimate tensile strength
(*d*) none of these

345. Softening of hardened steel is done by its
(*a*) normalising (*b*) tempering
(*c*) annealing (*d*) carburising

346. .... is not a case hardening process.
(*a*) Carburising (*b*) Nitriding
(*c*) Cyaniding (*d*) Annealing

347. Castability of aluminium is increased by the addition of .... in it.
(*a*) silicon (*d*) silicon
(*c*) bismuth (*d*) lead

348. A steel containing pearlite and cementite is
(*a*) hard (*b*) soft
(*a*) tough (*d*) both (*b*) & (*c*)

349. A steel containing ferrite & pearlite is'
(*a*) soft (*b*) hard
(*a*) tough (*d*) both (*b*) and (*c*)

350. A hardened steel essentially contains
(*a*) sorbite (*b*) troostite
(*c*) martensite (*d*) none of these

351. .... steel is widely used for the manufacture of motor car crankshafts.
(*a*) Silicon (*b*) Nickel
(*c*) Chrome (*d*) High speed

352. The temperature at which new grains are formed in the metal is called the .... temperature.
(*a*) eutectic (*b*) recrystallisation
(*c*) upper critical (*d*) lower critical

353. The increase in hardness of metal due to its cold working is termed as the .... hardening.
(*a*) work (*b*) age
(*c*) induction (*d*) flame

354. Cast iron and steel pipes are produced by the .... casting.
(*a*) die (*b*) investment
(*c*) slush (*d*) true centrifugal

355. High speed steel tools retain their hardness upto a temperature of .... °C.
(*a*) 500 (*b*) 750
(*c*) 900 (*d*) 1100

356. Addition of .... in steel can help in increasing the depth of hardness.
(*a*) nickel (*b*) chromium
(*c*) vanadium (*d*) tungsten

357. Case hardening of a material is
(*a*) followed by tempering or carburising.
(*b*) preceded by its tempering.
(*c*) done to get a soft ductile interior with a very hard surface.
(*d*) carried out to get extreme hardness in its core.

Answers

342. (d)	343. (a)	344. (b)	345. (b)	346 (d)	347. (a)
348. (a)	349. (a)	350. (c)	351. (c)	352. (b)	353. (a)
354. (d)	355. (c)	356.(d)	357. (c)		

358. The process by which steel is coated with a thin layer of phosphate to act as a base or primer for paints & enamels is called
(a) phosphorising (b) sheardising
(c) parkerising (d) spheroidising

359. Steel castings
(a) have poor endurance limit.
(b) can not withstand impact.
(c) are not weldable.
(d) are weldable.

360. During hardening of steel parts, higher ratio of surface to mass produces
(a) greater depth of hardening.
(b) non-uniformity in surface hardness.
(c) surface defects.
(d) smaller depth of hardening

361. Very high sulphur in pig iron makes
(a) its casting unsound.
(b) it hard and machinable.
(c) it brittle and malleable.
(d) all (a), (b), & (c).

362. Dies for wire drawing are generally made of
(a) mild steel (c) stainless steel
(c) carbides (d) high carbon steel

363. Surface hardness achieved by nitriding is of the order of .... VPN.
(a) 150 - 200 (b) 500 - 650
(c) 750 - 850 (d) 1000-1100

364. The ability of tool steel to resist softening at high temperatures is termed as the .... hardness.
(a) red (6) extended
(c) super (d) extreme

365. Coarse grained steels have
(a) low toughness.
(b) no tendency to distort.
(c) high density.
(d) very high toughness.

366. Fine grained steels have
(a) high brittleness.
(b) higher tendency to distort.
(c) high ductility.
(d) none of these.

367. Maximum hardenability of steel depends upon its
(a) chemical composition.
(b) carbon content.
(c) grain size.
(d) *alloying elements content*

368. Young's modulus of elasticity of a material is the slope of the initial linear portion of the stress-strain curve. It decreases appreciably by
(a) alloying additions
(b) heat treatment
(c) cold work
(d) increasing temperature

369. Pick out the wrong statement.
(a) Both annealing and normalising release the internal stresses of the material besides improving the mechanical properties.
(b) Low carbon steel does not respond to the heat treatment for hardening of the material, hence it is subjected to case hardening or surface hardening processes like cyaniding, carburising nitriding etc, which produces high carbon outer layers resulting in increase of surface hardness.
(c) Induction hardening and flame hardening techniques are also used for surface hardening.
(d) Martempering of a material is a hardening process.

Answers

358. (c)	359. (d)	360 (a)	361. (a)	362. (c)	363. (d)
364. (a)	365. (a)	366. (c)	367. (b)	368. (d)	369. (a)

370. Spheroidising of a material is a/an .... process.
   (*a*) normalising (*b*) annealing
   (*c*) tempering (*d*) hardening

371. Hardenability of a material
   (*a*) determines its toughness.
   (*b*) indicates its hardness.
   (*c*) determines the depth of hardness.
   (*d*) none of these.

372. Hydrogen embrittlement of metal results from the
   (*a*) corrosion fatigue.
   (*b*) stress corrosion.
   (*c*) pitting corrosion.
   (*d*) none of these.

373. Ceramic materials are
   (*a*) organic in a nature.
   (*b*) stronger in compression than in tension.
   (*c*) always amorphous in nature.
   (*d*) always bad heat conductors.

374. Common house hold glass *i.e. soda-lime glass* is a/an .... material.
   (*a*) full crystalline (*b*) partly crystalline
   (*c*) amorphous (*d*) none of these

375. China clay is a major raw material for the production of
   (*a*) glass
   (*b*) fireclay refractory bricks
   (*c*) porcelain
   (*d*) none of these

376. Cermets are..................... materials.
   (*a*) refractory (*b*) reinforced
   (*c*) abrasive (*d*) fully metallic

377. Pick out the correct statement.
   (*a*) Plastics are good conductors of heat & electricity.
   (*b*) All the polymers are highly crystalline in nature.
   (*a*) Polymers can be vaporised by heating to a very high temperature.
   (*d*) The liquid polymer becomes greasy, then waxy and finally solid on increasing the degree of polymerisation.

378. Polystyrene is a .... plastic at room temperature.
   (*a*) ductile (*b*) brittle
   (*c*) malleable (*d*) none of these

379. Buna-s is a .... polymeric material.
   (*a*) fibrous (*b*) plastic
   (*c*) resinous (*d*) rubbery

380. Unbreakable crockeries are made from .... polymers.
   (*a*) polystyrene (*b*) melamine
   (*c*) polyester (*d*) polyurathane

381. Crystallisation of polymers is an undesirable property. Crystallisation of celluloid is prevented by adding
   (*a*) glycerol (*b*) nitrocellulose
   (*c*) camphor (*d*) none of these,

382. Pick out the wrong statement.
   (*a*) A ferromagnetic material becomes paramagnetic above the 'Curie temperature'.
   (*b*) Permanent magnets are made of hard materials, whereas electro magnets require soft magnetic materials.
   (*c*) Soft magnetic materials (e.g., pure iron) have higher permeability and low hysterisis loss & coercive forces.
   (*d*) Tungsten steel and alnico are not hard magnetic materials.

Answers

370. (b)	371. (c)	372. (b)	373. (b)	374. (b)	375 (b)
376. (a)	377. (d)	378. (b)	379. (d)	380. (b)	381. (c)
382. (d)					

383. Dielectrics are electrical insulating materials having
(*a*) very narrow energy gap in its band structure.
(*b*) very large number of charge carriers.
(*c*) positive temperature co-efficient of resistance.
(*d*) great influence of temperature and frequency on its permittivity.

384. Which of the following is not a dielectric material?
(*a*) Cotton, silk, & paper.
(*b*) Asbestos, glass, porcelain, & mica.
(*c*) Rubber & polymeric resins.
(*d*) All refractory materials.

385. Thermistors are used in the .... devices.
(*a*) voltage measuring
(*b*) temperature measuring
(*c*) temperature compensating
(*d*) both (*b*) & (*c*)

386. Pick out the wrong statement.
(*a*) Electrical conductivities of semi-con ductors is of the order of 10^{-15} (ohm . cm $)^{-1}$
(*b*) Semiconduction is also possible in ceramic & organic materials.
(*c*) The conductivity of an extrinsic semi conductor decreases with the rise in temperature.
(*d*) The operation of a p-n junction, which is a rectifier, is affected by temperature.

387. Pick out the correct statement.
(*a*) Materials exhibiting high elasticity obey Hooke's law.
(*b*) The elastic behaviour of rubber under compression is the same as its behaviour under tension.
(*c*) The damping capacity of a material is due to its plastic deformation.
(*d*) The stress required to cause plastic flow in polycrystalline material is higher as compared to monocrystalline materials due to the presence of grains of different orientations.

388. In cold working of metal as compared to its hot working
(*a*) cracks & blow holes are eliminated.
(*b*) ductility & impact strength improves.
(*c*) appreciable strain hardening is produced.
(*d*) yield stress, hardness & fatigue strength is not at all affected.

389. With decrease in the grain size of a material, its creep resistance
(*a*) increases
(*b*) decreases
(*c*) remains constant
(*d*) either (*a*) or (*b*); depends on the material

390. Pick out the wrong statement.
(*a*) Ductile fracture of a stressed material, which exhibits a large plastic deformation, is commonly caused by the formation and coalescence of voids in the necked region.
(*b*) Brittle fracture is caused by the propagation of pre existing cracks in the material and involves minimum plastic deformation.
(*c*) Fatigue fracture of a material is always brittle in nature and takes place due to the existence of line imperfections.
(*d*) Brittle materials are generally tested in tension.

391. The stress at which a metal fails by fatigue lies
(*a*) near the fracture point of the stress-strain curve.
(*b*) in the plastic range.
(*c*) in the elastic range.
(*d*) none of these.

Answers

383. (d)	384. (d)	385. (d)	386. (c)	387 (d)	388. (c)
389. (b)	390. (d)	391. (c)			

392. Which of the following metals is malleable but not ductile?
   (a) Silver (b) Lead
   (c) Copper (d) None of these

393. Reduction in the grain size reduces the .... of the material.
   (a) fatigue resistance
   (b) tensile strength
   (c) creep resistance
   (d) all (a), (b), & (c)

394. The formation of oxide film on a metal due to atmospheric exposure reduces its
   (a) toughness (b) stiffness
   (c) creep limit (d) all (a), (b), & (c)

395. Pick out the wrong statement.
   (a) The X-rays can not be deflected by electric field unlike cathode rays.
   (b) The intensity of X-rays can be measured by ionisation current produced due to the ionisation of gas by X-rays.
   (c) The quality of X-rays can be controlled by varying the anode-cathode voltage.
   (d) Crystal structure of a material can be studied by an electron microscope.

396. Most of the common metals have the .... crystal lattice structure.
   (a) cubic (b) hexagonal
   (c) orthorhombic (d) none of these

397. Pick out the wrong statement.
   (a) Every crystal system has certain important planes and directions which are described by means of three numbers called Miller indices.
   (b) The co-ordination number of a given crystal is an indication of the closeness of the packing of atoms.
   (c) The Miller indices of all planes parallel to one another are not identical.
   (d) The Miller indices of a plane are proportional to the reciprocal of numerical parameters of the intercepts.

398. Rockwell number represents the .... of a substance.
   (a) hardness
   (b) hardenability
   (c) surface hardness
   (d) depth of hardness

399. Sudden immersion/dipping of red hot steel bar in water makes it
   (a) malleable & tough.
   (b) ductile & soft.
   (c) brittle & hard.
   (d) none of these.

400. Diamond does not conduct electricity, because.
   (a) its structure is very compact.
   (b) no free electrons are present.
   (c) it is of crystalline nature.
   (d) there are only carbon atoms present.

401. Graphite is a good
   (a) thermal & electric insulator
   (b) conductor of heat
   (c) conductor of electricity
   (d) both 'b'& 'c'

402. Out of the following, ... waves have the largest wavelength,
   (a) radio
   (b) light
   (c) x-ray
   (d) gamma-ray

Answers

392. (b)	393. (c)	394. (c)	395. (d)	396. (a)	397. (c)
398 (a)	399. (c)	400. (b)	401. (d)	402. (a)	

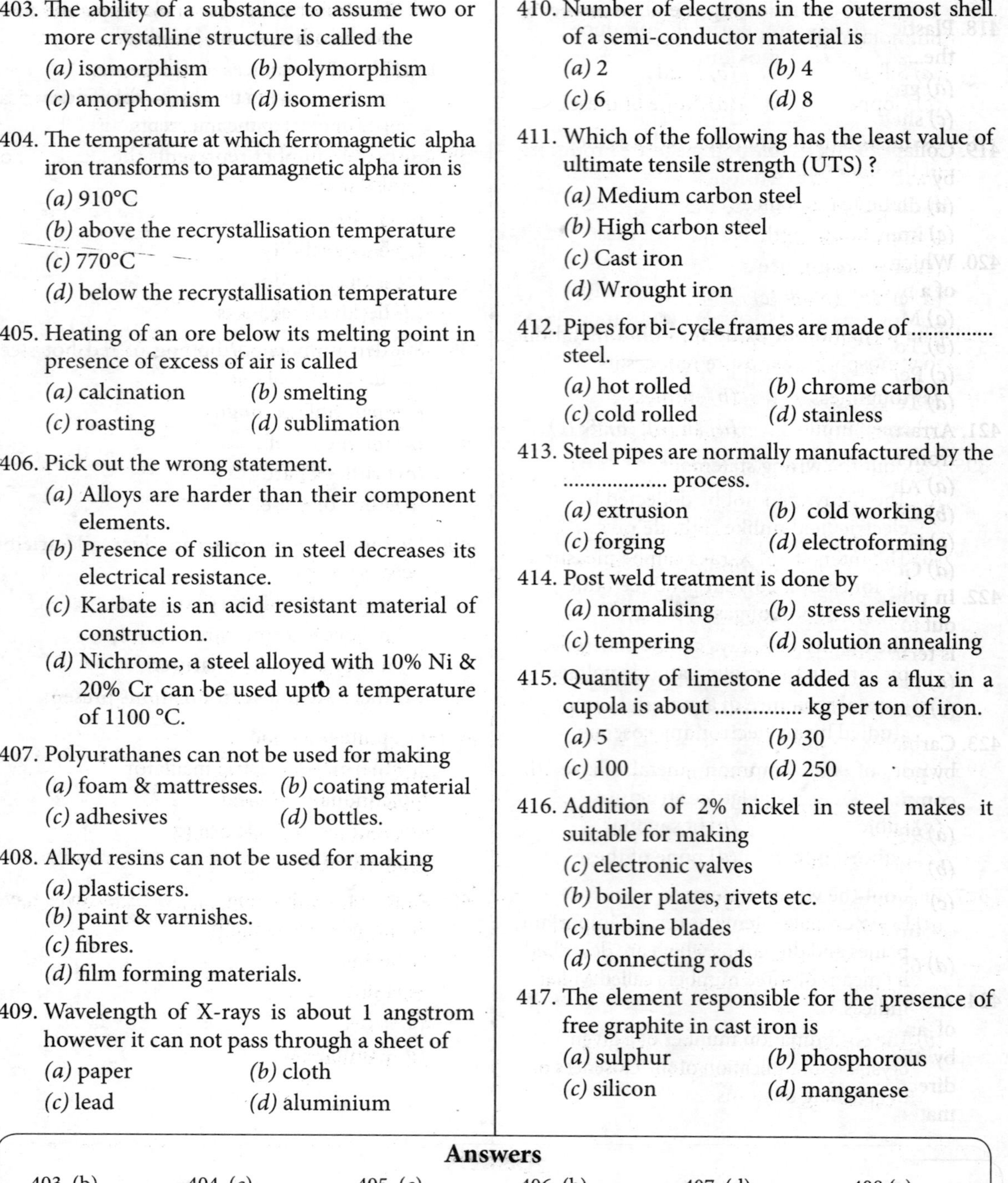

403. The ability of a substance to assume two or more crystalline structure is called the
    (*a*) isomorphism (*b*) polymorphism
    (*c*) amorphomism (*d*) isomerism

404. The temperature at which ferromagnetic alpha iron transforms to paramagnetic alpha iron is
    (*a*) 910°C
    (*b*) above the recrystallisation temperature
    (*c*) 770°C
    (*d*) below the recrystallisation temperature

405. Heating of an ore below its melting point in presence of excess of air is called
    (*a*) calcination (*b*) smelting
    (*c*) roasting (*d*) sublimation

406. Pick out the wrong statement.
    (*a*) Alloys are harder than their component elements.
    (*b*) Presence of silicon in steel decreases its electrical resistance.
    (*c*) Karbate is an acid resistant material of construction.
    (*d*) Nichrome, a steel alloyed with 10% Ni & 20% Cr can be used upto a temperature of 1100 °C.

407. Polyurathanes can not be used for making
    (*a*) foam & mattresses. (*b*) coating material
    (*c*) adhesives (*d*) bottles.

408. Alkyd resins can not be used for making
    (*a*) plasticisers.
    (*b*) paint & varnishes.
    (*c*) fibres.
    (*d*) film forming materials.

409. Wavelength of X-rays is about 1 angstrom however it can not pass through a sheet of
    (*a*) paper (*b*) cloth
    (*c*) lead (*d*) aluminium

410. Number of electrons in the outermost shell of a semi-conductor material is
    (*a*) 2 (*b*) 4
    (*c*) 6 (*d*) 8

411. Which of the following has the least value of ultimate tensile strength (UTS) ?
    (*a*) Medium carbon steel
    (*b*) High carbon steel
    (*c*) Cast iron
    (*d*) Wrought iron

412. Pipes for bi-cycle frames are made of............... steel.
    (*a*) hot rolled (*b*) chrome carbon
    (*c*) cold rolled (*d*) stainless

413. Steel pipes are normally manufactured by the .... process.
    (*a*) extrusion (*b*) cold working
    (*c*) forging (*d*) electroforming

414. Post weld treatment is done by
    (*a*) normalising (*b*) stress relieving
    (*c*) tempering (*d*) solution annealing

415. Quantity of limestone added as a flux in a cupola is about .... kg per ton of iron.
    (*a*) 5 (*b*) 30
    (*c*) 100 (*d*) 250

416. Addition of 2% nickel in steel makes it suitable for making
    (*c*) electronic valves
    (*b*) boiler plates, rivets etc.
    (*c*) turbine blades
    (*d*) connecting rods

417. The element responsible for the presence of free graphite in cast iron is
    (*a*) sulphur (*b*) phosphorous
    (*c*) silicon (*d*) manganese

Answers

403. (b)	404. (c)	405. (c)	406. (b)	407. (d)	408 (c)
409 (a)	410 (d)	411 (c)	412 (c)	413. (a)	414. (b)
415. (b)	416. (b)	417. (c)			

418. Plastic articles are normally produced by the........................ moulding.
(*a*) green sand (*b*) injection
(*c*) shell (*d*) dry sand

419. Collapsible tubes for toothpaste are produced by .... extrusion.
(*a*) direct (*b*) indirect
(*c*) impact (*d*) none of these

420. Which of the following mechanical properties of a material is most structure insensitive ?
(*a*) Modulus of elasticity (Young's modulus)
(*b*) Toughness
(*c*) Percentage reduction of area
(*d*) Tensile strength

421. Arrange copper, silver, gold and aluminium from lower to higher conductor of heat.
(*a*) Aluminium, gold, copper, silver
(*b*) Copper, gold, silver, aluminium
(*c*) Copper, gold, aluminium, silver
(*d*) Gold, copper, aluminium, silver

422. In powder metallurgy, the operation carried out to improve the bearing property of a bush is termed as the
(*a*) plating (*b*) heat treatment
(*c*) impregnation (*d*) infiltration

423. Carbide tipped cutting tools are manufactured by powder metallurgy techniques and have a composition of
(*a*) 90% tungsten carbide & 10% cobalt.
(*b*) 70% aluminium oxide & 30% silica.
(*c*) 30% nickel, 15% chromium & 55% tungsten.
(*d*) 65% tungsten & 35% zirconium.

424. An orthotropic material is a special class of anisotropic material, which is described by their properties in three perpendicular directions. An example of orthotropic material is
(*a*) stainless steel (*b*) cast iron
(*c*) wood (*d*) teflon

425. The temperature at which ferromagnetic material can no longer be magnetised by the outside forces is termed as the
(*a*) critical point.
(*b*) Curie temperature.
(*c*) inversion temperature.
(*d*) eutectic temperature.

426. A polymer is termed as an 'elastomer', if its percentage elongation is more than 100%. An elastomer is termed as 'rubber', if its percentage elongation is more than .... percent.
(*a*) 150 (*b*) 200
(*c*) 300 (*d*) 400

427. The maximum value, which the residual stress in a material can reach is the .... of the material.
(*a*) elastic limit (*b*) plastic limit
(*c*) yield stress (*d*) breaking stress

428. The maximum stress below which a material can withstand an infinite number of cycle of stress is termed as the
(*a*) fatigue strength
(*b*) creep strength
(*c*) resilience
(*d*) endurance limit

429. Wrought iron is never shaped by
(*a*) casting
(*b*) cold working
(*c*) forging
(*d*) welding

Answers

418. (b)	419. (a)	420. (a)	421. (a)	422 (d)	423 (a)
424 (c)	425 (b)	426 (b)	427. (a)	428. (d)	429. (a)

430. Grey cast iron (used for making underground water pipes & manhole covers) as compared to steel has higher
(*a*) ductility (*b*) impact strength
(*c*) machinability (*d*) melting point

431. Mho's scale of hardness, which consists of 10 standard minerals is used for the measurement of .... hardness.
(*a*) scratch (*b*) indentation
(*c*) dynamic (*d*) rebound

432. Which of the following hardness tests does not measure the indentation hardness of metals & alloys ?
(*a*) Vicker's hardness test.
(*b*) Shore sceleroscope test,
(*c*) Brinell hardness test.
(*d*) Rockwell hardness test,

433. Which of the following hardness tests uses the depth of penetration caused by the indentor as the parameter for arriving at the hardness value of a material ?
(*a*) Shore sceleroscope test.
(*b*) Vicker's hardness test.
(*c*) Brinell hardness test.
(*d*) Rockwell hardness test.

434. Which of the following non destructive tests for detection of flaws in metallic components can detect both surface as well as internal defects ?
(*a*) Magnetic particle inspection tests.
(*b*) Liquid penetration inspection test.
(*c*) Radiographic & ultrasonic inspection test.
(*d*) none of these.

435. Match the following forming/fabricating methods for polymeric materials with their characteristics.

List I

(a) A fabrication method in which a heated piston is forced into the shape of a mould cavity by internal pressure.
(b) A forming method for articles by fusing a plastic material in a chamber, and then forcing the mass into a hot mould, where it solidifies.
(*c*) A method of moulding a material, in a confined cavity by applying pressure and usually heat.
(*d*) A method of forming a material by forcing it in a fluid state under pressure through a runner system into the cavity of a closed mould.

List II

I. Transfer moulding
II Blow moulding
III Injection moulding
IV Compression moulding

436. Match the average percentage of metal in their ores.

List I	List II
(*a*) Gold ore	I 50
(*b*) Lead ore	II 30
(*c*) Aluminium ore	III 5
(*d*) Iron ore	IV 0.001

437. Match the crystal lattice structure of various materials.

List I	List II
(*a*) Zinc	I Body centred, cubic (bcc)
(*b*) Fluorspar	II Face centred cubic (fcc)
(*c*) Silver	III Simple cubic
(*d*) α-iron	IV Hexagonal close packed (hcp)

Answers

430. (c) 431. (a) 432. (b) 433. (d) 434. (c)
435. (a). -II (b). -I (c). IV (d). III 436. (a) -IV (b). -III (c). II (d) I 437. (a) -IV (b). -III (c). II (d). I

438. Match the manufacturing techniques used for producing various products.

List I

(*a*) Aluminium can for soft drink

(*b*) Aluminium brake shoe

(*c*) Stainless steel cups

(*d*) Crankshaft

List II

I. Sand casting

II. Pressure die casting

III. Deep drawing

IV. Impact extrusion

439. Silicon addition in plain carbon steel

(*a*) reduces its electrical resistance

(*b*) increases its magnetic permeability

(*c*) increases magnetostriction

(*d*) none of these

440. Tackiness is an adhesive property which means that the adhesive is changing from fluid to highly viscous phase. For good bond to develop, the .... of the adhesive used and the joining materials should be closely equal.

(*a*) impact strength

(*b*) co-efficient of expansion

(*c*) hardness

(*d*) tensile strength

441. The purpose of adding glass network modifiers to glass is to

(*a*) enhance its strength.

(*b*) decrease its viscosity in semi-liquid state.

(*c*) increase its toughness.

(*d*) increase its transparency.

442. Pick out the wrong statement.

(*a*) Curie temperature is required to be higher than the highest operating temperature, if the magnet is not to loose magnetism during operation.

(*b*) The hard magnetic materials used in the medical devices is 'alnico', which is an alloy of aluminium, iron & nickel.

(*c*) Metallic glasses are good ferromagnetic material because of its high permeability & non crystalline structure.

(*d*) Curie temperature of a ferromagnetic material like nickel increases with the addition of iron & cobalt, but decreases with the addition of silicon & gold.

443. A good ferromagnetic material must have low

(*a*) magnetic permeability.

(*b*) electrical resistivity.

(*c*) magnetostriction.

(*d*) none of these.

444. Which of the following alloying element added in aluminium increases its fluidity & strength ?

(*a*) Copper (*b*) Zinc

(*c*) Tin (*d*) Silicon

445. Hydrogen embrittlement occurs in electrolytic tough pitch (ETP.) copper due to Cu_20 reacting with H_2 at 400°C, which can be avoided by

(*a*) keeping its surface temperature always <400°C.

(*b*) keeping its surface always clean.

(*c*) adding phosphorous to ETP copper to form P_2O_5.

(*d*) continuous washing of its surface.

Answers

438. (a) -III (b). -I (c). IV (d). II	439. (b)	440. (b)	441. (b)	442. (b)
443. (c) 444. (d)	445. (c)			

446. Which of the following alloys of copper can be used against wear, fatigue & corrosion that is known for its golden color & strength ?
(*a*) Cartridge brass. (*b*) Aluminium bronze.
(*c*) Beryllium bronze (*d*) Phosphor bronze.

447. An alloy of cobalt, chromium, tungsten & carbon used for cutting metals at high speed & temperature is
(*a*) stellite (*b*) cemented carbide
(*c*) cermet (*d*) alnico

448. .... forbids the use of high melting point metals like tungsten & molybdenum against creep.
(*a*) Low tensile strength
(*b*) Easy oxidisability at elevated temperature
(*c*) Formation of weak unstable structure at elevated tempetature
(*d*) Brittleness at room temperature

449. Creep resistance can be improved by allowing the
(*a*) uniformly dispersed coarse particles precipitation in the metal matrix.
(*b*) directional solidification of alloys.
(*c*) grains to grow coarse.
(*d*) hard particles to precipitate along grain boundaries.

450. Pick out the wrong statment.
(*a*) Low melting point metals have high co-efficient of thermal expansion.
(*b*) Thermal co-efficient of expansion of metals is not related to its melting point.
(*c*) With increase in temperature, the linear thermal expansivity of all refractory oxides decreases steadily except for zirconia & silica.
(*d*) Compressive thermal stresses in metals cause cracks inclined at 45° to the surface.

451. Residual stresses in the welded joint are not reduced, if the welded structure is
(*a*) heated & cooled rapidly.
(*b*) shot penned on the weld & heat affected zone (HAZ).
(*c*) beaten by a hammer along the weld.
(*d*) plastically deformed and then load released.

452. Shell moulding employs a pattern made of
(*a*) plaster of paris
(*b*) wood
(*c*) metal
(*d*) wax

453. 'Transition temperature' of metals is concerned with its .... properties.
(*a*) creep (*b*) fatigue
(*c*) impact (*d*) tensile

454. Metals are good conductors, because their valance band is
(*a*) partially filled.
(*b*) completely filled.
(*c*) completely empty.
(*d*) either '*a*' or '*b*'.

455. .... failure comprises most of the service failures in engineering materials.
(*a*) Fatigue
(*b*) Ductile to brittle transition
(*c*) Creep
(*d*) Impact

456. Tungsten filaments are produced by .... technique.
(*a*) die casting.
(*b*) powder metallurgy.
(*c*) electrodeposition,
(*d*) forging.

Answers

446. (b)	447. (a)	448. (b)	449. (b)	450. (b)	451. (C)
452. (c)	453. (c)	454. (a)	455. (a)	456. (b)	

Section

A

2

Mineral Dressing

1. Shape factor for a cylinder whose length equals its diameter is
   (a) 1.5 (b) 0.5
   (c) 1 (d) 0
2. The ratio of the actual mesh dimension of Taylor series to that of the next smaller screen is
   (a) 2 (b) $\sqrt{2}$
   (c) 1.5 (d) $1/\sqrt{2}$
3. The opening of 200 mesh screen (Taylor series) is
   (a) 0.0074 cm (74 micron)
   (b) 0.0074 mm
   (c) 0.0047 cm
   (d) 0.074 cm
4. The ratio of the area of openings in cne screen (Taylor series) to that of the openings in the next smaller screen is
   (a) 1.5 (b) 1
   (c) $\sqrt{2}$ (d) none of these.
   assumption that all particles in a single
5. Equivalent diameter of a particle is the diameter of the sphere having the same as the actual particle.
   (a) ratio of surface to volume
   (b) ratio of volume to surface
   (c) volume
   (d) surface
6. For coarse reduction of hard solids , use
   (a) impact (b) attrition
   (c) compression (d) cutting
7. Soft & non-abrasive materials can be made into fines by
   (a) attrition (b) compression
   (c) cutting (d) none of these
8. Crushing efficiency is the ratio of the
   (a) surface energy created by crushing to the energy absorbed by the solid.
   (b) energy absorbed by the solid to that fed to the machine.
   (c) energy fed to the machine to the surface energy created by the crushing.
   (d) energy absorbed by the solid to the surface energy created by the crushing.
9. Rittinger's crushing law states that
   (a) work required to form a particle of any size is proportional to the square of the surface to volume ratio of the product.
   (b) work required to form a particle of a particular size is proportional to the square root of the surface to volume ratio of the product.
   (c) work required in crushing is proportional to the new surface created.
   (d) for a given machine and feed, the crushing efficiency is dependent on the sizes of feed & product.

Answers

1. (c)	2. (b)	3. (a)	4. (d)	5. (c)	6. (a)
7. (c)	8. (a)	9. (c)			

10. Bond crushing law
    (*a*) calls for relatively less energy for the smaller product particles than does the Rittinger's law.
    (*b*) is less realistic in estimating the power requirements of commercial crushers.
    (*c*) states that the work required to form particle of any size from very large feed is proportional to the square root of the volume to surface ratio of the product.
    (*d*) states that the work required for the crushing is proportional to the new surface created.

11. Work index is defined as the
    (*a*) gross energy (kWh/ton of feed) needed to reduce very large feed to such a size that 80% of the product passes a 100 micron screen.
    (*b*) energy needed to crush one tonne of feed to 200 microns.
    (*c*) energy (kWh/ton of feed) needed to crush small feed to such a size that 80% of the product passes a 200 mesh screen.
    (*d*) energy needed to crush one ton of feed to 100 microns.

12. The operating speed of a ball mill should be .... the critical speed.
    (*a*) less than (*b*) much more than
    (*c*) at least equal to (*d*) none of these

13. Wet grinding in a revolving mill ...compared to dry grinding.
    (*a*) gives less wear on chamber walls
    (*b*) requires more energy
    (*c*) increases capacity
    (*d*) complicates handling of the product

14. For the preliminary breaking of hard rock, we use a
    (*a*) gyratory crusher
    (*b*) ball mill
    (*c*) tube mill
    (*d*) squirrel-cage disintegrator

15. Fibrous material is broken by a
    (*a*) roll crusher
    (*b*) squirrel-cage disintegrator
    (*c*) ball mill
    (*d*) none of these.

16. As the product becomes finer, the energy required for grinding
    (*a*) decreases
    (*b*) increases
    (*c*) is same as for coarser grinding.
    (*d*) is 1.5 times that for coarser grinding.

17. Increasing the capacity of screen .... the screen effectiveness.
    (*a*) decreases (*b*) increases
    (*c*) does not affect (*d*) none of these.

18. Screen efficiency is
    (*a*) recovery/rejection
    (*b*) recovery
    (*c*) rejection
    (*d*) none of these.

19. Traces of solids are removed from a liquid in a
    (*a*) classifier
    (*b*) clarifier
    (*c*) sparkler filter
    (*d*) rotary vacuum filter.

20. As particle size is reduced
    (*a*) screening becomes progressively more difficult.
    (*b*) screening becomes progressively easier.
    (*c*) capacity & effectiveness of the screen is increased.
    (*d*) none of these.

Answers

10. (a)	11. (a)	12. (a)	13. (c)	14. (a)	15. (b)
16. (b)	17. (a)	18. (d)	19 (b)	20. (a)	

21. A screen is said to be blinded, when the
    (*a*) oversizes are present in undersize fraction.
    (*a*) undersizes are retained in oversize fraction.
    (*c*) screen is plugged with solid particles.
    (*d*) screen capacity is abruptly increased.

22. Size measurement of ultrafine particles can be best expressed in terms of
    (*a*) centimetre
    (*b*) screen size
    (*c*) mircon
    (*d*) surface area per unit mass

23. Ultrafine grinders operate principally by
    (*a*) slow compression
    (*b*) impact
    (*c*) attrition
    (*d*) cutting action.

24. Trommels separate a mixture of particles depending on their
    (*a*) size
    (*b*) density
    (*c*) wettability
    (*d*) electrical & magnetic properties.

25. The energy consumed by a ball mill depends upon
    (*a*) its speed
    (*b*) its ball load
    (*c*) the density of the material being ground
    (*d*) all (*a*), (*b*) & (*c*).

26. Grinding efficiency of a ball mill is of the order of..........percen*t*.
    (*a*) 1—5 (*b*) 40—50
    (*c*) 75—80 (*d*) 90—95

27. Screen capacity is expressed in terms of
    (*a*) tons/hr. (*b*) tons/m^2
    (*c*) both (*a*) & (*b*) (*d*) tons/hr.m^2

28. Which of the following screens has the maximum capacity ?
    (*a*) Grizzlies
    (*b*) Trommels
    (*c*) Shaking screens
    (*d*) Vibrating screens

29. In classification, particles are said to be equal settling, if they have the same terminal velocities in the
    (*a*) different fluids
    (*b*) same fluid
    (*c*) same field of force
    (*d*) both (*b*) & (*c*).

30. Separation of particles of various sizes, shapes & densities by allowing them to settle in a fluid is called
    (*a*) classification (*b*) froth floatation
    (*c*) thickening (*d*) none of these

31. For beneficiation of iron ore, the most commonly used method is
    (*a*) flocculation (*b*) froth floatation
    (*c*) jigging & tabling (*d*) none of these

32. Froth floatations is most suitable for treating
    (*a*) iron ores (*a*) sulphide ores
    (*c*) quartzite (*d*) none of these

33. In case of a revolving mill, wet grinding compared to dry grinding
    (*a*) requires more energy.
    (*b*) has less capacity.
    (*c*) complicates handling & classification of the product.
    (*d*) none of these.

34. Which is the most suitable for transportation of sticky material ?
    (*a*) Apron conveyor
    (*b*) Belt conveyor
    (*c*) Screw conveyor
    (*d*) Pneumatic conveyor.

Answers

21. (c)	22. (d)	23. (c)	24. (a)	25. (d)	26. (a)
27. (d)	28. (d)	29. (d)	30. (a)	31. (c)	32. (a)
33. (d)	34. (c)				

35. Ore concentration by Jigging is based on the .... of the particles.
(*a*) difference in specific gravities
(*b*) wettability
(*c*) shape
(*d*) none of these.

36. For sizing of fine materials, the most suitable equipment is a
(*a*) trommel (*b*) grizzly
(*c*) shaking screen (*d*) vibrating screen.

37. Which of the following cannot be recommended for transportation of abrasive materials ?
(*a*) Belt conveyor (*b*) Apron conveyor
(*c*) Flight conveyor (*d*) Chain conveyor

38. In froth floatation, chemical agent added to cause air adherence is called the
(*a*) collector (*b*) frother
(*c*) modifier (*d*) activator

39. Pine oil used in froth floatation technique acts as a/an
(*a*) collector (*b*) modifier
(*c*) frother (*d*) activator

40. The most efficient equipment for removal of sub-micronic dust particles from blast furnace gas is the
(*a*) venturi atomiser.
(*b*) gravity settling chamber.
(*c*) electro-static precipitator.
(*d*) cyclone separator.

41. Float and sink test determines the possibility of cleaning of coal by a process based on the
(*a*) gravity separation
(*b*) wettability
(*c*) particle shape
(*d*) none of these.

42. Washability curve based on float and sink test enables an assessment to be made of the possibility of cleaning a coal fraction based on the
(*a*) density separation.
(*b*) differential wettability.
(*c*) particle size.
(*d*) volatile matter content.

43. For efficient grinding, ball mills must be operated
(*a*) at a speed less than critical speed.
(*b*) at a speed more than critical speed.
(*c*) at a speed equal to critical speed.
(*d*) with minimum possible small balls.

44. For the transportation of ultrafine particles, the equipment used is the .... conveyor.
(*a*) belt (b) pneumatic
(*c*) screw (*d*) none of these

45. The material is crushed in a gyratory crusher by the action of
(*a*) impact (*b*) compression
(*c*) attrition (*d*) cutting

46. Mesh number indicates the number of holes per
(*a*) square inch (b) linear inch
(*c*) square foot (*d*) linear foot

47. For transporting pasty material, one will use the
(*a*) apron conveyor (*b*) belt conveyor
(*c*) screw conveyor (*d*) bucket elevator.

48. Ball mill is used for the
(*a*) crushing (*b*) coarse grinding
(*c*) fine grinding (*d*) attrition

Answers

35. (a)	36. (d)	37. (c)	38. (a)	39. (c)	40. (c)
41. (a)	42. (a)	43. (a)	44. (b)	45. (b)	46. (b)
47. (c)	48. (c)				

49. The capacity of a belt conveyor depends upon two factors. If one is the cross-section of the load, the other is the .... of the belt.
(a) speed (b) thickness
(c) length (d) none of these.

50. Dry powdery solid materials are transported by a.............conveyor.
(a) belt (6) bucket
(c) screw (d) none of these.

51. Grizzlies are used for separating the........ solids.
(a) coarse (6) fine
(c) any size of (d) abrasive

52. Cyclones are used primarily for separating the
(a) solids (b) solids from fluids
(c) liquids (d) solids from solids.

53. The main differentiation factor between tube mill and ball mill is the
(a) length to diameter ratio.
(b) size of the grinding media.
(c) final product size.
(d) operating speed.

54. Fick's law relates to the
(a) energy consumption
(b) final particles size
(c) feed size
(d) none of these.

55. Apron conveyors are used for the
(a) heavy loads & short runs.
(b) small loads & long runs.
(c) heavy loads & long runs.
(d) small loads & short runs.

56. Sphericity of a non-spherical particles is given by
(a)V/DS (b) DS/V
(c) 6V/DS (d) V/6DS

where, D = equivalent diameter of particles
S = surface area of one particle
V = volume of one particle

57. Sphericity of a cylindrical particles having equal length & diameter is
(a) 1 (b) > 1
(c) < 1 (d) 2

58. Sphericity of cube having its height as equivalent diameter is
(a) $1/\sqrt{3}$ (b) $\sqrt{3}$
(c) 1 (d) < 1

59. Sphericity of crushed coal particles is
(a) 1 (b) > 1
(c) < 1 (b) ∞

60. 200 mesh screen has 200 openings per
(a) inch (b) $inch^2$
(c) cm (d) cm^2

61. The mass flow rate of granular solid through a circular opening varies as the .... of the opening diameter.
(a) square (b) cube
(c) square root (d) fourth power

62. Crushing efficiency of a crusher ranges between......percent.
(a) 0.1—2 (b) 10—15
(c) 15—25 (d) 40—50.

63. The crushing energy required to create new surface is given by the..........law.
(a) Rittinger's (b) Fick's
(c) Fourier's (d) all (a), (b) & (c).

64. Which mode of size reduction is used by jaw crushers ?
(a) Cutting (b) Impact
(c) Attrition (d) Compression.

Answers

49. (a)	50. (c)	51. (a)	52. (b)	53. (a)	54. (a)
55. (a)	56. (c)	57. (a)	58. (c)	59. (c)	60. (a)
61. (b)	62. (a)	63. (a)	64. (d)		

65. Which of the following size reduction equipments can accept feed size > 25 cms ?
(*a*) Tube mill
(*b*) Ball mill
(*c*) Fluid energy mill
(*d*)Jaw crusher

66. Reduction ratio (feed dia/product dia) for fine grinders may be
(*a*) 5 (*b*) 15
(*c*) 50 (*d*) as high as 100

67. Temperature of solid particles during grinding
(*a*) increases
(*b*) decreases
(*c*) remains constant
(*d*) may increase or decrease ; depends on the nature of solid.

68. Large scale sedimentation is carried out in industries in
(*a*) cyclones (*b*) rotary filters
(*c*) thickeners (*d*) sorting classifiers

69. According to Bond crushing law, the work required to form particles of size d_p from very large feed is proportional to the .... of surface to volume ratio of the product.
(*b*) unit power (*b*) square
(*c*) square root (*d*) cube

70. Mechanical efficiency of a crusher is defined as the ratio of the
(*a*) energy supplied to the crusher to the energy created by crushing.
(*b*) surface energy created by crushing to the energy supplied to the crusher.
(*c*) energy absorbed by the solid to the energy supplied to the crusher.
(*d*) surface energy created by crushing to the energy absorbed by the solid.

71. The crushing efficiency of a crusher is the ratio of the
(*a*) energy supplied to the crusher to the surface energy created by crushing.
(*b*) surface energy created by crushing to the energy supplied to the crusher.
(*c*) energy absorbed by the solid to the energy supplied to the crusher.
(*d*) surface energy created by crushing to the energy absorbed by the solid.

72. The specific surface of spherical particles is ....the diameter of the particles.
(*a*) independent of
(*b*) proportional to the square of
(*c*) proportional to
(*d*) inversely proportional to

73. Which crusher is used for crushing coal blend for charging in coke ovens in steel plant ?
(*b*) Gyratory crusher
(*b*) Hammer crusher
(*c*) Jaw crusher
(*d*) Ball mills.

74. Coal middlings in thermal power plant are generally crushed by a
(*a*) hammer crusher
(*b*) ball mill
(*c*) jaw crusher
(*d*) gyratory crusher

75. Pick out the wrong statement.
(*a*) Grizzlies are used for coarse screening of large lumps and are of rugged construction.
(*b*) Trommels used for fairly large particles are rotating screens.
(*c*) Shaking & vibrating screens are used for fine sizing.
(*d*) all (*a*), (*b*) & (*c*).

Answers

65. (d)	66. (d)	67. (a)	68. (c)	69. (c)	70. (c)
71. (d)	72. (d)	73. (b)	74. (b)	75. (d)	

76. Which is a comminuting equipment ?
(*a*) Trommel (*b*) Crusher
(*c*) Classifier (*d*) Filter

77. Sorting classifiers separate particles of differing densities based on the .... method.
(*a*) sink and float
(*b*) differential settling
(*c*) both (*a*) & (*b*)
(*d*) neither (*a*) nor (b).

78. Operating speed of trommels is about... times the critical speed.
(*a*) 0.1 (*b*) 0.45
(*c*) 1.5 (*d*) 0.9

79. The critical speed of a trommel is ...... the diameter of the trommel.
(*a*) proportional to
(*b*) inversely proportional to
(*c*) proportional to the square root of
(*d*) inversely proportional to the square root of

80. The screen effectiveness....with increase in the capacity of the screen.
(*a*) decreases (*b*) increases
(*c*) increases linearly
(*d*) remains unaffected

81. Critical speed of a ball mill is
(*a*) $\frac{1}{2\pi}\sqrt{\frac{g}{R-r}}$ (*b*) $\frac{1}{2\pi}\sqrt{\frac{R-r}{g}}$
(c) $\frac{1}{2}\sqrt{\frac{g}{R-r}}$ (*d*) $\frac{1}{\pi}\sqrt{\frac{g}{R-r}}$

where, R and r are radii of ball mill and balls respectively

82. Angle of nip (2□) of a roll crusher is given by
(*a*) $\tan \square = \frac{dr+dp}{dr+df}$ (*b*) $\cos \square \; \frac{dr+df}{dr+dp}$
(*c*) $\cos \square = \frac{dr+d_p}{dr+df}$ (*d*) $\tan \square \; \frac{dr+df}{dr+dp}$

where d_r, dp and d_f are dia of crushing rolls. gap between the rolls and dia of feed particles respectively.

83. Which of the following is a rotating screen?
(*a*) Grizzly (*b*) Trommel
(*c*) Shaking screen (*d*) None of these

84. Run of mine (ROM) coal is crushed by a...... for use in domestic ovens.
(*a*) jaw crusher (*b*) hammer crusher
(*c*) ball mill (*d*) tube mill

85. During filtration operation, the filtrate encounters the resistance of the
(*a*) filter medium
(*b*) cake
(*c*) channel carrying the slurry to the upstream side of the cake and filtrate away from the filter medium.
(*d*) all (*a*), (*b*) & (*c*).

86. In case of a plate and frame filter press, filtrate flow through the cake follows .... flow.
(*a*) plug (*b*) turbulent
(*b*) laminar (*d*) none of these

87. Which one is a filter aid ?
(*a*) Convas fabric.
(*b*) Diatomaceous earth,
(*c*) Calcined lime.
(*d*) None of these.

88. The filtrate flow rate in a constant pressure filtration
(*a*) continuously increases.
(*b*) continuously decreases,
(*c*) remains constant throughout.
(*d*) may increase or decrease ; depends on the pressure.

Answers

76. (b)	77. (c)	78. (b)	79. (d)	80. (a)	81. (a)
82. (c)	83. (b)	84. (b)	85. (d)	86. (c)	87. (b)
88. (b)					

89. Raw materials are charged in the iron blast furnace using
(*a*) bucket elevator (*b*) skip hoist
(*c*) screw conveyor (*d*) none of these.

90. Screw conveyors are
(*a*) run at very high rpm.
(*b*) suitable for sticky materials.
(*c*) suitable for highly abrasive materials.
(*d*) all (*a*), (*b*) & (*c*).

91. A belt conveyor used for the transportation of materials can
(*a*) run upto 1 km.
(*b*) travel at speed upto 300 metres/minute.
(*c*) handle upto 5000 tons/hr.
(*d*) all (*a*), (*b*) & (*c*).

92. The maximum slope of a belt conveyor can be...........degrees.
(*a*) 15 (*b*) 30
(*c*) 45 (*d*) 60

93. Width and speed of a conveyor belt depends upon the.......of the material.
(*a*) lump size (*b*) bulk density
(*c*) both (*a*) & (*b*) (*d*) neither (*a*) nor (*b*).

94. Bucket elevators are not suitable for vertical lifting of the
(*a*) fine materials (*e.g.* -200 mesh size coal).
(*b*) sticky materials (*e.g.* clay paste).
(*c*) small lumpy materials (*e.g.* grains & sand).
(*d*) free flowing materials.

95. The capacity of a pneumatic conveying system depends upon the
(*a*) bulk density of materials.
(*b*) pressure of the conveying air.
(*c*) diameter of the conveying line.
(*d*) all (*a*), (*b*) & (*c*).

96. Which of the following must be stored in silos and not in open yard?
(*a*) Coke breeze
(*b*) High V.M. bituminous coal
(*c*) Sand
(*d*) B.F.coke

97. Reduction ratio of crushers is the
(*a*) ratio of feed opening to discharge opening.
(*b*) ratio of discharge opening to feed opening.
(*c*) determining factor for minimum dia of the feed & the product.
(*d*) none of these.

98. Grindability of a material does not depend upon its
(*a*) elasticity (*b*) hardness
(*c*) toughness (*d*) size

99. Which of the following gives the work required for size reduction of coal to -200 mesh in ball mill most accurately ?
(*a*) Rittinger's law (*b*) Kick's law
(*c*) Bond law (*d*) none of these

100. 200 mesh seive size corresponds to .... microns.
(*a*) 24 (*b*) 74
(*c*) 154 (*d*) 200

101. In case of grinding in a ball mill,
(*a*) wet grinding achieves a finer product size than dry grinding.
(*b*) its capacity decreases with increasing fineness of the products.
(*c*) grinding cost & power requirement increases with increasing fineness of the products.
(*d*) all (*a*), (*b*) & (*c*).

Answers

89. (b)	90. (b)	91. (d)	92. (b)	93. (c)	94. (b)
95. (d)	96. (b)	97. (a)	98. (d)	99. (a)	100. (b)
101. (d)					

102. Pick out the wrong statement pertaining to the roll crushers.
   (a) Maximum feed size determines the required roll diameter.
   (b) For hard materials's crushing, the reduction ratio should not exceed 4.
   (c) Both the rolls run necessarily at the same speed.
   (d) Reduction ratio and differential roll speed affect production rate & energy consumed per unit of surface product.

103. Horsepower required for a roll crusher is directly proportional to its
   (a) reduction ratio (b) capacity
   (c) both (a) & (b) (d) neither (a) nor (b).

104. In case of a hammer crusher,
   (a) crushing takes place by impact breaking.
   (b) maximum acceptable feed size is 30 cms.
   (c) reduction ratio can be varied by adjusting the distance from cage to hammers.
   (d) all (a), (b) & (c).

105. In case of a ball mill,
   (a) coarse feed requires a larger ball.
   (b) fine feed requires a larger ball.
   (c) operating speed should be more than critical speed.
   (d) none of these.

106. Which of the following is used for primary crushing of very hard lumpy materials ?
   (a) Toothed roll crusher.
   (b) Gyratory crusher.
   (c) Ball mill.
   (d) Tube mill.

107. In a roll crusher, both the rolls
   (a) have the same diameter.
   (b) are rotated towards each other.
   (c) run either at same or different speeds.
   (d) all (a), (b) & (c).

108. Power required to drive a ball mill with a particular ball load is proportional to
   (a) D (6) $1/D$
   (c) $D^{2.5}$ (d) $1/D^{2.5}$
   where, D = diameter of ball mill.

109. In case of a hammer crusher, the final product size depends upon the
   (a) feed rate
   (b) rotor speed
   (c) clearance between hammer & grinding plates
   (d) all (a), (b) & (c).

110. The grinding in hammer crusher takes place due to
   (a) attrition (b) impact
   (c) both (a) & (b)
   (d) neither (a) nor (b).

111. In case of a hammer crusher,
   (a) the feed may be highly abrasive (Mho's scale > 5).
   (b) minimum product size is 3 mm.
   (c) maximum feed size may be 50 mm.
   (d) rotor shaft carrying hammers can be vertical or horizontal.

112. Limestone is normally crushed in a
   (a) roll crusher (b) hammer crusher
   (c) ball mill (d) tube mill

113. Coal is finally pulverised to 200 mesh size for burning in boilers by a
   (a) hammer crusher. (b) ball mill
   (c) roll crusher (d) gyratory crusher

114. Which of the following comes in the category of primary crusher for hard & tough stone ?
   (a) Jaw crusher
   (b) Cone crusher
   (c) Gyratory crusher
   (d) None of these

Answers

102. (c)	103. (c)	104. (d)	105. (a)	106. (b)	107. (d)
108. (c)	109. (d)	110. (c)	111. (d)	112. (b)	113. (b)
114. (a)					

115. Which is a secondary crusher for a hard & tough stone ?
(*a*) Jaw crusher (*b*) Cone crusher
(*b*) Impact crusher (*d*) Toothed roll crusher

116. Filtrate flow rate in case of a rotary drum vacuum filter (in which $R_m << R_c$) is proportional to..........and the cycle time.
(*a*) $\sqrt{\mu}$ (*b*) $1/\sqrt{\mu}$
(*c*) $1/\mu$ (*d*) $1/\mu^2$
where, μ = filtrate viscosity
Rm = filter medium resistance
R_c = cake resistance

117. Out of the following size reduction equipments, the maximum feed size can be accepted by the
(*a*) tube mill (*b*) ball mill
(*c*) jaw crusher (*d*) jet pulveriser

118. Pick out the wrong statement.
(*a*) Hammer crushers operate by impact action.
(*b*) Standard screens have circular opening.
(*c*) With increase in mesh number of screens their diameter in microns decreases.
(*d*) 200 mesh screen has 200 openings per linear cm.

119. Trommels employ..........for screening of materials.
(*a*) fibrous cloth (*b*) woven wire screen
(*c*) punched plate (*d*) none of these

120. Electrical energy consumed by a jaw crusher is not a function of the
(*a*) average feed size
(b) average product size
(c) machine capacity
(d) none of these

121. In a size reduction crushing operation, feed size is 300 to 1500 mm, while the product size is 100 to 300 mm. This is a case of .... crushing.
(*a*) secondary (*b*) fine
(*c*) primary (*d*) ultrafine

122. 'Xanthates' are used in the froth floatation process as a/an
(*a*) conditioner (*b*) frother
(*c*) collector (*d*) activator

123. Screen capacity is not a function of the
(*a*) openings size
(*b*) screening mechanism
(*c*) screening surface
(*d*) atmospheric humidity

124. Gold ore concentration is mostly done using
(*a*) jigging (*b*) tabling
(*c*) froth floatation (*d*) elutriation

125. Activators are those chemicals which help buoying up one mineral in preference to the other in the froth floatation process. Which of the following is an activator?
(*a*) Cresylic acid
(*b*) Copper sulphate
(*c*) Calcium carbonate
(*d*) Sodium carbonate

126. Pick out the wrong statement.
(*a*) Magnetic separation method can be employed to treat both dry & wet ores.
(*b*) Reduction ratio in crushing operation is defined as the ratio of minimum feed size to the maximum product size.
(*c*) Gyratory crusher is used for the coarse crushing.
(*d*) Screens are of stationary, moving and vibratory types.

Answers

115. (b)	116. (b)	117. (c)	118. (d)	119. (c)	120. (d)
121. (c)	122. (c)	123. (d)	124. (b)	125. (b)	126. (b)

127. Trommels are revolving screens, which normally operate in the range of .... rpm.
(a) 1-2 (b) 15-20
(c) 40-50 (d) 60-75

128. Fluid medium used in the classification technique of mineral beneficiation is
(a) air
(b) water
(c) either (a) or (b)
(d) neither (a) nor (b)

129. Pine oil and cresylic acid are used as the in froth floatation process.
(a) frother (b) collector
(c) depressor (d) conditioner

130. Which of the following parts of a jaw crusher is subjected to maximum wear and tear during its operation ?
(a) Check plates (b) Jaw plates
(c) Toggles (d) Crush shaft

131. The constants (K_B, K_R & K_K) used in the laws of crushing (i.e. Bond's law, Rittinger's law & Kick's law) depend upon the
(a) feed material
(b) type of crushing machine
(c) both (a) & (b)
(d) neither (a) nor (b)

132. General mechanism of size reduction in intermediate and fine grinder is by
(a) cutting action
(b) compression
(c) compression & tearing
(d) impact & attrition

133. Length/diameter ratio of a ball mill is
(a) 1.5 (b) 1
(c) <1 (d) > 1

134. The crushed material received for separation is called feed or
(a) tailing (b) heading
(c) concentrate (d) middling

135. Screen capacity is proportional to
(a) S (b) 1/S
(c) S^2 (d) $\sqrt{S}$
where, S = screen aperture

136. Size reduction of asbestos and mica is done by
(a) hammer mills
(b) rod mills
(c) gyratory crushers
(d) crushing rolls

137. Tube mill compared to bail mill
(a) produces finer products.
(b) is larger in comparison with its diameter.
(c) uses smaller balls.
(d) all (a), (b) & (c).

138. Use of grinding aids results in
(a) enhanced production rate
(b) finer products
(c) both (a) & (b)
(d) neither (a) nor (b)

139. Size reduction of..........can be suitably done by ball mills, crushing rolls and rod mills.
(a) metailiferrous ores
(b) non-metallic ores
(c) basic slags
(d) asbestos and mica

140. ....mill fall in the category of tumbling mills.
(a) Ball and pebble
(b) Rod and tube
(c) Compartement
(d) all (a), (b) & (c)

Answers

127. (b)	128. (c)	129. (a)	130. (b)	131. (c)	132. (d)
133. (c)	134. (b)	135. (a)	136. (a)	137. (d)	138. (c)
139. (a)	140. (d)				

141. Toothed roll crushers achieve size reduction by
(*a*) tearing (shear) & compression
(*b*) impact & attrition
(*c*) both (*a*) & (*b*)
(*d*) neither (*a*) nor (*b*)

142. Colloid mills achieve size reduction mainly by
(*a*) impact (*b*) attrition
(*c*) cutting (*d*) compression

143. Ball mills and tube mills with flint or porcelain balls are used for the size reduction of
(*a*) asbestos
(*b*) rubber
(*c*) non-metallic ores
(*d*) limestone

144. Energy consumption in a crusher decreases with increase in the
(*a*) size of product (at constant feed size).
(*b*) capacity of the crushing machine.
(*c*) size of feed (at constant reduction ratio).
(*d*) all (*a*), (*b*) & (*c*).

145. Which of the following terminology is not used for the size reduction of materials to fine sizes or powders?
(*a*) Comminution (*b*) Dispersion
(*c*) Pulverisation (*d*) Compression

146. Size reduction does not occur due to compression in case of
(*a*) rod mills
(*b*) gyratory crushers
(*c*) jaw crushers
(*d*) smooth roll crushers

147. Basic slag is not ground in
(*a*) jaw crushers
(*b*) ball mills
(*c*) compartment mills
(*d*) tube mills

148. Specific surface area is the surface area of a unit...........of material.
(*a*) weight
(*b*) volume
(*c*) either (*a*) or (*b*)
(*d*) neither (*a*) nor (*b*)

149. What is the reduction ratio in a fine crushing operation having following feed & product sizes?

Parameters	*Unit*	*Maximum*	*Minimum*
Feed size	mm	20	10
Product size	mm	10	5

(*a*) 0.5 (*b*) 2
(*c*) 5 (*d*) 10

150. The term 'angle of nip' is concerned with the operation of..................crushers.
(*a*) jaw (*b*) roll
(*c*) gyratory (*d*) none of these

151. Which of the following achieves the least reduction ratio for a given feed size?
(*a*) Jaw crusher
(*b*) Roll crusher
(*c*) Cone crusher
(*d*) Gyratory crusher

152. Two particles are called to be equal settling, if they are of the same
(*a*) size
(*b*) specific gravity
(*c*) terminal velocities in the same fluid and in the same field of force.
(*d*) none of these.

Answers

141. (c)	142. (b)	143. (c)	144. (d)	145. (d)	146. (a)
147. (a)	148. (c)	149. (b)	150. (b)	151. (b)	152. (c)

153. .... is defined as the geometric mean of the relative rejections and the relative recoveries of two minerals.
    (a) Separation efficiency
    (b) Selectivity index
    (c) Concentration ratio
    (d) none of these

154. Which of the following minerals is not subjected to magnetic separation method?
    (a) Rutile (b) Galena
    (c) Chromite (d) Siderite

155. The study on washability of coal is done by using........................technique.
    (a) tabling
    (b) elutriation
    (c) heavy media separation
    (d) none of these

156. During size reduction by a jaw crusher, the energy consumed decreases with the
    (a) decreasing size of product at constant size of feed.
    (b) decreasing machine capacity.
    (c) increasing size of feed at constant reduction ratio.
    (d) none of these

157. Pick out the correct statement.
    (a) Removal of iron from ceramic material is necessitated (by magnetic separation method) so as to avoid discolouration of ceramic products.
    (b) The operating cost of shaking screen is more than that of a vibrating screen.
    (c) Screen capacity does not depend upon the specific gravity of the minerals.
    (d) Asphalt is best crushed using toothed roll crusher.

158. Sauter mean diameter is same as the .... mean diameter.
    (a) mass (b) arithmetic
    (c) volume-surface (d) geometric

159. Gummy & sticky materials like molasses, sugar etc. are best transported/handled by using a.............conveyor.
    (a) drag (b) ribbon
    (c) screw (d) slat

160. Hot, lumpy and abrasive materials are best transported by using a/an..........conveyor.
    (a) apron (b) belt
    (c) screw (d) flight

161. Which of the following is a coarse crusher?
    (a) Smooth roll crusher
    (b) Toothed roll crusher
    (c) Gyratory crusher
    (d) Tube mill

162. .... mills are termed as impactors.
    (a) Hammer
    (b) Cage
    (c) Rolling-compression
    (d) None of these

163. Which of the following is not a revolving/tumbling mill for size reduction?
    (a) Compartment mill
    (b) Pebble mill
    (c) Cage mill
    (d) Rod mill

164. Stamp mills are generally used for crushing
    (a) iron ores
    (b) gold ores
    (c) talc
    (d) diamond

Answers

153. (b)	154. (b)	155. (c)	156. (c)	157. (a)	158. (c)
159. (b)	160. (a)	161. (c)	162. (a)	163. (c)	164. (b)

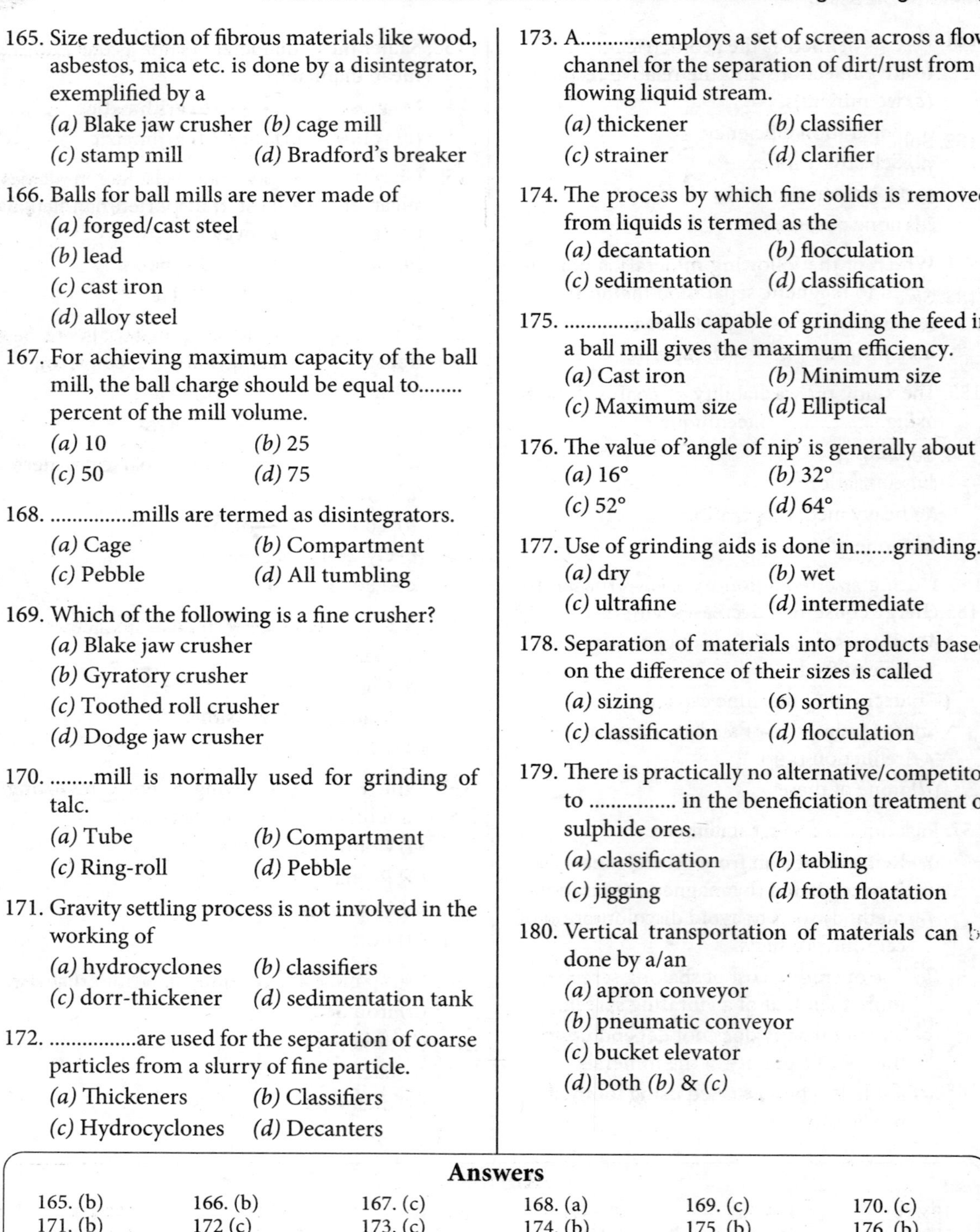

165. Size reduction of fibrous materials like wood, asbestos, mica etc. is done by a disintegrator, exemplified by a
    (*a*) Blake jaw crusher (*b*) cage mill
    (*c*) stamp mill (*d*) Bradford's breaker

166. Balls for ball mills are never made of
    (*a*) forged/cast steel
    (*b*) lead
    (*c*) cast iron
    (*d*) alloy steel

167. For achieving maximum capacity of the ball mill, the ball charge should be equal to........ percent of the mill volume.
    (*a*) 10 (*b*) 25
    (*c*) 50 (*d*) 75

168. ....mills are termed as disintegrators.
    (*a*) Cage (*b*) Compartment
    (*c*) Pebble (*d*) All tumbling

169. Which of the following is a fine crusher?
    (*a*) Blake jaw crusher
    (*b*) Gyratory crusher
    (*c*) Toothed roll crusher
    (*d*) Dodge jaw crusher

170. ....mill is normally used for grinding of talc.
    (*a*) Tube (*b*) Compartment
    (*c*) Ring-roll (*d*) Pebble

171. Gravity settling process is not involved in the working of
    (*a*) hydrocyclones (*b*) classifiers
    (*c*) dorr-thickener (*d*) sedimentation tank

172. ....are used for the separation of coarse particles from a slurry of fine particle.
    (*a*) Thickeners (*b*) Classifiers
    (*c*) Hydrocyclones (*d*) Decanters

173. A.............employs a set of screen across a flow channel for the separation of dirt/rust from a flowing liquid stream.
    (*a*) thickener (*b*) classifier
    (*c*) strainer (*d*) clarifier

174. The process by which fine solids is removed from liquids is termed as the
    (*a*) decantation (*b*) flocculation
    (*c*) sedimentation (*d*) classification

175. ....balls capable of grinding the feed in a ball mill gives the maximum efficiency.
    (*a*) Cast iron (*b*) Minimum size
    (*c*) Maximum size (*d*) Elliptical

176. The value of'angle of nip' is generally about
    (*a*) 16° (*b*) 32°
    (*c*) 52° (*d*) 64°

177. Use of grinding aids is done in.......grinding.
    (*a*) dry (*b*) wet
    (*c*) ultrafine (*d*) intermediate

178. Separation of materials into products based on the difference of their sizes is called
    (*a*) sizing (6) sorting
    (*c*) classification (*d*) flocculation

179. There is practically no alternative/competitor to .... in the beneficiation treatment of sulphide ores.
    (*a*) classification (*b*) tabling
    (*c*) jigging (*d*) froth floatation

180. Vertical transportation of materials can be done by a/an
    (*a*) apron conveyor
    (*b*) pneumatic conveyor
    (*c*) bucket elevator
    (*d*) both (*b*) & (*c*)

Answers

165. (b)	166. (b)	167. (c)	168. (a)	169. (c)	170. (c)
171. (b)	172 (c)	173. (c)	174. (b)	175. (b)	176. (b)
177. (a)	178. (a)	179. (d)	180. (d)		

181. ....conveyors are also called scrappers.
(a) Apron *(b)* Screw
(c) Helical flight *(d)* both *(b)* & *(c)*

182. Solid particles separation based on the difference in their flow velocities through fluids is termed as the
(a) clarification *(b)* classification
(c) elutriation *(d)* sedimentation

183. Sizing of very fine particles of the order of 5 to 10 microns is done by elutriation, which is a...............operation.
(a) clarification (b) sedimentation
(c) flocculation *(d)* classification

184. Separation of materials of the same density based on their sizes by using their different rates of flow is called
(a) sorting *(b)* sizing
(c) flocculation *(d)* elutriation

185. The equipment which helps in the removal of traces of solids from a liquid is called a
(a) classifier
(b) clarifier
(c) sedimentation tank
(d) flocculator

186. ....conveyor is the most suitable for long distance transportation of cold, non- abrasive granular/irregular shape/fine materials.
(a) Bucket *(b)* Belt
(c) Screw *(d)* Apron

187. A widely used size reduction equipment for.........is the Bradford breakers.
(a) talc *(b)* coal
(c) iron-ore *(d)* wheat

188. ....baffles are provided in ball mills.
(a) Horizontal *(b)* No
(c) Only two *(d)* Vertical

189. Separation of solid particles based on their densities is called
(a) sizing *(b)* sorting
(c) clarification *(d)* dispersion

190. Handling of ashes and similar materials can be done best by a .... conveyor.
(a) flight (6) drag or slat
(c) belt *(d)* ribbon

191. For spheres, the surface shape factor is given by
(a) $\pi(=A/D^2)$ *(b)* $\pi/6(= V/D^3)$
(c) AD/V *(d)* V/AD
where, A = area, V = volume and D - diameter

192. Pick out the wrong statement.
(a) Recycled coarse material to the grinder by a classifier is termed as the circulating load.
(b) Wear and tear in wet crushing is more than that in dry crushing of materials.
(c) Size enlargement (opposite of size reduction) is not a mechanical operation.
(d) A 'dust catcher' is simply an enlargement in a pipeline which permits the solids to settle down due to reduction in velocity of the dust laden gas.

193. If a force greater than that of gravity is used to separate solids and fluids of different densities, the process is termed as the
(a) sedimentation *(b)* flocculation
(c) dispersion *(d)* centrifugation

Answers

181. (d)	182. (b)	183. (d)	184. (b)	185. (b)	186. (b)
187. (b)	188. (a)	189. (b)	190. (b)	191. (a)	192. (c)
193. (d)					

194. The process opposite to 'dispersion' is termed as the
(a) flocculation (b) sedimentation
(c) filtration (d) clarification

195. For spheres, volume shape factor is given by
(a) ϖ (= A/D²) (b) 2ϖ(=2A/D²)
(a) ϖ/6 (=- V/D³) (d) AD/V

196. Size reduction of ice and gypsum can be accomplished suitably by a ....crusher.
(a) Blake jaw (b) toothed roll
(c) gyratory (d) none of these

197. For spheres, the specific surface shape factor is given by
(a) AD/V (b) D/V
(c) A/V (d) D/V²

198. In coal washeries, three products namely the valuable product (*i.e* clean/washed coal), discarded product (*i.e.*, mineral matter) and an additional concentrated product called the .... is produced.
(a) concentrate (b) tailing
(c) middling (d) none of these

199. Separation of solid suspended in liquid into a supernatant clear liquid and a denser slurry employs a process termed as the
(a) coagulation (b) flocculation
(c) sedimentation (d) clarification

200. Which of the following grinding mills has the horizontally arranged rods as the grinding elements, thereby delivering more uniform granular product with minimum fines ?
(a) Compartment mill (b) Rod mill
(c) Pebble mill (d) Tube mill

201. In ball mill operation, the feed size (D_f in metres) and the ball diameter (D_b in metres) are related as
(a) $D_b^2 = K.D_f^2$ (b) $D_b = K.D_f$
(c) $D_b^3 = K.D_f$ (d) $D_b^2 = K.D_f^2$
where, K = grindability constant (varying from 0.9 to 1.4 in increasing order of hardness)

202. The optimum moisture content in solids to be crushed/ground ranges from ...... percent.
(a) 3 to 4 (6) 8 to 10
(c) 10 to 15 (d) 15 to 20

203. Pick out the wrong statement.
(a) Gape is the greatest distance between the crushing surfaces or the jaws.
(b) The angle of nip (2α) is the angle between roll faces at the level where they will just take hold of a particle and draw it in the crushing zone.
(c) Crushing efficiency is the ratio of the energy absorbed by the solid to the surface energy created by crushing.
(d) Reduction ratio is the ratio of the maximum size of the particles in the feed to that in the product.

204. Critical speed of rotation, N (in rps-rotation per second) of a trommel is equal to
(a) $\frac{1}{2\pi}\sqrt{\frac{g}{r}}$ (b) $\frac{1}{\pi}\sqrt{\frac{g}{r}}$
(c) $\frac{1}{2}\sqrt{\frac{g}{r}}$ (d) $2\pi\sqrt{\frac{g}{r}}$

where, g = acceleration due to gravity = 9.81 m/sec² and, r = radius of trommel, metre

205. Energy consumed for crushing one ton of material ranges from .... kWh.
(a) 0.01 to 0.1 (b) 0.5 to 1.5
(c) 2 to 3.5 (d) 4 to 5

Answers

194. (a)	195. (c)	196. (b)	197. (a)	198. (c)	199. (c)
200. (b)	201. (a)	202. (a)	203. (b)	204. (a)	205. (b)

206. Capacity (in tons/hr) of jaw/gyratory crusher is equal to
(a) 0.01 ***L.S*** (b) 0.087 ***L.S***
(c) ***L.S*** (d) ***L.S***/0.087
where, L = length of the receiving opening, cm
S = greater width of the discharge opening, cm

207. .... mill is not a revolving mill.
(a) Pebble (b) Compartment
(c) Cage (d) Tube

208. A .... mill is a revolving mill divided into two or more sections by perforated partitions, in which preliminary grinding takes place at one end and the finishing grinding at the discharge end.
(a) compartment (b) tube
(c) rod (d) pebble

209. In closed circuit grinding as compared to open circuit grinding, the
(a) specific surface of product is more.
(b) product has lesser size uniformity.
(c) production rate at a given limiting size is lower.
(d) operation is economical.

210. Capacity of flight conveyor in tons/hr is given by
(a) 3.6 *W.D.V.p* (b) 3.6 *W.D.V*
(c) 3.6WVp (d) 3 . 6 D V p
where, *W* & *D* = width & depth of flight respectively in metre
V = speed of the conveyor, metre/second
ρ = bulk density of material, kg/m^3

211. A cottrel precipitator makes use of the .... for dusty air cleaning.
(a) electric spark
(b) corrona discharge
(c) alternating current
(d) low voltage

212. Gyratory crushers compared to the reciprocating jaw crushers
(a) have greater capacity per unit of discharge area.
(b) crush intermittently.
(c) are less suitable for coarse materials.
(d) have less steady power consumption.

213. Kick's law assumes that the energy required for size reduction is proportional to logarithm of the ratio between the initial & the final diameters. The unit of Kick's constant is
(a) kW.sec./kg (b) kWh/kg
(c) kWh/sec.kg (d) kg/sec

214. Theoretical capacity of crushing rolls in tons/hr is given by
(a) 3.6 V.W.Dr.p (b) 3.6V.W.p
(c) 3.6 W.Dr.p (d) 3.6 V.W.Dr/p
where, V = peripherial velocity, m/sec. W = width of rolls, m
D_r = distance between rolls
ρ = density of material to be crushed,kg/m^3
here, V = *n N.D* where,
N = speed of rolls in rotation per second
D = diameter of rolls, m

215. Vibrating screens are used for handling large tonnages of materials. The vibrating motion is imparted to the screening surface by means of
(a) electromagnets.
(b) cams or eccentric shafts.
(c) unbalanced flywheels.
(d) either 'a', 'b' or 'c'

Answers

206. (b)	207. (c)	208. (a)	209. (d)	210. (a)	211. (b)
212. (a)	213. (a)	214. (a)	215. (d)		

216. Pick out the correct statement.
 (*a*) The capacity and the effectiveness of a screen are the same.
 (*b*) The capacity and the effectiveness of a screen are opposing factors.
 (*c*) The screening surface of a 'reel' (a revolving screen used in flour mills) is made of silk bolting cloth supported by wire mesh.
 (*d*) both 'b' & 'c'

217. Wet seiving is employed, when the product contains .... materials.
 (*a*) abrasive
 (*b*) large quantity of very fine
 (*c*) coarse
 (*d*) non-sticky

218. The capacity of a classifier in 'tons of solid/hr' is given by
 (*a*) 3.6 AVS.p (*b*) 3.6A.V.p
 (*c*) 3.6 A.V.p (*d*) 3.6 AVS/p
 where, A = cross-sectional area, m^2
 V = rising velocity of fluid, m/sec
 S = percentage of solids in the suspension by volume
 □ = density of solids, kg/m^3

219. Which of the following is a batch sedimentation equipment ?
 (*a*) Dust catcher
 (*b*) Filter thickener
 (c) Dry cyclone separator
 (*d*) Rotary sprayer scrubber

220. Pick out the correct statement.
 (*a*) Angle of repose is always greater than the angle of slide.
 (b) A hopper is a small bin with a sloping bottom.
 (*c*) A silo is a short height vessel of very large diameter used for the storage of high volatile matter coal.
 (*d*) Pine oil is used as a 'modifying agent' (for activating or depressing the adsorption of filming agents) in froth floatation process.

221. Agglomeration of individual particles into clusters (floes.) is called flocculation. To prevent flocculation, the most commonly used dispersing agents are
 (*a*) carbonates
 (*b*) sulphates
 (*c*) silicates & phosphates
 (*d*) bi-carbonates

222. .... mean diameter of particles is given by sigma xi.Dpi .
 (*a*) Mass *b*) Volume
 (*c*) Arithmetic (*d*) Volume surface

223. Which of the following equations is Rittinger's crushing law ?
 (*a*) $P/m = K/\sqrt{D_P}$
 (*b*) $P/m = K.\ln(D_{sa}\sqrt{D_{sb}})$
 (*c*) $P/m = K'\left(\dfrac{1}{D_{sb}} - \dfrac{1}{D_{sb}}\right)$
 (*d*) none of these
 where, P = power required by the machine,
 m = feed rate, K = a constant
 $\overline{D_{sa}}$ & $\overline{D_{sb}}$ = volume surface mean diameter of feed & product respectively

224. General crushing equation is given by

$$d\left(\frac{p}{m}\right) = -K\frac{d\overline{D_s}}{D_s^n}$$

Bond's crushing law is obtained by solving this equation, for n = ... and feed of infinite size.
 (*a*) 1 (*b*) 1.5
 (*c*) 2 (*d*) 2.5

Answers

216. (d)	217. (b)	218. (a)	219. (b)	220. (b)	221. (c)
222. (b)	223. (c)	224. (b)			

225. Pick out the correct statement.
   (a) Plastic chips are called non-cohesive solids.
   (b) Kick's crushing law is,
   $\frac{P}{m} = K. \text{In} (\overline{D_{sa}} / \overline{D_b}$
   (c) Comminution is a generic term for the size enlargement operation.
   (d) Energy required in kWh per ton of product such that 80% of it passes through a 200 mesh screen is called Work index'.

226. Pick out the wrong statement.
   (a) More commonly used jaw crusher between Dodge jaw crusher and Blake jaw crusher is the later one.
   (b) There are only four methods namely compression, impact, attrition & cutting which the size reduction equipments employ.
   (c) Cutting machines mainly employ 'attrition' for size reduction of solids.
   (d) Operating principles of Dodge & Blake jaw crushers are combined in the working of universal jaw crushers.

227. Tabling process used for separating two materials of different densities by passing the dilute pulp over a table/deck, which is inclined from the horizontal surface at an angle of about
   (a) 1 to 2° (b) 2 to 5°
   (c) 5 to 10° (d) 10 to 15°

228. Grinding characteristic of a material is given by its
   (a) Hardgrove grindability index (HGI)
   (b) angle of repose
   (c) shatter index
   (d) abrasion index.

229. Which of the following relationships between co-efficient of friction between rock & roll and α (half of the angle of nip) of the particle to be crushed is correct ?
   (a) $\mu > \tan \alpha$ (b) $\mu \geq \tan \alpha$
   (c) $\mu > \tan 2\alpha$ (d) $\mu \leq \tan \alpha$

230. The mechanism of size reduction by a hammer mill is by impact & attrition between the
   (a) grinding element & housing.
   (b) feed particles.
   (c) both (a) & (b).
   (d) neither (a) nor (b).

231. Actual operating speed of a ball mill may vary from 65 to 80% of the critical speed. Which of the following duties would require the ball mill to be operated at maximum percentage of critical speed ?
   (a) Wet grinding in low viscous suspension.
   (b) Wet grinding in high viscous suspension.
   (c) Dry grinding of large particles (upto 1.25 cms).
   (d) Dry grinding of large particles in unbaffed mills.

232. A pebble mill
   (a) is a ball mill.
   (b) employs flints of ceramic pebbles as the grinding medium.
   (c) is a tube mill lined with ceramic or other non-metallic liner.
   (d) both (b) & (c).

233. Number of particles in a crushed solid sample is given by
   (a) m/p . Vp (b) m . p/Vp
   (c) m . Vp/p (d) Vp/m . p
   where, m = mass of particles in a sample, V_p = volume of one particle and p = density of particles.

Answers

225. (b)	226. (c)	227. (b)	228. (a)	229. (b)	230. (c)
231. (d)	232. (d)	233. (a)			

234. Introduction of slurry in a plate and frame filter press is done through a plate in each frame. The plate of this filter has a ....surface.
(a) plane (6) curved
(c) ribbed (d) either (a) or (b)

235. Which of the following is the most suitable filter for separation of abrasive solids suspended in a corrosive liquid ?
(a) Sand bed filter
(b) Plate and frame filter press
(c) Vacuum filter
(d) Batch basket centrifuge.

236. rpm of a trommel at critical speed is given by
(a) 76.65 / D (b) 76.65/√D
(c) 76.75 /D^2 (d) 76.75 √D
where, D = Diameter of trommel in ft.

237. Which of the following crushing laws is most accurately applicable to the fine grinding of materials ?
(a) Bond's crushing law
(b) Kick's law
(c) Rittinger's law
(d) None of these

238. Rittinger number, which designates the new surface produced per unit of mechanical energy absorbed by the material being crushed, depends on the
(a) state or manner of application of the crushing force.
(b) ultimate strength of the material.
(c) elastic constant of the material.
(d) all (a), (b) & (c).

239. Match the following pulverisers as per their characteristics.
List I
(a) It combines the action of hammer & attrition mills and is used for grinding plastic materials liable to be softened under warm mill conditions.
(b) It is a tumbling mill comprising of a cylinder divided into two or more section by a perforated partition, in which preliminary grinding takes place at one end and final grinding at the discharge end.
(c) It is an attrition mill with hard circular emery rock serving as grinding medium used for grinding special cereals and grains.
(d) This mill, in which the journals carrying the grinding rollers are stationary while the grinding ring rotates, is used for pulverising coal for boiler firing.
List II
I. Buhrstone mill
II. Bowl mill
III. Compartment mill
IV. Turbo pulveriser

240. Match the following conveying equipments with their field of application.
List I
(a) Skip hoist
(b) Bucket elevator
(c) Pneumatic conveyor
(d) Screw conveyor with ribbon flight
List II
I. Movement of blast furnace charge.
II. Handling of sticky materials.
III. Vertical movement of free flowing pulverised coal.
IV. Domestic dust cleaning by vacuum cleaner.

Answers

234. (c) 235. (c) 236. (b) 237. (c) 238. (d) 239. a-IV, b-III, c-I, d-II
240. a-I, b-III, c-IV, d-II

241. Match the type of size reduction equipments with their examples.

List I

(*a*) Shredder

(*b*) Jaw crusher

(*c*) Heavy duty impact mill

(*d*) Peripheral speed mill

List II

I. Hammer mill

II. Cage disintegrator

III. Buhrstone

IV. Dodge crusher

242. Match the proper size reduction equipment used for various applications.

List I

(*a*) size reduction of ice

(*b*) Pulverisation of coal midding

(*c*) Grinding of talc

(*d*) Crushing of gold ore.

List II

I. Stamp mill

II. Toothed roll crusher

III. Ball mill

IV. Ring roll mill

243. Match the typical transportation applications of various conveyors.

List I

(*a*) Transportation of sticky, pasty and dry powdery solid materials.

(*b*) Not fit for transportation of abrasive materials.

(*c*) Transportation of ultrafine materials.

(*d*) Not fit for vertical lifting of sticky materials.

List II

I. Flight conveyor

II. Screw conveyor

III. Bucket elevator

IV. Pneumatic conveyor

244. Match the crushing action involved in the operation of various size reduction equipments.

List I	*List II*
(*a*) Gyratory crusher	I. Cutting action
(*b*) Colloid mill	II. Compression
(*c*) Hammer crusher	III. Attrition
(*d*) Dicer or slitters ...	IV. Both attrition & impact

245. Match the crushers used for various applications.

List I

(*a*) Size reduction of fibrous materials like asbestos, mica etc.

(*b*) Primary crushing of hard and tough stone.

(*c*) Limestone crushing.

(*d*) Pulverisation of carbon black.

List I

I. Ball mill II. Cage mill

III. Jaw crusher IV. Hammer crusher

246. Match the techniques/principle employed in various separation processes.

List I

(*a*) Sorting (*b*) Sizing

(*c*) Classification (*d*) Clarification

List II

I. Separation of materials of same density based on their sizes by using their different rates of flow.

II. Removal of traces of solid from a liquid.

III. Separation of solid particles based on their densities.

IV. Separation of solid particles based on the difference in their flow velocities through fluids.

Answers

241. a-II, b-IV, c-I, d-III	242. a-II, b-III, c-IV, d-I	243. a-II, b-I, c-IV, d-III
244. a-II, b-III, c-IV, d-I	245. a-II, b-III, c-IV, d-I	246. a-III, b-I, c-IV, d-II

247. Match the symbols of various equipments used in mineral dressing.

List I

(a) Rake classifier *(b)* Vibrating screen

(c) Shaking table *(d)* Trommel

List II

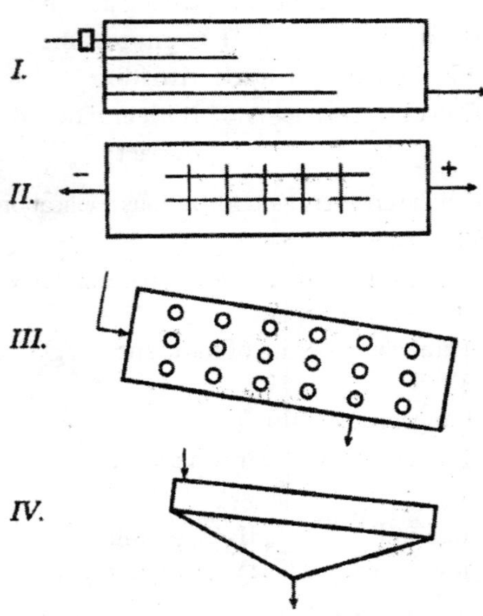

248. Match the ore treatment process suitable for ores having various principal constituents.

List I

(a) Copper sulphides *(b)* Clay

(c) Bituminous coal *(d)* Zinc carbonate or silicate

List II

I. Gravity concentration.

II. Heavy fluid separation, jigging, tabling.

III. Selective froth floatation.

IV. Dispersion and classification.

249. Match the various additive chemicals used in froth floatation process.

List I

(a) Modifiers *(b)* Frothers

(c) Activators *(d)* Collectors

List II

I. Pine oil, fusel oil and crcsylic acid.

II. Sulphuric acid, lime and sodium cyanide.

III. Potassium ethyl xanthate, oleic acid and palmitic acid.

IV. Copper sulphate.

250. Which of the following is not an industrial screening equipment ?

(a) Sharples centrifuge

(b) Vibrating screen

(c) Grizzly

(d) Trommel

251. The most suitable equipment for the transportation of 200 mesh size particles is a

(a) bucket elevator

(b) pneumatic conveyor

(c) screw conveyor

(d) belt conveyor

252. The resistance offered by the filter used in a bag filter is proportional to

(a) c/s *(b)* s/c

(c) s.c *(d)* 1/s.c

where, c = dust concentration,

and s = particle size

253. Which of the following crushers can be considered as a combination of a jaw crusher and a roller crusher ?

(a) Rod mill

(b) Fluid energy mill

(c) Gyratory crusher

(d) Ball mill

Answers

247. a-II, b-IV, c-I, d-III
248. a-III, b-IV, c-II, d-I
249. a-II, b-IV, c-I, d-III
250. (a)
251. (b)
252. (a)
253. (c)

254. What is the usual value of angle of nip for crushing of ordinary rock in smooth steel crushing rolls ?
(*a*) 16° (*b*) 32°
(*c*) 10° (*d*) 46°

255. Higher is the mesh number, smaller will be the aperture size of the screen. It means that the aperture size of a 200 mesh screen will be smaller than that of 20 mesh screen. This is valid for the .... standard screens.
(*a*) British
(*b*) German (DIN .1171).
(*c*) American (ASTM & Tylor).
(*d*) all (*a*), (*b*) & (*c*).

256. For Indian standard (IS) screens, the mesh number is equal to its aperture size expressed to the nearest deca-micron (0.01 mm). Aperture width of IS screen of mesh number 50 will be approximately .... microns.
(*a*) 5 (*b*) 50
(*c*) 500 (*d*) 5000

257. Rittinger's number designates the new surface created per unit mechanical energy absorbed by the material being crushed. Larger value of Rittinger's number of a material indicates its
(*a*) easier grindability.
(*b*) poor grindability.
(*c*) high power consumption in grinding.
(*d*) none of these.

258. Angle of nip of the crushing rolls does not depend upon the
(*a*) diameter of the rolls
(*b*) speed of the rolls
(*c*) product size
(*d*) feed size

259. In bag filters, filter fabrics are never made of the
(*a*) metallic wire woven mesh
(*b*) polyester fibres
(*c*) cotton fibres
(*d*) nylon fibres

260. In the cyclone separator used for the separation of dust from dust laden gas, the gas
(*a*) enters the cyclone from the top.
(*b*) is admitted tangentially at high velocity.
(*c*) develops a helical motion inside the chamber.
(*d*) both (*b*) & (*c*).

261. Dust collection efficiency of an electrostatic precipitator can be as high as 99.9%. Maximum temperature & pressure of dust laden gas that can be cleaned in an electrostatic precipitator is respectively.
(*a*) 200°C & 5 atm
(*b*) 1000°C & 10 atm
(*c*) 500°C & 50 atm
(*d*) 1000°C & 500 atm

262. Jigging is a technique by which different particles can be separated based on the particle
(*a*) size (*b*) density
(*c*) shape (*d*) wettability

263. The energy required per unit mass to grind limestone particles of very large size to 100 micro metre is 12.7 kWh/ton. An estimated (using Bond's law) consumption of the energy to grind the particles from a very large size to 50 micro metre is .... kWh/ton.
(*a*) 6.35 (*b*) 9.0
(*c*) 18 (*d*) 25.4

Answers

254. (b)	255. (d)	256. (c)	257. (a)	258. (b)	259. (a)
260. (d)	261. (b)	262. (a)	263. (c)		

264. Ball mill is used for
(a) crushing *(b)* coarse grinding
(c) fine grinding *(d)* attrition

265. Which of the following crushers can be considered as a combination of a jaw crusher and a roller crusher ?
(a) Rod mill *(b)* Fluid energy mill
(c) Gyratory crusher *(d)* Ball mill

266. Pulverised coal passing through a 200 mesh screen has a diameter of 0.074 mm. (74 microns). The same passing through a 50 mesh screen will have a diameter of ......mm.
(a) 0.0074 *(b)* 0.30
(c) 0.05 *(d)* 0.014

267. In a ball mill, the volume occupied by the balls (when the mill is stopped) is about .... percent of the volume of the mill.
(a) 35 *(b)* 50
(c) 70 *(d)* 85

268. The work index in Bond's law for crushing of the solids has the following dimensions.
(a) No units (dimensionless)
(b) kWh/ton
(c) kW/ton
(d) kWh.$m^{1/2}$/ton

269. The sphericity of a solid particle of cubical shape is
(a) ϖ *(b)* $(ϖ /6)^{1/3}$
(c) $(ϖ /6)^{1/2}$ *(d)* ϖ /3

270. Sphericity is the ratio of the surface area of a spherical particle having the same volume as the particle to the surface area of the particle. Which of the following has the maximum value of sphericity ?
(a) Sphere
(b) Cube
(c) Cylinder (L/D = 1)
(d) Raschig rings

271. Reciprocal of sphericity is termed as the
(a) Specific surface ratio
(b) Shape factor
(c) Sauter diameter
(d) Surface area per unit mass

272. Reciprocating screens are normally inclined at an angle of 5° with the horizontal and employ gyratory motion at feed end & reciprocating motion at the discharge end. They are not suitable for the screening of the
(a) light metal powder down upto 4 mesh size.
(b) dry chemicals.
(c) heavy tonnages of rocks or gravel.
(d) powdered food & granular materials.

273. Work index is the gross energy (KWh/tonne of feed) necessary to reduce a very large feed to such a size that 80% of product particles will pass through a 0.1 mm screen. The value of work index determined for wet grinding should be multiplied with .... to get the same for dry grinding.
(a) 1.0 *(b)* 0.5
(c) 1.34 *(d)* 4.34

274. The value of work index does not change materially from one equipment to another. If the value of work index determined for close circuit grinding is Wi, then the same for open circuit grinding will be
(a) 0.5 Wi *(b)* *Wi*
(c) 1.34 Wi *(d)* 3.34 Wi

275. The capacity of a gyratory crusher is .... that of a jaw crusher with the same gape, handling the same feed & for the same product size range.
(a) same as *(b)* 2.5 times
(c) 5 times *(d)* 10 times

Answers

264. (c)	265. (c)	266. (b)	267. (b)	268. (b)	269. (b)
270. (a)	271. (c)	272. (c)	273. (c)	274. (b)	275. (d)

276. Rod mills employed for grinding
  (*a*) employs a steel shell having L/D ratio 1.5 to 3.0.
  (*b*) is useful for handling sticky materials.
  (*c*) employ steel rods of 2-12 cms diameter extending over full length of the mill.
  (*d*) all 'a' 'b' & 'c'

277. Tubular bowl centrifuges as compared to disk bowl centrifuges.
  (*a*) operate at higher speed.
  (*b*) employ bowl of larger diameter.
  (*c*) can not be operated under pressure/ vacuum.
  (*d*) can't be used for separation of fine suspended solids from a liquid.

278. Ultracentrifuges running at speeds upto 100000 rpm is normally used for the
  (*a*) separation of isotopes based on their density or molecular weights difference.
  (*b*) concentration of rubber latex.
  (*c*) separation of cream from milk.
  (*d*) dewaxing of lubricating oil.

279. Filter aids like asbestos, Kieselguhr, diatomaceous earth etc. are used to increase the porosity of the final filter cake & reducing the cake resistance during filtration. Filter aid is
  (*a*) added to the feed slurry.
  (*b*) precoated on the filter medium prior to filtration.
  (*c*) separated from the cake by dissolving solids or by burning it off.
  (*d*) all 'a 'b' & 'c'

280. The sphericity of a cylinder of 1 mm diameter and length 3 mm is
  (*a*) 0.9 (*b*) 0.78
  (*c*) 0.6 (*d*) 0.5

281. For crushing of solids, the Rittinger's law states that the work required for crushing is proportional to
  (*a*) the new surface created.
  (*b*) the size reduction ratio.
  (*c*) the change in volume due to crushing.
  (*d*) none of these.

282. A suspension of glass beads in ethylene glycol has a hindered settling velocity of 1.7 mm/s, while the terminal setting velocity of a single glass bead in ethylene glycol is 17 mm/s. If the Richardson-Zaki hindered settling index is 4.5, the volume fraction of solids in the suspension is
  (*a*) 0.1 (6) 0.4
  (*c*) 0.6 (*d*) 0.9

283. Arrange the following size reduction equipment in the decreasing order of the average particle size produced by each of them.
  (*a*)Jaw crusher, Ball mill, Fluid energy mill.
  (*b*) Ball mill, Jaw crusher, Fluid energy mill.
  (*c*) Fluid energy mill, Jaw crusher, Ball mill.
  (*d*) Fluid energy mill, Ball mill, Jaw crusher.

284. What is the critical rotation speed in revolutions per second, for a ball mill of 1.2 m diameter charged with 70 mm dia balls.
  (*a*) 0.5 (*b*) 1.0
  (*c*) 2.76 (*d*) 0.66

285. A sand mixture was screened through a standard 10-mesh screen. The mass fraction of the oversize material in feed, overflow & underflow were found to be 0.38, 0.79 & 0.22 respectively. The screen effectiveness based on the oversize is
  (*a*) 0.50 (*b*) 0.58
  (*c*) 0.68 (*d*) 0.62

Answers

276. (d)	277. (a)	278. (a)	279. (d)	280. (a)	281. (a)
282. (c)	283. (a)	284. (d)	285. (a)		

286. Energy requirement (per unit mass of material crushed/ground) is the highest for
(*a*) jaw crusher (*b*) rod mill
(*c*) ball mill (*d*) fluid energy mill

287. In constant pressure filtration,
(*a*) resistance decreases with time.
(*b*) rate of filtration is constant.
(*c*) rate of filtration increases with time.
(*d*) rate of filtration decreases with time.

288. A 30% (by volume) suspension of spherical sand particles in a viscous oil has a hindered settling velocity of 4.44 μm/s. If the Richardson-Zaki hindered settling index is 4.5, then the terminal velocity of a sand grain is ----------- μm/s.
(*a*) 0.90 (*b*) 1
(*c*) 22.1 (*d*) 0.50

289. Match the sphericity of various materials/ particles.
List I
(*a*) Spherical particles
(*b*) Cubical particles
(*c*) Cylinderical (L/D = 1) particles
(*d*) Mica flakes
List II
I. 0.28
II. 0.874
III. 0.806
IV. 1

Answers

286. (d) 287. (d) 288. (b) 289. (iv) (iii) (ii) (i)

3 Iron Making

1. Finest dust particles present in blast furnace gas is removed by a/an
   (*a*) dust catcher
   (*b*) wet scrubber
   (*c*) electrostatic precipitator
   (*d*) hydrocyclone.
2. Which is a dry dust removal equipment used in the blast furnace gas cleaning system ?
   (*a*) Dust catcher
   (*b*) Scrubber
   (*c*) Venturi scrubber
   (*d*) None of these.
3. Dust content in cleaned blast furnace gas is less than .... mg/'Nm^3.
   (*a*) 1 (*b*) 10
   (*c*) 100 (*d*) 1000
4. Dust content in raw B.F. gas coming out from the blast furnace operating at high top pressure (say 1.5 kg/cm^2) is about .... gm/Nm^3 .
   (*a*) 1 — 2 (*b*) 10 — 15
   (*c*) 35 — 40 (*d*) 60 — 80
5. Dust content in raw blast furnace gas coming out from the blast furnace operating at low top pressure (say 0.2 kg/cm^2) is about......... gm/Nm^3.
   (*a*) 1-5 (*b*) 5-10
   (*c*) 10-15 (*d*) 30-35
6. Temperature of blast furnace gas (top gas) emanating from blast furnace is about .... °C.
   (*a*) 80- 100 (*b*) 200-250
   (*c*) 350 - 500 (*d*) 650 - 750
7. Hot blast temperature in Indian blast furnaces is in the range of .... °C.
   (*a*) 300 – 500 (*b*) 500 – 700
   (*c*) 800 – 1100 (*d*) 1300 – 1500
8. Maximum hot blast temperature so far used in modern blast furnaces (operating in Japan, U.S.S.R. etc) is about .... °C.
   (*a*) 1100 (*b*) 1250
   (*c*) 1500 (*d*) 1650
9. Function of coke in a blast furnace is to
   (*a*) produce heat.
   (*b*) act as a reducing agent.
   (*c*) act as a spacer in bosh region.
   (*d*) all (*a*), (*b*) & (*c*)
10. Main area of fuel economy in blast furnace operation is to reduce the
   (*a*) coke rate.
   (*b*) gaseous fuel consumption in stoves.
   (*c*) sensible heat loss from blast furnace gas.
   (*d*) coke dust losses in outgoing B.F. gas.

Answers

1. (c)	2. (a)	3. (b)	4. (b)	5. (d)	6. (b)
7. (c)	8. (b)	9. (d)	10. (a)		

11. Solid raw materials required to produce a ton of pig iron is about .... tons.
(a) 1 (b) 3
(c) 6 (d) 10

12. Air requried to produce a ton of pig iron is about .... tons.
(a) 1 (b) 4
(c) 10 (d) 15

13. Average iron content in Indian iron ore is about .... percent.
(a) 40 (b) 62
(c) 80 (d) 90

14. Which iron bearing mineral has the widest distribution in the world?
(a) Hematite (b) Magnetite
(c) Limonite (d) Siderite

15. Which has the highest % Fe ?
(a) Magnetite (b) Hematite
(c) Limonite (d) Iron pyrite

16. Black iron ore mainly contains
(a) hematite (b) magnetite
(c) limonite (d) none of these

17. Red or reddish black iron ore contains mainly
(a) hematite (b) magnetite
(c) limonite (d) carbonate

18. Brown iron ores are rich in
(a) magnetite (b) hematite
(c) limonite (d) carbonate

19. Argillaceous iron ores are rich in
(a) silica (b) alumina
(c) clayey matter (d) coaley matter

20. Fe content in hematite is about .... percent.
(a) 45 (b) 60
(c) 70 (d) 80

21. Indian iron ores in general are rich in
(a) hematite (b) magnetite
(c) limonite (d) ilmenite

22. Iron ore required to produce a ton of pig iron in Indian steel plant is about tons.
(a) 1 (b) 2
(c) 4 (d) 6

23. Chief source of sulphur in the blast furnace charge is
(a) coke (b) iron ore
(c) sinter (d) flux

24. Slag is
(a) always basic in nature,
(b) always acidic in nature.
(c) molten oxide product of smelting.
(d) heavier than iron.

25. Flux is added during iron ore smelting to reduce the
(a) softening point of gangue materials.
(b) viscosity of slag.
(c) activity of some of its components t make them stable or unstable in the slag phase.
(d) all (a), (b) & (c).

26. With decrease in the softening point of the slag, its viscosity at a given temperature
(a) increases.
(b) decreases.
(c) remains constant.
(d) may increase or decrease, depending on its chemical composition.

27. Limestone required as a flux for making a ton of pig iron is about .... ton.
(a) 0.1 (b) 0.5
(c) 1 (d) 3

Answers

11. (b)	12. (b)	13. (b)	14. (a)	15. (a)	16. (b)
17. (a)	18. (c)	19. (c)	20. (c)	21. (c)	22. (c)
23. (a)	24. (b)	25. (d)	26. (b)	27. (b)	

28. Function of coke in the blast furnace is to
    (*a*) provide heat for meeting the endothermic requirements of chemical reactions and melting of slag & metal
    (*b*) produce and regenerate reducing gases for the reduction of iron oxides.
    (*c*) provide an open permeable bed through which slag & metal pass down into the hearth and hot reducing gases pass upwards.
    (*d*) all (*a*), (b) & (*c*).

29. Average coke rate in Indian blast furnaces is about .... kg/ton of pig iron or hot metal.
    (*a*) 750 —950 (*b*) 250 — 450
    (*c*) 550 —750 (*d*) 950 —1150

30. Average coke rate in the modern blast furnaces in developing countries is about kg/ ton hot metal.
    (*a*) 100-200 (*b*) 400-500
    (*c*) 700-800 (*d*) 900-1000

31. Permeability of the charge in the bosh region of the blast furnace is maintained by
    (*a*) sinter (*b*) coke
    (*c*) iron ore (*d*) limestone flux

32. The main function of coke in the bosh region of blast furnace is to
    (*a*) act as a spacer
    (*b*) produce heat
    (*c*) generate reducing gases
    (*d*) melt the slag

33. A good blast furnace coke should be
    (*a*) of narrow size range.
    (*b*) strong & abrasive.
    (*c*) thermally stable at high temperature.
    (*d*) all (*a*), (*b*) & (*c*).

34. Each percent increase in the ash content of coke over a critical limit
    (*a*) decreases the blast furnace production by 3%.
    (*b*) increases the coke rate in blast furnace by 3%.
    (*c*) increases the slag volume per ton of hot metal.
    (*d*) all (*a*), (*b*) & (*c*).

35. Sulphur percentage in Assam coal (which is of good coking quality) is about
    (*a*) 1-2 (*b*) 4-6
    (*c*) 10 -12 (*d*) 15 - 20

36. Blast furnace alone consumes about .... percent of the total energy consumption in producing a ton of crude (ingot) steel.
    (*a*) 10—15 (*b*) 35—40
    (*c*) 55—60 (*d*) 75—80

37. Average productivity of Indian blast furnaces is about .... tons/m^3/day.
    (*a*) 0.1 (*b*) 1
    (*c*) 5 (*d*) 10

38. Productivity of modern blast furnaces working in developed countries is about .... tons/m^3/day.
    (*a*) 0.1 (*b*) 1
    (*c*) 2 (*d*) 5

39. Most of the iron ore deposits in India are located in
    (*a*) Jharkhand & West Bengal.
    (*b*) Jharkhand & Orissa.
    (*c*) Andhra Pradesh & West Bengal.
    (*d*) Madhya Pradesh & West Bengal.

Answers

28. (d)	29. (c)	30. (b)	31. (b)	32. (a)	33. (d)
34. (d)	35. (b)	36. (b)	37. (b)	38. (c)	39. (b)

40. Most of the coking coal reserves in India are located in
    (*a*) Jharkhand & West Bengal.
    (*b*) Jharkhand & Orissa.
    (*c*) Orissa & Madhya Pradesh.
    (*d*) Andhra Pradesh & Orissa.

41. A reduction of 0.1% sulphur in blast furnace coke will
    (*a*) reduce the coke rate by 2 %.
    (*b*) reduce the limestone consumption by 1.5—2%.
    (*c*) increase the B.F. productivity.
    (*d*) all (*a*), (*b*) & (*c*).

42. Washing of Indian metallurgical coal is done to reduce its ash content to about .... percent.
    (*a*) 8—10 (*b*) 18—20
    (*c*) 28—30 (*d*) 24—26

43. Percentage of prime coking coal in coal blend used for making metallurgical coke in Indian steel plant is about
    (*a*) 10 (*b*) 25
    (*c*) 50 (*d*) 80

44. Imported coal (from Canada and Australia) used in coal blend for B.F. coke making contains about .... percent ash.
    (*a*) 1 (*b*) 10
    (*c*) 50 (*d*) 17

45. Rate of burning of coke (which controls the rate of production from blast furnace) is directly proportional to the
    (*a*) area of coke exposed to the blast.
    (*b*) temperature & pressure of the blast.
    (*c*) affinity of the particular type of carbon for oxygen.
    (*d*) all (*a*), (*b*) & (*c*).

46. High reactivity of coke with CO_2
    (*a*) causes degradation of coke during its descent in the furnace due to solution loss reaction.
    (*b*) is further catalysed by the presence of alkalis.
    (*c*) is detrimental to blastfurnace operation.
    (*d*) all (*a*), (*b*) & (*c*).

47. Blast temperature in a blast furnace should be kept .... the ignition temperature of the coke.
    (*a*) more than (*b*) less than
    (*c*) equal to (*d*) either more or less.

48. Which of the following charge materials in the blast furnace burden has got maximum average size ?
    (*a*) Iron ore lump (*b*) Sinter
    (*c*) Coke (*d*) Lime

49. Size of the blast furnace coke is about .... mm.
    (*a*) 10—30 (*b*) 25—40
    (*c*) 25—80 (*d*) 100

50. +40 mm size in the B.F. coke should be about .... percent.
    (*a*) 20 (*b*) 40
    (*c*) 80 (*d*) 100

51. Mean size of the B.F. coke should be about .... mm.
    (*a*) 30—35 (*b*) 55—60
    (*c*) 75—80 (*d*) 20—25

52. Ratio of average size of coke to iron bearing burden in blast furnace should be about
    (*a*) 1 : 1 (*b*) 2 : 1
    (*c*) 4 : 1 (*d*) 8 : 1

Answers

40. (a)	41. (d)	42. (b)	43. (c)	44. (b)	45. (d)
46. (d)	47. (a)	48. (c)	49. (c)	50. (c)	51. (a)
52. (c)					

53. Percentage of+80 mm fraction in B.F. coke should be
    (*a*) 0 (*b*) 25
    (*c*) 10 (*d*) 80

54. Thermal stability of coke is favoured by
    (*a*) low chemical reactivity.
    (*b*) absence of large lumps in feed coke.
    (*c*) uniformity of coke texture.
    (*d*) all (*a*),(*b*) & (*c*).

55. Strength and hardness of coke depends upon the
    (*a*) quality & blending procedure of coal.
    (*b*) carbonisation time & temperature.
    (*c*) bulk density of coal & coke quenching method.
    (*d*) all (*a*), (*b*) & (*c*).

56. Shatter index of B.F. coke is a measure of its
    (*a*) strength (*b*) hardness
    (*c*) both (*a*) & (*b*)
    (*d*) neither (*a*) nor (*b*).

57. Abrasion index of coke is a measure of its
    (*a*) strength
    (*b*) hardness
    (*c*) both (*a*) & (*b*)
    (*d*) neither (*a*) nor *lb*).

58. Driving rate of blast furnaces is directly related to the
    (*a*) rate of reduction of iron oxide.
    (*b*) rate of heating of the burden.
    (*c*) quantity of blast that goes through the furnaces.
    (*d*) velocity of burden descent.

59. With increase in the ranges of sizes of particles in the burden bed inside the blast furnace, the
    (*a*) packing density increases.
    (*b*) permeability decreases.
    (*b*) driving rate reduces at a given pressure of air blowing.
    (*d*) all (*a*), (*b*) & (*c*).

60. Increased fines in the charge of blast furnace increases the
    (*a*) burden permeability.
    (*b*) pressure drop inside the furnace.
    (*c*) reduction rate.
    (*d*) driving rate at a given air blowing pressure.

61. The fraction of iron bearing materials in the blast furnace burden (*e.g.* sinter) is termed as 'fine', which has a size of less than .... mm.
    (*a*) 1 (*b*) 5
    (*c*) 10 (*d*) 25

62. The size range of iron ore charged in modern blast furnace is .... mm.
    (*a*) 5-15 (*b*) 10-20
    (*c*) 10-40 (*d*) 25-80

63. An ideal burden for blast furnace should have low
    (*a*) swelling tendency during reduction.
    (*b*) alumina/silica ratio.
    (*c*) percentage of fines and close size range.
    (*d*) all (*a*), (*b*) & (*c*).

64. An ideal burden for blast furnace should have high
    (*a*) softening temperature with a narrow temperature range of fusion,
    (*b*) bulk reducibility to obtain close equilibrium conditions between solid & gas phases in the stack.
    (*c*) iron & low gangue contents.
    (*d*) all (*a*), (*b*) & (*c*).

Answers

53. (a)	54. (d)	55. (d)	56. (a)	57. (b)	58. (c)
59. (d)	60. (b)	61. (b)	62. (c)	63. (d)	64. (d)

65. During iron ore washing, the ratio of alumina to silica in lumps
   (*a*) increases,
   (*b*) decreases,
   (*c*) remains constant.
   (*d*) may increase or decrease ; depends on the type of ore.

66. For efficient blast furnace operation and achieving optimum level of productivity, the iron ore should have
   (*a*) > 60% iron content.
   (*b*) $Al_2O_3/Fe < 0.5$.
   (*c*) $Al_2O_3/SiO_2 < 1$.
   (*d*) all (*a*),(*b*) & (*c*).

67. Blue dust
   (*a*) is generated during iron ore mining,
   (*b*) contains > 68% Fe.
   (*c*) is soft.
   (*d*) all (*a*), (*b*) & (*c*).

68. Size of sinter charged in the blast furnace is .... mm.
   (*a*) 5—10 (*b*) 5—50
   (*c*) 25—80 (*d*) +80

69. Maximum percentage of sinter in the blast furnace burden may be about
   (*a*)10 (*b*) 25
   (*c*) 50 (*d*) 85

70. Moisture percentage in sinter mix charge fed to the sintering machine is maintained (for optimum permeability) at
   (*a*) 1 (*b*) 7
   (*c*) 18 (*d*) 30

71. Temperature maintained in the ignition hearth of sinter machine is about .... °C.
   (*a*) 500 (*b*) 800
   (*c*) 1200 (*d*) 1800.

72. Specific coke consumption in sinter making in Indian sinter plant is about .... kg dry coke/ton sinter.
   (*a*) 25 (*b*) 50
   (*c*) 75 (*d*) 150

73. Sinter is cooled by
   (*a*) forcing up air from the bottom of the sinter bed.
   (*b*) forcing down air from the top of the sinter bed.
   (*c*) air suction through sinter bed on sinter machine itself.
   (*d*) water quenching.

74. Vacuum employed in the sinter machine is about .... mm water gauge.
   (*a*) 200 — 400 (*b*) 800 — 1200
   (*c*) 2000 — 3000 (*d*) 5000 — 6000

75. Sinter bed height in sinter machine is in the range of. .... mm.
   (*a*) 50-100 (*b*) 350-600
   (*c*) 800-1000 (*d*) 1500-2000

76. Specific heat consumption in sinter making is about .... Giga calorie/ton sinter.
   (*a*) 0.005 (*b*) 0.05
   (*c*) 0.5 (*d*) 1.5

77. Working principle of a sinter machine is similar to the
   (*a*)) top blown oxygen converter.
   (*b*) bottom blown oxygen converter.
   (*c*) puffing of a cigarette.
   (*d*) functioning of Bessemer converter.

78. Normal speed of the sinter machine is usually kept at about .... metres/minute.
   (*a*) 0.5—1 (*b*) 3—5
   (*c*) 10—15 (*d*) 20—25

Answers

65. (a)	66. (d)	67. (d)	68. (b)	69. (d)	70. (b)
71. (c)	72. (c)	73. (a)	74. (b)	75. (b)	76. (b)
77. (c)	78. (b)				

79. Fuel used for the ignition in sinter machine is
    (a) coke oven gas/mixed gas
    (b) pulverised coke
    (c) tar
    (d) furnace oil

80. Which of the following is stationary in the sinter machine ?
    (a) Wind box (b) Pallets
    (c) Sinter bed (d) None of these

81. Final mixing and balling of charge material in the sinter plant is done in a
    (a) drum cooler (b) balling drum
    (c) pug mill (d) kneading mixer

82. Specific power consumption in sinter making is about .... kWh/ton sinter.
    (a) 25 (b) 60
    (c) 150 (d) 250

83. Specific power consumption in pig iron production by blast furnace is about .... kWh/ton pig iron.
    (a) 5 (b) 25
    (c) 100 (d) 500

84. In a good quality sinter, the sinter return (*i.e.* < 5 mm size) generated should be less than .... percent.
    (a) 5 (b) 15
    (c) 25 (d) 40

85. Sinter is transported from sinter plant to blast furnace by .... conveyor.
    (a) screw (b) belt
    (c) pneumatic (d) chain

86. Optimum permeability of sinter mix charge is obtained,. when it is preheated in the balling drum to a temperature of. .... °C.
    (a) 35 (b) 60
    (c) 125 (d) 150

87. Optimum size of coke breeze used for sinter making is .... mm.
    (a) 0—3 (b) 5—15
    (c) 15—20 (d) + 5

88. The rate of pellet production in a drum pelletiser depends upon the drum
    (a) speed (b) diameter
    (c) inclination (d) all (a), (b) & (c).

89. Production rate in a drum pelletiser does not depend upon the .... of the feed.
    (a) nature, size & rate.
    (b) moisture content.
    (c) depth of material in residence.
    (d) none of these.

90. Size of iron ore fines suitable for pelletisation is minus .... mm.
    (a) 3 (b) 100
    (c) 1 (d) 5

91. Maximum temperature is generated in the blast furnace in the
    (a) front of tuyers (b) hearth
    (c) bosh (d) stack

92. For a blast furnace operating at a top pressure of 1.5 kg/cm^2, the blast pressure may be about .... kg/cm^2.
    (a) 0.5 (b) 1.5
    (c) 2.5 (d) 5.5

93. CO/CO_2 ratio in the blast furnace gas coming out of a very efficient modern blast furnace is about
    (a) 0.5 (b) 1
    (c) 2 (d) 2.5

Answers

79. (a)	80. (a)	81. (b)	82. (b)	83. (b)	84. (a)
85. (b)	86. (b)	87. (a)	88. (d)	89. (d)	90. (b)
91. (a)	92. (c)	93. (b)			

94. CO/CO_2 ratio in B.F. gas of Indian steel plants is about
(a) 0.5—0.8 (b) 1.1—1.4
(c) 2—2.5 (d) 3—3.5

95. Based on the volumetric composition of outgoing B.F. gas, the reduction efficiency of a blast furnace is expressed as
(a) $\frac{CO_2}{CO+CO_2}$ (b) $\frac{CO}{CO_2+CO}$
(c) $\frac{CO}{CO_2}$ (d) $\frac{CO_2}{CO}$

96. Raceway is the .... in the blast furnace.
(a) pear shaped coke combustion zone in front of tuyers
(b) heat source as well as the source of reducing gas
(c) both (a) & (b).
(d) neither (a) nor (b)

97. Coke containing a large amount of fines tends to
(a) raise the pressure in the raceway because of decreased permeability.
(b) create unblanced pressure near the raceway.
(c) create an unstable raceway shape.
(d) all (a), (b) & (c).

98. The main reaction in the tuyers area of blast furnace is the
(a) reduction of iron oxide.
(b) combustion of coke.
(c) solution loss reaction.
(d) melting of slag.

99. Reduction efficiency of blast furnace is increased, if iron ore
(a) is more porous and of uniform size.
(b) contains no fines.
(c) burden permeability is more.
(d) all (a), (b) & (c)

100. Higher % of CO in B.F. gas means
(a) lower reduction efficiency of blast furnace.
(b) unnecessary higher coke rate in the furnace.
(c) its higher calorific value.
(d) all (a), (b) & (c).

101. High top pressure in the blast furnace helps to increase the
(a) effective concentration of CO.
(b) reduction efficiency of blast furnace.
(c) contact time of reducing gases with the charge.
(d) all (a), (b) & (c).

102. The most efficient blast furnace operation is obtained, when .... of the reduction is achieved by solid carbon (*i.e.* direct reduction) and the remaining by gaseous reduction (*i.e.* indirect reduction).
(a) one fourth (b) one third
(c) two third (d) three fourth

103. Sensible heat of blast furnace gas is about .... percent of total heat released by coke consumption in the blast furnace.
(a) 1 (b) 5
(c) 25 (d) 50

104. 1% increase in ash content of coke decreases the effective carbon percentage in coke by about .... percent.
(a) 1 (b) 1.5
(c) 0.5 (d) 3

105. Slag rate in Indian blast furnaces is in the range of. .... kg/ton of pig ion.
(a) 150 – 250 (b) 300 – 450
(c) 250 – 350 (d) 1000 –1500

Answers

94. (b)	95. (a)	96. (c)	97. (d)	98. (b)	99. (d)
100. (d)	101. (d)	102. (b)	103. (b)	104. (b)	105. (b)

106. Alumina content in the blast furnace slag in India is in the range of. .... percent.
   (a) 8 - 12   (b) 18 - 22
   (c) 32 -36   (d) 2-6

107. Alumina content in the blast furnace slag in the modern blast furnaces abroad is in the range of .... percent.
   (a) 8-12   (b) 18-22
   (c) 32-36   (d) 2-6

108. High alumina in blast furnace slag
   (a) makes the slag highly viscous.
   (b) impairs the kinetics of sulphur & silicon transfer to the slag.
   (c) is undesirable.
   (d) all (a), (b) &.(c).

109. Flush slag basicity (CaO/SiO_2) in Indian blast furnace slag is about
   (a) 0.5   (b) 1
   (c) 1.7   (d) 2.5

110. Fluidity of B.F. slag can be increased (i.e. viscosity can be decreased) by
   (a) charging quartz in the blast furnace.
   (b) maintaining high MgO (10—12%) in the slag.
   (c) both (a) & (b).
   (d) neither (a) nor (b).

111. 'Salamander' (in a blast furnace)
   (a) is a large mass of iron accumulated at the bottom of the hearth.
   (b) is. removed by dynamite. blasting after shutting off the blast furnace.
   (c) does not get tapped during normal tapping operation.
   (d) all (a), (b) & (c).

112. Bustle pipe (in blast furnace) is
   (a) water cooled.
   (b) not lined with refractories.
   (c) meant for supplying hot blast at uniform pressure to all the tuyers.
   (d) same as blow pipe.

113. Gas cleaning efficiency of dust catcher in a blast furnace gas cleaning plant is about percent.
   (a) 15   (b) 30
   (c) 65   (d) 95

114. Gas cleaning efficiency of an electrostatic precipitator may be about .... percent.
   (a) 15   (b) 30
   (c) 65   (d) 99

115. Potential (*i.e.* pressure) energy of high top pressure blast furnace gas is used for
   (a) increasing the hot blast pressure.
   (b) operating the small bell in the charging system.
   (c) meeting the requirement of pressure drop of the gas during its long distance transportation.
   (d) generating electrical power in a top pressure recovery turbine (TRT).

116. Electrical power generation in top pressure recovery turbine (operating at about 1.5 kg/cm^2) is about .... kWh/ton pig iron produced.
   (a) 1   (b) 20
   (c) 100   (d) 250

117. About .... percent of total blast furnace gas generated is consumed in heating the blast in stoves.
   (a) 15   (b) 35
   (c) 60   (d) 80

Answers

106. (b)	107. (a)	108. (d)	109. (c)	110. (c)	111. (d)
112. (c)	113. (c)	114. (d)	115. (d)	116. (b)	117. (b)

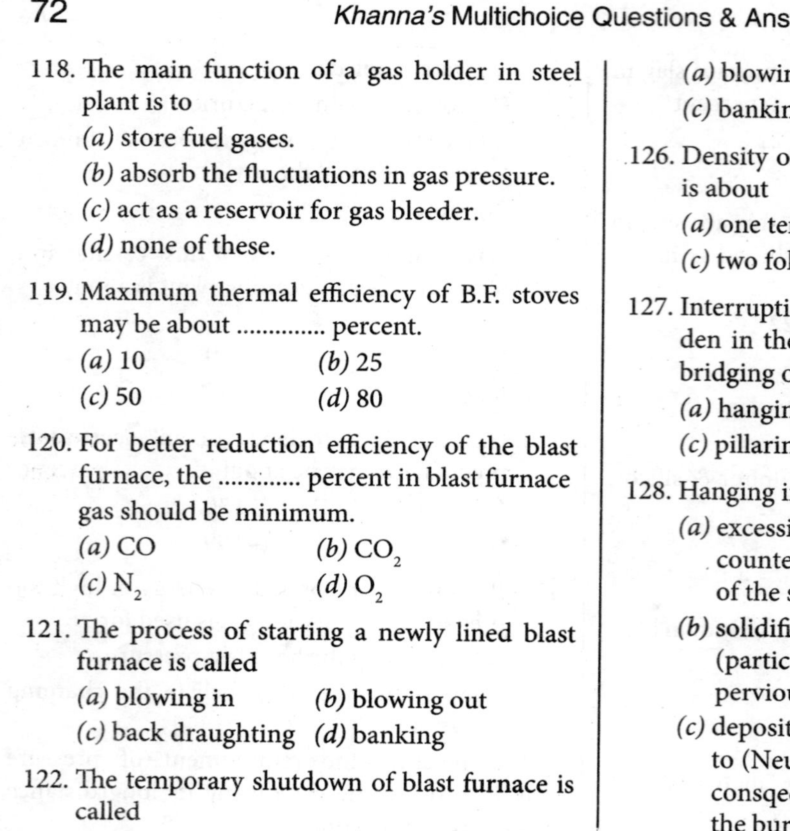

118. The main function of a gas holder in steel plant is to
(*a*) store fuel gases.
(*b*) absorb the fluctuations in gas pressure.
(*c*) act as a reservoir for gas bleeder.
(*d*) none of these.

119. Maximum thermal efficiency of B.F. stoves may be about .... percent.
(*a*) 10 (*b*) 25
(*c*) 50 (*d*) 80

120. For better reduction efficiency of the blast furnace, the .... percent in blast furnace gas should be minimum.
(*a*) CO (*b*) CO_2
(*c*) N_2 (*d*) O_2

121. The process of starting a newly lined blast furnace is called
(*a*) blowing in (*b*) blowing out
(*c*) back draughting (*d*) banking

122. The temporary shutdown of blast furnace is called
(*a*) blowing in (*b*) blowing out
(*c*) banking (*d*) back draughting

123. Heating of a newly lined blast furnace stove to optimum dome temperature requires about .... days.
(*a*) 1-2 (*b*) 15-20
(*c*) 60-90 (*d*) 120-150

124. The process of stopping the blast furnace operation at the end of its campaign is known as
(*a*) blowing in (*b*) blowing out
(*c*) banking (*d*) back draughting

125. 'Salamander' tapping in the blast furnace is done after .... the furnace.
(*a*) blowing in (*b*) blowing out
(*c*) banking (*d*) back draughting

126. Density of slag compared to that of pig iron is about
(*a*) one tenth (*b*) one third
(*c*) two fold (*d*) three fourth

127. Interruption of uniform descent of the burden in the blast furnace either by wedging/bridging of stock or by scaffolding is called
(*a*) hanging (*b*) slipping
(*c*) pillaring (*d*) channelling

128. Hanging in a blast furnace is caused by the
(*a*) excessive blast pressure resulting in counteracting the downward movement of the stock.
(*b*) solidification of previously fused slag (particularly in sinter) into a large impervious mass.
(*c*) deposition of large amount of carbon due to (Neumann reversion) in the voids and consqequent decrease in permeability of the burden.
(*d*) all (*a*), (*b*) & (*c*).

129. Hanging in a blast furnace can be eliminated by
(*a*) reducing the hot blast temperature & pressure.
(*b*) adding large size limestone, which on dissociation generates CO2, which in turn forces the solution loss reaction to take place and thereby improves the permeability of the bed.
(*c*) blowing down the furnace to bosh level and filling it with coke blank.
(*d*) all (*a*), (*b*) & (*c*).

Answers

118. (b)	119. (d)	120. (a)	121. (a)	122. (c)	123. (b)
124. (b)	125. (b)	126. (b)	127. (a)	128. (d)	129. (d)

130. Sticking of a large mass of material (apparently as a single block) to the blast furnace wall (particularly in the top portion of the bosh) is called
   (a) scaffolding   (b) slipping
   (c) pillaring   (d) breakout

131. Scaffolding in a blast furnace results in
   (a) increase in blast pressure, coke rate & dust losses.
   (b) decrease in gas yield & hot metal output.
   (c) reduction in the effective cross-section of the furnace for gas-solid interaction.
   (d) all (a), (b) & (c).

132. The tendency to scaffold formation in the blast furnace can be minimised by minimising the
   (a) refractoriness of the slag.
   (b) alkali content of the burden.
   (c) erratic working of furnace and shut off of blast furnace for any reason.
   (d) all (a), (b) & (c).

133. Scaffold in the blast furnace can be removed by
   (a) charging a few blanks of coke and reducing the blast pressure on the affected side.
   (b) making the burden more silicious on the affected side.
   (c)'jumping'the furnace, *i.e.* by putting off the blast suddenly to relieve pressure under the scaffold.
   (d) all (a), (b) & (c).

134. Sudden sinking of the stock in the blast furnace caused by collapse of scaffold, wedging, hanging etc. is called
   (a) pillaring   (b) slipping
   (c) breakout   (d) channelling

135. Formation of a cold central column of the stock with an annular hot zone all around it in a blast furnace (resulting from non-penetration of blast right upto the centre of furnace) is called
   (a) pillaring   (b) hanging
   (c) scaffolding   (d) breakout

136. Chilled hearth in blast furnace operation is caused because of
   (a) low coke input.
   (b) excessive steam injection in the blast.
   (c) water leaking in from tueyers.
   (d) all (a), (b) & (c).

137. Flooding in a blast furnace is caused by the
   (a) interruption in uniform blowing rate or keeping some tueyers blanked.
   (b) accumulation of metal & slag in the coke interstices in the bosh region.
   (c) impermeable coke bed in the bosh.
   (d) all (a), (b) & (c)

138. Flooding in the iron blast furnace can be minimised by
   (a) having a high voidage in bosh region.
   (b) using higher mean size of the coke.
   (c) using better quality coke.
   (d) all (a), (b) & (c).

139. Preferential flow of ascending gases through certain areas of burden, because of their relatively much better permeability is called
   (a) slipping   (b) pillaring
   (c) channelling   (d) hanging

140. Channelling in a blast furnace occurs due to the .... of charge in the furnace.
   (a) wide size range
   (b) improper distribution
   (c) both (a) & (b)
   (d) neither (b) nor (b)

Answers

130. (a)	131. (d)	132. (d)	133. (d)	134. (b)	135. (a)
136. (d)	137. (d)	138. (d)	139. (c)	140. (c)	

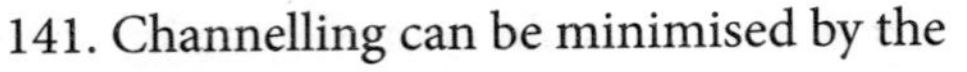

141. Channelling can be minimised by the
   (*a*) proper burden distribution.
   (*b*) use of a more uniformly sized burden,
   (*c*) both (*a*) & (*b*).
   (*d*) neither (*a*) nor (*b*).

142. Channelling in a blast furnace does not decrease its
   (*a*)productivity directly.
   (*b*)effective cross-section for gas-solid interaction.
   (*c*) both (*a*) & (*b*).
   (*d*) neither (*a*) nor (*b*).

143. Silicon and sulphur content in the blast furnace pig iron depends upon the
   (*a*) slag volume. (*b*) slag basicity.
   (*c*) operating temperature of the furnace.
   (*d*) all (*a*), (*b*) & (*c*).

144. Manganese content of pig iron depends upon the
   (*a*) MnO content of the burden.
   (*b*) slag basicity.
   (*c*) operating temperature of the blast furnace.
   (*d*) all (*a*), (*b*) & (*c*).

145. Manganese percentage in pig iron is in the range of
   (*a*) 0.001— 0.1 (*b*) 0.1 - 0.5
   (*c*) 0.5—1.5 (*d*) 3.5 - 4.2

146. Sulphur percentage in the pig iron is in the range of
   (*a*). 0.001-0.1 (*b*) 0.04 — 0..15
   (*c*) 0.5 —2 (*d*) 3.5 — 4.5

147. If the S & P contents are each below .... percent. the pig iron is called acid iron.
   (*a*) 0.001 (*b*) 0.04
   (*c*) 0.15 (*d*) 0.5

148. Basic pig iron is the one, which contains phosphorous more than .... percent.
   (*a*) 0.001 (*b*) 0.04
   (*c*) 0.5 (*d*) 1.5

149. Ideal silicon content in pig iron used for basic process of steel making is .... percent.
   (*a*) 0.01 (*b*) 0.5
   (*c*) 1.5 (*d*) 2

150. Phosphorous percentage in most of the world's pig iron is about
   (*a*) 0.15—0.35 (*b*) 1.5—2
   <0.015 (*d*) <0.01

151. Desulphurisation inside the blast furnace can be done by
   (*a*) increasing the slag volume.
   (*b*) having higher slag basicity.
   (*c*) raising the hearth temperature.
   (*d*) all (*a*), (*b*) & (*c*).

152. Higher alumina content in blast furnace burden necessitates high
   (*a*) furnace hearth temperature to keep the slag thin & free 11 owing.
   (*b*) slag basicity to obtain proper slag.
   (*c*) both (*a*) & (*b*).
   (*d*) neither (*a*) nor (*b*).

153. Bell-less top (BLT) burden charging system (which improves the burden distribution in the blast furnace) is a patent of the
   (*a*) Paul Worth Co.. Luxemburg.
   (*b*) Demag, Germany.
   (*c*) British Steel Corporation.
   (*d*) Nippon Steel, Japan.

Answers

141 (c)	142. (c)	143. (d)	144. (d)	145. (c)	146. (b)
147. (b)	148. (b)	149. (b)	150. (a)	151. (d)	152. (c)
153. (a)					

154. The performance of blast furnace is assessed in terms of its
   (*a*) productivity
   (*b*) coke rate
   (*c*) both (*a*) & (*b*))
   (*d*) neither (*a*) nor (*b*)

155. Hard driving of blast furnace means
   (*a*) blowing maximum volume of blast in the furnace.
   (*b*) charging hard coke (having low M_{10} value).
   (*c*) maintaining a highly permeable coke bed in the bosh region.
   (*d*) none of these.

156. Productivity of a blast furnace does not depend upon the
   (*a*) driving rate.
   (*b*) bulk permeability of the burden.
   (*c*) shape & size of the burden particles.
   (*d*) none of these.

157. With increase in every 10% sinter proportion of the blast furnace burden, the
   (*a*) output of B.F. increases by 3.5%.
   (*b*) coke rate reduces by 3%.
   (*c*) both (*a*) & (*b*).
   (*d*) neither (*a*) nor (*b*).

158. Increase in basicity of sinter beyond 1.5
   (*a*) increases its strength.
   (*b*) improves permeability in the furnace.
   (*c*) reduces its degradation during handling & charging,
   (*d*) all (*a*), (*b*) & (*c*).

159. Desirable value of shatter index of+ 10 mm sinter for blast furnace burden is about .... percent.
   (*a*) >50 (*b*) >84
   (*c*) >95 (*d*) >36

160. Porosity of sinter is in the range of.............. percent.
   (*a*) 10 —18 (*b*) 40 — 45
   (*c*) 63 —70 (*d*) 2—5

161. Porosity of pellets is in the range of.....percent,
   (*a*) 10 — 15 (*b*) 20 —30
   (*c*) 65—70 (*d*) 2 —5

162. Pellets compared to sinter
   (*a*) require finer (100 mesh) raw materials.
   (*b*) is less prone to hanging in the blast furnace.
   (*c*) give better bulk permeability of burden.
   (*d*) all (*a*), (*b*) & (*c*).

163. The output of a blast furnace is limited by the
   (*a*) appearance of flooding in the bosh.
   (*b*) hot blast temperature mainly,
   (*c*) top pressure mainly.
   (*d*) maximum hearth temperature attainable.

164. Production capacity of the largest blast furnace in the world is about .... tons pig iron/day.
   (*a*) 1000 (*b*) 6000
   (*c*) 12000 (*d*) 25000

165. Maximum useful volume of the largest blast furnace in the world is about .... m^3.
   (*a*) 1500 (*b*) 1000
   (*c*) 5000 (*d*) 10000

166. The hearth diameter of the largest blast furnace (10000 tons/day capacity) is about .... metres.
   (*a*) 1.5 (*b*) 5
   (*c*) 15 (*d*) 25

Answers

154. (c)	155. (a)	156. (d)	157. (c)	158. (d)	159. (b)
160. (a)	161. (b)	162. (d)	163. (a)	164. (c)	165. (c)
166. (c)					

167. Maximum top pressure used in the blast furnace abroad is about .... kg/cm^2.
(*a*) 0.5 (*b*) 1
(*c*) 1.5 (*d*) 3

168. Maximum top pressure used in the Indian blast furnace is about .... kg/cm^2 in Visakhapatnam steel plant.
(*a*) 2.5 (*b*) 0.5
(*c*) 1 (*d*) 2.2

169. With increase in top pressure of the blast furnaces, its
(*a*) reduction efficiency increases.
(*b*) coke rate decreases.
(*c*) gas velocity through the furnace decreases.
(*d*) all (*a*), (*b*) & (*c*).

170. With increase in the top pressure of the blast furnace,
(*a*) the dust content in B.F. gas decreases.
(*b*) solution loss reaction is stiffled (*i.e.* Boudouard equilibrium moves to the left).
(*c*) its output increases.
(*d*) all (*a*), (*b*) & (*c*).

171. Higher hot blast temperature in blast furnaces
(*a*) reduces coke rate.
(*b*) is used in conjunction with humidified blast or hydrocarbon oil/gas injection.
(*c*) may cause hanging of the burden.
(*d*) all (*a*), (*b*) & (*c*).

172. Maximum practical limit of oxygen enrichment of hot blast in the blast furnace is about .... percent oxygen.
(*a*) 4 (*b*) 10
(*c*) 15 (*d*) 25

173. Which of the following modern techniques used in the blast furnace operation has got least/marginal effect (*i.e.* decrease) on coke rate ?
(*a*) Oxygen enrichment of hot blast,
(*b*) Increase in top pressure.
(*c*) Increase in hot blast temperature.
(*d*) Natural gas/fuel oil injection.

174. One kilogram of furnace oil injected in blast furnace saves about .... kilogram of coke.
(*a*) 0.1 (*b*) 0.5
(*c*) 1.5 (*d*) 5

175. Alumina rich blast furnace slag in India results in
(*a*) greater viscosity of slag resulting in decreased desulphurisation ability,
(*b*) higher free running temperature leading to a high silicon content in pig iron.
(*c*) difficulties in skimming the slag from the metal and llushing of slag.
(*d*) all (*a*), (*b*) & (*c*).

176. Viscosity of the molten blast furnace slag is about .... poise.
(*a*) 0.5 (*b*) 4.5
(*c*) 100 (*d*) 500

177. With increase in the alumina percentage of B.F. slag, its
(*a*) basicity decreases.
(*b*) free running temperature increases.
(*c*) viscosity increases.
(*d*) all (*a*), (*b*) & (*c*).

178. Coke replacement ratio in the blast furnace by non coking coal dust injection is about
(*a*) 0.5 (*b*) 1.0
(*c*) 2.8 (*d*) 5.6

Answers

167. (d)	168. (d)	169. (d)	170. (d)	171. (d)	172. (a)
173. (a)	174. (c)	175. (d)	176. (b)	177. (d)	178. (b)

179. Specific energy consumption in modern Indian blast furnaces is about .... giga calories/ton of pig iron.
   (a) 0.5   *(b)* 1.5
   (c) 3.5   *(d)* 9.5

180. Share of blast furnace in total energy consumption in steel making is about .... percent.
   (a) 10   *(b)* 35
   (c) 50   *(d)* 75

181. With increase in the sinter bed height, the
   (a) specific output (tons/m^2/day) of sinter machine increases,
   (b) sinter strength increases.
   (c) coke rate in sintering reduces.
   (d) all *(a), (b)* & *(c).*

182. Purpose of blending medium coking/non-coking coal in prime coking coal in coke making is to
   (a) check its excessive swelling during carbonisation which may damage coke oven walls,
   (b) improve the yield & micum index of B.F. grade coke.
   (c) improve the reactivity of coke.
   (d) all *(a), (b)* & *(c).*

183. In selective crushing of coal.
   (a) coal blend is subjected to milder crushing.
   (b) harder & softer coals are separately crushed and then blended.
   (c) softer coal is subjected to milder crushing. whereas harder coal is subjected to intense grinding to finer sizes.
   (d) all *(a), (b)* & (c).

184. Selective crushing of coal (in coke making)
   (a) reduces the prime coking coal requirement in the blend.
   (b) improves (*i.e.* reduces) M_{10} value of coke (*i.e.* makes it more abrasive).
   (c) reduces coke rate and improves output of the blast furnace.
   (d) all *(a), (b)* & *(c).*

185. Stamped charging of coal in the coke ovens facilitates
   (a) increase in the bulk density of coal charge to 1 ton/m^3 oven volume and hence lower specific heat consumption in coke making,
   (b) improvement in micum index of coke & output of coke oven.
   (c) use of non coking coals upto 15—30% to produce satisfactory coke.
   (d) all *(a), (b)* & *(c).*

186. Preheating of coal charge (upto 150 – 200°C) before charging in the coke ovens
   (a) improves the micum index of coke.
   (b) improves its bulk density and permits the use of more non coking coals in the blend.
   (c) reduces the tar & gas yield on carbonisation.
   (d) all *(a), (b)* & *(c).*

187. Dry quenching of coke
   (a) is done by nitrogen to recover its sensible heat.
   (b) reduces the specific heat consumption in coke making by 40%.
   (c) cools the coke upto 100°C.
   (d) all *(a), (b)* & *(c).*

Answers

179. (c)	180. (b)	181. (d)	182. (d)	183. (d)	184. (d)
185. (d)	186. (d)	187. (d)			

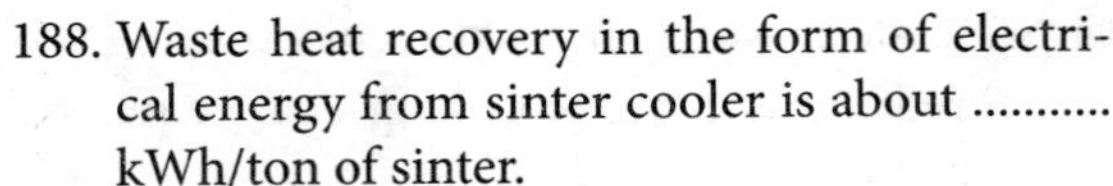

188. Waste heat recovery in the form of electrical energy from sinter cooler is about .... kWh/ton of sinter.
(*a*) 1— 2 (*b*) 10—15
(*c*) 100—200 (*d*) 800—1000

189. Secondary energy output in the form of blast furnace gas is about .... percent of the input energy in the blast furnace (mainly coke).
(*a*) 5 (*b*) 30
(*c*) 65 (*d*) 85

190. Coke rate in Indian blast furnaces is double than that of modern blast furnaces abroad due to
(*a*) higher ash content.
(*b*) poor coke quality (i.e. micum index).
(*c*) lack of use of advanced techniques for B.F. operation.
(*d*) all (*a*), ,(*b*) & (*c*).

191. Increase in slag rate by 100 kg/ton hot metal increases the coke rate by .... kg/ton hot metal within the range of slag rate 250-350 kg/ton hot metal.
(*a*) 1—5 (*b*) 15—25
(*c*) 80—100 (*d*) 180—200

192. High slag rate (300-450 kg/ton hot metal) in Indian blast furnaces is due to
(*a*) low top pressure operation of blast furnace.
(*b*) high ash content in coke.
(*c*) improper burden distribution facilities.
(*d*) low sinter % in burden.

193. For every 100 kg of limestone replaced from B.F. burden (by increasing the % sinter in burden), there is a decrease of coke rate by .... kg/ton pig iron and an increase of output of blast furnace by 2.4 percent.
(*a*) 5 (*b*) 30
(*c*) 100 (*d*) 200

194. With increase in hot blast temperature by 100°C (in the hot blast temperature range of 900 - 1250°C), the coke rate reduces by .... kg/ton pig iron.
(*a*) 1-5 (*b*) 30-40
(*c*) 80-100 (*d*) 180-200

195. Humidification of hot blast by steam in a blast furnace
(*a*) helps in controlling/reducing the hearth temperature,
(*b*) helps in controlling the silicon % in pig iron.
(*c*) is practised with the use of high hot blast temperature or oxygen enrichment of hot blast.
(*d*) all (*a*), (*b*) & (*c*).

196. Blast humidity in blast furnace is maintained in the range of. .... gm/Nm³ of blast.
(*a*) 1—5 (*b*) 35—65
(*c*) 150—200 (*d*) 500—750

197. Hydrocarbon fuel (*e.g.*, furnace oil) injection in the blast furnace must be practised with
(*a*) oxygen enrichment of hot blast.
(*b*) high top pressure operation of blast furnace.
(*c*) very low hot blast temperature.
(*d*) humidification of hot blast.

198. Temperature drop of hot metal during its transportation from blast furnace to mixer can be reduced by
(*a*) reducing its track time & turn around time of laddles.
(*b*) use of torpedo laddles.
(*c*) covering the B.F. runner and putting coke breeze in hot metal laddies.
(*d*) all (*a*), (*b*) & (*c*).

Answers

188. (b)	189. (b)	190. (d)	191. (b)	192. (b)	193. (b)
194. (b)	195. (d)	196. (b)	197. (a)	198. (d)	

199. During iron manufacture in the blast furnace, the reaction represented by, $2Fe_2O_3 + 3CO$ -- $3CO_2 + 4\ Fe$, is called the .... reaction.
   (*a*) indirect reduction
   (*b*) direct reduction
   (*c*) solution loss
   (*d*) none of these.

200. During iron manufacture in the blast furnace, the reaction, $Fe_20_3 + 3C - 3CO + 2Fe$, is called the .... reaction.
   (*a*) direct reduction (*b*) indirect reduction
   (*c*) solution loss (*d*) calcination

201. The reaction, $C + CO_2 = 2CO$, occuring in the blast furnace is called the .... reaction.
   (*a*) reduction (*b*) solution loss
   (*c*) Neumann's (*d*) none of these

202. During blast furnace operation, if the metal tapping hole is not opened as per schedule and liquid iron & slag is allowed to remain in the hearth, then the
   (*a*) blast pressure will increase.
   (*b*) burden descent rate will be reduced.
   (*c*) both (*a*) & (*b*).
   (*d*) neither (*a*) nor (*b*).

203. During tapping of hot metal from the blast furnace,
   (*a*) blast pressure falls.
   (*b*) burden descent is fast & irregular.
   (*c*) bosh gas distribution is affected.
   (*d*) all (*a*), (*b*) & (*c*).

204. Which of the following technologies has the least effect on the coke rate reduction in the blast furnace ?
   (*a*) Reduction in ash content in coke.
   (*b*) Oxygen enrichment of blast.
   (*c*) Coal dust injection in blast furnace.
   (*d*) Increase in top pressure.

205. With increase in the top pressure of the blast furnace, the
   (*a*) reduction rate of iron oxide increases.
   (*b*) contact time between solid & gas increases.
   (*c*) velocity of the gases through the furnace is reduced.
   (*d*) all (*a*), (*b*) & (*c*).

206. Oxygen enrichment of hot blast
   (*a*) is practically done up to maximum 25% oxygen in the blast.
   (*b*) requires humidification of hot blast.
   (*c*) requires natural gas/fuel oil injection in the blast furnace.
   (*d*) all (*a*), (*b*) & (*c*).

207. Humidification of hot blast
   (*a*) affects the temperature level in the combustion zone adversely.
   (*b*) controls the silicon content in pig iron.
   (*c*) provides extra reductant in the form of hydrogen.
   (*d*) all (*a*), (*b*) & (*c*).

208. Full advantages of auxiliary fuel injection (*e.g.* coal dust/natural gas/fuel oil) in the blast furnace is achieved, when it is accompanied with the use of
   (*a*) oxygen enrichment of blast.
   (*b*) humidification of blast.
   (*c*) low hot blast temperature.
   (*d*) low top pressure operation

209. In case of fuel oil injection in the blast furnace, the coke replacement ratio is about
   (*a*) 0.2 (*b*) 1.5
   (*c*) 2 (*d*) 3.5

Answers

199. (a)	200 (a)	201. (b)	202. (c)	203. (d)	204. (b)
205. (d)	206. (d)	207. (d)	208. (a)	209. (b)	

210. 1 Nm^3 of natural gas when injected in the blast furnace, reduces the coke consumption by .... kg.
(*a*) 0.1 (*b*) 0.75
(*c*) 2 (*d*) 3.5

211. Solution loss reaction (C + CO2 - 2CO) is also called the .... reaction.
(*a*) Boudourd (*b*) Neumann
(*c*) reduction (*d*) none of these

212. External desulphurisation (*i.e.* outside blast furnace) of pig iron is done by
(*a*) mixing lime with pig iron.
(*b*) oxygen lancing in transfer ladle.
(*c*) adding phosphoric acid in the pig iron.
(*d*) none of these.

213. External desiliconisation of pig iron is done by
(*a*) mixing lime with pig iron.
(*b*) oxygen lancing (using consumable mild steel pipe) in a transfer ladle.
(*c*) mixing CaC_2 with pig iron.
(*d*) none of these.

214. Maximum carbon percentage, which can be dissolved in pure iron at just above its melting point, is about
(*a*) 2.5 (*b*) 5
(*c*) 7.5 (*d*) 10

215. Presence of silicon in cast iron
(*a*) ranges from 1 to 3 percent.
(*b*) increases the formation of graphite by the decomposition of Fe_3C.
(*c*) controls its hardness to produce hard white iron to soft grey iron.
(*d*) all (*a*), (*b*) & (*c*).

216. Presence of phosphorous in cast iron
(*a*) decreases its fluidity.
(*b*) increases its melting point.
(*c*) reduces its shrinkage.
(*d*) increases its tensile strength, if it exceeds 0.5%.

217. Indian iron ores are mostly
(*a*) hematites (*b*) magnetites
(*c*) siderites (*d*) laterites

218. Total theoretical iron percentage in hematite is about
(*a*) 60 (*b*) 70
(*c*) 80 (*d*) 90

219. Hematite is
(*a*) bluish grey to dark red in colour.
(*b*) more easily reducible than magnetite.
(*c*) the world's main iron ore.
(*d*) all (*a*), (*b*) & (*c*).

220. Pick out the wrong statement.
(*a*) Sulphur present in iron ore upto 1% can be removed in the blast furnace in the form of slag.
(*b*) Sulphur present in iron ore fines can be reduced during its sintering.
(*c*) All the phosphorous present in iron ore can be removed in the blast furnace in the form of slag.
(*d*) The specific gravity of hematite & magnetite is about 5.

221. An iron ore is called 'acid ore', if its .... percent.
(*a*) sulphur content is > 1
(*b*) phosphorous content is < 0.05
(*c*) sulphur content is < 1
(*d*) phosphorous content is > 0.05

Answers

210. (b)	211. (a)	212. (a)	213. (b)	214. (b)	215. (c)
216. (c)	217. (a)	218. (b)	219. (d)	220. (c)	221. (b)

222. Al_2O_3/SiO_2 ratio in iron ore should be low, because
   (*a*) Al_2O_3 has a lower melting point than iron.
   (*b*) higher temperature will be required to be maintained in B.F.
   (*c*) B.F. can then operate at lower temperature, requiring less coke.
   (*d*) more quartzite will be needed to reduce the slag fluidity in B.F.

223. Sulphur in B.F. grade coke should not exceed .... percent.
   (*a*) 0.1 (*b*) 1
   (*c*) 3 (*d*) 5

224. For production of acid grade pig iron, phosphorous in coke should not exceed .... percent.
   (*a*) 0.01 (*b*) 0.03
   (*c*) 0.1 (*d*) 0.25.

225. Non-coking coal dust injection is done in blast furnace to reduce the coke rate. The replacement ratio of coke to coal is about (for the same ash contents)
   (*a*) 1 : 1 (*b*) 0.5 : 1
   (*c*) 1.5 : 1 (*d*) 2 : 1

226. Insoluble contents of blast furnace grade limestone should not exceed .... percent.
   (*a*) 3—5 (*b*) 8—12
   (*c*) 20—25 (*d*) 30—35.

227. Pick out the wrong statement.
   (*a*) B.F. slag is lighter than pig iron.
   (*b*) Blast furnace gas is used in heating of stoves without cleaning.
   (*c*) B.F. slag is hotter than pig iron, when tapped from B.F.
   (*d*) none of these.

228. In blast furnace operating on humidified blast, the reducing agent is
   (*a*) CO_2 (*b*) H_2
   (*c*) CO (*d*) both (*b*) & (*c*).

229. The calorific value of blast furnace gas is about .... of that of coke oven gas.
   (*a*) half (*b*) one third
   (*c*) one fifth (*d*) one eighth

230. In the pig iron, suitable for basic steel making .... percent.
   (*a*) silicon should be < 1
   (*b*) sulphur should be < 0.04
   (*c*) both (*a*) & (*b*)
   (*d*) neither (*a*) nor (*b*)

231. Foundry grade pig iron has
   (*a*) 2—3% silicon
   (*b*) < 0.06% sulphur
   (*c*) 0.5—1% phosphorous
   (*d*) all (*a*), (*b*) & (*c*).

232. Pick out the wrong statement pertaining to sulphur control in blast furnace operation.
   (*a*) Sulphur has a greater affinity for manganese than for iron, forming MnS which is readily soluble in the slag, whereas FeS is absorbed by the metal.
   (*b*) Sulphur can be controlled by increasing the basicity / slag ratio *i.e*, MgO/SiO_2 thus producing a more limey slag with greater affinity for sulphur.
   (c) Sulphur can be controlled by increasing the slag volume (so as to dilute the sulphur content) and by increasing the slag fluidity.
   (*d*) all (*a*), (*b*) & (*c*)

Answers

222. (c)	223. (b)	224. (b)	225. (a)	226. (b)	227. (b)
228. (d)	229. (c)	230. (c)	231. (d)	232. (d)	

233. Three major constituents of blast furnace slag are SiO_2, AI_2O_3 and
(a) CaO *(b)* FeO
(c) CaS *(d)* MnO

234. The difference between the slag temperature and the metal temperature in the blast furnace is about .... ^{0}C.
(a) 10 *(b)* 50
(c) 250 *(d)* 450

235. Which of the following is not injected in the blast furnace to reduce the coke rate ?
(a) Tar & furnace oil.
(b) Natural gas.
(c) Non-coking coal dust.
(d) Coking coal dust.

236. Largest capacity blast furnaces in India are installed at Visakhapatnam steel plant. The useful volume of each blast furnace there is .... m^3.
(a) 1500 *(b)* 2000
(c) 3200 *(d)* 4000

237. Tallest coke ovens (7 metres tall) are installed in India at .... steel plant.
(a) Bokaro *(b)* Salem
(c) Durgapur *(d)* Visakhapatnam

238. Maximum top pressure maintained in Indian B.F. is about 2.2 kg/cm^2. These blast furnaces are installed in the .... steel plant.
(a) Bokaro *(b)* Bhilai
(c) Visakhapatnam *(d)* Durgapur

239. .... tons of blast furnace gas is produced per ton of pig iron made in the blast furnace.
(a) 0.5 *(b)* 1
(c) 2.5 *(d)* 5.

240. Pick out the wrong statement pertaining to iron blast furnace operation.
(a) Acidic slag and low blast furnace temperature tend to produce low silicon, but high sulphur pig iron,
(b) Basic slag and a high furnace temperature tend to produce a low silicon, low sulphur pig iron.
*(a)*Slag which is basic enough to absorb sulphur, but not basic enough to reduce the silicon and a high furnace temperature tend to produce high silicon & low sulphur pig iron.
(d) none of these.

241. Pick out the correct statement.
(a) Double bell burden charging system employed in blast furnace is more effective than bell-less top (BLT) charging system,
(b) Normal speed of a sinter machine is about 3 metres/minute.
(c) Use of the super fluxed sinter in the blast furnace needs additional flux charging in the blast furnace.
(d) The basicity of super fluxed sinter is half than that of the slag produced in the blast furnace.

242. Percentage of moisture maintained in the sinter charge is about
(a) 2 *(b)* 7
(c) 10 *(d)* 15

243. Maximum temperature in the blast furnace is generated in front of the tuyers, which is about .... ^{0}C.
(a) 1500 *(b)* 2000
(c) 2500 *(d)* 3000

Answers

233. (a)	234. (b)	235. (d)	236. (c)	237. (d)	238. (c)
239. (c)	240. (d)	241. (b)	242. (b)	243. (b)	

244. Reduction reaction in the blast furnace is promoted by the .... of the charge.
(*a*) lower permeability
(*b*) non-uniform size
(*c*) lower porosity
(*d*) none of these

245. The thermal efficiency of blast furnace .... stoves may be about percent.
(*a*) 25 (*b*) 50
(*c*) 75 (*d*) 95

246. Blast furnace stoves are heated by
(*a*) waste flue gases.
(*b*) coke oven gas.
(*c*) B.F. gas or mixed gas.
(*d*) super heated steam.

247. Preheating of sinter charge mix in the balling drum is done upto a temperature of about .... °C.
(*a*) 40 (*b*) 65
(*c*) 100 (*d*) 150.

248. Preheating of sinter charge mix in the balling drum is done by .... in the balling drum.
(*a*) charging hot sinter return (which is at 80 - 90°C)
(*b*) burning coke oven gas
(*c*) injecting steam
(*d*) all (*a*), (*b*) & ic)

249. For every increase of 2 mm of sinter bed height, ... million calorie of heat can be saved.
(*a*) 1 (*b*) 5
(*c*) 10 (*d*) 20

250. For every increase of 25 mm of sinter bed height, a power saving of. .... kWh/ton of sinter results.
(*a*) 1 (*b*) 25
(*c*) 50 (*d*) 100

251. Insufficient strength of the sinter produced can be controlled by
(*a*) increasing the sinter machine speed.
(*b*) decreasing the coke breeze addition,
(*c*) increasing the bed height & suction,
(*d*) all (*a*), (*b*) & (*c*).

252. Iron (Fe) content in sinter is about .... percent.
(*a*) 38 (*b*) 53
(*c*) 67 (*d*) 77

253. Reduction degradation index (RDI) of sinter is improved (i.e. reduced) by increased .... content.
(*a*) MgO, SiO_2 & FeO
(*b*) AI2O.3
(*c*) alkali
(*d*) Al_2O_3/SiO_2 ratio in fine raw mix

254. High reducibility index of sinter (65%) is achieved by reducing the
(*a*) sinter bed height.
(*b*) FeO & SiO_2 in sinter.
(*c*) strength of sinter.
(*d*) suction in the sinter machine.

255. Preheating of sinter charge mix results in the improvement in
(*a*) sinter productivity.
(*b*) bed permeability.
(*c*) coke & electrical power consumption rate.
(*d*) all (*a*), (*b*) & (*c*).

Answers

244. (d)	245. (c)	246. (c)	247. (b)	248. (d)	249. (a)
250. (a)	251. (c)	252. (b)	253. (a)	254. (b)	255. (d)

256. MgO in sinter is maintained at about 2.5-3%. Higher MgO percentage in sinter results in increased .... of sinter.
(*a*) RDI (-3mm)
(*b*) reducibility
(*c*) cold strength
(*d*) all (*a*), (*b*) & (*c*)

257. FeO in Indian sinter is maintained at 8- 10%. Higher FeO percentage in sinter leads to .... of the sinter.
(*a*) poor reducibility
(*b*) higher RDI
(*c*) lower permeability in the B.F
(*d*) better high temperature properties

258. A good sinter should have higher .... content.
(*a*) basicity (1.6—2) & MgO (1.5—2.5%)
(*b*) $A1_2O_3$
(*c*) SiO_2 > 10%)
(*d*) FeO > 15%)

259. Hot air recovered from the sinter cooler is at a temperature of about .... °C
(*a*) 150 (*b*) 350
(*b*) 650 (*d*) 850

260. Specific coke breeze consumption in sinter making in Indian plants is about .... kg/ ton of skip sinter.
(*a*) 40 (*b*) 70
(*c*) 100 (*d*) 150

261. Specific electrical power consumption in Indian sinter plants is about .... kWh/ ton of gross sinter produced.
(*a*) 25 (*b*) 55
(*c*) 75 (*d*) 95

262. Tuyers in the blast furnace are located between
(*a*) iron notch and slag notch
(*b*) hearth and bosh zone
(*c*) bosh and stack
(*d*) big bell and small bell

263. Main constituents of blast furnace slag are
(*a*) SiO_2. FeO, CaO
(*b*) SiO_2, CaO, Al_2O_3
(*c*) CaO, MgO, SiO_2
(*d*) SiO_2, Al_2O_3, MgO

264. Sponge iron is produced, if iron oxide is reduced in the solid state with the help of a reducing gas like CO, H_2 etc. Sponge iron as compared to pig iron
(*a*) has lesser porosity.
(*b*) is intermixed with considerable slag.
(*c*) has higher hardness.
(*d*) none of these.

265. The passage through which molten pig iron flows the blast furnace to ladle is called
(*a*) iron notch (*b*) cinder notch
(*c*) runner (*d*) hearth

266. Pick out the wrong statement.
(*a*) Ferro-manganese can be manufactured in a blast furnace.
(*b*) Cawper stoves used for heating the blast (air) to be supplied to the blast furnace works on regenerative principle.
(*c*) Silica is a neutral fluxing material.
(*d*) Reduction efficiency of an efficient blast furnace may go upto 50%.

267. Tuyers in the blast furnace are cooled by .... circulation.
(*a*) water (*b*) air
(*c*) cooling oil (*d*) nitrogen

Answers

256. (c)	257. (a)	258. (a)	259. (b)	260. (b)	261. (b)
262. (b)	263. (b)	264. (b)	265. (c)	266. (c)	267. (a)

268. Which of the following does not involve 'agglomeration' technique?
(*a*) Sintering and pelletising
(*b*) Briquetting
(*c*) Nodullising
(*d*) None of these

269. Basicity of super-fluxed sinter is
(*a*) 0 (*b*) 1
(*c*) <1 (*d*) > 1

270. Which ofthe following is not removed from the charged material in the blast furnace?
(*a*) Sulphur (*b*) Gaunge
(*c*) Phosphorous (*d*) Alumina

271. Basicity of acidic sinter is
(*a*) > 4 (*b*) 1
(*c*) < 1 (*d*) > 1

272. Decomposition of limestone occurs in the .... region of the blast furnace.
(*a*) stack (*b*) bosh
(*c*) hearth (*d*) inter-bell space

273. External/ladle disiliconisation of pig iron is done by lancing with oxygen at 10 Kg/cm^2. To remove 0.1% silicon from the pig iron, the oxygen required is about .... m^3 .
(*a*) 1 (*b*) 17
(*c*) 93 (*d*) 257

274. .... iron is produced in cupola.
(*a*) Pig (*b*) Cast
(*c*) Wrought (*d*) High silicon

275. With one per cent of oxygen enrichment of hot blast in the blast furnace (upto the permissible limit), its productivity increases by about .... percent.
(*a*) 1 (*b*) 3
(*c*) 10 (*d*) 20

276. A blast furnace operating with high top pressure (say 1.5 - 2 Kg/cm^2) has higher
(*a*) burden descent velocity.
(*b*) top gas velocity.
(*c*) dust content in raw B.F. gas.
(*d*) none of these.

277. Which of the following factors does not favour desulphurisation in the blast furnace?
(*a*) Higher slag basicity
(*b*) Lower slag viscosity
(*c*) High FeO in slag
(*d*) None of these

278. FeO content in the slag is the lowest in case of .... slag.
(*a*) L.D. converter
(*b*) blast furnace
(*c*) open hearth furnace
(*d*) electric reduction furnace

279. Basicity of self-fluxed sinter is
(*a*) >5 (*b*) 1
(*c*) <1 (*d*) > 1

280. Pick out the correct statement.
(*a*) The diameter of the hearth of the blast furnace is smaller than that of the bosh.
(*b*) Hearth is the hottest part of the blast furnace.
(*c*) Sintering is agglomeration of iron & carbon bearing fine materials to produce a lumpy, hard and non-porous material by complete fusion.
(*d*) A blast furnace can not operate with 100% sinter charge in its burden.

Answers

268. (d)	269. (d)	270. (c)	271. (c)	272. (a)	273. (b)
274. (b)	275. (b)	276. (a)	277. (c)	278. (b)	279. (b)
280. (a)					

281. Pick out the wrong statement.
   (*a*) Reduction of alumina takes place in the iron blast furnace.
   (*b*) B.F. slag ladles are not lined with refractory.
   (*c*) B.F. slag is used as a raw material for cement manufacture.
   (*d*) Vizag steel plant has the largest capacity blast furnace in India (having 3200 m^3 useful volume).

282. Blast furnace produces .... iron.
   (*a*) wrought (*b*) cast
   (*c*) pig (*d*) none of these

283. Presence of. .... in molten iron affects its fluidity most.
   (*a*) sulphur (*b*) silicon
   (*c*) phosphorous (*d*) manganese

284. In one day of normal blast furnace operation, maximum tonnage of .... is produced.
   (*a*) pig iron
   (*b*) flue dust
   (*c*) blast furnace gas
   (*d*) B.F. slag

285. Siderite is the .... ore of iron.
   (*a*) oxide (*b*) carbonate
   (*c*) sulphide (*d*) sulphate

286. Metal tap hole of the blast furnace is closed by using
   (*a*) lime powder paste
   (*b*) clay
   (*c*) plaster of paris
   (*d*) alumina powder paste

287. Melting point of pure iron is .... °C.
   (*a*) 1259 (*b*) 1379
   (*c*) 1539 (*d*) 1739

288. Desulphurisation in the blast furnace is favoured by the .... of the slag.
   (*a*) low basicity
   (*b*) high basicity
   (*c*) high temperature
   (*d*) both (*b*) & (*c*)

289. .... favours the reduction of manganese oxide in the blast furnace.
   (*a*) Low temperature
   (*b*) High temperature
   (*c*) High slag viscosity
   (*d*) both (*b*) & (*c*)

290. .... is the cause of 'hanging' in the blast furnace.
   (*a*) High blast temperature
   (*b*) High blast pressure
   (*c*) Charging of lump iron ore
   (*d*) none of these

291. Puddling process is meant for the conversion of pig iron to
   (*a*) wrought iron
   (*b*) mild steel
   (*c*) high silicon (14%) iron
   (*d*) stainless steel

292. Addition of calcium fluoride to the slag .... the sulphur transfer rate from metal to slag.
   (*a*) decreases
   (*b*) increases
   (*c*) does not change
   (*d*) may increase or decrease, depending upon the slag composition

Answers

281. (a)	282. (c)	283. (c)	284. (c)	285. (b)	286. (b)
287. (c)	288. (d)	289. (a)	290. (b)	291. (a)	292. (b)

293. Which of the following chemical reactions occuring in blast furnace is a direct reduction reaction ?
(a) $FeO + CO \rightarrow Fe + C0_2$
(b) $FeO+C \rightarrow Fe+CO$
(c) $Fe_3 O_4 + 4CO \rightarrow Fe + 4C0_2$
(d) $Fe_2O_3+3CO + 3CO \rightarrow 2Fe + 3CO_2$

294. Basic iron contains
(a) high silicon
(fe) low silicon
(c) high phosphorous
(d) both (b) & (c)

295. Iron in the blast furnace melts at only 1200°C,because
(a) slag present inside the B.F., reduces its melting temperature,
(b) CO present lowers its melting point.
(c) impurities (e.g., Si, C, Mn, S, P etc) present in it, lowers its melting temperature.
(d) none of these.

296. Pick out the correct statement.
(a) Pig iron is softer than pure iron,
(b) Melting point of pig iron is higher than that of pure iron,
(c) Pig iron is termed acid or basic depending on its sulphur content.
(d) Pig iron is heavier than B.F. slag.

297. Acidic iron is produced in the blast furnace (by maintaining highly basic slag), which is rich in
(a) phosphorous (b) sulphur
(c) silicon (d) manganese

298. Which of the following materials present in the blast furnace charge burden, descends to the tuyere level in the solid form ?
(a) Sinter (b) Lump iron ore
(c) Coke (d) Limestone

299. Which of the following reactions occuring in the blast furnace is not endothermic ?
(a) $CaCO_3 = CaO + CO_2$
(b) $SiO_2 + 2C = Si + 2 CO$
(c) $FeO + C = Fe + CO$
(d) none of these

300. Basic iron ore is the one, which contains more than .... percent phosphorous.
(a) 0.005 (b) 0.01
(c) 0.05 (d) 0.5

301. Reduction of manganese oxide ($MnO + C = Mn + CO$) in the blast furnace starts above a temperature of. .... °C.
(a) 500 (b) 1000
(c) 1450 (d) 2000

302. Installation of bell less top(BLT) system in modern blast furnaces helps in improved
(a) burden distribution.
(b) quality of pig iron.
(c) hearth cooling.
(d) calorific value of B.F. gas.

303. Oxygen enrichment of hot blast in the blast furnace results in
(a) less B.F. gas yield.
(b) higher hearth temperature.
(c) larger concentration of CO in the furnace.
(d) all (a), (b) & (c).

304. High top pressure operation of blast furnace results in
(a) suppression of dissociation of carbon monoxide.
(b) increase in upward gas velocity in the furnace.
(c) more loss of iron ore, sinter and coke fines in outgoing B.F. gas.
(d) none of these.

Answers

293. (b)	294. (d)	295. (c)	296. (d)	297. (c)	298. (c)
299. (d)	300. (c)	301. (c)	302. (a)	303. (d)	304. (a)

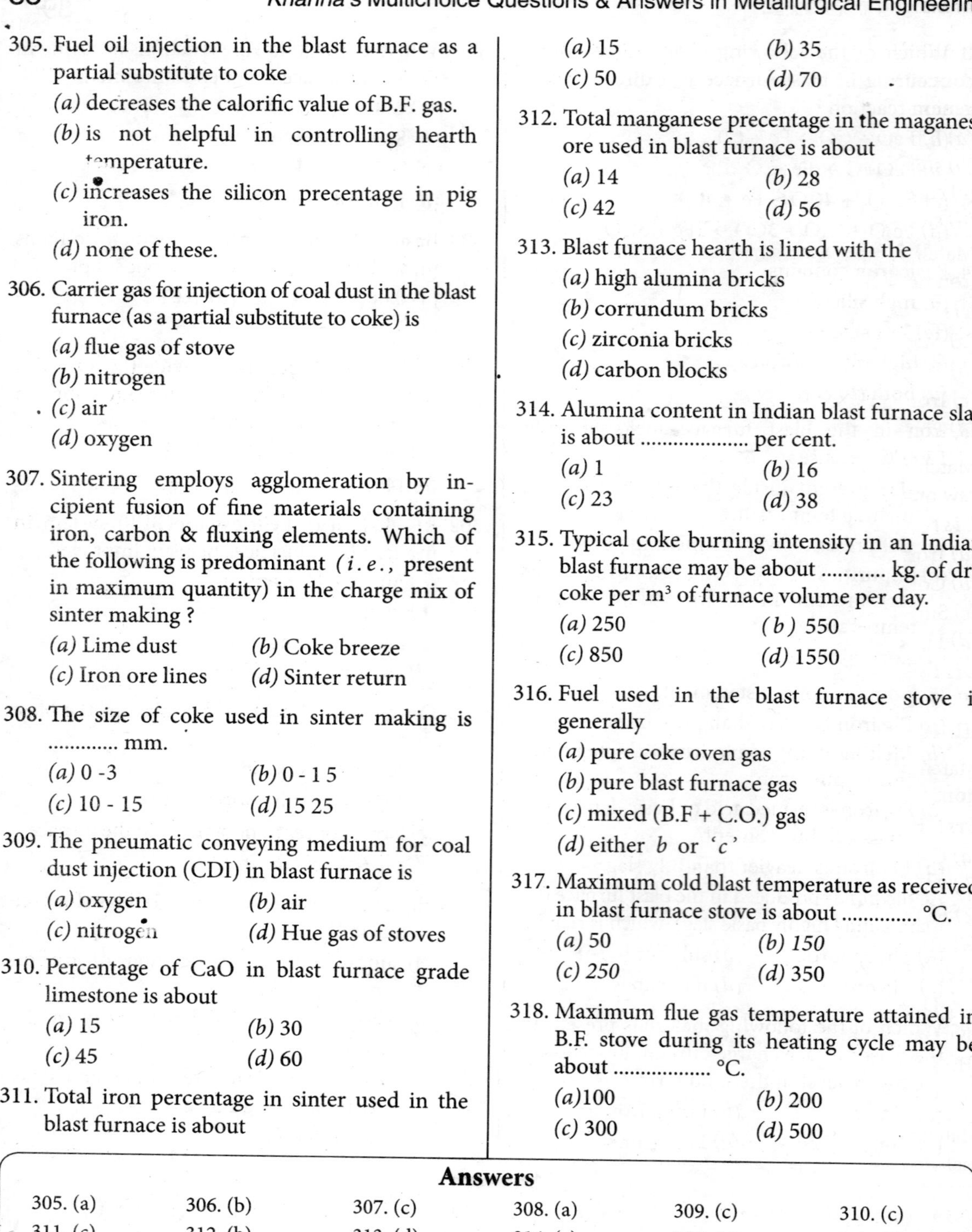

305. Fuel oil injection in the blast furnace as a partial substitute to coke
    (*a*) decreases the calorific value of B.F. gas.
    (*b*) is not helpful in controlling hearth temperature.
    (*c*) increases the silicon precentage in pig iron.
    (*d*) none of these.

306. Carrier gas for injection of coal dust in the blast furnace (as a partial substitute to coke) is
    (*a*) flue gas of stove
    (*b*) nitrogen
    (*c*) air
    (*d*) oxygen

307. Sintering employs agglomeration by incipient fusion of fine materials containing iron, carbon & fluxing elements. Which of the following is predominant (*i.e.*, present in maximum quantity) in the charge mix of sinter making ?
    (*a*) Lime dust (*b*) Coke breeze
    (*c*) Iron ore lines (*d*) Sinter return

308. The size of coke used in sinter making is .... mm.
    (*a*) 0 -3 (*b*) 0 - 1 5
    (*c*) 10 - 15 (*d*) 15 25

309. The pneumatic conveying medium for coal dust injection (CDI) in blast furnace is
    (*a*) oxygen (*b*) air
    (*c*) nitrogen (*d*) Hue gas of stoves

310. Percentage of CaO in blast furnace grade limestone is about
    (*a*) 15 (*b*) 30
    (*c*) 45 (*d*) 60

311. Total iron percentage in sinter used in the blast furnace is about
    (*a*) 15 (*b*) 35
    (*c*) 50 (*d*) 70

312. Total manganese precentage in the maganese ore used in blast furnace is about
    (*a*) 14 (*b*) 28
    (*c*) 42 (*d*) 56

313. Blast furnace hearth is lined with the
    (*a*) high alumina bricks
    (*b*) corrundum bricks
    (*c*) zirconia bricks
    (*d*) carbon blocks

314. Alumina content in Indian blast furnace slag is about .... per cent.
    (*a*) 1 (*b*) 16
    (*c*) 23 (*d*) 38

315. Typical coke burning intensity in an Indian blast furnace may be about .... kg. of dry coke per m^3 of furnace volume per day.
    (*a*) 250 (*b*) 550
    (*c*) 850 (*d*) 1550

316. Fuel used in the blast furnace stove is generally
    (*a*) pure coke oven gas
    (*b*) pure blast furnace gas
    (*c*) mixed (B.F + C.O.) gas
    (*d*) either *b* or '*c*'

317. Maximum cold blast temperature as received in blast furnace stove is about .... °C.
    (*a*) 50 (*b*) 150
    (*c*) 250 (*d*) 350

318. Maximum flue gas temperature attained in B.F. stove during its heating cycle may be about .... °C.
    (*a*)100 (*b*) 200
    (*c*) 300 (*d*) 500

Answers

305. (a)	306. (b)	307. (c)	308. (a)	309. (c)	310. (c)
311. (c)	312. (b)	313. (d)	314. (c)	315. (c)	316. (d)
317. (b)	318. (c)				

319. If the iron tapped out from blast furnace contains more than *2 %* silicon, then it is generally used for
(a) L.D. steel making
(b) steel ingot mould making
(c) casting pigs by pig casting machine
(d) steel ingot bottom plate making

320. Match the chemical nature of the various iron minerals.

List I	*List II*
(a) Magnetite	I. Carbonate
(b) Siderite	II. Hydroxide
(c) Iron pyrite	III. Sulphide
(d) Limonite	IV. Oxide

321. Match the ideal size (mm) ranges of various raw materials charged in a blast furnace.
List I
(a) Iron ore lump
(b) Coke
(c) Sinter
(d) Limestone
List II

I 5 to 50	II. 30 to 80
III. 10 to 35	IV. 25 to 80

322. Match the name of various chemical reactions taking place inside the blast furnace.
List I
(a) $C + H_2O = CO + H_2$
(b) $2\ Fe_2O_3 + 3CO = 3CO_2 + 4Fe$
(c) $Fe_20_3 + 3C = 3CO + 2Fe$
(d) $C + CO_2 = 2\ CO$
List II
1. Direct reduction reaction.
II. Gasification reaction.
III. Solution loss reaction.
IV. Indirect reduction reaction.

323. Match the operational parameters of a normal working blast furnace.
List I
(a) Al_2O_3 percentage in blast furnace slag.
(b) FeO percentage in sinter.
(c) N_2 percentage in blast furnace gas.
(d) Si percentage in pig iron.
List II

I. 9	II. 24
III. 0.8	IV. 55

324. Match the typical temperatures (°C) of various parameters encountered in a normal blast furnace operation.
List I
(a) Temperature of raw blast furnace gas.
(b) Temperature of cold blast.
(c) Temperature of blast furnace slag.
(d) Temperature of hot blast.
List II

I. 125	II. 1500
III. 1000	IV. 225

325. Match the reasons for various irregularities observed in the operation of an iron blast furnace.
List 1
(a) Scaffolding
(b) Chilling of hearth
(c) Channelling
(d) Hanging
List 11
1. Leaking of bosh & hearth cooling plates and sudden lowering of fixed carbon content of coke.
II. Action of volatile alkali metals, zinc, lead, tin & their compounds, low & wide softening temperature ranges of ores, presence of fines in sinter and carbon deposition.
III Increase in flame temperature caused due to increased hot blast temperature, decreased blast humidity and increased carbon content of coke.
IV. Due to non-uniform blast distribution at tuyere and drop in blast pressure.

Answers

319. (c)	320. (a)	321. (iv) (i) (iii) (ii)	322. (iii) (iv) (i) (ii)
323. (ii) (iv) (i) (iii)		324..(ii) (i) (iv) (iii)	325. (iv) (i) (ii) (iii)

326. Match the normal value (percentage) of various blast furnace operational parameters.

List I

(*a*) Thermal efficiency of blast furnace stove.

(*b*) Heat supplied by hot blast of the total heat requirement of the blast furnace.

(*c*) Maximum beneficial limit of oxygen enrichment of hot blast.

(*d*) Total Fe content in sinter charged in the biast furnace.

List II

I. 20-30 II. 75-80

III. 52-55 IV. 3-4

327. Match the techniques used for controlling the various blast furnace operational irregularities.

List I

(*a*) Hot hanging (*b*) Chocking of hearth

(*c*) Scaffolding (*d*) Channelling

List 11

I. By improving mechanical strength of coke, avoiding burning of tuyers, using more fluid bosh & hearth slag, giving blank charge of coke every 5-7 charges.

II. By lowering the hot blast temperature or using any coolant injectant (*e.g.* steam, fuel oil etc.), increasing biast humidity and decreasing carbon content of coke.

III. By reducing the blast pressure to 0.2 atm or by lowering hot biast temperature or by increasing the blast humidity.

IV. By decreasing the concentrations of alkali metals in 800-1100°C zone by increasing their exit in the slag to be achieved by decreasing the activity coefficient of the alkali oxides by lowering the $Ca0/SiO_2$ ratio (*i.e.*, basicity) of the slag.

328. Match the beneficial/deleterious effects of various operational parameters of blast furnace.

List *I*

(*a*) Oxygen enrichment of biast.

(*b*) High top pressure operation,

(*c*) Coal dust injection.

(*d*) Hearth chilling.

List II

I. Increased residence time of gas in the furnace, lower gas velocity and lower dust content in B.F. gas.

II. Increased productivity.

III. Reduced pig iron temperature.

IV. Increased top gas (B.F. gas) credit (MCal/ton pig iron).

329. Match the optimum temperatures (°C) required for various reactions occuring inside the blast furnace.

List I

(*a*) Decomposition of calcium carbonate of limestone.

(*b*) Carbon deposition reaction ($2CO = CO_2 + C$).

(*c*) Indirect reduction of magnetite to wustite (FeO),

(*d*) Hydrated (chemically bonded) water release from B.F. charge materials.

List II

450 - 600 700 - 900

300- 100 900 - 1100

330. Limestone is added in the blast furnace (during pig iron manufacture) to

(*a*) form slag by combining with impurities.

(*b*) supply heat by undergoing exothermic reaction with impurities.

(*c*) reduce the coke consumption.

(*d*) none of these.

Answers

326. (ii) (i) (iv) (iii) 327. (ii) (i) (iv) (iii) 328. (ii) (i) (iv) (iii)

329. (ii) (i) (iv) (iii) 330. (a)

331. Coke is added in iron blast furnace to
- *(a)* reduce & melt the iron ore.
- *(b)* increase the fluidity of the slag.
- *(c)* remove sulphur & phosphorous from molten metal.
- *(d)* none of these.

332. Blast furnace slag is mainly molten
- *(a)* sand
- *(b)* magnesium silicate
- *(c)* calcium silicate
- *(d)* iron pyrite

333. The main reducing agent in blast furnace is
- *(a)* air
- *(b)* carbon monoxide
- *(b)* hydrogen
- *(d)* limestone

334. Dust content in raw B.F. gas is about .... gm/Nm^3, hence it is cleaned to the level of about <10 mg/Nm^3.
- *(a)* 15-30
- *(b)* 1-2
- *(c)* 70-100
- *(d)* 150-200

335. The approximate height of a blast furnace having a useful volume of 2000 m^3 is about .... metres.
- *(a)* 20
- *(b)* 35
- *(c)* 60
- *(d)* 80

336. The height of blast furnace stoves meant for supplying hot blast at 1000°C to the blast furnace of 2000 m^3 useful volume is about .... metres.
- *(a)* 15
- *(b)* 25
- *(c)* 45
- *(d)* 65

337. Pick out the wrong statement
- *(a)* Gross calorific value of B.F gas is very low, because of its high inert content.
- *(b)* Blast furnace tuyers are made of bronzes.
- *(c)* Zinc vapor produced in the iron blast furnace damages its refractory lining, if zinc oxide is present in the iron ore.
- *(d)* Mould for the steel ingot is made of wrought iron.

338. The diameter of the blast furnace is the maximum at the
- *(a)* hearth
- *(b)* bosh
- *(c)* stack
- *(d)* throat

339. In the top portion of the stack of shaft of the blast furnace, the
- *(a)* burden material is completely in solid form.
- *(b)* melting of the burden (except coke) takes place.
- *(c)* gangue & flux combine to form the slag.
- *(d)* iron oxide reduction just starts.

340. The value of an iron ore is drastically reduced by the presence of
- *(a)* lime & magnesia
- *(b)* alumina
- *(c)* alkali oxides
- *(d)* both *b* & *c*

341. Blast furnace coke should contain a maximum of 2% of
- *(a)* volatile matter
- *(b)* phosphorous
- *(c)* ash
- *(d)* sulphur

342. Chief source of sulphur in the blast furnace is
- *(a)* iron ore
- *(b)* coke
- *(c)* sinter
- *(d)* limestone

343. Flux (*e.g.*, limestone/dolomite) added during smelting in blast furnace does not reduce the
- *(a)* coke consumption rate.
- *(b)* slag viscosity.
- *(c)* softening point of the gauge material.
- *(d)* activity of some components to make them stable or unstable in the slag phase.

Answers

331. (a)	332. (c)	333. (b)	334. (a)	335. (c)	336. (c)
337. (d)	338. (b)	339. (a)	340. (d)	341. (a)	342. (b)
343. (a)					

344. Pick out the wrong statement.
   (*a*) Burden composition of the blast furnace determines the type & amount of flux required.
   (*b*) The value of flux *e.g.,* limestone/dolomite used in B.F[1] is expressed in terms of available base and it should contain $Al_2O_3 < 3\%$.
   (*c*) The lower is the softening point of the slag, the lower is its viscosity at a given temperature.
   (*d*) In normal blast furnace operation, the basicity $(CaO + MgO)/SiO_2$, of the slag ranges from 3 to 3.5.

345. Average coke size to other burden size ratio in the blast furnace charge should be about
   (*a*) 1 : 1 (*b*) 2 : 1
   (*c*) 4 : 1 (*d*) 8 : 1

346. Maximum height of the modern blast furnace is decided mainly by the
   (*a*) strength of the coke available.
   (*b*) maximum wind rate to be blown
   (*c*) production capacity of the blast furnace.
   (*d*) maximum hot blast temperature to be used.

347. The production rate of the blast furnace is directly determined by the rate of reduction of iron oxide & rate of heating of the descending burden, which in turn depends on the degree & time of contact of the gases with the burden in the .... region which, is a diffusion controlled counter-current gas-solid reactor.
   (*a*) bosh (*b*) throat
   (*c*) hearth (*d*) stack

348. Pick out the wrong statement about blast furnace operation.
   (*a*) In double bell material charging system of a B.F. the big bell forms an angle of 50-53° from the horizontal.
   (*b*) True specific gravities of iron ore, limestone & coke are respectively 5-6, 3- 3.5 & 1.5, which implies that rolling tendency of coke particles is maximum & that of iron ore is minimum.
   (*c*) The reduction rate is drastically reduced with increasing percentage of fines in the blast furnace burden,
   (*d*) The lower is the angle of repose of B.F. burden materials, the more it has the tendency to form the ridges on charging in the B.F.

349. Optimum size range of iron ore lump to be charged in B.F is 10-30 mm. Charging of too large size ore lump will result in increased
   (*a*) productivity of B.F.
   (*b*) total demand of heat.
   (*c*) flow of unreduced material down the stack into the tuyere area.
   (*d*) both b & c.

350. Pick out the wrong statement pertaining to efficient blast furnace operation leading to its high productivity. For achieving optimum level of productivity in B.F., the
   (*a*) Al_2O_3/Fe ratio of burden should be < 0.05.
   (*b*) Al_2O_3/SiO_2 ratio of the burden should be <l.
   (*c*) iron content of the ore should be above 60%.
   (*d*) Sinter in the burden should be maximum 50%.

351. Which of the following is not an objective of sintering process ?
   (*a*) To incorporate flux in the burden.
   (*B*) To increase the size of iron ore fines to form a strong agglomerates with high bulk reducibility.
   (*c*) To convert Fe of iron ore to FeO, which has better reducibility.
   (*d*) To remove volatiles like CO_2 from carbonates, sulphur from sulphides & H_2O from hydroxides.

Answers

344. (d)	345. (c)	346. (a)	347. (d)	348. (d)	349. (d)
350. (d)	351. (c)				

352. Binder used for making stronger pellets of iron bearing material is
   (a) bentonite (b) tar
   (c) lime (d) pitch

353. Desulphurisation in blast furnace is not favored by
   (a) raising the hearth temperature.
   (b) making the slag acidic in nature.
   (c) having higher basicity of the slag.
   (d) increasing the slag volume.

354. Pick out the wrong statement about the use of pellets in the blast furnace charge.
   (a) The swelling properties of pellets is a result of transformation of hematite to magnetite during reduction.
   (a) Size degradation of pellets during transportation is much less than that of sinter.
   (c) Excessive swelling of pellets improves its crushing strength & permeability of the blast furnace burden.
   (d) Pellets with more than 40% swelling should not be used in B.F burden, however pellets with less than 20% swelling can be used upto 70% in the B.F. burden.

355. Silicon & sulphur content in the pig iron produced in the blast furnace does not depend upon the
   (a) furnace operating temperature
   (b) blast pressure
   (c) slag volume
   (d) slag basicity

356. For reasons of poor off take of pig iron to the steel melting shop because of its shut-down/breakdown, the production rate of blast furnace is reduced by reducing the blast input. If the blast is reduced to a value less than .... percent of the normal value. the technique is termed as the **fanning**.
   (a) 10 (b) 25
   (c) 50 (d) 75

357. Hanging in a blast furnace does not take place because of the
   (a) use of larger average size of the burden materials, particularly coke.
   (b) condensation of alkali oxide vapor in the upper part of the stack, thereby cementing the charge into an impervious mass.
   (c) carbon deposition (due to Neumann reversal reaction) in voids in the stack resulting in decrease in burden permeability.
   (d) solidification of previously fused slag in bosh particularly in sinter, into a large impervious mass.

358. Scaffold is a large mass of material that gets stuck to the blast furnace wall, a single, block particularly in the top portion of the bosh. Formation tendency of scaffold is not due to the
   (a) presence of high amount of alkali oxides in the burden.
   (b) presence of chlorides or other compounds of alkali metals in the burden,
   (c) sudden lowering of the fusion zone adjacent to the wall.
   (d) increasing refractoriness of the slag.

Answers

352. (a) 353. (b) 354. (c) 355. (b) 356. (b) 357. (a)
358. (b)

359. Scaffold formation tendency in the blast furnace can be reduced by reducing the
   (a) alkali content of the burden,
   (b) coke rate in the furnace,
   (c) flux content in the burden.
   (d) heavy burden charging.

360. Scaffolding in a blast furnace does not cause
   (a) uneven movement of the burden.
   (b) rise in blast pressure.
   (c) decrease in furnace output.
   (d) decrease in coke consumption rate.

361. Chilled hearth in the blast furnace is not caused by the
   (a) excessive humidification of the blast.
   (b) leakage of water from the tuyers.
   (c) lower coke input in the furnace.
   (d) higher blast pressure.

362. The ratio of the density of the pig iron to that of the blast furnace slag is
   (a) 1.5 (b) 3
   (c) 5.5 (d) 8.5

363. Rate of reduction of iron oxide in the blast furnace is
   (a) slower in case of hematite as compared to magnetite.
   (b) reduced by the presence of associated gangue oxide.
   (c) more, if the temperature of the out going blast furnace gas is more.
   (d) not a diffusion controlled gas-solid reaction.

364. Pick out the wrong statement pertaining to the use of sinter in the blast furnace.
   (a) Acid sinter does not contain flux at all.
   (b) In case of fluxed sinter, the basicity of the sinter mix is equal to that of the slag to be produced in the blast furnace.
   (c) In case of super fluxed sinter, the basicity of the sinter mix is more than that of the slag to be produced in the blast furnace.
   (d) A blast furnace can not operate on < 80 % of fluxed sinter in the burden.

365. Hot blast enters the blast furnace through the tuyers at a velocity of about .... metre/second at a gage pressure of 2—4 kg/cm^2.
   (a) 10 — 20 (b) 50 — 60
   (c) 200 — 300 (d) 1000— 1500

366. Backward direction of the solution loss reaction in a blast furnace *e.g.*, $2CO = C + CO_2$, is known as the .... reaction.
   (a) Boudouard
   (b) Neumann reversal
   (c) shift conversion
   (d) gasification

367. Sensible heat of the blast provides about .... percent of the total process heat required in a blast furnace.
   (a) 10 (b) 30
   (c) 60 (d) 80

368. Which of the following portion of blast furnace is not lined with fireclay bricks ?
   (a) Stack
   (b) Bosh
   (c) Hearth bottom
   (d) Throat

369. Volume of the blast furnace slag is about .... times that of the hot metal (pig iron) of the same weight.
   (a) 1.5 (b) 3
   (c) 6 (d) 8

Answers

359. (a)	360. (a)	361. (d)	362. (b)	363. (b)	364. (d)
365. (c)	366. (b)	367. (b)	368. (c)	369. (b)	

370. Hanging in a blast furnace can not be controlled by

(*a*) reducing the concentration of alkali chlorides in the burden.

(*b*) reducing the hot blast pressure & temperature.

(*c*) blowing down the furnace to the bosh level & filling it with coke blank.

(*d*) charging large size limestone to facilitate improvement of bed permeability by generating CO_2 upon its dissociation, which in turn forces the solution less reaction to take place.

371. Burden slipping in the blast furnace *i.e.* sudden sinking of the stock caused by collapse of the hanging/wedging/ scaffold etc., may

(*a*) be due to the bad bosh design.

(*b*) cause heart h chilling or even explosion in severe cases.

(*c*) be controlled by adjustment of blast pressure & temperature.

(*d*) all *a*, *b* & c.

372. Formation of cold central coloumn of the stock with an annular hot zone all around it in a B.F is known as **pillaring,** which can be eliminated by

(*a*) increasing the blast pressure, which will penetrate more & heat up the pillar.

(*b*) increasing the coke input in the furnace.

(*c*) increasing the wind blowing rate in the furnace.

(*d*) increasing the hot blast temperature

373. Preferential flow of ascending gases through certain areas of burden, because of their relatively much better permeability is termed as **chanelling,** which

(*a*) may cause coke ejection through metal tap hole.

(*b*) is caused by wide size range & improper distribution of the burden in the B.F.

(*c*) reduces the effective cross-section of the furnace for gas-solid interaction and thereby decreases the B.F. productivity directly.

(*d*) both b & c.

374. Pick out the wrong statement pertaining to blast furnace operation.

(*a*) An increase in the bosh gas flow prevents the liquid metal/slag flow downwards through the permeable coke bed causing it to accumulate in the coke interstices. If the weight ofthe liquid overcomes the upward thrust of the gases & descends suddenly into the hearth, the phenomenon is known as **flooding.**

(*b*) Flooding in the blast furnace can be minimised by using higher mean size ofthe coke.

(*c*) Coke ejection through the tap hole of the B.F. can be remedied by its uniform blowing.

(*d*) Blast furnace hearth breakout may be caused by the chilling of hearth due to low coke input.

375. Match the colour/appearance of various iron ores.

List I

(*a*) Siderite (carbonate ore)

(*b*) Hemetite (oxide ore)

(*c*) Magnetite (oxide ore)

(*d*) Limonite (hydroxide ore)

List II

I. Brown

II. Black

III. Red or reddish black

IV. Grey or whitish

Answers

370. (a) 371. (d) 372. (a) 373. (d) 374. (d) 375. IV, III, II, I

376. Pig iron production rate of the blast furnace is directly not determined by the rate of
(*a*) heating of the burden
(*b*) burden descent velocity
(*c*) reduction of iron oxide
(*d*) all *a*, *b* and c

377. All the solid material present in the blast furnace hearth and bosh is
(*a*) flux (*b*) iron ore lump
(*c*) coke (*d*) gaunge

378. While the process of stopping the blast furnace operation at the end of its com- paign is termed as **blowing out;** the temporary shut down of blast furnace is called
(*a*) back-draughting
(*b*) banking
(*c*) drying
(*d*) salamander tapping

379. In external ladle deseliconisation of big iron (hot metal) by oxygen lancing depends upon the amount of flux bath temperature and rate and quantity of oxygen blown. The find level of silicon achieved after ladle deseliconisation is .... percent.
(*a*) 0.1-0.15 (*b*) .0.30 - 0.50
(*c*) 0.60-0.80 (*d*) 0.80-1.0

380. The elastic limit of cast iron is close to its ultimate breaking strength. High silicon (14%) cast iron has an important property of high
(*a*) hardness (*b*) brittleness
(*c*) tenacity (*d*) plåsticity

Answers

376. (b)	377. (b)	378. (b)	379. (b)	380. (a)

Section

A

4

Steel Making

1. Carbon percentage in high carbon steel is
   (*a*) < 0.25 (*b*) 0.25—0.65
   (*c*) 0.65—1.70 (*d*) 1.7—2.5
2. Carbon percentage in medium carbon steel is
   (*a*) < 0.25 (*b*) 0.25—0.65
   (*c*) 0.65—1.70 (*d*) 1.7—2.5
3. Carbon percentage in low carbon steel or mild steel is
   (*a*) < 0.25 (*b*) 0.25—0.66
   (*c*) 0.65—1.70 (*d*) 1.7—2.5
4. Total alloying percentage in low alloy steel is
   (*a*) < 5 (*b*) 5—10
   (*c*) > 10 (*d*) > 20
5. Total alloying percentage in medium alloy steel is
   (*a*) <5 (*b*) 5—10
   (*c*) > 10 (*d*) > 5
6. Total alloying percentage in high alloy steel is
   (*a*) < 5 (*b*) >5
   (*c*) 5-10 (*d*) > 10
7. 18/8 stainless steel is a .... alloy steel.
   (*a*) high (*b*) medium
   (*c*) low (*d*) none of these.
8. Stainless steel is a .... steel.
   (*a*) plain carbon (*b*) high carbon
   (*c*) high alloy (*d*) none of these.
9. With increase in steel melting furnace temperature, the slag viscosity
   (*a*) increases.
   (*b*) decreases.
   (*c*) remains same.
   (*d*) may increase or decrease, depends on the slag composition.
10. In general, with increase in liquidus temperature, the viscosity of the slag (at the steel making temperatures.)
    (*a*) increases (*b*) decreases
    (*c*) remains same (*d*) none of these
11. Tap to tap time is maximum in the case of
    (*a*) L.D. converter
    (*b*) Kaldo process
    (*c*) Bessemer converter
    (*d*) Open hearth furnace
12. IRSID/CAFL method of steel making employs
    (*a*) open combustion.
    (*b*) suppressed combustion.
    (*c*) acidic brick lining in the furnace.
    (*d*) none of these.

Answers

1. (c)	2. (b)	3. (a)	4. (a)	5. (b)	6. (d)
7. (a)	8. (c)	9. (b)	10. (a)	11. (d)	12. (b)

13. Which is not an autogenous process (*i.e.* requires external heating)?
    (*a*) Bessemer process
    (*b*) L.D. converter
    (*c*) Open hearth furnace
    (*d*) Basic oxygen furnace

14. Electric arc furnaces for steel making
    (*a*) are usually used for the manufacture of costly alloy steels only due to high cost of electrical energy.
    (*b*) usually runs on cold charge.
    (*c*) is of small capacity.
    (*d*) all (*a*), (*b*) & (*c*).

15. Which is a basic oxygen furnace (BOF) ?
    (*a*) Bessemer converter.
    (*b*) Basic open hearth furnace.
    (*c*) L.D. converter.
    (*d*) None of these.

16. IRSID / CAFL method of steel making is a modified version of the
    (*a*) Open hearth furnace
    (*b*) L.D. converter
    (*c*) Bessemer converter
    (*d*) Kaldo vessel

17. Scrap charging rate in L.D. converters of Indian steel plants is about .... kg/ton of crude steel.
    (*a*)100 (*b*) 200
    (*c*)300 (*d*) 400

18. Which property of slag is unimportant in steel melting furnaces from the point of view of rate of slag-metal, slag-gas and slag-refractory reactions ?
    (*a*) Viscosity
    (*b*) Thermal conductivity
    (*c*) Surface tension
    (*d*) None of these

19. Slag basicity is the ratio of
    (*a*) SiO_2 to CaO
    (*b*) CaO to SiO_2
    (*c*) basic to acid oxides
    (*d*) both (*b*) & (*c*).

20. In basic steel making process, the slag can be highly basic and
    (*a*) highly oxidising (*b*) quite reducing
    (*c*) both (*a*) & (*b*)
    (*d*) neither (*a*) nor (*b*)

21. In acid steel making processes, the slag is
    (*a*) having practically no basicity.
    (*b*) highly oxidising.
    (*c*) both (*a*) & (*b*).
    (*d*) neither (*a*) nor (*b*).

22. Slag basicity (CaO/SiO_2) of L.D. slag is
    (*a*) 0.01—0.05 (*b*) 0.5—1
    (*c*) 2—3.5 (*d*) 4—6.5

23. Which one produces minimum slag basicity?
    (*a*) L.D. converter
    (*b*) Basic open hearth furnace
    (*c*) Basic Bessemer converter
    (*d*) Acid open hearth furnace

24. Viscosity of highly basic slag can be reduced by the addition of
    (*a*) sand (*b*) bauxite
    (*c*) both (*a*) & (*b*) (*d*) neither (*a*) nor (*b*)

25. Finished steel develops cold shortness, if phosphorous content in it is more than......... percent.
    (*a*) 0.01 (*b*) 0.05
    (*c*) 0.5 (*d*) 1.

Answers

13. (c)	14. (d)	15. (c)	16. (b)	17. (b)	18. (d)
19. (d)	20. (c)	21. (c)	22. (c)	23. (d)	24. (c)
25. (b)					

26. Finished steel develops hot shortness, if sulphur content in it is percent.
 (a) 0.01 (b) 0.5
 (c) 0.05 (d) 1

27. When silicon is the chief impurity to be eliminated from iron, and S & P need not be eliminated, then we use .... steel making process.
 (a) acid (b) basic
 (c) either (a) or (b) (d) neither (a) nor (b).

28. When S & P are the chief impurities to be removed from iron along with even silicon & manganese, we use .... steel making process.
 (a) acid (b) basic
 (c) either (a) or (d) neither (a) nor (b).

29. In a basic steel making furnace, the
 (a) refractory lining is basic in nature.
 (b) slag is highly acidic in nature.
 (c) flux used is necessarily basic.
 (d) none of these.

30. Soda lime is far superior than lime for dephosphorisation of hot metal in steel making furnaces; still it is not used due to its
 (a) prohibitively high cost.
 (b) severe corrosive action on furnace refratories.
 (c) poor thermal stability.
 (d) none of these.

31. Which of the following constituents of slag can be controlled to effect the required dephosphorisation of the steel melt ?
 (a) CaO
 (b) SiO_2
 (c) FeO
 (d) all (a), (b) & (c)

32. Higher silicon content in pig iron (in basic steel making process) requires higher lime consumption to
 (a) maintain the required basicity.
 (b) effectively dephosphorise it.
 (c) both (a) & (b)
 (d) neither (a) nor (b).

33. Which is the most stable oxide product during refining of pig iron ?
 (a) SiO_2 (b) MnO
 (c) CaO (d) MgO

34. Steel scrap charging rate in the L.D. converter is reduced, if the
 (a) pig iron temperature is low.
 (b) loss on ignition (LOl) value of lime is high.
 (c) silicon content in pig iron is low.
 (d) all (a), (b) & (c).

35. In steel making, sulphur transfer (removal) from metal to slag is favoured by the .... of the slag.
 (a) high basicity.
 (b) high temperature.
 (c) low oxidising potential (reducing condition)
 (d) all (a), (b) & (c).

36. Killing of steel means removal of dissolved / residual .... from it.
 (a) oxygen (b) slag
 (c) sulphur (d) phosphorous

37. Which is used most extensively as a deoxidiser producing killed steel in practice?
 (a) Aluminium (b) Bauxite
 (c) Nickel (d) Zirconia

Answers

26. (c)	27. (a)	28. (b)	29. (a)	30. (b)	31. (d)
32. (c)	33. (a)	34. (d)	35. (d)	36. (a)	37. (a)

38. Which is not at all deoxidised ?
(*a*) Killed steel (*b*) Semi-killed steel
(*c*) Rimming steel (*d*) None of these

39. Tapping temperature of refined steel depends upon the
(*a*) chemical composition of bath.
(*b*) holding time in the laddle and the teeming practice.
(*c*) amount of recarburiser, deoxidisers and alloying additions to be added in the laddle after tapping the steel.
(*d*) all (*a*), (*b*) & (*c*).

40. The process of pouring molten steel into ingot moulds is called
(*a*) teeming (*b*) tapping
(*c*) charging (*d*) none of these

41. Slide gate system is used for molten metal pouring in
(*a*) mixer
(*b*) teeming laddle
(*c*) hot metal charging laddle
(*d*) none of these

42. A special category of basic iron which contains more than 1.5% phosphorous is known as the .... iron.
(*a*) delta (*b*) Thomas
(*c*) wrought (*d*) none of these

43. Purity of oxygen used in L.D. converter blowing should be at least .... percent.
(*a*) 92 (*b*) 95
(*c*) 99.5 (*d*) 97

44. Consumption of oxygen in L.D. steel making in Indian steel plants is typically about .... Nm^3/ton steel.
(*a*) 40—45 (*b*) 65—70
(*c*) 100--150 (*d*) 200--250

45. Scrap charging rate in L.D. converter can be maximised by
(*a*) increasing the iron scrap to steel scrap ratio in the scrap charge.
(*b*) increasing the charging hot metal temperature.
(*c*) using scrap in place of lime as bath coolant when the opening bath temperature exceeds the steel tapping temperature.
(*d*) all (*a*), (*b*) & (*c*).

46. Flux is added during steel refining to reduce the
(*a*) softening temperature of gangue materials.
(*b*) slag viscosity.
(*c*) activity of some components to make it stable in the slag phase.
(*d*) all (*a*), (*b*) & (*c*).

47. Pick out the wrong statement.
(*a*) Lime is added during steel making to make the slag basic enough to retain P &S.
(*b*)Bauxite is added to decrease the viscosity of refining slags.
(*c*) both (*a*) & (*b*).
(*d*) neither (*a*) nor (*b*).

48. The refractory lining of tar dolomite bricks in L.D. converter in Indian steel plant lasts for about .... heats.
(*a*) 50 (*b*) 200
(*c*) 500 (*d*) .1000

49. Ferro-silicon is added in the molten steel laddle to .... it.
(*a*) carburise (*b*) deoxidise
(*c*) cool (*d*) solidify

Answers

38. (c)	39. (d)	40. (a)	41. (b)	42. (b)	43. (c)
44. (b)	45. (d)	46. (d)	47. (d)	48. (b)	49. (b)

50. Titanium when added to molten steel in the laddle
    (a) acts as a strong deoxidiser.
    (b) neutralises the effect of nitrogen on steel by bonding it into stable insoluble nitride compounds.
    (c) makes it useful for aircraft industry due to its lightness coupled with strength.
    (d) all (a),(b)) & (c).

51. Addition of which of the following does not help in the deoxidation of molten steel ?
    (a) Nickel (b) Chromium
    (c) Silicon (d) Manganese

52. Addition of vanadium (is done in structural, tool and spring steel) in steel
    (a) helps in its deoxidation.
    (b) increases its strength and plasticity.
    (c) imparts resistance to attrition and impact.
    (d) all (a), (b) & (c)

53. Which is the weakest deoxidiser of molten steel ?
    (a) Titanium (b) Vanadium
    (c) Tungsten (d) Chromium

54. Addition of tungsten (is done in tool and high speed steels) in steel increases its
    (a) hardness (b) strength
    (c) elasticity (d) all (a), (b) & (c)

55. Addition of molybdenum (is done in steel used for gears, rolls, shafts etc.) in steel
    (a) ensures uniform microcrystalline structure.
    (b) augments hardeneability.
    (c) eliminates post temper brittleness.
    (d) all (a), (b) & (c).

56. Zirconium when added to molten steel
    (a) acts as a deoxidiser.
    (b) decreases the deleterious effects of nitrogen & sulphur in steel.
    (c) both (a) & (b)
    (d) neither (a) nor (b)

57. Boron addition as an alloying element in molten steel
    (a) helps in its deoxidation.
    (b) enhances the mechanical properties.
    (c) improves the hardeneability.
    (d) all (a), (b) & (c).

58. Niobium addition to stainless steel and heat resisting steel does not improve their
    (a) plasticity
    (b) corrosion resistance
    (c) welding properties
    (d) none of these

59. Which of the following alloying elements are added to molten steel in elemental form ?
    (a) Silicon (b) Cobalt
    (c) Boron (d) Zirconium

60. Hot metal mixer
    (a) is heated with coke oven gas.
    (b) homogenises the hot metal received from different blast furnaces.
    (c) stores the metal as a buffer stock.
    (d) all (a), (b) & (c).

61. Hot metal (i.e., pig iron) temperature in mixer is maintained at about .... °C.
    (a) 700 (b) 900
    (c) 1100 (d) 1300.

62. Efficiency of a steel making process is assessed by the .... of steel.
    (a) refractory consumption per ton.
    (b) yield.
    (c) production rate & quality.
    (d) all (a), (b) & (c).

Answers

50. (d)	51. (a)	52. (d)	53. (d)	54. (d)	55. (d)
56. (c)	57. (d)	58. (d)	59. (b)	60. (d)	61. (d)
62. (d)					

63. Metallic loss during steel making is due to the
 (*a*) oxidation of metalloids.
 (*b*) rejection as slopping & sparking and skull formation on the mouth of the converter.
 (*c*) loss of iron as fume & dust and in slag (both chemical & mechanical).
 (*d*) all (*a*), (*b*) & (*c*).

64. Metallic yield of steel during steel making in Indian steel plant is about .... percent.
 (*a*) 67 (*b*) 77
 (*c*) 87 (*d*) 97

65. Steel yield can be increased by
 (*a*) charging purer hot metal in the converter/furnace.
 (*b*) charging more steel scrap in the converter.
 (*c*) keeping the slag volume low.
 (*d*) all (*a*), (*b*) & (*c*).

66. The annual production rate of a steel making furnace does not depend upon the .... time.
 (*a*) tap to tap (*b*) compaign
 (*c*) down (*d*) none of these

67. Total time required for converting a batch of raw materials into steel including handling time is called the .... time.
 (*a*) tap to tap or heat
 (*b*) compaign
 (*c*) down
 (*d*) blowing

68. Duration before the steel melting furnace is again due for relining is called the .... time.
 (*a*) heat (*b*) campaign
 (*c*) down (*d*) blowing

69. Interval between two consecutive campaigns of steel making furnaces is called the .... time.
 (*a*) down (*b*) blowing
 (*c*) heat (*d*) campaign

70. Production rate of steel making furnace can be increased by decreasing the .... time.
 (*a*) campaign & blowing
 (*b*) down
 (*c*) tap to tap
 (*d*) all (*a*), (*b*) & (*c*)

71. Typical tapping temperature of steel in L.D. converter is about .... °C.
 (*a*) 1300 (*b*) 1440
 (*c*) 1640 (*d*) 1840

72. Alloying elements in steel which tend to form carbides are
 (*a*) chromium & tungsten.
 (*b*) titanium & vanadium.
 (*c*) manganese & molybdenum.
 (*d*) all (*a*), (*b*) & (*c*).

73. Which alloying elements of steel tends to graphitise the carbide ?
 (*a*) Silicon and cobalt
 (*b*) Aluminium
 (*c*) Nickel
 (*d*) All (*a*), (*b*) & (*c*).

74. Alloying elements in steel which tend to stabilise austenite are
 (*a*) copper and cobalt
 (*b*) manganese
 (*c*) nickel
 (*d*) all (*a*), (*b*) & (*c*)

Answers

63. (d)	64. (c)	65. (d)	66. (d)	67. (a)	68. (b)
69. (a)	70. (d)	71. (c)	72. (d)	73. (d)	74. (d)

75. Which alloying elements in steel tend to stabilise ferrite ?
   (a) Chromium & tungsten
   (b) Molybdenum & vanadium
   (c) Silicon
   (d) All (a), (b) & (c)

76. Specific lime consumption in L.D. converter is about .... kg/ton crude steel.
   (a) 50 (b) 90
   (c) 180 (d) 320

77. Which has the most shallow bath ?
   (a) Open hearth furnace
   (b) L.D. converter
   (c) Kaldo rotary furnace
   (d) Bessemer converter

78. Open hearth furnaces can be heated by burning
   (a) oil (b) gas
   (c) pulverised coal (d) all (a), (b) & (c)

79. In laddle desiliconisation of hot metal, the amount of silicon oxidised depends upon
   (a) its down concentration.
   (b) bath temperature.
   (c) amount of flux added and rate & quantity of oxygen blown.
   (d) all (a), (b) & (c).

80. With increasing current in the circuit in electric arc furnace, the
   (a) arc voltage decreases.
   (b) power factor decreases.
   (c) power consumption increases upto a certain limit and then decreases.
   (d) all (a), (b) & (c).

81. Which steel making furnace has no oxidising atmosphere of its own?
   (a) L.D. converter
   (b) Open hearth furnace
   (c) Bessemer converter
   (d) Electric furnace

82. Alloy steels are usually produced in
   (a) Bessemer converter
   (b) electric furnace
   (c) acid open hearth furnace
   (d) none of these

83. L.D. process of steel making is
   (a) a basic process.
   (b) an acidic process.
   (c) a duplex process.
   (d) not suitable for pig iron containing < 2%- Si.

84. Temperature of hot metal charged in L.D. converter is about .... °C.
   (a) 900 (b) 1100
   (c) 1300 (d) 1500

85. Which steel making process can take up maximum metal charge as scrap ?
   (a) L.D. process
   (b) Electric furnace process
   (c) Acid Bessemer process
   (d) Kaldo furnace.

86. Oxygen blowing time in L.D. converter is about .... minutes.
   (a) 5-10 (b) 20-25
   (c) 45-50 (d) 60-70

87. Advantage of double deslagging practice in L.D. process is that the
   (a) main portion of blow is conducted with less slag volume of higher basicity.
   (b) slopping tendency is reduced.
   (c) wear on refractory lining is lesser, and a better control on FeO content of slag results in cleaner steel.
   (d) all (a), (b) & (c).

Answers

75. (d)	76. (b)	77. (a)	78. (d)	79. (d)	80. (d)
81. (d)	82. (b)	83. (a)	84. (c)	85. (b)	86. (b)
87. (d)					

88. Tapping temperature of steel depends upon the
   (*a*) chemical composition of the steel melt.
   (*b*) amount of additions *ie.g.* recarburisers, deoxidiser and alloying elements) to be made in the ladle.
   (*c*) holding time of the melt in the ladle before teeming starts and the teeming practice.
   (*d*) all (*a*), (*b*) & (*c*).

89. For dephosphorisation in the L.D. process, the slag should be
   (*a*) basic and of low viscosity
   (*b*) oxidising
   (*b*) at low temperature
   (*d*) all (*a*), (*b*) & (*c*)

90. Slopping in a L.D. vessel is caused by
   (*a*). lower lance position during blowing for long.
   (*b*) violent and intermittent reaction due to build up of FeO in the slag when carbon is high.
   (*c*) overcharging of vessel.
   (*d*) all (*a*), (*b*) & (*c*).

91. Permanent lining in L.D. converters is of .... bricks.
   (*a*) tar dolomite (*b*) magnesite
   (*c*) fireclay (*d*) silica

92. L.D. process can remove .... percent of the phosphorus in the hot metal pig iron.
   (*a*) 30 (*b*) 50
   (*c*) 10 (*d*) 90

93. L.D. process can satisfactorily handle hot metal containing silicon upto .... percent.
   (*a*) 0.1 (*b*) 0.5
   (*c*) 1.4 (*d*) 2.5

94. A high manganese content in the L.D. slag tends to
   (*a*) retard dephosphorisation of bath.
   (*b*) accelerate dephosphorisation of bath.
   (*c*) make the slag viscous.
   (*d*) none of these.

95. Maximum permissible percentage of sulphur in the pig iron to be processed in L.D. process is about .... percent.
   (*a*) 0.01 (*b*) 0.06
   (*c*) 0.5 (*d*) 1

96. Which is the flux normally added in L.D. converter for early formation of slag ?
   (*a*) Lime (*b*) Limestone
   (*c*) Dolomite (*d*) Iron scrap

97. Scrap in a L.D. converter is used as a/an .... agent.
   (*a*) early slag formation
   (*b*) cooling
   (*c*) slag viscosity reducing
   (*d*) phosphorous removal

98. To produce deep drawing type of steel, the phosphorous content in pig iron may be about .... percent.
   (*a*) <0.01 (*b*) <0.0-0.5
   (*c*) 0.12—0.15 (*d*) >0.5

99. The rate of dissolution of lime in L.D. slag can be increased by
   (*a*) retaining part of the previous heat's slag.
   (*b*) increasing FeO content of the slag.
   (*c*) adding a little dolomite in the beginning and increasing the reactivity of lime.
   (*d*) all (*a*), (*b*) & (*c*).

Answers

88. (d)	89. (d)	90. (d)	91. (b)	92. (d)	93. (c)
94. (a)	95. (b)	96. (a)	97. (b)	98. (c)	99. (d)

100. Elimination of. .... is completed first (in about six minutes after the starting of oxygen blow) in the L.D. process.
(*a*) silicon (*b*) manganese
(*c*) sulphur (*d*) phosphorous

101. Basic, thin and oxidising slag which is needed for dephosphorisation of hot metal in L.D. process is formed only after .... minutes of oxygen blowing.
(*a*) 5 (*b*) 15
(*c*) 20 (*d*) 25

102. Dephosphorisation in L.D. process is complete by the time carbon is down to .... percent.
(*a*) 3 (*b*) 2
(*c*) 0.07 (*d*) 0.7

103. In L.D. process, the relative rates of dephosphorisation and decarburisation is controlled by adjusting the
(*a*) lance height (*b*) flow rate of oxygen
(*c*) both (*a*) & (*b*) (*d*) neither (*a*) nor (*b*)

104. Rate of decarburisation in L.D. process is reduced by
(*a*) raising the height of the oxygen lance.
(*b*) decreasing the oxygen pressure,
(*c*) both (*a*) & (*b*).
(*d*) neither (*a*) nor (*b*).

105. Dephosphorisation reaction is increased by
(*a*) raising the lance height while maintaining the constant oxygen flow rate.
(*b*) decreasing the oxygen pressure keeping the lance height constant.
(*c*) both (*a*) & (*b*).
(*d*) neither (*a*) nor (*b*).

106. Desulphurisation of hot metal in L.D. process is favoured by the .... slag.
(*a*) increased FeO content of
(*b*) basic
(*c*) thin
(*d*) all (*a*), (*b*) & (*c*).

107. Double slag practice in L.D. steel making process decreases the
(*a*) steel yield (*b*) heat time
(*c*) blow time (*d*) none of these

108. Double slag practice is adopted in L.D.steel making, if .... content in hot metal is very high.
(*a*) silicon (*b*) phosphorous
(*c*) both(*a*) & (*b*) (*d*) neither (*a*) nor (*b*)

109. Delay in tapping a heat (*i.e.* extra holding time of heat) causes
(*a*) decrease in steel yield.
(*b*) wear of refractory lining.
(*c*) skull formation at the mouth of the converter.
(*d*) none of these.

110. With increase in the vessel size of the L. D. converter, its .... increases.
(*a*) tap to tap time
(*b*) blowing time
(*c*) specific oxygen consumption
(*d*) none of these

111. Pick out the wrong statement pertaining to L.D. steel making process.
(*a*) Slag composition in a L.D. vessel varies continuously during the blow (basicity increases as the blowing progresses).
(*b*) The iron content of the slag is on lower side, if high carbon steel is made.
(*c*) If phosphorous in the charge is more, intermediate slagging off is carried out 2-3 times during the blow.
(*d*) none of these.

Answers

100. (a)	101. (a)	102. (d)	103. (c)	104. (c)	105. (c)
106. (d)	107. (a)	108. (c)	109. (b)	110. (a)	111. (d)

112. Steel yield in L.D. steel making decreases, if the
  - (*a*) phosphorous content in hot metal is very high.
  - (*b*) silicon content in the charge is very high.
  - (*c*) iron is lost in the slag due to excess lime charging.
  - (*d*) all (*a*), (*b*) & (*c*).

113. In L.D. steel making process
  - (*a*) the residual element content is very low.
  - (*b*) phosphorous can be reduced from 0.15% to 0.012% in one slag operation. It can further be brought down (at the expense of some metallic yield) with a double slag practice,
  - (*c*) carbon and nitrogen content can be reduced down to 0.04% and 0.002% respectively,
  - (*d*) all (*a*), (*b*) &(*c*).

114. Use of multi hole oxygen lances (3-7 holes) compared to single hole nozzle lance helps in
  - (*a*) reducing the blowing time
  - (*b*) increasing lining life
  - (*c*) improving steel yield
  - (*d*) all (*a*), (*b*) & (*c*)

115. L.D. converter gas is recovered, cleaned and cooled for use as a gaseous fuel in .... process.
  - (*a*) IRSID/CAFL (*b*) OBM
  - (*c*) OLP (*d*) LDAC

116. 'Catch carbon technique' is employed in L.D. process to produce .... steel.
  - (*a*) high alloy (*b*) high carbon
  - (*c*) low carbon (*d*) killed

117. The main drawback of L.D. process is the
  - (*a*) lack of homogenisation in bath w.r.t. its carbon content and temperature due to inadequate mixing.
  - (*b*) slopping tendency.
  - (*c*) lower steel yield.
  - (*d*) high refractory consumption.

118. Oxygen lance in L.D. process is cooled with
  - (*a*) air (*b*) water
  - (*c*) oil (*d*) none of these

119. Kaldo process of steel making employs a/an .... vessel.
  - (*a*) vertical (*b*) horizontal
  - (*c*) inclined rotary (*d*) acidic brick lined

120. The rate of decarburisation is increased in Kaldo process by
  - (*a*) increasing the speed of rotation of the vessel.
  - (*b*) increasing the velocity of oxygen flow.
  - (*c*) lowering the lance further down.
  - (*d*) all (*a*), (*b*) & (*c*).

121. The rate of dephosphorisation in Kaldo process is increased by
  - (*a*) decreasing the speed of rotation of the vessel which builds up FeO content of the slag.
  - (*b*) decreasing the velocity of oxygen flow
  - (*c*) raising the lance further up.
  - (*d*) all (*a*), (*b*) & (*c*).

122. Converter gas at the mouth of the L.D converter is rich in
  - (*a*) CO (*b*) CO_2
  - (*c*) O_2 (*d*) N_2

123. Refractory lining life in Kaldo vessel is about .... heats.
  - (*a*) 60 – 100 (*b*) 200 – 250
  - (*c*) 10 – 25 (*d*) 500 – 600

Answers

112. (d)	113. (d)	114. (d)	115. (a)	116. (b)	117. (a)
118. (b)	119. (c)	120. (d)	121. (d)	122. (a)	123. (a)

124. Refractory lining life is minimum in
   (a) L.D. converter
   (b) Open hearth furnace
   (c) Kaldo vessel
   (d) Bessemer converter

125. Kaldo process of steel making is not practised widely mainly due to its
   (a) low yield of steel.
   (b) high refractory consumption.
   (c) high slopping tendency.
   (d) very high oxygen consumption.

126. High nitrogen' content in steel has deleterious effect on its
   (a) cold workability
   (b) weidability
   (c) both (a) & (b)
   (d) neither *(a)* nor *(b)*

127. Oxygen is blown from sides in
   (a) Kaldo vessel
   (b) L.D. converter
   (c) *IRSID*/CAFL process
   (d) basic oxygen furnace

128. Combined blowing process in which oxygen from top and inert gas like nitrogen or argon from the bottom are blown simultaneously to achieve relatively a calmer bath than top blowing alone, results in
   (a) lower FeO% in slag.
   (b) less slopping during blowing,
   (c) more scrap melting capability.
   (d) all *(a)*, *(b)* & *(c)*.

129. Which of the following can not be added in teeming laddle after the steel is tapped from L.D. converter ?
   (a) Deoxidisers
   (b) Recarburisers
   (c) Alloying elements
   (d) None of these

130. Modern steel teeming ladle are equipped with
   (a) stopper rod.
   (b) slide gate system.
   (c) audco valves in the discharge line.
   (d) none of these.

131. Ingot mould is coated with tar to
   (a) avoid the sticking of ingot to the mould.
   (b) accelerate the rate of solidification.
   (c) produce killed steel.
   (d) accelerate deoxidation of steel.

132. Ingot moulds are made of
   (a) stainless steel   *(b)* cast iron
   (c) killed steel   *(d)* none of these

133. Fully deoxidised steel is called .... steel.
   (a) semi-killed   *(b)* rimming
   (c) killed   *(d)* none of these

134. Hot top is generally provided in ingot mould carrying .... steel.
   (a) killed   *(b)* rimming
   (c) semi-killed   *(d)* none of these

135. Hot top is provided in ingot mould mainly for avoiding the
   (a) pipe formation in the steel ingots due to shrinkage on solidification.
   (b) oxidation of ingot.
   (c) chance of sticker generation.
   (d) cooling of steel ingot.

136. Bottom plates (on which steel ingot mould rests) are made of
   (a) stainless steel   *(b)* killed steel
   (c) cast iron   *(d)* none of these

Answers

124. (c)	125. (b)	126. (c)	127. (a)	128. (d)	129. (a)
130. (b)	131. (a)	132. (b)	133. (c)	134. (a)	135. (a)
136. (c)					

137. Life of ingot mould is about .... pours.
 (*a*) 10 (*b*) 50
 (*c*) 150 (*d*) 250

138. Ingot mould consumption is about .... kg/ton of crude steel.
 (*a*) 5—10 (*b*) 20 - 30
 (*c*) 80—100 (*d*) 200 —300

139. Tar or pitch coating in the ingot mould is done, when its surface temperature is in the range of .... C^0.
 (*a*) 95 – 230 (*b*) 40 – 60
 (*c*) 400 – 500 (*d*) 700 – 800

140. Rimming steels are produced, only if the carbon percentage in steel is
 (*a*) < 0.15 (*b*) 0.5 – 0.8
 (*c*) > 0.5 (*d*) very high

141. Rimming steels
 (*a*) require a lot of gas evolution'during solidification.
 (*b*) must contain enough dissolved oxygen.
 (*c*) is possible to e produced with low-carbon <0.15%) steel only.
 (*d*) all (*a*), (*b*) & (*c*).

142. Capped steel is another variety of the .... steel.
 (*a*) killed (*b*) semi-killed
 (*c*) rimming (*d*) none of these

143. Semi-killed steels
 (*a*) are partially deoxidised steels.
 (*b*) contain carbon in the range of 0.15— 0.30%.
 (*c*) are used for manufacture of structural shapes, plates and merchant products.
 (*d*) all (*a*), (*b*) & (*c*).

144. Killed steels
 (*a*) during solidification do not evolve gases.
 (*b*) contain > 0.3% carbon.
 (*c*) are cast in wide-end-up moulds with hot tops.
 (*d*) all (*a*), (*b*) & (*c*).

145. The problem of pipe formation is most serious in the case of .... steel.
 (*a*) capped (*b*) killed
 (*c*) rimming (*d*) semi-killed.

146. Volume contraction (during solidification of steel) resulting in pipe formation is about percent of the total .... apparent volume of the ingot.
 (*a*) 0.5—1 (*b*) 2.5—3
 (*c*) 5—10 (*d*) 10—15

147. Primary pipe formation in ingots
 (*a*) reduces the steel yield on rolling.
 (*b*) does not weld during rolling.
 (*c*) both (*a*) & (*b*).
 (*d*) neither (*a*) nor (*b*).

148. Blowholes (formed due to entrapment of gas evolved during solidification of steels) are not formed in .... steels.
 (*a*) capped (*b*) killed
 (*c*) semi-killed (*d*) rimming

149. Segregation (which can be minimised by prolonged soaking of ingots before working)
 (*a*) means departure from average composition.
 (*b*) increases with increasing time of solidification for an ingot.
 (*c*) tendency is more in large ingots compared to small ones.
 (*d*) all (*a*), (*b*) and (*c*).

150. Segregation is maximum in .... steels.
 (*a*) killed (*b*) semi-killed
 (*c*) capped (*d*) rimming

Answers

137. (b)	138. (b)	139. (a)	140. (a)	141. (d)	142. (c)
143. (d)	144. (d)	145. (d)	146. (b)	147. (c)	148. (b)
149. (d)	150. (d)				

151. Vacuum deoxidation of steel (for achieving high degree of cleanliness) is necessary for .... steels.

(*a*) ball bearing (*b*) die and tool

(*c*) aircraft (*d*) all (*a*), (*b*) and (*c*)

152. Nitrogen in steel

(*a*) makes steel strain brittle.

(*b*) poses problem in its cold working without intermediate annealing.

(*c*) must be < 0.002% in deep drawing type steel.

(*d*) all (*a*), (*b*) &(*c*).

153. Nitrogen content (in L.D. steel) is minimum in case of .... steel.

(*a*) killed (*b*) semi-killed

(*c*) rimming (*d*) capped

154. High hydrogen content in steel

(*a*) causes hair line cracking in fully killed high alloy steel.

(*b*) reduces its ductility.

(*c*) both (*a*)& (*b*).

(*d*) neither (*a*) nor (*b*).

155. Sources of hydrogen pick up in steel making are

(*a*) wet/rusty scrap and wet solid charge materials.

(*b*) steam fed into the furnace.

(*c*) wet refractory channel, runner, vessels etc.

(*d*) all (*a*), (*b*) & (*c*).

156. Vacuum degassing of steel is done to

(*a*) remove hydrogen from steel to avoid prolonged annealing treatment particularly in forging quality steels.

(*b*) improve cleanliness by removing oxygen in the form of CO gas.

(*c*) produce very low carbon steel (0.03%) by transferring part of the refining from the furnace to the.' degassing unit.

(*d*) all (*a*), (*b*) & (*c*).

157. Continuous casting of steel

(*a*) increases the yield of finished steel product.

(*b*) reduces the specific energy consumption in steel making.

(*c*) increases the production rate.

(*d*) all (*a*), (*b*) & (*c*).

158. For effective removal of phosphorous in basic steel making furnace, the

(*a*) slag should have high basicity.

(*b*) basicity of slag should not fall much low, otherwise phosphorous may revert back to the metal phase.

(*c*) both (*a*) & (*b*).

(*d*) neither (*a*) nor (*b*).

159. In Steel making processes, the ratio of phosphorous content in slag to that in metal of the slag.

(*a*) is called dephosphorising index.

(*b*) has a higher value for higher basicity.

(*c*) has a higher value for higher oxidising power

(*d*) all (*a*), (*b*) &(*c*).

160. Silicon content of pig iron should be as low-as possible during basic steel making process as it will reduce the

(*a*) lime consumption

(*b*) slag volume

(*c*) both (*a*) & (*b*)

(*d*) neither (*a*) nor (*b*)

Answers

151. (d)	152. (d)	153. (c)	154. (c)	155. (d)	156. (d)
157. (d)	158. (c)	159. (d)	160. (c)		

161. The solubility of oxygen in pure carbon during steel making at 1600°C & 1800°C are respectively
(*a*) 0.23% & 0.48% (*b*) 1.2% & 3.5%
(*c*) 2% & 0.48% (*d*) 3%. & 5%

162. Which is the most effective and stable oxide used for deoxidation of steel ?
(*a*) SiO_2 (*b*) MnO
(*c*) Al_2O_3 (*d*) TiO_2

163. A flux added during smelting and refining
(*a*) reduces the softening point of gangue material.
(*b*) reduces the viscosity of slag.
(*c*) decreases the activity of some components to make it stable in the slag phase.
(*d*) all (*a*), (*b*) & (*c*).

164. Which of the following is not used as a deoxidiser in steel making?
(*a*) Nickel (*b*) Silicon
(*c*)Aluminium (*d*) Manganese

165. The efficiency of steel making process is assessed by the .... of steel.
(*a*) yield & quality.
(*b*) production rate.
(*c*) refractory consumption per ton.
(*d*) all (*a*), (*b*) & (*c*).

166. Specific energy consumption in L.D. steel melting shop is about ....G.cal/ton of crude steel.
(*a*) 0.01 – 0.05 (*b*) 0.25 – 0.30
(*c*) 1 – 1.5 (*d*) 2 – 2.51

167. L.D. converters employing suppressed combustion are not used at .... steel plant.
(*a*) Vizag (*b*) Bokaro
(*c*) Salem (*d*) Bhilai

168. The lining life of steel melting converter is prolonged by doing hot repairs of the worn out areas. It is known as
(*a*) fettling or gunning
(*b*) ceramic welding
(*c*) fritting
(*d*) ramming

169. The time required for converting a batch of raw material into steel including handling time and time required for various repairs (*e.g.* fettling of the converter/furnace) is called the .... time.
(*a*) compaign (*b*) heat
(*c*) tap to tap (*d*) both (*b*) & (*c*)

170. The duration before the steel melting furnace is due again for relining is called the .... time.
(*a*) down (*b*) compaign
(*c*) track (*d*) tap to tap

171. The main function of a hot metal mixer is to
(*a*) homogenise the pig iron from various casts.
(*b*) heat up the pig iron,
(*c*) store the pig iron.
(*d*) conserve the heat of pig iron.

172. Phosphorous occurs in steel as
(*a*) Fe_3P (*b*) $Fe\ PO_4$
(*c*) $Fe_2(PO_4)_3$ (*d*) none of these.

173. Maximum permissible percentage of phosphorous in plain carbon steel is .... percent.
(*a*) 0.01 (*b*) 0.05
(*c*) 0.5 (*d*) 1

Answers

161. (a)	162. (c)	163. (d)	164. (a)	165. (d)	166. (b)
167. (c)	168. (a)	169. (d)	170. (b)	171. (a)	172. (a)
173. (b)					

174. Presence of manganese in steel
(*a*) indicates its good deoxidation.
(*b*) decreases its tensile strength.
(*c*) results in blow holes.
(*d*) produces coarse grained structure.

175. Alloying elements are added to steel for
(*a*) improving high temperature strength and magnetic properties.
(*b*) increasing abrasion/corrosion resistance and the depth of hardening.
(*c*) refinement of grain size and high toughness.
(*d*) all (*a*), (*b*) & (*c*).

176. Addition of aluminium to plain carbon steel
(*a*) increases its high temperature strength.
(*b*) helps in its deoxidation.
(*c*) restricts grain growth.
(*d*) both (*b*) & (*c*).

177. Addition of cobalt to plain carbon steel helps in
(*a*) making high speed steel.
(*b*) increasing its red hardness,
(*c*) both (*a*) & (*b*).
(*d*) neither (*a*) nor (*b*).

178. Which of the following alloying elements, when added to plain carbon steel, increases its corrosion/oxidation resistance ?
(*a*) Chromium (*b*) Cobalt
(*c*) Molybdenum (*d*) Tungsten

179. Which of the following alloying elements when added to plain carbon steel does not increase its red hardness ?
(*a*) Cobalt (*b*) Molybdenum
(*c*) Aluminium (*d*) Tungsten

180. Presence of silicon in plain carbon steel
(*a*) is in the form of SiO_2.
(*b*) is an indication of its poor deoxidation.
(*c*) is in the form of FeSi.
(*d*) increases the blow holes formation tendency.

181. Sulphur occurs in steel as
(*a*) FeS (*b*) MnS
(*c*) both (*a*) & (*b*) (*d*) neither (*a*) nor (*b*).

182. Addition of chromium to plain carbon steel increases its
(*a*) corrosion/oxidation resistance.
(*b*) hardeneability.
(*c*) high temperature strength.
(*d*) all (*a*), (*b*) & (*c*).

183. Addition of nickel & chromium as alloying element in the plain carbon steel improves its
(*a*) strength & ductility
(*b*) fatigue resistance
(*c*) shock resistance
(*d*) all (*a*), (*b*) & (*c*).

184. 'Temper brittleness' of nickel-chromium steel occurs when it is cooled in air from the tempering temperature thereby making it brittle. This can be eliminated by
(*a*) quenching the steel from the tempering temperature.
(*b*) small addition of molybdenum if air cooling is indispensable.
(*c*) both (*a*) & (*b*)
(*d*) neither (*a*) nor (*b*)

185. Pick out the wrong statement.
(*a*) Tungsten percentage in steel varies from 1 to 20.
(*b*) High speed steel after proper heat treatment possess red hardness.
(*c*) Main alloying elements in high speed steel are chromium and molybdenum.
(*d*) Main alloying elements in high speed steel are tungsten, chromium & vanadium.

Answers

174. (a)	175. (d)	176. (d)	177. (c)	178. (a)	179. (c)
180. (c)	181. (d)	182. (d)	183. (d)	184. (c)	185. (c)

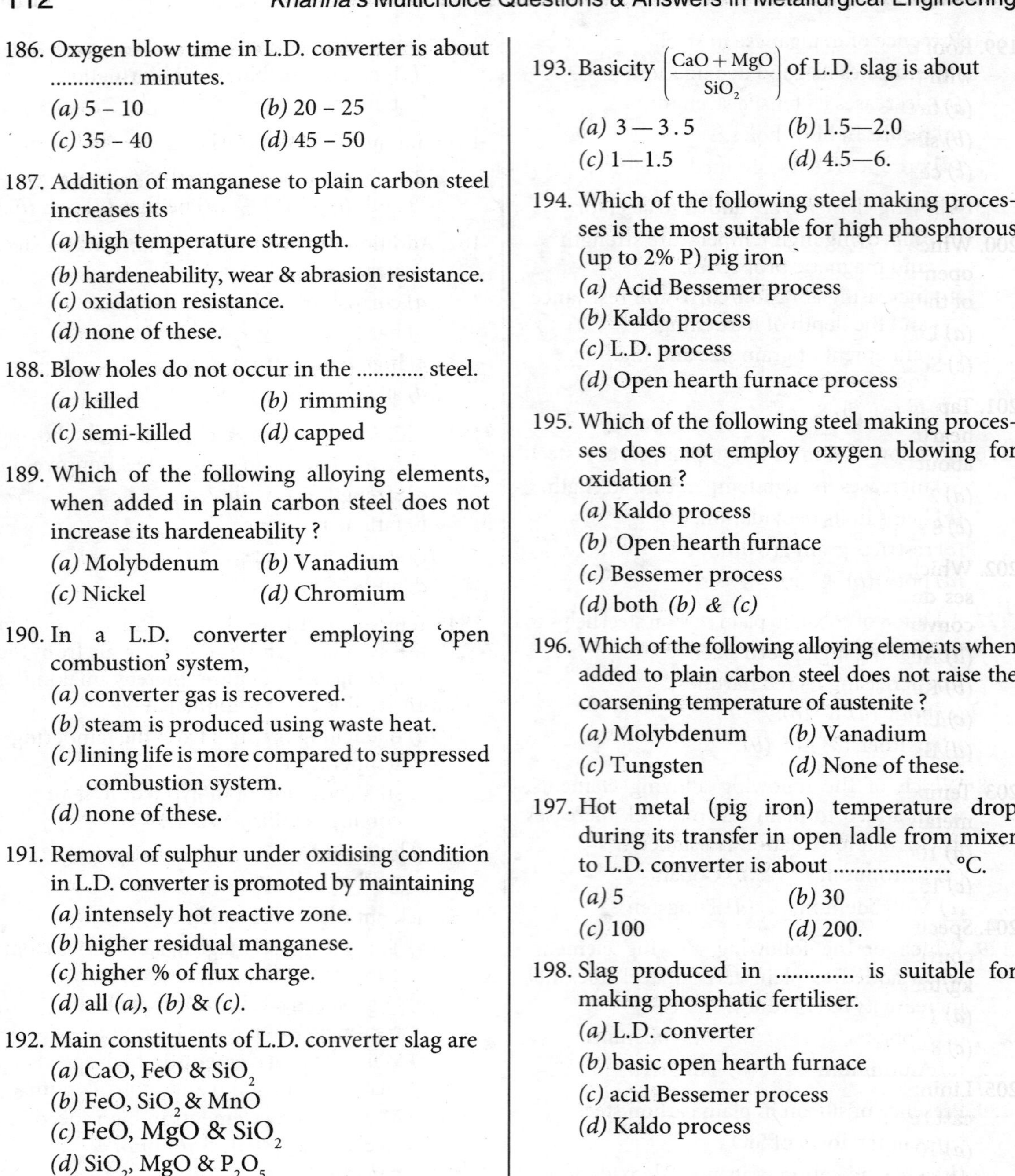

186. Oxygen blow time in L.D. converter is about .... minutes.
(*a*) 5 – 10 (*b*) 20 – 25
(*c*) 35 – 40 (*d*) 45 – 50

187. Addition of manganese to plain carbon steel increases its
(*a*) high temperature strength.
(*b*) hardeneability, wear & abrasion resistance.
(*c*) oxidation resistance.
(*d*) none of these.

188. Blow holes do not occur in the .... steel.
(*a*) killed (*b*) rimming
(*c*) semi-killed (*d*) capped

189. Which of the following alloying elements, when added in plain carbon steel does not increase its hardeneability ?
(*a*) Molybdenum (*b*) Vanadium
(*c*) Nickel (*d*) Chromium

190. In a L.D. converter employing 'open combustion' system,
(*a*) converter gas is recovered.
(*b*) steam is produced using waste heat.
(*c*) lining life is more compared to suppressed combustion system.
(*d*) none of these.

191. Removal of sulphur under oxidising condition in L.D. converter is promoted by maintaining
(*a*) intensely hot reactive zone.
(*b*) higher residual manganese.
(*c*) higher % of flux charge.
(*d*) all (*a*), (*b*) & (*c*).

192. Main constituents of L.D. converter slag are
(*a*) CaO, FeO & SiO_2
(*b*) FeO, SiO_2 & MnO
(*c*) FeO, MgO & SiO_2
(*d*) SiO_2, MgO & P_2O_5

193. Basicity $\left(\frac{CaO + MgO}{SiO_2}\right)$ of L.D. slag is about
(*a*) 3—3.5 (*b*) 1.5—2.0
(*c*) 1—1.5 (*d*) 4.5—6.

194. Which of the following steel making processes is the most suitable for high phosphorous (up to 2% P) pig iron
(*a*) Acid Bessemer process
(*b*) Kaldo process
(*c*) L.D. process
(*d*) Open hearth furnace process

195. Which of the following steel making processes does not employ oxygen blowing for oxidation ?
(*a*) Kaldo process
(*b*) Open hearth furnace
(*c*) Bessemer process
(*d*) both (*b*) & (*c*)

196. Which of the following alloying elements when added to plain carbon steel does not raise the coarsening temperature of austenite ?
(*a*) Molybdenum (*b*) Vanadium
(*c*) Tungsten (*d*) None of these.

197. Hot metal (pig iron) temperature drop during its transfer in open ladle from mixer to L.D. converter is about .... °C.
(*a*) 5 (*b*) 30
(*c*) 100 (*d*) 200.

198. Slag produced in .... is suitable for making phosphatic fertiliser.
(*a*) L.D. converter
(*b*) basic open hearth furnace
(*c*) acid Bessemer process
(*d*) Kaldo process

Answers

186. (b)	187. (b)	188. (a)	189. (c)	190. (b)	191. (d)
192. (a)	193. (a)	194. (b)	195. (c)	196. (d)	197. (b)
198. (d)					

199. Roof of a basic open hearth furnace is lined with the .... bricks.
(a) fireclay
(b) silica
(c) chrome magnesite
(d) none of these

200. Which of the following is added in basic-open hearth furnace to increase the fluidity of the slag?
(a) Limestone (b) Iron ore
(c) Scrap (d) Fluorspar

201. Tap to tap time in a modern basic open hearth furnace of 250 tons capacity may be about .... hours.
(a) 2 (b) 4
(c) 8 (d) 16

202. Which of the following steel making processes does not employ oxygen blowing in the converter ?
(a) Acid Bessemer converter
(b) Kaldo rotary converter
(c) L.D. converter
(d) Basic open hearth furnace

203. Temperature of pig iron maintained in hot metal mixer is about............ ^{0}C.
(a) 1000 (b) 1350
(c) 1550 (d) .1650.

204. Specific refractory consumption in L.D. converter steel making process is about kg/ton of steel.
(a) 1—1.5 (b) 3—5
(c) 8—10 (d) 15—20

205. Lining life of L.D. converter lined with fusion cast refractories is about .... heats.
(a) 100 (b) 400
(c) 1000 (d) 2000

206. Which of the following thermocouples is used for measuring the steel bath temperature in L.D. converter before tapping ?
(a) Pt-Pt+Rh
(b) Chromel-alumel
(c) Platinum-constantan
(d) None of these

207. Pick out the wrong statement.
(a) Oxygen blowing time in L.D. vessel is kept almost same irrespective of the size and capacity of the converter.
(b) Tap to tap time in L.D. converter is same irrespective of the size of the converter.
(c) Dephosphorisation in L.D. converter is favoured if the slag is basic, thin & oxidising and at a low temperature.
(d) Maximisation of scrap charging in L.D. converter increases the steel yield.

208. Top & bottom combined blowing in L.D. converter using oxygen & inert argon results in comparatively
(a) calmer bath than in top blowing alone.
(b) lower FeO in slag.
(c) lower slopping during blowing & more scrap melting capability.
(d) all (a), (b) & (c).

209. Pick out the correct statement.
(a) Ingot moulds are coated with tar or pitch before pouring steel in it.
(b) The steel which is fully deoxidised by strong deoxidisers is called 'rimming steel'.
(c) The steel which is only partially deoxidised is called 'killed steel'.
(d) Rimming steel is produced only if carbon is more than 0.15%.

Answers

199. (b)	200. (d)	201. (c)	202. (a)	203. (b)	204. (b)
205. (c)	206. (a)	207. (b)	208. (d)	209. (a)	

210. Desulphurisation and dephosphorisation during steel making is favoured
(a) by high slag basicity.
(b) at low temperature.
(c) by oxidising & reducing furnace atmosphere respectively.
(d) none of these.

211. Cosiderable amount of nitrogen present in steel makes it unsuitable for deep drawing purposes. Steel made by which of the following processes is prone to have maximum nitrogen in it ?
(a) L.D. process
(b) Bessemer process
(c) Open hearth process
(d) Electric arc furnace

212. Which of the following is a demerit of electric arc steel making ?
(a) Higher temperature can not be attained.
(b) Its capacity is low compared to other processes.
(c) It can not remove sulphur & phosphorous.
(d) All types of steel can not be made by this process.

213. Pick out the wrong statement.
(a) Electric arc furnace can be used to produce cast iron.
(b) Carbide slag formed by addition of lime and coke in the steel bath during refining is a reducing slag.
(c) Stainless steel can not be made using L.D. process.
(d) Carbon content in steel made by L.D. process can be reduced to as low as 0.05%.

214. Which of the following is the most comprehensive & complete definition of the basicity of the slag produced in L.D. steel making?
(a) $\frac{CaO}{SiO_2}$ *(b)* $\frac{CaO + MgO}{SiO_2}$
(c) $\frac{CaO + MgO}{SiO_2 + Al_2O_3}$ *(d)* $\frac{CaO + MgO}{SiO_2 + Al_2O_3 + P_2O_5}$

215. Ajax process of steel making is a modified .... process.
(a) open hearth
(b) L.D.
(c) electric arc furnace
(d) Bessemer

216. Which of the following gases is not present in steel in the dissolved state ?
(a) H_2 *(b)* O_2
(c) N_2 *(d)* SO_2

217. As per the Sievert's law for solubility of diatomic gases (e.. H2, N2, O2) in steel at a particular temperature, the percentage of dissolved gas is proportional to
(a) p (b) $\sqrt{P}$
(c) p^2 (d) 1/P
where p = gas pressure

218. The deoxidising effect of complex deoxidiser (in case of deoxidation of steel is much better than that of its constituents used separately. Which of the following is a complex deoxidiser of steel?
(a) Zirconium *(b)* Silica-calcium
(c) Aluminium *(d)* Manganese

219. Which of the following defects are found in steel ingots ?
(a) Segregation
(b) Slag or non-metallic inclusions
(c) Both *(a)* & *(b)*
(d) Neither *(a)* nor *(b)*

Answers

210. (a)	211. (b)	212. (b)	213. (c)	214. (d)	215. (a)
216. (d)	217. (b)	218. (b)	219. (c)		

220. Segregation is the process of selective crystallisation and discarding of particles or impurities from the steel melt during solidification. Which of the following has the tendency to segregate ?
(a) Carbon
(b) Sulphur
(c) Phosphorous
(d) All (a), (b) & (c)

221. The formation of larger grain size in the structure of steel ingot is called
(a) shrinkage defect
(b) ingotism
(c) negative segregation
(d) pipe defect

222. Cracks formation in steel ingots results from
(a) low pouring temperature.
(b) low pouring speed.
(c) high pouring temperature.
(d) none of these.

223. Coating of ingot moulds with coal tar prior to steel pouring in it is done for
(a) easy stripping of ingot.
(b) avoiding ingotism.
(c) avoiding pipe defect.
(d) eliminating blow holes.

224. Pick out the correct statement.
(a) Deoxidation of steel can not be done in steel making furnace.
(b) Deoxidation of steel is done in steel teeming ladle bath only.
(c) Kaldo converter for steel making rotates in different angles for charging, slagging and tapping operations.
(d) Operational cost of electric arc steel making furnace is lower than L.D. process.

225. .... steel is the most prone to loss of ductility due to ageing.
(a) Bessemer
(b) L.D.
(c) Open hearth
(d) Electric arc furnace

226. The main purpose of using oxygen in open hearth process of steel making is to
(a) promote dephosphorisation of steel.
(b) reduce nitrogen content of steel.
(c) improve operational efficiency by increasing the bath temperature.
(d) none of these.

227. In electric arc process of steel making, ... is reduced during the oxidation period of refining of steel.
(a) silicon (b) phosphorous
(c) sulphur
(d) all (a), (b) & (c)

228. Maximum temperature in electric arc process of steel making is attained
(a) during reducing period.
(b) at the end of oxidation period.
(c) at the time of metal tapping.
(d) during deoxidation period.

229. Ferro-manganese and ferro-silicon used in L.D. steel making acts as a
(a) desulphuriser (b) decarburiser
(c) desiliconiser (d) none of these

230. Which of the following is the quickest steel making process ?
(a) Open hearth process
(b) Electric arc process
(c) L.D. process
(d) Induction furnace process

Answers

220. (d)	221. (b)	222. (c)	223. (a)	224. (c)	225. (a)
226. (c)	227. (b)	228. (b)	229. (d)	230. (c)	

231. Which of the following is not an alloying element in steel ?
   (*a*) Tungsten (*b*) Lead
   (*c*) Silicon (*d*) Vanadium

232. During steel ingot casting, shrinkage cavity can be minimised by using
   (*a*) high teeming temperature
   (*b*) hot tops
   (*c*) exothermic powders
   (*d*) both (*b*) & (*c*)

233. Vaccum refinning of liquid steel aims at reducing the .... in it.
   (*a*) carbon content
   (*b*) dissolved gases
   (*c*) non-metallic inclusions
   (*d*) both (*b*) & (*c*)

234. In electric arc process of steel making, the slag during oxidation period has
   (*a*) low FeO content
   (*b*) low basicity
   (*c*) high FeO content
   (*d*) both (*b*) & (*c*)

235. Degree of deoxidation achieved is maximum in the case of .... steel.
   (*a*) killed
   (*b*) semi-killed
   (*c*) rimmed
   (*d*) all have the same degree of deoxidation

236. Presence of .... in steel is mainly responsible for the occurence of hair line crack.
   (*a*) phosphorous (*b*) sulphur
   (*c*) hydrogen (*d*) nitrogen

237. Ferro-tungsten and ferro-manganese is added to steel in electric arc furnace steel making process
   (*a*) as alloying elements.
   (*b*) to lower the steel melting temperature.
   (*c*) to enhance the slag fluidity.
   (*d*) all (*a*), (*b*) & (*c*).

238. With increase in temperature, the electrical conductivity of steel slag
   (*a*) increases
   (*b*) decreases
   (*c*) remains unaffected
   (*d*) may increase or decrease, depends on slag compsoition

239. Cracks in steel ingots during casting occurs due to
   (*a*) use of uncoated mould
   (*b*) high teeming speed
   (*c*) high teeming temperature
   (*d*) both (*b*) & (*c*)

240. Gas evolution is practically nil in case of casting of .... steel ingots.
   (*a*) rimmed (*b*) killed
   (*c*) semi-killed (*d*) all (*a*). (*b*) & (*c*)

241. In electric arc process of steel making, use of oxygen helps in accelerating the rate of
   (*a*) carbon removal (*b*) sulphur removal
   (*c*) slag oxidation (*d*) coke combustion

242. Aluminium shots added in liquid steel ladles acts as a
   (*a*) desulphuriser (*b*) grain refiner
   (*c*) deoxidiser (*d*) both (*b*) & (*c*)

243. All types of alloy steel can be produced by .... steel making process only.
   (*a*) open hearth (*b*) L.D.
   (*c*) Bessemer (*d*) electric arc

Answers

231. (b)	232. (d)	233. (d)	234. (d)	235. (a)	236. (c)
237. (a)	238. (a)	239. (d)	240. (b)	241. (a)	242. (d)
243. (d)					

244. The oxidation reaction is the slowest in the .... process of steel making.
(a) L.D. (b) Bessemer
(c) Kaldo (d) Open hearth

245. Steel made by the .... process is not suitable for deep drawing operation due to its high nitrogen content.
(a) electric arc furnace
(b) L.D.
(c) Bessemer
(d) open hearth

246. Slag generation rate in L.D. steel making is about .... kg/ton of crude steel.
(a) 50 (b) 150
(b) 350 (d) 550

247. Fuel used for heating the hot metal (pig iron) mixer is generally
(a) blast furnace gas
(b) coke oven gas
(c) furnace oil
(d) no fuel is used for heating the mixer

248. Refractory lining life of L.D. converter is reduced, if the .... is more.
(a) tap to tap time
(b) idle time of the converter
(c) silicon content in the pig iron
(d) capacity of the converter

249. Dephosphorisation during L.D. steel making is favoured, if the slag is
(a) highly viscous.
(b) very hot.
(c) formed within three minutes of blowing.
(d) basic in nature.

250. L.D. steel making using very high silicon and phosphorous containing iron necessitates the use of double slag practice, which increases the
(a) yield (b) production rate
(c) heat time (d) all 'a', 'b' & 'c

251. With increase in the size of L.D. converter, the .... time can be kept the same.
(a) tap to tap (b) blowing
(c) charging (d) tapping

252. Basicity of typical L.D. slag may be about
(a) 1 (b) 2
(c) 3.5 (d) 5

253. L.D. slag contains maximum percentage of
(a) CaO (b) FeO
(c) SiO_2 (d) MnO

254. Progress of a typical L.D. blow, showing the rate of removal of carbon, phosphorous, sulphur and manganese is shown below in figure.
0.08

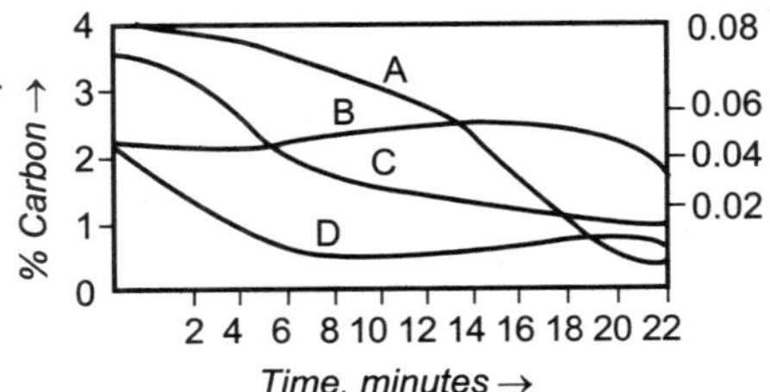

The rate of removal of phosphorous is illustrated by the curve
(a) C (b) A
(c) D (d) B

255. Curve 'A' in the above figure represents the rate of removal of
(a) managanese (d) sulphur
(c) carbon (d) none of these

Answers

244. (d)	245. (c)	246. (b)	247. (b)	248. (c)	249. (d)
250. (c)	251. (b)	252. (c)	253. (a)	254. (a)	255. (c)

256. Match the refractory lining life of various equipments used in steel making.

List I

(*a*) Pig iron mixer.

(*b*) L.D. steel teeming ladle.

(*c*) Kaldo vessel.

(*d*) L.D. converter having wear out lining of fusion cast magnesia carbon bricks.

List II

I. 60-100 heats

II. 9 lakh tons of pig iron pouring

III. 25 heats

IV. 600 heats

257. Match the tap to tap time (minutes) of various steel making processes.

List I

(*a*) L.D. converter (300 tons capacity).

(*b*) Kaldo process.

(*c*) Open hearth furnace (250 tons capacity).

(*d*) Bessemer converter.

List II

I. 35 II. 60

III. 80\ IV. 480

258. Match the various temperatures (°C) encountered in steel making processes.

List I

(*a*) Temperature of pig iron charged in L.D. converter.

(*b*) L.D. steel tapping temperature

(*c*) L.D. slag temperature

(*d*) Ferro-alloy heating temperature

List II

I. 1650 II. 1300

III. 100 IV. 1700

259. Match the operating parameters encountered in L.D. steel making in modern Indian steel plants. (TCS-tons of crude steel).

List I

(*a*) Overall specific energy consumption rate.

(*b*) Specific oxygen consumption rate.

(*c*) Specific lime consumption rate.

(*d*) Specific slag generation rate.

List II

I. 65 Nm^3/TCS

II. 80 kg/TCS

III. 150 kg/TCS

IV.7.5 Giga calorie/TCS

260. Match the beneficial effects of various alloying element addition to steel.

List I

(*a*) Vanadium (*b*) Tungsten

(*c*) Zirconium (*d*) Titanium

261. *List II*

I. Decreases the deleterious effects of nitrogen & sulphur in steel

II. Increases the strength, plasticity & impact resistance of steel.

III. Increases strength, hardness & elasticity.

IV. Imparts lightness coupled with strength to steel, making it suitable for aircraft industry.

262. Match the deleterious effect of high level of non-metallic inclusions in steel.

List I

(*a*) Sulphur (*b*) Phosphorous

(*c*) Nitrogen (*d*) Hydrogen

List II

I. Cold shortness of steel.

II. Makes steel grain brittle.

III. Decrease in ductility of steel plus hair line crack in fully killed high alloy steel.

IV. Hot shortness of steel.

Answers

256. (ii) (iii) (i) (iv) 257. (ii) (iii) (iv) (i) 258. (ii) (i) (iv) (iii)
259. (iv) (i) (ii) (iii) 260. (ii) (iii) (i) (iv) 261. (iv) (i) (ii) (iii)
262. (iii) (iv) (ii) (i)

263. Match the various types of steel ingot defects with their description.

List I

(a) Pipe

(b) Blowholes

(c) Segregation

(d) Ingot skin

List II

I. Chilling effect of mould forming a thin solid layer on the side faces and bottom of the ingot immediately after pouring.

II. Departure from average composition of ingot.

III. Volumetric contraction resulting from solidification of especially killed steel.

IV. Entrapment of gas evolved during solidification of all except killed steel.

264. Match the description of various surface defects of steel ingots.

List I

(a) Scab

(b) Lap

(c) Crazing

(d) Boot Leg

List II

I. A fold in the ingot skin caused by freezing of a slowly rising top surface of the metal in the mould before the pouring is over.

II. A projection on the side surface of an ingot caused by freezing of steel in a cavity in the mould wall or in a mould with uneven walls.

III. Sinking of an ingot top below the original level in the mould due to decreased evolution of gas.

IV. Freezing of steel in the large number of cracks present in mould walls giving rise to a network of fins on the ingot face.

265. Match the carbon percentage in various types of ingot steels.

List I

(a) Killed steel

(b) Rimming steel

(c) Semi-killed steel

(d) Capped steel

List II

I. 0.15

II. <0.15

III. 0.15 - 0.30

IV. > 0.3

266. Sulphur content in free cutting steels is about .... percent.

(a) 0.021 *(b)* 0.21

(c) 0.38 *(d)* 0.48

267. Sulphur content in tool steel is limited to a maximum of .... percent.

(a) 0.005 *(b)* 0.035

(c) 0.35 *(d)* 0.80

268. Increasing percentage of manganese in plain carbon steel decreases its

(a) weldability *(b)* tensile strength

(c) hardness *(d)* all 'a', *b* & 'c?

269. In acid steel making process, ... is not removed from pig iron.

(a) silicon

(b) carbon

(c) sulphur & phosphorous

(d) manganese

270. Grade of pig iron is decided by its .... content.

(a) silicon *(b)* manganese

(c) sulphur *(d)* carbon

Answers

263. (iii) (iv) (ii) (i) 264. (ii) (i) (iv) (iii) 265. (iv) (ii) (iii) (i)

266. (b) 267. (b) 268. (a) 269. (c) 270. (a)

271. Pick out the wrong statement pertaining to steel making.
   (*a*) The refinning reaction during steel making takes place at the slag-metal interface, but carbon reaction takes place at gas metal interface.
   (*b*) For effective dephosphorisation, use of basic flux to produce slag with basicity 2-4 decreases the activitv co - efficient of P_2O_5.
   (*c*) MnO is the least stable oxide product during refinning of pig iron.
   (*d*) With increase in the slag basicity, the activity co-efficient of MnO increases and the ratio [% MnO / % Mn| decreases ; however for the same basicity, this ratio increases with the oxidising power of slag.

272. L.D. process can not take more than .... percent of scrap in the metallic charge because of its thermal constraint (cooling effect), which makes it difficult to consume all the circulating scrap in an integrated steel plant with steel ingot route having no continuous casting facility for steel slabs.
   (*a*) 10- 15 (*b*) 25-30
   (*c*) 45 - 50 (*d*) 65 - 70

273. Which of the following is not a function of the flux (lime/dolo/ganister) used in the steel making process.
   (*a*) To reduce the slag viscosity.
   (*b*) To act as a deoxidiser.
   (*c*) To decrease the activity of some components to make it stable in the slag phase.
   (*d*) To reduce the softening point of the gangue.

274. Which of the following steel making process can take upto 100% cold metallic charge (*i.e.*, scrap)?
   (*a*) Bessemer process.
   (*b*) L.D. process.
   (*c*) Open hearth furnace.
   (*d*) None of these.

275. Which of the following is normally not used as a recarburiser in electric arc furnace for steel making ?
   (*a*) Limestone (*b*) Graphite
   (*c*) Coke breeze (*d*) Anthracite coal

276. During steel making in the L.D. converter, metallic loss ia not due to the
   (*a*) loss of iron in slag or as fume & dust.
   (*b*) penetration inside the refractory bricks.
   (*c*) ejection as slopping & sparking and skull formation on the converter mouth.
   (*d*) oxidation of metalloids.

277. Yield of steel in the steel making process will be more, if
   (*a*) high oxidising potential is maintained in the converter at the time of slag removal.
   (*b*) duplex process of steel making (*i.e.*, use of more than one steel making processes) is practised.
   (*c*) double slag practice is adopted resulting in slag volume more than necessary.
   (*d*) the steel scrap proportion of metallic charge is increased.

278. Oxygen is blown in the L.D. converter through a water cooled lance, whose tip is made of
   (*a*) copper (*b*) aluminium
   (*c*) brass (*d*) nickel

Answers

271. (c)	272. (b)	273. (b)	274. (c)	275. (a)	276. (b)
277. (d)	278. (a)				

279. Pick out the wrong statement pertaining to the steel making process.
   (a) Activity co-efficient of carbon increases in the presence of non-carbide forming elements.
   (b) During steel making, carbon oxidation reaction does not take place at the slag-metal interface, because of the difficulty in nucleating gas bubble.
   (c) It is easier to remove sulphur from pig iron than from steel, because the presence of unusual impurities in pig iron raises the activity co-efficient of sulphur.
   (d) Low slag basicity & high oxygen potential is essential for efficient removal of sulphur during steel making.

280. The L.D. converter/vessel top is a truncated frustrum of a cone staggered by nearly 8-10° to the main axis of the converter, whose height to diameter ratio is about
   (a) 1 (b) 1.5
   (c) 2 (d) 2.5

281. Maximum permissible cooling water temperature rise in the cooling of oxygen lance in L.D converter is about .... °C.
   (a) 5 (b) 20
   (c) 40 (d) 65

282. During oxygen blow in the L.D converter, the jet should be expanded to obtain maximum impact area at the steel bath surface, simultaneous with the maximum penetration of the bath; because the depth of penetration of the jet in steel bath varies inversely with the impact area at the bath surface. The oxygen lance height from\ the still bath level should be about .... times the nozzle diameter.
   (a) 10 - 15 (61 40-50
   (c) 100 - 200 (d) 500 - 750

283. During oxygen blowing in L.D. converter, the velocity of oxygen jet at the nozzle exit of the lance is supersonic & is about twice the velocity of sound. The exit temperature of the supersonic oxygen jet may be lowered by about .... °C, because of the Joule-Thomson effect.
   (a) 10-20 (b) 40-50
   (c) 100 - 120 (d) 200 - 300

284. The total weight of refractory used for lining the L.D converter is about .... times its steel production capacity; out of which 2/3rd is the wear out lining & 1/3 rd is the permanent refractory lining.
   (a) 1.5 (b) 3
   (c) 0.5 (d) 5

285. Pick out the wrong statement pertaining to L.D. steel making.
   (a) Oxidising condition of L.D. process is not at all conducive to desulphurisation necessitating lime addition to produce high basicity basic slag for effective desulphurisation.
   (b) L.D. process can remove more than 90% of phosphorous present in pig iron, bringing it down to as low as 0.025%.
   (c) Iligh manganese content in the slag tends to retard dephosphorisation of the steel bath ; manganese content in the range of 0.5 -1.1% are tolerable but 0.7% is considered better.
   (d) The temperature of the pig iron (hot metal) charged in the converter is about 1000° C.

Answers

279. (d) 280. (b) 281. (c) 282. (b) 283. (c) 284. (b)
285. (d)

286. Which of the following condition is not conducive for effective dephosphorisation in L.D. steel making ?
(*a*) Thin slag b) Oxidising slag
(*b*) Low temperature (*d*) Acidic slag

287. Which of the following is absent in the L.D, slag ; whose high percentage in B.F, slag is highly undesirable [9]
(*a*) CaO (*b*) SiO_2
(*b*) MgO (*d*) $Al2O_3$

288. Pick out the wrong statement pertaining to L.D. steel making.
(*a*) Energy recovery through converter gas is more in L.D. vessel employing open combustion system as compared to the similar vessel employing suppressed combustion system.
(*b*) Basic, thin & oxidising slag, which is required for effective dephosphorisation is formed in L.D. process alter the initial 4 - 6 minutes of blowing.
(*c*) Dephosphorisation rate increases in the L.D. vessel with the raising the oxygen lance height or decreasing tilt- pressure of oxygen and vice versa.
(*d*) Towards the end of oxygen blow in the L.D. vessel, when the temperature is high, the danger of phosphorous reversion is prevented by maintaining a high basicity of the slag.

289. V-ratio is defined as basic oxide/acidic oxide. The V-ratio for slag with maximum oxidizing power is
(*a*) 0.5 (*b*) 1
(*c*) 2 (*d*) 4

290. For facilitating the removal of phosphorous and sulphur from iron during refining (i.e., steel making), the V-ratio should be
(*a*) low
(*b*) high
(*c*) low at the beginning and high at the end of oxygen blow
(*d*) low throughout the oxygen blow

291. The deciding factor for adoption of either acid or basic process of steel making is the
(*a*) composition of slag
(*b*) nature of slag
(*c*) compostion of pig iron
(*d*) none of these

292. Favourable condition for efficient removal of sulphur is
(*a*) high basicity, high oxygen potential
(*b*) high basicity, low oxygen potential
(*c*) low basicity, low oxygen potential
(*d*) low basicity, high oxygen potential

293. In all the steel making processes, the rate determining steel in the oxidation reaction is the rate of supply of oxygen. The oxygen requirement of the bath in a basic open hearth is determined by the quantity of
(*a*) slag
(*b*) deoxidiser
(*c*) hot metal (ie, pig iron)
(*d*) none of these

294. Nose angle of LD. converter is
(*a*) 37° (*a*) 47°
(*c*) 67° (*d*) 77°

295. For each ton of steel produced, the refractory material consumption is about 3.5 kg. The total weight of refractory material required for lining a LD. converter is about .... times its nominal capacity.
(*a*) 1.5 (*b*) 3
(*c*) 4.5 (*d*) 5.5

Answers

286. (d)	287. (d)	288. (a)	289. (c)	290. (b)	291. (c)
292. (b)	293. (c)	294. (c)	295. (b)		

Section

A

5

Extractive Metallurgy (Non-Ferrous)

1. Which of the following operations use physical means to yield a product enriched in the metal bearing mineral ?
   (*a*) Hydrometallurgy.
   (*b*) Pyrometallurgy.
   (*c*) Electrometallurgy.
   (*d*) Mineral dressing.

2. In which of the following processes, the chemical combination between the metal sought and other elements is not broken up ?
   (*a*) Mineral dressing
   (*b*) Hydrometallurgy
   (*c*) Pyrometallurgy
   (*d*) Electrometallurgy.

3. Which one occupies the lowermost position in the electromotive series of metals ?
   (*a*) Aluminium.
   (*b*) Alkali metals (*e.g.*, K, Na, Li),
   (*c*) Noble metals (Ag, Pt, Au).
   (*d*) Zinc.

4. The metals occurring at the lowermost position in electromotive series
   (*a*) do not resist corrosion.
   (*b*) resist corrosion very strongly.
   (*c*) are very brittle.
   (*d*) are heat insulators.

5. Oxides of which of the following metals in electromotive series are not reducible by carbon in fuel fired furnaces ?
   (*a*) Ag & Hg (*b*) Cu&Sn
   (*c*) Pb & Fe (*d*) Zn & Cr

6. Oxides of which of the following metals decompose by heat alone ?
   (*a*) Ag & Hg (*b*) Cu&Sn
   (*c*) Pb & Fe (*d*) Zn & Cr

7. Which of the following does not offer an effective protective coating against corrosion of iron (as being evident from electromotive series of metals) ?
   (*a*) Zinc (*b*) Chromium
   (*c*) Tin (*d*) Nickel

8. Electromotive series of metals helps in
   (*a*) deciding the mode of its extraction from ore.
   (*b*) classifying the metals into different groups based on their reactivity towards oxygen, water etc.
   (*c*) both (*a*) & (*b*).
   (*d*) neither (*a*) nor (*b*).

Answers

1. (d)	2. (a)	3. (c)	4. (b)	5. (a)	6. (a)
7. (d)	8. (c)				

9. Which of the following metals occurs in nature in their free metallic state ?
(*a*) Au & Ag (*b*) Pb & Zn
(*c*) Cu & Ni (*d*) none of these

10. Extraction of which of the following metals is the most expensive ?
(*a*) Fe (*b*) Au
(*c*) Zn (*d*) Pb

11. Which of the following metals generally occurs in the form of oxides?
(*a*) Sn (*b*) Cu
(*c*) Pb (*d*) Zn

12. Which of the following metals generally occurs in the form of sulphides ?
(*a*) Fe (*b*) Sn
(*c*) A1 (*d*) Pb

13. Iron, tin & aluminium ores mostly occurs in the form of
(*a*) pyrites (*b*) sulphides
(*c*) chlorides (*d*) oxides

14. Copper, lead & zinc ores mostly occur in the form of
(*a*) oxides (*b*) carbonates
(*c*) sillicates (*d*) sulphides

15. Which is a metalliferrous mineral ?
(*a*) Chalcopyrite (*b*) Calcite
(*c*) Flourspar (*d*) Quartz

16. Which is non-metalliferrous mineral ?
(*a*) Cassiterite (*b*) Chalcopyrite
(*c*) Sphalerite (*d*) Flourspar

17. Chemical formula of chalcopyrite is
(*a*) $Cu_2S. Fe_2S_3$ (*b*) $Cu_2S.FeS$
(*c*) SnO_2 (*d*) CaF_2

18. Chemical formula of cassiterite is
(*a*) ZnS (*b*) SnO_2
(*c*) CaF_2 (*d*) SiO_2

19. Chemical formula of calcite is
(*a*) CaF_2 (*b*) $CaCO_3$
(*c*) $Cu_2S. Fe_2S_3$ (*d*) ZnS

20. Chemical formula of sphalerite is
(*a*) ZnS (*b*) SnO_2
(*c*) CaF_2 (*d*) ZnS

21. Chemical formula of flourspar is
(*a*) CaF_2 (*b*) SiO_2
(*c*) $Cu_2S. Fe_2S_3$ (*d*) ZnS

22. Percentage of gangue is maximum in the ore of
(*a*) iron (*b*) gold
(*c*) zinc (*d*) copper

23. Percentage metal in extractable ore is minimum in case of the .... ore.
(*a*) copper (*b*) iron
(*c*) gold (*d*) zinc

24. Function of flux in extractive metallurgy is to
(*a*) chemically combine with the gangue.
(*b*) make the gangue fusible.
(*c*) produce a new compound fusible at a temperature lower than that at which the gangue alone or flux alone would fuse.
(*d*) all (*a*), (*b*) & (*c*).

25. With increase in the gangue content of the ore .... increases.
(*a*) flux requirement
(*b*) smelting time
(*c*) smelting furnace size requirement
(*d*) all (*a*), (*b*) & (*c*).

Answers

9. (d)	10. (b)	11. (a)	12. (d)	13. (d)	14. (d)
15. (a)	16. (d)	17. (a)	18. (b)	19. (b)	20. (a)
21. (a)	22. (b)	23. (c)	24. (d)	25. (d)	

26. Metallic ores are subjected to physical methods of gangue removal before smelting mainly to
(*a*) reduce the cost of metal extraction,
(*b*) get high purity metals.
(*c*) get high quality metals.
(*d*) get metals with good workability.

27. Fe content in Indian iron ore is about .... percent.
(*a*) 20-25 (*b*) 40-45
(*c*) 60-65 (*d*) 90-95

28. Iron oxides percentage in Indian iron ore is about
(*a*) 20-25 (*b*) 40-45
(*c*) 60-65 (*d*) 90-95

29. Which method of metal extraction is applied to the metalliferrous ores ?
(*a*) Hydrometallurgical
(*b*) Pyrometallurgical
(*c*) Electrometallurgical
(*d*) All (*a*), (*b*) & (*c*)

30. The metals which stand higher in the electrochemical series (*i.e.* alkali & alkaline earth metals.) are usuaully extracted by the .... method.
(*a*) hydrometallurgical
(*b*) pyrometallurgical
(*c*) electrometallurgical
(*d*) all (*a*), (*b*) & (*c*)

31. Gold ores and most of the silver ores are normally subjected to .... method of metal extraction.
(*a*) hydrometallurgical
(*b*) pyrometallurgical
(*c*) electrometallurgical
(*d*) none of these

32. Which is the cheapest method of metal extraction ?
(*a*) Hydrometallurgical.
(*b*) Pyrometallurgical.
(*c*) Electrometallurgical.
(*d*) All are equally cheap.

33. Pyrometallurgical method of metal extraction is a/an .... method.
(*a*) dry (*b*) wet
(*c*) electrolytic (*d*) none of these

34. Leaching of ore is done in the .... method of metal extraction.
(*a*) pyrometallurgical
(*b*) hydrometallurgical
(*c*) electrometallurgical
(*d*) none of these

35. Roasting and smelting of ore is done in .... method of metal extraction.
(*a*) hydrometallurgical
(*b*) pyrometallurgical
(*c*) electrometallurgical
(*d*) none of these

36. Electrometallurgical method is widely employed for the extraction of
(*a*) Al & Mg (*b*) Fe
(*c*) Zn (*d*) Pb

37. Fluxes are most commonly
(*a*) oxides minerals (*b*) metal oxides
(*c*) metal sulphides (*d*) quartzitic materials

38. Melting point of slag is lower than that of
(*a*) flux
(*b*) gangue
(*c*) both (*a*) & (*b*)
(*d*) neither (*a*) nor (*b*)

Answers

26. (c)	27. (c)	28. (d)	29. (c)	30. (a)	31. (a)
32. (a)	33. (a)	34. (b)	35. (b)	36. (a)	37. (a)
38. (c)					

39. Decomposition of carbonates during calcination of dolomite, limestone & magnesite takes place at .... °C.
(*a*) < 300 (*b*) >500
(*c*) > 800 (*d*) > 1100

40. Roasting of an ore is done to
(*a*) fuse it without changing its chemical nature.
(*b*) change the ore chemically by devolatalisation.
(*c*) both (*a*) & (*b*).
(*d*) neither (*a*) nor (*b*).

41. Which is the most common form of ore roasting ?
(*a*) Oxidising roasting
(*b*) Chloridizing
(*c*) Suspension roasting
(*d*) Sulphatizing

42. Ore roasting in which the heat generated by the oxidation of sulphides is sufficient to propagate the reaction is called the
(*a*) flash roasting
(*b*) sulphatizing
(*c*) blast roasting
(*d*) autogenous roasting

43. Distillation is adopted in the treatment of
(*a*) volatile metals.
(*b*) zinc, cadmium & mercury.
(*b*) both (*a*) &(*b*).
(*d*) neither (*a*) nor (*b*).

44. Matte is
(*a*) an intermediate product during metal extraction from sulphide rich ores.
(*b*) produced when the gangue percentage in sulphide ore is very high.
(*c*) a heterogeneous mixture of metallic sulphides in which the metallic contents are concentrated.
(*d*) all (*a*), (*b*) & (*c*).

45. Matte is
(*a*) brittle.
(*b*) produced during extraction of copper & nickel.
(*c*) blue or grey and lustre is bright.
(*d*) all (*a*), (*b*) & (*c*).

46. Speiss is
(*a*) hetergeneous mixture of two or more metals arsenides.
(*b*) brittle.
(*c*) formed when arsenical ores are smelted.
(*d*) all (*a*), (*b*) & (*c*).

47. Secondary metals are those
(*a*) resulting from refining of scraps.
(*b*) resulting from reduction of their ores.
(*c*) which are having high level of impurities.
(*d*) which can be cold rolled.

48. The chemical composition of slag formed during metal extraction does not depend upon the .... of flux.
(*a*) nature
(*b*) amount
(*c*) both (*a*) & (*b*)
(*d*) neither (*a*) nor (*b*).

49. If the gangue in the ore to be fluxed is basic, then
(*a*) an acid flux is required.
(*b*) a basic flux is required.
(*c*) either acid or basic flux will do.
(*d*) its removal is very easy.

Answers

39. (c)	40. (b)	41. (a)	42. (d)	43. (c)	44. (d)
45. (d)	46. (d)	47. (a)	48. (d)	49. (a)	

50. The gangue material in ore consists of
(a) acid components only.
(b) basic components only.
(c) both acid & basic components but the acid components generally predominates.
(d) both acid & basic components but the basic components generally predominates.

51. Self fluxing ores are those
(a) which do not need external flux.
(b) in which acid & basic components are balanced.
(c) both (a) & (b).
(d) neither (a) nor (b).

52. Which is not an acid flux ?
(a) Sand (b) Gravel
(c) Quartz rock (d) None of these

53. Depending upon the condition of the slag, which of the following is capable of acting both as an acid or a base ?
(a) Lime (b) Silica
(c) Alumina (d) Magnesia

54. High MgO in slag
(a) increases its viscosity.
(b) lowers its fusion temperature.
(c) eliminates the formation of dibasic silicates.
(d) all (a), (b) & (c).

55. The main basic flux is
(a) limestone and dolomite.
(b) quartz and sand.
(c) alumina and gravel.
(d) magnesia and silica.

56. Neutral fluxes
(a) do not increase the acidity or basicity of the slag.
(b) make slag more fluid by forming double and easily fusible salts.
(c) are exemplified by flourspar (CaF_2).
(d) all (a), (b) & (c).

57. Which of the following is an undesirable property of slag ?
(a) Low fluidity & fusibility
(b) Low density
(c) Poor heat conductivity
(d) High chemical activity and solvent poor for impurities.

58. Main flux used in smelting of tin is
(a) limestone
(b) silica
(c) manganese ore
(d) flourspar

59. Cinder is
(a) same as slag.
(b) a fused mass formed by the action of flux on the gangue of the ore and fuel.
(c) a product resulting from the interaction of acid & basic oxides of high melting points.
(d) all (a), (b) & (c).

60. Pick out the wrong statement.
(a) In pyrometallurgy, all reduced elements join the metal and all oxidised ones go to form slag.
(b) The hearth temperature of blast furnace can be regulated by adjusting the slag composition,
(c) The slag furnishes a means for sulphur removal in iron blast furnace.
(d) none of these.

Answers

50. (c)	51. (c)	52. (d)	53. (c)	54. (b)	55. (a)
56. (d)	57. (a)	58. (a)	59. (d)	60. (d)	

61. Low specific gravity of slag is desirable mainly due to the fact that, it
(a) permits clean and better separation from metal/matte.
(b) makes the slag more fluid.
(c) protects the metal from overheating due to poor heat conductivity.
(d) is more fusible.

62. Flux required in Bessemerising of copper mattee is
(a) acid flux
(b) basic flux
(c) neutral flux
(d) none, as it is self fluxing

63. Slags are used for
(a) cement making.
(b) fertiliser production.
(c) insulation & railroad ballast.
(d) all *(a)*, *(b)* & *(c)*.

64. Copper is not
(a) ductile.
(b) malleable.
(c) a very good conductor of heat & electricity.
(d) none of these.

65. Presence of high quantity of impurities (like arsenic, antimony, oxygen, bismuth etc.) in copper greatly affects its
(a) thermal conductivity
(b) electrical conductivity
(c) tensile strength
(d) all *(a)*, *(b)* & *(c)*

66. Presence of high% of oxygen in copper (0.05% O_2 is considered desirable) increases its
(a) brittleness
(b) ductility
(c) both *(a)* & *(b)*
(d) neither *(a)* nor *(b)*

67. Presence of arsenic in copper greatly reduces its
(a) electrical conductivity.
(b) tenacity & hardness.
(c) malleability & ductility.
(d) none of these.

68. Presence of antimony in copper increases its
(a) electrical conductivity
(b) brittleness
(c) both *(a)* & *(b)*
(d) neither *(a)* nor *(b)*.

69. Presence of bismuth in copper
(a) causes brittleness.
(b) renders it unsuitable for wire-drawing,
(c) both *(a)* & *(b)*.
(d) neither *(a)* nor *(b)*.

70. Copper is used for the
(a) manufacture of electric transmission cable.
(b) alloy making.
(c) sheet, plate, rod & tube making.
(d) all *(a)*, *(b)* & *(c)*.

71. Which of the following is an oxide ore of copper ?
(a) Chalcocite *(b)* Chalcopyrite
(c) Malachite *(d)* Bornite

72. Which of the following is a sulphide ore of copper ?
(a) Cuprite *(b)* Azurite
(c) Malachite *(d)* None of these

Answers

61. (a)	62. (a)	63. (d)	64. (d)	65. (d)	66. (d)
67. (a)	68. (b)	69. (c)	70. (d)	71. (c)	72. (d)

73. Percentage of copper in pure mineral is maximum (about 89%) in
   (*a*) cuprite (*b*) chalcopyrite
   (*c*) chalcocite (*d*) malachite

74. Major Indian deposits of copper ore is in
   (*a*) Mosabani mines (Jharkhand)
   (*b*) Khetri (Rajasthan)
   (*c*) both (*a*) & (*b*)
   (*d*) neither (*a*) nor (*b*)

75. Copper percentage in Indian copper ore is about
   (*a*) 1-3 (*b*) 30-33
   (*c*) 60-64 (*d*) 85-89

76. Copper is extracted from its ore by the .... method.
   (*a*) pyrometallurgical
   (*b*) hydrometallurgical
   (*c*) both (*a*) & (*b*)
   (*d*) neither (*a*) nor (*b*)

77. More widely used process for extraction of metallic copper from its oxide ore is
   (*a*) by reduction smelting in blast furnace,
   (*b*) mixing it with sulphide ore and smelting the mixture in reverberatory furnace.
   (*c*) roasting it in blast furnace.
   (*d*) none of these.

78. Which of the following is not a pyrometallurgical operation involved in copper extraction from its sulphide ores ?
   (*a*) Roasting of concentrates and smelting of roasted product to matte.
   (*b*) Bessemerising of the matte.
   (*c*) Refining of blister copper.
   (*d*) None of these.

79. The main purpose of roasting sulphide copper ore is to
   (*a*) facilitate the existence of oxidising atmosphere in the reverberatory furnace.
   (*b*) eleminate some sulphur (by its oxidation to SO_2) from it.
   (*c*) produce copper rich (about 75%) matte.
   (*d*) all (*a*), (*b*) & (*c*).

80. Desirable percentage of copper in the matte resulting from smelting of sulphide copper ore is about
   (*a*) 10 (*b*) 40
   (*c*) 75 (*d*) 95

81. Grade of matte produced in a reverberatory furnace can be controlled by adjusting the
   (*a*) charge composition
   (*b*) smelting time
   (*c*) smelting temperature
   (*d*) none of these

82. A low grade matte
   (*a*) is rich in FeS.
   (*b*) is lean in Cu_2S.
   (*c*) requires large quantity of silica for fluxing.
   (*d*) all (*a*), (*b*) & (*c*).

83. In flash smelting process of copper extraction
   (*a*) no external fuel is required (preheated air and exothermic reactions serve as a source of heat).
   (*b*) high grade of matte is produced thereby-reducing blowing time in the converter,
   (*c*) smelting process is easier to control and has greater flexibility.
   (*d*) all (*a*), (*b*) & (*c*).

84. Fire refining of blister copper is done to produce .... copper.
   (*a*) tough
   (*b*) very high purity (99.99%)
   (*c*) very high electrical conductivity
   (*d*) electrolytic

Answers

73. (a)	74. (c)	75. (a)	76. (c)	77. (b)	78. (d)
79. (b)	80. (b)	81. (a)	82. (d)	83. (d)	84. (a)

85. Electrolytic refining of blister copper is done to produce copper with high
(*a*) purity (99.99%)
(*b*) electrical conductivity
(*c*) both (*a*) & (*b*)
(*d*) neither (*a*) nor (*b*).

86. Purity of copper produced by fire refining is about .... percent.
(*a*) 98.7 (*b*) 99.3
(*c*) 99.9 (*d*) 87.3

87. H_2SO_4 concentration in electrolyte used in refining of copper is about .... percent.
(*a*) 1—3 (*b*) 10 —16
(*c*) 46 —52 (*d*) 83 — 87.

88. Precious metals like gold, silver & platinum are recovered in electrolytic refining of copper from the
(*a*) mud at the bottom of the electrolytic cell.
(*b*) spent electrolyte.
(*c*) both (*a*) & (*b*).
(*d*) neither (*a*) nor (*b*).

89. Hydrometallurgical method of copper extraction
(*a*) is applied to very poor oxidised ores.
(*b*) is the cheapest of all methods.
(*c*) involves crushing, washing, leaching & precipitation of ore.
(*d*) all (*a*), (*b*) & (*c*).

90. Which is the cheapest and the most common leaching chemical used in hydrometallurgical method of copper extraction from its oxide ores ?
(*a*) Sulphuric acid
(*b*) Nitric acid
(*c*) Ammonium sulphate
(*d*) Brine

91. Of all the commercial metals, which is the most plentiful ?
(*a*) Copper (*b*) Aluminium
(*c*) Iron (*d*) Lead

92. Aluminium and its alloys are given excellent protection against corrosion by
(*a*) an anodising treatment.
(*b*) galvanising.
(*c*) lead coating.
(*d*) none of these.

93. Aluminium is produced by
(*a*) smelting its ore in a reverberatory furnace.
(*b*) electrolysis of a solution of alumina in fused cryolite (Na_3AlF_6).
(*c*) thermal decomposition of bauxite.
(*d*) none of these.

94. Alumina percentage in bauxite is about
(*a*) 10 (*b*) 25
(*c*) 60 (*d*) 90

95. Bauxite requirement to produce one ton of aluminium is about .... tons.
(*a*) 1.5 (*b*) 5.5
(*c*) 15.5 (*d*) 25.5

96. Melting point of pure alumina is .... °C.
(*a*) 1300 (*b*) 1700
(*c*) >2000 (*d*) 3400

97. The most important requirement of an aluminium plant is the
(*a*) cheap & abundant electrical power supply.
(*b*) availability of high purity bauxite.
(*c*) availability of high purity alumina.
(*d*) lower melting point of alumina.

Answers

85. (c)	86. (b)	87. (b)	88. (a)	89. (d)	90. (a)
91. (b)	92. (a)	93. (b)	94. (c)	95. (b)	96. (c)
97. (a)					

98. Bayer's process is used for purifying
   (*a*) bauxite (*b*) alumina
   (*c*) aluminium (*d*) none of these

99. Main impurities in bauxite are
   (*a*) $Fe_2 0_3$, $Si0_2$ & TiO_2.
   (*b*) $CaSiO_3$, $MgSiO_3$ & $Si0_2$.
   (*c*) CaO, MaO & $Si0_2$.
   (*d*) FeO, CaO & MgO.

100. In Bayer's process, powdered bauxite is digested (in a steam jacketted autoclave at 150°C for 2—4 hours) in
   (*a*) NaOH (*b*) NH_4Cl
   (*c*) H_2SO_4 (*d*) HCl

101. Lime added to bauxite during its digestion in Bayer's process
   (*a*) prevents the dissolution of silica.
   (*b*) causes the formation of insoluble calcium silicate.
   (*c*) prevents the loss of alumina.
   (*d*) all (*a*), (*b*) & (*c*).

102. Complete precipitation of $Al(OH)_3$ from sodium aluminate in settling tank (in Bayer's process) takes about .... hours.
   (*a*) 2 (*b*) 25
   (*c*) 60 (*d*) 72

103. $Al(OH)_3$ is converted to alumina (in Bayer's process) by
   (*a*) electrolysis (*b*) calcination
   (*c*) smelting (*d*) none of these

104. Bayer's process is suitable and economical, only when silica percentage of bauxite is
   (*a*) very high (*b*) <0.5
   (*c*) < 3 (*d*) > 5

105. Electric current used for conversion of calcined alumina to aluminium by electrolysis
   (*a*) keeps the electrolyte liquid by generation of heat.
   (*b*) causes electrolytic dissociation of alumina.
   (*c*) both (*a*) & (*b*).
   (*d*) neither (*a*) nor (*b*).

106. Conversion of calcined alumina to aluminium by electrolysis is done in a
   (*a*) carbon lined furnace.
   (*b*) concrete chamber.
   (*c*) wooden chamber.
   (*d*) stainless steel chamber.

107. Anode and cathode in the aluminium reduction cell are made respectively of
   (*a*) carbon and iron bar.
   (*b*) iron bar and carbon.
   (*c*) aluminium and carbon.
   (*d*) carbon and aluminium.

108. Reinforcement of annealed aluminium by steel core
   (*a*) increases its load carrying capacity.
   (*b*) does not affect its electrical conductivity.
   (*c*) facilitates its use for manufacture of cable and bus bar for transmission of electric power.
   (*d*) all (*a*), (*b*) & (*c*).

109. The most significant property of aluminium alloy is its
   (*a*) corrosion resistance.
   (*b*) high strength to weight ratio.
   (*c*) machinability without tearing.
   (*d*) ability for cold working.

110. Ageing of aluminium alloys causes increase in its
   (*a*) strength (*b*) hardness
   (*c*) both (*a*) & (*b*) (*d*) neither (*a*) nor (*b*).

Answers

98. (a)	99. (a)	100. (a)	101. (d)	102. (c)	103. (b)
104. (c)	105. (c)	106. (a)	107. (a)	108. (d)	109. (b)
110. (c)					

111. Tin can be severely cold worked without the necessity for annealing mainly due to its
(*a*) high malleability.
(*b*) high ductility.
(*c*) low recrystallisation temperature.
(*d*) high machinability.

112. Tin exists in .... allotropic forms.
(*a*) 2 (*b*) 3
(*c*) 4 (*d*) 5

113. Tin coating on metals is done by
(*a*) hot-dipping
(*b*) electro-deposition
(*c*) spraying
(*d*) all (*a*), (*b*) & (*c*)

114. Tin is used for
(*a*) corrosion resistant coating of steel & copper alloys.
(*b*) making soft solders.
(*c*) alloy making (*e.g.* bronzes, fusible alloys, white bearing metals, pewter type metals etc.)
(*d*) all (*a*), (*b*) & (*c*).

115. The most important ore of tin is
(*a*) chalcopyrite (*b*) flourspar
(*c*) cassiterite (*d*) calamine

116. Tin percentage in pure cassiterite is about
(*a*) 26.4 (*b*) 43.6
(*c*) 78.6 (*d*) 92.2

117. Tin percentage in its crude ore is about
(*a*) 2 (*b*) 44
(*c*) 78 (*d*) 89

118. Tin is recovered from its ore by the .... methods.
(*a*) pyrometallurgical
(*b*) hydrometallurgical
(*c*) electrometallurgical
(*d*) both (*b*) & (*c*)

119. Flux used in smelting of black tin concentrates to extract tin is
(*a*) silica (*b*) flourspar or lime
(*c*) iron ore (*d*) sand

120. The temperature required for reduction of tin oxide by carbon and carbon monoxide during smelting of black tin concentrate is about .... °C.
(*a*) 300 (*b*) 500
(*c*) 1000 (*d*) 1600.

121. Melting point of pure tin is .... °C.
(*a*) 116 (*b*) 232
(*c*) 696 (*d*) 900

122. Tin smelting is done in a .... furnace.
(*a*) shaft
(*b*) reverberatory
(*c*) both (*a*) & (*b*)
(*d*) neither (*a*) nor (*b*)

123. Use of reverberatory furnace compared to a shaft furnace for tin smelting
(*a*) produces purer tin from less pure extremely fine ores.
(*b*) retains a large quantity of tin in the bed of the furnace, which is recovered only when the smelting compaign is ended.
(*c*) produces slag richer in tin and the process is intermittent.
(*d*) all (*a*), (*b*) and (*c*).

124. Tin percentage in the slag resulting from the smelting of tin may be about
(*a*) 2 (*b*) 8
(*c*) 32 (*d*) 64

Answers

111. (c)	112. (a)	113. (d)	114. (d)	115. (c)	116. (c)
117. (a)	118. (a)	119. (b)	120. (c)	121. (b)	122. (c)
123. (d)	124. (c)				

125. Tin is present in the slag mostly as
(*a*) silicate (*b*) chloride
(*c*) oxide (*d*) sulphide

126. Flux used in tin slag melting for recovery of tin from it, is
(*a*) silica (*b*) limestone
(*c*) bauxite (*d*) tin oxide

127. Refining of impure tin is done
(*a*) electrolytically.
(*b*) in a reverberatory furnace.
(*c*) at a temperature just above the melting point of tin (*i.e.* 232°C).
(*d*) both (*b*) & (*c*).

128. Lead
(*a*) can't be work hardened.
(*b*) has a low recrystallisation temperature.
(*c*) can be made into rod/pipe by extrusion.
(*d*) all (*a*), (*b*) & (*c*).

129. Measure of purity of lead is the
(*a*) production of a very dull sound when struck with a hammer.
(*b*) darkness of its streak on white paper,
(*c*) both (*a*) & (*b*).
(*d*) neither (*a*) nor (*b*).

130. Most significant & important property of lead is its
(*a*) low melting point.
(*b*) shock absorbing capacity.
(*c*) high resistance to corrosion.
(*d*) high recrystaiiisation temperature.

131. Corrosion resistance of lead can be increased by
(*a*) work hardening.
(*b*) annealing.
(*c*) additions of Cu, Ni & Te.
(*d*) none of these.

132. Lead mattresses are used
(*a*) as shock absorbers between the foundations and steel framework of skyscrappers.
(*b*) as pigments.
(*c*) for electrical cable sheathing.
(*d*) none of these.

133. Which is not an ore of lead ?
(*a*) Galena (*b*) Anglesite
(*c*) Cerussite (*d*) Azurite

134. Which in its purest form contains maximum percentage of lead ?
(*a*) Galena (*a*) Anglesite
(*c*) Cerussite (*d*) Cassiterite

135. Lead percentage in pure 'Galena' is about
(*a*) 8.6 (*b*) 35.8
(*c*) 61.4 (*d*) 86.6

136. Lead percentage in pure Anglesite is about
(*a*) 4.2 (*b*) .6.8
(*c*) 68.3 (*d*) 89.3

137. Lead percentage in pure cerussite is about
(*a*) 6.5 (*b*) 18.5
(*c*) 44.5 (*d*) 77.5

138. Lead percentage in a typical lead ore may be about
(*a*) 4 (*b*) 24
(*c*) 64 (*d*) 84

139. The main/maximum use of lead is in the
(*a*) storage batteries manufacture.
(*b*) lining of chemical equipment.
(*c*) bearing material manufacture.
(*d*) paint manufacture.

Answers

125. (a)	126. (b)	127. (d)	128. (d)	129. (c)	130. (c)
131. (c)	132. (a)	133. (d)	134. (a)	135. (d)	136. (c)
137. (d)	138. (b)	139. (a)			

140. Most of the lead is produced in a
  (*a*) reverberatory furnace
  (*b*) blast furnace
  (*c*) rotary kiln
  (*d*) none of these

141. Extraction of lead from lead sulphide is done by
  (*a*) direct reduction with carbon.
  (*b*) direct reduction by CO.
  (*c*) first converting it into oxide (by roasting) before smelting.
  (*d*) All (*a*), (*b*) & (*c*).

142. The most common ore of lead is
  (*a*) galena (*b*) anglesite
  (*c*) cerussite (*d*) chalcocite

143. Cross-section of the blast furnace used for smelting of lead ores is
  (*a*) circular (*b*) rectangular
  (*c*) same as that used in case of copper ore
  (*d*) both (*b*) & (*c*).

144. Roasting of lead ore is done in
  (*a*) Dwight-Lloyd Sintering Machine.
  (*b*) blast furnace.
  (*c*) reverberatory furnace.
  (*d*) none of these.

145. Lead ores rich in gold & silver is charged raw in the blast furnace (*i.e.* not roasted) to
  (*a*) prevent their loss by volatalisation.
  (*b*) facilitate its complete melting.
  (*c*) achieve its better desulphurisation.
  (*d*) increase the yield of lead.

146. Slag from a lead blast furnace consists mainly of. .... of iron & calcium.
  (*a*) oxides (*b*) silicates
  (*c*) sulphates (*d*) chlorides

147. Coke consumption in lead blast furnace is about .... percent of the charge.
  (*a*) 2—4 (*b*) 12—15
  (*c*) 48—50 (*d*) 80—85

148. Temperature in the tuyere level in the lead blast furnace is about .... °C.
  (*a*) 500 (*b*) 1300
  (*c*) 1800 (*d*) 2200

149. Pattinson's process and Parke's process is used for the
  (*a*) removal of impurities like S, Cu & As from lead.
  (*b*) desilverisation of lead.
  (*c*) drossing.
  (*d*) none of these.

150. Litharge is chemically
  (*a*) PbS (*b*) PbO
  (*c*) Pb-Ag alloy (*d*) Pb-Ag-Au alloy

151. Cupellation is carried out for the
  (*a*) purification of gold,
  (*b*) purification of copper.
  (*c*) extraction of silver & gold from lead.
  (*d*) none of these.

152. Zinc is not
  (*a*) resistant to atmospheric corrosion.
  (*b*) very malleable or ductile at ordinary temperature.
  (*c*) rollable into thin sheets under any circumstances.
  (*d*) able to be drawn into wire in any case.

153. Bulk of the zinc is used mainly for the
  (*a*) protection of steel by galvanising.
  (*b*) production of zinc based die casting.
  (*c*) manufacture of pigments.
  (*d*) manufacture of brasses.

Answers

140. (b)	141. (c)	142. (a)	143. (d)	144. (a)	145. (a)
146. (b)	147. (b)	148. (b)	149. (b)	150. (b)	151. (c)
152. (b)	153. (a)				

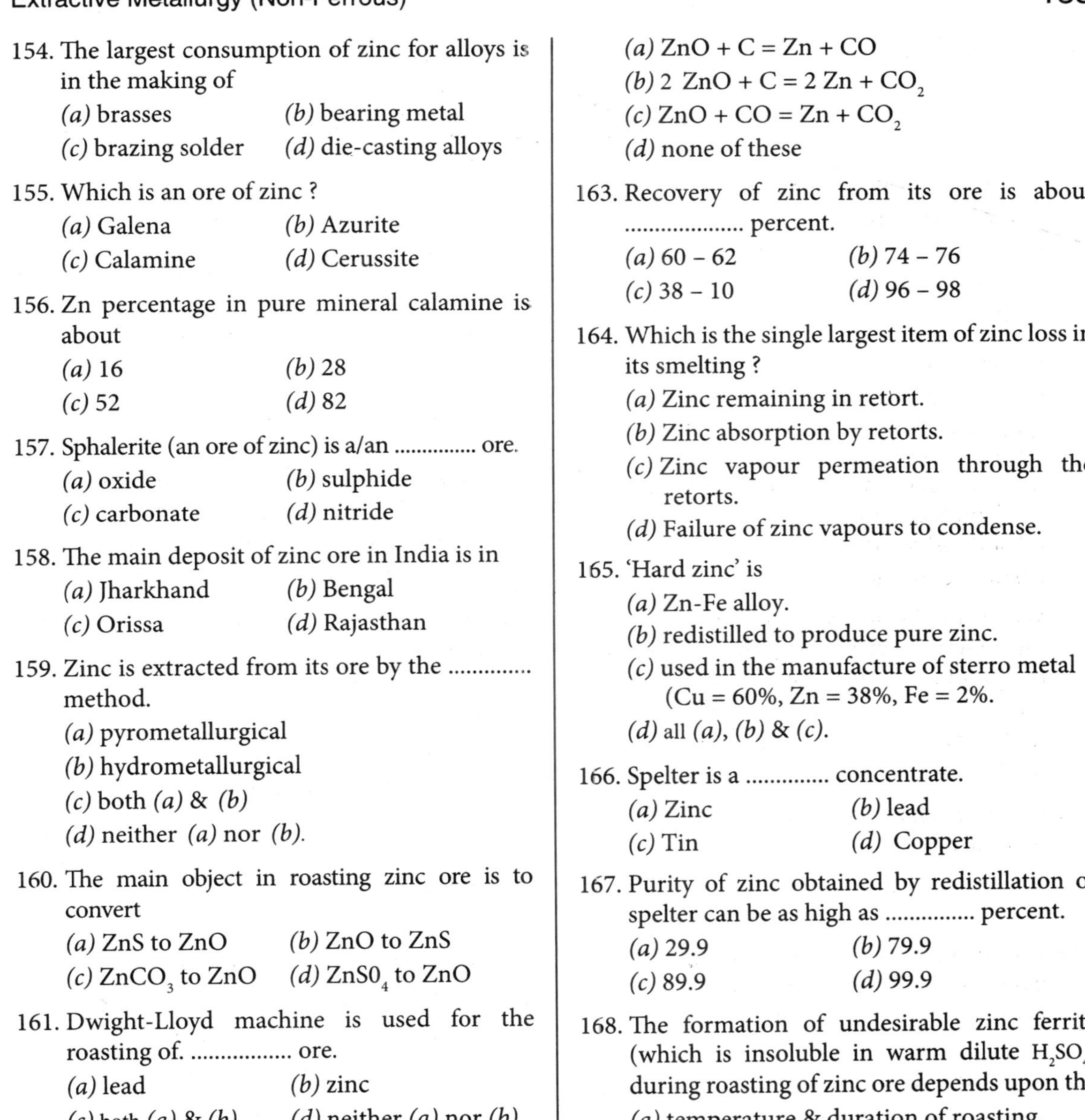

154. The largest consumption of zinc for alloys is in the making of
(*a*) brasses (*b*) bearing metal
(*c*) brazing solder (*d*) die-casting alloys

155. Which is an ore of zinc ?
(*a*) Galena (*b*) Azurite
(*c*) Calamine (*d*) Cerussite

156. Zn percentage in pure mineral calamine is about
(*a*) 16 (*b*) 28
(*c*) 52 (*d*) 82

157. Sphalerite (an ore of zinc) is a/an .... ore.
(*a*) oxide (*b*) sulphide
(*c*) carbonate (*d*) nitride

158. The main deposit of zinc ore in India is in
(*a*) Jharkhand (*b*) Bengal
(*c*) Orissa (*d*) Rajasthan

159. Zinc is extracted from its ore by the .... method.
(*a*) pyrometallurgical
(*b*) hydrometallurgical
(*c*) both (*a*) & (*b*)
(*d*) neither (*a*) nor (*b*).

160. The main object in roasting zinc ore is to convert
(*a*) ZnS to ZnO (*b*) ZnO to ZnS
(*c*) $ZnCO_3$ to ZnO (*d*) $ZnS0_4$ to ZnO

161. Dwight-Lloyd machine is used for the roasting of. .... ore.
(*a*) lead (*b*) zinc
(*c*) both (*a*) & (*b*) (*d*) neither (*a*) nor (*b*).

162. Which of the following reactions involved in the reduction of zinc from its oxide is exothermic ?
(*a*) ZnO + C = Zn + CO
(*b*) 2 ZnO + C = 2 Zn + CO_2
(*c*) ZnO + CO = Zn + CO_2
(*d*) none of these

163. Recovery of zinc from its ore is about .... percent.
(*a*) 60 – 62 (*b*) 74 – 76
(*c*) 38 – 10 (*d*) 96 – 98

164. Which is the single largest item of zinc loss in its smelting ?
(*a*) Zinc remaining in retort.
(*b*) Zinc absorption by retorts.
(*c*) Zinc vapour permeation through the retorts.
(*d*) Failure of zinc vapours to condense.

165. 'Hard zinc' is
(*a*) Zn-Fe alloy.
(*b*) redistilled to produce pure zinc.
(*c*) used in the manufacture of sterro metal (Cu = 60%, Zn = 38%, Fe = 2%.
(*d*) all (*a*), (*b*) & (*c*).

166. Spelter is a .... concentrate.
(*a*) Zinc (*b*) lead
(*c*) Tin (*d*) Copper

167. Purity of zinc obtained by redistillation of spelter can be as high as .... percent.
(*a*) 29.9 (*b*) 79.9
(*c*) 89.9 (*d*) 99.9

168. The formation of undesirable zinc ferrite (which is insoluble in warm dilute H_2SO_4) during roasting of zinc ore depends upon the
(*a*) temperature & duration of roasting.
(*b*) iron content of the ore.
(*c*) association of iron & zinc.
(*d*) all (*a*), (*b*) & (*c*).

Answers

154. (a)	155. (c)	156. (c)	157. (b)	158. (d)	159. (c)
160. (a)	161. (c)	162. (d)	163. (c)	164. (a)	165. (d)
166. (a)	167. (d)	168. (d)			

169. Zinc sulphide roasts at a temperature of .... °C.
(*a*) <600 (*b*) > 600
(*c*) 350 (*d*) 1300

170. Leaching of roasted zinc ore is done with
(*a*) dilute H_2SO_4
(*b*) concentrated (99%) H_2SO_4
(*c*) dilute HCl
(*d*) dilute HNO_3.

171. In zinc smelting blast furnace, ... of the furnace.
(*a*) preheated air is introduced both at the top & bottom
(*b*) zinc vapour & other gases are withdrawn from the centre
(*c*) both (*a*) & (*b*).
(*d*) neither (*a*) nor (*b*).

172. Which of the following materials requires maximum electrical power (kWh/ton of product), when produced by either electrolysis or in electric furnace ?
(*a*) Copper (*b*) Aluminium
(*c*) Pig iron (*d*) Zinc

173. Charge for the zinc smelting blast furnace is preheated to .... °C.
(*a*) 150 (*b*) 300
(*c*) 800 (*d*) 1100.

174. Percentage of zinc vapour in the gas coming out of zinc smelting blast furnace is about
(*a*) 1 (*b*) 5
(*c*) 25 (*d*) 45

175. Residual zinc percentage in the slag from zinc smelting blast furnace is about
(*a*) 1—5 (*b*) 20—25
(*c*) 40—45 (*d*) 60—65

176. Oxides of .... is not reducible by carbon in fuel fired furnace.
(*a*) lead (*b*) tin
(*c*) copper (*d*) aluminium

177. Which of following metals occurs mostly in the form of sulphides ?
(*a*) Tin (*b*) Aluminium
(*c*) Silver (*d*) Iron

178. Which of the following metals occurs mostly in the form of oxides?
(*a*) Lead (*b*) Zinc
(*c*) Tin (*d*) Copper

179. A mineral is called the ore of the metal, if the metal
(*a*) percentage in the ore is above 25'%.
(*b*) can be economically extracted from it.
(*c*) has a moderately low melting point.
(*d*) is not brittle.

180. Which of the following is not a mineral dressing operation ?
(*a*) Roasting
(*b*) Comminution
(*c*) Concentration
(*d*) Screening & classification

181. Which of the following ore concentration operations does not utilise the differences in densities ?
(*a*) Spiralling (*b*) Jigging
(*c*) Tabling (*d*) Froth floatation.

182. Sulphatising of ores is a .... process.
(*a*) smelting (*b*) roasting
(*c*) calcining (*d*) distillation

183. Heterogeneous mixture of arsenides of two or more metals (resulting from smelting of arsenical ores) is called the
(*a*) matte (*b*) speiss
(*c*) bimetal slag (*d*) none of these

Answers

169. (b)	170. (a)	171. (c)	172. (b)	173. (c)	174. (b)
175. (a)	176. (d)	177. (c)	178. (c)	179. (b)	180. (a)
181. (d)	182. (c)	183. (b)			

184. Which of the following is an acid flux ?
(a) MnO (b) SiO_2
(c) CaO (d) CaF_2

185. Neutral fluxes like fluorite increase the .... of the slag.
(a) basicity (b) acidity
(c) fluidity (d) none of these

186. Slag is not used for making
(a) cement (b) fertiliser
(c) road (d) buildings

187. .... tons of bauxite is required to produce one ton of aluminium.
(a) 1.5 (b) 3.5
(c) 5.5 (d) 8.5

188. Aluminium industry is a highly electrical energy intensive industry. Specific electrical energy consumption in aluminium making is about .... kWh/ton of aluminium.
(a) 7000 (b) 14000
(c) 21000 (d) 28000

189. Aluminium is produced by the electrolysis of a solution of alumina in fused cryolite (m.p. - 1000°C). The chemical formula of cryolite is
(a) Na_3AlF_6
(b) $Al_2O_3 . 2H_2O$
(c) Al_2F_6
(d) $NaAlO_2$

190. Bayer's process is meant for the
(a) reduction of $A1_20_3$ to Al.
(b) purification of bauxite.
(c) refining of aluminium.
(d) none of these.

191. During refining of aluminium in Hoope's cell, one ton of aluminium is produced from about .... tons of alumina.
(a) 2 (b) 4
(c) 6 (d) 8 .

192. Purity of electrical grade aluminium should be > .... percent.
(a) 80 (b) 85
(c) 95 (d) 99

193. Voltage employed in the aluminium reduction cell is about .... volts.
(a) 5 – 7 (b) 220 – 240
(c) 4000 – 5000 (d) 11000

194. Which of the following pure minerals of copper contains least % of copper ?
(a) Chalcopyrite (b) Chalcocite
(c) Cuprite (d) Malachite

195. The copper content in the Indian copper ores varies from .... percent.
(a) 1 to 3 (b) 7 to 10
(c) 10 to 15 (d) 15 to 20

196. Coke consumption in the blast furnace for the smelting of copper mattee is about .... percent of the charge.
(a) 5 (b) 15
(c) 30 (d) 50

197. Pick out the wrong statement.
(a) In pyrometallurgy, all reduced elements join the metal and all oxidised ones go to form a slag.
(b) The fused mass formed by the action of the flux on the gangue of the ore & the fuel is called 'cinder'.
(c) The alkali & alkaline earth metal are usually obtained by hydro-metallurgical method.
(d) Froth floatation is used mainly for concentrating sulphide minerals.

Answers

184. (b)	185. (c)	186. (d)	187. (c)	188. (c)	189. (a)
190. (b)	191. (a)	192. (d)	193. (a)	194. (a)	195. (a)
196. (b)	197. (c)				

198. The most extensive and major use of copper is in
(*a*) electrical industry
(*b*) making brasses
(*c*) making bronzes
(*d*) none of these

199. Purity of copper produced by electrolytic refining of blister copper is about .... percent.
(*a*) 90 (*b*) 95
(*c*) 98 (*d*) 99.99

200. Leaching operation is carried out in the .... method of metal extraction.
(*a*) hydrometallurgical
(*b*) pyrometallurgical
(*c*) electro-metallurgical
(*d*) all (*a*), (*b*) & (*c*)

201. Sphalerite is the main ore of
(*a*) cadmium (*b*) zinc
(*c*) lead (*d*) tin

202. Calamine is a zinc ore. Zinc percentage in calamine is about
(*a*) 38 (*b*) 52
(*c*) 67 (*d*) 87

203. Extraction of tin from the black tin concentrate (called black tin) is done by smelting in a
(*a*) blast furnace
(*b*) shaft furnace
(*c*) reverberatory furnace
(*d*) both (*b*) & (*c*)

204. Pick out the wrong statement pertaining to extraction of copper from sulphide ores.
(*a*) Sulphur has greater affinity for iron than for copper.
(*b*) Oxygen has greater affinity for iron than for copper.
(*c*) FeO combines readily with SiO_2 forming a fusible slag.
(*d*) CuS, FeS & other sulphides are all miscible and they form matte.

205. Which of the following metals is not extracted by hydrometallurgical method from its ore ?
(*a*) Tin (*b*) Zinc
(*c*) Copper (*d*) None of these

206. Which of the following metals is not extracted by pyrometallurgical method from its ore ?
(*a*) Lead (*b*) Copper
(*c*) Tin (*d*) Aluminium

207. Which of the following is not an ore of lead?
(*a*) Galena (*b*) Angelsite
(*c*) Cerussite (*d*) Azurite

208. Pick out the wrong statement pertaining to desilverisation of lead by Parke's process.
(*a*) Zinc has greater capacity for dissolving silver than lead.
(*b*) Lead and zinc are mutually miscible in the liquid state,
(*c*) Lead is much heavier than zinc.
(*d*) Lead has a lower freezing point than zinc.

209. Litharge is chemically represented as
(*a*) PbO (*b*) PbS
(*c*) $PbSO_4$ (*d*) $PbCO_3$.

210. Alkali and alkaline earth metals are usually obtained by the .... methods.
(*a*) pyrometallurgical
(*b*) electrometallurgical
(*c*) hydrometallurgical
(*d*) none of these

Answers

198. (a)	199. (d)	200. (a)	201. (b)	202. (b)	203. (d)
204. (a)	205. (a)	206. (d)	207. (d)	208. (b)	209. (a)
210. (b)					

211. Sphalerite is the main source of zinc having .... percent zinc in it.
(a) 47 (b) 52
(c) 67 (d) 87

212. Pick out the wrong statement.
(a)Bauxite and alumina are chemically same.
(b) Bauxite is the main ore of aluminium.
(c) Aluminium is not used for thermal insulation due to its high thermal conductivity.
(d) Production of vanadium uses aluminium as a reducing agent.

213. The pyrometallurgical process used for the extraction of magnesium is called the .... process.
(a) Hoope's (b) Pidgeon
(c) Bayer's (d) Mond

214. Which of the following processes is used for the extraction of nickel?
(a) Hoope's process (b) Harris process
(c) Mond's process (d) Pattinson's process

215. Copper matte comprises mainly of copper sulphide and
(a) iron oxide (b) iron sulphide
(c) iron sulphate (d) copper oxide

216. Copper ores are concentrated using
(a) jigging process
(b) magnetic separator
(c) floatation cell
(d) classification equipments

217. A .... furnace is used for smelting of lead ores.
(a) rotary (b) blast
(c) reverberatory (d) crucible

218. Electrolytic refining of aluminium is done by the .... process.
(a) Hoope's (b) Hall and Heroult
(c) Harris (d) Bayer's

219. Which of the following processes is used for the electrolytic reduction of alumina ?
(a) Hoope's process
(b) Hall and Heroult process
(c) Harris process
(d) none of these

220. Which of the following is a sulphide ore of zinc ?
(a) Zincite (b) Calamine
(c) Zinc blende (d) None of these

221. Matte obtained during smelting of nickel ore comprises of nickel sulphide and .... sulphide.
(a) iron (b) copper
(c) cadmium (d) calcium

222. A/an .... furnace is employed for smelting of roasted tin ore at 1200°C.
(a) rotary (b) blast
(c) reverberatory (d) electric resistance

223. Which of the following is an ore of tungsten?
(a) Wolframite (b) Wulfenite
(c) Carnellite (d) Pitchblende

224. Carnellite is an ore of
(a) tungsten (b) nickel
(c) magnesium (d) beryllium

225. What is the percentage of copper in the matte obtained after smelting in the extraction of copper ?
(a) 25 (b) 50
(c) 75 (d) 95

Answers

211. (c)	212. (a)	213. (b)	214. (c)	215. (b)	216. (c)
217. (b)	218. (a)	219. (b)	220. (c)	221. (a)	222. (c)
223. (a)	224. (c)	225. (b)			

226. Aluminium can not be produced by reduction of alumina by carbon due to the fact that
(*a*) aluminium has a very high affinity for oxygen upto 2000°C.
(*b*) alumina has a very high melting point.
(*c*) suitable furnace for reduction is not available.
(*d*) none of these.

227. Pick out the wrong statement.
(*a*) The largest source of magnesium is sea water.
(*b*) Cerium is not a rare earth metal.
(*c*) Presence of impurities in copper reduces its electrical conductivity.
(*d*) Poling is the process for refining of blister copper.

228. Zinc refining is done by
(*a*) liquation (*b*) distillation
(*c*) electrolysis (*d*) both (*b*) & (*c*)

229. Which of the following is an ore of molybdenum ?
(*a*) Sheelite (*b*) Pitchblende
(*c*) Wulfenite (*d*) Carnellite

230. Concentrated sulphide nickel ore is subjected to .... pressure leaching for the extraction of nickel.
(*a*) CO_2 (*b*) SO_2
(*c*) NH_3 (*d*) H_2SO_4

231. In Mond's process, nickel is obtained in the form of
(*a*) liquid (*b*) vapour
(*c*) shots (*d*) none of these

232. Which of the following is added during smelting of roasted tin ore ?
(*a*) Lime & coal powder
(*b*) Limestone
(*c*) Coke
(*d*) Dolomite.

233. When sulphide ores of copper is subjected to pressure leaching, the
(*a*) ore is roasted.
(*b*) ore does not need roasting.
(*c*) copper produced does not require refining.
(*d*) none of these

234. Which of the following is the major constituents of the slag produced during pyrometallurgical method of extraction of copper from its sulphide ore ?
(*a*) Iron oxide (*b*) Iron silicate
(*c*) Iron sulphide (*d*) Iron sulphate

235. Which of the following is the carbonate ore of copper?
(*a*) Cuprite (*b*) Chalcocite
(*c*) Malachite (*d*) Chalcopyrite

236. Percentage of copper in chalcopyrite is about
(*a*) 10 (*b*) 35
(*c*) 55 (*d*) 70

237. Percentage of copper in high grade copper matte may be about
(*a*) 20 (*b*) 40
(*c*) 60 (*d*) 80

238. Hydrometallurgical process of copper extraction is suitable for its .... ore.
(*a*) carbonate (*b*) oxide
(*c*) sulphide (*d*) none of these

239. Which of the following is a silicate ore of aluminium ?
(*a*) Alunite (*b*) Kaolin
(*c*) Cryolite (*d*) Bauxite

Answers

226. (a)	227. (b)	228. (d)	229. (c)	230. (c)	231. (c)
232. (a)	233. (b)	234. (b)	235. (c)	236. (b)	237. (b)
238. (b)	239. (b)				

240. Aluminium is not extracted from its oxide ore by pyrometallurgical process as in the case of iron ores, mainly because the
   (*a*) melting point of aluminium is very high (2000°C).
   (*b*) reduction of aluminium oxide to aluminium with carbon requires a temperature of 2500 °C while aluminium vaporises at 2000°C.
   (*c*) aluminium becomes brittle at very high temperature.
   (*d*) none of these.

241. Hoope's process in aluminium extraction from its ore is concerned with the
   (*a*) conversion of bauxite to alumina.
   (*b*) reduction of aluminium oxide to aluminium.
   (*c*) electrolytic refining of aluminium.
   (*d*) conversion of alumina to aluminium.

242. Which of the following is the carbonate ore of zinc ?
   (*a*) Troosite (*b*) Calamine
   (*c*) Zinc blende (*d*) Franklinite

243. Which of the following metals can not be extracted from its ore by both pyrometallurgical as well as hydrometallurgical process ?
   (*a*) Aluminium
   (*b*) Zinc
   (*c*) Copper
   (*d*) None of these

244. Dwight Lloyd machine is used for the
   (*a*) roasting of zinc sulphide ore.
   (*b*) electrolytic reduction of alumina.
   (*c*) distillation of zinc oxide.
   (*d*) electrolytic refinning of copper.

245. Percentage of copper in .... is about 97.5.
   (*a*) high grade copper matte
   (*b*) blister copper
   (*c*) electrolytically refined copper
   (*d*) none of these

246. 'Cryolite' used during the extraction of aluminium (Bayer's process) contains a mixture of flourides of calcium, sodium & aluminium. Percentage of aluminium flouride in cryolite is about
   (*a*) 20 (*b*) 40
   (*c*) 60 (*d*) 80

247. Which is the most important ore of mercury?
   (*a*) Galena (*b*) Cinnabar
   (*c*) Cassiterite (*d*) Argentite

248. .... is the main ore of silver.
   (*a*) Argentite (*b*) Sylvanite
   (*c*) Azurite (*d*) Cryolite

249. Pick out the wrong statement.
   (*a*) Rolled gold is not an alloy of gold.
   (*b*)Sulphur does not occur in free state in nature.
   (*c*) Copper occurs in native state also.
   (*d*) Copper ore reserves are found mainly in Noamundi (Jharkhand),

250. Temperature maintained during treatment of bauxite with alkali solution in Bayer's process is about .... °C.
   (*a*) 35 (*b*) 100
   (*c*) 300 (*d*) 500

251. Zinc oxide is reduced to zinc using .... in the retort process of zinc extraction.
   (*a*) coke (*b*) hydrogen
   (*c*) carbon dioxide (*d*) steam

Answers

240. (b)	241. (c)	242. (b)	243. (a)	244. (a)	245. (b)
246. (c)	247. (b)	248. (a)	249. (d)	250. (b)	251. (a)

252. Purity of aluminium achieved in Hoope's refinning process is about .... percent.
(*a*) 95 (*b*) 96.5
(*c*) 98 (*d*) 99.9

253. .... is the main ore of gold.
(*a*) Calaverite (*b*) Argentite
(*c*) Willemite (*d*) none of these.

254. Pick out the correct statement.
(*a*) Vertical retort process is not the widely used process for the reduction of zinc oxide during the extraction of zinc.
(*b*) Lead is such a soft metal, that it can be scratched by finger's nail.
(*c*) Copper ore is sintered in the reverberatory furnace.
(*d*) During roasting of ores, generally incipient fusion takes place.

255. Copper matte is produced during .... operation in course of extraction of copper from its ore.
(*a*) smelting (*b*) roasting
(*c*) bessemerisation (*d*) none of these

256. Iron .... is the most predominant constituent of the slag produced in the pyrometallurgical process for extraction of copper from its sulphide ore.
(*a*) silicate (*b*) phosphate
(*c*) aluminate (*d*) silicide

257. Roasting results in the production of metals in case of
(*a*) iron pyrites (*b*) galena
(*c*) bauxite (*d*) cinnabar

258. Roasting of zinc blende converts it into
(*a*) ZnO (*b*) Zn
(*c*) ZnS (*d*) $ZnSO_4$

259. Parke's process is used in the extraction of
(*a*) aluminium (*b*) zinc
(*c*) silver (*d*) copper

260. The cyanide process is used for the extraction of
(*a*) aluminium (*b*) gold
(*c*) copper (*d*) mercury

261. The electrolyte in the extraction of alumina is
(*a*) fused cryolite with flourspar.
(*b*) fused cryolite with felspar.
(*c*) pure alumina with bauxite and molten cryolite.
(*d*) pure alumina in molten cryolite.

262. Thermite (a mixture of one part aluminium powder and three parts of Fe_2O_3) process is used to extract metals, when their
(*a*) sulphides can not be converted into oxides by roasting.
(*b*) oxides can not be reduced by carbon.
(*c*) carbonates do not yield oxides by thermal decomposition.
(*d*) melting points are very high.

263. Copper is extracted from low grade ore by
(*a*) electrometallurgy
(*b*) hydrometallurgy
(*c*) pyrometallurgy
(*d*) all 'a', b' & 'c'

264. High purity copper metal is obtained by electrolytic reduction. Which of the following properties is not exhibited by copper ?
(*a*) Low electrical conductivity.
(*b*) Malleability.
(*c*) High thermal conductivity.
(*d*) Ductility.

265. 'Cinnabar' is an important ore of
(*a*) zinc (*b*) mercury
(*c*) lead (*d*) copper

Answers

252. (d)	253. (a)	254. (b)	255. (a)	256. (a)	257. (b)
258. (a)	259. (b)	260. (b)	261. (d)	262. (b)	263. (b)
264. (a)	265. (b)				

266. .... process for metal refining aims at the formation of a volatile halide of a metal at low temperatures by reaction with a halogen and its subsequent decomposition to produce very pure crystalline metal.

(a) Zone refinning (b) Poling
(c) Van Arkel (d) Parke's

267. Pidgeon process is meant for the extraction of

(a) aluminium (b) lead
(c) magnesium (d) tin

268. Which of the following processes is not involved in the extraction of aluminium from its ore ?

(a) Serpeck process
(b) Hall-Heroult process
(c) Bayer process
(d) Magnotherm process

269. .... can be extracted from its ore by employing both hydrometallurgy as well as pyrometallurgy.

(a) Copper (b) Aluminium
(c) Iron (d) none of these

270. Pick out the correct statement.

(a) In self fluxing ores, the acid and bases are so well balanced, that they need no external flux.
(b) 'Speiss' produced during smelting of arsenical ores is not brittle.
(c) 'Mattee' (or regulus) produced during smelting of sulphide ores is not brittle.
(d) In 'fire refining', molten metal is subjected to reducing conditions.

271. The maximum consumption of tin is in .... making.

(a) alloy (b) tin plate
(c) bearing material (d) roof sheet

272. Match the characteristics of various non-ferrous metals.

List I

(a) Metal whose extraction is the most expensive.
(b) Metal standing higher in electrochemical series.
(c) Metal whose oxide is not reducible by carbon in a fuel fired furnace.
(d) Metal whose oxide decomposes by heat alone.

List II

I. Aluminium
II. Gold
III. Silver
IV. Alkali & alkaline earth metals.

273. Match the name of ores of various non-ferrous metals.

List I	*List II*
(a) Molybdenum	I. Anglesite
(b) Tungsten	II. Carnellite
(c) Magnesium	III. Wolframite
(d) Lead	IV. Wulfenite

274. Match the types of ores of various non-ferrous metals.

List I

(a) Carbonate ore of lead.
(b) Oxide ore of tin.
(c) Sulphide ore of copper.
(d) Carbornate ore of zinc.

List II

I. Calamine II. Cerussite
III. Cassiterite IV. Chalcocite

Answers

266. (c) 267. (c) 268. (d) 269. (a) 270. (a) 271. (b)
272. (ii) (iv) (i) (iii) 273. (iv) (iii) (ii) (i) 274. (ii) (iii) (iv) (i)

275. Match the metal content present in pure ores of various metals.

List I	*List II*
(a) Cuprite	I. 78.6% Sn
(b) Cassiterite	II. 88.8% Cu
(c) Calamine	III. 68.3% Pb
(d) Anglesite	IV. 52% Zn

276. Match the purposes for which various processes are used in the extractive metallurgy of various non-ferrous metals.

List I

(a) Parke's process and Pattinson's process

(b) Hoope's process.

(c) Hall and Heroult Process.

(d) Mond's process.

List II

I. Electrolytic refining of aluminium.

II. Desilverisation of lead.

III. Extraction of nickel.

IV. Electrolytic reduction of alumina.

277. Match the temperature (°C) maintained in various processes employed in extractive metallurgy of non-ferrous metals.

List I

(a) Temperature in tuyere level in lead blast furnace.

(b) Preheating temperature of charge for zinc smelting blast furnace.

(c) Smelting of roasted tin ore in a reverberatory furnace

(d) Treatment of bauxite with alkali solution in Bayer's process.

List II

I. 800	II. 1300
III. 100	IV. 1200

278. Match the equipments used for various purposes during extraction of zinc from its ore.

List I

(a) Smelting of zinc ore.

(b) Roasting of zinc blende.

(c) Refinning of crude spelter.

(d) Leaching or roasted ore with dilute H_2SO_4.

List II

I. Pachuka vats.

II. Reverberatory furnace.

III. Dwight Lloyd sintering machine.

IV. Blast furnace.

279. Match the chemical formula.

List I	*List II*
(a) Cryolite	I. PbS
(b) Zinc ferrite	II. PbO
(c) Litharge	III. Na_3AlF_6
(d) Galena	IV. $ZnO.FeO_3$

280. Match the purposes for which various processes are used in the extractive metallurgy of non-ferrous metals.

List I

(a) Belgian process.

(b) Cyanide process.

(c) Hutington-Heberlein process.

(d) Pidgeon process.

List II

I. Extraction of magnesium.

II. Extraction of zinc.

III. Extraction of gold.

IV. Extraction of lead.

281. Match the chemical formula of various minerals/ores.

List I	*List II*
(a) Chalcocite	I. $ZnCO_3$
(b) Calamine	II. Cu_2S
(c) Anglesite	III. $PbSO_4$
(d) Cassiterite	IV. SnO_2

Answers

275. (ii) (i) (iv) (iii)	276. (ii) (i) (iv) (iii)	277. (ii) (i) (iv) (iii)
278. (iv) (iii) (ii) (i)	279. (iii) (iv) (ii) (i)	280. (ii) (iii) (iv) (i)
281. (ii) (i) (iii) (iv)		

282. The purpose of smelting an ore is to reduce it. Smelting involves reduction of metal oxide with
(a) C (b) CO
(c) A1 (d) Mg

283. Which of the following metals can not be extracted by carbon reduction process?
(a) Mercury (b) Zinc
(c) Lead (d) Aluminium

284. Van Arkel method of purification of metal involves converting the metal to a .... compound.
(a) volatile stable
(b) volatile unstable
(c) non-volatile stable
(d) non-volatile unstable

285. Heating pyrites to remove sulphur is called roasting. Roasting results in the production of metal in case of
(a) bauxite (b) galena
(c) cinnabar (d) both b and c

286. In which of the following minerals, aluminium is not present ?
(a) Fluorspar (b) Feldspar
(c) Cryolite (d) Mica

287. Beneficiation of the mineral $AI_2O_3 . 2H_2O$ is done by
(a) leaching
(b) liquation
(c) froth floatation
(d) magnetic separation

288. Bessemerisation is involved in the extraction of
(a) iron (b) aluminium
(c) copper (d) silver

289. Refinning of silver is done by
(a) poling (b) liquation
(c) cupellation (d) Van Arkel method

290. The salt which is least likely to be found in minerals is
(a) sulphates (b) nitrates
(c) chloride (d) carbonate

291. Which of the following metals is sometimes found native in nature ?
(a) Copper (b) Iron
(c) Magnesium (d) Aluminium

292. Which of the following metals can be purified by liquation?
(a) Aluminium (b) Tin
(c) Lead (d) both b and c

293. Alkali metals do not exist in free state in nature, because they are highly
(a) electro-negative (b) metallic
(c) volatile (d) reactive

294. Heating an ore in the absence of air below its melting point is called
(a) smelting (b) calcination
(c) roasting (d) leaching

295. Concentration of cassiterite (an ore of tin) is done by
(a) froth floatation
(b) levigation
(c) electromagnetic separation
(d) roasting

296. The most electropositive metals are isolated from their ores by
(a) thermal decomposition
(b) high temperature reduction with carbon
(c) electrolysis of fused ionic salts
(d) self reduction.

Answers

282. (a)	283. (d)	284. (a)	285. (d)	286. (a)	287. (a)
288. (c)	289. (c)	290. (b)	291. (a)	292. (d)	293. (d)
294. (b)	295. (c)	296. (c)			

297. Electrometallurgical method is involved into the extraction of
(*a*) copper (*b*) silver
(*c*) iron (*d*) sodium.

298. Liquation process is not used for the purification of
(*a*) copper (*b*) lead
(*c*) tin (*d*) any of these.

299. The anode mud obtained during electro refinning of copper does not contain
(*a*) silver (*b*) gold
(*c*) magnesium (*d*) any of these.

300. Pick out the wrong statement
(*a*) Alkali metals are generally extracted by electrolysis of their ores in fused state.
(*b*) Every mineral is an ore, but every ore is not a mineral,
(*c*) Aluminium is the most abundant metal in the earth's crust.
(*d*) During electro refinning of a metal, impure metal is made anode.

Answers

297. (d) 298. (a) 299. (c) 300. (b))

6 Mechanical Metallurgy

1. Hooke's law
   (*a*) applies to the elastic deformation.
   (*b*) applies beyond limit of proportionality in stress-strain curve.
   (*c*) states that stress is inversely proportional to strain up to elastic limit.
   (*d*) all (*a*), (*b*) & (*c*).
2. As the load on the tensile test specimen is increased beyond the elastic limit,
   (*a*) plastic deformation occurs.
   (*b*) work hardening sets in.
   (*c*) the shape of the subsequent portion of stress-strain diagram depends on the rate of work hardening.
   (*d*) all (*a*), (*b*) & (*c*).
3. 'Yield point' in the stress-strain curve of a metal is the state
   (*a*) after which deformation increases rapidly with very slight increase in load.
   (*b*) upto which the curve is linear.
   (*c*) at which plastic deformation ceases.
   (*c*) none of these.
4. Stress causing a deformation of .... percent is called the yield strength.
   (*a*) 0.01 (*b*) 0.2
   (*c*) 1 (*d*) 3.3
5. Stress continues to increase after the elastic limit of the metal has been passed before final fracture, due to the
   (*a*) loss of plasticity of metal during this region.
   (*b*) strain hardening of metal owing to cold working.
   (*c*) highly elastic nature of metals.
   (*d*) none of these.
6. The shape of stress-strain curve of a metal beyond elastic limit (but before final fracture) depends on the
   (*a*) increasing strength due to strain hardening.
   (*b*) diminution of the cross-sectional area suffered by the specimen.
   (*c*) both (*a*) & (*b*).
   (*d*) neither (*a*) nor (*b*).
7. Beyond elastic limit, until the point of maximum stress, the tensile test metal specimen uniformly
   (*a*) increases in strength.
   (*b*) decreases in cross-sectional area.
   (*c*) both (*a*) & (*b*).
   (*d*) neither (*a*) nor (*b*).

Answers

1. (a) 2. (d) 3. (a) 4. (b) 5. (b) 6. (c)
7. (c)

8. Percentage elongation and reduction of cross-sectional area of a metal are measures of its
   (*a*) ductility
   (*b*) formability
   (*c*) both (*a*) & (*b*)
   (*d*) neither (*a*) nor (*b*)

9. A high reduction of cross-sectional area of a metal (on application of tensile load) indicates that the material has
   (*a*) a low rate of hardening by deformation.
   (*b*) a high rate of hardening by deformation.
   (*c*) lost its ductility.
   (*d*) lost its malleability.

10. Materials (*e.g.* 18/8 steel, Hadfield manganese steel etc.) having a high elongation but a low reduction of cross-sectional area (on application of tensile load)
    (*a*) are tough.
    (*b*) harden very rapidly when deformed.
    (*c*) both (*a*) & (*b*).
    (*c*) neither (*a*) nor (*b*).

11. Cold working of metals
    (*a*) raises yield point & limit of proportionality.
    (*b*) reduces elongation.
    (*c*) increases ultimate tensile stress.
    (*d*) all (*a*), (*b*) & (*c*).

12. A skin pass is given to the steel sheets in a rolling mill prior to deep drawing operation to
    (*a*) eliminate stretcher strains (*i.e.* localised plastic deformation region in grains).
    (*b*) make the surface smooth,
    (*c*) impart ductility.
    (*d*) none of these.

13. Plastic deformation begins with the first loading and there is no elastic limit (*i.e.* stress-strain curve is a uniform curve with no straight portion) for
    (*a*) cast iron.
    (*b*) soft ductile metals (*e.g.* annealed copper),
    (*c*) both (*a*) & (*b*).
    (*d*) neither (*a*) nor (*b*).

14. Hardness of a material determines the resistance of materials to
    (*a*) scratching
    (*b*) wear & indentation
    (*c*) machinability
    (*d*) all (*a*), (*b*) & (*c*)

15. Brinnel Hardness Number (BHN) of annealed aluminium sheet may be about
    (*a*) 20 (*b*) 200
    (*c*) 300 (*d*) 400

16. The dia of ball in BHN test for steel and other hard materials is .... mm.
    (*a*) 5 (*b*) 10
    (*c*) 25 (*d*) 50

17. The load applied in BHN test is
    (*a*) 100 kg
    (*b*) 1000 kg
    (*c*) 3000 kg
    (*d*) different for soft & hard materials

18. The dia of ball in BHN test for soft metals (*e.g.* Al, Cu, etc.) is .... mm.
    (*a*) 5 (*b*) 10
    (*c*) 20 (*d*) 40

19. The time of loading in BHN test is
    (*a*) 1 second
    (*b*) 15 seconds
    (*c*) 5 minutes
    (*d*) not specified, it can be any thing between 30 to 45 minutes

Answers

8. (c)	9. (a)	10. (c)	11. (d)	12. (a)	13. (c)
14. (d)	15. (a)	16. (b)	17. (c)	18. (a)	19. (b)

20. The thickness of the test specimen in BHN test should not be less than .... times the depth of impression.
   (*a*) two (*b*) five
   (*c*) seven (*d*) ten

21. Piling up of metal round the edge of impression in BHN test indicates
   (*a*) strong capacity for strain hardening.
   (*b*) low capacity for hardening by deformation.
   (*c*) high ductility of metal.
   (*d*) its inability to be cold worked.

22. Tensile strength of steel in ton/inch^2 is obtained by multiplying the BHN by
   (*a*) 0.002 (*b*) 0.22
   (*c*) 22 (*d*) 2200

23. BHN test gives erratic result, when the
   (*a*) piling up of metal round the edge of impression occurs.
   (*b*) sinking of metal round the edge of impression occurs.
   (*c*) hardness of material under test is equal to or more than the hardness of Brinell steel ball itself.
   (*d*) none of these.

24. The depth of indentation in BHN test should be .... the thickness of the test piece.
   (*a*) < 1/10th of (*b*) twice
   (*c*) just less than (*d*) just more than

25. Centre of indentation of BHN test should not be .... its diameter from the edge of the test piece.
   (*a*) < 2.5 times (*b*) > 2.5 times
   (*c*) less than (*d*) none of these

26. Applications of Brinell test include
   (*a*) rapid control of chemical carbon determination during iron and steel making.
   (*b*) examining the effect of heat treatment.
   (*c*) examining the effect of cold working.
   (*d*) all (*a*), (*b*) & (*c*).

27. Vicker's Pyramid method, which measures hardness as Vicker's Pyramid Number (VPN) employs a
   (*a*) diamond pyramid indenter
   (*b*) load of 30 kg
   (*c*) both (*a*) & (*b*)
   (*d*) neither (*a*) nor (*b*)

28. The distinguishing feature between the Vicker's diamond pyramid method and Brinell hardness method is that the former
   (*a*) is used for soft materials only.
   (*b*) is used for very hard materials only.
   (*c*) gives a square impression at the surfaces, whereas the latter gives a circular one.
   (*d*) none of these.

29. Vicker's diamond pyramid method of hardness determination does not give accurate result in case employed for the
   (*a*) polished and hardened steel surfaces.
   (*b*) rough forging or heterogeneous materials like grey cast iron.
   (*c*) soft materials,
   (*d*) hard materials.

30. Rockwell hardness test is useful for
   (*a*) only soft materials.
   (*b*) only hard materials.
   (*c*) rapid routine tests on finished products.
   (*d*) none of these.

31. BHN of a high speed steel is 650. Its hardness in VPN may be about
   (*a*) 250 (*b*) 400
   (*c*) 600 (*d*) 750

Answers

20. (d)	21. (b)	22. (b)	23. (c)	24. (a)	25. (a)
26. (d)	27. (c)	28. (c)	29. (b)	30. (c)	31. (d)

32. In which of the hardness tests, the height of rebound of a small diamond pointed hammer is taken as the index of hardness ?
(*a*) Rockwell's test (*b*) Brinnel's test
(*c*) Vicker's test (*d*) Shore test

33. For steels, shore hardness value multiplied by .... gives roughly the BHN.
(*a*) 2 (*b*) 6
(*c*) 40 (*d*) 0.1

34. Izod value of material is the measure of its
(*a*) hardness (*b*) impact resistance
(*c*) creep strength (*d*) none of these

35. The ability of a material to withstand suddenly applied loads is called its
(*a*) hardness (*b*) impact resistance
(*c*) creep strength (*d*) none of these

36. Detection of faulty heat treatment of metals can be best done by the
(*a*) Brinell's test (*b*) Vicker's test
(*c*) Izod impact test (*d*) Rockwell's test

37. Which of the following is liable to creep even at room temperature ?
(*a*) Low carbon steel
(*b*) Copper
(*c*) Aluminium
(*d*) Lead

38. The ability of a material to withstand repeatedly applied stress is determined by the .... test.
(*a*) impact (*b*) creep strength
(*c*) fatigue (*d*) Brinnel's hardness

39. The fatigue limit of metals is decreased by
(*a*) carburising (*b*) decarburising
(*c*) nitriding (*d*) cold working

40. The usual ratio of fatigue limit to the ultimate tensile stress of metals is about
(*a*) 0.01—0.1 (*b*) 0.45--0.55
(*c*) 2—3 (*d*) 1.0-15

41. With increase in Izod value of metals, its fatigue limit
(*a*) increases.
(*b*) decreases.
(*c*) remains same.
(*d*) may increase or decrease ; there is no direct relationship between the two.

42. Penetrating liquid test is used for the
(*a*) detection of surface cracks.
(*b*) non-ferrous metals.
(*c*) both (*a*) & (*b*).
(*d*) neither (*a*) nor (*b*).

43. Magnaflux method is
(*a*) a non-destructive test.
(*b*) based on distortion of magnetic field by cracks & flaws.
(*c*) suitable for detection of cracks caused by quenching, fatigue, blowholes in castings, defects in welding and seams in rolled & forged products.
(*d*) all (*a*), (*b*) & (*c*).

44. Which of the following is not a non-destructive test ?
(*a*) Ultrasonic testing
(*b*) Magnaflux method
(*c*) Radiography
(*d*) Wohler fatigue test

45. Permanent deformation of a metal under stress occurs, when the nature of deformation is
(*a*) elastic (*b*) plastic
(*c*) both(*a*) & (*b*) (*d*) neither (*a*) nor (*b*)

Answers

32. (d)	33. (b)	34. (b)	35. (b)	36. (c)	37. (d)
38. (c)	39. (b)	40. (b)	41. (d)	42. (c)	43. (d)
44. (d)	45. (b)				

46. Which of the following has the least Young's modulus ?
    (a) Aluminium (b) Copper
    (c) Iron (d) Steel

47. Plastic deformation of a metal is not a function of the
    (a) applied stress (b) temperature
    (c) strain rate (d) none of these

48. Plastic deformation of metals is not carried out
    (a) to improve some mechanical properties of metals/alloys.
    (b) intentionally under any circumstances.
    (c) both (a) & (b).,
    (d) neither (a) nor (b).

49. Plastic deformation of single crystal occurs by
    (a) slip
    (b) twinning
    (c) both (a) & (b)
    (d) neither (a) nor (b)

50. All metals of similar crystal structure slip on the same crystallographic
    (a) planes (b) directions
    (c) both (a) & (b) (d) neither (a) nor (b)

51. Which of the following deforms by twinning?
    (a) Zinc (b) Tin
    (c) Iron (d) All (a), (b) & (c)

52. Pick out the wrong statement pertaining to slip phenomenon involved in metal deformation.
    (a) All atoms in one block move the same distance.
    (b) Slip appears as thin lines under microscope.
    (c) Slip usually requires a lower shear stress than twinning.
    (d) none of these.

53. Pick out the wrong statement pertaining to metal deformation by twinning.
    (a) Atoms in each successive plane within a block move different distances.
    (b) Twinning appears as broad lines/bands under microscope.
    (c) There is a markedly different lattice orientation in the twinned region.
    (d) none of these.

54. Deformation in polycrystalline cubic metals takes place by
    (a) slip (b) twinning
    (c) both (a) & (b) (d) neither (a) nor (b)

55. Tensile test is used for the .... materials.
    (a) ductile (b) hard
    (c) brittle (d) none of these

56. Compression test is mostly used for the .... materials.
    (a) ductile (b) brittle
    (c) hard (d) none of these

57. Stress-strain curve of a metal cannot be helpful in estimating the .... of materials.
    (a) yield point, yield strength & ultimate strength
    (b) % elongation, % reduction in area & Young's modulus of elasticity
    (c) ductility, stiffness, resilience & toughness
    (d) none of these

58. The elongation of metals that occurs at constant stress is called the
    (a) yield point elongation
    (b) elastic deformation
    (c) plastic deformation
    (d) none of these

Answers

46. (a)	47. (d)	48. (c)	49. (c)	50. (c)	51. (d)
52. (d)	53. (d)	54. (a)	55. (a)	56. (b)	57. (d)
58. (a)					

59. Work strain hardening of metals/alloys
    (*a*) reduces ductility (formability) & plasticity.
    (*b*) improves hardness & strength.
    (*c*) both (*a*) & (b).
    (*d*) neither (*a*) nor (*b*).

60. The time dependence of strain on loading and unloading is called the
    (*a*) elastic after effect
    (*b*) Cottrel theory
    (*c*) Luders band formation
    (*d*) none of these

61. The mechanism that produces elastic after effect is called
    (*a*) slip
    (*b*) twinning
    (*c*) internal friction
    (*d*) none of these

62. Bauschinger effect is
    (*a*) same as elastic after effect.
    (*b*) a phenomenon in plastic flow.
    (*c*) irreversible.
    (*d*) none of these.

63. Season cracking
    (*a*) results from combined effects of corrosion and internal stresses.
    (*b*) is usually applied to stress corrosion cracking of cold worked brasses.
    (*c*) may be eliminated by annealing strain hardened brass (at 200—300°C for several hours).
    (*d*) all (*a*), (*b*) & (*c*).

64. Preferred orientation
    (*a*) is a condition of polycrystalline aggregates in which crystal orientations are not random.
    (*b*) occurs when certain crystallographic planes tend to orient themselves in a preferred manner w.r.t. the direction of maximum strain.
    (*c*) depends on the slip and twinning system available for deformation.
    (*d*) all (*a*), (*b*) & (*c*).

65. Preferred orientation is
    (*a*) produced by working of metals.
    (*b*) determined by X-ray methods.
    (*c*) not affected by roll speed, reduction per pass etc.
    (*d*) all (*a*), (*b*) & (*c*).

66. Cold working of metals reduces its
    (*a*) ductility & plasticity.
    (*b*) hardness.
    (*c*) yield strength & ultimate strength.
    (*d*) electrical resistance.

67. While drawing metal wire through a die, the
    (*a*) metal work hardens.
    (*b*) resistance to further deformation increases.
    (*c*) power required to continue drawing increases.
    (*d*) all (*a*), (*b*) & (*c*).

68. Annealing of cold worked metal causes
    (*a*) recovery/stress relief
    (*b*) recrystallisation
    (*c*) grain growth
    (*d*) all (*a*), (*b*) & (*c*)

69. Stress relief annealing of cold worked metals
    (*a*) lowers the strength acquired during its cold working.
    (*b*) prevents its stress corrosion cracking.
    (*c*) does not release the internal stresses in rolled, drawn & extruded objects.
    (*d*) none of these.

Answers

59. (c)	60. (c)	61. (c)	62. (b)	63. (d)	64. (d)
65. (d)	66. (a)	67. (d)	68. (d)	69. (b)	

70. Recrystallisation
    (*a*) produces strain free crystals from the deformed metals.
    (*b*) results in the formation and growth of new grains.
    (*c*) proceeds at a higher temperature than recovery.
    (*d*) all (*a*), (*b*) & (*c*).

71. Recrystallisation of which of the following metals, can occur even at room temperature?
    (*a*) Iron (*b*) Copper
    (*c*) Nickel (*d*) Zinc

72. Recrystallisation temperature does not depend upon the
    (*a*) purity of metal.
    (*b*) annealing time.
    (*c*) grain size of cold worked metal.
    (*d*) none of these.

73. With increase in percentage of soluble impurities in metal, its recrystallisation temperature
    (*a*) increases.
    (*b*) decreases.
    (*c*) remains same.
    (*d*) may increase or decrease, depends on the type of metal.

74. In pure metals, recrystallisation occurs at a temperature of about .... T_m
    (Tm = melting point of metal , °K).
    (*a*) 0.01 (*b*) 0.3
    (*c*) 1.5 (*d*) 10

75. In alloys, recrystallisation occurs at a temperature of about .... T_m.
    (*a*) 0.01 (*b*) 0.5
    (*c*) 1.8 (*d*) 8.7

76. The greater is the degree of cold work on a metal, the higher is the
    (*a*) recrystallisation temperature.
    (*b*) grain size.
    (*c*) deformation necessary to cause recrystallisation.
    (*d*) none of these.

77. With increase in the annealing time of cold worked metals, its recrystallisation temperature
    (*a*) increases.
    (*b*) decreases.
    (*c*) remains same.
    (*d*) may increase or decrease, depends on the type of metal.

78. With decrease in the grain size of cold worked metal, the recrystallisation temperature
    (*a*) increases.
    (*b*) decreases.
    (*c*) remains same.
    (*d*) may increase or decrease, depends on the type of metal.

79. Recrystallisation of cold worked metals increases its
    (*a*) strength (*b*) hardness
    (*c*) ductility (*d*) none of these

80. Recrystallisation of cold worked metal causes disappearance of
    (*a*) distorted & elongated grains
    (*b*) slip
    (*c*) further internal stresses
    (*d*) all (*a*), (*b*) & (*c*)

Answers

70. (d)	71. (d)	72. (d)	73. (a)	74. (b)	75. (b)
76. (d)	77. (b)	78. (b)	79. (c)	80. (d)	

81. Grain growth (which is the third stage of annealing that follows recrystallisation)
    (*a*) means an increase in grain size.
    (*b*) decreases with increasing temperature or time.
    (*c*) is associated with gain in strength and hardness and a loss in ductility.
    (*d*) none of these.

82. The extent of grain growth depends on the
    (*a*) annealing time
    (*b*) annealing temperature
    (*c*) rate of heating
    (*d*) all (*a*), (*b*) & (*c*)

83. With increase of temperature, the grain size
    (*a*) increases.
    (*b*) decreases.
    (*c*) remains same.
    (*d*) may increase or decrease, depends on the metal.

84. As the annealing time increases, the grain size
    (*a*) decreases.
    (*b*) increases.
    (*c*) remains same.
    (*d*) may increase or decrease, depends on the annealing temperature.

85. Slower rate of heating during annealing of a cold worked metal will
    (*a*) form few nuclei
    (*b*) favour grain growth
    (*c*) result in coarse grain
    (*d*) all (*a*), (*b*) & (*c*).

86. High degree of deformation of metals prior to recrystallisation favours the production of
    (*a*) new nuclei (*b*) many nuclei
    (*c*) small grain size (*d*) both (*b*) & (*c*).

87. Pick out the correct statement pertaining to recrystallisation.
    (*a*) The greater the amount of insoluble impurities, the finer is the final grain size.
    (*b*) Alloying elements like nickel etc. limits grain growth during annealing of steel.
    (*c*) In general, the production of coarse grains by growth after recrystallisation is not desirable as it impairs the mechanical properties.
    (*d*) all (*a*), (*b*) & (*c*).

88. Mechanical working of metals does not change its
    (*a*) shape
    (*b*) properties
    (*c*) surface conditions
    (*d*) none of these

89. Plastic deformation of a metal above the recrystallisation temperature, but below the melting/burning point is called
    (*a*) hot working (*b*) cold working
    (*c*) forging (*d*) casting

90. Plastic deformation of a metal below its recrystallisation temperature is called
    (*a*) hot working (*b*) cold working
    (*c*) forging (*d*) deep drawing

91. Hot working of a metal
    (*a*) results in strain hardening.
    (*b*) does not cause its work hardening.
    (*c*) can be carried out below recrystallisation temperature.
    (*d*) none of these.

92. Cold working of a metal
    (*a*) is always carried out at room temperature for all the metals.
    (*b*) results in strain hardening.
    (*c*) is carried out above its recrystallisation temperature.
    (*d*) does not cause work hardening.

Answers

81. (a)	82. (d)	83. (a)	84. (b)	85. (d)	86. (d)
87. (d)	88. (d)	89. (a)	90. (b)	91. (b)	92. (b)

93. In cold working of metals, the
   (*a*) grains structure is not distorted.
   (*b*) amount of deformation without introducing other treatments is not limited.
   (*c*) stress required to cause deformation increases rapidly with the amount of deformation.
   (*d*) all (*a*),(*b*) & (*c*).

94. Hot working of metals
   (*a*) is preferred for large deformations.
   (*b*) uniformly disperses the non metallic impurities throughout the metal.
   (*c*) refines grain structure.
   (*d*) all (*a*), (*b*) & (*c*).

95. Surface finish of hot worked metals is not as good as cold worked metal mainly because of
   (*a*) higher internal stresses.
   (*b*) lower internal stresses.
   (*c*) oxidation and scaling.
   (*d*) work hardening.

96. Which of the following defects in metal does not get welded up during its hot working ?
   (*a*) Blowholes (*b*) Internal porosity
   (*c*) Cracks (*d*) None of these

97. Cold working of metal produces strain hardening, because
   (*a*) it takes place below recrystallisation temperature.
   (*b*) it takes place above recrystallisation temperature.
   (*c*) of low extent of deformation.
   (*d*) of uniform distribution of insoluble impurities in the metal.

98. Cold working compared to hot working of metals obtains
   (*a*) thinner material.
   (*b*) better surface finish & chemical properties.
   (*c*) greater dimensional accuracy.
   (*d*) all (*a*), (*b*) & (*c*).

99. Cold working of metals does not improve the
   (*a*) Izod value (*b*) residual stresses
   (*c*) yield strength (*d*) hardness

100. Cold working of metals increases its
   (*a*) ultimate tensile strength
   (*b*) ductility
   (*c*) % elongation
   (*d*) reduction of area

101. Pick out the wrong statement.
   (*a*) Prior to cold working, hot worked metals need not be acid pickled (by H_2SO_4/HCI) to make the surface clean & scale free.
   (*b*) In hot working of metals, surface decarburisation occurs.
   (*c*) Percentage elongation, reduction of area & impact strength are all improved by hot working of metals.
   (*d*) all (*a*), (*b*) & (*c*).

102. Which of the following is an exclusively cold working process ?
   (*a*) Rolling (*b*) Extrusion
   (*c*) Spinning (*d*) Deep drawing

103. Hot rolling is carried out at temperature above .... T_m. (where T_m = melting point of the metal, °K)
   (*a*) 0.1 (*b*) 0.01
   (*c*) 0.7 (*d*) 0.9

Answers

93. (c)	94. (d)	95. (c)	96. (d)	97. (a)	98. (d)
99. (a)	100. (a)	101. (a)	102. (d)	103. (c)	

104. Which is the most suitable for the production of very thin strip in hard materials ?
(*a*) Three high rolling mill
(*b*) Four high rolling mill
(*c*) Tandem rolling mill
(*d*) Cluster rolling mill

105. Drop/die forging
(*a*) enables the control of spread of metal during compression.
(*b*) is suitable for making connecting rods and crankshafts.
(*c*) uses steam or air hammers.
(*d*) all (*a*), (*b*) & (*c*).

106. Press forging
(*a*) employs a very fast squeezing action.
(*b*) does not allow time for flow of the metal.
(*c*) is used for forging only small sections.
(*d*) is faster than hammer forging.

107. Up set forging is
(*a*) employed to forge heads on bolts, rivets, valves, coupling etc.
(*b*) produced by a fast impact force.
(*c*) completed in only one impression.
(*d*) none of these.

108. Most of the energy spent in deforming metal by cold working appears as heat. The extent of cold work that a material will stand depends upon its
(*a*) ductility (*b*) purity
(*c*) carbon content (*d*) none of these

109. Cold work tool steel should have
(*a*) poor hardenability
(*b*) toughness
(*c*) high wear resistance
(*d*) none of these

110. A/an .... material has similar properties throughout its volume.
(*a*) isotropic (*b*) anisotropic
(*c*) homogenous (*d*) continuous

111. Stress concentration occurs, when a body is subjected to non-uniform stress distribution, and it is caused by the
(*a*) change in its shape.
(*b*) change in its cross-sectional area.
(*c*) hole or notch in the body.
(*d*) change in any dimension in the body.

112. In cold working of steel
(*a*) internal stresses disappear.
(*b*) strength and hardness increases.
(*c*) grain structure does not change.
(*d*) it is heated below recrystallisation temperature.

113. Pick out the wrong statement.
(*a*) In a three high rolling mill, three rolls are placed one above the other and the upper & bottom roll rotate in the same direction, whiie the middle roll rotates in opposite direction.
(*b*) In a four high rolling mill, the diameter of back up rolls is more than that of working rolls.
(*c*) Both the rolls of a two high rolling mill rotate at the same speed and in the same direction.
(*d*) Both the rolls of a two high rolling mill are of the same diameter.

114. During hot working of metals
(*a*) porosity is almost eliminated.
(*b*) scale is formed on the metal surface.
(*c*) mechanical properties improve due to refinement of grain structure.
(*d*) all (*a*), (*b*) & (*c*).

Answers

104. (d)	105. (d)	106. (d)	107. (a)	108. (a)	109. (c)
110. (c)	111. (c)	112. (b)	113. (c)	114. (d)	

115. Pick out the wrong statement about the definition of sheet, plate, strip and foil.
   (*a*) A plate is more than 5 mm thick, while a sheet is less than 5 mm thick.
   (*b*) A strip is a sheet less than 6 mm in thickness and 300 mm in width.
   (*c*) A foil is a thin strip having a maximum thickness of 1.5 mm and a maximum width of 300 mm.
   (*d*) A sheet is more than 5 mm thick, while a plate is less than 5 mm thick.

116. Which rolling mill can not reduce a slab directly to strip in one pass through the mill ?
   (*a*) Cogging mill (*b*) Planetary mill
   (*c*) Three high mill (*d*) None of these

117. Flange wrinkling is the defect found in
   (*a*) rolling (*b*) forging
   (*c*) bending (*d*) deep drawing

118. The recommended range of reductions in diameter for drawing stainless steel in first stage draw is .... percent.
   (*a*) 10 – 15 (*b*) 35 – 40
   (*c*) 55 – 60 (*d*) 70 – 75

119. Billets are heated in extrusion of steel to .... °C.
   (*a*) 700 – 800 (*b*) 1000 – 1100
   (*c*) 1200 – 1300 (*d*) 1400 – 1500

120. Work hardening of metals
   (*a*) decreases tensile strength
   (*b*) decreases ductility
   (*c*) increases ductility
   (*d*) does not affect ductility

121. Cold rolling compared to hot rolling
   (*a*) decreases the strength of the rolled product due to strain hardening.
   (*b*) produces sheet/strip with poor surface finish.
   (*c*) requires less power for rolling.
   (*d*) none of these.

122. When cold worked metal is heated, the dislocation loops rearrange themselves and the .... are progressively reduced.
   (*a*) internal stresses
   (*b*) hardness
   (*c*) electrical resistance
   (*d*) all (*a*),(*b*) & (*c*)

123. Brittleness in a metal can be caused by the
   (*a*) extreme cold working which leaves no planes unaffected by slip.
   (*b*) alteration of ratio of plastic yield strength to cohesive strength.
   (*c*) application of equal tension stresses in three directions at right angles at the same time.
   (*d*) all (*a*) , (*b*) & (*c*).

124. Cold working of metal decreases the
   (*a*) proof stress (*b*) electrical resistance
   (*c*) creep properties (*d*) all (*a*) , (*b*) & (*c*)

125. Pick out the wrong statement pertaining to rolling.
   (*a*) The rolling load is increased by providing front and back tensions.
   (*b*) Usually high speed, four high tandem cold rolling mills are provided with front and back tensions.
   (*c*) Uniformity of thickness across the width of the rolled sheet is maintained by providing front & back tensions.
   (*d*) none of these.

126. Swaging is a forging operation
   (*a*) used to force the end of a heated bar into a desired shape.
   (*b*) in which two halves of a rotating die open and close rapidly while impacting the end of the heated tube or shell.
   (*c*) for reducing the diameter of a bar and in the process making it longer.
   (*d*) none of these.

Answers

115. (d)	116. (c)	117. (d)	118. (b)	119. (d)	120. (c)
121. (a)	122. (d)	123. (d)	124. (c)	125. (a)	126. (b)

127. A hot steel ingot can not just be picked and rolled without soaking in a soaking pit, because of
(a) sharp temperature gradients throughout the ingot.
(b) its big size.
(c) high power consumption in rolling it.
(d) none of these.

128. Which of the following steel ingot need not be inverted before charging in soaking pit?
(a) Ingots of open top killed steel ingots
(b) Semi-killed steel ingots
(c) Rimming steel ingots
(d) None of these.

129. Soaking time for steel ingots at constant temperature (say 1260—1315°C) should ideally be about .... of total ingot heating time.
(a) one third *(b)* half
(c) one fourth *(d)* two third

130. Track time of steel ingot is defined as the time between
(a) steel teeming finish to ingot charging 'in soaking pit) finish.
(b) ingot stripping and charging in soaking pit.
(c) teeming finish to stripping of ingot.
(d) none of these.

131. Heating ratio of ingot is defined as the ratio of
(a) heating time to soaking time.
(b) heating time to track time,
(c) track time to heating time.
(d) soaking time to heating time.

132. Bare/naked time of steel ingot is defined as the time
(a) between teeming finish to stripping finish.
(b) from end of stripping to charging.
(c) between teeming finish to charging finish.
(d) none of these.

133. Scale loss in ingot heating in soaking pit under ideal conditions (*i.e.* no overheating and excess oxygen in flue gas etc.) may be about .... percent.
(a) 3 *(b)* 17
(c) 22 *(d)* 34

134. Underheating of steel ingots in soaking pit
(a) increases the power required for rolling it.
(b) causes surface crack in the ingot.
(c) increases time required for rolling.
(d) all *(a)*, *(b)* & *(c)*.

135. Burning of ingot, which means surface melting (washing) and causes checking, tearing & cracking, occurs due to
(a) direct flame impingement resulting in localised slagging of iron.
(b) presence of inclusion or low melting element (like copper) in steel.
(c) fast rate of heating particularly of cold ingots.
(d) all (a), *(b)* & *(c)*.

136. Normal range of heating ratio of hot steel ingot is about
(a) 2-3 *(b)* 6-8
(c) 10-12 *(d)* 15-17

Answers

127. (a)	128. (c)	129. (a)	130. (a)	131. (b)	132. (b)
133. (a)	134. (d)	135. (d)	136. (a)		

137. Burning of steel ingot during heating in soaking pit is promoted by
   (*a*) high amount of sulphur, tellerium, phosphorous, silicon or aluminium, which reduce the heat conductivity of steel.
   (*b*) no'fuel off period for ingots with very low track time (*i.e.* hotter ingots).
   (*c*) non-uniform expansion of ingots during fast heating of cold ingots.
   (*d*) all (*a*), (*b*) & (*c*).

138. Heating time required for heating the cold ingots is proportional to its
   (*a*) charging temperature
   (*b*) track time
   (*c*) size
   (*d*) shape

139. Which of the following objective is fulfilled by rolling of steel ingot in a slabbing/blooming/billet mill ?
   (*a*) Improvement of structural quality by breaking up the weak cast structure and converting it into a more uniform wrought one.
   (*b*) Improvement of structural quality : scaling off and rolling out of minor surface defects and surface conditioning while still hot immediately after rolling (by not machine scarfing).
   (*c*) Improvement of internal quality: cropping off of piped, heavily segregated ingot tops and also lowermost bottom parts, after the semi finished section is rolled (welding of blowholes).
   (*d*) all (*a*) , (*b*) & (*c*).

140. Reduction of thickness of ingots in each pass (called draft) during roiling is
   (*b*) same.
   (*b*) different.
   (*c*) same, if ingot weight is more than 18 tons.
   (*d*) different, only when the ingot weight is less than 12 tons.

141. The reduction in steel slab thickness due to hot machine scarfing may be about .... mm.
   (*a*) 0.1 (*b*) 3
   (*c*) 12 (*d*) 22

142. Loss of steel due to hot machine scarfing of slabs may be about .... percent.
   (*a*) 0.5 (*b*) 3
   (*c*) 7 (*d*) 12

143. To ensure proper internal welding, the ratio of ingots thickness to slab thickness should be at least
   (*a*) 3 : 1 (*b*) 7 : 1
   (*c*) 12 : 1 (*d*) 20 : 1.

144. A steel is termed "hot short", if it has a .... ratio.
   (*a*) low Mn/S (*b*) high Mn/S
   (*c*) low S/Mn (*d*) none of these

145. Steel ingot to slab yield may be about .... percent.
   (*a*) 75 (*b*) 85
   (*c*) 94 (*d*) 98

146. Which fuel gas is commonly used for hot machine scarfing of slabs?
   (a) Blast furnace gas (*b*) Coke oven gas
   (*c*) Acetylene gas (*d*) None of these

147. When metal is shaped at temperature above recrystallisation temperature, the process is called .... of metal.
   (*a*) hot machining
   (*b*) thermal treatment
   (*c*) high energy rate forming
   (*d*) none of these

Answers

137. (d)	138. (c)	139. (d)	140. (b)	141. (b)	142. (b)
143. (a)	144. (a)	145. (b)	146. (b)	147. (d)	

148. The proper temperature for forging plain carbon steel is .... °C.
(*a*) 768 (*b*) 910
(*c*) 1300 (*d*) 1550

149. The mechanical properties of the metal improves in hot working due to the
(*a*) refinement of grain size
(*b*) grain growth
(*c*) recovery of grains
(*d*) formation of columnar grains

150. A polished and etched surface of the cross-section of a hot worked product appears to the naked eye to have a
(*a*) fibre like structure.
(*b*) mirror like surface.
(*c*) micro constituent appearing at the surface.
(*d*) none of these.

151. .... can be hot worked, even when no heat is applied to the metal.
(*a*) Steels (*b*) Plastics
(*c*) Certain metals (*d*) No metal

152. Pick out the wrong statement.
(*a*) Melting point of any metal is almost double of its recrystallisation temperature.
(*b*) Hot working of metal is done above its recrystallisation temperature,
(*c*) Cold working of metal is done below its recrystallisation temperature,
(*d*) Forging is not a mechanical working process.

153. Which of the following is a mechanical working process ?
(*a*) Machining (*b*) Grinding
(*c*) Casting (*d*) Extrusion

154. Which of the following is not a mechanical working process ?
(*a*) Spinning (*b*) Cupping
(*c*) Piercing (*d*) Grinding

155. Plastic deformation involves
(*a*) permanent deformation in dimension.
(*b*) mechanical working of metal above its recrystallisation temperature only.
(*c*) mechanical working of metal below its recrystallisation temperature only.
(*d*) none of these.

156. Elastic deformation involves
(*a*) permanent deformation in dimension.
(*b*) Hooke's law.
(*c*) restoration of original dimension on removal of stress.
(*d*) both (*b*) & (*c*).

157. A material which possess identical properties in all directions is called a/an material.
(*a*) anisotropic (*b*) isotropic
(*c*) plastic (*d*) none of these

158. Pick out the correct statement.
(*a*) Metals having fine grained structure are more plastic than those having coarse grained structure.
(*b*) The ductility of metal decreases with increase in temperature generally.
(*c*) Hammering is not a mechanical working process.
(*d*) none of these.

159. Cold working of metal causes
(*a*) increase in its strength.
(*b*) decrease in its ductility.
(*c*) both (*a*) & (*b*).
(*d*) neither (*a*) nor (*b*).

Answers

148. (c)	149. (a)	150. (a)	151. (c)	152. (d)	153. (d)
154. (d)	155. (a)	156. (d)	157. (b)	158. (a)	159. (c)

160. A material which does not possess identical properties in all directions is called the .... material.
(a) isotropic (b) anisotropic
(c) plastic (d) homogeneous

161. Cold working reduces the
(a) ultimate tensile stress
(b) yield point
(c) limit of proportionality
(d) elongation

162. The fatigue limit of a metal is decreased by its
(a) carburising (b) cold working
(c) decarburising (d) none of these

163. In which of the following operations, the cross-sectional area of bar is increased due to its reduction in height ?
(a) Upsetting (b) Swaging
(c) Drawing (d) Edging

164. Residual stresses in steel forging may develop due to the
(a) non-uniform deformation.
(b) faulty heat treatment.
(c) quenching of steel forgings,
(d) all (a), (b) & (c).

165. Piercing is done for producing
(a) forged parts (b) seamless tubes
(c) cup-shaped parts (d) wires

166. Which of the following cold working operation involves 'squeezing'?
(a) Broaching (b) Swaging
(c) Spinning (d) Roll forming.

167. Which of the following cold working operation involves 'drawing'?
(a) Spinning (b) Impact intrusion
(c) Punching (d) Slitting

168. Honing can not correct
(a) local irregularities.
(b) malposition of the axis.
(c) non-parallelism of cylindrical features.
(d) waviness of axis.

169. Degree of piastic deformation does not depend upon the
(a) strain rate
(b) kind of metal
(c) temperature of working
(d) none of these

170. A planetary mill reduces the steel slab to strip in .... pass.
(a) one (b) two
(c) three (d) four

171. Peening is related to
(a) sheet metal process.
(b) extrusion process.
(c) wire drawing process.
(d) mechanical working of metals by hammer blow.

172. Which of the following cold working operation involves 'shearing'?
(a) Perforating (b) Embossing
(b) Rivetting (d) Angle bending

173. Pick out the wrong statement.
(a) Very thin strip are produced by cold rolling process.
(b) Back up rolls in four stand high mills are smaller than work rolls.
(c) In direct extrusion process, the force is applied on one side and the extruded products comes out from the other side of the die.
(d) The material subjected to cold working becomes harder than that subjected to hot working.

Answers

160. (b)	161. (d)	162. (c)	163. (a)	164. (d)	165. (b)
166. (b)	167. (a)	168. (b)	169. (d)	170. (a)	171. (d)
172. (a)	173. (b)				

174. The operation in which cutting of outside contour or a piece is done by pressing is called
(*a*) broaching (*b*) coining
(*c*) blanking (*d*) burnishing

175. Which of the following operations produces corrugations ?
(*a*) Crimping (*b*) Piercing
(*c*) Seaming (*d*) Embossing

176. Which of the following defects appears in forged products as a result of hot shortness ?
(*a*) Surface cracks (*b*) Scab
(*c*) Flash (*d*) Cold shut

177. .... extrusion process is used for making collapsible shaving cream tubes.
(*a*) Direct (*b*) Indirect
(*c*) Impact (*d*) none of these

178. Lead is subjected to hot working at
(*a*) 100°C (*b*) 200°C
(*c*) 300°C (*d*) room temperature

179. The fatigue strength of materials under repeated stress .... as the temperature is increased.
(*a*) decreases
(*b*) increases
(*c*) remains unchanged
(*d*) may increase or decrease ; depends on the material

180. .... is used to determine the yield strength of a material.
(*a*) Compression (*b*) Tension
(*c*) Fatigue (*d*) Hardness

181. Young's modulus of elasticity is the
(*a*) ratio of strain to stress.
(*b*) ratio of strain to elasticity,
(*c*) slope of inital linear portion of the stress-strain curve.
(*d*) slope of the plastic region of the stress-strain curve.

182. .... of material is found by the cupping test.
(*a*) Compression strength
(*b*) Tensile strength
(*c*) Ductility
(*d*) Hardness

183. Creep in material under stress can occur due to
(*a*) compression
(*b*) tension
(*c*) torsion & blending
(*d*) all (*a*), (*b*) & (*c*)

184. Which of the following is mainly responsible for tertiary stage of creep to occur ?
(*a*) Low stress & low temperature level
(*b*) High stress & high temperature level
(*c*) High stress & low temperature level
(*d*) Low stress & high temperature level

185. If the magnitude of stress is increased, keeping the temperature constant, the creep rate
(*a*) increases.
(*b*) decreases.
(*c*) remains unchanged.
(*d*) may increase or decrease : depends upon the material.

186. Pick out the wrong statement.
(*a*) Failure of specimen may or may not occur in stress-rupture test.
(*b*) The load used in creep test is lower than that used in stress rupture test.
(*c*) Creep is a plastic deformation process.
(*d*) The creep rate decreases in primary stage of creep,

Answers

174. (c)	175. (a)	176. (a)	177. (c)	178. (d)	179. (a)
180. (b)	181. (c)	182. (c)	183. (d)	184. (b)	185. (a)
186. (a)					

187. Deformation in .... of the test specimen is measured in the creep test.
 (*a*) volume (*b*) length
 (*c*) width (*d*) none of these

188. The maximum pressure build up in the rolling of plates is at .... of the plate.
 (*a*) both ends
 (*b*) mid centre
 (*c*) one third of the length
 (*d*) two third of the length

189. In .... a block of metal is reduced in cross-section by forcing it to flow through a die orifice.
 (*a*) extrusion (*b*) forging
 (*c*) deep drawing (*d*) none of these

190. .... rolls are used in two high reversing mill.
 (*a*) Four horizontal (*b*) Two vertical
 (*c*) Two horizontal (*d*) Four vertical

191. .... test measures the micro-hardness.
 (*a*) Rockwell (*b*) Brinnel
 (*c*) Knoop (*d*) Meyer

192. At .... temperature, the grains and the grain boundaries have equal strength.
 (*a*) Curie (*b*) equi-cohesive
 (*c*) recrystallisation (*d*) absolute zero

193. .... base alloys is susceptible for magnetic particle method of testing.
 (*a*) Aluminium (*b*) Magnesium
 (*b*) Nickel (*d*) Copper

194. Source of production of magnetic lines of flux in magnetic particle method of testing is
 (*a*) alternating current
 (*b*) direct current
 (*c*) permanent magnet
 (*d*) either (*a*), (*b*) or (*c*)

195. Cold working of metals results in increase in its
 (*a*) dislocation density
 (*b*) strength
 (*c*) ductility
 (*d*) magnetic permeability

196. .... of the cast structure does not occur during hot working.
 (*a*) Welding of unoxidised blow holes & minor cracks
 (*b*) Crystalline structure refinement
 (*c*) Elimination of orientation
 (*d*) Decrease in chemical inhomogeneities

197. .... does not affect the extrusion force in the extrusion process.
 (*a*) Deformation speed
 (*b*) Extrusion ratio
 (*c*) Material of orifice of die
 (*d*) Working temperature

198. Sinking process produces
 (*a*) sheets (*b*) tubes
 (*c*) rods (*d*) wires

199. From among the following .... steel possess the maximum ductility.
 (*a*) quenched (*b*) carburised
 (*c*) normalised (*d*) annealed

200. A flat sheet can be made into a bath tub (cup-shaped) by
 (*a*) deep drawing (*b*) casting
 (*c*) extrusion (*d*) none of these

201. Mechanical working of metals, if carried out in protective/inert atmosphere or vacuum will necessarily result in the
 (*a*) prevention of scale formation.
 (*b*) increase in metal strength,
 (*c*) increase in metal's softness.
 (*d*) prevention of warping of the metal.

Answers

187. (b)	188. (b)	189. (a)	190. (c)	191. (c)	192. (b)
193. (c)	194. (d)	195. (b)	196. (d)	197. (c)	198. (b)
199. (d)	200. (a)	201. (a)			

202. Cold working of a metal may not necessarily result in
    (*a*) good surface finish/texture
    (*b*) loss in its weight
    (*c*) decrease in its hardness
    (*d*) none of these

203. Pick out the correct statement.
    (*a*) Hooke's law is concerned with plastic deformation of metals.
    (*b*) Energy requirement for same degree of deformation of metals by hot working is more than that by cold working.
    (*c*) Extrusion of metals is necessarily a hot working operation.
    (*d*) Von Mises theory is applied to plastic deformation.

204. .... of the metallic object remains constant/unchanged in case of plastic deformation,
    (*a*) Volume
    (*b*) Length
    (*c*) Cross-sectional area
    (*d*) none of these

205. The operation of removal of scale from the steel slabs formed by hot rolling of ingots is called
    (*a*) shot blasting (*b*) pickling
    (*c*) scarfing (*d*) shearing

206. The operation of removal of scale from hot rolled steel strip coils prior to its cold rolling is called
    (*a*) slitting (*b*) pickling
    (*c*) shot blasting (*d*) shearing

207. Hadfield manganese steel is normally not subjected to mechanical working; rather it is generally cast mainly due to the fact that, it
    (*a*) can not achieve the desired strength & grain structure.
    (*b*) is highly ductile.
    (*c*) strain hardens to a very large extent.
    (*d*) none of these.

208. Systematic deformation of .... materials is practically impossible.
    (*a*) ductile (*b*) brittle
    (*c*) malleable (*d*) none of these

209. In which of the following metal forming technology, the metal flow is at right angle to the direction of compressive force ?
    (*a*) Shearing (*b*) Forging
    (*c*) Cupping (*d*) Extrusion

210. Scale formation does occur, if the object is heated in a
    (*a*) molten salt (say barium & sodium chloride) bath
    (*b*) muffle furnace
    (*c*) vacuum or inert/protective atmosphere
    (*d*) none of these

211. Shearing force required to cut steel strips is equal to
    (*a*) U.S (*b*) 0.8U.S
    (*c*) US^2 (*d*) $\sqrt{U}.S$
    where, U = ultimate strength of metals at the cutting temperature
    S = cross-sectional area of the strip.

212. Which of the following facilities is not required in an integrated steel plant having cent percent continuous casting of steel facility?
    (*a*) Hot rolling mill (*b*) Cold rolling mill
    (*c*) Slabbing mill (*d*) None of these

213. Fuel used in soaking pits for heating of steel ingots is generally
    (*a*) pure coke oven gas
    (*b*) pulverised coal
    (*c*) mixed (B.F + C.O) gas
    (*d*) pulverised pitch

Answers

202. (c)	203. (d)	204. (a)	205. (c)	206. (b)	207. (c)
208. (b)	209. (b)	210. (d)	211. (b)	212. (c)	213. (c)

214. Heating and soaking time for cold steel ingots in soaking pit may be typically about .... hours.
(a) 4 (b) 8
(c) 16 (d) 30

215. Bottom making material used in soaking pit is generally
(a) alumina
(b) coke breeze
(c) fuller's earth
(d) pulverised coking coal

216. To remove the surface defect of rolled steel slab, its hot scarfing is done by intense burning flame of c.o. gas with high pressure oxygen. Scarfing loss is about percent of the total weight of the steel ingot.
(a) 0.2 (b) 3
(c) 5 (d) 8

217. Rolling temperature of steel slabs coming out of reheating furnace is about .... °C.
(a) 1100 (b) 1250
(c) 1400 (d) 1000

218. The main advantage of walking beam type reheating furnace over pusher type furnace is that the
(a) steam generation by evaporative cooling system is increased very much.
(b) skid mark and mechanical damage of steel slabs is minimised.
(c) sticking and pile-ups of steel slabs is obviated.
(d) both 'b' & 'c'

219. Fuel used in steel slab reheating furnace is generally
(a) pure B.F. gas
(b) mixed (B.F. + C.O) gas
(c) pure coke oven gas
(d) pulverised pitch

220. Specific heat consumption in walking beam reheating furnace is about .... giga calorie/ton of steel slab.
(a) 0.05 (b) 0.3
(c) 0.7 (d) 1

221. Pickling of hot rolled steel coil is done to remove the iron oxide scales from the surface of steel strip prior to its cold rolling. The pickling medium used is generally
(a) alkaline solution
(b) sulphuric acid
(c) hydrochloric acid
(d) either b or 'c'

222. In general, the strongest form in which a metal can be used is produced by its
(a) forging (b) hot rolling
(c) cold rolling (d) extrusion

223. .... is an operation of forging.
(a) Plug rolling (b) Swaging
(c) Thread rolling (d) Stretch forming

224. Thread rolling is a type of
(a) hot rolling (b) cold rolling
(c) cold forging (d) cold drawing

225. Cold working of steel does not increase its
(a) recrystallisation temperature.
(b) hardness & strength.
(c) ductility & corrosion resistance.
(d) elasticity.

226. Forging of plane carbon steel is carried out at .... °C.
(a) 900 (b) 1100
(c) 1300 (d) 1500

Answers

214. (c)	215. (b)	216. (b)	217. (b)	218. (d)	219. (b)
220. (b)	221. (d)	222. (a)	223. (b)	224. (b)	225. (c)
226. (c)					

227. In hot rolling of metal
    (*a*) annealing operation is a must.
    (*b*) electrical power requirement is very high.
    (*c*) surface finish is not very good.
    (*d*) grain refinement is not possible.

228. Plug rolling is used to .... of the tubes.
    (*a*) reduce diameter
    (*b*) increase diameter
    (*c*) reduce wall thickness
    (*d*) both *b* &'*c*'

229. Steel billets for extrusion are heated to a temperature of about .... °C.
    (*a*) 900 (*b*) 1000
    (*c*) 1200 (*d*) 1350

230. With increase in the carbon percentage in steel, its .... decreases.
    (*a*) percentage elongation
    (*b*) hardness
    (*c*) tensile strength
    (*d*) none of these

231. The ultimate tensile strength of a steel whose BHN is 200 will be approximately about .... kgf/mm^2.
    (*a*) 70 (*b*) 105
    (*c*) 200 (*d*) 300

232. .... is not suitable for impact load.
    (*a*) Mild steel (*b*) Cast iron
    (*c*) High carbon steel (*d*) Chrome steel

233. Mild steel is a typical structural steel, which can be .... with considerable ease.
    (*a*) forged (*b*) welded
    (*c*) cold worked (*d*) all '*a*', '*b*' & '*c*'

234. Match the temperature (°C) maintained/ required in various operations.

    List I
    (*a*) Hot working temperature for lead.
    (*b*) Rolling temperature of steel slab out from reheating furnace.
    (*c*) Temperature of acid pickling bath for hot rolled steel strip.
    (*d*) Forging temperature for plain carbon steel.

    List II
    I. 1250 II. 90
    III. 1300 IV. room temperature

235. Match the ratios of various parameters normally maintained in various processes.

    List I
    (*a*) Ratio of soaking time to total heating time of steel ingot in soaking pit.
    (*b*) Ratio of heating time to track time of steel ingot.
    (*c*) Ratio of slab produced to ingot steel rolled.
    (*d*) Ratio of fatigue limit to ultimate tensile stress of metals.

    List II
    I. 0.45-0.55 II. 0.33
    III. 2-3 IV. 0.85

236. Match the terminologies used for various operations.

    List I
    (*a*) Mechanical working of metals by hammer blow.
    (*b*) Removal of scale from steel slab,
    (*c*) Production of seamless tubes.
    (*d*) Removal of scale from hot rolled steel, strip.

    List II
    I. Peening II. Piercing
    III. Scarfing IV. Pickling

Answers

227. (c)	228. (d)	229. (c)	230. (a)	231. (a)	232. (b)
233. (d)	234. (iv) (i) (ii) (iii)		235. (ii) (iii) (iv) (i)		236. (i) (iii) (ii) (iv)

237. Match the processes/applications of various metal forming operations.

List I

(*a*) Crimping

(*b*) Blanking

(*c*) Extrusion

(*d*) Broaching

List II

I. Cutting of outside contour by pressing.

II. Production of corrugations.

III. Gear teeth and keyways making by shearing of hole of contour.

IV. Reduction of cross-section of a block of metal by forcing it to flow through a die orifice.

238. Match the typical specific heat consumption values in various furnaces in steel rolling mills.

List I

(*a*) Batch hood annealing furnace.

(*b*) Soaking pit.

(*c*) Pusher type reheating furnace.

(*d*) Walking beam type reheating furnace

List II

I. 0.3 Giga calorie/ton of slab rolled

II. 0.4 Giga calorie/ton of annealed coil

III. 0.45 Giga calorie/ton of ingot charged

IV. 0.60 Giga calorie/ton of slab rolled

239. Hardness of mild steel is in the range of .... BHN.

(*a*) 40 - 60 (*b*) 110 - 150

(*c*) 300 - 400 (*d*) 700 - 800

240. Pick out the wrong statement.

(*a*) All materials obey Hooke's law upto their plastic limits.

(*b*) The ratio of the young's modulus of elasticity for mild steel determined in tension and compression will be 1.

(*c*) An elasticity is the property of material, when the recovery after unloading is complete but not instantaneous,

(*d*) Tensile strain of a material increases more rapidly, when the elastic limit is reached.

241. Roll forging

(*a*) is a preliminary process followed by press or hammer forging.

(*b*) is used to make axles, leaf springs, tapered levers etc.

(*c*) employs semi-cylindrical rolls.

(*d*) all '*a*', '*b*' & '*c*'.

242. Cold forging process

(*a*) reduces the strength & hardness in the component.

(*b*) produces dull & rough surface in the component.

(*c*) eliminates decarburisation & forging scale.

(*d*) none of these.

243. Extrusion process is used to manufacture

(*a*) rods (*b*) tubes

(*c*) channels, I, Z, T & other sections

(*d*) all *a*, *b* & c

244. Which of the following can not be both a hot working as well as a cold working process?

(*a*) Spinning (*b*) Extrusion

(*c*) Forging (*d*) None of these

245. Deep drawing of a flat sheet

(*a*) requires very high ductility.

(*b*) employs mechanically or hydraulically operated presses.

(*c*) produces cup shaped components whose length is more than its diameter.

(*d*) all (*a*), (*b*) & (*c*).

Answers

237. (ii) (i) (iv) (iii) 238. (ii) (iii) (iv) (i) 239. (b) 240. (a)
241. (d) 242. (c) 243. (d) 244. (d) 245. (d)

246. Good machinability is not associated with the
(*a*) removal of material with very high force.
(*b*) good surface finish.
(*c*) medium degree of tool abrasion.
(*d*) formation of rather small chips.

247. In rolling, when a material is severely deformed in a particular direction, it becomes
(*a*) ductile (*b*) anisotropic
(*c*) isotropic (*d*) homogenous

248. The collapsable toothpaste tube is produced by the .... extrusion.
(*a*) impact (*b*) direct
(*c*) indirect (*d*) tube

249. In drop forging process, the forging is done by dropping the .... at high velocity.
(*a*) hammer
(*b*) workpiece
(*c*) die with hammer
(*d*) weight on hammer

250. To get uniform thickness of the plate by rolling process, we provide
(*a*) antifriction bearings.
(*b*) offset on the rolls.
(*c*) camber on the rolls.
(*d*) hardening of the rolls.

251. During spinning, the mode of metal deformation is
(*a*) bending (*b*) stretching
(*c*) both '*a*' & '*b*' (*d*) rolling & stretching

252. If metals are worked above the recrystallisation temperature, it is referred to as hot working. Which of the following metals can not be hot worked at room temperature, because its recrystallisation temperature is above room temperature ?
(*a*) Zinc (*b*) Tin
(*c*) Lead (*d*) Aluminium

253. Which of the following is an advantage of forging ?
(*a*) Close tolerance.
(*b*) Improved physical properties.
(*c*) Low tooling cost.
(*d*) Good surface finish.

254. A metal finishing operation in which mechanical treatment of surface is done in such a manner that no appreciable metal is removed but the surface is smoothened, is termed as the
(*a*) planishing (*b*) burnishing
(*c*) honing (*d*) bonderising

255. Hot shortness causes development of .... in a forged product.
(*a*) surface cracks
(*b*) porous structure
(*c*) flash
(*d*) scab

256. Slow but progressive deformation of a material over a period of time under the action of a constant stress is a measure of its creep property. Creep in steel becomes important at a temperature above .... °C."
(*a*) 50 (*b*) 150
(*c*) 300 (*d*) 500

257. Cold working & hot working of a metal involves the .... deformation.
(*a*) elastic (*b*) visco-elastic
(*c*) plastic (*d*) isotropic

Answers

246. (a)	247. (b)	248. (b)	249. (c)	250. (c)	251. (c)
252. (d)	253. (b)	254. (b)	255. (a)	256. (c)	257. (c)

258. Pick out the wrong statement about hot working of a metal, which is carried out at above recrystallisation temperature.
(a) Good surface finish & close tolerance is obtained.
(b) Refinement of grain structure is achieved.
(c) Rolling power requirement is low.
(d) Porosity in the metal is largely eliminated.

259. A material used for extrusion purpose must possess the following mechanical property.
(a) Ductility (b) Plasticity
(c) Elasticity (d) Brittleness

260. The increase in hardness due to cold working of a metal is termed as the work hardening. Pick out the wrong statement about the effect of cold working of metal.
(a) Increase in hardness & yield strength is achieved.
(b) Grain structure remains unaffected.
(c) Improvement in metallic surface finish is obtained.
(d) Specific electrical power consumption compared to hot working is more.

261. .... operation aims at removing the flash or burr from the forged parts in drop forging.
(a) Burring (b) Coining
(c) Trimming (d) Swaging

262. Out of the following processes, which would produce the strongest metallic components ?
(a) Extrusion (b) Die casting
(c) Forging (d) Cold rolling

263. First stage diameter reduction in drawing of stainless steel should be about .... percent.
(a) 10 - 15 (b) 35-40
(c) 55 - 60 (d) 75 - 80

264. Match the various metal forming processes to similar processes.

List I	List II
(a) Coining	I. Piercing
(b) Cup drawing	II. Wire drawing
(c) Extrusion	III. Rolling
(d) Blanking	IV. Embossing

265. Match the task performed by various metal forming operations.
List I
(a) Blanking
(b) Slitting
(c) Planishing
(d) Forming
List II
I. Cutting of a metal sheet in straight line along the length
II. Bending of a metal sheet along a curved axis
III. Cutting of flat metal sheet to the desired shape
IV. Straightening of a curved metal sheet

266. Match the tasks performed in various metal forming techniques.
List 1
(a Bulging
(b) Piercing
(c) Roll forging
(d) Perforating.
List II
I. Cutting of non cylinderical hole in a metal sheet by punch & die.
II. Expansion of cylindrical/tubular parts.
III. Even shaped hole cutting in a regular pattern on a metal sheet.
IV. Lengthening of a bar by reducing its diameter.

Answers

258. (a) 259. (b) 260. (b) 261. (c) 262. (c) 263. (b)
264. (iv) (iii) (ii) (i) 265. (iii) (i) (iv) (ii) 266. (ii) (i) (iv) (iii)

267. Match the operations involved in various metal forming techniques.

List I

(*a*) Upsetting

(*b*)Embossing

(*c*) Notching

(*d*) Broaching

List II

I. Removal of metal by pulling/ pushing a cutting tool.

II. Removal of metal to the desired shape from the edge of a plate.

III. Impression of letters, figures & design on metal sheet or parts,

IV. Increase in the diameter, of a bar by reducing its length.

268. Match the various products manufactured by different forging processes.

List I

(*a*) Upset forging

(*b*) Swaging

(*c*) Drop or die forging

(*d*) Roll forging

List II

I. Neck of a cylinder.

II. Axles, leaf springs & tapered levers,

III. Heads on bolts & rivets.

IV. Connecting rods & crankshafts.

269. Cold working of a metal

(*a*) decreases the tensile strength of pure metals, but increases the yield strength of an alloy.

(*b*) increases the yield strength & tensile strength of solid solution alloy more than that of the pure metal.

(*c*) increases the tensile strength & yield strength of only alloys, but not of pure metals.

(*d*) increases the tensile strength & yield strength of only pure metals, but not of alloys.

270. Which of the following is a measure of the ductility of a metal ?

(*a*) Young's modulus.

(*b*) U.T.S.

(*c*) Yield strength.

(*d*) Percent elongation.

271. Creep failure of solids may occur at any temperature. Creep is a very important design consideration in case of

(*a*) boiler tubes ,

(*b*) furnace

(*c*) gas turbine blades

(*d*) crank shaft

272. Die casting of aluminium is not done because of the reason that, it is

(*a*) having tendency to react chemically with the die surface.

(*b*) strong & light weight.

(*c*) having high melting point & cooling time.

(*d*) prone to high expansion on solidification.

273. Initiation of fatigue crack from the surface of a material is not because of the reason that the surface of the material is

(*a*) exposed to corrosive environment.

(*b*) free of dislocation.

(*c*) having higher stress than at any other point inside.

(*d*) likely to carry a stress concentration.

Answers

267. (iv) (iii) (ii) (i) 268. (iii) (i) (iv) (ii) 269. (b) 270. (d)
271. (c) 272. (a) 273. (b)

274. Pick out the wrong statement about, jog formation.

A jog in the line of dislocation is formed, when

(*a*) an edge dislocation climbs partially.

(*b*) an edge dislocation is stopped by grain boundary.

(*c*) two screw dislocations perpendicular to each other, cross each other.

(*d*) two edge dislocations having their planes of glide perpendicular, cross each other.

275. Cracking & .... can be reduced by preheating of parts to be welded and slow cooling of the welded structure.

(*a*) incomplete fusion

(*b*) underfill

(*c*) incomplete penetration

(*d*) residual stress

276. More than one fatigue crack will not originate from the surface, which is

(*a*) exposed to corrosive atmosphere.

(*b*) smoothly ground.

(*c*) having high stress concentration.

(*d*) roughly turned.

277. Three elastic properties, measured as stress from stress-strain diagram of a material, in ascending order are proportional limit, elastic limit & yield point. Proof stress of a material corresponds to

(*a*) lower yield point

(*b*) higher yield point

(*c*) elastic limit

(*d*) a specified strain

278. Pick out the wrong statement about the change in strain rate applied to materials. Increasing strain rate in the tension test of a material

(*a*) increases both yield strength & ultimate tensile strength (U.T.S.).

(*b*) increases yield strength but decreases the ultimate tensile strength (U.T.S.).

(*c*) causes yield point to appear in those materials, which do not show the yield point at low strain rate.

(*d*) and decreasing the temperature tends to make the material brittle.

279. In stretch forming process,

(*a*) metal is deformed during plastic state.

(*b*) dies need not be used.

(*c*) metal deformation is only in the direction of stretch.

(*d*) force is applied only to stretch the material.

280. Pick out the correct statement about bending of materials.

(*a*) A cold bend part has no spring back, because the deformation is plastic

(*b*) All the three processes of bending namely air bending, bottom bending & three point bending, require the same force, if the angle of bend is the same.

(*c*) Parts are normally overbent slightly in cold bending process.

(*d*) The correct combination in cold bending process is - thicker metal : larger bend angle : smaller bend radius.

281. The correct sequence for performing blocking operation on a fresh forging is :

(*a*) trimming, planishing, coining.

(*b*) trimming, coining, planishing.

(*c*) planishing, trimming, coining.

(*d*) planishing, coining, trimming,

Answers

274. (b)	275. (d)	276. (b)	277. (d)	278. (b)	279. (a)
280. (c)	281. (a)				

282. The maximum limit of the hot working temperature of metals is
(*a*) the temperature at which excessive oxidation occurs.
(*b*) 50 °C below the melting point.
(*c*) both '*a*' & '*b*'
(*d*) neither "*a*' nor '*b*'

283. A strip is defined as the rolled metallic product, whose .... mm.
(*a*) thickness is < 6
(*b*) width is not > 600
(*c*) both '*a*' & '*b*'
(*d*) neither 'a' nor '*b*'

284. In rolling of metals, smaller diameter rolls as compared to larger diameter rolls have less
(*a*) strength
(*b*) rigidity
(*c*) power requirement for rolling
(*d*) all '*a*' '*b*'&'*c*''

285. Cold rolling is not used to produce
(*a*) strips (*b*) sheets
(*c*) plates (*d*) both *b* & '*c*'

286. The total metal thickness reduction by cold rolling ranges from .... percent.
(*a*) 10 to 30 (*b*) 30 to 50
(*c*) 50 to 90 (*d*) 95 to 98

287. Roll diameter of two-high reversing hot rolling mills generally ranges from .... metres.
(*a*) 0.2 to 0.5 (*b*) 0.6 to 1.4
(*c*) 1.5 to 2 (*d*) 2.5 to 3

288. The material delivery/rolling speed by a five stand continuous cold rolling mill can be as high as .... metres/second.
(*a*) 5 (*b*) 15
(*c*) 30 (*d*) 60

289. Rolling load is the force with which the rolls press against the metal. Rolling load decreases
(*a*) with roll diameter (D) at a rate greater than $\sqrt{D}$.
(*b*) as the sheet centering the roll becomes thinner.
(*c*) with increase in friction force.
(*d*) with decreasing back & front tension of the rolls.

290. Usually, hot working of a metal is carried out at a temperature above
(*a*) 0.2 T_m (*b*) 0.4 T_m
(*c*) 0.6 T_m (*d*) 0.8 T_m
where, T_m = melting temperature of the metal, ° C

291. In cold working as compared to hot working of metals.
(*a*) strain hardening is never eliminated.
(*b*) corrosion resistance increases.
(*c*) residual stresses are not produced.
(*d*) refinement of crystals occurs.

292. Pick out the wrong statement pertaining to forging operations.
(*a*) Forging hammers are rated by the weight of the ram.
(*b*) Metal flow is parallel to the direction of the die motion in both upsetting & extrusion.
(*c*) In open die forging, squeeze ratio should be < 1.3, while the bite ratio should not be less than 1/3 to minimise the inhomogenous deformation.
(*d*) Amount of deformation should be limited to 50 - 70% reduction in cross- section of a forging to achieve an optimum balance between the ductility in longitudinal & transverse directions.

Answers

282. (c)	283. (c)	284. (d)	285. (c)	286. (c)	287. (b)
288. (c)	289. (a)	290. (c)	291. (a)	292. (b)	

293. Which of the following is a direct compression process?
   (*a*) Bending (*b*) Extrusion
   (*c*) Stretch forming (*d*) Forging.

294. The strain in hot working is
   (*a*) 0.5-1 (*b*) 1 - 2
   (*c*) 2 - 4 (*d*) 6 - 8

295. Which of the following is not a forging operation?
   (*a*) Extrusion (*b*) Fullering
   (*c*) Edging (*d*) Swaging

296. Rotary swaging is used for
   (*a*) diameter reduction of round bars and tubes by rotating die, which open and close rapidly on the work,
   (*b*) making seamless pipes and tubes
   (*c*) providing desired contour to sheet metal
   (*d*) manufacturing bolt and rivets

297. Surface velocity of the rotating roll is .... the speed of incoming metal in rolling operation.
   (*a*) more than (*b*) less than
   (*c*) equal to (*d*) unpredictable

298. .... press is the best suited press for drawing operation.
   (*a*) Rack and pinion
   (*b*) Crank shaft and connecting rod
   (*c*) Knuckle joint
   (*d*) Toggle.

299. In stretch forming process
   (*a*) all the deformation takes place in the direction of stretch
   (*b*) advantage is taken of the plastic state induced due to stretch
   (*c*) no dies are used
   (*d*) all forces are applied in the direction of stretch.

300. Ring rolling is used to decrease the thickness of a ring, in which the
   (*a*) outer roller is larger in diameter, but inner roller is power driven.
   (*b*) outer roller is larger in diameter, and is power driven.
   (*c*) inner roller is larger in diameter, but outer is power driven .
   (*d*) both rollers are equal in diameter, but inner roller is power driven.

301. Drawing process does not belong to the group of
   (*a*) stamping (*b*) pressing
   (*c*) deep drawing (*d*) shallow drawing.

302. A crank shaft can not be produced by
   (*a*) bending (*b*) welding
   (*c*) casting (*d*) rolling.

303. Which of the following metals is not good for impact extrusion?
   (*a*) Low carbon annealed steel
   (*b*) Alloys of Al and Pb
   (*c*) Alloys of Sn and Zn
   (*d*) Stainless steel.

304. A cylindrical vessel with flat bottom can be deep drawn by
   (*a*) single action deep drawing
   (*b*) triple action deep drawing
   (*c*) shallow drawing
   (*d*) double action deep drawing.

305. A poppet valve can be produced by
   (*a*) stretching
   (*b*) rolling
   (*c*) swaging
   (*d*) combined forging and extrusion

Answers

293. (d)	294. (c)	295. (a)	296. (a)	297. (a)	298. (d)
299. (b)	300. (b)	301. (c)	302. (d)	303. (d)	304. (d)
305. (d)					

306. Which of the following materials is suitable for extrusion?
   (*a*) Brass casting
   (*b*) Stainless steel
   (*c*) Low carbon work hardened steel
   (*d*) Low carbon annealed steel.

307. A file removes metal during its forward stroke. When file is moved to and fro over the work, it is known as .... filing.
   (*a*) draw (*b*) cross
   (*c*) pull and push (*d*) none of these

Answers

306. (d) 307. (a)

Section

A

7

Physical Metallurgy

1. Which of the following is a line defect found in metal crystals ?
   (a) Grain boundries
   (b) Cracks
   (c) Edge dislocation
   (d) None of these

2. Vacancy in a metal crystal lattice occurs due to
   (a) imperfect packing during the original crystallization.
   (b) thermal vibrations of atoms at elevated temperature.
   (c) both (a) & (b).
   (d) neither (a) nor (b).

3. Impurity defect like slag inclusion in metals
   (a) give rise to compositional defects
   (b) are introduced into crystal structure as substitutional or interstitial atoms
   (c) may distort the lattice
   (d) all (a), (b) & (c)

4. Dislocation is a line defect in a crystal structure and is intimately connected with the phenomenon of
   (a) slip
   (b) strain hardening & creep
   (c) fatigue & brittle fracture
   (d) all (a), (b) & (c)

5. Dislocations rise in a crystal as a result of
   (a) thermal stresses and external stresses causing plastic flow
   (b) phase transformations
   (c) segregation of solute atoms causing mismatches
   (d) all (a), (b) & (c)

6. Which of the following properties of metals are not affected by imperfections in crystal structure ?
   (a) Plastic & creep characteristics
   (b) thermal stresses and external stresses causing plastic flow.
   (c) Flow, fracture & corrosion characteristics.
   (d) None of these.

7. Substitutional solid solution forms, when the
   (a) solute atoms are very small compared to solvent atoms.
   (b) solute & solvent atoms possess almost equal diameters.
   (c) solvent atoms are very small compared to solute atoms.
   (d) none of these.

Answers

1. (c)	2. (c)	3. (d)	4. (d)	5. (d)	6. (d)
7. (b)					

8. According to Hume Rothery's rule regarding substitutional solid solutions, the .... for their complete solubility.
   (*a*) crystal lattice structure of two metals should be the same
   (*b*) atomic diameters of the two metals should be different
   (*c*) chemical affinity of the two metals should be high
   (*d*) none of these.

9. Metals with large valence electrons .... with a lower valence metal.
   (*a*) have poor solubility
   (*b*) have very good solubility
   (*c*) are completely miscible
   (*d*) are completely miscible, only when the crystal lattice structure of the two metals are same

10. The line/surface in an equilibrium diagram, which indicates the temperature of the beginning of solidification or completion of melting is called the .... temperature.
    (*a*) solidus (*b*) liquidus
    (*c*) solidification (*d*) melting

11. The line/surface in an equilibrium diagram which indicates the temperatures of the completion of solidification or the beginning of melting is called the .... temperature.
    (*a*) liquidus (*b*) solidus
    (*c*) solidification (*d*) melting

12. Which of the following is a structural constituent of an alloy ?
    (*a*) Eutectic (*b*) Pearlite
    (*c*) Martensite (*d*) Sorbite

13. In the equilibrium diagram of a binary alloy system, liquidus is that line
    (*a*) where solidification completes.
    (*b*) where solidification starts.
    (*c*) above which the alloy is in liquid state.
    (*d*) both (*b*) & (*c*).

14. Solidus is that line (in an equilibrium diagram)
    (*a*) above which the alloy is in liquid state.
    (*b*) below which the alloy is in solid state.
    (*c*) where solidification completes.
    (*d*) both (*b*) & (*c*).

15. Which one involves a solid-solid transformation ?
    (*a*) Eutectic (*b*) Peritectic
    (*c*) Eutectoid (*d*) None of these

16. Which of the following metals does not have face centered cubic (f.c.c.) crystal lattice ?
    (*a*) Cu (*b*) Al
    (*c*) Pb (*d*) Na

17. Which one has body centered cubic (b.c.c.) lattice of crystals ?
    (*a*)Na (*b*) Zn
    (*c*) Ag (*d*) Pb

18. Which one does not have hexagonal close packed (h.c.p.) crystal lattice structure ?
    (*a*) Zn (*b*) Mg
    (*c*) Au (*d*) Cd

19. Pick out the wrong statement.
    (*a*) Pure molten metal soldifies at a constant and definite temperature.
    (*b*) Molten alloys solidify over a range of temepratures.
    (*c*) In ternery equilibrium diagram, liquidus & solidus temperatures are indicated by surfaces ; not by lines.
    (*d*) none of these.

Answers

8. (a)	9. (a)	10. (b)	11. (b)	12. (a)	13. (d)
14. (d)	15. (c)	16. (d)	17. (a)	18. (c)	19. (d)

20. Allotropic forms of metals have the same
    (*a*) physical properties
    (*b*) crystal structure
    (*c*) both (*a*) & (*b*)
    (*d*) neither (*a*) nor (*b*)

21. α – iron
    (*a*) is stable below 760 °C.
    (*b*) is magnetic and crystallises in b.c.c. lattice.
    (*c*) has only 0.025% solubility for carbon.
    (*d*) all (*a*), (*b*) & (*c*).

22. γ - iron
    (*a*) is stable above 910°C.
    (*b*) is non magnetic and has f.c.c. lattice,
    (*c*) can hold upto *2%* carbon in solution.
    (*d*) all (*a*), (*b*) & (*c*).

23. Solid solution alloys compared to parent metals
    (*a*) work harden more rapidly.
    (*b*) are more ductile.
    (*c*) have better formability.
    (*d*) none of these.

24. Materials with metallic bonds in atoms are necessarily
    (*a*) ductile under stress.
    (*b*) hard.
    (*c*) gases at room temperature.
    (*d*) having low electrical conductivity.

25. Solids having covalent bonds have
    (*a*) high electrical conductivity.
    (*b*) low electrical conductivity.
    (*c*) high hardness.
    (*d*) both (*b*) & (*c*).

26. Iron exists in more than one type of lattice structure (*e.g. b.c.c/f.c.c.*) depending upon its
    (*a*) carbon content (*b*) temperature
    (*c*) melting point (*d*) hardness

27. At room temperature iron is .... in lattice arrangement.
    (*a*) b.c.c. (*b*) f.c.c.
    (*c*) h.c.p. (*d*) none of these

28. Iron is non magnetic
    (*a*) above Curie point.
    (*b*) when its lattice sturcture is f.c.c.
    (*c*) when it is in γ - iron form.
    (*d*) all (*a*), (*b*) & (*c*).

29. During cooling of steel containing 0.8% carbon from 1000°C, pearlite occurs at...........°C.
    (*a*) 480 (*b*) 723
    (*c*) 768 (*d*) 910

30. Maximum percentage of carbon in ferrite, which is a solid solution of carbon in the solvent of α-iron is
    (*a*) 0.025 (*b*) 0.25
    (*c*) 0.8 (*d*) 1.7

31. Maximum percentage of carbon in austenite, which is a solid solution of carbon in gamma iron is about
    (*a*) 0.025 (*b*) 0.25
    (*c*) 0.8 (*d*) 1.7

32. Carbon percentage in eutectoid steel is
    (*a*) 0.025 (*b*) 0.25
    (*c*) 0.8 (*d*) 1.7

33. Carbon percentage in steel used for the manufacture of drop forging dies is
    (*a*) 0.025 – 0.25 (*b*) 0.25 – 0.40
    (*c*) 0.6 – 0.7 (*d*) 1.4 – 1.7

34. Carbon percentage in steel used for the manufacture of heavy duty leaf and coil springs is
    (*a*) 0.25 (*b*) 0.8
    (*c*) 1 (*d*) 1.5

Answers

20. (d)	21. (d)	22. (d)	23. (a)	24. (a)	25. (d)
26. (b)	27. (a)	28. (d)	29. (b)	30. (a)	31. (d)
32. (c)	33. (c)	34. (c)			

35. Which of the following is added to steel to improve its corrosion resistance ?
    (*a*) Copper (*b*) Columbium
    (c) Cobalt (*d*) None of these

36. Which of the following elements is essentially added to steel during heat treatment to inhibit grain growth in austenite ?
    (*a*) Tin (*b*) Aluminium
    (*c*) Copper (*d*) Brass

37. Recrystallisation temperature is characterised by the
    (*a*) growth of crystals.
    (*b*) change of allotropic form.
    (*c*) formation of spherical crystals from deformed ones, when a strained metal is heated.
    (*d*) formation of first crystal from the molten metal, when it is cooled.

38. What is the criterion used for the determination of hardness of steel ?
    (*a*) Contents of alloying elements
    (*b*) Carbon content
    (*c*) Method of steel production
    (*d*) Shape & distribution of carbide in iron

39. Which etchant is used in metallography of carbon steel ?
    (*a*) 2% HNO_3 in ethyl alcohol.
    (*b*) 2% HNO_3 in water.
    (*c*) 2% HCl in water.
    (*d*) none of these.

40. Which of the following carbide forming elements affects the T-T-T diagram to have two minima ?
    (*a*) Nickel (*b*) Aluminium
    (*c*) Chromium (*d*) Copper

41. Normalising can not improve
    (*a*) mechanical properties.
    (*b*) yield strength.
    (*c*) ductility.
    (*d*) or refine the coarse grain structure developed during hot forging.

42. Cyaniding process comprises of
    (*a*) keeping low carbon steel in a cyanide bath maintained at 800°C.
    (*b*) dipping low carbon steel in a cyanide bath at room temperature.
    (*c*) dipping cast iron in a cyanide bath at 400°C.
    (*d*) dipping a non-ferrous metal in a cyanide bath at 400°C.

43. Internal/external quenching cracks during heat treatment can not be prevented by
    (*a*) austempering.
    (*b*) martempering.
    (*c*) relieving stress through annealing.
    (*d*) making the part smooth without any projections.

44. Isothermal annealing is done for chromium steel by the transformation on supercooled austenite in a bath maintained at a temperature of. .... °C.
    (*a*) 400 – 500 (*b*) 600 – 750
    (*c*) 800 – 950 (*d*) 1000 – 1200.

45. In nitriding operation, some portion of the part may not require surface hardening. Such portion is protected from nitrogen by a layer of
    (*a*) copper (*b*) tin
    (*c*) asbestos (*d*) glass

Answers

35. (d)	36. (b)	37. (c)	38. (d)	39. (a)	40. (c)
41. (c)	42. (a)	43. (d)	44. (b)	45. (b)	

46. In bright hardening process, steel part is heated in salt bath and subsequently cooled in fused alkaline bath maintained at a temperature
   (*a*) below that of martensitic transformation temperature.
   (*b*) above that of martensitic transformation temperature,
   (*c*) 150°C above that of upper critical temperature.
   (*d*) none of these.

47. Though expensive and complex, still surface hardening of steel parts is done by nitriding operation mainly due to the fact that
   (*a*) high production rates can be achieved.
   (*b*) quenching is not involved.
   (*c*) only 800°C furnace temperature is required.
   (*d*) there is no distortion of the furnace part.

48. Which of the following materials get diffused in the surface layer of the steel parts subjected to nitriding operation for increasing its wear resistance property ?
   (*a*) Monoatomic nitrogen
   (*b*) Molecular nitrogen
   (*c*) Ammonia
   (*d*) None of these.

49. The amount of oxygen consumed in flame hardening (which consists of heating the surface of steel parts above the critical temperature by an oxy acetylene torch followed by sudden cooling) is .... the thickness of the hardened layer.
   (*a*) directly proportional to
   (*b*) inversely proportional to
   (*c*) directly porportional to square root of
   (*d*) inversely proportional to the square root of

50. A steel part was hardened and then tempered to remove brittleness; but subsequently it was found that the part has very poor hardness. It may be due to tempering done at .... °C.
   (*a*) 150 (*b*) 350
   (*c*) 550 (*d*) 800.

51. Surface hardening of steel parts in an electrolytic bath containing sodium carbonate is based on a physical phenomenon called
   (*a*) cathode effect (*b*) corona discharge
   (*c*) electrolysis (*d*) none of these

52. The work surface in surface hardening of steel parts in an electrolytic bath (using 3.5 ampere/cm^2 d.c. density through the electrolyte at 200—220 volts) is heated to .... °C.
   (*a*) 200—400 (*b*) 500—800
   (*c*) 900—1200 (*d*) 1200—20

53. A steel part is placed in a varying magnetic field which induces an eddy current to heat it in the induction hardening process. If the depth to which the part is to be hardened is limited to one tenth of its diameter (D), then the relation between frequency (*f*) in cycles per second and diameter (D) is
   (*a*) f = 2,50,000/D^2 (*b*) f = 1,50,000/D^2
   (*c*) f = 50,000/D^2 (*d*) f = 80,000/D^2

54. For a soaking temperature of 930°C in a case carburizing process requiring case hardening to be done for a thickness of 2 mm, the holding time of heating is .... hours.
   (*a*) 6 – 10 (*b*) 19 – 24
   (*c*) 30 – 45 (*d*) 60 – 96

Answers

46. (a)	47. (b)	48. (a)	49. (c)	50. (d)	51. (a)
52. (d)	53. (a)	54. (b)			

55. Case carburising is the most widely used method of case hardening of steel parts with
(*a*) high hardness
(*b*) low carbon content
(*c*) low tensile strength
(*d*) high corrosion resistance

56. In order to overcome warping (which is a problem caused during quenching of slender parts like taps, drills etc,)), the articles should be quenched in a/an
(*a*) vertical manner with cutting end touching first to the quenching medium.
(*b*) inclined manner with the cutting end touching last to the quenching medium.
(*c*) horizontal manner.
(*d*) none of these.

57. Cemented carbide tools (produced by powder metallurgy and used to cut non-ferrous metals) is an alloy of cobalt ,
(*a*) tungsten & carbon.
(*b*) vanadium & molybdenum.
(*c*) chromium & carbon.
(*d*) tungsten & chromium.

58. Austenite is
(*a*) of high tensile strength (10500 kg/cm^2).
(*b*) magnetic.
(*c*) very hard.
(*d*) very stable at room temperature.

59. Which is the softest structure that appears on iron-carbon equilibrium diagram ?
(*a*) Cementite (*b*) Ferrite
(*c*) Pearlite (*d*) Austenite

60. Which is the hardest structure that appears on iron-carbon equilibrium diagram ?
(*a*) Pearlite (*b*) Martensite
(*c*) Cementite (*d*) Austenite.

61. Cementite has
(*a*) an orthorhombic crystal structure.
(*b*) high tensile strength (8000 kg/cm^2).
(*c*) low compressive strength.
(*d*) both (*b*) & (*c*).

62. Ledeburite
(*a*) contains 4.3% carbon.
(*b*) is formed at 1130°C.
(*c*) is the eutectic mixture of austenite & cementite.
(*d*) all (*a*), (*b*) & (*c*).

63. Pearlite which consists of ferrite & cementite and is the product of austenite decomposition by an eutectoid reaction is formed at .... °C.
(*a*) 480 (*b*) 723
(*c*) 980 (*d*) 1130

64. Bainite
(*a*) consists of ferrite and carbide (a decomposition product of austenite).
(*b*) is the constituent produced in steel when austenite transforms at a temperature below that at which pearlite is produced and above that at which martensite is formed,
(*c*) is produced by austempering.
(*d*) all (*a*), (*b*) & (*c*).

65. Martensite has a .... crystal lattice structure.
(*a*) body centered tetragonal
(*b*) body centered cubic
(*c*) face centered cubic
(*d*) close packed hexagonal

Answers

55. (b)	56. (a)	57. (a)	58. (a)	59. (b)	60. (c)
61. (a)	62. (d)	63. (b)	64. (d)	65. (a)	

66. Martensite
   (*a*) is an interstitial supersaturated solid solution of carbon in α - iron.
   (*b*) possesses a needle like structure and is a product of quenching.
   (*c*) is formed as a result of shear type transformation with no diffusion.
   (*d*) all (*a*), (*b*) & (*c*).

67. Troostite is a mixture of
   (*a*) ferrite & austenite.
   (*b*) austenite & martensite.
   (*c*) ferrite & cementite.
   (*d*) none of these.

68. Tempering of martensite below 450°C produces
   (*a*) troostite (*b*) ferrite
   (*c*) cementite (*d*) martensite

69. Which of the following is not a mixture of ferrite & cementite ?
   (*a*) Pearlite (*b*) Sorbite
   (*c*) Troostite (*d*) None of these

70. Machinability tends to decrease with decrease in
   (*a*) hardness & tensile strength.
   (*b*) grain size.
   (*c*) carbon content of carbide.
   (*d*) strain hardening tendency.

71. Etching solution used for medium/high carbon steel and cast iron is
   (*a*) picral (5% picric acid and ethyl alcohol).
   (*b*) 50% NH_4 Oil solution.
   (*c*) nital (2% $HN0_3$ in ethyl alcohol).
   (*d*) 1% HF in water.

72. Etching solution used for aluminium is
   (*a*) 1% HF in water (*b*) 50% NH_4OH
   (*c*) nital (*d*) picral

73. Etching solution used for copper is
   (*a*) 50% NaOH in water
   (*b*) 1% HF in water
   (*c*) nital
   (*d*) picral

74. Carbon percentage in hypoeutectoid steel is
   (*a*) 0.8 – 3 (*b*) < 0.83
   (*c*) 0.83 – 2 (*d*) > 2

75. Carbon percentage in hypereutectoid steel is
   (*a*) 0.83 (*b*) < 0.83
   (*c*) 0.83 – 2 (*d*) > 2

76. Hypereutectoid steels have the structure of
   (*a*) pearlite alone
   (*b*) phases of ferrite & pearlite.
   (*c*) phases of cementite & pearlite.
   (*d*) none of these.

77. Pearlite steels have the structure of
   (*a*) pearlite alone.
   (*b*) phases of cementite & pearlite.
   (*c*) phases of ferrite & pearlite.
   (*d*) none of these.

78. Eutectoid steels have the structure of
   (*a*) pearlite alone.
   (*b*) phases of ferrite & pearlite.
   (*c*) phases of cementite & pearlite.
   (*d*) none of these.

79. Spheroidal is the structure obtained by quenching austenite and then heating into the range of .... °C.
   (*a*) 150 – 250 (*b*) 360 – 660
   (*c*) 660 – 700 (*d*) 980 – 1120

Answers

66. (d)	67. (c)	68. (a)	69. (d)	70. (b)	71. (a)
72. (a)	73. (a)	74. (b)	75. (c)	76. (c)	77. (a)
78. (a)	79. (c)				

80. Sorbite is the structure obtained by quenching austenite and then heating into the range of. .... °C.
    (*a*) 150 – 260 (*b*) 375 – 660
    (*c*) 800 – 1020 (*d*) 980 – 1120

81. Troostite is the structure obtained by quenching austenite and then heating into the range of .... °C.
    (*a*) 200 – 375 (*b*) 360 – 660
    (*c*) 800 – 1020 (*d*) 980 – 1120

82. Toughness of a material implies
    (*a*) strength (*b*) softening
    (*c*) machinability (*d*) stress relieving

83. Annealing does not
    (*a*) improve machinability.
    (*b*) remove coarseness of grain.
    (*c*) relieve internal stresses induced by uneven cooling
    (*d*) none of these.

84. Annealing treatment is not applied to the
    (*a*) forgings.
    (*b*) castings.
    (*c*) cold worked sheets & wire.
    (*d*) none of these.

85. Prolonged annealing of steel causes
    (*a*) decrease in ductility
    (*b*) increase in strength
    (*c*) grain growth
    (*d*) all (*a*), (*b*) & (*c*)

86. A steel part subjected to normalising as compared to full annealing will have improved
    (*a*) tensile strength (*b*) yield point
    (*c*) Izod value (*d*) all (*a*), (*b*) & (*c*)

87. Overheating of steel during annealing (*i.e.* heating well above the upper critical temperature) causes
    (*a*) improvement in ductility.
    (*b*) improvement in resistance to shock.
    (*c*) surface decarburisation
    (*d*) none of these.

88. For a metal to be forged, it should have
    (*a*) high tensile strength.
    (*b*) good malleability.
    (*c*) high machinability.
    (*d*) low Izod value.

89. Heat treatment of steel involves the change of
    (*a*) ferrite to austenite.
    (*b*) austenite to ferrite.
    (*c*) a body centered cubic iron lattice to a face centered cubic iron lattice.
    (*d*) none of these.

90. Hardenability of steel is a measure of
    (*a*) its carbon content.
    (*b*) its cementite content.
    (*c*) the depth to which the steel will harden on quenching.
    (*d*) none of these.

91. In iron-carbon equilibrium diagram, the eutectic point is at .... percent carbon.
    (*a*) 0.8 (*b*) 2.5
    (*c*) 4.3 (*d*) 1.6

92. Steel containing 0.4% carbon is
    (*a*) hypoeutectoid steel.
    (*b*) completely austenite above upper critical-temperature line.
    (*c*) composed of approximately half ferrite and half austenite by the time it is cooled to lower critical temperature.
    (*d*) all (*a*), (*b*) & (*c*).

Answers

80. (b)	81. (a)	82. (a)	83. (d)	84. (d)	85. (c)
86. (d)	87. (c)	88. (b)	89. (b)	90. (c)	91. (c)
92. (d)					

93. Carbon content in steel does not affect its
    (a) hardness & tensile strength.
    (b) machinability.
    (c) melting point.
    (d) none of these.

94. Nickel in steel
    (a) renders high chromium iron allcys austenitic.
    (b) increases distortion in quenching.
    (c) decreases toughness & resistance to impact.
    (d) unites with carbon.

95. Chromium addition to steel
    (a) adds to depth hardenability.
    (b) adds some strength at high temperature.
    (c) improves resistance to abrasion, wear, corrosion & oxidation.
    (d) all *(a)*, *(b)* & *(c)*.

96. Addition of molybdenum to steel
    (a) forms abrasion resisting particles and promotes hardenability.
    (b) counteracts tendency towards temper brittleness.
    (c) raises tensile & creep strength at high temperature.
    (d) all *(a)*, *(b)* & *(c)*.

97. Addition of vanadium to steel
    (a) resists tempering and causes secondary hardening.
    (b) imparts strength & toughness to heat treated steel.
    (c) promotes fine grains in steel and increases hardeneability (when dissolved).
    (d) all *(a)*, *(b)* & *(c)*.

98. Tungsten in steel
    (a) promotes coarse grain.
    (b) decreases red hardness.
    (c) increases strength at elevated temperature.
    (d) increases corrosion resistance.

99. Presence of manganese in steel
    (a) counteracts brittleness from sulphur,
    (b) increases ductility.
    (c) contributes to strength & hardness to a much greater degree than carbon.
    (d) increases weldability.

100. Copper (0.2 — 0.5%) when added to steel
    (a) increases resistance to acid corrosion.
    (b) increases resistance to atmospheric corrosion.
    (c) acts as a strengthening agent.
    (d) both *(b)* & *(c)*.

101. Boron in steel increases its
    (a) corrosion resistance
    (b) hardeneability
    (c) weldability
    (d) impact resistance

102. Addition of aluminium to steel
    (a) acts as a deoxidiser.
    (b) produces fine austenitic grain size.
    (c) helps promoting nitriding (if present upto 1%).
    (d) all *(a)*, *(b)* & *(c)*.

103. Addition of cobalt to steel
    (a) contributes to red hardness by hardening ferrite.
    (b) increases corrosion resistance.
    (c) increases weldability.
    (d) none of these

Answers

93. (d)	94. (a)	95. (d)	96. (d)	97. (d)	98. (c)
99. (a)	100. (d)	101. (b)	102. (d)	103. (a)	

104. Addition of chromium (0.15—1%) to cast iron decreases its
(*a*) wear resistance
(*b*) resistance to heat
(*c*) machinability
(*d*) hardness and strength

105. Silicon in steel
(*a*) improves oxidation resistance.
(*b*) strengthens low alloy steels.
(*c*) acts as a deoxidiser.
(*d*) all (*a*) , (*b*) & (*c*).

106. Titanium in steel
(*a*) causes localised depletion of chromium in stainless steels during long heating.
(*b*) accelerates formation of austenite in high chromium steels,
(*c*) reduces martensitic hardness & hardeneability in medium chromium steel.
(*d*) none of these.

107. Magnesium addition (0.04—0.08%) to cast iron increases its
(*a*) corrosion resistance
(*b*) ductility
(*c*) hardness
(*d*) heat resistance

108. T.T.T.(Time—Temperature—Transformation) diagram is also known as
(*a*) S-curve or C-curve.
(*b*) Bain's curve.
(*c*) isothermal transformation diagram.
(*d*) all (*a*), (*b*) & (*c*)

109. T.T.T. diagram is used more particularly for
(*a*) the assessment of decomposition of austenite in a heat treatable steel.
(*b*) knowing the extent of austenite decomposition under equilibrium condition.
(*c*) knowing the phases and resulting micro structure corresponding to equilibrium conditions.
(*d*) none of these.

110. The fine austenite grains size in steel increases the
(*a*) hardeneability
(*b*) impact toughness
(*c*) creep strength
(*d*) rough machinability

111. The coarse austenite grains size in steel
(*a*) improves machining finishes.
(*b*) improves tensile strength as normalized.
(*c*) mitigate quenching cracks, distortion in quenching and surface decarburisation.
(*d*) none of these.

112. Austenitic grain size in steel affects its
(*a*) hardening response.
(*b*) fabrication properties.
(*c*) both (*a*) & (*b*).
(*d*) neither (*c*) nor (*b*).

113. Addition of deoxidisers (like ferrosilicon, aluminium etc.) to molten steel does not
(*a*) reduce its FeO content.
(*b*) promote fine grain formation.
(*c*) kill (*i.e.* eliminate dissolved gases) it.
(*d*) none of these.

114. In alloy steels, which element is added to achieve fine grain structure and to avoid grain growth at high temperature ?
(*a*) Vanadium (*b*) Copper
(*c*) Nickel (*d*) Cobalt

Answers

104. (c)	105. (d)	106. (c)	107. (b)	108. (d)	109. (a)
110. (b)	111. (b)	112. (c)	113. (d)	114. (a)	

115. A burnt steel is characterised by the
   (*a*) presence of brittle iron oxide films.
   (*b*) fine grain size.
   (*c*) glossy & smooth surface.
   (*d*) darkening of the surface.

116. Which of the following is not a metal heat treatment process ?
   (*a*) Marageing
   (*b*) Austempering
   (*c*) Quench hardening
   (*d*) Calorising

117. Heat treatment processes of metal can not improve its
   (*a*) ductility, toughness, corrosion resistance, hardness & strength.
   (*b*) grain size.
   (*c*) electrical & mechanical properties.
   (*d*) none of these.

118. Pick out the correct statement pertaining to heat treatment processes of steel.
   (*a*) Heat treatment processes for steel basically involve decomposition/transformation of austenite.
   (*b*) Slow cooling from above the critical range in steel produces a pearlite (soft) structure.
   (*c*) Rapid cooling from above the critical range in steel produces a martensitic (hard) structure.
   (*d*) all (*a*), (*b*) & (*c*).

119. Spheroids annealing of steel compared to normal annealing produces steel with lower
   (*a*) hardness & tensile strength.
   (*b*) relative elongation.
   (*c*) reduction of area.
   (*d*) none of these.

120. In case of full annealing, the austentizing temperature for hypoeutectoid steel is in the range of. .... °C.
   (*a*) 723 – 910 (*b*) 910 – 1130
   (*c*) 467– 723 (*d*) 668 – 800

121. The austentizing temperature of hyper-eutectoid steels in case of full annealing is in the range of. .... °C.
   (*a*) 723 – 910 (*b*) 723 – 1.130
   (*c*) 910 –1130 (*d*) 467 – 723

122. Full annealing of steel does not improve its
   (*a*) formability (*b*) machinability
   (*c*) hardness (*d*) none of these

123. Normalizing produces microstructure consisting of pearlite .... steels.
   (*a*) for eutectoid.
   (*b*) and cementite for hypereutectoid.
   (*c*) and ferrite for hypoeutectoid.
   (*d*) all (*a*), (*b*) & (*c*).

124. Normalizing
   (*a*) produces harder & stronger steel than full annealing.
   (*b*) eliminates the carbide network at the grain boundaries of hypereutectoid steels.
   (*c*) produces enough strength & ductility in a steel that is too soft & ductile for machining.
   (*d*) all (*a*), (*b*) & (*c*).

125. The degree of hardness produced in steel by its quench hardening depends upon the
   (*a*) rate of cooling.
   (*b*) quenching temperature.
   (*c*) nature & properties of quenching medium.
   (*d*) all (*a*), (*b*) & (*c*).

Answers

115. (a)	116. (d)	117. (d)	118. (d)	119. (a)	120. (a)
121. (b)	122. (c)	123. (d)	124. (d)	125. (d)	

126. The purpose of tempering is to reduce
(*a*) residual stresses & % elongation.
(*b*) ductility & toughness.
(*c*) hardness.
(*d*) none of these.

127. Quench hardening of steel is usually followed by its tempering, because the quenceh hardened steels are
(*a*) brittle, hard & highly stressed.
(*b*) liable to crack & distort.
(*c*) liable to dimensional change due to some retained austenite (unstable phase).
(*d*) all (a). (b) & (c).

128. Low temperature tempering is done at .... °C.
(*a*) 150 – 250 (*b*) 350 – 550
(*c*) 550 – 750 (*d*) 750 – 950

129. High temperature tempering is done at .... °C.
(*a*) 350 – 450 (*b*) 500 – 650
(*c*) 850 – 1050 (*d*) 1050 – 1250

130. High temperature tempering
(*a*) imparts high ductility without reducing hardness.
(*b*) develops a martensite structure.
(*c*) is used for components like connecting rods, charts, gears etc.
(*d*) developes a troostite structure.

131. Martempering of steel
(*a*) employs two stage quenching (first rapid cooling in salt bath and then slower air cooling).
(*b*) produces martensitic structure with a high stress and distortion.
(*c*) decreases ductility.
(*d*) is suitable for very large section.

132. Austempering of steel
(*a*) employs interrupted quenching.
(*b*) is a hardening treatment.
(*c*) can be done for very large sections.
(*d*) none of these.

133. The structure produced by austempering of steel is
(*a*) bainite (*b*) martensite
(*c*) troostite (*d*) austenite

134. Which of the following alloying elements does not increase the hardeneability of steel ?
(*a*) Manganese (*b*) Molybednum
(*c*) Cobalt (*d*) Vanadium

135. Softening of high speed steel is done by annealing at .... °C.
(*a*) 550 (*b*) 850
(*c*) 1050 (*d*) 1250.

136. Hardening temperature for high speed steel is .... °C.
(*a*) 550 – 650 (*b*) 800 – 900
(*c*) 1150 –1350 (*d*) 750 –950.

137. Which of the following types of stainless steel can only be hardened by heat treatment ?
(*a*) Martensitic (*b*) Ferritic
(*c*) Austenitic (*d*) None of these

138. Which type of stainless steel has the highest corrosion resistance?
(*a*) Martensitic (*b*) Ferritic
(*c*) Austenitic (*d*) Low chromium

139. Austenitic stainless steel is
(*a*) magnetic in annealed condition.
(*b*) hardenable only by cold working.
(*c*) hardenable by heat treatment also.
(*d*) not able to be hot worked readily.

Answers

126. (c)	127. (d)	128. (a)	129. (b)	130. (c)	131. (a)
132. (a)	133. (a)	134. (c)	135. (b)	136. (c)	137. (a)
138. (c)	139. (b)				

140. Which stainless steel is magnetic in annealed condition ?
(*a*) Austenitic (*b*) Ferritic
(*c*) Martensitic (*d*) None of these

141. Ferritic stainless steels are superior to martensitic stainless steel in
(*a*) corrosion resistance
(*b*) machinability
(*c*) both (*a*) & (*b*)
(*d*) neither (*a*) nor (*b*).

142. Which stainless steel has the highest thermal conductivity ?
(*a*) Austenitic
(*b*) Ferritic
(*c*) Martensitic
(*d*) All have equal value

143. Martensitic stainless steel is the only type of stainless steels which is hardenable by heat treatment, because of its
(*a*) higher carbon to chromium ratio.
(*b*) lower carbon to chromium ratio.
(*c*) high thermal conductivity.
(*d*) magnetic nature.

144. Carburising (also called cementation) of low carbon steel part
(*a*) increases the carbon content at the surface by absorption & diffusion.
(*b*) is done at 870 —925°C.
(*c*) is a case hardening process.
(*d*) all (*a*), (*a*) & (*c*).

145. Carburising medium in pack carburising is
(*a*) fused salt
(*b*) bone charcoal
(*c*) hydrocarbon gases
(*d*) none of these

146. Which of the following case hardening processes for steel produces a hard case without quenching or any further heat treatment ?
(*a*) Nitriding
(*b*) Pack carburising
(*c*) Liquid carburising
(*d*) Gas carburising

147. Nitriding produces case hardness (Vicker's pyramid number) of the order of
(*a*) 150 (*b*) 550
(*c*) 11,50 (*d*) 2000

148. Pick out the correct statement pertaining to nitriding.
(*a*) The case of the nitrided part is brittle.
(*b*) Nitriding cycle time is small (2-4 hours).
(*c*) All types of steels can be nitrided.
(*d*) Surface hardness of nitrided part is retained, if it is accidently overheated.

149. In cyaniding
(*a*) carbon & nitrogen are introduced into the surface of steel, and then it is quench hardened.
(*b*) case depth is about 10 mm.
(*c*) distortion does not occur during heat treatment.
(*d*) molten potassium cyanide bath is used.

150. Low carbon steel (C = 0.2%), when subjected to cyaniding is heated in the cyanide bath to a temperature of. .... °C.
(*a*) 450 -620 (*b*) 800 -870
(*c*) 920 - 1280 (*d*) 1200 - 1400.

Answers

140. (a)	141. (c)	142. (c)	143. (a)	144. (d)	145. (b)
146. (a)	147. (c)	148. (a)	149. (a)	150. (b)	

151. Carbonitriding
   (*a*) produces cases/surfaces containing both carbon and nitrogen by the use of gas (like hydrocarbon, CO and NH_3) atmosphere,
   (*b*) is used for case hardening of high speed steels.
   (*c*) of steel produces high distortion.
   (*d*) of steel produces extremely high dimensional changes.

152. Carrier gas used in carbonitriding process is a mixture of
   (*a*) N_2, H_2 & CO (*b*) C_3H_8 & C_4H_{10}
   (*c*) C_3H_8 & NH_3 (*d*) CO & NH_3

153. Carbonitriding is carried out for low carbon steel (C = 0.2%) at .... °C.
   (*a*) 500 (*b*) 845
   (*c*) 1135 (*d*) 1320

154. Active gas in carbonitriding process comprises of
   (*a*) hydrocarbons & ammonia,
   (*b*) N_2, H_2 & CO.
   (*c*) natural gas, CO & N_2.
   (*d*) CO & H_2.

155. In flame hardening process for steel, the
   (*a*) surface becomes hard.
   (*b*) core remains soft & tough.
   (*c*) objects are heated by an oxyacetylene flame.
   (*d*) all (*a*), (*b*) & (*c*).

156. The heating time (seconds) in flame hardening of steels (C = 0.3 - 0.6%) is
   (*a*) *y* 2 (*b*) 33 y^2
   (*c*) 7 *y* 2 (*d*) 100 y^2
   where, *y* = depth of the hardened layer, mm.

157. Hardening temperature in induction hardening of medium carbon steel is about .... °C.
   (*a*) 450-500 (*b*) 750 - 800
   (*c*) 1000- 1050 (*d*) 1150- 1250

158. Which of the case hardening processes can not be adopted for steel having less than 0.4% carbon ?
   (*a*) Carbonitriding
   (*b*) Induction hardening
   (*c*) Cyaniding
   (*d*) None of these

159. Typical steel parts hardened by induction hardening include
   (*a*) piston rods and pump shafts,
   (*b*) spur gears and cams.
   (*c*) crankshafts and camshafts.
   (*d*) all (*a*), (*b*) & (*c*).

160. Which of the following steel components are flame hardened ?
   (*a*) Teeth of gears and shafts.
   (*b*) Mill rolls and spindles.
   (*c*) Valve stems, open end wrenches, pully and worms.
   (*d*) All (*a*), (*b*) & (*c*).

161. Hadfield's manganese steel contains .... percent manganese.
   (*a*) 0.3 – 0.8 (*b*) 2 – 3.5
   (*c*) 5 – 6.5 (*d*) 12 – 14

162. Tool steel should be first hardened and then
   (*a*) normalized (*b*) tempered
   (*c*) annealed (*d*) cooled

163. The purpose of austempering is to obtain greater
   (*a*) hardness (*b*) softness
   (*c*) brittleness (*d*) toughness

Answers

151. (a)	152. (a)	153. (b)	154. (a)	155. (d)	156. (c)
157. (b)	158. (b)	159. (d)	160. (d)	161. (d)	162. (b)
163. (d)					

164. The solid solution of carbon in gamma iron is the .... phase of steel.
(a) cementitic (b) austenitic
(c) pearlitic (d) ferritic

165. The steels, while hardening by heat treatment becomes too brittle due to
(a) bainite (b) pearlite
(c) martensite (d) cementite

166. Heating the steel above upper critical temperature and then cooling in air is the process of
(a) annealing (b) normalising
(c) tempering (d) hardening

167. The super saturated solution of carbon in .... iron is known as martensite.
(a) alpha (b) beta
(c) gamma (d) none of these

168. Heating the steels above critical temperature and then slowly cooling it in a furnace is the process of
(a) hardening (b) annealing
(c) tempering (d) normalizing

169. Quenching of heated steel in bath imparts
(a) austempering (b) hardening
(c) tempering (d) age hardening

170. Drawing the .... from the tool is tempering.
(a) carbon (b) strain
(c) stress (d) both (b) & (c)

171. Since the atomic spacings in crystals are of the order of a few angstrom, the crystal structure of metal are studied by
(a) ultrasonic method.
(b) X-ray technique.
(c) electron microscopy.
(d) diffusion techniques.

172. The temperature of crystallisation is the temperature at which a metal must be heated to .... new strain free grains.
(a) decompose (b) form and grow
(c) precipitate (d) orient

173. Among the non-ferrous group, ... is the best for forging and press working.
(a) brass (b) magnesium alloys
(c) duralumin (d) aluminium bronze

174. Precipitation is a/an .... process for steel.
(a) alloying (b) tempering
(c) softening (d) none of these.

175. .... diagram is of fundamental importance in heat treatment process ?
(a) Iron-carbon equilibrium
(b) Decarburisation
(c) TTT
(d) Continuous cooling transformation

176. Pick out the wrong statement.
(a) A pure metal melts at a particular temperature, while an alloy melts over a range of temperatures.
(b) If two metals are completely soluble in both liquid & solid state, then this alloy is called solid solution.
(c) For solid solution to be formed, both the solvent (i.e. parent metal) and solute (i.e. added metal) atoms should be chemically dissimilar in nature.
(d) For solid solution to be formed, the difference in atomic sizes of the solute and solvent atoms should be less than 15%.

Answers

164. (b)	165. (c)	166. (b)	167. (a)	168. (b)	169. (b)
170. (b)	171. (b)	172. (b)	173. (c)	174. (d)	175. (a)
176. (c)					

177. Pick out the correct statement.
   (*a*) In substitutional solid solution, the solute atoms occupy the position in between the solvent atoms.
   (*b*) For solid solution to be formed, both the solute and the solvent atoms should have similar crystal lattice.
   (*c*) In interstitial solid solution, the atoms of the solute replace the solvent atom of the lattice.
   (*d*) Copper-nickel alloy does not form a solid solution.

178. Solid solution of carbon in gamma iron is called
   (*a*) cementite (*b*) austenite
   (*c*) iron carbide (*d*) leduburite

179. A mixture of ferrite (alpha iron) and cementite (iron carbide, Fe_3C) is called
   (*a*) pearlite (*b*) leduburite
   (*c*) austenite (*d*) none of these

180. Steels containing more than 0.8% carbon are called .... steels.
   (*a*) hypo eutectoid (*b*) hypereutectoid
   (*c*) eutectoid (*d*) none of these

181. Steels containing less than 0.8% carbon are called .... steels.
   (*a*) eutectoid (*b*) hyper eutectoid
   (*c*) hypo eutectoid (*d*) none of these

182. Which of the following is not specified in heat treatment of metals/alloys ?
   (*a*) Heating temperature
   (*b*) Heating rate
   (*c*) Cooling rate
   (*d*) None of these

183. Heat treatment operation can not achieve the
   (*a*) modification in grain structure.
   (*b*) homogeneity of phases.
   (*c*) welding of minor crack.
   (*d*) none of these.

184. Annealing process does not
   (*a*) weld the minor cracks.
   (*b*) remove the coarseness of grains.
   (*c*) refine the structure and make it homogeneous.
   (*d*) none of these.

185. An eutectoid steel part is heated to austenising temperature and then put in a bath having temperature of 700°C. At 700°C, austenite will be transformed into
   (*a*) martensite (*b*) coarse pearlite
   (*c*) coarse bainite (*d*) none of these

186. In martempering process of steel,
   (*a*) martensite is produced in steel.
   (*b*) it is heated below the transformation range.
   (*c*) the heated steel is cooled in air.
   (*d*) very high residual stress exists in the steel part.

187. In austempering process of steel,
   (*a*) bainite is produced in steel.
   (*b*) it is heated below the transformation range.
   (*c*) the heated steel is cooled in air.
   (*d*) impact resistance of steel reduces.

188. What is the range of temperature (°C) maintained in carburising ?
   (*a*) 250- 350 (*b*) 550- 750
   (*c*) 900 -930 (*d*) 1200- 1300

189. Carburiser used in gas carburising is
   (*a*) nitrogen (*b*) natural gas
   (*c*) ammonia (*d*) acetylene

Answers

177. (b)	178. (b)	179. (a)	180. (b)	181. (c)	182. (d)
183. (d)	184. (d)	185. (b)	186. (a)	187. (a)	188. (c)
189. (b)					

190. In .... process, steel part is heated in a solid or liquid medium to diffuse carbon and nitrogen elements to the surface layer of steel.
(*a*) nitriding (*b*) carburising
(*c*) cyaniding (*d*) austempering

191. Which of the following is not a diffusion coating method, which provide surface hardness of steel ?
(*a*) Calorising (*b*) Galvanising
(*c*) Chromising (*d*) Siliconising

192. Which of the following is not a carbide forming element in steel ?
(*a*) Chromium (*b*) Cobalt
(*c*) Nickel (*d*) Molybdenum

193. Which of the following docs not have a *b.c.c.* lattice structure ?
(*a*) Molybdenum (*b*) Silver
(*c*) Potassium (*d*) Sodium

194. Which of the following has a *f.c.c.* lattice structure ?
(*a*) Zinc (*b*) Magnesium
(*c*) Copper (*d*) None of these

195. Which of the following does not have a *h.c.p.* lattice structure ?
(*a*) Zinc (*b*) Magnesium
(*c*) Gold (*d*) None of these

196. .... steel is normally subjected to tempering.
(*a*) Annealed
(*b*) Quenched
(*c*) Normalised
(*d*) Plastically deformed

197. Which of the following hardening operations is involved in ausforming of steel ?
(*a*) Solution hardening
(*b*) Strain hardening
(*c*) Age hardening
(*d*) None of these

198. Which of the following heat treatment operations is applied to steel castings ?
(*a*) Normalising
(*b*) Quenching
(*c*) Sub-critical annealing
(*d*) Full annealing

199. Pick out the wrong statement.
(*a*) The reason for the steel having fine austenite grain size having lower hardeneability is that the formation of pearlite in the finer grains steel is more rapid than in coarse grain steel.
(*b*) The dislocation line in screw dislocation is parallel to Burger's vector.
(*c*) Residual stresses are relieved, if a body undergoes non-uniform plastic deformation.
(*d*) With decrease in temperature, the number of vacant lattice sites in a metal increases rapidly.

200. With increase in temperature, the grain size produced by annealing
(*a*) increases
(*b*) decreases
(*c*) remains unchanged
(*d*) may increase or decrease

201. .... causes vacancy in a pure metal.
(*a*) Annealing
(*b*) Normalising
(*c*) Rapid quenching
(*d*) Tempering

Answers

190. (c)	191. (b)	192. (b)	193. (b)	194. (c)	195. (c)
196. (b)	197. (b)	198. (d)	199. (c)	200. (b)	201. (c)

202. Which of the following is used for hardening of teeth of spur gear?
   (a) Quenching
   (b) Dispersion hardening
   (c) Induction hardening
   (d) None of these

203. In order to achieve austenitic structure, high manganese steel is
   (a) normalised
   (b) cooled slowly
   (c) quenched rapidly
   (d) none of these

204. .... is the most suitable case hardening process for alloy steels containing molybdenum and chromium.
   (a) Nitriding
   (b) Carburising
   (c) Cyaniding
   (d) Induction hardening

205. .... is the most suitable case hardening process for plain carbon steel containing < 0.2 % carbon.
   (a) Nitriding (b) Carburising
   (c) Cyaniding (d) none of these

206. Solubility of one metal into other is most extensive, when their atomic sizes do not differ by more than .... percent.
   (a) 1 (b) 15
   (c) 25 (d) 45

207. Pick out the wrong statement.
   (a) Strain ageing is most common in high carbon steels.
   (b) Alternating current is normally employed in the induction hardening process.
   (c) Carburising process takes more than eight hours.
   (d) With increase in temperature, the creep rate increases.

208. .... structure is called accicular structure.
   (a) Dendritic (b) Needle like
   (c) Amorphous (d) Crystalline

209. Pick out the correct statement.
   (a) Dislocation is a surface defect.
   (b) Martempering is also called isothermal annealing.
   (c) In case of normalising, the grains are finer than that in annealing.
   (d) With increasing carbon content, the hardeneability decreases for the similar austenitic grain size.

210. Transformation of austenite into .... takes place in martempering of steel.
   (a) bainite (b) martensite
   (c) pearlite (d) none of these

211. Which of the following has the lowest tensile strength ?
   (a) Martensite (b) Ferrite
   (c) Cementite (d) Bainite

212. Point defect occurs, when any atom is missing at any point in the lattice of the crystal structure. Which of the following is not a point defect?
   (a) Interstitial defect.
   (b) Vacancy defect.
   (c) Impurity atom defect.
   (d) Negative dislocations.

Answers

202. (c)	203. (b)	204. (a)	205. (b)	206. (b)	207. (a)
208. (b)	209. (c)	210. (b)	211. (b)	212. (d)	

213. A measure of the magnitude and direction of a slip, which is equal to the amount of dislocation is called the
(*a*)Burger's vector
(*b*) twin defect
(*c*) dislocation density
(*d*)none of these

214. During solidification of a metal/alloy, radial arms of small nuclei are formed to form a fir type structure called dendritic structure. Presence of dendritic structure in a metal/alloy
(*a*) indicates homogeneity.
(*b*) indicates inadequate hot working subsequent to casting.
(*c*) can not be corrected by rolling/'forging/extrusion.
(*d*) can not be corrected by annealing.

215. What is the maximum solubility of carbon in iron at 1130°C?
(*a*) 0.5% (*b*) 1%
(*c*) 2% (*d*) 3.5%

216. From iron-carbon equilibrium diagram, we find that at 723°C and 0.8% carbon, ferrite and cementite form eutectoid mixture called
(*a*) leduburite (*b*) pearlite
(*c*) austenite (*d*) none of these

217. Metallic crystal structure is non-existent in the .... form.
(*a*) orthorhombic (*b*) trapezium
(*c*) triclinic (*d*) tetragonal

218. Pick out the wrong statement.
(*a*) The melting point of an alloy is higher than that of either of its constituent elements.
(*b*) A supersaturated solution of carbon in α - iron is called 'martensite', which has a body centred tetragonal structure.
(*c*) Steels containing 0.8% and > 0.8% carbon are respectively called eutectoid and hypereutectoid steel.
(*d*) All types of heat treatment processes do not strengthen & harden the steels.

219. Which of the following is the softest?
(*a*) Austenite (*b*) Ferrite
(*c*) Pearlite (*d*) Cementite

220. Which of the following exemplifies an age hardening alloy?
(*a*) Brass (*b*) Babbit metal
(*c*) Duralumin (*d*) Bronze

221. Recrystallisation annealing temperature is given by
(*a*) 0.1 T (*b*) 0.2 T
(*c*) 0.4 T (*d*) 0.6 T
where, T = melting temperature of the metal, °C

222. Sub-zero treatment of steel is most applicable for the
(*a*) high speed steel tools.
(*b*) steels containing < 0.2% carbon.
(*c*) steels not containing retained austenite.
(*d*) low carbon steels.

223. Hume-Rothery's rule predicts the existence of solid solubility in which .... factor is not included.
(*a*) crystal structure (*b*) atomic weight
(*c*) atomic size (*d*) electronegativity

224. During .... stage of annealing, polygonisation in plastically bent crystals occurs.
(*a*) recrystallisation (*b*) recovery
(*c*) grain growth (*d*) none of these

Answers

213. (a)	214. (b)	215. (c)	216. (b)	217. (b)	218. (a)
219. (b)	220. (c)	221. (c)	222. (a)	223. (b)	224. (b)

225. Pick out the wrong statement.
   (*a*) With increase in carbon content in steel, its hardeneability decreases.
   (*b*) Sub-zero (–30 to – 120 °C) treatment of steel reduces the retained austenite in steel by conversion of austenite into martensite.
   (*c*) Retained austenite reduces the hardness, wear resistance & thermal conductivity of steel and makes its dimension unstable.
   (*d*) The carbon content in cementite is about 2%.

226. With increase in shear strain, the shear stress required to produce slip continuously increases. This is due to the
   (*a*) work hardening.
   (*b*) increase in stacking fault energy.
   (*c*) polygonisation.
   (*d*) none of these.

227. Hardenability of steel is the measure of the degree of
   (*a*) transformation of austenite to pearlite.
   (*b*) the depth to which steel can be hardened,
   (*c*) its hardness.
   (*d*) none of these.

228. Which of the following processes produces the finest grains ?
   (*a*) Normalising
   (*b*) Annealing
   (*c*) Quenching
   (*d*) Sub-critical annealing

229. In a metal crystal, the atoms are arranged in a random .... dimensional pattern.
   (*a*) two (*b*) repeated two
   (*c*) three (*d*) repeated three

230. Cooling rate of the material in case of normalising operation is higher than that in
   (*a*) water quenching
   (*b*) oil quenching
   (*c*) annealing
   (*d*) all (*a*), (*b*) & (*c*)

231. Frank-Read source is concerned with the
   (*a*) dislocation (*b*) diffusion
   (*c*) age hardening (*d*) none of these

232. Which of the following relations is valid for temperature of cold working of a material (T) and the melting point of the the material to be cold worked (T_m)?
   (*a*) $T < 0.5T_m$ (*b*) $T < 0.75\ T_m$
   (*c*) $T < T_m$ (*d*) $T < 0.\ 1T_m$

233. Pick out the correct statement.
   (*a*) Ferrous alloys are not prone to age hardening phenomenon.
   (*b*) α - iron & γ - iron have same crystal structure, as they are polymorphic forms of the same metal.
   (*c*) Highly overheated and burnt steels can be reused after proper heat treatment.
   (*d*) Induction hardening of steel need not be followed by quenching essentially.

234. The .... form of iron is highly meagnetic and is the one that exists at atmospheric temperature.
   (*a*) alpha (*b*) beta
   (*c*) gamma (*d*) delta

235. Each corner atom in a body centred cubic (b.c.c.) structure is shared by .... unit cells.
   (*a*) two (*b*) four
   (*c*) eight (*d*) twelve

Answers

225. (d)	226. (a)	227. (b)	228. (c)	229. (d)	230. (c)
231. (a)	232. (a)	233. (d)	234. (a)	235. (c)	

236. Corrosion resistance & endurance limit is the highest for .... steels.
(*a*) nitrided
(*b*) carburised
(*c*) cyanided
(*d*) all have equal properties

237. .... is a thermo-mechanical treatment of metal.
(*a*) Annealing (*b*) Martempering
(*c*) Ausforming (*d*) Austempering

238. TD-nickel comprises of nickel oxide and .... oxide.
(*a*) aluminium (*b*) barium
(*c*) thorium (*d*) none of these

239. Detection of temper brittleness can be fairly done by the .... test.
(*a*) creep
(*b*) tensile
(*c*) notched-bar impact
(*d*) fatigue

240. Which of the following involves the slowest cooling rate?
(*a*) Oil quench (*b*) Water quench
(*c*) Brine quench (*d*) Ideal quench

241. 'Martempering' is done to harden especially alloy steels with a minimum distortion & residual stresses by reducing the temperature difference of the outside and the inside of the piece being quenched. It is also called .... quenching.
(*a*) stepped (*b*) isothermal
(*c*) interrupted (*d*) none of these

242. 'Austempering' is used for thin steel sections to obtain products free from cracks and with good impact resistance. It is also called .... quenching.
(*a*) stepped (*b*) isothermal
(*c*) interrupted (*d*) none of these

243. Microstructure of a 0.6% carbon containing steel, when cooled slowly from 900°C to room temperature, will contain ferrite and
(*a*) cementite (*b*) austenite
(*c*) pearlite (*d*) none of these

244. Point corresponding to 0.16% carbon and 1493°C in iron-carbon equilibrium diagram represents the .... point.
(*a*) eutectic (*b*) peritectic
(*c*) eutectoid (*d*) none of these

245. In iron-carbon equilibrium diagram, ... represents the eutectic point.
(*a*) 0.8% carbon and 723°C
(*b*) 4.3% carbon and 1147°C
(*c*) 6.67% carbon and 1130°C
(*d*) none of these

246. Cementite has got .... crystal lattice structure.
(*a*) cubic (*b*) hexagonal
(*c*) orthorhombic (*d*) triclinic

247. On cooling the hypereutectoid steel to room temperature, the microstructure contains
(*a*) cementite & pearlite
(*b*) cementite & ferrite
(*c*) pearlite & ferrite
(*d*) none of these

248. Decomposition of austenite to pearlite, at eutectoid composition occurs at a constant temperature of .... °C.
(*a*) 1130 (*b*) 723
(*c*) 910 (*d*) 768

Answers

236. (a)	237. (c)	238. (c)	239. (c)	240. (a)	241. (c)
242. (b)	243. (a)	244. (b)	245. (b)	246. (c)	247. (a)
248. (b)					

249. Which is the only phase present in cast iron containing 4.3% carbon at room temperature?
(*a*) Ledeburite (*b*) Pearlite
(*c*) Ferrite (*d*) Bainite

250. The co-ordination number of a metal crystallising in a hexagonal closed packed (hcp) structure is
(*a*) 4 (*b*) 8
(*c*) 10 (*d*) 12

251. .... is an interstitial solid solution of carbon in FCC iron and is stable from 912 to 1394°C.
(*a*) α - ferrite (*b*) Austenite
(*c*) δ - ferrite (*d*) none of these

252. Recrystallisation is the nucleation and growth of new, strain free crystals from the cold worked metals. The recrystallisation rate .... with rise in temperature.
(*a*) increases linearly
(*b*) decreases exponentially
(*c*) increases exponentially
(*d*) decreases linearly

253. Pick out the wrong statement pertaining to the law of recrystallisation.
(*a*) The higher is the amount of cold work, the higher is the recrystallisation temperature.
(*b*) The finer is the initial grain size, the lower is the recrystallisation temperature.
(*c*) Increasing amount of cold work and decreasing initial grain size produce finer recrystallised grains.
(*d*) Higher is the temperature of cold working, the less is the strain energy stored in the metal and consequently, higher is the recrystallisation temperature.

254. Recovery, recrystallisation and grain growth processes refer to the microstructural changes that take place on heating a cold worked metal. In all these phenomenon, a change in .... is involved.
(*a*) crystal structure
(*b*) composition
(*c*) both '*a*' & '*b*'
(*d*) neither '*a*' nor '*b*'

255. Pick out the wrong statement.
(*a*) Nucleation and growth of new, strain free crystals from the cold worked metal is termed as 'recovery'.
(*b*) Grain growth is the increase in the average grain size following recrystallisation.
(*c*) Reheating of martensite to higher temperatures is called 'tempering'.
(*d*) Bainite is an extremely fine mixture of ferrite and carbide.

256. In case hardening of steel, its
(*a*) surface is made wear resistant,
(*b*) core becomes hard,
(*c*) core becomes brittle.
(*d*) surface is made soft.

257. A steel contains 0.65% carbon. Its surface hardening will not be generally done by
(*a*) induction hardening.
(*b*) carburising.
(*c*) flame hardening.
(*d*) any of these.

258. Which of the following processes achieves the shallowest surface hardening ?
(*a*) Laser hardening,
(*b*) Flame hardening.
(*c*) Induction hardening.
(*d*) All achieve same hardening.

Answers

249. (a)	250. (d)	251. (b)	252. (c)	253. (a)	254. (d)
255. (a)	256. (a)	257. (b)	258. (a)		

259. The hardness of steel increases with the appearance of
(a) pearlite (b) martensite
(c) austenite (d) cementite

260. Which of the following elements does not form carbide in steel ?
(a) Chromium (b) Nickel
(c) Molybdenum (d) Manganese

261. .... is the best suited heat treatment process for leaf and coil spring of steel,
(a) Case hardening (ft) Sub zero treatment
(c) Austempering (d) Annealing

262. Sub zero treatment of steel
(a) induces temper brittleness.
(b) reduces the retained austenite in hardened steel.
(c) makes it usable in sub zero temperature region.
(d) none of these.

263. Which of the following defects is responsible for the phenomenon of slip, by which most of the metal deform plastically ?
(a) Twinning
(b) Fracture
(c) Strain hardening
(d) Dislocation

264. Stress concentration occurs, when a body is subjected to non-uniform stress distribution. It may be caused by .... in the body.
(a) a hole or a notch
(b) dimensional change
(c) cross-sectional area change
(d) none of these

265. Hardening of steel can not be done unless it is heated above the .... critical point.
(a) highest (b) lowest
(c) middle (d) second & third

266. Steel containing 0.8% carbon and 100% pearlite is termed as the .... steel.
(a) eutectoid (b) hyper-eutectoid
(c) austenitic (d) none of these

267. The solid solution of carbon in gamma iron formed, when a steel part containing 0.2% carbon is heated above third critical point, is termed as the .... steel.
(a) austenitic (b) cementitic
(c) eutectoid (d) critical

268. A steel part containing ferrite and pearlite is
(a) hard (b) soft
(c) ductile (d) both b & c'

269. Fine grained steels are
(a) brittle (b) lighter
(c) ductile (d) prone to distortion

270. The solid solution of carbon in alpha iron formed when a steel part containing 0.2% carbon is cooled after heating above the third critical point is termed as the .... steel.
(a) cementitic (b) ferritic
(c) austenitic (d) pearlitic

271. Coarse grained steels are
(a) not prone to distortion
(b) lighter
(c) denser
(d) less tough

272. For any type of steel, the maximum hardenability depends upon the
(a) grain size
(b) carbon content
(c) amount of alloying elements
(d) none of these

Answers

259. (b)	260. (b)	261. (c)	262. (b)	263. (d)	264. (a)
265. (b)	266. (a)	267. (a)	268. (b)	269. (c)	270. (b)
271. (d)	272. (b)				

273. A steel part containing 0.2% carbon which has been heated above third critical temperature is cooled to the first critical point. In this case, the austenite remaining in the solution is transformed to a new structure called
(*a*) ferrite (*b*) martensite
(*c*) pearlite (*d*) cementite

274. .... is the essential component of any hardened steel.
(*a*) Austenite (*b*) Cementite
(*c*) Martensite (*d*) Pearlite

275. Materials having .... lattice structure are usually the most ductile.
(*a*) fcc (*b*) hcp
(*c*) bcc (*d*) none of these

276. Brass is an example of
(*a*) intermetallic compound
(*b*) interstitial solid solution
(*c*) substitutional solid solution
(*d*) none of these

277. Carbon in iron is an example of
(*a*) substitutional solid solution
(*b*) interstitial solid solution
(*c*) intermetallic compound
(*d*) none of those

278. Case hardening can not be done by
(*a*) nitriding
(*b*) electroplating
(*c*) cyaniding
(*d*) induction hardening

279. The size of 'coarse' grain in metals is more than .... mm.
(*b*) 0.005 (*b*) 0.05
(*c*) 0.5 (*d*) 1

280. The crystal structure of a material can be studied by a/an
(*a*) electron microscope.
(*b*) X-ray diffraction.
(*c*) X-ray microanalyser.
(*d*) all '*a*', *b* & *c*,

281. In plane carbon steel, with increase in carbon percentage from 0.2 to 0.8%, its .... decreases.
(*a*) tensile strength
(*b*) BHN
(*c*) percentage elongation
(*d*) none of these

282. Hardenability of steel is the
(*a*) property which determines the depth of the hardened zone induced by quenching.
(*b*) depth of penetration obtained by Vicker's hardness test.
(*c*) ability of the steel to resist abrasion, wear & penetration.
(*d*) none of these.

283. Hardness of tempered martensite (lower bainite) is about
(*a*) RC 48 (*b*) RC 57
(*c*) RC 65 (*d*) RC 80

284. .... is the softest and the least strong micro constituent of steel.
(*a*) Cementite (*b*) Ferrite
(*c*) Austenite (*d*) Pearlite

285. Pure iron contains
(*a*) 100% pearlite
(*b*) pearlite & ferrite
(*c*) 100% ferrite
(*d*) cementite & pearlite

Answers

273. (c)	274. (c)	275. (a)	276. (b)	277. (b)	278. (b)
279. (b)	280. (b)	281. (c)	282. (a)	283. (b)	284. (b)
285. (c)					

286. Hypoeutectoid steels containing < 0.85% carbon transforms on cooling below 690°C, into cementite and

(a) austenite (b) ferrite
(c) pearlite (d) none of these

287. Match the various types of defects found in crystals.

List I

(a) Vacant lattice sites.
(b) Defect present in metallic, ceramic & polymeric materials.
(c) Screw dislocation.
(d) Stacking fault.

List II

I. Line defects
II. Point defect
III. Surface defect
IV. Line defect

288. Match the crystal lattice patterns in which various metals crystallise.

List I	*List II*
(a) Copper	I. Hexagonal closed packed (hcp)
(b) Molybdenum	II. Body centred cubic (bcc)
(c) Zinc	III. Face centred cubic (fcc)
(d) Iron	IV. Both bcc and fcc

289. Match the carbon percentage present in various iron & steel.

List I	*List II*
(a) Hypoeutectoid steel	I. 0.83 - 2
(b) Hypoeutectoid cast iron	II. < 0.83
(c) Hypereutectoid steel	III. 2-4.3
(d) Hypereutectoid cast iron	IV. > 4.3

290. Match the relevant terminologies in the two lists.

List I

(a) Hardest structure appearing on iron-carbon equilibrium diagram.
(b) Eutectic mixture of austenite & cementite.
(c) Structure produced by austempering.
(d) Has body centred tetragonal crystal lattice structure.

List II

I. Martensite
II. Bainite
III. Ledeburite
IV. Cementite

291. Match the effects of various alloying elements addition in steel.

List I

(a) Boron (b) Cobalt
(c) Tungsten (d) Silicon

List II

I. Increases oxidation resistance.
II. Increases strength at elevated temperature.
III. Increases red hardness.
IV. Increases hardenability,

292. .... can be expected to have a hardness value of about 1400 BHN.

(a) Ferrite (b) Pearlite
(c) Cementite (d) None of these

293. Hardening by carburising is limited to a maximum depth of .... mm.

(a) 0.2 (b) 2
(c) 5 (d) 8

Answers

286. (b) 287. (ii) (i) (iv) (iii) 288. (iii) (ii) (i) (iv) 289. (ii) (iii) (i) (iv)
290. (iv) (iii) (ii) (i) 291. (iv) (iii) (ii) (i) 292. (c) 293. (b)

294. Development of austenitic grain size during heat treatment of steel imparts it
(a) lower internal stresses.
(b) higher toughness,
(c) lower hardenability.
(d) none of these.

295. Which property is expected to be imparted to a steel part, when it is heated to 730°C and cooled in the furnace itself, at the slowest possible rate?
(a) Hardness (b) Toughness
(c) Softness (d) None of these

296. Which of the following may not be a possible reason for a steel part developing insufficient hardening after quenching ?
(a) Use of too slow rate of cooling.
(b) Maintaining too low hardening temperature.
(c) Presence of internal stresses in the part before subjecting it to heating.
(d) Use of insufficient holding time at the hardening temperature.

297. Which property a hardened steel part will acquire, if it is heated to 300°C and then cooled in oil ?
(a) Toughness (b) Hardness
(c) Softness (d) None of these

298. Warping of steel part during heat treatment may result from the
(a) presence of internal stresses in the part before heating.
(b) non-uniform cooling.
(c) non-uniform heating.
(d) none of these.

299. .... is not a heat treatment process.
(a) Cyaniding (b) Parkerising
(c) Austempering (d) Martempering

300. Austempering of steel requires it to be heated to 875° C followed by sudden cooling to 250 - 525° C, thereby changing austenite to
(a) sorbite (b) bainite
(c) martensite (d) troostite

301. .... of a material results, when its strength is increased and ductility is decreased by heating at relatively lower temperature after cold working.
(a) Solid solution hardening
(b) Screw dislocation
(c) Strain ageing
(d) Twinning

302. Fatigue strength of a material increases by
(a) having notches in the specimen.
(b) increasing the temperature.
(c) under stressing the specimen.
(d) having scratches on the surface.

303. Nitriding of steel is a process for the
(a) case hardening. (b) spheroidising.
(c) normalising. (d) annealing.

304. Pick the odd man out of the following
(a) Calorising (b) Pack carburising
(c) Nitriding (d) Cyaniding

305. Hardening of steel is not possible, unless it is heated .... critical point.
(a) above the highest
(b) above the lowest
(c) between the first & second
(d) between the second & third

306. Which of the following heat treatment processes is used for softening the hardened material ?
(a) Normalising (b) Tempering
(c) Annealing (d) None of these

Answers

294. (c)	295. (c)	296. (d)	297. (c)	298. (c)	299. (b)
300. (c)	301. (a)	302. (c)	303. (a)	304. (b)	305. (b)
306. (b)					

307. Sheardising process is used for the
(a) surface coating.
(b) heat treatment of high speed steel.
(c) cold working on metals.
(d) none of these.

308. The heat treatment process applied to cold formed parts is
(a) normalising (b) tempering
(c) artificial ageing (d) solution annealing

309. The heat treatment process to which castings & the steel balls produced by cold heading are subjected, is
(a) tempering (b) normalising
(c) annealing (d) none of these

310. To improve the machinability of steel, it is generally subjected to
(a) spheroidismg (b) tempering
(c) normalising (d) annealing

311. Work hardening of a material
(a) decreases its tensile strength.
(b) decreases its ductility.
(c) increases its ductility.
(d) does not affect its ductility.

312. Eutectic reaction for iron-carbon system occurs at a temperature of .... °C
(a) 723 (b) 959
(c) 1147 (d) 1493

313. Eutectoid reaction for iron-carbon system occurs at a temperature of .... °C.
(a) 597 (b) 959
(c) 1147 (d) 1493

314. During peritectic solidification, one liquid
(a) solidifies into two different solids.
(b) forms one solid.
(c) forms one solid & another liquid.
(d) combines with one solid to form a second new solid.

315. Pearlite comprises of
(a) 87% cementite & 13% Fe.
(b) 87% ferrite & 1.3% cementite.
(c) 93.33% ferrite & 6.67% C.
(d) 87% ferrite & 13% C.

316. Hardenability of steel that determines its capacity to harden throughout the section when quenched, is determined by the .... test.
(a) Sceleroscope
(b) Rockwell
(c) Jimmy hardenability
(d) Maxwell

317. Annealing of metal does not
(a) precede its wire drawing.
(b) induce softness and removes stresses & dissolved gases.
(c) refine the crystalline structure & produce a definite microstructure.
(d) alter toughness, ductility and magnetic & electrical or other physical properties.

318. Steels developing finer austenitic grain sizes on heat treatment as compared to those developing coarse grain sizes will have
(a) less austenite retention.
(b) lower toughness & higher hardenability.
(c) higher internal stresses & more susceptibility to grinding cracks.
(d) higher warping/cracking tendency on quenching.

Answers

307. (a)	308. (d)	309. (b)	310. (a)	311. (b)	312. (a)
313. (d)	314. (a)	315. (b)	316. (c)	317. (a)	318. (a)

319. Microstructure of steel is completely pearlitic, when the percentage of carbon in it is 0.8% *i.e.*, it contains maximum percentage of pearlite, which results in attainment of maximum tensile strength of the order of .... N/mm^2.
 (*a*) 250 (*b*) 450
 (*c*) 850 (*d*) 1500

320. A work hardening operation for strengthening steel tubes usually carried out on gun barrels is termed as the
 (*a*) autofrettage
 (*b*) black annealing
 (*c*) burnishing
 (*d*) barffing

321. Normalising of an object does not
 (*a*) refine coarse grain structure obtained during hot working.
 (*b*) improve ductility.
 (*c*) improve yield strength,
 (*d*) improve mechanical properties.

322. Machine tool guideways are usually hardened by
 (*a*) electroplating with hard metal.
 (*b*) flame hardening.
 (*c*) vacuum hardening.
 (*d*) induction hardening.

323. Hypotectoid steel in stress relieving process is heated
 (*a*) above A_3 line and air cooled resulting in slight hardening.
 (*b*) above A3 line and cooled very slowly in furnace so as to refine old grain structure.
 (*c*) below A_1 line and cooled slowly resulting in removal of internal stresses.
 (*d*) below A_1 line with a view to making steel ductile for cold working,

324. Which of the following processes follows the hardening process for reducing the hardening strains & increasing the toughness of the steel part ?
 (*a*) Anodising (*b*) Tempering
 (*c*) Carburising (*d*) Annealing

325. Oxide layer formed on the non-ferrous metal surface after its annealing is
 (*a*) removed by acid pickling.
 (*b*) hammered into the surface.
 (*c*) removed with coarse emery cloth.
 (*d*) left as such to protect the surface.

326. What is the range of tempering temperature (°C) for most of the materials ?
 (*a*) 75 – 100 (*b*) 200 – 300
 (*c*) 350 – 450 (*d*) 500 – 600

327. Which of the following contains the least percentage of carbon?
 (*a*) Pig iron (*b*) Wrought iron
 (*c*) Cast iron (*d*) Malleable iron

328. Pick out the correct statement.
 (*a*) Hot worked materials are subjected to annealing to remove internal stresses.
 (*b*) Annealing of steel hardens it slightly.
 (*c*) Normalising of a material induces stresses.
 (*d*) Tempering of a material improves ductility & toughness but reduces hardness & brittleness.

Answers

319. (c)	320. (a)	321. (b)	322. (b)	323. (d)	324. (b)
325. (a)	326. (b)	327. (b)	328. (d)		

329. Match the Brinnel hardness numbers.

List I	*List II*
(a) Ferrite	I. 80
(b) Cementite	II. 200
(c) Annealed aluminium sheet	III. 20
(d) High speed steel	IV. 650

330. Match the Vickers hardness number (VHN) of various materials.

List I	*List II*
(a) Lead	I. 8400
(b) Plain Carbon steel	II. 12
(c) Tungsten carbide	III. 210
(d) Diamond	IV. 2100

331. Match the Brinell hardness number (BHN) of various materials.

List I	*List 11*	
(a) Lead	I.	25-40
(6) Zinc	II.	115-200
(c) Cast iron	III.	450-700
(d) High speed steel	IV.	4-8

332. Match the characteristics of various micro-structural constituents of heat treated steels.

List I	*List 11*
(a) Austenite	I. It appears white under microscope in white cast iron.
(b) Martensite	II. It comprises of alternate layers of cementite and ferrite.
(c) Pearlite	III. It. is extremely hard and brittle phase.
(d) Cementite	IV. It exists only above temperature of 723° C.

333. Match the following

List I

(a) Microstructure produced by oil cooling of austenite.

(b) Microstructure produced by reheating martensite upto 300°C.

(c) Microstructure produced by very slow cooling of austenite.

(d) Essential component of any hardened steel.

List II

I. Troostite

II. Martensite

III. Very fine pearlite

IV. Spheroidite

334. Increase in volume of a fixed mass of pure iron on cooling from 1000° C to room temperature, is because of the reason that the

(a) inter-atomic distance between iron atoms increases at lower temperature.

(b) iron has fcc crystal structure at 1000° C and changes to bcc at room temperature.

(c) iron has bcc crystal structure at 1000°C and changes to fcc at room temperature.

(d) atomic radius of iron atom increases at lower temperature.

335. The number of grains (N) per inch^2 at a magnification of 100 X is given by

(a) $2^n - 1$ *(b)* 2^{n-1}

(c) 2^{n+1} *(d)* $2^n + 1$

where, n = grain size number

336. Yield strength of a yielding material will increase, if the grains are

(a) of irregular shape.

(b) fine.

(c) coarse.

(d) mixture of both fine & coarse.

Answers

329. (i) (ii) (iii) (iv) 330. (ii) (iii) (iv) (i) 331. (iv) (i) (ii) (iii)

332. (iv) (iii) (ii) (i) 333. (iii) (i) (iv) (ii) 334. (b) 335. (b)

336. (b)

337. Pick out the wrong statement.
    (*a*) The precipitation hardening of Al-Cu alloy is because of the cluster formation of Cu atoms in Al matrix, thus impeding dislocation slide.
    (*b*) Annealing of a cold worked material results in recrystallisation at low-temperature and will maintain its directional deformation of grains,
    (*c*) An elasticity is the time dependent stress relaxation.
    (*d*) Isothermal straining occurs, if load is applied at low strain rate.

338. Heat treatment of steel does not
    (*a*) change the percentage of carbon & silicon in the bulk.
    (*b*) change its composition on the surface.
    (*c*) alter the grain size.
    (*d*) remove undesirable residual stresses.

339. Out of the following, the fastest cooling rate is achieved, when the steel is quenched in
    (*a*) fused salt (*b*) water
    (*c*) brine (*d*) oil

340. Pearlite phase in steel comprises of the alternate layers of
    (*a*) bainite & cementite.
    (*b*) ferrite & cementite.
    (*c*) martensite & cementite.
    (*d*) martensite & ferrite.

341. Eutectoid transformation reaction occuring at 723° C with 0.77% carbon transforms the austenite into
    (*a*) ferrite + cementite.
    (*b*) bainite + martensite.
    (*c*) ferrite + pearlite.
    (*d*) ferrite + martensite.

342. Out of the following, the slowest cooling rate is achieved, when the steel is quenched in
    (*a*) oil-water emulsion
    (*b*) fused salt
    (*c*) air
    (*d*) water

343. Number of grains observed per inch^2 at a magnification of 100 X, for ASTM grain size number 5 is
    (*a*) 8 (*b*) 16
    (*c*) 32 (*d*) 64

344. .... phase in steel is metastable and hence is not encountered, if the equilibrium cooling takes place.
    (*a*) Austenitic (*b*) Martensitic
    (*c*) Cementitic (*d*) Ferritic

345. Which of the following surface hardening treatment increases the carbon content in the surface of the material ?
    (*a*) Surface heating by gas flame.
    (*b*) Induction hardening & cyaniding.
    (*c*) Dipping the material in fused mixture of chlorides, carbonates & cyanides.
    (*d*) None of these.

346. Correct arrangement of the quenching media in order of increasing cooling rate is
    (*a*) air, fused salt, oil, water, brine
    (*b*) air, oil, fused salt, brine, water
    (*c*) air, brine, fused salt, oil, water
    (*d*) fused salt, air, water, brine, oil

Answers

337. (b)	338. (a)	339. (c)	340. (b)	341. (a)	342. (c)
343. (b)	344. (b)	345. (c)	346. (a)		

347. Pick out the wrong statement.
   (a) The pearlite content in plain carbon steel increases with carbon content upto 0.8% and then decreases.
   (b) Steel containing 0.4% carbon is less hardenable than that containing 0.8% carbon and its critical cooling rate is higher than that for 0.8% carbon steel.
   (c) The pearlite content in plain carbon steel increases with increasing carbon content upto 1.2%.
   (d) Martensite is marked with high hardness and has carbon dissolved in it, so that its structure is no more bcc.

348. Debye in his theory of specific heat of crystalline solids proposed that the
   (a) lattice of crystalline solid vibrates only in one direction with single characteristic frequency.
   (b) atoms do not vibrate independently but the lattice of crystalline solid vibrates in three direction with single characteristic frequency.
   (c) atoms vibrate in three directions with respective frequencies.
   (d) atoms vibrate independently in three directions with same frequency.

349. Pick out the wrong statement.
   (a) Yield point in low carbon steel reappears, if steel is unloaded after yielding and allowed to rest for several hours.
   (b) The reason for yielding of mild steel is that the carbon atoms in low carbon steel (0.2%C) anchor the dislocation lines by migrating just below the line of dislocations.
   (c) Yield point in mild steel occurs because carbon atoms form Fe_3C with iron.
   (d) The reduction of fatigue strength due to 'same stress concentration & geometry in samples of different materials is not the same.

350. Which of the following alloying elements of steel is not a ferrite stabiliser ?
   (a) Nickel (b) Chromium
   (c) Tungsten (d) Molybdenum

351. Which of the following alloying elements of steel is an austenite stabiliser ?
   (a) Vanadium (b) Copper
   (c) Manganese (d) Both 'b: & 'c'

352. Pick out the wrong statement
   (a) Most of the ferrite stabiliser alloying elements of steel have bcc crystal structure & they are more soluble in α-iron than in gamma iron.
   (b) Addition of ferrite stabiliser reduces/removes the austenitic phase field and raises the eutectoid temperature.
   (c) Addition of austenite stabiliser restrict the separation of carbide and lowers the eutectoid temperature.
   (d) Addition of austenite & ferrite stabilisers respectively lowers & raises the carbon content of eutectoid composition.

353. Addition of alloying elements in steel can not improve its
   (a) residual internal stresses.
   (b) hardenability & high temperature properties.
   (c) elastic ratio & endurance strength.
   (d) resistance to softening on tempering.

354. Internal stresses/residual stresses/locked- in stresses existing within a body do not
   (a) cause micro-structural changes.
   (b) develop during solidification of castings.
   (c) enhance warpage & dimensional instability tendency.
   (d) cause reduction in its fatigue strength.

Answers

347. (c)	348. (b)	349. (c)	350. (a)	351. (d)	352. (d)
353. (a)	354. (a)				

355. Internal stresses in a body do not develop during
   (*a*) machining.
   (*b*) phase transformation.
   (*c*) hot dip galvanising.
   (*d*) electroplated coatings.

356. Annealing is classified as full annealing, partial annealing & sub-critical annealing depending on the
   (*a*) phase transformation features.
   (*b*) heat treatment temperature,
   (*c*) holding time.
   (*d*) final properties of the material desired.

357. Which of the following phenomenon does not occur in sub-critical annealing ?
   (*a*) Recovery, recrytallisation & grain growth.
   (*b*) Phase transformation
   (*c*) Agglomeration of carbides.
   (*d*) Softening.

358. Which of the following is not an objective of normalising process ?
   (*a*) Enhancement in toughness & strength.
   (*b*)Improvement in machinability.
   (*c*) Grain refinement.
   (*d*) Enhanced softness.

359. Hardening of steel does not increase its
   (*a*) tensile strength.
   (*b*) yield strength.
   (*c*) residual internal stresses.
   (*d*) corrosion resistance.

360. Tempering of hardened steel improves its
   (*a*) tensile & yield strength
   (*b*) wear resistance
   (*c*) hardness
   (*d*) internal stresses

361. A good quenchant should have high
   (*a*) viscosity.
   (*b*) latent heat of vaporisation.
   (*c*) specific heat & thermal conductivity.
   (*d*) all '*a*' '*b*' & 'c'.

362. Pick out the wrong statement about the quenchants used in various heat treatment processes.
   (a) The temperature difference between the core & the case of the workpiece is less for oil quenching than for water quenching.
   (b) Most of the quenching oils used are generally paraffin base mineral oils.
   (c) Inverted solubility of the most widely used synthetic quenchant i.e. poly-alkylene glycol enables a thin film of glycol to wet the workpiece, as it is quenched.
   (d) Cooling/quenching rate achieved by-water quenching is much more than that achieved by brine quenching.

363. Which of the following is the most common & convenient method of determination of hardenability ?
   (*a*) Jominy end quench test.
   (*b*) Grossman's critical method.
   (*c*) Chemical composition method.
   (*d*) Fracture test.

364. Match the following.

List I	*List II*
(*a*) Frankel defect	I. Separation of crystals
(*b*) Schottky defect	II. Missing atomic plane
(*c*) Dislocation	III. Cation vacancy associated with anion vacancy.
(*d*) Grain boundary	IV. Cation vacancy associated with intersti tial cation.

Answers

355. (c)	356. (b)	357. (b)	358. (d)	359. (d)	360. (d)
361. (c)	362. (d)	363. (a)	364. (iv) (iii) (ii) (i)		

365. Match the temperature (°C) for various changes.

List I

(*a*) 210

(*b*) 723

(*c*) 770

(*d*) 1148

List II

I Eutectoid transformation reaction.

II Change of magnetic property of cementite.

III Eutectic transformation reaction.

IV Loss of ferromagnetism of ferrite.

366. Match the characteristics/heating temperature of steel in various annealing processes.

List I	*List II*
(*a*) Process annea ling	I. Heating of steel be tween A_1 & A_3 temperatures.
(*b*) Full annealing	II. Heating of hypoeutectoid steel above A_3 & that of eutectoid steel to above A_1 temperature followed by rapid cooling.
(*c*)Partial/incomplete annealing	III Heating of steel below A_1 temperature.
(*d*) Isothermal annealing	IV. Heating of hypoeutectoid toid steel to 30 - 50° C above upper critical temperature (A3) & that of eutectoid steel to 30 – 50°C above lower critical temperature followed by slow cooling.

367. Match the characteristics of various gaseous quenchants.

List I	*List II*
(*a*) Hydrogen	I. Prohibitively expen sive
(*b*) Helium	II. Unsafe to use (risk of explosion)
(*c*) Nitrogen	III. Lowest cooling ef ficiency of all the gaseous quenchants.
(*d*) Argon	IV. Most used gaseous quenchant.

368. Austenite contains a maximum carbon percentage of

(*a*) 0.48 (*b*) 0.80

(*c*) 1.20 (*d*) 1.70.

369. Which of the following elements in steel directly affects the critical temperature of the steel to be heat treated?

(*a*) Manganese (*b*) Carbon

(*c*) Phosphorous (*d*) Sulphur

370. Ductile-brittle transition temperature of steel can be lowered by

(*a*) Sub-Zero treatment

(*b*) Lowering carbon content

(*c*) heat treatment

(*d*) cold rolling.

Answers

365. (ii) (i) (iv) (iii) 366. (iii) (iv) (i) (ii) 367. (ii) (i) (iv) (iii)

368. (d) 369. (b) `370. (b)

Section

A

8

Materials Science

1. Iron alloyed with carbon upto 2% is called
   (*a*) pig iron (*b*) wrought iron
   (*c*) high carbon steel (*d*) none of these

2. The softest material in Mho's scale (for measuring hardness) is
   (*a*) talc (*b*) gypsum
   (*c*) rubber (*d*) none of these

3. Karbate is
   (*a*) a mixture of iron dust & saw dust.
   (*b*) carbon impregnated with resin (usually bakelite).
   (*c*) acid resistant material.
   (*d*) both (*b*) & (*c*).

4. Duralumin is an alloy of aluminium,
   (*a*) copper and manganese
   (*b*) nickel and silicon
   (*c*) and nickel
   (*d*) none of these

5. Corrosion resistance of steel is increased by the addition of
   (*a*) phosphorus and tungsten.
   (*b*) nickel and chromium.
   (*c*) lead and vanadium.
   (*d*) molybdenum and tungsten.

6. White metal is an alloy of
   (*a*) lead, tin & cadmium
   (*b*) copper, tin & zinc
   (*c*) copper & lead
   (*d*) none of theae

7. Presence of cobalt in steel improves its
   (*a*) cutting ability
   (*b*) corrosion resistance
   (*c*) tensile strength
   (*d*) none of these

8. Carbon content in steel is .... percent.
   (*a*) 0.1—2 (*b*) 4—6
   (*c*) <1 (*d*) < 0.5

9. Brine (15% concentration) can be stored in a vessel made of
   (*a*) monel (*b*) karbate
   (*c*) cast iron (*d*) none of these

10. Most suitable material of construction for the storage of concentrated nitric acid is
   (*a*) cast iron.
   (*b*) monel.
   (*c*) karbate.
   (*d*) aluminium or chromium alloy, (Cr > 18% for cold acid)

Answers

1. (d)	2. (a)	3. (d)	4. (a)	5. (b)	6. (d)
7. (a)	8. (a)	9. (a)	10. (d)		

11. Babbit metal (used for making bearings) comprises of
    (a) mainly tin (85%) & lead.
    (c) saw dust & iron dust mixture.
    (c) zinc & aluminium.
    (d) copper & aluminium.

12. Most common stainless steel type 316, which is highly resistant to corrosion contains
    (a) 16-18% chromium, 10-14% nickel & 2- 3% molybdenum.
    (b) 20-22% chromium, & 8-10% nickel.
    (c) 2-4% chromium, 22% nickel & 2-4% molybdenum.
    (d) none of these.

13. Brass is an alloy of
    (a) nickel & tin
    (b) copper & zinc
    (c) tin & lead
    (d) copper, nickel & zinc

14. Gun metal is an alloy of
    (a) nickel, tin & copper.
    (b) copper, tin & zinc.
    (c) copper, phosphorus & nickel.
    (d) 'maganese, phosphorus & nickel.

15. Balls in a ball mill are usually made of
    (a) steel (b) cast iron
    (c) stainless steel (d) bronze

16. Bronze is an alloy of copper and
    (a) lead (b) tin
    (c) nickel (d) zinc

17. Tank furnace used for the melting of glass is made of
    (a) mild steel (b) cast iron
    (c) refractory blocks (d) stainless steel

18. Stainless steel contains
    (a) chromium & nickel
    (b) copper
    (c) aluminium
    (d) vanadium

19. Monel metal is an alloy of
    (a) molybdenum & nickel.
    (b) nickel & copper.
    (c) molybdenum & aluminium.
    (d) molybdenum & zinc.

20. German silver in an alloy of
    (a) copper, nickel & zinc.
    (b) copper, aluminium & silver.
    (c) silver, zinc & aluminium.
    (d) silver, nickel & zinc.

21. Chromel (Nichrome) is an alloy of
    (a) chromium & molybdenum.
    (b) nickel & chromium.
    (c) molybdenum & nickel.
    (d) chromium & aluminium.

22. Aqueous nitric acid is stored in a
    (a) steel drum.
    (b) stainless steel vessel.
    (c) cast iron vessel lined with acid-proof masonry brick.
    (d) cast iron vessel.

23. Hydrochloric acid is .... corrosive to common metals.
    (a) least (b) not
    (c) mildly (d) highly

24. Valves in pipe sizes of 2" and under are normally made of
    (a) wrought iron (b) brass
    (c) bronze (d) monel

Answers

11. (a)	12. (a)	13. (b)	14. (b)	15. (a)	16. (b)
17. (c)	18. (a)	19. (b)	20. (a)	21. (b)	22. (b)
23. (d)	24. (b)				

25. The most resistant material to alkaline corrosion is
    (*a*) duriron (*b*) nickel
    (*c*) aluminium (*d*) karbate

26. Presence of nickel in steel improves its
    (*a*) corrosion resistance
    (*b*) cutting ability
    (*c*) wear resistance
    (*d*) all (*a*), (*b*) & (*c*)

27. An alloy of iron containing 4% carbon is called
    (*a*) high carbon steel
    (*b*) wrought iron
    (*c*) mild steel
    (*d*) none of these

28. Bell metal is an alloy of copper and
    (*a*) zinc (*b*) tin
    (*c*) nickel (*d*) chromium

29. Inconel is an alloy of
    (*a*) tin, zinc & nickel
    (*b*) iron, nickel & chromium
    (*c*) copper & nickel
    (*d*) copper & tin

30. Hastelloy comprises of
    (*a*) copper & tin
    (*b*) copper & nickel
    (*c*) molybdenum & nickel
    (*d*) lead & tin

31. Ability of a material to absorb energy in deformation in the plastic range is characterised as its
    (*a*) ductility (*b*) toughness
    (*c*) creep (*d*) resilience

32. Slow and progressive deformation of a material with time under constant stress is called
    (*a*) creep (*b*) erosion
    (*c*) resilience (*d*) none of these

33. Wrought iron is
    (*a*) high carbon iron.
    (*b*) highly resistant to acid corrosion.
    (*c*) malleable & ductile : hence used for making chain links, hooks and couplings.
    (*d*) an alloy of iron, chromium & carbon.

34. Which of the following constituents of cast iron is mainly responsible for imparting it an anti-corrosive property ?
    (*a*) Silicon (*b*) Phosphorus
    (*c*) Sulphur (*d*) None of these

35. Mild steel is
    (*a*) a low carbon steel (0.05 to 0.3% carbon).
    (*b*) highly resistant to corrosion as much as stainless steel).
    (*c*) a high carbon steel (0.5 to 1.5% carbon).
    (*d*) very poor in strength and ductility.

36. Presence of manganese in alloy steel improves its
    (*a*) corrosion resistance.
    (*b*) cutting ability.
    (*c*) abrasion resistance & toughness.
    (*d*) elasticity & creep resistance.

37. H_2SO_4 (50% concentration) is corrosive to
    (*a*) aluminium, mild steel, stainless steel, concrete & tin.
    (*b*) copper, cast iron & high silicon iron (14% Si).
    (*c*) rubber (butyl and hard), silicon rubber & teflon.
    (*d*) glass, graphite, porcelain & stoneware.

Answers

25. (b)	26. (a)	27. (d)	28. (b)	29. (b)	30. (c)
31. (b)	32. (a)	33. (c)	34. (a)	35. (a)	36. (c)
37. (a)					

38. 18/8 stainless steel means that it contains
(a) 18% chromium & 8% nickel.
(b) 18% chromium & 8% molybdenum.
(c) 18% nickel & 8% chromium.
(d) 18% molybdenum & 8% chromium.

39. Stainless steel is not corroded by
(a) hydrochloric acid (10%)
(b) nitric acid (10%)
(c) sulphuric acid (10%)
(d) saturated brine

40. Perspex is nothing but an
(a) acrylic sheet. (b) elastomer.
(c) alloy of lead & tin (d) epoxy resin.

41. The most commonly used resin for making reinforced plastic is
(a) unsaturated polyester
(b) polypropylene
(c) polyurathane
(d) nylon-6

42. Copper is dissolved from its ore by H_2SO_4 in a percolation tank made of
(a) wood
(b) stainless steel
(c) reinforced concrete lined with lead
(d) high silicon iron (14% Si)

43. Rotary driers are generally made of
(a) cast iron
(b) mild steel
(c) high silicon iron (14% Si)
(d) stainless steel

44. Zircalloy used as a fuel cladding material in a nuclear reactor (thermal) is an alloy of zirconium
(a) tin, nickel, iron & chromium.
(b) and graphite.
(c) and copper.
(d) none of these.

45. Which of the following is an alloy of nickel and copper ?
(a) Hastelloy (b) Duriron
(c) Monel (d) Inconel

46. Thermal shield used in high powered nuclear reactor to protect the walls of the reactor from radiation damage is made of
(a) concrete (b) steel
(c) zircalloy (d) graphite

47. Biological shield in a nuclear power reactor is made of
(a) concrete (b) steel
(c) cadmium (d) zircalloy

48. Muntz metal is an alloy of
(a) magnesium & tin
(b) ferrous materials
(c) lead & tin
(d) copper & zinc

49. Presence of nitrogen in steel
(a) makes it brittle.
(b) increases its strength.
(c) increases its hardness.
(d) is desirable.

50. Invar used in thermocouples is an alloy of .... and nickel.
(a) iron (b) copper
(c) chromium (d) lead

51. Constantan used in thermocouples is an alloy of
(a) iron & copper.
(b) copper & aluminium.
(c) lead & tin.
(d) none of these.

Answers

38. (a)	39. (b)	40. (a)	41. (a)	42. (c)	43. (b)
44. (a)	45. (c)	46. (b)	47. (a)	48. (d)	49. (a)
50. (a)	51. (d)				

52. Hardest material so far found is
    (*a*) diamond (*b*) graphite
    (*c*) carborundum (*d*) pumice stone

53. Lead alone is not used as a material of construction in storing equipment, because it
    (*a*) is very costly.
    (*b*) corrodes easily.
    (*c*) has low creep strength.
    (*d*) none of these.

54. With increase in the carbon percentage in the steel, its
    (*a*) hardness increases.
    (*b*) ductility reduces.
    (*c*) tensile strength increases.
    (*d*) all (*a*), (*b*) & (*c*).

55. Low carbon steels are those, in which the carbon percentage is around
    (*a*) 0.1 to 0.35 (*b*) 0.4 to 0.7
    (*c*) 0.8 to 1.5 (*d*) 2 to 4

56. Carbon percentage in medium carbon steel is around
    (*a*) 0.1 to 0.35 (*b*) 0.35 to 0.5
    (*c*) 0.8 to 1.5 (*d*) 1 to 1.5.

57. High carbon steel contains more than .... percent carbon.
    (*a*) 0.5 (*b*) 1
    (*c*) 1.5 (*d*) 2

58. Which of the following has the poorest weldability ?
    (*a*) Low carbon steel
    (*b*) Mild steel
    (*c*) Wrought iron
    (*d*) High carbon steel

59. Ferritic stainless steels compared to austenitic stainless steels
    (*a*) have lower corrosion resistance.
    (*b*) are harder to fabricate.
    (*c*) are less ductile and hence less suitable for cold pressing.
    (*d*) all (*a*), (*b*) & (*c*).

60. The material used in the filament of electric bulbs is
    (*a*) nichrome (*b*) tungsten
    (*c*) constantan (*d*) German silver

61. Ability of a material to undergo large permanent deformation in tension is called
    (*a*) plasticity (*b*) malleability
    (*c*) ductility (*d*) none of these

62. Ability of a material to undergo large permanent deformations in compression is called
    (*a*) malleability (*b*) ductility
    (*c*) plasticity (*d*) none of these

63. Machining properties of steel are improved by adding
    (*a*) silicon & chromium.
    (*b*) nickel & vanadium.
    (*c*) sulphur, lead & phosphorous.
    (*d*) aluminium & titanium.

64. Solder is an alloy comprising of
    (*a*) tin & copper (*b*) tin & lead
    (*c*) lead & copper (*d*) none of these

65. Improvement in wear resistance, cutting ability and toughness of steel is achieved by adding.
    (*a*) chromium (*b*) tungsten
    (*c*) cobalt (*d*) nickel

Answers

52. (a)	53. (c)	54. (d)	55. (a)	56. (b)	57. (a)
58. (d)	59. (d)	60. (b)	61. (c)	62. (a)	63. (c)
64. (b)	65. (a)				

66. Addition of .... in steel refines grain size, produces less tendency to carburisation and improves corrosion & heat resistant properties.
(*a*) chromium (*b*) tungsten
(*c*) cobalt (*d*) nickel

67. Improvement in cutting ability and reduction in hardeneability of steel is achieved by the addition of
(*a*) cobalt (*b*) chromium
(*c*) nickel (*d*) tungsten

68. Which of the following imparts ductility, toughness, tensile strength and anti corrosion properties to steel ?
(*a*) Cobalt (*b*) Chromium
(*c*) Nickel (*d*) Tungsten

69. Bell metal comprises of
(*a*) 75% Cu + 25% Sn
(*b*) 75% Cu + 25% Zn
(*c*) 75% Cu + 25% Ni.
(*d*) 75% Ni + 25% Zn

70. White metal contains
(*a*) nickel & copper
(*b*) lead, tin & cadmium
(*c*) copper & zinc
(*d*) nickel & zinc

71. Dow metal contains mainly .... and aluminium.
(*a*) magnesium, copper
(*b*) copper
(*c*) nickel
(*d*) zinc

72. Aluminium bronze contains mainly
(*a*) aluminium (90%) & copper.
(*b*) copper (90%) & aluminium,
(*c*) magnesium (90%) & copper.
(*d*) nickel (90%) & copper.

73. Iron alloyed with carbon in percentages greater than 2% is called
(*a*) steel
(*b*) mild steel
(*c*) high carbon steel
(*d*) cast iron

74. Admirality gun metal contains mainly
(*a*) copper (88%), tin (10%) & rest zinc.
(*b*) nickel (88%) & rest copper.
(*c*) zinc (88%) & rest tin.
(*d*) none of these.

75. A measure of toughness of a material is
(*a*) the area under strain-stress diagram till fracture.
(*b*) its yield strength.
(*c*) its ultimate strength.
(*d*) its percentage elongation.

76. With increase in carbon percentage, the ultimate strength of mild steel
(*a*) decreases.
(*b*) increases.
(*c*) remains same.
(*d*) may increase or decrease, depends on the range of carbon content.

77. The tensile stress-strain diagram for cast iron shows
(*a*) large deformation before fracture.
(*b*) no yield point at all.
(*c*) a pronounced yield point.
(*d*) linear relationship upto point of fracture.

78. With increase in carbon percentage, toughness of mild steel
(*a*) increases (*b*) decreases
(*c*) remains same (*d*) changes randomly

Answers

66. (b)	67. (a)	68. (c)	69. (a)	70. (b)	71. (a)
72. (b)	73. (d)	74. (a)	75. (a)	76. (c)	77. (d)
78. (c)					

79. Cast iron has high .... strength.
(*a*) tensile (*b*) compressive
(*c*) shear (*d*) fatigue

80. Which of the following is a suitable light duty bearing material ?
(*a*) Monel metal
(*b*) Phosphorous bronze
(*c*) White metal
(*d*) Nimonic alloys

81. Magnet steels contain .... of cobalt.
(*a*) high % (*b*) low %
(*c*) traces (*d*) carbide

82. Which is a suitable material for heavier duty bearings ?
(*a*) White metal (*b*) Phosphor bronze
(*c*) Monel metal (*d*) Nimonic alloys

83. Which is the most used non ferrous metal/ alloy ?
(*a*) Aluminium & its alloys.
(*b*) Copper & its alloys.
(*c*) Tin & its alloys.
(*d*) Zinc & its alloys.

84. Which is the low melting point (300 °C) non ferrous metal/alloy?
(*a*) Zinc & its alloys (*b*) Tin & its alloys
(*c*) Silver & its alloys (*d*) Copper & its alloys

85. Which is the high melting point (2000°C) non ferrous metal ?
(*a*) Tungsten (*b*) Berellium
(*c*) Uranium (*d*) Germanium

86. 18/8 stainless steel means that it contains
(*a*) 18% chromium & 8% nickel.
(*b*) 18% chromium & 8% molybdenum.
(*c*) 18% nickel & 8% chromium.
(*d*) 18% molybdenum & 8% chromium.

87. Stainless steel is not corroded by the
(*a*) hydrochloric acid (10%)
(*b*) nitric acid (10%)
(*c*) sulphuric acid (10%)
(*d*) saturated brine

88. Polythene (low or high density) containers are not corroded by
(*a*) sulphuric acid (10%) at room temperature.
(*b*) nitric acid (95%) at room temperature.
(*c*) sulphur trioxide at 60°C.
(*d*) any of these.

89. Dielectric
(*a*) is an electrical insulating material.
(*b*) of low resistivity is preferred.
(*c*) should have high thermal conductivity.
(*d*) need not defy the corrosive action of chemicals.

90. Glass is corroded by the
(*a*) fluorine (dry or wet.)
(*b*) sulphuric acid
(*c*) phosphoric acid
(*d*) none of these

91. Graphite is corroded by the
(*a*) sulphuric acid (10%)
(*b*) sea water
(*c*) hydrochloric acid
(*d*) none of these

92. Wood is corroded by the
(*a*) hydrochloric acid
(*b*) SO_2 (dry or wet)
(*c*) chlorine (dry or wet)
(*d*) all (*a*), (*b*) & (*c*)

Answers

79. (b)	80. (b)	81. (a)	82. (a)	83. (a)	84. (b)
85. (a)	86. (a)	87. (b)	88. (a)	89. (a)	90. (a)
91. (d)	92. (d)				

93. A material capable of undergoing large permanent deformation, when subjected to compression is termed as
(a) malleable (b) ductile
(c) brittle (d) none of these

94. Hastelloy C is a good material of construction in metallurgical process industry, since it is
(a) highly corrosion resistant and is readily fabricated.
(b) relatively inexpensive, although it can be fabricated with some difficulty.
(c) corrosion resistant to most alkalis, particularly because of its copper content.
(d) light and resists attack by acids.

95. Plastic tubes and pipes are generally made by .... moulding.
(a) injection (b) extrusion
(c) transfer (d) compression

96. Thermosetting plastics
(a) are cross-linked molecules
(b) are solvent soluble
(c) soften on heating
(d) none of these

97. Ability of a material to resist the applied forces is purely a measure of its
(a) strength properties
(b) ductility
(c) toughness
(d) none of these

98. .... does not contain tin.
(a) Brass (b) Pewter
(c) Solder (d) Babbit metal

99. Which of the following low melting alloy containing bismuth and lead is used for making electric fuse ?
(a) Wood's metal (b) Babbit metal
(c) Monel metal (d) Duralumin

100. Which of the following is the lowest cost plastic commercially available ?
(a) Polythene (b) Teflon
(c) Bakelite (d) PVC

101. Which of the following metals does not form an amalgam ?
(a) Cobalt (b) Silver
(c) Iron (d) Zinc

102. Nickel (76%) and chromium (34%) alloys are termed as
(a) monel
(b) inconel
(c) aluminium bronzes
(d) brass

103. The ratio of thermal & electrical conductivity is same for all the metals at the same temperature. This ratio is proportional to .... at around room temperature.
(a) T (b) 1/T
(c) T^2 (d) $1/T^2$

104. Cast iron has very high
(a) compressive strength
(b) ductility
(c) shock resistance
(d) resistance to brittleness

105. Nickel is a constituent of
(a) bronze (b) solder
(c) duralumin (d) monel metal

106. Light weight metallic alloy used in aircraft industry is
(a) aluminium (b) high silicon iron
(c) duralumin (d) phosphor bronze

Answers

93. (a)	94. (d)	95. (b)	96. (a)	97. (a)	98. (a)
99. (a)	100. (a)	101. (c)	102. (b)	103. (a)	104. (a)
105. (d)	106. (c)				

107. .... metal melts in boiling water.
   (*a*) Gun (*b*) Wood
   (*c*) Bell (*d*) Monel

108. Ductility of a material with work hardening
   (*a*) increases
   (*b*) decreases
   (*c*) remains unchanged
   (*d*) unpredictable

109. Constantan is an alloy of
   (*a*) Cu (55%) & Sn (45%)
   (*b*) Cu (55%) & Ni (45%)
   (*c*) Pt (95%) & Rh (10%)
   (*d*) Fe (80%) & Ni (20%)

110. For platinum thermocouples, lead wires are made of
   (*a*) copper & copper-nickel alloys.
   (*b*) copper & iron.
   (*c*) iron & nickel.
   (*d*) tin & nickel.

111. Ceramic recuperators are generally made of
   (*a*) silicon carbide
   (*b*) calcium carbide
   (*c*) fireclay bricks
   (*d*) high alumina bricks

112. The main purpose of galvanising iron sheets is to
   (*a*) harden the surface.
   (*b*) increase its glossiness & lustre.
   (*c*) prevent the action of water.
   (*d*) prevent the action of oxygen.

113. Which has the maximum melting point out of the following ?
   (*a*) Steel (*b*) Cast iron
   (*c*) Stainless steel (*d*) Wrought iron

114. The presence of impurities in iron
   (*a*) increases its melting point.
   (*b*) does not affect its melting point,
   (*c*) decreases its melting point.
   (*d*) makes it malleable.

115. Carbon percentage is the same in cast iron and
   (*a*) wrought iron
   (*b*) pig iron
   (*c*) mild steel
   (*d*) high silicon (14%) iron

116. Which of the following has the least carbon percentage ?
   (*a*) Low carbon steel (*b*) Mild steel
   (*c*) Wrought iron (*d*) White cast iron

117. Metalloid is
   (*a*) highly electronegative in nature,
   (*b*) an element which exhibits the properties of both metal and non metal.
   (*c*) an alloy.
   (*d*) all (*a*), (*b*) & (*c*) .

118. Iron rust is
   (*a*) ferrous oxide
   (*b*) hydrated ferric oxide
   (*c*) powdered iron
   (*d*) ferric sulphide

119. Which one has the maximum tensile strength out of the following?
   (*a*) Nodular cast iron
   (*b*) Pig iron
   (*c*) White cast iron
   (*d*) Grey cast iron

120. Which is not a thermoplastic material ?
   (*a*) Polystyrene (*b*) Acrylics
   (*c*) Polypropylene (*d*) None of these

Answers

107. (b)	108. (b)	109. (b)	110. (a)	111. (a)	112. (c)
113. (d)	114. (c)	115. (b)	116. (c)	117. (b)	118. (b)
119. (a)	120. (d)				

121. Presence of .... in steel decreases its hardeneability.
(a) manganese (b) cobalt
(c) silicon (d) molybdenum

122. The properties of materials which opposes the creation of magnetic flux in them is called
(a) resistance (b) reluctance
(c) dielectric (d) permeance

123. Carbon content is the highest in
(a) mild steel
(b) eutectic alloy
(c) hypoeutectic alloy
(d) hypereutectic alloy.

124. Manganese in steel affects its
(a) ductility (b) tensile strength
(c) hardness (d) none of these

125. The ability of a material to absorb energy in the elastic range is a measure of its
(a) toughness (b) resilience
(c) malleability (d) brittleness

126. The ability of a material to offer resistance to scratching or indentation is a measure of its
(a) brittleness (b) toughness
(c) hardness (d) resilience

127. The maximum strength which a specimen can tolerate without failure, when the stress is repeated for a specified number of cycle is called the
(a) compressive strength
(b) endurance limit
(c) creep limit
(d) tensile strength

128. Crystals of the material during cold working attains
(a) hardness (b) brittleness
(c) conductivity (d) toughness

129. Chiesels are usually made of ....steel.
(a) low carbon
(b) mild
(c) high carbon
(d) none of these

130. A material is able to retain the deformation permanently by virtue of its
(a) elasticity (b) plasticity
(c) ductility (d) malleability

131. Stiffness of a material is its ability to resist the
(a) fracture due to high impact load.
(b) deformation under stress,
(c) external force.
(d) all (a), (b) & (c).

132. Cast iron is a .... material.
(a) brittle (b) ductile
(c) tough (d) malleable

133. A steel alloy having 36% nickel is called
(a) invar
(b) high speed steel
(c) cutting tool steel
(d) none of these

134. High silicon steel is usually used for making
(a) single point cutting tools.
(b) connecting rods,
(c) crankshafts.
(d) generators/transformers in the form of laminated cores.

135. In high speed steel, tungsten can be replaced by
(a) vanadium (b) molybdenum
(c) silicon (d) cobalt

Answers

121. (b)	122. (b)	123. (d)	124. (b)	125. (b)	126. (c)
127. (b)	128. (b)	129. (c)	130. (b)	131. (b)	132. (a)
133. (a)	134. (d)	135. (b)			

136. Ball bearings are normally made of .... steel.
(*a*) high speed (*b*) high carbon
(*c*) chrome (*d*) silicon

137. Free cutting steels
(*a*) have free carbon.
(*b*) require very low cutting force.
(*c*) are used where rapid machining is required.
(*d*) none of these.

138. Shock resisting steels must have
(*a*) low hardness
(*b*) toughness
(*c*) low tensile strength
(*d*) none of these

139. Formation of iron sulphide in steel is reduced by the addition of
(*a*) tungsten (*b*) vanadium
(*c*) manganese (*d*) cobalt

140. Residual magnetism in steel for magnets is increased by the addition of
(*a*) nickel (*b*) cobalt
(*c*) tungsten (*d*) chromium

141. Which is an amorphous material ?
(*a*) Glass (*b*) Mica
(*c*) Brass (*d*) Cast iron

142. An example of mesomorphous material is
(*a*) lead (*b*) cast iron
(*c*) mica (*d*) none of these

143. Babbit metal used for bearings is a .... base alloy.
(*a*) tin (*b*) lead
(*c*) aluminium (*d*) copper

144. Pick out the most suitable material of construction for bearings subjected to light loads.
(*a*) Monel (*b*) White metal
(*c*) Phosphor bronze (*d*) None of these

145. Addition of lead & bismuth to aluminium improves its
(*a*) castability (*b*) machinability
(*c*) hardness (*d*) none of these

146. Addition of silicon to aluminium improves its
(*a*) castability (*b*) machinability
(*c*) ductility (*d*) none of these

147. Addition of manganese to aluminium improves its
(*a*) castability (*b*) machinability
(*c*) corrosion (*d*) none of these

148. Addition of silicon to copper improves its
(*a*) machinability
(*b*) ductility
(*c*) hardness & strength
(*d*) none of these

149. Addition of nickel to copper improves its
(*a*) machinability (*b*) ductility
(*c*) strength (*d*) hardness

150. A steel containing both cementite & pearlite is
(*a*) tough (*b*) soft
(*c*) hard (*d*) hard & tough

151. Addition of zinc to copper improves its
(*a*) machinability
(*b*) corrosion resistance
(*c*) hardness
(*d*) ductility and strength

Answers

136. (c)	137. (c)	138. (b)	139. (c)	140. (b)	141. (a)
142. (c)	143. (a)	144. (c)	145. (b)	146. (a)	147. (c)
148. (c)	149. (d)	150. (c)	151. (d)		

152. White cast iron is produced by the .... of grey cast iron.
(*a*) slow heating (*b*) rapid heating
(*c*) slow cooling (*d*) rapid cooling

153. Steel produced from phosphatic iron is
(*a*) tough (*b*) hard
(*c*) brittle (*d*) soft

154. Invar contains highest percentage of
(*a*) vanadium (*b*) iron
(*c*) tungsten (*d*) cobalt

155. Ferroelectric materials exhibit
(*a*) thermoelectric effect
(*b*) hysterisis
(*c*) piezoelectric
(*d*) both (*a*) and (*b*)

156. Which of the following is essentially present in hard steel ?
(*a*) Austenite (*b*) Ferrite
(*c*) Martensite (*d*) Sorbite

157. A ferromagnetic substance at Curie temperature changes to
(*a*) paramagnetic (*b*) diamagnetic
(*c*) insulator (*d*) none of these

158. .... materials tend to obey Curie law.
(*a*) Metallic (*b*) Non-metallic
(*c*) Diamagnetic (*d*) Paramagnetic

159. The number of degrees of freedom at the eutectic point of a mixture of two metals is
(*a*) 0 (*b*) 1
(*c*) 2 (*d*) 3

160. The electrical conductivity of a/an .... decreases with rise in temperature.
(*a*) alloy or metal (*b*) semiconductor
(*c*) dielectric (*d*) none of these

161. Interchain forces are the weakest in case of
(*a*) plastics
(*b*) fibres
(*c*) elastomers
(*d*) both rubbers and elastomers.

162. Creep is not exhibited at low temperature by
(*a*) rubber (*b*) acrylics
(*c*) lead (*d*) plastics

163. A material is called ductile, if it can be
(*a*) drawn into wires,
(*b*) hammered to a thin sheet.
(*c*) fractured without deformation.
(*d*) made lustrous by heating it.

164. The process of coating steel sheets by zinc to improve its corrosion resistance is called
(*a*) annealing (*c*) galvanising
(*c*) zincification (*d*) tempering

165. Pick out the wrong statement.
(*a*) Bronze is an alloy of copper & tin,
(*b*) Brass is an alloy of copper & zinc.
(*c*) The alloy named 'German silver' does not contain any silver.
(*d*) The ability of material to fracture without appreciable deformation is called its ductility.

166. Which of the following does not have a sharp melting point ?
(*a*) Thoria (*b*) Ice
(*c*) Pig iron (*d*) Glass

167. Pick out the correct statement.
(*a*) Stainless steel is nothing but chromium coated steel.
(*b*) Hardening of a soft metal can be done by alloying it with another metal or non-metal.
(*c*) Plastic deformation of a material is its temporary distortion under the action of applied stress.
(*d*) Rusting of iron is not electrochemical in nature.

Answers

152. (b)	153. (c)	154. (b)	155. (d)	156. (c)	157. (a)
158. (c)	159. (a)	160. (a)	161. (d)	162. (c)	163. (a)
164. (b)	165. (d)	166. (d)	167. (b)		

168. Which of the following is a low melting point metal ?
    (*a*) Stainless steel
    (*b*) Wrought iron
    (*c*) Tin
    (*d*) Copper

169. Babbit metal used for bearings is a .... base alloy.
    (*a*) tin (*b*) lead
    (*c*) aluminium (*d*) copper

170. Wood metal is not used for making
    (*a*) storage tanks for storing brine and caustic soda.
    (*b*) fusible safety plug for pressure cooker.
    (*c*) casting for dental work.
    (*d*) soft solder.

171. Cast iron as compared to steel is better in
    (*a*) ductility
    (*b*) fluidity & castability
    (*c*) strength
    (*d*) malleability

172. Electromotive force series of the metals gives an indication of the
    (*a*) chemical activity of metals.
    (*b*) metals which tend to displace hydrogen from an acid.
    (*c*) formation of corrosion retarding films on metal surface.
    (*d*) all (*a*), (*b*) & (*c*).

173. Reduction in volume of steel (*i.e.* shrinkage), when it solidifies from molten state is about .... percent by volume.
    (*a*) 1 to 1.8 (*b*) 3.5 to 3.8
    (*c*) 6.9 to 7.4 (*d*) .13.5 to 14.8

174. Which of the following is not an alloy of nickel ?
    (*a*) Inconel (*b*) Nimonics
    (*c*) Hastealloy (*d*) Gun metal

175. Nickel (56%) and molybdenum (17%) alloys are called
    (*a*) monel (*b*) inconel
    (*c*) bronzes (*d*) hastealloy C

176. The ball indentor used in the Brinell hardness test is usually made of
    (*a*) mild steel (*b*) cast iron
    (*c*) hardened steel (*d*) brass

177. Presence of manganese in cast iron beyond what is required to form MnS
    (*a*) improves soundness of its castings.
    (*b*) reduces its hardness.
    (*c*) reduces its tensile strength.
    (*d*) all (*a*), (*b*) & (*c*).

178. Presence of high phosphorous in cast iron increases its
    (*a*) fluidity (*b*) melting point
    (*c*) shrinkage (*d*) tensile strength

179. Silicon percentage in iron castings to have maximum tensile strength should be about
    (*a*) 0.5 (*b*) 0.8
    (*c*) 1.2 (*d*) 1.8

180. Silicon percentage in iron castings to have maximum hardness should be about
    (*a*) 0.3 (*b*) 0.5
    (*c*) 0.8 (*d*) 2.5

181. The most distinguished property of malleable cast iron is its high
    (*a*) ductility (*b*) hardness
    (*c*) malleability (*d*) none of these

Answers

168. (c)	169. (a)	170. (a)	171. (b)	172. (d)	173. (c)
174. (d)	175. (d)	176. (c)	177. (a)	178. (a)	179. (d)
180. (c)	181. (a)				

182. Which of the following exercises most powerful influence on the mechanical properties of steel ?
(*a*) Manganese (*b*) Carbon
(*c*) Phosphorous (*d*) Silicon

183. Presence of silicon in steel
(*a*) is in the form of free Si.
(*b*) indicates that steel is not deoxidised properly.
(*c*) decreases the blow holes formation tendency in castings.
(*d*) none of these.

184. Which of the following contains maximum percentage of carbon ?
(*a*) Boiler plate steel
(*b*) Rail steel
(*c*) Saws for cutting steel
(*d*) Railway spring steel

185. Which of the following alloy steels is the most suitable for making gun barrels and projectiles ?
(*a*) Tungsten steel
(*b*) Molybdenum steel
(*c*) Nickel-chrome steel
(*d*) Cobalt steel

186. Which of the following alloy steels has the energy saving properties making it suitable for use in electrical machines ?
(*a*) Silicon steel
(*b*) Chromium steel
(*b*) Molybdenum steel
(*d*) None of these

187. Presence of chromium in steel
(*a*) reduces its hardness.
(*b*) is in the form of carbide.
(*c*) seriously affects its ductility.
(*d*) decreases its elastic limit & tensile strength.

188. Temper brittleness (makes the alloy steel brittle when cooled in air from the tempering temperature) is a characteristic defect associated especially with .... steels.
(*a*) nickel-chromium
(*b*) chromium
(*c*) tungsten
(*d*) molybdenum

189. Pure nickel is
(*a*) ferromagnetic above its Curie point (*i.e.*, 415°C).
(*b*) having h.c.p. crystal lattice.
(*c*) ferromagnetic at room temperature.
(*d*) not resistant to oxidation at high temperature.

190. Austenitic manganese steel used for making jaws of crushing machines contains about .... percent manganese.
(*a*) 1.5-2 (*b*) 3.5-4.5
(*c*) 7-9 (*d*) 12-14

191. Tungsten steel possess the property of'red hardness' because tungsten obstructs the austenite to pearlite transformation and act as 'a fixing agent' for martensite which resists tempering. Tungsten percentage in steel varies from
(*a*) 1 to 5 (*b*) 1 to 10
(*c*) 1 to 15 (*d*) 1 to 20

192. Molybdenum steels containing 0.15 to 0.40% molybdenum
(*a*) are prone to temper brittleness.
(*b*) are used for automobile parts like gears, axle, shaft, bearings etc.
(*c*) have very poor fatigue resistance.
(*d*) do not contain nickel or chromium or both.

Answers

182. (b)	183. (c)	184. (c)	185. (c)	186. (a)	187. (b)
188. (a)	189. (c)	190. (d)	191. (d)	192. (b)	

193. Chrome vanadium steels usually contain 0.8 to 1.1% Cr, 0.25 to 0.35% C & 0.25% V. It is
(*a*) used for making axle & shafts of aeroplanes, automobiles and locomotives.
(*b*) having poor fatigue resistance.
(*c*) having poor toughness.
(*d*) not helpful in producing cleaner steel due to oxidising action of vanadium.

194. Silico-mangancse steels (containing 0.7% Mn, 0.5% C and 2% Si)
(*a*) do not require heat treatment for any of its applications.
(*b*) are used for leaf and coil springs.
(*c*) do not have good resilience properties.
(*d*) is tempered at 850°C.

195. Straight silicon steel (containing Si from 0.5 to 5%) because of their low hysterisis loss and high magnetic permeability are used for electrical appliances. Which of the following electrical appliances made of silicon steel contains maximum percentage of silicon (about 4%) ?
(*a*) Armature of small motors/generators.
(*b*) Generators and small motors.
(*c*) Induction motors.
(*d*) High frequency transformers.

196. 'Hypersil' is a .... steel.
(*a*) high speed (*b*) silicon
(*c*) molybdenum (*d*) tungsten

197. Nickel base alloys with .... are known as monel metal.
(*a*) zinc (*b*) tin
(*c*) copper (*d*) molybdenum

198. Spark plugs, ignition tubes & electrodes are made of nickel .... alloys.
(*a*) beryllium (*b*) manganese
(*c*) copper (*d*) iron

199. Which of the copper base alloys with nickel contains zinc in addition to Cu and Ni ?
(*a*) Cupro-nickel (*b*) German silver
(*c*) Monel (*d*) Constantan

200. Monel alloy is
(*a*) the costliest of all the alloys of nickel.
(*d*) not suitable for making kitchen utensils.
(*c*) used for making steam valves & turbine blades as it resists steam attack at high temperature.
(*d*) not used in cast and wrought forms.

201. Which of the following monel alloy contains aluminium in addition to Cu & Ni; and has good corrosion & oxidation resistance besides high creep strength at elevated temperature ?
(*a*) K-monel (*b*) R-monel
(*c*) H-monel (*d*) S-monel

202. Which of the following is not an alloy of nickel & chromium ?
(*a*) Inconel (*b*) Hastealloy
(*c*) Nimonic alloys (*d*) Duralumin

203. Because of their law creep rates and high resistance to high temperature oxidation, 'super alloys' are used for making baldes and rotors of jet engines which operate at above 900°C for prolonged periods. They are .... based alloys.
(*a*) cobalt (*b*) iron
(*c*) nickel (*d*) none of these

204. Hastealloy
(*a*) is not subjected to precipitation hardening.
(*b*) does not contain molybdenum,
(*c*) is not used in cast form.
(*d*) has high strength and oxidation & corrosion resistance.

Answers

193 (a)	194. (b)	195. (d)	196. (b)	197. (c)	198. (b)
199. (b)	200. (c)	201. (a)	202. (d)	203. (d)	204. (d)

205. Which of the following alloying components is absent in 'marageing steels' ? .
(*a*) Nickel (*b*) Aluminium
(*c*) Cobalt (*d*) Molybdenum

206. Which of the following is not a copper based alloy with nickel ?
(*a*) German silver (*b*) Monel metal
(*c*) Inconel (*d*) Constantan

207. Which of the following commercial metals is most abundantly found in India ?
(*a*) Copper (*b*) Aluminium
(*c*) Nickel (*d*) Zinc

208. Aluminium as a material of construction suffers from the disadvantage of
(*a*) very high cost.
(*b*) low tensile strength.
(*c*) low strength to weight ratio.
(*d*) scarce availability.

209. Aluminium alloy is one of the most suitable materials of construction for air crafts mainly due to its
(*a*) high strength to weight ratio.
(*b*) low temperature strength properties.
(*c*) its ability to be cast, rolled, forged and stamped.
(*d*) high strength and corrosion & oxidation resistance at elevated temperature.

210. Pick out the wrong statement.
(*a*) Aluminium ternery alloys are susceptible to "ageing" *i.e.* increase in strength and hardness with time.
(*b*) Aluminium alloys are stronger than pure aluminium.
(*c*) Aluminium alloys compared to pure aluminium have better hardness & impact resistance and higher fatigue strength.
(*d*) Aluminium alloys can not be used in both the cast and wrought form.

211. Aluminium-magnesium alloys (containing upto 1% magnesium)
(*a*) can not be welded.
(*b*) are weaker than pure aluminium.
(*c*) develop extra hardness by cold working.
(*d*) are suitable for very high temperature applications.

212. Age hardening of duralumin is due to
(*a*) copper
(*b*) magnesium
(*c*) both (*a*) & (*b*)
(*d*) neither (*a*) nor (*b*)

213. Age hardening of aluminium alloys
(*a*) may not take place at room temperature.
(*b*) at higher temperature accelerates the rate of hardening.
(*c*) is also termed as temper-hardening.
(*d*) both (*b*) &(*c*).

214. Pick out the wrong statement.
(*a*) Titanium exits in two allotropic forms *i.e.* α-form (h.c.p.-stable at room temperature) and β-form (b.c.c. - stable from 885 to 1675°C).
(*b*) Aluminium is an important alloying element for titanium alloys as it acts as α (alpha)-form stabiliser.
(*c*) β-phase of titanium is much stronger but more brittle than the α-phase.
(*d*) Titanium is heavier than both aluminium as well as steel.

215. Which of the following is the lightest of engineering metals ?
(*a*) Magnesium (*b*) Aluminium
(*c*) Titanium (*d*) Tin

Answers

205. (b)	206. (c)	207. (b)	208. (b)	209. (a)	210. (d)
211. (c)	212. (c)	213. (d)	214. (d)	215. (a)	

216. Magnesium alloys are
    (*a*) ferromagnetic (*b*) highly anodic
    (*c*) non-inflammable (*d*) not extruded

217. Copper has very low
    (*a*) malleability.
    (*b*) ductility.
    (*c*) tensile strength.
    (*d*) thermal & electrical conductivity.

218. German silver used for decorative purposes contains maximum percentage of
    (*a*) silver (*b*) copper
    (*c*) zinc (*d*) nickel

219. Pick out the wrong statement.
    (*a*) Crystal structure of magnesium is h.c.p., hence room temperature ductility and cold workability is poor.
    (*b*) Magnesium alloys are suitable for very high temperature usage (500°C).
    (*c*) Magnesium alloys can be cast by sand, permanent mould and die casting methods.
    (*d*) Magnesium alloys are not subjected to wire drawing operation.

220. A prism of brittle material will break in compression
    (*a*) by forming a bulge.
    (*b*) by shearing along oblique plane.
    (*c*) in a direction perpendicular to the application of load.
    (*d*) none of these.

221. Wrought iron is having very high
    (*a*) hardness
    (*b*) strength
    (*c*) corrosion resistance
    (*d*) susceptibility to changes in its properties by heat treatment.

222. Cast iron has very high
    (*a*) proximity between its elastic limit and ultimate breaking strength.
    (*b*) ductility.
    (*c*) tensile strength.
    (*d*) all (*a*), (*b*) & (*c*).

223. Which of the following is not an alloy of copper and tin ?
    (*a*) Bell metal
    (*b*) Gun metal
    (*c*) Admiralty gun metal
    (*d*) Monel metal

224. As the impurities are oxidised, the melting point of iron
    (*a*) decreases
    (*b*) remains unchanged
    (*c*) increases
    (*d*) may increase or decrease ; unpredictable

225. Maximum consumption of zinc is in
    (*a*) alloying
    (*b*) galvanising
    (*c*) utensil manufacture
    (*d*) electrical industry

226. Zinc is highly .... at room temperature.
    (*a*) ductile
    (*b*) atmospheric corrosion resistant
    (*c*) malleable
    (*d*) brittle

227. Zinc is not used
    (*a*) for producing zinc base die casting alloys.
    (*b*) in its oxide form as pigments for paints.
    (*c*) as anode for corrosion prevention in boiler.
    (*d*) as an alloying element in various bronze.

Answers

216. (b)	217. (c)	218. (b)	219. (b)	220. (b)	221. (c)
222. (a)	223. (d)	224. (c)	225. (b)	226. (b)	227. (d)

228. The largest consumption of zinc for alloys is in the making of
(*a*) bearing metal (*b*) brasses
(*c*) brazing solder (*d*) die casting alloy

229. Pick out the wrong statement.
(*a*) Tin can be readily and very easily drawn into very fine wire.
(*b*) Tin can be severally cold worked without the necessity of annealing due to its low recrystallisation temperature.
(*c*) Tin exists in two allotropic forms.
(*d*) The predominant use of tin is in the form of coating for steel and copper alloys.

230. Which of the following is not an alloy of tin?
(*a*) White-bearing metal
(*b*) Pewter type metal
(*c*) Soft solder
(*d*) German silver

231. Tin coating is not done by
(*a*) vaporising
(*b*) hot dipping
(*c*) electro deposition
(*d*) spraying

232. Tin in its pure form is used for making
(*a*) cans for food packing.
(*b*) collapsible tubes for toilet preparations.
(*c*) foil for wrapping cheese.
(*d*) all (*a*), (*b*) & (*c*).

233. Among the tin containing alloys, ... consumes the maximum amount of tin.
(*a*) solder (*b*) white metal
(*c*) pewter metal (*d*) bronzes

234. Lead
(*a*) is the hardest metal in common use.
(*b*) is the lightest metal in common use.
(*c*) can not be scratched by finger nails.
(*d*) can not be work hardened due to its low recrystallisation temperature.

235. Maximum consumption of lead is in the manufacture of
(*a*) storage batteries
(*b*) solder alloys
(*c*) electric cable sheathing
(*d*) lead lined vessels

236. Lead is
(*a*) not resistant to corrosion.
(*b*) used as a cathodic material.
(*c*) not used as pigment (in its compound forms) for paints.
(*d*) used as shock absorber (in mattress form) between the foundation and steel framework of skyscrapper buildings.

237. Some specific solder alloys may contain little bit (1 to 2%) of .... in addition to lead and tin.
(*a*) aluminium (*b*) antimony
(*c*) zinc (*d*) copper

238. A bearing metal should not have
(*a*) enough plasticity
(*b*) low thermal conductivity
(*c*) low co-efficient of friction
(*d*) high toughness

239. Presence of 0.3 to 0.5% arsenic in copper increases its
(*a*) ductility
(*b*) malleability
(*c*) tenacity and hardness
(*d*) electrical conductivity

Answers

228. (b)	229. (a)	230. (d)	231. (a)	232. (d)	233. (a)
234. (d)	235. (a)	236. (d)	237. (b)	238. (b)	239. (c)

240. Presence of even 0.005% of .... in copper makes it highly brittle, rendering it useless for wire-drawing.
(*a*) oxygen (*b*) arsenic
(*c*) antimony (*d*) bismuth

241. Maximum consumption of copper is in
(*a*) utensil manufacture.
(*b*) electrical industry.
(*c*) industrial equipment manufacture.
(*d*) small scale industries.

242. Presence of even very minute amount (say 0.1%) of. .... in copper, reduces its electrical conductivity very seriously.
(*a*) bismuth (*b*) arsenic
(*c*) antimony (*d*) both (*b*) and (*c*)

243. Which of the following is not an alloy of copper ?
(*a*) Muntz metal
(*b*) Admirability gun metal
(*c*) German silver
(*d*) Pewter metal

244. Babbit metals used for making bearings are
(*a*) tin or lead based alloys.
(*b*) short of antifrictional properties.
(*c*) having homogenous microstructure.
(*d*) yellow metals.

245. In lead base bearing metal alloys, if tin is not added to lead, then
(*a*) lubricating oil will attack the lead.
(*b*) its corrosion resistance will decrease.
(*c*) its solidus temperature will be more.
(*d*) all (*a*), (*b*) & (*c*).

246. Tin base and lead base alloys are used for bearing materials. Tin compared to lead has lower
(*a*) price
(*b*) corrosion resistance
(*c*) thermal conductivity
(*d*) abundant availability

247. Babbit metal is not a
(*a*) tin base alloy (*b*) white metal
(*c*) lead base alloy (*d*) pure metal

248. Which of the following is produced by condensation polymerisation?
(*a*) Bakelite
(*b*) Polythene
(*c*) Poly vinyl chloride
(*d*) Polystyrene

249. Which of the following is produced by addition polymerisation ?
(*a*) Polybutadiene
(*b*) Phenol formaldehyde
(*c*) Urea-formaldehyde
(*d*) Dacron

250. Thermosetting resins are those polymers, which
(*a*) do not increase in plasticity with rise in temperature.
(*b*) allow slip to occur between molecules.
(*c*) do not take on a permanent set.
(*d*) none of these.

251. Thermoplastic resins are those polymers, which
(*a*) has decreased plasticity at increasing temperature.
(*b*) can not be moulded.
(*c*) easily deform under pressure at high temperature.
(*d*) none of these.

Answers

240. (d)	241. (b)	242. (d)	243. (d)	244. (a)	245. (d)
246. (d)	247. (d)	248. (a)	249. (a)	250. (a)	251. (c)

252. Combination of metals and ceramics are called
(*a*) metalloy
(*b*) cermets
(*c*) metalloid
(*d*) non-crystalline ceramics

253. Cermets are used for making
(*a*) cutting tools
(*b*) abrasives
(*c*) both (*a*) & (*b*)
(*d*) neither (*a*) nor (*b*)

254. Ceramic materials
(*a*) are exemplified by clay and mica
(*b*) are poor electrical conductors due to lack of free electrons.
(*c*) never exhibit polymorphism.
(*d*) all (*a*), (*b*) & (*c*).

255. Which of the following is not used as the ceramic material in 'cermets' ?
(*a*) Silicides (*b*) Oxides
(*c*) Carbides (*d*) Nitrides

256. Carbide ceramics in 'cermets' are not bonded by
(*a*) Cu (*b*) Ni
(*c*) Co (*d*) Fe

257. Spark plugs are made of
(*a*) high alumina ceramic
(*b*) metallic carbides
(*c*) corundum
(*d*) carborundum

258. Ceramic materials fabrication can not be done by
(*a*) welding (*b*) slip casting
(*c*) extrusion (*d*) plastic pressing

259. Vycor, a widely used material for making "thermal wells' in temperature measurement. is a
(*a*) cermet
(*b*) glass
(*c*) thermosetting plastic
(*d*) metallic alloy

260. Presence of high amount of silicon in cast iron
(*a*) suppresses the decomposition of Fe_3C.
(*b*) increases the formation of graphite,
(*c*) increases the contraction of the metal,
(*d*) reduces its fluidity.

261. The softest material just next to talc in the Mho's scale is
(*a*) quartz (*b*) gypsum
(*c*) feldspar (*d*) fluorite

262. The hardest material just prior to diamond in "Mho's scale' of hardness is
(*a*) topax (*b*) carborundum
(*c*) corrundum (*d*) calcite

263. Which of the following alloys does not contain nickel ?
(*a*) Chlorimet-2 alloy
(*b*) Monel
(*c*) Inconel
(*d*) Babbit metal

264. Wrought iron is a suitable material of construction for handling .... solutions.
(*a*) dilute acidic
(*b*) concentrated acidic
(*c*) alkalis & alkaline
(*d*) none of these

265. Mild steel is a/an .... steel.
(*a*) low carbon (*b*) medium carbon
(*c*) high carbon (*d*) high alloy

Answers

252. (b)	253. (c)	254. (d)	255. (d)	256. (a)	257. (a)
258. (a)	259. (b)	260. (b)	261. (b)	262. (c)	263. (d)
264. (c)	265. (a)				

266. Mercury is transported in metal containers made of
(*a*) aluminium (*b*) iron
(*c*) lead (*d*) nickel

267. Which of the following is not a nickel based alloy ?
(*a*) Inconel (*b*) Hastelloys
(*c*) Nimonics (*d*) Babbit metal

268. .... alloys have the highest specific-strength of all the structural materials out of the following.
(*a*) Magnetic steel (*b*) Titanium
(*c*) Chromium (*d*) Magnesium

269. .... moulding process can not be used for thermoplastic plastics.
(*a*) Compression (*b*) Extrusion
(*c*) Injection (*d*) none of these

270. A material no longer behaves elastically beyond
(*a*) plastic limit (*b*) limiting load
(*c*) elastic limit (*d*) breaking load

271. Fatigue failure of a material results from .... stress.
(*a*) tensile (*b*) compressive
(*c*) fluctuating (*d*) none of these

272. High speed steel should have high
(*a*) toughness (*b*) wear resistance
(*c*) hardeneability (*d*) both *b* &'c'

273. High degree of toughness is a must for .... steel.
(*a*) high speed (*b*) shock resisting
(*c*) cold work tool (*d*) none of these

274. Cold work tool steel should have
(*a*) high toughness
(*b*) poor hardenability
(*c*) high wear resistance
(*d*) all *'a', b & 'c'*

275. Chilled iron castings are
(*a*) easily machinable
(*b*) soft on surface
(*c*) highly wear resistant
(*d*) none of these

276. Bush bearings are made of
(*a*) mild steel
(*b*) phosphorous bronze
(*c*) white metal alloys
(*d*) none of these

277. Steel balls for ball bearings are generally made of .... steel.
(*a*) cast (*c*) stainless
(*c*) free carbon (*d*) carbon chrome

278. Dielectric strength of a material is its capacity to
(*a*) store energy.
(*b*) undergo magnetisation,
(*c*) resist the flow of current.
(*d*) withstand high voltage.

279. Diamagnetic materials are magnetised
(*a*) strongly.
(*b*) only slightly.
(*c*) wifh eddy currents only.
(*d*) in a direction opposite to that of the applied held.

280. The phenomenon in which slow extension of material takes place with the time at constant load, is called
(*a*) plasticity (*b*) creep
(*c*) elasticity (*d*) ductility

Answers

266. (b)	267. (d)	268. (b)	269. (a)	270. (c)	271. (c)
272. (b)	273. (d)	274. (b)	275. (c)	276. (c)	277. (b)
278. (d)	279. (d)	280. (b)			

281. Compressive strength of wood is usually more along the grains. The tensile strength of wood as compared to that of steel is about
(*a*) half (*b*) one third
(*c*) one fifth (*d*) one twelfth

282. .... is a donor impurity for semi conductors.
(*a*) Boron (*b*) Antimony
(*c*) Gallium (*d*) None of these

283. .... does not contain copper as an alloy ing element.
(*a*) Nichrome (*b*) Manganin
(*c*) Perminivar (*d*) Monel metal

284. Steel rods are used in reinforced concrete to increase its .... strength.
(*a*) shear (*b*) tensile
(*c*) compressive (*d*) none of these

285. Thermosetting polymers are
(*a*) injection moulded
(*b*) cast molded
(*c*) extruded
(*d*) none of these

286. Sulphuric acid pickling tanks are lined with
(*a*) nickel (*b*) rubber
(*c*) aluminium (*d*) glass

287. A metallic alloy in general as compared to their pure metal components has
(*a*) higher corrosion resistantce.
(*b*) higher melting point.
(*c*) lower hardness, reactivity & fusibility.
(*d*) lower electrical conductivity.

288. A metallic alloy in which one of the constituent metal is .... is called an 'amalgam'.
(*a*) zinc (*b*) mercury
(*c*) lead (*d*) tin

289. Maximum carbon content in any variety of steel can be upto 1.5%. Steel containing 0.4% carbon is not used for making
(*a*) nuts (*b*) bolts
(*c*) chiesels (*c*) steel

290. .... is added in stainless steel to prevent inter crystal corrosion,
(*a*) Vanadium (*b*) Niobium
(*c*) Chromium (*d*) Nickel

291. Pick out the wrong statement.
(*a*) Hardening makes the steel more brittle.
(*b*) High alloy steels contain more than 50% alloying element.
(*c*) 18/8 stainless steel is a magnetic steel.
(*d*) both (*b*) & (*c*).

292. Alinco, which is a aluminium-cobalt-nickel steel, is used for making
(*a*) surgical instruments.
(*b*) powerful permanent magnets,
(*c*) chemical equipments.
(*d*) boiler tubes.

293. Condenser tubes are not made of
(*a*) cartridge brass
(*b*) muntz brass
(*c*) wood metal
(*d*) aluminium brass alloy

Answers

281. (c)	282. (b)	283. (a)	284. (b)	285. (b)	286. (b)
287. (a)	288. (b)	289. (c)	290. (b)	291. (b)	292. (c)
293. (c)					

294. The addition of antimony in tin-based alloys improves its
(*a*) rupture strength & hot hardness.
(*b*) impact strength & bonding strength.
(*c*) deformation resistance.
(*d*) wear resistance.

295. Match the trade names of various polymeric materials of construction.

List I	*List II*
(*a*) Ebonite	I. Polytetraflouroethylene (PTFE).
(*b*) Teflon	II. Hard rubber.
(*c*) Bakelite	III. Polymethyl methacrylate (PMMA).
(*d*) Perspex	IV. Phenol formal dehyde.

296. Match the materials used for the manufacture of various parts/equipments.

List I	*List II*
(*a*) Babbit metal	1. Safety plug for cooker
(*b*) Wood metal	II. Bearings
(*c*) Cartridge	III. Automobile gears brass
(*d*) Nickel steel	IV. Condenser tube

297. Match the main alloying elements present in various alloys.

List I	*List II*
(*a*) Nimonics	I. Cu, Zn
(*b*) Monel	II. Ni, Cr
(*c*) Muntz metal	III. Cu, Ni
(*d*) Durimet 20	IV. Fe, Ni, Cr

298. High resistance electrical heating elements/wire are not made of
(*a*) marageing steel (*b*) nimonic alloys
(*c*) kanthal wire (*d*) nichrome

299. Galvanising is generally done on
(*a*) non-ferrous metals
(*b*) low carbon steel
(*c*) stainless steel
(*d*) cast iron

300. Percentage elongation of a material is a measure of its
(*a*) ductility (*b*) brittieness
(*c*) toughness (*d*) malleability

301. Ability of a material to .... is indicated by its damping capacity.
(*a*) withstand compression
(*b*) absorb vibration
(*c*) absorb shock
(*d*) none of these

302. A measure of toughness of a material is its
(*a*) percentage elongation.
(*b*) yield strength.
(*c*) ultimate strength.
(*d*) area under stress-strain diagram.

303. .... does not contain tin as an alloying material.
(*a*) Brass (*b*) Pewter
(*c*) Solder (*d*) Babbit metal

304. Trough of an apron conveyor is made of
(*a*) lead lined concrete vessel
(*b*) wood
(*c*) metal
(*d*) either (*b*) or (*c*)

305. Metal at the top of electromotive series is the
(*a*) least active (*b*) most stable
(*c*) most active (*d*) most noble

Answers

294. (c) 295. (II) (I) (IV) (III) 296. (II) (IV) (I) (III) 297. (II) (III) (I) (IV)
298. (a) 299. (b) 300. (a) 301. (b) 302. (d) 303. (a)
304. (d) 305. (c)

306. Nickel is a constituent of
(a) bronze (b) solder
(c) duralumin (d) monel metal

307. 18/8 steel is a/an .... stainless steel.
(a) austenitic (b) ferritic
(c) martensitic (d) none of these

308. 18-4-1 high speed steel contains 18%, 4% and 1% respectively of
(a) tungsten, vanadium & chromium.
(b) tungsten, chromium & vanadium.
(c) vanadium, chromium & tungsten.
(d) chromium, tungsten & vanadium.

309. Age hardening is concerned with
(a) copper (b) brass
(c) duralumin (d) silver

310. Addition of lead, sulphur & phosphorous to low carbon steels help in improving its
(a) corrosion resistance.
(b) tensile strength.
(c) machinability.
(d) compressive strength.

311. Pick out the wrong statement.
(a) The toughness of a material decreases, when it is heated.
(b) Crane hooks are normally made of wrought iron.
(c) Cold working of a metal decreases its fatigue strength.
(d) The temperature at which new grains are formed in a metal is known as the recrystallisation temperature.

312. Failure of a material is termed as fatigue failure, if it fails below the yield point. The resistance to fatigue failure of a material is measured by the
(a) ultimate tensile strength
(b) endurance limit
(c) elastic limit
(d) none of these

313. The most resistant material to alkaline corrosion is
(a) karbate (b) nickel
(c) aluminium (d) duriron

314. Which of the following is the most elastic material ?
(a) Plastic (b) Glass
(c) Steel (d) Rubber

315. Fatigue failure of a material may occur, when it is subjected to the .... stress.
(a) fluctuating (b) tensile
(c) compressive (d) torsion

316. Pick out the wrong statement.
(a) Pure rubber is as useless as pure gold as a material of construction.
(b) Thermal conductivity of aluminium is higher than that of copper, hence it is increasingly used for utensil making.
(c) Copper has poor weldability.
(d) Reinforced plastics are made from both the thermoplastics as well as thermosetting plastics.

317. In hot wire anemometer used for gas flow measurement, the wire is normally made of
(a) copper (b) platinum
(c) constantan (d) chromium

318. Which of the following has the highest modulus of elasticity (about 7×10^6) kg/cm^2 ?
(a) High speed steel.
(b) Stainless steel.
(c) Tungsten carbide.
(d) Superalloys.

Answers

306. (d)	307. (a)	308. (b)	309. (c)	310. (c)	311. (c)
312. (b)	313. (b)	314. (c)	315. (a)	316. (b)	317. (b)
318. (c)					

319. Fibre reinforced plastic (FRP) are
(*a*) made of glass fibre & thermoplastic resins.
(*b*) anisotropic.
(*c*) made of thermosetting resin & glass fibre.
(*d*) both *b* & 'c'.

320. Material having maximum density is
(*a*) plutonium (*b*) osmium
(*c*) titanium (*d*) uranium

321. Alloys having more than 80% copper are generally more
(*a*) abrasion resistant.
(*b*) corrosion resistant.
(*c*) prone to season cracking.
(*d*) machinable.

322. Brass parts with high residual tensile stress at the surface are susceptible to season cracking (*i.e.* spontaneous cracking occuring on exposure to atmospheric corrosion), if its zinc percentage is more than
(*a*) 5 (*b*) 10
(*c*) 20 (*d*) 35

323. 'Ageing' of steel refers to the
(*a*) degradation of steel by corrosion.
(*b*) failure of steel structure by overstress- ing.
(*c*) return of sharp yield point in temper rolled low carbon sheet after a period of time.
(*d*) decarburisation of steel by overheating.

324. Presence of sulphur in steel reduces the
(*a*) hardness
(*b*) grain size
(*c*) tensile strength
(*d*) toughness & transverse ductility

325. Addition of .... to steel can not impart it high strength at elevated temperature.
(*a*) nickel (*b*) magnesium
(*c*) silicon (*d*) manganese

326. Main alloying element in Elinvar used in precision instruments, hair springs for watches etc. is
(*a*) chromium (*b*) nickel
(*c*) zinc (*d*) aluminium

327. Out of the following, ... iron has the best capability to bear sudden & excessive shocks.
(*a*) white (*b*) cast
(*c*) wrought (*d*) pig

328. Tungsten in high speed steel is responsible for its
(*a*) toughness & wear resistance.
(*b*) cold hardness.
(*c*) hot hardness.
(*d*) sharp cutting edge.

329. Diamagnetic materials
(*a*) are magnetised by eddy currents only.
(*b*) are non-magnetic and can not be magnetised.
(*c*) are magnetised in direction opposite to the magnetic held
(*d*) can be magnetised in one direction only.

330. Out of the following, the most ductile material is
(*a*) mild steel (*b*) copper
(*c*) nickel (*d*) aluminium

331. Out of the following, the most malleable material is
(*a*) copper (*b*) aluminium
(*c*) wrought iron (*d*) lead

Answers

319. (d)	320. (b)	321. (b)	322. (c)	323. (c)	324. (d)
325. (b)	326. (b)	327. (c)	328. (c)	329. (c)	330. (b)
331. (d)					

332. Out of the following, which will fracture most readily, when hit with a hard hammer ?
(*a*) Brass (*b*) Bronze
(*c*) German silver (*d*) Cast iron

333. What is the percentage of chromium in 18-4-1 high speed steel ?
(*a*) 1 (*b*) 4
(*c*) 18 (*d*) 8

334. Alloying elements present in Hayness satellite which has superior performance than high speed steel, are
(*a*) nickel, cobalt & chromium.
(*b*) tungsten, chromium & cobalt.
(*c*) maganese, chromium & cobalt.
(*d*) tungsten, molybdenum & nickel.

335. Exposure to .... accelerates the degradation of plastics.
(*a*) ultraviolet radiation
(*b*) high atmospheric temperature
(*c*) high ambient temperature
(*d*) damp atmosphere

336. Primary designation of steel is based on its
(*a*) hardness & tensile strength.
(*b*) carbon content,
(*c*) iron content.
(*d*) alloying elements content.

337. Addition of beryllium to copper is done mainly to produce a/an .... alloy.
(*a*) non-sparking & non-magnetic
(*b*) high strength
(*c*) precipitation hardenable
(*d*) corrosion resistant

338. In case of brasses, with decreasing zinc percentage and increasing copper percentage, its .... increases.
(*a*) percentage elongation
(*b*) Brinell hardness
(*c*) tensile strength
(*d*) all '*a*, *b* & 'c'

339. Metal shots used in shot blasting are made of
(*a*) stellites (*b*) steel
(*c*) bronze (*d*) lead

340. Addition of .... to steel does not impart hardness.
(*a*) nickel (*b*) copper
(*c*) chromium (*d*) silicon

341. A material is capable of resisting softening at high temperature because of its property termed as the
(*a*) hot temper (*b*) hot hardness
(*c*) fatigue (*d*) creep

342. Which of the following is universally employed as the low expansion metal in the bimetallic thermometer, which is an iron-nickel alloy containing 36% nickel and has very low co-efficient of expansion (l/20th of ordinary metals) ?
(*a*) Invar (*b*) Constantan
(*c*) Chromel (*d*) Alumel

343. The high expansion metal normally used in the bimetallic thermometer is .... temperature measurement.
(*a*) brass.for low
(*b*) nickel alloys for high
(*c*) both '*a*' & *b*
(*d*) aluminium for all

Answers

332. (d)	333. (a)	334. (b)	335. (a)	336. (d)	337. (a)
338. (a)	339. (b)	340. (b)	341. (b)	342. (a)	343. (c)

344. Thermometer bulb & capillary in case of mercury filled pressure spring expansion thermometer is normally made of .... as mercury does not amalgamate with it.
(*a*) copper (*b*) copper alloys
(*c*) stainless steel (*d*) monel

345. In a plate type heat exchanger, heat transfer plates are never made of
(*a*) stainless steel (*b*) cast iron
(*c*) titanium (*d*) hastealloy

346. German silver is an alloy of copper, nickel and
(*a*) zinc (*b*) silver
(*c*) tin (*d*) lead

347. Softness of silver can be converted into hardness by alloving it with small quantity of
(*a*) copper & nickel
(*b*) zinc
(*c*) aluminium
(*d*) tin

348. Match the composition of various alloys of copper
List I
(*a*) Bell metal
(*b*) Muntz metal
(*c*) Admirability gun metal
(*d*) Aluminium bronze
List II
I. 90% Cu + 10% A1
II. 88% Cu + 10% Sn + 2% Zn
III. 60% Cu + 40% Zn
IV. 75% Cu + 25% Sn

349. Machinability of a material is the ease with which it can be cut or removed by cutting tools in machining operations. Match the machinability characteristics of various materials.
List I
(*a*) 18/8 stainless steel
(*b*) Low carbon steel
(*c*) Malleable cast iron and gun metal
(*d*) AI alloys and Zn alloys
List II
I. Good machinability
II. Excellent machinability
III. Poor machinability
IV. Not machinable

350. Match the uses of various materials.
List I
(*a*) Babbit alloy
(*b*) Constantan
(*c*) Hadfield manganese steel
(*d*) Low carbon steel
List II
I. Thermocouple
II. Bulldozer blades
III. Wire nails
IV. Bearing

351. Match the major characteristic constituent of various cutting tool materials.
List I,
(*a*) Diamond, (*b*) Stellite,
(*c*) High speed steel,
(*d*) Coated carbide tool,
List II
I. Nitride
II.Molybdenum
III. Cobalt
IV. Carbon

Answers

344. (c) 345. (b) 346. (a) 347. (a) 348. (iv) (iii) (ii) (i)
349. (iv) (iii) (i) (ii) 350. (iv) (i) (ii) (iii) 351. (iv) (iii) (ii) (i)

352. Match the field of applications of various materials.

List I

(*a*) Fibre reinforced plastics (FRP)

(*b*) Boron

(*c*) Engineering Ceramics

(*d*) Synthetic carbon

List II

I. Bearings

II. Aircraft industry

III. Nuclear reactor control rods

IV. Electrical insulators

353. Match the various tests conducted to determine the different properties of materials.

List 1

(*a*) Hardness test

(*b*) Impact test

(*c*) Tension test

(*d*) Fatigue test

List II

I. Endurance limit

II. Ductility

III. Toughness

IV. Penetration resistance

354. Match the applications of various polymeric materials.

List I

(*a*) Polyurathane

(*b*) Araldite

(*c*) Bake lite

(*d*) Neoprene

List II

I. Oil seal

II. Electrical switches

III. Adhesive

IV. Thermal insulator

355. Match the field of application of various materials.

List I

(*a*) Aluminium oxide

(*b*) Silicon nitride

(*c*) Silicon carbide

(*d*) Tungsten carbide

List II

I. Pipes for conveying liquid metals

II. Drawing dies

III. Abrasive wheels

IV. Heating elements

356. Match the properties required in a material application for various requirements.

List I

(*a*) Strength for no plastic deformation under static load

(*b*) Rigidity

(*c*) Reliability and safety

(*d*) Strength to withstand overload without fracture

List II

I. Toughness and impact resistance

II. Endurance limit and yield point

III. Elastic modulus and yield strength

IV. Yield point

357. Match the percentage of carbon present in steel used for making various products

List I

(*a*) Heavy duty leaf and coil springs

(*b*) Dies, taps and drills

(*c*) Crank shafts, axles & large forgings

(*d*) Cold rolled steel sheet

List II

I. 0.1　　II. 1

III. > 1　　IV. 0.45 – 0.55

Answers

352. (ii) (iii) (iv) (i)　353. (iv) (iii) (ii) (i)　354. (iv) (iii) (ii) (i)

355. (iii) (i) (iv) (ii)　356. (iv) (iii) (ii) (i)　357. (ii) (iii) (iv) (i)

358. In super conducting state, the materials are .. .. in nature.
   (*a*) paramagnetic (*b*) diamagnetic
   (*c*) ferromagnetic (*d*) none of these.

359. The ratio of cobalt, aluminium, nickel and iron in alnico (an alloy used for making permanent magnet) is
   (*a*) 10:20:20: 50 (*b*) 10:20:30:40
   (*c*) 20 : 30 : 10 : 40 (*d*) 50 : 20 : 20 : 10

360. Vanadium in high speed steels
   (*a*) increases toughness
   (*b*) forms very hard carbides and thus increases wear resistance
   (*c*) provides high hot hardness
   (*d*) promotes retention of austenite

361. .... steel has almost zero temperature co-efficient of thermal expansion
   (*a*) Invar (*b*) Stainless
   (*c*) Cobalt (*d*) Platinum

362. Slow plastic deformation of metals under a constant stress is known as creep, stress relaxation is a phenomenon in which ....on the part.
   (*a*) no loading is done
   (*b*) stress remains constant on increasing load
   (*c*) stress reduces on increasing the load
   (*d*) deformation tends to loosen the joint and produces a stress reduction.

363. Which of the following metals or combination of metals does not form the bases of the alloys known as superailoys?
   (*a*) Nickel base (*b*) Iron-nickel base
   (*c*) Cobalt base (*d*) Carbon base.

364. Which of the following is not a composite material?
   (*a*) Cermet
   (*b*) Fiber reinforced plastic (FRP)
   (*c*) Galvanised steel
   (*d*) Wood.

365. Pick out the wrong statement.
   (*a*) Cutting tool materials contain tungsten and vanadium.
   (*b*) Calorising is done to protect the steel exposed to oxidation at low temperature.
   (*c*) For easy magnetisation and demagnetisation, silicon is added to low carbon steel.
   (*d*) Two modes of plastic deformation are

366. Float of a rotameter can be made of
   (*a*) steel (*b*) wood
   (*c*) Plastic (*d*) all of these

367. Which of the following classes of stainless steel does not contain nickel?
   (*a*) Ferritic (*b*) Austinitic
   (*c*) Martensitic (*d*) None of these.

368. Monel contains copper and nickel in the ratio of
   (*a*) 1 : 1 (*b*) 1 : 2
   (*c*) 2 : 1 (*d*) 4 : 1

369. Inconel typically contains
   (*a*) 76% Fe, 7% Cr, .15%. Ni
   (*b*) 15% Fe, 7% Ni, 76% Cr
   (*c*) 76% Ni, 7% Fe, 15% Cr
   (*d*) None of these.

370. In addition to nickel, chromium and iron, Hastealloy contains
   (*a*) silicon (*b*) aluminium
   (*c*) niobium (*d*) molybdenum

Answers

358. (b)	359. (a)	360. (b)	361. (a)	362. (d)	363. (d)
364. (c)	365. (b)	366. (a)	367. (a)	368. (c)	369. (c)
370. (d)					

371. Which of the following metals is called metallic glass?
(*a*) Lead (*b*) Molybdenum
(*c*) Tantalum (*d*) Titanium

372. Which of the following stainless steels variety has the highest corrosion resistance?
(*a*) Type 304 (*b*) Type 304 L
(*c*) Type 321 (*d*) Type 316

373. The rate of corrosion of mild steel by sulfuric acid is unacceptably high for
(*a*) dilute acid
(*b*) acid at concentrations above 70%
(*c*) acid of intermediate concentrations
(*d*) both (*a*) and (*b*)

374. Material of construction for heat exchanger tubes handling sea water is
(*a*) mild steel
(*b*) stainless steel
(*c*) cupro-nickel alloys (70% Cu, 30% Ni)
(*d*) aluminium.

375. Which one of the following materials contains copper ?
(*a*) SS 321 (*b*) Inconel 0.05%
(*c*) Monel (*d*) Karbate

376. The corrosion resistance of stainless steel is due to the
(*a*) formation of a thin oxide film on the surface.
(*b*) formation of graphitic carbon on the surface.
(*c*) formation of an inert Fe-C-Fe type complex chain.
(*d*) none of these.

377. Stainless steel is particularly susceptible to stress corrosion cracking in the presence of
(*a*) dilute H_2SO_4
(*b*) concentrated HNO_3
(*c*) dilute phosphoric acid.
(*d*) chloride solution.

378. Under oxidizing environment, the use of mild steel as material of construction is limited to temperatures below .... °C.
(*a*) 100 (*b*) 200
(*c*) 500 (*d*) 1000

379. Impellers of 98% sulfuric acid pump are made of
(*a*) mild steel
(*b*) cast iron
(*c*) lead
(*d*) illium (a nickel-chrome alloy).

380. For transportation of 98% sulfuric acid at temperatures above 50°C .... pipes are preferred.
(*a*) cast iron
(*b*) mild steel
(*c*) galvanized iron.
(*d*) (*a*), (*b*), (*c*) are equally useful.

381. Material of construction of vessels handling boiling nitric acid of any concentration is
(*a*) lead (*b*) cast iron
(*c*) mild steel (*d*) high-silicon iron.

382. Karbate is
(*a*) an alloy of copper and nickel.
(*b*) chemically impregnated carbon.
(*c*) high-silicon iron.
(*d*) an alloy of copper, tin and zinc.

383. The corrosion resistance of titanium is due to
(*a*) the formation of an oxide film as in stainless steel.
(*b*) inert titanium, which is not attacked by chemicals.
(*c*) the formation of titanium carbide, which is intert
(*d*) some hitherto unknown reason.

Answers

371. (c)	372. (d)	373. (d)	374. (c)	375. (c)	376. (a)
377. (d)	378. (c)	379. (d)	380. (a)	381. (d)	382. (b)
383. (a)					

384. Which of the following components present in stainless steel is called an austinite former?
(*a*) Niobium (*b*) Chromium
(*c*) Molybdenum (*d*) Nickel

385. Which of the following is called ferrite former ?
(*a*) Nickel (*b*) Niobium
(*c*) Molybdenum (*d*) Chromium

386. When the carbon content of iron-chromium alloys is increased, the chromium content above which the steel no longer becomes austenitic on heating
(*a*) increases.
(*b*) decreases.
(*c*) remains unchanged.
(*d*) unpredictable.

387. Sensitization in stainless steel is the combining of
(*a*) chromium with carbon along grain boundaries to form chromium carbides which give rise to intergranular corrosion.
(*b*) carbon with hydrogen to produce methane which reduces the ductility.
(*c*) nickel with carbon to form nickel carbide that improves the corrosion resistance of stainless steel in the presence of chloride ions.
(*d*) none of these.

388. Presence of ammonia can readily cause stress corrosion cracking of
(*a*) nickel. (*b*) copper.
(*c*) stainless steel 321.
(*d*) zirconium.

389. Which of the following performs best under reducing conditions?
(*a*) Stainless steel
(*b*) Aluminium
(*c*) Titanium
(*d*) Copper-base alloys

390. Copper-base alloys are resistant to
(*a*) nitric acid of any concentration.
(*b*) chromic acid of any concentration.
(*c*) most alkaline solutions at room temperatures.
(*d*) aqueous ammonia solutions.

391. Which of the following materials will **not** corrode aluminium?
(*a*) 90% sulfuric acid.
(*b*) 20% hydrochloric acid.
(*c*) 20% caustic soda solution.
(*d*) 90% nitric acid.

392. The alloy of composition 70% nickel, 28% molybdenum and 2% iron is called
(*a*) Monel alloy No. 400
(*b*) Hastealloy - B
(*c*) Carpenter alloy 20
(*d*) Hastalloy C

393. Hastalloy B is generally suitable as a material of construction for .... conditions.
(*a*) strongly oxidizing (*b*) reducing
(*c*) both oxidizing and reducing
(*d*) none of these

394. Hastelloy C has a composition of
(*a*) 60% nickel, 17% molybdenum, 15% chromium, 5% tungsten and 5% iron.
(*b*) 60% nickel, 40% zirconium.
(*c*) 30% nickel, 40% molybdenum, 30% iron.
(*d*) 60% vanadium, 17% molybdenum, 15% chromium, 8% iron.

Answers

384. (d)	385. (d)	386. (a)	387. (a)	388. (b)	389. (d)
390. (c)	391. (d)	392. (b)	393. (b)	394. (a)	

395. KBI-40 is an alloy of
(a) chromium and nickel.
(b) molybdenum and vanadium.
(c) columbium and iron.
(d) tantalum and columbium.

396. Of the two alloys, Hastelloy B and Hastelloy C,
(a) Hastelloy C has better resistance under reducing conditions.
(b) Hastelloy C has better resistance oxidizing conditions.
(c) Hastelloy B has better resistance under oxidizing conditions.
(d) None of these.

397. Addition of columbium to stainless steel improves its
(a) oxidation resistance.
(b) creep strength.
(c) machinability.
(d) pitting resistance.

398. Which of the following chemicals attacks equipment made of borosilicate glass?
(a) 10% caustic soda solution at 80°C.
(b) 85% phosphoric acid at 110°C.
(c) Hydrofluoric acid.
(d) All (a), (b) and (c).

399. Which of the following materials is the best in handling hydrochloric acid as well as caustic soda solution of all concentrations and at temperature upto boiling point ?
(a) Lead (b) Titanium
(c) Tantalum (d) Zirconium

400. The susceptibility of stainless steel to stress corrosion cracking is very dependent on the nickel content and is a maximum at .... percent nickel.
(a) 2 (b) 8
(c) 20 (d) 40

401. Which of the following stainless steels is considered immune to stress corrosion cracking under most circumstances ?
(a) Stainless steel Type 304.
(b) Stainless steel Type 316.
(c) Alloy 20. (d) Alloy 825.

402. Which of the following materials is/are resistant to moist sulfur dioxide at 25 °C ?
(a) Hastelloy alloy B-2.
(b) Stainless Steel Alloy 216.
(c) Monel.
(d) All (a), (b) and (c)

403. Material of construction for processes involving hydrogen under high pressures and temperatures should be selected carefully. For processes involving hydrogen under high pressures, carbon steel has severe limitation above about 200°C. The most suitable alloying elements offering resistance to high temperature hydrogen attack (under elevated pressures) to stainless steel are
(a) nickel.
(b) molybdenum/chromium.
(c) silicon
(d) All (a), (b) and (c).

404. The well known "silver steel" is a/an
(a) carbon steel containing 0.1% carbon.
(b) stainless steel containing 50% silver.
(c) carbon steel containing 0.9% carbon.
(d) alloy containing 90% silver, 1% carbon and 9% iron.

405. As the carbon content in steel increases, welding
(a) becomes progressively more difficult.
(b) becomes progressively easier.
(c) carbon content has nothing to do with the ease of welding.
(d) unpredictable.

Answers

395. (d)	396. (c)	397. (b)	398. (d)	499. (d)	400. (b)
401. (d)	402. (b)	403. (b)	404. (c)	405. (a)	

406. Creep strength is
   (*a*) another name of tensile strength.
   (*b*) another name of yield strength.
   (*c*) time dependent strain occurring under stress.
   (*d*) yield strength at elevated temperatures, greater than 1000°C.

407. Which of the following materials cannot be used to handle liquefied natural gas at 160°C?
   (*a*) Carbon steel.
   (*b*) Aluminium-magnesium alloys.
   (*c*) Austenitic stainless steels.
   (*d*) Copper alloys.

408. For situations where sea water is used as a coolant, heat exchanger tubes are made of
   (*a*) carbon steel.
   (*b*) stainless steel type 316
   (*c*) cupro-nickel alloys (70 Cu, 30 Ni).
   (*d*) aluminium.

409. Polytetrafluoroethylene, known under the trade names Teflon and Fluon, is resistant to all chemicals, except
   (*a*) molten alkalies and fluorine.
   (*b*) moderately concentrated sulfuric and nitric acid.
   (*c*) formic, acetic and higher carboxylic acids.
   (*d*) saturated brines at temperatures of 50 to 80°C.

410. The creep strength of a material is usually reported as the stress to cause rupture in .... hours at the test temperature.
   (*a*) 1 (*b*) 100
   (*c*) 1000 (*d*) 100,000

411. Brittle materials are
   (*a*) weak in tension but strong in compression.
   (*b*) strong in tension but weak in compression.
   (*c*) weak in tension as well as in compression.
   (*d*) strong in tension as well as in compression.

412. Stiffness is a measure of a material's resistance to
   (*a*) wear.
   (*b*) bending and buckling.
   (*c*) crack propagation.
   (*d*) both (*a*) and (*c*).

413. In sulfuric acid manufacture, sulfur trioxide is absorbed in 98-99% sulfuric acid in a packed absorber. The material of construction of the absorber is usually
   (*a*) copper and copper base alloys.
   (*b*) aluminium.
   (*c*) brick-lined steel or cast iron.
   (*d*) nickel and nickel base alloys.

414. The packing particles in sulfur trioxide absorber are usually made of
   (*a*) carbon steel.
   (*b*) stainless steel Type 316.
   (*c*) ceramic material.
   (*d*) alloy B-2.

415. Which of the following materials creeps under its own weight ?
   (*a*) Aluminium (*b*) Zinc
   (*c*) Copper (*d*) Lead

416. Concentrated hydrochloric acid is transported and stored in .... tanks.
   (*a*) carbon steel.
   (*b*) stainless steel Type 316.
   (*c*) rubber-lined.
   (*d*) aluminium.

Answers

406. (c)	407. (a)	408. (c)	409. (a)	410. (d)	411. (a)
412. (b)	413. (c)	414. (c)	415. (d)	416. (c)	

417. Chemical inhibitor that minimizes caustic embrittlement of industrial boilers is
(*a*) sodium nitrate.
(*b*) sodium sulfate.
(*c*) calcium chloride.
(*d*) ferric sulfate.

418. Admiralty brass is an alloy of 70% copper, 29% zinc and 1% tin. The alloy also contains about 0.5% arsenic. The presence of arsenic in Admiralty brass
(*a*) improves its corrosion resistance.
(*b*) improves its mechanical properties.
(*c*) prevents dezincification.
(*d*) prevents fatigue failure.

419. The addition of zinc to copper
(*a*) improves its mechanical properties.
(*b*) lowers its corrosion resistance in most chemical environments.
(*c*) improves its pitting resistance.
(*d*) improves its corrosion resistance in most chemical environments.

420. Which of the following pairs are correctly matched ?
1. Resilience .... Resistance to deformation.
2. Malleability .... shape change.
3. Creep .... Progressive deformation.
4. Plasticity .... Permanent deformation.
Select the correct answer using the codes given below the lists:
Codes :
(*a*) 2, 3 and 4 (*b*) 1, 2 and 3
(*c*) 1, 2 and 4 (*d*) 1, 3 and 4

421. Which of the following pairs are correctly matched ?
1. Silicon steels .... Transformer stamping.
2. Duralumin .... Cooking utensils.
3. Gun metal .... Bearings.
Select the correct answer using the codes given below :
Codes
(*a*) 1, 2 and 3 (*b*) 1 and 2
(*c*) 1 and 3 (*d*) 2 and 3

422. Cubic boron nitride is used
(*a*) as lining material in induction furnace
(*b*) for making optical quality glass
(*c*) for heat treatment
(*d*) for none of the above.

423. Which one of the following material is used as the bonding material for grinding wheel?
(*a*) Silicon carbide (*b*) Sodium silicate
(*c*) Boron carbide (*d*) Aluminium oxide

424. The fatigue failure of a material can be avoided by
(*a*) coating the surface
(*b*) shot peening
(*c*) nitriding
(*d*) any of the above

425. Presence of sulphur in cast iron
(*a*) promotes formation of graphite
(*b*) reduces fluidity during melting
(*c*) increases strength
(*d*) reduces oxidation.

426. Which of the following alloys does not contain copper ?
(*a*) Phosphor bronze (*b*) Invar
(*c*) Brass (*d*) Muntz metal

427. With increasing percentage of carbon in steel, its .... decreases.
(*a*) ductility
(*b*) ultimate strength
(*c*) hardness
(*d*) percentage of pearlite

Answers

417. (a)	418. (c)	419. (a)	420. (d)	421. (a)	422. (d)
423. (b)	424. (d)	425. (b)	426. (b)	427. (a)	

428. A ductile fracture of material is usually not preceded by
   (a) noise
   (b) deformation
   (c) plastic flow
   (d) large amount of non-recoverable energy absorption.

429. Which of the following engineering materials is the most suitable candidate for hot chamber die casting ?
   (a) Low carbon steel
   (b) Titanium
   (c) Copper
   (d) Tin

430. The material property which depends only on the basic crystal structure is the
   (a) fatigue strength *(b)* work hardening
   (c) fracture strength *(d)* elastic constant

Answers

428. (d) 429. (b) 430. (d)

9 Fuel Technology

1. When coal is heated in absence of air, it is called its
   (a) gasification (b) deoxidation
   (c) coalification (d) carbonization

2. Higher grindability index of coal implies its
   (a) easy pulverising nature.
   (b) difficult pulverising nature.
   (c) higher power consumption in grinding.
   (d) both (b) & (c).

3. Washing of coal
   (a) reduces its sulphur & ash content.
   (b) improves its caking properties & calorific value.
   (c) controls its ash fusibility.
   (d) all (a), (b) & (c).

4. Low temperature oxidation of coal resulting from bad storage conditions does not decrease its
   (a) caking power
   (b) calorific value
   (c) hydrogen content
   (c) oxygen content

5. Spontaneous combustion of coal on storage results due to
   (a) inadequate ventilation.
   (b) low temperature oxidation.
   (c) storage in large heaps with small surface to volume ratio.
   (d) ail (a) (b)& (c).

6. Fusion point of coal ash increases with increase in its .... content.
   (a) iron carbonate
   (b) iron sulphide
   (c) lime and magnesia
   (d) alumina and silica

7. Proximate analysis of coal determines, its moisture. .... content.
   (a) ash. sulphur & volatile matter.
   (b) ash, fixed carbon, & volatile matter.
   (c) sulphur, nitrogen, & fixed carbon.
   (d) phosphorous, sulphur & nitrogen.

8. Ultimate analysis of coal determines, its carbon, ... content.
   (a) hydrogen, nitrogen, sulphur, & oxygen
   (b) ash, sulphur, nitrogen, & calorific value
   (c) sulphur, volatile matter, & ash
   (d) sulphur, volatile matter, ash & moisture.

Answers

1. (d)	2. (a)	3. (d)	4. (d)	5. (d)	6. (d)
7. (b)	8. (a)				

9. A good quality coal should have
   (*a*) high ash content.
   (*b*) low fusion point of its ash.
   (*c*) high sulphur content.
   (*d*) none of these.

10. Fixed carbon in coal is defined as
   (*a*) that present in volatile matters.
   (*b*) the total quantity of carbon present in coal.
   (*c*) hundred minus the percentage of volatile matter, ash & moisture.
   (*d*) the one which is present in the residue after combustion.

11. Coalification means
   (*a*) the process of conversion of lignite into anthracite.
   (*b*) underground gasification of coal,
   (*c*) complete combustion of coal.
   (*d*) direct hydrogenation of coal.

12. Calorific value of bituminous coal is about .... kcal/kg.
   (*a*) 2000 (*b*) 5000
   (*c*) 7000 (*d*) 10000.

13. Anthracite coal
   (*a*) contains more volatile matter than bituminous coal.
   (*b*) ignites more easily than bituminous coal.
   (*c*) is essentially a coking coal.
   (*d*) burns with short bluish, yellow tipped flame producing very little smoke.

14. Bituminous coal
   (*a*) ignites less easily than anthracite.
   (*b*) is generally coking.
   (*c*) burns with smoky yellow flame.
   (*d*) both (*b*) & (*c*).

15. Which of the following petrographic constituents is responsible for bright & lustrous black band of bituminous coal?
   (*a*) Vitrain (*b*) Clarain
   (*c*) Durain (*d*) Fussain

16. A coal containing very high percentage of durain is called .... coal.
   (a.) bright (*b*) splint
   (*c*) non-banded (*d*) boghead

17. To avoid fire by spontaneous combustion of coal due to its low temperature oxidation, it should be stored in
   (*a*) shallow & small piles.
   (*b*) fine sizes without the presence of any lump.
   (*c*) closed space without any ventilation facility.
   (*d*) large heaps with small surface to volume ratio.

18. A coal having high amount of'volatile matter
   (*a*) requires smaller combustion chamber.
   (*b*) produces less tar & gas on carbonisation.
   (*c*) ignites easily & burns with long smoky flame.
   (*d*) all (*a*), (*b*) & (*c*).

19. A coal with high .... content would ignite most easily.
   (*a*) fixed carbon (*b*) volatile matter
   (*c*) ash (*d*) oxygen

20. A coal with high ash content is undesirable, as
   (*a*) it is abrasive to the coal pulveriser (*e.g.* ball mill) and the combustion chamber.
   (*b*) the ash in molten condition gets absorbed in the pores of the refractory lining of the furnace & causes its spalling.
   (*c*) the ash retains the S & P and thus affects the quality of products in metallurgical furnaces.
   (*d*) all (*a*), (*b*) & (*c*).

Answers

9. (d)	10. (c)	11. (a)	12. (c)	13. (d)	14. (d)
15. (a)	16. (b)	17. (a)	18. (d)	19. (b)	20. (d)

21. Caking coal is desirable for
    (a) burning on travelling grate
    (b) burning on fireballs
    (c) coke making
    (d) injection in B.F.

22. As the equilibrium moisture content of coal increases, its
    (a) calorific value increases.
    (b) swelling number increases.
    (c) swelling number decreases.
    (d) bulk density decreases.

23. Sulphur in metallurgical coal
    (a) contributes to its heating value.
    (b) affects the quality of steel produced as cracks develop on the surface while hot rolling the steel (hot shortness).
    (c) both (a) & (b).
    (d) neither (a) nor (b).

24. Presence of phosphorous in metallurgical coal
    (a) is not undesirable.
    (b) reduces its calorific value.
    (c) badly affects (causes cold shortness) the quality of steel.
    (d) increases its caking power.

25. Nitrogen in coal
    (a) is present upto 1 - 3%.
    (b) comes from protein in parent vegetable matter.
    (c) is recovered as NH_3 during its carbonization.
    (d) all (a), (b) & (c).

26. Pick out the correct statement.
    (a) Oxygen content decreases from lignite to bituminous coal, as the coalification increases.
    (b) The less the oxygen content better is the coal, as it reduces the calorific value.
    (c) With increase in oxygen content, moisture holding capacity of coal increases and the caking power decreases.
    (d) all (a), (b) & (c).

27. Pick out the wrong statement.
    (a) Pulverised fuel can be completely burnt with less percentage of excess air compared to lump coal.
    (b) Generally, high volatile matter coals are more suitable for making pulverised fuel.
    (c) Regulation of furnace temperature & atmosphere (oxidising or reducing) is easily possible with pulverised fuel firing.
    (d) all (a), (b) & (c).

28. Anthracite is used for
    (a) recarbonising steel.
    (b) making carbon electrodes.
    (c) blending with prime coking coal.
    (d) all (a), (b) & (c).

29. Which is the most matured coal ?
    (a) Lignite (b) Bituminous
    (c) Semi-anthracite (d) Anthracite

30. Which will have the least volatile matter content and, hence will be the most difficult to ignite ?
    (a) Bituminous coal.
    (b) Anthracite.
    (c) Coke.
    (d) Sub-bituminous coal.

31. A good coking coal should have high .... content.
    (a) ash (b) S & P
    (c) moisture (d) none of these

Answers

21. (c)	22. (c)	23. (c)	24. (c)	25. (d)	26. (d)
27. (d)	28. (d)	29. (d)	30. (c)	31. (d)	

32. A coal having high volatile matter content will
    (*a*) give less yield of tar & gas on carbonization.
    (*b*) burn with a small non-smoky flame.
    (*c*) have a very high calorific value.
    (*d*) none of these.

33. Caking index of the coal blend used for making metallurgical coke is about
    (*a*) 12 (*b*) 21
    (*c*) 37 (*d*) 49

34. High ash coals
    (*a*) are soft & friable.
    (*b*) require larger time of carbonization as ash offers resistance to heat transfer.
    (*c*) produce larger quantity of coke oven gas.
    (*d*) none of these.

35. Swelling index of coal is a measure of its
    (*a*) abradability (*b*) grindability
    (*c*) agglutinating properties
    (*d*) swelling characteristics

36. Caking index of coal is a measure of its
    (*a*) reactivity.
    (*b*) agglutinating (binding) properties.
    (*c*) porosity.
    (*d*) abradability.

37. Calorific value of coal middling generated in coal washeries during washing of coking coal is about .... kcal/kg.
    (*a*)2000 (*b*) 4000
    (*c*) 7500 (*d*) 10000

38. Gross & net calorific value of a fuel will be the same,
    (*a*) if its ash content is zero.
    (*b*) if its carbon content is very low.
    (*c*) if its hydrogen/hydrogen compounds content is zero.
    (*d*) under no circumstances.

39. With increase in moisture content of coal, its
    (*a*) calorific value increases sometimes.
    (*b*) bulk density always decreases.
    (*c*) clinkering tendency during combustion increases.
    (*d*) none of these.

40. Low temperature oxidation and spontaneous combustion of freshly mined coal is accentuated, if
    (*a*) it contains large amount of volatile matter.
    (*b*) it is stored in tall heaps.
    (*c*) smaller fines are stored in large quantity.
    (*d*) all (*a*), (*b*) & (*c*).

41. With increase in the oxygen content of coal, its .... decreases.
    (*a*) calorific value
    (*b*) caking power
    (*c*) both (*a*) & (*b*)
    (*d*) neither (*a*) nor (*b*)

42. Ignition temperature decreases progressively from anthracite to lignite, because
    (*a*) volatile matter content increases.
    (*b*) carbon content decreases.
    (*c*) moisture content increases.
    (*d*) ash content increases.

43. Coke compared to the coal from which it has been made contains
    (*a*) less volatile matter.
    (*b*) more carbon.
    (*c*) greater percentage of ash.
    (*d*) all (*a*), (*b*) & (*c*).

Answers

32. (d)	33. (b)	34. (b)	35. (d)	36. (b)	37. (b)
38. (c)	39. (c)	40. (d)	41. (d)	42. (a)	43. (d)

44. Which is not a manufactured fuel ?
    (a) Furnace oil (b) Bagasse
    (c) Semi-coke (d) Kerosene

45. Most of the coking coals are
    (a) anthracite coal (b) bituminous coal
    (c) lignite (d) none of these

46. Highly caking coals
    (a) produce weak coke.
    (b) produce strong coke.
    (c) may damage the coke oven walls due to excessive swelling during carbonisation.
    (d) both (b) & (c).

47. Presence of free moisture in coal during its high temperature carbonization
    (a) reduces the coking time.
    (b) protects the volatile products from pyrolysis (cracking) in the presence of hot coke & oven walls.
    (c) increases the loss of fine coal dust from the ovens during charging.
    (d) none of these.

48. Higher ash in coal for metallurgical coke making
    (a) increases the hardness of coke.
    (b) decreases the abrasion resistance of coke.
    (c) causes brittleness in steel.
    (d) none of these.

49. With increase in carbonisation temperature of coal,
    (a) gas yield increases.
    (b) tar yield increases.
    (c) hydrogen percent in coke oven gas decreases.
    (d) methane percent in coke oven gas increases,

50. Increase in ash content of blast furnace coke
    (a) reduces its specific consumption in the furnace.
    (b) increases its specific consumption in the furnace.
    (c) does not affect its specific consumption in the furnace.
    (d) decreases its hardness & abrasion resistance.

51. With increase in the time of carbonisation of coal at a particular temperature (say 1000°C), the ....percentage in coke oven gas increases.
    (a) hydrogen
    (b) methane
    (c) unsaturated hydrocarbons
    (d) all (a), (b) & (c)

52. Which is the most reactive fuel ?
    (a) Blast furnace coke.
    (b) Low temperature coke.
    (c) Anthracite.
    (d) Sub-bituminous coal.

53. Net Calorific value (C.V.) of coke oven gas is about .... kcal/Nm^3.
    (a) 900 (b) 4200
    (c) 1900 (d) 7500

54. Percentage of hydrogen in coke oven gas is around
    (a) 1 (b) 5
    (c) 25 (d) 55

55. Percentage of methane in coke oven gas is about
    (a) 1 (b) 10
    (c) 25 (d) 55

56. Coke oven gas compared to blast furnace gas is
    (a) more explosive & inflammable.
    (b) less poisonous.
    (c) lighter.
    (d) all (a) , (b) &(c) .

Answers

44. (b)	45. (b)	46. (d)	47. (b)	48. (a)	49. (a)
50. (b)	51. (a)	52. (b)	53. (b)	54. (d)	55. (c)
56. (d)					

57. The main product of high temperature carbonisation of coal is
   (*a*) coke (*b*) coke oven gas
   (*c*) tar (*d*) ammonia

58. Quantity of coke oven gas produced by high temperature carbonisation of one ton of dry coal is about .... Nm^3.
   (*a*)30 (*b*) 300
   (*c*) 3000 (*d*) 750

59. High temperature carbonization of coal takes place at .... °C.
   (*a*) 2100 (*b*) 600
   (*c*) 1100 (*d*) 1600.

60. Low temperature carbonisation of coal takes place at .... °C.
   (*a*)300 (*b*) 1100
   (*c*) 700 (*d*) 1500.

61. High temperature carbonisation of coal compared to its low temperature carbonisation produces
   (*a*) inferior coke (*b*) less of gases
   (*c*) less of tar (*d*) none of these

62. Metallurgical coke should
   (*a*) have very high porosity.
   (*b*) be brittle.
   (*c*) have low fusion point of its ash.
   (*d*) contain large quantity of volatile matter, moisture & phosphorous.

63. Which has the least percentage of ash ?
   (*a*) Petroleum coke (*b*) Beehive coke
   (*c*) Foundry coke (*d*) Metallurgical coke

64. Gray King Assay & Roga index of coal is a measure of its .... properties.
   (*a*) abradability
   (*b*) grindability
   (*c*) weathering
   (*d*) caking & swelling

65. Sometimes, water is sprayed during coal charging in the coke oven, which helps in
   (*a*) prevention of clinker formation in the oven.
   (*b*) controlling the dust nuisance, while charging the coal.
   (*c*) reducing the cracking of hydrocarbons in the coke oven gas.
   (*d*) all (*a*), (*b*) & (*c*).

66. Low temperature carbonisation of coal
   (*a*) is mainly for the production of smokeless domestic coke.
   (*b*) is meant for the production of metallurgical coke.
   (*c*) produces higher quantity of coke oven gas than high temperature carbonisation.
   (*d*) produces less quantity of tar than high temperature carbonisation.

67. In high temperature carbonisation as compared to low temperature carbonization of coal,
   (*a*) yield of ammonia is less.
   (*b*) aromatic content of tar is low.
   (*c*) H_2 content in the coke oven gas is more.
   (*d*) calorific value of the coke oven gas is lower.

68. During the high temperature carbonisation of coal,
   (*a*) all tar is evolved at < 600°C.
   (*b*) evolution of H_2 and formation of CH4 & aromatics occurs at > 700°C.
   (*c*) hard semi-coke starts shrinking at 600°C.
   (*d*) all (*a*), (*b*) & (*c*).

Answers

57. (a)	58. (b)	59. (c)	60. (c)	61. (c)	62. (c)
63. (a)	64. (d)	65. (d)	66. (a)	67. (d)	68. (d)

69. The cooling medium used in dry quenching of coke is
   (a) nitrogen (b) air
   (c) carbon dioxide (d) phenolic water

70. Width of the coke oven towards coke side is slightly more than that on pusher side to
   (a) facilitate easy discharging of coke as it swells during coking.
   (b) facilitate uniform heating of coke oven.
   (c) increase the output of the coke.
   (d) none of these.

71. Traces of tar fog present in the coke oven gas is removed by a/an
   (a) cyclone separator.
   (b) wet packed scrubber.
   (c) electrostatic precipitator.
   (d) washing with monoethanol amine (MEA).

72. High ash containing coke
   (a) produces more slag in the blast furnace.
   (b) has poor strength & hardness.
   (c) is desirable in producer gas manufacture.
   (d) is very soft.

73. High amount of sulphur & phosphorous in coke causes,
   (a) decrease in its calorific value.
   (b) increase in its strength.
   (c) brittleness of steel made by using it.
   (d) none of these.

74. High rate of heating of coke ovens
   (a) may damage its wall, due to abrupt excessive swelling of coal.
   (b) produces larger size coke,
   (c) increases the time of carbonization.
   (d) none of these.

75. Tar yield in high temperature carbonization of coal is .... percent of dry coal charged.
   (a) 1 (b) 3
   (c) 7 (d) 11

76. Coke yield in high temperature carbonisation is .... percent of dry coal charged.
   (a) 17 (b) 95
   (c) 77 (d) 37.

77. Ammonia content of raw coke oven gas is about .... gm/Nm^3 .
   (a) 1 (b) 15
   (c) 150 (d) 1500

78. Calorific value of coke oven gas produced by low temperature carbonisation of coal is about .... kcal/Nm^3.
   (a) 4000 (b) 2500
   (c) 6500 (d) 10000

79. Yield of coke oven gas in low temperature carbonisation of coal is about .... Nm^3 /ton dry coal charged.
   (a) 60 (b) 160
   (c) 500 (d) 750

80. Tar yield in low temperature carbonisation of coal is .... percent of dry coal charged.
   (a) 3 (b) 5
   (c) 10 (d) 20

81. Coke ovens in steel plants are heated
   (a) electrically (b) by coke oven gas
   (c) by mixed gas (d) either (b) or (c)

82. Micum-10 (M_{10}) of coke is the measure of its
   (a) hardness (b) strength
   (c) porosity (d) size

Answers

69. (a)	70. (a)	71. (c)	72. (a)	73. (c)	74. (a)
75. (b)	76. (c)	77. (b)	78. (c)	79. (b)	80. (c)
81. (d)	82. (a)				

83. Micum-40 (M_{40}) of coke is the measure of its
    (a) hardness *(b)* strength
    (c) reactivity *(d)* none of these

84. Coke oven gas is a better fuel than blast furnace gas, because of its higher
    (a) calorific value, cleanliness and relatively low distribution cost (due to its low specific gravity).
    (b) adiabatic flame temperature.
    (c) heat release rate (thus requiring smaller combustion chamber).
    (d) all *(a)*, *(b)* & *(c)*.

85. Density of coke oven gas is about .... kg/Nm^3, which implies that it is lighter than air.
    (a) 0.45 *(b)* 1.3
    (c) 0.01 *(d)* 2.5

86. Ash content in the coke produced from coking coal having 20 percent ash may be around .... percent.
    (a) 5 *(b)* 11
    (c) 17 *(d)* 24

87. Coking time in a commercial high temperature by product coke oven is about .... hours.
    (a) 6 *(b)* 18
    (c) 28 *(d)* 36

88. Coking time in beehive coke oven is about
    (a) 12 hours *(b)* 2-3 days
    (c) one week *(d)* two weeks

89. Calorific value of blast furnace gas is about .... kcal/Nm^3.
    (a) 800 *(b)* 1800
    (b) 4200 *(d)* 6500

90. Blast furnace gas is a very poisonous gas due to its high .... content.
    (a) CO_2 *(b)* CO
    (c) N_2 *(d)* H_2S

91. Percentage of carbon monoxide in blast furnace gas is about
    (a) 1 *(b)* 22
    (c) 1 *(d)* 55

92. Calorific value of L.D. converter gas is about .... kcal/Nm^3.
    (a) 800 *(b)* 1800
    (c) 3600 *(d)* 6200

93. Average carbon monoxide content in converter gas may be about .... percent.
    (a) 35 *(b)* 65
    (c) 95 *(d)* 10

94. Which of the following is the most poisonous gas ?
    (a) Converter gas *(b)* Blast furnace gas
    (c) Coke oven gas *(d)* Natural gas

95. Yield of blast furnace gas is about .... Nm^3/ton pig iron.
    (a) 290—350 *(b)* 2000 —2200
    (c) 3500—4000 *(d)* 5000 —7000

96. Which of the following fuel gases is heavier than air ?
    (a) Coke oven gas
    (b) Natural gas
    (c) Blast furnace gas
    (d) Water gas

97. Blast furnace gas burns with a bluish flame, because of the presence of
    (a) CO *(b)* CO_2
    (c) CH_4 *(d)* N_2.

Answers

83. (b)	84. (d)	85. (a)	86. (d)	87. (b)	88. (b)
89. (a)	90. (b)	91. (b)	92. (b)	93. (b)	94. (a)
95. (b)	96. (c)	97. (a)			

98. Coke oven gas burns with an yellowish flame because of the presence of
(a) CO_2 (b) CH_4
(c) H2 (d) N_2

99. Blast furnace gas constitutes mainly of
(a) N_2 & CH4 (b) N_2& H_2
(c) N_2 & CO (d) CH4 & CO_2

100. Volatile matter content of metallurgical coke may be around .... percent.
(a) 0.5 – 2 (b) 6 – 7
(c) 10 – 15 (d) 16 – 20.

101. Fuel gases containing hydrocarbons (*e.g.* coke oven gas) are not preheated before burning, mainly because
(a) the hydrocarbons crack, thereby chok-ing & fouling the heat transfer surface by carbon soot.
(b) it reduces its calorific value tremen-dously.
(c) it reduces its flame temperature tremendously.
(d) there are chances of explosion during preheating.

102. Coke having higher porosity has
(a) lower bulk density.
(b) lower strength,
(c) higher reactivity.
(d) all (a), (b) & (c).

103. Preheating of coal charge for the coke ovens reduces the
(a) coking time.
(b) yield of gas & tar.
(c) fuel consumption in coking.
(d) all (a), (b) & (c).

104. Beehive coke oven
(a) facilitates by products recovery.
(b) take 2-3 days for coking, but requires no external fuel for heating .
(c) gives larger yield of coke (about 85%) as compared to by-product ovens..
(d) produces coke with very poor strength.

105. Size of blast furnace grade coke is .... mm.
(a) 0—15 (b) 15—25
(c) 25—80 (d) >80

106. Micum index of coke is a direct measure of its
(a) reactivity
(b) porosity
(c) bulk density
(d) strength & hardness

107. Desirable 'Micum Index' values of metal-lurgical coke are
(a) $M_{40} > 78$ & $M_{10} < 10$.
(b) $M_{40} > 4$ & $M_{10} < 80$.
(c) $M_{40} > 10$ & $M_{10} < 78$.
(d) $M_{40} > 98$ & $M_{10} < 2$.

108. In high temperature carbonisation as com-pared to low temperature carbonization of coal,
(a) coke oven gas yield is more.
(b) tar yield is less but free carbon in tar is more.
(c) calorific value of coke oven gas is less.
(d) all (a), (b) & (c).

109. Shatter index of coke is the measure of its
(a) strength (b) hardness
(c) bulk density (d) reactivity

Answers

98. (b)	99. (c)	100. (a)	101. (a)	102. (d)	103. (d)
104. (b)	105. (c)	106. (d)	107. (a)	108. (d)	109. (a)

110. Shatter index of metallurgical coke on 2 inches & 0.5 inch screen should be respectively about .... percent.
(*a*) 80 & 97 (*b*) 95 & 100
(*c*) 40 & 70 (*d*) 25 & 97

111. Abrasion index of blast furnace coke should be around .... percent.
(*a*) 20 (*b*) 35
(*c*) 55 (*d*) 80

112. Height of the coke oven is limited (say maximum upto 7 metres) mainly by the
(*a*) problem of uniform heating along its height.
(*b*) structural strength of silica bricks.
(*c*) problem in door cleaning.
(*d*) buckling of ram of pusher car at the time of coke pushing.

113. lonisation potential applied across the electrodes of electrostatic tar precipitator is around
(*a*) 230 VAC (*b*) 60 k VDC
(*c*) 230 V DC (*d*) 60 k VAC

114. Benzol comprises mainly of
(*a*) benzene, toluene & xylene.
(*b*) tar & creosote.
(*c*) ammonia & phenol.
(*d*) anthracite & phenol.

115. Which fertiliser is produced in the by-products recovery plant attached to a steel plant (from coke oven gas) ?
(*a*) Ammonium phosphate
(*b*) Ammonium nitrate
(*c*) Ammonium sulphate
(*d*) Calcium ammonium nitrate (CAN)

116. Naphthalene is removed from coke oven gas by
(*a*) scrubbing with wash oil.
(*b*) adsorbing on bog iron bed.
(*c*) absorbing in vetro coke solution.
(*d*) adsorption on silica gel.

117. Coke oven gas is stripped of its H2S and mercaptans content by
(*a*)absorbing it in wash oil.
(*b*)adsorption on bog iron bed.
(*c*) bubbling it through dilute sulphuric acid.
(*d*) adsorption on silica gel.

118. Which is not a product of tar distillation ?
(*a*) Phenol & naphthalene
(*b*) Benzol & pitch
(*c*) Anthracene & creosote
(*d*) Petrolatum

119. Benzene is used
(*a*) as a motor fuel.
(*b*) as an explosive.
(*c*) for making insecticides (*e.g.* DDT, BHC etc.) detergent & rubber (SBR).
(*d*) as a perfume.

120. Which is used for making the explosive "TNT" (Tri-nitro toluene) ?
(*a*) Benzol (*b*) Toluene
(*c*) Pyridine (*d*) Creosote.

121. Hard pitch is used for making
(*a*) pulverised fuel (*b*) perfumes
(*c*) insecticides (*d*) plastics

122. Yield of pitch from distillation of high temperature tar is about .... percent.
(*a*) 10 (*b*) 25
(*c*) 65 (*d*) 90

Answers

110. (a)	111. (d)	112. (a)	113. (b)	114. (a)	115. (c)
116. (a)	117. (b)	118. (d)	119. (c)	120. (b)	121. (a)
122. (c)					

123. Benzol is used
   (*a*) as a motor fuel blend.
   (*b*) for producing benzene, toluene & xylene by its distillation.
   (*c*) both (*a*) & (*b*).
   (*d*) neither (*a*) nor (*b*).

124. Naphthalene is used for making
   (*a*) insecticides (*e.g.* moth balls).
   (*b*) unsaturated polyesters.
   (*c*) drug intermediates (*e.g.* β-naphthol).
   (*d*) all (*a*), (*b*) & (*c*).

125. Volatile matter content in coking coal blend is about .... percent.
   (*a*) 1 (*b*) 10
   (*c*) 23 (*d*) 46

126. Calorific value of tar is about .... kcal/kg.
   (*a*) 4000 (*b*) 9000
   (*c*) 13500 (*d*) 20000

127. Calorific value of furnace oil is about .... kcal/kg.
   (*a*) 5000 (*b*) 10000
   (*c*) 15000 (*d*) 20000

128. Tar is a better fuel than furnace oil, because of its
   (*a*) high calorific value.
   (*b*) lower sulphur content.
   (*c*) higher emissivity (0.8-0.9) resulting in higher radiation heat transfer rate.
   (*d*) both (*b*) & (*c*).

129. Largest constituent of blast furnace gas is
   (*a*) N_2 (*b*) CO
   (*c*) CO_2 (*d*) H_2

130. Largest constituent of coke oven gas is
   (*a*) N_2 (*b*) H_2
   (*c*) CH_4 (*d*) CO_2

131. Blast furnace gas compared to coke oven gas has
   (*a*) lower ignition temperature.
   (*b*) narrower explosion limit.
   (*c*) higher calorific value.
   (*d*) lower theoretical flame temperature.

132. L.D. converter gas comprises mainly of
   (*a*) CO & CO_2 (*b*) CO_2 & H_2
   (*c*) CO & O_2 (*d*) CO_2 & O_2

133. Pick out the wrong statement.
   (*a*) Cokes of high reactivity are obtained from weakly coking coals.
   (*b*) Cokes of high reactivity are obtained from strongly coking coals.
   (*c*) Reactivity of coke is inversely proportional to its absolute density.
   (*d*) Abrasion index of the coke is a measure of its hardness.

134. Which of the following is not a by-product fuel ?
   (*a*) Producer gas (*b*) Blast furnace gas
   (*c*) Coke oven gas (*d*) Reformed gas

135. Dust content in cleaned blast furnace gas is less than .... mg/Nm^3.
   (*a*) 10 (*b*) 100
   (*c*) 1000 (*d*) 1

136. Explosion limit/limit of inflammability of coke oven gas is .... percent gas in gas-air mixture.
   (*a*) 5 – 32 (*b*) 37 – 71
   (*c*) 4 – 74 (*d*) 12 – 75

137. Limit of inflammability of blast furnace gas is .... percent gas in-gas-air mixture.
   (*a*) 37 – 71 (*a*) 5 – 32
   (*c*) 12 – 74 (*d*) 4 – 74

Answers

123. (c)	124. (d)	125. (c)	126. (b)	127. (b)	128. (d)
129. (a)	130. (b)	131. (d)	132. (a)	133. (a)	134. (a)
135. (a)	136. (a)	137. (a)			

138. Which of the following is the lightest fuel gas?
(*a*) Coke oven gas (*b*) Producer gas
(*c*) Blast furnace gas (*d*) Gobar gas

139. When steam is passed over coal resulting in the endothermic gasification, reaction C + $H_20 = CO + H_2$, it is called .... of coal.
(*a*) carbonisation (*b*) oxidation
(*c*) gasification (*d*) coalification

140. A coal containing high amount of volatile matter will have
(*a*) low ignition temperature.
(*b*) very little ash content.
(*c*) high fusion point of its ash.
(*d*) low adiabatic flame temperature.

141. High sulphur content in the fuel .... of the flue gases.
(*a*) decreases the dew point
(*b*) increases the dew point
(*c*) reduces combustion efficiency by limiting the permissible temperature reduction
(*d*) both (*b*) & (*c*)

142. Which constituent of coal is the most important in the production of coke ?
(*a*) Moisture (*b*) Ash
(*c*) Volatiles (*d*) Carbon

143. Coke oven gas comprises mainly of
(*a*) H2 & CH4 (*b*) CO&CO2
(*c*) H_2 & CO (*d*) CH4 & CO

144. Washing medium in coal washries comprises of water and
(*a*) carbon tetrachloride
(*b*) sand/hematite iron ore
(*c*) mineral oil
(*d*) wash oil

145. The difference between gross & net calorific value will be minimum in case of
(*a*) coke oven gas (*b*) water gas
(*c*) natural gas (*d*) blast furnace gas

146. Main constituent of natural gas is
(*a*) CH_4 (*b*) C_2H_6
(*c*) H_2 (*d*) C_3H_8

147. The calorific value of natural gas is about .... kcal/Nm³.
(*a*) 10000 (*b*) 2500
(*c*) 25000 (*d*) 5000

148. In flue gas analysis by Orsat's apparatus, carbon monoxide is absorbed by
(*a*) cuprous chloride.
(*b*) potassium hydroxide.
(*c*) alkaline pyragallol solution.
(*d*) bromine water.

149. Removal of hydrogen from coke oven gas
(*a*) increases its calorific value.
(*b*) decreases its calorific value.
(*c*) does not alter its calorific value.
(*d*) is not possible on commercial scale.

150. In flue gas analysis by Orsat's apparatus, oxygen is absorbed by
(*a*) KOH
(*b*) Cu_2Cl_2
(*c*) alkaline pyragallol solution
(*d*) none of these.

151. In Orsat's apparatus, CO2 is absorbed by
(*a*) KOH (*b*) K2CO3
(*c*) CU2CI2 (*d*) bromine water

152. The sequence of absorption in flue gas analysis by Orsat's apparatus is respectively
(*a*) CO_2, O_2, CO (*b*) CO, O_2, CO_2
(*c*) CO_2, CO, O_2 (*d*) O_2, CO_2, CO

Answers

138. (a)	139. (c)	140. (a)	141. (d)	142. (b)	143. (a)
144. (b)	145. (d)	146. (a)	147. (a)	148. (a)	149. (a)
150. (c)	151. (a)	152. (a)			

153. Bomb calorimeter determines the
 (a) higher calorific value at constant volume.
 (b) gross calorific value at constant pressure.
 (c) lower calorific value at constant pressure.
 (d) net calorific value at constant volume.

154. Incomplete combustion of a fuel is characterised by the
 (a) smoke formation.
 (b) high flue gas temperature.
 (c) high % of O_2 in flue gas.
 (d) high % of CO in flue gas.

155. For maximum discharge through a chimney, its height should be
 (a) 200 metres.
 (b) infinitely long.
 (c) more than 105 metres.
 (d) equal to the height of hot gas column producing draught.

156. Oxygen percentage in air is .... by weight.
 (a) 79 (b) 21
 (c) 23 (d) 29

157. Stack height in a big thermal power plant is primarily dictated by the
 (a) pollution control aspect.
 (b) draught to be created.
 (c) limitation of constructional facility.
 (d) none of these.

158. Emission of a dense white smoke out of the chimney of a thermal power plant, is an indication of the use of
 (a) less air for combustion.
 (b) correct air for combustion.
 (c) too much air for combustion.
 (d) pulverised coal in boilers.

159. Which of the following accounts for maximum energy loss in a boiler ?
 (a) Flue gases.
 (b) Ash content in the fuel.
 (c) Incomplete combustion.
 (d) Unburnt carbon in flue gases.

160. The main function of primary air in pulverised coal fired burner is to
 (a) burn CO to CO2.
 (b) dry & transport the coal.
 (c) have proper combustion by supplying it around the burner.
 (d) preheat the tertiary air used for complete combustion of CO to CO2.

161. Orsat apparatus is meant for
 (b) gravimetric analysis of the flue gases.
 (a) finding out the combustion efficiency.
 (c) direct determination of nitrogen in the flue gas by absorbing it in ammonical cuprous chloride.
 (d) none of these.

162. The maximum adiabatic flame temperature in air as compared to that in pure oxygen is
 (a) much lower.
 (b) much higher.
 (c) same.
 (d) either lower or higher; depends on the type of fuel.

163. With increase in the calorific values of fuels, their adiabatic flame temperatures
 (a) increase.
 (b) decrease.
 (c) remain unchanged.
 (d) may increase or decrease ; depends on the quantity of products of combustion.

Answers

153. (a)	154. (d)	155. (d)	156. (c)	157. (b)	158. (c)
159. (a)	160. (b)	161. (d)	162. (a)	163. (d)	

164. If the specific heat of gaseous products of combustion of a fuel is high, the adiabatic flame temperature will be
(*a*) low
(*b*) high
(*c*) very high, if the fuel is of low calorific value.
(*d*) none of these.

165. Use of excess air in combustion of fuels results in
(*a*) heat losses.
(*b*) long flame.
(*c*) condensation of water vapour from flue gases.
(*d*) none of these.

166. A gaseous fuel in order to develop luminousity on burning must contain
(*a*) CO (*b*) H_2
(*c*) O2 (*d*) hydrocarbons

167. Preheating of gaseous fuels results in increased
(*a*) flame length
(*b*) flame temperature
(*c*) quantum of flue gas
(*d*) ignition temperature.

168. Which of the following constituents of a fuel does not contribute to its calorific value on combustion ?
(*a*) H (*b*) S
(*c*) C (*d*) O

169. Dry air required to burn 1 kg of carbon completely is about .... kg.
(*a*) 11.5 (*b*) 2.67
(*c*) 16 (*d*) 32

170. Dry air for burning 1 Nm^3 of CO to $C0_2$ is about .... Nm^3.
(*a*) 2.4 (*b*) 1.75
(*c*) 0.87 (*d*) 11.4

171. High excess air in combustion of fuels results in the
(*a*) increased fuel consumption.
(*b*) incomplete combustion.
(*c*) smoky flame.
(*d*) none of these.

172. Atomising steam to fuel oil ratio in a burner should be about
(*a*) 0.5 (*b*) 2
(*c*) 3 (*d*) 5

173. For every 10 percent increase in the excess air, the fuel consumption increases by .... percent.
(*a*) 0.1 (*b*) 2
(*c*) 5 (*d*) 10

174. The difference between the net and gross calorific value is maximum in case of
(*a*) pitch (*b*) fuel oil
(*c*) blast furnace gas (*d*) bituminous coal

175. Which fuel gas will require maximum amount of combustion air for 1 Nm^3 gas ?
(*a*) Blast furnace gas
(*b*) Natural gas
(*c*) Producer gas
(*d*) Water gas

176. The combustion reaction, $C + O_2 = CO_2$, is
(*a*) exothermic (*b*) endothermic
(*c*) auto catalytic (*d*) none of these

177. Junkar's calorimeter is used to determine the calorific value of
(*a*) pulverised coal (*b*) gaseous fuels
(*c*) fuel oil (*d*) both *b* & c

Answers

164. (a)	165. (a)	166. (d)	167. (b)	168. (d)	169. (a)
170. (a)	171. (a)	172. (a)	173. (a)	174. (b)	175. (b)
176. (a)	177. (b)				

178. Preheating of combustion air is done to
   (a) increase the adiabatic flame temperature.
   (b) increase the calorific value of the fuel.
   (c) complete the combustion of fuel.
   (d) reduce its requirement for complete combustion of fuel.

179. The average molecular weight of a flue gas having the composition by volume as CO_2 = 25%, O_2 = 25%, N_2= 50% will be
   (a) 27.6 (b) 23
   (c) 47.3 (d) 42.9

180. A fuel with high heat release rate will
   (a) require smaller combustion chamber.
   (b) have high calorific value.
   (c) have high adiabatic flame temperature
   (d) have long flame.

181. Presence of nitrogen in combustion air does not reduce the
   (a) amount of heat liberated.
   (b) flame temperature.
   (c) flue gas temperature.
   (d) any of the above.

182. Net C.V. is the gross C.V. minus the .... heat of water in the combustion products, when cooled to 15.5°C.
   (a) sensible
   (b) latent
   (c) sensible and latent
   (d)none of these.

183. Natural draught produced by chimney depends on the
   (a) density of flue gas
   (b) height of the chimney.
   (c) both (a) & (b).
   (d) neither (a) nor (b).

184. Artificial draught produced by a fan can be controlled by the
   (a) speed of the fan.
   (b) damper.
   (c) variation of the pitch of the fan blades.
   (d) all (a), (b) & (c).

185. A liquid/gaseous fuel containing hydrocarbons & high amount of sulphur is burnt with 40% excess air. The flue gas still contains large amount of carbon monoxide. This may be due to the
   (a) presence of large quantity of hydrogen in the fuel.
   (b) low C.V. of the fuel.
   (c) lack of thorough mixing of fuel with air.
   (d) high sulphur content in the fuel.

186. Which of the following would require least amount of secondary air for combustion ?
   (a) Coke breeze containing 25% ash & 2% volatile matter.
   (b) Anthracite containing 10% volatile matter & 8% ash.
   (c) Bituminous coal containing 20% ash & 25% volatile matter.
   (d) Semi-bituminous coal containing 25% ash & 20% volatile matter.

187. Pick out the wrong statement.
   (a) Theoretical flame temperature is the temperature attained by the products of combustion, when the fuel is burned without loss or gain of heat.
   (b) Burning the fuel with theoretically required amount of pure oxygen results in attainment of maximum adiabatic flame temperature.
   (c) Burning the fuel with excess pure oxygen results in attainment of maximum theoretical flame temperature.
   (d) Adiabatic flame temperature of actual combustions are always less than the maximum values.

Answers

178. (a)	179. (b)	180. (a)	181. (a)	182. (c)	183. (c)
184. (d)	185. (c)	186. (c)	187. (c)		

188. Preheating of .... the flame temperature.
(*a*) gaseous fuel before combustion decreases
(*b*) combustion air decreases
(*c*) either the fuel or the air or both increases
(*d*) either the fuel or the air does not affect

189. A gaseous fuel in order to develop luminosity on burning must contain
(*a*) CO_2 (*b*) N_2
(*c*) O_2 (*d*) hydrocarbons

190. Which fuel gas will generate maximum amount of flue gas on complete combustion of 1 Nm^3 of fuel gas ?
(*a*) Carburetted water gas
(*b*) Blast furnace gas
(*c*) Natural gas
(*d*) Producer gas

191. The maximum adiabatic flame temperature in air is .... that in pure oxygen.
(*a*) lower than (*b*) higher than
(*c*) same as (*d*) not related to

192. The optimum percentage of excess air for combustion depends upon the .... of the fuel.
(*a*) type (solid / liquid / gaseous)
(*b*) calorific value
(*c*) sulphur content
(*d*) ignition temperature

193. Which of the following fuels will require maximum amount of percent excess air for complete combustion ?
(*a*) Coke oven gas (*b*) Furnace oil
(*c*) Pulverised coal (*d*) Lump coal

194. Efficiency of combustion is judged by the .... the flue gas.
(*a*) % of CO_2 in (*b*) % of O_2
(*c*) temperature of (*d*) colour of

195. Which has the highest flame speed ?
(*a*) CO (*b*) H_2
(*c*) CH4 (*d*) C_2H_2

196. Gross C.V. & net C.V. will be equal for
(*a*) H_2 (*b*) CH_4
(*c*) CO (*d*) C_2H_6

197. Combustion requires
(*a*) supply of oxygen.
(*b*) combustible fuel.
(*c*) source of heat energy.
(*d*) all(*a*) , (*b*) & (*c*).

198. The lowest temperature at which a solid fuel produces enough vapours to support continuous combustion is called the
(*a*) flash point.
(*b*) kindling temperature.
(*c*) burning temperature.
(*d*) none of these.

199. The hottest part of the flame lies in its .... zone.
(*a*) blue (*b*) luminous
(*c*) non-luminous (*d*) unburn gases

200. How much Nm^3 of oxygen is required for the complete theoretical combustion of 1 Nm^3 of methane ?
(*a*) 1 (*b*) 2
(*c*)1.5 (*d*) 3

201. Fuel combustion is never cent percent efficient due to the
(*a*) incomplete combustion
(*b*) dry gas loss or stack loss
(*c*) moisture loss
(*d*) all (*a*), (*b*) & (*c*)

Answers

188. (c)	189. (d)	190. (c)	191. (a)	192. (a)	193. (d)
194. (b)	195. (b)	196. (c)	197. (d)	198. (b)	199. (c)
200. (b)	201. (d)				

202. When incomplete combustion loss is high, the flue gas analysis shows large amount of

(*a*) CO_2 (*b*) CO

(*c*) O_2 (*d*) N_2

203. If O2 content in the flue gas rises too high, fuel is being wasted by the

(*a*) incomplete combustion

(*b*) dry gas or stack loss

(*c*) moisture loss

(*d*) none of these.

204. The ratio of % total carbon obtained in the ultimate analysis of coke and % fixed carbon obtained in the proximate analysis is always

(*a*) 1 (*b*) < 1

(*c*) > 1 (*d*) ∞

205. C.V. (kcal/Nm^3) of gaseous fuels .... with increase in molecular weight.

(*a*) decreases

(*b*) increases

(*c*) remains constant

(*d*) may increase or decrease ; depends on % combustibles.

206. The dew point of the flue gases

(*a*) decreases with decrease in C/H ratio of a fuel.

(*b*) decreases with increase in C/H ratio of a fuel.

(*c*) is independent of C/H ratio of a fuel

(*d*) decreases with decrease in excess air.

207. Actual flame temperature is always lower than the adiabatic flame temperature, because there is

(*a*) no possibility of obtaining complete combustion at high temperature.

(*b*) always loss of heat from the flame.

(*c*) both (*a*) & (*b*).

(*d*) neither (*a*) nor (*b*).

208. The ratio of maximum adiabatic flame temperature in air to that in pure oxygen is always

(*a*) 1 (*b*) < 1

(*c*) > 1 (*d*) unpredictable

209. A sooty flame can be made non-luminous by

(*a*) admitting more fuel gas.

(*b*) admitting more air.

(*c*) shutting air supply.

(*d*) shutting fuel gas supply.

210. C/H ratio (bv weight) is maximum in case of

(*a*) coal (*b*) furnace coil

(*c*) natural gas (*d*) naphtha

211. C/H ratio (by weight) is minimum in case of

(*a*) furnace oil (*b*) natural gas

(*c*) coal (*d*) naphtha

212. Emission of dense white smoke out of the chimney of a thermal power plant is an indication of the use of

(*a*) less air for combustion.

(*b*) correct amount of air for combustion.

(*c*) too much air for combustion.

(*d*) pulverised coal in boilers.

213. As time passes, the calorific value of stored coal

(*a*) decreases.

(*b*) increases,

(*c*) remains unaltered.

(*d*) may increase or decrease ; depends on the method of storage.

214. Fischer-Tropsch method aims at the

(*a*) gasification of coal.

(*b*) synthesis of gasoline (from water gas).

(*c*) hydrogenation of coal to produce gasoline.

(*d*) none of these.

Answers

202. (b)	203. (b)	204. (c)	205. (b)	206. (b)	207. (c)
208. (b)	209. (b)	210. (a)	211. (b)	212. (c)	213. (a)
214. (b)					

215. Steam is intermittantly admitted into the fuel bed during the production of producer gas to
(*a*) convert CO to CO_2.
(*b*) increase the combustion rate.
(*c*) increase the gas production rate.
(*d*) minimise the chances of clinker formation.

216. Combustion of pulverised coal as compared to that of lump coal
(*a*) develops a non-luminous flame.
(*b*) develops a low temperature flame.
(*c*) can be done with less excess air.
(*d*) provides a lower rate of heat release.

217. Which has the highest gross calorific value?
(*a*) Blast furnace gas.
(*b*) Coke oven gas.
(*c*) Carburetted water gas.
(*d*) Oil refinery gas.

218. Rate of low temperature oxidation of coal due to bad storage conditions
(*a*) decreases with increase in surface area.
(*b*) does not vary with increase in surface area.
(*c*) is more for low volatile coals compared to high volatile coals.
(*d*) is accelerated by storage in large heaps with small surface to volume ratio.

219. Pulverized coal used in boiler firing need not have
(*a*) less moisture content.
(*b*) high fusion point of ash.
(*c*) high bulk density.
(*d*) lower ash content.

220. A coal having high ratio of volatile matter to fixed carbon
(*a*) is less liable to spontaneous combustion on storage.
(*b*) is more difficult to ignite & produces a shorter flame,
(*c*) requires smaller combustion space & less secondary air.
(*d*) none of these.

221. Presence of free moisture in coal is most disadvantageous during
(*a*) its pulverisation (as it requires more power).
(*b*) combustion of fine slacks on the grate,
(*c*) handling (*e.g.* when emptying wagons).
(*d*) none of these.

222. Which is not a binder for coal/coke briquettes ?
(*a*) Coal tar (*b*) Bitumen
(*c*) Molasses (*d*) None of these

223. Which has the highest calorific value ?
(*a*) Lignite
(*b*) Sub-bituminous coal
(*c*) Anthracite
(*d*) Peat

224. Washing of coal does not reduce its
(*a*) adhering impurities
(*b*) mineral matter content
(*c*) ash content
(*d*) inherent impurities

225. A carbonaceous fuel (containing no H_2 or hydrocarbons) is burnt and the resulting flue gas contains 21% CO_2. It means that
(*a*) 21% excess air has been used for combustion.
(*b*) 21% excess oxygen has been used for combustion,
(*c*) complete combustion of fuel has taken place.
(*d*) no excess air has been used for combustion.

Answers

215. (d)	216. (c)	217. (d)	218. (d)	219. (b)	220. (d)
221. (a)	222. (d)	223. (d)	224. (d)	225. (c)	

226. Lignite
   (a) is a high rank coal.
   (b) is a coking coal.
   (c) can be used for gasification & steam generation.
   (d) is a black banded coal, which burns with a non-smoky yellowish flame.

227. Froth floatation is used for washing
   (a) fine coal dust (0.5 mm size).
   (b) lump coal (80 mm size).
   (c) ash from the coal based on difference in specific gravity.
   (d) all (a), (b) & (c).

228. Combustion of pulverised fuel
   (a) requires larger combustion chamber than rich gaseous fuel.
   (b) in a metallurgical furnace does not contaminate the product in the furnace by ash from fuel.
   (c) suppresses the discharge of fine dust and grit into atmosphere.
   (d) all (a), (b) & (c).

229. Producer gas comprises mainly of
   (a) CO & N_2 (b) CO & H_2
   (c) CO_2 & N_2 (d) CO_2 & H_2.

230. Washing of coal
   (a) reduces its ash & sulphur content.
   (b) improves its coking properties.
   (c) increases the fusion point of its ash by removing chlorine compounds.
   (d) all (a), (b)& (c).

231. The gasification reaction, represented by, $C + H_2O = CO + H_2$, is a/an .... reaction.
   (a) exothermic   (b) endothermic
   (c) catalytic   (d) autocatalytic

232. Producer gas containing least amount of tar is produced by the
   (a) partial combustion of coal.
   (b) partial combustion of < -50 mm size coke.
   (c) high pressure gasification of coal.
   (d) atmospheric pressure gasification of coal.

233. Calorific value of a typical dry anthracite coal is about .... kcal/kg.
   (a) 1000   (b) 4000
   (c) 8000   (d) 15000

234. During combustion of gaseous fuels, deficiency of air
   (a) lengthens the flame.
   (b) shortens the flame.
   (c) does not affect the flame length.
   (d) increases flame temperature.

235. Calorific value of dry wood may be about .... kcal/kg.
   (a) 1000   (b) 2500
   (c) 3000   (d) 4500.

236. The calorific value of liquified petroleum gas (LPG) comprising of 75% propane & 25% butane is about .... kcal/Nm^3.
   (a) 2500   (b) 25000
   (c) 10000   (d) -50000.

237. The maximum adiabatic flame temperature is attained, when the fuel is burnt with
   (a) theoretically required amount of air.
   (b) more than theoretically required amount of air.
   (c) less than theoretically required amount of air.
   (d) theoretically required amount of oxygen.

Answers

226. (c)	227. (a)	228. (a)	229. (a)	230. (d)	231. (b)
232. (b)	233. (c)	234. (a)	235. (d)	236. (b)	237. (d)

238. Too much of excess air in combustion results in high
   (*a*) fuel consumption for the same heat load.
   (*b*) stack losses
   (*c*) percentage of oxygen in flue gases.
   (*d*) all (*a*), (*b*) & (*c*).

239. LPG stands for
   (*a*) liquid petroleum gas.
   (*b*) liquefied petrol gas.
   (*c*) liquid petrol gas.
   (*d*) liquefied petroleum gas.

240. Sweetening of petroleum product means the removal of
   (*a*) sulphur & its compounds.
   (*b*) water.
   (*c*) organic impurities.
   (*d*) wax.

241. Octane number of gasoline is a measure of its
   (*a*) knocking tendency.
   (*b*) ignition delay,
   (*c*) ignition temperature.
   (*d*) smoke point.

242. Cetane number of a diesel fuel is the measure of its
   (*a*) ignition delay (*b*) smoke point
   (*c*) viscosity (*d*) oxidation stability

243. Kerosene should have
   (*a*) low smoke point
   (*b*) high smoke point
   (*c*) high aromatics content
   (*d*) low paraffins content

244. The octane number of aviation gasoline may be
   (*a*) 10 (*b*) >50
   (*c*) >100 (*d*) none of these.

245. Viscosity index
   (*a*) is the measure of flash point.
   (*b*) is the measure of variation of viscosity with temperature.
   (*c*) of a lubricating oil should be low.
   (*d*) none of these.

246. Reid vapour pressure of gasoline is the measure of its
   (*a*) pour point
   (*b*) cloud point
   (*c*) vapour locking tendency
   (*d*) carbon residue

247. The main aim of cracking is to produce
   (*a*) gasoline (*b*) lube oil
   (*c*) petrolatum (*d*) coke

248. Stabilisation of gasoline (petrol) means
   (*a*) removal of dissolved gases from it.
   (*b*) increasing its oxidation stability.
   (*c*) stabilising its lead susceptibility.
   (*d*) increasing its vapour pressure.

249. Which of the following is the most important property for a jet fuel ?
   (*a*) Cloud point (*b*) Pour point
   (*c*) Colour (*d*) Freezing point

250. Flash point of motor gasoline is around .... °C.
   (*a*) 5 (*b*) 40
   (*c*) 120 (*d*) 160

251. Petroleum coke is used mainly in the
   (*a*) discoloration of yellow glycerine.
   (*b*) sugar refining.
   (*c*) manufacture of carbon electrode.
   (*d*) blast furnace for reduction of iron ore.

Answers

238. (d)	239. (d)	240. (a)	241. (a)	242. (a)	243. (b)
244. (c)	245. (b)	246. (c)	247. (a)	248. (a)	249. (d)
250. (b)	251. (c)				

252. Cetane number of diesel used in trucks may be about
(*a*) 5 (*b*) 15
(*c*) 35 (*d*) 85

253. Liquefied Petroleum Gas (LPG) used for household cooking comprises mainly of
(*a*) propane & butane.
(*b*) butane & ethane.
(*c*) methane & ethane.
(*d*) methane & carbon monoxide.

254. Which of the following is not a manufactured fuel ?
(*a*) LPG (*b*) Coal briquettes
(*c*) Tar (*d*) Colloidal fuels

255. Which of the following is not a by-product fuel ?
(*a*) Pitch (*b*) Blast furnace gas
(*c*) Petrol (*d*) Refinery gas

256. Yield of charcoal in high temperature carbonisation of wood is about .... percent.
(*a*) 10 (*b*) 25
(*c*) 50 (*d*) 75.

257. The most matured coal out of the following is
(*a*) lignite (*b*) semi-anthracite
(*c*) sub-bituminous (*d*) bituminous

258. Coking coals are invariably
(*a*) lignites (*b*) bituminous coals
(*c*) semi-anthracites (*d*) anthracites

259. A coal having higher volatile matter content has lower
(*a*) smoking tendency on burning.
(*b*) coke oven gas yield on carbonisation.
(*c*) chance of catching fire during storage in open space.
(*d*) ignition temperature.

260. A coal with high ash content
(*a*) has higher calorific value.
(*b*) is harder & stronger.
(*c*) is not subjected to washing.
(*d*) has low quantity of mineral matter.

261. The difference between % of total carbon & fixed carbon will be minimum in case of
(*a*) lignite.
(*b*) bituminous coal.
(*c*) anthracite.
(*d*) high temperature coke.

262. Assam coals suffers, mainly from the disadvantage of high .... content.
(*a*) ash (*b*) volatile matter
(*c*) sulphur (4-6%) (*d*) moisture

263. Washing of coal does not reduce its
(*a*) S & P content
(*b*) heating value
(*c*) fusion point of ash
(*d*) none of these

264. Low temperature oxidation of coal during storage does not decrease its
(*a*) oxygen content (*b*) caking power
(*c*) calorific value (*d*) none of these

265. Mineral matter content (M) and ash content (A) in coal are approximately related as
(*a*) M = 1.1A (*b*) M=A
(*c*) M = 1.5/A (*d*) M = 2A

266. Hardgrove grindability index of four coal samples are given below. Which of them is the easiest to grind ?
(*a*) 50 (*b*) 70
(*c*) 85 (*d*) 100

Answers

252. (c)	253. (a)	254. (c)	255. (c)	256. (b)	257. (b)
258. (b)	259. (d)	260. (b)	261. (d)	262. (c)	263. (d)
264. (d)	265. (a)	266. (d)			

267. 'Fat coals' are those coals, which have very high
   (*a*) caking capacity.
   (*b*) volatile matter content.
   (*c*) fusion point of its ash.
   (*d*) inherent moisture content.

268. Bulk density of pulverised coal may be about .... kg/m^3.
   (*a*) 100 (*b*) 500
   (*c*) 1000 (*d*) 1500

269. In low temperature carbonisation of coal,
   (*a*) yield of coke oven gas is 290 Nm^3/ton dry coal.
   (*b*) volatile matter in coke is zero.
   (*c*) temperature maintained is 700°C.
   (*d*) yield of tar is about 3% of dry coal.

270. Combustion reaction of fuels is a/an .... reaction.
   (*a*) auto catalytic (*b*) exothermic
   (*c*) endothermic (*d*) none of these

271. Which of the following gaseous fuels has the highest net calorific value (kcal/Nm^3) ?
   (*a*) Producer gas.
   (*b*) Carburetted water gas.
   (*c*) Natural gas.
   (*d*) Liquified petroleum gas.

272. .... has the widest inflammability limit (explosion limit).
   (*a*) Hydrogen (*b*) Carbon monoxide
   (*c*) Acetylene (*d*) Methane

273. Gobar gas is produced by the .... of gobar'.
   (*a*) hydrolysis (*b*) fermentation
   (*c*) oxidation (*d*) dehydration

274. 'Wobbe index' is a characteristic of the
   (*a*) solid fuels (*b*) gaseous fu«ls
   (*c*) liquid fuels (*d*) fat coals

275. Calorific value of gobar gas (containing CH_4 = 60%, H_2 = 10% & $C0_2$ = 30%) may be about .... kcal/Nm^3.
   (*a*) 1800 (*b*) 3200
   (*c*) 5400 (*d*) 10200

276. Main component of sewage gas produced during aneorobic decomposition of organic waste (by suitable bacteria) during sewage disposal is
   (*a*) H_2 (*b*) CH_4
   (*c*) CO_2 (*d*) N_2

277. Which of the following has the highest heat of combustion ?
   (*a*) H_2 (*b*) CO
   (*c*) CH_4 (*d*) C_2H_6

278. About .... Nm^3 of air will be required for the complete combustion of 2 Nm^3 of CO.
   (*a*) 2 (*b*) 4
   (*c*) 5 (*d*) 1

279. Pick out the wrong statement.
   (*a*) High concentration of oxygen in flue gas means high stack loss.
   (*b*) Gaseous fuels require the least % excess air for complete combustion.
   (*c*) The ratio of fixed carbon to volatile matter percentage in coal is called its 'fuel ratio'.
   (*d*) Calorific value of natural gas is more than that of LPG.

280. Pick out the wrong statement.
   (*a*) Main Constituents of LPG are propane & butane.
   (*b*) C.V. of LPG is more than that of natural gas.
   (*c*) C.V. of LPG is about 26000 kcal/Nm^3 (11500 kcal/kg).
   (*d*) L.P.G. is lighter than air.

Answers

267. (a)	268. (b)	269. (c)	270. (b)	271. (d)	272. (c)
273. (b)	274. (b)	275. (c)	276. (b)	277. (d)	278. (c)
279. (d)	280. (d)				

281. Which of the following is unsuitable fuel for producer gas manufacture ?
(*a*) Coke
(*b*) Anthracite
(*c*) Coal having low fusion point of its ash
(*d*) Coal having high fusion point of its ash

282. Which of the following has the least calorific value (kcal/Nm^3) ?
(*a*) Blast furnace gas
(*b*) Coke oven gas
(*c*) Sewage gas
(*d*) Natural gas

283. Which of the following has the highest theoretical flame temperature ?
(*a*) Blast furnace gas
(*b*) Hydrogen
(*c*) Acetylene
(*d*) Coke oven gas

284. Supply of excess air for complete combustion of fuel is necessitated to facilitate
(*a*) its thorough mixing with air.
(*b*) attainment of chemical equilibrium.
(*c*) attainment of high temperature.
(*d*) none of these.

285. The function of secondary combustion air is to
(*a*) pneumatically convey the pulverised coal.
(*b*) completely burn the volatile matter.
(*c*) burn the lumpy coal.
(*d*) none of these.

286. The main advantage of forced draft over natural draft is that
(*a*) combustion of fuel is complete.
(*b*) smaller height chimney can be used.
(*c*) furnace control is easier.
(*d*) all (*a*), (*b*) & (*c*).

287. Short/intense flame is produced during combustion of gaseous fuel by using
(*a*) high amount/current of combustion air.
(*b*) low amount/current of combustion air.
(*c*) preheated secondary air.
(*d*) very little excess air.

288. Ignition temperature of a fuel in air is .... that in pure oxygen.
(*a*) more than
(*b*) less than
(*c*) equal to
(*d*) either more or less ; depends on the type of fuel

289. Explosion limit of blast furnace gas is 37 to 71% gas in gas-air mixture. It means that blast furnace gas will explode, when burnt in a confined space, if its concentration in the gas-air mixture is .... percent.
(*a*) < 37
(*b*) > 71
(*c*) in between 37 & 71
(*d*) none of these.

290. In general, the limit of inflammability/explosion limit of fuel gases in widened by the
(*a*) increase in gas pressure.
(*b*) increase in gas temperature (*i.e.* preheating).
(*c*) use of oxygen instead of air for combustion .
(*d*) all (*a*), (*b*) & (*c*).

291. Long/loose flame is produced during combustion of gaseous fuels by using
(*a*) high amount of combustion air.
(*b*) preheated primary air.
(*c*) low amount/current of combustion air.
(*d*) preheated secondary air.

Answers

281. (c)	282. (a)	283. (c)	284. (a)	285. (b)	286. (d)
287. (a)	288. (a)	289. (c)	290. (d)	291. (c)	

292. Turndown ratio of a gas burner is equal to the
(*a*) maximum to minimum heat input ratio.
(*b*) maximum to minimum permissible gas flow rate.
(*c*) both (*a*) & (*b*).
(*d*) minimum to maximum heat input ratio.

293. Bunsen burner is an example of a/an .... type burner.
(*a*) inside mixing/premix
(*b*) outside mixing/diffusion flame/nozzle mix
(*c*) rotary cup
(*d*) submerged combustion

294. Advantages of fluidised bed combustion are
(*a*) reduced NOx formation in flue gas.
(*b*) lower furnace operating temperature.
(*c*) higher heat transfer rate.
(*d*) all (*a*), (*b*) & (*c*).

295. Main use of soft coke is as the .... fuel.
(*a*) domestic (*b*) blast furnace
(*c*) foundary (*d*) none of these

296. .... process is meant for the direct hydrogenation of coal to produce liquid fuels.
(*a*) Fischer-Tropsch (*b*) Bergius
(*c*) Lurgi (*d*) none of these

297. Calorific value of both the solid & liquid fuels can be determined by the .... calorimeter.
(*a*) Junker's (*b*) Bomb
(*c*) Boy's (*d*) none of these

298. Air/gas ratio for complete combustion will be the highest for
(*a*) LPG
(*b*) gobar gas
(*c*) coke oven gas
(*d*) carburetted water gas

299. Nitrogen present in the flue gas is determined in the Orsat apparatus by absorbing it in
(*a*) potassium hydroxide
(*b*) ammoniacal cuprous chloride.
(*c*) pyrogallol solution.
(*d*) none of these.

300. A travelling grate stoker is meant for efficient burning of .... coal.
(*a*) caking (*b*) pulverised
(*c*) non-caking (*d*) high ash

301. Volumetric ratio of N_2 to O_2 in dry atmospheric air is
(*a*) 0.2.1 (*b*) 3.76
(*c*) 4.76 (*d*) 0.79

302. Which of the following is the maximum coal producing state in India ?
(*a*) Orissa (*b*) West Bengal
(*c*) Jharkhand (*d*) Assam

303. During combustion of a geseous fuel, the presence of a non-luminous flame is an indication of the
(*a*) incomplete combustion.
(*b*) complete combustion.
(*c*) high oxygen in flue gas.
(*d*) none of these.

304. Pick out the wrong statement.
(*a*) Carburetted water gas is also called blue gas.
(*b*) Coals are divided in four species according to their carbon content in Seylor's classification.
(*c*) Carbonisation time in a by-product coke oven is about 18 hours.
(*d*) Gross & net calorific value of a fuel is the same, if it does not contain hydrogen or hydrocarbons.

Answers

292. (d)	293. (a)	294. (d)	295. (a)	296. (b)	297. (d)
298. (a)	299. (a)	300. (a)	301. (a)	302. (a)	303. (a)
304. (a)					

305. Coal is pulverised before burning in large capacity boiler furnaces mainly to
(*a*) ensure its complete combustion.
(*b*) facilitate easy ash removal,
(*c*) enhance its calorific value.
(*d*) provide trouble free operation.

306. Which of the following coal gasification processes will produce gas having maximum methane content ?
(*a*) Winkler process.
(*b*) Lurgi process.
(*c*) Kopper-Totzek process.
(*d*) All can produce same methane content in the gas.

307. The main reason for making the copper calorimeter (used in Bomb calorimeter) silvery white & shining/polished is to
(*a*) minimize its corrosion.
(*b*) avoid radiation heat loss.
(*c*) make it look attractive.
(*d*) none of these.

308. High pressure coal gasification is employed in the .... process.
(*a*) Lurgi (*b*) Kopper-Totzek
(*c*) Winkler (*d*) none of these

309. By-products are .... recovered in 'Beehive ovens'.
(*a*) fully (*b*) partially
(*c*) not at all (*d*) negligibly

310. .... of the coal is the basis for Seylor's coal classification,
(*a*) Proximate analysis
(*b*) Ultimate analysis
(*c*) Caking index
(*d*) Calorific value

311. By-products recovery process from coal carbonisation is termed as direct, indirect & semi-direct process, depending upon the method of recovery of
(*a*) tar (*b*) benzol
(*c*) ammonia (*d*) naphthalene

312. Out of the following, which is the most important parameter for the blast furnace grade coke ?
(*a*) CSR & CRI
(*b*) Ash content
(*c*) Moisture content
(*d*) Volatile matter content

313. Undercharging of coal in the by-product coke ovens results in
(*a*) decrease in the C.V. of coke oven gas.
(*b*) increase in its throughput.
(*c*) increase in the C.V. of coke oven gas.
(*d*) no change in the C.V. of coke oven gas.

314. Coke made in narrower by-product coke ovens (as compared to wider ovens) is
(*a*) less reactive
(*b*) stronger
(*c*) smaller in size
(*d*) all (*a*), (*b*) & (*c*)

315. Which of the following fuels is generally not used in thermal power plant boiler firing ?
(*a*) Furnace oil, light diesel oil & tar/PCM.
(*b*) Pulverised coking coal.
(*c*) Coal middling from washeries.
(*d*) B.F. gas, coke oven gas & L.D. converter gas.

316. Optimum preheating temperature for pitch creosote mixture (PCM) which is also termed as C.T.F-200, before atomisation through burners is
(*a*) 200°F (*b*) 200°C
(*c*) 200°K (*d*) 200°R

Answers

305. (a)	306. (b)	307. (b)	308. (a)	309. (c)	310. (b)
311. (c)	312. (a)	313. (a)	314. (d)	315. (b)	316. (a)

317. Ash content in the middling coal (a by product of coal washeries) in India ranges between .... percent.
(a) 15 to 25 *(b)* 35 to 45
(c) 55 to 75 *(d)* 75 to 85

318. Calorific value of PCM is about .... Kcal/kg.
(a) 12800 *(b)* 8800
(c) 4800 *(d)* 26000

319. Which adsorbant is used for removing sulphur compounds (SO_2, H_2S, RSH etc.) from coke oven gas in by-products recovery plant ?
*(a)*Silica gel.
(b) Diatomaceous earth.
(c) Basalt.
(d) Bog iron (ferric hydroxide).

320. Of the total tar present in raw coke oven gas, the tar recovered in primary gas cooler is about ....percent.
(a) 5 *(b)* 25
(c) 55 *(d)* 75

321. 'Mixed gas' used in steel plants is a mixture of
(a) B.F gas & coke oven gas.
(b) Coke oven gas & converter gas.
(c) Coke oven gas & L.P.G.
(d) none of these

322. Higher fuel combustion efficiency can not be achieved by
(a) preheating of fuel gases & combustion air.
(b) reducing sulphur content in the fuel.
(c) adopting proper fuel firing technique & fuel preparation.
(d) supplying correct amount of combustion air.

323. Higher efficiency in the combustion of solid fuel can not be achieved by
(a) proper fuel preparation.
(b) keeping the flue gas exhaust temperature very high.
(c) adopting efficient fuel firing technique and equipment.
(d) supplying correct quantity of combustion air.

324. 'Wind loss' resulting from unscientific storage of coal may be of the order of about .... percent.
(a) 2.5 *(b)* 7.5
(c) 10.5 *(d)* 14.5

325. The main drawback of supplying more excess air in the combustion of fuel is the
(a) excessive power requirement of air blower.
(b) enhanced sensible heat loss in the flue gas.
(c) intermittent & uncontrolled combustion of the fuel.
(d) high exit flue gas temperature from the furnace.

326. A certain thickness of the coal, if stored unscientifically on soft (katcha) ground having no metallic/concrete flooring, gets sunk into the ground, which is termed as the 'carpet loss'. The carpet loss may be of the order of. .... cms.
(a) 1 to 2 *(b)* 3 to 4
(c) 5 to 15 *(d)* 20 to 40

327. Furnace oil consumption in a furnace for a given duty employing preheated combustion air (at 300°C) can be reduced by about .... percent as compared to use of atmospheric combustion air.
(a) 5 *(b)* 10
(c) 20 *(d)* 35

Answers

317. (b)	318. (b)	319. (d)	320. (b)	321. (a)	322. (b)
323. (b)	324. (a)	325. (b)	326. (c)	327. (c)	

328. Correct viscosity of furnace oil at the burner tip for proper atomisation is about 25 centistokes. To reduce the viscosity of high viscosity furnace oil (250 centistokes) to the correct atomisation viscosity (*i.e.*, 25 cst), it should be preheated to about .... °C.

(*a*) 70 (*b*) 85
(*c*) 105 (*d*) 145

329. Presence of .... in a dry gaseous fuel does not contribute to its calorific value.

(*a*) sulphur (*b*) oxygen
(*c*) hydrogen (*d*) carbon

330. Which of the following oil gasification processes is non-catalytic ?

(*a*) Semet-Solvay process.
(*b*) Segas process.
(*c*) Onia-Gegi process.
(*d*) Steam reforming of naphtha.

331. Number of macrocomponents present in coal according to Stopes are

(*a*) four (*b*) five
(*c*) six (*d*) two

332. .... present is coal is not determined in its ultimate analysis.

(*a*) Fixed carbon (*b*) Total carbon
(*c*) Nitrogen (*d*) Sulphur

333. Improper storage condition results in weathering of coal & spontaneous combustion, which increases its

(*a*) caking index.
(*b*) yield of carbonised products.
(*c*) friability & oxygen content.
(*d*) calorific value.

334. Coal is heated in presence of air to a temperature of about .... °C, while determining its ash content for proximate analysis.

(*a*) 500 (*b*) 750
(*c*) 950 (*d*) 1100

335. During its calorific value determination by bomb calorimeter, coal is combusted by

(*a*) air
(*b*) oxygen
(*c*) oxygen enriched air
(*d*) none of these

336. Which of the following is a measure of agglutinating (*i.e.*, binding) property of coal ?

(*a*) Thickness of plastic layer
(*b*) Caking index
(*c*) Swelling index
(*d*) Gray-king index

337. Which of the following is a single stage, fixed bed high pressure coal gasification process ?

(*a*) Winkler process
(*b*) Kopper-Totzek process
(*c*) Lurgi Process
(*d*) None of these

338. Soft coke is not

(*a*) produced by low temperature carbonisation of coal.
(*b*) a domestic fuel.
(*c*) used in blast furnaces.
(*d*) none of these.

339. Which of the following is not a by-product recovered in a high temperature coal carbonisation plant ?

(*a*) Benzol
(*b*) Pitch-creosote mixture (PCM)
(*c*) Naphthalene
(*d*) Ethylene

Answers

328. (c)	329. (b)	330. (a)	331. (a)	332. (a)	333. (c)
334. (b)	335. (b)	336. (b)	337. (c)	338. (c)	339. (d)

340. Coal is heated in absence of air for 7 minutes at a temperature of about .... °C, during its volatile matter determination for proximate analysis,
(*a*) 550 (*b*) 750
(*c*) 950 (*d*) 1150

341. Pick out the wrong statement.
(*a*) Percentage of ash in coke produced from medium coking coal is more than that in coal.
(*b*) The calorific value (KCal/Nm³) of coke oven gas reduces on removal of hydrogen from it by cryogenic method.
(*c*) Ash is normally removed as 'flyash' in kopper-Totzek process of coal gasification.
(*d*) Coal based fertiliser plants in India at Talcher (Orissa) & Ramagundan (A.P.) employ Kopper-Totzek process of coal gasification.

342. In situ theory & drift theory are related to the
(*a*) origin of petroleum oil.
(*b*) origin of coal.
(*c*) coalification.
(*d*) variation of coal quality with depth.

343. Coal containing large quantity of .... are difficult to wash.
(*a*) ash
(*b*) interent mineral matter
(*c*) free impurities
(*d*) volatile matter

344. Main use of hard coke produced by the high temperature carbonisation of coal is in
(*a*) iron blast furnace
(*b*) cupola in foundaries
(*c*) sinter making
(*d*) domestic ovens

345. Compound coke ovens are those, which can be heated by
(*a*) both steam & electricity.
(*b*) lean gas (*e.g.*, B.F. gas).
(*c*) rich gas (*e.g.*, coke oven gas).
(*d*) both (*b*) & (*c*).

346. Degree of carbonisation of coal during coke making can be roughly judged by the .... of the coke produced.
(*a*) colour (*b*) moisture content
(*c*) ash content (*d*) volatile matter

347. Highly reactive coke have lower
(*a*) porosity.
(*b*) coke reactivity index (CRI).
(*c*) critical air blast (CAB) value.
(*d*) both (*b*) & (*c*).

348. The temperature at which plastic layer formation takes place during carbonisation of coal ranges from .... °C.
(*a*) 100 to 150 (*b*) 350 to 450
(*c*) 550 to 650 (*d*) 700 to 850

349. Pick out the wrong statement.
(*a*) Indian coals on an average contain 25 –30% ash as against 10 – 12% ash in imported coking coal.
(*b*) Ammonia is recovered in the form of ammonium sulphate in direct process of by-product recovery.
(*c*) A high swelling index number of coking coal indicates poor caking properties of coal.
(*d*) Wash oil is used for scrubbing benzol from coke oven gas.

350. Which of the following combustibles is absent in blast furnace gas ?
(*a*) H_2 (*b*) CH_4
(*c*) CO (*d*) none of these

Answers

340. (c)	341. (c)	342. (b)	343. (b)	344. (a)	345. (d)
346. (d)	347. (c)	348. (b)	349. (c)	350. (b)	

351. Which of the following is a lean fuel gas?
(a) Coke oven gas (b) LPG
(c) Blast furnace gas (d) Natural gas

352. Which of the following is a rich fuel gas ?
(a) Producer gas.
(b) Coal gas from underground gasification of coal.
(c) Refinery gases.
(d) Blast furnace gas.

353. The reaction, $C + CO_2 = 2CO$, taking place during coal gasification is called the .... reaction.
(a) Neumann reversal
(b) shift conversion
(c) Boudouard
(d) reduction

354. The shift conversion reaction taking place during water gas manufacture is given by
(a) $C + H_2O = CO + H_2$
(b) $C + 2H_2O = CO_2 + 2H2$
(c) $CO + H_2O = CO_2 + H_2$
(d) $CO + 1/2\,O_2 = CO_2$

355. Bomb calorimeter is used to determine the
(a) GCV at constant pressure.
(b) GCV at constant volume.
(c) NCV at constant pressure.
(d) NCV at constant volume,
where, GCV-Gross Calorific Value
NCV-Net Calorific Value

356. Which of'the following is a primary fuel ?
(a) Blast furnace coke
(b) Gasoline
(c) Natural gas
(d) Wood charcoal

357. Laboratory gas is obtained by the cracking of
(a) gasoline (b) diesel
(c) fuel oil (d) kerosene

358. Stoichiometric combustion of 12 kg of carbon requires .... of oxygen.
(a) 1 kg mole (b) 22.4 Nm^3
(c) 32kg (d) all (a), (b) & (c)

359. Presence of free moisture in coal is most disadvantageous during its
(a) transportation & handling
(b) washing
(c) pulverisation
(d) storage

360. Gasification of the solid fuel converts its organic part into combustible by interaction with air/oxygen and steam so as to obtain a secondary gaseous fuel of high calorific value having no ash. Gasification reactions are normally carried out at about .... °C.
(a) 400 – 500 (b) 900 – 1000
(c) 1400 – 1500 (d) 1700 – 1800

361. Light oil whose major component is benzol, is obtained by the distillation of crude tar in the temperature range of .... C.
(a) 80 – 170 (b) 200 – 300
(c) 250 – 270 (d) 280 – 300

362. Match the calorific values (Kcal/Nm^3) of various gaseous fuels.

List I	*List 11*
(a) Liquefied Petroleum gas (LPG).	I. 4300
(b) Gobar gas.	II. 2650
(c) Coke oven gas.	III. 26000
(d) Coal gas from Kopper-Totzek process.	IV. 5400

Answers

351. (c)	352. (c)	353. (c)	354. (c)	355. (b)	356. (c)
357. (d)	358. (a)	359. (c)	360. (b)	361. (a)	362. (iii) (iv) (i) (ii)

363. Match the various classified types of fuels.

List I	List II
(b) Rocket fuel	I. Coke breeze
(b) By-product fuel	II. Liquefied petroleum gas (LPG)
(b) Manufactured fuel	III. Acetvlene
(b) Illuminating fuel	IV. Hydrazine

364. Match the typical value of percent excess air required for complete combustion of various fuels.

List I	List II
(a) Lumpy coal	I.30
(b) Pulverised coal	11.15
(c) Furnace oil	III. 10
(b) Coke oven gas	IV. 20

365. Match the amount of theoretical air required for the complete combustion of various fuels.

List I	List 11
(a) Coking bituminous coal	I. 4.20
(b) Furnace oil	II. 7.35
(c) Coke oven gas	III. 10.25
(d) LPG	IV. 26.00

(Units – Nm^3/Kg for solid & liquid fuels and Nm^3/Nm^3 for gaseous fuels.)

366. Match the explosion limit (limit of inflammability i.e., % of gas in gas-air mixture) of various fuel gases.

List I	List II
(a) Hydrogen	I. 2 - 9
(b) Acetylene	II. 4-75
(c) LPG	III. 37-71
(d) B.F. gas	IV. 2.5 - 80

367. Match the spontaneous ignition temperature (°C) of various fuels in air.

List I	List II
(a) Hard coke	I. 255
(b) Kerosene	II. 500 - 650
(c) Blast furnace gas	III. 380-420
(d) Bituminous coal	IV. 680

368. Match the major constituent of various fuels.

List I	List II
(a) Natural gas	I. CO
(b) Coke oven gas	II. CH_4
(c) Lurgi gas	III. H_2, CH_4
(d) L.D. converter gas	IV. H_2, CO

369. Match the C/H ratio of various fuels.

List I	List II
(a) Coal	I. 16.5 : 1
(b) Petroleum	II. 15 : 1
(c) Pitch Creosote mixture (PCM/ C.T.F - 200)	III. 8 : 1
(d) Pitch	IV. 11:1

370. The weathering index of a coal

(a) gives an idea of the fusion temperature of ash.

(b) is related to its calorific value.

(c) is a measure of its size stability, when stored & exposed to weather.

(d) is a measure of its caking tendency.

371. Softening temperature of coal ash is a measure of the .... of coal.

(a) caking tendency

(b) coking tendency

(c) clinkering tendency

(d) size stability

372. Fossil fuels mean

(a) solid fuels.

(b) liquid fuels.

(c) those fuels, which are found in the crust of earth.

(d) premature fuels with low calorific value.

Answers

363. (iv) (i) (ii) (iii)
364. (i) (iv) (ii) (iii)
365. (ii) (iii) (i) (iv)
366. (ii) (iv) (i) (iii)
367. (ii) (i) (iv) (iii)
368. (ii) (iii) (iv) (i)
369. (ii) (iv) (iii) (i)
370. (c)
371. (a)
372. (c)

373. A good metallurgical coke should have very-low
   (a) sulphur & phosphorous content.
   (b) porosity.
   (c) fusion point of its ash.
   (d) hardness & strength.

374. Kopper-Totzek coal gasifier (installed in a coal based nitrogeneous fertiliser plant) employs a/an .... bed gasifier.
   (a) entrained *(b)* moving
   (c) fixed *(d)* fluidised

375. Calorific values of both the solid as well as the liquid fuels can be determined by the .... calorimeter.
   (a) bomb *(b)* Boy's
   (c) Junkar's *(d)* none of these

376. Gobar gas is generated by the .... of cow dung.
   (a) fermentation *(b)* oxidation
   (c) hydrogenation *(d)* hydrolysis

377. Main constituent of the gas produced from a gobar gas plant is
   (a) CO_2 *(b)* CH_4
   (c) H_2 *(d)* CO

378. The lowest temperature at which a solid fuel produces enough vapors to support continuous combustion is called the
   (a) fire point
   (b) smoke point
   (c) burning temperature
   (d) kindling temperature

379. The only commercial Fischer-Tropsch plant for producing liquid hydrocarbons fuel from coal is located at the
   (a) SASOL (in South Africa).
   (b) Redcar (U.K.).
   (c) Los-Angeles (U.S.A.).
   (d) Trombay (India).

380. Incomplete combustion of fuel in the furnace is judged by high .... of the flue gas.
   (a) CO content *(b)* dew point
   (c) CO_2 content *(d)* O_2 content

381. In case of liquid petrofuels, momentary combustion is observed at its
   (a) flash point.
   (b) flame temperature.
   (c) fire point.
   (d) preheating temperature corresponding to viscosity of 25 centistokes.

382. Which of the following is not increased by using preheated combustion air in place of ordinary air at room temperature ?
   (a) Calorific value of the fuel
   (b) Flame temperature achieved
   (c) Speed of combustion of fuel
   (d) Heat transfer rate to the stock in the furnace

383. The terminology BTX used in coal tar distillation industry refers to
   (a) benzol - toluol - xylol.
   (b) benzole - toluene - xylene.
   (c) benzene - toluol - xylol.
   (d)benzene - toluene - xylene.

384. Low temperature carbonisation of coal produces
   (a) metallurgical coke.
   (b) soft coke.
   (c) very low C.V. coke oven gas.
   (d) no by-products.

Answers

373. (a)	374. (a)	375. (a)	376. (a)	377. (b)	378. (c)
379. (a)	380. (a)	381. (a)	382. (a)	383. (d)	384. (b)

385. Gross calorific value of the L.D. converter gas analysing, CO = 65%, $C0_2$=15%, N_2 = 20% is about 2000 kcal/Nm^3. Its net calorific value will be .... Kcal/Nm^3.
(*a*) 1800 (*b*) 2000
(*c*) 2200 (*d*) 1500

386. Low grade solid fuels have low .... content.
(a) carbon (*b*) moisture.
(*c*) ash (*d*) none of these

387. Gross & net calorific value is the same for
(*a*) blast furnace gas.
(*b*) coke oven gas.
(*c*) L.D. converter gas.
(*d*) none of these.

388. Which of the following fuels has the highest calorific value per unit mass (Kcal/Kg) ?
(*a*) Coal (*b*) Kerosene
(*c*) Natural gas (*d*) Furnace oil

389. The gas which contributes maximum to the heating value of natural gas is
(*c*) CO (*b*) C_2H_6
(*c*) H_2 (*d*) CH_4

390. Synthesis gas is a mixture of
(a) CO & H_2 (b) N_2&H_2
(c) H_2 CH_4 & CO (d) CO_2 & H_2

391. For which pair of the fuel gases, calorific value of one fuel is almost double that of the other on the volume basis (i.e., Kcal/Nm^3) ; while the C.V. is same on weight basis (i.e., Kcal/Kg) ?
(*a*) Propane & acetylene.
(*b*) Propane & LPG.
(*c*) Sewage gas & Gobar gas.
(*d*) B.F. gas & C.O. gas.

392. The excess air required for the combustion of pulverised coal is of the order of about .... percent.
(*a*) 5 (*b*) 15
(*c*) 30 (*d*) 45

393. A coal with high .... content will ignite most easily.
(*a*) hydrogen (*b*) oxygen
(*c*)volatile matter (*d*)fixed carbon

394. Which of the following is not a by-product fuel ?
(*a*) Sewage gas (*b*) Refinery gas
(*c*) Producer gas (*d*) Bagasse

395. Hard coke is manufactured from
(*a*) lignite (*b*) bituminous coal
(*c*) semi-anthracite (*d*) anthracite

396. Temperature of preheated air used for the transportation of pulverised coal through pipes to the burner of a boiler furnace is restricted to a maximum limit of about 300° C, to avoid the
(*a*) requirement of large volume combustion chamber.
(*b*) risk of explosion.
(*c*) chances of clinker formation.
(*d*) incomplete combustion of coal.

397. Coals used for the generation of producer gas should have
(*a*) high caking index.
(*b*)low fusion point of ash.
(*c*) high volatile matter content.
(*d*) very low ash content (8-10%).

Answers

385. (b)	386. (d)	387. (c)	388. (c)	389. (d)	390. (a)
391. (a)	392. (b)	393. (c)	394. (a)	395. (b)	396. (b)
397. (c)					

398. Fusion point of coal ash generally, varies from 1000 to 1700°C. Ash having fusion point less than .... ^{0}C is liable to form clinker.
(*a*) 1100 (b) 1250
(*c*) 1350 (*d*) 1400

399. In low temperature carbonisation of coal as compared to high temperature carbonisation .... produced is less.
(*a*) difference in gross & net calorific value of the coke oven gas
(*b*) free carbon content in tar
(*c*) yield percentage of coke
(*d*) yield of ammonia present in coke oven gas

400. Pick out the wrong statement.
(*a*) Angle of repose of coal increases with its increasing size.
(*b*) Specific gravity of coal increases with its increasing maturity.
(*c*) Refractive index & reflectance of coal increases with the increasing rank of coal.
(*d*) Specific heat of coal decreases with increase in the volatile matter & decrease in the C/H ratio of coal.

401. Naphthalene recovered from coke oven gas is used
(*a*) as moisture proof coating on fibres.
(*b*) as moth balls (insecticides).
(*c*) as a fuel in furnaces.
(*d*) for making electrodes.

402. Coal tar fuel - 200 (CTF 200) is a mixture of
(*a*) pitch & creosote oil.
(*b*) light oil & pitch.
(*c*) anthracene & creosote oil.
(*d*) solar oil & pitch.

403. Coal tar fuels (CTF) as compared to petroleum based fuel oils have higher
(*a*) calorific value.
(*b*) higher C/H ratio.
(*c*) sulphur content.
(*d*) difference in gross & net calorific value.

404. .... prohibits the use of alcohol directly in petrol engines.
(*a*) Low octane number
(*b*) High cost and availability
(*c*) Low flash point
(*d*) Low calorific value

405. Power alcohol as compared to straight run gasoline has lower
(*a*) calorific value (*b*) octane number
(*c*) specific gravity (*d*) viscosity

406. Match the typical gas yield (Nm^3/ton) in various gaseous fuels producing processes.
List I
(*a*) High temperature carbonisation of dry wood
(*b*) Gasification of dry peat
(*c*) Low temperature carbonisation of dry coal
(*d*)Fermentation of gobar to produce gobar gas
List 11
1. 2500
II. 30
III. 850
IV. 150

407. Match the characteristics/usage of various fuels.

List I	List II
(*a*) Gobar gas	I. Butane & Propane
(*b*) Coal middling	II. Calorific value is 10500 Kcal/Kg
(*c*)Light diesel oil (LDO)	III. Fuel for thermal power plant
(*d*) Liquefied Petroleum gas (LPG)	IV.Carbon dioxide & methane

Answers

398. (a)	399. (c)	400. (d)	401. (b)	402. (a)	403. (b)
404. (b)	405. (a)	406. (iii) (i) (iv) (ii)		407. (iv) (iii) (ii) (i)	

408. Pure carbon is completely burnt in oxygen. The flue gas analysis is 70% CO_2, 20% CO and 10% O_2. The percent excess oxygen used is
(*a*) 20 (*b*) 12.5
(*c*) 0 (*d*) 10

409. Which of the following gaseous fuels requires the highest amount of theoretical combustion air?
(*a*) LPG (*b*) Coke oven gas
(*c*) Natural gas (*d*) Lurgi gas

410. Differences of gross calorific value and net calorific value of a fuel indicates the .... content in the fuel.
(*a*) carbon (*b*) hydrogen
(*c*) moisture (*d*) volatile matter

411. Suitable atomising viscosity of fuel oil for use in burner is .... Redwood-I seconds.
(*a*) 100 (*b*) 400
(*c*) 800 (*d*) 1200

412. Which parameter of coal gives a rough estimate of its calorific value?
(*a*) Ash (*b*) Hydrogen
(*c*) Fixed carbon (*d*) Total carbon

413. Gross calorific value of fuel oil, diesel and petrol is of the order of.... MJ/kg.
(*a*) 20 (*b*) 30
(*c*) 40 (*d*) 50

414. Gross calorific value of a good coal is 26000 MJ/ton, which is equivalent to .... Kcal/kg.
(*a*) 621 (*b*) 6210
(*c*)62100 (*d*) 621000

415. Calorific value of grade B Indian coal is in the range of .... Kcal/kg.
(*a*) 3600 - 4000 (*b*) 4500 - 5000
(*c*) 5600 - 6200 (*d*) 6500 - 7000

416. In a flue gas, the theoretical CO_2 is 15.5% and measured CO2 is 11% by the volume. The percentage of excess air will be
(*a*) 40.9 (*b*) 20.5
(*c*) 80.8 (*d*) 145.2

417. In a oil fired furnace, preheating of combustion air by 20°C will save about .... per cent of fuel.
(*a*) 0.2 (*b*) 1
(*c*) 5 (*d*) 10

418. Which of the following is not measured in ultimate analysis of coal?
(*a*) Hydrogen (*b*) Sulphur
(*c*) Ash (*d*) Total carbon

419. With increase in the percentage of excess air for combustion of fuel oil, the percentage of CO_2 in flue gas.
(*a*) increases (b) decreases
(*c*) remain same (*d*) unpredictable

420. Which of the following requires the least amount of oxygen/kg for complete combustion?
(*a*) Hydrogen (*b*) Methane
(*c*) Sulphur (*d*) Carbon

421. Out of the following, the lowest excess air will be required by the .... burner.
(*a*) coal (*b*) high pressure oil
(*c*) high pressure gas (*d*) low pressure oil

422. The purpose of oil atomisation in oil fired burner is to
(*a*) increase the surface area of oil .
(*b*) increase the excess air.
(*c*) reduce the power consumption.
(*d*) reduce the flue gas temperature.

Answers

408. (c)	409. (a)	410. (c)	411. (a)	412. (d)	413. (c)
414. (b)	415. (c)	416. (a)	417. (b)	418. (c)	419. (b)
420. (c)	421. (c)	422. (a)			

423. Pick out the wrong statement.
(a) Proximate analysis of coal has no connection with the word "approximate".
(b) Volatile matter content of coal influences the secondary air requirement for its combustion.
(c) Moisture present in coal aids in radiation heat transfer.
(d) High sulphur Assam coals contain a maximum of 2% sulphur.

424. Losses associated with the storage of coal include oxidation loss, wind loss and carpet loss. A 1% oxidation of coal has the same effect as ..... % ash in coal.
(a) 0.1 (6) 1
(c) 3 (d) 5.

425. Combustion of large size and irregular lump coal does not result in higher
(a) excess air requirement
(b) stack loss
(c) unburnts in the ash
(d) thermal efficiency.

426. Match the proper size of coal to be used (List II) for various coal firing systems (List I).
List I
P. Fluidiesed bed boiler
Q. Pulverised fuel fired boiler
R. Natural draft hand firing
S. Spreader stoker firing
List II
1. 80% - 200 mesh (74 micron)
2. > 10 mm
3. 15 - 25 mm
4. 25 - 75 mm
(a) P-2, Q-1, R-4, S-3
(b) P-1, Q-2, R-4, S-3
(c) P-2, Q-1, R-3, S-4
(d) P-1, Q-2, R-3, S-4

427. Natural gas
(a) is heavier than air
(b) produces heavy smoke and soot on burning
(c) practically contains no sulphur
(d) contains about 10% hydrogen

428. When 1 kg of carbon burns completely to $C0_2$ (C + O2 = CO2), it releases 8084 kcal of heat; while 1 kg of carbon when burns incompletely to produce CO(C + 1/2 O_2 = CO), it releases much less quantity (kcal) of heat. Each kg of CO thus formed means a loss of .... kcal of heat.
(a) 2430 (b) 5654
(c) 4042 (d) 7044

429. Turn down ratio of a burner is the ratio of the maximum fuel input to minimum fuel input of the burner without affecting the excess air level. The optimum turn down ratio of an oil burner is
(a) 1 : 2 (b) 1 : 5
(c) 1 : 10 (d) 1 : 15

430. Typically, Indian coals have ash fusion temperature as low as 1100°C, which results in clinker formation. Once clinker is formed, it has a tendency to
(a) stick to a hot surface rather than a cold one
(b) grow
(c) stick to a rough surface rather than a smooth one
(d) all 'a', 'b' and 'c'

Answers

423. (d) 424. (b) 425. (d) 426. (a) 427. (c) 428. (b)
429. (b) 430. (d)

431. Pick out the wrong statement about combustion of fuels.
   (*a*) Gaseous fuels require the least % excess air for combustion out of solid, liquid and gaseous fuels.
   (*b*) Spreader stoker grater firing of coal is thermally more efficient than puiverised coal firing in boilers.
   (*c*) Combustion of 1 kg of bituminous coal requires about 11 kg of theoretical air producing about 11 kg of flue gas.
   (*d*) Combustion of 1 kg of fuel oil requires about 14 kg of theoretical air producing about 14 kg of flue gas.

432. The heat loss dependent on the chemical characteristic of liquid fuel is
   (*a*) radiation loss (*b*) blowdown loss
   (*c*) hydrogen loss (*d*) unburnt loss.

433. Excess air can be derived by measuring the % of O_2 in the flue gases. In a flue gase, the theoretical CO2 is 15 % and the measured CO_2 is 12 % by volume. The % excess air is
   (*a*) 25 (b) 50
   (*c*) 100 (*d*) 150

434. Fluidised bed combustion takes place in the temperature range of.... °C.
   (*a*) 500-600 (*b*) 800-900
   (*c*) > 1000 (*d*) < 5000

435. Complete combustion of 8 kg of methane produces .... kg of CO_2.
   (*a*) 16 (*b*) 22
   (*c*) 44 (*d*) 56

436. When 1000 kg of fuel containing 60 % carbon is burnt with theoretical air, the weight of CO2 produced will be .... kg.
   (*a*) 220 (*b*) 320i
   (*c*) 440 (*d*) 4400

437. Which of the following fuels will have maximum viscosity at the same temperature?
   (*a*) Light diesel oil (LDO)
   (*b*) high speed diesel (HSD)
   (*c*) Furnace oil
   (*d*) LSHS (low sulphur heavy stock)

438. The lowest limit of exit flue gas temperature of furnace oil fired system is dependent on .... content in fuel.
   (*a*) carbon (*b*) sulphur
   (*c*) ash (*d*) moisture

439. Which of the following fuels requires maximum theoretical/stoichiometirc combustions air ?
   (*a*) Coal (*a*) Furnace oil
   (*c*) Bagasse (*d*) Hydrogen

440. The viscosity of LSHS (low sulphur heavy stock) will be maximum at which of the following temperature (°C)?
   (*a*) 35 (*b*) 65
   (*c*) 95 (*d*) 115

441. LSHS contains about .... per cent sulphur.
   (*a*) 0.5 (*b*) 2
   (*c*) 4 (*d*) 5.5

442. When pure hydrogen is burnt with stoichiometirc air, percentage nitrogen on volume basis in flue gas on dry basis will be
   (*a*) 21 (*b*) 79
   (*c*) 100 (*d*) <21

443. The country which has the largest share of coal reserve is
   (*a*) India (*b*) China
   (*c*) U.S.A (*d*) Russia

Answers

431. (b)	432. (a)	433. (c)	434. (a)	435. (b)	436. (b)
437. (a)	438. (d)	439. (b)	440. (d)	441. (a)	442. (a)
443. (c)					

444. A factory substitutes expensive fuel oil and starts firing rice husk in a boiler. The following scenario is most likely to happen.
(a) Higher boiler efficiency
(b) Higher energy consumption
(c) Higher fuel cost
(d) Less particulate emission in the chimney

445. Changing from furnace oil firing to natural gas firing in a boiler furnace will result in
(a) decreased % of wet flue gas loss.
(b) decreased SO_2 emission.
(c) increased CO2 emission.
(d) none of these.

446. With increase in excess air for combustion, which of the following will result in flue gas?
(a) %$C0_2$ decreases.
(b) %O_2 decreases,
(c) Both '*a*' and '*b*'
(d) Neither '*a*' nor

447. Sulphur percentage is furnace oil
(a) improves viscosity
(b) does not add to heat value
(c) forms soot
(d) sets lower flue gas temperature.

448. Theoretical combustion air to fuel ratio for burning 1kg of carbon is
(a) 3.76 *(b)* 4.76
(c) 11.6 *(d)* 44

449. Excess air can be derived by measuring percentage of.... in flue gases.
(a) CO (b) CO_2
(c) NO_2 *(d)* N_2

450. Carpet loss occurs in
(a) coal combustion
(b) coal storage
(c) furnace operation
(d) oil atomisation

451. For coal fired system, flame length is dictated by
(a) fixed carbon
(b) ash content
(c) volatile matter
(d) moisture

452. In pure stoichiometirc combustion of furnace oil, which of the following will be absent in flue gas?
(a) SO_2 *(b)* CO2
(c) N_2 *(d)* O_2

453. The factor that influences the atomisation of fuel oil the most is its
(a) pour point *(b)* density
(c) flash point *(d)* viscosity

454. Which of the following will require minimum excess air for combustion?
(a) Pulverised coal fired boiler
(b) Manually fired boiler
(c) Fluidised bed boiler
(d) Spreader stoker boiler

455. In oil firing boiler, primary air is used for .... while secondary air is used for completion of combustion.
(a) creating turbulence
(b) atomisation of oil
(c) cooling of oil
(d) transportation of oil

456. Which of the following cannot be used as fuel for gas turbine?
(a) LPG *(b)* Natural gas
(c) LSHS *(d)* Naphtha

Answers

444. (b)	445. (b)	446. (a)	447. (d)	448. (c)	449. (b)
450. (b)	451. (c)	452. (d)	453. (d)	454. (a)	455. (b)
456. (c)					

457. The content of solid left in the oven after volatile matter is distilled off (while analysing the composition of coal in laboratory is)
(*a*) only ash
(*b*) only moisture
(*c*) only fixed carbon
(*d*) mix of fixed carbon and ash

458. Which of the following fuels uses the lowest amount of excess air during combustion process?
(*a*) Bagasse (*b*) Fuel oil
(*c*) Natural gas (*d*) Pulverised coal

459. Moisture in combustion air .... in the fuel gas.
(*a*) contributes to latent heat loss but not sensible heat loss
(*b*) does not contributes to latent heat loss but contributes to sensible heat loss
(*c*) does not contributes to latent heat loss and sensible heat loss
(*d*) contributes to both the latent heat loss and sensible heat loss

460. Quality of waste heat in flue gas refers to its
(*a*) dust content
(*b*) temperature
(*c*) moisture content
(*d*) corrosive gases content

461. Chances of NOx formation is the least in
(*a*) pulverised coal fired boiler
(*b*) spreader stoker fired boiler
(*c*) chain grate stoker fired boiler
(*d*) fluidised bed combustion boiler

462. kg of CO2 produced on complete combustion of 1 kg of furnace oil containing 84% carbon is
(*a*) 2.4 (*b*) 3.1
(*c*) 3.8 (*d*) 4.2

463. Pour point of LSHS is .... °C.
(*a*) 2 6 (*b*) 52
(*c*) 72 (*d*) 85

464. In a flue gas theoretical CO_2 is *15%* and the measured $C0_2$ is 12 % by volume. The % excess air is.
(*a*) 10 (b) 50
(*c*) 25 (*d*) 125

465. 1 kg of wood contains 20% moisture and 5% hydrogen by weight. How many gms of water is evaporated during complete combustion of 1 kg of wood?
(*a*) 250 (*b*) 450
(*c*) 650 (*d*) 800

466. The largest heat loss in the heat balance of a coal fired boiler is due to
(*a*) radiation loss
(*b*) dry flue gas loss
(*c*) moisture in the air
(*d*) hydrogen in the fuel.

467. Which loss is the highest in a typical reheating furnace operating at a furnace temperature of 1250 °C ?
(*a*) Wall loses
(*b*) Necessary furnace opening losses
(*c*) Cooling water loss
(*d*) Flue gas loss.

468. Three "T"s for good combustion are
(*a*) thorough mixing, total air and temperature,
(*b*) time, temperature and turbulence
(*c*) total air, true fuel and turbulence
(*d*) total air, time and temperature.

Answers

457. (d)	458. (c)	459. (b)	460. (b)	461. (d)	462. (b)
463. (c)	464. (c)	465. (b)	466. (d)	467. (b)	468. (b)

469. The balance draft furnace is the one using .... for exiting the flue gas out of the furnace.
(a) forced draft fan and chimney
(b) induced draft fan and chimney
(c) induced draft fan and forced draft fan
(d) only chimney.

470. Velocity of fluidised air in a AFBC (atmospheric fluidised bed combustion) boiler is .... metre/second.
(a) 1.2-3.7 *(b)* 3.7-9.5
(c) 9.5 -12 *(d)* 12-15.5.

471. In a coal fired AFBC boiler, clinker formation occurs when bed temperature exceeds .... °C.
(a) 750 *(b)* 850
(c) 950 *(d)* 1050.

472. The coal size for over bed feeding in a AFBC boiler is mm.
(a) 3 (b) 6-10
(c) 10-15 *(d)* 15.

473. A PFBC boiler system can drive .... turbine.
(a) steam
(b) gas
(c) both 'a' and 'b'
(d) neither 'a' nor 'b'

474. Excess air quantity required in a FBC boiler is of the order of .... percent.
(a) 10-15 (b) 20-25
(c) 30 *(d)* 5-10

475. Pulverised coal is crushed upto .... percent below 200 mesh (74 microns).
(a) 45 *(b)* 55
(c) 75 *(d)* 95.

476. The A' grade Indian coals have calorific value .... kcal/kg.
(a) > 6200 *(b)* < 6200
(c) < 1300 *(d)* > 1300

477. The smallest size of coal (200 mesh = 74 micron) is required in .... coal firing.
(a) hand *(b)* stoker
(c) pulverised *(d)* fluidised bed

478. Under optimum combustion condition of fuel oil, the % of oxygen in fuel gas will be in the range of
(a) 0.1-05 *(b)* 2-3
(c) 5-8 *(d)* 10-12

479. Heat generated in combustion on per kg basis, is the highest for
(a) carbon *(b)* hydrogen
(c) sulphur *(d)* oxygen

480. Which of the following is the most prominent loss component for fuel oil fired boiler?
(a) Losses due to radiation and convection
(b) Loss due to hydrogen in fuel
(c) Loss due to moisture in fuel
(d) Loss due to dry flue gas

481. The calorific value of 'D' grade Indian coal varies between .... keal/kg.
(a) 1500 – 2500 *(b)* 2000 – 3000
(c) 4000 – 5000 *(d)* 6000 – 8000

482. Air fuel ratio (AFR) for coal ranges between
(a) 20-24 *(b)* 15-16
(c) 8-10 *(d)* 2-4

483. An oil fired boiler is operating at 5% oxygen in the flue gas. The percentage excess air supplied to the boiler is approximately .... per cent.
(a) 11 *(b)* 20
(c) 31 *(d)* 44

Answers

469. (c)	470. (a)	471. (c)	472. (b)	473. (b)	474. (c)
475. (a)	476. (c)	477. (c)	478. (b)	479. (b)	480. (d)
481. (c)	482. (b)	483. (b)			

484. Controlled wetting of coal (during coal preparation) prior to its combustion would result in the.
    (*a*) decrease in the percentage of unburnt carbon
    (*b*) reducation in the flue gas exit temperature
    (*c*) increase in the fines of coal
    (*d*) improper combustion

485. The presence of nitrogen in the products of combustion ensures that
    (*a*) complete combustion of fuel takes place
    (*b*) incomplete combustion of fuel occurs
    (*c*) dry products of combustion are analysed
    (*d*) air is used for the combustion

486. Consider the following statements. The maximum temperature produced by the combustionof a unit mass of fuel depends upon
    1. LCV
    2. ash content
    3. mass of air supplied
    4. pressure in the furnace
    of these statements:
    (*a*) 1 alone is correct
    (*b*) 1 and 3 are correct
    (*c*) 2 and 4 are correct
    (*d*) 3 and 4 are correct

487. Consider the following statements:
    1. Pulversied fuel gives up high and controlled burning rate.
    2. Insufficient air causes excessive smoking of exhaust.
    3. Excess air is provided to control the flue gas temperature.
    4. Effect of sulphur in fuel is to give high heat transfer rate.
    Which of following statements are correct?
    (*a*) 3 and 4 (*b*) 2 and 3
    (*c*) 1 and 2 (*d*) 1 and 4

488. Consider the following statements:
    1. For the combustion of pulverised coal, 5 to 10% excess air is required.
    2. Air contains 21% oxygen by weight.
    3. The flue gases from a coal fired furnace. contain around 70% nitrogen by volume.
    4. In the combustion of liquid fuels, the number of moles of the products are invariably greater than the number of moles of reactants:
    of these statements:
    (*a*) 1,2 and 4 are correct
    (*b*) 1,3 and 4 are correct
    (*c*) 2,3 and 4 are correct
    (*d*) 1 and 3 are correct

489. Consider the following statements:
    1. Motor gasoline is a mixture of various hydrocarbons with a major proportion being aromatic hydrocarbons.
    2. Compressed natural gas is mainly composed of methane.
    3. Producer gas has a predominant component of hydrogen with lesser proportion of carbon monoxide.
    4. Cetane number of fuel used in diesel engines in India is in the range of 80 to 90.
    Which of these statements are correct?
    (*a*) 1 and 2 (*b*) 1 and 3
    (*c*) 2, 3 and 4 (*d*) 1, 2, 3 and 4

Answers

484. (a) 485. (d) 486. (b) 487. (c) 488. (d) 489. (a)

10

Furnace Technology

1. An example of shaft furnace is
   (a) L.D. converter (b) Glass melting tank
   (c) Blast furnace (d) Soaking pit
2. Hearth furnaces are not used for
   (a) Roasting (b) Melting
   (c) Reheating (d) None of these
3. Which is not a hearth furnace ?
   (a) Glass tank furnace
   (b) Open hearth furnace
   (c) Cupola
   (d) Reheating furnace
4. An example of indirectly heated furnace is
   (a) Soaking pit (b) Muffle furnace
   (c) Reheating furnace (d) None of these
5. If the flame is produced under the hearth and then sweeps up into the heating chamber, this is called .... furnace.
   (a) Sidefired (b) Underfired
   (c) Overfired (d) Recirculating
6. Which is a regenerative furnace ?
   (a) Coke oven heating chamber
   (b) Open hearth furnace
   (c) Both (a) & (b)
   (d) Neither (a) nor (b)
7. Tunnel kilns are usually used for the
   (a) Roasting of pyrites
   (b) Drying of refractories
   (c) Reheating of steel slabs
   (d) Melting of glass
8. An example of recuperative furnace is
   (a) Soaking pit
   (a) Open hearth furnace
   (a) Coke ovens
   (d) None of these
9. Which is a continuous furnace ?
   (a) Coke ovens
   (b) Soaking pits
   (c) Glass tank furnace
   (d) None of these
10. Which furnace employs natural draft ?
   (a) Coke oven
   (b) High pressure boiler
   (c) Rotary kilns
   (d) L.D. converter
11. Which one is not an induced draught furnace ?
   (a) Blast furnace stove.
   (b) Sintering furnace,
   (c) High pressure boilers.
   (d) None of these.

Answers

1. (c)	2. (d)	3. (c)	4. (b)	5. (b)	6. (c)
7. (b)	8. (a)	9. (c)	10. (a)	11. (a)	

12. Which one shows the diagrammatic heat balance in a furnace ?
    (a) Sankey diagram (b) Cox chart
    (c) Ostwald chart (d) Reingleman chart
13. Specific heating capacity of a furnace is ex pressed as
    (a) weight heated/hr.
    (b) weight heated/furnace volume.
    (c) weight heated/hr/furnace volume.
    (d) none of these.
14. If fuel and air are mixed ahead of the burner, it is called a/an .... burner.
    (a) premix (b) outside mixing type
    (c)rotary (d) diffusion
15. Which of the following variables does not affect the furnace capacity?
    (a) Size of furnace.
    (b) Gas velocity in furnace.
    (c) Ratio of wall surface to stock surface.
    (d) None of these.
16. In salt bath furnaces, heat is transferred to the charge mainly by
    (a) conduction (b) convection
    (c) radiation (d) none of these
17. Reheating furnace (pusher type) is used for heating steel
    (a) ingots (b) slabs
    (c) coils (d) sheets
18. Maximum thermal efficiency of steam boiler may be about .... percent.
    (a) 10 (b) 25
    (c) 65 (d) 90
19. Soaking pits are meant for heating steel
    (a) ingots (c) coils
    (c) sheets (d) strips
20. Ostwald charts are meant for
    (a) computing the excess/deficiency of combustion air.
    (b) calculation of flue gas temperature,
    (c) computation of flue gas analysis.
    (d) none of these.
21. In which of the waste heat recovery equipment, the flow of flue gas & air is alternately reversed after a fixed interval ?
    (a) Recuperator (b) Regenerator
    (c) Waste heat boiler (d) None of these
22. Furnace pressure is normally controlled by regulating the
    (a) air pressure (b) fuel gas pressure
    (c) speed of I.D. fan (d) damper
23. Dampers are located
    (a) before the I.D. fan
    (b) after the I.D. fan.
    (c) near the top of chimney.
    (d) anywhere after the I.D. fan.
24. Furnace aerodynamics is related to .... in the furnace.
    (a) movement of gases
    (b) measurement of flue gas volume
    (c) temperature control
    (d) pressure adjustment
25. Main gas valve in a gaseous fuel burner is a .... valve.
    (a) gate (b) butterfly
    (c) globe (d) needle
26. Blast furnace furnace stoves are meant for heating
    (a) air (b) blast furnace gas
    (c) steam (d) none of these

Answers

12. (a)	13. (c)	14. (a)	15. (c)	16. (a)	17. (b)
18. (d)	19. (a)	20. (a)	21. (b)	22. (d)	23. (a)
24. (a)	25. (a)	26. (a)			

27. fuel used in B.F. stove is
    (a) pulverised coal.
    (b) furnace oil.
    (c) blast furnace gas/mixed gas.
    (d) coke oven gas.

28. Coke ovens in steel plant are heated by
    (a) electricity
    (b) blast furnace gas/mixed gas
    (c) coke oven gas
    (d) both (b) & (c).

29. Reverberatory furnace is used for
    (a) roasting/reducing of ores
    (b) annealing of steel coil
    (c) temperature control
    (d) steel melting

30. Which is the most thermally efficient furnace ?
    (a) Reheating furnace
    (b) Reverberatory furnace
    (c) Rotary kilns
    (d) Boiler furnace

31. Thermal efficiency of open hearth furnace may be about............percent.
    (a) 5 *(b)* 20
    (c) 50 *(d)* 80

32. Which furnace employs preheating, heating & soaking zones ?
    (a) Soaking pit
    (b) Reheating furnace
    (c) Open hearth furnace
    (d) Cupola

33. Rate of heat release in a furnace which is the measure of heat intensity, is defined as
    (a) Kcal/hr/m^3 combustion space
    (b) Keal/m^3 combustion space
    (c) Kcal/hr
    (d) none of these.

34. Unit of furnace loading is
    (a) Ton stock/hr/m^2 hearth area.
    (b) Ton stock/hr,
    (c) Ton stock/ m2 hearth area
    (d) Both (a) & (b).

35. In a heating process, a heat flow diagram in which the quantities of heat in the various items of a heat balance are represented by the width of a band is called the
    (a) Ostwald chart.
    (b) Cox chart.
    (c) Sankey diagram.
    (d) None of these.

36. In order to maintain an oxidising atmosphere in a furnace, it should have
    (a) more of excess air.
    (b) less of excess air.
    (c) more of CO in flue gas.
    (d) more or CO_2 in flue gas.

37. In order to maintain a reducing atmosphere in a furnace, it should have
    (a) no O_2 in flue gas.
    (b) CO in flue gas.
    (c) high ecxess air.
    (d) none of these.

38. Which of the following is not an additive for flue gases from furnace to reduce its dew-point ?
    (a) Ammonia.
    (b) Fine dolomite.
    (c) Alkaline powders.
    (d) None of these.

Answers

27. (c)	28. (d)	29. (a)	30. (d)	31. (b)	32. (b)
33. (a)	34. (c)	35. (c)	36. (a)	37. (b)	38. (d)

39. The purpose of controlling the furnace atmosphere is to prevent surface attack by
    (a) oxidation/scaling.
    (b) decarburising.
    (c) sulphur penetration
    (d) either of the above.

40. Pick out the wrong statement .
    (a) Addition of methane to the furnace atmosphere reduces decarburising by hydrogen.
    (b) Nitrogen in presence of steam decarburises high carbon steel whereas hydrocarbon gases carburise the surface of steel at annealing temperatures.
    (c) Active nitrogen (formed by cracking ammonia at the metal surface) causes nitride formation with increase in surface hardness.
    (d) none of these.

41. Protective gas used in the annealing furnace for steel coil comprises of
    (a) 95% N_2 + 5% H_2
    (b) 5% H_2 + 95% N_2
    (c) 100% CO
    (d) 100% H_2

42. Very high pressure boilers are usually .... boilers.
    (a) fire tube (b) water tube
    (c) waste heat (d) horizontal

43. The function of economiser in a boiler is to preheat
    (a) feedwater (b) combustion air
    (c) pulverised coal (d) fuel gases

44. In which of the following furnaces, gases are sucked through the stock bed from below ?
    (a) Soaking pit (b) Sintering machine
    (c) Beehive oven (d) None of these

45. In producer gas making furnace, steam is added along with air to mainly control the
    (a) fusion of coal ash and clinnker formation.
    (b) C.V. of producer gas.
    (c) tempreture of producer gas.
    (d) tar content in producer gas.

46. A refractory wall separating the stock and the source of heat is provided in a
    (a) updraft kiln
    (b) muffle furnace
    (c) continuous furnace
    (d) none of these

47. Size of combustion chamber of a furnace depends upon the
    (a) heat release rate of the fuel.
    (b) preheat tempreture of fuel & air.
    (c) method of mixing the fuel & air.
    (d) all (a), (b) (c)

48. Which furnace employs an I.D. fan for the removal of flue gases from the furnace ?
    (a) Coke ovens heating chamber.
    (b) Blast furnace stoves.
    (c) Beehive coke oven.
    (d) High pressure boiler.

49. The electric furnace in which heat is produced by a combination of induced current & skin effect is called the.....................furnace.
    (a) arc
    (b) resistance
    (c) low frequency induction
    (d) high frequency induction

Answers

39. (d)	40. (d)	41. (a)	42. (b)	43. (a)	44. (b)
45. (a)	46. (b)	47. (d)	48. (d)	49. (d)	

50. An electric furnace producing heat by means of an electric arc struck between each of three electrodes & the charge is called the............ furnace.
   (a) resistance
   (b) arc
   (c) low frequency induction
   (d) none of these

51. The resistance furnace produces heat by the
   (a) electric are struck between electrodes & the charge
   (b) flow of current through a heating element
   (c) combination of induced current & skin effect
   (d) none of these

52. An example of periodic furnace is the
   (a) blast furnace stoves
   (b) blast furnace.
   (c) coke ovens.
   (d) rotary kilns

53. Which of the following accounts for maximum loss of energy in a coal fired boiler ?
   (a) Incomplete combustion of fuel
   (b) Outgoing flue gases.
   (c) Radiation heat losses.
   (d) Unburnt carbon in ash.

54. A balanced draught boiler furnace employs .... draught fan.
   (a) induced *(b)* forced
   (c) both 'a' & 'b' *(d)* neither 'a' nor 'b'

55. Which of the following relatively employs highest heat transfer surface area in a high pressure boiler ?
   (a) Air preheater.
   (b) Economiser
   (c) Steam superheater.
   (d) furnace water wall.

56. Which of the following absorbs maximum heat in a high pressure boiler ?
   (a) Economiser
   (b) Steam superheater
   (c) Air preheater
   (d) Furnace water wall.

57. Draught produced by a tall chimney due to density difference between hot flue gas inside the chimney & atmospheric air outside the chimney is more appropriately known as the ....draught.
   (a) natural *(b)* induced
   (c) forced *(d)* chimney

58. The delivery pressure of boiler feed water pump compared to the boiler steam pressure is
   (a) same *(b)* slightly less
   (c) slightly more *(d)* much more

59. Oxygen percentage in the flue gas coming out of a gaseous fuel fired furnace should be.........percent.
   (a) < 2 (b) 5
   (c) 0 *(d)* > 10

60. The thermal efficiency of an air/fuel gas preheating recuperator may be as high as..........percent.
   (a) 50 (b) 65
   (c) 85 (d) 100

61. The thermal efficiency of a steel slab reheating furnace (pusher type) may be about .... percent.
   (a) 10 (b) 30
   (c) 65 *(d)* 85

Answers

50. (b)	51. (b)	52. (a)	53. (b)	54. (c)	55. (a)
56. (d)	57. (a)	58. (a)	59. (a)	60. (c)	61. (b)

62. Ceramic recuperators compared to metallic recuperators for the same duty
 (a) are lighter.
 (b) occupy more space.
 (c) are less costly.
 (d) have higher pressure differential between flue gas & air side.

63. Regenerators compared to the recuperators for the same duty
 (a) occupy more space.
 (b) are less costly.
 (c) are of smaller size.
 (d) all (a), (b) & (c).

64. Air infiltration in a furnace
 (a) reduces its thermal efficiency.
 (b) is indicated by flame sting out.
 (c) increases the flue gas temperature.
 (d) none of these.

65. .... atmosphere is maintained inside an iron blast furnace.
 (a) Oxidising (b) Reducing
 (c) Inert (d) Decarburising

66. Inert atmosphere is maintained in a
 (a) L.D. converter.
 (b) Reheating furnace.
 (c) Steel annealing furnace.
 (d) Boiler furnace.

67. Which of the following is a heat treatment furnace ?
 (a) Muffle furnace
 (b) Annealing furnace
 (c) Reheating furnace
 (d) Rotary kiln.

68. Calcination of limestone to produce lime is done in a .... kiln.
 (a) tunnel (b) rotary
 (c) vertical shaft (d) both (b) & (c).

69. Turndown ratio of a burner gives an idea of the .... in the furnace.
 (a) range of fuel firing rates
 (b) volume of the combustion chamber
 (c) maximum heat input rate only
 (d) minimum heat input rate only

70. Cement kiln is a
 (a) rotary kiln
 (b) tunnel kiln
 (c) natural draft furnace
 (d) batch furnace

71. Higher furnace temperature can not be achieved by the use of lean fuel gas in the furnace by
 (a) increasing the draft in the furnace.
 (b) preheating the fuel gas.
 (c) oxygen enrichment of combustion air.
 (d) preheating the combustion air.

72. Heat transfer rate to the stock/charge in the furnace does not depend upon the
 (a) emissivity of the refractory walls.
 (b) size of the furnace.
 (c) use of waste heat recovery equipments.
 (d) thickness of the stock,

73. Which of the following furnaces will have maximum thermal efficiency ?
 (a) Soaking pits.
 (b) Walking beam reheating furnace.
 (c) Boiler furnace.
 (d) Rotary kilns.

74. Thermal efficiency of blast furnace stoves used for heating blast (air) may be about .... percent.
 (a) 20 (b) 40
 (c) 60 (d) 80

Answers

62. (a)	63. (a)	64. (b)	65. (b)	66. (c)	67. (b)
68. (d)	69. (a)	70. (a)	71. (a)	72. (c)	73. (c)
74. (d)					

75. Regenerators are used for waste heat recovery in the
   (*b*) by-product coke ovens.
   (*b*) beehive coke ovens,
   (*c*) blast furnace stoves.
   (*d*) soaking pits.

76. Recuperator is
   (*a*) not used for fuel gas preheating.
   (*b*) an intermittant waste heat recovery equipment.
   (*c*) a continuous waste heat recovery equipment.
   (*d*) heavier than regenerator meant for the same duty.

77. Acid dew point temperature (ADT) of the flue gas produced by the combustion of a fuel containing *1%* sulphur may be about .... °C.
   (*a*) 80 (*b*) 130
   (*c*) 180 (*d*) 250

78. Which of the following is the most suitable equipment for preheating combustion air above 650°C ?
   (*a*) Regenerator.
   (*b*) Metallic recuperator.
   (*c*) Ceramic recuperator
   (*d*) None of these.

79. Low stack loss indicates the
   (*a*) use of sub-stoichiometric air for combustion.
   (*b*) efficient use of fuel.
   (*c*) inefficient use of fuel.
   (*d*) oxygen enrichment of combustion air.

80. Oxygen enrichment of combustion air does not result in increase of the
   (*a*) flame temperature.
   (*b*) oxygen in the flue gas.
   (*c*) stack loss.
   (*d*) heat transfer rate

81. An improved & fuel efficient version of the pusher type reheating furnace is the .... furnace.
   (*a*) walking beam (*b*) shaft
   (*c*) tunnel (*d*) rotary hearth

82. Which of the following is a batch furnace ?
   (*a*) Cupola
   (*b*) Reheating furnace
   (*c*) Glass tank furnace
   (*d*)None of these

83. Which of the following is a directly fired intermittant furnace ?
   (*a*) Walking beam reheating furnace.
   (*b*) Reverberatory furnace.
   (*c*) Tower furnace.
   (*d*) Tunnel kiln.

84. Overall thermal efficiency of a lumpy coal fired suitably designed heating furnace, if operated & maintained properly may be about .... per cent.
   (*a*) 10-15 (*b*) 25-30
   (*c*) 45-50 (*d*) 65-70

85. Turn down ratio of a burner
   (*a*) should be as low as possible i.e., 1 : 2.
   (*b*) should be 1:1 for a batch type furnace.
   (*c*) is the ratio of maximum to minimum permissible heat input rates.
   (*d*) is normally much more for a continuous furnace as compared to a batch furnace.

86. In practical operation of any furnace, zero oxygen percentage or theoretical CO_2 percentage in flue gas is rarely achieved, because of the
   (a) use of non-preheated combustion air.
   (b) use of pulverised solid fuels.
   (*c*) imperfect mixing of fuel & air and infiltration of air.
   (*d*) use of excessive positive draft in the furnace.

Answers

75. (a)	76. (c)	77. (b)	78. (c)	79. (b)	80. (b)
81. (a)	82. (a)	83. (b)	84. (b)	85. (c)	86. (c)

87. Thermal efficiency of a limestone calcination rotary kiln may be around .... per cent.
(a) 20 (b) 40
(c) 65 (d) 80

88. Metallic recuperators are not used for waste heat recovery, if the hot flue gas temperature is above .... °C. because corrosion prevails at higher temperatures,
(a) 350 (b) 750
(c) 1050 (d) 1250

89. Ceramic recuperators (made of SiC) is economical, only when used for preheating combustion air above .... °C.
(a) 250 (b) 650
(c) 850 (d) 1000

90. Flue gas outlet temperature from the chimney of any furnace should be ideally about .... °C.
(a) 50 (b) 100
(c) 150 (d) 250

91. Combustion of furnace oil in a furnace (soaking pit) with preheated combustion air at 400°C results in saving of about .... percent furnace oil as compared to its combustion with atmospheric air, if the flue gas outlet temperature from the soaking pit is 1200°C.
(a) 5 (b) 10
(c) 20 (d) 40

92. Out of the following fuels used in a furnace exhausting flue gas at a temperature of 600°C, the percentage stack loss will be maximum in case of complete combustion of
(a) furnace oil with air.
(b) furnace oil with oxygen.
(c) blast furnace gas with air.
(d) blast furnace gas with oxygen.

93. In furnaces operating at very high temperature (say 1250°C, *e.g.*, soaking pit), the maximum heat transfer takes place by
(a) conduction (b) natural convection
(c) radiation (d) forced convection

94. .... furnace is not an electric furnace.
(a) Arc
(b) Induction
(c) Pot
(d) Resistance

95. Pot furnace is either regenerative or recuperative and is used in the manufacture of
(a) glass
(b) stainless steel
(c) potteries
(d) refractory bricks

96. While the fuel fired furnace can be used upto a maximum temperature of about 1700°C, the electric furnace can be used upto a temperature of about .... °C.
(a) 2000 (b) 3000
(c) 4500 (d) 6000

97. Test specimen for determination of refractoriness under load (RUL) of a refractory is heated in a
(a) furnace employing neutral atmosphere.
(b) carbon resistance furnace.
(c) reducing atmosphere.
(d) furnace employing negative pressure.

98. Use of pulverised coal in boilers provides
(a) higher calorific value.
(b) better combustion.
(c) smokeless burning
(d) less erosion on furnace wall.

Answers

87. (b)	88. (b)	89. (b)	90. (c)	91. (c)	92. (c)
93. (c)	94. (c)	95. (a)	96. (c)	97. (b)	98. (b)

99. Overfire burning in a furnace is a phenomenon characterised by
    (a) supply of excess fuel.
    (b) supply of excess air.
    (c) burning carbon monoxide & other incombustible in upper zone of furnace by supplying more air.
    (d) none of these.

100. Which of the following is the most important deterrants to an extended use of pulverised coal in boiler firing ?
    (a) Ash disposal problem.
    (b) Excessive flyash discharge from the stack.
    (c) Higher power consumption in its transportation.
    (d) Erosion of induced draft fan blades.

101. Which of the following is a rotary furnace?
    (a) Lime kiln (b) Soaking pit
    (c) Stoves *(d)* Annealing furnace.

102. Match the type of atmosphere prevailing in various furnaces.
    List I
    (a) Blast furnace
    (b) Steel annealing furnace
    (c) L.D Converter
    (d) Boiler furnace
    List II
    I Oxidising atmosphere
    II Reducing atmosphere furnace
    III. Decarburising atmosphere
    IV. Neutral/annealing furnace

103. Match the examples of various classified type of furnaces.
    List I
    (a) Shaft furnace
    (b) Regenerative furnace
    (c) Recuperative furnace
    (d) Natural draft furnace
    List II
    I. By- product coke oven
    II. Blast furnace
    III. Blast furnace stove
    IV. Reheating furnace

104. Match the various furnaces with their fields of application.
    List I
    (a) Annealing furnace.
    (b) Tunnel kilns
    (c) Walking beam furnace
    (d) Electric arc furnace
    List II
    I. Scrap melting
    II. Stress relieving of metals
    III. Drying of refractories
    IV. Reheating of steel slabs

105. Match the temperature (°C) normally maintained in various furnaces.

List I	List II
(a) Soaking pits	1.1650
(b) Open hearth furnace	II. 1300
(c) Steel annealing furnace	III. 1800
(d) Blast furnace	III. 700

106. Match the various waste heat recovery equipments installed in various furnaces.
    List I
    (a) Glass melting tank furnace
    (b) Blast furnace stoves
    (c) Reheating furnace
    (d) Rotary lime kiln

Answers

99. (c) 100. (a) 101. (a) 102. (ii) (iv) (iii) (i) 103. (ii) (i) (iv) (iii)
104. (ii) (iii) (iv) (i) 105. (ii) (i) (iv) (iii) 106. (iv) (i) (ii) (iii)

List II

I. Combustion air prehealer

II. Recuperator

III. Waste heat boiler

IV. Regenerator

107. Match the examples of various types of furnaces.

List I

(a) Pusher type furnace

(b) Periodic furnace

(c) Pulverised coal fired

(d) Batch furnace

List II

I. Coke ovens

II. Steel slab reheating furnace

III. Cupola furnace

IV. Boiler

108. Match the fuels normally used in various furnaces.

List I

(a) Blast furnace

(b) Glass melting tank furnace

(c) Industrial boilers

(d) B.F. stoves

List II

I. Furnace oil

II. Coke

III. Blast furnace gas

IV. Pulverised coal

109. Which of the following is not a continuous waste heat recovery equipment from the flue gases going out of the furnace ?

(a) Economiser

(b) Regenerator

(c) Ceramic recuperator

(d) Waste heat boiler

110. Combustion of fuel in a furnace with oxygen enriched air results in higher

(a) flue gas volume

(b) flame temperature

(c) fuel consumption

(d) stack loss

111. Ceramic coating material for furnace refractory, which increases its emissivity & thus the radiation heat transfer rate in the furnace, comprises of

(a) graphite powder (b)thoria

(c) zircon powder *(d)* berrylium

112. Which of the following advantages is not associated with the use of preheated air for combustion of a fuel in the furnace ?

(a) Increase in calorific value of the fuel.

(b)Increase in flame temperature.

(c) Reduction in scale losses.

(d) Saving in fuel consumption.

113. Power required in case of forced draft furnace as compared to induced draft (for the same draught produced) is

(a) same (b) less

(c) more

(d) either more or less ; depends on the flue gas density

114. The method of Lobo and Evans is concerned with the calculation of the heat

(a) absorption in the radiant section of a furnace.

(b) transfer by forced convection in a reboiler.

(c) transfer in vacuum condensers.

(d) none of these.

Answers

107. (ii) (i) (iv) (iii)	108. (ii) (i) (iv) (iii)	109. (b)	110. (b)
111. (c) 112. (a)	113. (b)	114. (a)	

115. Flue gases pass through the following boiler accessories in a high pressure natural circulation boiler.

A. Induced draft fan

B. Economiser

C. Air preheater

D. Steam superheater

Which of the following is the correct sequence of flow of the flue gases through the above mentioned boiler accessories ?

(a) A, C, B, D
(b) D, B, C, A
(c) D, C, B, A
(d) D, A, B, C

116. The most commonly used combustion system manufactured in India for the thermal power plant boilers is the .... combustion system.

(a) pulverised fuel
(b) travelling grate stoker firing
(c) circulating fluidised bed
(d) fluidised bed

117. What is the effect on the performance of the economiser in the thermal power plant boiler, if the feed water temperature is very low due to preheaters being out for repairs or starving of steam supply ?

(a) It results in corrosion due to the condensation on the flue gas side of the economiser tubes due to handling of low temperature water.
(b) Economiser tube life increases due to handling of low temperature water.
(c) Heat transfer efficiency of economiser increases due to picking up high heat from the flue gases.
(d) Scale formation rate on economiser tubes increases.

118. A fire tube boiler is limited to a maximum steam pressure of about .... Kg/cm².

(a) 6
(b) 18
(c) 38
(d) 52

119. Function of feed check valve in a boiler is to

(a) facilitate circulation of high pressure feed water.
(b) check the water level in the boiler drum.
(c) prevent the back flow of water.
(d) regulate the water flow to boiler.

120. Match the location of various accessories of a boiler plant.

List I

(a) I.D. fan
(b) Air preheater
(c) Economiser
(d) Steam superheater

List II

(I) Between boiler furnace and economiser
(II) Between steam superheater and air preheater
(III) Between economiser and the chimney
(IV) Between the pressure controlling damper and the base of the chimney

121. Which property of ceramic material coating influences the energy savings in furnaces?

(a) convective heat transfer co-efficient
(b) Coating thickness
(c) Emissivity
(d) Conductivity

Answers

115. (b) 116. (a) 117. (a) 118. (b) 119. (c)
120. (iv) (iii) (ii) (i) 121. (c)

122. Heat loss through openings in furnaces is directly proportional to the .... furnace temperature.
(a) fourth power of
(b) square of absolute
(c) fourth power of absolute
(d) absolute

123. Thermal efficiency of a typical fluidised bed combustion (FBC) boiler will be around... per cent.
(*a*) 50 (*b*) 70
(*c*) 80 (*d*) 90

124. Fluidised bed temperature exceeding 950°C in a FBC boiler may result in
(a) ash carryover.
(b) melting of limestone.
(c) overheating.
(d) ash fusion.

125. In a ciruclating fluidised bed combustion boiler (CFBC), the capture and recycling of bed materials is accomplished by
(a) venturi scrubber
(b) bag filter
(c) cyclone separator
(d) gravity settling chamber.

126. In coal fired FBC boilers, the percentage bottom ash to total ash is in the range of .... per cent.
(*a*)20 (*b*) 30-40
(*c*) 60-70 (*d*) 90-95

127. Ceramic recuperators can withstand gas side temperature upto .... °C.
(*a*)500 (*b*) 1000
(*c*) 1300 (*d*) 1600.

128. The waste heat recovery equipment in a combustion system will be more economical, when the exit flue gases are at a temperature of .... °C.
(*a*) 300 (*b*) 500
(*c*) 800 (*d*) 1000

129. Which of the following loss components is the main contributor to total energy losses?
(a) Heat loss through openings.
(b) Sensible heat loss in flue gas.
(*c*) Skin loss.
(*d*) Loss due to evaporation of moisture in fuel.

130. Which of the following will not conserve energy in a furnace?
(a) Minimising excess air.
(b) Preheating of stock/charge.
(*c*) Preheating of combustion air.
(*d*) Addition of more burners

131. Pick out the wrong statement about the measures to reduce energy costs in a furnace by substitution of a fuel. Fuel switching
(*a*) may reduce energy efficiency.
(*b*) may reduce energy cost.
(*c*) may improve energy efficiency.
(*d*) always reduces energy consumption.

132. A company switches from a low ash and low moisture fuel to a less expensive high ash and high moisture fuel by retrofitting a furnace. The most likely impact of this measure may be
(*a*) Increase in both energy consumption and energy fuel costs.
(*b*) Reduction in both fuel costs and energy consumption.
(*c*) Reduction in fuel cost but increase in energy consumption.
(*d*) Reduction in fuel consumption but increase in fuel cost.

Answers

122. (c)	123. (d)	124. (d)	125. (c)	126. (c)	127. (c)
128. (c)	129. (b)	130. (d)	131. (d)	132. (c)	

133. Scale losses in reheating furnaces will increase with
(a) Decrease in furnace temperature
(b) Increase in excess air
(c) Decrease in excess air
(d) None of these.

134. Which of the following heat losses will be the highest in case of a typical oil fired reheating furnace operating at 1300°C?
(a) Cooling water loss.
(b) Wall loss.
(c) Flue gas loss.
(d) Necessary furnace opening loss,

135. Pick out the wrong statement about furnace operation. The thermal efficiency of a furnace increases by
(a) minimising the unburnt carbon mooxide loss.
(b) reducing the surface heat loss rate.
(c) increasing the excess air flow rate.
(d) preheating the combustion air.

136. Air infiltration into a continuous furnace can be most effectively avoided by
(a) closing all the furnace openings.
(b) reducing negative pressure inside the furnace.
(c) operating the furnace above 90% capacity.
(d) increasing the chimney height.

137. The low combustion temperature (850-950°C) in FBC boilers results in minimal formation of
(a) CO (b) SO_x
(c) NOx (d) NOx and SOx.

138. SO_x emission in a FBC boiler fired with high sulphur Assam coal are controlled by adding .... to the bed.
(a) lime (b) limetone
(c) sand (d) magnesia

139. Following energy efficiency measure may not give desired result in case of a batch type furnace operation.
(a) Increasing the hot face insulation of the furnace.
(b) Increasing the insulation at the outer surface of the furnace.
(c) Reducing the excess air level.
(d) Preheating the combustion air.

140. Most of the heat transfer in circulating fluidised bed (CFB) combustion boiler takes place
(a) in the bed tubes.
(b) outside the combustion zone.
(c) inside the combustion zone.
(d) in the superheater tubes.

141. Ceramic recuperator as a waste heat recovery equipment is used mainly in
(a) gas turbine.
(b) high pressure boiler .
(c) soaking pits and reheating furnace.
(d) none of these.

142. Maximum thermal efficiency of boilers occurs at about
(a) full load.
(b) 25% of full load.
(c) two third of full load.
(d)half of full load.

143. Radiation recuperators should not be employed with exhaust flue gases, which is at a temperature less than .... °C.
(a) 500 (b) 800
(c) 1000 (d) 1200

Answers

133. (b)	134. (c)	135. (c)	136. (b)	137. (c)	138. (b)
139. (b)	140. (a)	141. (c)	142. (c)	143. (b)	

144. SO_2 formation can be greatly minimised by addition of limestone for high sulphur coals (e.g., Assam coal containing 4-6 % sulphur) used in fluidised bed combustion boilers. For every 1% sulphur in coal feed, limestone required is .... per cent.
(*a*) 1 (*b*) 3
(*c*) 6 (*d*) 9.

145. Which of the following data is not required to calculate the thermal efficiency of boiler by indirect method?
(*a*) Ambient temperature.
(*b*) Blow down rate.
(*c*) Flue gas temperature.
(*d*) Calorific value of fuel.

146. In a furnace, high emissivity coatings are most effective on
(*a*) inner surface of furnace
(*b*) outer surface of furnace
(*c*) burner nozzle
(*d*) furnace charge.

147. High percentage of carbon monoxide presence in the flue gas of boiler is an indicator of
(*a*) high excess air
(*b*) low excess air
(*c*) good control of pollutants
(*d*) complete combustion.

148. Parallel flow arrangement instead of counter flow arrangement of two streams in a metallic recuperator is chosen because the
(*a*) temperature difference between the two streams is constant along the flow direction.
(*b*) pressure drops in the two streams are less in this arrangement,
(*c*) combustion air also serves the purpose of cooling the duct carrying the flue gases.
(*d*) combustion air can easily enter the recuperator in the flow direction of the flue gases.

149. For equal capacity, boiler size is smallest for .... boiler.
(*a*) AFBC (*b*) CFBC
(*c*) PFBC (*d*) pulverised coal fired

150. Preheating of combustion air by 100°C will save about .... per cent of fuel.
(*a*) 0.5 (*b*) 5
(*c*) 20 (*d*) 35

151. Increase in feed water temperature by 30°C for an oil fired boiler results in a saving of about .... per cent of fuel.
(*a*) 0.5 (*b*) 5
(*c*) 15 (*d*) 25

152. Which of the energy saving measure will not be applicable for a heat treatment furnace?
(*a*) Waste heat recovery from flue gas.
(*b*) Heat recovery from furnace openings.
(*c*) Optimum capacity utilisation.
(*d*) Complete combustion with minimum excess air.

153. For the same size and rating, efficiency of continuous furnaces will be more than batch type furnaces mainly because of
(a) reduced flue gas losses in continuous furnaces.
(b) reduced opening losses in continuous furnaces.
(*c*) increased flue gas losses in batch furnaces.
(*d*) loss of energy stored in walls of batch furnaces.

Answers

144. (b)	145. (b)	146. (a)	147. (b)	148. (c)	149. (c)
150. (b)	151. (b)	152. (b)	153. (d)		

154. A reheating furnace is operating at 10 tons/hr and consuming furnace oil @460 kg/hr (GCV of 10500 kcal/kg) for reheating the material (of sp. heat 0.13 kcal/kg) from 40° to 1100°C. The thermal efficiency of reheating furnace will be .... per cent.

(a) 14 (b) 29

(c) 42 *(d)* 50

155. Furnace wall heat loss does not depend on the

(a) temperature of external wall surface.

(b) velocity of air around the furnace.

(c) material of stock to be heated.

(d) thermal conductivity of wall brick.

156. NO_x (oxides of nitrogen) formation in fluidised bed combustion (FBC) boilers is eliminated because of the

(a) higher velocity of flue gas in combustion chamber.

(b) higher pressure of the air supplied,

(c) lower temperature in the bed and combustion chamber.

(d) higher contact time of solid particles in the flue gas.

157. The furnace temperature maintained in a batch type steel re-rolling mill furnace is .... °C.

(a) 800 *(b)* 950

(c) 1200 *(d)* 1500.

158. The temperature maintained in forging furnaces is in the range of.......................°C.

(a) 650-750 *(b)* 850-950

*(c)*1150-1250 *(d)* 1450-1500

159. Batch furnace for a given load should be operated for .... to retain the heat stored in furnace wails.

(a) 8 hours per day

(b) 48 hours every six days

(c) 24 hrs every third day

(d) 80 hours every ten days.

160. In large glass industry, the waste heat recovery equipment connected with glass melting furnace for preheating the combustion air is a

(a) regenerator.

(b) recuperator.

(c) shell and tube heat exchanger.

(d) heat wheels.

161. The operating temperature which separates ovens from furnace is .... °C.

(a) 175 *(b)* 275

(c) 370 *(d)* 475.

162. Use of higher excess air in an oil fired furnace would result in

(a) increased furnace temperature,

(b) reduced flame temperature.

(c) increased heating rate.

(d) increased furnace production rate.

163. The overall conversion efficiency of a PFBC boiler in co-generation mode is higher by .... per cent.

(a) 1 - 3 *(b)* 5 - 8

(c) 10 - 12 *(d)* 18-20.

164. In a combined cycle power plant consisting of gas turbine and waste heat boiler, the exhaust gas temperature is around .... °C.

(a) 75 *(b)* 150

(c) 350 *(d)* 500

165. The overall efficiency of combined cycle co-generation is of the order of .... per cent.

(a) 35-45 *(b)* 55-65

(c) 70-80 *(d)* 90 - 95

Answers

154. (b)	155. (c)	156. (c)	157. (c)	158. (c)	159. (a)
160. (a)	161. (c)	162. (b)	163. (b)	164. (b)	165. (c)

166. Heat to power ratio of the combined cycle co-generation is in the range of
(*a*) 0.1-0.5 (*b*) 1-1.7
(*c*) 4 - 6 (*d*) 8 - 10.

167. Clinker has a tendency to stick to .... surfaces.
(*a*) hot & rough (*b*) hot & smooth
(*c*) cold & rough (*d*) cold & smooth.

168. Excess air level in gas burner is of the order of about .... per cent.
(*a*) 1 (*b*) 10
(*c*) 5 (*d*) 20

169. Controlled wetting of coal during coal preparation would result in
(*a*) reduction in flue gas exit temperature.
(*b*) improper combustion.
(*c*) decrease in the percentage of unburnt carbon.
(*d*) increase in the fines of coal.

170. A modern recuperator used for preheating combustion air by heat recovery from flue gas can save energy upto .... per cent.
(*a*) 5 (*b*) 10
(*c*) 30 (*d*) 50

171. The typical waste gas temperature from glass melting furnace is about .... °C.
(*a*) 750 -850 (*b*) 850- 950
(*c*) 950 - 1050 (*d*) 1050 - 1550

172. In a low to medium temperature waste heat recovery system, which of the following devices is the most suitable ?
(*a*) Air preheater (*b*) Economiser
(*c*) Heat wheel (*d*) Recuperator.

173. An oil fired packaged boiler is converted to wood fired. Which of the following will be true?
(*a*) Reduction in dust emissions.
(*b*) Reduction in steam generation rate.
(*c*) Reduction in flue gas flow rate.
(*d*) Reduction in excess air rate.

174. For optimum fuel consumption, the pressure at which furnaces operate should be
(*a*) slightly positive (*b*) slightly negative
(*c*) highly positive (*d*) highly negative

175. The thermal efficiency of a furnace is decreased by
(*a*) reducing the surface heat loss
(*b*) minimising the carbon monoxide (CO) loss & unburnt loss
(*c*) increasing the excess air flow rate
(*d*) increasing the furnace loading

176. The normal operating temperature of a steel rolling mill furnace is about................. °C.
(*a*) 800 (*b*) 1000
Ic) 1200 (*d*) 1400

177. Pick out the correct statement with regard to the energy efficiency performance of a furnace.
(*a*) Infiltration and exfiltration of air have no impact on furnace efficiency.
(*b*) Infiltration of air in the furnace is more harmful than exfiltration of air.
(*c*) Exfiltration of air is more harmful than infiltration of air.
(*d*) Infiltration and exfiltration of air are equally harmful.

178. In industrial operations, fluid with temperature less than .... °C are set as the limit for waste heat recovery, because of the risk of condensation of corrosive liquids.
(*a*) 75 (*b*) 100
(*c*) 125 (*d*) 200

179. An economiser is provided to reuse flue gas heat for preheating the
(*a*) fuel (*b*) stock
(*c*) combustion air (*d*) boiler feed water

Answers

166. (b)	167. (a)	168. (c)	169. (c)	170. (c)	171. (d)
172. (c)	173. (b)	174. (a)	175. (c)	176. (c)	177. (b)
178. (b)	179. (d)				

11 Refractory Technology

1. High refractoriness of refractory bricks means, that it has a
   (a) High spalling resistance.
   (b) Low spalling resistance.
   (c) High resistance to fusion.
   (d) Low porosity.
2. Capacity of a refractory brick to withstand sudden changes in temperature is denoted by the property called
   (a) Spalling resistance.
   (b) Refractoriness.
   (c) Refractoriness under load.
   (d) None of these.
3. Permeability of bricks is a measure of the
   (a) Refractoriness.
   (b) Melting point.
   (c) Rate at which a fluid will pass through the pores.
   (d) Expansion during heating.
4. Pyrometric cone equivalent (PCE) of a refractory is the measure of its
   (a) Spalling resistance.
   (b) Fusion point.
   (c) Resistance to slag penetration
   (d) Resistance to carbon monoxide attack.
5. Fusion temperature of pure silica (SiO_2) is .... °C.
   (a) 1350 (c) 1715
   (c) 2570 (d) 2800.
6. Fusion point of an acidic refractory material is
   (a) Increased by the addition of basic oxides.
   (b) Reduced by the addition of basic oxides.
   (c) Not affected by the addition of basic oxides.
   (d) Always more than 2500°C.
7. Fusion point of a basic refractory material is
   (a) Reduced by the addition of acid oxides.
   (b) Increased by the addition of acid oxides,
   (c) Not affected by the addition of acid oxides.
   (d) Always less than 1000°C.
8. Refractoriness under load (RUL) is the most important property for the refractory bricks used in the
   (a) Hearth bottom of the furnace on which stock is placed.
   (b) Furnace insulation.
   (c) Roof of the furnace.
   (d) None of these.
9. Refractories subjected to alternate cycles of heating & cooling are liable to loose their resistance to
   (a) Thermal spalling. (b) Slag attack.
   (c) Fusion under load. (d) CO attack.

Answers

1. (c)	2. (a)	3. (c)	4. (b)	5. (b)	6. (b)
7. (b)	8. (a)	9. (a)			

10. Crushing strength of a refractory
    (*a*) Increases with rise in service temperature.
    (*b*) Decreases with rise in service temperature.
    (*c*) Is unaffected with change in service temperature.
    (*d*) Decreases with increase in porosity.
11. Thermal conductivity of refractory bricks
    (*a*) increases with decrease in porosity.
    (*b*) decreases with decrease in porosity.
    (*c*) is independent of its porosity and is max imum for insulating bricks.
    (*d*) increases with amount of air entrapped in pores.
12. Pick out the wrong statement.
    (*a*) Electrical conductivity of refractory is not important, when these are to be used in electrical furnaces.
    (*b*) Graphite & metals are the good electrical conductor among the refractories and others are all electrical insulators.
    (*c*) Refractories used for lining electrical furnaces should ordinarily have very low electrical conductivity.
    (*d*) Electrical conductivity of porous refractory materials is low.
13. Segar cones are used for the determination of the .... of refractories.
    (*a*) softening temperature.
    (*b*) spalling resistance.
    (*c*) electrical conductivity.
    (*d*) resistance to slag attack.
14. Test piece for determination for RUL of a refractory is heated in a/an
    (*a*) oxidising atmosphere.
    (*b*) reducing atmosphere.
    (*c*) electric furnace.
    (*d*) neutral atmosphere.
15. Panel test determines the.........of refractories.
    (*a*) fusion point
    (*b*) spalling resistance
    (*c*) slag penetration resistance
    (*d*) refractoriness under load
16. Spray test determines the.........of refractories.
    (*a*) resistance to slag penetration
    (*b*) resistance to CO attack
    (*c*) RUL
    (*d*) permanent linear change.
17. Which is an acidic refractory ?
    (*a*) Magnesite (*b*) Dolomite
    (*c*) Fireclay (*d*) Chrome magnesite
18. Which is a basic refractory ?
    (*a*) Fireclay
    (*b*) Silica
    (*c*) Chrome magnesite
    (*d*) None of these
19. Which is a neutral refractory ?
    (*a*) Graphite (*b*) Magnesite chrome
    (*c*) Silica (*d*) Magnesia
20. Which is not an acidic refractory ?
    (*a*) Silica
    (*b*) Fireclay
    (*c*) High alumina refractory
    (*d*) Carbon block.
21. Which is not a basic refractory ?
    (*a*) Chrome magnesite
    (*b*) Magnesite
    (*c*) Dolomite
    (*d*) Silicon carbide

Answers

10. (b)	11. (a)	12. (a)	13. (a)	14. (c)	15. (b)
16. (a)	17. (c)	18. (c)	19. (a)	20. (d)	21. (d)

22. Insulating refractories should have
    (*a*) high porosity
    (*b*) low thermal conductivity
    (*c*) both (*a*) & (*b*)
    (*d*) neither (*a*) nor (*b*).

23. Which is not an alumino-silicate refractory?
    (*a*) Fireclay bricks
    (*b*) Mullite bricks
    (*c*) Tar dolomite bricks
    (*d*) High alumina bricks.

24. Water content is ground refractory material to be shaped into bricks by hand moulding is about .... percent.
    (*a*) 5 (*b*) 20
    (*c*) 40 (*d*) 55.

25. Hollow refractory bricks are made by
    (*a*) slip casting
    (*b*) hand moulding
    (*c*) pressing/machine moulding
    (*d*) extrusion

26. Refractories are dried in the
    (*a*) rotary kilns (*b*) tunnel kilns
    (*c*) sun (*d*) none of these

27. Which refractory must have controlled atmosphere (temperature & humidity) for its safe storage ?
    (*a*) Tar bonded dolomite bricks.
    (*b*) Fireclay bricks.
    (*c*) Mullite bricks.
    (*d*) Magnesite bricks.

28. Which brick undergoes maximum shrinkage on drying ?
    (*a*) Tar bonded dolomite bricks.
    (*b*) Fireclay bricks.
    (*c*) Magnesite bricks.
    (*d*) Chromite bricks.

29. An indication of degree of firing in silica brick is its
    (*a*) specific gravity (*b*) fusion point
    (*c*) RUL (*d*) none of these.

30. Mixing of ground refractory material and water is done in a......................mill.
    (*a*) pug (*b*) ball
    (*c*) tube (*d*) rod

31. Colour of fireclay bricks is
    (*a*) light buff to reddish buff
    (*b*) yellow
    (*c*) black
    (*d*) none of these.

32. Porosity of fireclay refractories is................ percent.
    (*a*) 5—10 (*b*) 10—25
    (*c*) 25—35 (*d*) 35—50

33. Fireclay refractories have
    (*a*) low co-efficient of thermal expansion.
    (*b*) poor thermal spalling resistance.
    (*c*) tendency to expand unduly high during firing.
    (*d*) very high cost.

34. With increase in porosity, thermal spalling resistance of fireclay brick
    (*a*) increases
    (*b*) decreases
    (*c*) remains same
    (*d*) may increase or decrease

35. Refractoriness under load (RUL) of fireclay bricks (under a load of 2 kg/cm^2) is.............°C.
    (*a*) 500 (*b*) 100
    (*c*) >1350 (*d*) >2000

Answers

22. (c)	23. (c)	24. (b)	25. (a)	26. (b)	27. (a)
28. (b)	29. (d)	30. (a)	31. (a)	32. (a)	33. (b)
34. (a)	35. (c)				

36. Cold crushing strength (C.C.S.) of fireclay bricks is about..............kg/cm^2.
(*a*) 50—100 (*b*) 100—150
(*c*) 200—400 (*d*) 500—1000.

37. Fireclay bricks are not used in the
(*a*) blast furnace (*b*) hot blast stove
(*c*) cupola (*d*) walls of coke oven.

38. Fireclay bricks are used in the lining of the
(*a*) coke ovens regenerator.
(*b*) L.D. converter.
(*c*) hearth of blast furnace.
(*d*) none of these.

39. Maximum alumina content in high alumina refractory can be as high as.................percent.
(*a*) 30 (*b*) 50
(*c*) 70 (*d*) 90

40. High alumina refractory compared to fireclay bricks have
(*a*) less load bearing capacity.
(*b*) less resistance to slag attack.
(*c*) low refractoriness.
(*d*) high resistance to thermal shock & creep.

41. Sillimanite is a.......................refractory.
(*a*) basic (*b*) neutral
(*c*) high alumina (*d*) insulating

42. With increase of alumina content, the refractoriness of high alumina refractories
(*a*) increases
(*b*) decreases
(*c*) remains same
(*d*) may increase or decrease.

43. Refractoriness under high loads (RUL.) is quite close to the fusion temperature (PCE) for.....................bricks.
(*a*) fireclay (*b*) silica
(*c*) dolomite (*d*) very low alumina.

44. Si percentage in silica refractories used in the walls of coke oven is about
(*a*) 45 (*b*) 60
(*c*) 80 (*d*) 95

45. Silica refractories are not used in the
(*a*) coke oven walls.
(*b*) beehive coke ovens.
(*c*) dome and upper portion of stoves.
(*d*) open hearth furnace roof.

46. Dolomite bricks have good resistance to attack by
(*a*) molten steel (*b*) iron oxide
(*c*) lime slag (*d*) none of these

47. Periclase refractory contains mainly
(*a*) CaO (*b*) MgO
(*c*) Al_2O_3 (*d*) SiO_2.

48. Firing temperature of magnesite bricks is about.....................°C.
(*a*) 800-1000 (*b*) 1000-1200
(*c*) 1600-1800 (*d*) 2400-2600.

49. Maximum apparent porosity of magnesite bricks is about....................percent.
(*a*) 8 (*b*) 24
(*c*) 44 (*d*) 58

50. Which of the following impurities reduces the refractoriness of magnesite bricks ?
(*a*) Al_2O_3
(*b*) CaO
(*c*) SiO_2
(*d*) all (*a*), (*b*) & (*c*).

51. Magnesite bricks have poor resistance to attack by........................slag.
(*a*) lime (*b*) basic
(*c*) acid (*d*) none of these

Answers

36. (c)	37. (d)	38. (a)	39. (c)	40. (d)	41. (c)
42. (a)	42. (d)	43. (b)	44. (d)	45. (b)	46. (a)
47. (b)	48. (c)	49. (b)	50. (d)	51. (c)	

52. Slide gate in teeming ladle used for steel pouring in ingot mould is lined with the
    (a) bakelite impregnated or fused periclase.
    (b) silica bricks.
    (c) semi-silica bricks.
    (d) none of these.

53. Chrome magnesite bricks are
    (a) acidic in nature.
    (b) neutral in nature.
    (c) having higher RUL than silica bricks.
    (d) made by mixing 30% chromite & 70% periclase.

54. Chrome magnesite bricks are used in the
    (a) side walls of soaking pits & arc furnaces.
    (b) copper smelting furnaces & converters.
    (c) burner block of glass tanks.
    (d) all (a), (b) & (c).

55. Chrome magnesite is not used in the
    (a) inner lining of L. D. converter.
    (b) aluminium melting furnaces.
    (c) wear out lining of steel melting furnaces.
    (d) all (a), (b) & (c).

56. Magnesite chrome refractories .
    (a) have better spalling resistance than chrome magnesite refractories.
    (b) have very low thermal co-efficient of expansion.
    (c) are not at all resistant to the corrosive action of iron oxide.
    (d) have very low (50 kg/cm^2) cold crushing strength (C.C.S.) and cannot be used in 'metalcase' form.

57. Magnesite chrome bricks are used for the
    (a) roof lining of basic open hearth and other basic furnaces.
    (b) reheating furnaces.
    (c) soaking pits.
    (d) all (a), (b) & (c).

58. The most common bonding material for carbon refractory made from graphite is
    (a) tar/pitch (b) clay
    (c) lime (d) sodium silicate

59. Carbon refractories
    (a) do not burn/oxidise when exposed to air on heating.
    (b) are not attacked by slags as they are not wetted by melts.
    (c) do not resist temperature fluctuations.
    (d) have extremely low thermal & electrical conductivities.

60. Carbon refractories are exclusively used in the
    (a) hearth of blast furnace.
    (b) walls of coke oven.
    (c) regenerators of coke oven.
    (d) side wall of soaking pits.

61. Skull is not formed on the carbon blocks in the hearth of a blast furnace, when it becomes cold because of its
    (a) non-wetting characteristic.
    (b) high thermal conductivity.
    (c) high crushing strength.
    (d) none of these.

62. Which furnace consumes maximum refractory annually in an integrated steel plant ?
    (a) Soaking pit (b) Blast furnace
    (c) L.D. converter (d) Coke ovens

63. The main raw material for the manufacture of silicon carbide refractories is
    (a) corrundum (b) carborundum
    (c) bauxite (d) periclase

Answers

52. (a)	53. (c)	54. (d)	55. (c)	56. (a)	57. (d)
58. (a)	59. (b)	60. (a)	61. (a)	62. (b)	63. (b)

64. Silicon carbide refractories have very low
    (*a*) refractoriness (<1700°C).
    (*b*) thermal conductivity.
    (*c*) thermal shock resistance.
    (*d*) none of these.

65. Zirconia refractory
    (*a*) does not react with basic slags.
    (*b*) is produced from baddeleyite.
    (*c*) cannot be used as an insulator.
    (*d*) has poor electrical conductivity at high temperature.

66. Zirconia refractories are not used in
    (*a*) making sheaths for thermocouple.
    (*b*) lining high temperature ceramic kilns.
    (*c*) furnaces subjected to fluctuating temperature.
    (*d*) high frequency induction furnaces in the form of inductors.

67. Dry mix hydraulic compositions of refractory aggregates with suitable bonding materials are called refractory
    (*a*) mortars (*b*) cements
    (*c*) castables (*d*) none of these.

68. Dynamometer is used for the determination of.............................. of refractories.
    (*a*) Modulus of rupture.
    (*b*) Permanent linear change.
    (*c*) Resistance to CO attack.
    (*d*) RUL

69. A steel member used in furnace construction to take the thrust of the brick work is called the
    (*a*) buckstay (b) breast wall
    (*c*) armouring (*d*) baffle

70. Canister is a source of
    (*a*) silica (*b*) periclase
    (*c*) lime (*d*) none of these.

71. Thermal diffusivity of a refractory brick is high, when its.........................is high.
    (*a*) density
    (*c*) specific heat
    (*c*) thermal conductivity
    (*d*) none of these.

72. SiO_2 in firebrick is about...................percent.
    (*a*) 35—40 (*b*) 55—60
    (*c*) 80—85 (*d*) > 94

73. Which one contains minimum SiO_2 ?
    (*a*) Firebrick
    (*b*) Sillimanite
    (*c*) Semi-silica
    (*d*) Aluminous firebrick.

74. Which one contains maximum Al_2O_3 ?
    (*a*) Firebrick
    (*b*) Sillimanite
    (*c*) Magnesite
    (*d*) Aluminous firebrick

75. Which is the most stable crystalline form of silica at room temperature ?
    (*a*) Quartz (*b*) Cristobalite
    (*c*) Tridymite (*d*) None of these

76. Rate of slag attack on refractories increases with rise in temperature due to
    (*a*) decreased viscosity of slag.
    (*b*) increased thermal conductivity of brick.
    (*c*) oxidizing condition in the furnace.
    (*d*) none of these.

77. Hot metal runner in the blast furnace are lined with the.....................bricks.
    (*a*) silica (*b*) carborundum
    (*c*) fireclay (*d*) magnesite

Answers

64. (d)	65. (b)	66. (c)	67. (c)	68. (b)	69. (a)
70. (a)	71. (c)	72. (b)	73. (b)	74. (b)	75. (a)
76. (a)	77. (c)				

78. Hot blast main (carrying air at 1000°C) in blast furnace are lined with the..............bricks.
    (*a*) silica (*b*) fireclay
    (*c*) magnesite (*d*) zirconia

79. Mullite is chemically represented as
    (*a*) $Al_2O_3.2SiO_2$ (*b*) $3Al_2O_3.2SiO_2$
    (*c*) $Al_2O_3.SiO_2$ (*d*) $2Al_2O_3.3SiO_2$.

80. Which is not a natural insulating material?
    (*a*) Diatomaceous earth/kieselgur.
    (*b*) Asbestos.
    (*c*) Vermiculite.
    (*d*) None of these.

81. Which property of refractories is the most important for top section of the blast furnace ?
    (*a*) Resistance to abrasion.
    (*b*) Resistance to slag penetration.
    (*c*) Stability of volume at high temperature.
    (*d*) Resistance to corrosion by slag.

82. Lower part of hot metal mixer are lined with the.........................bricks.
    (*a*) superduty fireclay
    (*b*) high alumina
    (*c*) silica
    (*d*) carborundum.

83. Upper portion of hot metal mixer are lined with mullite bricks which helps in resisting the
    (*a*) washing action of metal.
    (*b*) corrosion by layer of slag.
    (*c*) CO attack on bricks.
    (*d*) none of these.

84. Bottom of basic open hearth furnace are constructed of the
    (*a*) dead burnt magnesite ramming mass.
    (*b*) porous fireclay bricks.
    (*c*) semi-silica bricks.
    (*d*) silicon carbide bricks.

85. Roof of a basic electric furnace is made of the.............bricks.
    (*a*) superduty fireclay
    (*b*) silica
    (*c*) chromite
    (*d*) none of these.

86. Lower courses of soaking pits are made of.................bricks to avoid the action of molten slag & scale.
    (*a*) chrome or magnesite
    (*b*) silicon carbide
    (*c*) silica
    (*d*) low duty fireclay

87. Walls, roofs & combustion chambers of annealing furnaces are made of the.........................bricks.
    (*a*) high duty fireclay
    (*b*) silica
    (*c*) mullite
    (*d*) carborundum

88. Which is the stable form of silica below 870°C ?
    (*a*) Tridymite (*b*) Cristobalite
    (*c*) Quartz (*d*) None of these

89. Which is the stable form of silica upto 1470°C ?
    (*a*) Quartz (*b*) Cristobalite
    (*c*) Tridymite (*d*) None of these

90. Which is the stable form of silica between 1470°C and the melting point 1713°C ?
    (*a*) Cristobalite (*b*) Tridymite
    (*c*) Quartz (*d*) None of these

91. Grog addition in fireclay during brick manufacture is done to
    (*a*) reduce its shrinkage on heating.
    (*b*) impart greater spalling resistance.
    (*c*) enhance the strength of fired refractories.
    (*d*) all (*a*), (*b*) & (*c*).

Answers

78. (b)	79. (b)	80. (d)	81. (a)	82. (d)	83. (a)
84. (a)	85. (b)	86. (a)	87. (a)	88. (c)	89. (c)
90. (a)	91. (d)				

92. With increasing alumina content, the fusion point of high alumina refractories
   (*a*) increases
   (*b*) decreases
   (*c*) remains constant
   (*d*) may increase or decrease

93. Chromite refractories are
   (*a*) acidic refractory.
   (*b*) neutral refractory.
   (*c*) basic refractory.
   (*d*) fired at 600°C only.

94. Carbon refractory blocks
   (*a*) are wetted by molten iron.
   (*b*) are used in the hearth of blast furnace.
   (*c*) are acidic in nature.
   (*d*) all (*a*), (*b*) & (*c*).

95. Periclase is
   (*a*) crystalline form of MgO.
   (*b*) $MgCO_3$.
   (*c*) ZrSO„.
   (*a*) amorphous magnesite.

96. Magnesite bricks are used in those parts of furnaces, which are
   (*a*) subjected to temperature fluctuation.
   (*b*) required to resist corrosive basic slag.
   (*c*) high load.
   (*d*) none of these.

97. Highest melting (m.p. = 3070°C) oxide refractory is
   (*a*) alumina (*b*) thoria
   (*c*) zirconia (*d*) magnesia

98. 'Super refractories' are made from pure
   (*a*) carbides (*b*) oxides
   (*c*) borides (*d*) nitrides

99. Carborundum used for making crucibles for melting non-ferrous metals is chemically
   (*a*) silicon carbide
   (*b*) silicon nitride
   (*c*) crystalline magnesia
   (*d*) zirconium sulphate

100. Chemically, mullite refractories is
   (*a*) Al_2O_3. $2SiO_2$ (*b*) Al_2O_3
   (*c*) $ZrSO_4$ (*d*) ThO_2.

101. ....bricks are used in the burning zone of a cement rotary kiln.
   (*a*) High alumina (*b*) Fireclay
   (*c*) Thoria (*d*) Silicon carbide

102. Roof of basic open hearth furnace is lined with the................bricks.
   (*a*) silica (*b*) fireclay
   (*c*) dolomite (*d*) magnesite.

103. Fireclay bricks is not used for lining the
   (*a*) cupola
   (*b*) gas producer
   (*c*) bottom of hot metal mixer
   (*d*) roof of open hearth furnace

104. Chrome magnesite brick is not used for lining the
   (*a*) hearth of soaking pits.
   (*b*) bottom hearth of reheating furnace.
   (*c*) coke oven regenerator.
   (*d*) burning zone of limestone rotary kilns.

105. Porosity of silica bricks varies from................ percent.
   (*a*) 5 to 10 (*b*) 20 to 30
   (*c*) 45 to 60 (*d*) 60 to 75

106. Which of the following bricks should not be used, if the furnace is to be used intermittantly ?
   (*a*) Firebricks
   (*b*) Silica bricks
   (*c*) Silicon carbide bricks.
   (*d*) Sillimanite

Answers

92. (a)	93. (b)	94. (b)	95. (b)	96. (b)	97. (b)
98. (b)	99. (a)	100. (a)	101. (a)	102. (a)	103. (c)
104. (c)	105. (b)	106. (b)			

107. Which of the following has the lowest electrical resistivity ?
(*a*) Graphite (*b*) Fireclay
(*c*) Alumina (*d*) Zircon

108. Refractoriness/fusion points of 'superduty' refractories is.................°C.
(*a*) 1520- 1630 (*b*) 1630- 1670
(*c*) 1730 (*d*) 2000

109. With decrease in porosity, the.......... ..of the refractories decreases.
(*a*) strength.
(*b*) thermal conductivity.
(*c*) spalling resistance.
(*d*) none of these.

110. ....is the measure of the strength of refractory under the combined effect of temperature & load.
(*a*) Porosity.
(*b*) RUL.
(*c*) Specific gravity.
(*d*) Thermal conductivity.

111. Spalling tendency of refractories is reduced by increasing its
(*a*) porosity.
(*b*) specific gravity.
(*c*) thermal conductivity.
(*d*) strength.

112. Refractories used in/for................should have low thermal conductivity.
(*a*) coke ovens (*b*) insulation
(*b*) regenerators (*d*) muffle furnaces

113. Except..............., all other refractories are bad conductors of electricity (i.e., have low electrical conductivity).
(*a*) fireclay (*b*) carborundum
(*c*) graphite (*d*) chromite

114. Maximum water percentage in refractory mix meant for hand moulding may be as high as.................precent.
(*a*) 2-3 (*b*) 5-7
(*c*) 7-12 (*d*) 14-20

115. Machine moulding of dry mixture of refractories requires a pressure of the order of..........................kg/cm^2.
(*a*) 10 (*b*) 500
(*c*) 100 (*d*) 1000

116. Maximum shrinkage in volume occuring during burning/firing of dried refractories may be as high as......................percent.
(*a*) 10 (*b*) 15
(*c*) 20 (*d*) 30

117. Which of the following does not occur during firing/burning of refractories ?
(*a*) Removal of water of hydration.
(*b*) Vitrification.
(*c*) Decrease in crushing strength.
(*d*) Development of stable mineral form.

118. Firing temperature is minimum (1250 - 1400°C) for the...................bricks.
(*a*) fireclay (*b*) direct bonded basic
(*c*) silica (*d*) magnesite

119. Cold crushing strength of ordinary fireclay brick is about 950 Kg/cm^2 . On exposure to a temperature of about 1500°C, its crushing strength may come down to as low as.............. Kg/cm^2.
(*a*) 450 (b) 250
(*b*) 150 (*d*) 65

120. Maximum safe working temperature for fireclay bricks is about......................°C.
(*a*) 1150 (*b*) 1300
(*c*) 1450 (*d*) 1550

Answers

107. (a)	108. (c)	109. (c)	110. (b)	111. (a)	112. (b)
113. (c)	114. (d)	115. (d)	116. (c)	117. (a)	118. (d)
119. (d)	120. (d)				

121. The maximum linear expansion of silica bricks during firing is about..............percent.
(*a*) 0.5 (*b*) 1
(*c*) 2 (*d*) 3.5

122. 10 to 30% magnesite is added to chromite to produce chrome-magnesite refractories. Magnesite addition is mainly done to improve the.....................of chromite.
(*a*) spalling resistance
(*b*) refractoriness
(*c*) crushing strength
(*d*) resistance to slag attack

123. The highest melting pure oxide (mp > 3000°C) is
(*a*) thoria (*b*) alumina
(*c*) beryllia (*d*) zirconia

124. Which is not a high alumina refractory ?
(*a*) Mullite (*b*) Corrundum
(*c*) Bauxite (*d*) Dolomite

125. 'Spinel', a refractory mineral is chemically represented as
(*a*) $MgAl_2O_4$ (*b*) Al_2O_3
(*c*) $MgSO_4$ (*d*) $MgAl_2O_3 . 2H_2O$

126. In panel test for spalling resistance, the average face temperature of panel assembly is maintained at.................°C for 24 hours.
(*a*) 700 (*b*) 1000
(*c*) 1600 (*d*) 2000

127. With increase in the alumina content in fire bricks, its fusion point (refractoriness)
(*a*) decreases linearly
(*b*) remains unchanged
(*c*) increases
(*d*) decreases

128. Refractoriness of a typical silica brick corresponds to Segar cone number 32, which is equivalent to a temperature of..................°C.
(*a*) 1380 (*b*) 1520
(*c*) 1710 (*d*) 1915

129. Which form of silica has the highest specific gravity ?
(*a*) Quartz
(*b*) Cristobalite
(*c*) Tridymite
(*d*) All have same specific gravity

130. The linear thermal expansion of.................. bricks upto 1000°C is very low of the order of < 0.5%.
(*a*) fireclay (*b*) silica
(*c*) magnesite (*d*) corundum

131. Silica bricks have low spalling resistance below 600°C, due to their
(*a*) very high thermal conductivity.
(*b*) high co-efficient of expansion upto this temperature.
(*c*) high thermal diffusivity.
(*d*) low refractoriness.

132. Electrical resistor bars are made of
(*a*) silicon carbide (*b*) alumina
(*c*) zirconia (*d*) graphite

133. Silica bricks are never used for lining the
(*a*) beehive coke ovens.
(*b*) by-product coke ovens.
(*c*) dome of blast furnace stoves.
(*d*) roof of open hearth furnace.

134. Rotary kilns meant for calcination of limestone are lined with chrome magnesite bricks in the.....................zone.
(*a*) preheating (*b*) cooling
(*c*) burning (*d*) all

Answers

121. (a)	122. (a)	123. (d)	124. (a)	125. (c)	126. (c)
127. (c)	128. (a)	129. (a)	130. (b)	131. (a)	132. (a)
133. (c)	134. (c)				

135. Bauxite calcining rotary kins are lined with the.....................bricks.
(*a*) fireclay (*b*) carbon
(*c*) 85% alumina (*d*) corundum

136. Vacuum steel degassing units are lined with the.......................bricks.
(*a*) silica (*b*) low duty fireclay
(*c*) high alumina (*d*) graphite

137. ....nozzles are used in continuous casting of steel.
(*a*) Zircon (*b*) Thoria
(*c*) Carborundum (*d*) Beryllia

138. Cold crushing strength of a refractory does not depend upon its
(*a*) shape
(*b*) composition
(*c*) firing temperature
(*d*) texture

139. High thermal conductivity of a refractory material is not important, when it is to be used in the
(*a*) coke oven regenerators
(*b*) muffle furnace
(*c*) blast furnace
(*d*) recuperators

140. Basic bricks are not made of
(*a*) fireclay (*b*) magnesite
(*c*) foresterite (*d*) chromite

141. ....of carbon blocks in the hearth of blast furnace helps in avoiding skull formation, when it becomes cold.
(*a*) High thermal conductivity
(*b*) Low porosity
(*c*) Non-wetting characteristic
(*d*) High density

142. Magnesite refractories are generally not used in the lining of
(*a*) electric furnace walls.
(*b*) steel melting furnace.
(*c*) open hearth furnace.
(*d*) burning zone of cement kilns.

143. Pick out the wrong statement.
(*a*) Insulating refractories used in place of regular refractory bricks are usually called light weight refractories and they have similar composition as heavy bricks.
(*b*) Graphite refractories are also called plumbago refractories.
(*c*) Superduty fireclay bricks correspond to a pyrometric cone equivalent of 26- 28.
(*d*) Calcined magnesite is also called dead burnt magnesite.

144. Pure bauxite is the best raw material for the manufacture of high alumina refractories in which maximum alumina content can be as high as................percent.
(*a*) 55 (*b*) 70
(*c*) 80 (*d*) 90

145. Silliminite ($Al_2O_3.SiO_2$) refractory, which is a neutral refractory, is not used in the
(*a*) pottery furnace
(*b*) glass melting furnace
(*c*) crucibles
(*d*) gas retorts

146. Which of the following is not a neutral refractory ?
(*a*) Silicon carbide (6) Magnesite
(*c*) Chromite (*d*) Graphite

147. Graphite or carbon refractories
(*a*) exhibit wetting characteristics.
(*b*) should be used in neutral or reducing atmosphere.
(*c*) exhibit high shrinkage on thermal treatment.
(*d*) are not resistant to corrosion by slag.

Answers

135. (c)	136. (a)	137. (a)	138. (c)	139. (a)	140. (c)
141. (b)	142. (c)	143. (d)	144. (d)	145. (b)	146. (b)
147. (b)					

148. ....bricks should not be used in the oxidising atmosphere.
(*a*) Tar dolomite (*b*) Carbon
(*c*) Silica (*d*) Fireclay

149. Beryllia (which is used in making crucibles for melting uranium & thorium) is superior to alumina in all respects for high temperature (1900°C) use, except
(*a*) cost.
(*b*) electrical conductivity,
(*c*) thermal conductivity.
(*d*) fusion point.

150. Most Cermets, which normally have high thermal conductivity & high thermal shock resistance, comprises of ceramic & metallic components of.............percent respectively.
(*a*) 80 & 20 (*b*) 20 & 80
(*c*) 50 & 50 (*d*) 60 & 40

151. Pick out the wrong statement.
(*a*) Fluxing material like lime is added in clay to reduce the vitrification temperature.
(*b*) Main constituents of clay are alumina & silica.
(*c*) Addition of sand in ceramic materials makes it non-plastic, increases its fusion point & reduces its shrinkage on burning.
(*d*) Vitrification of fireclay material is done to increase its porosity.

152. Because of its very high refractoriness of the order of.....................°C, silicon carbide refractories are used in zinc smelting furnace, muffle furnace & for supporting the wares in tunnel kilns.
(*a*) 1800 (b) 2200
(*c*) 2400 (*d*) 2700

153. Match the refractories used in various furnaces.

List I,	List II
(*a*) Roof of basic open hearth furnace.	I. Chrome magnesite brick.
(*b*) Burning zone of lime kiln.	II. High alumina brick,
(*c*) Dome of blast furnace stove.	III. Carbon block.
(*d*) Blast furnace hearth.	IV. Silica brick.

154. Match the values of refractoriness under load (RUL) at 2 Kg/cm^2 for various refractory bricks.

List I	List II
(*a*) MHD'firebrick	I. 1500°C
(*b*) Zirconia brick	II. 1300°C
(*c*) Carbon block	III. 1670°C
(*d*) Silica brick	IV. 1750°C

155. Match the examples of various classified type of bricks.

List I	List II
(*a*) Acidific refractory	I. Dolomite
(*b*) Basic refractory	II. Carbon
(*c*) Neutral refractory	III. Ceramic oxide + metallic alloys
(*d*) Cermet refractory	IV. Fireclay

156. Match the bulk density (gm./c.*c*) of various refractory bricks.

List I	List II
(*a*) Chrome brick	I. 1.7 - 1.9
(*b*) Zirconia brick	II. 2.7 - 3.0
(*c*) Magnesite brick	III. 4 - 4.6
(*d*) Silica brick	IV. 2.8 - 3.5

157. Match the refractory bricks used in various applications.

List I	List II
(*a*) Wear out lining of L.D converter.	I. Fireclay bricks,
(*b*) Regenerator of coke ovens.	II. Tar dolomite brick.
(*c*) Walls of coke ovens.	III. High alumina brick.
(*d*) Combustion chamber of B.F.stoves.	IV. Silica brick.

Answers

148. (b) 149. (a) 150. (a) 151. (d) 152. (d)
153. (iv) (i) (ii) (iii) 154. (ii) (i) (iv) (iii) 155. (iv) (i) (ii) (iii)
156. (iv) (iii) (ii) (i) 157. (ii) (i) (iv) (iii)

158. Spalling of a refractory means its
(*a*) softening.
(*b*) fracture due to uneven expansion at high temperature.
(*c*) resistance to compressive loads.
(*d*) resistance to chemical action of gases & molten fluxes.

159. An ideal refractory should have
(*a*) high spalling rate.
(*b*) high fusion point.
(*c*) shrinkage ability.
(*d*) none of these.

160. Refractory bricks having lower porosity have
(*a*) high insulating properties.
(*b*) low heat capacity.
(*c*) low thermal conductivity.
(*d*) greater strength.

161. The largest annual consumer of refractories is the....................industry.
(*a*) cement (*b*) metallurgical
(*c*) fertiliser (*d*) power

162. Which of the following consumes the maximum tonnage of refractories annually in an integrated steel plant ?
(*a*) Soaking pits (*b*) Reheating furnace
(*c*) L.D. converter (*d*) Rotary lime kiln

163. Ceramic recuperators used for the waste heat recovery from high temperature flue gas going out of the furnace, is made of
(*a*) fireclay (*b*) silicon carbide
(*c*) corrundum (*d*) siliceous fireclay

164. PCE value (Segar cone) of superduty refractories is more than 33, which is equivalent to the temperature of.................°C.
(*a*) 1520 (*b*) 1630
(*c*) 1670 (*d*) 1730

165. Cermets are combination of ceramic and metallic materials due to which they have high strength & resistance to high temperature. Cermets are used in the
(*a*) hearth of blast furnace.
(*b*) nuclear reactors, missiles & space crafts.
(*c*) insulation of high temperature furnaces.
(*d*) roof of electric furnaces.

166. Pick out the wrong statement.
(*a*) Refractories used in muffle furnace should have low thermal conductivity.
(*b*) The electrical resistivity of refractories drops rapidly with rise in temperature,
(*c*) For reducing spalling tendency, the refractory should be well fired and its porosity should be more.
(*d*) Refractoriness under load (RUL) of a refractory is always less than its refractoriness.

167. Pick out the wrong statement.
(*a*) A ceramic material which becomes fluid upon heating and can be moulded in liquid/viscous state is termed as glass.
(*b*) Ceramic materials do not undergo vitrification on heating.
(*c*) Ceramic materials are brittle in nature.
(*d*) Non-oxide ceramic materials generally act as a semi-conductor.

168. Refractory materials are never used in the construction of
(*a*) segar cones (*b*) orton cones
(*c*) pressure vessels (*d*) ovens & retorts

169. Spalling resistance of a refractory can not be increased by
(*a*) increasing its porosity.
(*b*) using a coarser grog during its manufacture.
(*c*) decreasing its thermal co-efficient of expansion.
(*d*) making it denser.

Answers

158. (b)	159. (b)	160. (d)	161. (b)	162. (c)	163. (b)
164. (d)	165. (b)	166. (a)	167. (b)	168. (c)	169. (d)

170. High density refractory bricks have lower
   (*a*) spalling resistance.
   (*b*) thermal conductivity.
   (*c*) fusion point.
   (*d*) slag penetration resistance.

171. Silica bricks are attacked by basic slags at high temperature. Which of the following is not used solely as a binding material ?
   (*a*) aluminium phosphate
   (*b*) water
   (*c*) lime
   (*d*) plaster of paris

172. Carbon bricks are not used in the lining of the
   (*a*) combustion chamber of blast furnace stoves.
   (*b*) electric furnaces.
   (*c*) highly chemical resistant equipments.
   (*d*) blast furnace hearth.

173. ....is not a single oxide refractory.
   (*a*) Zirconia (*b*) Silicon carbide
   (*c*) Magnesia (*d*) none of these

174. Pure oxide refractories are generally monocrystalline in nature and are self bonded.................bricks are generally used as moderator in nuclear reactors.
   (*a*) Beryllia (*b*) Carborundum
   (*b*) Corundum (*d*) Thoria

175. Thoria is an expensive refractory material and is radioactive in nature. Thorium oxide is used in the manufacture of
   (*a*) segar cones.
   (*b*) muffles for muffle furnaces.
   (*c*) insulating bricks.
   (*d*) crucibles used for melting of high purity metals.

176. An insulating refractory brick should have high porosity & low thermal conductivity. Which of the following is not used for inducing porosity in the insulating refractory bricks during its manufacture ?
   (*a*) Cork
   (*b*) Saw dust
   (*c*) Sand
   (*d*) Chemically prepared foam

177. Ferromagnetic ceramic material is not used in the
   (*a*) thermal insulation.
   (*b*) transformers.
   (*c*) magnetic switches.
   (*d*) television sets.

178. Spalling of silica bricks occurs due to abrupt volume changes, when it is cooled below a temperature of....................................°C.
   (*a*) 770 (*b*) 570
   (*b*) 270 (*d*) 70

179. Pick out the wrong statement.
   (*a*) Sand & coke is the main raw material for the manufacture of silicon carbide.
   (*b*) Carbon refractories can not be used in the furnaces operating under reducing conditions.
   (*c*) Mullite can be obtained by the heating of alusite, kyanite or silimanite.
   (*d*) Silica occurs in nature in all cellular, amorphous or crystalline form.

180. Match the percentage of various constituents in different typical refractory bricks.

List I	List II
(*a*) Percentage of SiO_2 in fireclay bricks	I. 38-42
(*b*) Percentage of Al_2O_3 in silica bricks	II. 20-35
(*c*) Percentage of Cr_2O_3 in magnesite bricks	III. 0.5-2.0
(*d*) Percentage of CaO in tar dolomite bricks	IV. 50-70

Answers

170. (a)	171. (b)	172. (a)	173. (b)	174. (a)	175. (d)
176. (c)	177. (a)	178. (c)	179. (b)	180. (iv) (iii) (ii) (i)	

181. Match the fusion point (°C) of various pure refractory oxides.

List I	List II
(a) SiO_2	I. 2050
(b) Al_2O_3	II. 2570
(c) ThO_2	III. 1715
(d) CaO	IV. 3070

182. Match the specific gravities of various pure refractory oxides.

List I	List II
(a) SiO_2	I. 9.69
(b) Al_2O_3	II. 5.56
(c) ThO_2	III. 2.32
(d) ZrO_2	IV, 3.97

183. Match the refractory used in various equipments used in non-ferrous metal industries.

List I	List II
(a) Roaster used in the extrac- extraction of zinc	I. Magnesite
(b) Blast, furnace (at tuyere level) used in the extraction of lead.	II. Dense fireclay
(c) Flash smelter used in the extraction of nickel	III. Phosphate bonded bricks
(d) Melting & holding furnaces in aluminium industry,	IV. Fusion cast mag- chrome bricks

184. The co-efficient of thermal expansion of refractory material should be
(a) low *(b)* high
(b) Medium *(d)* Very high

185. Alumina is a.....................type of refractory.
(a) Neutral *(b)* Basic
(c) Acidic *(d)* None of these

186. The emissivity of a refractory material .... with increasing temperature.
(a) increases
(b) decreases
(c) remains unchanged
(d) increases exponentially

187. An increase in the bulk density of a given refractory increases its
(a) slag penetration resistance.
(b) heat capacity.
(c) volume stability.
(d) all 'a', 'b' & 'c'.

188. Cold crushing strength (CCS) of refractory is most useful for assessing its
(a) ability to withstand the rigors of transport.
(b) adequacy of firing,
(c) abrasion resistance.
(d) bulk density and porosity.

189. Refractory of low thermal conductivity is required for use in
(a) muffles *(b)* regenerators
(c) recuperators *(d)* back up insulation

190. Which of the following does not fall in the category of monolithic refractories ?
(a) Ramming *(b)* Mouldables
(c) Castables *(d)* Fused cast.

191. Out of the following, which is the most important property for an insulating brick?
(a) Hot crushing strength.
(b) Cold crushing strength.
(c) Chemical property.
(d) Porosity.

192. The most common refractory used in general purpose furnace is.........................brick.
(a) Silica *(b)* Fireclay
(c) Zirconia *(d)* Magnesite

193. The term blanket refers to
(a) castables
(b) ramming mass
(c) ceramic fibre
(d) insulating bricks.

Answers

181. (iii) (i) (iv) (ii)		182. (iii) (iv) (i) (ii)		183. (ii) (i) (iv) (iii)	184. (a)
185. (b)	186. (d)	187. (a)	188. (d)	189. (d)	190. (d)
191. (b)	192. (c)	193. (b)			

194. Which of the following is added to aluminosilicate to help reduce the shrinkage level of ceramic fibre ?
(*a*) Al_2O_3
(*b*) ZrO_2
(*c*) SiO_2
(*d*) $CaSO_4$.

195. Zirconia is a useful refractory for glass melting furnaces, primarily because it is
(*a*) having high thermal conductivity
(*b*) not easily wetted by molten glass
(*c*) very cheap
(*d*) not attacked by basic slag.

196. The use of ceramic fibre in a batch type furnace results in fuel economy is due to
(*a*) thermal shock resistance.
(*b*) lower heat storage.
(*c*) chemical resistance.
(*d*) lower maintenance.

197. Which property is the most important for an insulating brick ?
(*a*) Porosity
(*b*) Chemical resistance
(*c*) Compact strength
(*d*) Mechanical strength

198. Waste heat recovery device in which high thermal conductivity refractory bricks are used for storing heat is
(*a*) thermocompressor
(*b*) heat pump
(*c*) regenerator
(*d*) heat pipe

199. Which is not a property of ceramic fibre insulation ?
(*a*) Thermal shock resistant.
(*b*) Low thermal conductivity.
(*c*) Light weight.
(*d*) High heat storage.

200. Ceramic fibre gives maximum savings when used in......................furnace.
(*a*) batch
(*b*) continuous
(*c*) arc
(*d*) induction

201. The insulation material suitable for low temperature application to prevent heat gain is
(*a*) fibre glass
(*b*) polyurathane
(*c*) silica
(*d*) mineral fibre.

202. The co-efficient of thermal expansion of refractory material should be
(*a*) low
(*b*) medium
(*c*) high
(*d*) very high.

203. Ceramic fibre is manufactured by blending and melting
(*a*) silica and chromite.
(*b*) alumina and silica.
(*c*) alumina and magnesia.
(*d*) silica and magnesia.

204. The heat stored in a ceramic fibre lining system is generally in the range of.................. Kcal/m^2.
(*a*) 270-400
(*b*) 2700-4000
(*c*) 27000-40000
(*c*) 27-40

Answers

194. (b)	195. (b)	196. (a)	197. (c)	198. (d)	199. (a)
200. (b)	201. (a)	202. (b)	203. (b)	204. (b)	

205. A refractory property 'PCE' stand for
    (a) preheated cone equivalant.
    (b) pyrometric cone equivalant.
    (c) pressurise cone equilibrium.
    (d) pressurised cone equivalant.

206. Out of the following, cold crushing strength is highest for...................bricks.
    (a) fireclay (b) silica
    (b) high alumina (d) dolomite.

207. Which of the following is an organic insulating material?
    (a) Mineral wool
    (b) Expanded polystyrene
    (c) Calcium silicate
    (d) Ceramic fibre.

Answers

205. (b) 206. (b) 207. (b)

12 Foundry Metallurgy

1. The bottom most part of flask in sand moulding is called
   (*a*) cope (*b*) drag
   (*c*) cheek (*d*) none of these.
2. The middle portion of flask in sand moulding is called
   (*a*) cheek (*b*) drag
   (*c*) cope (*d*) none of these
3. The top most part in sand moulding is called the
   (*a*) drag (*b*) cheek
   (*c*) cope (*d*) none of these.
4. Wax pattern is used in the
   (*a*) investment casting.
   (*b*) shell moulding,
   (*c*) die casting.
   (*d*) none of these.
5. Draft on pattern for casting is
   (*a*) identification number embossed on it.
   (*b*) shrinkage allowance.
   (*c*) taper to facilitate its removal from mould.
   (*d*) none of these.
6. Wood for patterns is considered dry, when its moisture content is............................per cent.
   (*a*) 0 (*b*) 1.5
   (*c*) < 15 (*d*) < 5.
7. Which is the best moulding process for production of large intricate castings weighing > 100 tons ?
   (*a*) Pit moulding
   (*b*) Ceramic moulding
   (*c*) Green sand moulding
   (*d*) Cement moulding
8. Loose piece patterns are
   (*a*) a type of split patterns.
   (*b*) similar to core patterns.
   (*c*) used when the pattern cannot be drawn from the mould.
   (*d*) none of these.
9. Main function of a riser is to
   (*a*) permit escape of the hot gases.
   (*b*) feed the metal to the casting, as it shrinks during solidification.
   (*c*) help for flow of metal towards the mould cavity.
   (*d*) none of these.

Answers

1. (b)	2. (a)	3. (c)	4. (a)	5. (c)	6. (c)
7. (a)	8. (c)	9. (b)			

10. Shrinkage allowance is made by
    (*a*) subtracting from external & internal dimensions.
    (*b*) adding to external & internal dimensions.
    (*c*) adding to external dimensions and subtracting from internal dimensions.
    (*d*) none of these.

11. Large floor and pit moulds are packed with sand by the help of
    (*a*) jolt machines (*b*) sand slingers
    (*c*) simple squeezers (*d*) jolt squeezer

12. The function of gate is to
    (*a*) feed molten metal from pouring basin to gate.
    (*b*) act as a reservoir of molten metal.
    (*c*) feed the casting at a rate consistent with the rate of solidification.
    (*d*) none of these.

13. Riddle is
    (*a*) used to scoop sand deep in the mould.
    (*b*) a round sieve.
    (*c*) used to make/repair corners in a mould.
    (*d*) none of these.

14. Lifter is
    (*a*) used to scoop sand deep in the mould.
    (*b*) a round sieve.
    (*c*) a long flat metal fitted with an offset handle.
    (*d*) none of these.

15. The position of the runner and ingate for ferrous metals should be as follows :
    (*a*) Runner in the drag and ingate in the cope.
    (*b*) Both runner & ingate in the cope.
    (*c*) Runner in the cope and the ingate in drag.
    (*d*) None of these

16. Purpose of riser is to
    (*a*) help feed the casting until all solidification takes places.
    (*b*) act as a reservoir for molten metal.
    (*c*) feed molten metal from pouring basin to gate.
    (*d*) none of these.

17. Plastic goods are usually produced by the .... moulding.
    (*a*) shell (*b*) injection
    (*c*) dry sand (*d*) pit

18. Cores in a centrifugal casting are made of
    (*a*) cast iron (*b*) steel
    (*c*) hard sand (*d*) none of these

19. Slick is
    (*a*) used to make/repair corners in a mould.
    (*b*) a round sieve.
    (*c*) used to scoop sand deep in the mould.
    (*d*) none of these.

20. Trowel is
    (*a*) a long flat metal plate fitted with an offset handle.
    (*b*) a round sieve.
    (*c*) used to make/repair corners in a mould.
    (*d*) none of these.

21. According to Chvarinov's rule, metal solidification time (t) is proportional to
    (*a*) V^2/A^2 (*c*) A^2/V^2
    (*c*) A/V (*d*) V/A

    where, *V* - volume of the metal,
    A - surface area

Answers

10. (c)	11. (b)	12. (c)	13. (b)	14. (a)	15. (c)
16. (a)	17. (b)	18. (d)	19. (a)	20. (a)	21. (a)

22. Function of sprue is to
    (*a*) feed molten metal from pouring basin to gate.
    (*b*) act as a reservoir of molten metal.
    (*c*) feed the casting at a rate compatible with the rate of solidification.
    (*d*) none of these.

23. Accuracy of shell moulding is of the order of .... mm/mm.
    (*a*) 0.003 - 0.005 (*b*) 0.1-0.2
    (*c*) 0.3 -0.8 (*d*) 1.6-2.5

24. Freezing ratio or relative freezing time which is used in Caine's equation is given by
    (*a*) $\frac{V_2 / A_2}{V_1 / A_1}$ (*b*) $\frac{V_2 / A_1}{V_2 / A_2}$
    (*c*) $\frac{V_1 A_1}{V_2 A_2}$ (*d*) $\frac{V_2 A2}{V_1 A_1}$

    where V_1 and A_1 = Volume and area of riser respectively
    V_2 and A_2 = Volume and area of casting respectively

25. The function of pouring basin is to
    (*a*) feed molten metal from pouring basin to gate.
    (*b*) act as a reservoir for molten metal.
    (*c*) feed the casting at a rate compatible with the rate of solidification
    (*d*) none of these.

26. Volumetric shrinkage of grey cast iron is about
    (*a*) 6-8.5% (*b*) 3.5-4.5%
    (*c*) 2.5-3% (*d*) 1.9% to negative

27. The internal hot tear defect in castings can be revealed by the........................inspection.
    (*a*) dimensional (*b*) radiographic
    (*c*) visual (*d*) none of these

28. Investment casting is used for the
    (*a*) shapes, which are very complex & intricate and can't be cast by any other method.
    (*b*) shapes, which are made by difficulty using complex patterns in sand casting.
    (*c*) mass production.
    (*d*) none of these.

29. Centrifugal method of casting is used to
    (*a*) obtain high density and pure castings.
    (*b*) cast symmetrical shapes.
    (*c*) ensure purity & density at extremities of a casting.
    (*d*) all (*a*), (*b*) & (*c*).

30. Semi-centrifugal casting is used to
    (*a*) ensure purity & density at extremities of a casting.
    (*b*) cast symmetrical object.
    (*c*) obtain high density & purity castings.
    (*d*) none of these.

31. True centrifugal casting is used to
    (*a*) obtain high density & purity castings.
    (*b*) ensure purity & density at extremities remities of a casting.
    (*c*) cast symmetrical objects.
    (*d*) none of these.

32. In slush casting process, molten metal is
    (*a*) poured into die cavity, and at a predetermined time the mould is inverted to permit a part of the metal still in molten state to flow out of cavity.
    (*b*) fed into the cavity in metallic mould by gravity.
    (*c*) forced into mould under high pressure.
    (*d*) none of these.

Answers

22. (a)	23. (a)	24. (a)	25. (b)	26. (d)	27. (b)
28. (a)	29. (b)	30. (a)	31. (c)	32. (a)	

33. In die casting process, molten metal is
    (a) fed into the cavity in metallic mould by gravity.
    (b) forced into mould under high pressure.
    (c) poured into dies cavity and the mould is inverted to permit a part of the metal still in molten condition to flow out of cavity.
    (d) none of these.

34. In permanent mould casting method, molten metal is
    (a) fed into the cavity in metallic mould by gravity.
    (b) forced into mould under high pressure.
    (c) forced into cavity by inserting a core/plunger.
    (d) none of these.

35. In hot chamber method of die casting
    (a) die is kept hot by electrical heating.
    (b) only low melting point metal can be cast.
    (c) the melting pot is separate from die casting machine.
    (d) high melting point metals can be cast.

36. In cold chamber method of die casting
    (a) die is kept hot by electrical heating.
    (b) high melting point metals can be cast.
    (c) the melting pot is separate from die casting machine.
    (d) both *(b)* & *(c)*.

37. Cold shots are casting defects, which is
    (a) due to some sand shearing from cope surface.
    (b) caused by two streams of metals, that are too cold to fuse properly.
    (c) due to discontinuities in metal casting resulting from hindered contraction.
    (d) none of these.

38. Scabs/buckles are the casting defects caused by
    (a) some sand shearing from the cope surface.
    (b) two streams of metals that are too cold to fuse properly.
    (c) discontinuity in metal casting resulting from hindered contraction.
    (d) none of these.

39. Blow holes are casting defects caused by
    (a) some sand shearing from the cope surface.
    (b) excessive gaseous substances not able to escape.
    (c) discontinuity in casting resulting from hindered contraction.
    (d) two stream of metals that are too cold to fuse properly.

40. Hot tears are casting defects caused by
    (a) discontinuity in metal casting resulting from hindered contraction.
    (b) two streams of metals that are too cold to fuse properly,
    (c) some sand shearing from the cope surface.
    (d) none of these.

41. Chilled castings (used for making cast iron rolls, railway carriage wheels etc.) are commonly produced by the..................moulding.
    (a) dry sand *(b)* injection
    (c) extrusion *(d)* green sand

42. Chills are used while casting, when.................. at a section.
    (a) additional strength is required
    (b) a void is desired
    (c) localised cooling is required
    (d) a metallic insertion is required

Answers

33. (b)	34. (a)	35. (b)	36. (d)	37. (b)	38. (a)
39. (b)	40. (a)	41. (a)	42. (b)		

43. The purpose of baking core is to
   (*a*) remove moisture.
   (*b*) harden the binder.
   (*c*) fuse silica grains.
   (*d*) burn carbonaceous matter.

44. When the change in casting sections is too abrupt, it causes the..........................defect.
   (*a*) shrinkage (*b*) hot tear
   (*c*) misrun (*d*) none of these

45. The most detrimental defect of cupola is, that it
   (*a*) is not possible to make high alloy addition during melting.
   (*b*) can not be used to refine metals.
   (*c*) is not possible to take sample of molten metal for the purpose of quality control.
   (*d*) none of these.

46. In true centrifugal casting processes, the .... of the mould & casting are the same.
   (*a*) size (*b*) shape
   (*c*) plane (*d*) axis

47. Small precision castings are preferably made by the............................process.
   (*a*) centrifugal (*b*) shell moulding
   (*c*) lost wax (*d*) die casting

48. Turbulent flow of metals in the casting of non-uniform sections causes the .... defect.
   (*a*) hot tear (*b*) shrinkage
   (*c*) cold shut (*d*) none of these

49. The shell of a shell moulding process is made from a mixture of sand and
   (*a*) water (*b*) oil
   (*c*) resin (*d*) dextrir

50. Which of the following is used for drawing the mould from the pattern ?
   (*a*) Sand slinger
   (*b*) Squeezing machine
   (*c*) Stripper plate machine
   (*d*) Jolt machine

51. Which of the following is used for producing uniform packing of sand in the mould ?
   (*a*) Diaphragm moulding machine
   (*b*) Sand slinger
   (*c*) Squeezing machine
   (*d*) Jolt machine

52. A...................machine is used to give uniform sand hardness throughout the mould.
   (*a*) squeezing
   (*b*) diaphragm moulding
   (*c*) jolt
   (*d*) stripper plate

53. A.......................machine is used for ramming the sand harder at the back of the mould and softer on the pattern face.
   (*a*) Jolt
   (*b*) Squeezing
   (*c*) Stripper plate
   (*d*) Diaphragm moulding

54. Which of the following is used to deliver molten metal from pouring basin to gate ?
   (*a*) Core (*b*) Sprue
   (*c*) Riser (*d*) chill

55. Sand used on the face of the pattern before moulding is called the......................sand.
   (*a*) green (*b*) loam
   (*c*) parting (*d*) oil

56. .... type binder is commonly used in making oil sand.
   (*a*) Organic (*b*) Inorganic
   (*c*) Clay (*d*) either

Answers

43. (b)	44. (a)	45. (b)	46. (a)	47. (c)	48. (c)
49. (b)	50. (c)	51. (b)	52. (b)	53. (b)	54. (b)
55. (c)	56. (b)				

57. Which of the following is used to ram the sand harder at the pattern face with decreasing hardness towards the back of mould ?
(*a*) Sand slinger
(*b*) Jolt machine
(*c*) Squeezing machine
(*d*) None of these

58. .... sand is used for making core.
(*a*) Dry (*b*) Green
(*c*) Oil (*d*) Loam

59. A casting defect, which results in mismatching of the top & bottom part of the casting is known as the
(*a*) swell (*b*) scab
(*c*) shift (*d*) sand wash

60. A casting defect, which results in general enlargement of the casting is known as the
(*a*) blow hole (*b*) scab
(*c*) sand wash (*d*) swell

61. A casting defect, which occurs near the ingates as rough lumps on the surface of a casting is known as the
(*a*) swell (6) blowhole
(*c*) sand wash (*d*) scab

62. Core in centrifugal casting method is made of
(*a*) sand (*b*) steel
(*c*) copper (*d*) none of these

63. .... casting method is adopted for ornaments and toys of non-ferrous alloys,
(*a*) Centrifugal (*b*) Slush
(*c*) Die (*d*) Investment

64. Casting method in which the molten metal is fed in the cavity of a metallic mould by gravity is known as the..................casting method.
(*a*) slush
(*b*) permanent mould
(*c*) centrifugal
(*d*) die

65. .... casting is used for producing steel/cast iron pipes.
(*a*) Investment (*b*) True/centrifugal
(*c*) Die (*d*) Slush

66. The molten metal is forced into mould under high pressure in a...................casting method.
(*a*) slush (*b*) die
(*c*) centrifugal (*d*) permanent mould

67. The molten metal is poured & allowed to solidify, while the mould is removing, in the .... casting method.
(*a*) permanent mould
(*b*) die
(*c*) slush
(*d*) none of these

68. .... core is used when a hole/ cavity to be cored is not in line with the parting surface.
(*a*) Drop (*b*) Balanced
(*c*) Horizontal (*d*) Vertical

69. The property of the sand due to which it clings to the sides of a moulding box is called
(*a*) adhesiveness (*b*) cohesiveness
(*c*) permeability (*d*) collapsibility

70. The property of the sand due to which the sand grains stick together is called the
(*a*) permeability (*b*) collapsibility
(*c*) adhesiveness (*d*) cohesiveness

71. A mixture of 50% sand and 50% clay is called...................sand.
(*a*) green (*b*) dry
(*c*) loam (*d*) none of these

Answers

57. (c)	58. (c)	59. (d)	60. (c)	61. (d)	62. (b)
63. (b)	64. (b)	65. (b)	66. (d)	67. (a)	68. (b)
69. (a)	70. (d)	71. (c)			

72. Which is used for moistening the sand around the edge before removing the pattern ?
(*a*) Slick (*b*) Swab
(*c*) Lifter (*d*) Riddle

73. Which is used for repairing and finishing the mould ?
(*a*) Slick (*b*) Swab
(*c*) Lifter (*d*) Riddle

74. .... is used for smoothing and cleaning out depression in the mould.
(*a*) Slick (*b*) Swab
(*c*) Lifter (*d*) Riddle

75. .... is used for reinforcement of sand in the top part of the moulding box.
(*a*) Riddle (*b*) Swab
(*c*) Lifter (*d*) Slick

76. A taper provided on the pattern for its easy & clean withdrawal from the mould is known as the .... allowance.
(*a*) shrinkage (*b*) draft
(*c*) distortion (*d*) machining

77. Cores do not
(*a*) form internal cavities in the casting.
(*b*) improve mould surface.
(*c*) form a part of a green sand mould.
(*d*) none of these.

78. On an average, the tolerance of the order of +0.2 mm can be produced by the.................. method.
(*a*) shell moulding
(*b*) investment casting
(*c*) both (*a*) & (*b*)
(*d*) neither (*a*) nor (*b*)

79. Blowhole is a casting defect, which occurs due to
(*a*) poor permeability of moulding sand.
(*b*) shearing of some sand particles from the mould cavity.
(*c*) two streams of metals that are too cold to fuse properly.
(*d*) none of these.

80. .... (a casting defect) occurs due to some shearing away of sand from the cope surface.
(*a*) Blowhole (*b*) Inclusion
(*c*) Scab (*d*) Honey

81. The cold chamber die casting method can be used for metals having
(*a*) low melting point (*b*) high melting point
(*c*) low fluidity (*d*) none of these

82. Loose piece patterns are
(*a*) same as core boxes.
(*b*) similar to split pattern.
(*c*) used for very small casting.
(*d*) none of these.

83. Shrinkage allowance for cast iron is about .... mm/metre.
(*a*) < 1 (*b*) 10
(*c*) 20 (*d*) <20.

84. Wood pattern used for the production of metal pattern is termed as the.............pattern.
(*a*) duplicate (*b*) solid
(*c*) master (*d*) skeleton

85. A long metal plate filled with offset handle for smoothing the mould is called the
(*a*) slick (*b*) lifter
(*c*) riddle (*d*) none of these

86. In die casting process, the molten metal is fed by
(*a*) allowing it to flow by gravity.
(*b*) forcing it into mould under pressure.
(*c*) forcing it by inserting the plunger into the cavity.
(*d*) none of these.

Answers

72. (b)	73. (c)	74. (c)	75. (a)	76. (b)	77. (d)
78. (a)	79. (a)	80. (c)	81. (b)	82. (d)	83. (b)
84. (c)	85. (d)	86. (b)			

87. For continuous casting method, generally, graphite moulds are used, due to fact that
   (*a*) they are cheaper.
   (*b*) they are self lubricating.
   (*c*) the metal wets the mould slightly.
   (*d*) a very thin film of lubricant is required.

88. Heavy & large castings are usually made by the
   (*a*) green sand moulding
   (*b*) dry sand casting
   (*c*) pit moulding
   (*d*) pressure moulding.

89. Casting of ferrous alloy is normally done by the .... machine.
   (*a*) direct blow (*b*) hot chamber
   (*c*) cold chamber (*d*) none of these

90. The core in the centrifugal casting is made of
   (*a*) sand (*b*) steel
   (*c*) plastic (*d*) none of these

91. The trowel is used to
   (*a*) blow the excess sand from the mould,
   (*b*) repair corners in a mould.
   (*c*) flatten & smoother the sand during moulding.
   (*d*) scoop sand deep in the mould.

92. .... patterns are used, when several patterns are to be mounted at a time,
   (*a*) Skeleton (*b*) Loose-piece
   (*c*) Sweep (*d*) Match plate.

93. Cast iron & steel pipes are cast by the .... casting.
   (*a*) investment
   (*b*) true centrifugal
   (*c*) continuous
   (*d*) die

94. ....is cast using hot chamber die casting method.
   (*a*) Aluminium
   (*b*) Magnesium
   (*c*) Brass
   (*d*) Alloys of lead, tin & zinc

95. Match plate pattern is used in the .... moulding.
   (*a*) pit (*b*) machine
   (*c*) green sand (*d*) bench

96. What is the material of construction of mould used for casting ferrous materials in continuous casting process ?
   (*a*) Copper
   (*b*) Mild steel
   (*c*) High carbon steel
   (*d*) Low carbon steel

97. In order to facilitate easy withdrawal of pattern, the
   (*a*) draft is provided on the pattern.
   (*b*) allowances are made on the pattern.
   (*c*) pattern is made smooth.
   (*d*) none of these.

98. The mould is housed in a
   (*a*) cope (*b*) drag
   (*c*) moulding box (*d*) flask

99. Draft on pattern for casting is
   (*a*) for machining allowance.
   (*b*) for shrinkage allowance.
   (*c*) tapered to facilitate its removal from mould.
   (*d*) none of these.

100. The middle part of flask in sand moulding is called the
   (*a*) drag (*b*) cope
   (*c*) cheek (*d*) none of these

Answers

87. (b)	88. (b)	89. (d)	90. (d)	91. (c)	92. (d)
93. (b)	94. (d)	95. (b)	96. (a)	97. (a)	98. (d)
99. (c)	100. (c)				

101. Sometimes, cow dung is used in the .... moulding.
(*a*) bench (*b*) green sand
(*c*) dry sand (*d*) none of these

102. Which of the following casting processes is used for making water pipes of large length & diameter ?
(*a*) Semi-centrifugal casting
(*b*) Sand spun process
(*c*) De Lavand method
(*d*) Watertown method

103. .... is not a casting process.
(*a*) Shell moulding (*b*) Slush process
(*c*) Extrusion (*d*) none of these

104. The function of trowel in a foundry is to
(*a*) blow the excess sand.
(*b*) swab the edges & the mould cavity.
(*c*) flatten & smoothen the sand during moulding.
(*d*) ram the moulding sand,

105. In true centrifugal casting, the impurities
(*a*) are thrown out.
(*b*) get collected at the inner surface.
(*c*) get collected at the outer surface.
(*d*) get collected at the middle portion.

106. Pick out the correct statement.
(*a*) Wood is used for making disposable patterns.
(*b*) Sweep patterns eliminate the need for three dimensional patterns.
(*c*) Loose piece patterns are used, when the pattern can be drawn from the mould.
(*d*) none of these.

107. Antioch process is
(*a*) for making porous moulds.
(*b*) continuous casting process.
(*c*) for brazing.
(*d*) none of these.

108. The draft on casting is generally of the order of .... mm/m.
(*a*) 5-10 (*b*) 10-15
(*c*) 15-25 (*d*) 25-35

109. Which of the following is a non-destructive test conducted for castings ?
(*a*) Tensile test (*b*) Fatigue test
(*c*) Hardness test (*d*) Radiographic test

110. Which of the following is a destructive test conducted for castings?
(*a*) Ultrasonic test.
(*b*) Dye-penetration test,
(*c*) Wear resistance test.
(*d*) Magnetic particle test.

111. Internal stresses may develop in a casting due to
(*a*) hydroblasting (*b*) shot blasting
(*c*) flame cutting (*d*) all (*a*), (*b*) & (*c*)

112. Ornaments, toys and statues are made using
(*a*) slush casting (*b*) continuous casting
(*c*) antioch process (*d*) none of these

113. Which of the following is used for increasing the distance between two risers in any mould ?
(*a*) Horn gate (*b*) Chill
(*c*) Skin bob (*d*) Strainer core

114. Strainer core, which is fixed in the bottom of the sprue, is used to
(*a*) minimise shrinkage cavity.
(*b*) provide conical hole in the casting.
(*c*) filter molten metal entering the mould cavity.
(*d*) none of these.

Answers

101. (b)	102. (c)	103. (c)	104. (c)	105. (b)	106. (b)
107. (a)	108. (b)	109. (d)	110. (c)	111. (d)	112. (a)
113. (b)	114. (c)				

115. Which of the following is used to compact sand in the mould ?
   (a) Gagger (b) Rammer
   (c) Trowel (d) none of these

116. Gaggers are used to
   (a) reinforce the sand in mould.
   (b) withdraw pattern from the mould during finishing.
   (c) remove big lumps of sand & foreign particles from the sand.
   (d) none of these

117. Drawspike is used to
   (a) make a hole in the casting.
   (b) withdraw pattern from the mould during finishing.
   (c) compact sand in the mould.
   (d) none of these.

118. Which of the following is not a pattern making material ?
   (a) Thermoplastic plastic
   (b) Thermosetting plastic
   (c) Plaster of Paris
   (d) None of these

119. .... pattern is used for making circular castings.
   (a) Match plate (b) Sweep
   (c) Segmental (d) both

120. Which is the most commonly used foundry clay ?
   (a) Fuller's earth (b) Bentonite
   (c) Limonite (d) Kaolinite

121. Silica and iron oxide is mixed as an additive in the moulding sand. It provides
   (a) good surface & mouldability property.
   (b) edge hardness & dry strength in the mould.
   (c) increased hot strength of mould thereby minimising metal penetration defect.
   (d) none of these

122. Moulds in a continuous casting process is made of
   (a) aluminium (b) copper
   (c) stainless steel (d) bronze

123. What is the melting point (°C) of pure silica sand (SiO_2 = 99.8%) ?
   (a) 750 (b) 950
   (c) 1150 (d) 1450

124. Pick out the wrong statement.
   (a) Finer sand has comparatively lower fusion point than that of coarser grain sand.
   (b) With the increase in moisture content, the green compression strength of the moulding sand decreases.
   (c) Bentonite is used in moulding sand to improve the hot strength of the mould.
   (d) The permeability of moulds made by finer sand is less than that made by coarser grain sand.

125. The use of....................enhances the efficiency of riser.
   (a) chills (b) chaplcts
   (c) bottom gate (d) none of these

126. Out of the following, the pouring rate is the slowest for .... casting.
   (a) aluminium (b) cast iron
   (c) mild steel (d) zinc

127. Visual inspection of casting can not detect the .... defect.
   (a) blow holes
   (b) shifts
   (c) metal penetration
   (d) none of these

Answers

115. (b)	116. (a)	117. (b)	118. (d)	119. (d)	120. (b)
121. (c)	122. (b)	123. (c)	124. (c)	125. (a)	126. (a)
127. (a)					

128. .... process is used for casting dental alloys.
   (*a*) True centrifugal
   (*b*) Investment casting
   (*c*) Dry sand
   (*d*) None of these

129. For steel castings, pouring time (*t*) in seconds is given by
   (*a*) $K.x^{1/2}$ (*b*) $K/x^{1/2}$
   (*c*) K.x (*d*) $K.x^2$
   where, x = weight of the casting, lbs k = a constant = 1.2 (for castings weighing < 100 lbs) and 0.4 (for castings weighing upto 1 lakh lbs).

130. To make a hole in the casting, ... core is used.
   (*a*) cover (*b*) wing
   (*c*) balanced (*d*) none of these

131. Pick out the correct statement.
   (*a*) Shrinkage allowance provided in cast iron is more than that in cast steel products.
   (*b*) Mouldings sand is also termed as base sand.
   (*c*) Fine sand and thermo-setting resin is used in shell moulding.
   (*d*) Plaster is never used for making patterns.

132. Rapping operation is meant for the
   (*a*) chaplet fixing.
   (*b*) pattern stripping after moulding.
   (*c*) core making.
   (*d*) none of these.

133. .... allowance is provided in the pattern to facililate its easy stripping from the moulds.
   (*a*) Draft (*b*) Shrinkage
   (*c*) Rapping (*d*) Warp

134. .... moulds are most suitable for satisfactory casting of reactive metals like titanium alloys.
   (*a*) Dry sand (*b*) Plaster of paris
   (*c*) Graphite (*d*) Loam sand

135. .... castings are normally made using true centrifugal casting process,
   (*a*) Hollow cylinderical
   (*b*) Alloy steel
   (*c*) Small solid
   (*d*) Complicated shape

136. .... pattern is used in the case of shell moulding.
   (*a*) Cold (*b*) Hot
   (*c*) No (*d*) Plaster of paris

137. Mould made of....................is used in centrifugal casting.
   (*a*) plaster of paris (*b*) wax
   (*c*) frozen mercury (*d*) metal

138. .... is used for making chaplets.
   (*a*) Plaster of paris (*b*) Metal
   (*c*) Wood (*d*) Wax

139. .... of sand is found by shatter test.
   (*a*) Toughness
   (*b*) Grain fineness number
   (*c*) Sintering temperature
   (*d*) None of these

140. Mould wash is used to...............of the moulds.
   (*a*) increase the hot strength
   (*b*) increase the green compressive strength
   (*c*) decrease the permeability
   (*d*) none of these

Answers

128. (b)	129. (a)	130. (b)	131. (b)	132. (a)	133. (c)
134. (c)	135. (a)	136. (b)	137. (d)	138. (b)	139. (a)
140. (d)					

141. With increase in its moisture content, the permeability number of moulding sand
(a) increases.
(b) first increases and then decreases
(c) decreases.
(d) first decreases and then increases

142. The purpose of providing relief sprues or skin bobs in gating system is to
(a) collect slag or dross present in the metal.
(b) minimise the erosion of downsprue.
(c) avoid the use of runner and step gate.
(d) none of these.

143. Cover core is used
(a) to cover kiss cores
(b) on the top of core
(c) in investment casting mould
(d) none of these.

144. Blow hole formation in castings is not promoted by the use of
(a) semi-dried moulds
(b) non-preheated ladles
(c) organic binder for core making
(d) inorganic binder as moulding material.

145. High pressed/density moulds are made on machines applying a pressure of .... kgf/cm^2 onwards.
(a) 2 (b) 5
(c) 10 (d) 25

146. Addition of....................is done to achieve grain refinement of aluminium melt.
(a) sodium (b) titanium
(c) chromium (d) boron

147. 'Dross' floating on the molten bath of aluminium is nothing but aluminium
(a) carbide (b) oxide
(c) sulphate (d) chloride

148. Pick out the wrong statement.
(a) Even without doing any heat treatment. low carbon steel castings may be used.
(b) Rough coarse grained surface is produced in shrinkage cavity in a cast product.
(c) The moulding materials obtained after shaking out the silica sand moulds are reused.
(d) In mercast process, the pattern is made of plaster of paris.

149. Cracks in casting is not caused due to the
(a) failure of tempering just after quenching.
(b) high pouring temperature,
(c) residual/internal stresses.
(d) use of improper quenching medium.

150. Skim bob is attached to
(a) ingates (b) pouring cup
(c) sprue (d) risers

151. 'Fin' in a casting is a
(a) void formation
(b) thin unwanted projection of metal
(c) crack formed due to internal/residual stresses
(d) defect due to misrun of metal.

152. In a foundry, a/an.......................furnace is not used for melting of steel.
(a) induction (b) acid electric arc
(c) rocking arc (d) none of these

153. Height of coke bed in a cupola is measured from the
(a) slag tapping hole
(b) tuyers level,
(c) metal tapping hole
(d) none of these.

Answers

141. (b)	142. (a)	143. (d)	144. (d)	145. (c)	146. (c)
147. (b)	148. (d)	149. (b)	150. (c)	151. (b)	152. (c)
153. (c)					

154. Riser of minimum size will be obtained from the............................shape.
(*a*) cubical (*b*) cylindrical
(*c*) spherical (*d*) semi-spherical

155. Which of the following is a non-baking binder used for core making?
(*a*) Phenol-furfural resin
(*b*) Bakelite
(*c*) Urea-formaldehyde
(*d*) None of these

156. Pick out the correct statement pertaining to centrifugal casting process.
(*a*) It is used for making small solid castings.
(*b*) Higher speed of rotation resulting in increased energy consumption is compensated by good quality of casting.
(*c*) Impurities and foreign materials present in the metal collect at the inner surface of the casting.
(*d*) None of these

157. Presence of........................in silica sand lowers the fusion point of the moulding sand.
(*a*) calcium phosphate
(*b*) alkali oxides
(*c*) titanium oxide
(*d*) none of these

158. The ratio of surface area to volume of casting to that of riser is equal to the
(*a*) relative freezing time for complete solidification of a casting
(*b*) maximum diameter of a core used in a mould
(*c*) volume of a chill used in the mould
(*d*) none of these.

159. Temperature & pressure of wax, at which it is injected into mould in investment casting, is respectively
(*a*) 20°C & 10 -15 kgf/cm^2.
(*b*) 75°C & 6.5 - 40 kgf /cm^2
(*c*) 150°C & 70-85 kgf /cm^2
(*d*) 175°C & 90-125 kgf /cm^2

160. In carbon dioxide process of moulding, CO_2 is passed into the mould through a pipe. It uses sodium silicate sand. The mould is hardened due to the formation of
(*a*) silica gel (*b*) sodium carbonate
(*c*) silicon (*d*) none of these

161. Which of the following is not a casting defect?
(*a*) Scab (*b*) Hot tear
(*c*) Ingate (*d*) Fin

162. Which of the following is a permanent mould process?
(*a*) Centrifugal casting process.
(*b*) Floor moulding process.
(*c*) Plaster moulding process.
(*d*) Sodium silicate moulding process.

163. Moulding sand comprises of sand, moisture and
(*a*) sea coal
(*b*) clay
(*c*) exothermic powder
(*d*) degassing agent

164. In foundry practice, a mixture of aluminium and iron oxide acts as a/an
(*a*) grain refiner (*b*) exothermic powder
(*c*) deoxidiser (*d*) degassing agent

165. ....gas is passed in sodium silicate moulding process.
(*a*) Nitrogen (*b*) Argon
(*c*) Carbon dioxide (*d*) Hydrogen

Answers

154. (c)	155. (a)	156. (c)	157. (b)	158. (a)	159. (b)
160. (a)	161. (c)	162. (a)	163. (b)	164. (c)	165. (b)

166. During casting of a metal, shrinkage occurs due to reduction in the volume of the metal, when it goes from molten to solid state. The volume contraction for steel is about................ per cent during solidification.
   (*a*) 2 (*b*) 7
   (*c*) 14 (*d*) 20

167. Facing sand used in foundry work comprises of
   (*a*) silica & clay.
   (*b*) silica & alumina.
   (*c*) clay & alumina.
   (*d*) none of these.

168. In casting of ferrous metals, the runners & ingates are located respectively in
   (*a*) drag & cope (*b*) cope & drag
   (*c*) both in cope (*d*) both in drag

169. ....moulding process will be the most suitable for casting a large intricate heavy casting (weighing in tons).
   (*a*) Green sand (*b*) Machine
   (*c*) Pit (*a*) Ceramic

170. Rammers are used to
   (*a*) flatten and smoothen the sand during moulding.
   (*b*) scoop the sand deep in the mould.
   (*c*) ram the sand around the pattern & over the mould.
   (*d*) none of these.

171. The impurities in centrifugal casting are
   (*a*) collected in the centre of casting.
   (*b*) present in the middle section of casting.
   (*c*) forced outside the surface.
   (*d*) none of these.

172. Graphite moulds are used for continuous casting process in order to impart
   (*a*) non wetting characteristic.
   (*b*) chilling effect.
   (*c*) self lubricating quality.
   (*d*) quick solidifying property.

173. Sweep pattern is used for moulding parts having........................shape.
   (*a*) elliptical
   (*b*) uniform symmetrical
   (*c*) rectangular
   (*d*) complicated

174. Hot tear are casting defects, which occur due to discontinuity in metal casting resulting from hindered contraction. The test carried out for inspection of castings to detect internal hot tear defects is the..................test.
   (*a*) damping
   (*b*) radiography
   (*c*) flouroscent penetrant
   (*d*) hardness

175. Shrinkage of metal on cooling is dependent on the
   (*a*) temperature of the melt.
   (*b*) material being cast.
   (*c*) design & size of casting.
   (*d*) all '*a*', '*b*' & '*c*'.

176. Toys and statues are cast by....................... casting process.
   (*a*) slush (*b*) die
   (*c*) centrifugal (*d*) continuous

177. .... allowance is not provided in the disposable pattern.
   (*a*) Distortion (*b*) Shrinkage
   (*c*) Draft (*d*) none of these

Answers

166. (b)	167. (a)	168. (b)	169. (c)	170. (c)	171. (a)
172. (c)	173. (b)	174. (b)	175. (d)	176. (a)	177. (c)

178. Large diameter long water pipes are made by the..........................casting.
    (*a*) continuous
    (*b*) sand
    (*c*) semi-centrifugal
    (*d*) none of these

179. Loose piece patterns are
    (*a*) similar to core prints.
    (*b*) used, when the pattern can not be drawn from the mould.
    (*c*) a type of split patterns.
    (*d*) not used in foundry operations.

180. Match the functions of various tools used in foundary practice.
    List I
    (*a*) Rammer
    (*b*) Swap
    (*c*) Slick
    (*d*) Bellow
    List II
    I. It is used to pack the sand in mould.
    II. It is a hand tool used for repairing moulds.
    III. Used for blowing loose sand from cavities and mould surface.
    IV. It is used for moistening the sand.

181. Match the significance and indications provided by various colors on a pattern.
    List I
    (*a*) Yellow color
    (*b*) Black broken lines
    (*c*) Green color
    (*d*) Yellow stripes
    List II
    I. Seats for loose core prints.
    II. Surface to be machined.
    III. Core prints for machined openings on periphery.
    IV. Fillets.

182. Match the relevant terms in the two columns.
    List I
    (*a*) Vent wire
    (*b*) Gagger
    (*c*) Gating system
    (*d*) Fillet
    List II
    I. A concave connecting surface or the rounding of an internal corner caused by two intersecting surfaces.
    II. The passage way of bringing molten metal to the mould cavity.
    III. A L-shaped metal accessory used in floor moulds to support hanging bodies of sand in cape.
    IV. A sharp pointed wire used in foundary for punching holes through the rammed sand.

183. Match the uses of various patterns used in foundary practice.
    List I
    (*a*) Sweep pattern
    (*b*) Match plate pattern
    (*c*) Metal pattern
    (*d*) Wax pattern
    List II
    I. For investment casting.
    II. Used for large scale production of castings.
    III. Casting of circular shape objects.
    IV. Production of large number of small castings.

184. Draft or taper allowances is provided in patterns on all the.....................of the pattern.
    (*a*) vertical surfaces (*b*) corners
    (*c*) sides and round surfaces
    (*d*) horizontal surfaces

Answers

178. (c) 179. (b) 180. (i) (iv) (ii) (iii) 181. (ii) (iv) (i) (iii)
182. (iv) (iii) (ii) (i) 183. (iii) (iv) (ii) (i) 184. (a)

185. The grain size of moulding sand is expressed in terms of the
   (*a*) seive number
   (*b*) grain fineness number
   (*c*) average grain diameter
   (*d*) grain density

186. Fetling of casting is done to remove the
   (*a*) extra material from castings.
   (*b*) internal stresses from the castings.
   (*c*) void by welding.
   (*d*) surface roughness.

187. In die casting process
   (*a*) heavy sections can be very easily cast.
   (*b*) surface finish of the casting is poor.
   (*c*) high production rates can not be achieved.
   (*d*) only certain non-ferrous alloys can be cast economically.

188. In centrifugal castings, cores are made of
   (*a*) plastic (*b*) wax
   (*c*) plaster of paris
   (*d*) none of these (as cores are not used in this process)

189. .... moulding machine is used for obtaining uniform sand hardness throughout the mould.
   (*a*) Sand slinger (*b*) Diaphragm
   (*c*) Stripper plate (*d*) Jolt

190. Use of synthetic sand in foundary shop is advantageous due to the fact that
   (*a*) it is non hygroscopic.
   (*b*) it is cheaper.
   (*c*) its properties are easily controllable.
   (*d*) it has high clay percentage.

191. Wood used in pattern making is considered dry, when the moisture content in it is less than....................per cent.
   (*a*) 5 (*b*) 10
   (*c*) 15 (*d*) 20

192. Water pipes of large length & diameter are made by semi-centrifugal casting. The impurities in centrifugal casting are
   (*a*) collected in the centre of casting.
   (*b*) uniformly distributed.
   (*c*) forced outside the surface.
   (*d*) in the middle section of the casting.

193. Which of the following pairs is not correctly matched ?
   (*a*) Lost wax process : Investment moulding.
   (*b*) Plaster moulding : Mould prepared in gypsum.
   (*c*) Pit moulding : For large jobs.
   (*d*) None of these.

194. An ideal gating system should facilitate the
   (*a*) entry of molten metal to the mould cavity with as high a velocity as possible.
   (*b*) prevention of absorption of air/gases from surrounding in the molten metal while flowing through it.
   (*c*) complete filling of the mould cavity.
   (*d*) both '*b*' & '*c*'

195. Defect developed by improper pouring condition such as splashing of steel in the moulds is termed as the
   (*a*) pipes (*b*) scab
   (*c*) segregation (*d*) laps

196. If volume to surface area of one riser is 8 & that of another riser is 6, then the ratio of their solidification time will be
   (*a*) 8 (*b*) 6
   (*c*) 4 (*d*) 2

Answers

185. (b)	186. (a)	187. (d)	188. (d)	189. (b)	190. (c)
191. (c)	192. (a)	193. (d)	194. (d)	195. (b)	196. (c)

197. If a cylinder & a sphere having the same volume are cast simultaneously, then the solidification time of sphere as compared to that for cylinder will be
   (*a*) more
   (*b*) less
   (*c*) same
   (*d*) either more or less; depends on the metal being cast

198. Match the characteristics of various casting processes.
   List I
   (*a*) Investment casting
   (*b*) Sand casting
   (*c*) Shell mould casting
   (*d*) Plaster mould casting
   List II
   I. Parts produced have soft skin but hard interior.
   II. It is suitable for non-ferrous metals only.
   III. Requires minimum post casting processing.
   IV. Parts produced have a tendency to warp.

199. Match the products produced by various moulding processes.

List I	*List II*
(*a*) Blow moulding.	I. Turbine blade.
(*b*) Slush moulding.	II. Plastic water bottle.
(*c*) Dry sand moulding.	III. Hollow castings.
(*d*) Shell moulding.	IV. Cylinder block.

200. Match the products produced by various casting processes,

List I	*List II*
(*a*) Die casting	I. Connecting rods
(*b*) Centrifugal casting	II. Turbine blades
(*c*) Investment casting	III. Cast iron pipes
(*d*) Drop forging	IV. Body of the car burettor

201. Match the working principle of various types of casting processes.

List I	*List II*
(*a*) Centrifugal casting	I. Metal solidifies while the mould is rotating
(*b*) Continuous casting	II. Molten metal is forced into die under pressure
(*c*) Centrifuging	III. Axis of rotation does not coincide with the axis of mould
(*d*) Die casting	IV. Molten metal is continuously poured in the mould

202. Match the products made/materials used in various processes.

List I	*List II*
(*a*) Centrifugal casting	I. Sodium silicate
(*b*) Die casting	II. Cast iron pipes
(*c*) CO_2 moulding	III. Phenolformaldehyde
(*d*) Shell moulding	IV. Non-Ferrous alloys

203. The purpose of inoculation is to .... of the cast metal.
   (*a*) decrease the melting temperature
   (*b*) alter the chemical composition
   (*c*) modify the structure and properties
   (*d*) improve the finishing.

Answers

197. (a)	198. (iii) (iv) (i) (ii)	199. (ii) (iii) (iv) (i)	200. (iv) (iii) (ii) (i)
201. (iii) (iv) (i) (ii)	202. (ii) (iv) (i) (iii)	203. (c)	

204. In pattern making, blue colour on pattern is
   (*a*) for machined surfaces
   (*b*) never used
   (*c*) for unfinished surfaces
   (*d*) for loose piece pattern

205. Surfaces to be machined are marked with........................colour on the pattern.
   (*a*) green (*b*) yellow
   (*c*) red (*d*) black

206. Slag inclusion in a casting is a .... defect.
   (*a*) internal (*b*) surface
   (*c*) superficial (*d*) none of these.

207. Chilled surfaces are marked on the pattern by
   (*a*) oblique red strips
   (*b*) oblique red strips on yellow
   (*c*) yellow cross strips
   (*d*) black cross strips on yellow.

208. Sprue in casting refers to
   (*a*) riser (*b*) runner
   (*c*) gate (*d*) vertical passage.

209. Tolerances produced by the investment casting process are of the order of .... mm.
   (*a*) ±0.05 (*b*) ±1
   (*c*) ±2.5 (*d*) ±5

210. Wood flour is added to the moulding sand to improve
   (*a*) permeability (*b*) collapsibility
   (*c*) hot strength (*d*) green strength.

211. Linseed oil is used in the core sand as a
   (*a*) binder (*b*) flux
   (*c*) catalyst (*d*) parting agent

212. Hexamethylene tetramine is added to the resin sand as a
   (*a*) dispersant (*b*) binder
   (*c*) catalyst (*d*) releasing agent

213. Nut and bolts are produced by .... process.
   (*a*) extrusion (*b*) hot piercing
   (*c*) upsetting (*d*) cold peening

214. Seamless tubes areproduced by .... process.
   (*a*) casting (*b*) forging
   (*c*) extrusion (*d*) piercing

215. Which of the following pairs is not correctly matched ?
   (*a*) Aluminium alloy piston.............Pressure die casting.
   (*b*) Jewellery.............Lost wax process.
   (*c*) Large pipes.............Centrifugal casting.
   (*d*) Large bells.............Loam moulding.

216. A sand casting mould assembly is shown in the figure given below. The elements marked A and *B* are respectively.

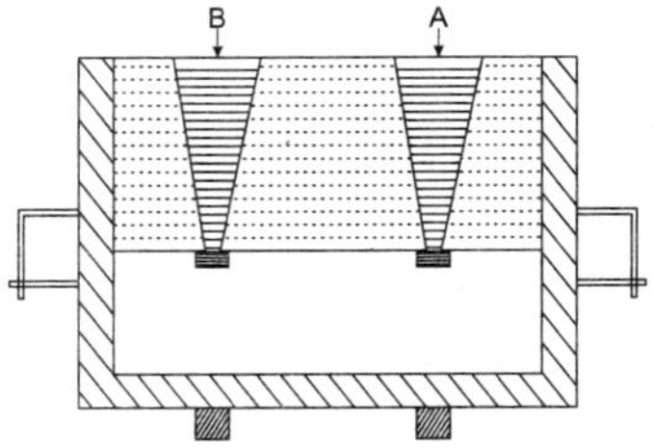

   (*a*) sprue & riser (*b*) ingate & riser
   (*c*) drag & runner (*d*) riser & runner

217. The height of the downsprue is 175 mm and its cross-sectional area at the base is 200 mm^2. The cross-sectional area of the horizontal runner is also 200 mm^2. Assuming no losses, indicate the correct choice for the time (in seconds) required to fill a mould cavity of volume 10^6 mm^3. (Use g = 10 m/sec^2)
   (*a*) 2.67 (*b*) 8.45
   (*c*) 26.72 (*d*) 84.50

Answers

204. (b)	205. (c)	206. (b)	207. (b)	208. (d)	209. (a)
210. (b)	211. (a)	212. (c)	213. (b)	214. (d)	215. (b)
216. (a)	217. (a)				

218. Match List I with List II and select the correct answer from the codes given below the list:

List I

(*a*) Aluminium brake shoe
(*b*) Plastic water bottle
(*c*) Stainless steel cups
(*d*) Soft drink can (aluminium)

List II

1. Deep drawing
2. Blow moulding
3. Sand casting
4. Centrifugal casting
5. Impact extrusion
6. Upset forging

Codes:

	A	B	C	D
(*a*)	1	2	3	4
(*b*)	4	3	2	1
(*c*)	3	2	5	1
(*d*)	3	6	5	4

219. Match the following moulding/casting processes with the product and select the correct answer from the codes given below the lists :

Moudling / casting process

A. Slush casting
B. Shell moulding
C. Dry sand moulding
D. Centrifugal casting

Product

1. Turbine blade
2. Machine tool bed
3. Cylinder block
4. Hollow casting like lamp shades
5. Rain water pipe
6. Cast iron shoe brake

Codes

	A	B	C	D
(*a*)	4	1	3	5
(*b*)	3	4	2	6
(*c*)	2	1	3	4
(*d*)	1	2	3	4

220. List 1 gives a number of processes and List II gives a number of products. Match List I with List II and select the correct answers.

List I

A. Investment casting
B. Die casting
C. Centrifugal casting
D. Die forging

List II

1. Turbine rotors
2. Turbine blades
3. Connecting rods
4. Galvanised iron
5. Cast iron pipes
6. Carburettor body

Codes

	A	B	C	D
(*a*)	1	2	5	6
(*b*)	2	6	5	3
(*c*)	3	4	5	6
(*d*)	4	5	2	6

221. Misrun is a casting defect, which occurs due to
(*a*) very high pouring temperature of the metal
(*b*) insufficient fluidity of the molten metal
(*c*) absorption of gases by liquid metal
(*d*) improper alignment of the mould flasks

Answers

218. (c) 219. (a) 220. (b) 221. (b)

222. Grey cast iron blocks 200 × 100 × 10 mm are to be cast in sand moulds. Shrinkage allowance for pattern making is l %.The ratio of the volume of pattern to that of the casting will be

(a) 0.97 (b) 0.99
(c) 1.01 (d) 1.03.

223. A mould has a downsprue whose length is 20 cms and the cross-sectional area at the base of the downsprue is 1 cm^2. The downsprue feeds a horizontal runner leading into the mould cavity of volume 1000 cm^3. The time required to fill the mould cavity will be........................seconds.

(a) 4.05 (b) 5.05
(c) 6.05 (d) 7.25

224. An expandable pattern is used in

(a) slush casting
(b) squeeze casting
(c) centrifugal casting
(d) investment casting.

225. In a sand casting operation, the total liquid head is maintained constant such that it is equal to the mould weight. The time taken to fill the mould with a top gate is 'tA' . If the same mould is filled with a bottom gate, then the time taken is *'tB'*- Ignore the time required to fill the runner and frictional effects. Assume atmospheric pressure at the top molten metal surfaces. The relation between t_A and t_B is

(a) $t_B = \sqrt{2.t_A}$ (b) $t_B = 2t_A$

(a) $t_B = t_A / \sqrt{2}$ (b) $t_B = 2\sqrt{2}\, t_A$

226. Volume of a cube of side T and volume of a sphere of radius V are equal. Both the cube and the sphere are solid and of same material. They are being cast. The ratio of the solidification time of cube to the same of the sphere is

(a) $\left(\frac{4\pi}{6}\right)^3 \left(\frac{r}{1}\right)^6$ (b) $\left(\frac{4\pi}{6}\right)\left(\frac{r}{1}\right)^2$

(a) $\left(\frac{4\pi}{6}\right)^2 \left(\frac{r}{1}\right)^3$ (b) $\left(\frac{4\pi}{6}\right)^2 \left(\frac{r}{1}\right)^4$

227. A 200 mm long downsprue has an area of cross-section of 650 mm2 where the pouring basin meets the downspure (*i.e.,* at the beginning of the downspure). A constant head of molten metal is maintained by the pouring basin. The molten metal flow rate is 6.5 x 10^5 mm^3/s. Considering the end of downsprue to be open to atmosphere and an acceleratoin due to gravity of 10^4 mm/s^2, the area of the downsprue in mm^2, at its end (avoiding aspiration effect) should be

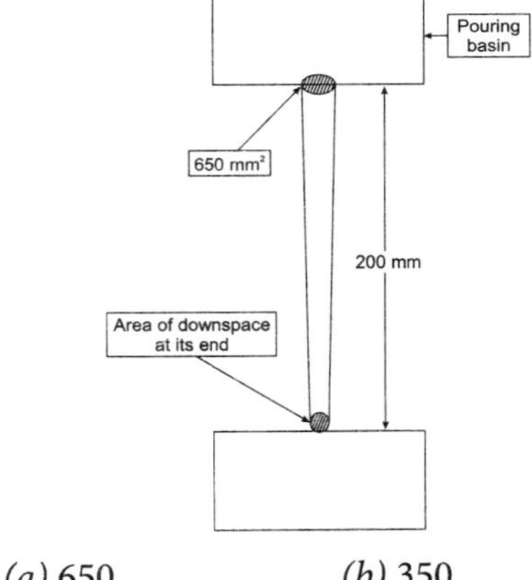

(a) 650 (b) 350
(c) 290.7 (d) 190.

228. While cooling a cubicle casting of side 40 mm undergoes 3%, 4%, and 5% volume shrinkage during the liquid state, phase transition and solid state respectively. The volume of metal compensated from the riser is...........per cent.

(a) 2 (b) 7
(c) 8 (d) 9

Answers

222. (d) 223. (b) 224. (d) 225. (c) 226. (d) 227. (c)
228. (b)

229. Two streams of liquid metal, which are not hot enough to fuse properly, result into a casting defect known as

(*a*) cold shut (*b*) swell
(*c*) sand wash (*d*) scab.

230. Match the items in coloumn I and II.

Coloumn I

P. Metallic chills
Q. Metallic chaplets
R. Riser
S. Exothermic padding

Coloumn II

1. Support for the core
2. Reservoir of the molten metal
3. Control cooling of critical section
4. Progressive solidification

(*a*) P-l, Q-3, R-2, S-4
(*b*) P-l, Q-4, R-2, S-3
(*c*) P-3, Q-4, R-2, S-l
(*d*) P-4, Q-l, R-2, S-3

231. In a gating system, the ratio 1:2:4 represents

(*a*) sprue base area : runner area : ingate area
(*b*) pouring basin area: ingate area : runner area
(*c*) sprue base area : ingate area : casting area
(*d*) runner area : ingate area : casting area

Answers

229. (a) 230. (d) 231. (a)

13 Welding Metallurgy

1. Oxygen to acetylene ratio in case of oxidising flame is
   (a) 1 : 1 (b) 1.5 : 1
   (c) 2 : 1 (d) 2.5 : 1

2. Temperature generated in arc welding is of the order of .... °C.
   (a) 1500 (b) 2500
   (c) 5500 (d) 8500

3. Voltage used for heating in resistance welding is .... volts.
   (a)10 (b) 230
   (c) 100 (d) 500

4. The thickness of material upto which plane and butt welds can be used is about .... mm.
   (a) 100 (b) 75
   (c) 50 (d) 25

5. Too slow welding speed in arc welding results in
   (a) excessive piling up of weld metal and overlapping without penetration of edges.
   (b) under cutting along edges, irregular, deposit and wasted electrodes.
   (c) too small bead, weld metal and wasted electrodes.
   (d) poor penetration and wasted electrodes.

6. Carburising flame is used to weld metals like
   (a) copper.
   (b) steel.
   (c) aluminium, stainless steel & nickel.
   (d) none of these.

7. Two sheets of same materials but different thicknesses can be butt welded by
   (a) control of pressure & current.
   (b) changing the size of one electrode.
   (c) adjusting the current.
   (d) none of these.

8. In arc welding, the arc length should be
   (a) d (b) 0.5 d
   (c) 2d (d) 2.5 d
   where, d = electrode rod diameter.

9. Thermit welding is
   (a) done by maintaining molten metal pool between plates,
   (a) a process in which arc is maintained under a blanket of flux.
   (a) a process which uses a mixture of granular aluminium & iron oxide.
   (a) none of these.

Answers

1. (b)	2. (c)	3. (a)	4. (d)	5. (a)	6. (c)
7. (b)	8. (a)	9. (c)			

10. Neutral flame is used to weld
    (*a*) copper & brass.
    (*b*) steel.
    (*c*) nickel, aluminium & stainless steel.
    (*d*) none of these.

11. Function of flux in brazing is to
    (*a*) prevent oxide formation both on base metal & brazing material,
    (*b*) avoid thermal distortion & cracking.
    (*c*) dissolve surface oxide coating formed prior to brazing.
    (*d*) facilitate the wetting by reducing the viscosity of the melt.

12. Oxygen to acetylene ratio in a carburising flame is
    (*a*) 0.9 : 1 (b) 1 : 2
    (*c*) 1 : 2 (*d*) 1 : 2.5.

13. In forehand gas welding, the angle between the rod and work is kept about .... degrees.
    (*a*) 80 (*b*) 60
    (*c*) 45 (*d*) 30

14. In electroslag welding,
    (*a*) arc is maintained under a blanket of flux.
    (*b*) hot molten metal pool is maintained between the plates.
    (*c*) a mixture of iron oxide and granular aluminium is used.
    (*d*) none of these.

15. Two sheets of different materials but same thickness can be spot welded by
    (*a*) changing the size of one electrode.
    (*b*) adjusting the current.
    (*c*) time duration of current.
    (*d*) none of these.

16. If arc is too long in arc welding, the result is
    (*a*) electrode sticking to the base metal.
    (*b*) arc extinction.
    (*c*) formation of large globules in an irregular pattern because of wandering of arc leading to poor fusion with base metal.
    (*d*) none of these.

17. In carbon arc welding
    (*a*) arc is maintained under blanket of flux.
    (*b*) hot molten metal pool between plates is maintained.
    (*c*) a mixture of iron oxide and granular aluminium is used.
    (*d*) none of these.

18. Seam welding is
    (*a*) multi spot welding process.
    (*b*) used to form mesh.
    (*c*) continuous spot welding process.
    (*d*) none of these.

19. For welding aluminium in inert gas arc welding, we use
    (*a*) combustible electrode and argon with alternating current.
    (*b*) non combustible electrode with helium and direct current.
    (*c*) straight polarity direct current.
    (*d*) CO_2 and high speed.

20. The distance between two metallic surfaces to be brazed properly (*i.e.*, for wetting action to take place) should be .... mm.
    (*a*) < 0.01
    (*b*)> 1
    (*c*) between 0.025 & 0.75
    (*d*) > 5.

Answers

10. (b)	11. (b)	12. (a)	13. (d)	14. (b)	15. (a)
16. (c)	17. (d)	18. (c)	19. (a)	20. (c)	

21. Plug weld joint is used
    (*a*) to join two metal pieces in the same manner as rivets join metals.
    (*b*) where longitudinal shear is present.
    (*c*) where severe loading is encountered.
    (*d*) none of these.

22. A.C. arc welding always employs coated electrodes, because it
    (*a*) has low leakage loss.
    (*b*) has high efficiency and low spatter.
    (*c*) uses less current.
    (*d*) all (*a*), (*b*) & (*c*).

23. Projection welding is
    (*a*) continuous spot welding process.
    (*b*) multispot welding process.
    (*c*) used to form mesh.
    (*d*) none of these.

24. Oxygen to acetylene ratio in neutral flame is
    (*a*) 1:1 (*b*) 1 : 5
    (*c*) 1 : 15 (*d*) 1 : 0.8.

25. In arc welding, too low welding current apart from wasted electrodes, results in
    (*a*) too small bead and weak weld.
    (*b*) excessive piling up of weld metal & over lapping without penetration of edges.
    (*c*) excessive piling up of weld metal & poor penetration.
    (*d*) excessive splatter, undercutting along edges & irregular deposits.

26. Pressure in resistance welding is released
    (*a*) just at the time of passing of current.
    (*b*) just after completion of current.
    (*c*) after the weld cools.
    (*d*) none of these.

27. Flux used for brazing cast iron is
    (*a*) boric acid, borax or fluoride with a wetting agent.
    (*b*) chlorides & fluorides mixed with water.
    (*c*) mixture of boric acid & a wetting agent.
    (*d*) none of these.

28. In straight polarity welding, electrode holder is
    (*a*) earthed and the work is positive.
    (*b*) connected to the negative and the work to positive.
    (*b*) connected to the positive and the work to negative.
    (*b*) none of these.

29. Cross wire welding is
    (*a*) multispot welding process.
    (*b*) used to form mesh.
    (*c*) continuous spot welding process.
    (*d*) none of these.

30. Too short arc in arc welding results in
    (*a*) electrode sticking to the base metal, base metal not melting and bead resting on top of work leading to poor fusion and gas & slag holes.
    (*b*) formation of large globules in an irregular pattern because of wandering of arc leading to poor fusion with base metal.
    (*c*) arc extinction, (*d*) none of these.

31. In arc welding employing coated electrodes
    (*a*) rod melts first and then coating melts.
    (*b*) which melts first depends on polarity.
    (*c*) coating melts first and then rod melts.
    (*d*) both rod & coating melts simultaneously.

Answers

21. (a)	22. (d)	23. (b)	24. (a)	25. (c)	26. (c)
27. (a)	28. (b)	29. (b)	30. (a)	31. (a)	

32. T-joint weld is used
    (*a*) to join two metallic pieces in the same manner as rivets join metals.
    (*b*) where longitudinal shear is present,
    (*c*) where severe loading is encountered.
    (*d*) none of these.

33. Apart from wastage of electrodes, too fast welding speed in arc welding results in
    (*a*) too small bead and weak weld.
    (*b*) undercutting along edges and irregular deposits.
    (*b*) excessive piling up of weld metal and poor penetration.
    (*d*) overlapping without penetration of edges.

34. In reverse polarity welding, electrode holder is
    (*a*) earthed and work is positive.
    (*b*) connected to the negative and work to positive.
    (*c*) connected to the positive and work to negative.
    (*d*) none of these.

35. Half corner weld is used, where
    (*a*) metals are joined similar to rivetting.
    (*b*) severe loading is encountered.
    (*c*) longitudinal shear is present.
    (*d*) none of these.

36. For welding stainless steel, copper or cast iron by inert gas arc welding, we use
    (*a*) straight polarity direct current.
    (*b*) combustible electrode and argon in combination with A.C.
    (*c*) non combustible electrode with helium and D.C.
    (*d*) CO_2 and high speed.

37. Joggled welded joint are used, where
    (a) longitudinal shear is present.
    (b) metals are joined similar to rivetting.
    (*c*) severe loading is encountered (the upper surface of both pieces must be in the same plane).
    (*d*) none of these.

38. Flux used for brazing brass, copper bronze and low carbon steel is
    (*a*) chlorides mixed with water.
    (*b*) fluorides mixed with water.
    (*c*) mixture of boric acid, borax & a wetting agent.
    (*d*) none of these.

39. Dissimilar materials or materials of different thicknesses can be butt welded by
    (*a*) time duration of current.
    (*b*) adjusting initial gap.
    (*c*) controlling pressure & current.
    (*d*) none of these.

40. Magnetic arc blow is
    (*a*) a phenomenon of occurrence of splatter due to magnetic field created in D.C. arc welding.
    (*b*) used to weld hard materials.
    (*c*) used to weld brittle materials.
    (*d*) none of these.

41. Temperature attained in soldering of metals is about .... °C.
    (*a*) 150-300 (*b*) 750- 1000
    (*c*) 1000 - 1500 (*d*) 1500 - 2000

42. Pick out the wrong statement.
    (*a*) Soldering and brazing cannot produce fluid tight joints.
    (*b*) Welding involves high temperature and produces stronger joints compared to soldering.
    (*c*) Soldering/brazing involves addition of a low melting point material radically different in composition from the base metal.
    (*d*) Weld joint is less liable to corrode preferentially compared to soldered joints.

Answers

32. (b)	33. (a)	34. (c)	35. (d)	36. (a)	37. (c)
38. (c)	39. (c)	40. (a)	41. (a)	42. (a)	

43. The most common soldering material 'tinman's solder'
    (a) is an eutectic mixture of 62% tin & 38% lead.
    (b) has a melting point of 180°C.
    (c) both (a) & (b)
    (d) neither (a) nor (b).

44. Brazing
    (a) is a high temperature soldering process.
    (b) uses silver or copper alloys as the jointing material.
    (c) is done using oxy-acetylene welding torch.
    (d) all (a), (b) & (c).

45. Pick out the wrong statement.
    (a) Brazing stands intermediate between soldering & welding for both mechanical properties and corrosion resistance.
    (b) Non-ferrous metals can not be joined by brazing,
    (c) Borax flux is normally used in brazing.
    (d) none of these.

46. Carbon arc welding is
    (a) useful for thinner materials,
    (b) most suitable for welding stainless steel without flux.
    (c) useful for thinner materials.
    (d) cheaper than gas welding.

47. Welding current does not depend on the
    (a) type of metal (b) thickness of metal
    (c) electrode used (d) none of these.

48. In resistance welding
    (a) no flux is used.
    (b) metal surfaces need not be very clean.
    (c) aluminium cannot be welded.
    (d) none of these.

49. Ultrasonic welding
    (a) is useful for joining section of greatly differing thickness,
    (b) is cheapest of all the welding processes.
    (c) is not used in lap form of construction.
    (d) cannot join materials of differing metallurgical characteristics.

50. Stud welding
    (a) produces excellent joint in stainless steel.
    (b) produces strongest joint in aluminium & copper as compared to other welding process.
    (c) cannot draw current from motor generator or transformer-rectifier system.
    (d) none of these.

51. Which is the most weldable of all metals ?
    (a) Stainless steel. (b) Plain carbon steel.
    (c) Aluminium. (d) Brass.

52. High carbon tool steel (C = 1-1.7%) are best joined by
    (a) stud welding.
    (b) metallic arc welding.
    (c) resistance welding.
    (d) brazing with a low temperature silver alloy.

53. Which is not suitable for welding thin sheets?
    (a) Oxyacetylene blow pipe,
    (b) Carbon arc.
    (c) Tungsten arc with argon gas shield.
    (d) None of these.

54. Which is not suitable for welding heavy plates ?
    (a) Metallic arc process using flux coated wires or CO_2 shield,
    (b) Submerged arc using a fused granular flux.
    (c) The electroslag process.
    (d) None of these.

Answers

43. (c)	44. (d)	45. (b)	46. (a)	47. (d)	48. (a)
49. (a)	50. (a)	51. (b)	52. (d)	53. (d)	54. (d)

55. In welding process, zinc present in base metal
   (*a*) vaporises & causes porous weld metal.
   (*b*) makes the weld very strong.
   (*c*) reduces the toughness of weld metal.
   (*d*) none of these.

56. Pick out the wrong statement.
   (*a*) Gaseous oxides formed during welding causes blow holes.
   (*b*) Soluble oxides in the molten metal reduces the strength & toughness of weld.
   (*c*) Metal of high thermal expansion & low thermal conductivity is not subjected to high cooling stresses in welding.
   (*d*) Carbon present in appropriate form after heat treatment changes its form & structure and thus welded portion becomes weak till heat treatment is carried out again and so on.

57. Which of the following parameters is not important for consideration before starting the welding of a particular metal ?
   (*a*) The type & gauge of electrodes.
   (*b*) Need to preheat to a particular temperature, if necessary.
   (*c*) Current & voltage rating.
   (*d*) None of these.

58. The amount of heat generated in the work piece in resistance welding depends on the
   (*a*) welding current only.
   (*b*) time the current is allowed to flow only.
   (*c*) resistance of the current conducting path only.
   (*d*) none of these.

59. Most weldable of the following steels is the
   (*a*) high carbon steel
   (*b*) mild steel
   (*c*) 0.08% carbon steel
   (*d*) 1% carbon steel.

60. The most useful flame for gas welding of brasses is the .... flame.
   (*a*) neutral (*b*) oxidising
   (*c*) reducing (*d*) none of these

61. Best process for joining of two stainless steel foils (0.075 mm thick) would be .... welding process.
   (*a*) TIG (*b*) arc
   (*c*) plasma arc (*d*) MIG

62. The protective gaseous atmosphere around the arc and weld pool in the case of arc welding is due to the
   (*a*) formation of metal vapours.
   (*b*) burning of flux.
   (*c*) vaporisation of electrode metal.
   (*d*) none of these.

63. Electrode diameter in case of electric arc welding mainly depends on the
   (*a*) composition of the materials to be welded.
   (*b*) current & voltage from power source.
   (*c*) thickness of the materials to be welded.
   (*d*) none of these.

64. Which is the most suitable for welding heavier sections subjected to severe load conditions ?
   (a) Butt joint.
   (b) Double V-butt joint.
   (*c*) Open square butt joint.
   (*d*) Close square butt joint.

65. In the electroslag welding method, the heat is created due to the resistance to electrical flow through the
   (*a*) molten slag.
   (*b*) contact resistance between metal & electrode.
   (*c*) flux.
   (*d*) none of these.

Answers

55. (a)	56. (c)	57. (d)	58. (d)	59. (c)	60. (b)
61. (c)	62. (b)	63. (c)	64. (b)	65. (c)	

66. Gases used in the case of tungsten inert gas welding are
(*a*) argon and neon.
(*b*) argon and helium.
(*c*) neon and helium.
(*d*) carbon dioxide and neon.

67. The electrode in tungsten inert gas welding
(*a*) melts the metal for welding.
(*b*) is not consumed.
(*c*) is consumed.
(*d*) none of these.

68. TIG is especially useful in welding
(*a*) stainless steel (*b*) cast iron
(*c*) aluminium (*d*) titanium

69. In MIG welding, the metal is transferred in the form of
(*a*) grains (*b*) droplets
(*c*) weld pool (*d*) a fine spray

70. Gates and risers are needed in the .... welding method.
(*a*) thermit (*b*) electroslag
(*c*) submerged arc (*d*) none of those

71. Metal cutting is possible by oxy-acetylene flame through a gas torch due to the
(a) oxidation of metal.
(b) intensive oxidation.
(*c*) burning of metal.
(*d*) evaporation of metal.

72. Which is used in brazing ?
(*a*) Copper rod (*b*) Spelter
(*c*) Gas rod (*d*) Solder

73. Electrode is consumed in the .... welding.
(*a*) gas (b) thermit
(*c*) arc (*d*) tungsten inert gas

74. Grey cast iron is mostly welded by .... welding, because it facilitates graphitisation.
(*a*) TIG (*b*) MIG
(*c*) submerged arc (*d*) oxy-acetylene flame

75. The arc created between the electrode & the work for welding purpose is due to the
(*a*) contact resistance
(*b*) applied voltage
(*c*) current flow
(*d*) none of these

76. Material used for coating the welding electrode is known as the
(a) binder
(b) slagging ingredients
(*c*) flux
(*d*) deoxidiser

77. Pressure applied in electric resistance welding is in the range of .... kg/cm^2.
(*a*) 15-65 (*b*) 65-150
(*c*) 250-550 (*d*) 600-850

78. Tip of electrodes used in spot welding are made of
(*a*) mild steel (*b*) copper
(*c*) soft solder (*d*) aluminium.

79. .... arc welding employs consumable electrodes.
(*a*) MIG (*b*) TIG
(*c*) Submerged (*d*) Carbon

80. Thermit (used in thermit welding) com prises of
(*a*) charcoal & iron oxide
(*b*) aluminium & iron oxide
(*c*) charcoal & aluminium
(*d*) none of these

Answers

66. (b)	67. (b)	68. (c)	69. (d)	70. (a)	71. (b)
72. (b)	73. (c)	74. (d)	75. (a)	76. (c)	77. (c)
78. (b)	79. (a)	80. (b)			

81. Oxy-acetylene welding mostly employs the .... flame.
(*a*) oxidising (b) reducing
(*c*) carburising (*d*) neutral.

82. .... welding is usually used to join the ends of two pipes of uniform cross-section.
(*a*) Upset butt (*b*) Flash butt
(*c*) Spot (*d*) Projection

83. Edges of plates do not require bevelling, if their thicknesses are .... mm.
(*a*) < 1 (*b*) < 5
(*c*) > 10 (*d*) > 15

84. If t=the thickness of the plates to be spot welded, then the electrode tip diameter should be equal to
(*a*) t (*b*) $\sqrt{t}$
(*c*) $2\sqrt{t}$ (*d*) 2t

85. Edges of the plates should have a double U or V-groove on both sides for welding, if their thicknesses are .... mm.
(*a*) < 5 (*b*) > 5
(*c*) <10 (*d*) >12.5

86. Mild steel shanks are joined to high speed drills by ....welding.
(*a*) upset butt (b) flash butt
(*c*) seam (*d*) spot

87. The spacing between two spot weld in spot welding should be
(*a*) D (*b*) >3 D
(*c*) < 3D (*d*) < D
where, D = electrode tip diameter.

88. The desirable pressure range of acetylene at the welding torch in oxy-acetylene gas welding is about .... kg/cm2.
(*a*) 0.07-1.03 (*b*) 0.7-2.5
(*c*) 3.8-6.2 (*d*) 5.8-9.4.

89. The desirable pressure range of oxygen at the welding torch in oxy-accetylene gas welding is about .... kg/cm^2.
(*a*) 0.7-2.8 (*b*) 2.8-3.6
(*c*) 0.07-0.1 (*d*) 3.8-5.2.

90. The temperature of oxy-hydrogen flame is
(*a*) 1200°C
(*b*) 2400°C
(*c*) 3600°C
(*d*) more than that of oxy-acetylene flame

91. Neutral flame can not be used for welding
(*a*) cast iron
(*b*) mild steel
(*c*) copper and aluminium
(*d*) none of these

92. Carburising flame is normally used to weld
(*a*) copper alloys
(*b*) stainless steel
(*c*) phosphor bronzes
(*d*) hard surfacing materials like stellite

93. The angle between the welding torch & the work in back-hand welding as compared to that in fore-hand welding is
(*a*) less (*b*) more
(*c*) same (*d*) either (*a*) or (*b*).

94. The angle between the welding rod & the work in fore hand welding as compared to that in back-hand welding is
(*a*) more (*b*) less
(*c*) same (*d*) either (*a*) or (*b*)

95. The distance between the nearest edge of the plate & the centre of weld, in spot welding should be
(*a*) D (*b*) $D^{1/2}$
(*c*) 1.5 D (*d*) 3.5 D
where, D = electrode tip diameter

Answers

81. (d)	82. (a)	83. (b)	84. (b)	85. (d)	86. (b)
87. (b)	88. (a)	89. (a)	90. (b)	91. (d)	92. (d)
93. (a)	94. (c)	95. (c)			

96. Lap joints are preferred for plates having thickness .... mm.
(*a*) > 12.5 (*b*) > 5
(*c*) < 3 (*d*) 5—10

97. A pool of molten metal is used in .... arc welding.
(*a*) MIG (*b*) TIG
(*c*) submerged (*d*) carbon

98. Arc is produced between a .... electrode & the work in submerged arc welding.
(*a*) bare metal (b) carbon
(*c*) metal (*d*) none of these

99. Normally, thermit welding is used for
(*a*) joining truck & locomotive frames.
(*b*) joining rails.
(*c*) replacing broken gear tooth.
(*d*) all (*a*), (*b*) & (*c*).

100. Use of more of acetylene with less volume of oxygen in a oxy-acetylene torch produces a / an ....flame.
(*a*) neutral (*b*) oxidising
(*c*) carburising (*d*) none of these.

101. Which type of flame is used for welding brasses/bronzes ?
(*a*) Carburising flame.
(*b*) Oxidising flame.
(*c*) Neutral flame.
(*d*) None of these.

102. The weld can be made either from left to right or from right to left in .... welding.
(*a*) fore-hand (*b*) back-hand
(*c*) vertical (*d*) none of these

103. Two sheets of the same thickness but of different metals can be spot welded by
(*a*) bevelling the edges.
(*b*) increasing the current.
(*c*) changing the size of one of the electrodes.
(*d*) no means.

104. During arc welding, the current intensity depends upon the
(a) thickness of the plate.
(b) electrode diameter.
(*c*) method of edge preparation.
(*d*) nature of metal.

105. Plates of dissimilar materials of different thicknesses can be spot welded by
(*a*) bevelling the edges,
(*b*) increasing the gap between the plates.
(*c*) controlling the pressure & current.
(*d*) no means.

106. During coated electrode arc welding, the rod compared to the coating melts
(*a*) earlier
(*b*) later
(*c*) simultaneously
(*d*) either earlier or later depending on the polarity used during welding.

107. In which type of welding a mixture of iron oxide and powdered aluminium is ignited by heating ?
(a) Carbon arc welding.
(b) Thermit welding,
(*c*) Resistance welding.
(*d*) None of these.

108. Carbon arc welding is done with D.C, only (not A.C.), because
(*a*) no fixed polarity can be maintained in A.C.
(*b*) D.C. can melt the electrodes easny.
(*c*) The carbon electrode can work on both the polarity.
(*d*) none of these.

Answers

96. (c)	97. (c)	98. (a)	99. (d)	100. (c)	101. (b)
102. (c)	103. (c)	104. (b)	105. (c)	106. (a)	107. (b)
108. (c)					

109. Stainless steel plates can be best gas welded by using .... flame.
(*a*) oxidising (*b*) neutral
(*c*) carburising (*d*) either (*a*), (*b*) or (*c*).

110. Fore-hand welding is suitable for plate thicknesses .... mm.
(*a*) <6 (*b*) > 12
(*c*) >6 (*d*) > 25

111. Back-hand welding is suitable for plate thicknesses .... mm.
(*a*) <6 (*b*) >6
(*c*) < 2 (*d*) > 12

112. Electric arc welding can be used to easily weld the stainless steel plates by using the electrodes rich in .... either in its core or the coating.
(*a*) chromium (*b*) nickel
(*c*) carbon (*d*) vanadium.

113. Coated welding electrodes are mainly recommended for use, because they
(*a*) join all thicknesses of metals.
(*b*) are cheap.
(*c*) give better finish.
(*d*) can work on both the polarities.

114. Straight polarity is recommended for welding .... in A.C. arc welding.
(*a*) aluminium (*b*) bronze
(*c*) mild steel (*d*) none of these.

115. .... welding is the most suitable for butt welding of two mild steel plates (each 30 cm thick).
(*a*) Electroslag (*b*) TIG
(*c*) Gas (*d*) Submerged arc

116. In .... welding, two pieces to be joined are overlapped & placed between two electrodes.
(*a*) seam (*b*) spot
(*c*) butt (*d*) projection

117. Which of the following is the most suitable for joining two stainless steel foils of 0.1 mm thickness ?
(*a*) Electroslag welding
(*b*) Gas welding
(*c*) MIG welding
(*d*) TIG welding

118. The acetylene cylinder is filled with a material saturated with
(*a*) calcium carbide
(*b*) calcium oxide
(*c*) acetone
(*d*) none of these.

119. Metals like copper and brass are welded using .... flame.
(*a*) carburising (*b*) neutral
(*c*) oxidising (*d*) none of these.

120. Solder essentially has a .... base.
(*a*) tin-lead (*b*) tin-silver
(*c*) silver-lead (*d*) bismuth-lead

121. Oxygen cylinders are
(*a*) subjected to proper heat treatment.
(*b*) fabricated by casting.
(*c*) seamless steel cylinders.
(*d*) fabricated by forging.

122. The distance between two metallic surfaces to be brazed properly *i.e.* for wetting action to take place should be .... mm.
(*a*) between 0.025 to 0.075
(*b*) <0.025
(*c*) > 0.5
(*d*) > 0.25

Answers

109. (a)	110. (a)	111. (b)	112. (a)	113. (d)	114. (c)
115. (a)	116. (b)	117. (b)	118. (c)	119. (c)	120. (a)
121. (c)	122. (a)				

123. Steel is welded using .... flame.
(*a*) carburising (*b*) neutral
(*c*) oxidising (*d*) none of these.

124. The oxidising flame compared to carburising flame is
(a) less luminous (b) more luminous
(*c*) equally luminous (*d*) unpredictable

125. .... flame is the most commonly used flame in gas welding.
(a) Neutral (*b*) Carburising
(*c*) Oxidising (*d*) none of these

126. Electrodes get consumed in .... welding.
(a) thermit (b) gas
(*c*) resistance (*d*) arc

127. Which of the following is the most suitable for butt welding of two mild steel plates of 20 cms thickness for boiler drum ?
(a) Electroslag welding.
(b) Electrical resistance welding.
(c) Submerged arc welding.
(*d*) MIG welding.

128. The melting point of the filler metal in brazing should be above .... °C.
(a) 420 (*b*) 650
(*c*) 850 (*d*) 1050

129. TIG welding is the most suitable for welding
(*a*) aluminium (b) stainless steel
(*c*) mild steel (*d*) silver.

130. What is the angle of the electrode holder with the direction of welding in the mannual TIG welding ?
(*a*) 15 (b) 70°
(c) 90° (*d*) 25°.

131. If oxygen to acetylene ratio is 0.9 : 1 in the inner cone of the flame, then the flame is
(a) oxidising (b) carburising
(c) neutral (*d*) none of these.

132. Filler material need not be necessarily used in case of
(a) soldering (b) brazing
(*c*) welding (*d*) both (b) & (*c*)

133. Temperature above which brazing is done is .... °C.
(a) 125 (b) 225
(c) 325 (*d*) 425

134. Seamless pipes have .... strength & corrosion resistance as compared to welded pipes.
(*a*) lower (b) higher
(*c*) same
(*d*) either (a) or (b); depends on the pipe material

135. Weld decay may occur in .... steel, when heated to 500-800 °C.
(*a*) marageing (a) stainless
(a) high speed (*d*) none of these

136. Joint strength will be minimum in case of
(a) soldering (b) welding
(*c*) brazing
(*d*) all(a), (*b*) and (*c*) can have the same joint strength

137. Pick out the correct statement.
(a) Weldability of aluminium is very poor, hence it is subjected to brazing.
(b) Those metals which are capable of forming a continuous series of solid solutions with each other exhibit good weldability.
(c) Highest temperature is achieved, in oxy-acetylene welding, in outer portion of the flame.
(d) none of these.

Answers

123. (b)	124. (a)	125. (a)	126. (d)	127. (a)	128. (a)
129. (a)	130. (b)	131. (b)	132. (c)	133. (b)	134. (b)
135. (b)	136. (a)	137. (b)			

138. Porosity in the welded joint is not caused by .... gas.
(*a*) hydrogen (*b*) oxygen
(*c*) nitrogen (*d*) none of these

139. .... based alloys are used as filler material in brazing.
(*a*) Cadmium (*b*) Copper
(*c*) Tin (*d*) none of these

140. Which of the following may not essentially be a non-pressure type welding process?
(*a*) Submerged arc welding.
(*b*) Thermit welding.
(*c*) Metal arc welding.
(*d*) None of these.

141. Which of the following metallurgical processes is not involved in a welding process?
(*a*) Heat treatment
(b) Change in grain structure
(*c*) Fusion
(*d*) None of these

142. Pick out the wrong statement.
(*a*) Good weld is as strong as the base metal.
(*b*) Both similar as well as dissimilar metals and alloys can be joined by welding.
(*c*) Heat developed in thermit welding is due to the exothermic reaction between equal parts of iron oxide and aluminium powder.
(*d*) Oxy-acetylene flame can be used both for cutting as well as welding of steel.

143. Braze welding is
(*a*) same as soldering.
(*b*) a fusion welding process.
(*c*) a combination of brazing and welding operation.
(*d*) neither brazing nor welding operation.

144. Pick out the correct statement.
(*a*) Because of its low melting point, magnesium metal pieces are always joined by either soldering or brazing.
(*b*) Soldering may he done at a temperature of 500°C.
(*c*) The highest temperature in oxy-acetylene welding is attained in the inner portion of the flame.
(*d*) none of these.

145. Filler metal is used in .... welding.
(*a*) butt (*b*) seam
(*c*) spot (*d*) none of these

146. Eye protection is needed during arc welding against .... rays.
(*a*) ultra-violet (*b*) infra red
(*c*) both 'a' & 'b' (*d*) neither 'a' nor '*b*'

147. Maximum thickness of materials upto which plain and butt welding may be done is about .... mm.
(*a*) 10 (*b*) 25
(*c*) 50 (*d*) 75

148. Welding of .... essentially requires preheating.
(*a*) aluminium (b) cast iron
(*c*) stainless steel (*d*) high speed steel

149. Arc welding of all types of metals can not be done with transformer set, because of the reason that, it does not have the provision and facility for
(a) changing polarity.
(b) controlling current.
(*c*) regulating voltage.
(*d*) none of these.

Answers

138. (a)	139. (b)	140. (b)	141. (d)	142. (c)	143. (b)
144. (c)	145. (d)	146. (c)	147. (b)	148. (b)	149. (a)

150. Spot welding of two different materials of same thickness can be done by changing the
(a) pressure applied
(b) current
(c) size of one electrode
(d) time duration of current

151. Widest application of laser welding is in the .... industry.
(a) electronic (b) heavy
(c) process (d) structural work

152. Butt welding of two sheets of same material of different thickness can be done by changing the
(a) pressure applied.
(b) current.
(c) size of one electrode.
(d) time duration of current.

153. Long arc length in welding would require
(a) low voltage (b) low current
(c) high voltage (d) both 'b' & 'c'

154. In resistance welding, the current required ranges from .... kVA/cm^2
(a) 1.5 to 2.5 (b) 4.5 to 6.2
(c) 7 to 8.5 (d) 10 to 15

155. When welding is in progress, the arc voltage ranges from .... volts.
(a) 5 to 10 (b) 20 to 40
(c) 40 to 95 (d) 230 to 300

156. Thermit welding is a form of .... welding.
(a) arc (b) gas
(c) fusion (d) resistance.

157. Match the temperature (°C) encountered in various welding processes.
List I
(a) Electric arc welding using carbon electrodes.
(b) Brazing temperature from gas torches.
(c) Temperature of tungsten filament for emission of electrons (in case of electron beam welding).
(d) Temperature of oxy-acetylene flame.
List II
I. >430 II. 6100
III. 3300 IV. 1100

158. Match the relevant terminologies in the two columns.
List I
(a) Tinman's solder (b) Brazing alloy
(c) Soldering flux (d) Brazing flux
List II
I. Eutectic mixture of $ZnCl_2$ & NH_4Cl
II. Borax.
III. Silver solder.
IV. An alloy of tin & lead.

159. Match the characteristics/typical application of various welding processes.
List I
(a) High frequency welding
(b) MIG welding
(c) Stud welding .
(d) Gas welding
List II
I. Cheap welding method preferred for thinner section welding.
II. Mass production of longitudinally welded tubes from flat strip.
III. Employs consumable wire electrode and is suitable for welding of heavy aluminium and copper sections.
IV. Attachment of fixing screw for structural work inside/outside chemical process vessels.

Answers

150. (c) 151. (a) 152. (c) 153. (d) 154. (b) 155. (b)
156. (c) 157. (I) (II) (III) (IV) 158. (IV) (III) (I) (II) 159. (II) (III) (IV) (I)

160. Match the characteristics of various welding processes.

List I

(*a*) Seam welding.

(*b*) Atomic hydrogen welding,

(*c*) Resistance welding.

(*d*) Cross wire welding.

List II

I. Used to form mesh.

II. No flux is used,

III. An A.C. arc is struck between two tungsten electrodes.

IV. Continuous spot welding process.

161. Match the relevant terminologies in the two columns.

List I

(*a*) Plasma welding.

(*b*) Laser welding.

(*c*) Solid state welding.

(*d*) Bronze welding.

List II

I. Used for joining dissimilar metals e.g., cast iron and steel at comparatively lower temperature.

II. Ultrasonic welding employing temperature as low as 35% of the melting temperature of the metal being welded.

III. Fusion achieved by highly concentrated beam of light.

IV. Heat generated for fusion by an electric arc highly intensified by the injection of a gas into the arc stream.

162. The soldering iron is mainly used to melt the

(*a*) metal (*b*) flux

(*c*) solder (*d*) none of these

163. Chromium molybdenum steel is generally not welded using .... welding.

(*a*) resistance (*b*) submerged arc

(*c*) thermit (*d*) oxy-acetylene

164. Molten metal is poured for joining the metals in case of thermit welding, which is different from other welding methods in the way that, it

(*a*) does not require electrodes.

(*b*) does not use heat.

(*c*) is less time consuming.

(*d*) employs exothermic chemical reaction for developing high temperature.

165. The thickness of material that can be gas welded with the welding rods of 30 mm diameter is .... mm.

(*a*) 3 (*b*) 6

(*c*) 9 (*d*) 12

166. Weld spatter is a welding defect resulting from the use of

(*a*) too low welding current.

(*b*) too high welding current.

(*c*) low voltage.

(*d*) high voltage.

167. The welding current requirement increases with the increasing thickness of the material to be welded by arc welding. The current required for welding two 12 mm plates with the help of 8 gauge electrodes will be .... amperes.

(*a*) 50 (*b*) 100

(*c*) 150 (*d*) 200

168. For gas welding of two 20 mm thick plates, the appropriate size of the welding rod to be selected out of the following will be of .... mm.

(*a*) 1.5 (*b*) 3

(*c*) 6 (*d*) 12

Answers

160. (IV) (III) (II) (IV) 161. (-IV) (-III) (-II) (-I) 162. (c) 163. (a)
164. (d) 165. (a) 166. (a) 167. (b) 168. (c)

169. Higher porosity of a welded joint results from the use of
(*a*) wrong electrode size.
(*b*) poor base metal quality.
(*c*) high welding current.
(*d*) low welding current.

170. Frequency range employed in ultrasonic welding is .... cycles/second.
(*a*) 500-1000 (b) 4000-20000
(*c*) 20000-50000 (*d*) 50000-100000

171. Tack welding is
(a) a preparatory job before welding.
(b) a type of weld.
(c) used exculsively for non-ferrous materials.
(d) a process of TIG welding.

172. The type of welding normally used in automobile body work is ....welding.
(*a*) seam (*b*) gas
(*c*) arc (*d*) spot

173. Electron beam welding is carried out in
(a) atmospheric air.
(b) vacuum.
(c) shielding gas environment.
(d) pressurised inert gas chamber.

174. In mannual arc welding, the equipment should have drooping characteristics in order to maintain
(a) weld pool red hot.
(b) constant temperature in the arc.
(c) constant voltage when arc length changes.
(*d*) constant current when arc length changes.

175. The maximum heat in the resistance welding is at the
(a) top surface of the plate at the time of electric contact with the electrode,
(b) interface between the two plates being joined.
(c) tip of the negative electrode.
(b) tip of the positive electrode.

176. D.C. reverse polarity is used in arc welding to bear greater advantage in
(a) flat welding of butt joint.
(b) flat welding of lap joint.
(c) edge welding,
(d) overhead welding.

177. If volumetric proportion of oxygen & acetylene are kept the same (i.e., 50% oxygen and oxygen *50%* acetylene), in oxy- acetylene welding, then the nature of flame produced will be
(a)oxidising (b) reducing
(c)carburising (*d*) neutral

178. MIG welding process uses
(a) D.C. power supply & consumable electrode.
(b) A.C. power supply & non consumable electrode,
(c) D.C. power supply & non consumable electrode.
(*d*) A.C. power supply & consumable electrode.

179. Which of the following is the deciding factor for the current value to be used in the arc welding ?
(a) Length of the portion to be welded
(b) Thickness of the plate to be welded
(c) Electrode size
(d) Voltage across the arc

Answers

169. (b)	170. (b)	171. (b)	172. (d)	173. (b)	174. (c)
175. (b)	176. (b)	177. (c)	178. (a)	179. (c)	

180. The filler metal in brazing is melted and deposited at the point, where the weld is to be made. The melting point of the filler metal used in braze welding should be more than .... °C.
(a) 220 (b) 420
(c) 620 (d) 820

181. In .... process, the part to be welded are heated to the welding temperature and then mechanical force is applied at the ends to join the metallic parts.
(a) forge (b) thermit
(c) plasma (d) TIG

182. Out of the following, the material which is best weldable with itself is
(a) aluminium (b) stainless steel
(c) mild steel (d) cast iron

183. Tungsten electrodes with small addition of .... is used for improvement in the arc starting & stabilisation.
(a) cobalt (b) thoria
(c) nickel (d) copper

184. The distinguishing feature between soldering & brazing is that, in case of soldering the
(a) filler metal used is known as spelter.
(b) process is carried out below the melting point of filler metal.
(c) strength of joint produced is relatively low.
(d) joining of dissimilar metal pipes & fittings is done.

185. Vacuum maintained in the 'electron beam welding machine is about .... torr. (1torr = 1 mm Hg coloumn).
(a) 10^{-1} (b) 10^{-3}
(c) 10^{-5} (d) 10^{-8}

186. Welding of stainless steel poses problem because of the formation of
(a) oxide film.
(b) rust.
(c) chromium carbide.
(d) cracks.

187. High pressure (160-200 kg/cm^2) oxygen cylinders are .... steel cylinders.
(a) welded (b) seamless
(c) cast (d) heat treated

188. Cutting of metal by oxy-acetylene flame is because of its intensive oxidation. Out of the following, the material, which is best cut by oxy-acetylene cutting process is
(a) mild steel (b) stainless steel
(c) brass (d) copper

189. Pick out the wrong statement about the function of flux-in brazing. Flux used in brazing helps in the
(a) wetting process by reducing the viscosity of the melt to give filler metal the fluidity to wet the joint surface completely.
(b) prevention of oxide formation on both the brazing material as well as the base metal.
(c) prevention of thermal distortion & cracking.
(d) dissolution of surface oxide coatings formed prior to brazing.

190. Brazing is the process of hard soldering using brass spelter. The flux used in the brazing process is in .... form.
(a) liquid (b) powder
(c) paste (d) either 'a', 'b' or 'c'

Answers

180. (b)	181 (a)	182. (c)	183. (b)	184. (c)	185. (c)
186. (a)	187. (b)	188. (a)	189. (c)	190. (d)	

191. Match the fillers used in various metal joining processes.

List I

(*a*) Soldering
(*b*) Brazing
(*c*) TIG welding of aluminium
(*d*) Braze welding

List II

I. Copper & tin alloy
II. Iron oxide & aluminium powder
III. Copper, zinc & silver alloy
IV Lead & antimony welding alloy

192. Match the applications of various welding processes.

List I

(*a*) Gas welding
(*b*) Submerged arc welding
(*c*) TIG welding
(*d*) Electron beam welding

List II

I. Deep penetration precision weld
II. Joining of high speed steel drill bit to carbon steel shank
III Welding of aluminium alloys
IV Ship welding

193. Arc stability is better in .... welding.

(*a*) AC
(*b*) DC
(*c*) rectified electric supply
(*d*) specially designed wave form

194. Pick out the wrong statement.

AC arc welding always employs coated electrodes, because it

(*a*) has low leakage loss and splatter
(*b*) uses less input current
(*c*) is safest for the operator.
(*d*) protects the newly form weld against atmosphere.

195. Pick out the wrong statement about spot welding.

(*a*) Spot welding is used for welding top joints in plates having thickness ranging from 0.25 to 1.25 mm.
(*b*) A continuous spot welding process is percussion welding.
(*c*) Spacing between two spot welds should not be less than 3d (d = electrode tip diameter.)
(*d*) Number of zones of heat generation in spot welding are five.

196. Round bars are joined by ....welding process.

(*a*) seam (*b*) projection
(*c*) spot (*d*) none of these

197. The welding process in which the metal to be joined are heated to the plastic state and then forged together by external pressure without the addition of filler material, is known as.................. welding

(*a*) projection (*b*) thermit
(*c*) fusion (*d*) plastic

198. If two pieces of different metals are to be welded by projection welding, then the projection should be made on the metal piece having higher conductivity. The diameter of projection as compared to the thickness of the sheet is about .... times in projection welding.

(*a*) 0.5 (*b*) 1.5
(*c*) 2 (*d*) 3

Answers

191. (-IV) (-III) (-II) (-I)	192 (-III) (-IV) (-II) (-I)	193. (a)	194. (c)
195. (b) 196. (a)	197. (d) 198. (b)		

199. Plain and butt welds may be used on materials upto approximately .... mm thick.
(*a*) 25 (*b*) 50
(*c*) 60 (*d*) 75

200. Arc is created between the electrode and the work, in arc welding by
(*a*) material characteristics
(*b*) voltage
(*c*) flow of current
(*d*) contact resistance

201. The name key hole welding refers to the .... welding.
(*a*) plasma arc (*b*) percussion
(*c*) projection (*d*) pulsed TIG

202. Arc length in electric arc welding is the distance between the tip of the electrode and the
(*a*) work piece
(*b*) work piece minus electrode diameter
(*c*) centre of crater
(*d*) bottom of crater

203. The major problem in welding of stainless steel is
(*a*) formation of oxide film
(*b*) formation of chromium carbide
(*c*) high electrical resistance
(*d*) poor thermal conduction

204. Both preheating and post heating is essential in welding
(*a*) nickel
(*b*) low carbon steel
(*c*) medium carbon steel
(*d*) high carbon steel

205. The voltage current characteristics of a dc generator for arc welding is a straight line between an open circuit voltage of 80 V and short circuit current of 300 A. The generator settings for maximum arc power will be
(*a*) 80 V and 150 A (*b*) 400 V and 300 A
(*c*) 40 V and 150 A (*d*) 80 V and 300 A

206. The correct sequence of the given materials in ascending order of their weldability is
(*a*) MS, Copper, Cast iron, Aluminium
(*b*) Cast iron, MS, Aluminium, Copper
(*c*) Copper, Cast iron, Aluminium
(*d*) Aluminium, Copper, Cast iron, MS

207. Which of the following statements is correct?
(*a*) No flux is used in gas welding of mild steel.
(*b*) Borax is the commonly used flux coating on welding electrodes,
(*c*) Laser beam welding employs a vacuum chamber and thus avoids use of a shielding method.
(*d*) AC can be used for GTAW process.

208. In d.c. welding, the straight polarity (electrode negative) results in
(*a*) lower penetration
(*b*) lower deposition rate
(*c*) less heating of work piece
(*d*) smaller weld pool.

Answers

199. (a)	200 (d)	201. (c)	202. (c)	203. (b)	204. (d)
205. (c)	206. (a)	207. (a)	208. (a)		

209. Match List I with List II and select the correct answer from the codes given below the lists:

List I

A. Welding of aluminium alloy

B. Ship building

C. Joining of HSS drill bit to carbon steel shank

D. Deep penetration precision welds

List II

1. Submerged arc welding.
2. Electron beam welding
3. TIG welding
4. Gas welding

Codes:

	A	B	C	D
(a)	5	2	3	1
(b)	5	1	3	2
(c)	3	5	1	2
(d)	4	5	2	3

210. For butt welding 40 mm thick steel plates, when the expected quantity of such jobs is 5000 per month over a period of 10 years, choose the best suitable welding process out of the following available alternatives.

(a) Submerged arc welding

(b) Oxy-acetylene gas welding

(c) Electron beam welding

(d) MIG welding

211. Resistance spot welding is performed on two plates of 1.5 mm thickness with 6 mm diameter electrodes using 15000 A current for a time duration of 0.25 seconds. Assuming the interface resistance to be 0.001 the heat generated to form the weld is .... watt, sec.

(a) 5625 *(b)* 8437

(c) 22500 *(d)* 33750

212. In an arc welding process, the voltage and current are 25 V and 300 A respectively. The arc heat transfer efficiency is 0.85 and welding speed is 8 mm/sec. The net heat input (in J/mm) is

(a) 64 *(b)* 797

(c) 1103 *(d)* 79700.

213. Which of the following is a solid state joining process?

(a) Gas tungsten arc welding

(b) Resistance spot welding

(c) Friction welding

(d) Submerged arc welding

214. A direct current welding machine with a linear power source characteristic provides open circuit voltage of 80 V and short circuit current of 800 A. During welding with the machine, the measured arc current is 500 A corresponding to an arc length of 5 mm and the measured arc current is 460 A corresponding to an arc length of 7 mm. The linear voltage (E) - arc length (L) characteristic of the welding arc can be given as (where, E is in volt and L is in mm)

(a) E = 20 + 2L (b) E = 20 + 8L

(c) E = 80 + 2L *(d)*. E = 80 + 8L

215. Two 1 mm thick sheets are to be spot welded at a current of 5000 A, Assuming effective resistance to be 200 micro-ohms and current flow time of 0.2 second, heat generated during the process will be .... Joules.

(a) 0.2 *(b)* 1

(c) 5 *(d)* 1000

Answers

209. (b) 210 (b) 211. (a) 212. (b) 213. (c) 214. (a)
215. (a)

216. The strength of a brazed joint
   (a) decreases with increase in the gap between the two joining surfaces.
   (b) increases with increases in the gap between the two joining surfaces.
   (c) decreases upto a certain gap between the two joining surfaces beyond which it increases.
   (d) increases upto a certain gap between the two joining surfaces beyond which it decreases.

217. Current (amperage) setting, which depends upon size, position, electrode type and metal thickness, should be adjusted
   (a) during the welding operation with a half used electrode.
   (b) during the welding operation with a new electrode.
   (c) using E6011 electrode on a 5 mm mild steel plate.
   (d) welding is not being done.

218. Increasing sub-arc voltage will
   (a) use more flux *(b)* increase undercut
   (c) use less flux *(d)* increase undercut

219. Pick out the wrong statement.
   (a) An irregular, wide and splattered weld results when welding current is too high or the arc length is too great.
   (b) When welding with a direct current, straight polarity, the electrode lead is connected to the negative terminal.
   (c) AC welding machines house transformers to reduce input amperage and increase output voltage.
   (d) The flux coating on electrode provides a gas shield that protects the molten metal.

220. Potassium and sodium silicate are used in the coatings of arc welding electrode for
   (a) the formation of slag and gas shield.
   (b) deoxidation of weld metal.
   (c) binding the particle together and causing them to adhere to the core wire.
   (d) introducing alloy in weld metal.

221. Pick out the wrong statement.
   (a) No electrode is consumed in TIG welding.
   (b) Continuous wire feeding and inert gas shielding is required in GMAW.
   (c) Flux dia/core dia ratio for medium coated electrode is generally 1.45.
   (d) An electric arc welding machine output is generally expressed as voltage.

222. Metal .... is not important when selecting the type of welding electrode to use.
   (a) thickness *(b)* type
   (c) weight *(d)* welding position

223. Polarity is the direction of current flow in a welding circuit. Changing the polarity in electric arc welding is useful because, it
   (a reduce splatter
   (b) controls heat and penetration depths
   (c) reduces slag deposits
   (d) reduces ultraviolet & infrared light rays

224. Plasma is a type of ionised gas. In plasma arc welding process, shielding gas used is
   (a) argon
   (b) oxygen
   (c) carbon dioxide
   (d) carbon monoxide

Answers

216. (d)	217 (d)	218. (a)	219. (c)	220. (c)	221. (d)
222. (c)	223. (b)	224. (a)			

225. Welds made with high heat input will show reduction in
    (a) ductility
    (b) toughness
    (c) fatigue strength
    (d) mechanical strength

226. Welding cable is used in electric arc welding due to its high current carraying capacity. No load voltage of a welding machine is generally .... volts.
    (a)30-50 (b) 60-80
    (c) 90-110 (d) 120-150

227. The welding .... helps to control the penetration and characteristics of the metal deposit.
    (a) type
    (b) polarity
    (c) speed
    (d) current type (AC or DC)

228. Pick out the wrong statement.
    (a) The fourth digit in the electrode number designates the tensile strength.
    (b) In order to obtain maximum penetration with a high deposition rate, two different electrodes may be used on heavy welds.
    (c) Hard facing of an equipment is a reclamation job and the hard facing electrodes is a separate type of electrode (neither SS, MS or low hydrogen electrode)
    (d) If an electrode is classified as ER 4211X, the digit '4' denotes the strength characteristics.

229. In an electric arc welding, electric current required for 4 mm electrode is approximately about .... amperes.
    (a) 100 (b) 160
    (c) 220 (d) 280

230. Which type of SMAW welding machine offers both AC and DC current when welding?
    (a) Plasma arc
    (b) Generator
    (c) Transformer/rectifier
    (d) Transformer

231. Which of the following does not cause undercutting?
    (a) High amperage
    (b) Arc too long
    (d) Low amperage
    (d) Electrode held at 90° angle.

232. In SAW process, flux is
    (a) not required
    (b) coated at the outer surface of the electrode
    (c) inside the electrode
    (d) fed on weld pool from outside

233. Which of the following causes excessive splatter, fiery arc, spalling of flux coating and porosity in weld metal?
    (a) Moisture absorption by the electrode
    (b) Defective welding machine
    (c) Loose electrical connection
    (d) Low voltage

Answers

225. (b)	226 (b)	227. (b)	228. (a)	229. (b)	230. (c)
231. (c)	232. (d)	233. (a)			

14

Metallurgy of Nuclear Metals

1. Heat is generated in a nuclear reactor (thermal) by
   (*a*) combustion of *a* nuclear fuel *e.g.*, uranium.
   (*b*) fusion of atoms of uranium.
   (*c*) absorption of neutrons in uranium atoms.
   (*d*) fission of U-235 by neutrons.

2. Which of the following undergoes fission reaction easily ?
   (*a*) U-235 (6) U-238
   (*c*) Th-232 (*d*) none of these

3. Percentage of U-238 in natural uranium is around
   (*a*) 0.71 (*b*) 99.29
   (*c*) 0.015 (*d*) 29.70.

4. Number of secondary neutrons emitted on fission of an atom of U-235 on slow neutron bombardment is
   (*a*) 3 (*b*) 235
   (*c*) 200 (*d*) 1

5. Thermal neutrons which are used to cause the fission of U-235 have energy .... eV.
   (*a*) < 0.025 (*b*) > 1
   (*c*) 1—25 (*d*) > 200.

6. Velocity of the thermal neutron (< 0.025 eV) used for fission of U-235 may be around .... m/sec.
   (*a*) 1 (*b*) 2200
   (*c*) 3 x 1010 (*d*) 9 x 1021

7. A fertile material is the one, which can be
   (*a*) converted into fissile material on absorption of neutron.
   (*b*) fissioned by slow (thermal) neutrons.
   (*c*) fissioned by fast neutrons.
   (*d*) fissioned by either slow or fast neutrons.

8. Which of the following is not a naturally occurring nuclear fuel ?
   (*a*) Uranium-238 (*b*) Thorium-233
   (*c*) Plutonium-239 (*d*) None of these

9. Thermal shield is used in high powered nuclear reactors to
   (a) protect the walls of the reactor from radiation damage.
   (b) absorb the fast neutrons.
   (*c*) slow down the secondary neutrons.
   (*d*) protect the fuel element from coming in contact with the coolant.

Answers

1. (d)	2. (a)	3. (b)	4. (a)	5. (a)	6. (b)
7. (a)	8. (c)	9. (a)			

10. 'Light water' used as a coolant in nuclear reactor is nothing but .... water.
    (a) ordinary
    (b) mildly acidic (pH = 6)
    (c) mildly alkaline (pH = 8)
    (d) none of these.

11. Enriched uranium means that, it contains
    (a) > 0.71% of U-235.
    (b) only fertile material.
    (c) only fissile material.
    (d) no impurities.

12. A homogeneous nuclear reactor is the one, in which the
    (a) fissile atoms are evenly distributed throughout the mass of nuclear reactor.
    (b) same substance *(e.g.;* heavy water) is used as moderator & coolant.
    (c) fuel & the moderator are mixed to form a homogeneous material.
    (d) all *(a)*, *(b)* & *(c)*.

13. A heterogeneous nuclear reactor is the one in which the moderator & the
    (a) coolant are different materials.
    (b) coolant are present in different phases *(e.g.* heavy water & graphite).
    (c) fuel are present in different phases *(e.g.* uranium & heavy water).
    (d) none of these.

14. Candu reactor is a .... reactor.
    (a) A natural uranium fuelled and heavy water cooled & moderated
    (b) highly enriched uranium (85% U-235) fuelled
    (c) homogeneous
    (d) fast breeder

15. In a pressurized water reactor (PWR), the
    (a) fuel is natural uranium and heavy water acts both as moderator & coolant.
    (b) coolant water boils in the core of the reactor.
    (c) coolant water is pressurised to prevent bulk boiling of water in the core.
    (d) use of moderator is not required.

16. A boiling water reactor is the one, in which the
    (a) coolant water is allowed to boil in the core of the reactor.
    (b) coolant water, after being heated in the reactor core, generates steam in a boiler,
    (c) pressurised water is pumped into the core.
    (d) fuel & the coolant are thoroughly mixed to form a homogeneous solution.

17. Liquid metal *(e.g.,* molten sodium) is preferred as a coolant in case of a/an.................. reactor.
    (a) homogeneous
    (b) graphite moderated
    (c) fast breeder
    (d) enriched uranium (3% U-235) fuelled

18. Which of the following does not need a moderator ?
    (a) Candu reactor
    (b) Fast breeder reactor
    (c) Homogeneous reactor
    (d) Pressurised water reactor.

19. A fast breeder reactor
    (a) utilises fast neutrons for causing nuclear fission.
    (b) converts fertile material *(e.g.,* U-238) into fissile material (Pu-239).
    (c) normally employs molten sodium as coolant.
    (d) all *(a)*, *(b)* & *(c)*.

Answers

10. (a)	11. (a)	12. (c)	13. (c)	14. (a)	15. (c)
16. (a)	17. (c)	18 (b)	19. (d)		

20. Thorium can be converted into U-233 in a .... reactor.
   (*a*) liquid metal cooled
   (b) fast breeder
   (*c*) thermal
   (*d*) swimming pool

21. Which of the following is not a fertile material ?
   (*a*) Th-232 (*b*) U-238
   (*c*) U-233 (*d*) none of these

22. A nuclear reactor can't be used for
   (*a*) the production of radioisotopes.
   (*b*) supplying intense fields or beams of neutron for scientific experiments.
   (*c*) marine ship propulsion.
   (*d*) none of these.

23. Moderating material used in a thermal reactor should be a
   (a) good absorber of neutrons.
   (b) solid substance.
   (*c*) poor absorber of neutrons.
   (*d*) none of these.

24. Ordinary water is not used as a moderator, because it
   (*a*) has a low absorption cross-section.
   (b) has a low scattering cross-section.
   (*c*) absorbs neutrons.
   (*d*) does not absorb neutrons.

25. Uranium is recovered from its ore (Pitchblende) by
   (*a*) froth floatation techniques.
   (*b*) leaching with sulphuric acid.
   (*c*) smelting in a furnace.
   (*d*) dissolving in water.

26. The main purpose of control rod in a nuclear reactor is to control the
   (*a*) chain reaction & hence the power output by regulating the number of secondary neutrons for fission.
   (*b*) emission of hazardous radiation.
   (*c*) conversion of fissile material into fertile material.
   (*d*) velocity of the secondary neutrons.

27. An ideal coolant for a nuclear reactor should be
   (*a*) a good absorber of neutrons.
   (*b*) capable of attaining high temperature, only when it is pressurised.
   (*c*) of high density, but of low heat transfer coefficient.
   (*d*) free from radiation damage & non-corrosive.

28. Which of the following may not need a control rod ?
   (*a*) Liquid metal cooled reactor
   (*b*) Fast breeder reactor
   (*c*) Candu reactor
   (*d*) None of these

29. Heavy water is preferred over ordinary water as a coolant, because it
   (*a*) acts both as an efficient coolant as well as a moderator.
   (*b*) can be heated to a higher temperature without pressurising.
   (*c*) is less prone to radiation damage.
   (*d*) all (*a*), (*b*) & (*c*).

30. Molten sodium (as a coolant in fast breeder reactor)
   (*a*) can't attain high temperature at normal pressure.
   (*b*) is not at all corrosive even at a higher temperature.
   (*c*) is highly radioactive at elevated temperature & can cause explosion when it comes in contact with air or water.
   (*d*) none of these.

Answers

20. (b)	21. (c)	22. (d)	23. (c)	24. (c)	25. (b)
26. (a)	27. (d)	28. (d)	29. (d)	30. (c)	

31. Uranium ore is currently mined & concentrated at
(*a*) Jadugoda (Jharkhand).
(*b*) Ghatsila (Jharkhand).
(*c*) Khetri (Rajasthan).
(*d*) Alwaye (Kerala).

32. Which of the following is a fuel for a fusion reactor (thermo nuclear reactor) ?
(*a*) Deuterium & tritium
(*b*) U-233
(*c*) Thorium
(*d*) Heavy water

33. Commercial power generation for fusion reactor is not yet possible, because
(*a*) it is difficult to control fusion reaction.
(*b*) the fuel required (*e.g.*, deuterium & tritium) is scarce.
(*c*) it is difficult to initiate fusion reaction.
(*d*) quantity of fuel required for initiating the fusion reaction is prohibitively high.

34. Biological shield is provided in a nuclear power reactor to protect the .... from radiation damage.
(*a*) fuel elements(*b*) operating personnels
(*c*) walls of the reactor
(*d*) none of these

35. Percentage of heavy water in ordinary water is around
(*a*) 0.015 (*b*) 7.54
(*c*) 0.71 (*d*) 32.97

36. The main ore of uranium is
(*a*) pitchblende (*b*) monazite sand
(*c*) cassiterite (*d*) chalcopyrite .

37. The main ore of thorium is
(*a*) pitchblende (*b*) monazite
(*c*) limonite (*d*) galena

38. Heavy water (D_2O) in a nuclear reactor serves as a
(*a*) coolant (*b*) moderator
(*c*) both (*a*) & (*b*) (*d*) neutron absorber

39. Which of the following is a moderating material used in nuclear reactor ?
(*a*) graphite
(*b*) cadmium
(*c*) zircalloy (an alloy of zirconium and aluminium)
(*d*) stainless steel

40. The function of moderators in nuclear reactor is to
(*d*) slow down the secondary neutrons.
(*d*) absorb the secondary neutrons.
(*d*) control the chain reaction.
(*d*) none of these.

41. Which of the following nuclear materials is fissile ?
(*a*) Uranium-238 (*b*) Thorium-232
(*c*) Plutonium-239 (*d*) None of these

42. Quantity of fissionable material (*i.e.*, U-235) in natural uranium is .... per cent.
(*a*) 0.71 (*b*) 6.31
(*c*) 99.29 (*d*) 12.73

43. "Critical mass" is the minimum mass of nuclear fissile material required for the
(*a*) sustainment of chain reaction.
(*b*) power generation on commercial scale.
(*c*) economic power generation.
(*d*) none of these.

44. Fast breeder reactors are most usable inIndia, because of our largest .... deposits.
(*a*) thorium (*b*) plutonium
(*c*) uranium (*d*) none of these

Answers

31. (a)	32. (a)	33. (a)	34. (b)	35. (a)	36. (a)
37. (b)	38. (c)	39. (a)	40. (a)	41. (c)	42. (a)
43. (a)	44. (a)				

45. Which is not a fissile nuclear material ?
    (*a*) U-233 (*b*) U-235
    (*c*) U-238 (*d*) Pu-239

46. Which is a fertile nuclear fuel ?
    (*a*) U-233 (*b*) U-235
    (*c*) Pu-239 (*d*) Th-232

47. Main source of............... is monozite sand.
    (*a*) uranium (*b*) polonium
    (*c*) hafnium (*d*) thorium

48. Indian monazite sand contains....... % $Th0_2$.
    (*a*) 10 (*b*) 25
    (*c*) 60 (*d*) 90

49. Disintegration rate of a radioactive element
    (*a*) progressively increases.
    (*b*) progressively decreases.
    (*c*) remains constant throughout.
    (*d*) may increase or decrease with time.

50. Decay product of tritium which is a Beta emitter is
    (*a*) hydrogen (*b*) helium
    (*c*) deuterium (*d*) none of these

51. The molecular weight of heavy water is
    (*a*) 10 (*b*) 18
    (*c*) 20 (*d*) 36

52. Radioactive decay is a.......... order reaction.
    (*a*) zero (*b*) first
    (*c*) second (*d*) third

53. Thorium metal
    (*a*) resembles steel in appearance.
    (*b*) is less hard (in the range of silver).
    (*c*) is highly ductile.
    (*d*) *all (a), (b)* & *(c)*.

54. Plutonium
    (*a*) is recovered from spent fuel from thermal nuclear reactor.
    (*b*) has much lower melting point (640°C) compared to thorium (1690°C).
    (*c*) both (*a*) & (*b*).
    (*d*) neither (*a*) nor (*b*).

55. Ceramic nuclear metals compared to solid nuclear metals have
    (*a*) higher melting points.
    (*b*) low strength at high temperature.
    (*c*) more chances of radiation damage.
    (*d*) poorer corrosion resistance.

56. Which is used as a coolant in nuclear reactor due to high neutron capture cross-section ?
    (*a*) H_2 (6) N_2
    (*c*) He (*d*) $C0_2$

57. The type of pump used for recirculation of molten sodium coolant in liquid metal cooled reactor is the.......................... pump.
    (*a*) electromagnetic (*b*) reciprocating
    (*c*) centrifugal (*d*) volute

58. H_2 is a better coolant than CO2 due to its relatively
    (*a*) lower density.
    (*b*) higher specific heat.
    (*c*) non reactivity to uranium.
    (*d*) lower neutron capture cross-section.

59. Which of the following is not an ore of uranium ?
    (*a*) Kyanite (*b*) Pitchblende
    (*c*) Carnotite (*d*) Rescolite

60. Percentage of natural uranium in uranium ore mined at Jadugoda (Jharkhand) is about
    (*a*) 0.001-0.005 (*b*) 0.05-0.10
    (*c*) 1-2 (*d*) 15-20

Answers

45. (c)	46. (d)	47. (d)	48. (a)	49. (b)	50. (b)
51. (c)	52. (b)	53. (d)	54. (c)	55. (a)	56. (a)
57. (a)	58. (b)	59. (a)	60. (b)		

61. Extraction of uranium from its ore is done by.......................... method.
   (*a*) chemical
   (*b*) physical beneficiation
   (*c*) electrometallurgical
   (*d*) pyrometallurgical

62. Out of the following places, heavy water plant is not located at
   (*a*) Baroda (*b*) Ramagundam
   (*c*) Talcher (*d*) Tuticorin

63. Which of the following is not a moderator used in nuclear reactors?
   (*a*) Boron hydride (b) Ordinary water
   (*c*) Molten sodium (*d*) Beryllium

64. A fast breeder reactor does not
   (*a*) use thorium-232 as fissile fuel.
   (*b*) convert fertile material to fissile material.
   (*c*) use fast neutron for fission.
   (*d*) use gaseous coolant.

65. Net efficiency of an ordinary (light) water cooled nuclear reactor is about .... per cent.
   (*a*) 32 (*b*) 14
   (*c*) 52 (*d*) 78

66. Safety rod provided in nuclear reactors to guard against accident in case of earthquake is made of
   (*a*) molybdenum
   (*b*) zircalloy
   (*c*) boron or cadmium
   (*d*) high carbon steel

67. Out of the following places, a nuclear power plant is not located at
   (*a*) Talcher (Orissa)
   (*b*) Kaiga (Karnataka)
   (*c*) Rawatbhata (Rajasthan)
   (*d*) Kalpakkam (Tamilnadu)

68. The half life period of a radioactive element depends upon its
   (*a*) temperature (b) pressure
   (*c*) amount (*d*) none of these

69. The decay product of tritium (a beta embitter) is
   (*a*) lithium (*b*) helium
   (*c*) deuterium (*d*) hydrogen

70. 'Heavy water' is termed as 'heavy', because it is
   (*a*) denser than water.
   (*b*) an oxide of deuterium.
   (*c*) a heavy liquid.
   (*d*) an oxide of heavier isotope of oxygen.

71. Which one is radioactive in nature ?
   (*a*) Helium (*b*) Deuterium
   (*c*) Heavy hydrogen (*d*) Tritium

72. An element having large number of......... is most easily subjected to nuclear fission.
   (*a*) protons (*b*) electrons
   (*c*) neutrons (*d*) nucleons

73. Which of the following isotopes is not present in natural uranium?
   (*a*) U-238 (6) U-234
   (*c*) U-235 (*d*) U-232

74. .... nuclear reactor does not require a heat exchanger to supply steam to power turbine.
   (*a*) Molten sodium cooled
   (*b*) Helium cooled
   (*c*) Boiling water
   (*d*) Pressurised water

Answers

61. (a)	62. (b)	63. (c)	64. (a)	65. (a)	66. (c)
67. (a)	68. (b)	69. (b)	70. (b)	71. (d)	72. (d)
73. (d)	74. (c)				

75. Energy produced in the nuclear fission is of the order of .... MeV.
(a) 20 *(b)* 200
(c) 1000 *(d)* 2000

76. Critical energy should be........ the neutron binding energy of the atom in order to initiate a nuclear fission.
(a) equal to *(b)* less than
(c) more than *(d)* either more or less.

77. The time required for half of the........ of a radioactive isotope to decay is called its half life.
(a) nuclei *(b)* electrons
(c) protons *(d)* neutrons

78. Artificially produced radioactive isotopes are used for
(a) power generation.
(b) treatment of certain diseases.
(c) initiating nuclear fission & fusion.
(d) all *(a)*, *(b)* & *(c)*.

79. Which of the following nuclear reactors is the most efficient thermodynamically while operating between the same temperature & pressure limits of the reactor ?
(a) Molten sodium cooled reactor.
(b) CO2 gas cooled reactor.
(c) Pressurized water reactor.
(d) Boiling water reactor.

80. Radioactivity of an isotope is expressed in
(a) Barns *(b)* MeV
(c) Curie *(d)* Ergs

81. Molten sodium is used as a coolant in fast breeder reactor, because of its
(a) excellent moderating properties.
(b) neutron breeding capability.
(c) faster heat removal capability from core.
(d) capability to increase the reaction rate in the core.

82. Which of the following gases has very high neutron capture cross-section, making it unsuitable as a coolant in nuclear reactors?
(a) Nitrogen *(b)* Helium
(c) Carbon dioxide *(d)* Hydrogen

83. Pressure in a PWR is used for
(a) maintaining constant pressure in the primary cooling circuit under varying loads.
(b) superheating the steam.
(c) pressurising the water in the primary coolant circuit.
(d) none of these.

84. Gas cooling as compared to water cooling of nuclear reactors
(a) can not attain a high temperature.
(b) is more efficient as gas has a higher specific heat.
(c) can produce only saturated steam for feeding to power turbine.
(a) none of these.

85. Which of the following nuclear reactors is the most prone to radioactive hazards ?
(a) Pressurised water reactor.
(b) Gas cooled reactor.
(c) Molten sodium cooled reactor.
(d) Boiling water reactor.

86. Coolant used in boiling water reactor is
(a) hydrogen gas
(a) water
(c) steam
(d) a mixture of water & steam

Answers

75. (b)	76. (c)	77. (a)	78. (b)	79. (d)	80. (c)
81. (c)	82. (a)	83. (a)	84. (d)	85. (d)	86. (c)

87. Function of control rod in a nuclear reactor- is to control the
    (a) temperature
    (b) pressure
    (c) fuel consumption
    (d) absorption of neutrons

88. Nuclear power reactor located at Tarapur is of the .... type.
    (a) fast breeder (b) pressurised water
    (c) boiling water (d) gas cooled

89. Nuclear reactors are provided with shield to guard against the emission of mainly .... rays.
    (a) X
    (b) □ & β
    (c) neutron & gamma
    (d) infrared

90. Heavy water has the maximum density at .... °C.
    (a) 4 (b) 11.6
    (c) 0 (d) 18.6

91. Nuclear fuel usually used in a BWR is
    (a) plutonium (b) enriched uranium
    (c) natural uranium (d) thorium

92. Which of the following radiation is the least penetrating ?
    (a) □-rays (b) β-rays
    (c) y-rays (d) X-rays.

93. Which of the following is the most harmful for the human being ?
    (a) β-rays (b) X-rays
    (c) y-rays (d) ultra violet rays

94. What is the average life of a radioactive atom having a 'half life period' of T ?
    (a) 1.44T (b) 0.144T
    (c) 14.4T (d) 2T

95. The time taken for a radioactive element to reduce to 50°/ of its original weight is .... years, if its half life period is 12 years.
    (a) 24 (b) 18
    (c) 6 (d) 36

96. Which of following ores contains maximum percentage of uranium?
    (a) Carnotite (b) Thorium
    (c) Rescolite (d) Pitchblende

97. Uranium percentage in monazite sand is about
    (a) 0.01 (b) 0.25
    (c) 1.2 (d) 7

98. Specific gravity of uranium & plutonium is about
    (a) 9 (b) 13
    (c) 19 (d) 27

99. Which of the following is artificially produced, as it does not occur in nature ?
    (a) Uranium-235 (b) Uranium-233
    (c) Plutonium-239 (d) Both (b) & (c)

100. Thorium-232 (a fertile material) on absorption of a neutron gets converted into .... which is a fissile material.
    (a) thorium-233 (b) uranium-235
    (c) uranium-233 (d) plutonium-239

101. Which of the following accounts for the maximum energy release in the nuclear fission process ?
    (a) Kinetic energy of fission products.
    (b) Radioactive decay of fission products.
    (c) Instantaneous release of y-rays.
    (d) Kinetic energy of neutrons.

Answers

87. (d)	88. (c)	89. (c)	90. (b)	91. (b)	92. (a)
93. (c)	94. (a)	95. (b)	96. (d)	97. (b)	98. (c)
99. (d)	100. (c)	101. (a)			

102. The velocity of thermal (slow) neutrons triggering nuclear fission reaction (having energy equal to 0.025 eV) is about .... metres/second.
(*a*) 1100 (*b*) 2200
(*c*) 3300 (*d*) 4400

103. Thorium-232 is converted into uranium- 233 in a .... nuclear reactor.
(*a*) thermal
(*b*) fast breeder
(c) heavy water moderated
(*d*) enriched uranium

104. Which of the following is a non-fissile material ?
(*a*) Plutonium-239 (*b*) Uranium-235
(*c*) Uranium-233 (*d*) Thorium-232

105. Thermal nuclear reactors using 'enriched uranium' as fuel contain a maximum of .... percent fissile material *i.e.,* U-235.
(*a*) 1 (b) 2
(*c*) 3 (*d*) 7

106. Which of the following is not used as a nuclear fuel cladding material?
(*a*) Zircalloy (*b*) Cadmium
(*c*) Ceramics (*d*) Stainless steel

107. .... moderator is used in a fast breeder reactor.
(*a*) Graphite (b) Heavy water
(*c*) Beryllium (*d*) No

108. Enrichment of uranium is done to increase the concentration of .... in the natural uranium.
(*a*) U-238 (*b*) U-233
(*c*) U-235 (*d*) Pu-239

109. In the gaseous diffusion process of uranium enrichment, the natural uranium is converted into gaseous uranium
(*a*) oxides. (6) hexaflouride.
(*c*) carbide. (*d*) sulphate.

110. Nuclear power generation capacity in India is
(*a*) 5000 MW
(*b*) 15000 MW
(*c*) more than hydroelectric power generation capacity.
(*d*) more than thermal power generation capacity.

111. The first underground nuclear test was conducted by India at
(*a*) Jaisalmer (*b*) Pokharan
(c) Kalpakkam (*d*) Narora

112. The disintegration rate of a radioactive element
(*a*) progressively increases.
(*b*) progressively decreases.
(*c*) remains constant throughout.
(*d*) may increase or decrease: depends on the material.

113. Unit of radioactivity is
(*a*) barn (*b*) fermi
(*c*) angstrom (*d*) Curie

114. If 4 gm of a radioisotope has a half life period of 10 days, the half life of 2 gm of the same isotope will be .... days.
(*a*) 5 (*b*) 10
(*c*) 20 (*d*) 30

115. A radioactive substance does not emit
(*a*) □-ray (6) proton
(*c*) positron (*d*) β-ray

Answers

102. (b)	103. (b)	104. (d)	105. (c)	106. (b)	107. (d)
108. (c)	109. (b)	110. (a)	111. (b)	112. (b)	113. (d)
114. (b)	115. (b)				

116. Positron is a/an
   (a) nucleus having two protons.
   (b) helium nucleus.
   (c) electron with positive charge.
   (d) nucleus with one proton & one neutron.

117. Atoms with the same number of neutrons, but different number of nucleons are called
   (a) isotones *(b)* isobars
   (c) isotopes *(d)* isoters

118. Pick out the wrong statement.
   (a) Atoms with the same number of nucleons but different number of protons are called isobars.
   (b) Atoms with same number of protons but different number of nucleons are called isobars.
   (c) Out of □, β & γ - rays, the one having maximum penetration power are γ- rays.
   (d) The product formed by emission of an □ particle has mass number less by 4 units than the parent nuclide.

119. The radioisotope used to study the thyroid gland is
   (a) iodine-131 *(b)* cobalt-60
   (c) iridium-192 *(d)* carbon-14

120. The most abundant isotope of natural uranium is
   (a) ${}_{92}U^{238}$ *(b)* ${}_{92}U^{235}$
   (c) ${}_{92}U^{234}$ *(d)* ${}_{92}U^{233}$

121. Which of the following is the best nuclear fuel?
   (a) N_p-239 *(b)* *U*-234
   (c) Pu - 239 *(d)* Th - 236

122. The ${}_{92}U^{238}$ emits an □-particle. The product will be
   (b) ${}_{90}U^{234}$ *(b)* ${}_{90}U^{238}$
   (c) ${}_{90}U^{236}$ *(d)* ${}_{92}U^{236}$

123. Match the coolants used in various nuclear power reactors.
   List I
   (a) Calder Hall reactor
   (b) Candu reactor
   (c) Fast breeder reactor
   (d) Pressurised water reactor
   List II
   I. Ordinary water
   II. Carbon dioxide
   III. Heavy water
   IV. Molten sodium

124. Match the fuel used in various nuclear power reactors.
   List I
   (a) Fast breeder reactor.
   (b) Pressurised water reactor.
   (c) Candu reactor.
   (d) Calder Hall gas cooled reactor.
   List II
   I. Slightly enriched uranium.
   II. Natural uranium.
   III. Highly enriched uranium.
   IV. Plutonium.

125. Match the materials used in nuclear power reactor for various duties.
   List I
   (a) Biological shield.
   (b) Moderator for fast breeder reactor
   (c) Control rod for thermal reactors.
   (d) Fuel cladding material.
   List II
   I. No material is used
   II. Concrete
   III. Zircalloy
   IV. Hafnium

Answers

116. (c) 117. (a) 118. (b) 119. (a) 120. (a) 121. (c)
122. (a) 123. (II) (III) (IV) (I) 124. (IV) (I) (II) (III) 125. (II) (I) (IV) (III)

126. Match the following nuclear facilities/installations in India with their place of location.

List I

(*a*) Fast breeder test reactor (FBTR).

(*b*) Nuclear fuel complex.

(*c*) Apsara reactor.

(*d*) Uranium mining & concentration plant.

List I

I. Bhabha atomic research centre (BARC), Trombay.

II. Kalpakkam. (Tamilnadu).

III. Hyderabad.

IV. Jadugoda (Jharkhand).

127. Fuel for a nuclear reactor (thermal) is

(*a*) uranium (*b*) plutonium

(*c*) radium (*d*) none of these

128. Fuel for a fast breeder reactor is

(*a*) plutonium (*b*) uranium

(*c*) radium (*d*) neptunium

129. A thermal nuclear reactor compared to a fast breeder reactor

(*a*) uses slower neutrons for fission.

(*b*) uses faster neutrons for fission.

(*c*) gives higher power density.

(*d*) requires less fuel to run at the same power level.

130. Which of the following factors is the most important in the site selection for a nuclear power plant ?

(*a*) Absence of earthquake prone zone in nearby areas.

(*b*) Abundant availability of water,

(*c*) Remotely located from residential areas

(*d*) Proximity to fuel source.

131. Graphite is used in nuclear reactor as

(*a*) insulation lining of the reactor.

(*b*) fuel.

(*c*) lubricant.

(*d*) retarder of neutron velocity.

132. Which of the following may be used to measure the rate of nuclear disintegration?

(*a*) Geiger-Muller counter.

(*b*) Cyclotron.

(*c*) Cloud chamber.

(*d*)Mass spectrograph.

133. The radioactive isotope of hydrogen is

(*a*) deuterium (*b*) tritium

(*c*) protium (*d*) none of these

134. Pick out the wrong statement about a fast breeder reactor. A fast breeder reactor

(a) uses molten sodium as moderator.

(b) converts fertile fuel into fissile fuel.

(*c*) requires highly enriched fuel.

(*d*) produces more fuel than it consumes.

135. Which of the following pairs is not correctly matched ?

(*a*) Uranium - 233 : Fertile material.

(*b*) Binding energy : Mass defect.

(*c*) Scattering : Absorption cross-section.

(*d*) Number of nucleons : Mass number.

136. If a nuclear reactor produces more fissile nuclear fuel than it consumes, then it is called a .... reactor.

(*a*) critical (6) breeder

(*c*) fertile (*d*) heterogeneous

Answers

126. (II) (III) (I) (IV) 127. (a) 128. (d) 129. (a) 130. (a) 131. (d)
132. (a) 133. (d) 134. (a) 135. (a) 136. (b)

137. The half life period of a radioactive substance is best determined by counting the number of alpha particles emitted per second in a Geiger Muller counter from its known quantity. If the half life period of a radioactive substance is one month, then

(a) 3/4th of it will disintegrate in two months.

(b) it will completely disintegrate in two months.

(c) it will completely disintegrate in four months.

(d) l/8th of it will remain intact at the end of four months.

138. Match the following pertaining to nuclear power reactors.

(a) Fuel Cladding material	I Carbon dioxide
(b) Control rod material	II Hafnium
(c) Poison	III Cadmium
(d) Coolant	IV Zircalloy

139. Enriched uranium is required as a fuel in a nuclear reactor. Light water is used as moderator & coolant, because light water has

(a) high neutron absorption cross-section.

(b) low moderating efficiency.

(c) high neutron scatter cross-section.

(d) low neutron absorption cross-section.

140. Match List I and List II and select the correct answer using the codes given below the lists:

List I	List II
P. Prepared fuel	1. Uranium-235
Q. Primary fuel	2. Graphite
R. moderator	3.Uranium-233
S. Control rod	4.Cadmium

(a) P-l, Q-3, R-2, S-4

(b) P-3, Q-l, R-4, S-2

(c) P-3, Q-l. R-2, S-4

(d) P-l, Q-3, R-4, S-2

141. Match List I (Material) with List II (Use) and select the correct answer using the codes given below the lists:

List I (Material)

P. Graphite

Q. Thorium-233

R. Molten sodium

S. Plutonium-239

List II (Use)

1. Coolant

2. Moderator

3. Fissionable material

4. Fissile material

(a) P-l, Q-4, R-2, S-3

(b) P-2, Q-4. R-l, S-3

(c) P-2, Q-3, R-l, S-4

(d) P-l, Q-3, R-2, S-4

Answers

137. (a) 138. (IV) (III) (II) (I) 139. (c) 140. (c) 141. (b)

Section A 15

Electrometallurgy and Corrosion

1. Corrosion does not include the destruction of metals
   (a) in all type of atmospheres & liquids at any temperature.
   (b) by dissolution of material in the environment.
   (c) by metallurgical interaction between the environment & material.
   (d) none of these.

2. Which of the following can withstand corrosive attack in all environments ?
   (a) Stainless steel
   (b) Glass
   (c) Wood and concrete
   (d) None of these.

3. Corrosion rate of metals are not affected by the
   (a) surface films and gases adsorbed on the metal surface.
   (b) presence of impurities (e.g., dust, dirt, foreign matter) and other ions in the solution.
   (c) concentration & temperature of corrodant applied stresses.
   (d) none of these.

4. In case of corrosion of a metal by gas, the surface film grows .... with time
   (a) linearly.
   (b) parabolically.
   (c) logarithmically.
   (d) either of (a), (b) & (c) depending upon the type of metal and temperature.

5. Which of the following does not obey the linear law of oxidation in case of corrosion of metal by a gas ?
   (a) Iron at >900°C.
   (b) Titanium between 650 to 950°C.
   (c) Impure calcium at 500°C.
   (d) Copper at 800°C.

6. Which of the following obeys parabolic law of oxidation ?
   (a) Copper at 800°C.
   (b) Iron at > 900°C.
   (c) Iron at < 200°C.
   (d) Aluminium at room temperature.

7. Which of the following follows logarithmic law of oxidation ?
   (a) Iron at < 200°C.
   (b) Zinc at < 400°C.
   (c) Aluminium at room temperature
   (d) All (a), (b) and (c).

Answers

1. (d)	2. (d)	3. (d)	4. (d)	5. (d)	6. (a)
7. (d)					

8. Iron at > 200°C temperature, follows the........ law of oxidation.
   (*a*) linear (*b*) logarithmic
   (*c*) parabolic (*d*) none of these
9. Atmospheric corrosion is
   (*a*) frequent in non ferrous materials.
   (*b*) due to dry atmosphere.
   (*c*) based on oxygen adsorption mechanism.
   (*d*) all (*a*), (*b*) & (*c*).
10. Pitting corrosion is
   (*a*) a uniform corrosion.
   (*b*) a result of electrochemical reaction.
   (*c*) a result of chemical reaction.
   (*d*) due to homogeneities in metals.
11. Intergranular corrosion is
   (*a*) a uniform corrosion which can be revealed by microscopic examination.
   (*b*) not detrimental to the strength of the alloy.
   (*c*) susceptible in case of 18/8 stainless steel, aluminium alloys & copper alloys.
   (*d*) none of these.
12. Stress corrosion
   (*a*) is more likely in cold worked alloys/metals subjected to high tensile stress in corrosive environment.
   (*b*) can not be checked by elimination of tensile stress from the component part.
   (*c*) involves electrochemical reaction,
   (*d*) none of these.
13. Which is an example of stress corrosion ?
   (*a*) Season cracking of brass.
   (*b*) Caustic embrittlement of steel.
   (*c*) Both (*a*) & (*b*).
   (*d*) Neither (*a*) nor (b).
14. Corrosion fatigue is
   (*a*) due to the combined action of corrosion and repeated loading (stresses).
   (*b*) same as stress corrosion.
   (*c*) encountered in metal gas reaction only.
   (*d*) none of these.
15. Corrosion fatigue can be minimised by
   (*a*) surface protection of the metal.
   (*b*) nitriding of steel.
   (*c*) treatment of the corroding medium.
   (*d*) all (*a*), (*b*) & (*c*).
16. Fretting corrosion is
   (*a*) a purely chemical phenomenon.
   (*b*) a purely mechanical phenomenon.
   (*c*) liable to occur, where close fitting machine parts are subjected to vibrational stresses.
   (*d*) a uniform corrosion.
17. Fretting corrosion can be minimised by
   (*a*) avoiding vibrations.
   (*b*) raising the hardness of mating surfaces or nitriding, hardening, chromium plating etc.
   (*c*) introducing compressive stresses in the surface by shot peening or surface rolling.
   (*d*) all (*a*), (*b*) & (*c*).
18. Erosion corrosion is
   (*a*) caused by the combined effects of mechanical abrasion & chemical corrosion.
   (*b*) a uniform corrosion.
   (*c*) due to pure electrochemical reactions.
   (*d*) none of these.
19. Erosion corrosion is encountered in
   (*a*) centrifugal pumps & steam turbines.
   (*b*) clamped/press fits.
   (*c*) vibrating bearings.
   (*d*) alkaline environment only.

Answers

8. (c)	9. (c)	10. (b)	11. (c)	12. (a)	13. (c)
14. (a)	15. (d)	16. (c)	17. (d)	18. (a)	19. (a)

20. Crevice corrosion
   (a) occurs at the junction of two metals exposed to a corrosive environment.
   (b) is a uniform corrosion.
   (c) occurs in the area having oxygen concentration.
   (d) none of these.

21. Microbiological corrosion is caused
   (a) in only aerobic (oxygen rich) environment.
   (b) in only anerobic environment.
   (c) by the metabolic activity of various microorganisms.
   (d) none of these.

22. Metal corrosion can not be controlled by the use of
   (a) high purity metals.
   (b) alloy additions,
   (c) special heat treatments.
   (d) none of these.

23. Which is the wrong statement pertaining to control of corrosion by proper design of metallic structure ?
   (a) Structure should be stress free and possess good surface finish.
   (b) If two dissimilar metals have to be used, they should be as far as apart to each other in the galvanic series.
   (c) Welded joints rather than brazed, soldered or bolted joints may be used in order to avoid crevices.
   (d) The anodic material should have large area, whereas the cathodic material should have much smaller area.

24. With increase in impurities in metals, their corrosion resistances
   (a) decrease.
   (b) increase
   (c) remain same
   (d) may increase or decrease ; depends on the type of metal

25. Corrosion resistance of metals can not be improved by
   (a) increasing its purity.
   (b) stress relieving treatment after cold working.
   (c) its heat treatment that leads to homogenization of solid solutions.
   (d) none of these.

26. Which is the most effective method for complete corrosion prevention ?
   (a) Use of inhibitor
   (b) Cathodic protection
   (c) Use of surface coating
   (d) Heat treatment.

27. Sacrificial anode used in cathodic protection .... in the electro chemical series than the metal to be protected.
   (a) should be a metal that is higher
   (b) should be lower
   (c) can be either higher or lower
   (d) none of these

28. Which is an example of cathodic protection?
   (a) Galvanised iron.
   (b) Painted steel.
   (c) Case hardened steel.
   (d) Lead lined reactor.

29. Inhibitors are chemical compound added to an electrolyte, which protectively coat the anode/cathode and stop corrosion. An example of anodic inhibitor as is
   (a) nickel *(b)* zinc salts
   (c) phosphate *(d)* magnesium

Answers

20. (a)	21. (c)	22. (d)	23. (b)	24. (a)	25. (d)
26. (b)	27. (a)	28. (a)	29. (c)		

30. An example of cathodic inhibitor is
    (a) calcium bicarbonates in hard water
    (b) chromates
    (c) phosphates
    (d) none of these

31. Corrosion rate can not be lowered by reducing the .... of corroding medium.
    (a) temperature (b) velocity
    (c) concentration (d) none of these.

32. Passivation as applied to corrosion means
    (a) formation of corrosion retarding films on metal surface.
    (b) cathodic protection of metals.
    (c) reduction in corrosion rate by improving the purity of metals.
    (d) corrosion due to abrasive action coupled with chemical action.

33. Which is most prone to corrosive attack ?
    (a) Metal highest in the emf series.
    (b) Metal lowest in the emf series.
    (c) Metals with high level of impurities.
    (d) Metals with slag inclusions.

34. Which of the following acts as cathodic areas in case of concentration/galvanic cells ?
    (a) Stagnant solution.
    (b) Dilute electrolyte.
    (c) Deaerated solution.
    (d) Metals high in emf series.

35. Which of the following acts as cathodic areas in case of concentration/galvanic cells ?
    (a) Concentrated electrolyte.
    (b) Aerated solution.
    (c) Dilute electrolyte.
    (d) Metals low in emf series.

36. Stress corrosion rate is not influenced by the
    (a) residual stress level.
    (b) amount of plastic strain.
    (c) time & temperature of exposure.
    (d) none of these.

37. Dezincification is a form of corrosion, which specifically applies to
    (a) brasses
    (b) austenitic stainless steel
    (c) plain carbon steel
    (d) babbit metal

38. Erosion corrosion is not caused
    (a) by turbulent flow of fluids.
    (b) by stagnant solution.
    (c) by cavitation.
    (d) in impellers of pumps & turbines.

39. Addition of chromium (maximum upto 30%) in steel improves its resistance of oxidation by way of
    (a) making its surface non-wetting.
    (b) formation of a passive film of chromium oxide.
    (c) case hardening.
    (d) none of these.

40. The most common inorganic corrosion inhibitor used in automobile cooling system is
    (a) borax (b) oleic acid
    (c) calgon (d) morpholine

41. Dry corrosion of metal results
    (a) at gas/metal interface.
    (b) due to high temperature oxidation.
    (c) both (a) & (b).
    (d) at metal/solution interface.

Answers

30. (a)	31. (d)	32. (a)	33. (a)	34. (a)	35. (c)
36. (d)	37. (a)	38. (b)	39. (b)	40. (a)	41. (c)

42. Cavitation corrosion is categorised as the .... corrosion.
 (*a*) galvanic (*b*) erosion
 (*c*) stress (*d*) uniform

43. The phenomenon in which a metal fails due to the presence of corrosion and cyclic stress is called
 (*a*) stress corrosion cracking.
 (*b*) corrosion fatigue.
 (*c*) fretting corrosion.
 (*d*) crevice corrosion.

44. The phenomenon in which metal deterioration occurs due to the combined action of corrosion and tensile stress is called
 (*a*) corrosion fatigue,
 (*b*) crevice corrosion.
 (*c*) stress corrosion cracking.
 (*d*) none of these.

45. Oxidation and erosion of the valves of internal combustion engine by the incandescent products of combustion is an example of the .... corrosion.
 (*a*) electrochemical (*b*) chemical
 (*c*) stress (*d*) galvanic

46. Electrochemical corrosion occurs
 (*a*) where the liquid electrolytes containing free ions are present.
 (*b*) only in alloys.
 (*c*) only in pure metals.
 (*d*) due to direct chemical action of metal with oxygen.

47. Corrosion can be prevented by
 (*a*) alloying of a metal.
 (*b*) drying the moist polluted air (with SO_2/SO_3).
 (*c*) covering the pipeline with soil.
 (*d*) coating the metal with varnish.

48. Galvanising (*i.e.* zinc coating) of steel sheet is done for improving its
 (*a*) corrosion resistance.
 (*b*) lustre.
 (*c*) high temperature strength.
 (*d*) tensile stress.

49. Electroplating of zinc on a metal to improve its corrosion resistance is called cold galvanising. In electroplating, the
 (*a*) object to be plated is made the anode of the cell.
 (*b*) cathode may be the metal to be deposited.
 (*c*) electrolyte contains the salt of metal to be deposited.
 (*d*) none of these.

50. Tin plated iron and galvanised iron are generally produced by
 (*a*) vapourising (*b*) hot dipping
 (*c*) electroplating (*d*) spraying .

51. Zinc and aluminium is plated on metal by striking their vapour on the metallic surface with which they alloy. The deposition of zinc by this process is called
 (*a*) sherardising (*b*) calorising
 (*c*) electroplating (*d*) hot dipping .

52. Deposition of aluminium vapour on metallic surface is called
 (*a*) sherardising (*b*) calorising
 (*c*) electroplating (*d*) hot dipping .

53. Coating of iron with zinc protects it against corrosion, because
 (*a*) zinc is less noble than iron (in electro chemical series of metal).
 (*b*) iron is less noble than zinc.
 (*c*) zinc has a higher melting point than iron.
 (*d*) iron has a higher melting point than iron.

Answers

42. (b)	43. (b)	44. (c)	45. (b)	46. (a)	47. (a)
48. (a)	49. (c)	50. (b)	51. (a)	52. (b)	53. (a)

54. Tin coating on iron protects it from............ damage.
   (*a*) mechanical.
   (*b*) electrochemical corrosion.
   (*c*) both (*a*) & (*b*).
   (*d*) neither (*a*) nor (*b*).

55. Pick out the wrong statement.
   (*a*) Aluminium coating on steel is done by hot dipping.
   (*b*) Aluminium offers galvanic protection (like zinc) by dissolving as anode thus protecting base metal steel, which forms cathode.
   (*c*) If a metal contains punched rivet holes, it corrodes rapidly near holes, because of the potential difference between the cold worked metal and the adjoining metal which has not been cold worked.
   (*d*) Rivets should have the different composition as the metal, otherwise the contact will corrode rapidly.

56. Galvanic corrosion is.
   (*a*) also called bi-metallic corrosion.
   (*b*) caused when a pure metal is dipped in corrosive solution.
   (*c*) caused when there is no potential difference between two dissimilar metals in contact with a corrosive solution.
   (*d*) none of these.

57. Caustic embrittlement of boiler tubes is an example of
   (*a*) stress corrosion.
   (*b*) dezincification.
   (*c*) crevice corrosion.
   (*d*) erosion-corrosion.

58. Dezincification is a form of corrosion applicable to
   (*a*) copper-zinc alloys (*e.g.*, brass).
   (*b*) contact between metallic & non-metallic surfaces (*e.g.*, gasket).
   (*c*) alkali attack in stressed metals.
   (*d*) two metals corrosion.

59. Corrosion of some metals can be prevented by its
   (*a*) purification (*b*) alloying
   (*c*) painting (*d*) all (*a*), (*b*) & (*c*).

60. The rate of corrosion is expressed as the loss in weight of metal per unit
   (*a*) surface area
   (*b*) time
   (*c*) either (*a*) & (*b*)
   (*d*) surface area per unit time

61. Pick out the wrong statement.
   (*a*) Pure copper is not susceptible to stress corrosion.
   (*b*) Galvanic corrosion is applicable to corrosion of single metal.
   (*c*) In galvanic corrosion, the metal near the top in electrochemical series of metals get corroded preferentially compared to those at the bottom.
   (*d*) none of these.

62. Ships operating in sea water are guarded against corrosion by
   (*a*) painting the outer surface
   (*b*) cathodic protection
   (*c*) ceramic coating
   (*d*) anodic protection

63. The noble metals
   (*a*) are exemplified by Na, K & Li.
   (*b*) do not resist corrosion very strongly.
   (*c*) are the lowermost in the electro-chemical series of metals.
   (*d*) none of these.

Answers

54. (c)	55. (d)	56. (a)	57. (a)	58. (a)	59. (d)
60. (d)	61. (b)	62. (b)	63. (c)		

64. Though tin occurs lower than iron in the electrochemical series, yet it is coated on steel for corrosion protection, because of its
    (*a*) lower cost.
    (*b*) abundant availability.
    (*c*) resistance to vegetable acids as the tin plate is commercially used in 'canning' industry.
    (*d*) none of these.

65. The most important requirement for aluminium industry is the availability of cheap
    (*a*) ore
    (*b*) electrical power
    (*c*) labour
    (*d*) and abundant water availability.

66. Hoope's cell is predominantly used for the electrolytic refinning of
    (*a*) zinc (*b*) copper
    (*c*) aluminium (*d*) tin .

67. Electrolytic reduction cell used for conversion of calcined Al_2O_3 to Al is a carbon lined furnace operating at 800-900°C. The purpose of electric current supplied to the furnace is to
    (*a*) achieve very high purity of aluminium (99.9%).
    (*b*) keep the electrolyte in liquid condition by the generation of heat.
    (*c*) electrolytically dissociate alumina.
    (*d*) both (*b*) & (*c*).

68. Which of the following metals is not subjected to electrolytic refinning/purification?
    (*a*) Copper (b) Zinc
    (*c*) Tin (*d*) Aluminium

69. Electrometallurgy is not involved in the extraction of .... from its ore.
    (*a*) aluminium (*b*) tin
    (*c*) lead (*d*) both (b) & (*c*)

70. Coating thickness in case of galvanising of steel sheet generally corresponds to deposition of .... gms, of zinc per m^2 of steel strip.
    (*a*) 1 – 5 (b) 25 – 50
    (*c*) 120 – 150 (*d*) 350 – 500

71. Which of the following metals is not used for coating of the base metal to guard it against corrosion?
    (*a*) Lead (*b*) Cadmium
    (*c*) Magnesium (*d*) Aluminium

72. Pick out the wrong statement.
    (*a*) Soft water is more corrosive than that carrying calcium salts.
    (*b*) The corrosion of brass condenser tubes handling brine solution is also termed as 'dezincification'.
    (*c*) Non-annealed brass is highly susceptible to season cracking (i.e., inter-crystalline corrosion due to internal stresses).
    (*d*) Pure copper is highly susceptible to stress corrosion.

73. Which of the following does not come under the category of electrochemical corrosion?
    (*a*) Corrosion in organic liquids.
    (*b*) Atmospheric corrosion in moist gas or air.
    (*c*) Soil corrosion as in the case of underground pipes.
    (*d*) Corrosion in sea water & molten metals.

74. Galvanic corrosion
    (*a*) occurs in a bimetal couple.
    (*b*) occurs in highly stressed metallic parts only.
    (*c*) is exemplified by caustic embrittlement of boiler tubes.
    (*d*) is also called dezincification.

Answers

64. (c)	65. (b)	66. (c)	67. (d)	68. (c)	69. (d)
70. (c)	71. (c)	72. (d)	73. (a)	74. (a)	

75. Which of the following metals can not be produced by aqueous electrolysis ?
(*a*) Copper (*b*) Zinc
(*c*) Aluminium (*d*) Tin

76. Production of reactive metals like........ requires the electrolysis of its fused salt at high temperature.
(*a*) manganese (*b*) sodium
(*c*) potassium (*d*) both '*b*' & '*c*'

77. In electrometallurgy, the electrolyte (either an aqueous solution or a mixture of fused salt of the metal) must not
(*a*) have low ionic conductivity.
(*b*) have thermal stability at the operating temperature.
(*c*) be chemically inert towards the electrodes, the container material & the electrolysis products.
(*d*) all (*a*), (*b*) & (*c*).

78. According to Faraday's first law of electrolysis, the metal output as a result of electrolysis is independent of the
(*a*) total current passed through the cell.
(*b*) cell voltage.
(*c*) both (*a*) & (*b*).
(*d*) neither (*a*) nor (*b*).

79. Cathodic protection against corrosion is exemplified by
(*a*) a painted surface.
(*b*) alloying of a metal with a metal higher in emf series.
(*c*) galvanised iron.
(*d*) none of these.

80. Intergranular corrosion of stainless steel can be minimised by
(*a*) use of extremely high carbon grade of stainless steel.
(*b*) avoiding any type of heat treatment after its welding.
(*c*) use of titanium, coiumbium and/or tantalum additions to stabilise carbides.
(*d*) none of these.

81. In connection with the corrosion of metals, passivation is the process that
(*a*) intensifies deterioration.
(*b*) changes the composition of the metal,
(*c*) inhibits further deterioration.
(*d*) none of these.

82. The quantity of electricity required to deposit one gm equivalent of any substance is one
(*a*) ampere
(*b*) coulumb
(*c*) micro faraday
(*d*) Faraday

83. Electrolytes conduct electric current by the movement of
(*a*) ions
(*b*) molecules
(*c*) electrons through them from cathode to anode
(*d*) electric charges through them

84. Which metal is protected against corrosion by the layer of its own oxide ?
(*a*) Iron (*b*) Silver
(*c*) Calcium (*d*) Aluminium

85. Metals at the top of the electromotive series is the
(*a*) least active (*b*) most stable
(*c*) most active (*d*) most noble

Answers

75. (c)	76. (d)	77. (a)	78. (b)	79. (c)	80. (c)
81. (c)	82. (d)	83. (a)	84. (d)	85. (c)	

86. Electrochemical corrosion can occur, only if,...... is present in contact with the metal.
   (*a*) air (*b*) oxygen
   (*c*) liquid medium (*d*) gaseous medium

87. Cathodic protection is widely used in controlling corrosion in
   (*a*) high pressure vessels.
   (*b*) marine structures and underground pipelines.
   (*c*) overhead fuel gas pipelines.
   (*d*) overhead water pipelines.

88. Rate of corrosion in metal is accelerated by the
   (*a*) flow of alternating current through it.
   (*b*) applied or residual stresses in it.
   (*c*) decrease in its temperature.
   (*d*) reduction in current flow through it.

89. Inorganic corrosion inhibitors suffer from the drawback that they are.
   (*a*) ineffective in alkaline environment.
   (*b*) prohibitively costly.
   (*c*) ineffective in acidic environment.
   (*d*) active after a very long induction time.

90. Which of the following is an invalid cathode-anode combination ?
   (*a*) Titanium-nickel (*b*) Zinc-silver
   (*c*) Iron-zinc (*d*) Tin-iron

91. Oxygen absorption can cause corrosion, if the electrolyte is
   (*a*) acidic (*b*) alkaline
   (*c*) neutral (*d*) either '*b*' or '*c*'

92. The reason for practically zinc becoming anodic to aluminium, inspite of the fact that pure aluminium is anodic to zinc, is that
   (*a*) aluminium forms a surface adhering protective oxide film.
   (*b*) zinc reacts with all the electrolytes.
   (*c*) zinc is never found in 100% pure form.
   (*d*) zinc gets coated with surface adhering protective oxide film.

93. When an iron surface is in contact with water and oxygen of different concentrations, the area with
   (*a*) no oxygen concentration acts as the cathode.
   (*b*) no oxygen concentration acts as the anode.
   (*c*) low oxygen concentration acts as the cathode.
   (*d*) low oxygen concentration acts as the anode.

94. Which of the following conditions need not be satisfied for the electrochemical corrosion to occur ?
   (*a*) The metallic surface must be surrounded by a reactive gas medium.
   (*b*) Both the solution & the metal must be conductor of electricity.
   (*c*) To make the current flow, an electrical potential must exist.
   (*d*) Two spatially separated metallic surface areas must act as electrodes.

95. The region of a car body very close to or under the rubber lining of wind screen
   (*a*) has low oxygen supply.
   (*b*) is likely to anode.
   (*c*) both 'a' & 'b'.
   (*d*) neither 'a' nor 'b'.

Answers

86. (c)	87. (b)	88. (b)	89. (c)	90. (b)	91. (d)
92. (a)	93. (a & d)	94. (a)	95. (c)		

96. In case of the corrosion by mechanism of oxygen absorption, the anodic area is......... the cathodic area.
    (*a*) equal to (*b*) longer than
    (*c*) smaller than (*d*) not related to

97. In case of the corrosion by mechanism of hydrogen evolution, the anodic area is .... the cathodic area.
    (*a*) equal to (*b*) larger than
    (*c*) smaller than (*d*) not related to

98. Stress corrosion occurs due to the presence of .... stress.
    (*a*) shear
    (*b*) compressive
    (*c*) tensile
    (*d*) corrosive environment coupled with tensile

Answers

96. (c) 97. (b) 98. (d)

Section A

16

Metallurgical Testing & Inspection

1. Which of the following metals appear silvery white on visual inspection ?
   (*a*) Lead (*b*) Magnesium
   (*c*) Zinc (*d*) None of these
2. Which of the following metals does not produce ringing sound, when it is hammered?
   (*a*) Aluminium (b) Lead
   (*c*) Magnesium (*d*) Copper
3. A material subjected to .... exemplifies dynamic load state.
   (*a*) compression (*b*) tension
   (*c*) bending (*d*) impact
4. If a material suffers permanent deformation due to application of stress/load, it is called
   (*a*) elastic deformation
   (*b*) plastic deformation
   (*c*) rupture stress
   (*d*) ultimate stress
5. With the increasing load in any metal specimen, a point comes when the metal fractures. This is called the
   (*a*) yield point (*b*) ultimate stress
   (*c*) rupture stress (*d*) both (*b*) & (*c*)
6. Metallography of a metal or alloy involves its microscopic examination. It does not give any idea about the
   (a) grain size
   (b) segregation
   (*c*) non-metallic inclusion
   (*d*) none of these
7. Microscopic examination of fractures in a metal revealed the presence of slag particles & blowholes. It may be due to the .... of the metal.
   (*a*) faulty manufacture
   (*b*) incorrect pouring temperature
   (*c*) faulty heat treatment
   (*d*) none of these
8. Hardness of a safety razor blade is best measured by the
   (*a*) Brinell hardness test.
   (*b*) Vicker's diamond pyramid test.
   (*c*) Shore or rebound hardness test.
   (*d*) Rockwell superficial hardness test.
9. Plastic deformation in metals is due to the presence of dislocations in crystals. This is a .... defect.
   (*a*) point (b) surface
   (*c*) line (*d*) none of these

Answers

1. (b)	2. (b)	3. (d)	4. (b)	5. (d)	6. (d)
7. (d)	8. (d)	9. (c)			

10. The defects in the crystalline structure of material affects its properties. Which of the following is not a structure sensitive property of a material ?

    (a) Ductility (b) Tensile strength
    (c) Density *(d)* Yield strength

11. Most common reason for fatigue failure of materials is due to

    (a) electrochemical corrosion.
    (b) application of repeated stress.
    (c) its less elasticity.
    (d) its high yield point.

12. .... can be detected by visual inspection.

    (a) Internal shrinkage in casting
    (b) Blow holes on casting surface
    (c) Sub-surface defects
    (d) None of these

13. What type of stress is applied to the specimen in the creep test ?

    (a) Biaxial tensile stress.
    (b) Uniaxial compressive stress.
    (c) Uniaxial tensile stress.
    (d) Alternating stress.

14. The depth of indentation in Brinell hardness test should not be greater than........................ percent of the thickness of test piece.

    (a) 1 *(b)* 10
    (c) 25 *(d)* 40

15. In Brinell hardness test, load applied is 3000 kg for testing both the hard as well soft materials, but the diameter of the hardened indenting steel balls are different. The diameter of steel balls used for hardness testing of soft (e.g., Cu & Al) & hard (e.g., steel) materials are respectively................. mm

    (a) 5 & 10 *(b)* 10 & 15
    (c) 15 & 20 *(d)* 15 & 25

16. Ability of a material to withstand vibratory stress by converting mechanical energy into heat is called its

    (a) endurance limit *(b)* damping capacity
    (c) toughness *(d)* brittleness

17. The resistance of a material to fatigue failure is called the

    (a) damping capacity
    (b) endurance limit
    (c) toughness
    (d) malleability

18. Which of the following is not a non-destructive test ?

    (a) Penetrating liquid test.
    (b) Magnaflux method.
    (c) Ultrasonic testing .
    (d) Creep test.

19. Ultrasonic testing of materials uses ultrasonic waves for flaw detection in materials. It is

    (a) similar to sound wave, but is inaudible to human ear.
    (b) of very very low frequency.
    (c) reflected back with distortion, when transmitted into sound defect free materials.
    (d) none of these.

20. Y-radiographs compared to *X-ray* radiographs are more usable for materials inspection, due to the fact that the

    (a) X-rays can not penetrate all materials which are opaque to ordinary light waves.
    (b) production of X-rays is. much castlier than that of Y-rays.
    (c) X-ray units above 300 KV are not portable.
    (d) none of these.

Answers

10. (c)	11. (b)	12. (b)	13. (c)	14. (b)	15. (d)
16. (b)	17. (b)	18. (d)	19. (a)	20. (c)	

21. Centre of indentation in Brinell hardness test should not be less than .... times its diameter from the edge of the test piece.
(*a*) 2.5 (*b*) 5
(*c*) 7.5 (*d*) 10

22. Microscopic examination of fractures in a metal revealed the presence of coarse grains. It may be due to the .... of the metal.
(*a*) faulty heat treatment
(*b*) incorrect pouring temperature
(*c*) both (*a*) & (*b*)
(*d*) faulty manufacture

23. The edge dislocation and the screw dislocation found in metallic crystals is a.............. defect.
(*a*) point (*b*) line
(*c*) surface (*d*) none of these

24. Stacking fault in metallic crystals occurs, if a plane of atoms is stacked out of sequence in a closed packed structure. It is a .... defect.
(*a*) line (*b*) point
(*c*) surface (*d*) none of these

25. In magnetic particle testing method for detection of flaws/sub-cracks.
(*a*) ferreso-ferric oxide particles are used.
(*b*) non-magnetic material can only be tested
(*c*) magnetic particles flock at the edge of the defect free specimen.
(*d*) none of these.

26. For testing the hardness of very hard materials (having B.H.N > 500), the material of construction of the indentor to be used should be .... ball.
(*a*) case hardened steel
(*b*) diamond or tungsten carbide
(*c*) monel
(*d*) stainless steel

27. Which of the following is not a destructive test?
(*a*) Radiography
(*b*) Impact test
(*c*) Tensile strength test
(*d*) Brinell hardness test

28. Magnaflux non-destructive testing method is not meant for detecting .... of iron & steel products.
(*a*) interior/sub-surface defects
(*b*) cracks caused by quenching & by fatigue
(*c*) blow holes in casting & defects in welding
(*d*) seams in rolled & forged products

29. In Brinell hardness test, the thickness of the specimen should not be less than......... times the depth of the impression.
(*a*) 5 (*b*) 10
(*c*) 20 (*d*) 25

30. Fluorescent dye method of testing is used for detecting the .... of an object.
(*a*) surface crack
(*b*) interior defects
(*c*) structural characteristics
(*d*) flaws/sub-cracks in a magnetic material only

31. Sulphur printing of metallic steel specimen is done to
(*a*) find the percentage of sulphur (in the form of sulphides) in steel.
(*b*) detect segregation of sulphur in steel.
(*c*) remove sulphide form of sulphur from steel.
(*d*) none of these.

Answers

21. (a)	22. (c)	23. (b)	24. (c)	25. (a)	26. (b)
27. (a)	28. (a)	29. (b)	30. (a)	31. (b)	

32. Material of construction of the indentor ball in the Brinell hardness test for hard metals is
    (a) stainless steel (b) diamond
    (c) tungsten carbide (d) high carbon steel

33. Etching of the test specimen is proceeded by its
    (a) tempering
    (b) grinding & polishing
    (a) quenching
    (d) recrystallisation

34. The ability of a material to withstand occasional stresses above the yield stress without fracturing it, is called its
    (a) hardness (b) toughness
    (c) damping capacity (d) resilience.

35. Pick out the wrong statement.
    (a) In non-destructive testing of materials, the test specimens are neither harmed nor impaired for actual service.
    (b) Creep test for material is a nondestructive test.
    (c) Universal hydraulic testing machine is used for both tension as well as compression test.
    (d) The slope of the initial linear portion of the stress strain curve is equal to the Young's modulus of elasticity.

36. A fibrous fracture is a characteristic of the
    (a) metals with non-metallic inclusion.
    (b) wrought iron.
    (c) stainless steel.
    (d) none of these.

37. Which of the following is a pre-requisite for the test specimen before it is subjected to examination under metallurgical microscope for studying its metallographic structure ?
    (a) Plastic deformation.
    (b) Elastic deformation.
    (c) Etching.
    (d) Heating & annealing.

38. Pick out the wrong statement.
    (a) Micrography of test specimen can not detect the presence of non-metallic inclusion in it.
    (b) The smallest distance that can be separately resolved is called the resolution of a microscope.
    (c) X-ray testing is suitable for the detection of both surface as well as interior defects.
    (d) none of these.

39. Magnetic particle method of flaw testing/detection in a specimen can not be applied for
    (a) iron (b) austenitic steel
    (c) alloys (d) none of these

40. The position of flaw in ultrasonic method of testing is located by measuring
    (a) T.V (b) T/2V
    (c) TV/2 (d) 2/TV
    where, T = the time taken for the return journey V, = velocity of sound in that medium

41. .... is the term used for the study of microstructure of metal & alloys by metallurgical microscope.
    (a) Micrography (b) Macrography
    (c) Fractography (d) none of these

42. Which of the following defects can be detected accurately by X-ray testing ?
    (a) Foreign inclusion.
    (b) Surface fine cracks as well as sub-surface defects.
    (c) Segregation defects.
    (d) All (a), (b) & (c).

Answers

32. (c)	33. (b)	34. (b)	35. (b)	36. (b)	37. (c)
38. (a)	39. (b)	40. (c)	41. (a)	42. (b)	

43. Tensile strength of steels in tons/inch^2 is given approximately by multiplying the B.H.N by
(*a*) 0.01 (*b*) 0.0022
(*c*) 0.22 (*d*) 0.85

44. The process in which fluorescent screen is used to get the shadow of the object during X-ray testing is called the
(*a*) fluoroscopy (*b*) radiography
(*c*) microscopy (*d*) none of these

45. Source of production of Y-rays in Y-radiology is
(*a*) impact of fast electrons on a metallic object.
(*b*) radio isotopes.
(*c*) passing high current through tungsten filament.
(*d*) none of these.

46. Source of production of X-ray in X-ray testing is
(*a*) impact of fast electrons on a metallic object.
(*b*) radio isotopes.
(*c*) passing high current through tungsten filament.
(*d*) none of these.

47. In shore or rebound hardness test used for testing steel rolls etc., the B.H.N, of steels is roughly given by multiplying the shore value by
(*a*) 2 (*b*) 6
(*c*) 12 (*d*) 20

48. Load applied in Brinell Hardness Test for non-ferrous and soft metal/alloys is about .... kg.
(*a*) 500 (*b*) 1000
(*c*) 2000 (*d*) 3000

49. The indentor used in Rockwell hardness test (B scale) of soft steels & non-ferrous metal/alloys is
(*a*) diamond cone.
(*b*) 1/16" hardened steel ball.
(*c*) 10 mm. dia tungsten carbide ball.
(*d*) diamond square pyramid.

50. The load applied in Vickers diamond pyramid hardness testing method used for hardened steel and nitrided & case hardened parts is .... kg.
(*a*) 3000 (*b*) 100
(*c*) 30 (*d*) 500

51. Pick out the wrong statement.
(*a*) Ultrasonic means a sound of too high a pitch to be detected by human ear.
(*b*) Ultrasonic and supersonic have the same frequencies.
(*c*) Ultrasonic testing can not be employed for thickness measurement for the specimen.
(*d*) both 'b' & 'c'.

52. Penetrating liquid test
(*a*) is not suitable for detection of surface cracks.
(*b*) is not applied to non-ferrous metals.
(*c*) is applied to non-ferrous materials which can not be tested by magnaflux method.
(*d*) does make use of a flouroscent liquid.

53. Distortion of magnetic field by cracks and flaws in the test specimen is the principle involved in
(*a*) radiography
(*b*) ultrasonic testing
(*c*) magnaflux method
(*d*) supersonic method

Answers

43. (c)	44. (a)	45. (b)	46. (a)	47. (b)	48. (d)
49. (b)	50. (c)	51. (d)	52. (c)	53. (c)	

54. Pick out the wrong statement.
   (a) With decrease in grain size, the fatigue strength of non-ferrous metals increase.
   (b) Microscope is used for measuring the extension in length of the test specimen in creep test.
   (c) Young's modulus of elasticity is a dimensionless quantity.
   (d) In magnetic particle method of testing, ferroso-ferric oxide particles are used.

55. The curve plotted in the stress rupture test is strain Vs.
   (a) time (b) temperature
   (c) stress (d) none of these

56. What type of test is the 'sceleroscope test' ?
   (a) Fatigue test (b) Hardness test
   (c) Tension test (d) None of these

57. In Brinell hardness test, the measurement of the..........................is most prone to error.
   (a) diameter of indentor ball
   (b) load applied
   (c) surface area of indentor ball
   (d) diameter of indentation

58. In penetrating liquid test for flaw detection, the flouroscent penetrant is applied to the test specimen at
   (a) 50°C (b) 100°C
   (c) 200°C (d) room temperature

59. Material of construction of the pyramid in Vicker hardness test is
   (a) stainless steel (b) diamond
   (c) silicon carbide (d) wrought iron

60. .... hardness test employs a brale indentor.
   (a) Brinell (b) Rockwell
   (c) Vicker pyramid (d) none of these

61. ....furnace is employed in the creep test.
   (a) Reheating (b) Muffle
   (c) Vertical tube (d) none of these

62. Pick out the correct statement.
   (a) The fatigue strength of non-ferrous materials is not affected by changes in its temperature,
   (b) Vicker hardness number is given by the applied load divided by the surface area of the indentor ball.
   (c) The resistance to scratch is measured in Brinell hardness test.
   (d) Softer materials are normally tested on B-scale in Rockwell hardness test.

63. Stress rupture test of a material gives idea about its
   (a) strain rate.
   (b) maximum strain rate at a given stress level.
   (c) maximum strain rate at a given temperature under constant stress.
   (d) none of these.

64. Which of the following is the most vulnerable point for notch brittleness to occur in a metallic bar specimen ?
   (a) Points of stress concentration,
   (b) Corner points.
   (c) Mid-point of the specimen.
   (d) A distance of three fourth of the length of the specimen.

65. Which of the following is expected to have the highest fatigue strength among the following ?
   (a) Tin (b) Mild steel
   (c) Copper (d) Aluminium

Answers

54. (c)	55. (d)	56. (b)	57. (d)	58. (d)	59. (b)
60. (b)	61. (c)	62. (d)	63. (d)	64. (a)	65. (b)

66. Which of the following is the most widely used hardness testing method on account of being relatively free from human error and speed of determination ?
(a) Vicker pyramid method.
(b) Brinell hardness test,
(c) Rockwell hardness method.
(d) Meyer hardness test.

67. Pick out the wrong statement.
(a) Surface and sub-surface defects of both the magnetic as well as non-magnetic materials can be detected by magnetic particle method.
(b) Ultrasonic frequency is of the order of mega cycle per second.
(c) The fatigue strength of a specimen having rough surface is less than that having a smooth surface.
(d) The tensile strength of annealed steel is less than that of the normalised steel.

68. The hardness of steel strip/sheet is recommended to be tested by the .... hardness test.
(a) Rockwell *(b)* Vicker pyramid
(c) Brinell *(d)* both *(b)* & *(c)*

69. Load applied in Rockwell hardness test (scale-G) is .... kg.
(a) 50 *(b)* 100
(c) 150 *(d)* 300

70. "Geometrically similar test bars deform similarly". This statement follows from the
(a) Barba's law
(b) definition of endurance limit
(c) Bragg's law
(d) Kirkendal effect

71. 'Parallel length' in the tension test (which is the marked length over which elongation measurements are made) is .... the gauge length.
(a) equal to *(b)* less than
(c) more than *(d)* also called

72. Decrease in stress at constant deformation under creep conditions is termed as the
(a) stress relief *(b)* stress relaxation
(c) proof stress *(d)* residual stress

73. Pick out the wrong statement.
(a) Creep rate is more in metals having high stacking fault energy.
(b) Magnetic particle method can detect the sub-surface crack present in a brass specimen.
(c) For detection of smaller flaws by ultrasonic testing, use of higher frequency is needed.
(d) X-ray diffraction technique is used to find the spacing of the atomic planes of a crystal.

74. Loading time in BHN test is................ seconds.
(a) 5 (6) 15
(c) 200 *(d)* 50

75. In .... hardness test, the height of rebound of a small diamond pointed hammer is taken as the index of hardness.
(a) Shore *(b)* Brinell's
(c) Vicker's *(d)* none of these

76. .... of a material is designated by its Izod value.
(a) Impact resistance (b) Tensile strength
(c) Creep strength *(d)* Hardness

Answers

66. (c)	67. (a)	68. (d)	69. (c)	70. (a)	71. (b)
72. (b)	73. (c)	74. (b)	75 . (a)	76. (a)	

77. For tensile test of a material, the metal specimen is
   (a) either flat or round
   (b) of standard dimension
   (c) both 'a' & '*b*'
   (d) neither 'a' nor 'b'

78. .... is not a non-destructive test for a material.
   (a) Radiography test
   (b) Magnaflux method
   (c) Ultrasonic test
   (d) Wohler fatigue test

79. Compression test is mostly used for the .... materials.
   (a) hard *(b)* ductile
   (c) brittle *(d)* none of these

80. The fatigue limit of metals .... with increase in Izod value.
   (a) decreases.
   (b) increases.
   (c) remains unaltered.
   (d) may increase or decrease ; there is no direct relationship between the two.

81. Rockwell hardness test is useful for
   (a) only hard metals.
   (b) only soft metals.
   (c) rapid routine tests on finished products.
   (d) both 'b' & 'c'

82. Magnaflux method is
   (a) not a non-destructive test.
   (b) based on distortion of magnetic field by cracks & flaws.
   (c) suitable for detecting cracks caused by quenching, fatigue, blowholes in castings, defects in welding and seams in forged & rolled products.
   (d) both 'b' & 'c'.

83. Shore hardness value multiplied by .... gives roughly the Brinell hardness number (BHN) in case of steels.
   (a) 3 (b) 6
   (c) 0.6 *(d)* 20

84. On increasing the load on tensile test specimen beyond the elastic limit,
   (a) work hardening sets in.
   (b) plastic deformation occurs.
   (c) the shape of the subsequent portion of stress-strain diagram depends on the rate of work hardening.
   (d) all 'a', 'b' & 'c'.

85. BHN of annealed aluminium sheet maybe about
   (a) 2 *(b)* 20
   (c) 100 *(d)* 250

86. The load applied in BHN test is 3000 kg. The diameter of ball used in BHN test for soft metals like aluminium, copper etc. is .... mm.
   (a) 5 (b) 10
   (c) 20 *(d)* 30

87. Penetrating liquid test is used for the
   (a) ferrous metals.
   (b) detection of surface cracks.
   (c) non-ferrous metals.
   (d) both 'b' & 'c'

88. Tensile test is mostly used for the.................. materials.
   (a) hard *(b)* ductile
   (c) brittle *(d)* none of these

Answers

77. (c)	78. (d)	79. (c)	80. (d)	81. (c)	82. (d)
83. (b)	84. (d)	85. (b)	86 . (a)	87. (d)	88. (b)

89. The thickness of the test specimen in BHN test should not be less than .... times the depth of impression.
    (*a*) three (*b*) six
    (*c*) ten (*d*) fifteen

90. Beyond elastic limit, until the point of maximum stress, the tensile test metal specimen uniformly
    (*a*) decreases in strength.
    (*b*) increases in strength.
    (*c*) decreases in cross-sectional area.
    (*d*) both '*b*' & '*c*'

91. Piling up of metal round the edge of impression in BHN test indicates
    (*a*) strong capacity for strain hardening.
    (*b*) high ductility of metal.
    (*c*) low capacity for hardening by deformation.
    (*d*) its inability to be cold worked.

92. The centre of indentation in BHN test should not be .... its diameter from the edge of the test piece.
    (*a*) > 2.5 times (*b*) just less than
    (*c*) < 2.5 times (*d*) > 3.5 times

93. Faulty heat treatment of metals can be best detected by the .... test.
    (*a*) Vicker's (*c*) Izod impact
    (*b*) Rockwell's (*d*) Brinell's

94. Erratic results are obtained by BHN test, when the,
    (*a*) piling up of metal round the edge of impression occurs.
    (*b*) hardness of the metal specimen is equal to or more than the hardness of Brinell steel ball itself.
    (*c*) sinking of metal round the edge of impression occurs.
    (*d*) all '*a*' '*b*'&'*c*'.

95. .... test determines the ability of a material to withstand repeatedly applied stress.
    (*a*) Creep strength (*b*) Impact
    (*c*) Fatigue (*d*) Brinell's hardness

96. VPN is determined by Vicker's pyramid method employing a
    (a) load of 50 kg.
    (b) diamond pyramid indenter.
    (*c*) load of 30 kg.
    (*d*) both '*b*' & 'c'.

97. The depth of indentation in BHN test should be less than .... of the thickness of the metal specimen (i.e., test piece).
    (*a*) 1/2 (*b*) 1/5
    (*c*) 1/10 (*d*) 1/3

98. VPN of a particular grade of steel is 750. Its hardness in BHN may be about
    (*a*) 800 (*b*) 650
    (*c*) 1000 (*d*) 1200

99. Vicker's diamond pyramid method of hardness determination does not give accurate result, when employed for the
    (*a*) polished & hardened steel surface,
    (*b*) hard materials,
    (*c*) soft materials.
    (*d*) rough forging or heterogeneous materials like grey cast iron.

100. The distinguishing features between Brinell's hardness test method & Vicker's diamond pyramid method is that the former
    (*a*) is used for very hard materials only.
    (*b*) is used for very soft materials only.
    (*c*) gives a circular impression at the surfaces while the latter gives a square one.
    (*d*) both 'b'&'c'.

Answers

89. (c)	90. (d)	91. (c)	92. (c)	93. (b)	94. (b)
95. (c)	96. (d)	97. (c)	98 . (b)	99. (d)	100. (c)

101. Steel balls used in ball bearings are hardened to about...................... VPN.
(*a*) 100 (*b*) 300
(*c*) 500 (*d*) 700

102. Brinell hardness number (BHN) of mild steel ranges from
(*a*) 25 to 75 (*b*) 110 to 150
(*c*) 250 to 350 (*d*) 350 to 450

103. Hardness of a glass sheet is best determined by the
(*a*) Brinell hardness tester.
(*b*) Vickers hardness tester.
(*c*) Shore scleroscope.
(*d*) Rockwell hardness tester.

104. .... test is generally conducted to check the performance of a material in deep drawing operation.
(*a*) Tensile (*b*) Cupping
(*c*) Izod (*d*) Charpy

105. While determining hardness of aluminium by Brinell hardness testing machine,
(*a*) load on the indentor is reduced.
(*b*) time of loading is reduced.
(*c*) P/D^2 ratio of 25 is used.
(*d*) indenting ball of smaller diameter is used.

106. Match the load (kg) applied in various hardness testing methods.
List I
(*a*) Brinell test for hardened steel.
(*b*) Vicker's diamond pyramid method for hardened steel.
(*c*) Rockwell hardness test (scale-*c*).
(*d*) Rockwell hardness test (scale-A)
List I
I.30 II. 3000
III.60 IV. 150

107. Match the indentor used in various hardness testing methods.
List I
(*a*) Rockwell hardness test (scale-*B*)
(*b*) Shore or rebound hardness test.
(*c*) Rockwell hardness test (scale-C and *D*).
(*d*) Brinell hardness test.
List II
I Small diamond pointed hammer.
II 1/16" hardened steel ball.
III Hardened steel ball of 10 mm dia.
IV Diamond cone.

108. Match the relevant figures for the ratio of various parameters.
List I
(*a*) Ratio of tensile strength of steel (in tons/inch^2) to its BHN.
(*b*) Ratio of BHN to shore value of a typical steel.
(*c*) Ratio of the depth of indention to the thickness of test specimen in BHN test.
(*d*) Ratio of BHN to VPN of a particular steel (having BHN = 650).
List II
I. 6 II. 0.22
III.0.86 IV. 0.1

109. Match the Brinell hardness number (BHN) of various metals.

List I	List II
(a) Mild steel	I. 21
(b) Aluminium	II. 47
(c) Copper	III. 100
(d)Wrought iron	IV. 120

Answers

101. (d) 102. (b) 103. (c) 104. (b) 105. (a) 106. (II) (I) (IV) (III)
107. (II) (I) (IV) (III) 108. (II) (I) (IV) (III) 109. (IV) (I) (II) (III)

110. Vicker hardness tester employs a diamond square based pyramid indentor, which has an angularity of
(*a*) 88° (*b*) 108°
(*c*) 136° (*d*) 152°

111. Which of the following is a non-destructive test for a metallic specimen ?
(*a*) Charpy test
(*b*) X-ray test or ultrasonic test
(*c*) Cupping test
(*d*) Tensile test

112. Tensile test conducted on ductile material does not determine the
(*a*) impact strength.
(*b*) yield point stress.
(*c*) ultimate tensile strength.
(*d*) Young's modulus of elasticity.

113. Charpy test
(*a*) is suitable for highly ductile material.
(*b*) is not an impact test.
(*c*) employs the test specimen bar as a cantilever.
(*d*) is a destructive test.

114. Hardness of diamond is about .... VPN.
(*a*) 200 (*b*) 2200
(*c*) 4200 (*d*) 8400

115. If a soft ball is used for indentation in Brinell hardness tester, then the
(*a*) correct measurement of the depth of indentation is not possible.
(*b*) ball may deform.
(*c*) indentation will not be circular.
(*d*) surface of indentation will be rough.

116. Steel balls for ball bearings are generally made of carbon chrome steel, and they are hardened to a hardness value of about .... VPN.
(*a*) 100-150 (*b*) 250-350
(*c*) 400-550 (*d*) 700-800

117. Pick out the wrong statement.
(*a*) A meehanite ball is used as an indentor in Rockwell hardness test on B scale.
(*b*) Rockwell scale C is used for the hard metals.
(*c*) The hardness in Brinell hardness test is measured by the surface area of depression.
(*d*) Brinell hardness tester fails to give accurate & reliable result in case of very hard surfaces, because the ball deforms on hard surfaces.

118. The hardness is measured by the .... of depression, in case of Rockwell hardness test.
(*a*) surface area (*b*) diameter
(*c*) width (*d*) depth

119. Tension testing machine measures the.......... of the materials.
(*a*) strength (*b*) ductility
(*c*) both 'a' & 'b' (*d*) neither 'a' nor 'b'

120. Dead weight loading is used (in load producing mechanism of tension testing machines) in materials during .... testing.
(*a*) creep (*b*) rupture
(*c*) both 'a' & 'b' (*d*) neither 'a nor 'b'

121. Extensiveness are used (during tension testing of materials) for the direct measurement of the
(*a*) strain
(*b*) extension of specimen
(*c*) stress
(*d*) none of these

Answers

110. (c)	111. (b)	112. (a)	113. (d)	114. (d)	115. (b)
116. (d)	117. (a)	118. (d)	119 . (c)	120. (c)	121. (b)

122. Lapped packed ram type tension testing machine is a............... type of machine.
(*a*) hydraulic (*b*) mechanical
(*c*) screw (*d*) none of these

123. Non-ferrous metals & hard steels do not exhibit a definite yield point, when pulled in a tension testing machine, and hence for these cases, a better measure of their elastic properties is defined by the .... stress.
(*a*) ultimate (*b*) yield
(*c*) proof (*d*) none of these

124. .... is used for determining the tensile strength of steels.
(*a*) Hydraulic press
(*b*) Universal testing machine
(*c*) Mechanical press
(*d*) None of these

125. The difference in one unit of Rockwell hardness number corresponds to a difference in the depth of indentation of .... mm.
(*a*) 0.001 (*b*) 0.01
(*c*) 0.002 (*d*) 0.02

126. A square diamond pyramid indentor is used in determining the .... hardness number.
(*a*) Vicker's (*b*) Rockwell
(*c*) Brinell (*d*) None of these

127. .... hardness testing is the best for measuring the hardness of the surface of a multi-impression forging die.
(*a*) Brinell (b) Scleroscope
(*c*) Rockwell (*d*) Vicker's

128. One of the phases observed in a given steel test specimen by metallurgical microscope using a magnification of 100 X is Fe_3C, which is also known as
(*a*) austenite (*b*) cementite
(*c*) martensite (*d*) ferrite

129. Square pyramid diamond indentor used in Vicker's method of hardness testing has an optical angle of 13°. Vicker's hardness number is given by
(*a*) P/0.5393 d (*b*) 0.5393 P/d
(*c*) 0.5393 P.d (*d*) Pd/0.5393
where, P = impressed load in kg
and, d = diagonal of indentation in mm

130. .... hardness tester is the most suitable for measuring the hardness of lathe bed material.
(*a*) Brinell
(*b*) Rockwell
(*c*) Shore scelerescope
(*d*) Vicker's

131. Etching of glass is done by .... acid
(*a*) hydroflouric (*b*) nitric
(*c*) sulphuric (*d*) hydrochloric

132. The crystal structure of a material can be examined by the
(*a*) electron microscope.
(*b*) optical microscope.
(*c*) gamma radiology.
(*d*) X-rays and electron diffraction.

133. Pick out the wrong statement.
(a) Rockwell hardness number is calculated by a formula, because it is not recorded directly on a dial gauge.
(*b*) Brinell hardness number is calculated by a formula.
(*c*) Rockwell hardness test employs Brale indentor.
(*d*) Knoop test measures the micro-hardness.

Answers

122. (a)	123. (c)	124. (b)	125. (c)	126. (a)	127. (b)
128. (b)	129. (a)	130. (c)	131 . (a)	132. (d)	133. (a)

134. .... test is a non-destructive test.
   (*a*) Cupping (*b*) Charpy
   (*c*) Radiography (*d*) Tensile

135. Polished & etched surface of the cross-section of a hot worked test specimen will have
   (*a*) carbon precipitated at the surface.
   (*b*) surface like a mirror.
   (*c*) a fibre like structure.
   (*d*) flaky carbon at the surface.

136. Inspection of surface & sub-surface cracks in non-magnetic alloys is done by the .... testing method.
   (*a*) dye penetrant
   (*b*) ultrasonic
   (*c*) X-ray
   (*d*) magnetic particle inspection

137. Inspection of surface cracks in the weldment of non-magnetic alloys is done by the .... testing method.
   (*a*) fluorescent (*b*) ultrasonic
   (*c*) X-ray (*d*) magnaflux

138. Which of the following is a non-destructive test?
   (*a*) Fatigue test
   (*b*) Torsion test
   (*c*) Bend test
   (*d*) Ultrasonic inspection

139. Hardness test of a material does not give any information about
   (a) its machinability.
   (b) its wear resistance.
   (*c*) the effectiveness of heat treatment given to the castings.
   (*d*) its ultimate tensile strength (UTS).

140. The same universal tensile testing machine can not be used for carrying out the................. test on the material.
   (*a*) fatigue (*b*) bend
   (*c*) tensile (*d*) compression

141. Tensile test of a material does not help in determining its
   (*a*) modulus of elasticity.
   (*b*) creep strength.
   (*c*) yield strength.
   (*d*) % elongation & % reduction in area.

142. Rockwell hardness test as compared to Brinell hardness test
   (*a*) is faster.
   (*b*) needs no surface preparation (polishing) of the test specimen.
   (*c*) employs smaller indentor & the load resulting in smaller & shallower indentation on the test specimen.
   (*d*) all a, b & c.

143. y-ray radiography as compared to X-ray radiography
   (*a*) can detect defects in thicker castings.
   (*b*) are more suitable for examining objects of varying thickness.
   (*d*) is a slower method but can inspect a number of castings at one time.
   (*d*) all *a*, *b* & c.

144. Match the testing machine used for determination of various properties of a material.

List I	List II
(*a*) Fatigue strength	I. Izod testing machine.
(*b*) Bending strength	II. Universal testing machine.
(*c*) Tensile strength	III. Three point loading machine.
(*d*) Impact strength	IV. Rotating bending machine.

Answers

134. (c)	135. (c)	136. (a)	137. (a)	138. (d)	139. (d)
140. (a)	141. (b)	142. (d)	143 . (d)	144. (IV) (III) (II) (I)	

145. Match the etching solution used for various metallic test specimens.

List I

(a) Low carbon steel and welds

(b) Copper

(c) Cast iron, medium & high carbon steel

(d) Aluminium

List II

I. 50% NH_4OH and 50% water

II. 1% hydroflouric acid in water

III. Nital - 2% nitric acid in ethyl alcohol

IV. Picral - 5% picric acid and ethyl alcohol

146. Match the applications of various metal lurgical tests.

List I

(a) Cupping test

(b) Durometer test

(c) Flattening test

(d) Huey test

List II

I. Rapid assesement of corrosion resistance.

II. Bend test for tube.

III. Hardness testing for rubber and plastics.

IV. Ductility test for sheet metal.

147. Match the applications of various metal-lurgical inspection processes & tests.

List I

(a) Strain gauging

(b) Drop test

(c) Electrography

(d) Jominy test

List II

I. Quality control operation usually applied to finished component.

II. Quality control test to give a measure of hardenability.

III. Process used to identify hard spots on soft materials or soft spots on hard materials.

IV. Inspection technique to micro show porosity in metallic or non- metallic coating.

148. .... hardness testing method suffers from the defects of ridging & sinking in/on the test specimens,

(a) Rockwell *(b)* Vickers

(c) Brinell (d) Both *b & c*

149. The load applied on the indentor of a Brinell hardness tester is such that the ratio of load (kg) to the .... is a constant equal to 30.

(a) d *(b)* d^2

(c) $\sqrt{d}$ *(d)* $d^{3/2}$

where, *d* - ball diameter, mm

150. Pick out the wrong statement pertaining to the tension test of materials. In a tension test, the

(a) elongation at the time of fracture is localised in the region of necking.

(b) cup and cone fracture in the test specimen is due to tearing of metal in the centre followed by shearing of planes at 45^0 to the axis.

(c) U.T.S (ultimate tensile strength) decreases & percent elongation increases on increasing the strain rate.

(d) maximum load corresponds to the true strain equal to strain hardening component.

151. Which of the following hardness testing-requires indentor to be coated with dye before taking impression ?

(a) Rockwell C on hardened material.

(b) Brinell on an annealed material.

(c) Brinell on hardened material.

(d) Rockwell B on an annealed material.

Answers

145. (III) (I) (IV) (II) 146. (IV) (III) (II) (I) 147. (III) (I) (IV) (II)

148. (d) 149. (b) 150 (c) 151 (b)

152. Pick out the wrong statement about Knoop hardness testing method. In Knoop hardness testing method, the
   (*a*) basic phenomenon involved in the hardness determination is the scratching of the test specimen.
   (*b*) diamond pyramidal indentor is employed.
   (*c*) surface of the test specimen must have a metallographic finish.
   (d) Knoop hardness nominal value tends to increase, if the load applied reduces.

153. A universal testing machine
   (*a*) subjects the test specimen to tensile load.
   (*b*) is a hydraulic machinery.
   (*c*) subjects the test specimen to the tensile force through the screw motion.
   (*d*) tests the specimens under tension, bending & shear loads.

154. In Rockwell hardness testing, the thickness of the plate specimen must be at least .... times the depth of indentation.
   (*a*) five (*b*) ten
   (*c*) twenty (*d*) thirty

155. Which hardness number is independent of load?
   (*a*) Rockwell (b) Vicker's
   (*c*) Mho's (*d*) Brinnel

156. Match List I with List II and select the correct answer from the codes given below:
   List 1
   I. The measure of strength under fatigue loading.
   II. The reduced strength of material under repeated or fluctuating cyclic stress.
   III. The highest stress that a material can stand for a specified length of time without rupture.
   IV. The highest stress that a material can stand for a specified length of time without excessive deformation.
   List II
   A. Creep strength
   B. Creep rupture strength
   C. Fatigue
   D. Endurance limit
   (*a*) I-D, II-C, III-B, IV-A
   (b) I-C, II-D, III-B, IV-A
   (*c*) I-D, II-C, III-A, IV-B
   (*d*) I-A, II-B, III-C, IV-D

157. Under torsion, mild steel fails in shear, while cast iron fails in tension. Torsion testing is done to determine the
   (*a*) Poisson's ratio
   (*b*) Modulus of rigidity
   (*c*) Modulus of elasticity
   (*d*) bulk modulus

158. In compression test, cast iron fails in shear. For a successful compression test, the ratio of the test piece length to diameter should not be
   (*a*) < 8 and > 10 (*b*) < 4
   (*c*) >5 (*d*) < 2

159. Bend test is performed on a metal to determine its toughness. Wrap around bend test is generally used for
   (*a*) round bars (b) tubes
   (*c*) sheet metal (*d*) wires

Answers

152. (a)	153. (d)	154. (b)	155. (c)	156. (a)	157. (b)
158. (a)	159. (d)				

160. Cupping test is performed to determine the .... of sheet metal.
(a) malleability (b) drawability
(c) plasticity (d) elasticity

161. When the tensile test is performed on mild steel, its initial fracture takes place in shear, while cast iron fails in tension. The rate of loading in case of tensile testing, at any stage should not be .... kg/cm^2 second.
(a) < 75 and > 100 (b) > 125
(c) >150 (d) <50

162. Silky edge in a tensile test failure of mild steel indicates the failure by
(a) separation (b) shear
(c) sliding (d) either 'b' or 'c'

163. Dull grained appearance in a tensile test failure of mild steel indicates
(a) ductile failure
(b) brittle failure
(c) failure by separation
(d) both b & c.

164. Screwed ends are used for performing tensile test on materials made of
(a) copper (b) aluminium
(c) grey cast iron (d) mild steel.

165. Minor load in case of Rockwell hardness test is 10 kg. If the Rockwell hardness of a material on scale B is 40, then it shall be written as
(a) 40 HRB (b) HRB 40
(c) 40 RHB (d) RHB 40

166. The minor load in case of superficial Rockwell hardness test is 3 kg. Hardness of razor blades may be determined by the......................... hardness test.
(a) Rockwell
(b) Rockwell superficial
(c) Vicker's
(d) Brinell

167. One point of hardness of Rockwell hardness corresponds to 2 micron. Scale A' of Rockwell hardness test is used for
(a) annealed steels
(b) case hardened steels
(c) annealed case-hardened steels
(d) non-ferrous metals

168. In case of scale 'C' of Rockwell hardness test, the angle of the diamond cone indentor is 120°. One point of hardness of superficial Rockwell hardness corresponds to .... microns.
(a) 1 (b) 4
(c) 6 (d) 8

169. Match List I (hardness testing method) with List II formula for calculation of hardness) and select the correct answer from the codes given below :

List I
A. Rockwell hardness of A' scale.
B. Rockwell hardness of B' scale.
C. Rockwell superficial hardness.
D. Brinell hardness

List II
I. 2F/\nD(D – √D^{2-} d^2)]
II. 100 - (d_1/0.001)
III. 130 - (d_1/0.002)
IV. 100 - (d_1/0.002)
where, d_l = depth of penetration in mm.
F = Load in kg
D = Steel ball diameter
d = diameter of indentation
(a) A- IV, B - III, C - II, I) D -1
(b) A - III, B - IV, C - II, D -1
(c) A - IV, B - III, C -1 , D - II
(d) A -1, B - II, C(- III, D - IV

Answers

160. (b)	161. (a)	162. (d)	163. (d)	164. (c)	165. (b)
166. (b)	167. (b)	168. (a)	169 . (a)		

170. The unit of Brinell hardness is kg/mm^2. The relation between Brinell hardness(HB) and Rockwell 'B' scale hardness (HRB) is given by
(a) HB = (130 -HRB)/ 7300
(b) HB = (100-HRB)/7300
(c) 1/HB = (130- HRB)/7300
(d) l/HB = (100 - HRB)/ 7300

171. Which of the following relationships between Brinell hardness (HB) and tensile strength (σ_t kg/cm^2) is correct ?
(a) HB = $\sigma_t/20$ (b) HB = $\sigma_t/30$
(c) HB = $\sigma_t/35$ (d) HB = $\sigma_t/50$

172. In case of Rockwell hardness test, the measure of hardness is the depth of penetration. Indentation method of hardness measurement is based upon the .... deformation of material.
(a) plastic
(b) visco-plastic
(c) elastic
(d) both plastic and elastic

173. Rebound hardness test is based upon the ....deformation of material.
(a) visco-plastic
*(b)*visco-elastic
(c) elastic
(d) both elastic and plastic

174. In case of Vicker's hardness, the hardness is proportional to
(a) D *(b)* 1/D
(c) D^2 *(d)* $1/D^2$
where, *D* = Diagonal of projected indentation.

175. Impact strength is measured in terms of kg.m. Beam un-notched impact test is performed on
(a) grey cast iron (b) mild steel
(c) aluminium *(d)* cast iron

176. The function of notch in impact testing is to localise the rupture and concentrate the stress. The speed of striking in charpy impact test is .... metres / second.
(a) 2 *(b)* 5
(c) 8 *(d)* 10

177. While the depth of the charpy V-notch is 2 mm, the same in case of charpy U-notch is .... mm.
(a) 1 *(b)* 5
(c) 8 *(d)* 10

178. While the cross-section of a standard charpy test piece is 10 x 10 mm^2, the length of the test piece is .... mm.
(a) 35 *(b)* 55
(c) 75 *(d)* 100

179. The speed of striking in Izod impact test should not be .... metres/second.
(a) <3 *(b)* >4
(c) <2 *(d)* >8

180. The plastic deformation which persists after the removal of applied load is called the permanent set. 'Necking' in tensile test occurs.
(a) before yielding *(b)* after yielding
(c) before fracture *(d)* after fracture

Answers

170 (c)	171. (c)	172. (a)	173. (d)	174. (d)	175. (a)
176. (b)	177. (b)	178. (b)	179 . (c)	180 . (b)	

17 Metallurgical Analysis

1. Coning and quartering is a well known process of the sampling of
   (a) ore, coal and coke
   (b) molten metal
   (c) solid metallic scrap
   (d) fuel gases
2. Samples for the analysis of solid metallic scrap & alloys are normally taken by
   (a) grinding
   (b) drilling
   (c) cutting
   (d) coning and quartering
3. A sampling tube known as a 'thief, which has two concentric and close fitting tubes with matching holes or opening, is used for the sampling of
   (a) molten metal and alloys
   (b) sand and other fine materials.
   (c) ores containing metallics.
   (d) blast furnace grade coke.
4. Fuel gas and flue gas samples are most conveniently collected in a
   (a) test tube *(b)* football bladder
   (c) glass tube *(d)* plastic bottle
5. For complete analysis of iron ore, the size of the dried sample should be
   (a) (about 100 microns)
   (b) 200 mesh (about 74 microns)
   (c) 3 mm
   (d) –10 mm
6. Estimation of total iron in oxide ore comprises of three basic steps namely (i) extraction of iron into solution (ii) oxidation for volumetric analysis or precipitation for gravimetric analysis, and (iii) titration in volumetric analysis or ignition and weighting in gravimetric analysis. Which of the following concentrated acid is used for the first step i.e. extraction of iron into solution ?
   (a) $HN0_3$ *(b)* H_2S0_4
   (c)HCl *(d)* either *(a)*, (b) or *(c)*
7. During dissolution of iron from iron ore in acid solution, the rate of dissolution is accelerated by the addition of little amount of ....which reduces ferric iron to ferrous state.
   (a) $BaCl_2$ *(b)* $CaCl_2$
   (c) $SnCl_2$ *(d)* NaCl

Answers

1. (a) 2. (b) 3. (b) 4. (b) 5. (a) 6. (c)
7. (c)

8. Determination of alumina (Al_2O_3) in iron ore is based on the selective formation of $AlPO_4$ from its acidic solution (with conc. HCl) by the addition of
(a) phosphoric acid.
(b) calcium phosphate.
(c) barium phosphate.
(d) sodium or ammonium phosphate.

9. Loss on ignition (LOI) of iron ore is determined by finiding the loss in weight of the sample by heating it first to 500°C for 5 minutes followed by its heating at .... °C, for 30 minutes.
(a) 250 *(b)* 750
(c) 950 *(d)* 1150

10. Combined carbon content in iron and steel is determined by treating the sample with dilute HNO_3, so that combined carbon forms a soluble brown nitrated compound, the intensity of which is matched with that of a standard sample of similar composition treated under similar condition. The intensity of color produced is proportional to the carbon content of iron of steel sample. This method of combined carbon determination is called the

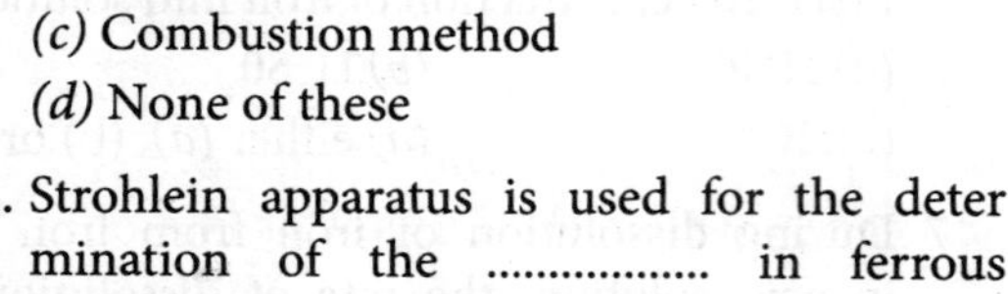

(a) Strohlein method
(b)Eggertz method
(c) Combustion method
(d) None of these

11. Strohlein apparatus is used for the deter mination of the .... in ferrous materials *e.g.*, iron or steel.
(a) graphitic carbon (b) total carbon
(c) sulphur *(d)* both (b) and (c)

12. Determination of phosphorous content of iron & steel is done by bringing phosphorous into solution and then finding it by the .... method.
(a) combustion *(b)* gravimetric
(c) volumetric *(d)* both*(b)* and *(c)*

13. Bomb calorimeter is used for determining the .... sulphur of coal.
(a) pyritic *(b)* sulphate
(c) total *(d)* organic

14. Eschka process is used for determining the .... sulphur of coal.
(a) pyritic *(b)* sulphate
(c) total *(d)* organic

15. For moisture determination during proximate analysis of coal, -72 mesh coal (2-10 gms) is heated between 105 to 110°C
(a.) for 30 minutes.
(b) for one hour.
(c) till constant weight is achieved.
(d) either (b) or *(c)*.

16. Determination of volatile matter in coal, during its proximate analysis requires heating of coal sample.
(a) weighing 25 gms (as received basis) in a muffle furnace.
(b) in a platinum crucible.
(c) weighing 1 gms (air dried).
(d) in a fuel oil fired furnace.

17. During volatile matter content determination, air dried coal sample is heated for 7 minutes at a constant temperature of .... °C.
(a) 725 *(b)* 925
(c) 1050 *(d)* 1175

18. Organic sulphur present in coal is found out by
(a) bomb calorimeter.
(b) Eschka process.
(c) deducting the inorganic sulphur from total sulphur.
(d) Kjeldahl method.

Answers

8. (d)	9. (c)	10. (b)	11. (d)	12. (d)	13. (c)
14. (c)	15. (d)	16. (c)	17. (b)	18. (c)	

19. Determination of ash content of coal during its proximate analysis, the temperature of weighed coal sample is raised to 400 - 500°C in half an hour and maintained at the same level for further half an hour.The temperature is then raised to .... °C at which the sample is heated for one more hour before weighing it.
   (*a*) 775 ± 25 (*b*) 875 ± 25
   (*c*) 975 ± 25 (*d*) 1075 ± 25

20. Fixed carbon of coal during its proximate analysis is determined by
   (*a*) Eschka process
   (*b*) Kjeldahl process,
   (*c*) complete combustion of coal in oxygen
   (*d*) substracting the sum of percentages of volatile matter and ash of dry coal from 100,

21. Temperature used in the determination of volatile matter of coal is .... °C.
   (*a*) 750 (*b*) 850
   (*c*) 950 (*d*) 1050

22. For determination of ash in coke
   (*a*) few drops of benzene is added to avoid the oxidation of coal.
   (*b*) temperature maintained is same as that for coal i.e. 775 ± 25°C.
   (*c*) muffle furnace is not used
   (*d*) temperature maintained is 1050°C.

23. Pick out the wrong statement.
   (*a*) Total carbon content of coal as determined in its ultimate analysis is more than the fixed carbon content as determined in its proximate analysis.
   (*b*) Both the Eschka process and the bomb calorimeter process determine the total sulphur content of coal.
   (*c*) Sulphate sulphur in coal is never present as $CaSO_4 2H_2O$ and is 0.5%.
   (*d*) Pyritic sulphur in coal is present between 0.2 to 0.3% and is generally present as FeS_2.

24. Eschka mixture used in total sulphur determination in coal comprises of a mixture of .... in 2:1 ratio by weight.
   (*a*) calcined MgO and anhydrous Na_2CO_3
   (*b*) calcined CaO and anhydrous Na_2C0_3
   (*c*) calcined MgO and anhydrous $MgCO_3$
   (*d*) calcined CaO and anhydrous $MgCO_3$

25. In the bomb calorimeter used for the determination of calorific value of coal,
   (*a*) high pressure oxygen at 25 atm is used for its combustion.
   (*b*) the bomb is made of copper.
   (*c*) net calorific value is determined directly.
   (*d*) ignition of coal sample is done by heat development due to compression by high pressure oxygen.

26. Conradson method and Ramsbottom method is used for the determination of.... of the fuel oil.
   (a) sediment content
   (b) carbon residue
   (*c*) ash content
   (*d*) flash point

27. Dean and stark apparatus is used in the determination of .... content in liquid fuels.
   (*a*) sediment (*b*) moisture
   (*c*) ash (*d*) sulphur

28. During Orsat analysis of coke oven gas, the unsaturated hydrocarbons (represented by $C2._5H_5$) is absorbed by the
   (*a*) cuprous chloride
   (*b*) fuming sulphuric acid
   (*c*) bromine water
   (*d*) either (*b*) or (*c*)

Answers

19. (a)	20. (d)	21. (c)	22. (b)	23. (c)	24. (a)
25. (a)	26. (b)	27. (b)	28. (d)		

29. Potassium hydroxide sulution used as an absorbent in orsat apparatus is capable of absorbing
(*a*) CO_2 (b) H_2S
(*c*) SO_2 (*d*) all (*a*), (*b*) and (*c*)

30. For hydrogen analysis in blast furnace gas by Orsat apparatus, it is completely oxidised in contact with CuO to H_2O at a temperature of about .... °C.
(*a*) 150 (*b*) 280
(*c*) 500 (*d*) 800

31. Determination of methane present in coke oven gas by orsat apparatus, it is completely oxidised in contact with CuO to $C0_2$ and H_2O at a temperature of, ... °C.
(*a*) 350 (*b*) 500
(*c*) 800 (*d*) 1050

32. Calorific value of liquid fuels can be determined by calorimeter.
(*a*) Boy's (*b*) Junker's
(*c*) bomb (*d*) none of these

33. Cuprous chloride solution is used in orsat apparatus for selectively absorbing carbon monoxide. However, it is capable of absorbing .... also.
(*a*) O_2 (*b*) $C0_2$
(*c*) C_2H_4 (*d*) all (*a*), (*b*) and (*c*)

34. Calorific value of fuel gases is not determined by the .... calorimeter,
(*a*) bomb (*b*) Boy's
(*c*) Junker's (*d*) none of these

35. Which of the following sulphur compounds present in coal that can not be reduced by washing of coal, is not determined directly ?
(*a*) Pyritic sulphur
(*b*) Organic sulphur
(*c*) Sulphate sulphur
(*d*) Total sulphur

36. Carrier agent used in gas analysis by chromatagraph for carrying of gas sample in adsorption coloumn is generally
(*a*) acetylene (*b*) helium
(*c*) nitrogen (*d*) either (*b*) or (*c*)

37. Qualitative determination of H2S in gases is done by wetting a filter paper with....and keeping it in the gas stream. The extent of brown or black colouration of the filter paper gives a rough estimate of H2S present in the gas.
(*a*) lead acetate
(*b*) methyl orange
(*c*) phenolphthalein
(*d*) ammonium hydroxide

38. Ammonia present in coke oven gas is determined by passing its measured volume through the standard solution of
(*a*) sulphuric acid and then titrating it with alkali.
(*b*) hydrochloric acid and then weighing the precipitate.
(*c*) nitric acid and then weighing the precipitate.
(*d*) none of these.

39. Match the solutions used for the selectivre absorption of various components of flue gases in orsat apparatus.

List I	List II
(*a*) Bromine water	I. Carbon dioxide
(*b*) Potassium hydroxide	II. Oxygen
(*c*) Alkaline pyragallol	III. Carbon monoxide
(*d*) Cuprous chloride	IV. Unsaturated hydrocarbons

Answers

29. (d)	30. (b)	31. (c)	32. (c)	33. (d)	34. (a)
35. (b)	36. (d)	37. (a)	38. (a)	39. (IV) (I) (II) (III)	

40. Match the equipments/method used for various metallurgical analysis processes.

List I

(*a*) Total sulphur content of liquid fuel

(*b*) Fuel gas analysis

(*c*) Calorific value of gaseous fuels

(*d*) Fuel gas analysis

List I

I. Boy's calorimater

II. Bomb calorimeter

III.Haldane apparatus

IV.Orsat apparatus

41. Match the equipments/processes used for various determinations/analysis.

List I

(*a*) Sigma calorimeter

(*b*) Thermal conductivity meter.

(*c*) Use of paramagnetic properties or zirconia probe.

(*d*) Dean and Stark apparatus.

List II

I. Oxygen content in flue gas (continuous measurement.)

II. Continuous determination of C.V. of fuel gases.

III. Measurement of CO_2 content in a gaseous mixture.

IV. Determination of moisture content in liquid fuel.

Answers

40. (II) (III) (I) (IV)

41. (II) (III) (I) (IV)

18 Physics of Metals

1. Which of the following is not explained by the electron theory of metals?
   (*a*) Binding in solids.
   (*b*) Ferromagnetism and electrical conductivity.
   (*c*) Elasticity and crystal structure.
   (*d*) None of these.

2. Classical free electron theory of metals assumes that the
   (*a*) electron moves in a periodic field provided by the lattice.
   (*b*) free electrons obey quantum laws.
   (*c*) metals contain free electrons obeying the law of classical mechanics.
   (*d*) none of these.

3. Which of the following electron theory of metals state that the electrons move in a periodic field provided by the lattice ?
   (*a*) Zone theory.
   (*b*) Quantum free electron theory.
   (*c*) Classical free electron theory.
   (*d*) None of these.

4. Solids retain their shapes due to .... forces only.
   (a) cohesive (b) repulsive
   (c) both (a) & (b) (d) neither (a) nor (b)

5. High cohesive forces in solid metals produce low
   (*a*) coefficient of thermal expansion
   (*b*) melting points
   (c) elastic constant
   (*d*) all (a), (b) & (c)

6. Metals with high cohesive forces have high
   (*a*) mechanical strength
   (*b*) melting point
   (*c*) elastic constant
   (*d*) all (a), (b) &(c)

7. The classical free electron theory of metals does not
   (a) assume that the electrons behave towards each other as uncharged particles.
   (b) assume a completely uniform potential field due to ions.
   (c) ignore mutual repulsion between electrons.
   (d) none of these.

8. Which of the following properties of metals can not be explained by free electron theory of metals ?
   (a) Thermal conductivity
   (b) Optical properties.
   (c) Electrical conductivity.
   (d) Specific heat.

Answers

1. (d)	2. (c)	3. (a)	4. (c)	5. (a)	6. (d)
7. (c)	8. (d)				

9. Wiedemann Franz Law, which states that the ratio of the electrical & thermal conductivities at the same temperature should be same for all metals is explained by the .... theory.
   (a) zone
   (b) classical free electron
   (c) quantum free electron
   (d) none of these.

10. Valence electrons in a metal
    (a) occupy the highest available energy states.
    (b) occur in the outermost orbit.
    (c) occupy tne lowest available energy state.
    (d) both (a) & (b).

11. An electron in the innermost shell of metal atom does not possess .... energy.
    (a) centrifugal (b) magnetic
    (c) potential (d) none of these

12. Electrical resistance of a metal is not affected by the
    (a) surface condition of materials.
    (b) extent to which the material is illuminated.
    (c) physical states and dimensions of materials.
    (d) none of these.

13. With increase in temperature of the metal, which of the following decreases ?
    (a) Thermal oscillations of ions in lattice.
    (b) Frequency of collisions between the valence electrons and ions.
    (c) Free time between collisions.
    (d) Electrical resistivity.

14. Intensity of magnetisation '*J*' of a metal placed in a magnetic field of strength '*H*' is given by
    (a) J = K.II (b) J = K/H
    (c) J = H/K (d) none of these
    where, *K* = susceptibility of material

15. For diamagnetic materials, the value of susceptibility, 'K' is
    (a) small & positive (*b)* small & negative
    (c) large & positive (d) large & negative

16. For paramagnetic materials, the value of *K* is
    (a) small & positive.
    (b) small & negative.
    (c) large & positive.
    (d) large & negative.

17. For ferromagnetic materials, the value of *K* is
    (a) small & negative. (b) small & positive.
    (c) large & positive. (d) large & negative.

18. Which of the following materials can retain their magnetism even after the removal of magnetising field (*i.e.* they can become permanent magnets) ?
    (a) Diamagnetic materials.
    (b) Ferromagnetic materials.
    (c) Paramagnetic materials.
    (d) None of these.

19. Non metallic crystals compared to metallic crystals are more
    (a) ductile
    (b) brittle
    (c) conductive electrically
    (d) none of these

20. Which of the following is not an amorphous material ?
    (a) Plastics (b) Glass
    (c) Rubbers (d) Carbon

21. Properties of polycrystalline materials do not depend on the .... of the crystals.
    (a) size
    (b) arrangement (random or orderly).
    (c) impurities present
    (d) none of these

Answers

9. (b)	10. (d)	11. (d)	12. (d)	13. (c)	14. (a)
15. (b)	16. (a)	17. (c)	18. (b)	19. (b)	20. (d)
21. (d)					

22. The atomic packing factor (*i.e.*, volume of atoms/volume of cell) for f.c.c. crystal structure is
(a) 0.68 (b) 0.74
(c) 0.89 (d) 0.16

23. The atomic packing factor for b.c.c. crystal structure is
(a) 0.68 (b) 0.74
(c) 0.83 (d) 0.24

24. The atomic packing factor for which of the following crystals lattice structures are same ?
(a) h.c.p. & f.c.c. (b) h.c.p. & b.c.c.
(b) f.c.c. & b.c.c. (d) none of these

25. Bragg equation for X-ray diffraction through a metal crystal structure is given by
(a) $n\lambda = 2d \sin Ø$
(b) $nd = 2\lambda \sin Ø$
(c) $\lambda d = 2n \sin Ø$
(d) $n\lambda = 2d \cos Ø$
where, d = spacing of atomic planes;
λ = wavelength of X-ray
$Ø$ = angle of reflection ;
n = an integer.

26. With the increase of interatomic bond strength of substance, their
(a) melting & boiling points decreases.
(b) melting point increases.
(c) boiling point increases.
(d) boiling & melting point increases.

27. Which of the following is not a primary bond?
(a) Ionic (electrovalent or heteropolar) bond.
(b) Covalent bond.
(c) Molecular (or Vanderwaals) bond.
(d) Metallic bond.

28. Metallic bonds in metals
(a) are unidirectional, *i.e.* do not act equally strong in all directions.
(b) leads to highly co-ordinated close packed structures accounting for plastic properties of metals and their ability to form alloys.
(c) have the valence electrons bound to a particular pair of atoms.
(d) are same as electrovalent bonds.

29. Ionic bond is
(a) a primary bond.
(b) a secondary bond,
(c) mostly found in steel alloys.
(d) non-existent between a metallic atom and a non-metallic atom.

30. Vander wall bond
(a) is a primary bond.
(b) involves the transfer of electrons between the atoms.
(c) involves sharing of electrons between the atoms.
(d) is a weak bond.

31. Which of the following is not an inter-molecular (Van der wall) bond?
(a) Dispersion bond (b) Dipole bond
(c) Hydrogen bond (d) None of these

32. Self diffusion
(a) is the migration of atoms in pure materials.
(b) occurs in binary metal alloys.
(c) involves atomic migration along grain boundaries.
(d) involves atomic migration along the surface of a phase.

Answers

22. (b)	23. (a)	24. (a)	25. (a)	26. (d)	27. (c)
28. (b)	29. (a)	30. (d)	31. (d)	32. (a)	

33. Diffusion of one component through the lattice of the other (e.g., Ni-Cu system) is called the .... diffusion.
(a) surface (b) inter
(c) grain boundary (d) self

34. Atomic migration through bulk of material is called the .... diffusion.
(a) volume (b) self
(c) inter (d) none of these

35. The unit of co-efficient of diffusion is
(a) sec/m^2 (b) m^2/sec.
(c) m^2sec (d) m^3/sec.

36. The inequality of diffusion of components in a binary solution of alloys is given by the
(a) Kirkendall effect (b) Fick's law
(c) Graham's law (d) none of these

37. Kirkendall effect is important in
(a) sintering
(b) metal to metal bonding
(c) creep
(d) all (a), (b) & (c)

38. Which mechanism of diffusion in solids is confirmed by Kirkendall effect ?
(a) Interstitial mechanism.
(b) Vacancy mechanism.
(c) Two atom, intercharge mechanism.
(d) Four atoms, ring intercharge mechanism.

39. Vacancy mechanism of diffusion occurs in diffusion of
(a) substitutional solid solution elements.
(b) ions in metals.
(c) ions in ceramics.
(d) all (a), (b) & (c).

40. Which diffusion mechanism is involved in the diffusion of carbon in γ-iron ?
(a) Interstitial mechanism
(b) Vacancy mechanism
(c) Two atoms interchange mechanism
(d) Four atoms interchange mechanism

41. Diffusion coefficient in case of solid-solid diffusion is not affected by the
(a) pressure
(b) temperature
(c) crystal structure
(d) none of these.

42. Diffusion co-efficient in solid diffusion depends upon the
(a) grain size.
(b) concentration.
(c) distortion by plastic deformation.
(b) all (a), (b) & (c).

43. Solid diffusion is not involved in the
(a) metal cladding.
(b) annealing of castings.
(c) recovery & recrystallisation.
(d) none of these.

44. Various metals differ in their properties like ductility, melting point, thermal conductivity etc. due to differences in their
(a) atomic structure (b) composition
(c) crystals structure (d) none of these

45. Energy required to remove a valance (*i.e.*, outermost) electron out of the atom is called its
(a) kinetic energy (b) ionisation potential
(c) Fermi energy (d) none of these

46. The bond between the atoms of a solid metal is called the .... bond.
(a) ionic (b) metallic
(c) electrovalent (d) none of these

Answers

33. (b)	34. (a)	35. (b)	36. (a)	37. (d)	38. (b)
39. (a)	40. (d)	41. (d)	42. (d)	43. (d)	44. (a)
45. (b)	46. (b)				

47. Bragg's law is related to the
(a) energy level of an atom
(b) reflection of X-ray within a crystal.
(c) degenerate energy systems.
(d) stereographic projection

48. Energy (E) and wave number (K) of an electron are related as $E = h^2 k^2/2M$
where, M = mass of the electron
and N = planck's constant.
This is the relation according to the
(a) Sommerfield theory.
(b) De Broglie equation.
(c) Fick's second law of diffusion
(d) Bragg's law

49. Diffusion co-efficient of a metal in a solid solution, depends upon its
(a) temperature (b) composition
(c) both (a) & (b) (d) neither (a) nor (b)

50. Which of the following metals has unstable nucleus?
(a) Alkali metals
(b) Radioactive metals
(c) Rare earth metals
(d) None of these

51. Various isotopes of a metal have
(a) different atomic numbers.
(b) same atomic numbers.
(c) different atomic weights.
(d) both (b) & (c).

52. What type of bond is present in diatomic molecules (*e.g.*, O_2, H_2 etc.)?
(a) Vanderwall bond (b) Ionic bond
(c) Metallic bond (d) Covalent bond

53. Which of the following has the most stable atomic structure?
(a) Rare earth metals.
(b) Alkali metals.
(c) Rare gases.
(d) Transition elements.

54. Which of the following is a diffusionless phenomenon?
(a) Phase change (b) Precipitation
(c) Twinning (d) None of these

55. Pick out the wrong statement.
(a) Mass spectrometer is used for measuring the mass of nucleus of an atom.
(b) Carbide tools can not be made by powder metallurgy techniques.
(c) Hysteresis and eddy current losses are low in soft magenetic materials.
(d) Atoms of a dimagnetic materials have zero magenetic moment.

56. Net magentic dipole moment is zero in an antiferromagnetic material like
(a) nickel (b) iron
(c) copper sulphate (d) none of these

57. Diffusion in solid solution is given by the
(a) Kirkendal effect.
(b) Fick's law
(c) De Broglie expression
(d) Schrodinger equation

58. What type of bond exists between sodium and chlorine ions when they form sodium chloride?
(a) Covalent bond (b) Vanderwall bond
(c) Ionic bond (d) None of these

Answers

47. (b)	48. (a)	49. (c)	50. (b)	51. (d)	52. (d)
53. (c)	54. (c)	55. (b)	56. (c)	57. (b)	58. (c)

59. Pick out the correct statement.
   (a) The energy associated with an electron at absolute zero temperature is minimum and is called ionisation energy.
   (b) The valance electrons in a metal are free to move in orbits of an atom as per sommerfield free electron theory.
   (c) There are three states of equal energy in degenerate states.
   (d) none of these.

60. Silicon is termed as a/an
   (a) super conductor (b) semi-conductor
   (c) insulator (d) none of these

61. On being heated up from 0°K, the electrical resistivity of a metal
   (a) increases.
   (b) decreases.
   (c) remains unchanged.
   (d) may increase or decrease ; depends on the metal.

62. Diameter of nucleus of an atom is
   (a) 10^{-12} cm (b) 10^{-12} mm
   (c) 10^{-12} m (d) 10^{-8} mm.

63. .... method is used for finding the diffusion co-efficient of a binary alloy.
   (a) Bragg's (b) Montanto's
   (c) Hess's (d) none of these

64. Energy of an electron as per Sommerfield theory is given by
   (a) $\frac{K^2 .h^2}{2M}$ (b) $\frac{K^2 .h^2}{M}$
   (c) $\frac{K^2 .h^2}{3M}$ (d) $\frac{2M}{K^2 . h^2}$

65. .... method is used for determining the phase composition of alloys and the releative arrangements of atoms in metals.
   (a) Powder (b) Bragg's
   (c) Rotating crystal (d) Sommerfield's

66. Mathematical expression, $n\lambda = 2d \sin \emptyset$, is called the
   (a) De Broglie equation.
   (b) Bragg's law.
   (c) Fick's law.
   (d) Pauli exclusion principle.

67. .... states that "no two electrons in an atom are in the same energy state as defined by four quantum numbers".
   (a) Pauli exclusion principle
   (b) Bragg's law
   (c) Sommerfiled theory
   (d) none of these

68. Pick out the wrong statement.
   (a) The distance between two nearest neighbouring atoms in body centred cubic structure is $\sqrt{2/3}$.
   (b) Vander wall's bonds provide weaker electrostatic attraction force than ionic bonds.
   (c) The mass of an electron at rest is 9.1×10^{-28} gm.
   (d) Debye – Scherrer method of X-ray diffraction technique is also called powder method.

69. Rotating crystal method used for determining the relative arrangement of atoms in metals and the phase composition of alloys uses
   (a) polarised light.
   (b) monochromatic light
   (c) monochromatic X-rays.
   (d) none of these

Answers

59. (c)	60. (b)	61. (a)	62. (a)	63. (b)	64. (a)
65. (c)	66. (b)	67. (a)	68. (a)	69. (c)	

70. 'Stereographic projection' is a way of mapping crystallographic planes and directions in two dimensions. It helps in finding the
(a) crystal orientation.
(b) slip lines.
(c) twin crack & etch pits.
(d) all (a), (b) & (c)

71. Kirkendal effect states that the
(a) Product of two uncertainties (i.e., position and momentum) of an electron can not be smaller than Planck's constant.
(b) each of the two atoms move with a different velocity in a binary solid solution.
(c) atoms of a solid vibrate with larger amplitude with increase in temperature.
(d) diffusivity is a function of concentratiun.

72. The megnitude of energy (E) that an electron gains by increase in the absolute temperature (T) is given by
(a) $E = K\sqrt{T}$ (b) $E = K.T$
(c) $E = K.T^2$ (d) $E = K/T$
where, K = Boltzman's constant.

73. Energy associated with an electron at absolute zero temperature (which is the maximum value for that electron) is called the .... energy.
(a) ionisation (b) Fermi
(c) degenerate (d) electron

74. Energy of an electron is given by the
(a) Sommerfield theory.
(b) theory of relativity.
(c) Heisenberg's principle.
(d) De-Broglie equation.

75. The specimen of crystal is in the form of a .... in the powder method of X-ray diffraction technique.
(a) thin rod (b) cube
(c) powder (d) water suspension

76. The necessary and sufficient condition for Bragg's reflection to occur is that,
(a) $\lambda \le 2d$ (b) $\lambda \le 4d$
(c) $\lambda \le d^2$ (d) $\lambda \le 1/2d$
where, λ = wavelength of X - ray & d = spacing of atomic planes

77. Pick out the correct statement.
(a) The primitive cell of an atom is also termed as unit cell.
(b) Neutrons and electrons can not be diffracted.
(c) The intensity of scattered wave is minimum, when the troughs & crests of X-ray beams coincide.
(d) The net enthalpy change due to mixing of components is zero in real solution.

78. The maximum value of magnetisation of iron (called saturation magnetisation) is obtained by increasing the field strength. It occurs at
(a) – 273°K (b) – 273°C
(c) – 763°C (d) Curie temperature

79. According to free electron theory, the nature of the plot of density of energy state Vs. energy is
(a) hyperbolic (b) elliptical
(c) parabolic (d) none of these

80. Electrons can move through lattice, only when the .... is satisfied as per free electon theory.
(a) Sievert's law
(b) Sommerfield theory
(c) Bragg's law
(d) Pauli exclusion principle

Answers

70. (d)	71. (b)	72. (b)	73. (b)	74. (a)	75. (a)
76. (a)	77. (c)	78. (b)	79. (c)	80. (c)	

81. Electrons are assumed to move in a region of .... potential, as per zone theory.
    (a) periodic (b) very low
    (c) constant (d) very high

82. The so called 'dielectric' materials are
    (a) semi-conductors
    (b) super-conductors
    (c) insulators
    (d) normal conductors

83. Pick out the wrong statement
    (a) Kirkendal effect can be explained by vacancy mechanism of diffusion.
    (b) Zerner ring mechanism of diffusion can not explain the diffusion in f.c.p & h.c.p crystals.
    (c) The constituent atoms in a binary solid solution diffuse at different rates at a particular temperature.
    (d) Lower melting point constituent of a binary solid solution diffuses at a faster rate compared to the higher melting point constituent.

84. With increase in the potential difference across a metal, the intensity of continuous X-ray spectrum of the metal .... at all wavelengths.
    (a) increases
    (b) decreases
    (c) remains constant
    (d) may increase or decrease; depends on the metal

85. Energy in the most stable state of an atom is
    (a) zero
    (b)minimum
    (c) maximum
    (d) in between minimum & maximum

86. Co-ordination number (*i.e.* an atom possessing the number of nearest neighbours) in a b.c.c structure is
    (a) 4 (b) 8
    (c) 10 (d) 12

87. Co-ordination number in both f.c.c & h.c.p structure is
    (a) 4 (b) 8
    (c) 12 (d) 14

88. .... is used for tracing stereographic projection.
    (a) Ordinary graph paper
    (b) Wulf net
    (c) Logarithmic graph paper
    (d) Ternery phase diagram

89. An electrograph is used for the study of the
    (a) composition of gases.
    (b) phase transformation in metals/alloys.
    (c) intensity of electron beams.
    (d) none of these.

90. Pick out the correct statement.
    (a) Yield strength and tensile strength of metals/alloys are high at low temperatures.
    (b) Atomic diameter of metals is equal to 10^{-11} meters.
    (c) Electrical conduction is not possible in a solid, comprising of overlapping zone.
    (d) Diamagnetic materials attract magnetic lines of force, while paramagnetic materials repel them.

91. A dipole is induced into the atom on application of an external electrical field to it. The induced dipole moment is given by
    (a) a/E (b) a . E
    (c) $\sqrt{a.E}$ (d) $\sqrt{a / E}$

Answers

81. (a)	82. (c)	83. (a)	84. (a)	85. (b)	86. (b)
87. (c)	88. (b)	89. (b)	90. (a)	91 . (b)	

92. Sum of the nucleons of an atom is called the
(a) mass number.
(b) atomic number.
(c) Avagadro's number.
(d) none of these.

93. In a face centred cubic cell, an atom at the face centre is shared by .... unit cells.
(a) 2 (b) 4
(c) 6 (d) 8

94. Number of nearest neighbours in an atom or in an octahedral hole of a close packed structure is
(a) 6 (b) 8
(c) 10 (d) 12

95. Total number of atoms required to form a bcc lattice will be
(a) 4 (b) 6
(c) 8 (d) 12

96. Which of the following does not crystallise in hcp structure ?
(a) Berrelium (b) Iron
(c) Magnesium (d) Molybednum

97. Which of the following crystallises in bcc structure ?
(a) Chromium (b) Lead
(c) Zinc (d) Aluminium

98. The properties of all materials are a function of the size, shape and orientation of the structured units in the material. Which of the following properties of a material is a structure sensitive property, (i.e., they are markedly influenced by grain boundaries)?
(a) Specific heat
(b) Density
(c) Tensile strength
(d) Coefficient of thermal expansion

99. Which of the following properties of a material is a structure-insensitive property (*i.e.*, almost not at all affected by the grain boundaries or crystal imperfections)?
(a) Hardness
(b) Yield strength
(c) Impact strength
(d) Coefficient of thermal expansion

100. α – iron as compared to δ-iron has the same
(a) atoms
(b) crystal structure
(c) physical properties
(d) mechanical properties

101. Pick out the wrong statement.
(a) Glass is a crystalline solid having a sharp melting point.
(b) A change in temperature or pressure may cause a solid to change its internal arrangement of atoms.
(c) Nearly, all commercial materials are polycrystalline.
(d) Amorphous materials as compared to a crystalline material are characterised by a relatively disordered and random spacing of atoms.

102. Match the relevant terminologies pertaining to material properties.
List I
(a) Slip (b)Stiffness
(c) Ductility (d) Toughness
List II
I. Percentage reduction of cross-section area of a metal.
II. Ability of a material to absorb energy upto the point of rupture.
III. Plastic deformation of metals.
IV. Resistance to elastic deformation of metals.

Answers

92. (a)	93. (a)	94. (a)	95. (b)	96. (b)	97. (a)
98. (c)	99. (d)	100. (a)	101. (a)	102 . (III) (IV) (I) (II)	

103. Match the characteristic crystal lattice structure possessed by various materials.

List I	List II
(a) Ferritic stainless steel	I. hcp
(b) Delta iron	II. fcc
(c) Cadmium	III. not hcp
(d) Austenitic stainless steel	IV. bcc

104. Atomic packing factor (APF) is the ratio of the total volume of atoms in a unit cell to the total volume of unit cell and is independent of the atom size. Match the A.P.F. of various crystal lattice structure.

List I	List 11
(a) Simple cubic	I. 0.74
(b) bcc	II. 0.52
(c) fcc	III. 0.68
(d) both (b) & (c)	IV. 0.74

105. Match the co-ordination number (i.e., an atom possessing the number of nearest neighbour).

List I	List II
(a) bcc structure	I. 6
(b) NaCI crystals	II. 12
(c) fcc	III. 12
(d) both fcc & hcp	IV. 8

106. Match the examples of various crystal lattice structure possessed by materials.

List I	List II
(a) bcc	I. Tin
(b) fcc	II. Zinc
(c) hcp	III. Gold
(d) not fcc	IV. Molybdenum

107. Atomic packing factor in case of copper crystals is
(a) 1.66 (b) 0.52
(c) 0.78 (d) 0.64.

108. .... is responsible for holding together the protons & neutrons present in the nucleus of an atom.
(a) Continuous exchange of sub-nucleonic particles
(b) Centrifugal forces
(c) Electromagnetic force
(d) Electrostatic force.

109. A crystal (is composed of ions & molecules) structure is a regular three dimensional pattern of atoms or ions in space. The lattice constant of a bcc unit cell with atomic radius of 1.24 A is .... °A.
(a) 0.062 (b) 1.24
(c) 2.864 (d) 1.424

110. Which of the following is not a characteristic of the metallic bond ?
(a) Directionality (b) High conductivity
(c) Opacity (d) Ductility

111. The lattice energy of ionic compounds increases with the
(a) decreasing number of bonding electrons & increasing size of the ions.
(b) increasing number of bonding electrons.
(c) decreasing size of the ions.
(d) both 'b' & 'c'.

112. Coordination number in hcp, bcc & fcc unit cell are respectively
(a) 12, 8 & 12 (b) 6,8 &12
(c) 8, 12 & 12 (d) 12, 12 & 8

113. Electrons moving through the metals do not obey the law of mechanics, because
(a) they move with very high velocity.
(b) they are very small in size.
(c) its density changes during movement.
(d) they are reflected & interact with others of the same kind.

Answers

103. (IV) (III) (II) (I) 104. (II) (III) (IV) (I) 105. (IV) (I) (II) (III) 106. (IV) (III) (II) (I)
107. (b) 108. (a) 109. (c) 110. (a) 111. (d) 112. (a)
113. (d)

114. While in case of a bcc unit cell, one central atom is surrounded by eight identical atoms; in case of a fcc unit cell, a central atom is in contact with .... identical atoms.
(a) four (b) six
(c) ten (d) twelve

115. .... of an atom implies, its capability to attract electrons to itself.
(a) Electronegativity
(b) Electropositivity
(c) Oxidation number
(d) Electronic defect

116. X-ray photons possess larger energy compared to the visible light photons because of its comparatively higher
(a) wavelengths (b) frequency
(c) velocity (d) Planck's constant

117. Aluminium atoms occupy .... percent of the volume of the cube in a unit cell of aluminium.
(a) 68 (b) 74
(c) 80 (d) 94

118. Mixed-ionic-covalant bonds are normally found in the
(a) high strength metals.
(b) semi-conductors.
(c) thermal insulators.
(d) high electrical conductors.

119. Lattice constant of fcc or bcc is equal to the length of the
(a) line having the highest linear density.
(b) line having the lowest linear density.
(c) cube edge.
(d)body diagonal.

120. Pick out the wrong statement.
(a) X-ray is electromagnetic waves having frequency greater than that of ultraviolet rays.
(b) X-ray striking a semi-conductor reduces its electrical conductivity.
(c) The reason why the target in the X-ray tube is generally embedded in copper body is that copper being good conductor of heat carries away the heat from the target.
(d) X-rays do not refract through any medium & do not form images through lenses.

121. A solid solution is an alloy of
(a) two non-metals.
(b) two metals.
(c) a metal & a non-metal.
(d) both 'b' & 'c'.

122. Pick out the correct statement.
(a) Vacancies in a crystalline solid can form & multiply at any temperature due to diffusion process.
(b) A vacancy can exist in a solid solution, only when a solute atom does not fill the interstitial.
(c) Heterogeneous nucleation occurs in pure metal at several points of the melt independently.
(d) In an interstitial solid solution, an impurity atom reacts with the solvent atoms to form a compound.

123. Schottky and Frankel imperfections are .... crystals.
(a) vacancies in ionic
(b) dislocations in ionic
(c) vacancies in covalant
(d) grain boundaries in covalant

Answers

114. (d)	115. (a)	116. (b)	117. (b)	118. (b)	119. (c)
120. (b)	121. (d)	122. (a)	123. (a)		

124. For ASTM grain size 8, the number of grains observed per inch^2 at 200 X is
(a) 32 (b) 48
(c) 64 (d) 128

125. Substitutional diffusion becomes possible, if the
(a) vacancies are present in a substitutional solid solution.
(b) solid solution temperature is very high.
(c) atoms are significantly different in size.
(d) activation energy for vacancy formation is low.

126. Pick out the wrong statement.
(a) Electrons emission from a metallic surface results, when it is struck by a light beam having wavelength less than some threshold value.
(b) A semi-conductor, when struck by photons, becomes better conductor of electricity.
(c) Work function of the metal is the energy required to release one electron.
(d) Frequency of X-rays is equal to that of the ultraviolet rays.

127. Photoelectric emission is the emission of electrons from a metallic surface, when it is struck by
(a) X-rays (b) photons
(c) electrons (d) none of these

128. The area of (111) plane in FCC in unit cell is
(a) $\sqrt[4]{3R^2}$ (b) $\sqrt[4]{6R^2}$
(c) $\sqrt[8]{2R^2}$ (d) $\sqrt[8]{6R^2}$.
where, R = radius of the atom

Answers

124. (a) 125. (a) 126. (d) 127. (b) 128. (a)

Section

A

19

Metallurgical Thermodynamics

1. Which of the following is an extensive property of a system ?
   (a) Heat capacity
   (b) Molal heat capacity
   (c) Pressure
   (d) Concentration

2. Which of the following is a thermodynamic property of a system ?
   (a) Concentration *(b)* Mass
   (c) Temperature *(d)* Entropy

3. First law of thermodynamics is mathematically stated as
   (a) dQ - dE + dW *(b)* dQ = dE - dW
   (c) dE = dQ + dW *(d)* dW = dW + dE

4. First law of thermodynamics deals with the
   (a) direction of energy transfer.
   (b) reversible processes only.
   (c) irreversible processes only.
   (d) none of these.

5. An irreversible process,
   (a) is the analog of linear frictionless motion in machines.
   (b) is an idealised visualisation of behaviour of a system.
   (c) yields the maximum amount of work.
   (d) yields an amount of work less than that of a reversible process.

6. In an adiabatic process,
   (a) heat transfer is zero.
   (b) temperature change is zero.
   (c) work done is a path function.
   (d) enthalpy remains constant.

7. Enthalpy 'H' is defined as
   (a) H = E - PV *(b)* H =F -TS
   *(c)*H-E = PV *(d)* H=F + TS

8. Efficiency of a heat engine working on Carnot cycle between two temperature levels depends upon the
   (a) two temperatures only.
   (b) pressure of working fluid.
   (c) mass of the working fluid.
   (d) mass and pressure both of the working fluid.

9. What is the degree of freedom for a system comprising liquid water in equilibrium with its vapour ?
   (a) 0 *(b)* 1
   (c) 2 *(d)* 3

Answers

1. (a)	2. (d)	3. (a)	4. (a)	5. (d)	6. (a)
7. (c)	8. (c)	9. (a)			

10. Efficiency of a Carnot engine working between temperatures T_1 and T_2 $(T_1 < T_2)$ is
(a) $\frac{T_2 - T_1}{T_2}$ (b) $\frac{T_2 - T_1}{T_1}$
(c) $\frac{T_1 - T_2}{T_2}$ (d) $\frac{T_1 - T_2}{T_1}$

11. For a constant pressure reversible process, the enthalpy change (ΔH) of the system is
(a) $C_v.dT$ (b) $C_p.dT$
(c) $\int c_p.dT$ (d) $\int C_v.dT$.

12. Internal energy of an ideal gas
(a) increases with increase in pressure.
(b) decreases with increase in temperature.
(c) is independent of temperature.
(d) none of these.

13. Equation which relates pressure, volume and temperature of a gas is called the
(a) equation of state
(b) Gibbs-Duhem equation
(c) ideal gas equation
(d) Maxwell's equation

14. Isobaric process means a constant .... process.
(a) temperature *(b)* pressure
(c) volume *(d)* entropy

15. Isentropic process means a constant .... process.
(a) enthalpy *(b)* pressure
(c) volume *(d)* none of these

16. Throttling (Joule-Thomson effect) process is a constant .... process.
(a) enthalpy *(b)* entropy
(c) pressure *(d)* volume

17. The point at which all three (solid, liquid & gaseous) phases co-exist is known as the .... point.
(a) freezing *(b)* triple
(c) boiling *(d)* Boyle

18. C_v for an ideal gas is independent of the
(a) temperature.
(b) pressure.
(c) volume.
(d) pressure & volume both.

19. For an isothermal process, the internal energy of a gas
(a) increases
(b) decreases
(c) remains unchanged
(d) data insufficient, can't be predicted.

20. PV^{γ} = Constant (where, $\gamma = C_p/C_v$) is valid for the ....process.
(a) isothermal *(b)* isentropic
(c) isobaric *(d)* adiabatic

21. For an isothermal reversible compression of an ideal gas
(a) only $\Delta E = 0$ (b) only $\Delta H = 0$
(c) $\Delta E = \Delta H = 0$ (d) $dQ = dE$

22. As the time is passing, entropy of the universe
(a) is increasing.
(b) is decreasing.
(c) remains constant.
(d) data insufficient, can't be predicted.

23. Boyle's law for gases states that,
(a) $P \propto \frac{1}{V}$, when temperature is constant.
(b) $P \propto \frac{1}{V}$, when temperature & mass of the gas remains constant.
(c) $P \propto V$, at constant temperature & mass of the gas.
(d) $\frac{P}{V}$ = constant, for any gas.

Answers

10. (a)	11. (c)	12. (d)	13. (a)	14. (b)	15. (c)
16. (a)	17. (b)	18. (d)	19. (c)	20. (d)	21. (c)
22. (a)	23. (b)				

24. The equation, PV = nRT, is best obeyed by gases at
    (*a*) low pressure & high temperature.
    (*b*) high pressure & low temperature.
    (*c*) low pressure & low temperature.
    (*d*) high pressure & high temperature.

25. Compressibility factor of a gas is
    (*a*) not a function of its pressure.
    (*b*) not a function of its nature.
    (*c*) not a function of its temperature.
    (*d*) unity, if it follows, PV = nRT.

26. Critical compressibility factor for all substances
    (a) are more or less constant (vary from 0.2–0.3).
    (b) vary as square of the absolute temperature.
    (*c*) vary as square of the absolute pressure.
    (*d*) none of these.

27. Reduced pressure of a gas is the ratio of its
    (a) pressure to critical pressure.
    (b) critical pressure to pressure.
    (*c*) pressure to pseudocritical pressure.
    (*d*) pseudocritical pressure to pressure.

28. Number of components (C), phases (P) & degrees of freedom (F) are related by Gibbs phase rule as
    (a) P + F-C= 2 (b) C=P-F+ 2
    (c) F = C-P-2 (d) P=F-C-2

29. Degrees of freedom at triple point will be
    (*a*) 0 (*b*) 1
    (*c*) 2 (*d*) 3

30. $C_p - C_v = R$, is valid for .... gases.
    (*a*) ideal (*b*) all
    (*c*) high pressure. (*d*) low temperature.

31. An isolated system can exchange .... with its surroundings.
    (*a*) matter
    (*b*) energy
    (*c*) neither matter nor energy
    (*d*) both matter & energy

32. Heat of formation of an element in its standard state is
    (*a*) 0
    (*b*) < 0
    (*c*) >0
    (*d*) a function of pressure.

33. Heat of reaction is
    (*a*) dependent on pressure only.
    (*b*) dependent on temperature only.
    (*c*) dependent on both pressure & temperature.
    (*d*) independent of temperature changes.

34. Second law of thermodynamics is concerned with the
    (*a*) amount of energy transferred.
    (*b*) direction of energy transfer.
    (*c*) irreversible processes only.
    (*d*) non-cyclic processes only.

35. The absolute entropy for all crystalline substances at absolute zero temperature is
    (*a*) zero (*b*) negative
    (*c*) positive (*d*) indeterminate

36. Joule-Thomson coefficient is defined as
    (a) $\mu = (\partial P / \partial T)_H$ (b) $\mu = (\partial T / \partial P)_H$
    (c) $\mu = (\partial E / \partial T)_H$ (d) $\mu = (\partial E / \partial T)_H$

37. Mollier diagram is a plot of
    (*a*) temperature vs. enthalpy
    (b) temperature vs. entropy
    (c) entropy vs. energy
    (d) temperature vs. internal energy.

Answers

24. (a)	25. (a)	26. (a)	27. (a)	28. (a)	29. (a)
30. (a)	31. (c)	32. (a)	33. (c)	34. (b)	35. (a)
36. (b)	37. (c)				

38. Near their critical temperatures, all gases occupy volumes .... that of the ideal gas.
(*a*) less than (*b*) same as
(*c*) more than (d) half

39. Charles' law for gases states that
(a) V/T= Constant (b) $V = \propto \frac{1}{T}$
(c) $V = \propto \frac{1}{P}$ (d) PV/T = Constant.

40. Absolute zero temperature signifies the
(*a*) minimum temperature attainable.
(*b*) temperature of the heat reservoir to which a Carnot engine rejects all the heat that is taken in.
(*c*) temperature of the heat reservoir to which a Carnot engine rejects no heat.
(*d*) none of these.

41. Entropy is a measure of the .... of a system.
(a) disorder
(b) orderly behaviour
(*c*) temperature changes only
(*d*) none of these

42. For spontaneous changes in an isolated system (S=entropy)
(a) ds = 0 (b) ds < 0
(c) ds > 0 (d) ds = constant.

43. For equilibrium process (*i.e.* reversible) in an isolated system
(a) ds = 0 (b) ds < 0
(c) ds > 0 (d) ds = constant.

44. The four properties of a system viz. P, V, T, S are related by the
(*a*) Gibbs-Duhem equation
(*b*) Gibbs-Helmholtz equation
(*c*) Maxwell's equations
(*d*) Antonie's equation

45. For a constant volume process
(a) $DE = C_p dT$ (b) $dE = C_v dT$
(c) DQ = dE + pdV (d) dW = pdV

46. In a reversible process
(a) Tds = dE + dW (b) dE-dW = Tds
(c) dW - dE = Tds (d) Tds-dW + dE< 0

47. In an irreversible process
(a) Tds = dE - dW = 0
(b) dE – dW - Tds = 0
(a) Tds – dE - dW > 0
(d) Tds – T.dE+dW<0

48. C_v is given by
(a) $(\partial E/\partial T)_V$ (b) $(\partial E/\partial T)_T$
(c) $(\partial E/\partial P)_V$ (d) $(\partial V/\partial T)_P$

49. Third law of thermodynamics is concerned with the
(*a*) value of absolute entropy.
(*b*) energy transfer.
(*c*) direction of energy transfer.
(*d*) none of these.

50. Which of the following equations is obtained on combining 1st & 2nd law of thermodynamics, for a system of constant mass?
(a) dE = Tds – PdV
(b) $dQ = C_v dT + PdV$
(c) $dQ = C_p dT + Vdp$
(d) Tds = dE – PdV

51. The equation, Tds = dE – PdV, applies to
(*a*) single phase fluid of varying composition.
(*b*) single phase fluid of constant composition.
(*c*) open as well as closed systems.
(*d*) both (*b*) & (*c*).

Answers

38. (a)	39. (a)	40. (c)	41. (a)	42. (c)	43. (a)
44. (c)	45. (b)	46. (a)	47. (c)	48. (a)	49. (a)
50. (a)	51. (d)				

52. For an exothermic reaction,
    (*a*) only enthalpy change (ΔH) is negative.
    (b) only internal energy change (ΔE) is negative.
    (c) both (ΔH) & (ΔE) are negative.
    (d) enthalpy change is zero.

53. "If different processes are used to bring about the same chemical reaction, the enthalpy change is same for all of them". This is the .... law.
    (*a*) Hess's (*b*) Kirchoff's
    (*c*) Lavoisier (*d*) Laplace

54. Change of heat content when one mole of compound is burnt in oxygen at constant pressure is called
    (*a*) calorific value
    (*b*) heat of reaction
    (*c*) heat of combustion
    (*d*) heat of formation

55. Melting of wax is accompanied with the .... entropy.
    (*a*) increase in (*b*) decrease in
    (c) constant (*d*) none of these

56. Helmholtz free energy (*A*) is defined as
    (*a*) A = H-TS (*b*) A = E – TS
    (*c*) A = H+TS (*d*) A = E + TS

57. Gibbs free energy (F) is defined as
    (*a*) F = E-TS (*b*) F = H – TS
    (*c*) F = H+TS (*d*) F = E + TS

58. Gibbs-Helmholtz equation is.
    (*a*) $\Delta F = \Delta H + T\left[\frac{\partial(\Delta F)}{\partial T}\right]_P$
    (b) $\Delta F = \Delta H - T\Delta T$
    (c) d(E-TS)T,V< 0
    (d) $\frac{dp}{dT} = \frac{\Delta H_{vap}}{T\,\Delta V_{vap}}$

59. For a reversible process involving only pressure-volume work
    (a) $(dF)_{T.P} < 0$ (b) $(dF)_{T.P} = 0$
    (c) $(dF)_{T.P} > 0$ (d) $(dA)_{T.v} < 0$.

60. For an irreversible process involving only pressure-volume work
    (a) $(dF)_{T.P} < 0$ (b) $(dF)_{T.P} = 0$
    (c) $(dF)_{T.P} > 0$ (d) $(dA)_{T.v} < 0$

61. Pick out the correct equation relating *'F'* and 'A'.
    (*a*) F =A +PV (*b*) F = E+A
    (*c*) F = A - TS (*d*) F =A + TS

62. A chemical reaction will occur spontaneously at constant pressure & temperature, if the free energy is
    (*a*) zero (*b*) positive
    (*c*) negative (*d*) infinity

63. Clayperon equation deals with the
    (*a*) rate of change of vapour pressure with temperature.
    (*b*) effect of an inert gas on vapour pressure.
    (*c*) calculation of ΔF for spontaneous phase change.
    (*d*) temperature dependence of heat of phase transition.

64. In any spontaneous process,
    (a) only F decreases
    (b) only A decreases
    (c) both F & A decreases
    (d) both F & A increases.

Answers

52. (c)	53. (a)	54. (c)	55. (a)	56. (b)	57. (b)
58. (a)	59. (c)	60. (a)	61. (a)	62. (c)	63. (a)
64. (c)					

65. Pick out the Claussius-Clayperon equation from among the following.
   (a) $\frac{dp}{dT} = \frac{\Delta H}{T \Delta V}$
   (b) $\text{In } P = \frac{\Delta H}{RT} + \text{Constant}$
   (c) $\Delta F = \Delta H + T\left[\frac{\partial(\Delta F)}{\partial T}\right]_p$
   (d) none of these.

66. Free energy changes for two reaction mechanism 'X' and 'Y' are respectively – 15 and - 5 units. It implies that X is
   (a) slower than Y
   (b) faster than Y
   (c) three times slower than Y
   (d) three times faster than Y

67. Partial molal quantities are important in the study of
   (a) ideal gases
   (b) ideal solutions
   (c) non-ideal mixtures
   (d) a pure component

68. Fugacity and pressure are numerically equal, when the gas is
   (a) in standard state
   (b) at high pressure
   (c) at low temperature
   (d) in ideal state

69. The relation connecting the fugacities of various components in a solution with one another and to composition at constant temperature & pressure is called the .... equation.
   (a) Gibbs-Duhem (b) Van Laar
   (c) Gibbs-Helmholtz (d) Margules.

70. Which of the following is not affected by temperature changes ?
   (a) Fugacity (b) Activity co-efficient
   (c) Free energy (d) None of these

71. The activity of an ideal gas is numerically .... its pressure.
   (a) more than
   (b) less than
   (c) equal to
   (d) data insufficient to predict.

72. The point at which both liquid & gaseous phases are identical, is called the .... point,
   (a) critical *(b)* triple
   (c) freezing *(d)* boiling

73. Equilibrium constant of a reaction varies with the
   (a) initial concentration of the reactant.
   (b) pressure.
   (c) temperature.
   (d) none of these.

74. For an ideal solution, the value of activity coefficient is
   (a) 0 (b) 1
   (c) < 1 (d) >1.

75. Fugacity co-efficient of a substance is the ratio of its fugacity to
   (a) mole fraction (b) activity
   (c) pressure (d) activity coefficient

76. Equilibrium constant
   (a) decreases as the temperature increases, for an exothermic reaction.
   (b) decreases as the temperature decreases, for an exothermic reaction.
   (c) will decrease with increasing temperature for an endothermic reaction.
   (d) none of these.

Answers

65. (b)	66. (b)	67. (c)	68. (d)	69. (a)	70. (d)
71. (c)	72. (a)	73. (c)	74. (b)	75. (c)	76. (a)

77. As the entropy of the universe is increasing, day by day, the work producing capacity of a heat engine is
   (a) not changed
   (b) decreasing
   (c) increasing
   (d) data insufficient, can't be predicted

78. Work done in adiabatic process between two states depends on the
   (a) rate of heat transmission
   (b) initial state only
   (c) end states only
   (d) none of these.

79. Extensive properties of a thermodynamic system depend upon the .... of the system.
   (a) specific volume (b) temperature
   (c) mass (d) pressure

80. Measurement of thermodynamic property of temperature is facilitated by the .... law of thermodynamics.
   *(a)*first *(b)* zeroth
   (c) third *(d)* none of these.

81. In case of steady flow compression polytropic process (PV^n= Constant), the work done on air is the lowest, when
   (a) $n = \gamma = 1.4$ (b) $n = 0$
   (c) $n= 1$ (d) $n = 1.66$.

82. The change in entropy (for a gas following $PV = nRT$) is
   (a) $S_2 - S_1 = - R \ln (P_1/P_2)$
   (b) $S_2 - S_1 = - R \ln (V_2/V_1)$
   (c) $S_2 - S_1 = - R \ln (V_1/V_2)$
   (d) $S_2 - S_1 = - R \ln (P_2/P_1)$.

83. Which of the following is not an intensive property ?
   (a) Volume (b) Density
   (c) Temperature (d) Pressure

84. Enthalpy of a gas depends upon the
   (a) temperature (b) mass
   (c) volume (d) pressure

85. In case of reversible process (following pv^n = constant), work obtained for trebling the volume (v_1=1 m^3 and v_2 = 3 m^3) is maximum, when the value of 'n' is
   (a) 0 (b) 1
   (c) $y = 1.44$ (d) 1.66.

86. Which law of the thermodynamics provides basis for measuring thermodvnamic property ?
   (a) First law (b) Zeroth law
   (c) Third law (d) Second law

87. First law of thermodynamics is nothing but the law of conservation of
   (a) momentum (b) mass
   (c) energy (d) none of these

88. Entropy change in case of a reversible adiabatic process is
   (a) minimum (b) zero
   (c) maximum (d) indeterminate

89. Critical temperature is defined as the temperature, above which a gas will
   (a) not liquefy (barring exceptions).
   (b) immediately liquefy.
   (c) never liquify however high the pressure may be.
   (d) none of these.

90. Work done in case of free expansion is
   (a) indeterminate (b) zero
   (c) negative (d) infinity

Answers

77. (b)	78. (c)	79. (c)	80. (b)	81. (c)	82. (a)
83. (a)	84. (a)	85. (a)	86. (b)	87. (c)	88. (b)
89. (c)	90. (b)				

91. In case of a close thermodynamic system, there is
   (*a*) no heat & mass transfer across the boundaries.
   (*b*) no mass transfer but heat transfer across the boundaries may take place.
   (*c*) mass & energy transfer across the boundaries.
   (*d*) none of these.

92. In polytropic process (PV^n = constant), if n = 1 ; it means a/an .... process.
   (a) adiabatic (b) reversible
   (c) isothermal (d) none of these.

93. Pick out the extensive property out of the following.
   (*a*) Surface tension (b) Free energy
   (*c*) Specific heat (*d*) Refractive index

94. If two gases have the same reduced temperature & reduced pressure, they will have the same
   (*a*) volume
   (*b*) mass
   (*c*) critical temperature
   (*d*) none of these.

95. "Law of corresponding states" states that
   (*a*) two different gases behave similarly if their reduced properties (*i.e.*, P,V& T) are same.
   (b) the surface of separation (*i.e.* the meniscus) between liquid and vapour phase disappears at the critical temperature.
   (*c*) no gas can be liquefied above the critical temperature, howsoever high the pressure may be.
   (*d*) the molar heat of every gas at constant volume should be nearly constant (about 3 calories).

96. If heat contents of CH_4, C_2H_4 and C_3H_8 are -17.9, 12.5 and -24.8 kcal/mole respectively; then AH for the reaction $CH_4(g) + C_2H_4(g) \rightleftharpoons C_3H_8(g)$ will be .... kcal.
   (*a*) –19.4 (*b*) – 30.2
   (*c*) – 55.2 (*d*) – 24.8.

97. "When a gas is expanded from high pressure region to low pressure region ; temperature change occurs". This phenomenon is related with the
   (*a*) Gibbs-Duhem equation.
   (*b*) Gibbs-Helmholtz equation.
   (*c*) Third law of thermodynamics.
   (*d*) Joule-Thomson effect.

98. During adiabatic expansion of a gas, the
   (*a*) pressure remains constant.
   (*b*) pressure is increased.
   (*c*) temperature remains constant.
   (*d*) none of these.

99. "When a system in equilibrium is subjected to a change in temperature, pressure or concentration, the equilibrium is displaced in a direction, which tends to undo the effect of the change." This is called the
   (*a*) Le-Chatelier principle
   (*b*) Kopp's rule
   (*c*) Law of corresponding state
   (*d*) Arrhenius hypothesis

100. Choose the condition that must be maintained in order to liquefy CO_2 (triple point for CO_2 is - 57°C and 5.2 atm.).
   (*a*) Pressure must be kept below 5.2 atm.
   (*b*) Temperature must be kept above-57°C.
   (*c*) Pressure must be kept below 5.2 atm. and temperature must be kept above-57°C.
   (*d*) Pressure and temperature must be kept below 5.2 atm. and - 57°C respectively.

Answers

91. (b)	92. (c)	93. (b)	94. (d)	95. (a)	96. (a)
97. (d)	98. (d)	99. (a)	100. (d)		

101. Free energy change at equilibrium is
    (a) zero (b) positive
    (c) negative (d) indeterminate.

102. Pick out the correct statement.
    (a) A real gas on expansion in vacuum gets heated up.
    (b) An ideal gas on expansion in vacuum gets cooled.
    (c) An ideal gas on expansion in vacuum gets heated up.
    (d) A real gas on expansion in vacuum cools down, whereas ideal gas remains unaffected.

103. For a spontaneous process, the free energy
    (a) is zero
    (b) increases
    (c) decreases, whereas the entropy increases
    (d) and entropy both decreases.

104. Variation of equilibrium pressure with temperature for any two phases of a given substance is given by the .... equation.
    (a) Gibbs-Duhem *(b)* Maxwell's
    (c) Clayperon *(d)* Margules

105. Solid and liquid phases of a substance are in equilibrium at its
    (a) critical temperature
    (b) melting point
    (c) freezing point
    (d) both (b) & (c)

106. At triple point (for one component system), vapour pressure of solid as compared to that of liquid will be
    (a) more
    (b) less
    (c) same
    (d) more or less ; depends on the system

107. For a multicomponent system, the term chemical potential is equivalent to the
    (a) molar concentration difference.
    (b) molar free energy.
    (c) partial molar free energy.
    (d) molar free energy change.

108. Fugacity is most helpful in
    (a) representing the actual behaviour of real gases.
    (b) representing the actual behaviour of ideal gases.
    (c) the study of chemical equilibria involving gases at atmospheric pressure.
    (d) none of these.

109. For a given substance at a specified temperature, activity is .... to fugacity.
    (a) directly proportional
    (b) inversely proportional
    (c) equal
    (d) none of these

110. For an ideal gas, the activity coefficient is .... pressure.
    (a) directly proportional to
    (b) inversely proportional to
    (c) unity at all
    (d) none of these

111. "At the absolute zero temperature, the entropy of every perfectly crystalline substance becomes zero". This follows from the
    (a) third law of thermodynamics.
    (b) second law of thermodynamics.
    (c) Nernst heat theorem.
    (d) Maxwell's relations.

Answers

101. (a)	102. (d)	103. (c)	104. (c)	105. (d)	106. (c)
107. (c)	108. (a)	109. (a)	110. (c)	111. (a)	

112. As the temperature is lowered towards the absolute zero, the value of $\partial(\Delta F)/\partial T$ approaches
(a) unity
(b) zero
(c) that of the heat of reaction
(d) infinity.

113. At absolute zero temperature, all substances have the same
(a) heat capacity
(b) crystalline structure
(c) entropy
(d) none of these

114. In the reaction ; $N_2 + O_2 = 2NO$, increasing the pressure will result in
(a) shifting the equilibrium towards right.
(b) shifting the equilibrium towards left.
(c) no change in equilibrium condition.
(d) none of these.

115. In the reaction, $H_2 + I_2 \rightleftharpoons 2HI$, addition of an inert gas will
(a) increase the partial pressure of H_2.
(b)decrease the partial pressure of I_2.
(c) increase the total pressure, and hence will shift the equilibrium towards the right.
(d) not affect the equilibrium conditions.

116. In case of the decomposition of hydroiodic acid ($2HI \rightleftharpoons H_2 + I_2$), addition of H_2 (at equilibrium condition) will
(a) increase the partial pressure of I_2.
(b) decrease the partial pressure of HI.
(c) diminish the degree of dissociation of HI.
(d) none of these.

117. In the decomposition of PCI_5, ($PCI_5 \rightleftharpoons PCI_3 + Cl_2$), decrease in the pressure of the system will .... the degree of dissociation of PCI_5.
(a) increase *(b)* decrease
(c) not alter *(d)* none of these

118. Pick out the wrong statement pertaining to the decomposition of PCI_5 represented by ($PCI_5 \rightleftharpoons PCI_3 + CI_2$).
Degree of dissociation of PCI_5 will
(a) decrease on addition of CI_2.
(b) increase on addition of an inert gas at constant pressure.
(c) decrease on increasing the pressure of the system.
(d) none of these.

119. In the reaction, $N_2 + 3H_2 \rightleftharpoons 2NH_3 + 22.4$ kcal, the formation of NH_3 will be favoured by the
(a) high temperature.
(b) low pressure.
(c) low temperature only.
(d) low temperature & high pressure,

120. In the chemical reaction, represented by $2S0_2 + 0_2 \rightleftharpoons 2S0_3$, ΔH= - 42 kcal ; the forward reaction will be favoured by the
(a) low temperature
(b) high pressure
(c) both *(a)* & (b)
(d) neither *(a)* nor *(b)*

121. Henry's law is closely obeyed by a gas, when its .... is extremely high.
(a) pressure *(b)* solubility
(c) temperature *(d)* none of these

Answers

112. (b)	113. (a)	114. (c)	115. (d)	116. (c)	117. (a)
118. (d)	119. (d)	120. (c)	121. (d)		

122. At a given temperature, the volume of a gas dissolved in a solvent..................with increase in pressure.
    (a) increases
    (b) decreases
    (c) is unchanged
    (d) may increase or decrease ; depends on the gas.

123. Pick out the wrong statement.
    (a) At constant pressure, the solubility of a gas in a liquid diminishes with rise in temperature.
    (b) Normally, the gases which are easily liquefiable, are more soluble in common solvent.
    (c) The gases which are capable of forming ions in aqueous solution, are much more soluble in water than in other solvents.
    (d) At constant pressure, the solubility of a gas in a liquid increases with rise in temperature.

124. Solubility of a substance which dissolves with an increase in volume and liberation of heat will be favoured by .... temperature.
    (a) low pressure & high
    (b) low pressure & low
    (c) high pressure & low
    (d) high pressure & high

125. Degree of freedom of a system consisting of ice-water-vapour will be
    (a) 0 (b) 1
    (c) 2 (d) 3

126. Degree of freedom of a system consisting of a gaseous mixture of H_2 & NH_3 will be
    (a) 0 (b) 1
    (c) 2 (d) 3

127. The minimum number of phases that can exist in a system is
    (a) 0 (b) 1
    (c) 2 (d) 3

128. The temperature at the eutectic point of the system is the .... temperature that can be attained in the system.
    (a) lowest (b) highest
    (c) average (d) none of these.

129. If the vapour pressure at two temperatures of a solid phase in equilibrium with its liquid phase are known, then the latent heat of fusion can be calculated by the
    (a) Maxwell's equation.
    (b) Claussius-Clayperon equation.
    (c) Van Laar equation.
    (d) Nernst Heat Theorem.

130. If two pure liquid constituents are mixed in any proportion to give an ideal solution, there is no change in the
    (a) volume (b) enthalpy
    (c) both (a) & (b) (d) neither (a) nor (b)

131. Fugacity is a measure of the
    (a) escaping tendencies of the same substance in different phases of a system.
    (b) relative volatility of a mixture of two miscible liquids.
    (c) behaviour of ideal gases.
    (d) none of these.

Answers

122. (c)	123. (a)	124. (b)	125. (a)	126. (d)	127. (b)
128. (a)	129. (b)	130. (c)	131. (a)		

132. "The fugacity of a gas in a mixture is equal to the product of its mole fraction and its fugacity in the pure state at the total pressure of the mixture". This is the statement as per
   (a) Gibbs-Helmholtz equation.
   (b) Lewis-Randall rule.
   (c) Henry's law.
   (d) Raoult's law.

133. The standard state of a gas (at a given temperature) is the state in which the fugacity is equal to
   (a) unity (b) activity
   (c) both (a) & (b) (d) neither (a) nor (b).

134. The amount of heat required to decompose a compound into its elements is .... the heat of formation of that compound from its elements.
   (a) less than (b) more than
   (c) same as (d) not related to

135. In reactions involving solids & liquids (where change in volume is negligible), the heat of reaction at constant pressure as compared to that at constant volume is
   (a) more
   (b) less
   (c) same
   (d) unpredictable ; depends on the particular reaction.

136. Critical solution temperature (or the consolute temperature) for partially miscible liquids *(e.g.,* phenol-water) is the minimum temperature, at which
   (a) a homogeneous solution (say of phenol-water) is formed.
   (b) mutual solubility of the two liquids shows a decreasing trend.
   (c) two liquids are completely separated into two layers.
   (d) none of these.

137. Pick out the wrong statement.
   (a) Systems (of partially miscible liquid pairs), in which the mutual solubility increases with rise in temperature, are said to possess an upper consolute temperature.
   (b) Systems in which the mutual solubility increases with decrease in temperature, are said to possess lower consolute temperature.
   (c) Nicotine-water system shows both an upper as well as a lower consolute temperature, implying that they are partially miscible between these two limiting temperatures.
   (d) none of these.

138. Those solutions in which there is no volume change upon mixing the components in the liquid state and which, when diluted do not undergo any heat change *(i.e.* heat of dilution is zero), are called the .... solutions.
   (a) ideal (b) real
   (c) isotonic *(d)* buffer

139. "The rate at which a substance reacts is proportional to its active mass and the rate of chemical reaction is proportional to the product of active masses of the reacting substances". This is the statement of the
   (a) Lewis-Randall rule
   (b) Vant-Hoff equation
   (c) Le-Chatelier's principle
   (d) Kistyakowasky equation

Answers

132. (b) 133. (c) 134. (c) 135. (c) 136. (a) 137. (d)
138. (a) 139. (d)

140. The quantitative effect of temperature on chemical equilibrium is given by the
   (a) Vant-Hoff equation
   (b) Le-Chatelier's principle
   (c) Arrhenius equation
   (d) Antonie's equation

141. If we increase the pressure on a substance (which is at its triple point), then the triple point
   (a) increases.
   (b) decreases.
   (c) remains unchanged.
   (d) may increase or decrease ; depends on the substance.

142. The value of C_p & C_v respectively for monoatomic gases are .... Kcal kg.mole.°K.
   (a) 5 & 3 (b) 3.987 & 1.987
   (c) 1.987 & 0.66 *(d)* 0.66 & 1.987

143. For an ideal gas, the enthalpy
   (a) increases with pressure rise.
   (b) decreases with pressure rise.
   (c) is independent of pressure.
   (d) is a path function.

144. If the pressure of 100 c.c.air is halved, then its volume (at the same temperature) would be .... c.c.
   (a) 100 *(b)* 50
   (c) 250 *(d)* 200

145. A gas has a volume 27.3 c.c. at 0°C. Its volume at 10°C (if pressure remains unchanged) will be .... c.c.
   (a) 2.73 *(b)* 28.3
   (c) 273 *(d)* 283

146. The fusion of a crystalline solid at its melting point to form a liquid at the same temperature is accompanied by
   (a) decrease in enthalpy corresponding to evolution of heat.
   (b) decrease of entropy.
   (c) increase in enthalpy corresponding to absorption of heat.
   (d) no change in enthalpy.

147. The value of gas constant 'R' is
   (a) 1.987 cal/gm mole °K
   (b) 1.987 BTU/lb mole °R
   (c) both *(a)* & (b)
   (d) neither *(a)* nor (b).

148. The extensive properties are
   (a) volume, mass, entropy & number of moles.
   (b) free energy, entropy & enthalpy.
   (c) both *(a)* & *(b)*.
   (d) none of these.

149. The intensive properties are
   (a) molar volume, density, viscosity & boiling point.
   (b) refractive index & surface tension.
   (c) both *(a)* & (b).
   (d) none of these.

150. A thermodynamic system is said to be isopiestic, if there is no .... change.
   (a) temperature (b) pressure
   (c) volume *(d)* enthalpy

151. For a cyclic process, a fixed ratio between heat and work
   (a) always exists
   (b) may exist
   (c) never exists
   (d) is difficult to predict

Answers

140. (a)	141. (c)	142. (a)	143. (c)	144. (d)	145. (b)
146. (c)	147. (c)	148. (c)	149. (c)	150. (b)	151. (a)

152. The variation of heat of reaction with temperature at constant pressure is given by the .... law.
(a) Kelvin's (b) Antonie's
(c) Kirchoff 's *(d)* Amgat

153. Joule-Thomson co-efficient for a perfect gas is
(a) zero (b) positive
(c) negative *(d)* none of these

154. Pick out the wrong statement.
(a) An ideal liquid or solid solution is defined as the one in which each component obeys Raoult's law.
(b) If Raoult's law is applied to one component of a binary mixture; Henry's law or Raoult's law is applied to other also.
(c) Henry's law is rigorously correct in the limit of infinite dilution.
(d) none of these.

155. Vapour which is at a pressure smaller than the saturation pressure for the temperature involved is called the .... vapour.
(a) superheated
(b) desuperheated
(c) non-condensable
(d) saturated

156. Activity coefficient is a measure of the
(a) departure from ideal solution behaviour.
(b) departure of gas phase from ideal gas law.
(c) vapour pressure of liquid.
(d) none of these.

157. Fugacity and pressure are numerically not equal for gases at
(a) low temperature & high pressure.
(b) standard state.
(c) both (a) & (b).
(d) ideal state.

158. The number of degrees of freedom at triple point of water is
(a) 0 *(b)* 1
(c) 2 *(d)* 3

159. Number of molecules present in 1 gm mole of a gas is called the
(a) atomic number *(b)* Avogadro number
(c) vapor density *(d)* none of these

160. A substance above its critical temperature exists as a
(a) saturated vapour *(b)* gas
(c) liquid *(d)* solid

161. C_p and C_v are equal at
(a) 0°R *(b)* 0°K
(c) 0°C *(d)* 0°F

162. The specific heat of saturated water vapour at 100°C is
(a) zero *(b)* negative
(c) positive *(d)* infinity,

163. Cp/C_v for an ideal monoatomic gas is
(a) 1 *(b)* 1.67
(c) 1.27 *(d)* 1.45

164. The internal energy of an ideal gas does not change in a/an reversible .... process.
(a) isothermal *(b)* isobaric
(c) adiabatic *(d)* isometric

165. Mollier chart is a plot of the
(a) temperature Vs. entropy
(b) enthalpy Vs. entropy
(c) pressure Vs. volume
(d) pressure Vs. enthalpy,

Answers

152. (c)	153. (a)	154. (d)	155. (a)	156. (a)	157. (c)
158. (a)	159. (b)	160. (b)	161. (b)	162. (b)	163. (b)
164. (a)	165. (b)				

166. The temperature at which a real gas obeys the ideal gas laws over a wide range of pressure is called the .... temperature.
(*a*) inversion (*b*) critical
(*c*) Boyle (*d*) reduced

167. Claussius-Clayperon equation is applicable to the........................process.
(*a*) sublimation (*b*) melting
(*c*) vaporisation (*d*) all (*a*), (*b*) & (*c*)

168. The unit of fugacity is same as that of the
(a) pressure (b) energy
(c) temperature (d) concentration

169. Fugacity coefficient of a real gas is
(*a*) *0* (b) *1*
(*c*) <1 (*d*) >1

170. Which is not dimensionless ?
(*a*) Fugacity (*b*) Fugacity coefficient
(*c*) Activity (*d*) Activity coefficient

171. Which is unity for an ideal liquid solution?
(*a*)Activity (b) Activity coefficient
(*c*)Fugacity (*d*) Fugacitycoefficient

172. With increase in reduced temperature, the fugacity coefficient of a gas at constant reduced pressure
(*a*) increases (b) increases linearly
(*c*) decreases (*d*) remains unchanged

173. With increase in temperature, the equilibrium constant for an exothermic reaction
(a) increases (b) increases linearly
(*c*)decreases (d) remains unaffected

174. With increase in temperature, the equilibrium constant for an endothermic reaction
(a) increases (b) increases linearly
(c) decreases (d) remains unchanged

175. PV^{γ}= constant, holds good for an isentropic process, which is
(*a*) reversible & isothermal.
(*b*) isothermal & irreversible.
(*c*) reversible & adiabatic.
(*d*) adiabatic & irreversible.

176. During the phase transition, ... changes.
(*a*) pressure (*b*) volume
(*c*) temperature (*d*) all (*a*), (*b*) & (*c*)

177. Kopp's rule is used to calculate the heat capacity of the
(*a*) solids (*b*) liquids
(c) gases (*d*) all (*a*), (*b*) & (*c*)

178. Translational kinetic energy of molecules of an ideal gas is proportional to
(a) T (b) $\sqrt{T}$
(*c*) T^2 (d) $1/\sqrt{T}$.
where, T = absolute temperature of the gas

179. At .... point, all the three phases (*i.e.*, solid, liquid & gaseous) co-exist.
(*a*) eutectic (*b*) triple
(*c*) plait (*d*) critical

180. Boiling of liquid is accompanied with increase in the
(*a*) vapour pressure
(*b*) specific Gibbs free energy
(*c*) specific entropy
(*d*) all (*a*), (*b*) & (*c*).

181. The value of Joule-Thomson co-efficient, in case where cooling occurs after the throttling process is
(*a*) 0 (*b*) ∞
(*c*) +ve (*d*) –ve

Answers

166. (c)	167. (d)	168. (a)	169. (b)	170. (a)	171. (b)
172. (a)	173. (c)	174. (a)	175. (c)	176. (b)	177. (a)
178. (a)	179. (b)	180. (a)	181. (c)		

182. Linde gas liquefaction process employs cooling
(*a*) by throttling
(*b*) by expansion in an engine
(*c*) at constant pressure
(*d*) none of these

183. In the equation PV^n= constant, if the value of $n = \pm \infty$,then it represents a reversible .... process.
(*a*) adiabatic (*b*) isometric
(*c*) isentropic (*d*) isothermal.

184. Pick out the wrong statement.
(*a*) Activity coefficient is dimensionless.
(*b*) In case of an ideal gas, the fugacity is equal to its pressure,
(*c*) In a mixture of ideal gases, the fugacity of a component is equal to the partial pressure of the component.
(*d*) The fugacity coefficient is zero for an ideal gas.

185. What happens in a reversible adiabatic compression ?
(*a*) Heating occurs
(*b*) Cooling occurs
(*c*) Pressure is constant
(*d*) Temperature is constant

186. In an ideal solution, the activity of a component equals its
(*a*) mole fraction.
(*b*) fugacity at the same temperature & pressure.
(*c*) partial pressure.
(*d*) none of these.

187. Requisites of a reversible process is that the
(*a*) system & surroundings pressure be equal.
(*b*) friction in the system should be absent.
(*c*) system & surroundings temperature be equal.
(*d*) none of these.

188. With increases in temperature, the atomic heat capacities of all solid elements
(a) increases
(*b*) decreases
(*c*) remains unchanged
(*d*) decreases linearly.

189. Pick out the wrong statement.
(*a*) The conversion for a gas phase reaction increases with decrease in pressure. if there is an increase in volume accompanying the reaction.
(b) With increase in temperature, the equilibrium constant increases for an exothermic reaction,
(*c*) The equilibrium constant of a reaction depends upon temperature only.
(*d*) The conversion for a gas phase reaction increases with increase in pressure, if there is a decrease in volume accompanying the reaction.

190. A/an .... system is exemplified by a vessel containing a volatile liquid in contact with its vapour.
(*a*) isolated (*b*) closed
(*c*) open (*d*) none of these

191. $1/V\ (\partial V/\ \partial T)p$ is the mathematical expression for the
(*a*) Joule-Thompson coefficient.
(*b*) specific heat at constant pressure (C_p).
(*c*) coefficient of thermal expansion.
(*d*) specific heat at constant volume (C_v).

Answers

182. (a)	183. (b)	184. (d)	185. (a)	186. (a)	187. (b)
188. (a)	189. (b)	190. (c)	191. (d)		

192. What is the degree of freedom for two miscible (non-reacting) substances in vapor-liquid equilibrium forming an azeotrope ?
(*a*) 0 (b) 1
(*c*) 2 (*d*) 3

193. Gibbs free energy at constant pressure & temperature under equilibrium conditions is
(a) infinity (b) *zero*
(c) maximum (*d*) minimum

194. (del T/del P)$_{(H)}$ *is* the mathematical expression for the
(a) specific heat at constant pressure (C_p).
(b) specific heat at constant volume (C_v).
(c) Joule-Thompson coefficient.
(d) none of these.

195. The enthalpy change when ammonia gas is dissolved in water is termed as the heat of
(a) solution (*b*) formation
(*c*) dilution (*d*) combustion

196. Pick out the correct statement.
(*a*) In an isothermal system, irreversible work is more than reversible work.
(b) Under reversible conditions, the adiabatic work is less than isothermal work.
(*c*) Heat, work, enthalpy & entropy are all 'state functions'.
(*d*) Matter and energy can not be exchanged with the surroundings in a closed system.

197. An ideal gas is taken around the cycle ABCA as shown in P-V diagram :The work done by the gas during the cycle is equal to
(*a*) 12 P_1V_1 (*b*) 6 P_1V_1
(*c*) 3 P_1V_1 (*d*) P_1V_1

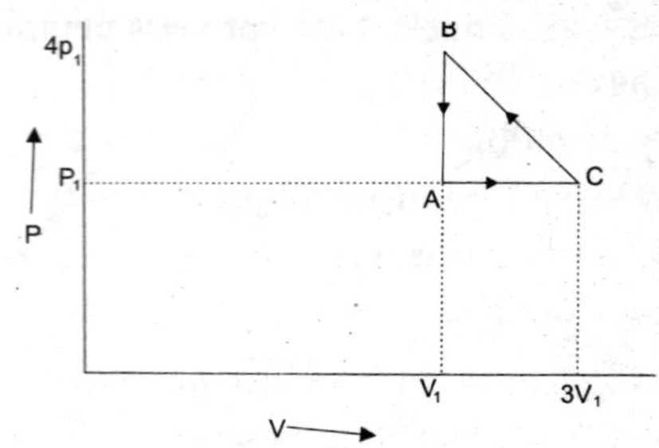

198. A thermodynamic system is taken from state A to B along ACB and is brought back to A along BDA as shown in the P-V diagram.The net work done during the complete cycle is given by the area covered by

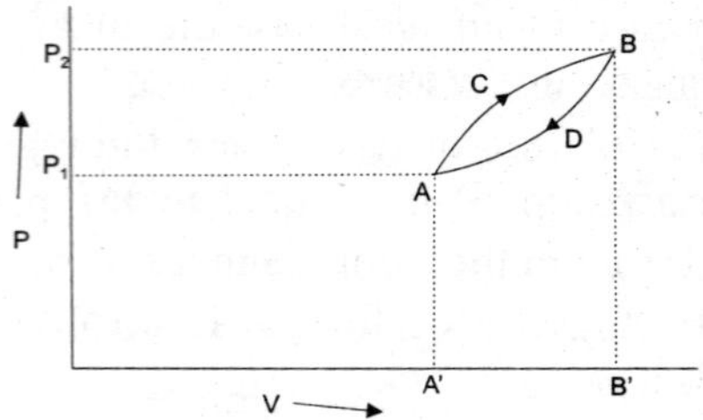

figure

(*a*) $P_1ACB\ P_2\ P_1$ (*b*) *ACB B' A'A*
(*c*) *ACBDA* (*d*) *ADBB'A'A*

199. An ideal monoatomic gas is taken round the cycle *ABCDA* as shown in the *P-V* diagram The net work done during the cycle is given by

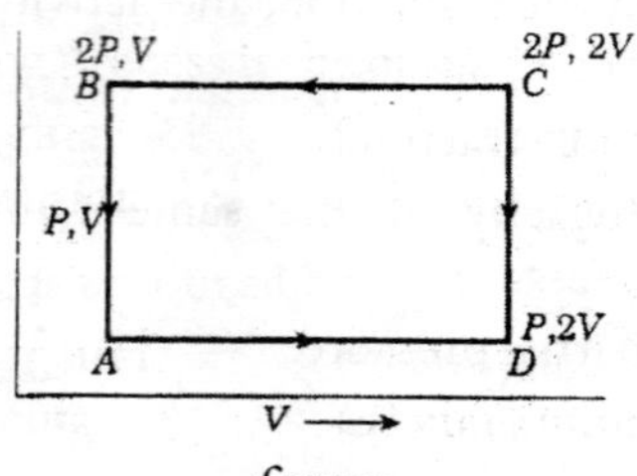

figure

(*a*) PV (b)2PV
(*c*) PV/2 (d)zero

Answers

192. (c)	193. (d)	194. (c)	195. (a)	196. (b)	197. (c)
198. (c)	199. (a)				

200. Heating of water under atmospheric pressure is an .... process.
(a) isochoric (b) isobaric
(c) adiabatic (d) isothermal

201. A gas performs the maximum work, when it expands
(a) non-uniformly (b) adiabatically
(c) isobarically (d) isothermally

202. Which of the following has the least efficiency ?
(a) Steam engine (b) Carnot engine
(c) Diesel engine (d) Otto engine

203. When a gas in vessel expands, its internal energy decreases. The process involved is
(a) reversible (b) irreversible
(c) isothermal (d) adiabatic

204. .... explains the equilibrium constant for any chemical reaction.
(a) Henry's law
(b) Law of mass action
(c) Hess's law
(d) none of these

205. Standard temperature & pressure (S.T.P) corresponds to
(a) 0°C & 760 mm Hg,
(b) 15°C & 760 mm Hg.
(c) 0°C & 1 kgf/cm^2.
(d) 15°C & 1 kgf/cm^2.

206. Normal temperature & pressure (N.T.P) corresponds to
(a) 0°C & 760 mm Hg,
(b) 15°C & 760 mm Hg.
(c) 0°C & 1 kgf/cm^2.
(d) 15°C & 1 kgf /cm^2.

207. Pick out the wrong statement.
(a) C_p of monoatomic gases such as metallic vapor is about 5 kCal / kg atom.
(b) The heat capacity of solid inorganic substance is exactly equal to the heat capacity of the substance in the molten state.
(c) There is an increase in entropy, when a spontaneous change occurs in an isolated system.
(d) At absolute zero temperature, the heat capacity for many pure crystalline substances is zero.

208. For metal-oxide system, Ellingham diagram does not give any idea about the
(a) kinetics of oxidation reaction.
(b) reduction of metal oxides.
(c) oxidation of metals.
(d) all (a), (b) & (c).

209. For metal sulphide system in Ellingham diagram, the position of the formation of calcium sulphide line is .... ; hence calcium oxide can be used to desulphurise many metals.
(a) in the middle (b) very low
(c) very high (d) none of these

210. Metal chlorides can not be reduced by carbon (as inferred from the metal- chloride system), because of the fact that
(a) unstable metal carbide are formed.
(b) unstable carbon tetrachloride is formed.
(c) reactions require very high temperature.
(d) reactions requires very low (sub-zero) temperature.

Answers

200. (b)	201. (c)	202. (a)	203. (a)	204. (b)	205. (b)
206. (a)	207. (b)	208. (a)	209. (a)	210. (b)	

211. As per the chemical reaction,
[S] + [Fe] + (O^{--})=(S^{--}) + (FeO); with increase in the basicity of slag, sulphur transfer from metal to slag will be
(a) unaffected
(b) increased
(c) decreased
(d) unpredictable from the reaction

212. Raoult's law is not obeyed in case of dissolution of chromium in the pure iron due to the fact that chromium .... the pure iron.
(a) reacts with
(b) is only partially soluble in
(c) has lower activity co-efficient than
(d) is completely soluble in

213. Dephosphorisation reaction, in oxygen steel making processes (*e.g.* L.D. process),......... with the presence of high percentage of carbon in iron.
(a) decreases
(b) increases
(c) remains unaffected
(d) unpredictable

214. Aluminium oxide is used as a/an .... agent due to the fact that it has a high negative free energy of formation.
(a) oxidising *(b)* reducing
(c) both *(a)* & *(b)* *(d)* neither *(a)* nor *(b)*

215. Pick out the wrong statement.
(a) The expansion of gas in vacuum is an irreversible process.
(b) An isometric process is a constant pressure process.
(c) Entropy change for a reversible adiabatic process is zero.
(d) Free energy change for a spontaneous process is negative.

216. The free energy change for a chemical reaction is given by
(a) RTlnK *(b)* – RT InK
(c) –R InK *(d)* T InK .

217. The equilibrium constant for a chemical reaction at two different temperatures is given by
(a) $\frac{Kp_2}{Kp_1} = \frac{\Delta H}{R}\left(\frac{1}{T_2} - \frac{1}{T_1}\right)$
(b) $\frac{Kp_2}{Kp_1} = \frac{\Delta H}{R}\left(\frac{1}{T_2} - \frac{1}{T_1}\right)$
(c) $\frac{Kp_2}{Kp_1} = -\Delta H\left(\frac{1}{T_2} - \frac{1}{T_1}\right)$
(d) $\frac{Kp_2}{Kp_1} = -\frac{-1}{R}\left(\frac{1}{T_2} - \frac{1}{T_1}\right)$

218. Which of the following represents the Virial equation of state?
(a) $T = \frac{RT}{V - b} - \frac{}{\sqrt{T \cdot V(V + b)}}$
(b) $\frac{PV}{RT} = 1 + B/V + C/V^2 + ...$
(c) $\left(P + \frac{á}{V^2}\right)(V - b) = RT$
(d) $n_1\, \mu_2 + n_2\, \mu_1 = 0$

219. Pick out the wrong statement.
(a) Phase rule variables are intensive properties.
(b) Heat and work are both state function.
(c) The work done by expansion of a gas in vacuum is zero.
(d) C_p and C_v are state function.

Answers

211. (b)	212. (b)	213. (b)	214. (b)	215. (b)	216. (b)
217. (a)	218. (b)	219. (b)			

220. In an ideal gas mixture, fugacity of a species is equal to its
(a) vapor pressure
(b) partial pressure
(c) chemical potential
(d) none of these

221. Pick out the correct statement.
(a) Entropy and enthalpy are path functions.
(b) In a closed system, the energy can be exchanged with the surrounding, while matter can not be exchanged.
(c) All the natural processes are reversible in nature.
(d) Work is a state function.

222. Which of the following is Clausius-Clayperon equation for vaporisation of an ideal gas under the condition that the molar volume of liquid is negligible compared to that of the vapor ?
(a) $\frac{d \text{ In } P}{dT} = \frac{H_{vap}}{RT_2}$
(b) $\frac{d \text{ In } P}{dT} = \frac{RT_2}{H_{vap}}$
(c) $\frac{dP}{dT} = \frac{RT_2}{H_{vap}}$
(d) $\frac{dP}{dT} = \frac{H_{vap}}{RT_2}$

223. The ratio of the equilibrium constants (Kp_2/Kp_1) at two different temperatures is given by
(a) $\frac{R}{\Delta H}\left[\frac{1}{T_1} - \frac{1}{T_2}\right]$
(b) $\frac{\Delta H}{R}\left[\frac{1}{T_1} - \frac{1}{T_2}\right]$
(c) $\frac{\Delta H}{R}\left[\frac{1}{T_2} - \frac{1}{T_1}\right]$
(d) $\frac{1}{R}\left[\frac{1}{T_1} - \frac{1}{T_2}\right]$

224. dW and dq are not the exact differential, because q and W are
(a) state functions
(b) path functions
(c) intensive properties
(d) extensive properties

225. The expression nRT In (P_1/P_2), is for the .... of an ideal gas.
(a) compressibility
(b) work done under adiabatic condition
(c)work done under isothermal condition
(d) co-efficient of thermal expansion

226. The equation relating E, P, V & T, which is true for all substances under all conditions, is given by $(\partial E/\partial V)_T = T \cdot (\partial P l \partial T)_y - P$. This equation is called the
(a) Maxwell's equation
(b) thermodynamic equation of state
(c) equation of state
(d) Redlich-Kwong equation of state

227. The expression for entropy change given by, the expression,
$\Delta S = -nR \text{ In } (V_2/V_1) + n\, Cv \text{ In } (T_2/T_1)$,
is valid for the
(a) reversible isothermal volume change.
(b) heating of a substance.
(c) cooling of a substance.
(d) simultaneous heating & expansion of an ideal gas.

Answers

220. (b)	221. (c)	222. (a)	223. (b)	224. (b)	225. (c)
226. (b)	227. (d)				

228. The expression for entropy change given by, $\Delta S = -nR \ln (P_2/P_1)$, holds good for the
(a) expansion of a real gas.
(b) reversible isothermal volume change.
(c) heating of an ideal gas.
(d) cooling of a real gas.

229. The expression, $\Delta G = nRT. \ln (P_2/P_1)$, gives the free energy change
(a) with pressure changes at constant temperature.
(b) under reversible isothermal volume change.
(c) during heating of an ideal gas.
(d) during cooling of an ideal gas.

230. Pick out the correct statement.
(a) Compression ratio of an Otto engine is comparatively higher than a diesel engine.
(b) Efficiency of an Otto engine is higher than that of a diesel engine for the same compression ratio.
(c) Otto engine efficiency decreases with rise in compression ratio due to decrease in work produced per quantity of heat.
(d) Diesel engine normally operates at lower compression ratio than an Otto engine for an equal output of work.

231. 4 kg moles of an ideal gas expands in a vacuum spontaneously. The work done is
(a) 4 J (b) ∞
(c) 0 (d) 8 J

232. Which of the following exemplifies an adiabatic process ?
(a) Melting of ice.
(b) Condensation of alcohol vapor.
(c) Sudden bursting of a cycle tube.
(d) Evaporation of water.

233. Match the following thermodynamic equations.
List I
(a) Redlich-Kwong equation.
(b) Virial equation of state.
(c) Gibbs-Helmholtz relation.
(d) Gibbs Duhem equation.
List II

I. $\sum_{i=1}^{n} xi \left(\partial G_i / \partial x_i\right) P, T = 0$

II. $\left[\partial (\Delta G / T) / \partial T\right]_{p, ni} = -\Delta H / T_2$

III. $pV / RT = 1 + B / V + C / V^2 + D / V^3 + ...$

IV. $p = \dfrac{RT}{v - b} - \dfrac{}{\sqrt{T . V (V + b)}}$

234. Match the value of n in polytropic process equation, $P. V^n$ = constant, for ideal gas for various thermodynamic processes.
List I
(a) Constant volume process.
(b) Constant pressure process.
(c) Constant temperature process.
(d) Reversible adiabatic process.
List II
I. n=0 II. n= ∞
111.n=y $=C_p/C_v$ IV. $n=1$

235. Rotary lime kiln is an example of a/an .... system.
(a) closed
(b) open
(c) isolated
(d) non thermodynamic

Answers

228. (b)	229. (a)	230. (b)	231. (c)	232. (c)	233. (IV) (III) (II) (I)
234. (II) (III) (IV) (I)		235. (b)			

236. Melting of ice exemplifies a/an
(a) adiabatic process.
(b) endothermic reaction.
(c) exothermic reaction.
(d) process involving a chemical reaction.

237. Addition of a non-volatile solute to a pure solvent.
(a) increases its freezing point
(b) increases its boiling point
(c) decreases its freezing point
(d) both 'b' & 'c'

238. At higher temperatures, molal heat capacity for most of the gases (at constant pressure) .... with increase in temperature.
(a) varies linearly (b) increases
(c) decreases (d) does not vary

239. During Joule Thomson expansion of gases, the .... remains constant.
(a) enthalpy (b) entropy
(c) temperature *(d)* none of these

240. For a single component, two phase mixture, the number of independent variable properties are
(a) 0 (b) 1
(c) 2 *(d)* 3

241. Ideal gas law is applicable at
(a) low T & low P
(b) high T & high P
(c) low T& high P
(d) high T & low P

242. The Maxwell's relation derived from the differential expression for the Helmholtz free energy (dA) is
(a) $(\partial T/\partial V)_s = -(\partial P/\partial S)_v$
(b) $(\partial S/\partial P)_T = -(\partial V/\partial T)_P$
(c) $(\partial V/\partial S)_P = (\partial T/\partial P)_s$
(d) $(\partial s/\partial V)_T = (\partial p/\partial T)_v$

243. Cryogenics is concerned with the producation & utilisation of low temperature in the range of
(a) 0°K to 123° K
(b) 0° C to - 123° C
(c) 0° F to - 123° F
(d) 0° *R* to - 123°R

244. Which of the following is not an equation of state ?
(a) Bertholet equation
(b) Clausius-Clayperon equation
(c) Beathie-Bridgeman equation
(d) Vanderwall's equation

245. When pressure is applied on the system, ice $\rightleftharpoons$ water, then
(a) equilibrium can not be established.
(b) more ice will be formed.
(c) more water will be formed.
(d) evaporation of water will take place.

246. .... does not change during phase transformation processes like sublimation, melting & vaporisation.
(a) Entropy
(b) Gibbs free energy
(c) Internal energy
(d) all *'a' 'b'* & 'c'

247. Raoult's law is not applicable to the
(a) solutes which desassociate or associate in the particular solution.
(b) concentrated solutions.
(c) both 'a' & *'b'*.
(d) solutions containing non-volatile solute.

Answers

236. (b)	237. (d)	238. (b)	239. (a)	240. (b)	241. (a)
242. (d)	243. (a)	244. (c)	245. (b)	246. (c)	247. (c)

248. All gases except .... shows a cooling effect during throttling process at atmospheric temperature & pressure.
(*a*) oxygen (*b*) nitrogen
(*c*) air (*d*) hydrogen

249. Triple point of water is
(*a*) 0° F (*b*) 492° R
(c) 0° K (*d*) - 273° C

250. The thermodynamic law PV^{γ} = constant, is not applicable, in case of
(a) ideal compression of air.
(*b*) free expansion of an ideal gas.
(*c*) adiabatic expansion of steam in a turbine.
(*d*) adiabatic compression of a perfect gas.

251. Filling of gas from a high pressure cylinder into small bottles is an example of a/an .... process.
(*a*) equilibrium (*b*) adiabatic
(*c*) steady (*d*) unsteady

252. For a thermodynamic system undergoing a process, which of the following pairs best expresses the relationship similar to that expressed by the pressure-volume plot ?
(*a*) Temperature – entropy
(*b*) Temperature – pressure
(*c*) Enthalpy – entropy
(*d*) Enthalpy – pressure

253. A gas having a negative Joule - Thompson coefficient, when throttled, will
(a) become hotter.
(b) become cooler.
(*c*) remain unchanged.
(*d*) liquefy.

254. Superheated vapour behaves like
(*a*) dry steam
(*b*) gas approximately
(*c*) an ordinary vapor
(*d*) none of these.

255. Absolute zero pressure will occur, when the molecular momentum of the system becomes zero. A liquid will cease to exist as liquid at
(*a*) high vacuum (*b*) zero pressure
(*c*) 0°K (*d*) earth's centre.

256. What is the value of entropy at 273° K ?
(*a*) 0 (*b*) *1*
(*c*) ∝ (*d*) none of these

257. Maxwell's thermodynamic relations applies to the
(*a*) chemical systems in equilibrium.
(*b*) mechanical systems in equilibrium.
(*c*) irreversible thermodynamic processes.
(*d*) reversible thermodynamic processes.

258. Thermodynamic cycle involved in the working of a thermal power plant is the .... cycle.
(*a*) Joule (*b*) Carnot
(*c*) Rankine (*d*) Brayton

259. Carnot cycle is also termed as the constant .... cycle in thermodynamics.
(*a*) entropy (*b*) pressure
(*c*) volume (*d*) heat

260. The value of y = Cp/Cv at < 500° C for air & most common gases can be safely assumed to be
(*a*) 0.8 (*b*) 1
(*c*) 1.4 (*d*) 1.8

Answers

248. (d)	249. (b)	250. (b)	251. (d)	252. (a)	253. (a)
254. (b)	255. (b)	256. (a)	257. (a)	258. (c)	259. (d)
260. (c)					

261. In case of compression of one kg of air, the work done will be the least, when the value of polytropic index 'n' is

(a) 1 *(b)* 1.4

(c) 1.5 *(d)* $y = C_p/C_v$,

262. Joule-Thomson coefficient is the ratio of

(a) pressure change to temperature change occuring during adiabatic compression of a gas.

(b) pressure change to temperature change occuring during adiabatic throttling of a gas.

(c) temperature change to pressure change occuring during adiabatic compression of a gas.

(d) temperature change to pressure change occuring during adiabatic throttling of a gas.

263. Claude's liquefaction process employs cooling of gases by

(a) expansion in an engine.

(b) following a constant pressure cycle.

(c) throttling.

(d) none of these.

264. Dryness fraction of wet steam is defined as the ratio of mass of vapor in the mixture to the mass of mixture. .... calorimeter is not used for measuring the dryness fraction of steam.

(a) Bomb *(b)* Separating

(c) Bucket *(d)* Throttling

265. Heat evolved/absorbed during conversion of a substance from one allotropic form to another is termed as the heat of

(a) fusion *(b)* vaporisation

(c) transition *(d)* none of these

266. In which of the following reaction equilibria, the value of equilibrium constant K_p will be more than K_c ?

(a) $2HI \rightleftharpoons H_2 + I_2$

(b) $N_2O_4 \rightleftharpoons 2NO_2$

(c) $2SO_2 + O_2 \rightleftharpoons 2SO_3$

(d) none of these

267. Partial molal free energy of an element A in solution is same as its

(a) chemical potential

(b) activity

(c) fugacity

(d) activity coefficient

268. Pick out the wrong statement.

(a) Enthalpies of all elements in their standard states are assumed to be zero.

(b) Combustion reactions are never endothermic in nature.

(c) Heat of reaction at constant volume is equal to the change in internal energy.

(d) Claussius-clayperon equation is not applicable to melting process.

269. The equation $dU=Tds-PdV$ is applicable to infinitesimal changes occuring in

(a) an open system of constant composition.

(b) a closed system of constant composition.

(c) an open system with changes in composition.

(d) a closed system with changes in composition.

270. The molar excess Gibbs free energy,g^E ,for a binary liquid mixture at T and P is given by, $(g^E/RT) = A.x_1x_2$, where A is a constant. The corresponding equation for In yi, where yi is the activity co-efficient of component 1, is

(a) Ax_2^2 (b) Ax_1

(c) Ax_2 *(d)* Ax_1^2

Answers

261. (a)	262. (d)	263. (a)	264. (a)	265. (c)	266. (b)
267. (a)	268. (d)	269. (d)	270. (a)		

271. Air enters an adiabatic compressor at 300 °K. The exit temperature for a compression ratio of 3, assuming air to be an ideal gas ($y = Cp/Cv = 7/5$) and the process to be reversible, is
    (a) 300 ($3^{2/7}$) (b) 300 ($3^{3/5}$)
    (c) 300 ($3^{3/7}$) *(d)* 300 ($3^{5/7}$)

272. When dilute aqueous solutions of two salts are mixed, the process is associated with
    (a) decrease in temperature.
    (b) increase in temperature,
    (c) no change in temperature.
    (d) change in temperature which is a function of composition,

273. One mole of nitrogen at 8 bar and 600 °K is contained in a piston-cylinder arrangement. It is brought to 1 bar isothermally against a resisting pressure of 1 bar. The work done by the gas is .... Joules.
    (a) 30554 *(b)* 10373
    (c) 4988.4 *(d)* 4364.9

274. A gas can be liquefied bv pressure alone, when its temperature is
    (a) higher than its critical temperature.
    (b) equal to its Boyle temperature.
    (c) equal to its critical temperature.
    (d) lower than its critical temperature.

275. As the pressure approaches zero, the ratio of fugacity to pressure (f/P) for a gas approaches
    (a) zero
    (b) unity
    (c) infinity
    (d) an indeterminate value

276. A gas obeys P(V-*b*) = RT. The work obtained from reversible isothermal expansion of one mole of this gas from an initial molar volume vi to a final molar volume vf is
    (a) RT In (vf/vi) *(b)* $\text{RT In}\dfrac{\text{vf} - \text{b}}{\text{vi}}$
    (c) $\text{RT In}\dfrac{\text{vf}}{\text{vi} - \text{b}}$ *(d)* $\text{RT In}\dfrac{\text{vf} - \text{b}}{\text{vi} - \text{b}}$

277. Match the item in the left column with the appropriate item in right column.
    P. dH 1.TdS-P.dV
    Q. dG 2. T.dS + V.dP
    (G - Gibbs free energy) 3. - PdV - S.dT
    4. C.dP - S.dT
    (a) P – 2, Q – 4 (b) P –2 , Q –1
    (c) P – 2 , Q – 3 (d) P – 3 . Q –4

278. Match the item in the left column with the appropriate item in right column.
    P. For any cyclic process 1. Δ U = 0
    Q. For any adiabatic process 2. Q = 0
    3.W = 0
    4. ΔU, Q and W all are zero
    (a) P – 1, Q – 3 (b) P – 1,Q – 2
    (c) P – 2, Q – 4 (d) p – 3, Q – 1

279. Pick out the correct statement.
    (a) If an insoluble gas is passed through a volatile liquid placed in a perfectly insulated container, the temperture of the liquid will increase.
    (b) A process is irreversible as long as ΔS for the system is greater than zero.
    (c) The mechanical work done by a system is always equal to ∫p.dv.
    (d) The heat of formation of a compound is defined as the heat of reaction leading to the formation of the compound from its reactants.

Answers

271. (a)	272. (b)	273. (b)	274. (b)	275. (b)	276. (d)
277. (a)	278. (b)	279. (d)			

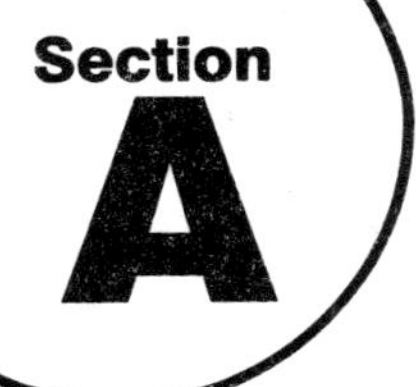

20
Metallurgical Kinetics

1. Rate of a chemical reaction is independent of the concentration of reactants for the .... reaction.
   (*a*) zero order (*b*) third order
   (*c*) consecutive (*d*) none of these

2. Which of the following is not a unit of reaction rate ?
   (*a*) Moles formed / (surface of catalyst) (time)
   (*b*) Mole formed / (volume of reactor) (time)
   (*c*) Mole formed / (volume of catalyst (time)
   (*d*) None of these.

3. If 'n' is the order of reaction, then the unit of rate constant is
   (*a*) $\dfrac{1}{(\text{time})(\text{concentration})^{n-1}}$
   (*b*) $(\text{time})^{-1}$ $(\text{concentration})^{n-1}$
   (*c*) $(\text{time})^{n-1}$ (concentration)
   (*d*) none of these

4. Which of the following is a controlling factor in very fast hetrogeneous reaction?
   (*a*) Heat & mass transfer effects
   (*b*) Pressure
   (*c*) Temperature
   (*d*) Composition of reacant

5. Variables affecting the rate of homogeneous reactions are
   (*a*) pressure & temperature only.
   (*b*) temperature & composition only.
   (*c*) pressure & composition only.
   (*d*) all pressure, temperature & composition.

6. Rate determining step in a reaction consisting of a number of steps in series is the .... step.
   (*a*) fastest
   (*b*) slowest
   (*c*) intermediate
   (*d*) data insufficient, can't be prediacted

7. Velocity of a chemical reaction
   (*a*) decreases with the increase in temperature.
   (*b*) increases with the increase of pressure of reactants for all reactions.
   (*c*) decreases with the increase of reactant concentration.
   (*d*) none of these.

8. Sum of the powers of the concentration terms in the rate equation is called the .... of the reaction.
   (*a*) order (*b*) overall order
   (*c*) molecularity (*d*) none of these.

Answers

1. (a)	2. (d)	3. (a)	4. (a)	5. (d)	6. (b)
7. (a)	8. (b)				

9. Molecularity of a reaction
   (*a*) is always equal to the overall order of a reaction.
   (*b*) may not be equal to the order of reaction.
   (*c*) can't have a fractional value.
   (*d*) both (*b*) & (*c*).

10. Concentration of the limiting reactant (with initial concentration of 'a' moles/litre) after time t is $(a - x)$. Then 't' for a first order reaction is given by
    (*a*) $k \cdot t = \text{In} \dfrac{a}{a - x}$
    (*b*) $k \cdot t = \text{In} \dfrac{x}{a\,(a - x)}$
    (*c*) $k \cdot t = \text{In} \dfrac{(a - x)}{a}$
    (*d*) $k \cdot t = \text{In} \dfrac{a\,(a - x)}{x}$

11. Half life period of a chemical reaction is
    (*a*) the time required to reduce the concentration of the reacting substance to half its initial value.
    (*b*) half of the space time of a reaction.
    (*c*) half of the residence time of a reaction.
    (*d*) none of these.

12. Half life period for a first order reaction is .... the initial concentration of the reactant.
    (a) directly proportional to
    (*b*) inversely proportional to
    (*c*) independent of
    (*d*) none of these

13. In a first order reaction, the time required to reduce the concentration of reactant from1 mole / litre to 0.5 mole/litre will be .... that required to reduce it from 10 moles/litre to 5 moles/litre in the same volume.
    (*a*) more than
    (*b*) less than
    (*c*) same as
    (*d*) data insufficient; can't be predicted.

14. Specific rate constant for a second order reaction
    (*a*) is independent of the temperature.
    (*b*) varies with the temperature.
    (*c*) depends on the nature of the reactants.
    (*d*) both (*b*) & (*c*).

15. The reaction in which the rate equation corresponds to a stoichiometric equation, is called a/an .... reaction.
    (*a*) elementary (*b*) non-elementary
    (*c*) parallel (*d*) autokinetic

16. Equilibrium of a chemical reaction as viewed by kinetics is a .... state.
    (*a*) dynamic steady
    (*b*) static steady
    (*c*) dynamic unsteady
    (*d*) none of these

17. For a zero order reaction, concentration of product increases with the
    (*a*) increase of reaction time.
    (*b*) increase in initial concentration.
    (*c*) increase in total pressure.
    (*d*) decrease in total pressure.

18. Arrhenious equation shows the variation of the .... with temperature.
    (*a*) reaction rate
    (*b*) rate constant
    (*c*) energy of activation
    (*d*) frequency factor

Answers

9. (d)	10. (a)	11. (a)	12. (c)	13. (c)	14. (d)
15. (a)	16. (a)	17. (a)	18. (b)		

19. Rate constant 'k' and absolute temperature T are related by collision theory (for bimolecular reaction) as
    (a) $k \propto T^{1.5}$ (b) $k \propto \exp(E/RT)$
    (c) $k \propto T^{1/2}$ (d) $k \propto T$

20. Transition state theory relates the above quantities as
    (a) $k \propto \exp(-E/RT)$ (b) $k \propto T.e(-E/RT)$
    (c) $k \propto \sqrt{T}$ (d) $k \propto T^{1.5}$.

21. Reactions with high activation energy are
    (a) very temperature sensitive.
    (b) temperature insensitive.
    (c) always irreversible.
    (d) always reversible.

22. In autocatalytic reactions,
    (a) one of the reactants acts as a catalyst.
    (b) one of the products acts as a catalyst.
    (c) catalysts have very high selectivity.
    (d) no catalyst is used.

23. With increase in temperature, the equilibrium conversion of a reversible exothermic reaction
    (a) decreases.
    (b) increases.
    (c) remains unaffected.
    (d) decreases linearly.

24. With decrease in temperature, the equilibrium conversion of a reversible endothermic reaction
    (a) decreases
    (b) increases
    (c) remains unaffected
    (d) increases linearly with temperature

25. The equilibrium constant of a chemical reaction .... in the presence of a catalyst.
    (a) increases
    (b) decreases
    (c) remains unaffected
    (d) can either increase or decrease, depends on the type of catalyst.

26. Conversion increases with increase in the temperature in case of a /an .... reaction.
    (a) autocatalytic
    (b) irreversible
    (c) reversible endothermic
    (d) reversible exothermic

27. The heat of reaction
    (a) depends on the reaction.
    (b) depends on the mechanism of reaction only.
    (c) depends on both the pressure & mechanism of reaction.
    (d) is independent of the mechanism of reaction

28. According to Arrhenius equation of temperature dependency of rate constant for an elementary reaction
    (a) $k \propto \sqrt{T}$ (b) $k \propto e^{-E/RT}$
    (c) $k \propto Te^{-E/RT}$ (d) $k \propto e^{E/RT}$

29. With increase in temperature, the rate constant obeying Arrhenius equation
    (a) increases
    (b) decreases
    (c) decreases exponentially
    (d) can either increase or decrease, depends on the frequency factor.

30. Fluidised bed reactor is characterised by
    (a) uniformity of temperature.
    (b) comparatively smaller equipment.
    (c) very small pressure drop.
    (d) absence of continuous catalyst regeneration facility.

Answers

19. (b)	20. (b)	21. (a)	22. (b)	23. (a)	24. (a)
25. (c)	26. (c)	27. (d)	28. (b)	29. (a)	30. (a)

31. A second order reaction of the form $A+B \rightarrow C$ is called pseudo-first order reaction, when
(*a*) $C_{A0} = C_{Bo}$ (*b*) $C_{Ao} > C_{Bo}$
(*c*) $C_{Bo} > C_{AO}$ (*d*) $C_{Bo} >> C_{Ao}$

32. The concentration of A in a first order reaction, $A \rightarrow B$, decreases
(*a*) linearly with time.
(*b*) exponentially with time.
(*c*) very abruptly toward the end of the reaction.
(*d*) logarithmically with time.

33. Pick out the wrong statement.
(*a*) A catalyst does not alter the final position of equilibrium in a reversible reaction.
(*b*) A catalyst initiates a reaction,
(*c*) A catalyst is specific in action.
(*d*) A catalyst remains unchanged in chemical composition at the end the reaction.

34. Catalyst is a substance which .... chemical reaction.
(*a*) increases the speed of a
(*b*) decreases the speed of a
(*c*) can either increase or decrease the speed of a
(*d*) alters the value of equilibrium constant in a reversible

35. Catalyst carriers
(*a*) have very high selectivity.
(*b*) increase the activity of a catalyst.
(*c*) provide large surface area with a small amount of active material.
(*d*) inhibit catalyst poisoning.

36. A catalyst promoter
(*a*) improves the activity of a catalyst.
(*b*) acts as a catalyst support.
(*c*) itself has very high activity.
(*d*) all (*a*), (*b*) & (*c*).

37. A catalyst inhibitor
(*a*) lessens its selectivity.
(*b*) may be useful for suppressing undesirable side reactions.
(*c*) is added in small quantity during the catalyst manufacture itself.
(*d*) all (*a*), (*b*) & (*c*).

38. A reaction $A \rightarrow B$ is conducted in a constant pressure vessel. Starting with pure A, the volume of the reaction mixture increases 3 times in 6 minutes. The fractional conversion is
(*a*) 0.33 (*b*) 0.5
(*c*) 1 (*d*) 0.66

39. BET apparatus is used to determine the
(*a*) specific surface of a porous catalyst.
(*b*) pore size distribution.
(*c*) pore diameter.
(*d*) porosity of the catalyst bed.

40. If pore diffusion is the controlling step in a solid catalysed reaction, the catalyst
(*a*) porosity is very important.
(*b*) porosity is of less importance.
(*c*) internal surface area is utilised efficiently
(*d*) none of these.

41. BET apparatus
(*a*) measures the catalyst surface area directly.
(*b*) operates at very high pressure,
(*c*) is made entirely of stainless steel.
(*d*) none of these.

42. Helium-mercury method can be used to determine the .... of the catalyst particle.
(*a*) pore volume (*b*) solid density
(*c*) porosity (*d*) all (*a*), (*b*) & (*c*)

Answers

31. (d)	32. (b)	33. (b)	34. (c)	35. (c)	36. (a)
37. (d)	38. (c)	39. (a)	40. (b)	41. (d)	42. (d)

43. If the time required to complete a definite fraction of reaction varies inversely as the concentration of the reactants; then the order of reaction is
   (a) 0 (b) 1
   (c) 2 (d) 3

44. If the time required for half change is inversely proportional to the square of initial concentration and the velocity constant depends on the units in which the concentration term is expressed; then the order of reaction is
   (a) 1 (b) 2
   (c) 3 (d) 0

45. A catalyst loses its activity due to
   (a) loss in surface area of the active component.
   (b) agglomeration of metal particles caused by thermal sintering of the solid surface.
   (c) covering of the catalytic active sites by a foreign substance.
   (d) all (a), (b) & (c).

46. Higher free energy of activation of a chemical reaction (at a given temperature) implies the
   (a) slower rate of reaction.
   (b) higher rate of reaction.
   (c) higher equilibrium conversion.
   (d) both (b) & (c).

47. A reaction which is catalysed by a base is catalysed by all substances, which have a tendency to
   (a) lose a proton (b) gain a proton
   (c) gain an electron (d) none of these

48. If, n = overall order of a chemical reaction, a = initial concentration of reactant, t = time required to complete a definite fraction of the reaction. Then, pick out the correct relationship.
   (a) $t \propto \frac{1}{a^n}$ (b) $t \propto \frac{1}{a^{n-1}}$
   (c) $t \propto \frac{1}{a^{n+1}}$ (d) $t \propto a^n$

49. The rate of reaction does not decrease appreciably as the reaction proceeds, in case of an .... reaction.
   (a) autocatalytic (b) exothermic
   (c) endothermic (d) autothermal

50. The rate of homogeneous reaction is a function of
   (a) temperature & pressure only.
   (b) temperature & composition only.
   (c) pressure & composition only.
   (d) all temperature, pressure & composition.

51. A catalyst is said to be a negative catalyst, if it
   (a) retards the rate of reaction.
   (b) reduces the value of equilibrium constant.
   (c) does not initiate the reaction.
   (d) all (a), (b) & (c).

52. When the reaction occurs in the diffusion controlled region, the apparent activation energy as measured is .... the true value.
   (a) twice (b) half
   (c) equal to (d) none of these.

53. When the reaction is dominated by intraparticle diffusion, the apparent order of reaction (n_D) as measured is related to the true order (n) as
   (a) $n_D = \frac{n+1}{2}$ (b) $n_D = \frac{n}{2}$
   (c) $n_D = n+1$ (d) $n_D = n-1$

Answers

43. (c)	44. (c)	45. (d)	46. (a)	47. (b)	48. (b)
49. (d)	50. (d)	51. (a)	52. (b)	53. (a)	

54. Half life period of a first order irreversible reaction, A → B, is
(a) $k/2$ (b) In $k/2$
(c) In $2/k$ (d) In $0.5/k$.

55. Half life period of a chemical reaction is proportional to CA_0, if the reaction is of the .... order.
(a) first (b) zero
(c) second (d) third

56. A consecutive reaction, $A \xrightarrow{k_1} B \xrightarrow{k_2} C$ is characterised by the
(a) maxima in the concentration of A.
(b) maxima in the concentration of B.
(c) maxima in the concentration of C.
(d) high exothermicity.

57. In a consecutive reaction system, $A \xrightarrow{E_1} B \xrightarrow{E_2} C$; when E_1 is much greater than E_2, the yield of *B* increases with the
(a) increase of temperature.
(b) decrease of temperature.
(c) increase in initial concentration of A.
(d) decrease in initial concentration of *B*.

58. The knowledge of initial concentration and rate constant is necessary to determine the half life time of a reaction of the .... order.
(a) zero (b) first
(c) second (d) none of these

59. If ΔG (free energy change) for a chemical reaction is very large and negative, then the reaction is
(a) not feasible.
(b) just feasible.
(c) very much feasible.
(d) unpredictable as, ΔG is not a measure of the feasibility of a reaction.

60. An imbalanced chemical equation is against the law of
(a) multiple proprtions
(b) conservation of mass
(c) constant proportion
(d) none of these.

61. Chemical reactions involve the participation of
(a) proton (b) neutron
(c) electron (d) none of these

62. A chemical reaction occurs, when the energy of the reacting molecules is .... the activation energy of the reaction.
(a) less than (b) equal to
(c) more than (d) either (b) or (c)

63. Calcination of limestone represented by $CaC0_3 \rightleftharpoons CaO + CO_2$, proceeds to completion in rotary lime kilns, because
(a) CO_2 escapes.
(b) of high temperature.
(c) CaO is more stable than $CaCO_3$.
(d) CaO and CO2 do not combine.

64. The excess energy of the reactants required to dissociate into products is known as the .... energy.
(a) thermal (b) activation
(c) threshold (d) binding

65. The rate at which a chemical substance reacts is proportional to its
(a) active mass
(b) equivalent weight
(c) molecular weight
(d) none of these.

66. The rate constant of a reaction depends on the
(a) time of reaction.
(b) temperature of the system.
(c) extent of reaction.
(d) initial concentration of the reactants.

Answers

54. (c)	55. (c)	56. (b)	57. (a)	58. (c)	59. (c)
60. (b)	61. (c)	62. (d)	63. (a)	64. (b)	65. (a)
66. (b)					

67. If a_0 is the quantity of reactant initially present, the quantity left after *n* half life periods will be
(a) $(a_0/2)^n$ *(b)* $(1/2)^n.a_0$
(c) $(a_0/2)^{1/n}$ *(d)* $(a_0)^{1/2n}$.

68. Liquid *B* decomposes by irreversible first order kinetics and the half life period of this reaction is 12 minutes. The time required for 75% conversion of B is .... minutes.
(a) 24 *(b)* 16
(c) 8 *(d)* 4

69. For the chemical reaction, $A \rightarrow B$, it is found that the rate of reaction doubles as the concentration of A is doubled. If rate is proportional to C_A^n, what is '*n*' for this reaction ?
(a) 1 *(b)* 2
(c) 3 *(d)* 0.

70. A reaction in which one of the products of reaction acts as a catalyst is called a/an .... reaction.
(a) catalytic *(b)* autocatalytic
(c) photochemical *(d)* none of these

71. The action of a catalyst follows its ability to change the
(a) activation energy
(b) equilibrium constant
(c) heat of reaction
(d) none of these

72. In a reversible chemical reaction, a catalyst increases the rate of forward reaction .... that of the backward reaction.
(a) only and not.
(b) to a greater extent than
(a) and decreases
(d) equally to

73. A catalyst
(a) increases the equilibrium concentration of the product.
(b) changes the equilibrium constant of the reaction.
(c) shortens the time to reach equilibrium.
(d) none of these.

74. If a solid-gas non catalytic reaction occurs at very high temperature, the rate controlling step is the .... diffusion.
(a) film *(b)* ash layer
(c) pore *(d)* none of these

75. The rate constant of a reaction increases by increasing the
(a) temperature
(b) pressure
(c) concentration of reactants
(d) none of these

76. When a catalyst increases the rate of reaction, the value of rate constant
(a) increases *(b)* decreases
(c) remains same *(d)* becomes infinite

77. As the chemical reaction proceeds, the rate of reaction
(a) increases
(b) decreases
(c) remains same
(d) may increase or decrease, depending on the type of reaction

78. The rate constant of a chemical reaction increases by 100 times, when the temperature is increased from 400° K to 500° K. Assuming transition state theory is valid, the value of E/R is .... ^{0}K.
(a) 8764 *(b)* 8987
(c) 9210 *(d)* 8621

Answers

67. (b)	68. (a)	69. (d)	70. (b)	71. (a)	72. (d)
73. (c)	74. (a)	75. (a)	76. (a)	77. (b)	78. (a)

79. Consider the 'n' th order irreversible liquid phase reaction, $A \to B$. Which one of the following plots, involving half life of the reaction ($t_{1/2}$) and the intial reactant concentration (CA_0) gives a straight line plot ?
   (*a*) In C_{A0} *Vs.* In $t_{1/2}$ (*b*) CAo Vs. $t_{1/2}$
   (*c*) In Ca_0 Vs $t t_{1/2}$ (*d*) CA_0 Vs. In $t_{1/2}$

80. Molecularity of an elementary reaction, $P + Q \to R + S$, is
   (*a*) 2 (*b*) 1
   (*c*) 4 (*d*) 3

81. In a reversible reaction, a catalyst increases the rate of forward reaction
   (*a*) only.
   (*b*) to a greater extent than that of the backward reaction.
   (*c*) and decreases that of the backward reaction.
   (*d*) and the backward reaction equally.

82. When a catalyst increases the rate of forward reaction, the value of rate constant
   (*a*) increases (*b*) decreases
   (*c*) remain same (*d*) becomes infinite

83. For reaction, $P + 2Q \to 3R$, molar rate of consumption of P is .... that of Q.
   (*a*) half of (*b*) same as
   (*c*) double of (*d*) 2/3rd of

84. In a chemical reaction, the .... are conserved.
   (*a*) ions (*b*) masses
   (*c*) atoms (*d*) both (*b*) & (*c*)

85. Pick out the correct statement.
   (*a*) A chemical reaction accompanied by absorption of heat is called an exothermic reaction.
   b) A chemical reaction accompanied by evolution of heat is called an endothermic reaction.
   (*c*) The rate constant for a first order reaction does not change on changing the concentration units.
   (*d*) Chemical equilibrium state is dynamic in nature.

86. For the chemical reaction $X \to Y$, it is observed that, on doubling the concentration of 'X', the reaction rate quadruples. If the reaction rate is proportional to C_x^n, then what is the value of 'n'
   (*a*) 4 (*b*) 1/4
   (*c*) 16 (*d*) 2

87. For the non-catalytic reaction of particles with surrounding fluid, the time needed to achieve the same fractional conversion for particles of different but unchanging sizes is proportional to the square of particle diameter, when the .... is the controlling resistance.
   (*a*) film diffusion
   (*b*) diffusion through ash layer
   (*c*) chemical reaction
   (*d*) either (*a*), (*b*) or (*c*).

88. A balanced chemical reaction equation conforms to the law of
   (*a*) conservation of mass
   (*b*) Avogadro's hypothesis
   (*c*) gaseous volumes
   (*d*) none of these

89. For a first order chemical reaction, the rate constant
   (*a*) changes on changing the concentration units.
   (*b*) is not a function of the unit of time.
   (*c*) has unit of time^{-1}.
   (*d*) none of these.

Answers

79. (c)	80. (c)	81. (d)	82. (a)	83. (a)	84. (d)
85. (c)	86. (a)	87. (b)	88. (a)	89. (c)	

90. Pick out the correct statement.
   (a) Reactions with high activation energies are very temperature sensitive.
   (b) Chemical equilibrium is a static state.
   (c) A photochemical reaction is catalysed by light.
   (d) A chemical reaction occurs when the energy of the reacting molecule is less than the activation energy of the reaction.

91. The rate of forward reaction, at chemical equilibrium is .... the rate of backward reaction.
   (a) more than *(b)* less than
   (c) equal to *(d)* equal to or less than.

92. For the reaction $2X + Y \xrightarrow{K} Z$, the rate of formation of Z is 0.2 gm mole/litre-hr. What is the rate of disappearance of X in gm mole/litre-hr ?
   (a) 0.4 *(b)* 0.1
   (c) 0.2 *(d)* 0.3

93. In a chemical reaction $X + Y + Z \xrightarrow{K} P$, it is observed that the
   (i) rate of formation of *'P'* is doubled on doubling the concentration of *X'*.
   (ii) rate of formation of *'P'* is quadrupled on doubling the concentration of *'Y*.
   (iii) doubling the concentration of *'Z'* does not affect the rate of formation of *'P'*. What is the order of the above chemical reaction ?
   (a) Zeroth order *(b)* First order
   (c) Second order *(d)* Third order

94. In a chemical reaction, $A \xrightarrow{K} B$, it is observed that, the rate of reaction increases by a factor of .... the concentration of the reactant.
   (i) 4 on doubling, (ii) 9 on trebling.
   Then the rate of the reaction is proportional to
   (a) C_A *(b)* C_A^2
   (c) C_A^3 *(d)* C_A^4
   where, C_A = concentration of the reactant

95. The irreversible reaction is a special case of reversible reaction, when the
   (a) equilibrium constant is 1.
   (b) conversion of the reactant at equilibrium condition is zero.
   (c) reactant concentration at equilibrium condition is zero.
   (d) all *(a)*, *(b)* & *(c)*.

96. Pick out the correct statement.
   (a) In catalytic reactions, the catalyst reacts with the reactants.
   (b) A catalyst initiates a chemical reaction .
   (c) A catalyst lowers the activation energy of the reacting molecules.
   (d) A catalyst can not be recovered chemically unchanged at the end of the chemical reaction.

97. For the irreversible elementary reactions in parallel *viz* $K \xrightarrow{K_1} Y ; X \xrightarrow{K_2} Z$ the rate or disappearance or is equal to
   (a) $C_A(K_1 + K_2)$ (b) $C_A (K_1 + K_2)/2$
   (c) $C_A . K_1/2$ (d) $C_A . K_2/2$.

98. For the irrreversible elementary first order reaction in parallel *viz* $K \xrightarrow{K_1} Y ; X \xrightarrow{K_2} Z$ a plot of Cy *vs.* Cz will give a straight line having a slope of
   (a) K_2/K_1 (b) K_1/K_2
   (c) $K_1 . K_2$ (d) $K_1 + K_2$

Answers

90. (a)	91. (c)	92. (a)	93. (d)	94. (b)	95. (c)
96. (c)	97. (a)	98. (b)			

99. Which of the following holds good for an elementary reaction, $X + 2Y \xrightarrow{K_1} 3Z$?
    (*a*) The rate of disappearance of 'Y' is equal to the rate of appearance of 'Z'.
    (*b*) The rate of disappearance of 'Y'is equal to the rate of disappearance of 'X'.
    (*c*) Three times the rate of disappearance of 'X' is equal to the rate of appearance of 'Z'.
    (*d*) The rate of disappearance of 'X' is equal to the rate of appearance of 'Z'.

100. Decomposition rate of a liquid 'X' which decomposes as per the reaction $X \xrightarrow{K_1} A$ $X \xrightarrow{K_2} B, X \xrightarrow{K_3} C$ is given by
    (*a*) $K_1 . C_x$ (*b*) $(K_1 + K_2 + K_3) C_x$
    (*c*) $(K_1 + K_2) C_x$ (*d*) $(K_2 + K_3) C_x$

101. Promotor is added to catalyst to improve its
    (*a*) porosity (*b*) sensitivity
    (*c*) surface area (*d*) fusion resistance

102. The unit of frequency factor in Arrhenious equation is
    (a) same as that of rate constant.
    (b) same as that of activation energy.
    (c) dimensionless.
    (d) none of these.

103. With an increase in pressure in gaseous chemical reactions, the fractional conversion .... when the number of moles decreases.
    (*a*) increases
    (*b*) decreases
    (*c*) remains unaffected
    (*d*) unpredictable from the data

104. With increase in initial concentration, the fractional conversion of a first order reaction in a given time
    (*a*) increases
    (*b*) decreases
    (*c*) remains constant
    (*d*) unpredictable.

105. A non-catalytic chemical reaction of the $A \xrightarrow{+Z,K_1} B \xrightarrow{+Z,K_2} C \xrightarrow{+Z,K_3} D$, is called a .... reaction.
    (*a*) parallel (*b*) series
    (*c*) series-parallel (*d*) none of these

106. .... explains the mechanism of catalysis.
    (*a*) Activated complex theory
    (*b*) Collision theory
    (*c*) Thermodynamics
    (*d*) none of these

107. Carrier in a catalyst increases its
    (*a*) surface area (*b*) activity
    (*c*) performance (*d*) fusion resistance

108. Pick out the wrong statement.
    (*a*) Use of different catalyst in a reversible catalytic chemical reaction does not change the equilibrium composition.
    (*b*) Alumina is added as a promoter to iron catalyst in ammonia synthesis reaction.
    (*c*) Activation energy for a reaction is obtained from the intercept of the Arrhenious plot.
    (*d*) Presence of inerts affects the equilibrium conversion of reactants in a chemical reaction.

109. For a reaction of the type, $V \xrightarrow{K_1} X \xrightarrow{K_2} Y \xrightarrow{K_3} Z$, the rate of reaction $(-r_x)$ is given by
    (*a*) $(K_1 + K_2) C_x$ (*b*) $(K_1 + K_2 + K_3) C_x$
    (*c*) $K_1 C_v - K_2 C_x$ (*d*) $(K_1 - K_2) C_x$

Answers

99. (c)	100. (c)	101. (b)	102. (a)	103. (a)	104. (c)
105. (c)	106. (a)	107. (a)	108. (c)	109. (c)	

110. Pick out the correct statement.
   (a) A catalyst speeds up the forward reaction but slows down the reverse reaction.
   (b) Addition of catalyst changes the equilibrium constant.
   (c) Pressure changes do not change the equilibrium concentrations.
   (d) The composition of equilibrium is changed by the catalyst.

111. What is the order of a chemical reaction whose rate is determined by the variation of one concentration term only?
   (a) Zero *(b)* First
   (c) Second *(d)* Third

112. The role of a catalyst in a chemical reaction is to change the
   (a) equilibrium constant
   (b) activation energy
   (c) final products
   (d) heat of reaction.

113. The increase in the rate of reaction with temperature is due to the
   (a) increase in the number of effective collisions.
   (b) decrease in activation energy.
   (c) increase in the average kinetic energy of the reacting molecules.
   (d) none of these.

114. Pick out the wrong statement.
   (a) In a multistep reaction, the slowest step is the rate determining step.
   (b) In general, the rate of a reaction becomes triple for every 10°C rise in temperature.
   (c) Slow chemical reactions have generally high values of activation energy.
   (d) Molecularity of a reaction can not be zero but order of a reaction can be zero.

115. Bulk diffusion in catalyst pore .... with increase in pressure.
   (a) increases
   (b) decreases
   (c) remains unchanged
   (d) increases exponentially

116. .... gas is normally employed in B.E.T. method of finding out the surface area of a catalyst.
   (a) N_2 *(b)* H_2
   (c) CO_2 *(d)* He

117. Pick out the wrong statement.
   (a) A particular chemical reaction is more temperature sensitive at low temperature.
   (b) A very high value of equilibrium constant, K (K » 1) indicates that the reaction is practically irreversible in nature.
   (c) The intercept of the Arrhenious plot is called the activation energy.
   (d) Chemical reactions with high activation energy are very temperature sensitive.

118. There is no correspondence between stoichiometry and the rate equation in case of a/an .... reaction.
   (a) elementary *(b)* multiple
   (c) auto catalytic *(d)* non-elementary

119. Rate of a gaseous phase reaction is given by, $-dp_a/dt = K \cdot P_A^2 C_x^n$ atm /hr. The unit of rate constant is
   (a) $(atm)^{-1}$ *(b)* $(hr)^{-1}$
   (c) $(atm)^{-1} . (hr)^{-1}$ *(d)* atm $(hr)^{-1}$

120. Radioactive decay follows the .... order kinetics.
   (a) first *(b)* second
   (c) third *(d)* zero

Answers

110. (c)	111. (b)	112. (b)	113. (b)	114. (b)	115. (b)
116. (a)	117. (c)	118. (d)	119 (c)	120. (a)	

121. The reactions of high molecularity are rare, because
   (*a*) of very large activation energy of many bodies.
   (*b*) of low probability of many body collisions.
   (*c*) many body collisions are not favoured energetically.
   (*d*) of the requirement of very large concentration for such reactions.

122. The reactions with low activation energy are
   (*a*) always spontaneous
   (*b*) slow
   (*c*) fast
   (*d*) non-spontaneous

123. The half life period of a first order reaction is
   (*a*) always the same irrespective of the reaction.
   (*b*) dependent on initial concentration of the reactants.
   (*c*) proportional to the initial concentration of reactants.
   (*d*) half the specific rate constant.

124. The rate of the reaction, $X \rightarrow Y$, quadruples, when the concentration of X is doubled. The rate expression for the reaction is, $r = K\, C_x^{\,n}$, the value of '*n*' in this case will be
   (*a*) 0 (*b*) 1
   (*c*) 2 (*d*) 3

125. The reaction rate almost gets doubled for 10°C rise in temperature. This is due to the fact that the
   (*a*) increased temperature reduces the activation energy.
   (*b*) fraction of molecules having threshold energy increases.
   (*c*) collision frequency increases.
   (*d*) value of threshold energy decreases.

126. Which of the following does not produce a change in the value of rate constant of a reaction ?
   (*a*) Pressure
   (*b*) Temperature
   (*c*) Concentration and catalyst
   (*d*) None of these

127. Pick out the wrong statement.
   (*a*) Visible radiation provides the necessary activation energy in photochemical reactions.
   (*b*) The order & molecularity of a complex reaction may not be the same.
   (*c*) For a second order reaction, the slope of the graph/plot between rate and $(\text{concentration})^2$ is equal to rate constant (k).
   (*d*) Molecularity of the reaction is always a whole number greater than zero.

128. In the reversible reaction of the type, $A+B \rightleftharpoons AB$, in general
   (*a*) both the forward & backward reactions will be exothermic.
   (*b*) neither of the reactions will be endothermic.
   (*c*) the combination reaction will be exothermic, while the dissociation reaction will be endothermic.
   (*d*) the combination reaction will be endothermic, while the dissociation reaction will be exothermic.

Answers

121. (b)	122. (c)	123. (b)	124. (c)	125. (b)	126. (c)
127. (b)	128. (c)				

129. In a reversible chemical reaction having two reactions in equilibrium, if the concentration of the reactants are doubled, then the equilibrium constant will
(a) remain same
(b) be halved
(c) also be doubled
(d) become one fourth.

130. In which of the following gaseous phase reactions, the product yield can not be increased by the application of high pressure?
(a) $N_2 + O_2 \rightleftharpoons 2\,NO$
(b) $PC1_3 + Cl_2 \rightleftharpoons PCI_5$
(c) $N_2 + 3H_2 \rightleftharpoons 2NH_3$
(d) $2SO_2 + O_2 \rightleftharpoons 2SO_3$

131. In which of the following reactions, the equilibrium will shift to the right, if the total pressure is increased ?
(a) $H_2 + Cl_2 \rightleftharpoons 2\,HCl$
(b) $N_2 + 0_2 \rightleftharpoons 2\,NO$
(c) $N_2 0_4 \rightleftharpoons 2\,NO_2$
(d) $H_2 + I_2 \rightleftharpoons 2HI$

132. In which of the following gaseous phase reactions, the equilibrium of the reaction remains unaffected by pressure changes ?
(a) $2\,O_3 \rightleftharpoons 3\,0_2$
(b) $N_2 + O_2 \rightleftharpoons 2NO$
(c) $2\,NO_2 \rightleftharpoons N_2O_4$
(d) $2\,SO_2 + O_2 \rightleftharpoons 2\,SO_3$

133. Reverse reaction in a chemical equilibrium is favoured by the
(a) removal of one of the products regularly.
(b) increase in the concentration of one of the products.
(c) increase in the concentration of one of the reactants.
(d) none of these.

134. Considering the endothermic dissociation of $CaCO_3$ in a closed vessel ($CaCO_3 \rightleftharpoons CaO + CO_2$), the pressure of CO2 increases, if
(a) a catalyst is added.
(b) the temperature is increased.
(c) an inert gas is pumped keeping the temperature constant.
(d) none of these.

135. In an exothermic chemical reaction, the reactants compared to the products have
(a) higher temperature
(b) more energy
(c) less energy
(d) same energy

136. In a zero order reaction, reactants concentration does not change with time and the
(a) time for half change is half the time taken for completion of the reaction.
(b) time for half change is independent of the initial concentration.
(c) time for completion of the reaction is independent of the initial concentration.
(d) reaction rate is trebled, when the initial concentration is trebled.

137. If the rate of a chemical reaction becomes slower at a given temperature, then the
(a) initial concentration of the reactants remains constant.
(b) free energy of activation is lower.
(c) entropy changes.
(d) free energy of activation is higher.

Answers

129. (a)	130. (a)	131. (b)	132. (b)	133. (b)	134. (b)
135. (b)	136. (a)	137. (d)			

138. Rate constant for a first order reaction does not depend upon the reaction time, extent of reaction & the initial concentration of the reactants, but it is a function of reaction temperature. In a chemical reaction, the time required to reduce the concentration of reactant from 100 gm moles/litre to 50 gm moles/litre is same as that required to reduce it from 2 gm moles/litre to 1 gm mole/litre in the same volume. Then the order of this reaction is
   (*a*) 0 (*b*) 1
   (*c*) 2 (*d*) 3

139. Arrhenious equation represents graphically the variation between the .... and temperature.
   (*a*) rate of reaction (*b*) frequency factor
   (*c*) rate constant (*d*) activation energy

140. Pick out the wrong statement.
   (*a*) For a first order consecutive reaction, a tubular flow reactor as compared to a stirred tank reactor provides higher overall selectivity.
   (*b*) For an ideal mixed reactor at steady state, the exit stream has the same composition as fluid within the reactor and the space time is equivalent to holding time for the constant density system.
   (*c*) Plug flow reactor (PFR) is always smaller than mixed reactor for all positive reaction orders for a.particular duty.
   (*d*) Reaction rate does not decrease appreciably as the reaction proceeds, in case of an autocatalytic reaction.

141. For a heterogeneous catalytic reaction.
   (*a*) free energy of activation is lowered in the presence of catalyst which remains unchanged at the end of reaction.
   (*b*) a relatively small amount of catalyst can cause the conversion of large amount of reactants, which does not mean that catalyst concentration is important,
   (*c*) the catalyst does not form an intermediate complex with the reactant.
   (*d*) the surface of the catalyst does not play an important role during reaction.

142. In Langmuir treatment of adsorption.
   (*a*) whole surface of the catalyst does not have the same activity for absorption and there is attraction between the adsorbed molecules.
   (*b*) whole surface of the catalyst is essentially uniform and the adsorbed molecule has no effect on the rate of adsorption per site.
   (*c*) all the adsorption does not take place by the same mechanism.
   (*d*) extent of adsorption is more than one complete monomolecular layer on the surface.

143. The rate of reaction of a/an .... reaction is not affected by temperature rise.
   (*a*) auto catalytic (*b*) photochemical
   (*c*) consecutive (*d*) zero order

Answers

138. (b)	139. (c)	140. (a)	141. (a)	142. (b)	143. (b)

144. Match the time Vs. concentration plot of different components X, Y and Z for various reactions.

List I

(*a*) Elementary reversible parallel reaction of the type, $X \rightleftharpoons Y$

$X \rightleftharpoons Z$

(*b*) Consecutive unimolar first order reaction of the type $X \xrightarrow{K1} Y \xrightarrow{K1} Z$

(*c*) Elementary reversible reaction of the type $X \rightleftharpoons Y \rightleftharpoons Z$

(*d*) Consecutive unimolecular first order reaction of the type $X \xrightarrow{K_2} Y \xrightarrow{K_2} Z$

List II

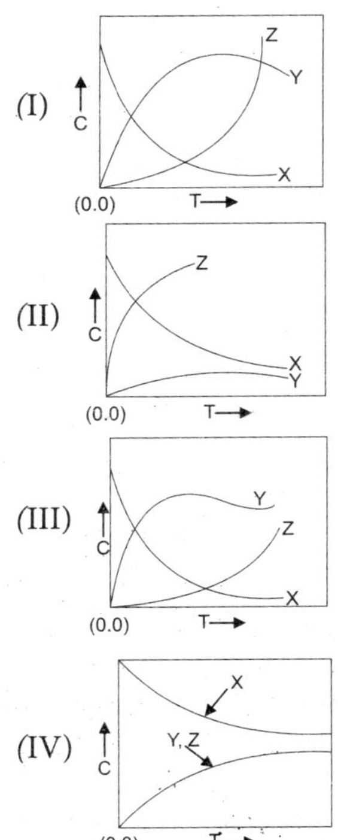

145. Match the relationship of rate constant 'k' as per various parameter/theories.

List I	*List II*
(*a*) Transition theory	I. $k = 0.693/t_{1/2}$
(*b*) Collision theory	II. $k \alpha T^{1/2}$
(*c*) Arrhenious equation	III. $k \alpha T^{e-E/RT}$
(*d*) Relation for first order reaction	IV. $k \alpha e^{-E/RT}$

146. Organic catalysts differ from the inorganic catalyst in the sense, that the former is

(*a*) active at cryogenic temperatures only.
(*b*) prohivitively costly.
(*c*) proteinous in nature.
(*d*) active at very high temperature only.

147. Photochemical reactions occur in the presence of

(*a*) sunlight
(*b*) darkness
(*c*) solid catalysts
(*d*) monochromatic radiation only

148. The unit of frequency factor in Arrhenious equation

(*a*) is the same as that of the rate constant.
(*b*) depends on the order of the reaction.
(*c*) depends on temperature & pressure of the reaction.
(*d*) is cycles per unit time.

149. The reaction rate constants at two different temperatures T_1 and T_2 are related by

(*a*) $\text{In}\,(k_2 / k_1) = \frac{E}{R}\left(\frac{1}{T_2} - \frac{1}{T_1}\right)$

(*b*) $\text{In}\,(k_2 / k_1) = \frac{E}{R}\left(\frac{1}{T_1} - \frac{1}{T_2}\right)$

(*c*) $\exp\,(k_2 / k_1) = \frac{E}{R}\left(\frac{1}{T_1} - \frac{1}{T_2}\right)$

(*d*) $\exp\,(k_2 / k_1) = \frac{E}{R}\left(\frac{1}{T_2} - \frac{1}{T_1}\right)$

Answers

144. (IV) (I) (III) (II) 145. (III) (II) (IV) (I) 146. (c) 147. (a)
148. (a) 149. (b)

150. According to the law of mass action, the rate of reaction is directly proportional to the
(*a*) equilibrium constant.
(*b*) volume of reaction vessel.
(*c*) nature of the reactants.
(*d*) molar concentration of the reactants.

151. What is the unit of rate constant in a chemical reaction in which 10% of the reactant decomposes in one hour, 20% in two hours, 30% in three hours and so on ?
(*a*) Litre/mole. second
(*b*) Moles/litre. second
(*c*) Litre/mole
(*d*) Litre/second

152. The order of a chemical reaction is
(*a*) an experimentally determined quantity.
(*b*) always equal to the total stoichiometrical number of reactants.
(*c*) never fractional
(*d*) none of these.

153. For a .... order reaction, the units of rate constant & rate of reation are the same.
(*a*) zero (*b*) first
(*c*) second (*d*) fractional

154. The minimum energy required to allow a chemical reaction to proceed is termed as the 'threshold energy'. Chemical reactions with low activation energy are
(*a*) always irreversible.
(*b*) insensitive to temperature changes.
(*c*) mostly irreversible.
(*d*) highly temperature sensitive.

155. In case of a/an .... reaction, conversion increases with rise in temperature.
(*a*) reversible exothermic
(*b*) irreversible exothermic
(*c*) irreversible endothermic
(*d*) reversible endothermic

156. What is the order of a chemical reaction, in which doubling the initial concentration of the reactants doubles the half life time of the reaction ?
(*a*) 0 (*b*) 1
(*c*) 2 (*d*) 3

157. Limiting reactant in a chemical reaction decides the
(*a*) rate constant
(*b*) conversion
(*c*) reaction speed
(*d*) equilibrium constant

158. Overall rate of reaction in a heterogeneous catalytic reaction depends upon the mass & energy transfer from the fluid to solid surface and its rate of reaction is usually .... the concentration of catalyst, if it does not entail a chain mechanism.
(*a*) proportional to
(*b*) independent of
(*c*) inversely proportional to
(*d*) proportional to the square of

159. To maximise the formation of *R* in the simultaneous reaction,
$A + B \rightarrow R, r_R = C_A^{0.5} \cdot C_B^2$
$A + B \rightarrow S, r_s = 1.5\ C_A \cdot C_B$
We should have
(*a*) low C_A, low C_B
(*b*) low C_A, high C_B
(*c*) high C_A, low C_B
(*d*) high C_A, high C_B

Answers

150. (d)	151. (a)	152. (a)	153. (a)	154. (b)	155. (d)
156. (a)	157. (b)	158. (a)	159. (b)		

160. The rate expression for a heterogenous catalytic reaction is given by, $-r_A = K.\ K_A\ P_A / (1+K_A.P_A+K_R.\ P_R)$, where K is surface reaction rate constant and K_A & K_R are absorption equilibrium constants of A & R respectively. If $K_R\ P_R >> (1 + K_A\ P_A)$, the apparent activation energy E_A is equal to (given E is the activation energy for the reaction and ΔH_R & ΔH_A are the activation energies of adsorption of R & A
   (a) E
   (b) $E + \Delta H_A$
   (c) $E + \Delta H_A - \Delta H_R$
   (d) $\Delta H_A + \Delta H_r$

161. When an exothermic reversible reaction is conducted adiabatically, the rate of reaction
   (a) continuously increases.
   (b) continuously decreases.
   (c) passes through a maximum.
   (d) passes through a minimum.

162. The decomposition of A into B is represented by the exothermic reaction, $A \rightarrow 2B$. To achieve maximum decomposition, it is desirable to carry out the reaction at
   (a) high P & high T.
   (b) low P & high T.
   (c) low P & low T.
   (d) high P & low T.

163. For a first order isothermal chemical reaction in a porous catalyst, the effectiveness factor is 0.3. The effectiveness factor will increase, if the catalyst size is
   (a) reduced or the catalyst diffusivity is reduced.
   (b) reduced or the catalyst diffusivity is increased.
   (c) increased or the catalyst diffusivity is reduced.
   (d) increased or the catalyst diffusivity is increased.

164. The rate of the heterogenous catalytic reaction $A(g) + B(g) \rightarrow C(g)$ is given by $-r_A = r.K_A.\ p_A.\ p_B/(1 + K_A.p_A + K_C.\ p_C)$, where K_A and K_C are the adsorption equilibrium constants. The rate controlling step for this reaction is
   (a) adsorption of A.
   (b) surface reaction between adsorbed A and absorbed B.
   (c) surface reaction between absorbed A & B in the gas phase.
   (d) surface reaction between A in the gas phase and absorbed B.

165. Second order consecutive irreversible reactions $A \xrightarrow{k_1} B \xrightarrow{k_2} C$ were carried out in a constant volume isothermal batch reactor with different initial feed compositions. Reactor temperature was same in all the cases. In experiments where the ratio of concentration of B to that of A in the initial feed was less than 0.5, the concentration of B increased first, reached a maximum and then declined with time. However, for all experiments where this concentration ratio was 0.5 or above, concentration of B decreased monotonically with time right from the beginning. What is the ratio of the two rate constants $(k_1 k_2)$?
   (a) 1/4 (b) 1/2
   (c) 2 (d) 4

166. For the reaction, $A + B \rightarrow 2B + C$,
   (a) $r_A = r_B$ (b) $r_A = -r_B$
   (c) $r_A = 2r_B$ (d) $r_A = r_B/2$

Answers

160. (c) 161. (c) 162. (c) 163. (b) 164. (c) 165. (a)
166. (b)

167. An endothermic second order reaction is carried out in an adiabatic plug flow reactor. The rate of heat generation is
(a) maximum at the inlet of the reactor.
(b) maximum at the exit of the reactor,
(c) maximum at the centre of the reactor.
(d) constant throughout the reactor.

168. For the reversible reaction $A \rightleftharpoons 2B$, if the equilibrium constant K is 0.05 mole/litre, starting from initially 2 moles of A and zero moles of B, how many moles will be formed at equilibrium ?
(a) 0.253 (b) 0.338
(c) 0.152 (d) 0.637

169. The experimentally determined overall order of the reaction, $A + B \rightarrow C + D$ is two, then the
(a) reaction is elementary with a molecularity of 2.
(b) molecularity of the reaction is 2, but the reaction may not be elementary.
(c) reaction may be elementary with molecularity of 2.
(d) reaction is elementary but the molecularity may not be 2.

170. The reaction $A \rightarrow B$ is conducted in an isothermal batch reactor. If the conversion of A increases linearly with holding time, then the order of the reaction is
(a) 0 (b) 1
(c) 1.5 (d) 2

171. In solid catalysed reactions, the diffusional effects are more likely to affect the overall rate of reaction for the
(a) fast reactions in catalyst of small pore diameter.
(b) fast reactions in catalyst of large pore diameter.
(c) slow reactions in catalyst of small pore diameter.
(d) slow reactions in catalyst of large pore diameter.

172. The first order series reaction $A \xrightarrow{k_1} B \xrightarrow{k_2} C$ is conducted in a batch reactor. The initial concentrations of A, B & C (C_{A_0}, C_{B_0}, C_{C_0} respectively) are all non-zero. The variation of C_B with reaction time will not show a maximum, if
(a) $k_2 C_{B0} > k_1 C_{Ao}$
(b) $k_1 C_{Ao} > k_2 . C_{Bo}$
(c) $C_{Bo} > C_{A0}$
(d) $C_{Ao} > C_{Bo}$

173. The conversion for a second order, irreversible reaction (constant volume), $A \xrightarrow{k_2} B$, in batch mode is given by
(a) $\dfrac{1}{1 + k_2 . C_{A0} . t}$ (b) $\dfrac{k_2 . C_{A0} . t}{1 + k_2 . C_{A0} . t}$
(c) $\dfrac{(k_2 . C_{A0} . t)^2}{(1 + k_2 . C_{A0} . t)}$ (d) $\dfrac{k_2 . C_{A0} . t}{(1 + k_2 . C_{A0} . t)^2}$

174. The first order gas phase reaction $A \xrightarrow{k_1} 2B$ is conducted isothermally in batch mode. The rate of change of conversion with time is given by
(a) $\dfrac{dx_A}{dt} = k_1 (1 - X_A)^2 (1 + 2 X_A)$
(b) $\dfrac{dx_A}{dt} = k_1 (1 - X_A) (1 + 0.5 X_A)$
(c) $\dfrac{dx_A}{dt} = k_1 (1 - X_A)$
(d) $\dfrac{dx_A}{dt} = \dfrac{k_1 (1 - X_A)}{(1 + X_A)}$

Answers

167. (a)	168. (b)	169. (a)	170. (a)	171. (c)	172. (a)
173. (b)	174. (c)				

175. For a vapor phase catalytic reaction (A + B → P), which follows the Ridel mechanism and the reaction step is rate controlling, the rate of reaction is given by (reaction rate is irreversible & product also absorbs)

(a) $-rA = \frac{k \cdot p_A^2 k_1 \cdot P_p}{1 + K_A \cdot P_A + K_p \cdot P_p}$

(b) $-rA = \frac{k \cdot p_A^2 k_1 \cdot P_p}{1 + K_A \cdot P_A + K_p \cdot P_p}$

(c) $-rA = \frac{k \cdot P_A \cdot P_B}{1 + K_{APA} + K_B \cdot P_B \; P_A + K_p \cdot P_p}$

(d) $-rA = \frac{k \cdot P_A \cdot P_B}{1 + K_A \cdot P_A}$

176. An exothermic reaction takes place in an adiabatic reactor. The product temperature .... reactor feed temperature.

(a) is always equal to
(b) is always greater than
(c) is always less than
(d) may be greater or less than

177. The extent of a reaction is

(a) different for reactants & products.
(b) dimensionless.
(c) depends on the stoichiometric co-efficient.
(d) all of the above.

178. For a series of reactions $A \rightarrow B \rightarrow C$ having $k_1 << k_2$, the reaction system can be approximated as K.

(a) $A \xrightarrow{k_1} B$ (b) $A \xrightarrow{k_2} B$

(c) $A \xrightarrow{k_2} C$ (d) $A \xrightarrow{k_1} C$

179. Find a mechanism that is consistent with the rate equation and reaction given below: $2A + B \rightarrow A_2B$, $(-rA) = k.C_A.C_B$

(a) $A + B \rightleftharpoons AB$; $AB + A \rightarrow A_2B$
(b) $A + B \rightarrow AB; AB + A \rightarrow A_2B \rightarrow$
(c) $A + A \rightarrow AA$; $AA + B \rightarrow A_2B$
(d) $A + A \rightleftharpoons AA$; $AA + B \rightarrow A_2B$.

180. A liquid phase reaction is to be carried out under isothermal conditions. The reaction rate as a function of conversion has been determined experimentally and is shown in the figure given below. What choice of reactor or combination of reactors will require the minimum overall reactor volume, if a conversion of 0.9 is desired ?

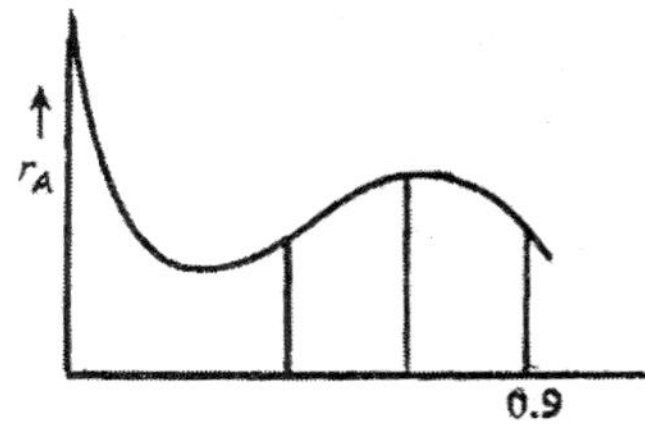

(a) CSTR followed by a PFR.
(b) PFR followed by a CSTR.
(c) CSTR followed by a PFR followed by a CSTR.
(d) PFR followed by a CSTR followed by a PFR.

Answers

175. (a)	176. (b)	177. (d)	178. (d)	179. (d)	180. (d)

181. Following isothermal kinetic data are obtained in a basket type of mixed flow reactor for a porous catalyst. Determine the role of pore diffusion and external mass transfer processes.

Pellet diameter	Leaving con centration of the reactant	Spinning rate of bas ket	(-r'A)
1	1	high	2
2	1	low	1
3	1	high	1

(*a*) Strong pore diffusion control and mass transfer not controlling.
(*b*) Both pore diffusion and mass transfer not controlling.
(*c*) Both pore diffusion and mass transfer controlling
(*d*) Mass transfer controlling.

182. For an ideal gas mixture undergoing a reversible gaseous phase chemical reaction, the equilibrium constant
(*a*) is independent of pressure.
(*b*) increases with pressure.
(*c*) decreases with pressure.
(*d*) increases/decreases with pressure depending on the stoichiometric co-efficients of the reaction.

183. The rate expression for the gaseous phase reaction, $CO + 2H_2 \rightleftharpoons CH_3OH$, is given by, $r = - K1\,P_{CO}^{\alpha} . P_{H2}^{\beta} - k2.\,P_{CH_3OH}^{\gamma}$. Which of the following is not possible ?
(*a*) $\alpha = 1$, $\beta = 1$, $r = 1$
(*b*) $\alpha = 1$, $\beta = 2$, $r = 1$
(*c*) $\alpha = 1/3$, $\beta = 2/3$, $r = 1/3$
(*d*) $\alpha = 1/2$, $\beta = 1$, $r = 1/2$

184. A first order gaseous phase reaction is catalysed by a non-porous solid. The kinetic rate constant and the external mass transfer coefficients are k and k_g respectively. The effective rate constant & k_{eff}) is given by
(*a*) $k_{eff} = k + kg$ (*b*) $k_{eff} = (k + kg)/2$
(*c*) $k_{eff} = (kkg)^{1/2}$ (*d*) $\frac{1}{K_{eff}} = \frac{1}{k} + \frac{1}{kg}$

185. A pollutant *P* degrades according to first order kinetics. An aqueous stream containing *P* at 2 k mole/m³ and volumetric flow rate 1 m³/hr requires a mixed flow reactor of volume *V* to bring down the pollutant level to 0.5 k mole/m³. The inlet concentration of the pollutant is now doubled and the volumetric flow rate is trippled. If the pollutant level is to be brought down to the same level of 0.5 kmole/m³, the volume of the mixed flow reactor should be increased by a factor of
(*a*) 7 (b) 6
(*c*) 3 (*d*) 7/3

186. The following gas phase reaction is taking place in a plug flow reactor.
$A + \frac{1}{2}B \rightarrow C$
A stoichiometric mixture of A & *B* at 300 °K is fed to the reactor. At 1 m along the length of the reactor, the temperature is 360 °K. The pressure drop is negligible and an ideal gas behaviour can be assumed. Identify the correct expression relating the concentration of A at the inlet (C_{A0}), concentration of A at 1 m (CA) and the corresponding conversion of A (X).
(*a*) $C_A = 1.2C_{Ao} \dfrac{(1-x)}{(1-0.33x)}$
(*b*) $C_A = 1.2\,C_{Ao} \dfrac{(1-x)}{(1-0.5\,x)}$
(*c*) $C_A = 0.83\,C_{Ao} \dfrac{(1-x)}{(1-0.33\,x)}$
(*d*) $C_A = 0.83\,C_{Ao} \dfrac{(1-x)}{(1-0.5\,x)}$

Answers

181. (a) 182. (b) 183. (a) 184. (d) 185. (a) 186. (c)

187. Pure A in gas phase enters a reactor. 50% of this A is converted to B through the reaction, $A \rightarrow 3B$. Mole fraction of A in the exit stream is

(a) 1/2 (b) 1/3
(c) 1/4 (d) 1/5

188. From among the following, choose one which is not an exothermic process.

(a) Methanol synthesis.
(b) Catalytic cracking.
(c) Ammonia synthesis.
(d) Oxidation of sulphur.

189. 6 kg of carbon is burnt with an amount of air containing 18 gm oxygen. The product contains 16.5 gms CO_2 and 2.8 gms CO basides other constituents. What is the degree of conversion (%) on the basis of disappearance of limiting reactant ?

(a) 100 (b) 95
(c) 75 (d) 20

190. At a given temperature, K_1, K_2 & K_3 are equilibrium constants for the following reactions 1, 2, & 3 respectively.

1. $CH_4 (g) + H_2O (g) \rightleftharpoons CO (g) + 3H_2 (g)$,
2. $CO (g) + H_2O (g) \rightleftharpoons CO_2(g) + H_2(g)$
3. $CH_4 (g) + 2H_2O (g) \rightleftharpoons CO_2 (g) + 4H_2 (g)$

Then K_1, K_2 and K_3 are related as :

(a) $K_3 = K_1 . K_2$
(b) $K_3 = (K_1 . K_2)^{0.5}$
(c) $K_3 = (K_1 + K_2)/2$
(d) $K_3 = (K_1 . K_2)^2$

191. Match the reactions in List I with their names in List II.

List I

P. $CO + H_2O \rightleftharpoons CO_2 + H_2$
Q. $CH_4 + H_2O \rightleftharpoons CO + 3H_2$

List II

1. Boudard reaction
2. Water gas shift reaction
3. Water gas reaction
4. Steam reforming

(a) P – 2, Q – 4 (b) P – 2, Q – 3
(c) P – 1, Q – 4 (d) P – 3, Q – 2

192. Match the item in List I to the appropriate item in List II.

List I

P. $A + B \rightarrow C$ Q. $A+B \rightarrow 2B + C$

List II

1. Rate of reaction of A shows a maximum with time.
2. C_B shows a maximum with time.
3. C_C shows a maximum with time.
4. C_B continuously decreases with time.

(a) P–2, Q–3 (b) P–2, Q–4
(c) P–1, Q–4 (d) P–4, Q–2

193. Pick out the wrong statement.

(a) In a first order reaction, A → products, the reaction becomes slower as it proceeds, because the concentration of A decreases and the rate is proportional to the concentration of A.
(b) Transition state theory approaches the problem of calculating reaction rates by concentrating on the idea of activated complexes.
(c) According to the penetration theory, the mass transfer co-efficient decreases, if the exposure time of an eddy to the solute decreases.
(d) If the rate of an irreversible reaction, $A + B \rightarrow 2C$, is $K.C_A.C_B$, then the reaction is always elementary.

Answers

187. (b) 188. (b) 189. (b) 190. (b) 191. (a) 192. (a)
193. (d)

194. Pick out the wrong statement.
   (a) For the same conversion, the holding time required in a batch reactor, is always equal to the space time required in a PFR.
   (b) Two mixed reactors of unequal size are available for producing a specified product, formed by a homogenous second order reaction. To achieve maximum production rate, the smaller reactor should be placed in series before the larger reactor.
   (c) Arrhenius equation describing the effect of temperature on rate constant is given by, $K = A.e^{-E/RT}$
   (d) The mechnism for the decomposition of CH_3CHO into CH_4 and CO in presence of I_2 is :
   $CH_3CHO + I_2 \rightarrow CH_3I + HI + CO$; slow
   $CH_3I + HI \rightarrow CH_4 + I_2$; fast
   Then the rate of disappearance of CH_3CHO is equal to $K.C_{CH_3I}.C_{HI}$ and acts as a catalyst.

Answers

194. (d)

21

Miscellaneous Questions

1. Atoms of the same element but having different masses are called
   (a) isobars (b) isotones
   (c) isotopes (d) isomers

2. Mass number of an atom is the sum of the numbers of
   (a) neutrons & protons.
   (b) protons & electrons.
   (c) neutrons & electrons.
   (d) both (a) and (c).

3. X - rays are
   (a) positively charged.
   (b) negatively charged.
   (c) neutral.
   (d) of higher wavelength than visible light.

4. The pH value is the measure of hydrogen ion concentration in a solution. The term pH stands for the .... in the solution.
   (a) power of hydrogen ion concentration
   (b) percentage of hydrogen
   (c) presence of hydrogen ions
   (d) none of these

5. Compositions expressed in mole per cent & volume per cent are exactly same for
   (a) ideal solutions
   (b) ideal gases only
   (c) colloidal solutions
   (d) any gas

6. A material lacking in toughness is usually termed as
   (a) plastic (b) malleable
   (c) brittle (d) ductile

7. .... cycle is the most efficient thermodynamic cycle.
   (a) Diesel (b) Otto
   (c) Rankine (d) Carnot

8. α-rays are
   (a) negatively charged
   (b) positively charged
   (c) neutral
   (d) same as X-rays

9. Volume occupied by 1 gm mole of gas at NTP is
   (a) 22,400 litres (b) 22.4 litres
   (c) 359 litres (d) 22.4 c.c.

10. The factor of safety for brittle materials (having elognation < 5%) is based on the .... stress.
    (a) yield (b) proof
    (c) ultimate (d) none of these

Answers

1. (b)	2. (d)	3. (c)	4. (a)	5. (b)	6. (c)
7. (d)	8. (b)	9. (b)	10. (c)		

11. Gas evolution does not occur during solidification in case of the .... steel.
(*a*) capped (*b*) killed
(*c*) semi-killed (*d*) rimmed

12. LASER
(*a*) stands for Light Amplification by Stimulated Emission of Radiation.
(*b*) has same intensity as visible light.
(*c*) can't be used for cutting/welding metals.
(*d*) none of these.

13. Depreciation calculated by which of the following methods is maximum ?
(*a*) Diminishing balance method.
(*b*) Straight line method.
(*c*) Sum of the years digits method.
(*d*) Sinking fund method.

14. The main ore of uranium is
(*a*) pitchblende (*b*) monazite sand
(*c*) cassiterite (*d*) chalcopyrite

15. The main ore of thorium is
(*a*) pitchblende (*b*) monazite sand
(*c*) limonite (*d*) galena

16. Hardening of aluminium is done by
(*a*) precipitation hardening.
(*b*) cold working.
(*c*) hot working.
(*d*) solution heat treatment.

17. .... has the highest melting point out of the following.
(*a*) Tungsten (*b*) Zirconium
(*c*) Molybdenum (*d*) Tantalum

18. Heavy water (D_2O) in a nuclear reactor serves as the
(*a*) coolant.
(*b*) moderator
(*c*) both coolant & moderator.
(*d*) neutron absorber.

19. Which of the following is a moderating material used in nuclear reactor ?
(*a*) Graphite.
(*b*) Cadmium.
(*c*) Zircalloy (an alloy of zirconium and aluminium).
(*d*) Stainless steel.

20. Fahrenheit & Centigrade temperature scales have the same readings at
(*a*) – 55° (*b*) – 40°
(*c*) – 33° (*d*) – 58°

21. Giga stands for
(*a*) 10^{9} (*b*) 10^{-12}
(*c*) 10^{12} (*d*) 10^{15}

22. 1 ata is equivalent to
(*a*) 1 atm (*b*) 10 torr
(*c*) 0.98 Pascal (*d*) 1 kgf/cm²

23. pH value of H_2SO_4 (5% concentration) is
(*a*) 5 (b) 7
(*c*) >7 (*d*) < 7

24. One torr is equal to
(*a*) 1 mm Hg
(*b*) 1 Pascal
(*c*) 1 ata
(*d*) 1 mm water column

25. Cobalt - 60 is used as a source of .... in medical therapy & industrial radiography.
(*a*) X-rays (*b*) γ-rays
(*c*) α-rays (*d*) β-rays

26. pH value of an alkaline solution is
(*a*) 7
(*b*) >7
(*c*) <7
(*d*) constant over a wide range

Answers

11. (b)	12. (a)	13. (a)	14. (a)	15. (b)	16. (b)
17. (a)	18. (c)	19. (a)	20. (b)	21. (a)	22. (d)
23. (d)	24. (a)	25. (b)	26. (b)		

27. Elements in a periodic table are arranged in order of their
    (a) atomic number
    (b) mass number
    (c) atomic weight
    (d) metallic characteristics

28. Which of the following metals reacts violently with water ?
    (a) Mercury (b) Sodium
    (c) Calcium (d) Magnesium

29. Size range of colloidal particles is
    (a) 5 – 200 milli-microns
    (b) 50 – 200 microns
    (c) 500 –1000 microns
    (d) 10 – 50 Angstrom

30. Dowtherm is a
    (a) high temperature heating medium (a petroleum product).
    (b) product of coal tar distillation.
    (c) very heat sensitive material.
    (d) none of these.

31. "Critical mass" is the minimum mass of nuclear fissile material required for the
    (a) sustainment of chain reaction.
    (b) power generation on commercial scale.
    (c) economic power generation.
    (d) none of these.

32. The viscosity of water at room temperature may be around one
    (a) centipoise (b) poise
    (c) stoke (d) both (b) & (c)

33. With increase in molecular weight, the vapour pressure of liquids (of similar chemical nature) at any particular temperature
    (a) increases
    (b) decreases
    (c) remains unaffected
    (d) none of these

34. A vapour whose partial pressure is less than its equilibrium vapour pressure is called a .... vapour.
    (a) superheated (b) saturated
    (c) supersaturated (d) none of these

35. In case of a solution (not of a solid in a liquid), whose total volume is less than the sum of the volumes of its components in their pure states, solubility is
    (a) increased by pressure rise.
    (b) lowered by pressure rise.
    (c) unaltered by pressure rise.
    (d) independent of temperature.

36. Solutions having the same osmotic pressure are called .... solutions.
    (a) isotonic (b) dilute
    (c) saturated (d) ideal

37. Permanent set in a material is produced by subjecting it to
    (a) triaxial loading.
    (b) fatigue loading.
    (c) stresses beyond the elastic limit.
    (d) tensile & compressive loads simultaneously.

38. Viscosity of atmospheric air may be around centipoise (cp).
    (a) 0.015 (b) 0.6
    (c) 1 (d) 100

39. Atmospheric pressure corresponds to a hydrostatic head of
    (a) 34 ft of water. (b) 13.6 cm of Hg.
    (c) 1 metre of water. (d) 100 metre of water.

Answers

27. (a)	28. (b)	29. (a)	30. (a)	31. (a)	32. (a)
33. (b)	34. (a)	35. (a)	36. (a)	37. (c)	38. (a)
39. (a)					

40. A pipe is generally made of circular cross-section, because a circular cross-section has
(*a*) higher structural strength.
(*b*) lowest surface area requirement for a given volume.
(*c*) both (*a*) & (*b*).
(*d*) none of these.

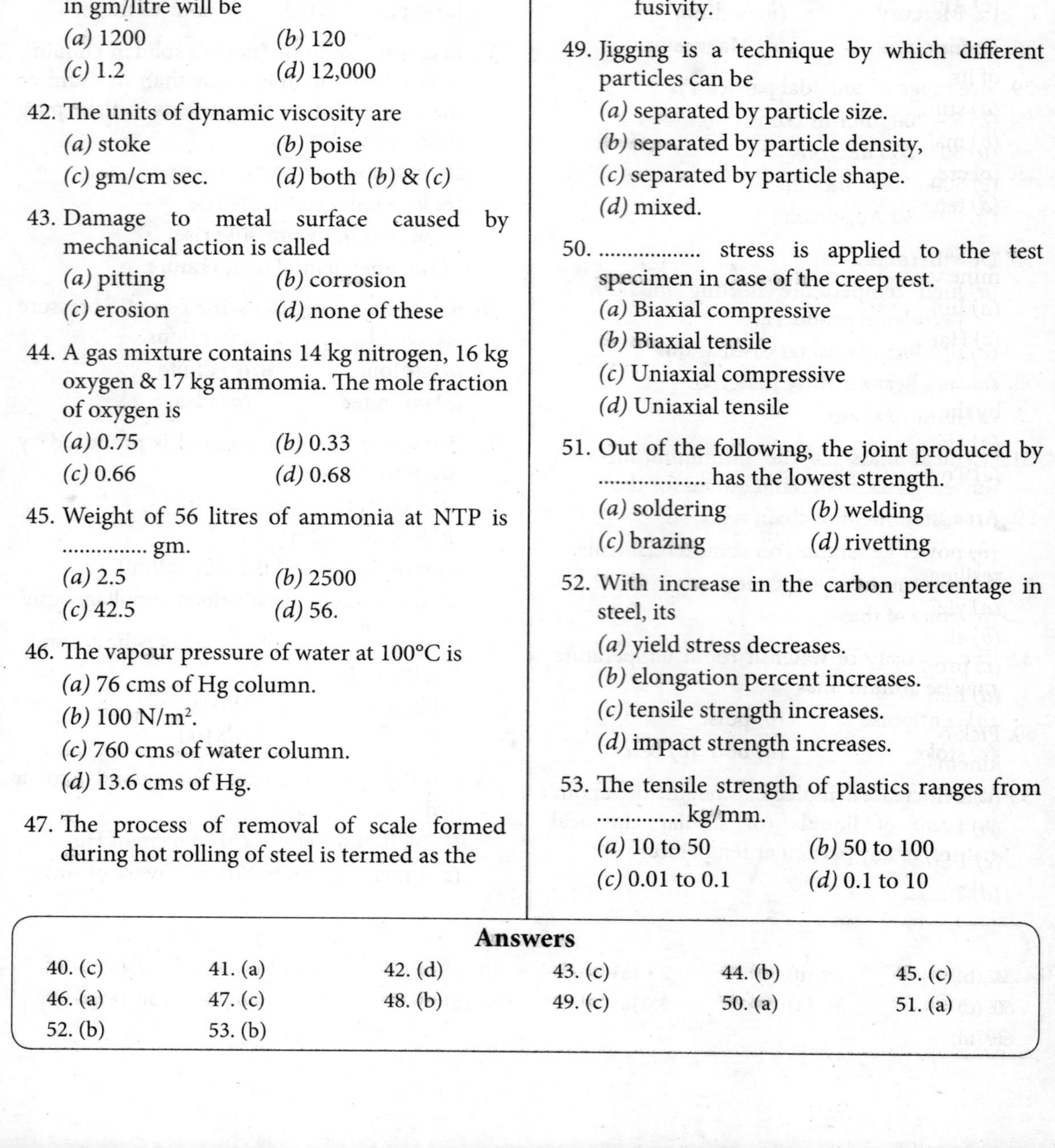

41. The density of a liquid is 1200 kg/m^3. Its value in gm/litre will be
(*a*) 1200 (*b*) 120
(*c*) 1.2 (*d*) 12,000

42. The units of dynamic viscosity are
(*a*) stoke (*b*) poise
(*c*) gm/cm sec. (*d*) both (*b*) & (*c*)

43. Damage to metal surface caused by mechanical action is called
(*a*) pitting (*b*) corrosion
(*c*) erosion (*d*) none of these

44. A gas mixture contains 14 kg nitrogen, 16 kg oxygen & 17 kg ammonia. The mole fraction of oxygen is
(*a*) 0.75 (*b*) 0.33
(*c*) 0.66 (*d*) 0.68

45. Weight of 56 litres of ammonia at NTP is .... gm.
(*a*) 2.5 (*b*) 2500
(*c*) 42.5 (*d*) 56.

46. The vapour pressure of water at 100°C is
(*a*) 76 cms of Hg column.
(*b*) 100 N/m^2.
(*c*) 760 cms of water column.
(*d*) 13.6 cms of Hg.

47. The process of removal of scale formed during hot rolling of steel is termed as the
(*a*) descaling (*b*) shot blasting
(*c*) pickling (*d*) skimming

48. Prandtl number is the ratio of the
(*a*) mass diffusivity to thermal diffusivity.
(*b*) momentum diffusivity to thermal diffusivity.
(*c*) thermal diffusivity to mass diffusivity.
(*d*) thermal diffusivity to momentum diffusivity.

49. Jigging is a technique by which different particles can be
(*a*) separated by particle size.
(*b*) separated by particle density,
(*c*) separated by particle shape.
(*d*) mixed.

50. .... stress is applied to the test specimen in case of the creep test.
(*a*) Biaxial compressive
(*b*) Biaxial tensile
(*c*) Uniaxial compressive
(*d*) Uniaxial tensile

51. Out of the following, the joint produced by .... has the lowest strength.
(*a*) soldering (*b*) welding
(*c*) brazing (*d*) rivetting

52. With increase in the carbon percentage in steel, its
(*a*) yield stress decreases.
(*b*) elongation percent increases.
(*c*) tensile strength increases.
(*d*) impact strength increases.

53. The tensile strength of plastics ranges from .... kg/mm.
(*a*) 10 to 50 (*b*) 50 to 100
(*c*) 0.01 to 0.1 (*d*) 0.1 to 10

Answers

40. (c)	41. (a)	42. (d)	43. (c)	44. (b)	45. (c)
46. (a)	47. (c)	48. (b)	49. (c)	50. (a)	51. (a)
52. (b)	53. (b)				

54. Thermoplastics are never moulded using
    (*a*) injection moulding,
    (*b*) blow moulding.
    (*c*) extrusion.
    (*d*) compression moulding.

55. Which of the following is not a ferromagnetic material ?
    (*a*) Nickel (*b*) Cobalt
    (*c*) Aluminium (*d*) Iron

56. Young's modulus of a material is the measure of its
    (*a*) stiffness
    (*b*) malleability
    (*c*) creep resistance
    (*d*) tensile strength

57. .... test is the appropriate test to determine whether a material is ductile or brittle.
    (*a*) Impact (*b*) Cupping
    (*c*) Hardness (*d*) Tensile

58. .... property of a material is determined by the Herbert Pendulum device.
    (*a*) Hardness (*b*) Tensile
    (*c*) Toughness (*d*) Compressive

59. Area under the stress-strain curve upto the .... is referred to as the modulus of resilience.
    (*a*) yield strength
    (*b*) elastic limit
    (*c*) proportional limit
    (*d*) maximum point

60. Pick out the wrong conversion of absolute & kinematic viscosities.
    (*a*) 1 Stoke = 1 cm^2/sec.
    (*b*) 1 N. sec/m^2 = 10 Poise
    (*c*) 1 Stoke = 1 m^2/sec.
    (*d*) 1 m^2/sec = 10^4 Stokes

61. Specific gravity of a metal which weighs 5 kg in air & 4 kg when submerged in water will be
    (*a*) 5 (*b*) 1.25
    (*c*) 2.5 (*d*) 3.75

62. The most important material handling system in an integrated steel plant is
    (*a*) belt conveyor
    (*b*) diesel locomotive
    (*c*) truck
    (*d*) over head hoists & cranes

63. The minimum carbon content in steel should be .... per cent for it to respond to a hardening treatment.
    (*a*) 0.2 (*b*) 0.4
    (*c*) 0.6 (*d*) 0.8

64. .... is not used as a material of construction in thermocouples.
    (*a*) Alumel (*b*) Rhodium
    (*c*) Constantan (*d*) Duralumin

65. Solidification time of metal in sand casting is proportional to
    (*a*) A/V (*b*) V/A
    (*c*) $(V/A)^2$ (*d*) $(A/V)^2$
    where, A = Cross-sectional area of the mould-interface
    and, V = Volume of the casting

66. Which of the following properties of a solid is not dependent on crystal imperfection ?
    (*a*) Ductility (*b*) Semiconductivity
    (*c*) Melting point (*d*) Yield stress

67. Pick out the wrong statement.
    (*a*) Reflectivity of gases is very low.
    (*b*) Absorptivity for metals is very high.
    (*c*) Reflectivity for non-conducting materials is low.
    (*d*) Reflectivity of polished surfaces is very high.

Answers

54. (b)	55. (b)	56. (a)	57. (a)	58. (a)	59. (c)
60. (c)	61. (a)	62. (b)	63. (a)	64. (b)	65. (d)
66. (a)	67. (c)				

68. All modes of heat transfer i.e., conduction, convection & radiation occur in case of the
    (*a*) condensation of steam in a partial condenser.
    (*b*) boiler furnace.
    (*c*) insulated pipe carrying high pressure steam.
    (*d*) none of these.

69. A spherical drop of molten metal of 2 mm radius solidifies in 5 seconds. The time required for the solidification of a similar drop of molten metal of radius 4 mm will be .... seconds.
    (*a*) 20 (*b*) 40
    (*c*) 5 (*d*) 10

70. .... is used as a material of construction for the blade of power saw.
    (*a*) Wrought iron (*b*) Stainless steel
    (*c*) Mild steel (*d*) High speed steel

71. The concentration of the dissolved gases in metal kept in a mould is proportional to
    (*a*) P^2 (*b*) P
    (*c*) $\sqrt{P}$ (*d*) 1/P
    where, P = Partial pressure of the gases above the metal in a mould.

72. The phenomenon of strain-ageing is generally experienced with the
    (*a*) cast iron
    (*b*) low carbon steel
    (*c*) high manganese steel
    (*d*) high carbon steel

73. The distribution given by microscopic analysis of powder is
    (*a*) number (*b*) length
    (*c*) area (*d*) volume

74. Which of the following welding processes employs the highest welding temperature?
    (*a*) Arc welding
    (*b*) Gas welding
    (*c*) Thermit welding
    (*d*) Resistance welding

75. Quartz is a .... material.
    (*a*) diamagnetic (*b*) ferromagnetic
    (*c*) ferroelectric (*d*) piezoelectric

76. Which of the following can be manufactured using powder metallurgy techniques?
    (*a*) Brake linings
    (*b*) Bearings
    (*c*) Carbide tool tips
    (*d*) all 'a', 'b' & 'c'

77. Scab is a defect developed during .... of metals.
    (*a*) forging (*b*) sand casting
    (*c*) machining (*d*) welding

78. Vanadium addition to steel improves its
    (*a*) high temperature oxidation resistance.
    (*b*) fatigue strength.
    (*c*) heat treatability by quenching.
    (*d*) hardenability.

79. Fatigue strength of a metal is improved by
    (*a*) overstressing.
    (*b*) scratching the surface.
    (*c*) raising the temperature.
    (*d*) understressing.

80. A cube, sphere & a thin circular plate (all having same mass and made of same material) are all heated to 300°C and allowed to cool in natural air. Which one will cool at the slowest rate ?
    (*a*) Cube
    (*b*) Plate
    (*c*) Sphere
    (*d*) All will cool at the same rate

Answers

68. (a)	69. (a)	70. (b)	71. (c)	72. (b)	73. (c)
74. (c)	75. (b)	76. (c)	77. (b)	78. (a)	79. (d)
80. (c)					

81. A tank painted with which of the following paints would heat up the maximum by radiation from sun ?
   (a) Yellow paint (b) White paint
   (c) Black paint (d) Grey paint

82. Roll separating force in the rolling process can be decreased by
   (a) providing back up rolls.
   (b) increasing the roll diameter.
   (c) decreasing the roll diameter.
   (d) increasing the friction between rolls & the metal.

83. Which of the following does not discharge the dust collected as a dry solid ?
   (a) Electrostatic precipitator
   (b) Wet scrubber
   (c) Bag filter
   (d) Gravity settling chamber

84. Laser is a device to produce
   (a) gamma rays
   (b) microwaves
   (c) a beam of white light
   (d) a beam of coherent light

85. High .... of cast iron makes it suitable for use in machine beds.
   (a) compressive strength
   (b) endurance strength
   (c) damping capacity
   (d) tensile strength

86. Compensation for solid contraction in solidification of metal during casting is provided by
   (a) properly placed risers.
   (b) promoting directional solidification.
   (c) oversize pattern.
   (d) chills.

87. Theoretical volume of oxygen required for the complete combustion of 1 Nm^3 of acetylene, in oxy-acetylene welding is .... Nm^3.
   (a) 0.5 (b) 1
   (c) 2 (d) 2.5

88. Strain hardening effect in a metal subjected to cold working is due to the .... mechanism.
   (a) fracture (b) dislocation
   (c) slip (d) twinning

89. The hardening process normally used for the hardening of machine tool guideways is
   (a) induction hardening.
   (b) martempering.
   (c) flame hardening
   (d) vacuum hardening

90. Which of the following parameters is actually measured during tensile testing of a specimen using Universal testing machine?
   (a) Load & elongation.
   (b) Engineering stress & strain.
   (c) True stress & strain.
   (d) Poisson's ratio & Young's modulus.

91. Atomic .... of an element is a whole number.
   (a) weight (b) number
   (c) volume (d) radius

92. One micron is equal to
   (a) 10^{-4} mm (b) 10^{-4} cm
   (c) 10^{-6} mm (d) both (b) & (c)

93. .... has a negative coefficient of linear expansion.
   (a) Iron (b) Copper
   (c) Rubber (d) Nickel.

Answers

81. (c)	82. (b)	83. (b)	84. (d)	85. (a)	86. (c)
87. (d)	88. (c)	89. (b)	90. (a)	91. (b)	92. (d)
93. (b)					

94. For the same size of casting, which of the following metals will require the largest size of riser ?
(*a*) Copper (*b*) Cast iron
(*c*) Aluminium (*d*) Steel

95. Addition of silicon to cast iron
(*a*) promotes graphite module formation.
(*b*) improves its ductility.
(*c*) does not promote graphite flakes formation.
(*d*) increases the fluidity of molten metal.

96. Pick out the wrong statement about powder metallurgy technique.
(*a*) Powder metallurgy techniques are applied at room temperature.
(*b*) Final product in powder metallurgy is not necessarily metallic powder.
(*c*) The final product in powder metallurgy is made from metallic powder.
(*d*) Silver and nickel when melted together do not mix to form an alloy ; however alloy of these constituents can be made by powder metallurgy technique.

97. Which of the following defects is caused in castings due to hard core, which restricts the castings during solidification ?
(*a*) Scabs (*b*) Hot tears
(*c*) Fissures (*d*) Fins

98. Babbit lining is used on brass/bronze bearings to increase the
(*a*) antifriction properties.
(*b*) compressive strength.
(*c*) bearing resistance.
(*d*) wear resistance.

99. The moment of inertia of a body comes into play in .... motion.
(*a*) curvilinear (*b*) linear
(*c*) rotational (*d*) none of these

100. Two wires of the same radius & material having length in the ratio of 1:2 are stretched by the same force. The strain produced in the two wires will be in the ratio of
(*a*) 1:1 (*b*) 1:2
(*c*) 2:1 (*d*) 1:4

101. Force between the molecules of the same substance is called the .... force.
(*a*) adhesive (*b*) cohesive
(*c*) moleular (*d*) Vanderwalls

102. Avogadro's number is the number of molecules in one .... of a gas.
(*a*) gram (*b*) kilogram
(*c*) gm. mole (*d*) litre

103. An ideal material for making cooking vessel should have
(*a*) high heat capacity
(*b*) low heat capacity
(*c*) high thermal conductivity
(*d*) both (*b*) and (*c*)

104. Acetylene should never be conveyed through .... pipe, as it forms an unstable compound that dissociates violently at the slightest shock.
(*a*) stainless steel (*b*) copper
(*c*) aluminium (*d*) brass

105. Cast iron detachable joint is used in .... pipes.
(*a*) steel
(*b*) cast iron
(*c*) asbestos cement
(*d*) reinforced cement concrete

Answers

94. (b)	95. (d)	96. (a)	97. (b)	98. (d)	99. (c)
100. (a)	101. (b)	102. (c)	103. (d)	104. (b)	105. (a)

106. .... are normally subjected to subzero treatment.
   (a) Highly stressed parts
   (b) Gauges or precision parts
   (c) Cold rolled sheets
   (d) Non machinable materials

107. Notched bar test is used for testing the .... of a material.
   (a) impact strength
   (b) endurance limit
   (c) machinability
   (d) corrosion resistance

108. Which of the following is prone to cup and cone fracture ?
   (a) Cast iron.
   (b) Soft brass.
   (c) Round specimen of ductile metal
   (d) Flat tensile specimen of ductile metal

109. Presence of small quantity of sulphur in low carbon steel improves its
   (a) formability (b) machinability
   (c) weldability (d) hardenability

110. "Flare tower' used in metallurgical industry is meant for
   (a) venting off (after burning) the excess inflammable/toxic gases at high pressure.
   (b) absorbing the pollutant gases from the furnace exhaust.
   (c) cooling of furnace exhaust gases.
   (d) none of these

111. .... is not used as the control rod material in a nuclear reactor.
   (a) Cobalt (b) Hafnium
   (c) Cadmium (d) Boron

112. The salt used in the salt bath for hardening at a temperature of 1100 – 1350°C is
   (a) barium chloride
   (b) sodium chloride
   (c) potassium chloride
   (d) calcium chloride

113. Out of the following substances, one ton of .... will store the maximum heat for a rise of 30°C in temperature.
   (a) steel
   (b) lignite
   (c) water
   (d) limestone

114. For parts with section thickness < 5 mm, the most suitable heat treatment process for achieving minimum distortion during quenching is
   (a) martempering
   (b) nitriding
   (c) flame hardening
   (d) tempering

115. Smoke is an example of
   (a) solid dispersed in liquid.
   (b) solid dispersed in gas.
   (c) liquid dispersed in gas.
   (d) gas dispersed in liquid.

116. The most suitable heat treatment process for achieving toughness in hardened components is
   (a) cyaniding (b) tempering
   (c) nitriding (d) flame hardening

117. The depth of indentation in Rockwell hardness test is measured by a
   (a) microscope
   (b) scale
   (c) micrometer screw gauge
   (d) dial gauge

Answers

106. (b)	107. (a)	108. (c)	109. (b)	110. (a)	111. (a)
112. (a)	113. (c)	114. (a)	115. (b)	116. (b)	117. (d)

118. A material expanding freely due to heating will develop .... stress.
(*a*) no (*b*) compressive
(*c*) tensile (*d*) thermal

119. Glass reacts with
(*a*) H_2SO_3 (*b*) HF
(*c*) HNO_3 (*d*) $K_2Cr_20_7$

120. Lead is most readily dissolved in the
(*a*) acetic acid (*b*) hydrochloric acid
(c) nitric acid (*d*) sulphuric acid

121. In which of the following surface hardening processes, quenching is not required ?
(*a*) Cyaniding
(*b*) Case carburising
(*c*) Nitriding
(*d*) Induction hardening

122. Tempering of quenched martensitic steel is necessary to improve the .... of the metal.
(*a*) corrosion resistance.
(*b*) ductility.
(*c*) surface texture.
(*d*) hardness.

123. A hot short metal is
(*a*) hard when hot
(*b*) brittle when cold
(*c*) brittle when hot
(*d*) brittle under all conditions

124. Commonly used flux for brazing is
(*a*) zinc chloride
(*b*) borax
(*c*) ammonium chloride
(*d*) rosin

125. Binding wire used to support the joints for soldering is made of soft iron. The 'bit' of a soldering iron is made of
(*a*) tin (*b*) copper
(*c*) mild steel (*d*) aluminium

126. Dilatometer is used to find out the .... of the moulding sand.
(*a*) fineness (*b*) hot strength
(*c*) permeability (*d*) compactness

127. .... can not be forged.
(*a*) Wrought iron (*b*) Mild steel
(*c*) Cast iron (*d*) High Carbon steel

128. .... allowance is the terminology used for the taper provided on the pattern for its easy & clean withdrawal from the mould.
(*a*) Distortion (*b*) Draft
(*c*) Taper (*d*) Pattern

129. There are one octahedral void and .... tetrahedral void in the closest packing of atoms.
(*a*) one (*b*) two
(*c*) three (*d*) none of these

130. The size of tetrahedral void in the closest packing of atoms is .... that of octahedral void.
(*a*) equal to
(*b*) greater than
(*c*) less than
(*d*) either (*a*), (*b*) or (*c*); depends on the size of the atom.

131. Protection of ship hulls against sea water corrosion is done by coating with zinc, because of the fact that zinc compared to iron is
(*a*) reduced preferentially.
(*b*) cheaper.
(*c*) oxidised preferentially.
(*d*) lighter.

Answers

118. (a)	119. (b)	120. (a)	121. (c)	122. (b)	123. (c)
124. (b)	125. (b)	126. (b)	127. (c)	128. (b)	129. (b)
130. (c)	131. (c)				

132. Location & detection of the surface & sub-surface cracks in non-magnetic alloys can be easily done by .... testing.
(a) dye-penetrant
(b) ultrasonic
(c) X-ray
(d) none of these

133. Tin & lead can be hot worked at room temperature. Improvement of mechanical properties of the metal in hot working is due to the
(a) formation of colomnar grains.
(b) refinement of grain size.
(c) grain growth.
(d) recovery of grains.

134. Mass production of metallic cans is usually done by the .... process.
(a) extrusion (b) coining
(c) spinning (d) embossing

135. Bigger rollers in four high rolling mill is called the .... rolls.
(a) working (b) back up
(c) guide (d) support

136. Fatigue strength of a material can not be increased by
(a) cold working (b) hot working
(c) shot peening (d) any of these

137. Out of the following, the rays having the lowest wavelength is the
(a) radio waves (b) ultrasonic waves
(c) infra red rays (d) X-rays

138. Out of the following, the materials having maximum magnetic permeability is
(a) silicon steel (Si = 4%)
(b) cast iron
(c) wrought iron
(d) mild steel

139. The highest stress that a material can withstand for a specified length of time without excessive deformation is called the .... strength.
(a) creep (b) endurance
(c) fatigue (d) none of these

140. When a metal or alloy is cold worked,
(a) it is worked above recrystallisation temperature.
(b) its hardness increases, but strength does not increase.
(c) it is worked below room temperature.
(d) its hardness & strength both increases.

141. Out of the following, the .... pipe has the least corrosion resistance.
(a) mild steel (b) copper
(c) cast iron (d) wrought iron

142. Metallic surveying tapes are made of .... which has a low coefficient of expansion & enough strength.
(a) invar (b) hastealloy
(c) duralumin (d) monel metal

143. Annealing of white cast iron produces .... iron.
(a) grey (b) nodular
(c) malleable (d) spheroidal

144. Carbon is present in the form of .... in grey cast iron.
(a) spheroids
(b) nodular aggregates of graphite
(c) flakes
(d) cementite

Answers

132. (a)	133. (b)	134. (b)	135. (b)	136. (b)	137. (a)
138. (a)	139. (a)	140. (b)	141. (d)	142. (a)	143. (c)
144. (c)					

145. Fibrous fracture is normally encountered in .... materials.
   (*a*) hard (*b*) elastic
   (*c*) ductile (*d*) brittle

146. Materials having .... lattice structure are usually the most ductile.
   (*a*) body centred cubic (bcc)
   (*b*) hexagonal close packed (hcp)
   (*c*) face centred cubic (fcc)
   (*d*) none of these

147. Failure of a material is attributed to .... if it fails below its yield point.
   (*a*) impact loading
   (*b*) fatigue
   (*c*) creep
   (*d*) straining

148. The difference between gross & net calorific value of a fuel is due to the
   (*a*) sensible heat carried away by the flue gases.
   (*b*) heat carried away by steam formed the moisture content of the fuel.
   (*c*) heat lost by radiation.
   (*d*) heat carried away by steam formed the combustion of hydrogen in the fuel.

149. The only suitable method for hardening the low carbon steel is case hardening. Which of the following is a case hardening process?
   (*a*) Cyaniding (*b*) Sheardising
   (*c*) Spherodising (*d*) None of these

150. Normalising does not .... of a metal.
   (*a*) improve machinability & tensile strength
   (*b*) remove internal stresses
   (*c*) refine the structure
   (*d*) remove strains caused by cold working

151. The product out from a cupola is called
   (*a*) wrought iron (*b*) pig iron
   (*c*) cast iron (*d*) none of these

152. Lead is poured into the joint between two .... pipes.
   (*a*) mild steel (*b*) concrete
   (*c*) cast iron (*d*) asbestos cement

153. Silicon percentage in the silicon steel used for electrical appliances/equipments is
   (*a*) 0.5 – 1 (*b*) 12 –14
   (*c*) 3 – 4 (*d*) 20 – 25

154. Wrought iron does not have the
   (*a*) uniform strength in all directions.
   (*b*) ability to hold protective coating.
   (*c*) high ductility & corrosion resistance.
   (*d*) easily weldable characteristics.

155. Match the typical temperature (°C) encountered in various metallurgical processes.

List I
   (*a*) L.D. slag temperature.
   (*b*) Heat treatment for hardening of steel.
   (*c*) Dome temperature of B.F. stove at the end of heating cycle.
   (*d*) Raw blast furnace gas (top gas) temperature.

List II
   I. 950
   II. 1700
   III. 250
   IV. 1300

156. Machinability of hard alloys & tool steels is improved by
   (*a*) spherodising (*b*) annealing
   (*c*) tempering (*d*) normalising

Answers

145. (c)	146. (c)	147. (b)	148. (d)	149. (a)	150. (b)
151. (c)	152. (c)	153. (c)	154. (a)	155. (II) (I) (IV) (III)	
156. (b)					

157. For hardening high alloy steel, it is to be heated slowly & uniformly to avoid
   (a) local hardening.
   (b) scale formation.
   (c) segregation.
   (d) warpage.

158. Out of the following, the best material capable of withstanding shock & vibration without the danger of cracking is
   (a) malleable iron.
   (b) grey cast iron,
   (c) chilled cast iron.
   (d) white cast iron.

159. Brittleness induced due to the presence of sulphur in steel can be reduced by adding
   (a) manganese (b) magnesium
   (c)vanadium (d) copper

160. Cold chiesel & hammers are made of
   (a) high speed steel
   (b) high carbon steel
   (c) forged steel
   (d) mild steel

161. Eutectoid composition of carbon steel at room temperature is
   (a) cementite (b) pearlite
   (c) martensite (d) ferrite

162. Which of the following liquid metals has the highest thermal conductivity ?
   (a) Molten sodium (b) Molten lead
   (c) Mercury (d) Molten potassium

163. .... is the most important element, which controls the physical properties of steel.
   (a) Manganese (b) Silicon
   (c) Carbon (d) Vanadium

164. Particle nature of cathode rays is proved due to the fact that they
   (a) travel in a straight line.
   (b) get deflected by magnetic & electric field.
   (c) produce fluorescence.
   (d) heat the exposed material.

165. The passage between the nozzle and the .... is called 'sprue' in case of injection moulding.
   (a) runner (b) cylinder
   (c) mold (d) none of these

166. Which of the following set of constituents is expected in equilibrium cooling of a hypereutectoid steel from austenitic state ?
   (a) Cementite & martensite.
   (b) Cementite & pearlite.
   (c) Ferrite & Pearlite.
   (d) Ferrite & bainite.

167. Welded joint efficiency in the design of metallurgical process equipment is taken as
   (a) 0.55 (b) 0.75
   (c) 0.85 (d) 0.95

168. .... addition increases the depth of hardness of steel.
   (a) Chromium (b) Nickel
   (c) Tungsten (d) Vanadium

169. Structural steel is not made of...... shape.
   (a) V (b) 0
   (c) T (d) I

170. .... is the hardest material out of the following.
   (a) Boron carbide
   (b) Tungsten carbide
   (c) Hardened steel
   (d) Silicon carbide

Answers

157. (d)	158. (a)	159. (a)	160. (b)	161. (b)	162. (a)
163. (c)	164. (b)	165. (a)	166. (b)	167. (c)	168. (a)
169. (a)	170. (a)				

171. Source of heat supply in cupola is
(*a*) coke (*b*) charcoal
(*c*) lumpy coal (*d*) pulverised coal

172. Which of the following is not a principal alloying element for the structural steel ?
(*a*) Molybdenum (*b*) Nickel
(*c*) Manganese (*d*) Chromium

173. Resilience of a material is important, when it is subjected to
(*a*) creep (*b*) fatigue
(*c*) shock loading (*d*) galvanising

174. Any shell opening greater than 5 cms for a storage tank must be reinforced for reason of
(*a*) preventing the local overstressing of the shell around the opening.
(*b*) reduction of discontinuity in shape at the junction.
(*c*) making the joint leak proof.
(*d*) none of these.

175. Consideration of the creep is the most important in case of
(*a*) blades of gas turbine.
(*b*) piston of I.C. engine.
(*c*) flywheel of steam engine.
(*d*) cycle chain.

176. Hardening of a small selected portion of a job can be best done by
(*a*) induction hardening or flame hardening.
(*b*) pack carburising.
(*c*) nitriding.
(*d*) cyaniding.

177. Fatigue resistance of a material is measured by the
(*a*) elastic limit.
(*b*) ultimate tensile strength.
(*c*) young's modulus.
(*d*) endurance limit.

178. Swell is a casting defect, which occurs due to the
(*a*) presence of moisture in the moulding sand.
(*b*) enlargement of the mould cavity by metal pressure causing partial or overall enlargement of the casting,
(*c*) loose ramming of sand.
(*d*) mismatching of mould at the parting line.

179. In metallurgical process equipments, the conical bottom heads used, usually has an apex angle of .... degrees.
(*a*) 20 (*b*) 40
(*c*) 60 (*d*) 80

180. Which of the following is not a polymer of two monomers?
(*a*) Teflon (*b*) Bakelite
(*c*) SBR (*d*) None of these

181. Thermal diffusivity of a substance is proportional to
(*a*) $1/k$ (*b*) k
(*c*) k^2 (*d*) $1/k^2$
where, k = Thermal conductivity

182. The testing pressure of storage tanks and pressure vessels designed as per Indian standard codes should be about .... times the design pressure.
(*a*) 1.5 to 2 (b) 3 to 4
(*c*) 4 to 5 (*d*) >5

183. Which of the following metals is the most prone to work hardening ?
(*a*) Brass (*b*) Aluminium
(*c*) Copper (*d*) Lead

Answers

171. (a)	172. (a)	173. (c)	174. (a)	175. (a)	176. (a)
177. (d)	178. (b)	179. (c)	180. (a)	181. (b)	182. (a)
183. (a)					

184. Vertical storage vessels are never supported by the .... support.
(a) bracket (b) saddle
(c) skirt (d) lug

185. Which of the following charge materials of iron blast furnace does not contain phosphorous ?
(a) Coke (b) Iron ore lump
(c) Sinter (d) Quartz

186. Which of the following does not occur inside the iron blast furnace ?
(a) Solution loss reaction.
(b) Oxidation of carbon.
(c) Reduction of alumina.
(d) Desulphurisation of charge materials.

187. Which of the following is not used as a coolant in L.D. process of steel making, when the opening steel bath temperature after finishing of oxygen blow is found to be more than the steel tapping temperature ?
(a) Coke breeze (b) Iron ore
(c) Steel scrap (d) Lime

188. Which of the following is an acidic constituent of B.F slag?
(a) SiO_2 (b) FeO
(c) MgO (d) CaO

189. Tin base white metals are used, where bearings are subjected to
(a) high pressure & load.
(b) low pressure & load.
(c) high temperature.
(d) large surface wear.

190. The ratio of the diameter of the back up rolls to that of the working rolls in a four high rolling mill is
(a) 1 (b) < 1
(c) > 1 (d) either 'b' or 'c'

191. Which of the following is not an oxide ore of its metal ?
(a) Cuprite (b) Chalcocite
(c) Magnetite (d) Bauxite

192. Which of the following places does not have a limestone mine ?
(a) Bhavnathpur (Jharkhand)
(b) Bilaspur (Chattisgarh)
(c) Jaisalmer (Rajasthan)
(d) Kiriburu (Jharkhand)

193. Which of the following is not a charge material for a cupola furnace ?
(a) Limestone (b) Iron scrap
(c) Iron ore (d) Pig iron

194. In the formation of cermets, the ratio of ceramic material to metallic material is usually 80 : 20. Which of the following is a cermet ?
(a) Zirconia (b) Alumina
(c) Bakelite (d) Tungsten carbide

195. Residual stresses in materials
(a) is always harmful.
(b) is independent of external loads.
(c) acts when external load is applied.
(d) becomes zero, when the external load is removed.

196. Which of the following materials can not be used for making patterns ?
(a) Lead (b) Mercury
(c) Wax (d) Aluminium

197. Duralumin is an alloy of aluminium & copper. Percentage of copper in duralumin is
(a) 38 (b) 26
(c) 16 (d) 4

Answers

184. (b)	185. (d)	186. (c)	187. (a)	188. (a)	189. (a)
190. (c)	191. (b)	192. (d)	193. (c)	194. (d)	195. (b)
196. (a)	197. (d)				

198. Material of construction of foundary crucible is
   (*a*) lead (*b*) stainless steel
   (*c*) graphite (*d*) cast iron

199. The pyramid indentor apex in the Vicker's hardness testing is
   (*a*) 163° (*b*) 151°
   (*c*) 136° (*d*) 118°

200. Grading of sand is done on the basis of its
   (*a*) permeability.
   (*b*) grain size & clay content.
   (*c*) source of origin.
   (*d*) strength.

201. Stellite is a trade name for the
   (*a*) cermets.
   (*b*) ferrous cast alloys.
   (*c*) non-ferrous cast alloys.
   (*d*) products made by powder metallurgy techniques.

202. Centrifugally cast iron pipe has poor machinability due to.
   (*a*) dense structure.
   (*b*) chilling.
   (*c*) high mould rotation speed.
   (*d*) segregation.

203. Which of the following pairs regarding the effect of alloying elements in steel are not matched correctly ?
   (a) Cobalt: It contributes to red hardness by hardening ferrite.
   (*b*) Molybdenum: It forms abrasion resisting particles.
   (*c*) Phosphorous : It improves machinability in free cutting steels.
   (*d*) Silicon : It reduces oxidation resistance.

204. Match the effects on the properties of materials of various heat treatment processes.

List I	*List II*
(*a*) Normalising	I. Improves the hardness of whole mass.
(*b*) Martempering	II. Increases surface hardness.
(*c*) Nitriding	III. Improves ductility.
(*d*) Annealing	IV. Refines grain structure.

205. Heating the hypoeutectoid steels to 30°C above the upper critical temperature line, soaking at that temperature and then cooling slowly to the room temperature to form a pearlite & ferrite structure is called the
   (*a*) tempering (*b*) hardening
   (*c*) annealing (*d*) normalising

206. Hot extrusion of aluminium is done in the temperature range of .... °C.
   (*a*) 550-600 (*b*) 450-500
   (*c*) 350-400 (*d*) 250-300

207. Which of the following is not an ore of aluminium ?
   (*a*) Troosite (*b*) Kaolin
   (*c*) Alunite (*d*) Cryolite

208. A material being tested for endurance strength is subjected to the .... load.
   (*a*) impact
   (*b*) completely reversed
   (*c*) dynamic
   (*d*) static & dynamic

209. The process of heating the cold pressed metallic powder is called .... in powder metallurgy.
   (*a*) precipitation (*a*) fusion
   (*c*) sintering (*d*) agglomeration

Answers

198. (c) 199. (c) 200. (b) 201. (c) 202. (b) 203. (b)
204. (IV) (I) (II) (III) 205. (c) 206. (c) 207. (a) 208. (b)
209. (c)

210. No cutting fluid is normally used, while machining
   (a) cast iron (b) aluminium
   (c) mild steel (d) stainless steel

211. Three elastic constants for a completely anisotropic elastic material, which follows the Hooke's law are Young's modulus, elastic limit stress & yield stress. Which of the following properties is the most sensitive to increase in strain rate ?
   (a) Tensile strength.
   (b) Elastic limit.
   (c) Proportional limit.
   (d) Yield strength.

212. Fullering is done to .... the material in the forging operation.
   (a) upset (b) draw out
   (c) extrude (d) bend

213. Increase in temperature, in general results in the
   (a) decrease in the viscosities of both liquids & gases.
   (b) increase in the viscosities of both liquids & gases.
   (c) increase in the viscosity of liquids & decrease in that of gases.
   (d) decrease in the viscosity of liquids & increase in that of gases.

214. Which of the following commonly used condenser tube materials has the lowest thermal conductivity ?
   (a) Admirability brass.
   (b) Stainless steel.
   (c) Aluminium brass.
   (d) Titanium.

215. Half life of a radioactive isotope corresponds to the time required for half of the .... to decay.
   (a) atoms (b) neutrons
   (c) nucleons (d) electrons

216. Factor of safety is the ratio of the .... stress to the working stress.
   (a) tensile (b) compressive
   (c) yield (d) bearing

217. Heating of steel above critical point and then quenching in water gives .... structure.
   (a) spherodite (b) acicular
   (c) bainite (d) sorbite

218. Property of a material to resist a tensile force without giving away is a measure of its
   (a) stiffness (b) toughness
   (c) resilience (d) tenacity

219. Green strength of sand is more for fine particles. High dry strength of the moulding sand produces the following defect.
   (a) Pipes (b) Blow hole
   (c) Scab (d) Hot tear

220. Skim bob in casting refers to the part of the pouring system. The volume of the riser is kept between .... per cent of the casting, depending on the metal being cast.
   (a) 60-80 (b) 25-55
   (c) 10-15 (d) 1-5

221. During solidification of molten metal in a mould, the equiaxed crystals are produced near the wall of the mould. Defect caused by two streams of metals that are too cold to fuse properly is termed as the
   (a) cold short
   (b) pour short
   (c) hot tear
   (d) none of these

Answers

210. (a)	211. (c)	212. (b)	213. (d)	214. (b)	215. (a)
216. (c)	217. (b)	218. (d)	219. (d)	220. (b)	221. (a)

222. .... is the process of coating the surface of steel with aluminium oxide, thereby imparting it increased resistance to corrosion & oxidation upto a temperature of 700°C.
(*a*) Vaneering (*b*) Galvanising
(*c*) Electroplating (*d*) Calorising

223. The forms in which various metals & non metals occur in nature are called minerals. Which of the following is not a metalliferrous mineral ?
(*a*) Cassiterite (*b*) Sphalerite
(*c*) Chalcopyrite (*d*) Calcite

224. Which of the following is the most effective inhibitor of grain growth, when added in small quantity ?
(*a*) Manganese (*b*) Vanadium
(*c*) Cobalt (*d*) Copper

225. Increasing sulphur content in pig iron tends to make it
(*a*) soft (*b*) hard
(*c*) tough (*d*) ductile & malleable

226. Addition of small amount of .... to grey cast iron is done to produce nodular grey cast iron.
(*a*) manganese (*b*) phosphorous
(*c*) magnesium (*d*) chromium

227. Frother is added in the froth floatation cell used in ore beneficiation to stabilise the air bubbles (*i.e.*, froth), which will hold the ore particles, but it does not affect the floatability of minerals. Which of the following is not used as a frother ?
(*a*) Cresylic (*b*) Xanthates
(*c*) Pine oil (*d*) All 'a', 'b' & 'c'

228. Distillation process is adopted in the extraction of volatile metals from its ore or concentrate heated to the boiling point and then condensing the vapor of metals in a suitable condenser. Which of the following metals is not volatile & hence distillation, is not employed in its extraction from ore?
(*a*) Zinc (*b*) Aluminium
(*c*) Mercury (*d*) Cadmium

229. Matte being bright & brittle is an intermediate product comprising of heterogenous mixture of metallic sulphides in which metallic contents are concentrated. Production of matte is an important step in the metallurgy of .... whose sulphide ore due to its large gaunge content runs the risk of loosing its precious metallic content in the large quantity of slag produced during smelting.
(*a*) zinc & cobalt
(*b*) copper & nickel
(*c*) lead & aluminium
(*d*) tin & silver

230. Pinch effect in welding is the result of expansion of gases in the arc. Arc blow is more common in
(*a*) A.C. welding.
(*b*) A.C. welding with bare electrodes.
(*c*) D.C. welding with bare electrodes.
(*d*) D.C. welding with straight polarity.

231. Large process vessels operating under extreme temperature & pressure conditions are normally lined with materials like stainless steel, aluminium, nickel & alloys, copper & alloys and titanium. The cladding thickness is generally about .... percent of the total thickness.
(*a*) 1 to 5 (*b*) 10 to 20
(*c*) 30 to 40 (*d*) 40 to 50

Answers

222. (d)	223. (d)	224. (b)	225. (b)	226. (c)	227. (b)
228. (b)	229. (b)	230. (b)	231. (b)		

232. If the melting ratio of a cupola is 8:1, then it means that coke requirement for one ton melt will be .... tons.
(*a*) 8 (*b*) 0.8
(*c*) 1/8 (*d*) 9

233. Regeneration of molecular sieve requires it to be heated to a temperature of about .... °C.
(*a*) 80-120 (*b*) 200-300
(*c*) 600-800 (*d*) 1000-1100

234. Hardness & tensile strength of austenitic stainless steel can be increased by
(*a*) martempering (*b*) cold working
(*c*) normalising (*d*) hardening

235. An electrostatic precipitator is normally used for separating particles from gases, when the
(a) gases contain very high concentration of solids.
(*b*) particle size is less than one micron,
(*c*) gases contain high concentration ofcarbon monoxide.
(*d*) particle size is more than one mm.

236. Which of the following is capable of acting both as an acid flux as well as a basic flux in the metal extraction from its ore ?
(*a*) Silica (*b*) Lime
(*c*) Alumina (*d*) Fluorite

237. Dip stick used for measuring the level of petro fuels (*e.g.*, petrol, diesel, fuel oil etc.) in the storage tank is usually made of
(*a*) brass (*b*) copper
(*c*) high carbon steel (*d*) aluminium

238. The function of neutral flux used in the pyrometallurgy of metal extraction is to increase the .... of the slag.
(*a*) fluidity (*b*) basicity
(*c*) acidity (*d*) viscosity

239. Which of the following is an undesirable property of slag produced during the pyrometallurgical method of metal extraction from its ore ?
(*a*) High fusibility & fluidity.
(*b*) High thermal conductivity.
(*c*) Low density.
(*d*) High chemical activity.

240. Tensile strength of a mild steel specimen can be roughly predicted from the result of the .... hardness test.
(*a*) Vicker's (*b*) Brinell
(*c*) Rockwell (*d*) Shore's scleroscope

241. The ultimate strength in tension for steel is .... times the ultimate strength in shear.
(*a*) 0.5 (*b*) 0.9
(*c*) 1.5 (*d*) 2.5

242. Work hardenability of a material is determined by the
(*a*) Shore's scleroscope.
(*b*) Herbert Pendulum.
(*c*) Vicker's hardness test.
(*d*) Rockwell hardness test.

243. Which of the following is the most suitable material of construction for the condenser tubes, where the cooling medium is brine (salty water) ?
(*a*) Aluminium (*b*) Copper
(*c*) Titanium (*d*) Stainless steel

244. Buffing is the process of
(*a*) providing aluminium coating on the metal.
(*b*) producing lustre on metal surface.
(*c*) cleaning of castings.
(*d*) preventing corrosion damage of metal.

Answers

232. (a)	233. (b)	234. (b)	235. (b)	236. (c)	237. (a)
238. (a)	239. (b)	240. (b)	241. (c)	242. (b)	243. (c)
244. (b)					

245. .... is the trade name assigned to a non-ferrous cast alloy composed of cobalt, chromium & tungsten.
(*a*) Stellite (*b*) High speed steel
(*c*) Cermet (*d*) Alnico

246. Carbon percentage in high speed steel tool material is 0.6 to 1.0. The hardness of carbon tool steels is increased, when it is alloyed with
(*a*) vanadium & chromium
(*b*) silicon
(*c*) tungsten
(*d*) manganese

247. A fluid energy mill is used for
(*a*) cutting (*b*) grinding
(*c*) ultra grinding (*d*) crushing

248. Broaching tools are usually made of
(*a*) cermets
(*b*) high speed steel
(*c*) tungsten carbide
(*d*) stellite

249. Which is the only metal that exists in liquid state at room temperature ?
(*a*) Sodium (*b*) Mercury
(*c*) Lithium (*d*) Strontium

250. Killed steels
(*a*) are free from oxygen.
(*b*) contain not even traces of sulphur & phosphorous.
(*c*) are produced by L.D. process only.
(*d*) contain minimum impurity level.

251. Match the tests done to determine the various material properties.
List I
(*a*) Penetration resistance
(*b*) Toughness
(*c*) Endurance limit
(*d*) Ductility
List II
I. Hardness test
II. Impact test
III. Fatigue test
IV. Tension test

252. Match the nature of strain produced with the elastic properties of an isotropic elastic material.
List I
(*a*) Volumetric strain
(*b*) Normal strain
(*c*) Transverse strain
(*d*) Shear strain
List II
I. Young modulus
II. Bulk modulus
III. Modulus of rigidity
IV. Poisson's ratio

253. Match the property conferred on steel by various alloying elements addition in steel.
List I
(*a*) Heat resistance
(*b*) Hardenability
(*c*) Magnetic permeability
(*d*) Corrosion resistance
List II
I. Nickel II. Silicon
III. Chromium IV. Tungsten

254. The same volume of all gases is representative of their
(*a*) specific gravities
(*b*) densities
(*c*) gas characteristic constants
(*d*) molecular weights

Answers

245. (a) 246. (a) 247. (c) 248. (b) 249. (b) 250. (a)
251. (I) (II) (III) (IV) 252. (II) (I) (IV) (III) 253. (IV) (I) (II) (III) 254. (d)

255. Holes in a plate for rivetted joints is preferably made by
(a) drilling (b) tapping
(c) reaming (d) punching

256. Which of the following material handling equipment is not suitable for moving materials in varying paths ?
(a) Hand trolley (b) Belt conveyor
(c) Crane (d) Truck

257. Gas carrying pipelines are best joined by
(a) nuts & bolts (b) rivetting
(c) welding (d) none of these

258. Suitability of steel for its use in cable is judged by its strength in
(a) tension (b) compression
(c) torsion (d) shear

259. Out of the following materials, the one which generally exhibits an yield point is
(a) hard bronze
(b) soft brass
(c) cold rolled mild steel
(d) hot rolled mild steel

260. The behaviour of a metal specimen, which when plastically strained in tension reduces its yield stress in compression & vice versa, is termed as the
(a) work hardening
(b) Baus chinger effect
(c) creeping effect
(d) stress recovery effect

261. An "orange peel" effect is observed
(a) as a defect during metal casting/welding.
(b) during deep drawing of metallic sheet having large grain size.
(c) during faulty heat treatment of steel.
(d) during extrusion of metal through a die.

262. Square steel key is normally strong in failure by shear & crushing. Keys are normally made of .... steel bars.
(a) hot rolled mild
(b) cold rolled mild
(c) forged
(d) machinable stainless

263. Viscoelastic behaviour is observed in .... materials.
(a) non-crystalline organic polymeric
(b) ceramic
(c) all crystalline
(d) all amorphous

264. The capacity of a material to absorb energy depends upon both its strength & stiffness. The term 'stiffness' of a material is the opposite of its
(a) flexibility (b) plasticity
(c) hardness (d) hardenability

265. Elimination of brittleness resulting from welding of saw blades is done by .... of the welded portion."
(a) annealing (b) toughening
(c) work hardening (d) tempering

266. .... property of steel increases by addition of large amount of silicon in it.
(a) Mechanical (b) Machining
(c) Magnetic (d) Refractory

267. Number of electrons in the outermost orbit of a semi-conductor is
(a) 2 (b) 4
(c) 6 (d) 8

268. The hardness of a grinding wheel is specified by the
(a) alphabetical letter
(b) Rockwell hardness number
(c) Brinell hardness number
(d) none of these

Answers

255. (a)	256. (b)	257. (c)	258. (a)	259. (d)	260. (b)
261. (b)	262. (b)	263. (a)	264. (a)	265. (a)	266. (c)
267. (b)	268. (a)				

269. Which of the following materials has the poorest electrical conductivity ?
(*a*) Carbon (*b*) Aluminium
(*c*) Silver (*d*) Stainless steel.

270. Toughness of a material is measured by the
(*a*) Vicker's hardness test
(*b*) Scratch test
(*c*) Universal testing machine
(*d*) Notched bar test

271. Luder lines (stretcher strains) are observed
(*a*) during extrusion of a metal through a die.
(*b*) during faulty heat treatment.
(*c*) during forming in the low carbon steel sheet due to sudden local elongation because of its sharp yield point characteristics.
(*d*) as casting defects while processing plastics.

272. Principal alloying elements of cast tool alloys which have very high wear resistance & high temperature strength are
(*a*) cobalt, chromium & tungsten
(*b*) cobalt, chromium & nickel
(*c*) molybdenum, tungsten & chromium
(*d*) cobalt, zirconium & molybdenum

273. Grain size of moulding sand is expressed in terms of grain fineness number. Toughness of sand is measured by the .... test.
(*a*) scratch (*b*) Charpy
(*c*) shatter (*d*) Sceleroscope

274. Pick out the wrong statement.
(*a*) Age hardening phenomenon is not observed in ferrous alloys.
(*b*) Precipitation of particles from supersaturated solution of the compound takes place in precipitation hardening.
(*c*) The atoms in a metal crystal are arranged in a regular, repeated three dimensional pattern considered to be tiny spheres of diameter 10 cm.
(*d*) The crystal structure of polymorphic metals changes at the transition temperature.

275. Aluminium shots are added in the steel teeming laddle as a/an
(*a*) flux
(*b*) alloying element
(*c*) deoxidiser
(*d*) grain refiner as well as deoxidiser

276. The mechanism involved in the removal of metal in drilling operation is by
(*a*) compression (*b*) extrusion
(*c*) shearing (*d*) both '*b*' & '*c*'

277. Which of the following is the most suitable abrasive for grinding high tensile strength materials ?
(*a*) Silicon carbide (*b*) Corundum
(*c*) Aluminium oxide (*d*) Boron carbide

278. Which of the following is not a natural abrasive ?
(*a*) Silicon carbide (*b*) Quartz
(*c*) Corundum (*d*) Emery

279. Pick out the wrong statement about the machinability of metals. Machinability of a metal
(*a*) decreases with increase in strain hardening tendencies.
(*b*) decreases with increase in hardness, in general.
(*c*) depends on the composition, microstructure and physical & mechanical properties.
(*d*) decreases with increase in tensile strength & decrease in grain size.

Answers

269. (a)	270. (c)	271. (c)	272. (a)	273. (c)	274. (a)
275. (d)	276. (d)	277. (c)	278. (a)	279. (d)	

280. Highest cutting speed is achieved by the .... tool material.
   (a) high speed steel
   (b) carbide
   (c) cast alloy
   (d) plain carbon steel

281. Pick out the wrong statement.
   (a) Processing of thermosetting plastics is ideally done by injection moulding.
   (b) Processing of thermoplastics is ideally done by compression moulding.
   (c) Mass production of seamless pipes are done by electrical resistance welding (ERW)
   (d) Stretch forming is a process of cold drawing.

282. .... is the process used for setting up compressive stresses in the surface of a metal to improve its fatigue strength,
   (a) Lancing (b) Shot peening
   (c) Slugging (d) Spinning

283. Yield strength of a material is determined by the .... test.
   (a) creep (b) tension
   (c) compression (d) endurance

284. Temper brittleness of a material can be fairly detected by the .... test.
   (a) fatigue
   (b) notched bar impact
   (c) tensile
   (d) hardness

285. A jig is used while .... a hole.
   (a) boring (b) counter boring
   (c) drilling (d) enlarging

286. The softness or hardness of a grinding wheel depends upon the type & amount of bonding material used. For general purpose cutter grinding, ... grinding wheel is normally used.
   (a) hard (b) soft
   (c) silicon carbide (d) aluminium oxide

287. Flange wrinkling defect is observed in
   (a) forging (b) deep drawing
   (c) bending (d) none of these

288. Grinding wheel made of silicon carbide is most suitable for grinding materials of low tensile strength (e.g. tough bronze). Grinding is a .... operation.
   (a) turning (b) dressing
   (c) forming (d) surface finishing

289. A dense structure of grinding wheel is not used for
   (a) ductile materials
   (b) hard materials
   (c) brittle materials
   (d) finishing cuts

290. A fine grained grinding wheel should not be used for grinding .... materials.
   (a) hard (b) brittle
   (c) soft & ductile (d) none of these

291. For grinding of softer materials, the grinding wheel should have .... grain size.
   (a) finer (b) coarser
   (c) medium (d) any type of

292. Aluminium oxide (an abrasive) is used for grinding
   (a) ceramics (b) high speed steel
   (c) cemented carbide (d) none of these

Answers

280. (b)	281. (c)	282. (b)	283. (b)	284. (b)	285. (c)
286. (d)	287. (b)	288. (d)	289. (a)	290. (c)	291. (b)
292. (b)					

293. Circular cross-section machine parts which are symmetrical about the axis of rotation are made by hot
(*a*) piercing (*b*) spinning
(*c*) drawing (*d*) extrusion

294. A spring material should have low
(*a*) elastic limit (*b*) deflection value
(*c*) fatigue resistance (*d*) none of these

295. Resilience of a bolt can be increased by increasing its
(*a*) length (*b*) shank diameter
(*c*) head diameter (*d*) none of these

296. Machine tool bodies are usually made of
(*a*) cast iron (*b*) mild steel
(*c*) wrought iron (*d*) copper alloys

297. Fatigue limit improvement by over stressing the metal by successively increasing the load is called coaxing. In fatigue failure, the material fails
(*a*) below the yield point
(*b*) above the yield point
(*c*) below the elastic limit
(*d*) at the elastic limit

298. Endurance is a measure of fatigue limit of a metal. Fatigue crack in nitrided machine parts originates
(*a*) at the surface
(*b*) within the core
(*c*) just below the surface
(*d*) anywhere on the surface

299. The yield point in fatigue loading compared to that in static loading is less. The ratio of ultimate tensile strength to endurance limit of a material subjected to fatigue loading is
(*a*) 0.5 (*b*) 1
(*c*) 1.5 (*d*) 2

300. The bolt is subjected to .... when the nut is tightened by putting the washer beneath it.
(*a*) tension (*b*) shear
(*c*) compression (*d*) bending & tension

301. Ultimate strength in tension as compared to that in shear for steel is
(*a*) half
(*b*) one third
(*c*) one and half times
(*d*) one fourth

302. Fatigue limit of a material should not be increased by
(*a*) shot peening
(*b*) surface decarburisation
(*c*) understressing
(*d*) cold working

303. Hollow shafts can be made as strong as solid shafts by making the twisting moments of both the shafts same. Shafts made by .... have residual stresses.
(*a*) forging (*b*) cold rolling
(*c*) hot rolling (*d*) casting

304. Springs are not made of
(*a*) phosphor bronze
(*b*) nickel steel
(*c*) monel
(*d*) carbon steel (C 0.5%)

305. Large diameter shafts are manufactured by
(*a*) hot rolling
(*b*) cold drawing
(*c*) casting & turning to size
(*d*) forging & turning to size

Answers

293. (b)	294. (d)	295. (a)	296. (a)	297. (a)	298. (b)
299. (d)	300. (a)	301. (c)	302. (b)	303. (b)	304. (d)
305. (d)					

306. A material subjected to .... must have high resilience.
(*a*) shock loading (*b*) vibration
(*c*) fatigue (*d*) tension

307. The kinematic viscosity (in stoke) and the absolute/dynamic viscosity (in poise) are the same for .... at room temperature.
(*a*) air (*b*) water
(*c*) mercury (*d*) alcohol

308. Hot dipping process is used for coating a low melting point metal (*ie.g.*, Pb, Sn, Zn) on iron, steel & copper having relatively higher melting points. Which of the following is not a hot dipping process ?
(*a*) Galvanising (*b*) Tinning
(*c*) Sheardising (*d*) none of these

309. Pick out the wrong statement about coating/electroplating of metals.
(*a*) Phosphate coating is used as an excellent base for paints & enamels.
(*b*) Phosphate coating is less corrosion resistant then chromate coating.
(*c*) The base metal to be electroplated is normally made the cathode of the electrolytic cell and coating metal is made the anode.
(*d*) Calorising is the process of coating steel with chromium.

310. Material having negative coefficient of linear expansion is
(*a*) copper (*b*) aluminium
(*c*) rubber (*d*) nickel

311. Which of the following materials has the least scrap value ?
(*a*) Aluminium (*b*) Stainless steel
(*c*) Brass (*d*) Copper

312. Out of the following surfaces, the emissivity will be minimum for
(*a*) aluminium foil
(*b*) concrete
(*c*) plaster
(*d*) smooth glass

313. Cladding of dissimilar metal is done by .... welding.
(*a*) explosive (*b*) seam
(*c*) forge (*d*) ultrasonic

314. Soldering iron is made of wedge shape in order to
(*a*) retain heat
(*b*) retain solder
(*c*) apply high pressure at edge
(*d*) facilitate molecular attraction

315. Molten iron is desulphurised by adding .... to the ladle.
(*a*) ferrosilicon (*b*) ferromanganese
(*c*) graphite (*d*) soda ash

316. Cold ducts are
(*a*) machine defect.
(*b*) casting defect due to two streams not able to fuse due to being cool.
(*c*) pores in the weld.
(*d*) forging defects caused by insufficient filling.

317. Zinc chloride is used in
(*a*) brazing (*b*) hardening
(*c*) tempering (*d*) soft soldering

318. The most notable precipitation hardenable alloys are those in which the base metal is
(*a*) nickel (*b*) magnesium
(*c*) copper (*d*) manganese

Answers

306. (a)	307. (b)	308. (c)	309. (d)	310. (c)	311. (b)
312. (b)	313. (d)	314. (a)	315. (a)	316. (d)	317. (d)
318. (b)					

319. Case hardening is the only method suitable for hardening .... steel
(*a*) high speed (*b*) high carbon
(*c*) low carbon (*d*) high alloy

320. The free carbon is distributed throughout the mass in the form of .... in ductile cast iron.
(*a*) crystals (*b*) flakes
(*c*) nodules (*d*) needles

321. Which of the following processes involves the preparation of objects from pressed powders?
(*a*) Shell moulding
(*b*) Permanent mould casting
(*c*) Semi-centrifugal casting
(*d*) None of these

322. The binding material for cementite carbide tool is
(*a*) nickel (*b*) chromium
(*c*) cobalt (*d*) iron

323. Trimming is the process associated with
(*a*) forging (*b*) machining
(*c*) press work (*d*) electroplating

324. Match the parameters related to the mechanical properties of the metals.
List I
(*a*) Resilience (*b*) Isotropy
(*c*) Hardness (*d*) Malleability
List *II*
I. Cold rolling
II. Impact load
III. Direction
IV. Indentation

325. Match the products made out of various types of steels.
List I
(*a*) Tool steel
(*b*) High carbon steel
(*c*) Mild steel
(*d*) Medium carbon steel
List II
I. Commercial beam
II. Screw driver
III. Blanking dies
IV. Crane hooks

326. Match the nature of products made out of various polymeric materials.
List I
(*a*) Foam
(*b*) Adhesive
(*c*) Film
(*d*) Fibre
List II
I. Epoxy resin
II. Polyurethane
III. Orion
IV. Polythene

327. Match the characteristics of following properties of material.
List I
(*a*) Hardness
(*b*) Toughness
(*c*) Endurance strength
(*d*) Deflection in a beam
List II
I. Moment area method
II. Fatigue loading
III. Energy absorbed before fracture in a tension test
IV. Abrasion resistance

Answers

319. (c) 320. (c) 321. (d) 322. (c) 323. (c) 324. (II) (III) (IV) (I)
325. (III) (II) (I) (IV) 326. (II) (I) (IV) (III) 327. (IV) (III) (II) (I)

328. Match the properties of various bearing materials.

List I

(*a*) Sintered powdered metal

(*b*) Bronze

(*c*) Cast iron

(*d*) Babbit metal

List II

I. Suitable for high loads and low speed

II. Porous

III. Good embeddability

IV. Runs well with cast iron journals

329. .... is the hardest oxide and is hence used where high wear resistance at high temperature is required.

(*a*) Beryllia (*b*) Zirconia

(*c*) Alumina (*d*) Magnesia

330. .... test is the most commonly applied test to steel of unknown quality for the identification purpose.

(*a*) Spark (*b*) Fracture

(*c*) Impact (*d*) Acid-etch

331. Out of the following processes, which one will produce the strongest component?

(*a*) Extrusion (*b*) Forging

(*c*) Die casting (*d*) Cold rolling

332. When metals are heated to high temperature, their .... increases in general.

(*a*) atomic vibration

(*b*) creep strength

(*c*) yield stress & ultimate tensile strength

(*d*) toughness and stiffness

333. While the thermosetting polymers are amorphous in nature, the thermoplastic polymers are either amorphous or crystalline. The crystalline polymers are characterised by their

(*a*) low impact strength

(*b*) high flexibility

(*c*) better finish and surface appearance

(*d*) high plastic deformation

334. Transformation range for ferrous material is the temperature interval during which .... is formed during its heating.

(*a*) cementite (*b*) austenite

(*c*) martensite (*d*) pearlite

335. Increasing the temperature of a material makes it

(*a*) harder (*b*) softer

(*c*) tougher (*d*) stronger

336. Materials having resistivity ranging from 1 to 100 ohm. cm is termed as

(*a*) conductor (*b*) insulator

(*c*) semi-conductor (*d*) none of these

337. Fibre glass is a composite material of

(*a*) glass fibre in a polymer matrix

(*b*) glass fibre in a metallic matrix

(*c*) polvmer fibre in a glassy matrix

(*d*) both 'b'& 'c'

338. Materials used for making permanent magnets containing iron and nickel are made by .... technique.

(*a*) powder metallurgy

(*b*) casting

(*c*) forging

(*d*) none of these

339. Which of the following materials does not form adherent oxide film on surface ?

(*a*) Copper (*b*) Nickel

(*c*) Aluminium (*d*) Gold and silver

Answers

328. (II) (I) (IV) (III)	329. (c)	330. (a)	331. (b)	332. (a)	333. (a)
334. (b)	335. (b)	336. (c)	337. (a)	338. (c)	339. (d)

340. Between 230 and 370° C, blue brittleness is caused in mild steel because of the
(a) immobility of dislocation
(b) strain-ageing
(c) increase in Young's modulus
(d) strain hardening

341. Which of the following relationships is correct for relating the three elastic constants of an isotropic elastic material ?
(a) $E = 2\ G\ (1 + v)$
(b) $E = G(1 + v)$
(c) $E = f(1 + v)$
(d) $E = 2\ G\ (1 + 2v)$
where, E = Young's modulus, G = Modulus of rigidity or shear modulus v =, Poisson's ratio

342. On increasing the yield strength of a steel from 450 to 650 MPa, its .... will increase.
(a) ultimate tensile strength (UTS)
(b) modulus of toughness
(c) percent elongation
(d) modulus of resilience

343. The two elements required to form substitutional solid solution should not have
(a) same crystalline structure
(b) same valancy
(c) widely differing electronegativity
(d) same atomic sizes

344. All of the following alloying elements of steel increase hardness but sacrifice ductility, except
(a) nickel (b) vanadium
(c) molybdenum (d) chromium

345. Non-ferrous alloys used for making cutting tools need not have high
(a) abrasion resistance
(b) toughness
(c) red hardness
(d) cutting speed

346. Plastics as a material of construction suffer from the drawback of low
(a) machinability
(b) density
(c) strength
(d) plastic deformation

347. Steel rods are normally used for concrete reinforcement because concrete and steel have almost equal
(a) tensile strength
(b) compressive strength
(c) Young's modulus
(d) thermal coefficient of expansion

348. Hard magnetic materials have high
(a) coercive force
(b) residual magnetic induction
(c) both 'a' & 'b'
(d) neither 'a' nor 'b'

349. Resistance of an electrical conductor is proportional to its
(a) A^2 (b) 1
(c) A^2 (d) both 'b' & 'c'
where, 1 = length & A = cross-sectional area of the conductor

350. Change in volume of metals from absolute zero temperature to their melting points is
(a) almost same for all metals.
(b) different for different metals.
(c) less for low melting point metals.
(d) less for high melting point metals.

Answers

340. (b)	341. (a)	342. (b)	343. (c)	344. (a)	345. (b)
346. (c)	347. (d)	348. (c)	349. (d)	350. (a)	

351. Thermal shock produced by cooling is more dangerous/deleterious than that produced by heating a surface, because of the reason that
(a) cooling produces tension on the surface.
(b) cooling produces compression on the surface.
(c) heating does not produces any thermal stress.
(d) heating produces compression on the surface.

352. Pick out the correct combination about the role of various additives used in polymers.
(a) Plasticiser : Increases the polymer strength.
(b) Heat stabiliser : Increases the maximum service temperature.
(c) Fillers : Does not affect the property of polymer but increases its weight.
(d) Lubricants : Increases the flexibility of polymers.

353. Maximum change in the hardness of martensite occurs in the carbon content range of .... percent in steel.
(a) 0.01 to 0.10 (b) 0.2 to 0.4
(c) 0.5 to 0.7 (d) 0.8 to 1.0

354. The transition temperature at which all the ferromagnetic materials become paramagnetic materials is termed as the 'Curie temperature'. Its value for cobalt is .... °C.
(a) 768 (b) 1127
(c) 1359 (d) 1480

355. Magnetic permeability of iron is increased by its
(a) decarburising
(b) alloying with cobalt
(c) purification
(d) alternate heating and cooling

356. A form of stress corrosion failure termed as 'season cracking' is generally observed in
(a) thermosetting polymers
(b) high carbon steels
(c) brasses
(d) borosilicate glasses

357. .... rubber is generally used for making 'O' rings used for vacuum sealings.
(a) Natural (b) Neoprene
(c) Butadiene (d) Nitrile

358. Pick out the wrong statement.
(a) copper is the metal having highest electronic conductivity.
(b) hardenability and weldability of metals are inversely related.
(c) covalent bonding formed by sharing of electrons is present in all semi-conductor materials.
(d) glass transition temperature applies to 'polymers' but not to 'glasses'.

359. The 'transition temperature' for ductile to brittle behaviour of steel increases with increase in the .... content in steel.
(a) carbon (b) manganese
(c) both *a* & *b* (d) neither 'a' nor 'b'

360. Measurement of the area of the indented surface is not involved in the .... hardness testing method.
(a) Vicker's (b) Rockwell
(c) Knoop (d) Brinell

361. Which of the following is not a characteristic observed in material failure by fatigue fracture ?
(a) Plastic deformation of material does not occur.
(b) Initiation of crack from below the surface does not occur.
(c) Initiation of crack occurs on the surface of the machine part.
(d) Presence of both rough and smooth zone with conchoidal markings in smooth zone of the surface.

Answers

351. (a)	352. (c)	353. (b)	354. (b)	355. (b)	356. (c)
357. (b)	358. (a)	359. (a)	360. (b)	361. (b)	

362. Mechanism/phenomenon involved in the improvement of fatigue strength of a material by shot peening is the
(*a*) induction of residual compressive stress on the surface.
(*b*) removal of stress concentration from the surface.
(*c*) grain refinement.
(*d*) surface polishing.

363. Pick out the wrong statement.
(*a*) Weldability of high carbon steel is poorer compared to low carbon steel.
(*b*) Invar is a magnetic alloy.
(*c*) Magnetic permeability of diamagnetic material is less than one.
(*d*) Martensitic transformation never goes to completion (*i.e.*, 100%) at room temperature.

364. Welding defects, which can be avoided by increasing the heat input and reducing the electrode travel speed are porosity and
(*a*) incomplete penetration
(*b*) slag inclusion
(*c*) underfill
(*d*) none of these

365. If .... then the pinhole porosity in a casting increases.
(*a*) mould sand contains high moisture
(*b*) mould sand contains hydrogen producing compounds
(*c*) sudden local chilling occurs
(*d*) metal is poured at high temperature

366. Thicker section of a casting as compared to thinner section
(*a*) contain coarser grains
(*b*) carry residual stresses
(*c*) solidify on the outside first and it takes longer time for the inner portion to solidify
(*d*) all '*a*', '*b*' and '*c*'

367. .... causes cuts and washes on the surface of castings.
(*a*) Erosion of mould sand by flowing metal
(*b*) Very high temperature of molten metal
(*c*) Entry of molten metal in the gaps between the grains of sand
(*d*) Movement of hydrogen gas through solidifying metal

368. A reversible change in the atomic structure of steel with corresponding change in the properties is known as the .... change.
(*a*) allotropic (*b*) atomic
(*c*) molecular (*d*) solidus

369. Electric power source used in an electric induction furnace is .... power.
(*a*) high frequency AC
(*b*) low frequency AC
(*c*) high frequency DC
(*d*) low frequency DC

370. In the double slag practice of steel making
(*a*) oxidising slag is made
(*b*) refinning is done under oxidising as well as reducing condition
(*c*) reducing slag is made
(*d*) sulphur is removed externally

371. Iron is
(*a*) paramagnetic (*b*) ferromagnetic
(*c*) ferroelectric (*d*) dielectric

372. In L.D. steel making, splashing is the least when the tip of the lance is just .... the slag surface.
(*a*) touching (*b*) above
(*c*) below (*d*) none of these

Answers

362. (a)	363. (b)	364. (a)	365. (d)	366. (d)	367. (a)
368. (a)	369. (b)	370. (b)	371. (b)	372. (c)	

373. The addition of lumpy iron ore during steel refining is known as .... which tends to make the slag foamy.
(a) deoxidising (b) fluxing
(c) oreing (d) none of these

374. Brass (an alloy of copper and zinc) is an example of
(a) substitutional solid solution
(b) intermetallic compounds
(c) interstitial solid solution
(d) none of these

375. Transverse cracking occurs due to the presence of
(a) nitrogen (b) phosphorous
(c) silicon (d) sulphur

376. Deoxidiser used in L.D. steel making is in the form of
(a) pellets (b) sinter
(c) pure metal (d) ferro-alloys

377. The refining reaction during steel making takes place at the .... interface.
(a) gas-gas (b) gas-slag
(c) gas-metal (d) slag-metal

378. In Hyl process of spong iron making, the ore is used in the form of pellets. The reducing agent used in Hyl process is
(a) coke
(b) graphite
(c) cooking coal
(d) reformed natural gas

379. Fluxes are used in welding in order to protect the molten metal and the surface to be joined from
(a) carburising
(b) oxidation
(c) distortion and warping
(d) unequal temperature distribution

380. Sulphur and silicon content in the iron blast furnace product (*i.e.*, pig iron) does not necessarily depend upon
(a) % sinter in burden
(b) slag volume
(c) slag basicity
(d) operating temperature

381. The correct order for descending order of machinability for following materials is:
P. Grey cast iron Q. Low carbon steel R. Wrought iron
(a) R P $\rightarrow$ Q (b) R Q $\rightarrow$ P
(c) P Q $\rightarrow$ R (d) P $\rightarrow$ R $\rightarrow$ Q

382. The degree of freedom for a system having equal number of components and phases will be
(a) 1 (b) 0
(c) 2 (d) 3

383. The reaction that yields two solid phases on cooling a single solid phase is called
(a) eutectoid (b) eutectic
(c) congruent (d) peritectoid

384. Proof stress corresponds to
(a) lower yield point
(b) elastic limit
(c) higher yield point
(d) a specified strain

385. The reaction that on heating one solid phase yields another solid phase together with one liquid phase is termed
(a) eutectic (b) peritectic
(c) eutectoid (d) peritectoid

Answers

373. (c)	374. (a)	375. (d)	376. (d)	377. (d)	378. (d)
379. (b)	380. (a)	381. (a)	382. (c)	383. (a)	384. (d)
385. (b)					

386. Recrystallisation temperature is the temperature at which recrystalisation just reaches to completion in .... hour.
(*a*) 0.5 (*b*) 1
(*c*) 1.5 (*d*) 2

387. The temperature at which a metal becomes superconductor is called the critical temperature (T_c). The transition temperature of most superconducting elements lie within the range of zero to .... °K.
(*a*) 10 (*b*) 25
(*c*) 50 (*d*) 150

388. In which of the following phases of steel, cementite is in particle form?
(*a*) Martensite (*b*) Ferrite
(*c*) Bainite (*d*) Pearlite

389. High elastic modulus in material arises from
(*a*) weak bonds
(*b*) shell potential well
(*c*) high strength of bond
(*d*) none of these.

390. Which of the following used as cutting material can sustain high temperature?
(*a*) High carbon steel alloy
(*b*) Cermets
(*c*) Composite of two metals
(*d*) None of these

391. In the phase diagram, a tie line is used for finding the
(*a*) composition of the alloy in the two phases in equilibrium.
(*b*) melting point of the alloy.
(*c*) solubility limit of the solute component of the alloy.
(*d*) proportional of the two phases in the alloy.

392. Iron-carbon alloy is ferromagnetic for compositions .... weight % of carbon.
(*a*) above 0.022 (*b*) below 0.022
(*c*) of all (*d*) above 6.7

393. In copper-nickel alloy,
(*a*) the solubility limit at 1000°C is 60 weight % nickel.
(*b*) there is no eutectic point.
(*c*) the alloy is in solid state below 1300°C for all compositions.
(*d*) a eutectic point exists for 65 wt% Ni at 1220°C.

394. Which of the following types of materials has the highest modulus of elasticity?
(*a*) Metals (*b*) Polymers
(*c*) Semi-conductors (*d*) Ceramics

395. Composite materials are made with strong fibres embedded in weaker and softer matrix to obtain strength better than strength of the matrix. Composite with molybdenum fibres and alumina matrix do not exhibit good strength, because molybednum
(*a*) dissolves in alumina.
(*b*) does not form any bond with alumina.
(*c*) fibres develop micro cracks.
(*d*) is weaker than alumina.

396. Brittle fracture occurs in material due to the
(*a*) low band strength
(*b*) presence of impurities
(*c*) presence of cracks
(*d*) rise in temperature

397. Time dependent deformation in which the material does not recover its original dimension is called
(*a*) plasticity (*b*) elasticity
(*c*) viscoelasticity (*d*) anelasticity

Answers

386. (b)	387. (a)	388. (c)	389. (c)	390. (b)	391. (a)
392. (b)	393. (b)	394. (d)	395. (c)	396. (c)	397. (c)

398. Materials that normally do not undergo creep failure are metals
    (a) having low melting points
    (b) having high melting point
    (c) at high temperature
    (d) none of these.

399. Creep curve is a plot of
    (a) strain *vs.* stress
    (b) elastic modulus *vs.* temperature
    (c) strain *vs.* time
    (d) strain *vs.* temperature

400. Ceramic materials are
    (a) basically crystalline oxides of metals
    (b) good conductors of electricity
    (c) inorganic compounds of metallic and non-metallic elements
    (d) none of these

401. Which of the following materials has the highest thermal coefficient of expansion?
    (a) Aluminium (b) Brass
    (c) Nylon (d) Fused silica

402. Which of the following materials is most suitable for thermal insulation?
    (a) Alumina (b) Aluminium
    (c) Glass (d) Perspex (PMMA)

403. The electrical conductivity of a metal is directly proportional to the
    (a) mean free path
    (b) density of free electrons
    (c) both *a* & *b*
    (d) probability of the collision of electrons within the lattice

404. Resistivity of a metal is increased by
    (a) purifying it
    (b) introducing imperfections in it
    (c) heating it
    (d) both *b* and c

405. n-type silicon is obtained by doping Si with
    (a) boron (b) arsenic
    (c) antimony (d) both *b* and c

406. Among the superconducting elements, the one with the highest transition temperature is
    (a) indium (b) lead
    (c) niobium (d) lanthanum

407. Diamagnetism is exhibited by
    (a) only those materials which possess permanent magnetic moments.
    (b) only those materials which do not possess permanent magnetic moments.
    (c) only those materials which have strong internal fields.
    (d) all materials.

408. Paramagnetic materials when placed in a magnetic field get
    (a) feebly attracted (b) feebly repelled
    (c) strongly repelled (d) strongly attracted

409. In high frequency applications, a ferrite is preferred to a ferromagnetic material because the ferrite has
    (a) high resistivity
    (b) high saturation magnetisation
    (c) high permeability
    (d) square hysterisis loop

410. Curie-Weiss law is obeyed by
    (a) paramagnetic materials
    (b) ferromagnetic materials above Curie temperature
    (c) antiferromagnetic materials
    (d) ferromagnetic materials below Curie temperature

Answers

398. (b)	399. (d)	400. (c)	401. (c)	402. (d)	403. (c)
404. (d)	405. (d)	406. (d)	407. (d)	408. (a)	409. (a)
410. (b)					

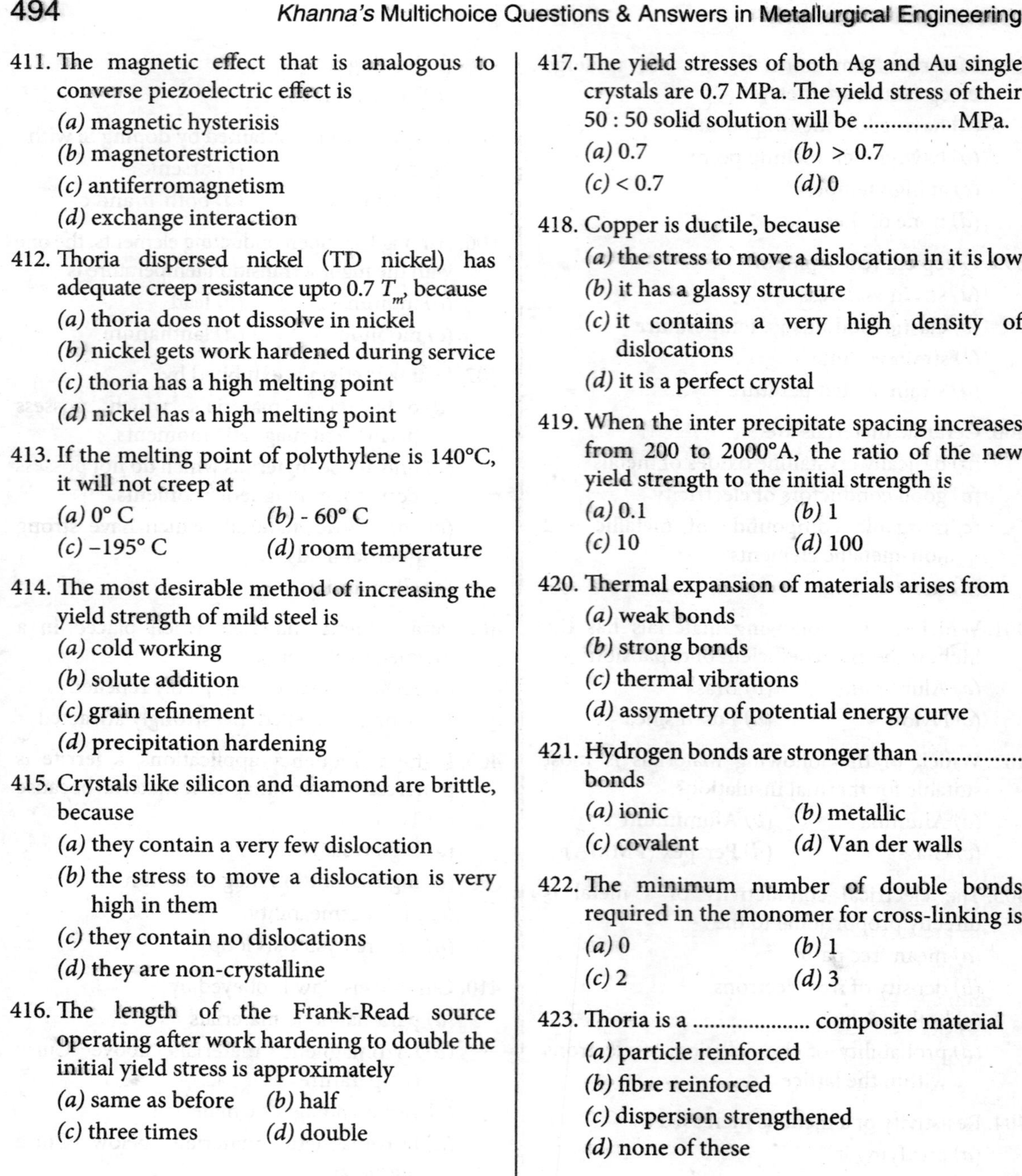

411. The magnetic effect that is analogous to converse piezoelectric effect is
   (*a*) magnetic hysterisis
   (*b*) magnetorestriction
   (*c*) antiferromagnetism
   (*d*) exchange interaction

412. Thoria dispersed nickel (TD nickel) has adequate creep resistance upto 0.7 T_m, because
   (*a*) thoria does not dissolve in nickel
   (*b*) nickel gets work hardened during service
   (*c*) thoria has a high melting point
   (*d*) nickel has a high melting point

413. If the melting point of polythylene is 140°C, it will not creep at
   (*a*) 0° C (*b*) - 60° C
   (*c*) –195° C (*d*) room temperature

414. The most desirable method of increasing the yield strength of mild steel is
   (*a*) cold working
   (*b*) solute addition
   (*c*) grain refinement
   (*d*) precipitation hardening

415. Crystals like silicon and diamond are brittle, because
   (*a*) they contain a very few dislocation
   (*b*) the stress to move a dislocation is very high in them
   (*c*) they contain no dislocations
   (*d*) they are non-crystalline

416. The length of the Frank-Read source operating after work hardening to double the initial yield stress is approximately
   (*a*) same as before (*b*) half
   (*c*) three times (*d*) double

417. The yield stresses of both Ag and Au single crystals are 0.7 MPa. The yield stress of their 50 : 50 solid solution will be .... MPa.
   (*a*) 0.7 (*b*) > 0.7
   (*c*) < 0.7 (*d*) 0

418. Copper is ductile, because
   (*a*) the stress to move a dislocation in it is low
   (*b*) it has a glassy structure
   (*c*) it contains a very high density of dislocations
   (*d*) it is a perfect crystal

419. When the inter precipitate spacing increases from 200 to 2000°A, the ratio of the new yield strength to the initial strength is
   (*a*) 0.1 (*b*) 1
   (*c*) 10 (*d*) 100

420. Thermal expansion of materials arises from
   (*a*) weak bonds
   (*b*) strong bonds
   (*c*) thermal vibrations
   (*d*) assymetry of potential energy curve

421. Hydrogen bonds are stronger than .... bonds
   (*a*) ionic (*b*) metallic
   (*c*) covalent (*d*) Van der walls

422. The minimum number of double bonds required in the monomer for cross-linking is
   (*a*) 0 (*b*) 1
   (*c*) 2 (*d*) 3

423. Thoria is a .... composite material
   (*a*) particle reinforced
   (*b*) fibre reinforced
   (*c*) dispersion strengthened
   (*d*) none of these

Answers

411. (b)	412. (a)	413. (c)	414. (c)	415. (b)	416. (b)
417. (b)	418. (a)	419. (a)	420. (d)	421. (d)	422. (c)
423. (c)					

424. Cold worked metal is .... susceptible to corrosion as compared to the same material in an annealed condition.
(*a*) more (*b*) less
(*c*) same (*d*) unpredictable

425. Contact-bend-stretch (CBS) process is a .... process.
(*a*) forging
(*b*) rolling
(*c*) stretch forming
(*d*) extrusion

426. In mechanical processing operation of a metal, its .... is not conserved.
(*a*) area (*b*) mass
(*c*) volume (*d*) none of these

427. Punching indentation into the edge of the sheet is called
(*a*) blanking (*b*) piercing
(*c*) notching (*d*) trimming

428. Poisson's ratio for a plastic material is
(a) 0.10 (b) 0.20
(c) 0.30 (d) 0.50

429. Recrystalisation temperature can not be lowered by
(*a*) purification of metal
(*b*) doping metal with impurities
(*c*) working at lower temperature
(*d*) grain refinement

430. Upgradation of hametite ore of size 0.1–1.5 mm can be best done by
(*a*) spiralling
(*b*) jigging
(*c*) tabling
(*d*) heavy media separation

431. A .... rolling mill can reduce the slab directly to strip in a single pass.
(*a*) two high (*b*) three high
(*c*) four high (*d*) planetary

432. Cutting and forming operations can be done in a single operation on .... die.
(*a*) compound (*b*) simple
(*c*) combination (*d*) progressive

433. Which type of bond is not formed during sintering?
(*a*) Covalent bond
(*a*) Diffusion or recrystalisation bond
(*c*) Solid state bond
(*d*) Slag or glass bond

434. Aluminium does not have
(*a*) good wear resistance
(*b*) very high modulus of elasticity
(*c*) very high fatigue strength
(*d*) good corrosion resistance

435. Casting process is preferred for producing parts having
(*a*) no detail
(*b*) a few details
(*c*) many details
(*d*) non-symmetrical shape

436. An ionic solid
(*a*) forms relatively hard and stable crystals
(*b*) is a poor electric conductor
(*c*) has a high vaporisation temperature
(*d*) all 'a' 'b' & 'c'

437. Ceramic compounds compared to metallic compounds
(*a*) are able to resist greater tensile stress
(*b*) crystallise more slowly
(*c*) have lower melting temperatures
(*d*) are better conductor of electricity

Answers

424. (a)	425. (b)	426. (a)	427. (c)	428. (d)	429. (b)
430. (a)	431. (d)	432. (c)	433. (a)	434. (a)	435. (c)
436. (d)	437. (d)				

438. An undesirable side effect of motional inductance of large conducting specimens is
(*a*) hysterisis (*b*) skin effect
(*c*) eddy currents (*d*) dielectric loss

439. Cold working process can be applied on the components having diameter upto .... mm.
(*a*) 10 (*b*) 25
(*c*) 50 (*d*) 70

440. Forging iron at a temperature below the recrystalisation temperature will increase the hardness of iron, because the
(*a*) increase in the density will cause pores & cavities in the metal to disappear.
(*b*) surface defects such as dislocations are formed.
(*c*) surface shrinkage from the forging operation puts pressure on the interior of the iron.
(*d*) internal stresses do not appear in the iron.

441. Match the following.
Group 1
P. Process used for making seamless tubes
Q. Process used to improve the fatigue resistance of a metal
R. Process used for making nuts and bolts
S. Parts of circular cross-section which are symmetrical
Group 2
1. Hot spinning
2. Hot piercing
3. Piercing
4. Cold peening
(*a*) P – 4, Q – 3, R – 2, S –1
(*b*) P – 3, Q – 2, R – 4, S – 1
(*c*) P – l, Q – 2, R – 4, S – 3
(*d*) P – 3, Q – 4, R – 2, S – 1

442. Chiesels used for metal cutting are never
(*a*) softened (*b*) hardened
(*c*) tempered (*d*) annealed

443. The crystal structure, which has the highest packing of atoms is .... lattice structure.
(*a*) bcc (*b*) hcp
(*c*) simple cubic (*d*) none of these

444. The nature of atomic bond found in diamond is
(*a*) ionic (*b*) covalent
(*c*) metallic (*d*) none of these

445. The Miller indices of the diagonal plane of a cube are
(*a*) 110 (*b*) 111
(*c*) 100 (*d*) 000

446. The crystal structure of most of the common metals is
(*a*) cubic (*b*) orthohombic
(*c*) hexagonal (*d*) none of these

447. Which of the following ceramic material is not isotropic?
(*a*) $Si0_2$ (*b*) $Si0_4$
(*c*) $A1_20_3$ (*d*) $MgSiO_3$

448. In rolling operation, the rolls rotate with a surface velocity .... the speed of the incoming metal.
(*a*) exceeding (*b*) lower than
(*c*) equal to (*d*) none of these

449. Swaging process is the opposite of
(*a*) forging (b) piercing
(*c*) extrusion (*d*) none of these

450. Needle is produced by
(*a*) extrusion (*b*) swaging
(*c*) forging (*d*) machining

451. Upsetting or cold heading is a .... process.
(*a*) bending (6) forging
(*c*) extrusion (*d*) rolling

Answers

438. (c)	439. (b)	440. (b)	441. (d)	442. (a)	443. (b)
444. (b)	445. (a)	446. (a)	447. (d)	448. (a)	449. (d)
450. (b)	451. (b)				

452. Electrical conductivities of semi-conductors is of the order of .... ohm.cm^{-1}.
(a) 10^{-3} (b) 10^{-5}
(c) 10^{10} (d) 10^{-10}

453. All atoms consist of electron, proton and neuton except .... atom.
(a) hydrogen (b) helium
(c) magnesium (d) boron

454. Ferromagnetic materials owe their properties to .... inner sub-shells.
(a) vacant
(b) completely filled
(c) partially filled
(d) equally filled all shells including

455. Magnetic permeability is maximum for .... materials.
(a) diamagnetic (b) paramagnetic
(c) ferromagnetic (d) none of these

456. Cobalt is a .... material.
(a) ferromagnetic (b) ferroelectric
(c) dielectric (d) none of these

457. Which of the following alloys has equal percentage of constituents?
(a) White metal (b) Gun metal
(c) Duralumin (d) Constantan

458. Which of the following effects is associated with cold forming?
(a) Surface discoloration
(b) Surface roughness
(c) Shrinkage
(d) Strain hardening

459. If there are bad effects of strain hardening on a cold formed part, the part must be
(a) hardenend (b) annealed
(c) normalised (d) tempered

460. Wood is a .... material.
(a) cellular (b) granular
(c) amorphous (d) homogenous

461. A toothpaste tube is produced by .... extrusion.
(a) solid forward (b) hollow forward
(c) solid backward (d) hollow backward

462. Which of the following metals is not good for impact extxusion?
(a) Stainless steel
(b) Low carbon annealed steel
(c) Alloys of Z_n & S_n
(d) Alloys of AI & Pb

463. Most commonly used n-type impurity doping material is
(a) phosphorous (b) silicon
(c) boron (d) arsenic/antimony.

464. Metal forming process which does not involve rotation of work piece is
(a) ring rolling (b) thread rolling
(c) upsetting (d) spinning

465. Which of the following materials is good and suitable for extrusion?
(a) Low carbon work hardened steel
(b) Brass casting
(c) Low carbon annealed steel
(d) Stainless steel

466. Which of the following has the highest thermal diffusivity?
(a) Wood (b) Iron
(c) Lead (d) Concrete

467. A semiconductor is damaged by a strong current because of .... electrons.
(a) lack of free (b) excess of
(c) decrease of (d) none of these

Answers

452. (a)	453. (a)	454. (c)	455. (c)	456. (a)	457. (d)
458. (d)	459. (b)	460. (a)	461. (d)	462. (a)	463. (d)
464. (c)	465. (c)	466. (c)	467. (b)		

468. For a given heat flow and for the same thickness, the temperature drop across the material will be maximum for
(*a*) copper (*b*) steel
(*c*) refractory brick (*d*) glass wood

469. Glass is an example of an amorphous solid which can be characterised as a
(*a*) good conductor
(*b*) very viscous fluid
(*c*) malleable solid
(*d*) crystal like structure

470. .... is harder than diamond
(*a*) Alumina ceramics
(*b*) Tantalum carbide
(*c*) Polycrystalline glass
(*d*) None of these

471. Piezoelectric is a material
(*a*) which becomes polarised when stressed.
(*b*) which changes dimensions due to applied field.
(*c*) in which magnetising force reduces when current flowing is increased.
(*d*) none of these.

472. Ferrites are
(*a*) ferro-magnetic (*b*) ferri-magnetic
(*c*) para-magnetic (*d*) anti-magnetic

473. As the carbon percentage increases in steel, its .... decreases.
(*a*) % of pearlite
(*b*) ultimate tensile strength
(*c*) ductility
(*d*) none of these

474. Which of the following is a copper free alloy?
(*a*) Muntz metal
(*b*) Phosphorous bronze
(*c*) Invar
(*d*) Brass

475. Which of the following forces does not act in case of fluids?
(*a*) Elastic force
(*b*) Vibratory force
(*c*) Centrifugal force
(*d*) Tensile force

476. Burgers vector of an edge dislocation is perpendicular to the dislocation line. The climb of dislocation occurs, when extra half plane moves .... by one atomic spacing.
(*a*) left or right (*b*) up
(*c*) down (*d*) up or down

477. A material having good damping property is likely to have
(*a*) greater defect density
(*b*) fine grained structure
(*c*) lesser defect density
(*d*) high tensile strength

478. Pick out the wrong statement.
(*a*) a single dislocation line produces several dislocation lines, when it is pinned between two obstacles.
(*b*) dislocations moving in parallel glide planes have tendency to attract each other and multiply.
(*c*) force acting upon a dislocation is the product of the shearing stress in the sliding plane of a dislocation and Burgers vector.
(*d*) the reason for occurence of yield point in mild steel is that the carbon atoms occupy the vacancy line just below the dislocation and anchor dislocation by reducing its energy.

Answers

468. (d)	469. (b)	470. (d)	471. (a)	472. (b)	473. (c)
474. (c)	475. (d)	476. (d)	477. (a)	478. (b)	

479. Match the type of defects normally encountered in various methods of processing of metals.

List I

(a) Welding

(b) Casting

(c) Forging

(d) Deep drawing

List II

I. Wrinkling and local necking

II. Cold shut or fold

III. Buckels and rat tails

IV.Cold and hot cracks

480. Match the property/characteristic of various types of cast iron.

List I

(a) Ni resist cast iron

(b) White cast iron

(c) Nodular cast iron

(d) Grey cast iron

List II

I. High toughness property

II. Carbon is present in the form of graphite flakes

III. Contains cementitic form of carbon

IV. Heat and corrosion resistant

481. Match the type of microstructure obtained, when a eutectoid steel is heated to just above $A\,i$ temperature and cooled in various quench media.

List I

(a) Coarse pearlite *(b)* Fine pearlite

(c) Very fine pearlite *(d)* Martensite

List II

I. Air cooling II. Water cooling

III. Furnace cooling IV. Oil quenching

482. Match the effect of increasing the content of various alloying element in steel on its property.

List I

(a) increase in nickel content in steel

(b) increase in carbon content in steel

(c) increase in nitrogen content in steel

(d) increase in manganese content in steel

List II

I. Ductility transition temperature of steel decreases.

II. Reduction in impact strength of steel.

III. Ductility transition temperature of steel increases.

IV. Increase in impact strength of steel.

483. Match the characteristics of elements/metals having various types of bonds.

List I

(a) Metallic bond *(b)* Ionic bond

(c) Covalant bond *(d)* Dipole

List II

I. High melting point and strength

II. Poor strength

III. High strength and ductility

IV. High strength and directionality

484. Match the percentage of copper and zinc in its various alloys.

List I

(a) Cartridge brass *(b)* Muntz metal

(c) Red brass *(d)* Admirability brass

List II

I. 60% Cu + 40% Zn

II. 71% Cu + 28% Zn + 1% Sn

III. 70% Cu + 30% Zn

IV. 85% Cu + 15% Zn

Answers

479. (IV) (III) (II) (I) 480. (IV) (III) (I) (II) 481. (III) (I) (IV) (II)

482. (IV) (III) (II) (I) 483. (III) (I) (IV) (II) 484. (III) (I) (IV) (II)

485. Match the composition of various alloys.

List I

(*a*) Gun metal

(*b*) Brazing alloy

(*c*) Vitallium

(*d*) Nimonic alloy

List II

I. 75% Ni + 21% Cr + 2.5% Ti + 1.2% AI + 0.04% C

II. 88% Cu + 10% Sn + 2% Zn

III. 50% Cu + 50% Zn

IV. 62% Co + 28% Cr + 2.5% Ni + 5.5% Mo + 1.7% Fe + 0.28% C

486. Match the melting points (°C) of various ferrous and non-ferrous alloys.

List I

(*a*) 75% Cu + 25% Pb alloy

(*b*) Babbit metal

(*c*) Cast iron

(*d*) Steel

List II

I. 1380-1500 II. 1200

III. 180-245 IV. 980

487. Match the composition of various alloys of high ohmic resistance.

List I

(*a*) Constantan

(*b*) German silver

(*c*) Nicealine

(*d*) Nichrome

List II

I. 65% Cu + 20% Zn + 15% Ni

II. 67.5% Ni + 16% Fe + 15% Cr + 1.5% Mn

III.58.5% Cu + 40% Ni + 1.5% Mn

IV. 54% Cu + 26% Ni + 20% Zn

488. Match the sources of defects with the type of defects developed during metal processing.

List I

(*a*) Moulding material

(*b*) Pouring metal

(*c*) Metallurgical defects

(*d*) Presence of hydrogen in the melt and moisture in the sand

List II

I. Pinholes and Blowholes

II. Hot spots and hot tear

III. Cold shut and misrun

IV. Rat tails and fusion

489. Match the type of defects encountered in various metal processing techniques.

List I

(*a*) Casting

(*b*) Forging

(*c*) Welding

(*d*) Deep drawing

List II

I. Cold shut or fold

II. Wrinkling and local necking

III. Buckles and rat rails

IV. Cold and hot cracks

490. A p-type germanium semi-conductor is doped with

(*a*) silicon (*b*) arsenic

(*c*) gallium (*d*) selenium

491. A polished and etched surface of the cross-section of a hot worked product will have

(*a*) carbon in the form of flakes

(*b*) carbon precipitated at the boundaries

(*c*) fibre like structure

(*d*) grain field like structure

Answers

485. (II) (III) (IV) (I) 486. (IV) (III) (II) (I) 487. (III) (I) (IV) (II)

488. (IV) (III) (II) (I) 489. (III) (I) (IV) (II) 490. (a) 491. (d)

492. Advantage of cold forming is that
   (a) grain refinement takes place
   (b) strength and hardness increases
   (c) no consequent heat treatment is needed
   (d) force required is relatively small

493. Bending process requires highest force in
   (a) bottom bending
   (b) three point bending
   (c) air bending
   (d) none of these

494. Which of the following glass is resistant to heat and chemicals?
   (a) Silica glass
   (b) Sodalime glass
   (c) Borosilicate glass
   (d) None of these

495. Out of the following .... thermocouple is capable of measuring the highest temperature.
   (a) platinum-rhodium
   (b) chromel-alumel
   (c) irridium-rhodium
   (d) iron-constantan

496. A fine powder of a refractory metal may be added to molten metal prior to solidification of the metal. The fine powder may act to
   (a) induce homogenous nucleation in the molten metal.
   (b) produce large numbers of moving dis-locations in the crystal.
   (c) reduce the average grain size of the final solidified metal.
   (d) none of these

497. Flattening is a process in which a metal strip is
   (a) bent over 18° and then pressed
   (b) pressed to remove wrinkles and kinks
   (c) pressed against rubber pad
   (d) bent to create a small kink

498. Silicon is present in
   (a) solder (b) glass
   (c) brass (d) none of these

499. Pick out the wrong statement.
   (a) Cold cracking of a weld is due to the presence of hydrogen gas in the weld.
   (b) True stress is given by, $\sigma = \sigma E\ (1 + \epsilon E)$, where σE and ϵE are engineering stress and engineering strain respectively.
   (c) Phosphorous can be easily recovered in the iron blast furnace.
   (d) High residual stress at the surface is beneficial for fatigue properties of a material.

500. Pick out the wrong statement.
   (a) In Parke's process, zinc is added to remove silver from lead.
   (b) In the Pidgeon process of producing magnesium, powdered ferro-silicon is used as the reducing agent.
   (c) Molten lead is used for shock cooling of the zinc vapours in the zinc blast furnace.
   (d) Pure copper electrodes are used as anode for electro-refinning of copper.

501. Pick out the wrong statement.
   (a) Lead can creep under its own weight at room temperature.
   (b) The electrical conductivity of gold is considerably reduced by alloying additions due to the decrease in electron movement.
   (c) Recrystallisation temperature decreases with decrease in strain in a cold worked metal.
   (d) With increase in temperature, the electrical conductivity of intrinsic semi-conductor will increase.

Answers

492. (b)	493. (b)	494. (c)	495. (a)	496. (a)	497. (a)
498. (b)	499. (c)	500. (d)	501. (c)		

502. Pick out the wrong statement.
   (*a*) Copper sulphate is used as collector in the froth floatation of sphalerite ore.
   (*b*) A combination of flash roasting and smelting is involved in the flash smelting of chalcopyrite concentrates.
   (*c*) A sessile edge dislocation can become glissile by dislocation glide.
   (*d*) FeO content of the slag in a top blown steel converter is less than that in a combined blown converter.

503. Bar of a section can be cold bent by
   (*a*) roll bending
   (*b*) rotating die and wiper block
   (*c*) series of rolls
   (*d*) matching shape wheels

504. Injection moulding is the ideal method of processing
   (*a*) thermosetting plastics
   (*b*) thermoplasics
   (*c*) non-ferrous metals
   (*d*) rubbery materials

505. Hot/cold rolling does not produce a/an .... section.
   (*a*) I- (*b*) T-
   (*c*) channel (*d*) hollow circular

506. Odd pair out of the following is
   (*a*) threading and boring
   (*b*) dimpling and flaring
   (*c*) welding and soldering
   (*d*) hobbing and swaging

507. Compression moulding is the ideal method of processing
   (*a*) thermoplastics
   (*b*) thermosetting plastics
   (*c*) non-ferrous metals
   (*d*) rubbery materials

508. Vacuum enviroment is required in .... welding.
   (*a*) electron beam (*b*) plasma arc
   (*c*) laser beam (*d*) ultrasonic

509. TTT diagram indicates the time and temperature transformation of
   (*a*) pearlite (*b*) cementite
   (*c*) austenite (*d*) ferrite

510. Pipe material not used with ammonia is
   (*a*) copper
   (*b*) stainless steel
   (*c*) wrought iron
   (*d*) all 'a', '*b*', and '*c*'

511. In a fillet welded joint, the weakest area of the weld is
   (*a*) root (*b*) toe
   (*c*) face (*d*) throat

512. The degree of freedom when FCC iron and BCC iron co-exist in equilibrium, are
   (*a*) –1 (*b*) 0
   (*c*) 1 (*d*) 5

513. The potential of a galvanic cell of copper (potential of + 0.34 V) and aluminium (potential of –1.66 V) is .... V.
   (*a*) 0 (*b*) 1.32
   (*c*) –1.32 (*d*) 2

514. The peak in corrosion rate of martensite occurs when
   (*a*) tempered at about 400°C
   (*b*) in untempered condition
   (*c*) in spherodized state
   (*d*) temperated at 900°C

Answers

502. (d)	503. (b)	504. (a)	505. (d)	506. (d)	507. (a)
508. (a)	509. (c)	510. (a)	511. (d)	512. (c)	513. (d)
514. (a)					

515. As we go down a coloumn of periodic table, the elastic modulus of the material
(*a*) increases (*b*) decreases
(*c*) remains constant (*d*) remains constant

516. The change in Young's modulus of ordinarily elastic materials between 0°K and the melting point is
(*a*) 10 – 20% decrease
(*b*) 10 – 20% increase
(*c*) 60 – 70% increase
(*d*) no change

517. In BCC iron, residual carbon occupies the mid points of
(*a*) < 100 > (*b*) < 110 >
(*c*) < 111 > (*d*) none of these

518. Which of the following is not equilibrium heat treatment?
(*a*) Annealing (*b*)Normalising
(*c*) Austenetising (*d*) Precipitation

519. Covalant and ionic solids are not suitable as structural components, because they
(*a*) are brittle
(*b*) have high elastic moduli
(*c*) are ductile
(*d*) have weak bonds

520. Pick out the wrong statement.
(*a*) Annealing temperature is less than the normalising temperature.
(*b*) Hardenability of steel is assessed by impact test.
(*c*) Martensite which is formed during quenching is too brittle.
(*d*) Austenite has FCC structure.

521. Whiskers are very thin
(*a*) copper wires (*b*) steel wires
(*c*) single crystals (*d*) none of these

522. The matrix phase of a fibrous composites may be
(*a*) ceramic
(*b*) polymer
(*c*) metal
(*d*) either of 'a', 'b' and 'c'

523. Cermets are example of .... composities.
(*a*) carbon–carbon (*b*) metal–matrix
(*c*) ceramic–metal (*d*) none of these

524. Fatigue failure of materials occur due to cyclic stress. The curve between strees *vs.* number of cycles to failure is plotted for .... test.
(*a*) shear
(*b*) hardness
(*c*) fatigue
(*d*) tensile

525. Rockwell hardness number represents the net
(*a*) increase in depth of impression
(*b*) size of diamond cone indentor used
(*c*) increase in diameter of impression
(*d*) increase in load

526. For metal-matrix composites the matrix is a
(*a*) ceramic
(*b*) polymer
(*c*) ductile metal
(*d*) none of these

527. Which of the following properties of materials are not determined by tensile test?
(*a*) Ductility
(*b*) Hardness
(*c*) Yield strength
(*d*) UTS

Answers

515. (b)	516. (a)	517. (c)	518. (d)	519. (a)	520. (a)
521. (c)	522. (d)	523. (c)	524. (c)	525. (a)	526. (c)
527. (b)					

528. Match the product in group I with the nature of reaction in group II.

Group I
P. Polythene
Q. Nylon
R. Polystyrene

Group II
1: Condensation polymerisation
2. Addition polymerisation

(*a*) P – 1, Q – 1, R – 2
(*b*) P – 2, Q – 2, R – 1
(*c*) P – 1, Q – 2, R – 1
(*d*) P – 2, Q – 1, R – 2

529. Aluminium is resistant to corrosion because, it is
(*a*) below hydrogen in the reactivity series
(*b*) rare
(*c*) protected by surface oxide layer
(*d*) inexpensive

530. What is the largest end use application of steel?
(*a*) Ship building
(*b*) Construction industry
(*c*) Automobile industry
(*d*) Oil & gas pipelines

531. The approximate density of steel is .... tonnes/cubic metre.
(*a*) 4.5 – 5 (*b*) 6.2 – 6.5
(*c*) 7.8 – 8.0 (*d*) 8.8 – 9.0

532. Stainless steel is defined as an alloy of carbon and iron that has a minimum chromium content of roughly .... %.
(*a*) 5 (*b*) 10
(*c*) 15 (*d*) 18

533. Approximate melting point range of carbon steel is .... °C.
(*a*) 1300 – 1400 (*b*) 1400 – 1450
(*c*) 1500 – 1550 (*d*) 1600 – 1620

534. Class of ores from which extraction of metals is difficult and expensive is the .... ores.
(*a*) halide (*b*) silicate
(*c*) carbonate (*d*) sulfide

535. What mass (in gm) of hematite ore that is 76% Fe_2O_3 contains 95 gm of Fe? Assuming no other form of Fe is present.
(*a*) 42.5 (*b*) 85.1
(*c*) 178.6 (*d*) 201.4

536. Tons of SO_2 produced by the roasting of 10 tons of sphalerite, which is 60% ZnS assuming that all ZnS is converted to ZnO and no other sulfur sources are present, will be
(*a*) 2.36 (*b*) 3.94
(*c*) 5.40 (*d*) 6.40

537. Pick out the wrong statement about the natural distribution of metals.
(*a*) Aluminium is the most abundant metal on earth.
(*b*) Metals make up only 25% of the earth's crust.
(*c*) The earth's core contains an accessible deposit of iron and nickel.
(*d*) Alkali and alkaline earth metals make up more than 1% of the earth's crust.

538. Pick out the wrong statement about copper.
(*a*) Copper has high electrical conductivity and hence is used to make electric wire.
(*b*) Copper is not recyclable, but is found in abundant deposits.
(*c*) Unlike many metals, copper can be found in its elemental form.
(*d*) Copper was one of the first metals to be isolated and used by humans.

Answers

528. (b)	529. (c)	530. (b)	531. (c)	532. (b)	533. (c)
534. (b)	535. (c)	536. (b)	537. (c)	538. (b)	

539. An alloy in which small, usually non-metallic atoms fit between the metallic atoms of a crystal is called .... alloy.
   (*a*) metallic (*b*) substitutional
   (*c*) interstitial (*d*) binary

540. The densest direction (highest linear density) in f.c.c. & b.c.c crystals are respectively
   (*a*) <110> and <111>
   (*b*) <111> and <110>
   (*c*) <110> and <110>
   (*d*) <111> and <111>

541. Fatigue strength from a group of samples of the same material will be the highest, if the samples are
   (*a*) polished and size is < 12 mm
   (*b*) fine turned and the size is < 12 mm
   (*c*) ground finished and the size is > 12 mm
   (*d*) drawn or forged and the size is > 12 mm

542. Two metallic plates are joined in bulk weld along length by a perfect weld. The jointed plate is likely to .... along the weld line.
   (*a*) elongate (*b*) contract
   (*c*) bend (*d*) warp

543. An eutectoid reaction is marked by
   (*a*) Solid phase → Solid phase A + Solid phase B
   (*b*) Solid phase → Solid phase A + Liquid phase B
   (*c*) Solid phase → 4 Liquid phase A + Liquid phase B
   (*d*) Solid phase → Solid phase B

544. The hardness of quenched martensite .... as carbon in steel increases.
   (*a*) increases
   (*b*) decreases
   (*c*) first increases and then remains constant
   (*d*) first increases and then decreases

545. Pick out the wrong statement.
   (*a*) Martensite has .a bcc crystal structure.
   (*b*) Austenite has a fcc crystal structure.
   (*c*) Retained austenite can be transformed into martensite.
   (*d*) Martensite is a solid solution of carbon in bcc iron.

546. Pick out the wrong statement.
   Martensite is harder than ferrite because martensite has
   (*a*) higher dislocation density
   (*b*) very fine structure
   (*c*) bcc crystal structure whereas ferrite has fcc structure.
   (*d*) distorted bcc crystal structure with carb-on on the edge whereas ferrite has bcc structure.

547. Pick out the wrong statement.
   (*a*) 60/40 brass can be easily cold worked.
   (*b*) Small amount of aluminium is added to brass to improve its properties.
   (*c*) Aluminium bronzes are not the alloys of aluminium
   (*d*) Corrosion resistance of cartridge brass is better than that of Muntz metal.

548. Aluminium oxide before being subjected to direct current electrolysis for extraction of aluminium is dissolved in
   (*a*) hot NaOH
   (*b*) molten NaF
   (*c*) water
   (*d*) both molten NaF and $A1F_3$

549. Polymers are not distinguished from metals in respect of
   (*a*) low density
   (*b*) compounds of carbon and other elements
   (*c*) amorphous structure
   (*d*) long chain molecules

Answers

539. (c)	540. (a)	541. (a)	542. (b)	543. (a)	544. (c)
545. (a)	546. (d)	547. (a)	548. (d)	549. (b)	

550. The resistivity of a superconductor
    (*a*) is very low at all temperatures.
    (*b*) is constant at very low level and is unaffected by temperature.
    (*c*) decreases uniformly to zero as temperature decreases to absolute zero (0°K).
    (*d*) suddenly becomes zero at a temperature very close to absolute zero, otherwise it increases with temperature.

551. In an intrinsic semi-conductor, number of holes is .... the number of conduction electrons.
    (*a*) less than (*b*) more than
    (*c*) equal to (*d*) not related to

552. Alloying of metals is not done to improve its
    (*a*) specific weight
    (*b*) tensile strength, elastic limit, ductility.
    (*c*) hardness and hardenability
    (*d*) weldability and wear resistance

553. Pick out the wrong statement.
    (*a*) Creep occurs in some metals at room temperature.
    (*b*) Ceramic materials mainly contain secondary bonding
    (*c*) Graphite is a good conductor of electricity.
    (*d*) Butyl rubber is used to make automobile tubes.

554. Which of the following is not a unique property of austenitic stainless steel?
    (*a*) Low corrosion resistance
    (*b*) High strength at elevated temperature
    (*c*) Good impact strength at low temperatures
    (*d*) High ductility and malleability

555. Pick out the wrong statement.
    (*a*) Magnesite is a refractory material.
    (*b*) The important property required in springs is resilience.
    (*c*) Intergranular corrosion is the most common type of corrosion is stainless steel.
    (*d*) Silicon carbide is a polymeric material.

556. Match the item in List I with that in List II.
    List I
    A. Carbide tips for machining
    B. Nitriding
    C.Griffith's criterion
    D. Dielectrics
    List II
    1. Ammonia
    2. Mica
    3. Powder metallurgy
    4. Brittle fracture

	A	B	C	D
(*a*)	4	3	2	1
(*b*)	2	1	4	3
(*c*)	3	1	4	2
(*d*)	1	3	4	2

557. Pick out the correct statement.
    (*a*) Wood is a natural composite material.
    (*b*) Non-crystalline materials have genrally cubic structure.
    (*c*) Twinning is possible only in hcp metals.
    (*d*) Machinability of grey cast iron is very good compared to that of high carbon steel.

558. Process of isothermal transformation to form bainite is
    (*a*) bright annealing
    (*b*) austempering
    (*c*) spheroidising
    (*d*) nitriding

Answers

550. (d)	551. (c)	552. (a)	553. (b)	554. (a)	555. (d)
556. (c)	557. (d)	558. (b)			

559. Strain hardening of metals results from its plastic deformation at temperature below recrystallisation temperature which increases its
   (a) corrosion resistance
   (b) fatigue strength
   (c) strength and hardness
   (d) brittleness

560. Pick out the correct statement.
   (a) Slip and twinning are important modes of elastic deformation.
   (b) Toughness of an alloy depends upon both its strength and ductility.
   (c) Mechanical working of lead at room temperature is a cold working process.
   (d) Rimmed steel is a deoxidised steel.

561. Match the items in List I with that in List II.
   List I
   A. Brass/Hydrogen embrittlement
   B. Hall-Petch equation
   C. Transition metals
   D. Cu-Be alloy
   List II
   1. Grain size dependence of yield stress
   2. Ferromagnetism
   3. Age hardening
   4. Stress corrosion cracking

	A	B	C	D
(a)	4	1	2	3
(b)	1	4	2	3
(c)	4	1	3	2
(d)	4	3	2	1

562. Pick out the wrong statement.
   (a) Polymers are good electrical conductors.
   (b) Aluminium and chromium are not suitable for high temperature.
   (c) For a protective oxide film, Pilling Bedworth (PB) ratio should be slightly more than one.
   (d) Plain carbon steel can be effectively nitrided.

563. Surface hardness of material is improved by
   (a) spheroidising
   (b) carburising
   (c) tempering
   (d) process annealing

564. Match the item in List I to that in List II.
   List I
   A. Composite material
   B. Amorphous polymer
   C. Piezoelectric effect
   D. Porcelain
   List II
   1. Quartz
   2. Atactic structure
   3. Insulator
   4. Concrete

	A	B	C	D
(a)	4	3	2	1
(b)	4	1	2	3
(c)	2	4	3	1
(d)	4	2	1	3

565. Pick out the wrong statement.
   (a) Graphite is a covalant bonded solid.
   (b) The property by virtue of which a material resists deformation is called toughness.
   (c) Fibreglass reinforced plastic (FRP) is a composite material.
   (d) Hacksaw blades are made of high carbon steel.

Answers

559. (c) 560. (b) 561. (a) 562. (a) 563. (b) 564. (d)
565. (b)

566. When a steel is cooled, austenite can transform to
(*a*) ferrite (*b*) pearlite
(*c*) martensite (*d*) either of (*a*),(*b*) & (*c*)

567. What must be done to a piece of metal to view its microstructure with an optical microscope?
(*a*) etch it
(*b*) polish it
(*c*) encapsulate in resin
(*d*) all (*a*), (*b*) & (*c*)

568. Supersaturated solid solutions are important for
(*a*) forming cementite
(*b*) precipitation hardening
(*c*) diffusion
(*d*) pearlite lamellae spacing

569. Which of the following is not important in alloying elements?
(*a*) Substitutions (*b*) Vacancies
(*c*) Intertitials (*d*) Solute atoms

570. Which of the following impedes dislocation motion?
(*a*) Voids
(*b*) Precipitates
(*c*) Voids & precipitates
(*d*) none of these

571. Stainless steel demand is mostly in the form of...... products.
(*a*) tubular (*b*) long
(*c*) flat (*d*) no particular form

572. Residuals refers to so *"tramp elements"*, which are typically introduced into the steel making process in the form of unsorted scrap. The presence of these elements can impair the final properties of the steel. Which of the following is the main *"tramp element"*?
(*a*) Lead (*b*) Copper
(*c*) Tin (*d*) Vanadium

573. The process preferred for producing grains along geometrical shape of a crank shaft will be
(*a*) welding (*b*) cold bending
(*c*) casting (*d*) forging

574. Pick out the correct statement.
(*a*) All thermoplastic polymers are crystalline.
(*b*) All thermosetting polymers are crystalline.
(*c*) All thermosetting and thermoplastic polymers are essentially amorphous.
(*d*) Thermoplastic polymers can be either amorphous or crystalline but thermosetting polymers are amorphous.

575. Pick out the correct statement.
(*a*) Metal, plastic and glass are amorphous.
(*b*) Metal and some plastics are crystalline.
(*c*) Glass and some plastics are amorphous.
(*d*) Glass and some metals are crystalline.

576. Pick out the wrong statement.
(*a*) The dielectric strength of a dielectric is the highest field (voltage per metre thickness) that can be withstood by the dielectric.
(*b*) The relative permittivity of dielectric increases with temperature but reduces with frequency.
(*c*) Polarisation of a dielectric on application of electrical field is orientation or displacement of electrons locally.
(*d*) An extrinsic semi-conductor is p-type when impurity atoms provide electron holes close to the conduction band.

Answers

566. (d)	567. (d)	568. (b)	569. (b)	570. (b)	571. (a)
572. (c)	573. (d)	574. (d)	575. (b)	576. (d)	

577. The ferromagnetic materials for detecting and transmitting small signals must have
   (a) low initial permeability.
   (b) low constant permeability through B-H curve.
   (c) high initial permeability.
   (d) constant permeability through B-H curve.

578. Work roll during hot rolling of steel slab experiences .... stress during one complete revolution.
   (a) no *(b)* cyclic
   (c) compressive *(d)* tensile

579. Defect caused during hot rolling of steel slab due to the effect of work roll flattening and thermal camber is a .... defect.
   (a) stip profile *(b)* shell
   (c) scale *(d)* surface

580. Which one of the following crystal structures has the fewest slip directions and therefore the metals with this structure are generally more difficult to deform at room temperature?
   (a) BCC. (b) FCC
   (c) HCP. (d) none

581. Grain boundaries are an example of which one of the following types of crystal structure defects?
   (a) dislocation. (b) Frenkel defect.
   (c) line defects. (d) surface defects.

582. Twinning is not
   (a) elastic deformation.
   (b) mechanism of plastic deformation.
   (c) more likely at high deformation rates.
   (d) more likely in metals with HCP structure

583. Polymers are not characterized by which of the following bonding type ?
   (a) adhesive (b) covalent
   (c) van der Waals. (d) none

584. Which one of the following is not the basic types of static stresses to which a material can be subjected ?
   (a) compression (b) shear
   (c) tensile (d) true stress

585. Which of the following is the correct definition of ultimate tensile strength, as derived from the results of a tensile test on a metal specimen?
   (a) the stress encountered when the stress-strain curve transforms from elastic to plastic behavior.
   (b) the maximum load divided by the final area of the specimen.
   (c) the maximum load divided by the original area of the specimen.
   (d) the stress observed when the specimen finally fails.

586. Which one of the following types of stress strain relationship best describes the behavior of brittle materials such as ceramics and thermosetting plastics?
   (a) elastic and perfectly plastic
   (b) elastic and strain hardening,
   (c) perfectly elastic
   (d) none of the above.

587. Which one of the following types of stress strain relationship best describes the behavior of most metals at room temperature:
   (a) elastic and perfectly plastic.
   (b) elastic and strain hardening.
   (c) perfectly elastic
   (d) none

Answers

577. (b)	578. (b)	579. (a)	580. (c)	581. (d)	582. (a)
583. (a)	584. (d)	585. (c)	586. (c)	587. (b)	

588. Which of the following types of stress strain relationship best describes the behavior of metals at temperatures above their respective recrystallization points:
(a) elastic and perfectly plastic.
(b) elastic and strain hardening.
(c) perfectly elastic
(d) none of these

589. Which one of the following materials has the highest hardness?
(a) alumina ceramic.
(b) gray cast iron.
(c) hardened tool steel.
(d) high carbon steel

590. Which one of the following metals has the lowest density?
(a) aluminum (b) copper
(c) magnesium (d) tin.

591. Which of the following materials has the highest specific heat?
(b) aluminum (b) concrete
(c) polyethylene (d) water

592. The mass diffusion rate *dm/dt* across a boundary between two different metals is not a function of which of the following variables ?
(a) concentration gradient *dc/dx*.
(b) contact area.
(c) density.
(d) temperature.

593. Which of the following pure metals is the best conductor of electricity?
(a) aluminum (b) copper
(c) gold (d) silver

594. Which of the metallic elements is the most abundant on the earth?
(a) aluminum (b) copper
(c) iron (d) magnesium

595. The predominant phase in the iron-carbon alloy system for a composition with 99% Fe at room temperature is which of the following?
(a)austenite (b) cementite
(c) delta (d) ferrite

596. A steel with 1.0% carbon is known as which of the following:
(a) eutectoid (b) hypoeutectoid
(c) hypereutectoid (d) wrought iron.

597. Which of the following is not a common alloying ingredient in steel?
(a) chromium (b) manganese
(c) nickel (d) zinc.

598. Which of the following is the most important cast iron commercially?
(a) ductile cast iron (b) gray cast iron
(c) malleable cast iron(d) white cast iron.

599. Which of the following metals has the highest density?
(a) gold (b) lead
(c) platinum (d) tungsten.

600. Which of the following has the lowest melting point?
(a) aluminum (b) lead
(c) tin (d) zinc.

601. Which one of the following is the most common element in the earth's crust?
(a) aluminum (b) calcium
(c) iron (d) oxygen

602. Glass products are based primarily on which one of the following minerals?
(a) alumina (b) corundum
(c) feldspar (d) silica.

Answers

588. (a)	589. (a)	590. (c)	591. (d)	592. (c)	593. (d)
594. (a)	595. (d)	596. (c)	597. (d)	598. (b)	599. (c)
600. (d)	601. (d)	602. (d)			

603. Which of the following ceramics are commonly not used as abrasives in grinding wheels ?
(a) aluminum oxide (b) silicon carbide
(c)silicon dioxide (d) none

604. Which one of the following is generally the most porous of the clay-based pottery ware?
(a) china clay (b) earthenware
(c) porcelain (d) stoneware.

605. Which of the following is fired at the highest temperatures?
(a) china day (b) earthenware
(c) porcelain (d) stoneware.

606. Which one of the following comes closest to expressing the chemical composition of day?
(a) Al_2O_3 (b) $Al_2(Si_2O_5)(OH)_4$,
(c) $3AL_2O_3$-$2SiO_2$ (d) MgO

607. Which one of the following materials is closest to diamond in hardness?
(a) aluminum oxide
(b) cubic boron nitride
(c) silicon dioxide
(d) tungsten carbide.

608. Which of the following best characterizes the structure of glass-ceramics?
(a) 95% polycrystalline
(b) 95% vitreous
(c) 50% polycrystalline.
(d) none of these

609. Properties and characteristics of the glass-ceramics does not include which of the following ?
(a) efficiency in processing
(b) electrical conductor
(c) strong, relative to other ceramics
(d) none of these

610. The specific gravity of graphite is closest to which one of the following?
(a) 1.0 (b) 2.0
(c) 4.0 (d) 8.0

611. Of the polymer types, which one is the most important commercially?
(a) thermoplastics (b)thermosets
(c) elastomers (d) synthetic fibres

612. Which one of the polymer types does not involve cross- linking?
(a) thermoplastics (b)thermosets
(c) elastomers (d) fibres

613. Degree of polymerization is which one of the following?
(a) average number of mers in the molecule chain.
(b) proportion of the monomer that has been polymerized.
(c) sum of the molecular weights of the mers in the molecule.
(d) none

614. As temperature of a polymer increases, its density
(a) increases
(b) decreases
(c) remains fairly constant
(d) may increase or decrease;depends on the polymer

615. As the temperature of an amorphous thermoplastic polymer is gradually reduced, the glass transition temperature Tg is indicated, when
(a) the polymer transforms to a crystalline structure.
(b) the coefficient of thermal expansion increases markedly.
(c) the polymer becomes stiff, strong, and elastic.
(d) the polymer solidifies from the molten state.

Answers

603. (d)	604. (b)	605. (c)	606. (b)	607. (b)	608. (a)
609. (b)	610. (b)	611. (a)	612. (a)	613. (a)	614. (b)
615. (c)					

616. Which of the following plastics has the highest market share?
(a) polyethylene (b) polypropylene
(c) polystyrene (d) polyvinylchloride.

617. Which of the following polymers are normally not thermoplastic?
(a) acrylics (b) cellulose acetate
(c) polychloroprene (d) polyurethane.

618. The fiber *rayon* used in textiles is based on which of the following polymers?
(a) cellulose (b) nylon
(c) polyester (d) polypropylene.

619. Among the thermosetting polymers, the most widely used commercially is which of the following?
(a) epoxies (b) phenolics
(c) silicones (d) urethanes.

620. Polyurethanes can be which of the following ?
(a) thermoplastic (b) thermosetting
(c) elastomeric (d) all a, b & c

621. The chemical formula for polyisoprene in natural rubber is which of the following:
(a) CH_2 (b) C_2H_4
(c) C_3H_6 (d) C_5H_8

622. The leading commercial synthetic rubber is which of the following ?
(a) butyl rubber
(b) isoprene rubber
(c) polybutadiene
(d) styrene-butadiene rubber

623. Anisotropic means which one of the following?
(a) composite materials with composition consisting of more than two materials.
(b) properties are the same in every direction.
(c) properties vary depending on the direction in which they are measured.
(d) strength and other properties is a function of curing temperature.

624. Which one of the following reinforcing geometries offers the greatest potential for strength and stiffness improvement in the resulting composite material?
(a) fibers. (b) flakes.
(c) particles. (d) infiltrated phase.

625. Which of the following materials are not used as fibers in fiber-reinforced plastics :
(a) aluminum oxide.
(b) boron.
(c) carbon/graphite.
(d) unsaturated polyester.

626. Which of the following metals are not commonly used as the matrix material in fiber-reinforced MMCs ?
(a) aluminum (b) copper
(c) titanium (d) zinc.

627. Which of the following metals is used as the matrix metal in nearly all WC cemented carbides?
(a) aluminum (b) chromium
(c) cobalt (d) tungsten

628. Ceramic matrix composites are designed to overcome which of the following weaknesses of ceramics ?
(a) compressive strength.
(b) hardness.
(c) modulus of elasticity.
(d) toughness.

Answers

616. (a)	617. (d)	618. (a)	619. (b)	620. (d)	621. (d)
622. (d)	623. (c)	624. (a)	625. (d)	626. (d)	627. (c)
628. (d)					

629. Which of the following polymer types are most commonly used in polymer matrix composites?
(a) elastomers (b) thermoplastics
(c) thermosets (d) fibres

630. Which one of the following is the most common reinforcing material in FRPs?
(a) Al_2O_3
(b) boron
(c) carbon
(d) SiO_2

631. Identify which of the following materials are not composites (more than one)?
(a) cemented carbide
(b) phenolic molding compound
(c) plywood
(d) 1020 steel.

632. In casting, a flask is which one of the following?
(a) beverage bottle for foundrymen.
(b) box which holds the cope and drag.
(c) container for holding liquid metal.
(d) metal which extrudes between the mold halves.

633. In foundry work, a runner is which one of the following?
(a) channel in the mold leading from the downsprue to the main mold cavity.
(b) foundryman who moves the molten metal to the mold.
(c) vertical channel into which molten metal is poured into the mold.
(d) none

634. Total solidification time is defined as which one of the following?
(a) time between pouring and complete solidification.
(b) time between pouring and cooling to room temperature.
(c) time between solidification and cooling to room temperature.
(d) time to give up the heat of fusion.

635. During solidification of an alloy when a mixture of solid and liquid metals are present, the solid-liquid mixture is referred to as which one of the following?
(a) eutectic composition.
(b) ingot segregation.
(c) liquidus.
(d) mushy zone

636. Chvorinov's Rule states that total solidification time is proportional to (a) $(A/V)^2$,
(b) Hf, (c) Tm, (d) $(V/A)^2$
where A = surface area of casting, Hf = heat of fusion, Tm = melting temperature, and V = volume of casting.

637. In a sand casting mold, the V/A ratio of the riser should be which one of the following relative to the V/A ratio of the casting itself ?
(a) equal.
(b) greater.
(c) smaller.
(d) unpredictable.

638. A riser that is completely enclosed within the sand mold and connected to the main cavity by a channel to feed the molten metal is called which of the following ?
(a) blind riser. (b) side riser.
(c) both a & b (d) top riser.

Answers

629. (c)	630. (d)	631. (d)	632. (b)	633. (a)	634. (a)
635. (d)	636. (a)	637. (b)	638. (c)		

639. Which one of the following casting processes is most widely used?
(a) centrifugal casting.
(b) die casting.
(c) investment casting.
(d) sand casting

640. For which of the following reasons is a *green mold* so named?
(a) green is the color of the mold.
(b) moisture is contained in the mold,
(c) mold is cured.
(d) mold is dry.

641. Given that W_m = weight of the molten metal displaced by a core and W_c = weight of the core, the buoyancy force is which one of the following?
(a) downward force = $W_m + W_c$.
(b) downward force = $W_m - W_c$.
(c) upward force = $W_m + W_c$.
(d) upward force = $W_m - W_c$.

642. Which of the following casting processes are not expendable mold operations (more than one)?
(a) investment casting.
(b) low pressure casting.
(c) sand casting.
(d) shell molding.

643. Shell molding is which one of the following ?
(a) casting operation in which the molten metal has been poured out after a thin shell has been solidified in the mold.
(b) casting operation used to make artificial sea shells.
(c) casting process in which the mold is a thin shell of sand binded by a thermosetting resin.
(d) sand casting operation in which the pattern is a shell rather than a solid form.

644. Investment casting is also known by which one of the following names?
(a) full-mold process.
(b) lost-foam process.
(c) lost pattern process.
(d) lost-wax process.

645. In plaster mold casting, the mold is made of which one of the following materials?
(a) Al_2O_3. (b) $CaSO_4$–H_2O.
(c) SiC. (d) SiO_2.

646. Which of the following qualifies as a precision casting process (more than one)?
(a) ingot casting. (b) investment casting.
(c) sand casting. (d) shell molding.

647. Which of the following casting processes are not permanent mold operations.
(a) centrifugal casting.
(b) die casting.
(c) low pressure casting.
(d) shell molding.

648. Which of the following metals would not typically be die casted ?
(a) aluminum. (b) cast iron.
(c) tin. (d) zinc.

649. Which of the following are not the advantages of die casting over sand casting ?
(a) better surface finish.
(b) higher production rates,
(c) larger parts can be casted.
(d) mold can be reused.

650. Cupolas are furnaces used to melt which of the following metals ?
(a) aluminum. (b) cast iron.
(c) steel. (d) zinc.

Answers

639 (d)	640. (b)	641. (d)	642. (b)	643. (c)	644. (d)
645. (b)	646. (b)	647. (d)	648. (b)	649. (c)	650. (b)

651. A misrun is which one of the following defects in casting?
(a) globules of metal becoming entrapped in the casting,
(b) metal is not properly poured into the downsprue,
(c) metal solidifies before filling the cavity.
(d) "pipe" formation.

652. Which one of the following casting metals is most important commercially?
(a) aluminum and its alloys.
(b) bronze.
(c) cast iron.
(d) cast steel.

653. Which one of the following terms refers to the glassy state of a material?
(a) crystalline.
(b) devitrified.
(c) polycrystalline.
(d) vitreous.

654. Besides helping to preserve the environment, the use of recycled glass as an ingredient of the starting material in glassmaking serves what other useful purpose ?
(a) adds coloring variations to the glass for aesthetic value.
(b) makes the glass easier to melt.
(c) makes the glass stronger.
(d) reduces odors in the plant.

655. Typical glass melting temperatures are in which of the following ranges?
(a) 400 to 500°C.
(b) 900 to 1000°C.
(c) 1500 to 1600°C.
(d) 2000 to 2200°C.

656. Which one of the following processes or processing steps is not applicable in glassworking?
(a) annealing. (b) pressing.
(c) quenching. (d) sintering.

657. Which one of the following processes is used to produce glass tubing?
(a) Danner process. (b) pressing.
(c) rolling. (d) spinning.

658. If a glass part with a wall thickness of 5 mm takes 10 minutes to anneal, how much time would a glass part of similar geometry but with a wall thickness of 7.5 mm take to anneal ?
(a) 10 minutes. (b) 15 minutes.
(c) 20 minutes. (c) 30 minutes.

659. The shear viscosity of a polymer melt is not affected by which of the following ?
(b) degree of polymerization.
(b) polymer type.
(c) density.
(d) temperature.

660. Which of the following is not a section of a conventional extruder barrel for thermoplastics?
(a) compression section.
(b) die section.
(c) feed section.
(d) shaping section.

661. Which of the following processes is not associated with the production of plastic sheet & film?
(a) blown-film extrusion process.
(b) chill-roll extrusion.
(c) doctor blade method.
(d) slit-die extrusion.

Answers

651 (c)	652. (c)	653. (d)	654. (b)	655. (c)	656. (d)
657. (a)	658. (c)	659. (c)	660. (b)	661. (c)	

662. Spinning in the production of synthetic fibers refers to which of the following ?
   (a) extrusion of polymer melt through small die openings.
   (b) drawing the strands to elongate and thin them.
   (c) both of the above.
   (d) none of the above.

663. Which is not a component of an injection molding machine ?
   (a) clamping unit.
   (b) hopper.
   (c) injection unit.
   (d) none

664. The ***parting line*** in injection molding is which one of the following ?
   (a) the lines formed where polymer melt meets after flowing around a core in the mold.
   (b) the narrow gate sections where the parts are separated from the runner,
   (c) where the clamping unit is joined to the injection unit in the molding machine,
   (d) where the two mold halves come together

665. The function of the ejection system is to
   (a) move polymer melt into the mold cavity.
   (b) open the mold halves after the cavity is filled.
   (c) remove the molded parts from the runner system after molding.
   (d) separate the part from the cavity after molding

666. A three-plate mold does not offer which of the following advantages when compared to a two-plate mold?
   (a) automatic separation of parts from runners.
   (b) gating is usually at the base of the part to reduce weld lines.
   (c) sprue does not solidify.
   (d) none of the above.

667. Which of the following defects or problems is not associated with injection molding ?
   (a) flash. (b) melt fracture.
   (c) short shots. (d) sink marks.

668. Use of a parison is associated with which one of the following plastic shaping processes?
   (a) bi-injection molding.
   (b) blow molding.
   (c) compression molding.
   (d) pressure thermoforming.

669. A thermoforming mold with a convex form is called
   (a) a die. (b) a negative mold.
   (c) a positive mold. (d) a three-plate mold.

670. The term ***encapsulation*** refers to which one of the following plastics shaping processes?
   (a) casting.
   (b) compression molding.
   (c) extrusion of hollow forms,
   (d) injection molding in which a metal insert is encased in the molded part

671. Which of the following terms does not apply to the processing of foam plastics ?
   (a) chemical blowing agents.
   (b) open cell structure.
   (c) powder injection molding.
   (d) sandwich molding,

672. The uncommon polymer foam is
   (a) polyacetal. (b) polystyrene.
   (c) polyurethane. (d) none.

Answers

662 (c)	663. (b)	664. (d)	665. (d)	666. (c)	667. (b)
668. (b)	669. (c)	670. (a)	671. (c)	672. (a)	

673. In which of the following property categories do plastic parts compare favorably with metals ?
(a) impact resistance.
(b) resistance to ultraviolet radiation.
(c) stiffness.
(d) strength.

674. Which of the following processes are generally limited to thermoplastic polymers ?
(a) blow molding.
(b) compression molding.
(c) reaction injection molding.
(d) transfer molding.

675. Which of the following processes would not be applicable to produce hulls for small boats ?
(a) blow molding.
(b) injection molding.
(c) rotational molding.
(d) vacuum thermoforming.

676. The chemical name of the ingredient recovered from the latex of the rubber tree is
(a) polybutadiene. (b) polyisobutylene.
(c) polyisoprene. (d) polystyrene.

677. Of the following rubber additives, which one would rank as the single most important?
(a) antioxidants.
(b) carbon black.
(c) clays and other hydrous aluminum silicates.
(d) plasticizers and softening oils

678. Which one of the following molding processes is the most important in the production of products made of conventional rubber?
(a) compression molding.
(b) injection molding.
(c) thermoforming.
(d) transfer molding.

679. Which of the following ingredients do not contribute to the vulcanizing process ?
(a) carbon black. (b) stearic acid.
(c) sulfur. (d) zinc oxide.

680. How many minutes are required to cure (vulcanize) a modern passenger car tire?
(a) 5. (b) 15.
(c) 25. (d) 45.

681. When is the tread pattern imprinted onto the circumference of the tire?
(a) during preforming.
(b) while building the carcass.
(c) during molding.
(d) during curing.

682. Which one of the following is the most common polymer type in fiber-reinforced polymer composites?
(a) elastomers. (b) thermoplastics.
(c) thermosets. (d) fibres.

683. Which of the following are normally used in the processing of thermoplastic elastomers ?
(a) blow molding.
(b) compression molding
(c) vulcanization.
(d) none

684. Most rubber products are not classified into which of the following categories ?
(a) elastomer reinforced with carbon black.
(b) fiber- reinforced composite.
(c) particle -reinforced composite.
(d) polymer matrix composite.

Answers

673 (a)	674. (a)	675. (b)	676. (c)	677. (b)	678. (a)
679. (a)	680. (b)	681. (c)	682. (c)	683. (a)	684. (b)

685. Filament winding involves the use of which one of the following fiber reinforcements?
(a) continuous filaments.
(b) fabrics.
(c) mats
(d) woven rovings.

686. Pultrusion is most similar to which one of the following plastic shaping processes?
(a) blow-molding.
(b) extrusion.
(c) injection molding.
(d) thermoforming.

687. For a given weight of metallic powders, the total surface area of the powders is not increased by
(a) larger particle size.
(b) smaller particle size.
(c) higher shape factor.
(d) none

688. Which of the following powder shapes would tend to have the lowest interparticle friction?
(a) acicular. (b) cubic.
(c) flakey. (d) spherical.

689. Which of the following statements is incorrect in the context of metallic powders ?
(a) porosity + packing factor = 1.0.
(b) packing factor = 1/porosity.
(c) packing factor = 1.0 – porosity.
(d) packing factor = bulk density/true density.

690. Impregnation does not refer to which of the following ?
(a) soaking oil by capillary action into the pores of a PM part.
(b) putting polymers into the pores of a PM part.
(c) filling the pores of the PM part with a molten metal
(d) none

691. In cold isostatic pressing, the mold is most typically made of which one of the following?
(a) rubber.
(b) sheet metal.
(c) tool steel.
(d) thermosetting polymer.

692. Which of the following processes does not combine pressing and sintering of the metal powders?
(a) metal injection molding.
(b) hot pressing.
(c) spark sintering.
(d) hot isostatic pressing.

693. Which of the following design features would not be difficult or impossible to achieve by conventional pressing and sintering (more than one)?
(a) side holes.
(b) threaded holes.
(c) outside rounded corners.
(d) vertical stepped holes.

694. The following equipment is used for crushing and grinding of minerals in the preparation of traditional ceramics raw materials. Which one of the pieces listed is used for grinding?
(a) ball mill. (b) hammer mill.
(c) jaw crusher. (d) roll crusher.

695. Which one of the following compounds becomes a plastic and formable material when mixed with suitable proportions of water?
(a) aluminum oxide.
(b) hydrogen oxide.
(c) hydrous aluminum silicate.
(d) silicon dioxide.

Answers

685 (a)	686. (b)	687. (a)	688. (d)	689. (b)	690. (c)
691. (a)	692. (a)	693. (d)	694. (a)	695. (c)	

696. At which one of the following water contents does clay become a suitably plastic material for the traditional ceramics plastic forming processes?
(a) 5%. (b) 10%.
(c) 20%. (d) 40%

697. Which of the following processes are not plastic forming methods used in the shaping of traditional ceramics ?
(a) extrusion. (b) jangling.
(c) jiggering. (d) none

698. Which one of the following processes for the new ceramic materials accomplishes shaping and sintering simultaneously?
(a) freeze drying.
(b) hot pressing.
(c) injection molding.
(d) isostatic pressing.

699. Which of the following is not the purpose of finishing operation used for parts made of the new ceramics (more than one)?
(a) apply a surface coating.
(b) improve surface finish,
(c) increase dimensional accuracy.
(d) remove material.

700. Which one of the following terms best describes what a cemented carbide is?
(a) ceramic. (b) cermet.
(c) composite. (d) metal.

701. Which of the following geometric features need not be avoided if possible in the design of structural components made of new ceramics ?
(a) complicated shapes.
(b) rounded inside corners.
(c) sharp edges.
(d) thin sections.

702. Which of the following are not the bulk deformation processes ?
(a) deep drawing. (b) extrusion.
(c) forging. (d) rolling.

703. Hot working of metals refers to which one of the following temperature regions relative to the melting point(T^m) of the given metal on an absolute temperature scale?
(a) room temperature.
(b) 0.2Tm.
(c) 0.4Tm.
(d) 0.6Tm.

704. Which of the following are not the advantages and characteristics of hot working relative to cold working?
(a) fracture of workpart less likely.
(b) increased strength properties.
(c) isotropic mechanical properties.
(d) less overall energy required,

705. Increasing strain rate tends to have which one of the following effects on flow stress during hot forming of metal?
(a) decreases flow stress.
(b) has no effect.
(c) increases flow stress.
(d) unpredictable

706. The coefficient of friction between the part and the tool in cold working tends to be which of the following relative to its value in hot working?
(a) higher. (b) lower.
(c) no effect. (d) unpredictable

Answers

696 (c)	697. (a)	698. (b)	699. (a)	700. (b)	701. (b)
702. (a)	703. (d)	704. (b)	705. (c)	706. (b)	

707. The maximum possible draft in a rolling operation does not depend on which of the following parameters?
(a) coefficient of friction between roll and work.
(b) roll diameter.
(c) roll velocity.
(d) none

708. Which of the following rolling mill types are not associated with relatively small diameter rolls in contact with the work ?
(a) cluster mill.
(b) continuous rolling mill.
(c) four-high mill.
(d) none

709. Production of pipes and tubes is not associated with which of the following bulk deformation processes?
(a) extrusion. (b) ring rolling.
(c) roll piercing. (d) tube sinking

710. Which of the basic bulk deformation processes use compression to effect shape change ?
(a) bar and wire drawing.
(b) extrusion.
(c) forging.
(d) all a, b & c

711. Which of the following are not classified as forging operations ?
(a) coining.
(b) fullering.
(c) impact extrusion.
(d) roll forging.

712. Theoretically, the maximum reduction possible in a wire drawing operation, under the assumptions of a perfectly plastic metal, no friction, and no redundant work, is which of the following ?
(a) zero. (b) 0.63.
(c) 1.00 (d) 2.72.

713. Which of the following bulk deformation processes are not involved in the production of nails for lumber construction ?
(a) bar and wire drawing.
(b) extrusion.
(c) forging.
(d) rolling.

714. Johnson's formula is associated with which of the four bulk deformation processes ?
(a) bar and wire drawing.
(b) extrusion.
(c) forging.
(d) rolling.

715. As sheet metal stock hardness increases, the clearance between punch and die should
(a) be decreased.
(b) be increased.
(c) be unaffected.
(d) unpredictable.

716. The cutting force in a blanking operation depends on which mechanical property of the sheet metal ?
(a) compressive strength.
(b) modulus of elasticity.
(c) shear strength.
(d) tensile strength

717. Sheet metal bending does not involve which of the following stresses and strains ?
(a) compressive. (b) shear.
(c) tensile. (d) all a, b & c.

Answers

707 (c)	708. (b)	709. (c)	710. (d)	711. (c)	712. (b)
713. (b)	714. (b)	715. (b)	716. (c)	717. (b)	

718. Which one of the following is the best definition of ***bend allowance***?

(a) amount by which the die is larger than the punch,

(b) amount of elastic recovery experienced by the metal after bending,

(c) safety factor used in calculating bending force,

(d) length before bending of the straight sheet metal section to be bent.

719. Which of the following are not the variations of sheet metal bending operations ?

(a) coining. (b) flanging.

(c) hemming. (d) tube forming.

720. Which one of the following stamping dies is the most complicated?

(a) blanking die.

(b) combination die.

(c) compound die.

(d) progressive die.

721. Which one of the following press types is usually associated with the highest production rates in sheet metal stamping operations?

(a) adjustable bed.

(b) open back inclinable.

(c) press brake.

(d) solid gap.

722. Which of the following processes are classified as high- energy-rate forming processes ?

(a) electrochemical machining.

(b) electromagnetic forming.

(c) electron beam cutting.

(d) hydroforming

723. A lathe is used to perform which of the following machining operations?

(a) broaching. (b) drilling.

(c) milling. (d) turning.

724. With which one of the following geometric forms is the drilling operation most closely associated?

(a) external cylinder. (b) flat plane.

(c) round hole. (d) screw threads .

725. Which of the three types of chip would be expected in a turning operation conducted at low cutting speeds on a brittle work material ?

(a) continuous.

(b) continuous with built-up edge.

(c) discontinuous.

(d) unpredictable.

726. Of the following cutting conditions, which one has the greatest effect on tool wear?

(a) cutting speed. (b) depth of cut.

(c) feed. (d) all have same effect.

727. As an alloying ingredient in high speed steel, tungsten does not serve which of the following functions?

(a) forms hard carbides to resist abrasion.

(b) improves strength and hardness.

(c) increases corrosion resistance.

(d) increases hot hardness.

728. Cast cobalt alloys typically contain which of the following main ingredient ?

(a) aluminum. (b) cobalt.

(c) chromium. (d) tungsten.

729. Which of the following is not a common ingredient in cemented carbide cutting tools?

(a) Al_2O_3. (b) Co.

(c) TiC. (d) WC.

Answers

718 (d)	719. (a)	720. (d)	721. (b)	722. (b)	723. (d)
724. (c)	725. (c)	726. (a)	727. (c)	728. (a)	729. (a)

730. An increase in cobalt content has which of the following effects on WC-Co cemented carbides?
(a) decreases transverse rupture strength.
(b) increases hardness.
(c) increases toughness.
(d) none

731. Steel cutting grades of cemented carbide are typically characterized by which of the following ingredients ?
(a) Co. (b) Ni.
(c) TiC. (d) WC.

732. Which of the following processes are used to provide the thin coatings on the surface of coated carbide inserts?
(a) chemical vapor deposition.
(b) electroplating.
(c) pressing and sintering.
(d) none

733. Which of the following materials has the highest hardness?
(a) aluminum oxide.
(b) cubic boron nitride.
(c) high speed steel.
(d) titanium carbide.

734. Which of the following are not the functions of a cutting fluid in machining ?
(a) improve surface finish on the workpiece.
(b) reduce friction at the tool-chip interface.
(c) remove heat from the process.
(d) all a , b & c.

735. Which of the following conventional machining processes is closest to grinding ?
(a) drilling. (b) milling.
(c) shaping. (d) turning.

736. Smaller grain size in a grinding wheel tends to do which one of the following?
(a) improve surface finish.
(b) have no effect on surface finish.
(c) degrade surface finish.
(d) none

737. Which of the following abrasive materials is most appropriate for grinding steel and cast iron ?
(a) aluminum oxide.
(b) cubic boron nitride.
(c) diamond.
(d) silicon carbide.

738. Which of the following abrasive materials is most appropriate for grinding hardened tool steel ?
(a) aluminum oxide.
(b) cubic boron nitride.
(c) diamond.
(d) silicon carbide.

739. Which of the following abrasive materials is most appropriate for grinding nonferrous metals ?
(a) aluminum oxide.
(b) cubic boron nitride.
(c) diamond.
(d) silicon carbide.

740. Which of the following will not help to reduce the incidence of heat damage to the work surface in Grinding ?
(a) frequent dressing or truing of the wheel.
(b) higher infeeds.
(c) higher work speeds.
(d) lower wheel speeds.

Answers

730 (c)	731. (c)	732. (a)	733. (b)	734. (a)	735. (b)
736. (a)	737. (a)	738. (b)	739. (d)	740. (b)	

741. Which of the following abrasive processes achieves the best surface finish?
(a) centerless grinding.
(b) honing.
(c) lapping.
(d) superfinishing.

742. Which of the following abrasive processes can not be used to finish a hole or internal bore ?
(a) centerless grinding.
(b) honing.
(c) cylindrical grinding.
(d) lapping.

743. The term ***deep grinding*** refers to which of the following ?
(a) alternative name for any creep feed grinding operation.
(b) external cylindrical creep feed grinding.
(c) grinding operation performed at the bottom of a hole.
(d) surface grinding which uses a large crossfeed.

744. Which of the following processes do not use mechanical energy as the principal energy source ?
(a) grinding.
(b) laser beam machining.
(c) milling.
(d) ultrasonic machining

745. Which one of the following is closest to the temperatures used in plasma arc cutting?
(a) 2750°C (b) 5500°C
(c) 8300°C (d) 11,000°C

746. Chemical milling is used in which of the following?
(a) drilling holes with high depth-to-diameter ratio.
(b) making intricate patterns in sheet metal.
(c) removing material to make shallow pockets in metal.
(d) cutting of plastic sheets.

747. Of the following processes, which one is noted for the highest material removal rates?
(a) electric discharge machining.
(b) electrochemical machining.
(c) laser beam machining.
(d) plasma arc cutting.

748. Which of the following processes would be appropriate to drill a hole with a square cross-section, 0.25 inch on a side and 1-inch deep in a steel workpiece ?
(a) abrasive jet machining.
(b) chemical milling.
(c) EDM.
(d) laser beam machining.

749. Which of the following are usually not the objective of heat treatment ?
(a) increase hardness.
(b) increase toughness.
(c) recrystallization of the metal.
(d) reduce density.

750. On which one of the following metals can the treatment called austenitizing be performed?
(a) aluminum alloys. (b) brass.
(c) copper alloys. (d) steel.

751. The treatment in which the brittleness of martensite is reduced is called
(a) annealing. (b) austenitizing.
(c) normalizing. (d) tempering.

Answers

741 (d)	742. (d)	743. (b)	744. (b)	745. (d)	746. (c)
747. (d)	748. (c)	749. (d)	750. (d)	751. (d)	

752. The Jominy end-quench test is designed to indicate which one of the following?
(a) cooling rate. (b) ductility.
(c) hardenability. (d) strength.

753. In precipitation hardening, the hardening and strengthening of the metal occurs in which one of the following steps ?
(a) aging.
(b) quenching.
(c) solution treatment.
(d) none

754. Which one of the following surface hardening treatments is the most common?
(a) carbonitriding. (b) carburizing.
(c) chromizing. (d) nitriding.

755. Which of the following chemicals is associated with alkaline cleaning ?
(a) borax. (b) sulfuric acid.
(c) trichlorethylene. (d) none

756. In sand blasting, which one of the following abrasives is used?
(a) Al_2O_3. (b) nylon beads.
(c) SiC. (d) SiO_2.

757. Calorizing is the same as which of the following?
(a) aluminizing. (b) doping.
(c) hot sand blasting. (d) siliconizing.

758. Carburizing involves which one of the following?
(a) acid pickling. (b) diffusion.
(c) tumbling. (d) vapor degreasing.

759. Which one of the following plate metals produces the hardest surface on a metallic substrate?
(a) cadmium. (b) chromium.
(c) copper. (d) nickel, or tin.

760. Which one of the following terms is used in connection with dip coating of lead onto a substrate such as sheet steel?
(a) aluminizing. (b) anodizing.
(c) galvanizing. (d) terneplating.

761. Which of the following processes involves electrochemical reactions?
(a) anodizing.
(b) chromate coatings.
(c) electroless plating.
(d) phosphate coatings.

762. With which one of the following metals is anodizing most commonly associated ?
(a) aluminum. (b) magnesium.
(c) steel. (d) titanium.

763. Sputtering is a form of which one of the following?
(a) chemical vapor deposition.
(b) defect in arc welding.
(c) diffusion.
(d) physical vapor deposition.

764. Which of the following gases is the most commonly used in sputtering and ion plating?
(a) argon. (b) chlorine.
(c) neon. (d) nitrogen.

765. The Mond process is used for which one of the following?
(a) chemical vapor deposition of silicon nitride onto silicon.
(b) an electroplating process,
(c) physical vapor deposition for coating tin onto cutting tools,
(d) reducing nickel carbonyl to metallic Ni.

Answers

752 (c)	753. (a)	754. (b)	755. (b)	756. (d)	757. (a)
758. (b)	759. (b)	760. (d)	761. (a)	762. (a)	763. (d)
764. (a)	765. (d)				

766. The principal methods of applying powder coatings are which of the following ?
(a) brushing.
(b) electrostatic spraying.
(c) immersion.
(d) roller coating.

767. Which of the following is an alternative name for thermal spraying ?
(a) flexible overlay process.
(b) hard facing.
(c) metallizing.
(d) none

768. Hard facing utilizes which one of the following basic processes?
(a) arc welding. (b) brazing.
(c) dip coating. (d) electroplating.

769. A fillet weld has a cross-sectional shape that is approximately
(a) rectangular. (b) round.
(c) square. (d) triangular.

770. Groove welds are most closely associated with which one of the following joint types:
(a) butt. (b) corner.
(c) edge. (d) lap.

771. A flange weld is most closely associated with which one of the following joint types:
(a) butt. (b) corner.
(c) edge. (d) lap.

772. For metallurgical reasons, it is desirable to melt the weld metal with minimum energy input. Which one of the following heat sources is most consistent with this objective?
(a) high power.
(b) high power density.
(c) low power.
(d) low power density.

773. The amount of heat required to melt a given volume of metal depends strongly on which of the following properties ?
(a) coefficient of thermal expansion.
(b) heat of fusion.
(c) modulus of elasticity.
(d) thermal conductivity.

774. Which of the following processes are not classified as fusion welding ?
(a) electrogas welding.
(b) electron beam welding.
(c) explosive welding.
(d) percussion welding.

775. Which of the following processes are classified as solid state welding?
(a) friction welding.
(b) resistance spot welding.
(c) thermit welding.
(d) upset welding,

776. MIG welding is a term sometimes applied when referring to which one of the following processes?
(a) FCAW. (b) GMAW.
(c) GTAW. (d) SMAW.

777. Shielding gases used for welding include which of the following?
(a) argon. (b) carbon monoxide.
(c) hydrogen (d) nitrogen.

778. The strength of a brazed joint is typically which one of the following relative to the filler metal out of which it is made.
(a) equal to. (b) stronger than.
(c) weaker than. (d) unpredictable

Answers

766. (b)	767. (c)	768. (a)	769. (d)	770. (a)	771. (a)
772. (b)	773. (b)	774. (c)	775. (a)	776. (b)	777. (a)
778. (b)					

779. Clearances between surfaces in brazing are which one of the following:
(a) 0.0025 to 0.025 mm
(b) 0.025 to 0.250 mm
(c) 0.250 to 2.50 mm
(d) 2.5 to 5.0 mm

780. Which of the following is not an advantage of brazing :
(a) dissimilar metals can be joined.
(b) less heat and energy required than fusion welding.
(c) multiple joints can be brazed simultaneously.
(d) stronger joint than welding.

781. Which of the following soldering methods are not used for brazing ?
(a) dip soldering.
(b) infrared soldering.
(c) soldering iron.
(d) torch soldering,

782. Which one of the following is not a function of a flux in brazing or soldering?
(a) chemically etch the surfaces to increase roughness for better adhesion of the filler metal.
(b) promote wetting of the surfaces.
(c) protect the faying surfaces during the process.
(d) remove or inhibit formation of oxide films.

783. Which type of soldering flux is preferred for electrical and electronics connections?
(a) inorganic fluxes such as zinc chloride.
(b) natural rosin fluxes,
(c) water-soluble organic fluxes.
(d) none

784. Adhesively bonded joints are strongest under which type of stresses?
(a) cleavage.
(b) peeling.
(c) shear.
(d) none

785. Which of the following are the mechanisms that does not operate in adhesive bonding ?
(a) chemical bonding, in which a primary chemical bond is formed between the adhesive & the parts being joined,
(b) mechanical interlocking,
(c) secondary bonding forces between atoms of opposing surfaces,
(d) surface tension of the fluid adhesive.

786. The correct order of the coordination number is SC, BCC, FCC and HCP unit cells is
(a) 12, 8, 12, 6.
(b) 6, 8, 12, 12.
(c) 8, 6, 12, 12.
(d) 6, 12, 12, 8.

787. The forbidden energy gap of carbon in diamond structure is
(a) 7.0 ev
(b) 1.0 ev
(c) 0.01 ev
(d) none

788. A suitable material for audio and TV transformers is
(a) Fe – 4% Si.
(b) Ferrite.
(c) Fe – 30% Ni.
(d) Pure Fe.

Answers

779. (b)	780. (d)	781. (c)	782. (a)	783. (a)	784. (c)
785. (d)	786. (b)	787. (a)	788. (b)		

789. Which of the following is not the function of oxide layer during IC (Integrating circuit) fabrication ?
(a) to increase the melting point of silicon.
(b) to mask against diffusion or ion implant.
(c) to insulate the surface electrically.
(d) to produce a chemically stable surface.

790. In normalizing, one of the following is not correct:
(a) it relieves internal stresses
(b) it produces a uniform structure.
(c) the rate of cooling is rapid
(d) the rate of cooling is slow.

791. Metallic bond is not characterized by
(a) ductility. (b) high conductivity.
(c) directionality. (d) opacity.

792. If the Fermi energy of silver at 0^0 K is 5 electron volt, the mean energy of electron in silver at 0^0 K is
(a) 6 electron volt.
(b) 12 electron volt.
(c) 1.5 electron volt.
(d) 3 electron volt.

793. The Fermi level in an n-type semiconductor at 0^0 K lies
(a) below the donor level.
(b) Half way between the bottom of conduction band and donor level.
(c) Exactly in the middle of hand gap.
(d) Half way between the top of valence band and the acceptor level.

794. Hard magnetic material is characterized by
(a) High coercive force and low residual magnetic induction.
(b) Low coercive force and high residual magnetic induction.
(c) Only low coercive force.
(d) High coercive force and high residual magnetic induction.

795. Piezoelectric effect is the production of electricity by
(a) chemical effect. (b) pressure.
(c) varying field. (d) temperature.

796. Zinc has hcp structure. In a unit cell of zinc, the zinc atoms occupy
(a) 74% of volume of unit cell.
(b) 80% of volume of unit cell.
(c) 68% of volume of unit cell.
(d) 90% of volume of unit cell.

797. Polarization in a dielectric on application of electric field is
(a) Displacement/separation of opposite charge centres.
(b) Passing of current through dielectric.
(c) Breaking of insulation.
(d) Excitation of electrons to higher energy level.

798. Which one of the following is not the purpose of full annealing ?
(a) refines grains
(b) induces softness.
(c) removes strains and stresses
(d) produces hardest material.

799. Which of the following elements is a covalently bonded crystal?
(a) aluminium (b) sodium chloride
(c) germanium (d) lead

800. Binary phase diagrams of two component systems are usually two dimensional plots of
(a) temperature and pressure.
(b) temperature and composition.
(c) pressure and composition.
(d) pressure, temperature and composition.

Answers

789. (a)	790. (d)	791. (c)	792. (d)	793. (b)	794. (d)
795. (b)	796. (a)	797. (a)	798. (d)	799. (c)	800. (b)

801. Imperfection arising due to the displacement of an ion from a regular site to an interstitial site maintaining overall electrical neutrality of the ionic crystal is called
(a) Frenkel imperfection
(b) Schottky imperfection
(c) Point imperfection
(d) Volume imperfection

802. The Fermi level is
(a) an average value of all available energy levels.
(b) an energy level at the top of the valence band.
(c) the highest occupied energy level at 0^0 C.
(d) the highest occupied energy level at 0^0 K.

803. Among the common dielectric materials, the highest dielectric strength is possessed by
(a) mica.
(b) polyethylene.
(c) PVC.
(d) transformer oil.

804. Annealing is generally done to impart .... to the material.
(a) hardness
(b) softness
(c) brittleness
(d) high conductivity

805. A ferromagnetic material is one in which neighbouring atomic magnetic moments are
(a) antiparallel and unequal.
(b) predominantly parallel.
(c) all randomly oriented.
(d) predominantly parallel in a small region of material.

806. In intrinsic semiconductor there are
(a) no mobile holes.
(b) no free electrons.
(c) as many free electrons as there are holes.
(d) neither free electrons nor mobile holes.

807. Mobility of electron is
(a) Average flow of electrons per unit field.
(b) Average applied field per unit drift velocity.
(c) Average drift velocity per unit field.
(d) Reciprocal of conductivity per unit charge.

808. The acceptor type impurity is formed by adding impurity of valency
(a) 6. (b) 5.
(c) 4. (d) 3.

809. Which of the following processes is used to harden a steel?
(a) Normalizing (b) Annealing
(c) Carburizing (d) Quenching

810. If the atomic number of an element is Z, and its atomic mass number is A, the number of protons in its nucleus is
(a) A. (b) Z.
(c) A – Z. (d) A / Z.

811. Miller indices of the diagonal plane of a cube are
(a) (200). (b) (111).
(c) (010). (d) (110).

812. The steady state conditions in diffusion are governed by
(a) Fick's second law.
(b) Fick's first law.
(c) both (a) and (b).
(d) none

Answers

801. (a)	802. (d)	803. (a)	804. (b)	805. (b)	806. (c)
807. (c)	808. (d)	809. (d)	810. (b)	811. (d)	812. (b)

813. Highest electrical resistivity exists in
(a) platinum wire.
(b) nichrome wire.
(c) silver wire.
(d) kanthal wire.

814. For high speed of reading and storing informations in a computer, use is made of
(a) ferrites.
(b) pyroelectrics.
(c) piezo electrics.
(d) ferromagnetics above 768^0 C.

815. Hall effect can be used to measure .... of semiconductors .
(a) mobility
(b) conductivity of semiconductors.
(c) resistivity
(d) all of these.

816. A pair of one cation and one anion missing in a crystal of the type AB is called
(a) Schottky defect.
(b) Frenkel defect.
(c) Pair of vacancies.
(d) None of these.

817. Pure silicon at zero K is an
(a) intrinsic semiconductor.
(b) extrinsic semiconductor.
(c) metal.
(d) insulator.

818. The dielectric strength of a material is the highest
(a) current which can pass through it.
(b) voltage that can be applied to it.
(c) field (voltage per meter thickness) that can be with-stood by it.
(d) current density that can be transmitted by it.

819. A Ge atom contains
(a) four protons
(b) four valence electrons
(c) six valence electrons
(d) only two electron orbits

820. The energy required to break a covalent bond in a semiconductor
(a) is equal to 1 eV
(b) is equal to the width of the forbidden gap
(c) is greater in Ge than in Si
(d) is the same in Ge and Si

821. An electron in the conduction band
(a) is located near the top of the crystal
(b) has no charge
(c) has a higher energy than an electron in the valence band
(d) is bound to its parent atom

822. At .0° K, all the valence electrons in an intrinsic semiconductor are in the
(a) valence band
(b) forbidden gap
(c) conduction band
(d) none of these

823. Malleability of a metal is the
(a) ability to withstand compressive stresses
(b) ability to withstand deformation under shear
(c) property by which a material can be cold-worked
(d) ability to undergo permanent deformation

824. Insulating material used in spark plug is
(a) rubber (b) porcelain
(c) mica (d) Polysterene

Answers

813. (b)	814. (a)	815. (d)	816. (a)	817. (d)	818. (c)
819. (b)	820. (b)	821. (c)	822. (a)	823. (c)	824. (b)

825. Which of the following has piezoelectric properties ?
(a) corundum (b) neoprene
(c) quartz (d) glass

826. For metallization, the property not desirable is
(a) reproducibility
(b) quick dissipation of heat
(c) low thermal conductivity
(d) high melting point

827. In a ferromagnetic material neighbouring atomic magnetic moments are predominantly
(a) parallel in small regions of material.
(b) parallel and unequal in small regions of material.
(c) equal and parallel through out the material.
(d) unequal and parallel throughout the material.

828. Which of the following is not the advantage of ion-implantation over diffusion doping ?
(a) it is a low temperature process.
(b) point imperfections are not produced.
(c) shallow doping is possible.
(d) gettering is possible.

829. The hardness of quenched Martensite .... as the carbon percentage increases.
(a) increases
(b) decreases
(c) first increases and then remains almost constant
(d) first increases and then decreases

830. Preheating of parts to be welded and slow cooling of the welded structure will reduce
(a) cracking and incomplete fusion
(b) cracking and residual stress.
(c) residual stress and incomplete penetration.
(d) cracking and underfill.

831. Missing of one cation & one anion in an ionic crystal (having charge neutrality) is called
(a) Frenkel imperfections.
(b) Compositional imperfections.
(c) Electronic imperfections.
(d) Schottky imperfections.

832. The probability of occupancy of electrons above Fermi level at T=0°K is .... percent.
(a) 0 (b) 25
(c) 50 (d) 100

833. In a ferroelectric material, the spontaneous polarization vanishes above
(a) Transition temperature.
(b) Debye temperature.
(c) Fermi temperature.
(d) Curie temperature.

834. P-type & N-type extrinsic semiconductors are formed by adding impurities of valency .... respectively.
(a) 5 & 3
(b) 5 & 4
(c) 3 & 5
(d) 3 & 4

835. Aluminium is not good for die casting because
(a) it is light and strong.
(b) it takes longer time to cool.
(c) it tends to react chemically with the die surface.
(d) its melting point is high and it expands on solidification.

Answers

825. (c)	826. (a)	827. (c)	828. (c)	829. (c)	830. (c)
831. (d)	832. (b)	833. (d)	834. (d)	835. (c)	

836. Time dependent permanent deformation is called
   (a) Plastic deformation
   (b) Elastic deformation
   (c) Creep
   (d) Anelastic deformation

837. Figure-out the odd point in the following :
   (a) Proportinal limit
   (b) Elastic limit
   (c) Yield point
   (d) Fracture point

838. If a material is subjected to two incremental true strains namely ε_1 and ε_2, then the total true strain is
   (a) $\varepsilon_1 * \varepsilon_2$ (b) $\varepsilon_1 - \varepsilon_2$
   (c) $\varepsilon_1 + \varepsilon_2$ (d) $\varepsilon_1 / \varepsilon_2$

839. Engineering stress-strain curve and True stress-strain curve are equal up to
   (a) Proportional limit
   (b) Elastic limit
   (c) Yield point
   (d) Tensile strength point

840. Value of Poisson's ratio for ionic solids in the range of
   (a) 0.1 (b) 0.2
   (c) 0.3 (d) 0.4

841. Hydrostatic stress results in the .... strain.
   (a) linear
   (b) shear
   (c) both linear & shear
   (d) None

842. High elastic modulus in materials arises from
   (a) High strength of bonds
   (b) Weak bonds
   (c) combination of bonds
   (d) None

843. Change in elastic modulus for ordinary materials between 0 K and melting point is
   (a) 10-20% increase
   (b) 10-20% decrease
   (c) 80-90% decrease
   (d) 80-90% increase

844. Bauschinger effect is about
   (a) Hysteresis loss during loading and unloading.
   (b) Anelastic deformation
   (c) Dependence of yield stress on path and direction
   (d) None

845. Shape of true stress-strain curve for a material depends on
   (a) Strain (b) Strain rate
   (c) Temperature (d) All

846. Toughness of a material is equal to area under .... part of the stress-strain curve.
   (a) Elastic (b) Plastic
   (c) Both (d) None

847. True stress-strain curve need to be corrected after
   (a) Elastic limit (b) Yield limit
   (c) Tensile strength (d) no need to correct

848. As compared with conventional stress-strain curve, the true stress-strain curve is
   (a) Above and right
   (b) Below and right
   (c) Above and left
   (d) Below and left

Answers

836. (c)	837. (d)	838. (c)	839. (c)	840. (b)	841. (d)
842. (a)	843. (b)	844. (c)	845. (d)	846. (c)	847. (c)
848. (c)					

849. According to distortion-energy criterion, yielding occurs when
(a) Distortion energy reaches a critical value
(b) Second invariant of the stress deviator exceeded some critical value
(c) Octahedral shear stress reaches a critical value
(d) All

850. Von Mises and Tresca criteria give different yield stress for
(a) Uni-axial stress
(b) Balanced bi-axial stress
(c) Pure shear stress
(d) All

851. Plastic deformation results from the following :
(a) Slip (b) Twinning
(c) Both (d) None

852. Time dependent recoverable deformation under load is called .... deformation.
(a) Elastic
(b) Anelastic
(c) Elastic after-effect
(d) Visco-elastic

853. Materials which can store electrical energy are called
(a) magnetic materials.
(b) semi conductors.
(c) dielectric materials.
(d) super conductors.

854. ACSR (Aluminium Conductor Steel Reinforced) are used as
(a) over head transmission lines.
(b) super conductors.
(c) fuse
(d) underground cables.

855. In n type semi conductor added impurity is
(a) pentavalent. (b) divalent.
(c) tetravalent. (d) trivalent.

856. The covers of electrical machines are made of
(a) soft magnetic materials.
(b) hard magnetic materials.
(c) super conductors.
(d) semiconductors.

857. The dielectric constant of air is practically taken as
(a) more than unity.
(b) unity.
(c) less than unity.
(d) zero.

858. n-type semiconductor is an example of
(a) extrinsic semiconductor.
(b) intrinsic semiconductor.
(c) super conductor.
(d) insulators.

859. Gold and silver are
(a) dielectric materials
(b) low resistivity conducting materials.
(c) magnetic materials.
(d) insulating materials.

860. Phenol and Formaldehyde are polymerised to a resultant product known as
(a) PVC. (b) bakelite.
(c) polyester. (d) teflon.

861. Dielectric materials are
(a) Insulating materials.
(b) Semiconducting materials.
(c) Magnetic materials.
(d) Ferroelectric materials.

Answers

849. (d)	850. (c)	851. (c)	852. (b)	853. (c)	854. (a)
855. (a)	856. (a)	857. (b)	858. (a)	859. (b)	860. (b)
861. (a)					

862. All semiconductors in their last orbit have .... electrons.
(a) 8 (b) 2
(c) 4 (d) 6

863. The material with lowest resistivity is
(a) constantan. (b) silver.
(c) manganin. (d) nichrome.

864. The property due to which the resistance of some metal or compound vanishes under certain conditions is
(a) Semi conductivity.
(b) Super conductivity.
(c) Curie point.
(d) Magnetostriction.

865. Ceramics are good
(a) insulators.
(b) conductors.
(c) superconductors.
(d) semiconductors.

866. Holes are majority carriers in
(a) P-type semiconductors.
(b) N-type semiconductors.
(c) Insulators.
(d) Superconductors.

867. Materials, which provide a path to the magnetic flux, are classified as
(a) insulating materials.
(b) semi conducting materials.
(c) magnetic materials.
(d) dielectric materials.

868. Germanium possesses .... valance electrons.
(a) one (b) two
(c) three (d) four

869. Ferrites are
(a) ferromagnetic material.
(b) ferrimagnetic materials.
(c) anti ferromagnetic material.
(d) diamagnetic materials.

870. The relative permeability of a paramagnetic substance is
(a) unity.
(b) slightly more than unity.
(c) zero.
(d) less than unity.

871. Hall effect may be used for which of the following?
(a) determining whether the semiconductor is p or n type.
(b) determining the carrier concentration.
(c) calculating the mobility.
(d) All the above.

872. Manganin is an alloy of
(a) copper, manganese and nickel.
(b) copper and manganese.
(c) manganese and nickel.
(d) manganese, aluminium and nickel.

873. Eddy current loss is proportional to the
(a) frequency.
(b) square of the frequency.
(c) cube of the frequency.
(d) square-root of the frequency.

874. A pure semiconductor under ordinary conditions behaves like
(a) a conductor.
(b) an insulator.
(c) a magnetic material.
(d) a ferro-electric material.

Answers

862. (c)	863. (b)	864. (b)	865. (a)	866. (a)	867. (c)
868. (d)	869. (a)	870. (b)	871. (d)	872. (a)	873. (b)
874. (b)					

875. In p-type semiconductor the majority carriers are
(a) holes. (b) electrons.
(c) positive ions. (d) negative ions.

876. Copper is completely miscible with
(a) nickel. (b) gold.
(c) hydrogen. (d) lead.

877. For germanium the forbidden gap is .... ev.
(a) 0.15 (b) 0.25
(c) 0.5 (d) 0.7

878. The dielectric strength of transformer oil should be
(a) 100 V. (b) 5 KV.
(c) 30 KV. (d) 132 KV.

879. Electrical resistivity of conductors is most affected by
(a) composition. (b) temperature.
(c) pressure. (d) current.

880. Mica is a .... material.
(a) Dielectric
(b) Insulating
(c) Magnetic
(d) Both insulating and dielectric

881. The conductivity of copper is less than that of silver by .... percent.
(a) 5 – 10 (b) 50 – 60
(c) 80 – 90 (d) 20 – 30

882. A ferrite core has less eddy current loss than an iron core because ferrites have.
(a) High resistance.
(b) Low resistance.
(c) Low permeability.
(d) High hysteresis.

883. Transformer cores are laminated with
(a) Low carbon steel.
(b) Silicon sheet steel.
(c) Nickel alloy steel.
(d) Chromium sheet steel.

884. For silicon the forbidden gap is .... ev.
(a) 1.1 (b) 0.25
(c) 0.5 (d) 0.7

885. Plastics are
(a) Good conductors of heat.
(b) Good conductors of electricity.
(c) Bad conductors of electricity.
(d) High density.

886. In order to obtain p-type germanium it should be doped with a
(a) Trivalent impurity.
(b) Tetravalent impurity.
(c) Pentavalent impurity.
(d) Any of the above will do.

887. Barrier potential in a P-N junction is caused by
(a) Thermally generated electrons and holes.
(b) Diffusion of majority carriers across the junction.
(c) Migration of minority carriers across the junction.
(d) Flow of drift current.

888. The most malleable, ductile low resistivity material is
(a) Copper. (b) Aluminium.
(c) Silver. (d) Iron.

889. Hard ferrites are used for making
(a) Transformer cores.
(b) Electrical machinery.
(c) High frequency equipment.
(d) Light weight permanent magnets.

Answers

875. (a)	876. (b)	877. (d)	878. (c)	879. (b)	880. (d)
881. (a)	882. (a)	883. (b)	884. (a)	885. (c)	886. (a)
887. (b)	888. (b)	889. (d)			

890. Micanite is a form of
   (a) Built up mica.
   (b) Hydrated potassium aluminium silicate.
   (c) Magnesium mica.
   (d) Calcium mica.

891. What is the type of bonding in silicon?
   (a) Ionic. (b) Covalent.
   (c) Metallic. (d) Ionic + Metallic

892. P-N junction is
   (a) a rectifier. (b) an amplifier.
   (c) an Oscillator. (d) a Coupler.

893. The conductivity of an extrinsic semiconductor .... with rise in temperature.
   (a) decreases
   (b) increases
   (c) remains constant
   (d) decreases and then increases

894. When a semiconductor is doped with a P-type impurity, each impurity atom will
   (a) acquire negative charge.
   (b) acquire positive charge.
   (c) remain electrically neutral.
   (d) give away one electron.

895. Which of the following material has the highest melting point?
   (a) Copper. (b) Aluminium.
   (c) Tungsten. (d) Gold.

896. A transistor has .... p-n junction.
   (a) One (b) Two
   (c) Four (d) Five

897. An example of high resistivity material is
   (a) Nichrome (b) Silver
   (c) Gold (d)Copper

898. Hard magnetic materials are used for making
   (a) Permanent magnets.
   (b) Temporary magnets.
   (c) Conductors.
   (d) Insulator.

899. Hall effect is associated with
   (a) Conductors. (b) Semiconductors.
   (c) Thermistors. (d) Solders.

900. Addition of trivalent impurity to a semiconductor creates many
   (a) holes.
   (b) free electrons.
   (c) valance electrons.
   (d) bound electrons.

901. Magnetic materials
   (a) provide path to magnetic flux.
   (b) are good insulators.
   (c) are semiconductors.
   (d) None.

902. In a ferromagnetic material the state of flux density .... when external magnetic field is applied to it.
   (a) Increased
   (b) Decreased
   (c) Remains unchanged
   (d) Becomes zero

903. Insulators have
   (a) A full valence band.
   (b) An empty conduction band.
   (c) A large energy gap.
   (d) All the above.

904. Hysteresis loss is least dependent on
   (a) Frequency
   (b) Magnetic field intensity.
   (c) Volume of the material.
   (d) Grain orientation of material.

Answers

890. (a)	891. (b)	892. (a)	893. (b)	894. (a)	895. (c)
896. (b)	897. (a)	898. (a)	899. (b)	900. (b)	901. (a)
902. (c)	903. (c)	904. (d)			

905. Semi-conductors have .... temperature coefficient of resistance.
(a) Negative
(b) Positive
(c) Both positive and negative
(d) None of the above

906. The material, which is different from the group is
(a) Constantan. (b) Manganin.
(c) Nichrome. (d) Brass.

907. The property, which is different from the group is
(a) Ductility (b) Resistivity
(c) Tensile strength. (d) Hardness.

908. Ferroelectric materials are characterised by
(a) Very high degree of polarisation.
(b) A sharp dependence of polarisation on temperature.
(c) Non-linear dependence of the charge Q on the applied voltage.
(d) All the above.

909. Decreasing grain size in a polycrystalline material
(a) Increases yield strength and corrosion resistance.
(b) Decreases yield strength and corrosion resistance
(c) Decreases yield strength but increases corrosion resistance
(d) Increases yield strength but decreases corrosion resistance.

910. When the temperature of a solid metal increases.
(a) Strength of the metal decreases but ductility increases
(b) Both strength and ductility of the metal decrease
(c) Both strength and ductility of the metal increase
(d) Strength of the metal increases but ductility decreases

911. Which one of the following factors is more relevant to represent complete solubility of two metals in each other?
(a) Chemical affinity
(b) Valency factor
(c) Crystal structure factor
(d) Relative size factor

912. Assertion (A): Elements are classified into metals and non-metals on the basis of their atomic weights.
Reason (R): The valence electron structures contribute to the primary bonding between the atoms to form aggregates.
(a) Both A and R are true and R is the correct explanation of A
(b) Both A and R are true but R is NOT the correct explanation of A
(c) A is true but R is false
(d) A is false but R is true

913. Assertion (A): Unlike in the case of ionic bonds, the co-ordination numbers for covalently bonded atoms are not controlled by the radii ratio.
Reason (R): A covalent bond has a specific direction of bonding in space.
(a) Both A and R are individually true and R is the correct explanation of A
(b) Both A and R are individually true but R is not the correct explanation of A
(c) A is true but R is false
(d) A is false but R is true

Answers

905. (a)	906. (d)	907. (b)	908. (d)	909. (a)	910. (a)
911. (c)	912. (d)	913. (c)			

914. Which of the following statement is true about brittle fracture?
   (a) High temperature and low strain rates favour brittle fracture
   (b) Many metal with HCP crystal structure commonly show brittle fracture
   (c) Brittle fracture is always preceded by noise
   (d) Cup and cone formation is characteristic for brittle materials

915. Magnesium is extruded and not rolled because
   (a) It has a low melting point
   (b) It has a low density
   (c) Its reactivity with roll material is high
   (d) It has a dose-packed hexagonal structure

916. Match the items in Column I and Column II.

Column I	Column II
P. Charpy test	1. Fluidity
Q. Knoop test	2. Micro hardness
R. Spiral test	3. Formability
S. Cupping test	4. Toughness
	5. Permeability

   (a) P – 4, Q – 5, R – 3, S – 2
   (b) P – 3, Q – 5, R – 1, S – 4
   (c) P – 2, Q – 4, R – 3, S – 5
   (d) P– 4, Q – 2, R – 1, S – 3

917. Assertion (A): The hardness test is a slow, expensive method of assessing the mechanical properties of materials.
   Reason (R): The hardness is a function of yield stress and the work hardening rate of material.
   (a) Both A and R are individually true and R is the correct explanation of A
   (b) Both A and R are individually true but R is not the correct explanation of A
   (c) A is true but R is false
   (d) A is false but R is true

918. A carbon steel having Brinell Hardness number 100 should have ultimate tensile strength closer to .... N/mm^2.
   (a) 100 (b) 200
   (c) 350 (d) 1000

919. Which of the following would you prefer for checking the hardness of very thin sections?
   (a) Hebert cloud burst test
   (b) Shore's Scleroscope
   (c) Knoop hardness test
   (d) Vickers hardness test

920. Herbert cloudburst Hardness test is conducted to know.
   (a) Uniformity of hardness over a surface
   (b) Softness of non-metallic components
   (c) Hardness of non-metallic components
   (d) Hardness at specified depth inside the surface.

921. The material property which depends only on the basic crystal structure is
   (a) Fatigue strength (b) Work hardening
   (c) Fracture strength (d) Elastic constant

922. For a Rhombohedral space lattice, which one of the following is correct?
   (a) $\alpha = \beta = \gamma = 90^0$ (b) $\alpha = \beta = \gamma \neq 90^0$
   (c) $\alpha = \gamma = 90^0 \neq \beta$ (d) $\alpha \neq \beta \neq \gamma \neq 90^0$

Answers

914. (b)	915. (a)	916. (d)	917. (b)	918. (c)	919. (c)
920. (a)	921. (c)	922. (b)			

923. Match List-I (Crystal Structure) with List-II (Example) and select the correct answer using the codes given below the Lists:

List-I (Crystal Structure)	List-II (Example)
A. Simple Cubic	1. Zinc
B. Body-centered Cubic	2. Copper
C. Face-centered Cubic at room temperature	3. Alpha iron
D. Hexagonal Close Packed	4. Manganese

Codes:

A B C D   A B C D
(a) 4 3 1 2   (b) 4 3 2 1
(c) 3 4 2 1   (d) 3 4 1 2

924. Match List-I (Name of the Element) with List-II (Crystal Structure) and select the correct answer using the codes given below the lists:

List I	List II
A. Fluorspar	1. Body-centered cubic
B. Alpha-Iron	2. Hexagonal closed packed
C. Silver	3. Simple cubic
D. Zinc	4. Face-centered cubic

Codes:

	A	B	C	D
(a)	3	2	4	1
(b)	4	1	3	2
(c)	4	2	3	1
(d)	3	1	4	2

925. Match List-I (Element) with List-II (Crystal Structure) and select the correct answer using the code given below the Lists:

List – I	**List - II**
A. Alpha Iron	1. Hexagonal closed packed
B. Copper	2. Body-centred cubic
C. Zinc	3. Amorphous
D. Glass	4. Face-centred cubic

Codes:

	A	B	C	D
(a)	2	3	1	4
(b)	1	4	2	3
(c)	2	4	1	3
(d)	1	3	2	4

926. In Zinc Blende structure, each atom is surrounded by four atoms of the opposite kind which are located at the comers of which one of the following?
(a) Tetrahedron (b) Hexahedron
(c) Cube (d) Orthorhombic

927. Consider the following temperature ranges:
1. Room temperature
2. 0 to 910°C
3. 910 to 1400°C
4. 1400°C to below melting point
In which of the above temperature ranges ferrite with body centered cubic structure is indicated in, the Fe-Fe3C phase diagram?
(a) 1, 2 and 4 (b) 2, 3 and 4
(c) 1 and 3 (d) 2 and 3

928. The microstructure composition of pearlite for a Fe_3C diagram consists of
(a) Carbon dissolved in alpha iron having a body centered cubic structure.
(b) Carbon dissolved in gama iron having a face centered cubic structure.
(c) A mixture of body centered alpha iron and face-centered gamma iron
(d) Carbon dissolved in body centered alpha iron and Fe_3C.

Answers

923. (b) 924. (d) 925. (c) 926. (a) 927. (a) 928. (d)

929. Assuming atoms to be perfect spheres, what is the value of the highest possible atomic packing factor (APF) in metals?
(a) 0.95 (b) 0.74
(c) 0.66 (d) 0.5

930. Match List-I (Crystal structure) with List-II (Atomic packing factor) and select the correct answer using the codes given below the Lists:

List-I	List-II
A. Simple cubic	1. 74%
B. Body-centered cubic	2. 74%
C. Face-centered cubic	3. 52%
D. Hexagonal close packed	4. 68%

Codes:

	A	B	C	D
(a)	1	2	3	4
(b)	4	3	2	1
(c)	3	4	1	2
(d)	4	3	1	2

931. Which one of the following is the correct ascending order of packing density for the given crystal structures of metals?
(a) Simple cubic – Face centred cubic – Body centred cubic
(b) Body centred cubic – Simple cubic – Face centred cubic
(c) Simple cubic – Body centred cubic – Face centred cubic
(d) Body centred cubic – Face centred cubic – Simple cubic

932. Consider the following statements about FCC and HCP crystal structure:
1. Both have same coordination number and atomic packing fraction.
2. Both represent closely packed crystal structures.
3. Both structures are generated by stacking of close packed plants on top of one another, but only the stacking sequence is different.

Which of the statements given above are correct?
(a) 1 and 2 (b) 2 and 3
(c) 1, 2 and 3 (d) 1 and 3

933. Which of the following properties of a solid are dependent on crystal imperfections?
1. Yield stress 2. Melting point
3. Semiconductivity 4. Ductility

Select the correct answer using the codes given below:
(a) 1 and 3 (b) 1, 3 and 4
(c) 2, 3 and 4 (d) 2 and 4

934. Which of the following is a point imperfection?
1. Vacancy
2. Interstitialcy
3. Frenkel imperfection
4. Schottky imperfection
(a) 1 and 2 only (b) 2 and 3 only
(c) 2, 3 and 4 only (d) 1, 2, 3 and 4

935. Which one of the following defects is 'Schottky defect'?
(a) Vacancy defect
(b) Compositional defect
(c) Interstitial defect
(d) Surface defect

936. Which one of the following pairs is not correctly matched?
(a) Point defect in crystal lattice : Self interstitials
(b) Linear defect in crystal lattice : Grain boundary
(c) Planar defect in crystal lattice : External surface
(d) Volume defect in crystal lattice : other phases

Answers

929. (b) 930. (c) 931. (c) 932. (d) 933. (b) 934. (d)
935. (a) 936. (b)

937. Assertion (A): Carbon forms interstitial solid solution when added to iron.

Reason (R): The atomic radius of carbon atom is much smaller than that of iron.

(a) Both A and R are individually true and R is the correct explanation of A
(b) Both A and R are individually true but R is not the correct explanation of A
(c) A is true but R is false
(d) A is false but R is true

938. Which of the following factors govern solubility of two non-ferrous metals both in liquid state, as well as in solid state?

1. Crystal structure
2. Relative size factor
3. Chemical-affinity factor
4. Relative valence factor

Select the correct answer using the codes given below:

Codes:

(a) 1, 2 and 3 (b) 2, 3 and 4
(c) 1 and 4 (d) 1, 2, 3 and 4

939. Which one of the following is correct for "Burger's vector" in screw dislocation?

(a) Perpendicular to the dislocation line
(b) Inclined to the dislocation line
(c) Parallel to the dislocation line
(d) Opposite to the dislocation line

940. Assertion (A): Refining the grain size of a polycrystalline material renders it harder and stronger.

Reason (R): Grain boundaries provide easy paths to dislocation motion.

(a) Both A and R are individually true and R is the correct explanation of A
(b) Both A and R are individually true but R is not the correct explanation of A
(c) A is true but R is false
(d) A is false but R is true

941. Chemicals attack atoms within grain boundaries preferentially because they have

(a) Lower energy than those in the grains
(b) Higher energy than those in the grains
(c) Higher number of atoms than in the grains
(d) Lower number of atoms than in the grains

942. What is a surface imperfection, which separates crystals of different orientations in a poly-crystalline aggregate, called?

(a) Edge dislocation
(b) Stacking fault
(c) Grain boundary
(d) Screw dislocation

943. What is the movement of block of atoms along certain crystallographic planes and directions, termed as?

(a) Glide (b) Twinning
(c) Slip (d) Jog

944. The B.C.C. and H.C.P. metals undergo plastic deformation by:

(a) Slip
(b) Twinning
(c) Edge dislocation
(d) Twinning in combination with slip

945. Which one of the following is correct for 'Climb'?

(a) Dislocation moves parallel to the slip plane
(b) Dislocation moves perpendicular to the slip plane
(c) Sliding of one plane of atoms over the other plane
(d) Dislocation moves from a slip plane to another slip plane

Answers

937. (a)	938. (d)	939. (c)	940. (c)	941. (b)	942. (c)
943. (c)	944. (d)	945. (b)			

946. Assertion (A): Plastic deformation in metals and alloys is a permanent deformation under load. This property is useful in obtaining products by cold rolling.
Reason (R): Plastic or permanent deformation in metal or alloy is caused by movement or dislocations.
(a) Both A and R are individually true and R is the correct explanation of A
(b) Both A and R are individually true but R is not the correct explanation of A
(c) A is true but R is false
(d) A is false but R is true

947. The true strain for a low carbon steel bar which is doubled in length by forging is
(a) 0.307 (b) 0.5
(c) 0.693 (d) 1.0

948. Consider the following statements about medium carbon steel:
1. It can be quench-hardened but not case-hardened.
2. It cannot be quench-hardened but case-hardening can be done.
3. It exhibits distinct yield point under tension test.
Which of the following statements given above are correct?
(a) 1 and 2 (b) 2 and 3
(c) 1 and 3 (d) 1, 2 and 3

949. Match List I with List II and select the correct answer using the codes given below the lists:

List I (Alloy)	List II (Use)
A. Low carbon steel	1. Bearing
B. Hadfield manganese steel	2. Thermocouple
C. Constantan	3. Wire nails.
D. Babbitt allo y	4. Bulldozer blades.

Code:

	A	B	C	D
(a)	1	2	3	4
(b)	3	4	1	2
(c)	3	2	1	4
(d)	3	4	2	1

950. Match List I (Steel) with List II (Application) and select the correct answer using the code given below the Lists:

List I	List II
A. Mild Steel	1. Ball bearing
B. Tool Steel	2. Cold chisels
C. High Carbon Steel	3. Shaft and axles
D. Medium Carbon Steel	4. Rolled steel sections

Codes:

	A	B	C	D
(a)	2	1	4	3
(b)	4	3	2	1
(c)	2	3	4	1
(d)	4	1	2	3

951. In case of power screws, what is the combination of materials used for the screw and the nut?
(a) Cast iron screw and mild steel nut
(b) Carbon steel screw and phosphor bronze nut
(c) Cast iron screw and cast iron nut
(d) Aluminium screw and alloy steel nut

Answers

946. (c) 947. (c) 948. (c) 949. (d) 950. (d) 951. (b)

952. Hot cracks occur in the weld and fusion zone as the metal solidifies. Which of the following are the causes for hot cracks?
   1. Presence of sulphur and phosphorus in the base metal
   2. High carbon or alloy content of the base metal
   3. Moisture in the joint or electrode
   4. Joint restraint

   Select the correct answer using the code given below:
   (a) 1, 2 and 4
   (b) 1, 2 and 3
   (c) 3 and 4
   (d) 1, 2, 3 and 4

953. Which of the following factors influence in a plain carbon steel?
   1. Percentage carbon
   2. Quenching media
   3. Work size

   Select the correct answer using the code given below:
   (a) 1 and 2 only
   (b) 2 and 3 only
   (c) 1 and 3 only
   (d) 1, 2 and 3

954. Consider the following statements:
   1. Strength of steel increases with carbon content.
   2. Young's Modulus of steel increases with carbon content.
   3. Young's Modulus of steel remains unchanged with variation of carbon content.
   4. Which of the statements given above is/are correct?

   (a) 1 only
   (b) 2 only
   (c) 1 and 2
   (d) 1 and 3

955. Presence of hydrogen in steel causes.
   (a) Reduced neutron absorption cross-section
   (b) improved weldability
   (c) Embrittlement
   (d) corrosion resistance

956. Match List-I (Fe-Fe_3C Phase Diagram Characteristic) with List-II (Phase) and select the correct answer using the code given below the Lists:

   List-I
   A. Alpha (α) iron
   B. Iron carbide having crystal lattice with 3 iron and 1 carbon atom
   C. BCC pure allotrope of iron is stable between 1388 °C and its melting point at 1535°C

   List-II
   1. δ iron
   2. Eutectic
   3. Ferrite
   4. Cementite

   Codes:

	A	B	C
(a)	4	2	1
(b)	3	4	1
(c)	4	2	1
(d)	2	1	3

Answers

952. (d) 953. (d) 954. (c) 955. (c) 956. (b)

957. Match List I with List II and select the correct answer:

List I (Phase diagram)

A. Isomorphous system

B. Eutectic system

C. Peritectic system

D. Monotectic system

List II (Characteristic)

1. One liquid decomposes into another liquid and solid
2. One liquid and another solid combine to form a new solid
3. Two metals are completely soluble in liquid state And completely insoluble in solid state
4. Two metals, soluble in solid and liquid state

Codes:

	A	B	C	D
(a)	2	3	4	1
(b)	4	1	2	3
(c)	2	1	4	3
(d)	4	3	2	1

958. According to Gibbs' phase rule, the number of degrees of freedom of an eutectic point in a binary system is

(a) 1 (b) 2

(c) 0 (d) 3

959. In a eutectic system, two elements are completely

(a) Insoluble in solid and liquid states

(b) Soluble in liquid state

(c) Soluble in solid state

(d) Insoluble in liquid state

960. Eutectic reaction for iron-carbon system occurs at

(a) 600°C (b) 723°C

(c) 1147°C (d) 1493°C

961. During peritectic solidification, one liquid

(a) Combines with one solid to form a second new solid

(b) Solidifies into two different solids

(c) Forms one solid

(d) Forms one solid and another liquid

962. Which one of the following is the correct statement?

Pearlite in iron-carbon system is a

(a) Phase consisting of ferrite and cementite at room temperature

(b) Mechanical mixture of ferrite and cementite at room temperature

(c) Eutectic mixture ferrite and cementite at room temperature

(d) All the above three are correct

963. Increase of ferrite phase in steel increases:

(a) Strength (b) Hardness

(c) Ductility (d) Brittleness

964. A 60 C-plain carbon steel has, approximately:

(a) 75% of pearlite and 25% of ferrite

(b) 25% of pearlite and 75% of ferrite

(c) 75% of cementite and 25% of ferrite

(d) 75% of pearlite and 25% of cementite

965. A given steel test specimen is studied under metallurgical microscope. Magnification used is 100 X. In that different phases are observed. One of them is Fe_3C. The observed phase Fe_3C is also known as

(a) Ferrite

(b) Cementite

(c) Austenite

(d) Martensite

Answers

957. (d)	958. (a)	959. (b)	960. (c)	961. (a)	962. (b)
963. (a)	964. (a)	965. (b)			

966. Which one of the following sets of constituents is expected in equilibrium cooling of a hypereutectoid steel from austenitic state?
(a) Ferrite and pearlite
(b) Cementite and pearlite
(c) Ferrite and bainite
(d) Cementite and martensite

967. Vibration damping in machinery is best achieved by means of base structures made of which one of the following materials?
(a) Low carbon steel
(b) Nodular iron
(c) Grey cast iron
(d) White cast iron

968. Which of the following materials is used in the manufacture of extrusion nozzles?
(a) Grey cast iron
(b) Malleable cast iron
(c) White cast iron
(d) Nodular cast iron

969. Which one of the following cast irons consists of carbon in rosette form?
(a) White cast iron
(b) Gray cast iron
(c) Malleable cast iron
(d) Nodular cast iron

970. Consider the following statements:
1. Cast Iron has poor ability to damp vibrations.
2. Cast Iron has higher compressive strength compared to that of steel.
3. Cast Iron parts are suitable where permanent deformation is preferred over fracture.

Which of the statements given above is/are correct?
(a) 1, 2 and 3 (b) 1 and 3
(c) 3 only (d) 2 only

971. Which of the following pairs are correctly matched? (Designation of Steel/Cast Iron) (Description)
1. Fe E 250 : Minimum tensile strength of 250 N/mm2
2. 40 C 8 : Percentage of Manganese is 0.7% – 0.9%
3. FG 200 : Grey cast iron with ultimate tensile strength of 200 N/mm^2

Select the correct answer using the code given below:
(a) 1 and 2 (b) 2 and 3
(c) 1 and 3 (d) 1, 2 and 3

972. Machine tool manufacturers prefer grey cast-iron grade 40 for producing machine columns and tables because grey cast-iron is
1. Heavy 2. Easily castable
3. Easily weldable
4. Having good damping capacity

Select the correct answer using the codes given below:
Codes:
(a) 1 and 2 (b) 2 and 4
(c) 1 and 3 (d) 3 and 4

973. Piston compression rings are made of which one of the following ?
(a) Cast iron (b) Bronze
(c) Aluminium (d) White metal

974. Nodular grey cast iron is obtained from the grey cast iron by adding a small amount of
(a) Manganese (b) Phosphorus
(c) Magnesium (d) Chromium

Answers

966. (b)	967. (b)	968. (c)	969. (d)	970. (d)	971. (c)
972. (b)	973. (a)	974. (c)			

975. Cast iron is used for machine beds because of its high
(a) Tensile strength
(b) Endurance strength
(c) Damping capacity
(d) none

976. Assertion (A): Cast iron is generally hard, brittle and wear resistant.
Reason (R): Cast iron contains more than 2% carbon and as such the percentage cementite in it is higher.
(a) Both A and R are individually true and R is the correct explanation of A
(b) Both A and R are individually true but R is not the correct explanation of A
(c) A is true but R is false
(d) A is false but R is true

977. Assertion (A): The notch sensitivity of cast iron component is zero.
Reason (R): Cast iron does not have a yield point.
(a) Both A and R are individually true and R is the correct explanation of A
(b) Both A and R are individually true but R is not the correct explanation of A
(c) A is true but R is false
(d) A is false but R is true

978. Consider the following statements:
Addition of silicon to cast iron
1. Promotes graphite module formation.
2. Promotes graphite flake formation.
3. Increases the fluidity of the molten metal.
4. Improves the ductility of cast iron.
Select the correct answer using the codes given below:
(a) 1and 4 (b) 2 and 3
(c) 1and 3 (d) 3 and 4

979. An iron-carbon binary alloy has 0.5% C by weight. What is this alloy called?
(a) Eutectoid alloy
(b) Eutectic alloy
(c) Hypo-eutectoid alloy
(d) Hypereutectoid alloy

980. Consider the following work materials:
1. Titanium 2. Mild steel
3. Stainless stee 4. Grey cast iron.
The correct sequence of these materials in terms of increasing order of difficulty in machining is
(a) 4,2,3,1
(b) 4,2, 1,3
(c) 2,4,3,1
(d) 2, 4, 1, 3.

981. Addition of magnesium to cast iron increases its
(a) Hardness
(b) Ductility and strength in tension
(c) Corrosion resistance
(d) Creep strength.

982. Assertion (A): Machine tool beds are generally made of grey cast iron.
Reason (R): Cast iron possesses good self-lubricating properties.
(a) Both A and R are individually true and R is the correct explanation of A
(b) Both A and R are individually true but R is not the correct explanation of A
(c) A is true but R is false
(d) A is false but R is true

Answers

975. (c)	976. (a)	977. (a)	978. (b)	979. (c)	980. (a)
981. (a)	982. (a)				

983. Which of the following pairs are correctly matched?
1. Lead screw nut .... Phosphor bronze
2. Piston .... Cast iron.
3. Cam .... EN-31 steel
4. Lead screw ....Wrought iron.
Select the correct answer using the codes given below:
Codes:
(a) 2, 3 and 4 (b) 1, 3 and 4
(c) 1, 2 and 4 (d) 1, 2 and 3

984. Assertion (A): Fracture surface of grey cast iron is dark.
Reason (R): Failure takes place along the weak cementite plates.
(a) Both A and R are individually true and R is the correct explanation of A
(b) Both A and R are individually true but R is not the correct explanation of A
(c) A is true but R is false
(d) A is false but R is true

985. Which of the following display properties similar to that of steel ?
1. Black-heart cast iron
2. White-heart cast iron
3. Gray cast iron
4. Pig iron
(a) 1 and 2 only (b) 3 and 4 only
(c) 2 and 4 only (d) 1 and 3 only

986. Which of the following metal shrinks most from molten state to solid state?
(a) Cast iron (b) Cast steel
(c) Brass (d) Admiralty metal

987. Consider the following statements:
1. From design considerations, it is always advantageous to place cast iron ribs on the tension side rather than on the compression side.
2. Cast iron is an excellent choice for machine tool guides and frames.
3. Cast iron parts have low notch sensitivity.
Which of these statements are correct?
(a) 1, 2 and 3 (b) 2 and 3
(c) 1 and 3 (d) 1 and 2

988. A cast iron specimen in a torsion test gives a
(a) Cup-and-cone fracture
(b) Fracture along a plane normal to the axis of the specimen
(c) Fracture along a helix of approximately 45°
(d) Fracture along a plane inclined at 60° to the axis

989. The alloying element mainly used to improve the endurance strength of steel materials is
(a) Nickel (b) Vanadium
(c) Molybdenum (d) Tungsten

990. Alloy steel which is work hardenable and which is used to make the blades of bulldozers, bucket wheel excavators and other earth moving equipment contain iron, carbon and
(a) Chromium (b) Silicon
(c) Manganese (d) Magnesium.

991. The alloy steel designated as 40 Cr18 Ni 2 by Bureau of Indian Standards contains
(a) 0.4% C, 18% Cr and 2% Ni
(b) 4.0% C, 1.8% Cr and 0.2% Ni
(c) 0.4% C, 1.8% Cr and 2% Ni
(d) 0.4% C, 1.8% Cr and 0.2% Ni

Answers

983. (d) 984. (a) 985. (a) 986. (b) 987. (b) 988. (b)
989. (b) 990. (c) 991. (a)

992. Match List I (Name of alloy) with List II (Major alloying elements) and select the correct answer using the codes given below the Lists:

List I	List II
A. Invar	1. Manganese
B. Hadfield steel	2. Chromium
C. Stellite	3. Nickel
D. Stainless steel	4. Tungsten
	5. Molybdenum

Codes:

	A	B	C	D
(a)	5	1	4	2
(b)	3	2	5	1
(c)	5	2	4	1
(d)	3	1	5	2

993. Match List I (Alloying Element) with List II (Effect on Steel) and select the correct answer using the code given below the Lists:

List I	List II
A. Vanadium	1. Increases endurance strength
B. Molybdenum	2. Improves creep properties
C. Silicon	3. Increases hardness
D. Chromium	4. Increases resistance to high temperature oxidation

Codes:

	A	B	C	D
(a)	2	1	3	4
(b)	1	3	2	4
(c)	2	1	4	3
(d)	1	2	3	4

994. Match List-I (Alloying element in steel) with List-II (Property conferred on steel by the element) and select the correct answer using the codes given below the lists:

List-I	List-II
A. Nickel	1. Corrosion resistance
B. Chromium	2. Magnetic permeability
C. Tungsten	3. Heat resistance
D. Silicon	4. Hardenability

Codes:

	A	B	C	D
(a)	4	1	3	2
(b)	4	1	2	3
(c)	1	4	3	2
(d)	1	4	2	3

995. Consider the following statements in respect of austenitic stainless steels:

1. Austenitic stainless steels are hardened and strengthened by cold working.
2. Austenitic stainless steels cannot be quenched and tempered.

Which of the statements given above is/are correct?

(a) 1 only (b) 2 only
(c) Both 1 and 2 (d) Neither 1 nor 2

996. Assertion (A): Austenitic stainless steel contains 18% chromium and 8% nickel. Since it retains its austenitic structure at room temperature, it is called austenitic stainless steel.

Reason (R): Chromium present in the steel improves its corrosion resistance by forming a thin film of chromium oxide on the surface.

(a) Both A and R are individually true and R is the correct explanation of A
(b) Both A and R are individually true but R is not the correct explanation of A
(c) A is true but R is false
(d) A is false but R is true

Answers

992. (d) 993. (d) 994. (a) 995. (c) 996. (a)

997. Match List I (Alloy) with List II (Major Constituent) and select the correct answer using the code given below the Lists:

List I	List II
A. Babbitt	1. Nickel
B. Invar	2. Tin and lead
C. Gun Metal	3. Aluminium
D. Duralumin	4. Copper

Code:

	A	B	C	D
(a)	2	4	1	3
(b)	3	1	4	2
(c)	2	1	4	3
(d)	3	4	1	2

998. Coefficient of Expansion is practically nil in a particular alloy. What is this alloy?
(a) Hadfield Manganese Steel
(b) Invar
(c) Vitallium
(d) Stellite

999. For improving the strength of steel at elevated temperatures, which one of the following alloying element is used?
(a) Copper (b) Tungsten
(c) Aluminium (d) Zinc

1000. Addition of which of the following improves machining of copper?
(a) Sulphur (b) Vanadium
(c) Tin (d) Zinc

1001. Which of the following pairs are correctly matched?
1. Silicon steels ....Transformer stampings
2. Duralumin ......Cooking utensils.
3. Gun metal................ Bearings

Select the correct answer using the codes given below.
Codes:
(a) 1, 2 and 3 (b) 1 and 2
(c) 1 and 3 (d) 2 and 3

1002. Consider the following statements: An increase in the cobalt content in the straight carbide grades of carbide tools
1. Increases the hardness.
2. Decreases the hardness.
3. Increases the transverse rupture strength
4. Lowers the transverse rupture strength.
Which of the statements given above are correct?
(a) 1 and 3 (b) 2 and 4
(c) 1 and 4 (d) 2 and 3

1003. Disruptive strength is the maximum strength of a metal, when
(a) Subjected to 3 principal tensile stresses at right angles to one another and of equal magnitude
(b) Loaded in tension
(c) Loaded in compression
(d) Loaded in shear

1004. Which one of the following elements is an austenitic stabilizer?
(a) Chromium
(b) Tungsten
(c) Nickel
(d) Molybdenum

1005. Which one of the following elements is a ferritic stabilizer?
(a) Nickel (b) Manganese
(c) Copper (d) Chromium

Answers

997. (c)	998. (b)	999. (b)	1000. (d)	1001. (c)	1002. (d)
1003. (a)	1004. (c)	1005. (d)			

1006. Which of the following elements given below determine(s) the maximum attainable hardness in steel?

1. Chromium 2. Manganese
3. Carbon 4. Molybdenum

select the correct answer using the code given below:

(a) 1 only (b) 1 and 2
(c) 3 only (d) 2 and 4

1007. Small percentage of boron is added to steel to

(a) Increases hardenability
(b) Reduce machinability
(c) Increases wear resistance
(d) Increase endurance strength

1008. Assertion(A): In high speed steels, alloying elements tungsten, chromium and vanadium are added to make them suitable to work at higher speeds than tool steel or low alloy steels.

Reason(R): Vanadium adds to the property of red hardness and tungsten and chromium add to high wear resistance.

(a) Both A and R are individually true and R is the correct explanation of A
(b) Both A and R are individually true but R is not the correct explanation of A
(c) A is true but R is false
(d) A is false but R is true

1009. Match List I (Alloying elements for tool steel) with List II (Improved mechanical property) and select the correct answer using the codes given below the Lists.

List I
(Alloying elements for tool steel)
A. Carbon
B. Manganese
C. Chromium
D. Vanadium

List II
(Improved mechanical property)
1. Hardness
2. Hot Hardness
3. Lower Critical Temp
4. Toughness

Codes:

	A	B	C	D
(a)	1	3	4	2
(b)	2	4	3	1
(c)	1	4	3	2
(d)	2	3	4	1

1010. Match List I with List II and select the correct answer using the codes given below the lists:

List I (Cutting tools)
A. Stellite B. H.S.S.
C. Ceramic D. DCON

List II (Major constituent)
1. Tungsten 2. Cobalt
3. Alumina 4. Columbium
5. Titanium

Codes:

	A	B	C	D
(a)	5	1	3	4
(b)	2	1	4	3
(c)	2	1	3	4
(d)	2	5	3	4

Answers

1006. (c) 1007. (a) 1008. (a) 1009. (c) 1010. (c)

1011. Cutting tool material 18-4-1 HSS has which one of the following compositions?
(a) 18% W, 4% Cr, 1% V
(b) 18% Cr, 4% W, 1% V
(c) 18% W, 4% Ni, 1% V
(d) 18% Cr, 4% Ni, 1% V

1012. The limit to the maximum hardness of a work material which can be machined with HSS tools even at low speeds is set by which one of the following tool failure mechanisms?
(a) Attrition
(b) Abrasion
(c) Diffusion
(d) Plastic deformation under compression.

1013. The blade of a power saw is made of
(a) Boron steel
(b) High speed steel
(c) Stainless steel
(d) Malleable cast iron

1014. The compositions of some of the alloy steels are as under:
1. 18 W 4 Cr 1 V
2. 12 Mo 1 W 4 Cr 1 V
3. 6 Mo 6 W 4 Cr 1 V
4. 18 W 8 Cr 1 V

The compositions of commonly used high speed steels would include
(a) 1 and 2 (b) 2 and 3
(c) 1 and 4 (d) 1 and 3

1015. Percentage of various alloying elements present in different steel materials are given below:
1. 18% W; 4% Cr; 1% V; 5% Co; 0.7% C
2. 8% Mo; 4% Cr; 2% V; 6% W; 0.7% C
3. 27% Cr; 3% Ni; 5% Mo; 0.25% C
4. 18% Cr; 8% Ni; 0.15% C

Which of these relate to that of high speed steel?
(a) 1 and 3
(b) 1 and 2
(c) 2 and 3
(d) 2 and 4

1016. Match List-I (Alloys) with List-II (Applications) and select the correct answer using the codes given below the lists:

List-I	List-II
A. Chromel	1. Journal bearing
B. Babbit alloy	2. Milling cutter
C. Nimonic alloy	3. Thermocouple wire
D. High speed steel	4. Gas turbine blades

Code:

	A	B	C	D
(a)	3	1	4	2
(b)	3	4	1	2
(c)	2	4	1	3
(d)	2	1	4	3

1017. The main alloying elements in high speed Steel in order of increasing proportion are :
(a) Vanadium, chromium, tungsten
(b) Tungsten, titanium, vanadium
(c) Chromium, titanium, vanadium
(d) Tungsten, chromium, titanium

1018. Assertion (A): The characteristic feature of High speed Steel is its red hardness.

Reason (R): Chromium and cobalt in High Speed promote martensite formation when the tool is cold worked.
(a) Both A and R are individually true and R is the correct explanation of A
(b) Both A and R are individually true but R is not the correct explanation of A
(c) A is true but R is false
(d) A is false but R is true

Answers

1011. (a)	1012. (a)	1013. (b)	1014. (d)	1015. (b)	1016. (a)
1017. (a)	1018. (b)				

1019. Which of the following processes can be used for production thin, hard. heat resistant coating at TiN, on HSS?

1. Physical vapour deposition.
2. Sintering under reducing atmosphere.
3. Chemical vapour deposition with post treatment
4. Plasma spraying.

Select the correct answer using the codes given below:

Codes:

(a) 1 and 3 (b) 2 and 3
(c) 2 and 4 (d) 1 and 4

1020. Match List I with List IT and select the correct answer using the codes given below the lists:

List - I (Cutting tool Material)

A. High speed steel
B. Stellite
C. Diamond
D. Coated carbide tool

List - I I(Major characteristic constituent)

1. Carbon
2. Molybdenum
3. Nitride
4. Columbium
5. Cobalt

Codes:

	A	B	C	D
(a)	2	1	3	5
(b)	2	5	1	3
(c)	5	2	4	3
(d)	5	4	2	3

1021. Which one of the following is the hardest cutting tool material next only to diamond?

(a) Cemented carbides
(b) Ceramics
(c) Silicon
(d) Cubic boron nitride

1022. The straight grades of cemented carbide cutting tool materials contain

(a) Tungsten carbide only
(b) Tungsten carbide and titanium carbide
(c) Tungsten carbide and cobalt
(d) Tungsten carbide and cobalt carbide.

1023. Which of the following given the correct order of increasing hot hardness of cutting tool material?

(a) Diamond, Carbide, HSS
(b) Carbide, Diamond, HSS
(c) HSS, carbide, Diamond
(d) HSS, Diamond, Carbide

1024. Match List-I with List-II and select the correct answer using the codes given below the Lists:

List I (Materials)	List II (Applications)
A. Tungsten carbide	1. Abrasive wheels
B. Silicon nitride	2. Heating elements
C. Aluminium oxide	3. Pipes for conveying liquid metals
D. Silicon carbide	4. Drawing dies

Code:

	A	B	C	D
(a)	3	4	1	2
(b)	4	3	2	1
(c)	3	4	2	1
(d)	4	3	1	2

Answers

1019. (a) 1020. (b) 1021. (d) 1022. (c) 1023. (d) 1024. (d)

1025. Cubic boron nitride
(a) Has a very high hardness which is comparable to that of diamond.
(b) Has a hardness which is slightly more than that of HSS
(c) Is used for making cylinder blocks of aircraft engines
(d) Is used for making optical glasses.

1026. Cubic boron nitride is used
(a) As lining material in induction furnace
(b) For making optical quality glass.
(c) For heat treatment
(d) For none of the above.

1027. Which one of the following is not a synthetic abrasive material?
(a) Silicon Carbide
(b) Aluminium Oxide
(c) Titanium Nitride
(d) Cubic Boron Nitride

1028. Which one of the following is not a ceramic?
(b) Alumina (b) Porcelain
(c) Whisker (d) Pyrosil

1029. Consider the following tool materials :
1. HSS 2. Cemented carbide
3. Ceramics 4. Diamond
The correct sequence of these materials in decreasing order of their cutting speed is
(a) 4, 3, 1, 2 (b) 4, 3, 2, 1
(c) 3, 4, 2, 1 (d) 3, 4, 1, 2

1030. Cermets are
(a) Metals for high temperature use with ceramic like properties
(b) Ceramics with metallic strength and luster
(c) Coated tool materials
(d) Metal-ceramic composites

1031. Assertion (A): Ceramic tools are used only for light, smooth and continuous cuts at high speeds.
Reason (R): Ceramics have a high wear resistance and high temperature resistance.
(a) Both A and R are individually true and R is the correct explanation of A
(b) Both A and R are individually true but R is not the correct explanation of A
(c) A is true but R is false
(d) A is false but R is true

1032. Consider the following tool materials:
1. Carbide
2. Cermet
3. Ceramic
4. Borazon.
Correct sequence of these tool materials in increasing order of their ability to retain their hot hardness is
(a) 1,2,3,4
(b) 1,2,4,3
(c) 2, 1, 3, 4
(d) 2, 1, 4, 3

1033. Assertion (A): Diamond tools can be used at high speeds.
Reason (R): Diamond tools have very low coefficient of friction.
(a) Both A and R are individually true and R is the correct explanation of A
(b) Both A and R are individually true but R is not the correct explanation of A
(c) A is true but R is false
(d) A is false but R is true

Answers

1025. (a)	1026. (d)	1027. (b)	1028. (d)	1029. (c)	1030. (d)
1031. (b)	1032. (a)	1033. (c)			

1034. Assertion (A): Non-ferrous materials are best machined with diamond tools.

Reason (R): Diamond tools are suitable for high speed machining.

(a) Both A and R are individually true and R is the correct explanation of A

(b) Both A and R are individually true but R is not the correct explanation of A

(c) A is true but R is false

(d) A is false but R is true

1035. Match. List I (Cutting tool materials) with List II (Manufacturing methods) and select the correct answer using the codes given below the Lists:

List I	List II
A. HSS	1. Casting
B. Stellite	2. Forging
C. Cemented carbide	3. Rolling
D. UCON	4. Extrusion
	5. Powder metallurgy

Codes:

	A	B	C	D
(a)	3	1	5	2
(b)	2	5	4	3
(c)	3	5	4	2
(d)	2	1	5	3

1036. Consider the following cutting tool materials used for metal-cutting operation at high speed:

1. Tungsten carbide
2. Cemented titanium carbide
3. High-speed steel
4. Ceramic

The correct sequence in increasing order of the range of cutting speeds for optimum use of these materials is

(a) 3,1,4,2 (b) 1,3,2,4

(c) 3,1,2,4 (d) 1,3,4,2

1037. Which of the following tool materials have cobalt as a constituent element?

1. Cemented carbide 2. CBN

3. Stellite 4. UCON

Select the correct answer using the codes given below:

Codes:

(a) 1 and 2 (b) 1 and 3

(c) 1 and 4 (d) 2 and 3

1038. At room temperature, which one of the following is the correct sequence of increasing hardness of the tool materials?

(a) Cast alloy-HSS-Ceramic-Carbide

(b) HH-Cast alloy-Ceramic-Carbide

(c) HSS-Cast alloy-Carbide-Ceramic

(d) Cast alloy-HSS-Carbide-Ceramic

1039. The coating materials for coated carbide tools, includes

(a) TiC, TiN and NaCN

(b) TiC and TiN

(c) TiN and NaCN

(d) TiC and NaCN

1040. The ductility of a material with work hardening

(a) Increases

(b) Decreases

(c) Remains unaffected

(d) Unpredictable

1041. Cast steel crankshaft surface is hardened by

(a) Nitriding

(b) Normalising

(c) Carburising

(d) Induction heating

Answers

1034. (b) 1035. (c) 1036. (c) 1037. (b) 1038. (d) 1039. (b)
1040. (b) 1041. (d)

1042. From the lists given below, choose the most appropriate set of heat treatment process and the corresponding process characteristics

Heat Treatment Process

P. Tempering

Q. Austempering

R. Martempering

Process Characteristics

1. Austenite is converted into bainite
2. Austenite is converted into martensite
3. Cementite is converted into globular structure
4. Both hardness and brittleness are reduced
5. Carbon is absorbed into the metal

(a) P–3 Q–1 R–5 (b) P–4 Q–3 R–2
(c) P–4 Q–1 R–2 (d) P–1 Q–5 R–4

1043. Hardness of steel greatly improves with
(a) Annealing (b) Cyaniding
(c) Normalising (d) Tempering

1044. On completion of heat treatment, the resulting structure will have retained Austenite if
(a) Rate of cooling is greater than the critical cooling rate
(b) Rate of cooling is less than the critical cooling rate
(c) Martensite formation starting temperature is above the room temperature
(d) Martensite formation finish temperature is below the room temperature

1045. Carburized machine components have high endurance limit because carburization
(a) Raises the yield point of the material
(b) Produces a better surface finish
(c) Introduces a compressive layer on the surface
(d) Suppresses any stress's, concentration produced in the component.

1046. During heat treatment of steel, the hardness of various structures in increasing order is
(a) Martensite, fine pearlite, coarse pearlite, spherodite
(b) Fine pearlite, coarse pearlite, spherodite, martensite
(c) Martensite, coarse pearlite, fine pearlite, spherodite
(d) Spherodite, coarse pearlite, fine pearlite, martensite

1047. The iron-carbon diagram and the TTT curves are determined under
(a) Equilibrium and non-equilibrium conditions respectively
(b) Non-equilibrium and equilibrium conditions respectively
(c) Equilibrium conditions for both
(d) Non-equilibrium conditions for both.

1048. TTT diagram indicates time and temperature transformation of
(a) Cementite (b) Pearlite
(c) Ferrite (d) Austenite

1049. Assertion (A): Lever Rule can be applied to determine relative amounts of phases present at any temperature.

Reason (R): Lever Rule is restricted to estimate relative phases, only if they are solid phases.
(a) Both A and R are true and R is the correct explanation of A
(b) Both A and R are true but R is NOT the correct explanation of A
(c) A is true but R is false
(d) A is false but R is true

Answers

1042. (c)	1043. (b)	1044. (c)	1045. (c)	1046. (d)	1047. (a)
1048. (d)	1049. (c)				

1050. Consider the following treatments:

1. Normalizing 2. Hardening
3. Martempering 4. Cold working

Hardness and tensile strength in austenitic stainless steel can be increased by

(a) 1, 2 and 3 (b) 1 and 3
(c) 2 and 4 (d) 4 alone

1051. Quenching is not necessary when hardening is done by

(a) Case carburizing
(b) Flame hardening
(c) Nitriding
(d) Any of the above processes

1052. Consider the following quenching media:

1. Oil 2. Water
3. Water + NaOH 4. Brine

The correct sequence of these media in order of increasing hardness of steel undergoing heat treatment is

(a) 1, 3, 2, 4 (b) 2, 1, 3, 4
(c) 1, 2, 3, 4 (d) 4, 3, 2, 1

1053. Match List-I (Effect of Cooling) with List-II (Cooling Medium) and select the correct answer using the code given below:

List -I	List - II
A. Martensite	1. Water quenched
B. Very fine pearlite	2. Air cooled
C. Fine pearlite	3. Furnace cooled
D. Coarse pearlite	4. Oil quenched

Code:

	A	B	C	D
(a)	1	4	2	3
(b)	2	3	1	4
(c)	2	3	4	1
(d)	1	2	3	4

1054. Which one of the following materials can be subjected to an age hardening process?

(a) HSS (b) Aluminium
(c) Pure iron (d) Stellite

1055. Tempering is a process of annealing

(a) Martensite at low temperatures
(b) Martensite at higher temperatures
(c) Bainite at low temperatures
(d) Bainite at higher temperatures

1056. Austempering is employed to obtain:

(a) 100% martensitic structure
(b) 100% bainitic structure
(c) 50% martensitic and 50% bainitic structure
(d) 100% pearlitic structure

1057. Consider the following pairs:

Heat treatment Effect on medium carbon steel

1. Normalizing : Grain refinement
2. Full annealing : Uniform grain structure
3. Martempering : Decreased ductility
4. Spheroidizing : Maximum softness

Which of the pairs given above are correctly matched?

(a) 1 and 2 (b) 2 and 3
(c) 3 and 4 (d) 1, 2, 3 and 4

1058. 'Tempering' of quenched martensitic steel is necessary to improve the .... of the metal.

(a) Hardness
(b) Surface texture
(c) Corrosion resistance
(d) Ductility

Answers

1050. (d)	1051. (d)	1052. (c)	1053. (a)	1054. (b)	1055. (a)
1056. (b)	1057. (d)	1058. (d)			

1059. Match List I with List II and select the correct answer using the codes given below the lists:

List I (Heat treatment)

A. Annealing

B. Nitriding

C. Martempering

D. Normalising

List II (Effect on the properties)

1. Refined grain structure
2. Improves the hardness of the whole mass
3. Increases surface hardness
4. Improves ductility

Codes:

	A	B	C	D
(a)	4	3	2	1
(b)	1	3	4	2
(c)	4	2	1	3
(d)	2	1	3	4

1060. The pattern known as Widmanstatten structure is encountered in:

(a) Tempering (b) Normalizing

(c) Spheroidizing (d) Annealing

1061. The complete phase recrystallization and fine grain structure is obtained in casting, forging and rolled parts by:

(a) Recrystallization annealing

(b) Normalizing

(c) Spheroidizing

(d) Austenising

1062. Heating the hypoeutectoid steels to 30oC above the upper critical temperature line, soaking at that temperature and then cooling slowly to room temperature to form a pearlite and ferrite structure, is known as

(a) Hardening (b) Normalizing

(c) Tempering (d) Annealing

1063. Which of the following statements are true of annealing of steels?

1. Steels are heated to 500 to 700°C.
2. Cooling is done slowly and steadily.
3. Internal stresses are relieved.
4. Ductility of steel is increased.

Select the correct answer using the codes given below:

Codes:

(a) 2, 3 and 4 (b) 1, 3 and 4

(c) 1, 2 and 4 (d) 1, 2 and 3

1064. Temperature required for full annealing in hyper-eutectoid steel is

(a) 50°C above upper critical temperature (AC3)

(b) 50°C below upper critical temperate (AC3)

(c) 50°C above lower critical temperature (AC1)

(d) 50°C below lower critical temperature (AC1)

1065. Assertion (A): Normalized steel will have lower hardness than annealed steel.

Reason (R): The pearlite of normalized steel is finer and has lower intermolecular space.

(a) Both A and R are individually true and R is the correct explanation of A

(b) Both A and R are individually true but R is not the correct explanation of A

(c) A is true but R is false

(d) A is false but R is true

1066. Globular form of cementite in the structure of steel is obtained through

(a) Normalizing (b) Malleabilising

(c) Spheroidizing (d) Carbonizing

Answers

1059. (a) 1060. (a) 1061. (a) 1062. (d) 1063. (a) 1064. (c)

1065. (d) 1066. (c)

1067. In case carburising Carbon is introduced to form a high carbon layer at the surface. The carbon is introduced in the form of

(a) Graphite flakes (b) Pearlite
(c) Cementite (d) Free carbon

1068. Match List I (Name of treatment) with List II (Media used) and select the correct answer using the codes given below the Lists.

List I
A. Pack carburizing
B. Gas carburizing
C. Cyaniding
D. Nitriding

List II
1. Ammonia gas
2. Sodium cyanide
3. Carburizing compound
4. Ethane

Codes:

	A	B	C	D
(a)	3	4	2	1
(b)	2	1	3	4
(c)	3	1	2	4
(d)	2	4	3	1

1069. Induction hardening is basically a

(a) Carburising process
(b) Surface hardening process
(c) Core-hardening process
(d) None of the above

1070. Guideways of lathe beds are hardened by

(a) Carburising (b) Cyaniding
(c) Nitriding (d) Flame hardening

1071. Which one of the following elements/ alloy exhibits season cracking?

(a) Iron (b) Brass
(c) Aluminium (d) Steel

1072. Which one among the following is the most effective strengthening mechanism of non-ferrous metal?

(a) Solid solution hardening
(b) Strain hardening
(c) Grain size refinement
(d) Precipitation hardening

1073. Which one of the following pairs is correctly matched?

(a) Solid solution strengthening .... Increasing density of dislocations
(b) Dispersion hardening .... Creating strained region in the crystal
(c) Strain-hardening .... Creating particles to resist the movement of dislocations
(d) Precipitation-hardening .... Creating particles by decreasing solubility of one phase in another

1074. Assertion (A): Carburizing is done on non-ferrous alloys to increase the surface hardness.

Reason (R): Precipitation hardening of non-ferrous alloys involves solution heat treatment followed by precipitation heat treatment.

(a) Both A and R are individually true and R is the correct explanation of A
(b) Both A and R are individually true but R is not the correct explanation of A
(c) A is true but R is false
(d) A is false but R is true

Answers

1067. (d)	1068. (a)	1069. (b)	1070. (d)	1071. (b)	1072. (d)
1073. (d)	1074. (d)				

1075. If the surface of a component is heavily stressed while the stresses in the core are of comparative small magnitude, which one of the following heat treatment methods is employed?

(a) Annealing (b) Tempering
(c) Quenching (d) Case hardening

1076. Major operations in the manufacture of steel balls used for Ball bearings are given below.

1. Oil lapping 2. Cold heading
3. Annealing 4. Hardening
5. Rough grinding

The correct sequence of these operations is

(a) 3,2,4,1,5 (b) 3,2,1,4,5
(c) 2,3,4,5,1 (d) 2,3,5,4,1

1077. Two plain carbon steel specimens having 0·8% carbon content are welded. If we observe the weldment under Metallurgical Microscope from centre towards either side, the following structures are observed at different zones:

1. Fine Pearlite
2. Coarse Pearlite
3. Martensite

Select the correct sequence using the codes given below:

Codes:

(a) 1, 2, 3 (b) 1, 3, 2
(c) 2, 1, 3 (d) 3, 1, 2

1078. Structure of a polymer is:

(a) Long chain
(b) Rhombic
(c) Cubic
(d) Closed pack hexagonal

1079. Assertion (A): Linear polymers are rigid at low temperatures but soft and mouldable at elevated temperatures.

Reason (R): Linear polymers are thermo-setting.

(a) Both A and R are individually true and R is the correct explanation of A
(b) Both A and R are individually true but R is not the correct explanation of A
(c) A is true but R is false
(d) A is false but R is true

1080. Teflon is a

(a) Thermosetting fluorocarbon polymer
(b) Thermo-plastic fluorocarbon polymer
(c) Inorganic compound of fluorine and carbon
(d) Laminated phenolic material

1081. Match List I with List II and select the correct answer using the code given below the Lists:

List I
(Material)
A. Fibre reinforced plastics
B. Acrylics
C. Phenolics
D. Butadiene rubber

List II
(Application)
1. Automobile tyres 2. Aircraft
3. Lenses 4. Electric switch cover

Code:

	A	B	C	D
(a)	1	4	3	2
(b)	2	3	4	1
(c)	1	3	4	2
(d)	2	4	3	1

Answers

1075. (c) 1076. (c) 1077. (d) 1078. (a) 1079. (c) 1080. (b)
1081. (b)

1082. Which of the following are fabricated using engineering plastics?

1. Surface plate
2. Gears
3. Guide ways for machine tools
4. Foundry patterns

Select the correct answer using the codes given below:

(a) 1, 2 and 3 (b) 1
(c) 2, 3 and 4 (d) 1, 2, 3 and 4

1083. Consider the following statements:

Polytetrafluoroethene is

1. A thermoplastic material
2. Having high friction coefficient
3. A thermosetting material
4. Having low friction coefficient
5. An electric insulator 6. Non sticking to surfaces

Which of the above statements are correct?

(a) 1, 2 and 5 (b) 2, 3 and 6
(c) 3, 4 and 5 (d) 3, 2 and 5

1084. The molecular weight of vinyl chloride is 62.5. Thus the molecular weight of a polyvinyl chloride with a degree of polymerization of 20000 is

(a) 20000 (b) 62.5
(c) 62.5× 20000 (d) 20000

1085. Consider the following pairs of plastics and their distinct characteristics:

1. Acrylics .... Very good transparency to light
2. Polycarbonate ..... Poor impact resistance
3. PTFE ... . Low coefficient of friction.
4. Polypropylene .... Excellent fatigue strength

Which of these pairs are correctly matched?

(a) 2 and 3 (b) 1 and 3
(c) 1 and 4 (d) 2 and 4

1086. Consider the following statements:

Thermosetting plastics are

1. Formed by addition polymerisation.
2. Formed by condensation polymerisation.
3. Softened on heating and hardened on cooling for any number of times
4. Moulded by heating and cooling.

Select the correct answer using the codes given below:

(a) 1 and 3
(b) 2 and 4
(c) 1 and 4
(d) 2 and 3

1087. Match List I with List II and select the correct answer

List I (Material)

A. Polyethylene
B. Polyurethane
C. Cyano-acrylate
D. Nylon

List II (Nature of product)

1. Adhesive
2. Film
3. Wire
4. Foam

Codes:

	A	B	C	D
(a)	2	4	3	1
(b)	4	2	3	1
(c)	2	4	1	3
(d)	4	2	1	3

Answers

1082. (c) 1083. (a) 1084. (c) 1085. (c) 1086. (b) 1087. (c)

1088. Consider the following statements:

Fibre Reinforced Plastics are

1. Made of thermosetting resins and glass fibre
2. Made of thermoplastic resins and glass fibre
3. Anisotropic
4. isotropic

Select the correct answer using the codes given below:

(a) 1 and 4

(b) 1 and 3

(c) 2 and 3

(d) 2 and 4

1089. Which of the following pairs of plastics and their modes of formation are correctly matched?

1. Polythene .... Condensation polymerization.
2. Polycarbonate .... Addition polymerisation.
3. Polystyrene .... .Addition polymerisation.
4. Polyamide .... Either by addition or by condensation polymerisation.

Select the correct answer using the codes given below:

(a) 1 and 2

(b) 2 and 4

(c) 1 and 4

(d) 3 and 4

1090. Polyamides are characterized by

(a) Flexible chain

(b) Rigid chain

(c) Amorphous structure

(d) Crystalline structure

1091. Polyesters can be defined as the condensation products of

(a) Dicarboxylic acids with dihydroxy alcohols

(b) Bisphenol-A and epichloro-hydrin

(c) Phenol and formaldehyde

(d) Benzene and toluene

1092. Assertion (A) In Addition Polymerization method, polymer is produced by adding a second monomer to the first, then a third monomer to this dimmer and so on.

Reason (R): There must exist at least one double bond in the monomer for Addition Polymerization reaction.

(a) Both A and R are individually true and R is the correct explanation of A

(b) Both A and R are individually true but R is not the correct explanation of A

(c) A is true but R is false

(d) A is false but R is true

1093. Assertion (A): Addition polymerization is a primary summation of individual molecules into long chains.

Reason (R): In addition polymerization, the reaction produces a small molecule as by-product.

(a) Both A and R are individually true and R is the correct explanation of A

(b) Both A and R are individually true but R is not the correct explanation of A

(c) A is true but R is false

(d) A is false but R is true

Answers

1088. (b) 1089. (d) 1090. (d) 1091. (a) 1092. (a) 1093. (c)

1094. Match List I (materials) with List II (applications) and select the correct answer using the codes given below the Lists:

List I

A. Engineering ceramics

B. Fibre reinforced plastics

C. Synthetic carbon

D. Boron

List II

1. Bearings

2. Control rods in nuclear reactors

3. Aerospace industry

4. Electrical insulator

Codes:

	A	B	C	D
(a)	1	2	3	4
(b)	1	4	3	2
(c)	2	3	1	4
(d)	4	3	1	2

1095. Match List I (Type of moulding) List II (Mechanism involved) and select the correct answer.

List I

A. Compression moulding

B. Injected moulding

C. Jet moulding

D. Extrusion moulding

List II

1. Mould cavity must be heated to cure the plastic forced into it.

2. Similar to Hydraulic extrusion

3. Analogous to the hot pressing of powered metals

4. Analogous to die casting of metals

Code:

	A	B	C	D
(a)	2	4	1	
(b)	3	1	4	2
(c)	2	1	4	3
(d)	3	4	1	2

1096. Match List-I with List-II and select the correct answer using the code given below the Lists:

List-I

(Article)

A. Disposable coffee cups

B. Large water tanks

C. Plastic sheets

D. Cushion pads

List-II

(Processing Method)

1. Rotomoulding

2. Expandable bead moulding

3. Thermoforming

4. Blow moulding

5. Calendering

Code:

	A	B	C	D
(a)	3	5	1	2
(b)	4	5	1	2
(c)	4	3	2	1
(d)	3	1	5	2

Answers

1094. (d) 1095. (d) 1096. (c)

1097. Which one of the following are the processes for thermosetting materials?
1. Compression
2. Transfer moulding
3. Injection moulding
4. Extrusion

Select the correct answer using the codes given below:
(a) 1 and 4 (b) 1 and 2
(c) 2 and 3 (d) 2, 3 and 4

1098. Which one of the following statements is not correct?
(a) In injection die moulding, exact amount of material to fill the cylinder is delivered
(b) Injection die moulding is generally limited to forming thermoplastic material
(c) Thermosetting plastics are more suitable for extrusion moulding
(d) Extrusion moulding process is used for giving shapes, such as rods, tubes, pipes, ropes etc.

1099. Match List I with List II and select the correct answer using the code given below the Lists:

List I
(Additive for Polymers)
A. Plasticizer B. Filler
C. Initiator

List II (Purpose)
1. Allows polymerization to begin
2. Colours the material
3. Acts as internal lubricants
4. Improves strength

Code:

	A	B	C
(a)	1	4	3
(b)	3	2	1
(c)	1	2	3
(d)	3	4	1

1100. Consider the following statements:
1. Thermoplastics possess a strong intermolecular bonding compared to that of thermosetting plastics.
2. Plastics have a high creep under continuous loading.
3. Embrittlement occurs in plastics at low temperature.

Which of these statements are correct?
(a) 1 and 2 (b) 2 and 3
(c) 1 and 3 (d) 1, 2 and 3

1101. Weldable type plastic(s) include (s)
(a) Thermosets alone
(b) Thermoplastics alone
(c) Both thermosets and thermoplastics
(d) Neither thermosets and thermoplast

1102. Match List- I (Name of moulding composition to prepare plastics) with List-II (Property of moulding composition) and select the correct answer using the codes given below the lists:

List- I
A. Binder B. Filler
C. Plasticize D. Lubricant

List – II
1. Reduce cost, shrinkage
2. Make the moulding of plastic easier
3. Cellulose derivatives
4. Accelerate condensation and polymerization
5. Toughness and resistance to temperature.

Codes:

	A	B	C	D
(a)	3	1	2	5
(b)	3	1	5	2
(c)	5	3	1	4
(d)	3	5	1	4

Answers

1097. (b) 1098. (c) 1099. (a) 1100. (b) 1101. (b) 1102. (b)

1103. Consider the following statements in respect of fabrication of plastic products:
1. Compression moulding is analogous to hot pressing of powdered metals.
2. Jet moulding is a modification of compression moulding.
3. Injection moulding is analogous to die casting of metals
4. Transfer moulding is similar to hydraulic extrusion.

Which of these statements are correct?
(a) 1 and 2 (b) 1 and 3
(c) 1, 2 and 4 (d) 2, 3 and 4

1104. Which of the following processes can be used for mass production of plastic containers (with lid) of 5 liter capacity?
1. Injection moulding
2. Jolt moulding
3. Blow moulding

Select the correct answer using the codes given below:
Codes:
(a) 1 and 2 (b) 2 and 3
(c) 1 and 3 (d) 1, 2 and 3

1105. To reduce the consumption of synthetic resins, the ingredient added is
(a) Accelerator (b) Elastomer
(c) Modifier (d) Filler

1106. Consider the following:
Which of the following properties are possessed by plastics?
1. Good resistance to corrosive atmosphere
2. Ease of fabrication into complex shapes.
3. Good resistance to shocks and vibrations.

Select the correct answer using the code given below:
(a) 1, 2 and 3 (b) 2 and 3 only
(c) 1 and 3 only (d) 1 and 2 only

1107. In the case of rubber, vulcanization refers to the process of producing a
(a) Linear polymer
(b) Branched polymer
(c) Cross-linked polymer
(d) Net-work polymer

1108. Consider the following statements relating to mechanical properties of ceramics:
1. Tensile strength is theoretically high but in practice quite low.
2. Compressive strength is many times lower than tensile strength.
3. Shear strength is high.
4. Transverse strength is easy to ascertain.

Which of the statements given above are correct?
(a) 1 and 3 (b) 1 and 4
(c) 2 and 3 (d) 2 and 4

1109. Match List I with List II and select the correct answer:
List I (Material)
A. Ceramics B. Refractory
C. Stones D. High silica glass
List II (Application)
1. Construction of chemical plants
2. Columns and pillars
3. Lining of furnaces
4. Tiles

Codes:

	A	B	C	D
(a)	4	3	2	1
(b)	2	1	4	3
(c)	4	1	2	3
(d)	2	3	4	1

Answers

1103. (c) 1104. (c) 1105. (d) 1106. (a) 1107. (c) 1108. (a)
1109. (d)

1110. Which one of the following is true?

(a) Structure of metallic materials consists of atoms having valence of 5, 6 or 7

(b) Ceramic materials have long range electron matrix bond

(c) Polymers are composed of long chain of repeating molecules

(d) Ceramics are weaker than metals because of weak electrostatic bond

1111. Which one of the following is correct? When "devitrification" of inorganic glasses is done,

(a) Glass transforms from crystalline to non-crystalline state

(b) Glass transforms into a fully transparent material

(c) Glass transforms from non-crystalline state to poly-crystalline state

(d) Glass is relieved of internal stresses

1112. Which of the following fibre materials are used for reinforcement in composite materials?

1. Glass 2. Boron carbide

3. Graphite

Select the correct answer using the codes given below:

(a) 1 and 2 (b) 1 and 3

(c) 2 and 3 (d) 1, 2 and 3

1113. Consider the following statements:

The strength of the fibre reinforced plastic product

1. Depends upon the strength of the fibre alone

2. Depends upon the fibre and plastic

3. Is isotropic

4. Is anisotropic

Which of these statements are correct?

(a) 1 and 3

(b) 1 and 4

(c) 2 and 3

(d) 2 and 4

1114. Wood is a natural composite consisting of which of the following?

(a) Lignin fibres in collagen matrix

(b) Lignin fibres in apatite matrix

(c) Cellulose fibres in apatite matrix

(d) Cellulose fibres in lignin matrix

1115. Nano composite materials are highly preferable in design consideration for their

(a) High resistance to crack propagation

(b) Vibration resistance

(c) Impact resistance

(d) High resilience

1116. Which one of the following materials is not a composite?

(a) Wood

(b) Concrete

(c) Plywood

(d) Sialon

1117. Which of the following composites are 'dispersion-strengthened composites'?

(a) Particulate composites

(b) Laminar composites

(c) Fiber reinforced composites

(d) Short-fiber discontinuous composites

Answers

1110. (c)	1111. (c)	1112. (d)	1113. (d)	1114. (d)	1115. (b)
1116. (d)	1117. (a)				

1118. Match List I with List II and select the correct answer using the code given below the Lists:

List I

(Component)

A. Blades of bulldozer

B. Gas turbine blades

C. Drill bit

D. Springs of automobiles

List II

(Required Property)

1. High wear resistance and high toughness
2. Low Young's modulus and high fatigue strength
3. High wear and abrasion resistance
4. High creep strength and good corrosion resistance

Code:

	A	B	C	D
(a)	3	2	1	4
(b)	1	4	3	2
(c)	3	4	1	2
(d)	1	2	3	4

1119. Match List-I (Composition) with List-II (Application) and select the correct answer using the code given below the Lists:

List- I

A. Commercial bronze (10% Zn)

B. Red brass (15% Zn)

C. Aluminium brass (22% Zn, 2% Al)

D. P-bronze (11 % tin) small amount of P)

List -II

1. Radiator
2. Spring metal
3. Forging and stamping
4. Power plant and chemical equipment

Codes:

	A	B	C	D
(a)	2	4	1	3
(b)	3	1	4	2
(c)	2	1	4	3
(d)	3	4	1	2

1120. Which one of the following possesses the property of nonsparking character?

(a) Hadfield's manganese steel

(b) Spring steel

(c) Stellite

(d) Invar

1121. Match List I (Material) with List II (Typical use) and the correct answer using the codes given below the Lists:

List I

A. Branched polyethylene

B. Polyester

C. Polyvinylidene chloride

D. Linear Polyethylene

List II

1. Bottles
2. Textile fibres
3. Films for packaging
4. Transparent film

Code:

A B C D

(a) 2 3 4 1

(b) 3 2 1 4

(c) 2 3 1 4

(d) 3 2 4 1

Answers

1118. (c) 1119. (b) 1120. (b) 1121. (b)

1122. Gunmetal, which is used in journal bearings, contains
(a) 88% Cu, 10% Sn, 2% Zn
(b) 80% Cu, 10% Zn, 10% Al
(c) 85% Cu, 5% Mg, 10% Al
(d) 85% Cu, 5% Sn, 10% Pb

1123. The correct composition of austenitic stainless steel used for domestic utensils is
(a) 0.08% C, 18% Cr, .8% Ni, 2% Mn, 1% Si
(b) 0.08% C, 24% Cr, 12% Ni, 2% Mn, 1% Si
(c) 0.15% C, 12% Cr, 0.5% Ni, 1% Mn, 1% Si
(d) 0.30% C, 12% Cr, 0.4% Ni, 1% Mn, 1% Si

1124. Babbit lining is used on brass/bronze bearings to
(a) Increase bearing resistance
(b) Increase compressive strength
(c) Provide anti-friction properties
(d) Increase wear resistance.

1125. Why are babbit alloys used for bearing material?
(a) They have excellent embeddability
(b) They are relatively stronger than other bearing materials
(c) They do not lose strength with increase in temperature
(d) They have high fatigue strength

1126. Match List-I with List-II and select the correct answer using the codes given below the Lists:

List-I
A. Neoprene
B. Bakelite
C. Foamed polyurethane
D. Araldite

List-II
1. Electric switches
2. Adhesive
3. Thermal insulator
4. Oil seal

Code:

	A	B	C	D
(a)	4	1	2	3
(b)	1	4	2	3
(c)	4	1	3	2
(d)	1	4	3	2

1127. Which material is used for bushes in the bushed-pin type of flexible coupling?
(a) Gun metal (b) Plastic
(c) Rubber (d) Aluminium

1128. The percentage of phosphorous in phosphor bronze is
(a) 0.1 (b) 1
(c) 11.1 (d) 98

1129. Tin base white metals are used where the bearings are subjected to
(a) Large surface wear
(b) Elevated temperatures.
(c) Light load and pressure
(d) High pressure and load.

1130. Machine tool frame should have
(a) High rigidity to weight ratio
(b) Graphite in the form of nodules
(c) Low hardness
(d) High work hardness

1131. Which of the following statement is incorrect about duralumin?
(a) It is prone to age hardening.
(b) It can be forged.
(c) It has good machining properties.
(d) It is lighter than pure aluminium.

Answers

1122. (a)	1123. (a)	1124. (c)	1125. (a)	1126. (c)	1127. (c)
1128. (a)	1129. (d)	1130. (a)	1131. (c)		

1132. Cartridge brass can be
(a) Cold rolled into sheets
(b) Drawn into wires
(c) Formed into tubes
(d) Any of the above.

1133. A riser in casting is not described by which of the following ?
(a) an insert in the casting that inhibits buoyancy of the core.
(b) metal that is not part of the casting,
(c) source of molten metal to feed the casting and compensate for shrinkage during solidification.
(d) waste metal that is usually recycled.

1134. Frankel and Schottky imperfections are
(a) dislocations in ionic crystals.
(b) Grain boundaries in covalent crystals.
(c) Vacancies in ionic crystals.
(d) Vacancies in covalent crystals.

Answers

1132. (d) 1133. (a) 1134. (c)

Section B

MULTIPLE CHOICE QUESTION (MCQ) & ANSWER

GATE (Graduate Aptitude Test in Engineering)

FROM 1991 to 2016

1

General Metallurgy

1. Powder metallurgy processing can be used for making
   (a) dispersion strengthened copper rod.
   (b) self lubricating bearing.
   (c) connecting rod.
   (d) cemented carbide.

(GATE 91)

2. For preparation of porous bearing by powder metallurgy, preferred particle is
   (a) spherical
   (b) nodular
   (c) irregular
   (d) no preferred shape.

(GATE 92)

3. Match the commonly used manufacturing process for the following products.
   A. Thoriated nickel wire
   B. Aircraft turbine blade made form a nickel based super alloy
   C. Automobile crank shaft made for ductile iron
   D. Silicon wafers used in the electronic in dustry
   1. Single crystal growing
   2. Sand casting
   3. Powder metallurgy
   4. Investment casting
   (a) A-1, B-4, C-2, D-3
   (b) A-4, B-3, C-2, D-1
   (c) A-1, B-4, C-3, D-2
   (d) A-2, B-4, C-3, D-1

[GATE 94]

4. Match the following features in tensile stress - strain curves.
   A. Yield drop
   B. Serrations.
   C. Increase in flow stress with plastic deformation
   D. 1000% uniform strain
   1. Strain ageing
   2. Superplasticity
   3. Dislocation pinning
   4. Dislocation multiplication
   (a) A-3, B-1, C-2, D-4
   (b) A-3, B-1, C-4, D-2
   (c) A-3, B-2, C-4, D-1
   (d) A-4, B-3, C- 2, D-1

[GATE 95]

Answers

1. (b & d) 2. (d) 3. (a) 4. (b)

5. The steady state temperature profile of a rectangular sheet of metal inside a furnace can be obtained by solving the following partial differential equation;

$$\frac{\delta^2 T}{\delta x^2} + \frac{\delta^2 T}{\delta y^2} = 0$$

The number of boundary conditions needed to solve this equation are :
   (a) one in x-direction, one in y-direction
   (b) two in x-direction, two in y-direction
   (c) two in any of the two directions
   (d) four in any of the two directions.

[GATE 96]

6. Increasing the mean stress influences the S-N curve as following : (S represents alternating stress)
   (a) shifts upwards
   (b) keeps unaltered
   (c) shifts downwards

[GATE 96]

7. Match the following materials with the method most commonly used for making their powders.
   A. Nickel
   B. Tungsten
   C. Silicon carbide
   D. Super alloys
   1. Carbothermic reduction
   2. Inert gas atomisation
   3. Oxide reduction
   4. Carbonyl process
   (a) A-2, B-3, C-4, D-1
   (b) A-1, B-3, C-4, D-2
   (c) A-2, B-3, C-1, D-4
   (d) A-4, B-3, C-2, D-1

[GATE 96]

8. Powder metallurgy is used to produce
   (a) high precision components with complex cavities and sharp features
   (b) components of large size
   (c) porosity free components
   (d) components of such alloys whose constituents do not form alloys readily

[GATE 98]

9. $\begin{bmatrix} 21 & 0 & 0 \\ 0 & 9 & 0 \\ 0 & 0 & 6 \end{bmatrix}$ MPa

The maximum shear stress available for the above stress tensor is ----MPa.
   (a) 9 (b) 6
   (c) 1.5 (d) 7.5

[GATE 98]

10. The minimum energy required to impose a plastic strain of ε to a metal having unit volume and a constant flow stress = σ is
   (a) σϵ b) 2 σϵ
   (c) σϵ/2 (c) 3 σϵ /4

[GATE 98]

11. A high cycle fatigue failure is identified by the presence of
   (a) dimples
   (b) beach marks or striations
   (c) slip lines
   (d) mirror like fractures.

[GATE 99]

12. Two samples A and B of a brittle material have crack lengths in the ratio 3:1 The ratio of the tensile strengths (measured normal to the cracks) of A and B will be in the ratio
   (a) 1 : 3 (b) √3 : 1
   (c) 1 : √3 (d) 1 : 9

[GATE 99]

Answers

5. (b)	6. (c)	7. (c)	8. (c)	9. (a)	10. (c)
11. (b)	12. (b)				

13. During solid state sintering of powder, the following mechanisms can be active.
    (a) Evaporation and condensation
    (b) Solid state diffusion process
    (c)Liquid formation in grain boundaries
    (d) Creation of more dislocations.
    [GATE 99]

14. One of the methods of purification of leach liquor is ion exchange which involves exchange between
    (a) two liquid phases
    (b) a gaseous phases and a liquid phase
    (c) a liquid phase and an organic resin phase
    (d) a solid phase and a gas phase.

15. The Larson- Miller parameter 'P' connecting the temperature 'T' and the rupture time ' ' is given as
    (a) P=T(log t_r + c) (b) P=log t_r –c/T
    (c) P=(c–T)/t_r (d) P=Tlog t_r
    [GATE 2000]

16. Alloy powders manufactured by the following process have spherical shapes:
    (a) Electro- chemical deposition
    (b) Gaseous reduction
    (c) Atomisation
    (d) Mechanical attrition.
    [GATE 2001]

17. Fick's second law of diffusion is stated as:
    (a) $\frac{\partial C}{\partial T} = D.\frac{\partial^2 C}{\partial x^2}$
    (b) $\frac{\partial C}{\partial x} = D.\frac{\partial^2 C}{\partial t^2}$
    (c) $\frac{\partial C}{\partial t} = \frac{\partial}{\partial x}\left(D.\frac{\partial C}{\partial t}\right)$
    (d) $\frac{\partial C}{\partial t} = \frac{\partial}{\partial t}\left(D.\frac{\partial C}{\partial x}\right)$
    [GATE 2001]

18. The percentage of ferrite and pearlite in annealed 0.5% carbon steel is approximately
    (a) 7.3% ferrite and 92.7% pearlite
    (b) 92.7% ferrite and 7.3% pearlite
    (c) 37.5% ferrite and 62.5% pearlite
    (d)63.5% pearlite and 37.5% ferrite.
    [GATE 2002]

19. Coating of zine over steel is known as
    (a) cladding (b) galvanising
    (c) anodising (d) passivation
    [GATE 2003]

20. Match the following.
    Group 1
    P. Grain refinement of aluminium
    Q. Improvement of fluidity of cast iron
    R. Refinement of graphite flakes in cast iron
    S. Removal of dissolved hydrogen from molten aluminium
    Group 2
    1. Magnesium
    2. Titanium
    3. Phosphorous
    4. Ferro-silicon
    5. Chlorine
    (a) P-2, Q-3, R-4, S-5
    (b) P-4, Q-3, R-1, S-2
    (c) P-2, Q-4, R-1, S-5
    (d) P-3, Q-5, R-4, S-2
    [GATE 2003]

21. An Fe/graphite diffusion couple is annealed at 1273 K. The carbon content (in mass %) on the Fe side of the Fe/graphite diffusion couple will be close to
    (a) 0.5 (b) 1
    (c) 1.6 (d) 6.7
    [GATE 2003]

Answers

13. (b)	14. (c)	15. (a)	16. (c)	17. (a)	18. (c)
19. (b)	20. (a)	21. (d)			

22. Cemented carbide cutting tools are
    P. made by casting
    Q. made of mainly WC and cobalt
    R. made of Fe_3C and cobalt
    S. made by liquid phase sintering
    (a) P, Q (b) Q, R
    (c) P, R (d) Q, S

[GATE 2004]

23. Identify the correct statements from the following :
    P. 0.2% yield strength of a material implies 0. 2% of the yield strength.
    Q. Von -Misses yield criterion implies that yielding occurs when the distortion energy reaches a critical value.
    R. Radius of the cylinderical Von-Misses yield surface increases as the grain size of a single phase material decreases.
    S. Tresca's yield criterion gives a circular cylinderical surface in the space of the three principal stresses.
    (a) P, Q (b) Q, R
    (c) P, R (d) Q, S

[GATE 2004]

24. For sintering of green powder compacts of copper, choose the correct statement:
    P. Sintering should be done in an inert or reducing atmosphere.
    Q. At a given sintering temperature, the rate of shrinkage will be higher for finer powder size.
    R. Full density parts can be produced in a finite time by solid state sintering
    S. Sintered compacts will have a higher strength than those made by metal working
    (a) P, Q (b) Q,R
    (c) P, R (d) Q, S

[GATE 2004]

25. Match the following :
    Group 1
    P. Dulong formula
    Q. Carbon
    R. Dwight-Lloyd machine
    S. Radiation
    Group 2
    1. Ultimate analysis
    2. Gray body
    3. Sintering
    4. Refractory
    (a) P-1, Q-2, R-3, S-4
    (b) P-2, Q-4, R-3, S-1
    (c) P-1, Q-4, R-3, S-2
    (d) P-3, Q-1, R-4, S-2

[GATE 2004]

26. Match the following:
    Group1
    P. Hall petch relation
    Q. Orowan mechanism
    R. Nabarro-Herring creep
    S. Griffith criterion
    Group 2
    1. Bulk diffusion between grain boundaries
    2. Fracture of brittle materials
    3. Grain boundary strengthening
    4. Dispersion strengthening
    (a) P-1, Q-2, R-3, S-4
    (b) P-2, Q-3, R-4, S-1
    (c) P-4, Q-1, R-2, S-3
    (d) P-3, Q-4, R-1, S-2

[GATE 2004]

27. The driving force for sintering of a powder compact is
    (a) strain energy
    (b) surface energy
    (c) volume energy
    (d) stacking fault energy

[GATE 2003]

Answers

22. (d) 23. (b) 24. (a) 25. (c) 26. (d) 27. (b)

28. In powder compacting of a monolithic component, it is generally advised to keep the ratio of thickness to width below a limit (say 2.0)this is essentially due to
    (a) difficulty in ejection of compact leading to breakage.
    (b) sidewall friction leading to non-uniform density.
    (c) difficulty in sintering.
    (d) difficulty in burn off.

[GATE 2005]

29. The engineering stress-strain curve for a ceramic material is
    (a) parabolic
    (b) exponential
    (c) logarithmic
    (d) linear

[GATE 2010]

30. In heterogrnous nucleation, the radius of the critical nucleus does not depends upon
    (a) contact angle
    (b) undercooling
    (c) the surface energy of the interface between the product and parent phases.
    (d) enthalpy change per unit volume of the product phase.

[GATE 2010]

31. Which of the following are not commercially manufactured by powder metallurgy?
    (a) Aircraft brake pads
    (b) Self-lubricating bearings
    (c) Tungsten carbide based cutting tools
    (d) Turbine blades

[GATE 2010]

32. In the powder metallurgy processing, the objective of pressing before sintering is to
    (a) squeeze out the moisture around the powder particles
    (b) further refine the grain size
    (c) break up the oxides around the particles
    (d) compacting the powder particles into mechanical and atomic closeness.

[GATE 2006]

33. Identify the combination of mechanisms which best describes sintering of pure metals in powder metallurgy.
    P. Grain boundary melting
    Q. Liquid metal freezing
    R. Interparticle melting
    S. Recrystallisation
    T. Grain growth
    U. Oxidation
    V. Reduction
    (a) S and I
    (b) P and Q
    (c) R and Q
    (d) U and V

[GATE 2006]

34. Transport mechanisms that do not contribute to densification during sintering are :
    P. Surface diffusion
    Q. Bulk diffusion
    R. Bulk diffusion
    S. Evaporation-condensation
    T. Viscous flow
    (a) P, Q
    (b) Q, S
    (c) Q, T
    (d) P, S

[GATE 2008]

Answers

28. (a)	29. (d)	30. (a)	31. (d)	32. (d)	33. (d)
34. (b)					

35. Match the particle morphologies in group 1 with the powder production methods in group 2.
Group 1
P. Superalloy powder with rounded morphology
Q. Monosized spherical Ta powder
R. Fe powder with onion peel structure
S. Irregularly shaped W powder
Group 2
1. Carbonyl process
2. Gas atomisation
3. Oxide reduction
4. Rotating electrode process
(a) P-2, Q-1, R-4, S-3
(b) P-1, Q-4, R-3, S-2
(c) P-2, Q-4, R-1, S-3
(d) P-4, Q-1, R-2, S-3
[GATE 2008]

36. Oil impregnated bronze bearings are manufactured using
(a) pressure die casting
(b) centrifugal casting
(c) solid-state sintering
(d) liquid phase sintering
[GATE 2008]

37. Which one of the following expands upon solidification?
(a) Low carbon steel
(b) High carboon steel
(c) White cast iron
(d) Gray cast iron
[GATE 2011]

38. When load is applied to a material, instantaneous strain develops with
(a) the speed of light
(b) half the speed of sound
(c) the speed of sound
(d) infinite speed
[GATE 2011]

39. When a zinc metal rod is immersed in dilute hydrochloric acid, it results in
(a) Evolution of hydrogen
(b) Evolution of chlorine
(c) Evolution of oxygen
(d) No evolution of any gas
[GATE 2012]

40. Copper can be reduced from acidic copper sulphate solution by
(a) Silver (b) Iron
(c) Carbon (d) Lead
[GATE 2012]

41. Which one is NOT an agglomeration process?
(a) Nodulizing (b) Briquetting
(c) Roasting (d) Pelletizing
[GATE 2012]

42. The temperature field of a slab is given by $T= 400 - 50z \exp(-t-x^2-y^2)$. The temperature gradient in y-direction is
(a) $100yz \exp(-t-x^2-y^2)$.
(b) $-100yz \exp(-t-x^2-y^2)$.
(c) $100xz \exp(-t-x^2-y^2)$.
(d) $-100xz \exp(-t-x^2-y^2)$.
[GATE 2012]

Answers

35. (b)	36. (c)	37. (d)	38. (c)	39. (a)	40. (b)
41. (c)	42. (a)				

43. Thin layer of material B (of total amount m) is plated on the end faces of two long rods of material A. These are then joined together on the plated side (see the figure below) and heated to a high temperature. Assuming the diffusion coefficient of B in A is D the composition profile CB along the rod axis x after a time t is described by

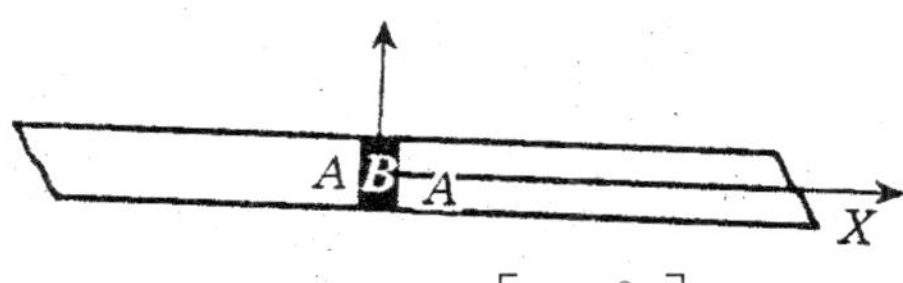

(a) $C_B = \frac{m}{2\sqrt{\pi Dt}} \exp\left[-\frac{x^2}{4Dt}\right]$

(b) $C_B = \frac{m}{2\sqrt{\pi Dt}} \mathrm{erf}\left[-\frac{x^2}{4Dt}\right]$

(c) $C_B = \frac{m}{2\sqrt{\pi Dt}} \left[1 - \mathrm{erf}\left(\frac{x^2}{4Dt}\right)\right]$

(d) $C_B = \frac{m}{2\sqrt{\pi Dt}} t$

44. A polymer matrix composite is reinforced with long continuous ceramic fibers aligned in one direction. The young's moduli of the matrix and fibers are Em and Er respectively and the volume fraction of the fibers is f. Assuming iso -stress condition, Young's modulus of the composite Ec in a direction perpendicular to the length of fibers, is given by the expression

(a) $Ec = (1-f) E_m + f E_f$

(b) Ec = f Em +(1−f) Er

(c) $\frac{1}{Ec} = \frac{(1-f)}{E_m} + \frac{f}{Ef}$

(d) $\frac{1}{Ec} = \frac{f}{E_m} + \frac{(1-f)}{Ef}$

[GATE 2012]

45. Match the processes in Group 1 with the objectives in Group ll.

Group 1

P. Vacuum Arc Degassing (VAD)

Q. LD

R. COREX

S.Blast Furnace

Group ll.

1. Primary iron making
2. secondary steel making
3.Direct smelting
4. Primary steel making

(a) P-3, Q-4, R-2, S-1

(b) P-4, Q-3, R-1, S-2

(c) P-3, Q-2, R-1, S-4

(d) P-2, Q-4, R-3, S-1

[GATE 2012]

46. The sulphide capacity (C_s) of liquid slag of composition 55 wt% CaO, 20 wt% SiO_2, 15 wt% Al_2O_3 and 10 wt% MgO is given by the following equation

$\log C_s = 3.44 (X_{CaO} + 0.1 X_{MgO} - 0.8 X_{Al2O3} - X_{SiO2}) - (9894/T) + 2.05$

where, X is mole fraction of the respective components. Atomic weights of Ca, Mg, Si, Al and O are 40, 24, 28, 27, and 16 respectively. The value of C_s at 1900 K is

(a) 0.0009 (b) 0.0009

(c) 0.09 (d) 0.9

[GATE 2012]

47. Fracture stress for a brittle material having a crack lenght of 1 μm is 200 MPa. Fracture stress for the same material having a crack lenght of 4 μm is

(a) 200 MPa (b) 150 MPa

(c) 100 MPa (d) 50 MPa

[GATE 2012]

Answers

43. (a) 44. (c) 45. (b) 43. (b) 47. (c)

48. Which one of the following can give information about the corrosion rate?
   (a) Pourbaix diagram
   (b) Polarization technique
   (c) EMF series
   (d) Galvanic series

[GATE 2013]

49. The total number of possible heat transfer mode (s) is
   (a) 1 (b) 2
   (c) 3 (d) 4

[GATE 2013]

50. If σ and ε are true stress and true strain, respectively, the maximum true uniform strain that can be imparted to a material obeying, $\sigma = 1050\ \varepsilon^{0.25}$ is
   (a) 0.15 (b) 0.25
   (c) 0.35 (d) 0.45

[GATE 2013]

51. The yield strenght of a polycrystalline metal increases from 100 MPa to 145 MPa on decreasing the grain size from 64 μm to 25 μm. The yield strenght of this metal (in MPa) having a grain size of 36 μm is
   (a)110 (b) 125
   (c) 140 (d) 165

[GATE 2013]

52. In a brittle material, the maximum internal crack is 8 μm. If Young's modulus is 400 GPa and surface energy is 3.14 J/m^2, the estimated theoretical fracture strenght (in MPa) is
   (a) 375 (b) 412
   (c) 327 (d) 447

[GATE 2013]

53. Match the powder production technique given in Group 1 with the corresponding shape listed in Group ll.

Group l

P. reduction Q. Gas Atomization
R. Milling S. Electrolysis

Group ll

1. Flaky 2. Spongy
3. Dendritic 4. Spherical

   (a) P-2, Q-4, R-1, S-3
   (b) P-1, Q-3, R-2, S-4
   (c) P-2, Q-3, R-4, S-1
   (d) P-3, Q-2, R-1, S-4

[GATE 2013]

54. For the following electrochemical reaction $Sn + 2H^+ = Sn^{2+} + H_2$ if the solution has Sn^{2+} concentration 10^{-2} M and pH 5 at 298 K, which of the following is true ?
Given : standard reduction potential for $Sn^{2+} + 2e^- \rightarrow Sn$ is - 0.136 V versus SHE; pH_2 = 1 atm
   (a) Sn undergoes oxidation
   (b) H^+ undergoes reduction
   (c) Sn^{2+} undergoes reduction
   (d) No net reaction

[GATE 2013]

55. The pilling-Bedworth ratio is defined as
   (a) the molar weight of an oxide divided by the molar weight of the metal consumed in oxide formation.
   (b) the volume of the oxide divided by the volume of the metal consumed in oxide formation.
   (c) the density of the oxide divided by the density of the metal consumed in oxide formation.
   (d) the molar Gibbs energy of the oxide divided by the Gibbs energy of the metal consumed in oxide formation.

[GATE 2014]

Answers

48. (b)	49. (c)	50. (b)	51. (b)	52. (d)	53. (a)
54. (c)	55. (b)				

56. What is the theoretical requirement of air (in m^3 at STP) for the complete combustion of 100 m^3 (at STP) of a fuel consisting of pure CH_4 ? Assume that air contains 21 vol.% of oxygen.
(a)386 (b) 488
(c) 805 (d) 952
[GATE 2014]

57. An electrolytic refining cell for copper consists of an alloy with activity of copper a_{Cu} = 0.8 as the anode, and pure copper as the cathode. What is the absolute value of the cell potential (in millivolts) at 25°C, given that copper is divalent ? Faraday constant is 96500 C/mol and the universal gas constant is 8.314 J/(mol.K).
(a) 2.9 (b)4.6
(c) 6.8 (d) 8.9
[GATE 2014]

58. A rod of a metal with Young's modulus of 200 GPa is pulled in tension to a stress of 400 MPa. What is the elastic strain (in %) that is recovered, when the rod is completely unloaded ?
(a) 0.1 (b) 0.2
(c) 0.3 (d) 0.4
[GATE 2014]

59. Match the operations listed in Group I with the type of processes listed in Group II.
Group I :
P. Blast Furnace Iron making
Q. BOF Steel making
R. Hall- Heroult Process
S. Bayer Process
Group II.:
1. Refining
2. Electrolysis
3. Smelting
4. Leaching
(a) P-1, Q-3, R-2, S-4
(b) P-3, Q-1, R-2, S-4
(c) P-4, Q-2, R-3, S-1
(d) P-3, Q-1, R-4, S-2
[GATE 2014]

60. The driving force for sintering a compact consisting of spherical particles of radius R_1 is ΔG_1. If the particle size is reduced to R_2=0.1 R_1, the corresponding driving force $\Delta G_2 = \alpha \Delta G_1$, where α is
(a) 2
(b) 5
(c) 10
(d) 15
[GATE 2015]

61. During the roasting of sulfide ore a metal **M.** the possible solid phased are M. MS. MO and MSO_4. Assuming that both SO_2 and O_2 are always present in the roaster. The solid phases that can co-exist at thermodynamic equilibrium are
(a) M, MS, MO, MSO_4
(b) M, MO, MSO_4
(c) MS, MO, MSO_4
(d) M, MSO_4
[GATE 2016]

Answers

56. (d) 57. (a) 58. (b) 59. (b) 60. (c) 61. (c)

62. A sub- lance is used to monitor composition and temperature in
    (a) BOF
    (b) Ladle refining furnace
    (c) Continuous casting mould
    (d) Blast furnace

[GATE 2016]

63. The lining of a box-type furnace is made up of a refractory layer and steel plate as shown in the figure. Steady state temperature at the surface of the refractory is 1273 K and that at the outer steel surface is 473 K. If the steady-state heat flux through the refractory-steel plate xomposite is 1600 W, M^{-2} m and heat flow is along x-direction, the thermal contact resistance (W^{-1} .m2 .K) betwwen refractory and steel is ....

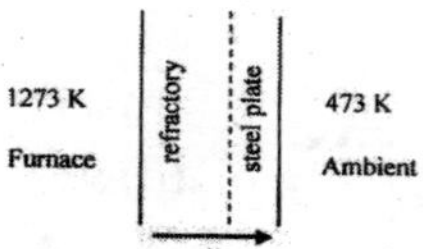

Given data:

Thermal conductivity of refractory = 1.2 w .m^{-1} K^{-1}
Thickness of refractory lining = 80 mm
Thermal conductivity of steel = 32 W .m^{-1} K^{-1}
Thickness of steel plate = 4mm

[GATE 2016]

Answers

62. (a)   63. (0.42 to 0.44)

2 Mineral Dressing

1. The following is a typical anionic collector used in floatation .
   (a) Ethyl dixanthogen
   (b) Tri methyl cetyl ammonium bromide
   (c) Potassium ethyl xanthate
   (d) Lauryl amine hydrochloride
   (GATE 94)

2. Match the following processes with their physical principles.
   Group 1
   P. Floatation
   Q. Jigging
   R. Heavy media separation
   S. Tabling
   Group 2
   1. Difference in specific gravity
   2. Difference in hydrophobicity
   3. Difference in lateral movements
   4. Differential initial accelaration
   (a) P-4, Q-2, R-1, S-3
   (b) P-2, Q-4, R-1, S-3
   (c) P-4, Q-3, R-2, S-1
   (d) P-1, Q-2, R-3, S-4
   (GATE 95)

3. In the floatation process, pine oil is used as
   (a) collector (b) activator
   (c) frother (d) depressant.
   [GATE 98]

4. A simple close grinding circuit may consist of a
   (a) ball mill and a classifier
   (b) rod mill and a hydrocyclone
   (c) jaw crusher and a hydrocyclone
   (d) ball mill and a froth floatation cell.
   [GATE 99]

5. Hydrocyclone is a
   (a) crusher (b) wet classifier
   (c) dry (d) magnetic separator.
   [GATE 2000]

6. Role of a "collector" in floatation is to
   (a) form a water repelling film on the mineral surface
   (b) create and stabilise the froth
   (c) act as a surfactant
   (d) collect the minerals according to their specific gravity.
   [GATE 2000]

7. Electrostatic separation of minerals from each other is based on their differences in the following property:
   (a) Densities
   (b) Magnetic permeabilites
   (c) Electrical conductivities
   (d) Hardness
   [GATE 2001]

Answers

1. (c) 2. (b) 3. (c) 4. (a) 5. (b) 6. (a)
7. (c)

8. Match the following:
   Group 1
   P. Floatation
   Q. Electrostatic concentration
   R. Comminution
   S. Heavy media separation
   Group 2
   1. Separation on the basis of density difference
   2. Reduction in size
   3. Surface charge is induced in particles
   4. Modification of surface tension
   (a) P-2, Q-1, R-4, S-3
   (b) P-1, Q-2, R-3, S-4
   (c) P-3, Q-4, R-1, S-2
   (d) P-4, Q-3, R-2, S-1
   [GATE 2003]

9. When the weight of the ball inside a ball mill is just balanced by the centrifugal force , the ball
   (a) abandons its circular path for a parabolic path
   (b) abandons its parabolic path for a circular path
   (c) continues to move on a circular path
   (d) continues to move on a parabolic path.
   [GATE 2004]

10. Identify the correct statements:
    P. A reagent could be a collector for one system and a frother for another.
    Q. Usage of a frother makes the bubble stable.
    R. Usage of a frother makes the mineral surface hydrophobic.
    S. Some mineral floatation processes will not require any collector.
    (a) P, Q (b) Q, R
    (c) P, S (d) P, R
    [GATE 2004]

11. Match the reagents in group 1 with their properties in group 2, in the context of froth floatation.
    Group 1
    P. Frother Q. Activator
    R. Regulator S. Depressant
    Group 2
    1. Alters the chemical nature of mineral surfaces so that they become hydrophobic due to action of collector
    2. Absorbs on air-water interface and reduces surface tension to stabilise the bubble.
    3. Increases the selectivity of floatation by rendering certain mineral hydrophilic.
    4. Modifies the action of collector either by intensifying or reducing its water repellant effect.
    (a) P-2, Q-1, R-4, S-3
    (b) P-1, Q-3, R-2, S-4
    (c) P-4, Q-2, R-3, S-1
    (d) P-3, Q-2, R-4, S-1
    [GATE 2005]

12. A 200 mesh screen made of steel wire having a diameter of 0.053 mm has an aperture size (in μm)
    (a) 21 (b) 36
    (c) 74 (d) 127
    [GATE 2006]

13. Cyclones are primarily used for
    (a) comminution (b) concentration
    (c) dewatering (d) classification.
    [GATE 2007]

Answers

8. (d) 9. (a) 10. (a) 11. (a) 12. (c) 13. (d)

14. A typical collector used in sulphide mineral floatation is
    (a) pine oil
    (b) potassium ethyl xanthate
    (c) oleic acid
    (d) polyacrylamide.

[GATE 2007]

15. Match the processes in group 1 with the physical principales in group 2.
    Group 1
    P. Floatation
    Q. Jigging
    R. Tabling
    S. Heavy media separation
    Group 2
    1. Differential initial accelaration
    2. Differential lateral movement
    3. Differential in density
    4. Modification of surface tension
    (a) P-4, Q-1, R-2, S-3
    (b) P-4, Q-1, R-2, S-2
    (c) P-2, Q-3, R-4, S-1
    (d) P-1, Q-3, R-4, S-2

[GATE 2008]

16. In froth floatation, collector refers to a reagent which primarily.
    (a) promoters bubble break-up and stabilises the foam.
    (b) absorbs on the surface of the minerals and makes it hydrophobic.
    (c) promotes separation of the particles from the froth.
    (d) absorbs on the unwanted minerals and makes it sink.

[GATE 2011]

17. Floatation beneficiation is based on the principle of
    (a) Mineral surface hydrophbicity
    (b) Gravity difference
    (c) Chemical reactivity
    (d) Partical size difference

[GATE 2012]

18. Match the unit operation in Group I with its corresponding principal in Group II :
    Group I :
    (P) Jigging
    (Q) Tabling
    (R) Heavy media separation
    (S) Flotation
    Group II :
    (1) Modification of surface tension
    (2) Difference in density
    (3) Differential initial acceleration
    (4) Differential lateral movement
    (a) P-3, Q-4, R-2, S-1
    (b) P-2, Q-3, R-1, S-4
    (c) P-4, Q-2, R-3, S-1
    (d) P-1, Q-1, R-2, S-4

[GATE 2013]

19. The aperture size (in μm)of a 200 mesh sieve having a wire diameter of 53 μm is
    (a) 37 (b) 53
    (c) 74 (d) 86

[GATE 2013]

Answers

14. (b)	15. (a)	16. (b)	17. (a)	18. (a)	19. (c)

20. The figure below shows a schematic of a water droplet placed on the surfaces of two mineral P and Q.

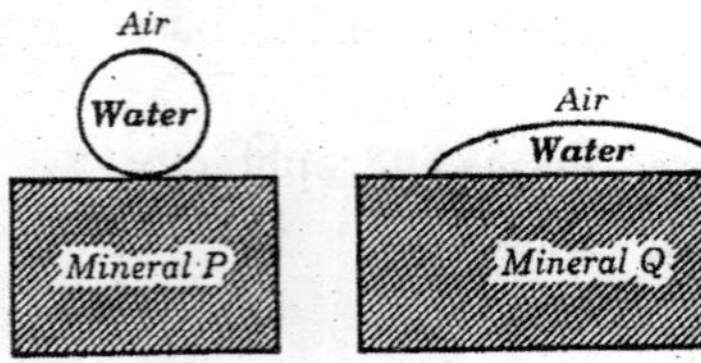

Given this information, pick the CORRECT statement from the following:

(a) Mineral P ascends with air bubbles preferentially over mineral Q.

(b) Mineral Q ascends with air bubbles preferentially over mineral P.

(c) Both minerals P and Q ascend with the air bubbles without preference.

(d) Both minerals P and Q sink to the bottom

[GATE 2015]

21. The figure below shows the cumulative size distribution of particles of a crushed mineral. 10 kg of this material is first passed through a sieve of size 400 micron and then through a sieve of size 300 micron. The weight of mineral (in kg) that is retained on the 300 micron sieve is ....

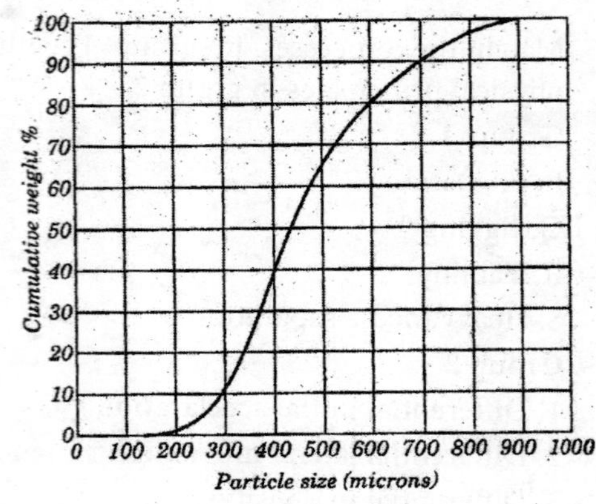

(a) 1 (b) 1.8

(c) 3 (d) 5.6

[GATE 2015]

Answers

20. (a) 21. (c)

Section

B

3 Iron Making

1. High top pressure in blast furnace
   (a) increases the silicon content in the hot metal.
   (b) increases the sulfur content in the hot metal.
   (c) increases the phosphorous content in the hot metal.
   (d) increases the manganese content in the hot metal.

[GATE 91]

2. The product of a commercial direct reduction process is
   (a) liquid iron
   (b) solid
   (c) sponge iron
   (d) iron saturated with carbon

[GATE 91]

3. Basicity [defined as (%CaO + %MgO)/%SiO_2] of the slag in the indian blast furnaces is in the range of
   (a) 0.7 – 1.0 (b) 1.1 – 1.4
   (c) 1.5 – 1.8 (d) 2.0 – 2.5

[GATE 95]

4. Match the following blast furnace reactions with the regions where they take place. (Note that the underlined elements are present in the hot metal and the species in the parenthesis belong to slag phase).
   P. $Fe_3O_4 + CO = 3FeO + CO_2$
   Q. $C + 0.5\ O_2 = CO$
   R. $y . Fe + c = Fe_yc$
   S. $C + (CaO) + S = (CaS) + CO$
   1. Hearth 2. Tuyer
   3. Stack 4. Bosh
   (a) P-3, Q-2, R-4, S-1
   (b) P-3, Q-2, R-1, S-4
   (c) P-2, Q-3, R-4, S-1
   (d) P-4, Q-3, R-2, S-1

[GATE 96]

5. Solution loss reaction occurs
   (a) in the blast furnace shaft
   (b) during pretreatment of hot metal
   (c) in the LD process
   (d) during deoxidation in the ladle.

[GATE 98]

6. The deadman's zone in the blast furnace consists of
   (a) closely packed central columm of coke, limestone and iron ore
   (b) gases only
   (c) closely packed central column of coke only
   (d) column of hot metal and slag

[GATE 99]

Answers

1. (a)	2. (c)	3. (b)	4. (a)	5. (a)	6. (c)

7. Desulfurisation of hot metal is favoured by
   (a) reducing and neutral slag
   (b) high activity co- efficient of sulphur in metal
   (c) oxidising and basic slag
   (d) reducing and basic slag

[GATE 99]

8. The Midrex process of iron making
   (a) uses a fluidised bed reactor
   (b) recycles a part of exit gas
   (c) uses both CO and H_2 for reduction
   (d) uses only H_2 for reduction

[GATE 99]

9. Pellets are not popular as a burden material as sinter in the iron blast furnace, because of their
   (a) shape
   (b) swelling tendency
   (c) poor reducibility
   (d) low mechanical strength

[GATE 2000]

10. Desulfurisation of molten pig iron outside the blast furnace is carried out by the
   (a) Aston process
   (b) Bayer process
   (c) Perrin process
   (d) Strategic- Udy process

[GATE 2000]

11. The chemical reserve zone in the iron blast furnace consists mainly of
   (a) lustite (b) magnetite
   (c) hematite (d) iron.

[GATE 2000]

12. Dephosphorisation of molten pig iron is favoured by
   (a) oxidising and basic slag
   (b) reducing and basic slag
   (c) high activity co-efficient of phosphorous in metal
   (d) oxidising and neutral slag

[GATE 2000]

13. High top pressure in the iron blast furnace
   (a) increases indirect reduction and decreases silicon content in the hot metal.
   (b) increases direct reduction and increases silicon content in the hot metal.
   (c) decreases coke rate and increases sulfur content in pig iron.
   (d) increases production rate and reduces the extent of solution loss reaction.

[GATE 2000]

14. Most of the phosphorous present in the blast furnace burden enters into
   (a) hot metal (b) flue gases
   (c) slag (d) refractory lining.

[GATE 2001]

15. The sintering of iron are is predominantly a
   (a) chemical process
   (b) combustion process
   (c) heat transfer process
   (d) none of these

[GATE 2002]

16. External desiliconisation of hot metal is carried out in the steel plant located at
   (a) Rourkela (b) Durgapur
   (c) Bokaro (d) Bhilai

[GATE 2002]

17. For efficient performance of a blast furnace, the extent of reduction of lustite should be
   (a) 100% indirect reduction
   (b) 100% direct reduction
   (c) 50 – 60% indirect reduction
   (d) 30 – 40% indirect reduction

[GATE 2002]

Answers

7. (c)	8. (c)	9. (b)	10. (a)	11. (d)	12. (a)
13. (b)	14. (a)	15. (a)	16. (b)	17. (c)	

18. Coke bed height in a cupola is measured from the
    (a) level of slag hole (b) bottom level
    (c) level of tap hole (d) level of tuyers.
    [GATE 2002]
19. Alkali metal oxide content in the blast furnace raw material
    (a) increases the strength of coke
    (b) increases the reactivity of coke
    (c) decreases the reactivity of coke
    (d) neither affects the strength nor the reactivity of coke.
    [GATE 2002]
20. Blast furnace is a
    (a) counter- current reactor
    (b) co-current reactor
    (c) cross- current reactor
    (d) combination of all the three.
    [GATE 2004]
21. In the blast furnace, as the oxygen percentage in air blast increases, the
    P. flame temperature increases
    Q. blast volume increases
    R. blast volume decreases
    S. flame temperature decreases
    (a) P, Q (b) P, R
    (c) R, S (d) Q, S
    [GATE 2004]
22. Select the correct statement for a blast furnace.
    (a) Hearth diameter is greater than bosh diameter.
    (b) Stack is water cooled.
    (c) Liquid slag comes out through tuyers.
    (d) Tuyers are located above the bosh.
    [GATE 2005]
23. In an iron blast furnace, the sinter is added in preference to iron ore lumps because, sinter
    (a) has better reducibility than ore
    (b) increases the raceway adiabatic flame temperature
    (c) decreases the carbon content of hot metal
    (d) helps in reducing the manganese content of hot metal
24. Group I is a list of technologies for alternate methods of producing iron. Group ll is a list of terms you come across in the context of these technologies. Match the items.
    Group 1
    P. MIDREX Q. COREX
    R. SL/RN S. HyI- I
    Group ll
    1. Retort
    2. Rotary kiln
    3. Smelting reduction
    4. Shaft furnace
    (a) P-3, Q-1, R-2, S-4
    (b) P-4, Q-2, R-1, S-3
    (c) P-1, Q-3, R-2, S-4
    (d) P-4, Q-3, R-2, S-1
    [GATE 2010]
25. Match the processes for alternate methods of producing iron in group 1 with the type of furnaces listed in group 2.
    Group 1
    P. MIDREX Q. SL/RN
    R. Echevarria S. FLOR
    Group 11
    1. Rotary kiln 2. Retort
    3. Shaft furnace 4. Fluidised bed
    (a) P-3, Q-1, R-2, S-4
    (b) P-2, Q-4, R-1, S-3
    (c) P-4, Q-1, R-2, S-3
    (d) P-1, Q-3, R-2, S-4
    [GATE 2006]

Answers

18. (d) 19. (b) 20. (a) 21. (b) 22. (b) 23. (a)
24. (a) 25. (a)

26. The primary source of heat in cupola melting is provided by the reaction:
   (a) $C + O_2 \rightarrow CO_2$
   (b) $C + H_2O \rightarrow CO_2 + H_2$
   (c) $C + CO_2 \rightarrow 2CO$
   (s) $CaCO_3 \rightarrow CaO + CO_2$

   [GATE 2007]

27. Identify the correct statement with reference to blast furnace iron making.
   (a) Hematite is reduced to magnetite in the lower part of the shaft.
   (b) Coke rate can not be improved by oil injection through tuyere.
   (c) High exit gas temperature may be an indication of "channeling".
   (d) Pressure drop in blast furnace can not be improved by proper burden distribution.

   [GATE 2007]

28. High top pressure in the blast furnace
   (a) decreases the time of contact between gas and solid
   (b) increases the time of contact between gas and solid
   (c) increases fuel consumption
   (d) increases the rate of solution loss reaction

   [GATE 2008]

29. Humidification of the blast in the iron blast furnace leads to
   (a) lowering of the raceway temperature
   (b) increase in the raceway temperature
   (c) difficulty in pulverised coal injection
   (d) decrease of oxygen content in the hot metal

   [GATE 2011]

30. In an ideal blast furnace, the input and output are as follows:

   Input :
   Ore, Pure Fe_2O_3, no gaunge :1357kg/THM*
   Coke, Pure C, no ash : 400 kg/ THM
   Blast air, dry : O_2 : 293 kg/HTM
   N_2 : 964 KG/THM
   Flux : nil
   Output :
   Hot metal : 5wt% C, rest iron
   Slag : nil
   Top gas : CO, CO_2, N_2

   * THM refers ton of hot metal (liquid pig iron) ; Atomic weights : C-12, O-16, Fe-56 The amount of oxygen in CO & CO_2 leaving with the top gas is .... kg.
   (a) 293 (b) 407
   (c) 700 (d) 1050

   [GATE 2011]

31. In Q. No. 30. The CO/CO_2 molar ratio in the top gas is
   (a) 0.9 (b) 1.0
   (c) 1.1 (d) 1.5

   [GATE 2011]

32. High top pressure in a blast furnace operation
   (a) favours the solution loss reaction
   (b) suppresses the solution loss reaction.
   (c) decreases gas-solid contact time.
   (d) increases coke rate.

   [GATE 2013]

33. The permeability of burden in an iron making blast furnace can be improved by using
   (a) fine charge.
   (b) agglomerated charge.
   (c) oxygen enriched air blast.
   (d) pulverized coal injection through the tuyeres.

   [GATE 2013]

Answers

26. (a)	27. (a)	28. (b)	29. (a)	30. (c)	31. (b)
32. (b)	33. (b)				

34. The important factors for producing low silicon pig iron in a blast furnace are
(a) higher temperature and higher basicity.
(b) lower temperature and lower basicity.
(c) lower temperature and higher basicity.
(d) higher temperature and lower basicity.
[GATE 2014]

35. An iron blast furnace produces hot metal containing 95% Fe. The iron ore charged into the furnace contains 95% Fe_2O_3 and the rest is gangue. Assume that all the iron in the ore goes to hot metal. The amount of iron ore (in kg) required for producing 1000 kg of hot metal is .... (Atomic weight of Fe = 56 g mol^{-1} and that of Fe_2O_3 = 160 g mol^{-1})
(a) 1025 (b) 1425
(c) 1675 (d) 2000
[GATE 2015]

36. The chemical formula of **Wustite** is
(a) FeS_2 (b) Fe_2O_3
(c) Fe_3O_4 (d) FeO
[GATE 2016]

37. Direct Reduced Iron (DRI) produced from a gas based process contains Fe, FeO, C and remainder being gangue. The chemical composition of DRI is: *Total Fe = 92 wt.% and Metallic Fe = 84 wt/ %*. The weight percent of FeO in DRI is ....
[GATE 2016]

Answers

34. (c) 35. (b) 36. (d) 37. (10.0 to 10.5)

Section B

4 Steel Making

1. Manganese recovery in steel making is aided by
   (a) low slag basicity
   (b) low silicon content in hot metal
   (c) high slag basicity and high silicon content in the hot metal
   (d) none of these

[GATE 92]

2. The usual energy consumption in electric are furnace steel making is ----k Wh/ton of steel.
   (a) 60 – 100 (b) 400 – 700
   (c) 1200 – 1500 (d) 2000 – 2300

[GATE 94]

3. A steel sample which has been deoxidised with Fe – Mn at 1600°C contains 0.51 wt%Mn. The equilibrium constant for dissolution of MnO in steel with 1 wt% standard state is 0.051. the residual oxygen level in the sample is ---- wt%
   (a) 0.51 (b) 0.1
   (c) 2.6×10^{-3} (d) 10

[GATE 95]

4. Hydrogen in liquid steels is dissolved
   (a) as tiny gas bubbles
   (b) in the atomic form
   (c) in the ionic form
   (d) in the molecular form

[GATE 99]

5. In a good rimming steel
   (a) both silicon and carbon should be low
   (b) silicon should be low but carbon should be high
   (c) both carbon and silicon should be high
   (d) silicon should be high but carbon should be low.

[GATE 99]

6. In the acid Bessemer process, the hot metal should have the following composition:
   (a) $S < 0.05\%$ and $P < 0.05\%$
   (b) $S < 0.05\%$ and $P < 1.5\%$
   (c) $S < 0.05\%$ and $P > 1.5\%$
   (d) $S > 1.5\%$ and $P < 0.05\%$

[GATE 2001]

7. The nozzle used in the lance of LD. steel making process is
   (a) convergent
   (b) convergent-divergent
   (c)divergent
   (d) divergent-convergent

[GATE 2002]

Answers

1. (c) 2. (b) 3. (c) 4. (a) 5. (a) 6. (a)
7. (b)

8. In continuous casting of liquid steel, the mould is made of
   (a) water cooled steel
   (b) water cooled copper
   (c) refractory oxide
   (d) silicon carbide

   [GATE 2003]

9. Match the following
   Group 1
   P. Aluminium addition
   Q. Argon stirring
   R. Oxygen injection
   S. Calcium carbide injection
   Group 2
   1. Homogenization 2. Deoxidation
   3. Decarburisation 4. Desulfurisation
   (a) P–2, Q–1, R–3, S–4
   (b) P–2, Q–1, R–4, S–3
   (c) P–4, Q–3, R–2, S–1
   (d) P–1, Q–2, R–3, S–4

   [GATE 2003]

10. Electromagnetic stirring of the liquid steel in continuous casting is primarily used to
    (a) decrease the extent of growth of columnar grains
    (b) improve the surface quality of steel slab.
    (c) minimise the internal cracks in steel slab
    (d) increase the rate of solidfication of steel

    [GATE 2004]

11. During deoxidation of steel, the sequence of addition of elements should be as follows:
    (a) Si, Mn, Al (b) Al, Mn, Si,
    (c) Mn, Si, Al (d) Al, Si, Mn

    [GATE 2004]

12. The oxidising power of a steel making slag is dependent on the concentration of
    (a) CaO, because Ca^{2+} ions are oxidising to the elements
    (b) CaO, because O^{2-} ions are oxidising to the elements
    (c) FeO, because Fe^{2+} ions are oxidising to the elements
    (d) FeO, because O^{2-} ions are oxidising to the elements

    [GATE 2004]

13. The purpose of injection of calcium into liquid steel is to
    P. replace Al_2O_3 inclusions by CaO inclusions
    Q. modify Al_2O_3 inclusions into calcium aluminate inclusions
    R. reduce sulfur content of steel
    S. reduce nitrogen content of steel
    (a) P, Q (b) P, R
    (c) Q, R (d) Q, S

    [GATE 2004]

14. Molten steel is kept in a ladle. Due in natural convection, mixing occurs inside the melt. A 1cm diameter sphere is held in the centre of the melt where the steel flows upward, so as to measure the force exerted by the melt on the sphere. The force, F exerted by the melt on the sphere is given by $F = f\,(\pi/8).\,(\mu^2/8)\,(Re)^2$.
    Data : Density of liquid steels
    $= 7100$ kg. m^{-3}
    Viscosity of liquid steel
    $= 6.5 \times 10^{-3}$ kg. $m^{-1}.s^{-1}$
    Reynolds number (Re) of the melt
    $= 5 \times 10^3$
    Friction factor (f) = 0.3
    The velocity of melt ($m.s^{-1}$) in the central portion of the ladle would be
    (a) 0.0046 (b) 0.46
    (c) 4.6 (d) 5

    [GATE 2004]

Answers

8. (b) 9. (a) 10. (a) 11. (c) 12. (d) 13. (c)
14. (b)

15. In Q. No. 14, the force exerted by steel on the sphere would be .... N.
(a) 0.018 (b) 0.18
(c) 0.5 (d) 18
[GATE 2004]

16. Necessary condition for removal of sulphur from liquid steel are
(a) high temperature, high viscosity, high FeO content of slag.
(b) high temperature, low viscosity, low FeO content of salg.
(c) low temperature, low viscosity, low FeO content of slag.
(d) high temperature, high viscosity, low FeO content of slag.
[GATE 2003]

17. Necessary conditions for removal of phosphorous from liquid steel are
(a) high temperature, high viscosity, low FeO content of slag
(b) low temperature, high viscosity, high FeO content of slag
(c) low temperature, low viscosity, and low FeO content of slag.
(d) high temperature, low viscosity, and high FeO content of slag
[GATE 2003]

18. If a steel has a to be deoxidised with silicon, manganese and aluminium in the ladle, the preferred sequence of addition is
(a) Silicon, manganese, aluminium
(b) aluminium, manganese, silicon
(c) manganese, silicon, aluminium
(d) aluminium, silicon, maganese
[GATE 98]

19. The main function of the RH process of steel treatment is to
(a) reduce dissolve nitrogen in steel
(b) improve ferro alloy recovery
(c) reduce sulfur in steel
(d) reduce carbon and hydrogen in steel
[GATE 2005]

20. In AOD process of stainless steel making, a mixture of argon and oxygen gas in blown through the bottom because the argon gas
(a) dissolve in steel to offer better corrosion protection
(b) acts as a catalyst for the reaction of carbon and oxygen
(c) lowers the partial pressure of CO gas in the bubble and hence helps to increase the decarburisation rate.
(d) Prevents the oxidation of expensive nickel in the stainless steel.
[GATE 2005]

21. In steel making the addition of bauxite is done to
(a) increase the activity of FeO in slag
(b) decrease the viscosity of slag
(c) improve phosphorous distribution ratio
(d) improve sulfur distribution ratio
[GATE 2005]

22. Liquid steel (300 metric tons) initially containing 0.03 mass% sulfur, is desulphurised with calcium (1000kg) according to the reaction, Ca + [S] = Ca S. The slag (1 metric ton) lying on top of metal contains (in mass%) 50% CaO, 25% SiO_2 and the rest is Al_2O_3. Only 10 mass% of the CaO injected reacts with the sulfur and the rest escapes as gas into the atmosphere.
What will be the final sulfur content, in mass% in liquid steel?
(a) 0.003 (b) 0.0010
(c) 0.0033 (d) 0.0066
[GATE 2005]

Answers

15. (a)	16. (b)	17. (c)	18. (c)	19. (d)	20. (d)
21. (c)	22. (d)				

23. In. Q. No. 22, the distribution ratio, defined as mass% sulfur in slag to mass% sulfur in metal is
(a) 2 ms (b) 10 ms
(c) 9.5 ms (d) 20 ms
[GATE 20050]

24. In a basic oxygen furnace, under appropriate conditions, which of the following statements is not correct?
(a) Carbon can be removed in preference to P and S.
(b) Phosphorous can be removed in preference to C and S.
(c) Sulfur can be removed in preference to C and P.
(d) Carbon and phosphorous can be removed in preference to S.
[GATE 2010]

25. In continuous casting of steel, the mould is subjected to vertical oscillations in order to
(a) allow easy floatation of inclusion
(b) ensure good casting homogeneity
(c) increase the heat transfer rate from the steel to the mould
(d) prevent the skin sticking to the mould
[GATE 2010]

26. Stainless steel is most commonly produced using the .... process.
(a) VOD (b) AOD
(c) EOF (d) EAF
[GATE 2006]

27. Match the facilities in a steel plant in group 1 with the associated terms listed in group 2.
Group 1
P. Electric arc furnace
Q. L-D converter
R. Continuous caster
S. Blast furnace
Group 2
1. High top pressure
2. Dummy bar
3. Slag splashing
4. Eccentric bottom tapping
(a) P-4, Q-l, R-2, S-3
(b) P-2, Q-4, R-l, S-3
(c) P-4, Q-3, R-2, S-l
(d) P-l, Q-3, R-2, S-4
[GATE 2006]

28. Identify the condition that is not beneficial for dephosphorisation of hot metal in an L-D converter.
(a) High slag basicity
(b) Low temperature
(c) High silicon in hot metal
(d) Highly oxidising slag.
[GATE 2006]

29. Identify the incorrect statement with reference to L.D. steel making
(a) The temperature of L.D. furnace is maintained at around 1600°C.
(b) The basicity of slag is maintained at around unity.
(c) Dephosphorisation and decarburisation should proceed simultaneously.
(d) High silicon hot metal may lead to sloping.
[GATE 2007]

30. Identify the correct statement about manganese in steels from the following:
(a) It decreases hardenability
(b) It makes the steel succeptible to hot-shortness
(c) It is a strong austenite stabiliser
(d) It decreases hardness of martensite
[GATE 2007]

Answers

23. (b)	24. (c)	25. (d)	26. (d)	27. (c)	28. (c)
29. (b)	30. (c)				

31. Manganese recovery in steel making is aided by
    P. Oxidising slag
    Q. Reducing slag
    R. High temperature
    S. Low temperature
    T. Acidic slag
    (a) P, Q (b) Q, S
    (c) R, T (d) Q, R

[GATE 2008]

32. Given the following Assertion A' and the Reason 'R'; the correct option is
    Assertion A: Phosphorous removal in steel making is favoured by basic slag.
    Reason R: Basic slag decreases the activity of P_2O_5 in the slag.
    (a) Both A and R are true and R is the correct reason for A.
    (b) Both A and R are false.
    (c) A is true but R is false.
    (d) Both A and R are true but R is not correct reason for A.

[GATE 2008]

33. Match the practices in group 1 with the reactors in group 2.
    Group 1
    P. Layered charging of coke and ore
    Q. Oxygen injection through supersonic nozzle
    R. Aluminium wire feeding
    S. Foamy slag practice
    Group 2
    1. Ladle furnace
    2. Electric are furnace
    3. Blast furnace
    4. L.D. converter
    (a) P-3, Q-l, R-2, S-4
    (b) P-2, Q-4, R-3, S-l
    (c) P-4. Q-3, R-2, S-l
    (d) P-3, Q-4, R-l, S-2

[GATE 2009]

34. VOD process is preferred over AOD process for making extra low carbon stainless steel, because
    (a) pco can be lowered to a much lower level in the VOD than in the AOD.
    (b) AOD does not have adequate stirring.
    (c) free board needed for such operation is not available in the AOD.
    (c) AOD refractory is not stable in contact with extra low carbon steel.

[GATE 2011]

35. During LD blow in steel making the impurity that gets removed first is
    (a) Carbon (b) Phosphorous
    (c) Manganese (d) Silicon

[GATE 2012]

36. In L-D steel making, the final slag can be best described as
    (a) oxidizing
    (b) basic
    (c) oxidizing and basic
    (c0 reducing and basic

[GATE 2013]

37. Ladle deoxidation of liquid steel is done at 1600°C by adding ferro-aluminium. By assuming Stokes law behaviour, time (in s) required for alumina particles of 50 μm diameter to float to the surface from a depth of 2 m would be
    (a) 1620 (b) 2158
    (c) 2630 (d) 4450
    [Given: density of steel = 7000 kg/m^3, density of alumina = 3650 kg/m^3, viscosity of steel = 6×10^{-3} kg/m/s]

[GATE 2013J

Answers

31. (d)	32. (c)	33. (d)	34. (a)	35. (d)	36. (c)
37. (c)					

38. Which one of the following factors is *NOT* desirable for effective phosphorus removal in BOF steel making process?
    (a) Higher FeO level in slag
    (b) Higher basicity
    (c) Higher temperature
    (d) Lower temperature

[GATE 2014]

39. Which one of the following processes is performed for inclusion modification in ladle metallurgy of steel making?
    (a) Aluminium wire injection
    (b) Calcium wire injection
    (c) Oxygen top blowing
    (d) Oxygen bottom blowing

[GATE 2014]

40. Determine the correctness or otherwise of the following Assertion(a) and Reason (r):

    Assertion (a) : Deoxidation and decarburisation of liquid steel are done by vacuum treatment process.

    Reason (r): Decreased partial pressure of CO(g) promotes deoxidation and decarburisation.

    (a) a is false but r is true.
    (b) a is true but r is false.
    (c) a is true and r is also true, but r is not the reason for a.
    (d) a is true and r is also true, and r is the reason for a.

[GATE 2014]

41. In BOF steelmaking, 5 metric ton of the lime containing 90 wt.% CaO is used to refine 100 metric ton of hot metal containing 93.2 wt.% Fe. The slag produced during refining contains 40 wt.% CaO and 22 wt.% FeO. Neglecting material losses, the yield of Fe (in%) is ....

[GATE 2016]

42. In vacuum degassing of steel, 14 ppm of dissolved nitrogen is in equilibrium with 1 mbar of nitrogen gas at 1873 K. At the same temperature, if the pressure is lowered to 0.7 mbar, the equilibrium nitrogen content (in ppm) is ....

[GATE 2016]

Answers

38. (c) 39. (b) 40. (d) 41. (97.5 to 98.3) 42. (11.6 to 11.8)

5 Extractive Metallurgy (Non-Ferrous)

1. In the commercial production of which of the following metals, metallothermic reduction is used?
   (a) Copper (b) Zinc
   (c) Zirconium (d) Nickel.
   [GATE 92]

2. The product(s) of roasting of a sulfide ore is (are)
   (a) oxide only
   (b) sulphate only
   (c) oxide and sulphate
   (d) dependent on temperature and partial pressures of oxygen and sulfur dioxide
   [GATE 95]

3. Match the following metals with their extraction processes.
   Group 1
   P. Gold Q. Titanium
   R. Magnesium S. Nickel
   Group 2
   1. Pidgeon process 2. Cyanidation
   3. Mond process 4. Kroll's process
   (a) P-2, Q-3, R-l, S-4
   (b) P-2, Q-3, R-4, S-l
   (a) P-4, Q-3, R- 2, S-1
   (b) P-1, Q-2, R-3, S-4
   [GATE 95]

4. Traditional copper converters are
   (a) side blown (b) top blown
   (c) bottom blown (d) both b and c
   [GATE 96]

5. The major advantage of using flash roasters over multi hearth roasters is
   (a) a significant increase in production rate.
   (b) a significant decrease in environmental pollution due to condensation of liquid sulfur.
   (c) possibility of complete automation.
   (d) no need for an additional fuel.
   [GATE 961

6. Match the following ores with their chemical formula.
   Group 1
   P. Sphalerite Q. Covellite
   R. Anglesite S. Chalcocite
   Group 2
   1. Cu_2S 2. CuS
   3. $PbSO_4$ 4. ZnS
   (a) P-4, Q-2, R-3. S-l
   (b) P-4, Q-3, R-2, S-l
   (c) P-4, Q-2, R-l, S-3
   (d) P-3, Q-2, R-l, S-4
   [GATE 96]

Answers

1. (b) 2. (a) 3. (a) 4. (a) 5. (a) 6. (a)

7. Match the following phenomena with the corresponding extraction related processes.
P. Anode effect
Q. Formation of scaffolds
R. Formation of white metal
S. Classification
1. Cu extraction
2. Blastfurnace
3. Hydrocyclone
4. Al extraction
(a) P-l, Q-2, R-3, S-4
(b) P-3, Q-4, R-l, S-2
(c) P-4, Q-3, R-2, S-l
(d) P-l, Q-2, R-4, S-3
[GATE 96]

8. Cadmium in zinc leach liquor is removed by cementation using
(a) zinc (b) iron
(c) nickel (d) copper.
[GATE 98]

9. In the Bayer process, bauxite is digested under pressure using
(a) H_2SO_4 (b) NaOH
(c) NH_3 (d) HCl.
[GATE 98]

10. Mattee smelting is used in the extraction of
(a) lead (b) zinc
(c) aluminium (d) copper
[GATE 98]

11. In Pidgeon process, the reducing agent used is
(a) carbon (b) carbon monoxide
(c) hydrogen (d) ferrosilicon.
[GATE 98]

12. Match the following:
Group I
P. Fire refinning
Q. Zone refinning
R. Refinning via carbonyl
S. Iodide refining
Group 2
1. Zirconium
2. Nickel
3. Germanium
4. Copper
(a) P-4, Q-3, R-2, S-l
(b) P-4, Q-3, R-l, S-2
(c) P-3, Q-4, R-2, S-l
(d) P-3, Q-4, R-l, S-2
[GATE 98]

13. Match the following:
Group 1
P. Flash smelting
Q. Blast furnace
R. Reduction by hydrogen
S. Fused salt electrowinning
Group 2
1. Tungsten 2. Aluminium
3. Copper 4. Lead
(a) P-3, Q-4, R-2, S-l
(b) P-4, Q-3, R-2, S-l
(c) P-3, Q-4, R-l, S-2
(d) IM, Q-2, R-3, S-4
[GATE 98]

14. Thiobacillus ferroxidans is used in
(a) bacterial leaching in acidic medium.
(b) bacterial leaching in alkaline medium.
(c) bacterial leaching of cuprite ore.
(d) pressure leaching using ammoniacal solution.
[GATE 99]

Answers

7. (d)	8. (a)	9. (b)	10. (d)	11. (d)	12. (a)
13. (c)	14. (a)				

15. The most abundant metal present in the earth's crust is
    (a) iron (b) copper
    (c) titanium (d) aluminium.
    [GATE 2001]

16. Monazite deposits constitute an important source for the following metal.
    (a) Thorium (b) Titanium
    (c) Molybdenum (d) None of the above.
    [GATE 2001]

17. The reductant used in the extraction of magnesium from calcinated dolomite via Pidgeon's process is
    (a) pure carbon (b) pure silicon
    (c) ferrosilicon (d) ferromanganese.
    [GATE 2001]

18. In "Imperial Smelting Process" for extraction of zinc, zinc vapour thus produced is quenched in the external condenser by the use of the following:
    (a) Jet of water at high pressure
    (b) Blast of air
    (c) Mixture of air and water
    (d) Rain of molten lead
    [GATE 2001]

19. Copper deposits are found in India at the following location:
    (a) Kudremukh (b) Kolar
    (c) Khetri (d) Ramagundam.
    [GATE 2001]

20. Speisses is a mixture of the following:
    (a) Arsenides of heavy metals
    (b) Antimonides of heavy metals
    (c) Arsenides and antimonides of heavy metals
    (d) Iron, cobalt and nickel.
    GATE 2001]

21. Phosphorous is added to copper melt to
    (a) remove hydrogen
    (b) deoxidise the metal
    (c) improve fluidity
    (d) reduce shrinkage.
    [GATE 2002]

22. The $A1_20_3$ content of cryolite in Hall-Heroult's cell is maintained between
    (a) 2 – 5% (b) 18 – 20%
    (c) 12 – 15% (d) 6 – 12%.
    [GATE 2002]

23. Titanium is produced by
    (a) electrolytic reduction of $TiCl_4$
    (b) calcium reduction of $TiCl_4$
    (c) magnesium reduction of $TiCl_4$
    (d) thermal dissociation of $TiCl_4$
    [GATE 2003]

24. A conventional (Pierce-Smith) copoer converter is
    (a) blown from both top and bottom
    (b) bottom blown
    (c) side blown with single tuyer
    (d) side blown with multiple tuyer.
    [GATE 2004]

25. Identify the correct statement.
    (a) Sphalerite is zinc oxide.
    (b) The first law of thermodynamics is stated as, $\delta E = \delta Q - \delta W$.
    (c) Lead can be produced in a blast furnace.
    (d) T. Ferroxidans is a fungus that can be used for leaching chalcopyrite.
    [GATE 2004]

Answers

15. (d)	16. (a)	17. (c)	18. (d)	19. (c)	20. (a)
21. (a)	22. (a)	23. (c)	24. (d)	25. (c)	

26. Identify the correct statement with reference to the extractive metallurgy of aluminium.
    (a) The electrolyte consists of molten Na_3AlF_6 with approximately 1 to 8% AI_2O_3.
    (b) Approximately 80% of the aluminium deposited on the cathode comes from cryolite.
    (c) Sodium is deposited along with aluminium, but is immediately vapourised.
    (d) Anode effect sets in when the cryolite concentration goes below 40%.

[GATE 2004]

27. Zone refinning is a process in which
    (a) a metal rod is dipped in a molten pool of liquid to dissolve slowly and the liquid is refined continuously.
    (b) liquid metal is refined in the zone of electric arc in the DC arc furnace.
    (c) a metal in the form of long rod is melted over a short length and molten region is made to traverse along the length of the rod.
    (d) a long metal rod is cut into small parts and each part is melted and refined separately.

[GATE 2005]

28. Match the metals in group 1 with the processes in group 2.

Group 1

P. Aluminium Q. Copper

R. Zinc S. Iron sponge

Group 2

1. Poling
2. Fused salt electrolysis
3. Rotary kiln process
4. Distillation

(a) P-l, Q-2, R-3, S-4

(b) P-2, Q-l, R-4, S-3

(c) P-4, Q-2, R-4, S-3

(d) P-3, Q-4, R-2, S-l

[GATE 2005]

29. Which of the following statements regarding Kroll's process is not correct?
    (a) Pure metal chloride serves as main raw material.
    (b) Reduction is done only by sodium.
    (c) Reduction chamber should be free of oxygen.
    (d) It is used for extraction of titanium and zirconium.

[GATE 2010]

30. Which of the following is a suitable method to remove hydrogen from molten Aluminium?
    (a) Expose flowing melt to vacuum.
    (b) Bubble humidified argon gas through the melt.
    (c) Increase melt temperature.
    (d) Cover melt surface with a flux.

[GATE 2010]

31. Match the metallurgical processes in group I with their corresponding reactors type in group II.

Group 1

P. Roasting of sulfide concentrate

Q. L.D. steel making

R. Dwight-Lloyd sintering

S. Zinc extraction

Group 2

1. Pneumatic reactor
2. Retort
3. Travelling grate reactor
4. Fluidised bed reactor

(a) P-3, Q-2, R-4, S-l

(b) P-4, Q-l. R-3, S-2

(c) P-3, Q-4, R-2, S-l

(d) P-4, Q-l, R-2, S-3

[GATE 2010]

Answers

26. (a) 27. (c) 28. (d) 29. (b) 30. (b) 31. (b)

32. Identify the metal that can not be produced by carbothermic reduction.
 (a) Iron (b) Lead
 (c) Tin (d) Silver.
 [GATE 2006]

33. Match the type of reagent used in floatation in group 1 with reagent in group 2.
 Group 1
 P. Collector Q. Regulator
 R. Activator 5. Frother
 Group 2
 1. Pine oil
 2. Copper sulphate
 3. Sodium ethyl xanthate
 4. Lime
 (a) P-2, Q-3, R-4, S-l
 (b) P-4, Q-2, R-3, S-l
 (c) P-3, Q-4, R-2, S-l
 (d) P-l, Q-3, R-2, S-4
 [GATE 2006]

34. Which metal is extracted by leaching?
 (a) Iron (b) Aluminium
 (c) Lead (d) Gold
 [GATE 2007]

35. Identify the incorrect statement with reference to unit processes in extractive metallurgy.
 (a) Selective distillation is a purification technique used in extractive metallurgy.
 (b) Coking of coal is carried out in a shaft furnace.
 (c) Precipitation is a hydrometallurgy route of purification.
 (d) Predominance area diagram is used to select the opening conditions of roastings
 [GATE 2007]

36. Match the process in group 1 to its description in group 2.
 Group 1
 P. COREX process Q. OBM process
 R. Carbonyl process S. AOD process
 Group 2
 1. Decarburisation of liquid steel
 2. Steel making using oxygen
 3. Nickel refmning
 4. Alternative route of liquid iron production
 (a) P-4, Q-2, R-3, S-l
 (b) P-3, Q-l, R-2, S-4
 (c) P-l, Q-4, R-3, S-l
 (d) P-2, Q-l, R-3, S-4
 [GATE 2007]

37. The process of cementation involves
 (a) separation of the desired metal by adding a more reactive metal.
 (b) elimination of a more reactive metal from molten metal by preferential oxidation.
 (c) refinning by preferential dissolution of the desired metal in a organic solvent.
 (d) extraction by selective dissolution of the desired metal in an inorganic solvent.
 [GATE 2007]

38. Cassiterite is an important source for
 (a) tin (b) titanium
 (c) molybdenum (d) thorium.
 [GATE 2008]

Answers

32. (d) 33. (c) 34. (d) 35. (b) 36. (a) 37. (b)
38. (a)

39. Match the unit processes in group 1 with the objectives in group 2.

Group 1

P. Leaching Q. Cementation

R. Roasting S. Converting

Group 2

1.Precipitation of metal in aqueous solution

2. Selective dissolution of metal

3. Conversion of mattee to metal

4. Conversion of sulfide to oxide

5. Separation of metal from slag

(a) P-2, Q-l, R-3, S-5

(b) P-2, Q-l, R-4, S-3

(c) P-3, Q-4, R-5, S-2

(d) P-4, Q-3, R-2, S-l

40. Match the following metals in group 1 with their production methods in group 2.

Group 1

P. Titanium Q. Nickel

R. Magnesium S. Zinc

Group 2

1. Mond's process

2. Pidgeon's process

3. Imperial smelting

4. Kroll's process

5. Cynidation

(a) P-5, Q-2, R-3, S-4

(b) P-3, Q-5, R-4, S-2

(c) P-4, Q-l, R-2, S-3

(d) P-4, Q-l, R-5, S-3

[GATE 2008]

41. Given the following assertion 'A' and the reason 'R' the correct option is

Assertion A: In Bayer's process, high pressure is used to dissolve alumina from bauxite.

Reason R: Pressure increases the boiling point of water.

(a) Both A and R are correct, but R is not the correct reason for A.

(b) Both A and R are false.

(c) Both A and R are correct and R is the correct reason for A.

(d) A is true but R is false.

[GATE 2008]

42. Match the extraction methods in group 1 with the metals in group 2.

Group 1

P. Roasting followed by carbothermic reduction

Q. Electrolysis of fused salt

R. Roasting followed by controlled oxidation

S. Halide process

Group 2

1. Ti 2. Pb

3. A1 4. Cu

5. Au

(a) P-2, Q-3, R-4, S-l

(b) P-5, Q-4, R-3, S-l

(c) P-2, Q-5, R-l, S-4

(d) P-3, Q-2, R-5, S-l

[GATE 2009]

43. Aluminium is NOT commercially produced by carbon-thermic reduction primarily, because

(a) aluminium metal will have excessive dissolved oxygen.

(b) it melts at too low a temperature.

(c) it does not vaporise at reasonable temperature.

(d) AI-AI_2O_3 line is too low in the Ellingham diagram' and needs excessively high temperature.

[GATE 2010]

Answers

39. (b) 40. (c) 41. (c) 42. (a) 43. (d)

44. Match the metal in group I with the corresponding ores in group II.

Group 1	Group 2
P. Lead	1. Columbite
Q. Zinc	2. Cassiterite
R. Uranium	3. Galena
S. Niobium	4. Pitchblende
	5. Sphalerite

(a) P-3, Q-5. R-2, S-l
(b) P-3, Q-2, R-5, S-4
(c) P-3, Q-5, R-4, S-l
(d) P-3, Q-4, R-5, S-2

[GATE 2011]

45. Match the processes given in Group-I with the corresponding metals in Group II.

Group 1
P. Matte smelting
Q. Cyanide leaching
R. Carbothermic reduction
S. Fused salt electrolysis

Group 2
1. Lead
2. Copper
3. Gold
4. Aluminium

(a) P-l, Q-2, R-l, S-4
(b) P-2, Q-3, R-l, S-4
(c) P-2, Q-l, R-3, S-4
(d) P-2, Q-3, R-4, S-l

[GATE 2012]

46. Identify the correct combination of the following statements.

P. Bessemer converter can be used in copper smelting.
Q. The Mond process for nickel involves reaction of metal with H_2 gas.
R. Roasted ZnS concentrates can be smelted in a blast furnace.
S. Magnesium metal can be produced by electrolysis of sea water.

(a) P, R and S (b) P, Q and R
(c) P and Q (d) ∩ and S

[GATE 2012]

47. In a roasting process, the set of conditions that favour sulphate formation from metal sulphide concentrates are.

P. high temperature
Q. high partial pressure of oxygen
R. use of excess air
S. high partial pressure of SO_3

(a) P, R and S (b) P, Q and R
(c)QandS (d) R and S

[GATE 2013]

48. Match the metal in Group I with its corresponding ore in Group II.

Group 1
P.Ni Q.Th
R. Pb S. Sn

Group 2
1 Monazite 2. Cassiterite
3. Penlandite 4. Galena

(a) P-l, Q-3, R-4, S-2
(b) P-4, Q-3, R-3, S-l
(c) P-3, Q-l, R-4. S-2
(d) P-2, Q-3, R-l, S-4

[GATE 2013]

Answers

44. (c) 45. (b) 46. (a) 47. (c) 48. (c)

49. Determine the correctness or otherwise of the following

Assertion: *For the extraction of metal values from* their sulphide concentrates by hydrometallurgical route, leaching with oxygen under high pressure is used. **Reason:** Presence of oxygen under high pressure causes roasting of sulphides, which helps in leaching of the values.

(a) a is true but r is false.

(b) a is false but r is true.

(c) both a and r are true, and r is the reason for a.

(d) both a and r are true, but r is not the reason for a.

[GATE 2013]

50. Which one of the following minerals is a source of titanium?

(a) Haematite (b) Magnetite

(c) Ilmenite (d) Pyrolusite

[GATE 2014]

51. Zn is commercially extracted from which of the following minerals?

(a) Sphalerite (b) Magnetite

(c) Chalcopyrite (d) Galena

[GATE 2015]

52. Match the metals listed in Group I with the most appropriate extraction routes listed in Group II.

Group 1

P. Al Q. Ti

R. Cu S. Fe

Group 2

1. Blast Furnace
2. Matte Smelting
3. Electrolysis of Fused Salts
4. Halide Metallurgy

(a) P-3, Q-2, R-4, S-l

(b) P-3, Q-4, R-2, S-1

(c) P-2, Q-4, R-3, S-l

(d) P-4, Q-l, R-3, S-2

[GATE 2015]

53. Match the entities in Column I with the corresponding processes in Column II.

Column 1

P. Xanthate salts

Q. Thiobacillus Ferrooxidans

R. Hydrocyclone

S. Anode effect

Column 2

1. Extraction of Al
2. Flotation
3. Classification
4. Bacterial Leaching

(a) P-4, Q-2, R-3, S-l

(b) P-2, Q-4, R-3, S-1

(c) P-3, Q-4, R-1, S-2

(d) P-4, Q-l, R-2, S-3

[GATE 2016]

Answers

49. (c) 50. (c) 51. (a) 52. (b) 53. (b)

6

Mechanical Metallurgy

1. Earing is a defect found in steels after the following metal working operation (s)?
   (a) Deep drawing
   (b) Rolling
   (c) Extrusion
   (d) Wire drawing

   [GATE 91]

2. Springback in sheet metal bending depends upon
   (a) elastic limit
   (b) bend radius
   (c) degree of bend
   (d) thickness of sheet.

   [GATE 91]

3. In extrusion of metals, which of the following statement(s) is/are true?
   (a) Speed of the extruded material is the same as that of ram speed.
   (b) Redundant work is a function of die angle.
   (c) Relative motion between the billet surface and the container wall is always present.
   (d) Hollow ram is used for indirect extrusion.

   [GATE 91]

4. The typical dislocation density (lines/cm^2) of a hot rolled material is
   (a) 10^2 (b) 10^{12}
   (c) 10^6 (d) 10^9

   [GATE 94]

5. The maximum axial compression stress during cold upsetting of a cylindrical rod of radius, 'r', occurs at
   (a) the outer edges of the rod
   (b) r/3 from the centre
   (c) r/2 from the centre
   (d) the centre

   [GATE 94]

6. As the % reduction increases, the flow stress during hot isothermal forging of a metal
   (a) increases linearly
   (b) decreases exponentially
   (c) decreases linearly
   (d) remains almost constant

   [GATE 94]

7. Fatigue strength of steel can be increased by
   (a) increasing tensile surface residual stresses
   (b) introducing hydrogen in steel
   (e) increasing the grain size
   (d) increasing the specimen size
   (e) increasing compressive surface residual stresses

   [GATE 94]

Answers

1. (a) 2. (b) 3. (b) 4. (b) 5. (d) 6. (b)
7. (e)

8. Match the following products to their method of manufacture.
   P. Seamless tubes
   Q. Aluminium toothpaste tubes
   R. Currency coins
   S. Car bodies
   1. Deep drawing
   2. Stamping
   3. Impact extrusion
   4. Roll piercing or extrusion
   (a) P-4, Q-3, R-2. S-l
   (b) P-4, Q-3. R-l, S-2
   (c) P-3, Q-4, R-l. S-2
   (d) P-2, Q-3, R-l, S-4

[GATE 95]

9. The most serious manufacturing defect from fracture toughness point of view is
   (a) surface roughness
   (b) pore
   (c) spherical inclusion
   (d) crack.

[GATE 95]

10. The co-efficient of friction alters by 5 times when a slab is hot rolled instead of cold rolling in a specific rolling mill. The maximum reduction in the thickness of the slab will increase by a factor of
    (a) 25 (b) 5
    (c) 0.04 (d) 1

[GATE 96]

11. Match the following components with their method of production.
    P. Automobile crankshaft
    Q. Boiler drum
    R. Tungsten
    S. Seamless tube
    1. Powder metallurgy
    2. Submerged arc welding
    3. Extrusion
    4. Forging
    (a) P-4, Q-2, R-3. S-l
    (b) P-4, Q-2, R-l, S-3
    (c) P-4, Q-3, R-2, S-l
    (d) P-3, Q-2, R-4, S-l

[GATE 96]

12. The rolling process can not be used to produce
    (a) plates (b) rods
    (c) tubes (d) wires

[GATE 98]

13. Amongst the following equipments, the productivity is the maximum for
    (a) gravity hammer
    (b) hydraulic press
    (c) mechanical press
    (d) screw press

[GATE 98]

14. Match the process with the property required.
    Process
    P. Cold rolling
    Q. Milling
    R. Brazing
    S. Powder metallurgy
    Property
    1. Wettability
    2. Ductility
    3. Comparability
    4. Machinability
    (a) P-2, Q-l, R-3, S-4
    (b) P-l, Q-2, R-4, S-3
    (c) P-2, Q-l, R-4, S-3
    (d) P-4, Q-3, R-2, S-l

[GATE 98]

Answers

8. (a) 9. (d) 10. (a) 11. (b) 12. (b) 13. (b)
14. (c)

15. Match the stress state with the corresponding forming process.
    P. Direct compressive
    Q. Indirect compressive
    R. Tensile
    S. Shear
    1. Wire drawing
    2. Sheet blanking
    3. Forging
    4. Stretch forming
    (a) P-3, Q-l, R-2, S-4
    (b) P-l, Q-2, R-3, S-4
    (c) P-4, Q-3, R-2, S-l
    (d) P-3, Q-l, R-4, S-2

    [GATE 98]

16. Often earing defects are found during deep drawing operation, because the
    (a) surface finish of the sheet is poor
    (b) sheet material has been given substantial spring back
    (c) starting sheet has planer anisotropy due to its texture
    (d) starting sheet has normal anisotropy due to its texture

    [GATE 99]

17. The spring back phenomenon in metal sheet bending can be compensated by
    (a) bending the part to a smaller than desired radius of curvature
    (b) bottoming the punch in a die
    (c) using low temperature bending
    (d) using high viscosity lubricant

    [GATE 99]

18. Flash in a close die forging.
    (a) is excess material that has been squirted out of the die
    (b) is to provide excess temperature to the material
    (c) helps in filling all the recesses in the die
    (d) is always of uniform cross-sectional area

    [GATE 99]

19. A planetary mill is used for
    (a) hot reduction of a slab directly to strip
    (b) grinding ceramic powders to a very fine mesh
    (c) reduction gear in a pulley system
    (d) rotary kiln assembly in cement production

    [GATE 2000]

20. In sheet metal forming, stretcher strains occur in
    (a) duralumin sheets
    (b) low carbon steel sheets
    (c) austenitic stainless steel sheet
    (d) Ni-base alloy sheet

    [GATE 2000]

21. In hot working, dynamic recovery occurs in
    (a) metals of low stacking fault energy.
    (b) metals of high stacking fault energy.
    (c) single crystal of Ni-based super alloys.
    (d) alpha iron.

    [GATE 2000]

22. Classification of metal forming processes into hot and cold working is based on the following parameter.
    (a) Stacking fault energy
    (b) Recrystallisation temperature
    (c) solidus temperature
    (d) transformation temperature.

    [GATE 2001]

Answers

15. (d)	16. (c)	17. (a)	18. (a)	19. (a)	20. (a)
21. (b & c)	22. (b)				

23. For the manufacture of thin foil of aluminium, the following rolling mill is used:
    (a) Three high rolling mill
    (b) Sendzimir mill
    (c) Four stand continuous mill
    (d) Planetary mill

[GATE 2001]

24. For the occurence of "bite" in rolling, the following condition should be satisfied.
    (a) The co-efficient of friction should exceed the tangent of the contact angle.
    (b) The roll separating force should reach a maximum value.
    (c) The friction co-efficient should be zerc.
    (d) The contact length should be minimum.

[GATE 2001

25. Extrusion pressure is proportional to
    (a) fractional reduction in area (r)
    (b) extrusion ratio (R)
    (c) natural logarithm of r
    (d) natural logarithm of R

[GATE 2002]

26. An ingot is hot forged to a 60% reduction in cross-section area. The percentage reduction in the volume for the above process is

(a) 0	(b) 60
(c) 30	(d) 20

[GATE 2003]

27. Match the following :
    Process

P. Forging	Q. Rolling
R. Extrusion	S. Deep drawing Product
1. Rails	2. Plane wires
3. Crankshaft	4. Toothpaste tube
5. LPG cylinder	

    (a) P-5, Q-3, R-l, S-4
    (b) P-5, Q-2, R-l, S-4
    (c) P-3, Q-l, R-4, S-5
    (d) P-2, Q-3, R-5, S-l

[GATE 2003]

28. A wedge shaped piece of copper is plastically deformed to a plate of uniform thickness. The finest recrystallised grains will be observed at
    (a) the end of the plate corresponding to the thinner part of the wedge
    (b) the end of the plate corresponding to the thicker part of the wedge
    (c) in the centre of the plate
    (d) at no particular location of the plate.

[GATE 2004]

29. A steel plate of 10 mm thickness is to be cold rolled. The mill has rolls of diameter 72 mm. If the co-efficient of friction $\mu = 0.3$ the maximum possible thickness reduction in the first pass will approximately be .... %.

(a) 10	(b) 20
(c) 30	(d) 40

[GATE 2004]

30. Match the following:
    Group 1
    P. Direct extrusion
    Q. Impact extrusion
    R. Tube drawing
    S. Hydrostatic extrusion
    Group 2
    1. Motion of ram and work piece in the same direction
    2. Motion of ram and work pieces in the opposite direction
    3. No containerwall friction
    4. Use of mandrel
    (a) P-l, Q-2, R-4, S-3
    (b) P-4, Q-2, R-3, S-l
    (c) P-2, Q-3, R-4, S-l
    (d) P-l, Q-2, R-2, S-4

[GATE 2004]

Answers

23. (c)	24. (a)	25. (d)	26. (a)	27. (c)	28. (b)
29. (c)	30. (a)				

31. Match the phenomenon with typically observed features

    P. Fatigue fracture

    Q. Yield point phenomenon

    R. Ductile fracture

    S. Deep drawing of a sheet

    1. Earing

    2. Dimple structure

    3. Beach marks

    4. Luder's band

    (a) P-3, Q-4, R-2, S-l

    (b) P-3, Q-4, R-l, S-2

    (c) P-4, Q-3, R-2, S-l

    (d) P-4, Q-l, R-2, S-3

    [GATE 94]

32. For successful rolling, the condition required is

    (a) friction angle should be greater than the angle of bite.

    (b) friction angle should be less than the angle of bite.

    (c) friction angle should be equal to the angle bite

    (d) sum of the friction angle and the angle of bite should be equal to the neutral angle

    [GATE 2005]

33. Patenting process is a

    (a) wire drawing process

    (b) heat treatment for drawn wires

    (c) coating process for wires

    (d) deep drawing process.

    [GATE 2005]

34. Match the items in group 1 with the manufacturing process in group 2.

    Group 1

    P. I-section for structural work

    Q. Base plates for heavy machines

    R. Self lubricating bearings

    S. Rotor for turbine

    Group 2

    1. Forging

    2. Rolling

    3. Casting

    4. Powder metallurgy

    5. Extrusion

    (a) P-2, Q-3, R-4, S-l

    (b) P-5, Q-4, R-2, S-3

    (c)P 1, Q-3, R-5, S-4

    (d) P-4, Q-l, R-3, S-5

    [GATE 2005]

35. Match the items in group 1 with the manufacturing processes in group 2.

    Group 1

    P. Upsetting operation

    Q. Port hole die

    R. Bull block

    S. Orange peel effect

    Group 2

    1. Extrusion

    2. Forging

    3. Rolling

    4. Wire drawing

    5. Deep drawing

    6. Sheet forming

    (a) P-2, Q-6, R-3, S-l

    (b) P-5, Q-2, R-l, S-6

    (c) P-4, Q-6, R-3, S-3

    (d) P-2, Q-l, R-4, S-5

    [GATE 2005]

Answers

31. (a) 32. (a) 33. (b) 34. (a) 35. (d)

36. A metallic sheet 1 m wide and 10 mm thick is to be rolled under a front tension force of 100 KN. A specimen from the sheet, tested under tension, shows a yield stress of 100 MPa. Assume plain strain condition during rolling and ignore the friction. If yielding is assumed to start at a roll pressure of $\sigma 3$ MPa, then the stress in the width direction (where strain is zero) is

(a) $2 + \sigma 3/2$ (b) $4 + \sigma 3/5$
(c) $5 + \sigma 3/2$ (d) $5 + \sigma 3/3$

[GATE 2005]

37. In Q. No. 36. according to Von Misses criterion of yielding, the minimum roll pressure at which rolling may be carried out is approximately .... MPa.

(a) 10.7 (b) 108.6
(c) 231.2 (d) 301.5

[GATE 2005]

38. The maximum possible reduction in a single pass for cold rolling of a 200 mm slab is (given the co-efficient of friction 0.1 and roll diameter is 400 mm) .... mm.

(a) 5 (b) 3
(c) 2 (d) 1

[GATE 2010]

39. Tungsten filament wire for lamp industry is commonly produced by

(a) powder metallurgy and deep drawing
(b) powder metallurgy and welding
(c) casting and welding
(d) casting and forging

[GATE 2006]

40. Match the following failure/fracture observations in group 1 with the likely deformation/process in group 2.

Group 1
P. Cup and cone
Q. Beach marks
R. Clevage
S. Triple point cracking

Group 2
1. Creep failure
2. Brittle failure
3. Normal ductile failure
4. Completely ductile failure
5. Fatigue failure

(a) P-4, Q-l, R-5, S-2
(b) P-3, Q-5, R-2, S-l
(c) P-5, Q-4, R-l, S-2
(d) P-3, Q-2, R-5, S-l

[GATE 2006]

41. For ideal defect free deep drawability of material, it should have

(a) low plastic anisotropy and low planer anisotropy
(b) low plastic anisotropy and high planer anisotropy
(c) high plastic anisotropy and low planer anisotropy
(d) high plastic anisotropy and high planer anisotropy

[GATE 2006]

42. Identify the statement that precisely describes the principle of hot working of metals and alloys.

(a) It is mechanical deformation carried out above the temperature of recrystallisation
(b) It is mechanical deformation carried out above the room temperature
(c) It is mechanical deformation carried out above the annealing temperature
(d) It is mechanical deformation carried out just below the melting temperature

[GATE 2006]

Answers

36. (c) 37. (a) 38. (c) 39. (a) 40. (b) 41. (d)
42. (a)

43. Identify the correct metal forming operations in group 1 with the product in group.

Group 1

P. Drawing Q. Cold forging
R. Extrusion S. Swaging

Group 2

1. Tailoring needles
2. Beverage cans
3. Radiator caps
4. Steel half domes

(a) P-l, Q-4, R-2, S-3
(b) P-l, Q-2. R-3, S-4
(c) P-2, Q-3. R-4, S-l
(d) P-1, Q-3, R-2, S-4

[GATE 2006]

44. Match the product in group 1 and the main manufacturing process in group 2

Group 1

P. Cement kilns Q. Gas pipelines
R. Cycle rims S. Seamless pipe

Group 2

1. Rolling and welding
2. Casting and welding
3. Forging and welding
4. Extrusion and welding
5. Forging and extrusion

(a) P-l. Q-2. R-3, S-5
(b) P-2, Q-l, R-4, S-5
(c) P-2, Q-5, R-4, S-l
(d) P-2, Q-4, R-5, S-l

[GATE 2006]

45. Match the fracture processes in group 1 to the fracture surface morphologies in group 2.

Group 1

P. Ductile fracture Q. Brittle fracture
R. Fatigue fracture

Group 2

1. Clevage 2. Dimples
3. Striations 4. Veins

(a) P-4, Q-2, R-3
(b) P-2, Q-l, R-3
(c) P-2, Q-3, R-l
(d) P-4, Q-4, R-3

[GATE 2007]

46. For ingot breakdown by hot rolling, the rolls are generally grooved paralled to the roll axis in order to

P. increase the angle of bite
Q. decrease the rolling load
R. achieve larger reduction
S. decrease roll flattening

(a) P, Q (b) Q, R
(c) P, R (d) Q, S

47. The pressure required to maintain flow during indirect extrusion

P. decreases with decreasing rate of the billet
Q. Increases with increasing extrusion ratio
R. is independent of the length of the billet
S. is independent of the extrusion ratio

(a) P, Q (b) Q. R
(c) R, S (d) P, S

[GATE 2007]

48. Match the forming methods in group I with the defects in group II.

Group I

P. Extrusion Q. Closed die forging
R. Rolling

Group II

1. Flash cracking 2. Fir-tree cracking
3. Alligatoring 4. Earing

(a) P-l. Q-2, R-3
(b) P-4, Q-3, R-l
(c) P-3, Q-l, R-4
(d) P-2, Q-l, R-3

[GATE 2007]

Answers

43. (b) 44. (b) 45. (b) 46. (d) 47. (a) 48. (d)

49. Match the terms in group 1 to their descriptions in group 2.
Group 1
P. Hall-Petch effect Q. Bauschinger effect
R. Cottreli atmosphere
Group 2
1. Solute dislocation interaction
2. Dislocation multiplication
3. Grain boundary strengthening
4. Barelling under compression
5. Mechanical hysterisis during plasticity
(a) P-3, Q-5, R-l
(b) P-l, Q-4, R-3
(c) P-5, Q-l, R-2
(d) P-3, Q-4, R-2
[GATE 2007]

50. The yield point phenomenon observed in annealed low carbon steels is due to the presence of
(a) silicon (b) chromium
(c) phosphorous (d) carbon.
[GATE 2008]

51. In a tensile test of ductile material, necking starts at
(a) lower yield stress
(b) upper yield stress
(c) ultimate tensile strength
(d) just before fracture
[GATE 2008]

52. Fatigue resistance of a steel is reduced by
(a) decarburisation
(b) polishing the surface
(c) reducing the grain size
(d) shot peening
[GATE 2008]

53. The defects that are not observed in extruded products are
P. Chevron cracking Q. Fold
R. Piping S. Surface cracking
T. Alligatoring
(a) P, Q (b) R, T
(c) P, S (d) Q, T
[GATE 2008]

54. Match the mechanical behaviour in group 1 with the terms in group 2.
Group 1
P. Low cycle fatigue Q. Creep
R. Impact toughness S. Stretcher strains
Group 2
1. Charpy test
2. Portevin-LeChatelier effect
3. Coffin-Manson equation
4. Larson - Miller parameter
5. Jominy test
(a) P-2, Q-4, R-l, S-5
(b) P-2, Q-l, R-5, S-3
(c) P-3. Q-4, R-l, S-2
(d) P Q-l, R-4, S-5
[GATE 2008]

55. Match the processes given in group 1 with corresponding typical defects in group 2.
Group 1
P. Forging Q. Rolling
R. Deep drawing S. Extrusion
Group 2
1. Alligatoring 2. Cold shut
3. Chevron Cracks 4. Wrinkles
(a) P-l, Q-2, R-3, S-4
(b) P-2, Q-l, R-4, S-3
(c) P-2, Q-l, R-3, S-4
(d) P-3, Q-l, R-4, S-2
[GATE 2009]

Answers

49. (a) 50. (d) 51. (d) 52. (a) 53. (d) 54. (c)
55. (b)

56. Match the loading conditions in group 1 with the characteristics in group 2.

Group 1

P. Tensile Q. Compressive

R. Fatigue S. Creep

Group 2

1. Barelling
2. Intergrannular cracking
3. Striations
4. Cup and cone
5. Earing

(a) P-4, Q-5, R-3, S-l
(b) P-4, Q-l, R-3, S-2
(c) P-5, Q-l, R-4, S-2
(d) P-l, Q-2, R-3, S-2

[GATE 2009]

57. In fracture toughness characterised by K_{IC} or J_{IC}, I in the subscript indicates loading by
(a) cracj opening mode
(b) forward shear mode
(e) parallel shear mode
(d) perpendicular shear mode

[GATE 2011]

58. Match the manufacturing process of group I to be used for producing the product for group II.

Group I

P. Drawing Q. Forging

R. Rolling S. Stretch forming

Group II

1. Large curved disc
2. Tube
3. Crank shaft
4. Plate

(a) P-2, Q-3, R-4, S-l
(b) P-l, Q-4, R-3, S-2
(c) P-3, Q-2, R-l S-4
(d) P-4, Q-l, R-2, S-3

[GATE 2011]

59. An aluminium billet of 300 mm diameter is extruded with an extrusion ratio of 16. What is the diameter of the final product ?
(a) 150 mm (b) 75 mm
(c) 59 mm (d) 19 mm

[GATE 2011]

60. In Q.No. 59, What is the ideal extrusion pressure, if the effective flow stress in compression is 250 MPa?
(a) 693 MPa (b) 346 MPa
(c) –346 MPa (d) –703 MPa

[GATE 2011]

61. The flow stress (σ) of an alloy varies with strain rate (ϵ) as $\sigma = 100\ (\epsilon)^{0.1}$ MPa. When the alloy is hot extruded from 10 cm diameter to 5 cm diameter at a speed of 2 cm/s, the flow stress is
(a) 1000 MPa (b) 105 MPa
(c) 150 MPa (d) 1050 MPs

[GATE 2012]

62. Determine the correctness or otherwise of the following. Assertion (a) and Reason (r).
Assertion : During rolling, front tension and (or) back tension are (is) employed to decrease rolling load.
Reason : Roll pressure decreases due to lowering of flow stress as a result of front tension/back tension.
(a) a is false but r is true.
(b) a is true and r is also true, but r is not the reason for a.
(c) a is true and r is also true, and r is the reason for a.
(d) a is true but r is false.

[GATE 2012]

Answers

56. (b) 57. (a) 58. (a) 59. (b) 60. (a) 61. (b)
62. (c)

63. Match the defects listed in Group I with the processes listed in Group II.

Group I

P. Cold Shut Q. Earing

R. Alligatoring S. Shrinkage porosity

Group II

1. Rolling 2. Forging

3. Deep drawing 4. Fusion welding

(a) P-2, Q-4, R-1. S-4

(b) P-2, Q-4, R-3, S-1

(c) P-2, Q-3, R-1, S-4

(d) P-4, Q-1, R-2, S-3

[GATE 2012]

64. The strain hardening behaviour of an annealed rod during cold rolling is given by $\sigma = 700\ (\epsilon)^{0.2}$ MPa. where σ is the flow stress at strain ϵ.

Flow stress after 50% reduction in area of the annealed rod on cold rolling is approximately

(a) 750 MPa (b) 650 MPa

(c) 609 MPa (d) 559 MPa

[GATE 2012]

65. In Q. 64, if a wire of 5 mm diameter is drawn from the above cold rolled rod of 10 mm diameter, the drawing stress, neglecting the effect of friction and redundant work, is approximately

(a) 650 MPa (b) 550 MPa

(c) 450 MPa (d) 400 MPa

[GATE 2012]

66. A 480 mm thick slab is hot-rolled using a roll of 720 mm diameter. For a coefficient of friction of 0.5, the maximum possible reduction (in mm) is

(a) 90 (b) 180

(c) 240 (d) 360

[GATE 2013]

67. Match the defects listed in Group I with the corresponding manufacturing process listed in Group II.

Group I

P. Orange-peel effect

Q. Chevron cracking

R. Flash

S. Undercut

Group II

1. Extrusion 2. Deep drawing

3. Arc welding 4. Forging

(a) P-1, Q-2, R-4, S-3

(b) P-2, Q-3, R-1, S-4

(c) P-3, Q-4, R-2, S-1

(d) P-2, Q-1, R-4, S-3

68. From a 2 m × 1.2 m sheet, squares are cut out from each of the four corners as shown in the figure and then the sides are bent to form an open box. The maximum possible volume (in m³) of the box is

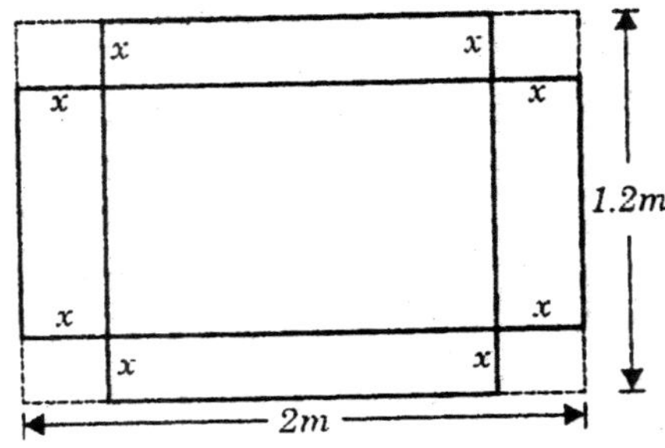

(a) 0.14 (b) 0.26

(c) 0.54 (d) 0.82

[GATE 2013]

69. Which one of the following metal forming processes is the earing defect associated with?

(a) Deep drawing (b) Rolling

(c) Forging (d) Wire drawing

[GATE 2014]

Answers

63. (c) 64. (b) 65. (b) 66. (a) 67. (d) 68. (b)
69. (a)

70. A metal powder with an apparent density of 2.5 g/cm^3 was compacted in a cylindrical die to a green density of 5.5 g/cm^3. If the height of the green sample was 12 mm, then the fill height of powder (in mm) would have been
(a) 26.4 (b) 18.6
(c) 42.8 (d) 56.8
[GATE 2014]

71. Which of the following metal working operations can be categorized as an indirect compression process?
(a) Forging (b) Wire drawing
(c) Extrusion (d) Stretch forming
[GATE 2015]

72. Which of the following is a typical rolling defect?
(a) Buckling (b) Edge cracking
(c) Cold shut (d) Porosity
[GATE 2015]

73. Which of the following metal forming techniques is used to produce soft drink cans from aluminium sheets?
(a) Rolling (b) Forging
(c) Deep drawing (d) Extrusion
[GATE 2015]

74. If a cylindrical billet of height 1.0 m and diameter 0.5 m is upset forged to form a cylindrical pancake of height 0.25 m, the diameter of the pancake (in m) is
(a) 0.4 (b) 0.6
(c) 1.0 (d) 1.8
[GATE 2015]

75. Railway tracks are typically manufactured using
(a) Forging (b) Extrusion
(c) Deep Drawing (d) Rolling
[GATE 2016]

76. Aluminium powder having an apparent density of 810 kg.m^{-3} is compacted in a cylindrical die at 600 Mpa. The density of the as-pressed aluminium compact is 1755 kg.m^{-3}. If the height of the as-pressed compact is 12 mm. the fill height (in mm) required is ....
[GATE 2016]

77. A rolling mill has a roll diameter of 200 mm. If coefficient of friction is 0.1. then the maximum possible reduction (in mm) during rolling of a 250 mm thick plate is ....
[GATE 2016]

78. Match the deformation processes in Column I with the corresponding stress states listed in Column II.
Column I
P. Wire Drawing
Q. Forging
R. Stretch Forming
S. Cutting
1. Direct Compression
2. Indirect Compression
3. Tension
4. Shear
(a) P-1, Q-2, R-3, S-4
(b) P-1, Q-2, R-4, S-3
(c) P-2, Q-l, R-3, S-4
(d) P-2, Q-1, R-4, S-3
[GATE 2016]

Answers

70. (a)	71. (c)	72. (b)	73. (c)	74. (c)	75. (d)
76. (25.5 to 26.5)	77. (0.9 to 1.1)	78. (c)			

7

Physical Metallurgy

1. In precipitation hardenable alloy like duralumin, intermediate precipitates can form due to
   (a) difficulty of nucleation of the final precipitate.
   (b) difficulty of the growth of the final precipitate.
   (c) ease of diffusion.
   (d) coherency strain.

[GATE 91]

2. Which of the following phenomenon/ phenomena is/are diffusion controlled?
   (a) Dislocation climb
   (b) Cross-slip
   (c) Twinning
   (d) Recrystallisation
   (e) all a, b, c, d

[GATE 91]

3. It is possible to form martensite
   (a) sometimes by cold working
   (b) sometimes without nucleation
   (c) sometimes by gas quenching
   (d) only in iron-based alloys

[GATE 91]

4. Martensitic transformation in plane carbon steel occurs by
   (a) diffusion of carbon
   (b) no diffusion of carbon
   (c) shear mechanism
   (d) diffusion of iron

[GATE 92]

5. A binary alloy system A-B shows the peritectic reaction, $L + \alpha\ (\rightarrow \beta)$. An alloy of the peritectic composition when cooled under natural cooling conditions will show coring in
   (a) α phase only
   (b) β phase only
   (c) both α and β phases
   (d) none of the phases

[GATE 92]

6. With increase in annealing temperature, the following defect density decreases.
   (a) Vacancy (b) Dislocation
   (c) Grain boundary (d) All of them

[GATE 92]

7. Solute atoms which cause yield point phenomenon in mild steel are/is
   (a) Aluminium (b) Boron
   (c) Carbon (d) Nitrogen

[GATE 92]

Answers

1. (d)	2. (e)	3. (d)	4. (b)	5. (b)	6. (d)
7. (c & d)					

8. Martensite in steels is
   (a) an interstitial solid solution of C in alpha iron
   (b) a supersaturated interstitial solution of C in BCT iron
   (c) a supersaturated solid solution of C in gamma iron
   (d) a very finely dispersed lamellar structure

[GATE 94]

9. Yield strength of a polycrystalline metal with an average grain size, d, is proportional to
   (a) $d^{1/2}$
   (b) $d^{-1/2}$
   (c) d
   (d) d^{-1}

[GATE 94]

10. A number of solid state phase transformations follow a sigmoidal pattern. In these cases, at any time the fraction transformed can be expressed as
   (a) $1 - \exp(-\alpha^2 t)$
   (b) $1 + \exp(-\alpha^2 t)$
   (c) $\exp(\alpha^2 t)$
   (d) $\exp(\alpha^2 t) - 1$

[GATE 94]

11. A peritectic reaction is
   (a) $\alpha + \beta \rightarrow \gamma$
   (b) $L + \alpha \rightarrow \beta$
   (c) $L_1 + L_2 \rightarrow \beta$
   (d) $L + \alpha + \beta \rightarrow y$

[GATE 94]

12. Eutectic Al-Si alloys can be modified by small additions of
   (a) Na
   (b) Mg
   (c) B
   (d) Cr
   (e) Cu

[GATE 94]

13. Other parameters remaining same, the recrystallisation temperature of an alloy is lowered, when
   (a) strain rate is increased
   (b) grain size is increased
   (c) prior cold deformation is increased
   (d) not affected by any of the above parameters.

[GATE 94]

14. Match the basic strengthening mechanisms in the following materials.

Group 1

P. 60:40 brass is stronger than 70:30 brass.

Q. An automobile gear made from 0.6% C steel and quenched in water from a temperature of 850°C.

R. Tungsten filament drawn at a temperature of 400°C.

S. A ball bearing heated in furnace in ammonia atmosphere.

Group 2

1. Case hardening
2. Strain hardening
3. Martensitic strengthening
4. Second phase strengthening

(a) P-4, Q-3, R-2, S-l
(b) P-l, Q-2, R-3, S-4
(c) P-4, Q-3, R-l, S-2
(d) P-3, Q-4, R-2, S-l

[GATE 94]

Answers

8. (b) 9. (b) 10. (b) 11. (b) 12. (c) 13. (a & c) 14. (a)

15. Match the microstructural state of the following steels with their respective heat treatments

Group 1

P. Eutectoid steel homogenised at 900°C and quenched in water at room temperature

Q. A 0.4% C steel soaked at 750°C and quenched in water at room temperature.

R. Normalised 1%C steel.

S. Eutectoid steel subjected to normalising

Group 2

1. Ferrite, martensite and retained austenite
2. Pearlite and carbide
3. Martensite and retained austenite
4. Pearlite

(a) P-4, Q-3, R-l, S-2
(b) P-3, Q-4, R-2, S-l
(c) P-4, Q-3, R-2, S-l
(d) P-l, Q-2, R-4, S-3

[GATE 94]

16. In a dilute solid solution of nickel and carbon in y-iron.

(a) $D_c > D_{NI}$ (b) $D_c < D_{NI}$
(c) $D_c > D_{Fe}$ (d) $D_c = D_{NI}$

[GATE 95]

17. The primary strengthening mechanism in 70:30 brass is

(a) solid solution strengthening
(b) precipitation hardening
(c) dispersion strengthening
(d) order hardening

[GATE 95]

18. Dislocation cross-slip is difficult in those materials which have

(a) large number of slip systems
(b) high work hardening rate
(c) coarse grain size
(d) low stacking fault energy

[GATE 95]

19. In a binary isomorphous system A-B, constitutional supercooling can occur in the

(a) metal with higher melting point
(b) metal with lower melting point
(c) solid solution
(d) all of the above

[GATE 95]

20. Malleabilisation heat treatment is performed on

(a) cast steel
(b) gray cast iron
(c) white cast iron
(d) spheroidal graphite cast iron

[GATE 95]

21. The following reactions are written such that the forward reaction occurs during cooling. Match the reactions to their names.

Group 1

P. $Y = \alpha + \beta$

Q. $L + \alpha = \beta$

R. $L_1 = L_2 + \alpha$

S. $\alpha + \beta = Y$

Group2

1. peritectoid
2. Eutectoid
3. Peritectic
4. Monotectic

(a) P-2, Q-3, R-4, S-l
(b) P-2, Q-3, R-2, S-4
(c) P-3, Q-2, R-4, S-l
(d) P-4, Q-3, R-2, S-l

[GATE 95]

Answers

15. (c) 16. (d) 17. (a) 18. (d) 19. (c) 20. (c)
21. (a)

22. A 0.4 wt% C steel is austenitised at 900°C for 1 hour and heat treated to get different inicrostructures. Match the following microstructures to the heat treatments.
    P. Martensite + Retained austenite
    Q. Ferrite + fine pearlite
    R. Ferrite + spheroidal cementite
    S. Ferrite + coarse pearlite
    1. Annealing
    2. Quenching
    3. Quenching + temperature at 600°C
    4. Normalising
    (a) P-4, Q-3, R-2, S-l
    (b) P-l, 0-2, R-3, S-4
    (c) P-2, Q-4. R-l, S-3
    (d) P-2, Q-4, R-3, S-l

    [GATE 95]

23. The low angle grain boundary occurs when the orientation difference between the adjacent grains is of the order of
    (a) 100° (b) 10°
    (c) 1° (d) none of these.

    [GATE 96]

24. The elastic strain energy of a unit length of an edge dislocation as compared to that of a screw dislocation is
    (a) more (b) equal
    (c) less (d) double.

    [GATE 96]

25. Substantial amount of bainite can form during continous cooling in
    (a) unalloyed low carbon steel
    (b) unalloyed high carbon steel
    (c) alloy steel
    (d) none of these

    [GATE 96]

26. A 0.2% C steel is equilibriated just above the eutectoid temperature and then quenched in iced brine. The room temperature microstructure will consist of
    (a) 77% ferrite and 23% martensite
    (b) 76% ferrite and 24% pearlite
    (c) 60% ferrite and 40% martensite
    (d) 97% ferrite and 3% pearlite

    [GATE 96|

27. Match the following fractographic features with the type of fracture with which they are associated.
    P. Striations
    Q. Separated grain facets
    R. Dimples
    S. Facets, river pattern
    1. Ductile fracture
    2. Brittle fracture
    3. Fatigue fracture
    4. Intergrannular fracture
    (a) P-3, Q-4. R-l, S-2
    (b) P-3, Q-4, R-2, S-l
    (c) P-4, Q-3, R-2, S-l
    (d) P-l, Q-2, R-3, S-4

    [GATE 96]

28. Match the following.
    P. Portevin-Le Chatlier effect
    Q. Bauschinger effect
    R. Elastic after effect
    S. Modulus effect
    1. Dislocation pile up
    2. Internal friction
    3. Precipitation hardening
    4. Dynamic strain ageing
    (a) P-4, Q-3, R-2, S-l
    (b) P-4, Q-2, R-l, S-3
    (c) P-4, Q-3, R-l, S-2
    (d) P-l, Q-2, R-3, S-4

    [GATE 96]

Answers

22. (d)	23. (c)	24. (a)	25. (c)	26. (b)	27. (a)
28. (c)					

29. Ellingham diagram for M - MO_x reaction is a plot of
(a) ΔG Vs . T (b) ΔG° Vs . T
(c) Δ G Vs . 1/T (d) Δ G° Vs 1/T
[GATE 98]

30. Spheroidal graphite is obtained in cast irons by
(a) inoculating the melt with ferrosilicon.
(b) adding nickel to the melt in the form of round shots.
(c) treating the melts of controlled compositions with magnesium.
(d) annealing of flake graphite cast iron.
[GATE 98]

31. The net effect of continuous cooling on the TTT curves is that the curves are shifted to the
(a) right and lowered
(b) left and lowered
(c) right and raised
(d) left and raised
[GATE 98]

32. In age hardenable alloys maximum ductility is obtained
(a) in as cast state
(b) immediately after solution treatment and subsequent quenching
(c) after optimum ageing
(d) after overageing.
(GATE 98]

33. The yield point phenomenon observed in annealed low carbon steel is due to the presence of the following element.
(a) silicon (b) carbon
(c) phosphorous (d) chromium
[GATE 98]

34. Match the following.
Group 1
P. Hardness of martensite in steels
Q. Hardness of pearlite in steels
R. Hardness of tempered martensite in steels
S. Hardness of α° brasses
Gruop 2
1. Extent of cold working
2. Particle size of the resulting carbide
3. Interlamellar spacing
4. Weight per cent of carbon
(a) P-4, Q-3, R-l, S-2
(b) P-4, Q-3, R-2, S-l
(c) P-3. Q-4, R-l, S-2
(d) P-l, Q-2, R-3, S-4
[GATE 98]

35. Match the following.
Gruop 1
P, Line defect
Q. Point defect
R. Area defect
S. Volume defect
Group 2
1. Vacancy
2. Grain boundary
3. Voids
4. Dislocations
(a) P-4, Q-l, R-2, S-3
(b) P-4, Q-2, R-l, S-3
(c) P-4, Q-3, R-2, S-l
(d) P-l, Q-2, R-3, S-4
[GATE 98]

Answers

29. (b) 30. (c) 31. (c) 32. (c) 33. (b) 34. (b)
35. (a)

36. Match the following.

Group 1

P. Hot shortness Q. Cold shortness

R. Season cracking S. Sensitization

Group 2

1. Overcome by adding excessive amount of manganese
2. Weld decay
3. Reduction in impact strength at low temperatures
4. Tendency of brasses to corrode due to high residual stress.

(a) P-l, Q-3, R-4, S-2 (b) P-4, Q-3, R-2, S-l

(c) P-l, Q-3, R-2, S-4 (d) P-l, Q-2, R-3, S-4

[GATE 98]

37. The relation between dendrite arm spacing (x) and the solidification time (t) is given by (A and *n* are constants)

(a) $x = At^n$ (b) $x = At^{-n}$

(c) $x = At^{-1/n}$ (d) $x = At^{1/n}$

[GATE 99]

38. Stretcher strains found in a low carbon sheet are associated with

(a) texture

(b) dislocation density

(c) yield point phenomenon

(d) thickness of the sheet

IGATE 99]

39. In an annealed metal, the density of dislocations is typically of the order of .... m^{-2}.

(a) 10^4 (b) 10^8

(c) 10^2 (d) 10^{13}

[GATE 99]

40. The hardenability of steels decreases with

(a) decrease in dislocation density

(b) increase in solutionising temperature

(c) increase in strength

(d) decrease in grain size.

[GATE 2000]

41. Movement of jogs can produce

(a) vacancies

(b) interstitial

(c) grain boundary sliding

(d) grain boundary migration

[GATE 2000]

42. Fine grain size in metallic materials will

(a) increase the yield strength

(b) increase the creep strength

(c) increase the fatigue strength

(d) decrease the impact strength.

[GATE 2000]

43. Internal cracks in drawn bars are due to

(a) secondary tensile stresses

{b) temperature gradient in the work piece

(c) heated dies and grips

(d) internal compressive residual stresses.

[GATE 2000]

44. Secondary hardening in steels arises out of the

(a) precipitation of fine alloy carbides at high temperature

(b) refinement of ferrite grain size by working

(c) decomposition of retained austenite upon heat treatment

(d) precipitation of complex inter-metallics upon heat treatment.

[GATE 2003)

45. During decarburising of a plane carbon steel, the thickness of ferrite layer growth is proportional to

(a) time (b) square root of time

(c) square of time (d) cube of time

[GATE 2004]

Answers

36. (a)	37. (c)	38. (c)	39. (c)	40. (d)	41. (a)
42. (a & c)	43. (a)	44. (c)	45. (b)		

46. The peritectic reaction in binary systems is given by
    (a) L = α + β (b) α = L + β
    (c) Y = α + β (d) L + α = β

[GATE 2C04]

47. The energetic driving force for grain growth is due to
    (a) dislocations in the matrix
    (b) grain boundary energy
    (c) residual strains in different grains
    (d) stacking fault energy.

[GATE 2004]

48. The self diffusion in FCC metals occurs by one of the following mechanisms.
    (a) Ring (b) Interstitial
    (c) Vacancy (d) Interstitialcy

[GATE 2004]

49. Which of the following microstructures of the eutectic is not observed in binary alloys?
    (a) Lamellar (b) Widmanstatten
    (c) Acicular (d) Nodular

[GATE 20041

50. An Fe – 3 wt % C – 1 wt % Si alloy is cooled very slowly from the liquid state to a temperature of 1023 K. Thereafter, it is cooled in air. The microstructure at room temperature will consist of
    (a) ferrite + graphite
    (b) pearlite + graphite
    (c) martensite + graphite
    (d) ferrite + pearlite

[GATE 2004]

51. A liquid alloy of the eutectic composition (L) undergoes the eutectic reaction to give α and β phases of compositions a and b respectively. The ratio of the relative amounts of α and β phases is
    (a) (b – L)/(L – a)
    (b) (L – a)/(b – L)
    (c) (b – LI)/(b – a)
    (d) (L – a)/(b – a)

[GATE 2004]

52. The interlamellar spacing, S, and the undercooling, ΔT below the eutectoid temperature in plane carbon steels are related as
    (a) S α ΔT
    (b) S α ΔT^2
    (c) S α ΔT^{-1}
    (d) S α $\Delta T^{-1/2}$

[GATE 2004]

53. The growth rate of pearlite, v, and the interlamellar spacing S are given by the following relation, $v = k.\frac{1}{S}\left(\frac{1 - S_c}{S}\right)$ where S_c is the critical spacing at which the growth rate is zero and k is a materials constant.The actual interlamellar spacing observed will correspond to
    (a) $S=S_C$ (6) $S = 2S_c$
    (c) $S = 3S_c$ (d) $S = S_C^2$

[GATE 2004]

54. A copper sample has been metallographically analysed for determining its mean grain size. It is found to have an ASTM grain size number of 5. The number of grains per mm^2 in the sample will be
    (a) 72 (b) 124
    (c) 248 (d) 496

[GATE 20041

Answers

46. (d)	47. (b)	48. (c)	49. (b)	50. (b)	51. (a)
52. (c)	53. (a)	54. (c)			

55. Choose the correct statements.
    P Interstitial atoms diffuse slower than substitutional atoms
    Q. In pure metals the vacancy concentration increases with temperature
    R. Martensitic transformation is athermal in nature
    S. Atoms in the grain boundary diffuse slower than in the bulk at relatively lower temperatures.
    (a) P. Q (b) Q, R
    (b) Q, S (d) R,S

[GATE 2004]

56. In the homogenous nucleation, as compared to heterogeneous nucleation
    (a) undercooling is more and the critical radius of the spherical surface of the nucleus r_{cr} is smaller
    (b) undercooling is more and r_{cr} is larger.
    (c) undercooling is smaller and r_{cr} is smaller
    (d) undercooling is smaller and r_{cr} is larger

[GATE 98]

57. Maraging steels derive their strength from the following mechanism:
    (a) A line, highly dislocated and strong martensite
    (b) Fine dispersions of intermetallics of Fe, Ni, Ti etc.
    (c) Fine dispersions of alloy carbides in a ferrite matrix
    (d) Fine dispersions of Fe_3C nucleated on dislocations in austenite.

[GATE 2001]

58. Metal A nucleates as a spheres in a melt. Assuming Y(liquid/solid surface energy) = 200mJ/m^2 and ΔGv (change in volume free energy) = -10^8 J/M^3 the critical radius (in nm) for stable nuclei is
    (a) 0.25 (b) 0.5
    (c) 2 (d) 4

[GATE 2004]

59. The plane strain fracture toughness (K_{IC}) and yield strength of a material are 100 MN$^{-3/2}$m and 500 MN^{-2}m respectively. The minimum plate thickness to determine K_{IC} is .... cm.
    (a) 1 (b) 10
    (c) 50 (d) 100

[GATE 95]

60. A 10 mm thick En-24 steel yielded a valid plane strain fracture toughness of 80 Mpa $\sqrt{m}$. The minimum yield strength of the steel is .... MPa.
    (a) 40 (b) 1265
    (c) 1000 (d) 126.5

[GATE 96)

61. Von Mises criterion for plastic yielding of a ductile metal predicts that the yield stress in uniaxial tension is related to that in pure torsion as
    (a) equal to each other
    \(b) $\sqrt{3}$ times
    (c) 2 times
    (d) one half

[GATE 2001]

62. An A1 4.5 mass% Cu alloy solutionising at 500°C, quenched to room temperature and aged at 200°C for 2 hrs is found to exhibit an increase in hardness. The primary mechanism contributing to increased hardness is
    (a) martensitic transformation due to quenching
    (b) solid solution strengthening
    (c) precipitation hardening
    (d) point defect strengthening

[GATE 2005]

Answers

55. (b)	56. (a)	57. (a)	58. (d)	59. (b)	60. (b)
61. (b)	62. (c)				

63. Which of the following is true for martensitic transformation in steels?
    (a) Symmetry of the product is same as that of the parent phase but the unit cell volume decreases.
    (b) Symmetry of the product is different from the parent phase and the unit cell column decreases.
    (c) Symmetry of the product is different from the parent phase but the unit cell volume increases.
    (d) Symmetry of the product is same as that of the parent phase but unit cell volume increases.

[GATE 2005]

64. Under application of stress, when a straight dislocation (radius of curvature, $r = \infty$) tries to bow out around precipitates of spacing L, there is an instability during changing of curvature at
    (a) $r = L$ (b) $r=L/2$
    (c) $r = L/3$ (d) $r = L/A$

[GATE 2005]

65. During load Vs load-line displacement measurement for the determination of J_{10} compliance is measured at intervals to estimate
    (a) strain energy
    (b) crack radius
    (c) crack length
    (d) crack opening displacement

[GATE 200]

66. For the same volume fraction, size and size distribution of precipitates, the highest strength in a precipitation-hardened material is obtained when the precipitates are
    (a) uniformly distributed in annealed matrix
    (b) distributed along grain boundaries
    (c) nucleated on dislocation substructure
    (d) nucleated at impurities and inclusions

[GATE 2005]

67. Upon addition of sodium oxide to molten silicate slag, the viscosity of the slag
    (a) decreases because the oxygen ions supplied by the sodium oxide break the chain structure of the silicate slag.
    (b) increases because the oxygen ion supplied by the sodium oxide further add to the oxygen network in the slag.
    (c) remains unaffected because fundamentally one can not distinguish between the oxygen ions supplied by the sodium oxide and those already present in the slag.
    (d) remains unaffected because the oxygen ions fill up the holes and vacancies present in the liquid slag.

[GATE 2005]

68. Stainless steel A has a stacking fault energy of 8 $mJ.m^{-2}$ and stainless steel B has a stacking fault energy of 45 $mJ. m^{-2}$. The correct statement out of the following are.
    P. A will have a wider stacking faults than B
    Q. A will strain harden more rapidly than B
    R. A and B will develop similar type of substructure
    S. A and B will have the same temperature dependence of flow stress.
    (a) P, Q, R (b) P, Q
    (c) P, R, S (d) P, S

[GATE 2005]

69. Gold has surface free energy of 1.5 $J.m^{-2}$ and grain boundary energy of 0.365 $J.m^{-2}$ The groove angle that forms where a grain boundary intersects the free surface in a well annealed sample is
    (a) 42° (b) 83°
    (c) 166° (d) 180°

[GATE 2005]

Answers

63. (c)	64. (b)	65. (d)	66. (c)	67. (a)	68. (b)
69. (c)					

70. In order to double its yield strength, the grain size of a steel of ASTM grain size number 2 should be changed to ASTM grain size number (neglect friction stress)

(a) 1 (b) 4
(c) 6 (d) 8

[GATE 2005]

71. Match the items in Group 1 with those in Group 2.

Group 1
P. Fatigue
Q. Hall-petch equation
R. Creep
S. Plastic deformation

Group 2
1. Cracking at extrusions and intrusions
2. Dislocation pile up
3. Slip and twinning
4. Grain boundary sliding
5. Angle of grain boundary
6. Young's modulus

(a) P-l, Q-4, R-5, S-6
(b) P-3, Q-5, R-2, S-6
(c) P-l, Q-2, R-4. S-3
(d) P-4, Q-2, R-3, S-l

[GATE 2005]

72. According to Griffith theory, which of the following are true for the fracture stress in a brittle material?

P. Increases with increasing surface energy
Q. Decreases with increasing elastic modulus.
R. Decreases with increasing length of pre-existing crack
S. Decreases with increasing number of pre-existing cracks of same size

(a) P, R, S (b) Q, R, S
(c) P, R (d) P, S

[GATE 2005]

73. Match the materials in group 1 with the possible microstructure adjacent to the fusion line in the heat affected zone, in group 2.

Group 1
P. Pure aluminium
Q. 0.6 mass% carbon steel
R. Aged Al 4.5 mass% Cu alloy
S. 18/8 stainless steel

Group 2
1. Martensite
2. Grain boundary precipitation
3. Coarse precipitates
4. Bainite
5. Coherent precipitates
6. Twins
7. Grain growth

(a) P-6, Q-l, R-5, S-2
(b) P-7, Q-l, R-2, S-4
(c) P-4, Q-3, R-2, S-6
(d) P-7, Q-l, R-3, S-2

[GATE 2005]

74. A steel contains 38.5 mass % pro-eutectoid ferrite and 61.3 mass % austenite at eutectoid temperature (723°C). The carbon content of steel, in mass %, is

(a) 0.35
(b) 0.40
(c) 0.45
(d) 0.50

[GATE 2005]

75. In Q. No, 74, the total ferrite content, in mass %, in the above steel at room temperature is

(a) 87.5
(b) 92. 5
(c) 95.5
(d) 97.5

[GATE 2005]

Answers

70. (c)	71. (c)	72. (b)	73. (b)	74. (d)	75. (a)

76. Driving force for grain growth after completion of recrystallisation is
   (a) stored energy of cold work
   (b) vacancy concentration
   (c) dislocation density in the crystal
   (d) grain boundary curvature

[GATE 2010]

77. Four eutectoid steel samples W, X, Y and Z are austenitized and then subjected to normalising, quenching, martempering and austempering treatments respectively. Which of the following statements is not correct?
   (a) The microstructure of sample W will be fully pearlitic
   (b) The microstructure of sample X will be untempered martensite
   (c) The microstructure of sample Y will be tempered martensite.
   (d) The microstructure of sample Z will be bainitic.

[GATE 2010]

78. Assertion: 'a': Hardenability of steel can be increased by adding certain alloying elements.
   Reason: 'r': The alloying elements can provide a fine dispersion of alloy carbides
   (a) Both 'a' and 'r' are true but 'r' is not a correct reason for 'a'
   (b) Both 'a' and 'r' are false
   (c) 'a' is true but 'r' is false
   (d) Both 'a' and 'r' are true and 'r' is a correct reason for 'a'

[GATE 2010]

79. At 910°C γ-Fe transforms to α - Fe resulting in a percentage volume expansion of
   (a) 5.6
   (b) 7.1
   (c) 7.6
   (d) 8.8

[GATE 2010]

80. In the hypothetical phase diagram, the melting point of each pure component is 1000 K and the eutectic temperature is 800 K. The eutectic is located at the equi- atomic composition. The maximum solid solubility in α-phase is given by the mole fraction, Ne = 0.1.
   The freezing range (in K) of the alloy with composition, Ne = 0.1 is

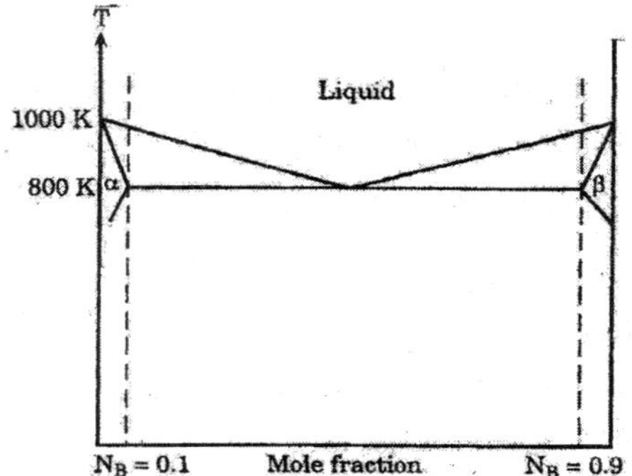

   (a) 100
   (b) 130
   (c) 160
   (a) 190

[GATE 2010]

81. In Q. No. 80, on cooling an alloy of composition $N_B = 0.2$, the fraction of proeutectic α phase at the eutectic temperature is
   (a) 0.75
   (b) 0.65
   (b) 0.55
   (d) 0.45

[GATE 2010]

Answers

76. (d)	77. (c)	78. (a)	79. (d)	80. (c)	81. (a)

82. The diffusion couple shown below is made from two A-B alloys. The initial compositions of the two alloys are indicated in the diagram. The centreline is at x = 0. The couple is held at an elevated temperature for 40 hours. Diffusivity, D = 3 x 10^{-11} m^{-2} s^{-1} Assume the diffusion couple to be infinitely long.

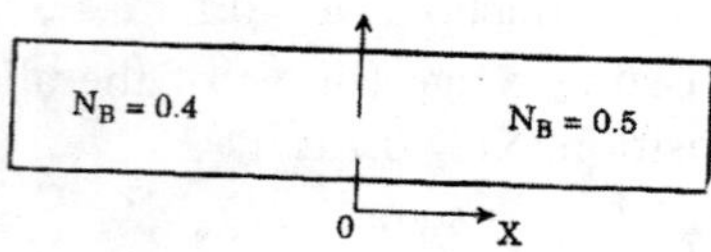

Which of the parameters give the composition profile in the following form?

$C(x, t) + c_1 + c_2 \text{ erf}\left(\frac{1}{2\sqrt{Dt}}\right)$

(a) $c_1 = 0.45$, $c_2 = 0.05$
(b) $c_1 = 0.5$, $c_2 = 0.4$
(c) $c_1 = -0.05$, $c_2 = 0.45$
(d) $c_1 = 0.1$, $c_2 = 0.9$

[GATE 2010]

83. In Q. No. 82, the composition at a distance, x = 2 mm is approximately [assuming *erf* (x) = x for small x]

(a) 0.3 (b) 0.474
(c) 0.524 (d) 0.7

[GATE 2010]

84. Match the associated primary change in group 2 with the corresponding phenomenon in group 1.

Group 1
P. Recovery
Q. Recrystallisation
R. Malleabilising
S. Texture

Group 2
1. change in dislocation density
2. change in the shape of a phase
3. change in vacancy concentration
4. change in grain orientation

(a) P-3, Q-4, R-l, S-2
(b) P-2, Q-4, R-l, S-3
(c) P-3, Q-l, R-2, S-4
(d) P-4, Q-3, R-l, S-2

[GATE 2006]

85. From an alloy, two specimens are machined and tested separately under tension and compression. The engineering stress-strain curves as well as the true stress- true strain curves for tension and compression are plotted in the same diagram. Identify the correct statements:

P. The engineering stress-strain curves in tension and compression are identical.
Q. The true stress-true strain curve in compression is lower than the true stress-true strain curve in tension due to Bauschinger effect.
R. The true stress-true strain curve in tension is above the corresponding engineering stress-strain curve.
S. The true stress-true strain curve in tension and compression are identical.

(a) P, Q
(b) Q, R
(c) Q, S
(d) R, S

[GATE 2006]

Answers

82. (a) 83. (b) 84. (c) 85. (b)

86. Match each phase in group 1 with a description in group 2.

Group 1

P. E- carbide Q. Sigma phase

R. E- ferrite S. Steadite

Group 2

1. A three component eutectic of iron, iron carbide, iron phosphate found in cast iron
2. An embrittling compound found in ferritic stainless steel
3. Obtained on tempering of hardened steels
4. Responsible for causing the weld deposit on austenitic stainless steels to be slightly magnetic.

(a) P-l, Q-3, R-4, S-2
(b) P-l, Q-2, R-4, S-3
(c) P-3, Q-2, R-l, S-4
(d) P-3, Q-2, R-4, S-l

[GATE 2007]

87. In Cu-Al phase diagram, the solubility of A1 in Cu at room temperature is about 10% and that of Cu in A1 is less than 1%. The Hume-Rothery rule that justifies this difference is

(a) size factor
(b) electro-negativity
(c) structure
(d) valancy

[GATE 2007]

88. The structure-sensitive properties are

P. Elastic modulus Q. Yield strength

R. Melting point S. Fracture strength

(a) P, S
(b) Q, S
(c) Q, R
(d) P, R

(GATE 2008]

89. Match the names in group 1 with invarient reactions in binary phase diagram in group 2.

Group 1

P. Eutectic

Q. Eutectoid

R. Peritectoid

S. Monotectic

Group 2

1. $S_1 = S_2 + S_3$
2. $L = S_1 + S_2$
3. $L_1 = L_2 + S$
4. $S_1 + S_2 = S_3$

(a) P-2, Q-l, R-3, S-4
(b) P-2, Q-l, R-4, S-3
(c) P-3, Q-4, R-2, S-l
(d) P-4, Q-3, R-l, S-2

[GATE 2008]

90. Match the heat treatment processes of steel in group 1 with the microstructural features in group 2.

Group 1

P. Quenching

Q. Maraging

R. Tempering

S. Austempering

Group 2

1. Bainite
2. Martensite
3. Intermetallic precipitates
4. Epsilon carbide

(a) P-2, Q-3, R-l, S-4
(b) P-l, Q-3, R-2, S-4
(c) P-2, Q-3, R-2, S-l
(d) P-3, Q-2, R-l, S-4

[GATE 2008]

Answers

86. (d) 87. (a) 88. (b) 89. (b) 90. (c)

91. Identify the attributes associated with dispersion hardened alloys.
    (a) Dispersoids do no dissolve in the matrix even at high temperature.
    (b) Dispersoids are coherent with the matrix.
    (c) Dispersoids impart creep resistance to the alloy.
    (d) Dispersoids improve the corrosion resistance of the alloy.
    (a) P, S (b) Q, R
    (c) Q, S (d) P, R
    [GATE 2008]
92. An annealed plain carbon steel, showing fully pearlitic microstructure, has a carbon content of .... wt%.
    (a) 0.01
    (b) 0.20
    (c) 0.77
    (d) 1.20
    [GATE 2009]
93. A defect that is bounded by two mirror planes is
    (a) twin (b) stacking fault
    (c) grain boundary (d) edge dislocation
    [GATE 2009]
94. Hardenability of steel does not depend upon the
    (a) alloy content
    (b) grain size
    (c) amount of carbon present
    (d) amount of cold work
    [GATE 2009]
95. Assertion : a : During hardening of steel, the component to be heat treated is strongly agitated in the quenching medium.
    Reason : r : The agitation breaks down the vapour barrier allowing the quench to proceed at a more rapid rate.
    (a) Both 'a' and 'r' are correct but 'r' is not correct reason of'a'
    (b) Both 'a' and 'r' are false
    (c) 'a' is true but 'r' is false
    (d) Both 'a' and 'r' are correct and 'r' is the correct reason for 'a'
    [GATE 2009]
96. As per TTT diagram, bainite will form in eutectoid plain carbon steel, when heated to 850° followed by
    (a) air cooling to room temperature
    (b) isothermal holding between eutectoid temperature and the nose
    (c) quenching to room temperature
    (d) isothermal holding between the nose and the Ms temperature
    [GATE 2009]
97. According to Hume-Rothery rules, extensive solid solubility between elements X and Y is promoted by the two factors in the following list.
    P. Same crystal structure of Z and Y.
    Q. Large atomic size difference (> 20%) between X and Y.
    R. Same valance of X and Y.
    S. Large difference in melting points of X and Y.
    (a) P, Q (b) P. R
    (c) Q, S (d) P, S
    [GATE 2009J
98. In the eutectoid steel, which one of the following structures DOES NOT form during continuous cooling?
    (a) Fully pearlitic (b)Pearlitic + bainitic
    (c) Fully bainitic (d) Martensitic
    [GATE 2011]

Answers

91. (d)	92. (c)	93. (a)	94. (d)	95. (b)	96. (d)
97. (b)	98. (c)				

99. Which one of the following is a ferrite stabiliser in steels?
(a) Ni (6) Cu
(c) Cr (d) Mn
[GATE 2011]

100. The angle between the line vector and the burgers vector of an edge dislocation is
(a) 0 degree (b) 90 degrees
(c) 120 degrees (d) 180 degrees
[GATE 2011]

101. With the increase in the degree of supercooling, the growth rate of a nucleus follows which one of the following trends?
(a) First increases and then decreases
(b) First decreases and then increases
(c) Only increases
(d) Only decreases
[GATE 2011]

102. Match the heat treatment process in group I with resultant microstructure group II.
Group I
P. Martempering
Q. Normalising
R. Sub-critical annealing for long time
S. Full annealing
Group II
1. Coarse pearlite
2. Fine pearlite
3. Tempered martensite
4. Spheroidised cementite in the matrix of ferrite
(a) P-l, Q-4, R-3, S-2
(b) P-2, Q-3, R-l, S-4
(c) P-4, Q-l, R-2, S-3
(d) P-3, Q-2, R-4, S-l
[GATE 2011]

103. A binary phase diagram of components P and Q displays an eutectoid reaction with terminal solid solutions α on the P rich side and β on the Q rich side. At the eutectoid temperature, the solubilities of Q in α and β are 5 and 90 wt% respectively. The densities of α and β phases are 9.5 and 2.49 gm/cm^3 respectively.
At the eutectoid point, the alloy has α and β in the weight ratio 1:1. The eutectoid point occurs at composition .... wt % Q.
(a) 46 (b) 47.5
(c) 50 (d) 52.5
[GATE 2011]

104. In Q.No. 103, at the eutectoid temperature, the ratio of α and β phases in the specimen observed under microscope is
(a) 0.50 (b) 0.40
(e) 0.25 (d) 0.20
[GATE 2011]

105. Which one of the following is an equilibrium defect?
(a) Vacancies (b) Dislocations
(c) Stacking faults (d) Grain boundaries
[GATE 2012]

106. During the solidification of a pure metal, it was found that dendrites are formed. Assuming that the liquid-solid interface is at the melting temperature, the temperature from the interface into the liquid
(a) Decreases
(b) Increases
(c) Remains constant
(d) Increases and then decreases
[GATE 2012]

Answers

99. (c)	100. (b)	101. (a)	102. (d)	103. (b)	104. (c)
105. (a)	106. (a)				

107. The property of a material that CANNOT be significantly changed by heat treatment is
    (a) Yield strength
    (b) Ultimate tensile strength
    (c) Ductility
    (d) Elastic modulus

[GATE 2012]

108. If d is the inter-planar spacing of the planes {h k 1}, the inter-planar spacing of the planes {nh nk nl}, n being an interger, is
    (a) d
    (b) d/n
    (c) nd
    (b) d/n^2

[GATE 2012]

109. Match the phases of steel in Group I with the crystal structures in Group II.

Group I
P. Martensite
Q. Cementite
R. Austenite
S. Ferrite

Group II:
1. bcc
2. fcc
3. bct
4. Orthorhombic

(a) P-3, Q-4, R-l, S-2
(b) P-2, Q-3, R-l, S-4
(c) P-3, Q-4, R-2, S-l
(d) P-4, Q-3, R-2. S-l

[GATE 2012]

110. Arrange the following in terms of increasing severity of quench.
P. Oil quenching
Q. Water quenching
R. Water quenching with agitation
S. Brine quenching

(a) $P < Q < R < S$
(b) $Q < R < P < S$
(c) $P < Q < S < R$
(d) $Q < P < R < S$

[GATE 2012]

111. Regarding recrystalisation, which one of the following statements is NOT correct?
    (a) Higher the amount of cold work, lower is the recrystalisation temperature.
    (b) Higher the recovery, higher is the recrystallization temperature.
    (c) Higher the temperature, of cold work, higher is the recrystallization temperature.
    (d) Finer the initial grain size, higher is the recrystallization temperature.

[GATE 2012]

112. Match the phenomena listed in Group I with the possible mechanisms in Group II.

Group I
P. Fatigue
Q. Creep
R. Strain hardening
S. Yield point phenomenon

Group II
1. Grain boundary sliding
2. Slip band extrusion and intrusion
3. Cottrell atmosphere
4. Dislocation interaction

(a) P-2, Q-3, R-4, S-l
(b) P-2, Q-4, R-3, S-l
(c) P-2, Q-l, R-4, S-3
(d) P-l, Q-2, R-4, S-3

[GATE 2012]

Answers

107. (d)	108. (b)	109. (c)	110. (a)	111. (d)	112. (c)

113. A material with grain size of ASTM No. 6 has a lattice frictional stress 100 MN/m^2 and locking parameter (Hall-Petch constant) 0.10 $MN/m^{3/2}$

Grain size of the material is approximately

(a) 45 μm (b) 35 μm

(c) 4.5 μm (d) 3.5 μm

[GATE 2012]

114. In Q. 113, yield strength of the material is approximately

(a)100MPa (b) 115 MPa

(c) 165 MPa (d) 215 MPa

[GATE 2012]

115. Which one of the following techniques does NOT require quenching to obtain final case hardness?

(a) Flame hardening

(b) Induction hardening

(c) Nitriding

(d) Carburizing

[GATE 2013]

116. Which one of the following elements is an austenite stabilizer?

(a) Nitrogen (b) Molybdenum

(c) Vanadium (d) Tungsten

[GATE 2013]

117. A 0.2 wt.% plain carbon steel sheet is heated and equilibrated in the inter-critical region followed by instant water quenching. The microstructure of the quenched steel sheet consists of

(a) fully martensite

(b) proeutectoid ferrite + martensite

(c) martensite + pearlite

(d) martensite + austenite

[GATE 2013]

118. Which one of the following does NOT improve fatigue life of steel component?

(a) Nitriding

(b) Decarburization

(c) Improving surface finish

(d) Shot-peening

[GATE 2013]

119. In the Ellingham diagram for oxides, C-CO line cuts the M-MO line at temperature T_1 and the M'-M'O line at a higher temperature T_2. At a temperature greater than T_1 and less than T_2, carbon can reduce

(a) MO

(b) both MO and M'O

(c) M'O

(d) neither MO nor M'O

[GATE 2013]

120. A steel specimen containing 0.2 wt.% C is carburized in an atmosphere that maintains a carbon content of 1.2 wt.% C at the surface of the specimen.

Given:

For carbon diffusion in austenite: $D_0 = 2.0 \times 10^{-5}$ m^2/s

Activation energy for diffusion, Q = 142 kj/mol

y	erf(y)
0.85	0.7707
0.90	0.7970
0.95	0.8209

What is the depth (in (.μm) from the surface of the specimen at which a composition of 0.4 wt.% C is obtained after carburizing at 870°C for 10 h?

(a) 15 (b) 84

(c) 113 (d) 875

[GATE 2013]

Answers

113. (a) 114. (b) 115. (c) 116. (a) 117. (b) 118. (b)
119. (a) 120. (d)

121. In Q. 120, how long (in hr) will it take to double the depth at which 0.4wt.% C is reached?
(a) 40 (b) 20
(c) 18 (d) 14
[GATE 2013]

122. Which one of the following can avoid weld decay of austenitic stainless steels?
(a) Reducing carbon content
(b) Increasing carbon content
(c) Eliminating strong carbide formers
(d)Decreasing chromium content
[GATE 2014]

123. Identify the INCORRECT statement with respect to grain growth.
(a) As the average grain size increases, the grain boundary energy per unit area deceases.
(b) The driving force for grain growth is the decrease in grain boundary energy per unit volume of the material.
(c) Higher the temperature, the faster is the grain growth.
(d) Impurity atoms segregated at grain boundaries can retard grain growth.
[GATE 2014]

124. Identify the type of the following invariant reaction:
liquid 1 + solid 1 $\rightleftharpoons$ solid 2
(a) Eutectic (b) Eutectoid
(c) Peritectic (d) Peritectoid
[GATE 2014]

125. Two Cu-Ni alloys, one with 47 wt.% Ni and the other with 53 wt.% Ni, are kept in separate crucibles and are allowed to equi-librate at the same temperature T_1 between the liquidus and the solidus. The two alloys have
(a) the same weight fraction of the liquid but the liquid compositions are different.
(b) the same composition of the liquid but the weight fractions of the liquid are different.
(c) the same weight fractions of the liquid and the solid.
(d) different compositions and weight fractions of the liquid.
[GATE 2014]

126. Assume that Sievert's law holds for solubility of hydrogen dissolved in a thin metal foil. If the partial pressure of hydrogen in contact with the foil is increased by a factor of 4, the solubility increases by a factor of
(a) 1 (b) 2
(c) 3.6 (d) 4.8
[GATE 2014]

127. Match the following strengthening methods (Group I) in metallic alloys with typical mechanisms responsible (Group II) for them.
Group I
P. Grain size strengthening
Q. Work hardening
R. Dispersion strengthening
S. Solid solution hardening
Group II
1. Orowan looping
2. Suzuki interaction
3. Lomer-Cottrell-barrier
4. Dislocation pile up at grain boundaries
(a) P-3, Q-4, R-2, S-l
(b) P-3, Q-2, R-l, S-4
(c) P-4, Q-3, R-l, S-2
(d) P-4, Q-l, R-3, S-2
[GATE 2014]

Answers

121. (a) 122. (a) 123. (a) 124. (c) 125. (b) 126. (b)
127. (c)

128. Determine the correctness or otherwise of the following Assertion (a) and Reason (r).
Assertion (a): Peak aging time for an Al-4 wt.% Cu alloy is indicated by a maximum in the hardness.
Reason (r): The maximum volume fraction of the θ precipitates is formed at the peak aging time.
(a) a is true but r is false
(b) a is false but r is true
(c) both a and r are true, and r is the reason for a
(d) both a and r are true, but r is nct the reason for a

[GATE 2014]

129. Match the elements is Group I with their corresponding crystal structures in Group II.
Group I
P. Tungsten Q. Nickel
R. Magnesium S. Polonium
Group II
1. HCP
2. Simple cubic
3. BCC
4. FCC
(a) P-2, Q-l, R-4, S-3
(b) P-3, Q-4, R-l, S-2
(c) P-4, Q-2, R-3, S-l
(d) P-3, Q-4, R-2, S-l

[GATE 2014]

130. In an Ellingham diagram, the standard free energy change $\Delta G°$ for the oxidation reaction of a metal M given by : $xM(s) + O_2(g) \rightarrow MxO_2(s)$, is plotted as a function of temperature. The slope of this line is positive because
(a) $\Delta S°$ is positive (b) $\Delta S°$ is negative
(c) $\Delta H°$ is positive (d) $\Delta H°$ is negative

[GATE 2015]

131. Which of the following manufacturing processes is NOT used for producing fine grained metals?
(a) Electrodeposition
(b) Czochralski method
(c) Equi-Channel Angular Pressing (ECAP)
(d) Sintering of milled powders

[GATE 2015]

132. The stress required for Orowon dislocation bypass is 200 MPa in an alloy when the interprecipitate spacing is 500 nm. In the same alloy, if the inter-precipitate spacing is reduced to 200 nm, the stress required (in MPa) is
(a) 100
(b) 250
(c) 500
(d) 750

[GATE 2015]

133. Match the names listed in Group I with the reactions listed in Group II.
Group I
P. Eutectic
Q. Peritectic
R. Peritectoid
S. Monotectic
Group II
1. $\gamma + \beta \rightarrow \alpha$ 2. $L \rightarrow \alpha + \beta$
3. $L_1 \text{—>} L_2 + \alpha$ 4. $L + \beta \rightarrow \alpha$
(a) P-2, Q-3, R-l, S-4
(b) P-3, Q-4, R-l, S-2
(c) P-2, Q-4, R-l, S-3
(d) P-4, Q-l, R-2, S-3

[GATE 2015]

Answers

128. (a) 129. (b) 130. (b) 131. (b) 132. (c) 133. (c)

134. Match the heat treatment for an eutectoid steel shown in the TTT diagram below (as P, Q, R and S) with the resulting microstructure listed below :

1. Fine pearlite
2. martensite
3. Bainite
4. Coarse pearlite

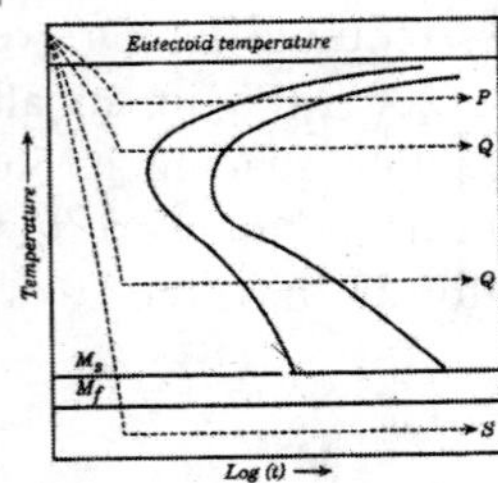

(a) P-l, Q-2, R-4, S-3
(b) P-4, Q-l, R-3, S-2
(c) P-2, Q-l, R-3, S-4
(d) P-l, Q-4, R-3. S-2

[GATE 2015]

135. An alloy of overall composition, X_B = 0.7 was equilibrated at temperature T_1. Microstructural analysis showed two phases, α and β, and that the phase fraction of β was 0.75. Given that the equilibrium composition of β at Ti is 0.9 as shown in the phase diagram below, the maximum solubility of B in α (in mole fraction) at this temperature is ....

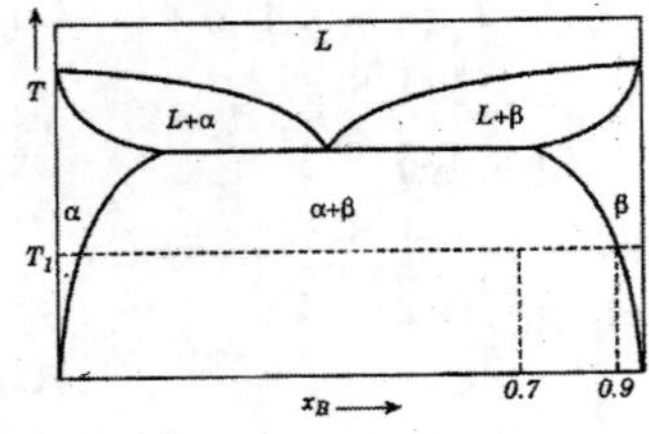

(a) 0.05 (b) 0.10
(c) 0.50 (d) 0.85

[GATE 2015]

136. Match the following fracture surface features listed in Group I with the fracture mechanisms listed in Group II.

Group I
P. Striations
Q. Dimples and microvoids
R. Flat facets and "river markings"
S. Jagged surface with grain-like features

Group II
1. Intergranular fracture
2. Cleavage fracture
3. Ductile fracture
4. Fatigue fracture

(a) P-l, Q-2, R-3, S-4
(b) P-l, Q-3, R-2, S-4
(c) P-4, Q-3, R-2, S-l
(d) P-2, Q-l, R-4, S-3

[GATE 2015]

137. Match the scientist pairs listed in Group I with phenomena listed in Group II.

Group I
P. Hall-Patch
Q. Nabarro-Herring
R. Lomer-Cottrell
S. Frank-Read Group

Group II
1. Dislocation reaction product
2. Diffusional creep
3. Dislocation source
4. Grain boundary strengthening

(a) P-l, Q-2, R-3, S-4
(b) P-l, Q-2, R-4, S-3
(c) P-4, Q-2, R-l, S-3
(d) P-4, Q-l, R-2, S-3

[GATE 2015]

Answers

134. (b) 135. (b) 136. (c) 137. (c)

138. Match the desired mechanical properties listed in Group I with the microstructural features listed in Group II.

Group I

P. Creep resistance

Q. Elastic modulus enhancement

R. Superplasticity

S. Increased strength

Group II

1. Fine grained two-phase microstructure
2. Single crystal
3. Coherent precipitates
4. Glass fibres in epoxy

(a) P-3, Q-4, R-2, S-l

(b) P-l, Q-2, R-3, S-4

(c) P-2, Q-4, R-l, S-3

(d) P-l, Q-4, R-2, S-3

[GATE 2015]

139. in a typical Ellingham diagram for the oxides, the $C + O_2 = CO_2$ line is nearly horizontal because

(a) The slope of the line is equal to the enthalpy change at standard state, which is approximately zero is this case.

(b) The slope of the line is equal to the entropy change at standard state, which is approximately zero in the case.

(c) CO_2 shows non-ideal behaviour.

(d) CO_2 is a gaseous oxide.

[GATE 2016]

140. Which of the following cooling curves (shown in schematic) in an eutectoid steel will produce 50% bainitic structure?

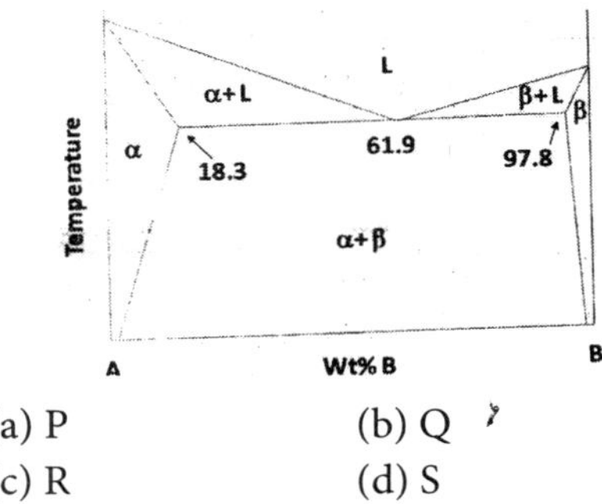

(a) P (b) Q

(c) R (d) S

[GATE 2016]

141. A plastically deformed metal crystal at low temperature exhibits wavy slip line pattern due to

(a) Dislocation pile-up

(b) Large number of slip systems

(c) Low stacking fault energy

(d) Dislocation climb

[GATE 2016]

142. Creep resistance decreases due to

(a) Small grain size

(b) Fine dispersoid size

(c) Low stacking fault energy

(d) High melting point

[GATE 2016]

143. During isothermal phase transformation (in solid-state), fraction transformed is measured at two different transformation times:

Transformation Time, t (s)	Fraction Transformed f,
75	0.11
150	0.37

Assuming Avrami kinetics [$f = 1 - exp(-ktn)$]. the fraction transformed in 300 seconds is ....

[GATE 2016]

Answers

138. (c)	139. (b)	140. (b)	141. (b)	142. (a)	143. (0.81 to 0.87)

144. The critical nucleus size (in mm) when copper melt is under-cooled by 100 K is

Given data:

Melting point:	1356 K
Density:	8900 kg. m^{-3}
Solid-liquid interfacial energy:	0.5 Jm^{-2}
Latent heat of freezing:	13000 J. mol^{-1}
Molar volume:	7×10^{-6} m^3 mol^{-1}

(a) 0.36 (b) 1.55
(c) 3.65 (d) 7.30

[GATE 2016]

145. A binary phase diagram is shown in the schematic.

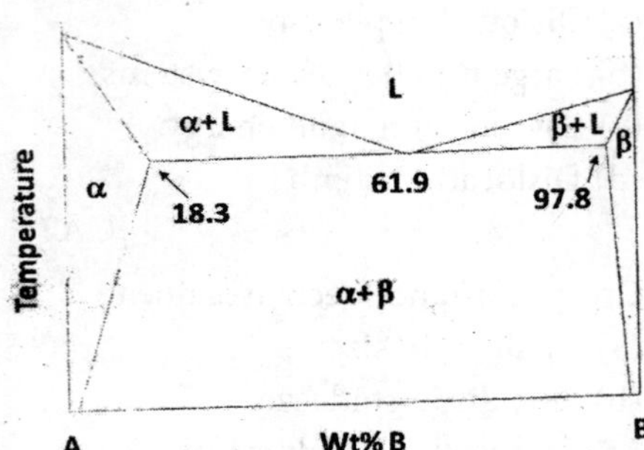

Upon complete solidification of a binary alloy system A-B, the fraction of pro-eutectic α-phase present is 0.50. The alloy composition in terms of wt%. B is ....

[GATE 2016]

146. A liquid phase sintered SiC-Ni composite has a solid-solid grain boundary energy ($\gamma_{SiC-SiC}$) of 0.80 $J.M^{-2}$ and a solid-liquid ($\gamma_{SiC-SiC}$) interfacial energy of 0.45 $J.m^{-2}$. For a SiC grain size of 20 μm, the average interparticle (SiC-SiC) neck size (in μm) is:

(a) 3.03 (b) 4.28
(c) 9.16 (d) 18.32

[GATE 2016]

Answers

144. (a) 145. (40.0 to 40.2) 146. (c)

8 Materials Science

1. Liquid nitrogen containers can be made, from
   (a) ferritic stainless steel
   (b) HSLA steel
   (c) Titanium
   (d) Austenitic stainless steel.

[GATE 91]

2. The strength of material increases with
   (a) increase in dislocation density
   (b) decrease in dislocation density
   (c) increase in grain size
   (d) decrease in grain size

[GATE 92]

3. Young's modulus of a material gives an idea about
   (a) toughness
   (b) stiffness
   (c) hardness
   (d) strength
   (e) electrical conductivity

[GATE 94]

4. An alloy of Fe-0.4% C is
   (a) cast iron
   (b) hypoeutectoid steel
   (c) hypereutectoid steel
   (d) eutectoid steel

[GATE 94]

5. The single most important requirement for a turbine blade material is
   (a) damping
   (b) resilience
   (c) creep resistance
   (d) DBTT

[GATE 94]

6. Match the following components to their material.
   P. Bush bearings
   Q. Incandescent lamp filaments
   R. Cutting tools
   S. Heavy duty electrical contacts
   1. Tungsten-silver
   2. Copper-tin
   3. Thoria doped tungsten
   4. Tungsten carbide-13% cobalt
   (a) P-2, Q-3, R-4, S-l
   (b) P-2, Q-3, R-l, S-4
   (c) P-4, Q-3, R-2, S-l
   (d) P-l, Q-2, R-3, S-4

[GATE 95]

Answers

1. (b) 2. (b & d) 3. (b) 4. (b) 5. (c) 6. (a)

7. Match the following materials to their specific applications.
   P. White cast irons
   Q. Age hardened Al alloys
   R. Grey cast irons
   S. Superalloys
   1. Damping capacity
   2. Wear resistance
   3. High temperature applications
   4. Aerospace structural applications
   (a) P-2, Q-4, R-l, S-3
   (b) P-4, Q-l, R-3, S-2
   (c) P-2, Q-4, R-3, S-l
   (d) P-l, Q-3, R-2, S-4

   [GATE 95]

8. The preferred alloying for low temperature application of steel is
   (a) Cr (b) N
   (c) Mo (d) Ni

   [GATE 96]

9. Lithium is a useful alloying addition to aluminium, because it
   (a) is cheap
   (b) imparts solid solution strengthening
   (c) lowers density and contributes to age-hardening
   (d) improves the corrosion resistance of aluminium

   [GATE 96]

10. Titanium is added to molten aluminium alloys before casting for the purpose of
    (a) grain refinement
    (b) increasing corrosion resistance
    (c) reducing porosity
    (d) improving fluidity.

    [GATE 96]

11. Match the material/ application in the following cases.
    P. Grey cast iron
    Q. Austenitic stainless steel
    R. High strength low alloy steel
    S. Nickel base superalloy
    1. Lathe bed
    2. Oil piplines
    3. Gas turbine engines components
    4. Cryogenic service
    (a) P-l, Q-2, R-4, S-3
    (b) P-l, Q-2, R-3, S-4
    (c) P-4, Q-3, R-2, S-l
    (d) P-l, Q-2, R-3, S-4

    [GATE 96]

12. Most important property of steels for use in automoblie bodies is
    (a) formability (b) yield strength
    (c) toughness (d) resilience.

    [GATE 98]

13. Match the cast irons with applications.
    P. Grey cast iron
    Q. White cast iron
    R. Nodular cast iron
    S. Ni resist cast iron
    1. S.G iron pipes
    2. Chemical processing equipments requiring high corrosion, erosion and heat resistance
    3. Bases or beds of industrial equipments
    4. Manufacture of malleable cast iron
    (a) P-3, Q-4, R-l, S-2
    (b) P-3, Q-4, R-2, S-l
    (c) P-4, Q-3, R-2, S-l
    (d) P-l, Q-2, R-4, S-3

    [GATE 98]

Answers

7. (a) 8. (d) 9. (d) 10. (d) 11. (a) 12. (c)
13. (a)

14. Match the following

P. Nickel-silvers Q. Nitralloys

R. Babbits S. Permalloy

1. Alloys of tin, antimony and copper usec as a bearing material.
2. Steel used for nitriding containing aluminium, chromium and molybed num.
3. Soft magnetic material used as electromagnet.
4. Ternery alloys of copper, nickel and zinc with good cold formability.

(a) P-4, Q-2, R-3, S-l

(b) P-4, Q-2, R-l, S-3

(c) P-4, Q-3, R-2, S-l

(d) P-l, Q-2, R-4, S-3

[GATE 98]

15. Silicon crystal can be converted to p-type semi-conductor by doping with

(a) phosphorous (b) nitrogen

(c) carbon (d) boron

[GATE 99]

16. On decreasing the grain size of a polycrys talline material, the property most likely to deteriorate is

(a) creep (b) toughness

(c) tensile strength (d) fatigue strength

[GATE 99]

17. Transformer cores are normally made from

(a) soft ferrites

(b) high purity iron

(c) grain oriented Fe-Si alloy

(d) AINiCo alloy

[GATE 99J

18. Which of the following materials do not respond well to induction hardening?

(a) Ferritic cast iron

(b) Low carbon steel

(c) Medium carbon steel

(d) Pearlitic cast iron

[GATE 991

19. A heating element that is resistant in air upto 1700°C is

(a) platinum-rhodium

(b) silicon carbide

(c) molybdenum disilicide

(d) tungsten

[GATE 2000]

20. Increasing the carbon content of steel

(a) reduces the upper shelf energy

(b) increases the ductility transition temperature

(c) decreases brittleness

(d) decreases hardness

[GATE 2000]

21. The fatigue resistance of material is improved by the following techique:

(a)Anodising (b) Carburising

(c) Ion nitriding (d) Shot peening

[GATE 2001]

22. Lead is added to 60-40 brass primarily to improve

(a) machinability

(b) corrosion resistance

(c) fluidity

(d) strength

[GATE 2002]

23. Which of the following materials is not suitable as a die material for wire drawing?

(a) Diamond (b) Tungsten carbide

(c) Tool steel (d) Bronze

[GATE 2004]

Answers

14. (b)	15. (d)	16. (a)	17. (c)	18. (d)	19. (b)
20. (b)	21. (d)	22. (a)	23. (d)		

24. To obtain super-plasticity, the alloy should
    P. have a fine grain structure that is also stable at high temperature
    Q. be deformed at low temperature
    R. be deformed at low strain rates
    S. be deformed at high strain rates
    (a) P, R (b) Q, S
    (c) Q, R (d) P, Q

[GATE 2004]

25. A soft magnetic material should have
    (a) high coercivity
    (b) high energy product
    (c) high permeability
    (d) high retentivity

[GATE 2005]

26. Match the materials in group 1 with the Young's modulus (GPa) in group 2.
    Group 1
    P. Copper Q. Ployvinyl cholride
    R. Steel S. Silicate glass
    Group 2
    1. 4 2. 37
    3. 70 4. 110
    5. 207 6. 700
    (a) P-4, Q-l, R-5, S-3
    (b) P-5, Q-l.R-4, S-2
    (c) P-4, Q-2, R-6, S-3
    (d) P-2, Q-3. R-5, S-6

[GATE 2005]

27. Match the properties in group 1 with materials in group 2.
    Group 1
    P. Diamagnetic Q. Paramagnetic
    R. Ferromagnetic S. Ferrimagnetic
    Group 2
    1. Aluminium
    2. Copper
    3. Yittrium iron garnet
    4. Cobalt
    (a) P-4, Q-3, R-2, S-l
    (b) P-2, Q-l, R-4, S-3
    (c) P-3, Q-4, R-2, S-l
    (d) P-2, Q-4, R-l, S-3

IGATE 2006]

28. Which one of the following alloy systems exhibits complete solid solubility?
    (a) Cu-Ni (b) Fe-Cu
    (c) Pb-Sn (d) Cu-Zn

[GATE 2007]

29. A small amount of thoria is doped into tungsten filament wires used in light bulbs. This is because thoria particles
    (a) decreases solute diffusivity
    (b) enhances the mobility of grain boundary
    (c) increases solute segregation to the grain boundary
    (d) are effective in limiting grain growth.

[GATE 2007]

30. Match each application in group 1 with a material in group 2.
    Group 1
    P. Cores for electric motor
    Q. stripe on credit cards
    R. Permanent magnets
    S. Multilayer capacitors
    Group 2
    1. Y-Fe_2O_3 particles
    2. Barium titanate
    3. COsSm intermetallic compound
    4. Grain oriented silicon steel
    (a) P-3, Q-l, R-2, S-4
    (b) P-4, Q-l, R-3, S-2
    (c) P-4. Q-l, R-2, S-3
    (d) P-2, Q-l, R-3, S-4

[GATE 2007]

Answers

24. (a) 25. (c) 26. (a) 27. (b) 28. (a) 29. (d)
30. (b)

31. Match the non-ferrous alloy in group 1 with their applications in group 2.
    Group 1
    P. Ti alloy   Q. Zr alloy
    R. Ni alloy   S, Cu alloy
    Group 2
    1. Nuclear reactors   2. Bells
    3. Dental implants   4. Gas turbines
    (a) P-3, Q-l, R-4, S-2
    (b) P-2, Q-3, R-4, S-l
    (c) P-2, Q-l, R-3, S-4
    (d) P-3, Q-4, R-l, S-2

    [GATE 2008]

32. Match the materials in group 1 with their functional applications in group 2.
    Group 1
    P. Nb_3Sn Q. Ga. As
    R. Fe - 4% Si alloy   S. $Si0_2$
    Group 2
    1. Dielectric   2. Soft magnet
    3. Super conductor   4. Semi-conductor
    (a) P-3, Q-l, R-4, S-2
    (b) P-l, Q-4, R-2, S-3
    (c) P-3, Q-2, R-4, S-l
    (d) P-3, Q-4, R-2, S-l

    [GATE 2008]

33. Match the cast irons in group 1 with the distinguishing microstructural features in group 2.
    Group 1
    P. Grey cast iron
    Q. Ductile cast iron
    R. Malleable cast iron
    S. White cast iron
    Group 2
    1. Temper graphite
    2. Pearlite
    3. Graphite flakes
    4. Massive cementite
    5. Nodular graphite
    (a) P-3, Q-5, R-4, S-2
    (b) P-l, Q-5, R-4, S-2
    (c) P-2, Q-4, R-5, S-3
    (d) P-3, Q-5, R-l, S-4

34. Super alloys are
    (a) Al-based alloys   (b) Cu-based alloys
    (c) Ni-based alloys   (d) Mg-based alloys

    [GATE 2009]

35. Wood is a naturally occuring .... material.
    (a) malleable   (b) composite
    (c) ceramic   (d) isotropic

    [GATE 2009]

36. p-type semi-conductor can be obtained by doping silicon with
    (a) antimony   (b) phosphorous
    (c) arsenic   (d) boron.

    [GATE 2009]

37. Match the properties in group 1 with the metals in group 2.
    Group 1
    P. Ferromagnetism
    Q. Superconductivity
    R. Diamagnetism
    S. Antiferromagnetism
    Group 2
    1. Nb   2. Fe
    3. Cu   4. Cr
    (a) P-2, Q-4, R-3, S-l
    (b) P-2, Q-l, R-3, S-4
    (c) P-3, Q-4, R-l, S-2
    (d) P-l, Q-2, R-3, S-4

    [GATE 2009]

Answers

31. (a)	32. (d)	33. (d)	34. (c)	35. (b)	36. (d)
37. (b)					

38. The engineering stress-strain curve for a ceramic material is
    (a) parabolic (b) exponential
    (c) logarithmic (d) linear
    [GATE 2010]

39. The material in which there is conduction primarily by holes is
    (a) conductor
    (b) insulator
    (c) p-type semi-conductor
    (d) n-type semi-conductor
    [GATE 2011]

40. As temperature increases, the electrical resistivities of pure metals (ρm) and intrinsic semiconductors (ρs) vary as follows.
    (a) Both ρm and ρs increase
    (b) Both ρm and ρs decrease
    (c) ρm increases and ρs decreases
    (d) ρm decreases and ρs increases
    [GATE 2012]

41. With respect to the matrix of Al-Cu alloys, G-P zones are
    (a) coherent
    (b) incoherent
    (c) semi-coherent
    (d) chemically indistinguishable
    [GATE 2013]

42. As compared to the engineering stress-engineering strain curve, the true stress-true strain curve for a given material
    (a) lies above and to the left.
    (b) lies below and to the right.
    (c) crosses the engineering stress-engineering strain curve.
    (d) is identical.
    [GATE 2013]

43. Which one of the following microstructures of a Ni-base superalloy imparts the highest creep resistance?
    (a) Fine grained equiaxed
    (b) Coarse grained equiaxed
    (c) Columnar
    (d) Single crystal
    [GATE 2014]

44. As mercury is cooled from room temperature to 3 K, its electrical behaviour changes from that of
    (a) conductor to insulator
    (b) semiconductor to insulator
    (c) conductor to superconductor
    (d) semiconductor to superconductor
    [GATE 2014]

45. Which one of the following refers to the ability of a material to absorb energy when deformed elastically?
    (a) Toughness (b) Fracture toughness
    (c) Resilience (d) Hardness
    [GATE 2014]

46. Match the types of corrosion in Group I with their examples in Group II.
    Group I.
    P. Galvanic corrosion
    Q. Filiform corrosion
    R. Stress corrosion cracking
    S. Crevice corrosion
    Group II.
    1. Season cracking in brass
    2. Nut-bolt assembly with gasket
    3. Brass fittings joined with lead-tin solder
    4. Painted food can
    (a) P-4, Q-l. R-2, S-3
    (b) P-3, Q-l, R-4, S-2
    (c) P-3. Q-4, R-l, S-2
    (d) P-4, Q-2, R-3, S-l
    [GATE 2014]

Answers

38. (d)	39. (c)	40. (c)	41. (a)	42. (a)	43. (d)
44. (c)	45. (c)	46. (c)			

47. Match the alloy names listed in Group I with the main elements present in them listed in Group II.

Group I

P. Babbit Q. Muntz metal

R. Invar S. Inconel Group II

1. Fe-Ni 2. Ni-Cr-Fe

3. Cu-Zn 4. Sn-Sb-Cu

(a) P-3, Q-l, R-4, S-2
(b) P-3, Q-4, R-l, S-2
(c) P-4, Q-l, R-2, S-3
(d) P-4, Q-3, R-l, S-2

[GATE 2014]

48. When boron (trivalent) is doped to silicon, the resulting material is
(a) a p-type semiconductor.
(b) an n-type semiconductor.
(c) a superconductor.
(d) an insulator.

[GATE 2015]

49. Select the INCORRECT statement related to the effect of a small amount of carbon addition on mechanical properties of iron.
(a) Ductile-Brittle-Transition-Temperature (DBTT) increases.
(b) Hardenability increases.
(c) Toughness increases.
(d) Yield point phenomenon occurs.

[GATE 2015]

50. In polymers such as epoxy, creep resistance can be enhanced by
(a) increasing the bulkness of side groups.
(b) increasing the cross-link density.
(c) addition of plasticizers.
(d) annealing.

[GATE 2015]

51. Match the materials listed in Group I with the most appropriate applications listed in Group II.

Group I.

P. Iron-silicon alloy Q. GaAs

R. Nichrome S. Quartz crystals

Group II.

1. Heating element
2. Ultrasonic generator
3. Transformer core
4. Light emitting diode

(a) P-3, Q-4, R-l, S-2
(b) P-2, Q-4, R-l, S-3
(c) P-l, Q-3, R-4, S-2
(d) P-3, Q-2, R-4, S-l

[GATE 2015]

52. For the tensile stress-strain curve of a material shown in the schematic, the resilience (in MPa) is ....

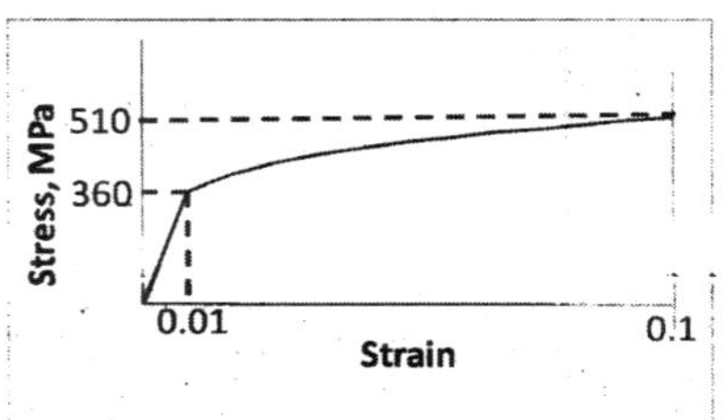

[GATE 2016]

53. A glass plate has two parallel cracks. One of them is an internal crack of length 5 μm and the other is a surface crack of length 3 μm. A tensile stress is applied perpendicular to the crack surfaces. The fracture stress (in MPa) is ....

Given data (for glass plate):

Young's Modulus = 70 GPa

Surface energy per unit area = 1 $J.m^{-2}$

[GATE 2016]

Answers

47. (d) 48. (a) 49. (c) 50. (b) 51. (a) 52. (1.7 to 1.9)
53. (115 to 125)

Section **B**

9 Fuel Technology

1. M_{10} index of coke indicates.
   (a) Compressive strength
   (b) hardness
   (c) abrasion resistance
   (d) impact stregth

[GATE 2002]

2. Natural gas is essentially
   (a) methane
   (b) propane + butane
   (c) 30% CO + 15% H_2 + 45% N2
   (d) 50% methane 50% H_2.

[GATE 2005]

Answers

1. (c) 2. (a)

10

Furnace Technology

1. Hot gases from a furnace are entering at the base of a 30m high tubular vertical chimney at 600°C. The density of air and furnace gases at respective temperatures are 1.165 and 0.405 kgm^{-3} respectively. The static draft produced by the chimney is .... Pa.

   (a) 2.3 (b) 22.8

   (c) 223.6 (d) 2193.3

   [GATE 95]

2. In a furnace, with heating element temperature at 1700°C, the dominant mechnaism of heat, transfer will be

   (a) conduction

   (b) radiation

   (c) natural convection

   (d) forced convection

   [GATE 2003]

3. Match the following:

   Group 1

   P. Blast furnace

   Q. Rotary kiln

   R. Continuous casting

   S. stoves

   Group 2

   1. Lime 2. Pig iron

   3. Heat exchanger 4. Billets

   (a) P-2, Q-1, R-4, S-3

   (b) P-1, Q-2, R-3, S-4

   (c) P-4, Q-3, R-2, S-1

   (d) P-3, Q-4, R-1, S-2

   [GATE 95]

Answers

1. (b) 2. (b) 3. (a)

11

Refractory Technology

1. L.D. Converters are generally lined with
   (a) magnesite
   (b) fireclay
   (c) chrome magnesite
   (d) silica

   [GATE 99]

2. The refractory brick which has good thermal shock resistance at high temnperature but cracks on cooling below 400°C is
   (a) magnesite (b) chrome
   (c) silica (d) fireclay

   [GATE 2000]

3. The refractory lining of the bottom in a basic electric are furnace in made of
   (a) silica (b) alumina
   (c) magnesia (d) fireclay

   [GATE 2003]

4. Arrange the following refractory materials in ascending order of their melting points
   P. Alumina
   Q. Silica
   R. Carbon
   S. Zirconia
   (a) R > P > Q > S (b) S > R > Q > P
   (c) R > S > P > Q (d) P > S > R > Q

   [GATE 2003]

5. Which of the following refractory materials is NOT used in the BOF (LD) working lining?
   (a) Tar bonded dolomite
   (b) Pitch bonded magnesite
   (c) Fired and pitch impregnated magnesite
   (d) Graphite-alumina composite

   [GATE 2011]

Answers

1. (c) 2. (c) 3. (c) 4. (a) 5. (d)

12 Foundary Metallurgy

1. Eutectic Al-Si castings are modified by
   (a) Na (b) P
   (c) S (d) none of these
   [GATE 92]

2. Mould oscillation is used in continuous casting of steels to
   (a) heal cracks formed on the surface of the casting
   (b) obtain good mixing of the liquid metal inside the mould
   (c) float out the inclusions.
   (d) avoid rhomboidity of the casting.
   [GATE 94]

3. Riser in a casting compensates for
   (a) liquid state shrinkage
   (b) solidification shrinkage
   (c) liquid state shrinkage, solidification shrinkage and solid state shrinkage
   (d) liquid state shrinkage and solidification shrinkage
   [GATE 95]

4. If two parts of a mould are not joined properly, the casting may have the following defect.
   (a) Scab (b) Blowhole
   (c) Gas porosity (d) Flash
   [GATE 98]

5. In the jolt-squeeze operation in green sand moulding, the nature of mould hardness distribution for the type of pattern shown in the figure will be

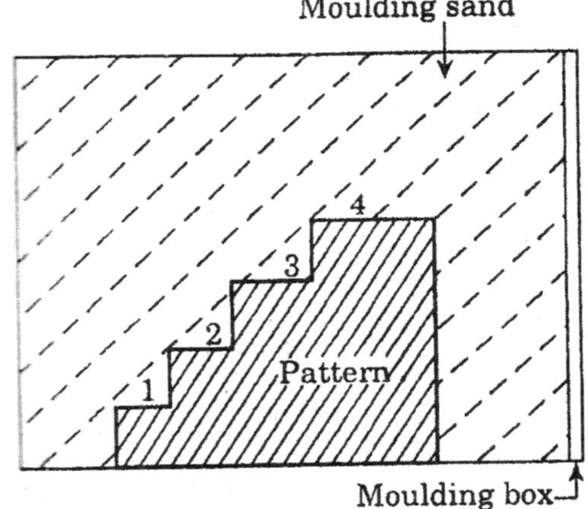

	1	2	3	4
(a)	25	35	60	85
(b)	80	75	75	85
(c)	75	60	50	45
(d)	75	80	80	75

[GATE 98]

Answers

1. (a) 2. (d) 3. (d) 4. (d) 5. (a)

6. Match the defect with appropriate cause.
   List I
   P. Hot tear
   Q. Porosity
   R. Sand inclusions
   S. Shrinkage cavity
   List II
   1. Inadequate riser
   2. Loosely rammed sand mould
   3. Mould restraint
   4. Gas entrapment
   (a) P-3. Q-2, R-4, S-l
   (b) P-3. Q-2, R-l, S-4
   (c) P-l, Q-4, R-2, S-3
   (d) P-4, Q-3, R-2, S-l

   [GATE 98]

7. Suggest which of the following is the correct practice of using chills.

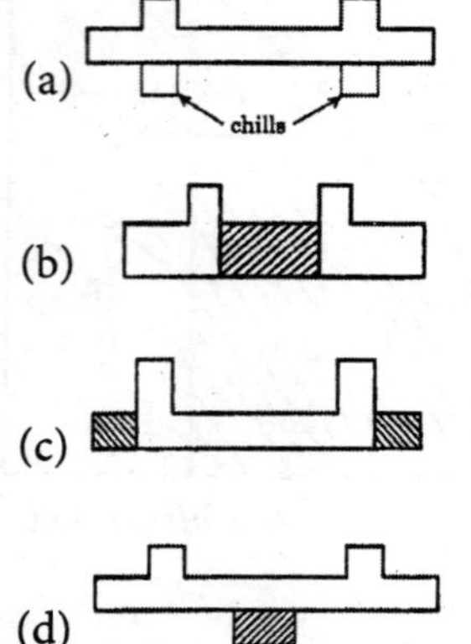

   [GATE 99]

8. In a casting, cold shut defect is caused by
   (a) low pouring temperature
   (b) melt-mould reaction
   (c) faulty gating system
   (d) very low mould permeability.

   [GATE 99]

9. Mould coating material which helps in grain refinement of metal casting is
   (a) cobalt aluminide (b) zinc
   (c) tellurium (d) boron

   [GATE 2000]

10. Solidification time of a thin plate casting in a sand mould increases by approximately 4 times with the doubling of
    (a) thickness
    (b) length
    (c) both width and length
    (d) length, width and thickness.

    [GATE 2002]

11. The solidification time for a cube casting having 5 cms side is 10 minutes. Under similar condition of heat transfer, the solidification time, in minutes, for a cube casting having 10 cm side will be
    (a) 10 (b) 20
    (c) 40 (d) 80

    [GATE 2003]

12. Draft allowance given to patterns is for
    (a) compensating the liquid state shrinkage
    (b) easy removal of pattern from the mould cavity
    (c) providing support for the core placement
    (d) compensating the solidification shrinkage

    [GATE 2004]

13. Two square shaped steel plates A and B have thickness of 4 and 8 cms respectively. They are sand cast under identical conditions. Plate A takes 10 minutes to solidify. Plate B would solidify in .... minutes.
    (a) 16 (b) 20
    (c) 30 (d) 40

    [GATE 2004]

Answers

6. (a)	7. (a)	8. (a)	9. (a)	10. (a)	11. (c)
12. (b)	13. (d)				

14. A vertical tapered sprue of 16 cms length is kept full during pouring. To just avoid any aspiration, the cross-sectional areas at the centre and the bottom of the sprue must be in the ratio of
 (a) 1:1 (b) $\sqrt{2}$: 1
 (c) 2:1 (d) 4:1
 [GATE 2004]

15. Casting alloys with a large freezing range tend to result in
 P. shrinkage cavity
 Q. distributed shrinkage porosity
 R. liquid metal with poorer fluidity
 S. liquid metal with superior fluidity
 (a) P, R (b) Q, R
 (c) P, S (d) Q, S.
 [GATE 2004]

16. A jet engine turbine blade is normally manufactured by
 (a) forging
 (b) shell moulding
 (c) investment casting
 (d) pressure die casting
 [GATE 99]

17. Match the defects in group 1 with possible causes in group 2.
 Group 1
 P. Misruns Q. Shifts
 R. Blowholes S. Scabs
 Group 2
 1. Sand expansion
 2. Molding materials
 3. Improper heating of metals
 4. Molding and core making
 (a) P-3, Q-4, R-2, S-l
 (b) P-l, Q-2, R-3, S-4
 (c) P-4, Q-3, R-2, S-l
 (d) P-2, Q-l, R-4, S-3
 [GATE 2005]

18. For designing a pressurised gating system in sand casting, the gating ratio should be
 (a) 1: 0.75 : 0.50 (b) 1:1:1
 (c) 1:2:4 (d) 1 : 3 : 3
 [GATE 2005]

19. During metal casting of a slab, the thickness of solid formed after time 't' is proportional to
 (a) $t^{1/3}$ (b) $t^{1/2}$
 (c) t (d) t^2
 [GATE 2010]

20. Match the requirement from group 1 with the suitable casting process from group 2.
 Group 1
 P. Good surface finish
 Q. Expandable mould
 R. Heavy casting
 S. Hollow ornamental casting
 Group 2
 1. Slush casting
 2. Pressure die casting
 3. Investment casting
 4. Sand casting
 (a) P-2, Q-3, R-4, S-l
 (b) P-4, Q-2, R-4, S-l
 (c) P-4, Q-3, R-2, S-4
 (d) P-2, Q-l, R-3, S-4
 [GATE 2010]

21. Feed heads, feeders and risers in casting serve to provide sources of molten metal to compensate for
 (a) misruns (b) cold shuts
 (c) hot tears (d) shrinkage.
 [GATE 2006]

Answers

14. (b)	15. (c)	16. (c)	17. (a)	18. (a)	19. (b)
20. (a)	21. (d)				

22. In a casting, shrinkage occurs
    (a) only after transformation from liquid to solid
    (b) only during transformation from liquid to solid
    (c) before, during and after transformation from liquid to solid
    (d) only when the metal is in the liquid state

[GATE 2006]

23. In foundary practice, the fluidity of an alloy does not increase with increasing
    (a) superheat
    (b) channel size
    (c) flow velocity
    (d) heat transfer coefficient

[GATE 2007]

24. Riser is NOT required for the casting of
    (a) grey cast iron
    (b) white cast iron
    (c) Al- 4% Cu
    (d) Al- 12% Si

[GATE 2008]

25. Match the casting defects in group 1 with cause given in group 2.

    Group 1

    P. Hot tear Q. Misrun
    R. Blister S. Rat tail

    Group 2

    1. Insufficient melt superheat
    2. High residual stresses
    3. Improper venting
    4. Expansion of sand

    (a) P-l, Q-2, R-3, S-4
    (b) P-3, Q-4, R-l, S-2
    (c) P-4, Q-3, R-2, S-l
    (d) P-2, Q-l, R-3, S-4

[GATE 2008]

26. The casting processes that require expendable moulds are

    P. Investment casting
    Q. Low pressure casting
    R. Shell moulding
    S. Slush casting

    (a) P, Q (b) Q, R
    (c) R, S (d) P, R

[GATE 2008]

27. In a sound casting the last liquid to solidify is in the
    (a) runner (b) riser
    (c) gate (d) vent

[GATE 2009]

28. Assertion 'a' : During casting of aluminium, grain refinement can be achieved by addition of certain alloying elements.

    Reason 'r' : The addition of the alloying elements may result in the formation of deoxidation products or intermetallic compounds which may act as nucleation sites for grain refinement.

    (a) Both 'a' and 'r' are true, but 'r' is not the reason for 'a'
    (b) Both 'a' and 'r' are true and 'r' is the reason for 'a'
    (c) 'a' is true but 'r' is false
    (d) 'a' is false but 'r' is true

[GATE 2011]

29. The riser is designed such that the melt in the riser solidifies
    (a) Before casting solidifies.
    (b) At the same time as casting solidifies.
    (c) After casting solidifies.
    (d) Irrespective of the solidification of the casting.

[GATE 2012]

Answers

22. (b)	23. (b)	24. (a)	25. (d)	26. (d)	27. (b)
28. (a)	29. (c)				

30. Riser are *Not* required for casting
(a) stainless steel (b) plain carbon steel
(c) grey cast iron (d) white cast iron
[GATE 2013]

31. In a sand casting of a rectangular slab, the thickness of the solidified layer after 2 minutes is known to be 3 cm. Based on Chvorinov's rule, the solidified thickness (in cm) after 4 minutes is
(a) 1.8 (b) 4.3
(c) 6.8 (d) 9.2
[GATE 2014]

32. At the mould exit of a continuous caster, the metal consisting of a solidified shell with a liquid metal core exits at the rate of 35 kg s^{-1}. Given that the latent heat of fusion is 3 x 10^5 J kg^{-1} and the total rate of heat removal by the mould is 4.2 x 10^6 W, the mass fraction of solid at the mould exit is ....
Assume that both while solid and liquid remain at the melting point, they are in the mould.
(a) 0.1 (b) 0.4
(c) 0.8 (d) 1.5
[GATE 2015]

33. Match the features observed in castings listed in Group I with the most appropriate reasons listed in Group II.
Group I
P. Macrosegregation
Q. Fine grained structure
R. Porosity
S. dendrites
Group II:
1. Inoculation
2. Gas evolution and shrinkage
3. Temperature gradients and supercooling
4. Density difference and convection currents
(a) P-l, Q-3, R-2, S-4
(b) P-4, Q-l, R-2, S-3
(c) P-2, Q-4, R-l, S-3
(d) P-4, Q-l, R-3, S-2
[GATE 2015]

34. The operation **NOT** associated with casting is
(a) Gating (b) Fettling
(c) Stack moulding (d) Calendaring
[GATE 2016]

35. Mould heat flux (q_m) for billet casters is expressed (in SI unit) as a function of distance below the meniscus (z)

$$q_m (z) = \left[2.67 - 0.33 \sqrt{\frac{z}{U_c}} \right] \times 10^6 \quad (0 \leq z \leq L_m)$$

if mould (L_m) is 0.8 m and casting speed (U_c) is 0.2 m.s^{-1}, the average mould flux (in mW. m^{-2}) is ....
[GATE 2016]

36. In a sand-mould, a sprue of 0.25 in height and a top cross-section area of 2.2 m^2 is provided to maintain the melt flow rate at 4 m^3 .s^{-1}. To prevent aspiration of molten metal, the maximum cross-section area (in m^2) at the base of the sprue is ....
[GATE 2016]

37. For casting a cylindrical aluminum bloom having a length of 1000 mm and diameter of 700 mm. the approximate solidification time (in minutes) estimated using Chvorinov's rule is
(a) 45 (b) 316
(c) 440 (d) 620
[GATE 2016]

Answers

30. (c) 31. (c) 32. (b) 33. (b) 34. (d) 35. (2.0 to 2.5)
36. (1.7 to 1.9) 37. (d)

Section

B

13

Welding Metallurgy

1. A weldment consists of
   (a) fused part of the weld
   (b) fused part and the heat affected zone (HAZ)
   (c) fused part + HAZ + base metal
   (d) HAZ

[GATE 94]

2. Stainless steel is welded using
   (a) oxy-acetylene flame
   (b) oxy-hydrogen flame
   (c) arc welding
   (d) inert gas arc welding

[GATE 95]

3. Preheating prior to welding is done for the purpose of
   (a) decreasing cooling rate
   (b) facilitating fusion of high melting metals
   (c) preventing hot cracking
   (d) ensuring full penetration

[GATE 96]

4. In an arc welding experiment, the current, voltage and electrode travel speed were 140 A, 22 V, and 15 cm/min respectively. The heat input per unit length of the weld is given by .... kJ/cm.
   (a) 24.64 (b) 4.62
   (c) 9.24 (d) 12.32

[GATE 96]

5. Thermit welding uses the following energy sources.
   (a) electrical energy
   (b) chemical energy
   (c) energy of high velocity electrons
   (d) heat generated by friction

[GATE 98]

6. Melting rate of electrodes in manual metal arc welding (MMAW) process is mainly governed by the
   (a) welding current
   (b) arc voltage
   (c) type of coating
   (d) length of the electrode

[GATE 99]

7. In multipass are welds shot peening is done after each pass to
   (a) close the surface porosity
   (b) break the continuity of columner grains
   (c) flatten the weldment
   (d) introduce texture in the weld.

[GATE 99]

Answers

1. (c)	2. (d)	3. (c)	4. (d)	5. (b)	6. (c)
7. (a)					

8. In TIG welding thoriated tungsten electrodes are used because it
   (a) has higher current carrying capacity
   (b) has better electron emissivity
   (c) is stronger than ordinary tungsten
   (d) is easy to prepare

[GATE 99]

9. "Cold cracking" in the heat affected zone of a high strength steel weld can take place because of
   (a) retained austenite
   (b) martensite formation
   (c) relatively high sulphur content in the base metal
   (d) sufficient hydrogen present in the welding arc

[GATE 2000]

10. Brazing filler metal used for joining steel plates
    (a) melts below the melting point of base metals
    (b) melts below 300°C
    (c) is copper-phosphorous alloy
    (d) is copper

[GATE 2000]

11. The titanium alloys are welded using the following process.
    (a) TIG welding
    (b) Submersible arc welding
    (c) Butt welding
    (d) Electron beam welding

[GATE 2001]

12. Lap joints are preferred over butt joints in soldering/brazing because
    (a) these are weaker in tension but stronger in shear.
    (b) these are weaker in shear but stronger in tension.
    (c) these are stronger in both shear and tension.
    (d) the lap joints are easily made.

[GATE 2002]

13. Preheating of steel plate during welding is required to
    (a) reduce the heat input
    (b) increase the heat input
    (c) increase the cooling rate
    (d) decrease the cooling rate

[GATE 2003]

14. Two sheets of iron and copper of equal thickness are welded together into one sheet. Iron is in contact with water at 298 K and copper is in contact with steam at 373 K. At steady state, the temperature profile within the sheet will be
    (a) parabolic
    (b) hyperbolic
    (c) combination of two separate linear profiles in copper and iron
    (d) combination of two separate parabolic profiles in copper and iron.

[GATE 2003]

15. Match the following:

    Process
    P. Ultarasonic welding
    Q. Plasma arc welding
    R. Thermit welding
    S. Shielded metal arc welding

    Characteristics
    1. Most commonly used process
    2. Use of explosives
    3. Highest temperature
    4. Solid state process
    5. Use of chemical energy

    (a) P - 4, Q - 3, R - 5, S - 1
    (b) P - 2, Q - 4, R - 5, S - 1
    (c) P - 1, Q - 2, R - 4, S - 3
    (d) P - 3, Q - 5, R - 2, S - 4

[GATE 2003]

Answers

8. (b) 9. (b) 10. (a) 11. (a) 12. (d) 13. (d)
14. (c) 15. (a)

16. The grain size in the heat affected zone (HAZ) of a weld is maximum
    (a) near the weld metal/HAZ interface
    (b) near the base metal/HAZ interface
    (c) in the middle of HAZ
    (d) at a location 1/3rd away from the weld/HAZ interface

[GATE 2004]

17. Steels with high carbon equivalent have poor weldability because in these steels during welding
    (a) carbon and other alloying elements get oxidised from the weld pool.
    (b) excessive ferrite forms in the heat affected zone leading to the poor toughness of the weld.
    (c) martensite forms in the heat affected zone leading to poor toughness/ductility of the weld.
    (d) segregation of carbon and other elements occurs in the weld pool leading to poor properties of the weld.

[GATE 2004]

18. *Match the following:*
    Group 1
    P. Thermit welding
    Q. Arc welding
    R. Welding in solid state
    S. Friction welding
    Group 2
    1. Globular metal transfer
    2. Mechanical energy converted to heat energy
    3. Exothermic process
    4. Diffusion bonding
    (a) P - 3, Q - 1, R - 4, S - 2
    (b) P - 3, Q - 2, R - 4, S -1
    (c) P - 2, Q -1, R - 3, S - 4
    (d) P -1, Q - 2, R - 3, S - 4

[GATE 2004]

19. In the consumable electrode shielded metal arc welding process, which of the following observations are correct ?
    P. The weld composition does not depend upon the basicity of the slag.
    Q. Basic flux is used to achieve low hydrogen content and low inclusion content in the weld metal.
    R. Reverse polarity of the direct current power source is used to weld thin sheets.
    S. Coated electrodes are classified by tensile strength of the weld metal deposit.
    (a) P, Q (b) P, R
    (c) Q, S (d) R, S

[GATE 2005]

20. Which of the following is NOT a solid state welding process?
    (a) Friction stir welding
    (b) Ultrasonic welding
    (c) Explosive welding
    (d) Flux cored arc welding

[GATE 2010]

21. Match the welding processes in group 1 with the principal mechanism in group 2.
    Group 1
    P. Seam welding Q. Explosive welding
    R. Gas welding S. TIG welding
    Group 2
    1. Ionisation and plasma
    2. Exothermic reaction
    3. Interfacial deformation
    4. Interfacial resistance heating
    5. Interfacial diffusion of atoms
    (a) P - 5, Q - 3, R - 1, S - 4
    (b) P - 5, Q - 3, R - 2, S - 1
    (c) P - 4, Q - 3, R - 2, S - 1
    (d) P -- 5, Q - 2, R - 1, S - 4

[GATE 2006]

Answers

16. (c) 17. (c) 18. (a) 19. (b) 20. (d) 21. (c)

22. For better resistance welding, the metal must have
    P. high electrical resistivity and low melting point
    Q. high thermal conductivity
    R. high electrical resistivity and high melting point
    S. low thermal conductivity
    (a) P, Q (b) P, S
    (c) R, Q (d) R, S
    [GATE 2007]

23. The order of decreasing weldability among the following steel is
    P. Fe-0.6%C Q. Fe 0.4% C
    R. HSLA
    (a) R → Q → P (b) P → Q → R
    (c) Q → P → R (d) Q → R → P
    [GATE 2008]

24. Match the welding process in group 1 with the source of heat in group 2.
    Group 1
    P. Ultrasonic welding
    Q. Spot welding
    R. SMAW
    S. Thermit welding
    Group 2
    1. Thermochemical 2. Electrical resistance
    3. Friction 4. Electric arc
    (a) P - 3, Q - 2, R - 1, S - 4
    (b) P - 4, Q - 3, R - 2, S - 1
    (c) P - 1, Q - 3, R - 4, S - 2
    (d) P - 3, Q - 2, R - 4, S - 1
    [GATE 2008]

25. Match the joining processes in group I with the filler material in group 2.

Group 1	Group 2
P. Soldering	1. Silver-titanium alloy
Q. Welding	2. Silver-tin alloy
R. Brazing	3. Mild steel
	4. Lead fluoride

    (a) P-2, Q-3, R-l
    (b) P-l, Q-2, R-3
    (c) P-3, Q-l, R-2
    (d) P-2, Q-4.R-1
    [GATE 2009]

26. Flux in welding process acts as
    (a) catalyst (b) protective agent
    (c) filler (d) heat generator
    [GATE 2009]

27. In brazing process, the liquid metal fills the gap by which one of the following means?
    (a) Capillary infiltration
    (B) Gravity infiltration
    (c) Pressure infiltration
    (d) Vacuum infiltration
    [GATE 2011]

28. The nature of submerged arc welding flux with basicity index of 0.5 is
    (a) neutral (b) basic
    (c) semi-basic (d) acidic
    [GATE 2011]

29. Which one of the following carbon equivalent in steel is considered good for weldability ?
    (a) 1.0 (b) 0.8
    (c) 0.6 (d) 0.4
    [GATE 2011]

30. Which of the following is NOT a fusion welding process?
    (a) Arc welding
    (b) Gas welding
    (c) Resistance welding
    (d) Friction stir welding
    [GATE 2012]

Answers

22. (b)	23. (c)	24. (d)	25. (a)	26. (b)	27. (a)
28. (d)	29. (d)	30. (d)			

31. For a good quality brazing, the molten filler alloy should have
    (a) low contact angle with the base metal.
    (b) low density.
    (c) high surface tension.
    (d) high viscosity.

[GATE 2013]

32. Arc welding is done using current, voltage and welding speed of 200 A, 20 V and 0.01 m/s, respectively. The heat input in KJ per unit length is
    (a) 100 (b) 200
    (c) 300 (d) 400

[GATE 2013]

33. Which one of the following welding processes uses a non-consumable electrode?
    (a) Gas tungsten arc welding
    (b) Gas metal arc welding
    (c) Submerged arc welding
    (d) Flux cored arc welding

[GATE 2014]

34. Determine the correctness or otherwise of the following Assertion (a) and Reason (r).
    Assertion (a): During welding of aluminium alloys by the gas tungsten arc welding process, direct current reverse polarity (DCRP) mode is used.
    Reason (r): DCRP facilitate breaking up of the oxide film on the workpiece surface.
    (a) a is true but r is false.
    (b) a is false but r is true.
    (c) both a and r are true, and r is the reason for a.
    (d) both a and r are true, and r is not the reason for a.

[GATE 2014]

35. Which of the following is NOT a solid state metal joining technique?
    (a) Ultrasonic welding
    (b) Friction welding
    (c) Diffusion bonding
    (d) Electroslag welding

[GATE 2015]

36. Determine the correctness or otherwise of the following Assertion [a] and the Reason [r].
    Assertion: In a pure metal weld, elastic modulus in the heat affected zone (HAZ) is the same as that in the base metal.
    Reason: Coarse grained microstructure in the HAZ results in lower hardness.
    (a) Both [a] and [r] are true and [r] is the correct reason for [a].
    (b) Both [a] and [r] are true and [r] is not the correct reason for [a],
    (c) Both [a] and [r] are false.
    (d) [a] is true but [r] is false.

[GATE 2015]

37. Of the following welding processes
    [P] Laser Beam Welding
    [Q] Submerged Arc Welding
    [R] Metal Inert Gas Welding
    the width of the heat-affected zone in decreasing order is
    (a) $P > Q > R$ (b) $Q > R > P$
    (c) $R > P > Q$ (d) $P > R > Q$

[GATE 2016]

Answers

31. (a)	32. (d)	33. (a)	34. (c)	35. (d)	36. (b)
37. (b)					

14

Electrometallurgy and Corrosion

1. A pipeline buried in soil is commonly protected from corrosion by
   (a) anodic protection
   (b) cathodic protection
   (c) using inhibitors
   (d) using a special alloy resistant to corrosion.
   [GATE 95]

2. Two electrolytic cells with $CuSO_4$ and $AgNO_3$ solutions are connected in series to a power source. If after 10 minutes the weight of copper (atomic weight 63.5) deposited in the first cell is 63.5 gms, the corrosponding weight of silver (atomic weight 108) deposited in the second cell is .... gms.
   (a) 108 (b) 54
   (c) 216 (d) 81
   [GATE 95]

3. Match the corrosion phenomenon with the material in which they are commonly observed.
   P. Season cracking
   Q. Pitting corrosion
   R. Galvanic corrosion
   S. Caustic embrittlement
   1. 70 : 30 brass
   2. GI water line fitted with a brass tap
   3. Duralumin
   4. Boiler steel
   (a) P-l, Q-2, R-3, S-4
   (b) P-4, Q-3. R-2, S-l
   (c) P-l, Q-2, R-4, S-3
   (d) P-3, Q-4, R-l, S-2
   [GATE 95]

4. The cathode in an electrochemical cell always carries
   (a) negative charge
   (b) positive charge
   (c) zero charge
   (d) positive or negative charge depending upon the nature of the cell.
   [GATE 96]

5. Match the following:
   P. Sensitization Q. Sacrificial anode
   R. Nickel plating S. Conversion coating
   1. Phosphating
   2. Anodic protection
   3. Cathodic protection
   4. Stainless steel
   (a) P-4, Q-3, R-2, S-l
   (b) P-l, Q-2, R-3, S-4
   (c) P-2. Q-3, R-4, S-l
   (d) P-3, Q-2, R-l, S-4
   [GATE 96]

Answers

1. (b) 2. (a) 3. (a) 4. (a) 5. (c)

6. Match the following :
   P. Dezincification
   Q. Flade potential
   R. Intergranular corrosion
   S. Tafel relationship
   1. Reactivation
   2. Stainless steel
   3. Polarisation
   4. Brasses
   (a) P-4, Q-l, R-3, S-2
   (b) P-4, Q-l, R-2, S-3
   (c) P-4, Q-3, R-2, S-l
   (d) P-l, Q-4, R-3, S-2

[GATE 98]

7. Match the following:
   P. Cathodic protection
   Q. Hydrogen attack
   R. Pilling Bedworth ratio
   S. Cavitation
   1. Decarburisation
   2. Erosion
   3. Galvanising
   4. Oxidation of metals
   (a) P-3, Q-l, R-4, S-2
   (b) P-3, Q-l, R-2, S-4
   (c) P-4, Q-2, R-l, S-3
   (d) p-4, Q-3, R-2, S-l

[GATE 98]

8. For oxide layer to be protective, the Pilling Bedworth ratio should be
   (a) close to 1 (b) less than 0.5
   (c) nearly 5 (d) greater than 10.

[GATE 99]

9. The following is employed as sacrificial anode to provide cathodic protection to ships, pipelines and storage tanks :
   (a) Iron (b) Copper
   (c) Nickel (d) Magnesium,

[GATE 2001]

10. The thickness of oxide film is 'y' at a time 't' If K_1, K_2 and K_3 are the temperature dependent constants, the parabolic law of oxidation is given by
    (a) $y^2 = 2K_1t + K_2$
    (b) $y = K_1 \log(K_2t + K_3)$
    (c) $y = K_1f + K_2$
    (d) $y = K_1t^3 + K_2$

[GATE 2003]

11. Identify the false statement:
    (a) Martensitic steel are less susceptible to pitting corrosion than austenitic steels.
    (b) Pitting corrosion is usually very localised.
    (c) Hydrogen embrittlement is facilitated by tensile stress.
    (d) Stress corrosion cracking is facilitated by tensile stress.

[GATE 2004]

12. The oxygen activity in a liquid metal can be measured by a concentration cell using stabilised zirconia electrolyte $PO_2^{(1)}$ | ZrO_2 based solid electrolyte | $PO_2^{(2)}$
    Where, $PO_2^{(1)}$ is the reference oxygen pressure and $PO_2^{(2)}$ is the oxygen pressure in equilibrium with liquid metal. If, F is Faraday constant, the EMF of the cell is equal to

    (a) $-\dfrac{RT}{2F} \cdot \dfrac{PO_2^{(2)}}{PO_2^{(1)}}$ (b) $-\dfrac{RT}{4F} \cdot \dfrac{PO_2^{(2)}}{PO_2^{(1)}}$

    (c) $-\dfrac{RT}{2F} \ln \dfrac{PO_2^{(2)}}{PO_2^{(1)}}$ (d) $-\dfrac{RT}{4F} \ln \dfrac{PO_2^{(2)}}{PO_2^{(1)}}$

[GATE 2004]

Answers

6. (b) 7. (a) 8. (a) 9. (d) 10. (a) 11. (a)
12. (d)

13. The difference in potential between two standard half cells is $\Delta E°$. If the half cell are galvanically coupled, the value of $\Delta E°$, with changes in concentration of ions in the solution, is such that
   (a) $\Delta E°$ decreases with increase in concentration of the ions.
   (b) $\Delta E°$ increases linearly with increase in concentration of the ions.
   (c) $\Delta E°$ increases exponentially with increase in concentration of the ions.
   (d) $\Delta E°$ is constant.

[GATE 2005]

14. Which of these metals can not be electroplated from aqueous electrolyte ?
   (a) Al (b) Cu
   (c) Ni (d) Zn

[GATE 2010]

15. The difference in reversible potential between oxygen reduction reaction and hydrogen evolution reaction of any pH in an aqueous electrolyte is .... volt.
   [Given, standard reduction potentials for hydrogen evolution reaction, $E^0 2H^+/H_2$ = OV, SHE and oxygen reduction reaction : $E°O_2/4OH$ = 0.4V, SHE. Also $PH_2 = PO_2 = 1$ atm]
   (a) 0 (b) 0.41
   (c) 0.82 (d) 1.23

[GATE 2010]

16. Identify the incorrect statement.
   (a) A concentration gradient in the electrolyte may lead to the formation of galvanic cell.
   (b) Chromium is added to improve the oxidation resistance of stainless steel.
   (c) Cathodic protection can be provided by applying a coating.
   (d) A steel bolt or nut is permissible on a large copper vessel.

[GATE 2007]

17. Intergrannular corrosion of light 18-8 stainless steel can NOT be prevented by
   (a) reducing the carbon content to less than 0.05%.
   (b) quenching it from high temperature to prevent chromium carbide precipitation.
   (c) adding strong carbide forming elements.
   (d) increasing the carbon content.

[GATE 2008]

18. Match applications in group 1 with the commonly used corrosion protection methods in group 2.
   Group 1
   P. Sea going vessel
   Q. Underground pipeline
   R. Electric traction tower
   S. Electric poles
   Group 2
   1. Inorganic coating
   2. Sacrificial anode
   3. Aluminium paint
   4. Impressed current
   5. Galvanising
   (a) P-2, Q-4, R-5, S-3
   (b) P-2, Q-3, R-5, S-l
   (c) P-l, Q-2, R-5, S-4
   (d) P-4, Q-3, R-l, S-2

[GATE 2008]

Answers

13. (c) 14. (a) 15. (d) 16. (d) 17. (d) 18. (a)

19. Match the form of corrosion in group 1 with the typical examples in group 2.

Group 1
P. Filiform corrosion
Q. Crevice corrosion
R. Galvanic corrosion
S. Stress corrosion cracking

Group 2
1. Austenitic stainless steel in chloride environment
2. Nut bolt with gasket
3. Painted food cans
4. Steel studs in copper plate

(a) P-3, Q-2, R-4, S-l
(b) P-l, Q-3, R-4, S-2
(c) P-3, Q-4, R-2, S-l
(d) P-2, Q-3, R-4, S-l

[GATE 2008]

20. From the list given below :

P. Cu Q. Mg
R. Ni S. Zn

Two metals which provide cathode protection to steel are:
(a) P, R
(b) R, S
(c) Q, R
(d) Q, S

[GATE 2009]

21. Zinc rod is immersed in dilute HCl (pure). If a very small amount of $FeCl_3$ is added to the solution, the corrosion rate of zinc
(a) decreases
(b) increases
(c) remains constant
(d) is zero (possivation)

[GATE 2011]

22. A metal .is electrochemically polarised to a potential which is higher than the standard reduction potential of the metal. The overvoltage will be
(a) zero
(b) negative
(c) positive
(d) initially negative, then positive

[GATE 2011]

23. $Al_2O_3 + 6H^+ = 3H_20 + 2Al, \Delta G° = 897.3$ kJ, where hydrogen ion concentration is unity. The reaction potential of the above reaction under standard state will be
(a) – 155 V
(b) – 1.40 V
(c) 1.65 V
(d) 1.75 V

[GATE 2011]

24. Match the principle given in Group I with corresponding corrosion terminology in Group II.

Group I:
P. Electrode polarization
Q. Passivity
R. Selective leaching
S. Grain boundary precipitation

Group II:
1. Dezincification
2. Intergranular attack
3. Over voltage
4. Surface oxide film

(a) P-3, Q-4, R-l, S-2
(b) P-3, Q-4, R-2, S-l
(c) P-4, Q-2, R-l, S-3
(d) P-2, Q-l, R-4, S-3

[GATE 2012]

Answers

19. (a)	20. (d)	21. (d)	22. (c)	23. (a)	24. (a)

25. Identify the correct combination of the following statements.
    P. Hydrogen electrode is a standard used to measure redox potentials.
    Q. Activation polarization refers to electrochemical process controlled by reaction sequence at metal-solution interface.
    R. Potential-pH diagrams can be used to predict corrosion rates of metals.
    S. Cathodic protection can use sacrificial anodes such as magnesium.
    (a) P, Q, and R (b) Q, R and S
    (c) P, Q and S (d) P, R and S
    [GATE 2012]

26. Based on the standard galvanic series, select the CORRECT sequence of metals in the increasing order of anodic behaviour:
    (a) Zn, Fe. Pt, Cu
    (b) Pt, Zn, Cu, Fe
    (c) Fe. Pt, Cu, Zn
    (d) Pt, Cu, Fe, Zn
    [GATE 2015]

27. In electrolytic refining of nickel (Ni), the anode is Cu-10 atom% Ni and the cathode is pure Ni. Assuming the Cu-Ni solution to be ideal, the absolute value of the minimum voltage (in mv) required for refining is.
    Given, Faraday's constant = 96490 C $mole^{-1}$,
    Temperature = 300 K, R = 8.314 J. mol^{-1}. K^{-1}
    (a) 10 (b) 15
    (c) 20 (d) 30
    [GATE 2015]

28. The passive film in stainless steel forms above the
    (a) Primary passive potential
    (b) Breakdown potential
    (c) Trans-passive potential
    (d) Pitting potential
    [GATE 2016]

Answers

25. (c) 26. (d) 27. (d) 28. (a)

15 Metallurgical Testing and Inspection

1. The technique (s) which can be used for the direct observation of dislocations is (are)
   (a) scanning electron microscopy
   (b) transmission electron microscopy
   (c) field-ion microscopy
   (d) electron probe micro analysis

[GATE 91]

2. Which of the following are considered applications of ultrasonic testing ?
   (a) Determination of elastic constants
   (b) Detection of defects in metals
   (c) Measurement of material thickness
   (d) none of these

[GATE 92]

3. The best method for determining the average hardness of an aluminium casting is
   (a) Rockwell A
   (b) Rockwell C
   (c) Knoop
   (d) Brinnel
   (e) Vickers

[GATE 94]

4. A pearlitic steel is observed under an optical microscope which has a numerical aperture (N.A) of 1.5 and uses the radiation of 4500A. The minimum lamellar spacing (in A) which can be resolved using this microscope is
   (a) 500 (b) 750
   (c) 1000 (d) 1500

[GATE 94]

5. Match the examination/inspection technique(s) in the following cases :
   P. Internal crack developed in a component due to faulty metal flow conditions during its forging.
   Q. Segregation in steel casting.
   R. Hydride formation in a Tia Boy component due to its exposure to H_2 at high temperature.
   S. Surface cracks formed on a car body during its manufacture.
   1. X-ray diffraction
   2. Dye penetrant test
   3. Sulphur printing
   4. Radiography
   (a) P-4, Q-3, R-l, S-2
   (b) P-3, Q-4, R-l, S-2
   (c) P-4, Q-3, R-2, S-l
   (d) P-2, Q-l, R-3, S-4

[GATE 94]

Answers

1. (b) 2. (b) 3. (b) 4. (b) 5. (a)

6. The ASTM grain size number N for a structural steel which shows 65 grains per square inch at a magnification of 100 X is
   (a) 1 (6)3
   (c) 5 (d) 7
   [GATE 95]

7. Brinell hardness measurement made with a 10 mm diameter steel ball at a load of 1000 kg gives an indentation of diameter 4 mm in a material. The BHN of this material is approximately
   (a) 75 (b) 100
   (c) 300 (d) 600
   [GATE 95]

8. Match the following:
   P. Rockwell hardness
   Q. Knoop hardness
   R. Vickers hardness
   S. Meyer hardness
   1. Unrecovered projected area of indentation
   2. Surface area of indentation
   3. Depth of indentation
   4. Projected area of indentation
   (a) P-3, Q-l, R-2, S-4
   (b) P-3, Q-l, R-4, S-2
   (c) P-3, Q-4, R-2, S-l
   (d) P-2, Q-3, R-l, S-4
   [GATE 96]

9. In the X-ray radiography technique the tube voltage for thicker plates as compound to the thin plates, should be
   (a) higher as it gives higher wavelength
   (b) lower as it gives higher wavelength
   (c) higher as it gives shorter wavelength
   (d) lower as it gives shorter wavelength.
   [GATE 98]

10. By X-ray diffraction, it is not possible to determine
    (a) preferred orientation
    (b) residual localised stresses
    (c) particle size
    (d) ordered arrangement of magnetic moments
    [GATE 98]

11. The swift cup test evaluates the following property of a sheet metal:
    (a) Stretchability (b) Drawability
    (c) Bendability (d) none of these.
    [GATE 98]

12. The tensile load-elongation curve of a metal does not describe
    (a) work hardening
    (b) yield stress
    (c) anisotropy index
    (d) necking strain
    [GATE 98]

13. Match the technique for proper distinction between the two :
    P. Soft and hard magnets
    Q. Amorphous and crystalline materials
    R. n and p-type semi-conductors
    S. Cast and wrought alloys
    1. Optical microscopy
    2. Hall effect measurement
    3. X-ray diffraction
    4. Magnetic hysteresis
    (a) P-4, Q-3, R-l, S-2
    (b) P-l, Q-3, R-4, S-2
    (c) P-4, Q-3, R-2, S-l
    (d) P-2. Q-l, R-3, S-4
    [GATE 98]

Answers

6. (d)	7. (d)	8. (a)	9. (c)	10. (b)	11. (b)
12. (c)	13. (c)				

14. Fatigue life is expected to increase by
(a) increasing the size of the sample
(b) smooth polishing of the surface of the sample
(c) having compressive residual stresses at the surface
(d) having tensile residual stresses at the surface

[GATE 99J

15. Soft materials are tested on Rockwell
(a) C scale
(b) B scale
(c) with 1.6 mm steel ball and a 100 kg major load
(d) with diamond indentor and a 150 kg major load

[GATE 2000J

16. A brale indentor used in Rockwell hardness test has the following geometry and material.
(a) Square based diamond pyramid with 136° included angle.
(b) Conical shape steel with 120° apex angle.
(c) Conical shaped diamond with 120° apex angle.
(d) 10 mm diameter hardened steel ball.

[GATE 2001]

17. Hydrogen embrittled steels samples can exhibit
(a) only clevage fracture
(b) clevage, dimple and intergranular fracture
(c) both clevage and dimple fracture
(d) only intergranular fracture

[GATE 2002]

18. Ultrasonic testing can be used for (choose the correct combination of the following statements P, Q, R, S):
P. Quantitative analysis of phases
Q. Determination of elastic constants
R. Determination of endurance limit
S. Detection of internal defects
(a) P, R (b) P, Q
(c) Q, S (d) R, S

[GATE 2003]

19. Radiographic appearance of inclusions resembles
(a) dark patches as compared to the background.
(b) bright patches as compared to the background.
(c) dark or bright patches depending on radiation energy.
(d) dark or bright patches depending on relative density.

[GATE 2003]

20. The Brinell hardness measurement of a material is made by applying a 500 kg load with a 10 mm ball indentor. If the indentation diameter is 6 mm, the Brinell hardness (kg/mm^2) will be
(a) 31.8 (b) 45
(c) 60 (d) 159

[GATE 2003]

21. Match the NDT methods with the type of defects detected.
NDT methods
P. Ultrasonic Q. X-ray
R. Eddy current S. Liquid penetrant
Type of defects detected
1. Internal 2. Most
3. External 4. Surface breaking
(a) P-l, Q-2, R-3, S-4
(b) P-3, Q-l, R-2, S-4
(c) P-2, Q-l, R-3, S-4
(d) P-2, Q-4, R-l, S-3

[GATE 2004]

Answers

14. (b & c)	15. (b & c)	16. (c)	17. (d)	18. (c)	19. (a)
20. (a)	21. (a)				

22. Match the following :

Group 1

P. Ultrasonic

Q. Radiography

R. Eddy current

S. Magnetic particle

Group 2

1. change in accoustic impedence
2. change in thermal conductivity
3. change in electrical conductivity
4. change in density
5. change in magnetic flux leakage

(a) P-l, Q-3, R-2, S-5

(b) P-2, Q-3, R-4, S-l

(c) P-l, Q-4, R-3, S-5

(d) P-5, Q-l, R-4, S-3

[GATE 2004]

23. A tensile specimen with 20 mm gauge length was pulled with a nominal strain rate of 2.2 x 10^{-3}/sec. The true strain rate of the specimen at 2 mm extension is

(a) 1.1 x 10^{-2}/sec

(b) 1.1 x 10^{-3}/sec

(c) 2.0 x 10^{-3}/sec

(d) 2.2 x 10^{-3}/sec

[GATE 96]

24. Match the defects with the most suitable NDT method for examining them.

Group 1

P. Porosity

Q. Lack of fusion in welds

R. Surface cracks in aluminium

S. Microcrack initiation in austenitic stainless steel tubing

Group 2

1. Accoustic emission
2. X-ray radiography
3. Ultrasonic test casting
4. Dye-penetrant inspection

(a) P-3, Q-2, R-4, S-l

(b) P-3, Q-2, R-l, S-4

(c) P-2, Q-3, R-4, S-l

(d) P-4, Q-3, R-2, S-l

[GATE 96]

25. It is observed that the tensile flow stress of an aluminium alloy decreases on increasing the strain rate during the test. The possible magnitude for the strain rate sensitivity of the alloy is

(a) 0

(b) 0.01

(c) 0.1

(d) −0.06

[GATE 98]

26. The resolution of an optical microscope is of the order of

(a) 1 nm

(b) lμm

(c) 1 mm

(d) 1 cm

[GATE 2005]

27. The indentor used in the vicker's harness test is

(a) 10 mm dia steel ball

(b) 120° diamond cone with a slightly rounded point

(c) 3.2 mm dia steel ball

(d) square base diamond pyramid (included angle 136° between opposite faces)

[GATE 2005]

Answers

22. (c)	23. (d)	24. (c)	25. (c)	26. (b)	27. (d)

28. In the context of tensile test of a ductile alloy, the following observations were made. Choose the correct ones from the following observations:
    P. Per cent elongation is independent of the dimensions of the test specimens.
    Q. The number of necking instabilities are different in cylindrical and sheet specimens.
    R. The range of strain rates to be used for 'static tensile test' is 10^{-1} to 10^{-5} s^{-1}.
    R. The yield stress is independent of the temperature of the test.
    (a) Q, R, S (b) P. Q, R
    (c) Q, R (d) R, S

[GATE 2005]

29. A 150 mm long cylinderical rod having a diameter of 100 mm contains six microvoids which are situated at radial distances of 5 mm, 8 mm, 12 mm. 20 mm, 30 mm, 36 mm respectively, on a given r-z plane. Along the length (z-direction) of the rod, the voids are spaced uniformly. The specimen is to be tested by an ultrasonic cylinderical probe, 7.5 mm in diameter. The maximum number of flows, that can be detected in a single measurement by placing the probe at one end of the rod (flat surface) is
    (a) 1 (b) 2
    (c) 3 (d) 4

[GATE 2005]

30. Match the items in group 1 with the measurement methods in group 2.
    Group 1
    P. Heat of fusion Q. Grain size
    R. Hardness S. Internal craks
    Group 2
    1. Optical microscope
    2. Dye penetrant test
    3. Calorimetry
    4. Viscosity meter
    5. Ultrasonic technique
    6. Brinell test
    (a) P-3, Q-l, R-6, S-5
    (b) P-l, Q-4, R-2, S-3
    (c) P-4, Q-2, R-5, S-l
    (d) P-3, Q-4, R-2, S-6

[GATE 2005]

31. The energy dispersive spectrometer (EDS) in an electron microscope does chemical analysis by analysing the energy of
    (a) secondary electrons
    (b) characteristics X-rays
    (c) auger electrons
    (d) back-scattered electrons.

[GATE 2010]

32. Match the defects given in group I with the suitable non-destructive evaluation technique from group II.
    Group I
    P. Crack in a flat aluminium slab
    Q. Sub-surface porosity in a bronze casting
    R. Surface cracks in a steel tool
    S. Internal porosity in a ceramic block
    Group II
    1. Radiography
    2. Eddy current technique
    3. Ultrasonic technique
    4. Magnetic particle technique
    (a) P-3. Q-4, R-l, S-2
    (b) P-2, Q-4, R-l, S-3
    (c) P-4, Q-2, R-l, S-3
    (d) P-3, Q-2, R-4, S-l

[GATE 2010]

Answers

28. (c) 29. (a) 30. (a) 31. (b) 32. (d)

33. The tensile test of a sheet material exhibits 20% elongation in length and 10% decrease in width. The plastic strain ratio is
   (a) 2.37
   (b) 1.37
   (c) 1.17
   (d) 0.87

[GATE 2010]

34. Crack propagation in metallic materials is detected by the NDT method :
   (a) Eddy current testing
   (b) Magnetic particle inspection
   (c) Accoustic emission testing
   (d) Ultrasonic testing

[GATE 2006]

35. Rockwell hardness on the 'C' scale is measured using an indentor with a
   (a) 120° diamond cone with a slightly rounded tip
   (b) square base diamond pyramid
   (c) 10 mm diameter steel ball
   (d) 3 mm diameter steel ball

[GATE 2006]

36. In fracture control design procedures, *NDT* plays an important role because it primarily enables to
   (a) predict the time it will take for a given defect to grow to a critical size.
   (b) evaluate accurately the defect type, loca tion and size that exist in the material.
   (c) measure the mechanical properties of the design materials.
   (d) measure the grain sizes of the micro-structures in the materials.

[GATE 2006]

37. Match the *NDT* method in group 1 with the items in group 2.

   Group 1
   P. Ultrasonic testing
   Q. Dye penetrant testing
   R. Magnetic particle inspection
   S. Radiography
   Group 2
   1. Core shift in castings
   2. Surface defects in HSLA welds
   3. Fillet welds
   4. Hot crack in austenitic SS welds
   (a) P-3, Q-2, R-4, S-l
   (b) P-3, Q-4, R-2, S-l
   (c) P-l, Q-4, R-2, S-3
   (d) P-l, Q-2, R-3, S-4

[GATE 2006]

38. A suitable technique for monitoring a growing crack in an alloy is
   (a) accoustic emission
   (b) radiography
   (c) magnetic particle technique
   (d) liquid penetrant test

[GATE 2007]

39. The *NDT* technique used to detect deep lying defects in a large sized casting is
   (a) liquid penetrant inspection
   (b) magnetic particle inspection
   (c) ultrasonic inspection
   (d) eddy current inspection

[GATE 2008]

40. A property that can not be obtained from a tensile test is
   (a) Young's modulus
   (b) Yield strength
   (c) Ultimate tensile strength
   (d) Endurance limit

[GATE 2009]

Answers

33. (b)	34. (b)	35. (a)	36. (b)	37. (b)	38. (c)
39. (c)	40. (d)				

41. Match the properties in Group 1 with the testing techniques in Group 2,

Group 1

P. Electrical conductivity

Q. Impact energy

R. Thermal expansion

S. Specific heat

Group 2

1. Jominy test

2. Izod test

3. Dilatometry

4. Four probe technique

5. Differential scanning calorimetry

(a) P-4, Q-2, R-5, S-l

(b) P-5, Q-3, R-2, S-l

(c) P-2, Q-l, R-3, S-4

(d) P-4, Q-2, R-3, S-5

[GATE 2009]

42. For a given ductile materia, which one of the following tensile properties obtained with non-standard specimen is *NOT* comparable to that obtained with standard specimen?

(a) Elongation to fracture

(b) Tensile strength

(c) Uniform elongation

(d) Yield strength

[GATE 2011]

43. Match the hardness test methods in group 1 with the indentor used in Group 2.

Group 1

P. Brinnel hardness

Q. Vickers hardness

R. Rockwell C hardness

S. Rockwell B hardness

Group 2

1. Brale indentor

2. Square base diamond pyramid

3. 10 mm diameter steel ball

4. 1.6 mm diameter steel ball

(a) P-l, Q-2, R-3. S-4

(b) P-3, Q-2, R-l, S-4

(c) P-l, Q-4, R-3, S-2

(d) P-l, Q-2, R-4, S-3

[GATE 2011]

44. Match those listed in group 1 with the *NDT* methods listed in group 2.

Group 1

P. Penetrameter

Q. Differential coil probe

R. Piezo-electric probe

S. Developer

Group 2

1. Ultrasonic test

2. Dye-penetrant test

3. X-ray radiography

4. Accoustic emission test

(a) P-3, Q-4, R-l, S-2

(b) P-2, Q-l, R-3, S-4

(c) P-l, Q-2, R-4, S-3

(d) P-4, Q-3, R-2, S-l

[GATE 2011]

45. Radiography technique of detecting defects is based on the principle of

(a) Diffraction

(b) Reflection

(c) Interference

(d) Absorption

[GATE 2012]

Answers

41. (d) **42. (c)** **43. (b)** **44. (a)** **45. (d)**

46. Match the suitability of non-destructive testing method in Group I for the detection of defects listed in Group II.

Group I:

P. Magnetic particle inspection

Q. X-ray radiography

R. Dye penetrant test

S. Ultrasonic testing

Group II :

1. Surface crack in martensitic stainless steels
2. Surface crack in austenitic stainless steels
3. Hairline crack in aluminium
4. Inclusion in steels

(a) P-2, Q-4, R-3, S-l
(b) P-4, Q-2, R-l, S-3
(c) P-3, Q-l, R-2, S-4
(d) P-l, Q-4, R-2, S-3

[GATE 2013]

47. Which one of the following *NDT* techniques *CANNOT* be used to detect an internal crack in a steel shaft?

(a) Liquid penetrant inspection
(b) Radiography
(c) Ultrasonic testing
(d) X-ray tomography

[GATE 2014]

48. A tensile specimen was deformed to a true strain of 0.405. The change in the gauge length was 5 mm. The final gauge length of the specimen, in mm, is

(a) 10 (b) 12
(c) 15 (d) 18

[GATE 2014]

49. Which of the following techniques are not applicable for detecting internal flows in a ceramic material?

1. Liquid penetration test
2. Radiography
3. Ultrasonic testing
4. Eddy current method

(a) 1 and 3 (b) 3 and 4
(c) 2 and 4 (d) 1 and 4

[GATE 2015]

50. A brittle material is mechanically tested in medium P in which it has surface energy γs = 0.9 J m^{-2}. This material has a fracture strength of 300 MPa for a given flaw size. The same solid containing the same flaws is then tested in medium Q in which γs = 0.1 J m^{-2}. The fracture strength (in MPa) in medium Q based on Griffith's theory is

(a) 50 (b) 100
(c) 180 (d) 360

[GATE 2015]

51. For dye-penetrant test, identify the **CORRECT** statement.

(a) Pre-and post-cleaning of parts are not required
(b) Internal defects can be detected
(c) Surface oxides helps in crack indentification
(d) Dye with low contact angle is required

[GATE 2016]

Answers

46. (d) 47. (a) 48. (c) 49. (d) 50. (b) 51. (d)

16 Physics of Metals

1. At room temperature F.C.C. crystal structure is observed in
   (a) lead (b) cupro-nickel
   (c) titanium (d) tungsten
   [GATE 91]

2. When the wavelength of the incident X-ray increases the angle of diffraction
   (a) decreases
   (b) increases
   (c) remains constant
   (d) shows no systematic variation.
   [GATE 92]

3. Slip plane in copper is
   (a) (100) (b) (110)
   (c) (111) (d) (0001)
   [GATE 94]

4. In a single crystal of copper (lattice parameter 3.615°A). The distance between (111) planes (in °A) is
   (a) 1.807 (b) 2.087
   (c) 2.556 (d) 3.615
   [GATE 94]

5. The co-ordination number in simple cubic structure is
   (a) 4 (b) 6
   (c) 8 (d) 12
   [GATE 94]

6. In a cubic lattice, the direction [123] is contained in the
   (a) plane (222)
   (b) plane (123)
   (c) neither a nor b
   (d) both a and b
   [GATE 95]

7. In a powder diffraction photograph of copper (fcc structure with a lattice parameter of 0.3608 nm) taken with a *cu-ka* radiation (γ = 0.154 nm), the Bragg angle for the first line is
   (a) 12.3° (b) 17.6°
   (c) 21.7° (d) 25.3°
   [GATE 95]

8. In a hcp single crystal, slip on the basal plane may occur; if the tensile axis is along
   (a) <0001> (b) <1120>
   (c) <1100> (d) <1121>
   [GATE 95]

9. The lowest angle reflection in a powder diffraction experiment or an fcc metal using *Cu-Ka* radiation (γ = 1.54°A) occured at an angle of 19°30'. The lattice parameter of the metal is .... A.
   (a) 2.8 (b) 3.2
   (c) 3.6 (d) 3.99
   [GATE 96]

Answers

1. (a)	2. (a)	3. (c)	4. (b)	5. (b)	6. (d)
7. (c)	8. (b)	9. (d)			

10. Match the phenomenon with their specific effects on X-ray diffraction patterns

P. Macroscopic residual stress
Q. Microscopic residual stress
R. Ordering reaction
S. Preferred orientation

1. Additional diffraction lines
2. X-ray line shift
3. Debye rings of non-uniform intensity
4. X-ray line broadening

(a) P-2, Q-4, R-3, S-l
(b) P-2, Q-4, R-l, S-3
(c) P-4, Q-2, R-3, S-l
(d) P-3, Q-4, R-2, S-l

[GATE 96]

11. Match the following metals and alloys with their slip planes for room temperature deformations.

P. Austenitic stainless steel
Q. Molybednum
R. Cadmium
S. Tin

1. None
2. (110)
3. (111)
4. (0001)

(a) P-2, Q-l, R-4, S-3
(b) P-l, Q-2, R-4, S-3
(c) P-4, Q-3, R-2, S-l
(d) P-2, Q-l, R-3, S-4

[GATE 96]

12. The elastic strain in copper is due to

(a) motion of dislocations
(b) stretching of atomic bonds
(c) breakage of atomic bonds
(d) none of these

[GATE 98]

13. The direction [011] in cubic crystals

(a) is perpendicular to the (111) plane
(b) lies within the (111) plane
(c) makes an angle of 30° with the normal to (111) plan
(d) Inakes an angle of 45° with the normal to (111) plane.

[GATE 2001]

14. The following type of bonding is strongly directional in solids:

(a) Vanderwalls (b) Ionic
(c) Metallic (d) Covalant

[GATE 20011

15. Fermi level of an atom refers to the

(a) highest energy level occupied by the electron at absolute zero R.
(b) energy level with a 50 % probability of occupation.
(c) energy of the outer most electrons.
(d) none of the above.

[GATE 2002]

16. The majority charge carriers in p-type silicon are

(a) Free electrons
(b) ions
(c) conduction electrons
(d) holes

[GATE 2004]

17. If the drift velocity of holes under a field gradient of 150 V/m is 7.5 m/s, their mobility (in SI units) is

(a) 0.05 (b) 0.5
(c) 0.75 (d) 50.

[GATE 2004]

18. The condition of diffraction from a crystal is given by

(a) $n\gamma = 2d \sin \theta$ (b) $\gamma = d \sin 2\theta$
(c) $\Upsilon = 2d \sin 2\theta$ (d) $n\gamma = d \sin \theta$

[GATE 2004]

Answers

10. (a)	11. (b)	12. (a)	13. (c)	14. (d)	15. (a)
16. (d)	17. (a)	18. (a)			

19. Identify the false statement.
    (a) Burger vector and the dislocation line are parallel to each other for screw dislocations.
    (a) Burger vector and the dislocation line are perpendicular to each other for edge dislocations.
    (a) Screw dislocations glide parallel to its Burger vector.
    (d) Edge dislocations glide parallel to its Burger Vector.

[GATE 2004]

20. The Fe. A1 intermetallic phase has a disordered bcc structure at high temperatures. This first four Bragg reflections will be
    (a) (100), (110), (200), (211)
    (b) (100), (110), (111), (200)
    (c) (110), (200), (220), (311)
    (d) (110), (200), (211), (220)

[GATE 2004]

21. The electrical conductivity of pure silicon at 300 K is $4.34 \times 10^{-4} (\Omega m)^{-1}$ and at 500 K, it is $2.15(\Omega M)^{-1}$. Given that the Boltzman constant, R = 8.6×10^{-5} eV $.K^{-1}$, the band energy gap is .... eV.
    (a) 0.65
    (b) 1.10
    (c) 1.5
    (d) 2.2

[GATE 2004]

22. The Fe.Al intermetallic phase is an ordered bcc structure with Fe atoms at the corner and Al atoms at the body centered positions. Each Fe atom is surrounded by
    (a) 6 Fe atoms and 8 Al atoms
    (b) 8 Fe atoms and 8 Al atoms
    (c) 6 Fe atoms and 6 Al atoms
    (d) 4 Fe atoms and 8 Al atoms

[GATE 2004]

23. The grain size of pure copper after annealing for 10^3 S is 10 μm. If the time exponent is 0.5, the total time of annealing required to obtain a grain size of 40 μm is .... S.
    (a) 4×10^3
    (b) 8×10^3
    (c) 16×10^3
    (d) 64×10^3

[GATE 2004]

24. A dislocation line in a fcc crystal dissociates into two partials which have their Burger vectors as $\frac{a}{6}[2\bar{1}\bar{1}]$ and $\frac{a}{6}[11\bar{2}]$, Indicate the correct statement.
    (a) Burgers vector of the undissociated dislocation line is $\frac{a}{2}[10\bar{1}]$
    (b) Burgers vector of the undissociated dislocation line is $\frac{a}{2}[1\bar{1}0]$
    (c) Energy of each partial is proportional to $a^2/3$
    (d) Energy of the undissociated dislocation line is lesser than the sum of energy of two partials

[GATE 20041

Answers

19. (d) 20. (d) 21. (b) 22. (a) 23. (c) 24. (a)

25. An elliptical dislocation loop is gliding on plane ABCD of a single crystal of BCC material as shown in the following figure. It is gliding along the burger vector 'b'. Identify the correct statement.

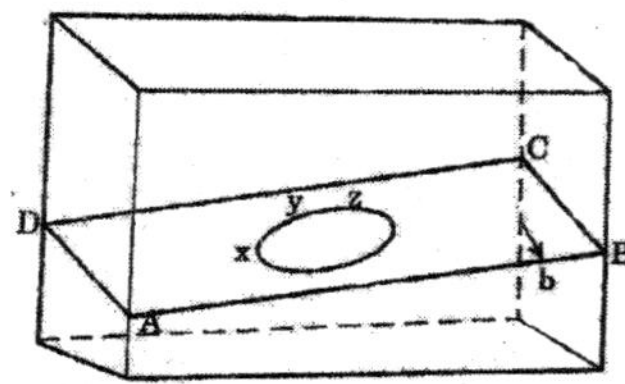

P. Plane ABCD belongs to the family of (110).
Q. Burger vector 'b' belong to <110> direction.
R. Dislocation has the edge character at point x and the screw character at point z.
S. Dislocation has the screw character at point x

(a) P, Q (b) Q, R
(c) P, R (d) P, S

[GATE 2004]

26. In an ionic crystal if the valancies of cation and anion are Vc and VA and the coordination numbers of cation and anions are Nc and NA respectively, then the necessary condition for obtaining a stable crystal structure is

(a) $V_C/N_C = V_A/N_A$
(b) $V_C.V_A = N_C.N_A$
(c) $V_C/N_A = V_A/N_C$
(d) $V_C/V_A = \sqrt{N_C}/\sqrt{N_A}$

[GATE 2005]

27. For dislocation with Burger vector b, the energy is

(a) independent of b
(b) proportional to b
(c) inversely proportional to b^2
(d) proportional to b^2

[GATE 2005]

28. Match the crystal structure in group I with the atomic packing factor in group II.

Group I
P. Body centred cubic
Q. Diamond cubic
R. Face centred cubic
S. Hexagonal close packed

Group II
1. 0.30 2. 0.34
3. 0.60 4. 0.68
5. 0.74 6. 0.80

(a) P-4, Q-2, R-5, S-5
(b) P-3, Q-1, R-5, S-5
(c) P-4, Q-3, R-5, S-6
(d) P-3, Q-5, R-4, S-2

[GATE 2005!

29. Which of the following is true for metals?

(a) Twin boundary energy > Grain boundary energy > Surface energy
(b) Grain boundary energy > Twin boundary energy > Surface energy
(c) Twin boundary energy > Surface energy > Grain boundary energy
(d) Surface energy > Grain boundary energy > Twin boundary energy

(GATE 20051

30. If the carrier charge concentration (n) and mobility (μa) in a metallic conductor are respectively
1.602×10^{-19}c, 8.50×10^{-3} m^2. V^{-1} S$^{-1,}$ the resistivity of the material in [Ω m] is

(a) 1.719×10^{-6}
(b) 1.719×10^{-7}
(c) 1.719×10^{-8}
(d) 1.719×10^{-9}

[GATE 2005]

Answers

25. (d) 26. (d) 27. (d) 28. (a) 29. (b) 30. (c)

31. The band gap of CdS is 2.45 eV. The wavelength that will be absorbed on incidence of visible light is
(a) 5057 Å (b) 5300 Å
(c) 5800 Å (d) 6800 Å
[GATE 2005]

32. Match the planes in group 1 with the interplaner spacings in group 2 for a cubic crystal having a lattice parameter of 'α'.

Group 1	Group 2
P. (100)	1. $\alpha/\sqrt{5}$
Q. (110)	2. $\alpha/\sqrt{3}$
R. (111)	3. $\alpha/\sqrt{2}$
S. (210)	4. α
	5. α/2
	6. α/3

(a) P-4, Q-3, R-2, S-5
(b) P-4, Q-3, R-5, S-l
(c) P-4, Q-2, R-5, S-6
(d) P-4, Q-3, R-2, S-l
[GATE 2005]

33. The outer electron configurations of CO, Fe and Ni atoms are respectively $3d^74s^2$, and $3d^2$, $3d^84s^2$. Which of the following ions has the potential to produce the highest magnetic moments?
(a) CO^{2+} (b) Fe^{2+}
(c) Ni^{2+} (d) Fe^{3+}
[GATE 2005]

34. Along which crystallographic direction, load must be applied in fcc single crystals so that slip starts simultaneously only on six slip planes?
(a) [110] (b) [111]
(c) [001] id) [011]
[GATE 2005]

35. Nickel and copper have fcc structures with lattice parameters 0.352 and 0.361 nm respectively. The density of Ni in $kg.m^{-3}$ is
(a) 8.94×10^{-1} (b) 8.94×10^{1}
(c) 8.94×10^{2} (d) 8.94×10^{3}
[GATE 2005]

36. In Q. No. 35, in powder XRD patterns obtained independently for Cu and Ni using monochromatic X-rays of the same wavelength
(a) The (200) reflection is the first reflection for both Cu and Ni.
(b) All reflections are such that $h + k + 1$ - odd.
(c) The first Cu reflection occurs at a larger 20 value compared to the first Ni reflection.
(d) The (111) reflection is the first reflection for both Cu and Ni.

37. An electron beam is accelarated by a potential and the velocity of electrons in the beam is 1.8754×10^8 m. s^{-1}. The accelarting potential in kilovolts is
(a) 80
(b) 100
(c) 120
(d) 200
[GATE 2005]

38. In Q. No. 37, the above electron beam is used in anelectron diffraction study of a sample. The parallel planes in the sample with interplaner spacing of 2 A diffract when the planes are
(a) nearly perpendicular to the beam
(b) nearly parallel to the beam
(c) about 30° to the beam
(d) about 60° to the beam
[GATE 2005]

Answers

31. (a)	32. (d)	33. (a)	34. (b)	35. (d)	36. (c)
37. (b)	38. (a)				

39. The miller indices of the direction common to the planes (111) and (110) in a cubic system is .
(a) [111] (b) [110]
(c) [110] (d) [111]
[GATE 2010]

40. The third peak in the X-ray diffraction pattern of a polycrystalline bcc metal is
(a) (111) (b) (110)
(c) (211) (d) (220)
[GATE 2010]

41. Number of slip systems in an ideal close packed hexagonal structure is
(a) 3 (b) 12
(c) 24 (d) 48
[GATE 2010]

42. The theoretical density of an fcc metal with atomic radius and atomic weight of 0.144 nm and 197 gm.mol^{-1} respectively, is approximately (in kg.m^{-3})
(a) 18110 (b) 18300
(c) .19360 id) 19890
[GATE 2010]

43. Silicon is doped with arsenic (concentration 10^{20} atoms.m^{-3}). At room temperature, the electron and hole mobilities in Si are 0.14 m^2.v^{-1}.s^{-1} and 0.05 m^2.v^{-1}.s^{-1} respectively. The conductivity in $(\Omega\,\mathrm{m})^{-1}$, at room temperature for Si doped with As is .
(a) 0.11 (b) 0.96
(c) 2.24 (d) 2.72.
[GATE 2010]

44. In one Fcc unit cell, there are
(a) 4 tetrahedral and 8 octahedral sites
(b) 8 tetrahedral and 4 octahedral sites
(c) 12 tetrahedral and 4 octahedral sites
(d) 4 tetrahedral and 4 octahedral sites.
[GATE 2007]

45. The atomic packing factor the diamond cubic structure is
(a) 0.74 (b) 0.68
(c) 0.34 (d) 0.25
[GATE 2007]

46. Match each material in group I with a bond type in group II.
Group I
P. Silicon Q. Copper
R. Sodium chloride
Group II
1. Metallic bonding
2. Covalant bonding
3. Ionic bonding
4. Vander-Wallls bonding
(a) P-2, Q-l, R-4,
(b) P-l, Q-3, R-l
(c) P-4, Q-l, R-3
(d) P-2, Q-l, R-3
[GATE 2007]

47. Which of the following statements are true about edge dislocations?
P. Edge dislocations do not have an extra half plane associated with them.
Q. The Burger vectors is perpendicular to the direction.
R. Edge dislocations can avoid obstacles by cross-slip.
S. Depending on geometry, parallel edge dislocations of opposite signs can attract or repel one another.
(a) R
(b) P, Q, S
(c) Q, S
(d) Q, R
[GATE 2007]

Answers

39. (c)	40. (c)	41. (a)	42. (a)	43. (c)	44. (b)
45. (c)	46. (d)	47. (c)			

48. Match the energy gaps in group II with the materials in group I.

Group I

P. Diamond Q. Silicon

R. Grey tin

Group II

1. 0.1 eV 2. 0.7 eV

3. 1.1 eV 4. 6.0 eV

(a) P-l, Q-3, R-4 (b) P-2, Q-4, R-l

(c) P-3. Q-l, R-2 (d) P-4, Q-3, R-l

[GATE 2007]

49. The planer density for (111) plane in a fcc crystal is

(a) 0.68 (b) 0.74

(c) 0.85 (d) 0.91.

[GATE 2008]

50. In an ideal HCP packing, the c/a ratio is

(a) 1.225 (b) 1.414

(c) 1.633 (d) 1.732.

[GATE 2009]

51. For a simple cubic unit cell with unit vectors, i, j and k, the angle between the lattice vector [100] and [111] in degrees is

(a) 35.2 (b) 54.7

(c) 60 (c) 90

[GATE 2011]

52. For a fcc unit cell, the ratio of the number of tetrahedral voids to the number of atoms is

(a) 2:1 (b) 3:1

(c) 4:1 (d) 5:1

[GATE 2011]

53. For a cubic metal with lattice parameter of 3.92°A, the first four diffraction peaks from the X-ray powder diffraction pattern taken with Cu-Ka radiation (λ = 1.5405°A) occur at 2θ values of 39.7, 46.2, 67.5 and 81.3 degrees. The crystal structure of the metal is

(a) simple cubic (b) fcc

(c) bcc (d) diamond cubic

[GATE 2011]

54. The largest size of immobilised segment of dislocation in a Frank Read (FR) source contained in a polycrystalline material is of the order of grain size. In a metal of 10 μm grain size, the shear stress required to operate such a FR source is 100 MPa. If the grain size in the same metal is reduced to 10 μm, the shear stress required to operate such FR source would be .... MPa.

(a) 10^2 (b) 10^3

(c) 10^5 (d) 10^6

[GATE 2011]

55. Which one of the following reactions is fcc/bcc crystals with lattice parameter 'a' is energetically favourable?

(a) $\frac{a}{2}\left[\bar{1}10\right]+\frac{a}{2}\left[0\bar{1}\bar{1}\right]$

(b) $\frac{a}{2}\left[111\right]+\frac{a}{2}\left[11\bar{1}\right]$

(c) $\frac{a}{2}\left[111\right]+\frac{a}{2}\left[11\bar{1}\right]$

(d) $\frac{a}{2}\left[111\right]+\frac{a}{2}\left[111\right]$

[GATE 2011]

56. Shear modulus of copper is 45 GPa. Lattice parameter of copper is 3.61°A. The magnitude of burgers vector in copper is .... °A.

(a) 2.54 (b) 2.39

(c) 2.20 (d) 2.18

[GATE 2011]

57. In Q.No. 56, the elastic strain energy per unit length of dislocation line in copper is .... N.

(a) 34.8×10^{-10} (b) 28.8×10^{-10}

(c) 24.8×10^{-10} (d) 14.5×10^{-10}

[GATE 2011]

Answers

48. (d)	49. (b)	50. (c)	51. (b)	52. (a)	53. (b)
54. (c)	55. (a)	56. (a)	57. (d)		

58. A peak in the X-ray diffraction pattern is observed at $2\theta = 78°$, corresponding to [311] planes of an fcc metal, when the incident beam has a wavelength of 0.154 nm. The lattice parameter of the metal is approximately
(a) 0.6 nm (b) 0.4 nm
(c) 0.3 nm (d) 0.2 nm
[GATE 2012]

59. At equilibrium spacing in a crystalline solid, which of the following is true for net inter-atomic force (F) and potential energy (U)?
(a) F is zero and U is Zero
(b) F is zero and U is minimum
(c) F is minimum and U is zero
(d) F is minimum and U is minimum
[GATE 2012]

60. A unit dislocation splits into two partial dislocations,. The correct combination of the Burgers vectors of the partial dislocations for a given unit dislocation having Burgers vector $\frac{a}{2}[1\bar{1}0]$

(a) $\frac{a}{6}[2\bar{1}1]$ and $\frac{a}{6}[1\bar{2}\bar{1}]$

(b) $\frac{a}{6}[1\bar{1}2]$ and $\frac{a}{6}[\bar{1}\bar{2}1]$

(c) $\frac{a}{6}[11\bar{2}]$ and $\frac{a}{6}[2\bar{1}\bar{1}]$

(d) $\frac{a}{6}[211]$ and $\frac{a}{6}[12\bar{1}]$

[GATE 2012]

61. In a binary system A-B, ε_{AA}, ε_{BB} and ε_{AB} correspond to A-A, B-B and A-B bond energies respectively . The miscibility gap will occur, if
(a) $\varepsilon_{AB} > 1/2\ (\varepsilon_{AA} + \varepsilon_{BB})$
(b) $\varepsilon_{AB} < 1/2\ (\varepsilon_{AA} + \varepsilon_{BB})$
(c) $\varepsilon_{AB} = 1/2\ (\varepsilon_{AA} + \varepsilon_{BB})$
(d) $\varepsilon_{AB} < 1/4\ (\varepsilon_{AA} + \varepsilon_{BB})$
[GATE 2013]

62. The atomic packing fraction of diamond cubic structure is
(a) 0.34 (b) 0.56
(c) 0.68 (d) 0.82
[GATE 2013]

63. A unit cell of an element has maximum linear density along the [110] direction. The packing density of its (100) plane is
(a) 0.68 (b) 0.74
(c) 0.79 (d) 0.91
[GATE 2013]

64. For an FCC metal, the ratio of interplanar spacing obtained from the first two peaks of the X-ray diffraction pattern is
(a) 1.91 (b) 1.63
(c) 1.41 (d) 1.15
[GATE 2013]

65. Saturation magnetization of an FCC metal with lattic parameter 0.2 nm is 600 kA/m. The net magnetic moment per atom is given by (in Bohr magnetron)
(a) 8.08×10^{57} (b) 2.02×10^{57}
(c) 0.517 (d) 0.129
[GATE 2013]

66. The density and associated crystallinity for two polypropylene samples are as follows:

density, g/cm^3	crystallinity, %
1.20	50
1.44	80

Density of totally amorphous polypropylene is
(a) 0.64 (b) 0.74
(c) 0.84 (d) 0.94
[GATE 2013]

Answers

58. (b)	59. (b)	60. (a)	51. (a)	62. (a)	63. (c)
64. (d)	65. (d)	66. (d)			

67. In Q.66, the percent crystallinity of polypropylene sample having a density of 1.3 g/cm³ is
(a) 54 (b) 64
(c) 74 (d) 84
[GATE 2013]

68. An edge dislocation is present in α-Fe. Atomic diameter of iron atom is 0.25 nm and its shear modulus is 70 GPa. Modulus of the Burgers vector (in nm) is
(a) 0.125 (b) 0.25
(c) 0.50 (d) 0.625
[GATE 2013]

69. In Q.68, energy (in J/m) of the dislocation is
(a) 0.5×10^{-9} (b) 1.1×10^{-9}
(c) 2.2×10^{-9} (d) 4.4×10^{-9}
[GATE 2013]

70. The respective units for dislocation density and stress intensity factor are
(a) m^2 and MPa.m
(b) m^2 and MPa.$m^{1/2}$
(c) m^{-2} and MPa.$m^{1/2}$
(d) m^{-2} and MPa.m
[GATE 2014]

71. Which one of the following signals from a specimen is used in a scanning electron microscope to get quantitative elemental analysis?
(a) Secondary electrons
(b) Backscattered electrons
(c) X-rays
(d) Transmitted electrons
[GATE 2014]

72. Dissociation of a dislocation into two particals in an FCC metal is given by the following equation.

$$\frac{a}{2}\left[1\bar{1}1\right] \rightarrow \frac{a}{6}\left[2\bar{1}1\right] + \frac{a}{6}\left[1\bar{2}\bar{1}\right]$$

On which plane do these two partial dislocations lie?
(a) $\bar{1}11$ (b) (1 1 1)
(c) $(1\bar{1}1)$ (d) $(11\bar{1})$
[GATE 2014]

73. The atomic radius of aluminium is 1.431Å. The interplanar spacing of (111) planes in aluminium, in Å, is
(a) 0.65 (b) 1.25
(c) 2.35 (d) 4.25
[GATE 2014]

74. In a conventional unit cell of a crystal a = b ≠ c and a = β = γ = 90°. This crystal belongs to which of the followings systems?
(a) Cubic (b) Tetragonal
(c) Orthorhombic (d) Triclinic
[GATE 2015]

75. In an X-Ray powder pattern of a simple cubic crystal, the 2nd peak corresponds to
(a) (111) (b) (100)
(c) (200) (d) (110)
[GATE 2015]

76. In an FCC crystal, the strain energy per unit length of a dislocation with Burgers vector $\frac{a}{2}$<110> is .... times that of a $\frac{a}{6}$ <112> dislocation.
(a) 1 (b) 2
(c) 3 (d) 4
[GATE 2015]

77. The lattice parameter of face-centered cubic iron (γ-Fe) is 0.3571 nm. The radius (in nm) of the octahedral void in γ-Fe is ....
[GATE 2016]

Answers

67. (b)	68. (b)	69. (a)	70. (c)	71. (c)	72. (d)
73. (c)	74. (b)	75. (d)	76. (c)	77. (0.045 to 0.060)	

78. For an ideal hexagonal-closed packed structure, the c/a ratio and packing efficiency respectively are.

(a) 1.633 and 52% (b) 1.633 and 74%
(c) 1.733 and 68% (d) 1.733 and 74%

[GATE 2015]

79. A schematic of X-ray diffraction pattern of a single phase cubic polycrystal is given below. The miller idices of **Peak A** is

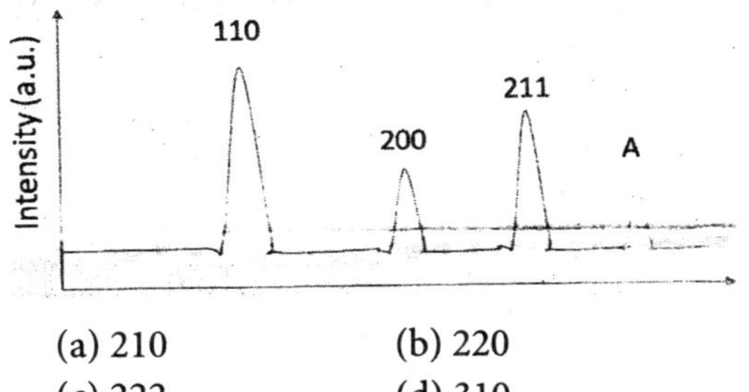

(a) 210 (b) 220
(c) 222 (d) 310

[GATE 2016]

80. The Burger's vector of a dislocation in a cubic crystal (with lattice parameter a) is $\frac{a}{2}$ [110] and dislocation line is along [112] direction. The angle (in degrees) between the dislocation line and its Burger's vector is ....

[GATE 2016]

81. A simplified energy band-diagram of an intrinsic semiconductor at thermal equilibrium (300 K) is shown. In the accompanying table, which one of the four cloumns correctly represents the listed parameters? Assume same effective mass for electrons and holes.

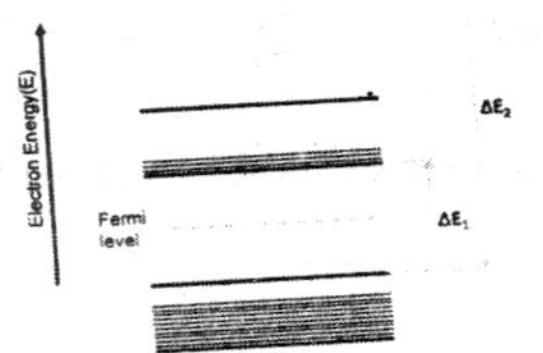

Parameter	Energy-difference			
	Column 1	**Column 2**	**Column 3**	**Column 4**
Band-gap	ΔE_2	ΔE_1	ΔE_2	ΔE_1
Electron affinity	$\Delta E_1/2$	$\Delta E_2-\Delta E_1$	$\Delta E_2/2$	$\Delta E_2-(\Delta E_1/2)$
Work function	$\Delta E_1+\Delta E_2$	$\Delta E_2-(\Delta E_1/2)$	$\Delta E_1-\Delta E_2/2$	$\Delta E_2+(\Delta E_1/2)$

(a) Column 1 (b) Column 2
(c) Column 3 (d) Column 4

[GATE 2016]

Answers

78. (b) 79. (b) 80. (54.0 to 55.5) 81. (b)

Section B

17

Metallurgical Thermodynamics

1. Chemical potential of a component 1 in a binary solution can be defined as

(a) $\left(\frac{\partial A}{\partial n_1}\right)_{T, V, n_2}$ (b) $\left(\frac{\partial U}{\partial n_1}\right)_{S, V, n_2}$

(c) $\left(\frac{\partial H}{\partial n_1}\right)_{S, V, n_2}$ (d) $\left(\frac{\partial G}{\partial n_1}\right)_{S, p, n_2}$

[GATE 91]

2. The activity of pure hydrogen gas at 1000°C and 5 atm. pressure.

(a) is always less than 1

(b) is always greater than 1

(c) can be 5

(d) depends on the choice of the standard state.

[GATE 91]

3. Heat capacity at constant pressure is defined by

(a) $\left(\frac{\partial E}{\partial T}\right)_p$ (b) $\left(\frac{\partial G}{\partial T}\right)_e$

(c) $\left(\frac{\partial H}{\partial T}\right)_p$ (d) $\left(\frac{\partial (E + Pv)}{\partial T}\right)_p$

4. The limiting condition for the appearance of the miscibility gap in a binary solution is

(a) $\left(\frac{\partial^2 G}{\partial X^2}\right) = 0$

(b) $\left(\frac{\partial^3 G}{\partial X^3}\right) = 0$

(c) $\left(\frac{\partial^2 G}{\partial X^2}\right) + \left(\frac{\partial^3 G}{\partial X_3}\right) = 0$

(d) $\left(\frac{\partial^2 G}{\partial X^2}\right) - \left(\frac{\partial^3 G}{\partial X^3}\right) = 0$

Where, G is the Gibbs free energy and X is the mole fraction of component Z.

[GATE 92]

5. For an ideal gas, Cp – Cv is

(a) R (b) –R

(c) 0 (d) $\left(\frac{3}{2}\right)$ R

[GATE 94]

6. The entropy change for a spontaneous process is

(a) > 0 for the system

(b) < 0 for the system

(c) > 0 for the system and the surrounding

(d) < 0 for the system and the surrounding

[GATE 94]

Answers

1. (d)	2. (d)	3. (c)	4. (a)	5. (a)	6. (d)

7. For a binary solution A–B, the α-function is given by $\alpha = \frac{\exp(x) - 1}{X}$ Where, X is the mole fraction of component A. The limiting value of a when X approaches zero, is
   (a) 1 (b) infinity
   (c) indeterminate (d) Zero
   [GATE 94]

8. For a spontaneous natural process at constant temperature and pressure, the free energy of the system always
   (a) increases
   (b) decreases
   (c) remains constant
   (d) increases to a maximum before decreasing
   [GATE 95]

9. Consider an ideal solution of component A and B. The entropy of mixing per mole of an alloy containing 50 wt% B, is
   (a) R In 2 (b) – R In 2
   (c) 3R In 2 (d) – 3R In 2.
   [GATE 95]

10. Match the following:
    Group 1
    P. Volume Q. Temperature
    R. Work done S. Molar entropy
    Group 2
    1. State function
    2. Intensive property
    3. Extensive property
    4. Path dependent
    (a) P-2, Q-3, R-4, S-l
    (b) P-3, Q-2, R-l, S-4
    (c) P-4, Q-3, R-2, S-l
    (d) P-3, Q-2, R-4, S-l
    [GATE 95]

11. The activity co-efficient of the solute in a dilute solution
    (a) decreases with increase of concentration of the solute.
    (b) increases with increase of concentration of the solute.
    (c) remains constant.
    (d) is unity at infinite dilution.
    [GATE 96]

12. The fugacity of liquid water at 298 K is approximately 3171 Pa. Considering the ideal heat of vaporisation as 43723 J/gm. mole, its fugacity at 300 K would be .... Pascal.
    (a)3171 (b) 3567
    (c) 1.01×10^5 (d) 5000
    [GATE 96]

13. The change in Gibbs free energy for the change of standard state Zn pure solid=Zn (1 wt% solution in Cu) at 298 K is given by
    (a) RT In Y_{Cu}
    (b) Zero
    (c) $\frac{\text{Molecular weight of } C_u}{100 \times \text{molecular weight of Zn}} y^0 Zn$
    (d) $\frac{\text{Molecular weight of } C_u}{100 \times \text{molecular weight of Zn}} y^0 Zn$

14. Match the following:

Group 1	Group 2
P. Henry's law	1. Extensive Property
Q. Disorder	2. Chemical potential
R. Activity	3. Entropy
S. Enthalpy	4. Dilute solution

    (a) P-4, Q-3', R-2, S-l
    (b) P-4, Q-3. R-2, S-l
    (c) P-3, Q-4, R-2, S-l
    (d) P-l, Q-2, R-3, S-4
    [GATE 96]

Answers

7. (d)	8. (a)	9. (a)	10. (d)	11. (c)	12. (b)
13. (d)	14. (a)				

15. Match the following:
    Group 1
    P. Debye temperature
    Q. State property
    R. Thermodynamic probability
    S. Activation energy
    Group 2
    1. Rate constant
    2. Specific heat
    3. Internal energy
    4. Microstate in a macrostate
    (a) P-4, Q-3, R-l, S-2 (b) P-4, Q-3, R-2, S-l
    (c) P-3, Q-4, R-2, S-l (d) P-3, Q-4, R-l, S-2
    [GATE 96]

16. In a totally irreversible isothermal expansion process for an ideal gas, $\Delta E = 0$, $\Delta H = 0$, and the ΔQ & ΔS will be
    (a) $\Delta Q = 0, \Delta S = 0$
    (b) $\Delta Q = 0, \Delta S = +$ ve
    (c) $\Delta Q = 0, \Delta S = +$ ve
    (d) $\Delta Q = +$ ve, $\Delta S = +$ ve
    [GATE 98]

17. If a solid is compressed adiabatically in its elastic range, its
    (a) internal energy remains constant
    (b) enthalpy remains constant
    (c) entropy remains constant
    (d) temperature remains constant.
    [GATE 98]

18. The chemical potential of a component 1 in a solution is given by, $\mu i =$
    (a) $\left(\frac{\delta E}{\delta n_1}\right) T, P, n_2, n_3 \ldots$
    (b) $\left(\frac{\delta H}{\delta n_1}\right)_{S, V, n_2, n_3 \ldots}$
    (c) $\left(\frac{\delta A}{\delta n_1}\right)_{T, V, n_2, n_3 \ldots}$
    (d) $\left(\frac{\delta G}{\delta n_1}\right)_{P, V, n_2, n_3 \ldots}$

19. The accepted sign conventions for the direction of heat and work transferred to a system are:

	Heat transferred to a system	Work transfered to a system
(a)	+ve	–ve
(b)	+ve	+ ve
(c)	–ve	– ve
(d)	–ve	+ve

[GATE 98]

20. At absolute zero temperature, for any reaction involving condensed phases
    (a) $\Delta G° = 0, \Delta H° = 0$
    (b) $\Delta H° = 0, \Delta S° = 0$
    (c) $\Delta S° = 0, \Delta E° = 0$
    (d) $\Delta S° = 0, \Delta C_p° = 0$
    [GATE 2004]

21. $AG_{(s)} = Ag_{(l)}$, T = 1234K; $\Delta H = 11300$ J.mol^{-1}
    $\Delta C_p = Cp_{(l)} - Cp_{(s)} = 0$
    When one mole of super cooled liquid silver freezes at an ambient temperature of 1000 K, the total entropy change of the system (Ag) and the surroundings is .... J.K^{-1}.
    (a) 11.3 (b) 9.16
    (c) 0 (d) 2.14
    [GATE 200.4]

22. pure nickel melts at 1726 K at 10^5 Pa (1 atm.) pressure, ΔV (solid-liquid) $= 0.26 \times 10^{-6}$ m^3 . mol^{-1} and the heat of fusion, $\Delta Hf = 18000$ J.mol^{-1}. The melting point of pure nickel, when acted upon by a pressure of 10^8 Pa is .... K.
    (a) 1701.5 (b) 1723.5
    (c) 1728.5 (d) 1751. 5
    [GATE 2004]

Answers

15. (b)	16. (b)	17. (c)	18. (d)	19. (a)	20. (c)
21. (d)	22. (b)				

23. In thermodynamics, the law of conservation of energy is expressed in the form of
    (a) zeroth law of thermodynamics
    (b) first law of thermodynamics
    (c) second law of thermodynamics
    (d) third law of thermodynamics.

[GATE 20C5]

24. The effect of change in temperature on the entropy of formation of an ideal binary solution, $\Delta S^{M,id}$, is such that $\Delta S^{M,id}$
    (a) increases with temperature
    (b) decreases with temperature
    (c) is always zero
    (d) is independent of temperature

[GATE 2035]

25. If a binary system exihibits a miscibility gap in the solid state, then the enthalpy of mixing in the solid state, ΔH^{mix} should necessarily be
    (a) zero
    (b) positive
    (c) negative
    (d) equal to the free energy of mixing

[2005]

26. For ideal gas, the difference between heat capacities per mole at constant pressure and at constant volume, C_P-C_V=R, is true because
    (a) $\left(\frac{\partial U}{\partial V}\right)_T \cdot \left(\frac{\partial V}{\partial T}\right)_P = 0$
    (b) $\left(\frac{\partial U}{\partial V}\right)_T \cdot \left(\frac{\partial V}{\partial T}\right)_P >> 0$
    (c) $\left(\frac{\partial U}{\partial V}\right)_T \cdot \left(\frac{\partial V}{\partial T}\right)_P >> 0$
    (d) $\left(\frac{\partial U}{\partial V}\right)_T \cdot \left(\frac{\partial V}{\partial T}\right)_P \to \infty$

[2005]

27. Uphill diffusion means diffusion from
    (a) a lower concentration and lower chemical potential to higher concentration and higher chemical potential.
    (b) a higher concentration to lower concentration.
    (c) a lower chemical potential to higher chemical potential,
    (d) a lower concentration and higher chemical potential to higher concentration but lower chemical potential.

[GATE 2005]

28. The vapour pressure of a solid M varies with temperature T in K, as

    $$\text{In } p(\text{atm}) = \frac{-33{,}200}{T} - 0.85 \text{ In } T + 21.46$$

    The vapour pressure of liquid M varies with temperature T, in K, as

    $$\text{In } p\ (\text{atm})\ \frac{-33{,}200}{T} = 0.85 \text{ In } T + 20.31$$

    The temperature of the triple point of M, in K is
    (a) 57 (b) 96
    (c) 579 (d) 957

[GATE 2005]

29. In Q. No. 28, heat of transformation for M, for liquid to vapour state transformation, ΔHi = v in K.J. mol^{-1}, at the temperature corresponding to the triple point, is
    (a) 27 (b) 62
    (c) 269 (d) 629.

[GATE 2005]

30. In a homogenous system (with C as the number of components) in equilibrium, the total number of independent intensive thermodynamic variable is
    (a) C – 1 (b) C
    (c) C+1 (d) C + 2

[GATE 2010]

Answers

23. (b)	24. (b)	25. (c)	26. (a)	27. (a)	28. (d)
29. (d)	30. (c)				

31. Which of the following partial derivative is equal to $\left(\frac{\partial S}{\partial P}\right)_T$?

(a) $-\left(\frac{\partial V}{\partial T}\right)_P$ (b) $-\left(\frac{\partial S}{\partial V}\right)_T$

(c) $\left(\frac{\partial V}{\partial T}\right)_S$ (d) $-\left(\frac{\partial S}{\partial V}\right)_T$

[GATE 2010]

32. In a binary system, the difference in chemical potentials of two components (μi–μj) is equal to

(a) $\frac{dG_m}{dxj}$ (b) 0

(c) $(1 - xj)\frac{dG_m}{dxj}$ (d) $-\frac{dG_m}{dxj}$.

[GATE 2010]

33. At 1200°C, the standard Gibbs energy of thermal decomposition of one mole of wustite into Fe and O_2 is 168 J.

The corresponding dissociation pressure (in atm.) is

(a) 2.51×10^{-15} (b) 1.22×10^{-12}

(c) 5.0×10^{-8} (d) 1.13×10^{-8}

[GATE 2010]

34. In Q. No. 33, given for the reaction $2CO + O_2 \leftrightarrow 2CO_2$, the standard Gibbs energy is 310 KJ. What is the equivalent (Pco/Pco_2)?

(a) 0.03 (b) 1.01

(c) 1.85 (d) 2.89.

[GATE 2010]

35. In a multicomponent heterogeneous system at thermodynamic equilibrium, identify the option that need not be true.

(a) Uniform pressure
(b) Uniform temperature
(c) Uniform chemical potential
(d) Uniform composition.

[GATE 2006]

36. The intensive thermodynamic variables among the following are:

Group 1

P. Pressure Q. Volume
R. Temperature S. Enthalpy

Group 2

(a) P, Q (b) P, R
(c) R, S (d) Q, R

[GATE 2008]

37. Intensive thermodynamic variables are:

(a) independent of the number of moles in the system
(b) dependent on the volume of the system
(c) dependent on the mass of the system
(d) independent of the temperature of the system

[GATE 2009]

38. At constant temperature and pressure, two phase α and β will be in equilibrium, when

(a) chemical potential of each component is same in α and β.
(b) partial molar free energy of each component is NOT the same in α and β.
(c) Gibbs free energy of mixing is minimum.
(d) enthalpy of mixing is zero.

[GATE 2009]

39. If two system P & Q are in thermal equi librium with a third system M, then P and Q will also be in thermal equilibrium with each other. This is following .... law of thermodynamics.

(a) first (b) second
(c) third (d) zeroth

[GATE 2011]

40. One mole of element P is mixed with one mole of element Q. The entropy of mixing at 0°K is

(a) 0 (b) – R In 0.5
(c) infinity (d) – R In 2

[GATE 2011]

Answers

31. (a)	32. (d)	33. (b)	34. (d)	35. (d)	36. (c)
37. (a)	38. (a)	39. (d)	40. (b)		

41. G = U + PV − TS. Then which one of the following is CORRECT?

(a) $\left(\frac{\partial V}{\partial T}\right)_P = \left(\frac{\partial S}{\partial P}\right)_T$ (b) $\left(\frac{\partial V}{\partial T}\right)_P = -\left(\frac{\partial S}{\partial P}\right)_T$

(c) $\left(\frac{\partial V}{\partial T}\right)_P = \left(\frac{\partial P}{\partial S}\right)_T$ (d) $\left(\frac{\partial V}{\partial T}\right)_P = -\left(\frac{\partial P}{\partial S}\right)_T$

[GATE 2011]

42. The hydrogen content of steel in equilibrium with hydrogen gas at 1 bar pressure is 28 ppm at some temperature. Hydrogen content in the metal at the same temperature gets reduced at 1 ppm, when the equilibrium PH_2 changes to .... bar.

(a) 28 (b) 1/28

(c) $(1/28)^{1.5}$ (d) $(1/28)^2$

[GATE 2011]

43. As the concentration of point defects in a crystal increases, its configurational entropy

(a) does not change

(b) decreases

(c) increases

(d) initially increases and then decreases

[GATE 2013]

44. Critical value of the Gibbs energy of nucleation at equilibrium temperature is

(a) zero (b) Infinite

(c) positive (d) negative

[GATE 2013]

45. When two phases α and β in an alloy are in thermodynamic equilibrium , then

(a) $c_p^{\alpha} = c_p^{\beta}$ (b) $V_m^{\alpha} = V_m^{\beta}$

(c) $G_m^{\alpha} = G_m^{\beta}$ (d) $G_i^{\alpha} = G_i^{\beta}$

[GATE 2013]

46. Isothermal compressibility of a material is given by

(a) $-\frac{1}{P}\left(\frac{\partial V}{\partial P}\right)_T$ (b) $\frac{1}{P}\left(\frac{\partial V}{\partial P}\right)_T$

(c) $-\frac{1}{V}\left(\frac{\partial V}{\partial P}\right)_T$ (d) $\frac{1}{V}\left(\frac{\partial V}{\partial P}\right)_T$

[GATE 2013]

47. Integral enthalpy of mixing (in J/mol) of liquid (Cu,Zn) solution can be approximated by $\Delta H_m^{mix} = -19250\, X_{Cu} X_{Zn}$

The corresponding partial molar enthalpy of mixing (in J/mol) for Cu is

(a) $19250\, x_{zn}^2$

(b) $-19250\, x^2 cu$

(c) $38500\, x_{Zn} - 19250 x_{zn}^2 - 19250$

(d) $-19250\, x_{zn}^2$

[GATE 2013]

48. In Q.47, assuming regular solution behaviour, the solution parameter (in J/mol) is

(a) −19250 (b) −9625

(c) 13.75 (d) 2315.4

[GATE 2013]

49. Which one of the following is NOT an intensive property?

(a) temperature (b) pressure

(c) volume (d) refractive index

[GATE 2014]

50. The activity coefficient of Q in a liquid Q-R alloy is represented by the following equation at a given temperature.

$\ln \gamma_Q = 0.6x_R^2 - 0.2.x_R^3$

What is the activity of Q in an alloy of composition XR = 0.6 at the same temperature?

(a) 0.16 (b) 0.24

(c) 0.32 (d) 0.48

[GATE 2014]

Answers

41. (b)	42. (b)	43. (d)	44. (b)	45. (d)	46. (c)
47. (d)	48. (a)	49. (c)	50. (d)		

51. Two phases α and β are in thermodynamic equilibrium. If x and y are the only components present in the phase α and β and μ_i^j represents the chemical potential of component *i* in phase *j*, then the condition for equilibrium is

(a) $\mu_x^\alpha = \mu_y^\beta$ and $\mu_x^\beta = \mu_y^\alpha$

(b) $\mu_\beta^x = \mu_\alpha^y$ and $\mu_\alpha^x = \mu_\beta^y$

(c) $\mu_x^\alpha = \mu_y^\beta$ and $\mu_x^\alpha = \mu_y^\beta$

(d) $\mu_x^\alpha = \mu_y^\alpha$ and $\mu_x^\beta = \mu_y^\beta$

52. The specific heat (C_p) of pure iron expressed in J/(mol.K) as a function of temperature T (in K) is given as :

$C_p = 17.49 + 24.77 \times 10^{-3}$ T

What is the change in the enthalpy of pure iron (in J/mol) when it is heated from 25°C to 700°C?

(a) 11200 (b) 22400
(c) 28600 (cl) –32500

[GATE 2014]

53. Which of the following properties is Intensive?

(a) Volume
(b) Gibbs free energy
(c) Chemical potential
(d) Entropy

[GATE 2015]

54. Select the correct plot of Gibbs free energy (G) Vs. Temperature (T) for a single component system from the following:

(a) P (b) Q
(c)R (d) S

[GATE 2015]

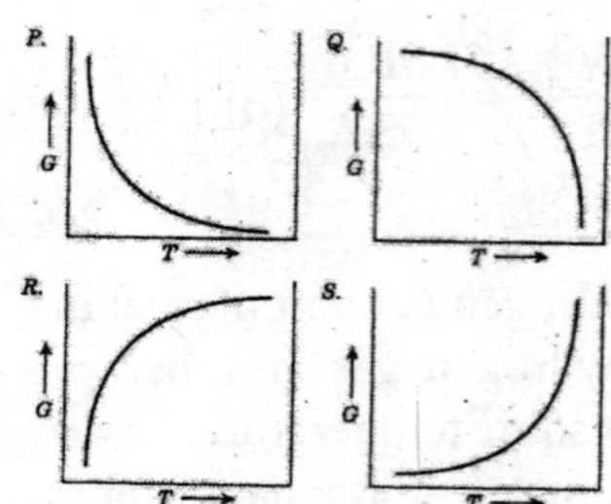

55. Configurational entropy due to ideal mixing in a binary A-B system is expressed as:

$\Delta S_{mix} = -R(X_A \ln X_A + X_B \ln X_B)$,

where X_A and X_B are mole fractions of A and B respectively.

ΔS_{mix} is maximum at X_A = ...

(a) 0.1 (b) 0.2
(c) 0.5 (d) 0.9

[GATE 2015]

56. Melting point of a metal is 1356 K. When the liquid metal is undercooled to 1256 K, the free energy change for solidification, $\Delta G^{L-S} = -1000$ J mol^{-1}. On the other hand, if the liquid metal is undercooled to 1200 K, the free energy change (in J mol^{-1}) for solidification is

(a) –950 (b) –1150
(c) –1550 (d) –1950

[GATE 2015]

57. The first law of thermodynamics can be stated as:

(a) $dE = \delta Q - \delta W$ (b) $dQ = dE - \delta W$
(c) $\delta W = dQ + dE$ (d) $dW = \delta Q - \delta E$

[GATE 2016]

Answers

51. (c)	52. (b)	53. (c)	54. (b)	55. (c)	56. (c)
57. (a)					

58. The change of standard state from pure liquid to 1 wt.% for Si dissolved in liquid Fe at 1873 K is expressed as

$$Si\ (liq.) = Si\ (1\ wt.\%)$$

Given that the activity coefficient of Si at infinite dilution in Fe is 10^{-3}, the standard Gibbs free energy change (in kJ) for this equilibrium is ....

[GATE 2016]

59. The following experimental data are available for a hypothetical binary liquid system **A-B** at 1073 K

Atom fraction of A	0.2	0.4	0.5	0.7	1.0
Partial pressure of A (bar)	0.01	0.04	0.06	0.07	0.08

When the atom fraction of **A** is 0.4 the activity of **A** in the liquid is ....

[GATE 2016]

60. Zinc oxide is reduced at a constant temperature in a closed reactor using ZnO (s) and C(s) as the only starting materials. The following reactions are assumed to be at thermodynamic equilibrium:

$ZnO(s) + C(s) = Zn(g) + CO(g)$

$2CO(g) = CO2(g) + C(s)$

Assume ideal gas behaviour. Based on mole balance, the relationship applicable to the system at equilibrium is

(a) $Pzn = Pco + 2Pco_2$

(b) $Pzn = 2pco + Pco_2$

(c) $Pzn = pco + Pco_2$

(d) $Pzn = 0.5\ Pco + 2Pco_2$

[GATE 2016]

Answers

58. (168.7 to 168.1) 59. (0.499 to 0.501) 60. (a)

18

Metallurgical Kinetics

1. The order of a chemical reaction
   (a) can be determined only expedrimentally
   (b) can be determined from the stoichiometry of the reaction
   (c) can not be zero
   (d) can be fractional

[GATE 91]

2. In which of the following reactions, application of vaccum will help?
   (a) Number of gas molecules in the product side is more than that on the reactant side.
   (b) Number of gas molecules in the reactant side of more than that on the product side.
   (c) Number of gas molecules are the same on both the reactant and product side.
   (d) None of the above

[GATE 92]

3. In BOF, desiliconisation is a first order reaction. So, the silicon content of the metal decreases
   (a) linearly with time
   (b) exponentially with time
   (c) logarithmically with time
   (d) in proportion to the square root of the time.

[GATE 94]

4. For a first order chemical reaction the concentration of the reactant decreases
   (a) linearly with time
   (b) exponentially with time
   (c) logarithmically with time
   (d) inversely with time.

[GATE 95]

5. The units of the rate constant for a second order reaction are
   (a) sec^{-1}. moles (b) $moles^{+1}$. sec^{-1}
   (c) $moles^{-1}$. sec^{-1} (d) $moles^{1}$. sec^{1}

[GATE 96]

6. For any given partial pressure of CO over liquid steel at a constant temperature, the activities of carbon and oxygen in the metal are related to a constant β as
   (a) $a_c . a_0 = \beta$ (b) $a_c/a_0=\beta$
   (c) $a_c = \beta . a^2_0$ (d) $a_c^2 = \beta.a^0$

[GATE 96]

7. The weight % of nitrogen (N_2) in liquid iron can be expressed as:
   (a) $K(PN_2)^{1/2}$ (b) $\sqrt{K} .PN_2$
   (c) $(K . PN_2)^{1/2}$ (d) $K . PN_2$

[GATE 2001]

Answers

1. (b) 2. (d) 3. (a) 4. (a) 5. (c) 6. (d)
7. (a)

8. The effect of temperature and pressure on chemical equilibrium can be predicted by
(a) Vant Hoff equation
(b) Le-Chatlier's principle
(c) Law of mass action
(d) Clausius-Clayperon equation
[GATE 2002]

9. In a dilute solution of elements X, Y etc. in liquid iron, the effect of Y on the activity co-efficient (*fx*) and the activity (*hx*) of X with respect to the Henrian 1 wt% standard state is taken into account by the activity interaction co-efficient e^{y}_{x}, which is

(a) $e_x^y = \frac{\partial \log hx}{\partial [\%Y]}$ (b) $e_x^y = \frac{\partial \log fx}{\partial [\%Y]}$

(c) $e_x^y = \frac{\partial hx}{\partial [\%Y]}$ (d) $e_x^y = \frac{\partial fx}{\partial [\%Y]}$
[GATE 2004]

10. For the irreversible reaction, Ca + C = CaC_2, ΔH°_{298} = – 60,000 J.mol–1. If a system, initially containing 2 moles of calcium, 3 moles of carbon and one mole of calcium carbide, is allowed to react to completion, the heat evolved, at 298 K will be .... J.
(a) 30,000 (b) 60,000
(c) 90,000 (d) 240,000
[GATE 2004]

11. Liquid steel contains initially 0.05 mass% P and this has to be reduced to 0.01 mass% using a basic slag. The equilibrium distribution ratio of P between slag and metal is, $Lp = \frac{(\%P)slag}{(\%P)metal}$, = 80; Assuming that in itially the slag does not contain any phosphorous (P), then the minimum weight of slag (ton) required per ton of steel is
(a) 0.025 (b) 0.05
(c) 0.075 (d) 0.10
[GATE 2004]

12. FeO(s) + CO(g) = Fe(s) + CO_2(g); K = 0.435 at 1173 K. At equilibrium, what will be the number of moles of CO gas required to reduce one mole of FeO at 1173K ?
(a) 1 (b) 1.3
(c) 2.3 (d) 3.3
[GATE 2004]

13. The diffusion co-efficient of Ni in Cu at 1000K is $1.93 \times 10^{-16} m^2.s^{-1}$ & it is 1.94×10^{-14} $m^2.s^{-1}$ at 1200K. The activation energy (in KJ. mol^{-1}) for the diffusion of Ni in Cu, is
(a)130 (b) 180
(c) 230 (d) 250
[GATE 2004]

14. Solution of nitrogen in liquid iron may be assumed to follow Sievert's law. Nitrogen content of liquid iron at 1873K in equilibrium with 1 atm pressure of nitrogen is measured as 0.044 (mass %). What will be the equilibrium nitrogen content in liquid iron (mass%), if nitrogen pressure is reduced to 0.25 atm?
(a) 0.011 (b) 0.022
(c) 0.088 (d) 0.176
[GATE 2004]

15. Consider the equilibrium A(g)+ B(g) = AB(g). When the partial pressure of A is 10^{-2} atm; the partial pressure of B is 10^{-3} atm and the partial pressure of AB is 1 atm; the equilibrium constant K is
(a) 10 atm^{-1}
(b) 10^5 atm^{-1}
(c) 10 (dimensionless)
(d) 10^5 (dimensionless)
[GATE 2004]

Answers

8. (a)	9. (b)	10. (c)	11. (b)	12. (d)	13. (c)
14. (b)	15. (b)				

16. For the reaction, A = X + Y, the respective concentrations are C_A, C_X and C_Y. The forward reaction rate constant is k_f and the backward reaction rate constant is k_b.Choose the correct statement from the following :

P. At equilibrium, $k_f C_A > k_B C_X C_Y$

Q. If the reaction is irreversible than, $k_b C_X C_Y = 0$

R. The backward reaction rate will essentially be first order, if the forward reaction rate is first order.

S. Activation energy for the first order forward reaction will be independent of temperature.

(a) P, Q (b) Q, R
(c) R, S (d) Q, S

[GATE 2004]

17. Roasting of a metallic sulfide MS in pure oxygen is carried out at a particular temperature where the probable stable solid phases are M, MS, MO and MSO_4. In case the system attains equilibrium and thus satisfies phase rule, it is possible to obtain a combination of the following solid phases in the roasted product.

P. M, MS and MO

Q. M, MS, MO and $MS0_4$

R. $MS0_4$, MS and MO

S. MS and MO

(a) P or Q (b) P or, Q or, R
(c) Q or, R or, S (d) P or, R or, S

[GATE 2004]

18. Which one of the following statements about the order of a reaction is false?

(a) Order of a reaction can be a positive integer.

(b) Order of a reaction can be a fraction.

(c) Order of a reaction can be zero.

(d) Order of a reaction can not change as the reaction proceeds.

[GATE 2005]

19. The rate of sequential multistep reaction, in a chemically controlled process, is expressed by the Arrhenius equation, $k = Ae^{(-Q/RT)}$. During an infinitesimallv small time step,this rate refers to the

(a) rate of the fastest step in the reaction

(b) rate of the slowest step in the reaction

(c) average rate of the fastest and the slowest step

(d) average rate of all the steps in the reaction.

[GATE 2005]

20. In reaction equilibria occuring between pure condensed phases and a gas phase, the equilibrium constant, K

(a) can be written solely in terms of those species which occur only in the gas phase.

(b) is always independent of the species that occur in the gas phase.

(c) can be written solely in terms of those species which occur only in the pure condensed phases.

(d) depends only on the concentration of pure species present in the mixture of condensed phases.

[GATE 2005]

21. The flux balance equation for the reaction 2[Al] + 3[O] = Al_2O_3, where J is the flux in $mol.S^{-1}.m^{-2}$, is

(a) $1/2 . J_{Al} = 1/3 = J_0 . J_{Al_2O_3}$

(b) $1/3 . J_{Al} = 1/3 = J_0 . J_{Al_2O_3}$

(c) $3J_{Al} = 2J_0 = 1/5\ J_{Al_2O_3}$

(d) $2J_{Al} = 3J_0 = 1/5 . J_{Al_2} . O_3$

[GATE 2005]

Answers

16. (b)	17. (b)	18. (b)	19. (b)	20. (a)	21. (a)

22. During the reduction of an oxide of hydrogen gas, it is observed that the rate of reaction increases by almost three fold due to a slight increases in temperature. The most likely rate controlling step is
    (a) inward mass transfer of hydrogen gas
    (b) outward mass transfer of H_2O
    (c) chemical reaction at gas metal interface
    (d) combined mass transfer of H_2 and H_2O
    [GATE 2005]

23. Decomposition of calcium carbonate occurs as:
    $CaCO_3 = CaO + CO_2$:
    $\Delta G^0_T = 17710 - 158T$ J. mol^{-1}
    What will be the decomposition temperature (in K) of $CaCO_3$, if the activity of CaO is 0.5 and the pressure is 3 atmospheres?
    (a) 142. 7 (b) 1120.9
    (c) 1424.8 (d) 2549.6
    [GATE 2005]

24. Reduction of nickel from a solution of hydrogen takes places as:
    $Ni^{2+} + H_2 = Ni + 2H^+$
    The maximum increase in the reduction rate can be obtained by
    (a) increasing pH of solution, increasing hydrogen pressure and keeping nickel concentration constant.
    (b) decreasing pH and nickel ion concentration and increasing hydrogen pressure.
    (c) decreasing hydrogen pressure, increasing pH and removing the nickel formed continuously.
    (d) increasing nickel ion concentration, hydrogen pressure and pH.
    [GATE 2005]

25. A recrystallisation process is 20% complete after 45 sec. and 85% complete after 75 sec. Assuming Avrami Kinetics, the value of Avrami exponent 'n' is
    (a) 4.19 (b) 3.12
    (c) 2.42 (d) 1.34
    [GATE 2010]

26. Reaction between A and B results in an intermediate complex AB* which leads to the final product AB as, A + B → AB* → AB. Collision rate theory views the rate as dependent on
    (a) frequency of breakdown of intermediate product.
    (b) frequency of energetic interaction to form AB.
    (c) frequency of formation of AB from AB*.
    (d) frequency of breakdown of AB.
    [GATE 2006]

27. The activation energy for a reaction is 100 kJ/mole. The approximate increase in temperature required for doubling the rate of reaction from that at 25°C, is .... °C.
    (a) 5 (b) 10
    (c) 15 (d) 20
    [GATE 2007]

28. For a reaction A → B, if the rate of change in concentration of A (CA), can be written as $\frac{-dCA}{dt} = k.\, C_A^2$, then the change in concentration with time from initial concentration of A, Cao is given by
    (a) $(1/C_A) - (1/C_{AO}) = k.t$
    (b) $(C_{AO} - C_A) = k.t$
    (c) $(C_{AO}^2 - C_A^2) = k.t$
    (d) In $(C_{A0}/C_A) = k.t$
    [GATE 2011]

Answers

22. (d) 23. (a) 24. (a) 25. (c) 26. (a) 27. (b)
28. (a)

29. If, $Y = k_1\left[1 - \exp\left(\frac{k_2\Delta x}{k_3 x}\right)\right]$, where k_1, k_2 and k_3 are constants. If $k_2\Delta X < \Delta X$, the value of Y up to first order of approximation would be

(a) $Y = k_1\left[1 - \frac{k_2\Delta x}{k_3 x}\right]$

(b) $Y = k_1\left[1 + \frac{k_2\Delta x}{k_3 x}\right]$

(c) $Y = k_1\ 1 - \frac{k_2\Delta x}{k_3 x}$

(d) $Y = -k1\ 1\frac{k_2\Delta x}{k_3 x}$

[GATE 2011]

30. If k is the rate constant for a reaction and T is the absolute temperature in the given figure, the activation energy for the reaction is .... J/K mole.

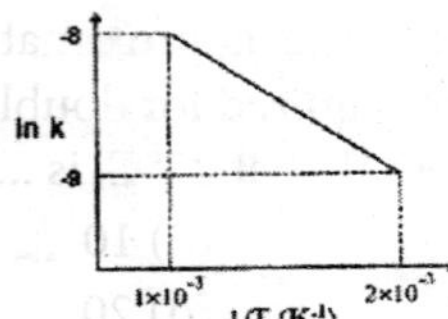

(a) 1000 (b)2000
(c) 4155 (d) 8314

[GATE 2011]

31. $2Cu(s) + 0.5\ O_2(g) = Cu_2O(s)$;
$\Delta G^0 = -162200 + 69.24\ T$,J
$2Cu(l) + 0.5\ 0_2(g) = CU_20(s)$;
$\Delta G^0 = -1188300 + 88.48\ T$,J
The molar free energy/ change at 1300 K for the transformation of solid Cu to liquid Cu will be .... J.

(a) 1050 (b) 960
(c) 544 (d) 445

[GATE 2011]

32. For the following reactions, the standard free energy change is given at 1773 K as follows:
$2/3\ Cr_2O_2(s) = 4/3\ Cr(s) + O_2\ (g)$
$\Delta G° = 447800$ J
$2H_2(g) + O_2(g) = 2H_2O(g)$
$\Delta G° = -297000$ J
If chromium oxide powder has to be reduced by hydrogen in a fludised bed, the minimum PH_2/PH_0 ratio that has to be maintained at the exit of the reactor is

(a) 8.5 (b) 10.6
(c) 100.2 (d) 166.5

[GATE 2011]

33. In case of homogenous nucleation, the critical edge length for a cubed shaped nucleus is (Y:Energy per unit area of interface between the product and the parent phase; Δg:Gibbs free energy change per unit volume)

(a) −4Y/Δg (b) −42/Δg
(c) Y/Δg (d) −3Y/Δg

[GATE 2011]

34. Consider a reaction with activation energy of 8.314 kJ/mol that takes place at 300 K. If the reaction rate is to be tripled, the temperature of the reaction should be

(a) 174.5 K (b) 447.5 K
(c) 600.5 K (d) 847.5 K

[GATE 2012]

35. The reduction of FeO with CO gas in co-current flow is given by the following equation:
$FeO+CO=Fe + CO_2$, $\Delta G° = 8120$ J at 1173K
The ratio of pco/pco_2 for this reaction at 1173 K is

(a) 0.0 (b) 0.25
(c) 0.44 (d) 2.3

[GATE 2012]

Answers

29. (c) 30. (d) 31. (c) 32. (d) 33. (a) 34. (b)
35. (d)

36. If 'Δx' represents adherent oxide layer thickness and 't' is time, which of the following curves represents diffusion-controlled oxidation kinetics?

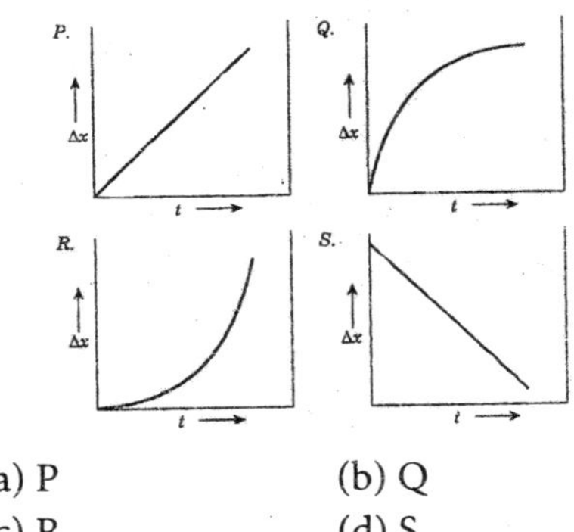

(a) P (b) Q
(c) R (d) S

[GATE 2015]

37. $C(s) + CO_2(g) \rightleftharpoons 2CO(g)$ is an important reaction in iron making.

Given ΔH^0_{298} = 172000 joules per mole of CO_2, which of the following conditions will favour the forward reaction?

(a) Increasing both temperature & pressure.
(b) Decreasing temperature & increasing pressure.
(c) Decreasing both temperature & pressure,
(b) Increasing temperature & decreasing pressure.

[GATE 2015]

38. Consider the reaction:

Fe_3O_4 (solid, pure) + CO (gas, 1 atm) → 3FeO (solid, pure) + CO_2 (gas, I atm)

For this reaction, ΔG_{1200} = −8000 joules per mole of CO and R = 8.314 J mol^{-1} K^{-1}.

The equilibrium ratio, p_{CO_2}/p_{CO} reaction at 1200 K and 1 atm is ....

(a) 0.8 (b) 1.8
(c) 2.2 (d) 4.6

[GATE 2015]

39. Activation energy of a chemical reaction, homogeneous or heterogeneous, is graphically estimated from a plot between

(a) K versus T (b) 1/k versus T
(c) 1/k versus In T (d) In k versus 1/T

[GATE 2016]

Answers

36. (b) 37. (d) 38. (c) 39. (d)

19

Miscellaneous Questions

1. The velocity at which individual particles from a fluidised bed are carried away by the fluid passing through it is defined as
   (a) minimum fluidisation velocity
   (b) terminal velocity
   (c) elutriation velocity
   (d) superficial velocity

   [GATE 95]

2. In fluid flow, heat and mass transfer, one encounters (*i*) kinematic viscosity μ (*ii*) molecular diffusivity (Ø) and (*iii*) thermal diffusivity (α) the units of these quantities are :
   (a) μ, α and Ø all have units of m/s
   (b) μ α and Ø all have units of m^2/s
   (c) μ and Ø have units of m^2/s while μ has units of m/s
   (d) μ and Ø have units of m/s while μ has units of m^2/s.

   [GATE 96]

3. Reynolds number is the ratio of
   (a) inertia forces to viscous forces
   (b) inertia forces to buoyancy forces
   (c) viscous forces to buoyancy forces
   (d) viscous forces to surface tension

   [GATE 97]

4. Match following:
   Group 1
   P. Ergun equation
   Q. Hagen-poiseulle equation
   R. Bernoulli equation
   S. Fourier equation
   Group 2
   1. Tube flow
   2. overall energy balance
   3. packed bed
   4. Heat flow
   (a) P-l. Q-3, R-2, S-4
   (b) P-4, Q-3, R-2, S-l
   (c) P-3, Q-l, R-4. S-2
   (d) P-3, P-l, R-2, S-4

   [GATE 98]

5. Bernoulli equation in the absence of electrical and magnetic field suggests
   (a) Mechanical energy = Potential energy + Kinetic energy
   (b) Potential energy = Kinetic energy + Friction energy
   (c) Mechanical energy = Potential energy + Kinetic energy + Pressure energy + Friction energy
   (d) Potential energy = Mechanical energy + Kinetic energy + Pressure energy

   [GATE 99]

Answers

1. (c) 2. (c) 3. (a) 4. (d) 5. (c)

6. Unit of viscosity in CGS system is
   (a) gm. cm^{-1}, sec^{-1} (b) gm. cm^{-2}, sec^{-1}
   (c) gm. cm^{-2}, sec^{-1} (d) gm. cm. sec^{-1}

   [GATE 2000]

7. If Reynolds number is greater than 1, then the
   (a) Viscous force is larger than the inertia force.
   (b) inertia force is larger than the viscous force.
   (c) inertia force is larger than the surface tension force.
   (d) inertia force is larger than the gravitational force.

   [GATE 2004]

8. For a steady state two dimensional incompressible flow of liquid iron, the x component of the velocity is given by $\mu = 5xy^2 - xy$ where x and y are rectangular co-ordinates. Using the continuity equation, the y component of the velocity would be
   (a) $5y - y^2$
   (b) $\frac{y^2}{3} 2.5Y^2 +$ C where C is an arbitrary constant
   (c) $y^3 - 5y + C$ where C is an arbitrary constant
   (d) $\frac{y^2}{3} - 5y^2$

   [GATE 2004]

9. A particle is settling in a liquid under Stokesian conditions. The free settling velocity of the particle is proportional to
   (a) $\sqrt{\text{particle diameter}}$
   (b) particle diameter
   (c) $(\text{particle diameter})^2$
   (d) $(\text{particle diameter})^3$

   [GATE 98]

10. Boundary layer thickens at solid-fluid interface
    (a) increases with increasing fluid density
    (b) decreases with increasing fluid density
    (c) independent of fluid flow condition.
    (d) decrease with increasing fluid velocity

    [GATE 2000]

11. In Laminar flow, the friction factor
    (a) increases with Reynolds number
    (b) decreases with Reyonds number
    (c) depends only upon the velocity of the fluid and increases with the velocity of the fluid
    (d) depends only upon the density of the fluid and increases with the density of the fluid.

    [GATE 2005]

12. Two fluids of density ρ_1 and ρ_2 are flowing at velocity V_1 and V_2 respectively through smooth pipes of identical diameter and pressure per unit length. When the friction factor is same, the ratio ρ_1 and ρ_2 is equal to
    (a) V_1/V_2 (b) $(V_1/V_2)^2$
    (c) $(V_2/V_1)^2$ (d) $(V_2/V_1)^{1/2}$

    [GATE 2010]

13. The velocity at which particles from a fluidised bed are carried away by the fluid passing through it, is known as
    (a) elutriation velocity
    (b) terminal velocity
    (c) minimum fluidisation velocity
    (d) superficial velocity

    [GATE 2008]

14. One face of a furnace wall is at 1030°C and the othere is exposed to the room temperature (30°C). If the thermal conductivity of the furnace wall is 3W. m^{-1} k^{-1} and the wall thickness is 0.3 m the maximum heat loss (in W/m^2) is
    (a) 100 (b) 900
    (c) 9000 (d) 10000

    [GATE 1994]

Answers

6. (a)	7. (b)	8. (a)	9. (b)	10. (d)	11. (b)
12. (c)	13. (a)	14. (d)			

15. The heat released by cooling one mole of copper from 400 K to room temperature (300 K) is (assume : Cp of copper is 23 J.K^{-1} mol^{-1}) .... J.
    (a) 2300 (b) 4600
    (c) 230 (d) 2.3 x 10^6
    [GATE 1995]

16. A thermally thin body is the one for which
    (a) Biot number is less than 0.1
    (6) Galileo number is less than 0.1
    (c) Fourier number is less than 0.1
    (d) Fourier number is greater than 0.1
    [GATE 1998]

17. Nusselt number/Biot number varies
    (a) inversely with thermal conductivity
    (b) directly with heat transfer co-efficient
    (c) directly with thermal conductivity
    (d) inversely with dimension of the solid
    [GATE.2000]

18. A furnace wall is made of three materials (I, II and III) of equal thickness and having thermal conductivity k_1 k_2 and k_3 respectively. The steady state temperature profile inside each material is shown in the figure. Thermal conductivities of the materials will vary as

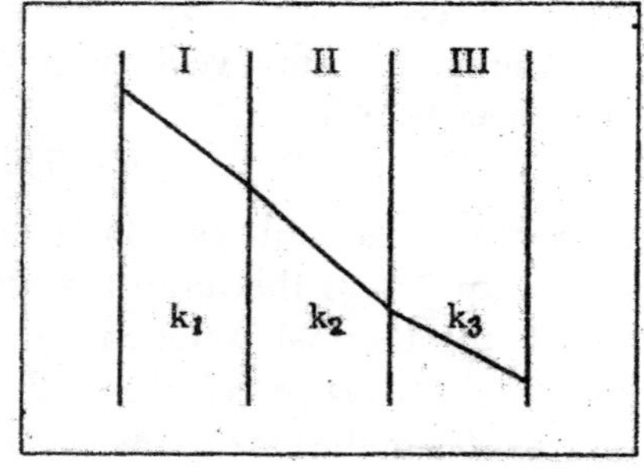

    (a) $k_1 > k_2 > k_3$ (b) $k_2 > k_1 > k_3$
    (c) $k_3 > k_1 > k_2$ (d) $k_3 > k_2 > k_1$
    [GATE 2004]

19. Match the following.
    Group 1
    P. Heat transfer coefficient
    Q. Thermal diffusivity
    R. Mass transfer coefficient
    S. Viscosity
    Group 2
    1. m^{2}S^{-1}
    2. W . m^{-2} K^{-1}
    3. kg . m^{-1}s^{-1}
    4. m.s^{-1}
    (a) P-2, Q-l, R-3, S-4
    (b) P-2, Q-l, R-5, S-4
    (c) P-4, Q-l, R-5, S-3
    (d) P-2, P-l, R-4, S-3
    [GATE 2004]

20. Select the best material, which can be used for the applications mentioned below using the following data:

Material	k(W.) m^{-1}. K^{-1}	ρ (kg.m^{-3})	Cp (J.kg^{-1}. K^{-1})
I	105	2320	687
II	425	10,500	234
III	238	2700	917
IV	2320	3500	519
V	63	2250	711

Choose a material which can be used as a heat reservior (to hold that)
    (a) III (b) II
    (c) V (d) IV
    [GATE 2004]

21. In Q. 20, choose a material which can be used as heat sink (to remove that).
    (a) I (b) III
    (c) V (d) IV
    [GATE 2004]

Answers

15. (a) 16. (a) 17. (a) 18. (c) 19. (b) 20. (c)
21. (d)

22. A spherical coordinate (r = radius, T = temperature and t = time), the heat conduction equation in steady state is

(a) $\frac{\partial}{\partial r}\left(r^2.\frac{\partial T}{\partial r}=0\right)$ (b) $\frac{\partial}{\partial T}\left(r^2.\frac{\partial T}{\partial r}=0\right)$

(c) $\frac{\partial}{\partial r}\left(T^2.\frac{\partial T}{\partial r}=0\right)$ (d) $\frac{\partial}{\partial t}\left(r^2.\frac{\partial T}{\partial r}=0\right)$

[GATE 20C4]

23. At steady state and when the inner and outer walls of a long hollow cylinder are kept at two different temperatures, the unidirectional temperature variation along the thickness of the wall is
(a) linear
(b) parabolic
(c) logarithmic
(d) constant

[GATE 2010]

24. The temperature of a gas flowing in a long duct as measured by thermocouple (having an emissivity of 0.5) is 800 K. The internal wall surface of the duct is at a temperature of 500 K. The convective heat transfer coefficient between the gas and the tip of the thermocouple is 100 $W.^{-1}$. $K.^{-1}$ The actual gas temperature is actually .... K.
(a) 400
(6) 500
(c) 820
(d) 900

[GATE 2010]

25. The infinitely long and wide parallel plates A and B are at temperature Ta and Tb respectively. The energy transferred from relatively hotter plate A to plate B is proportional to
(a) $T_a - T_b$
(b) $T_a^2 - T_b^2$
(c) $(T_a - T_b)(T_a^2 + T_b^2 - T_a.T_b)$
(d) $T_a^2 - T_a^2$

[GATE 2006]

26. Match the properties in group I with the units in group II.

Group I
P. Thermal conductivity
Q. Heat transfer coefficient
R. Specific heat
S. Diffusivity

Group II
1. $J.m^{-2}s^{-1}k^{-1}$
2. $J.m^{-1}s^{-1}k^{-1}$
3. $m.s^{-1}$
4. $J.mol^2s^{-1}$

(a) P-l, Q-2, R-4, S-3
(b) P-2, Q-3, R-l, S-4
(c) P-2, Q-l, R-4, S-3
(d) P-2, Q-4, R-3, S-l

[GATE 2008]

27. A furnace wall consists of four layers of different materials, M_1, M_2, M_3 and M_4. If the layers are of equal thickness and the steady state temperature profiles is as shown below, then the material with the lowest thermal conductivity is

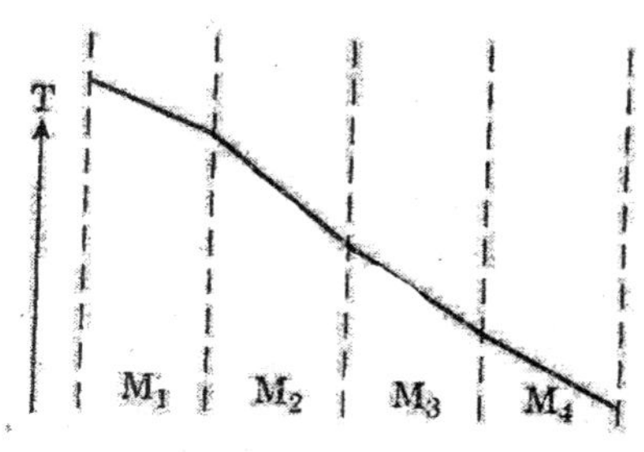

(a) M_1 (b) M_2
(c) M_3 (d) M_4

[GATE 2009]

Answers

22. (a) 23. (c) 24. (c) 25. (a) 26. (c) 27. (a)

28. A furnace wall consists of two layers. The inside layer of 450 mm is made of light weight bricks of thermal conductivity lW/m.K and outside layer of 900 mm is made of refractory of thermal conductivity 2W/m.K. The hot face of the inside layer is at temperature 1300 K and the cold face of the outer layer is at 400 K. The temperature at the interface between the two layers is .... K.
(a) 1000 (b) 850
(c) 700 (d) 600
[GATE 2009]

29. The ratio of the shear stress to the principal stress on a principal plane is
(a) 0 (b) 1
(c) 1/2 (d) 1/3
[GATE 94]

30. A material is loaded elastically under plane stress condition given by the following tensor:

$$\begin{bmatrix} 12 & 0 \\ 0 & 10 \end{bmatrix}$$

The modulus of rigidity is 25 MPa. The maximum elastic engineering strain is
(a) 0.48 (b) 0.40
(c) 0 (d) none of these.
[GATE 94]

31. The bulk modulus of a material with a poisson's ratio of 0.5 is equal to
(a) 3 x Young's modulus
(b) Young's modulus
(c) infinity
(d) zero
[GATE 95]

32. A metal having a Poisson's ratio = 0.3 is elastically deformed under uniaxial tension. If the longitudnal strain = 0.8, then the magnitude of thickness strain is
(a) – 0.8
(b) 0.8
(c) – 0.8
(d) –0.24
[GATE 98]

33. A cylindrical rod subjected to a tensile strain within the elastic limit undergoes a volume change. If the volume strain is equal to half the applied tensile strain then the poisson's ratio of the rod is
(a) 0.00
(b) 0.33
(c) 0.44
(d) 0.25
[GATE 99]

34. Thin walled pressure vessels are to be made either from Material A and B. If the applied stress is 1.2 times higher when material A is used and $(K_{IC})_A = 0.6(K_{IC})_B$ (where, K_{IC} denotes plain strain fracture toughness). The ratio of critical crack lengths in material A and B will be
(a) 1/2
(b) 1/4
(c) 1/6
(d) 1/8
[GATE 2004]

Answers

28. (d) 29. (a) 30. (a) 31. (c) 32. (d) 33. (c)
34. (b)

35. Engineering stress-strain curve for a metal under two conditions. A and B, are shown is the following figures. Identify the correct statement.

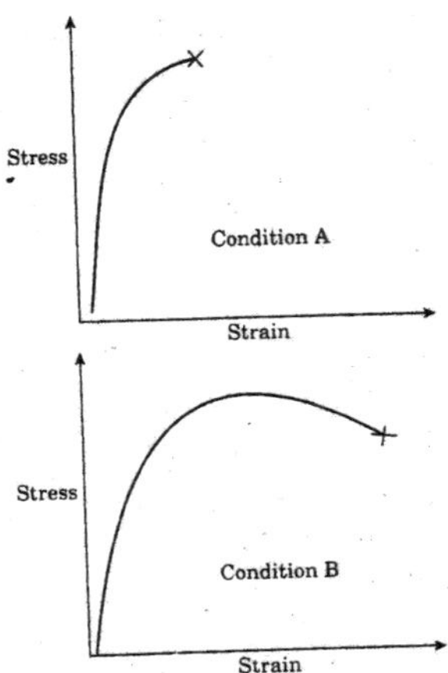

(a) The resilience of the material is the same in conditions A and B.
(b) The resilience of the material is higher in condition B than in condition A.
(c) The toughness of the material is higher in condition A than in condition B.
(d) The toughness of the material is higher in condition B than in conditon A.

[GATE 2004]

36. Graphite fibres are used to make un-directional Al-matrix composites. Young's modulus of graphite and aluminium are 400 GPa respectively. If a composite contains 60 volume % of fiber and is loaded along the fiber direction, the Youngs Modulus of the composite is .... GPa.
(a) 60
(b) 164
(c) 264
(d) 364

[GATE 2004]

37. Stress analysis of a mechanically loaded structure gives the state of stress as shown below. The stress tensor, σu of the above structure is given as

(a) $\begin{bmatrix} 15 & -30 & 0 \\ -30 & 200 & 0 \\ 0 & 0 & -80 \end{bmatrix}$ (b) $\begin{bmatrix} 200 & -0 & 0 \\ 0 & 150 & 0 \\ 0 & -30 & -80 \end{bmatrix}$

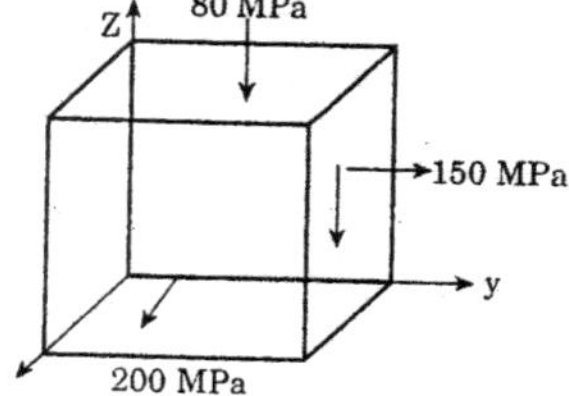

(c) $\begin{bmatrix} -80 & 30 & 0 \\ 30 & 150 & 0 \\ 0 & 0 & 200 \end{bmatrix}$ (d) $\begin{bmatrix} 200 & 0 & 0 \\ 0 & 150 & 30 \\ 0 & 0 & -80 \end{bmatrix}$

[GATE 2004]

38. In Q.No. 37, structure will undergo plastic yielding when the stress is close to .... MPa.
(a) 168
(a) 268
(c) 326
(a) 468

[GATE 2004]

39. If the volume of a material does not change during deformation, then the Poisson's ratio should be
(a) 0.25
(b) 0.50
(c) 0.67
(d) 1.00

[GATE 2005]

Answers

35. (d) 36. (c) 37. (b) 38. (b) 39. (d)

40. If the choice of axes changes for an externally loaded solid material, the correct statements are:

    P. The components of stress tensor changes.

    Q. The octahedral shear stress does not change.

    R. The distortion energy changes.

    S. The yield loci in the space of prinicpal stresses do not change.

    (a) P, Q, S (b) P,R, S

    (c) P, Q, R (d) Q, R

    [GATE 2005]

41. A glass reinforced polymer composite comprises of 35 volume % glass fiber and 65 volume % of polymer resin. The elastic moduli of glass and polymer resin are respectively 70 GPa and 3 GPa. The elastic modulus of the composite (in GPa) under iso-strain state is

    (a) 4.51 (b) 25.45

    (c) 26.45 (d) 27.45

    [GATE 2005]

42. In Q.No. 41, the elastic modulus (in GPa) of the composite under iso-stress state is

    (a) 3.51 (b) 4.51

    (c) 6.51 (d) 26.45.

    [GATE 2005]

43. A square of 9 mm^2 area is subjected to simple shear displacement $\sqrt{3}$ mm along x-direction, as shown below: The strain imparted will be

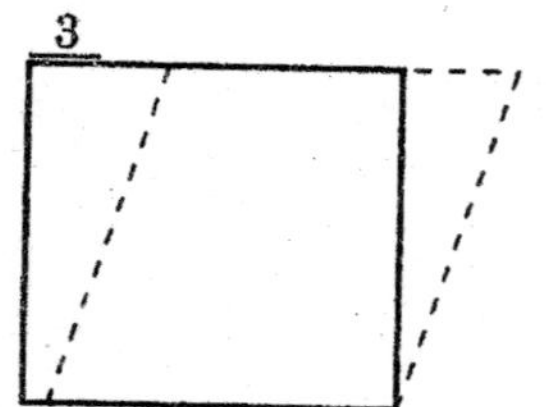

    (a) 1/3 (b) $1/\sqrt{3}$

    (c) $\sqrt{3}$ (d) 3

    [GATE 2010]

44. If true stress-true strain curve of ductile material is represented by the equation, $\sigma = 1100^{0.20}\epsilon$, the ultimate tensile strength (engineering) will be .... MPa.

    (a) 853

    (b) 753

    (c) 653

    (d) 556

    [GATE 2010]

45. An aluminium alloy rod of diameter 15 mm and length 120 mm is subjected to a tensile load of 35,000 N along its axis. The Young's modulus and Poisson's ratio for aluminium are 70 GPa and 0.33 respectively.

    (a) 0.11 (b) 0.014

    (c) 0.018 (d) 0.021

    [GATE 2010]

46. In Q.No. 45, the elastic strain energy is approximately .... KJ.m^{-3}.

    (a) 200

    (b) 240

    (c) 280

    (d) 320

    [GATE 2008]

47. Generalised Fick's law for a component i in a multicomponent solution can be written as

    (a) $Ji = \sum_k D_{ik} \cdot D_{kl} \nabla C_i, \nabla C_k$

    (b) $Ji = \sum_k D_{ik} \nabla C_k$

    (c) $Ji = D_{ii} \cdot \Delta C_i$

    (d) none of these

    [GATE 1992]

Answers

40. (a)	41. (c)	42. (b)	43. (b)	44. (c)	45. (b)
46. (c)	47. (d)				

48. If density and diffusion coefficient are assumed constant, then governing equation for mass transfer of A dissolved in solid B is

(a) $\frac{\partial C_A}{\partial t} = D_{AB}. \Delta C_A$

(b) $\frac{\partial C_A}{\partial t} = D_{AB}. \Delta_2 C_A$

(c) $\frac{\partial C_A}{\partial x} = D_{AB}. \Delta_2 C_A$

(d) $\frac{\partial C_A}{\partial x} = D_{AB}. \Delta_2 C_A$

[GATE 2003]

49. The two surfaces of a 5 mm thick plate of carbon steel are maintained at a carbon concentration of 10^4 mol. m^{-3} on one side and zero concentration on the other side. Taking the diffusion coefficient of carbon in austenite to be 10^{-11} $m^{-2}.s^{-1}$, the steady state flux (mol.$m^{-3}.s^{-1}$) through the thickness of the steel is

(a) 10^{-5} (b) 2×10^{-5}

(c) 4×10^{-5} (d) 6×10^{-5}

[GATE 2004]

50. The relative contribution of molecular diffusion to overall mass transfer is highest in

(a) stagnant liquid under natural convection

(b) liquid in laminar flow

(c) liquid in turbulent flow

(d) solid

[GATE 2005]

51. Match the items in group 1 with units/ dimensions in group 2.

Group 1

P. Diffusivity

Q. Surface tension

R. Dislocation density

S. Mass Transfer coefficient

Group 2

1. Lt^{-1}

2. $L^{-2}t^{-1}$

3. JL^{-2}

4. J $mol^{-1}.K^{-1}$

5. L^{-2}

(a) P-3, Q-4, R-2, S-5

(b) P-5, Q-3, R-3, S-l

(c) P-3, Q-2, R-l, S-l

(d) P-2, Q-3, R-5, S-l

[GATE 2005]

52. Fick's first law relates

(a) flux of atoms and the concentration gradient.

(b) amount of gas dissolved in the molten metal and the partial pressure.

(c) applied normal stress and the orientation of slip system.

(d) heat flux and the temperature gradient.

[GATE 2009]

53. Consider the following collection of polymer chains.

No. of molecules	10	5	4	2	1
Molecular weight gm.mol^{-1}	2800	3000	1200	3600	1000

Mer unit is ethylene. Atomic weights : carbon (12) and hydrogen (1). Calculate number average degree of polymerisation.

(a) 32.32 (b) 90.91

(c) 106.61 (d) 116.13

[GATE 2010]

54. Most of the thermosetting polymers are

(a) amorphous (b) semi-crystalline

(c) crystalline (d) transcrystalline

[GATE 2006]

Answers

48. (b) 49. (b) 50. (a) 51. (d) 52. (a) 53. (b)
54. (c)

55. Which one of the following methods is NOT used for numerically solving an ordinary differential equation (ODE)?
(a) Eulers method
(b) Runge-Kutta method
(c) Adam-Bashforth method
(d) Newton-Raphson method
[GATE 2011]

56. The inflection point of a non-linear function U(r) is at
(a) U = 0 (b) In U=0
(c) dU/dr = 0 (d) $d^2U/dr^2 = 0$
[GATE 2011]

57. A box contains 5 white balls and 3 red balls. Two balls are withdrawn from the box randomly, one after another (without replacement). The probability that the two balls withdrawn are of different colour is
(a) 15/64 (b) 25/64
(c) 25/56 (d) 30/56
[GATE 2011]

58. During fully developed laminar flow in a circular pipe, the velocity profile is parabolic, and symmetric around the axis. The velocity at the tube wall is zero. The ratio of the average velocity to the maximum velocity is
(a) 1/64 (b) 1/2
(c) 2/3 (d) 3/4
[GATE 2011]

59. A large set of data for a given measurement has been found to be normally distributed around a mean 'μ' with standard deviation 'σ'. Which of the following limits would have about 95% of the data points around the mean and rest outside?
(a) $\mu - 0.5\sigma$ and $\mu + 0.5\sigma$
(b) $\mu - \sigma$ and $\mu + \sigma$
(c) $\mu - 2\sigma$ and $\mu + 2\sigma$
(d) $\mu - 3\sigma$ and $\mu + 3\sigma$
[GATE 2011]

60. The following thermo couple may be used for measuring temperature upto 1873 K.
(a) chromel-alumel
(b) copper-constantan
(c) platinum-platinum rhodium
(d) iron-constant
[GATE 2003]

61. A fluid is flowing with a velocity of 0.5 m/s on a plate moving with a velocity of 0.01 m/s in the same direction. The velocity at the interface of the fluid and plate is
(a) 0.0 m/s (b) 0.01 m/s
(c) 0.255 m/s (d) 0.50 m/s
[GATE 2012]

62. Hot metal at 1700 K is poured in a sand mould that is open at the top. Heat loss from the liquid metal takes place by
(a) Radiation only
(b) Radiation and conduction only
(c) Radiation and convection only
(d) Radiation, conduction and convection
[GATE 2012]

63. A steel ball (density ρsteel = 7200 kg/m^3) is placed in an upward moving liquid AI (density ρal = 2360 kg/m^3, viscosity μAl = 1 x 10^{-3} Pa.s and Reynolds number = 5 x 10^5). The force (F) exerted on the steel ball is expressed as
$F = f\pi R^2 (\rho_{Al} V^2/2)$
where, f is friction factor (= 0.2), v is the velocity of liquid Al and R is the radius of steel ball.
The force exerted on the steel ball is
(a) 8.32 N (b) 6.70 N
(c) 1.67 N (d) 0.52 N
[GATE 2012]

Answers

55. (d)	56. (d)	57. (d)	58. (b)	59. (c)	60. (c)
61. (b)	62. (d)	63. (a)			

64. In Q.63, the terminal velocity of a fine spherical steel particle having diameter dp, in μm range, if allowed to fall in a quiescent liquid Al bath, is
(a) $5.2 \times 10^6 d_p^2$ m/s
(b) $2.6 \times 10^6 d_p^2$ m/s
(c) $1.3 \times 10^6 d_p^2$ m/s
(d) $6.6 \times 10^5 d_p^2$ m/s

[GATE 2012]

65. For the above stress cycle, stress ratio is
(a) 4 (b) 2
(c) –2 (d) –4

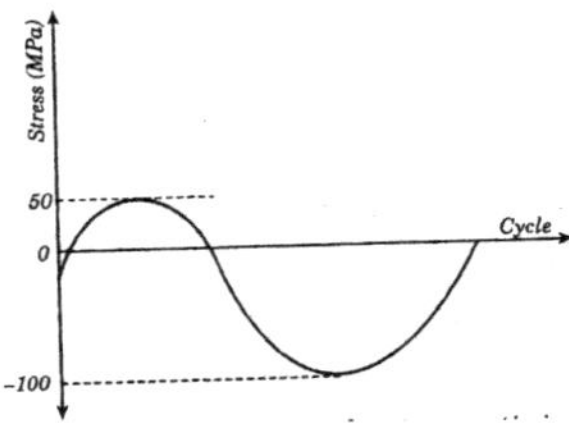

[GATE 2012]

66. In Q.65 Amplitude ratio is
(a) 3 (b) 1/3
(c) –1/3 (d) -3

[GATE 2012]

67. There are 150 gearwheels in a box, out of which 112 are within the required tolerance, 21 are below and rest are above the required tolerance. If the selection is done without replacement, the combined probability of randomly selecting a gearwheel below the tolerance and then a second one above the tolerance is
(a) 0.016 (b) 0.032
(c) 0.492 (d) 0.984

[GATE 2013]

68. The critical internal crack length (in mm) in a steel having K_{1c} of 45 MPa $\sqrt{m}$ to support a Mode-I stress of 400 MPa is
(a) 2 (b) 4
(c) 6 (d) 8

[GATE 2013]

69. Which one of the following plots relating shear stress with strain rate represents Newtonian behaviour of a fluid?

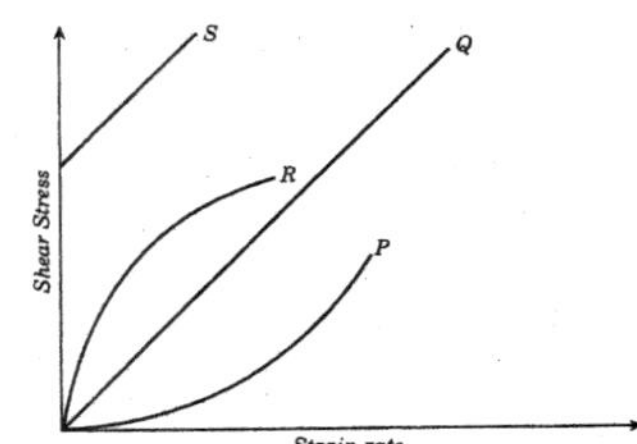

(a) P (b) Q
(c) R (d) S

[GATE 2013]

70. A tensile specimen was deformed at a constant cross head speed of 6 mm/min. The strain rate at the start of the testing was 5×10^{-3} s^{-1}. The initial gauge length of the specimen (in mm) was
(a) 10 (b) 20
(c) 35 (d) 50

[GATE 2014]

71. Two identical engineering components are made from alloys X and Y. The K_{1c} of alloy X is greater than that of alloy Y. If these components are subjected to the same stress during operation under mode-I loading, which one of the following statements is TRUE?
(a) Component made from alloy X can tolerate a larger flaw size compared to that made from Y.
(b) Component made from alloy Y can tolerate a larger flaw size compared to that made from X.
(c) Both can tolerate same flaw size.
(d) None of the above.

[GATE 2014]

Answers

64. (b)	65. (c)	66. (d)	67. (a)	68. (d)	69. (b)
70. (b)	71. (a)				

72. A steel rod was subjected to a fluctuating stress cycle that varied between a maximum of400 MPa in tension to a minimum of 300 MPa in compression. What is the stress amplitude, in MPa?
(a) 150 (b) 250
(c) 350 (d) 450
[GATE 2014]

73. A body starts yielding when it is subjected to stress state with principal stresses of 250 MPa, 50 MPa and - 50 MPa. What is the yield strength of the material, in MPa, if Tresca yield criterion is obeyed?
(a) 100 (b) 200
(c) 300 (d) 400
[GATE 2014]

74. Consider a steady heat flux across a rectangular slab composed of two layers of equal width as shown in the figure below. The thermal conductivities are in thratio of $\frac{k2}{k1}$ = 10. If the first layer experiences a temperature drop (T; - Tz) of 50 K, what is the temperature drop (T-2 - T3J, in K, across the second layer?

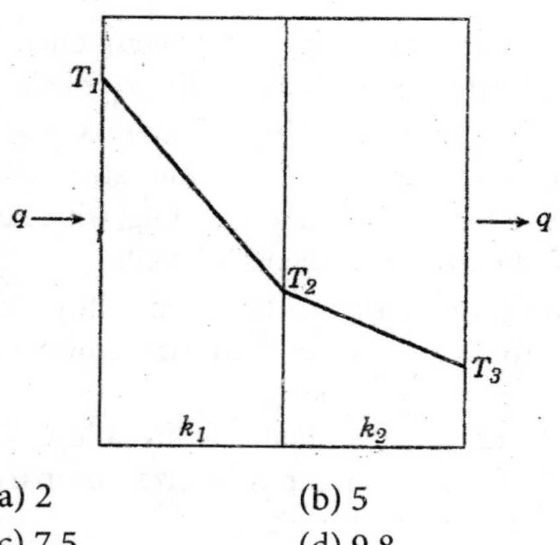

(a) 2 (b) 5
(c) 7.5 (d) 9.8
[GATE 2014]

75. A glass fibre reinforced epoxy composite is made with 60 wt.% unidirectional continuous glass fibres. The elastic moduli of the glass fibre and the epoxy matrix are 72.5 and 2.4 GPa, respectively. What is the elastic modulus of this composite along the fibre direction, in GPa? The densities of glass fibre and epoxy are 2.58 g/cm^3 and 1.14 g/cm^3, respectively
(a) 10.5 (b) 20.5
(c) 30.5 (d) 4.5
[GATE 2014]

76. Which of the following oxide addition results in polymerisation (e.g., network formation) in a silicate slag?
(a) CaO (b) MgO
(c) P_2O_5 (d) Na_2O

77. Self supporting arches for furnace roofs can be fabricated using silica bricks but not using magnesia bricks. Why?
(a) Silica has a significantly lower thermal expansion coefficient than magnesia at high temperatures.
(b) Silica has a significantly higher thermal conductivity than magnesia at high temperatures.
(c) Silica has a significantly lower melting point than magnesia.
(d) Silica is significantly more acidic than magnesia.
[GATE 2015]

78. A species can diffuse through the lattice (diffusion coefficient, D_L), along grain boundaries (diffusion coefficient, D_{GB}), and along free surfaces (diffusion coefficient, Ds). Which of the following relations is CORRECT?
(a) Dl > DGB > Ds (b) Ds > DL > DGB
(c) DGB > Ds > Dl (d) Ds > DGB > DL
[GATE 2015]

Answers

72. (c) 73. (c) 74. (b) 75. (c) 76. (c) 77. (a)
78. (d)

79. Which of the following Mohr's circles of a plane-stress condition corresponds to equi-biaxial tension

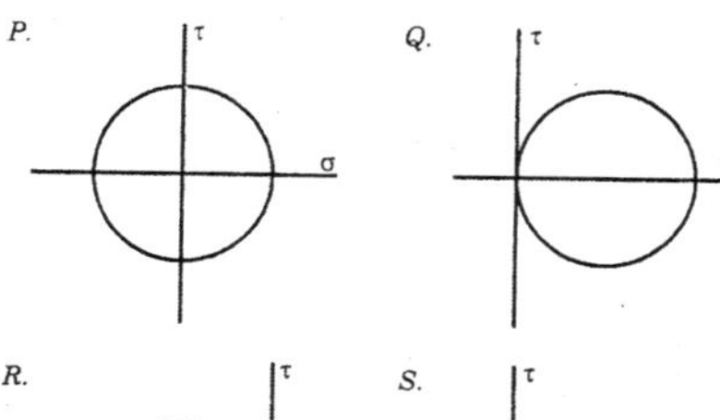

(a) P (b) Q
(c) R (d) S

[GATE 2015

80. The figure below shows water flowing through a pipe. The pressure difference between points P and Q measured using a water-over-mercury manometer is

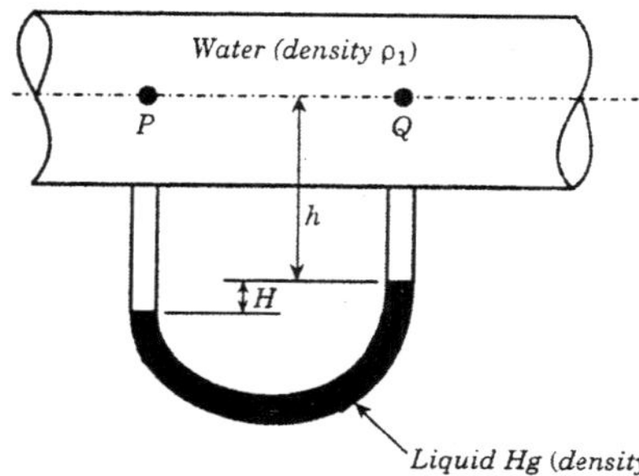

(a) $\rho_2 gH$
(b) ρ_1 gH
(c) $(\rho_2 - \rho_1)$ gH
(d) $(\rho_2 - \rho_1)$ gh

[GATE 2015]

81. It takes 10 hours to homogenize an alloy at 1273 K. The time required (in hours) to achieve the same extent of homogenization at 1373 K is ....

Given: Diffusivity, D_{1373} K = 10^{-18} m^2 s^{-1} and D_{1273} K = 10^{-19} m^2 s^{-1}

(a) 0.5 (b) 1
(c) 2.5 (d) 4

[GATE 2015]

82. Match entities listed in **Column I** with their correct dimensions given in **Column II**.

Column I

P. Drag coefficient
Q. Mass transfer coefficient
R. Viscosity
S. Mass flux

Column II

1. $ML^{-1} T^{-1}$
2. $ML^{-2} T^{-1}$
3. $M^0 L^0 T^0$
4. $M^0 LT^{-1}$

(a) P-3, Q-4, R-1, S-2
(b) P-3, Q-1, R-2, S-4
(c) P-1, Q-4, R-2, S-3
(d) P-4, Q-3, R-1, S-2

[GATE 2016]

83. The height of a liquid metal column in a cylindrical vessel is 3.2 m. At time t=0, liquid metal is drained out form the vessel through a small nozzle located at the base of the vessel. Neglecting frictional losses. the initial mass flow rate (in $kg.s^{-1}$) through the nozzle is ....

Given data

Density of liquid metal	= 7000 kg.m−3
Nozzle diameter	= 30 mm
Nozzle discharge coefficient	= 0.80

[GATE 2016]

Answers

79. (d) 80. (c) 81. (b) 82. (a) 83. (1 to 2)

84. The density and corresponding crystallinity of two poly-propylene material are given below.

Density.kg.m^{-3}	Crystallinity, %
904	62.8
895	54.4

The density of totally amorphous poly-propylene (in kg.m^{-3}) is:

(a) 723 (b) 841

(c) 905 (d) 956

[GATE 2016]

85. Fatigue behaviour of an aluminium alloy is shown is the S-N plot. A piston rod made of this material is subjected to: (i) 1000 cycles at 420 MPa. followed by (ii) 1000 cycles at 300 MPa. Using Miner's rule of cumulative damgae, the remaining fatigue life (in terms of number of cycles) at stress of 250 MPa is..................

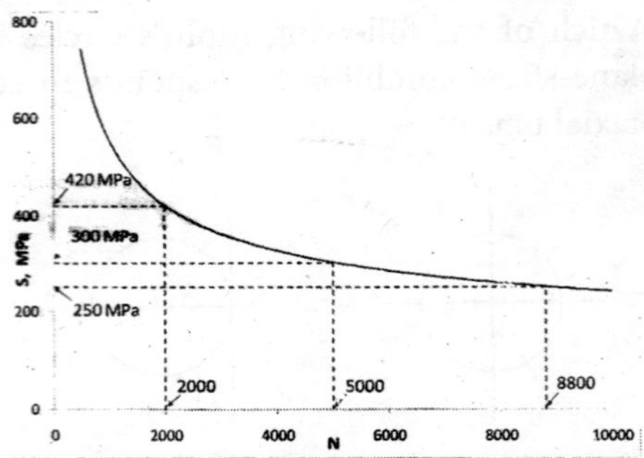

[GATE 2016]

86. For a polymer reinforced with 40 vol.% glass fiber, the elastic modulus (in GPa) along the transverse direction is

(a) 5.6 (b) 8.1

(c) 30.1 (d) 43.4

[$E_{\text{glass fiber}}$ = 70 Gpa; E_{polymer} = 6.5 GPa]

[GATE 2016]

Answers

84. (b) 85. (2630 to 2650) 86. (a)

Section

B

20

GATE–2017

1. For the matrix; $A = \begin{bmatrix} 1 & 1 & 2 \\ 2 & 1 & 1 \\ 1 & 1 & 2 \end{bmatrix}$ AAT is

(a) $\begin{bmatrix} 6 & 5 & 6 \\ 5 & 6 & 6 \\ 6 & 5 & 6 \end{bmatrix}$ (b) $\begin{bmatrix} 6 & 5 & 6 \\ 5 & 6 & 6 \\ 5 & 5 & 6 \end{bmatrix}$

(c) $\begin{bmatrix} 6 & 5 & 6 \\ 5 & 6 & 5 \\ 6 & 6 & 6 \end{bmatrix}$ (c) $\begin{bmatrix} 6 & 5 & 6 \\ 5 & 6 & 5 \\ 6 & 5 & 6 \end{bmatrix}$

Question Number : 2

Correct : 1 Wrong : –0.33

2. The mean of a numerical data-set is X and the standard deviation is S. If a number K is added to each term in the data-set then the mean and standard deviation become:

(a) $\overline{X}$ S (b) $\overline{X}$ + K, S

(c) $\overline{X}$, S + K (c) $\overline{X}$ + K, S K

Question Number : 3

Correct : 1 Wrong : –0.33

3. If (x) = $e^{|x|}$ then at x= 0, the function f(x) is

(a) continuous and differentiable.

(b) continuous but not differentiable.

(c) neither continuous nor differentiable.

(d) not continuous but differentiable.

Question Number : 4

Correct : 1 Wrong : –0.33

4. the pressure (P) versus volume (V) diagram given below represents reversible isothemal curves at temperatures, T1: T2 and T3.

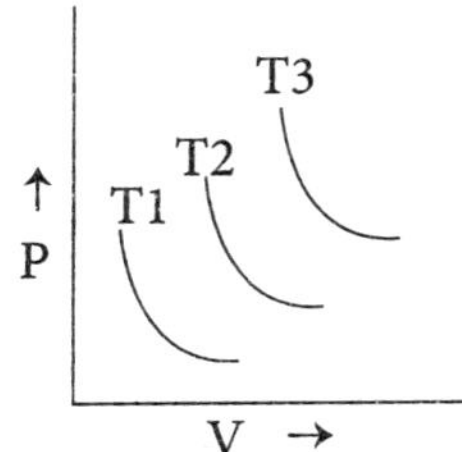

Considering one mole of ideal gas for the three isothermal processes, which one of the following is TRUE?

(a) T1>T2>T3 (a) T2>T3>T1

(c) T3>T1>T2 (d) T2>T1>T3

Question Number : 5

Correct : 1 Wrong : –0.33

Answers

1. (d) 2. (b) 3. (b) 4. (b)

5. For the electrochemical reaction, Cu^{2+} + Zu=Zn^{2+} + Cu : the standard cell potential at 25°C and 1 atm pressure is:
(Given: E° (Cu^{2+}/Cu) = 0.337 V and E°(Zn^{2+}/Zn) = 0.763 V)
(a) −0.426V (b) 0.426V
(c) 0.55V (d) 1.1V

Question Number : 6
Correct : 1 Wrong : −0.33

6. The rate of dissolution of Al particles in liquid steel is proportional to concentration differene (ΔC) ΔC is definied by:
(Given: (i) C_b = bulk concentration of dissolved Al in liquid steel, (ii) C* = saturation concentration of Al in liquid steel at the given temperature, (iii) C^m = Density of Al/Atomic weight of Al.)
(a) $C^* - C_b$ (b) $C_b - C_m$
(c) $C^* - C_m$ (d) $\sqrt{C^*}\ C_m{-}C_b$

Question Number : 7
Correct : 1 Wrong : 0

7. Hydrogen dissolves in Pd by the reaction H_2=2[H]. At 300°C and P_{H2} = 1 atm, the solubility of hydrogen i Pd is 1.64×10^4 mm^3 (STP) per kg of Pd is ________ (answer up to one decimal place).

Question Number : 8
Correct : 1 Wrong : 0

8. The sieve analysis of ground quartz particles is given in the table below.

Steve size (mm)	*Mass fraction of ground product retained on each steve*
4.76	0.0
3.36	0.2
2.38	0.4
1.68	0.3
1.19	0.08
<1.19	0.02

The cumulative mass fraction of particles of size less than 1.68 mm is__________(answer up to two decimal places)

Question Number : 9
Correct : 1 Wrong : −0.33

9. The sequence of precipitation to reach stable equilibrium during ageing of Al-4.5 wt.% Cu alloy is:
(a) GP zone →Ø' → Ø" → Ø
(b) GP zone → Ø" → Ø' → Ø
(c) GP zone → Ø' → Ø" → Ø'
(d) GP zone → Ø" → Ø → Ø'

Question Number : 10
Correct : 1 Wrong : 0

10. Tungsten powder is pressed at 150 MPa to a green density of 55%. After sintering, the compact attains 86.5% of its theoretical density. Assuming uniform shrinkage, the linear shrinkage (in%) is __________ (answer up to two decimal places)

Question Number : 11
Correct : 1 Wrong : 0

11. For a FCC metal, radius of the largest sphere that can fit in the tetrahedral void (in mm) is __________ (answer up to three decimal places).
(Given: lattice parameter = 0.401mm)

Question Number : 12
Correct : 1 Wrong : 0

12. In an iron-carbon alloy containing 0.35 wt.% C, the mass fraction of pearlite just below the eutectoid temperature is __________ (answer up to two decimal places)
(Given: eutectoid compostion = 0.8 wt.% carbon; and carbon content in ferrite is 0.225 wt. %)

Question Number : 13
Correct : 1 Wrong : 0

Answers

5.(d)	6. (a)	7. (4900 to 4940)	8. (0.09 to 0.11)	9. (b)
10. (13.30 to 15.50)		11. (0.030 to 0.034)	12. (0.36 to 0.44)	

13. A cubic metal has a density of 19000 kg. m^{-3}, lattice parameter of 0.4 mm and atomic weight of 183. The effective number of aroms in an unit cell of this metal is _______.

Question Number : 14

Correct : 1 Wrong : 0

14. Primary mechanisms of accommodating plastic strain at low temperatures in crystalline metals are:
(a) twimning and dislocation-slip
(b) dislocation-climb and dislocation-slip
(c) dislocation-slip and diffusion
(d) viscous-flow and dislocation-slip

Question Number : 15

Correct : 1 Wrong : 0

15. Spherical a phase particles are depicted in the hypothetical microstructure section shown below. Using the superimposed grid on the microstructure, the estimated volume fraction of a phase is ___________ (answer up to three decimal places).

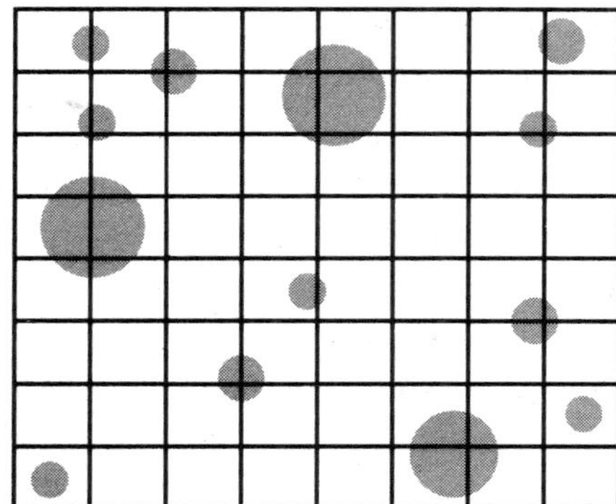

Question Number : 16

Correct : 1 Wrong : 0

16. A brittle material (Young's modulus = 60 GPa and surface energy = 0.5 $J.m^{-2}$) has a surface crack of length 2 μm. The fracture strength (in MPa) of this material is ________ (answer up to two decimal places)

Question Number : 17

Correct : 1 Wrong : -0.33

17. Both creep resistance and tensile strength of a metal can be enhanced by
(a) increase in the grain size
(b) decrease in the grain size
(c) addition of dispersoids
(d) annealing

Question Number : 18

Correct : 1 Wrong : -0.33

18. Stress required to operate a Frank- Read source of length L is approximately given by:

(a) $\frac{Gb}{L}$ (b) $\frac{Gb^2}{L}$

(c) $\frac{Gb^2}{L^2}$ (d) $\frac{Gb^2}{2L^2}$

Question Number : 19

Correct : 1 Wrong : 0

19. The second peak in the powder X-ray diffraction pattem of a FCC metal occurs at a Bragg angle Ø (in degrees) = _______ (answer up to two decimal places)
(Given: λ_{CuKa} = 0.154mn; lattice parameter of the metal = 0.36mn)

Question Number : 20

Correct : 1 Wrong : 0

20. A rod is elastically deformed by a uniaxial stress resulting in a strain of 0.02. If the Poisson's ration is 0.3, the volumetric strain is _________ (answer up to three decimal places)

Question Number : 21

Correct : 1 Wrong : -0.33

21. Four alloys, C1, C2, C3, C4, shown in the phase diagram are poured at temperature. T1 in a mold. During solidfication, which one of these alloys is expected to have the highest fluidity?

Answers

13. (4 to 4) 14. (a) 15. (0.130 to 0.160) 16. (95.00 to 125.00) 17. (c)
18. (a) 19. (24.00 to 26.00) 20.(0.006 to 0.010) 21.(c)

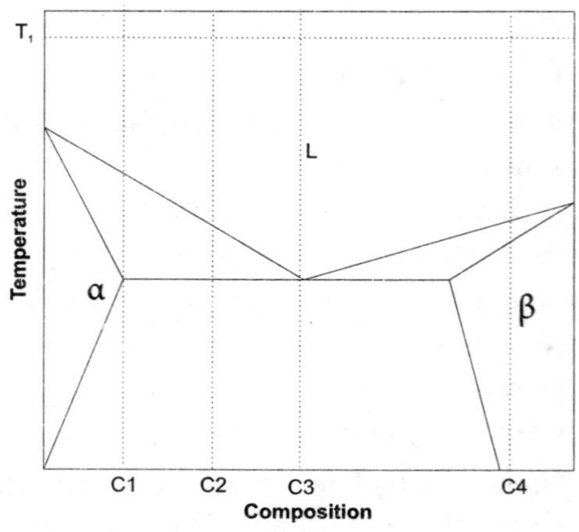

(a) C1 (b) C2
(c) C4 (d) C4

Question Number : 22
Correct : 1 Wrong : 0

22. A material, which shows power law behavior, $\bar{\sigma} = 50\varepsilon^{-0.3}$, being wire drawn. The maximum strain per pass in annealed condition (assume ideal work and efficiency η =1) is ________ (answer up to two decimal places)

Question Number : 23
Correct : 1 Wrong : –0.33

23. Schematic diagram shows rolling of a slab. P and Q are points on the surface of the workplace near entrance exit, respectively. With reference to the work piece, which one of the following statements of TRUE?

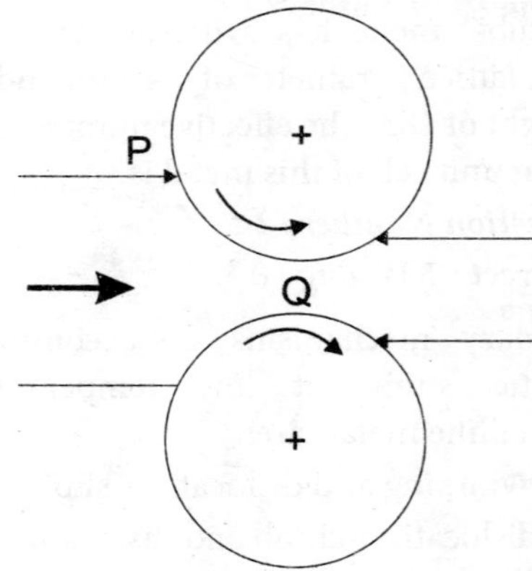

(a) Frictional force is along rolling direction both P and Q.
(b) Frictional force is opposite to rolling direction at both P and Q.
(c) Frictional force is along rolling direction at P and opposite to rolling direction at Q.
(d) Frictional force is opposite to rolling direction and P and along rolling direction at Q.

Question Number : 24
Correct : 1 Wrong : –0.33

24. Which one of the following manufacturing techniques is used for making window glass?
(a) Investment casting
(b) Patenting
(c) Spray forming
(d) Float-bath method

Question Number : 25
Correct : 1 Wrong : –0.33

25. Dye penetrant test is based on the principle of
(a) polarized sound waves in liquid.
(b) magnetic domain.
(c) absorption of X-rays.
(d) capillary action.

Answers

22. (1.20 to 1.40) 23. (c) 24. (d) 25. (d)

Question Number : 26
Correct : 1 Wrong : 0

26. Assume that the probability of South Africa winning against India is 1/3. If South Africa plays a 3 march cricket series against India, the probability that South Africa wins only one match is (assume that no matches are tied or abandoned) __________ (answer up to three decimal places)

Question Number : 27
Correct : 1 Wrong : 0

27. The function $f(x) = X^3 - 3x$ has a minimum at x = __________.

Question Number : 28
Correct : 1 Wrong : 0

28. The definite integral, $\int_0^1 e^{-x2}$ dx is to be evaluated numerically. Divide the integration interval into exactly 2 subintervals of equal length. Applying the trapezoidal rule, the approximate value of the integral is ______ (answer up to two decimal places)

Question Number : 29
Correct : 1 Wrong : -0.66

29. For the second order linear ordinary differential equation.

$$\frac{d^2y}{dx^2} + P\frac{dy}{dx} + qy = 0,$$

the following function is a solution:

$y = e^{\lambda x}$

Which one of the following statements is NOT TRUE?

(a) λ has two values: one complex and one real
(b) $\lambda^2 + p\lambda + q = 0$
(c) λ has two real values
(d) λ has two complex values

Question Number : 30
Correct : 2 Wrong : 0

30. Using the bisection method, the root of the equation $x^2 + x - 1 = 0$ after three iterations is________ (answer up to two decimal places)
(Assume starting values of x = –1 and +1)

Question Number : 31
Correct : 2 Wrong : -0.66

31. T_1 and T_2 are the melting points of pure metal A and pure stoichiometric oxide AO_2 respectively, and $T_1 < T_2$. The stoichiometric metal oxidation reaction A(s) + O2 (g) = AO2 (s) is in equilibrium at 1 atm pressure at temperature less thatn T1. If the temperature increases, which schematic represents the correct standard free energy change versus temperature plot?

(a) (b)
(c) (d)

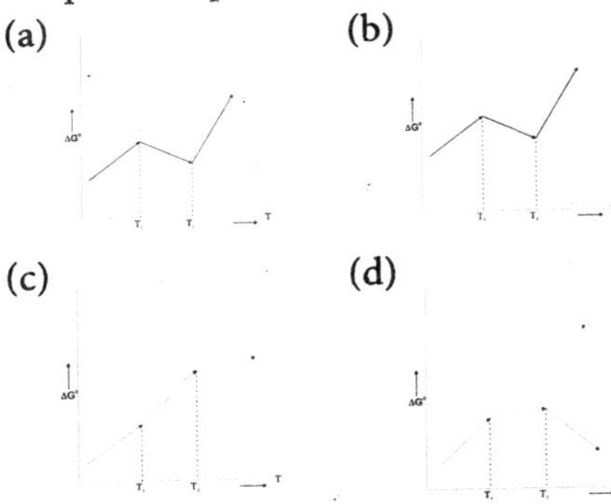

Question Number : 32
Correct : 2 Wrong : 0

32. A continuous cast steel slabm 1 m × 1 m × 0.1m, at 1298 K cools in air. The initial rate of heat loss (in kW) from the top surface of slab by radiation and convection is __________ (answer up to two decimal places)
Given: (1) Amblent temperature = 293K, (ii) emissiviy of steel = 0.8 (iii) convective heat transfer coefficient = 4.6 $Wm^{-2}m\ K^{-1}$, (iv)

Answers

26. (0.400 to 0.500)	27. (1 to 1)	28. (0.70 to 0.80)	29. (a)
30. (0.74 to 0.76)	31. (c)	32. (130.00 to 135.00)	

stefan-Boltzmam constant (σ) = 5.7×10^{-5} $W.m^{-2}.K^{-4}$

Question Number : 33

Correct : 2 Wrong : –0.66

33. The Pourbaix plot of the reaction AP^+ + $2H_2O$ – AlO_2 + $4H^+$ in potential (E) versus pH diagram is:

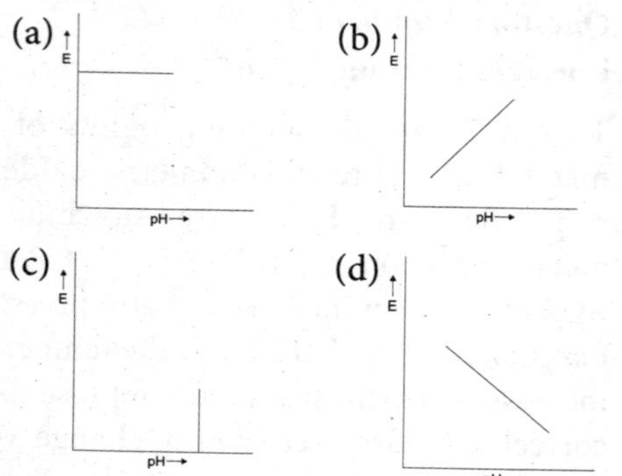

Question Number : 34

Correct : 2 Wrong : 0

34. During the end blow period in LD steelmaking the de-carburization rate is expressed by the equation: $\frac{dC}{dt} = -(C - C^*)$. Here, c and C* are the instantaneous and equilibrum concentration of carbon in steel respectively, in units of wt,% Given that C*= 0.04 wt.% and c (t=0mm) = 0.4 wt.% the concentration of carbon in steel (in wt.%) at t=1 min is ________ (answer up to three decimal places).

Question Number : 35

Correct : 2 Wrong : 0

35. $CaCO_3$(s) dissociates in a closed system according to the reaction.
$CaCO_3$(s) CaO(s) + CO_2 (g)
Assuming the reaction is in thermodynamic equilibrium, the degree (s) of freedom, F = ________

Question Number : 36

Correct : 2 Wrong : –0.66

36. A ladle containing molten steel is being discharged. The relevant forces are listed in ***Column I.*** Match them with their corresponding expressions in ***Column II.***

Column I	***Column II***
(P) Pressure force	(1) μUL
(Q) Inertial force	(2) ρgL^3
(R) Gravity force	(3) ρU^2L^3
(S) Viscous force	(4) PL^2

μ = viscosity, U= characteristic velocity, L= characteristic length.

g = acceleration due to gravity, P = pressure.

(a) P-4, Q-3, R-2,; S-1
(b) P-1, Q-3, R-2, S-4
(c) P-2, Q-3, R-4, S-1
(d) P-4, Q-3, R-1, S-2

Question Number : 37

Correct : 1 Wrong : 0

37. In primary steelmaking, dissolved oxygen ($\underline{O}$) reacts with carbon ($\underline{C}$) to produce CO (g), at 1 atm pressure according to the reactionL $\underline{C} + \underline{O}$ = CO (g). The equilibrium constant for this reaction is: $\log K = \frac{1160}{T} + 2.003$, where T is in Kelvin. Assuming Henrian activity coefficient of both $\underline{O}$ and $\underline{C}$ to be unity, the dissolved oxygen content (in wt. %) of a plain carbon steel melt with 0.7 wt.% C at 1600 °C is ______ (Answer up to four decimal places)

Question Number : 38

Correct : 1 Wrong : 0

38. A storchiometric mixture of CO and pure oxygen at 1 atm and 25°C flows into a combstion reactor. The molar flow rate of CO entering the reactor in 1 kg-mol, h^{-1}. The adiabatic flame remperature (in K) for

Answers

33.(c) 34. (0.170 to 0.175) 35. (1 to 1) 36. (a)
37. (0.0010 to 0.0050) 38. (6650.00 to 6750.00)

the combustion of CO with stoichiometric oxygen is __________ (answer up to two decimal places)

Given : ΔH°_{298} ($CO \rightarrow CO_2$) = −282000 kJ. (kg-mol co)−1, C_p (CO_2) = 44 kJ. (kg-mol $K)^{-1}$.

Question Number : 39
Correct : 1 Wrong : 0

39. A solution contains 10^{-3} M of Fe^{3+} at 25°C. The solubility product of Fe $(OH)_3$ is 10^{-19}, Assuming activity equals concentration, the minumum pH at which Fe^{3+} will precipitate as Fe $(OH)_3$ is _______ (answer up to two decimal places)

Question Number : 40
Correct : 2 Wrong : 0

40. A zinc electrowinning cell is being at a current of 400 A, voltage of 3.5 V, and a cathodic current efficiency of 90%. The specific energy consumption (in kJ. kg^{-1}zinc) is ________ (answer up to two decimal places)

(Atomic weight of Zn = 65)

Question Number : 41
Correct : 2 Wrong : −0.66

41. Pure metals A and B form two real binary solid solutions a and β at temperature T and P. The free energy versus composition plots for both the solutions are shown below.

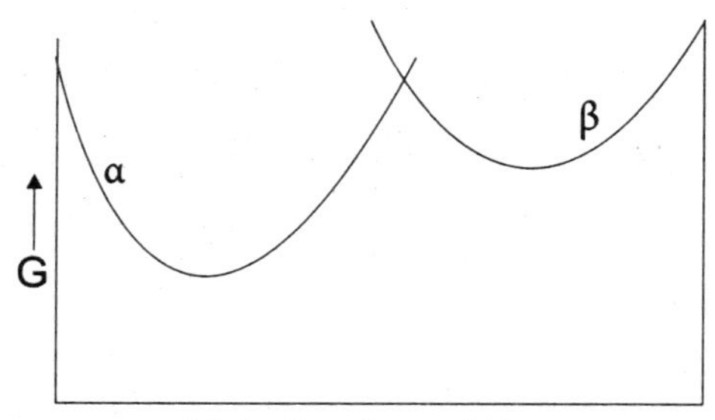

The condition for chemical equlibrium is:

(a) Mole fraction of A in a = mole fraction of A in β and mole fraction of B in a = mole fraction of B in β

(b) Mole fraction of B in a = mole fraction of A in β and mole fraction of A in a = mole fraction of B in β

(c) Activity of A in a = avtivity of A in β and activity of B in a = activity of B in β

(d) Activity of A in a = activity of B in β and activity of B in a = activity of A in β

Question Number : 42
Correct : 2 Wrong : 0

42. Pure orthorhombic sulfur transforms to stable monoclinic sulfur above 368.5K. Applying Third law of thermodynamies, the value of entropy (m J.K^{-1}) of transformation at 368.5 K is ___________(answer up to two decimal places)

Given:

i. Entropy change associated with hearing orthorhombic sulfur from 0 K to 368.5 K is 36.86 J. K^{-1}.

ii. Entropy change associated with cooling monocline sulfur from 368.5 K to 0 K is −37.8 J.$^{K-1}$.

Question Number : 43
Correct : 2 Wrong : 0

43. For homogeneous nucleation of solid in a liquid of a pure metal, the critical edge length (in mm) of a cube shaped nucleus is ______ (answer up to two decimal places).

(Given: surface energy y = 0.177 J.m^{-2}; change in volume free energy ΔGy = −208×102 J.m^{-3})

Question Number : 44
Correct : 2 Wrong : 0

44. Assuming the solid phases to be pure, the slope of line BC in the predominance area diagram schematically shown below is _______ (answer up to two decimal places)

Answers

39. (1.80 to 2.20) 40. (11535.00 to 11580.00) 41. (c) 42. (0.92 to 0.96)
43. (2.50 to 2.60) 44. (−0.51 to −0.49)

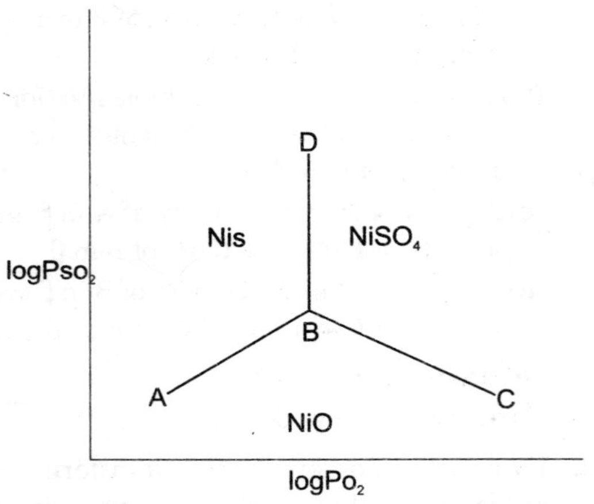

Question Number : 45

Correct : 2 Wrong : –0.66

45. For each of the crystallographic system listed in Group-I, match the corresponding minimum symmetry in Group-II.

Group-I	***Group-II***
(P) Tetragonal	(1) 1 two-fold rotation
(Q) Cubic	(2) 1 three-fold rotation
(R) Monoclinic	(3) 4 three-fold rotation
(S) Rhombohedral	(4) 1 four-fold rotation

(a) P-3, Q-4, R-2, S-3
(b) P-4, Q-3, R-2, S-1
(c) P-1, Q-2, R-4, S-3
(d) P-4, Q-3, R-1, S-2

Question Number : 46

Correct : 2 Wrong : –0.66

46. Arrange the magnetic moment of neighboring atoms in a one-dimensional lattice in Group-I to the corresponding magnetic material in Group-II.

Group - I

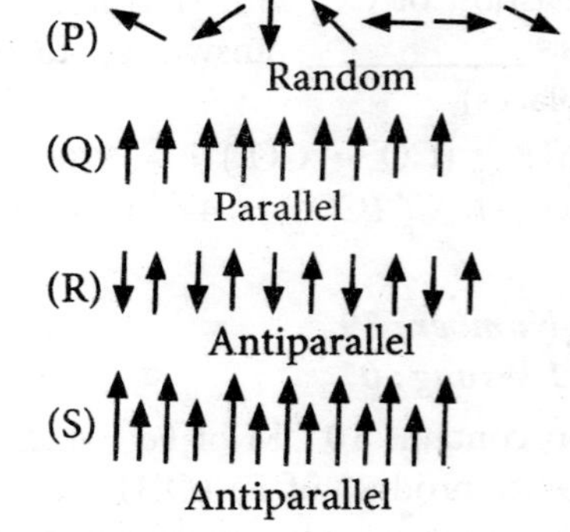

Group - II

(1) Antiferromagnetic
(2) Ferrimagnetic
(3) Paramagnetic
(4) Ferromagnetic

(a) P-4, Q-1, R-3, S-2
(b) P-3, Q-1, R-1, S-2
(c) P-2, Q-4, R-1, S-3
(d) P-1, Q-2, R-3, S-4

Question Number : 47

Correct : 2 Wrong : –0.66

47. For an intrinsic semiconductor, the room temperature electrical conductivity is 10^{-6} Ω^{-1}, m^{-1}. if the electron and hole mobilities are 0.75 and 0.06 $m^2\ V^{-1}\ S^{-1}$ respectively, the intrinsic carrier concentration (per m^3) at room temperature is:

(a) 5.1×10^{12} (b) 7.7×10^{12}
(c) 8.3×10^{12} (d) 1.1×10^{14}

Question Number : 48

Correct : 2 Wrong : 0

48. A steel is subjected to fatigue loading: σ (maximum) = 200 MPa, σ (minimum) –0. The component has an initial crack length of 1 mm. Propagation of crack is governed by

$$\frac{da}{dN} = 10^{-12}\ (\Delta K)^3.$$

Answers

45. (d)	46. (b)	47. (b)	48. (0.009 to 0.015)

where, the crack length α is in meters, N is the number of cycles and ΔK is in MPa. $M^{1/2}$. The length of the crack (in m) after one million cycles will be __________ (answer up to three decimal places)

Question Number : 49

Correct : 2 Wrong : 0

49. During heat treatment of a cold worked metal, recrystallization is 20% complete after 100s. The transformation (in%) 400s is ______ (answer up to two decimal places) (Assume avrami expnent, η=2)

Question Number : 50

Correct : 2 Wrong : –0.66

50. At low temperature, two parallel edge dislocations lying on parallel slip planes are shown in different configurations below. Match the following:

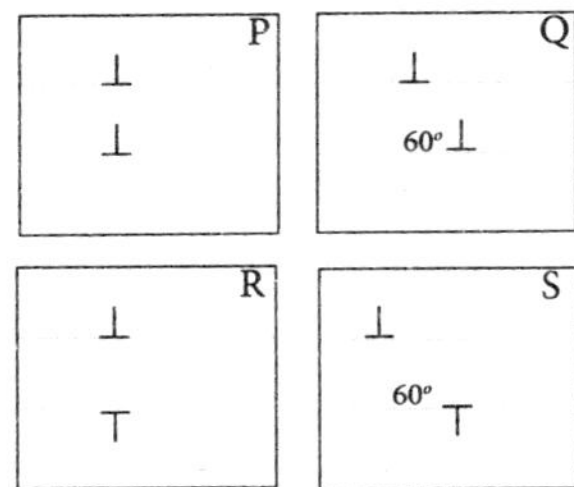

(P) Configuration (Q) Configuration
(R) Configuration (S) Configuration
(1) Dislocations repel
(2) Dislocations attract
(3) Dislocations are in stable equlibrium
(4) Dislocations are in unstable equilibrium
(a) P-3, Q-2, R-4, S-1
(b) P-4, Q-1, R-3, S-2
(c) P-1, Q-3, Q-2, R-1
(d) P-2, Q-4, R-1, S-3

Question Number : 51

Correct : 2 Wrong : –0.66

51. A single crystal of an FCC metal is subjected to a sufficiently large tensile stress along the [110} direction to activate some of the slip. Which one of the following slip systems will be activated:

(a) $\frac{a}{2}[\bar{1}10](111)$ (b) $\frac{a}{2}[011](11\bar{1})$

(a) $\frac{a}{2}[011](1\bar{1}\bar{1})$ (b) $\frac{a}{2}[110](\bar{1}1\bar{1})$

Question Number : 52

Correct : 2 Wrong : 0

52. A perfectly elastic-plastic material has a yield stress of 450 MPa and fractures at a strain of 0.45. The ratio of resilience to toughness for this material is __________ (answer up to three decimal places)
(Given the Young's modulus E= 4.3 GPa)

Question Number : 53

Correct : 2 Wrong : 0

53. Total time for solidification of a cubic casting of dimensions 5.0 cm × 5.0cm × 5.0 =cm is 1.6 min. A cylindricla riser with diameter to height ration 0.5 is required so that the time for solidification of riser is 3.2 min. Applying Chvorinov's rule, the height of the riser (in cm) is ________ (answer up to two decimal places)
(Assume that exponent (η) in chvorinov's equation is 2)

Question Number : 54

Correct : 2 Wrong : 0

54. A 250 mm thick slab of a nickle alloy is subjected to cold rolling using a roll of diameter 450 mm. If the angle of bite

Answers

49. (96.00 to 98.00) 50. (a) 51. (b) 52. (0.110 to 0.140)
53. (11.00 to 12.50)

during rolling is 10°, the maximum possible reduction (in mm) during rolling is _______ (answer up to two decimal places)

Question Number : 55
Correct : 2 Wrong : –0.66

55. W-Ni compact is prepared by liquid phase sintering at 1500°C. If the size of tungsten grains is 40 μm and the interfacial ningsten-mngsten and rungsten-nickel energies are 0.52 and 0.30 $J.m^{-2}$ respectively, the predicted average neck size (in μm) of sintered tungsten grain is:
(Melting points of tungsten and nickel are 3410°C and 1453°C, respectively)
(a) 10 (b) 15
(c) 20 (d) 25

Question Number : 56
Correct : 2 Wrong : –0.33

56. The ninth and the tenth of this month are Monday and Tuesday _________.
(a) figuratively (b) retrospectively
(c) respectively (d) rightfully

Question Number : 57
Correct : 1 Wrong : –0.33

57. It is _________ to read this year's textbook _________ the last year's.
(a) easier, than (b) most easy, than
(c) easier, from (d) easiest, from

Question Number : 58
Correct : 1 Wrong : –0.33

58. A rule states that in order to drink beer, one must be over 18 years old. In a bar, there are 4 people. P is 16 years old, Q is 25 years old, R is drinking milkshake and S is drinking a beer. What must be checked to ensure that the rule is being followed?
(a) Only P's drink
(b) Only P's drink and S's age
(c) Only S's age
(d) Only P's drink, Q's drink and S's age

Question Number : 59
Correct : 1 Wrong : –0.33

59. Fatima starts from point P, goes North for 3 km, and then East for 4 km to reach point Q. She then turns to face point P and goes 15 km in that direction. She then goes North for 6 km. How far is she from point P, and in which direction should she go to reach point P?
(a) 5 km. East (b) 12km. North
(c) 6km. East (d) 10km. North

Question Number : 60
Correct : 1 Wrong : –0.33

60. 500 students are taking one or more courses out of Chemistry, Physics, and Mathematics. Registration records indicate course enrolment as follows: Chemistry (329), Physics (186), Mathematic (295), Chemistry and Physics (83), Chemistry and Mathematics (217), and Physics and Mathematics (63). How many students are taking all 3 subjects?
(a) 37 (b) 43
(c) 47 (d) 53

Question Number : 61
Correct : 2 Wrong : –0.66

61. "If you are looking for a history of India, or for an account of the rise and fall of te British Raj, or for the reason of the cleaving of the subcontinent into two matually antagonistic parts and the effects this mutilation will have in the respective sections, and ultimately on Asia, you will not find it in these page; for though I have spent a lifetime in the country.

Answers

54. (6.60 to 7.40) 55. (c) 56. (c) 57. (a) 58. (b)
59. (a) 60. (d)

I lived too near the seat of eventsm and was too intimately associated with the actors, to get the perspective needed for the impartial recording of these matters."

Which of the following statements best reflects the author's opinion?

(a) An intimate association does not allow for the necessary perspective.

(b) Matters are recorded with an impartial perspective.

(c) An intimate association offers an impartial perspective.

(d) Actors are typicall associated with the impartial recording of matters.

Question Number : 62

Correct : 2 Wrong : –0.66

62. Each of P, Q, R, S, W, X, Y, and Z has been married at most once. X and Y are married and have two children P and Q, Z is the grandfather of the daughter S of P. Further Z and W are married and are parents of R. Which one of the following must necessarily be FALSE?

(a) X is the mother-in-law of R

(b) P and R are not married to each other

(c) P is a son of X and Y

(d) Q cannot be married to R

Question Number : 63

Correct : 2 Wrong : –0.66

63. 1200 men and 500 women can build a bridge in 2 weeks, 900 men and 250 women will take 3 weeks to build the same bridge. How many men will be needed to build the bridge in one week?

(a) 3000 (b) 3300

(c) 3600 (d) 3900

Question Number : 64

Correct : 2 Wrong : 0.66

64. The number of 3-digit numbers such that the digit 1 is never to the immediate right of 2 is.

(a) 781 (b) 791

(c) 881 (d) 891

Question Number : 65

Correct : 2 Wrong : 0.66

65. A contour line joins locations having the same height above the mean sea level. The following is a contour plot of a geographical region. Contour lines are shown at 25m intervals in this plot.

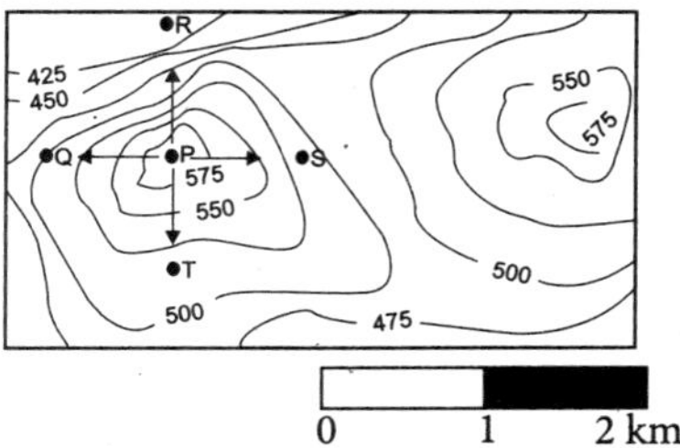

Which of the following is the steepest path leaving from P?

(a) P and Q (b) P to Q

(c) P to S (d) P to T

Answers

61. (a) 62. (d) 63. (c) 64. (c) 65. (b)

Section C

SHORT QUESTIONS & ANSWERS IN METALLURGICAL ENGINEERING

Section C
Short Questions & Answers in Metallurgical Engineering

Ques.1: The elements listed in the Periodic Table can be divided into three categories. What are these categories and give an example of each?

Answer: The three types of elements are metals (e.g., aluminum), non metals (e.g., oxygen), and semi metals (e.g., silicon).

Ques.2: Which elements are the noble metals?

Answer: The noble metals are copper, silver, and gold.

Ques.3: What is the difference between primary and secondary bonding in the structure of materials?

Answer: Primary bonding is strong bonding between atoms in a material, for example to form a molecule; while secondary bonding is not as strong and is associated with attraction between molecules in the material.

Ques.4: What is the difference between crystalline and noncrystalline structures in materials?

Answer: The atoms in a crystalline structure are located at regular and repeating lattice positions in three dimensions; thus, the crystal structure possesses a long-range order which allows a high packing density. The atoms in a noncrystalline structure are randomly positioned in the material, not possessing any repeating, regular pattern.

Ques.5: What are some common point defects in a crystal lattice structure?

Answer: Some of the common point defects are: (1) vacancy - a missing atom in the lattice structure; (2) ion-pair vacancy (Schottky defect) - a missing pair of ions of opposite charge in a compound; (3) interstitialcy - a distortion in the lattice caused by an extra atom present; and (4) Frenkel defect - an ion is removed from a regular position in the lattice and inserted into an interstitial position not normally occupied by such an ion.

Ques.6: Define the difference between elastic and plastic deformation in terms of the effect on the crystal lattice structure.

Answer: Elastic deformation involves a temporary distortion of the lattice structure that is proportional to the applied stress. Plastic deformation involves a stress of sufficient magnitude to cause a permanent shift in the relative positions of adjacent atoms in the lattice. Plastic deformation generally involves the mechanism of slip - relative movement of atoms on opposite sides of a plane in the lattice.

Ques.7: How do grain boundaries contribute to the strain hardening phenomenon in metals?

Answer: Grain boundaries block the continued movement of dislocations in the metal during straining. As more dislocations become blocked, the metal becomes more difficult to deform; in effect it becomes stronger.

Ques.8: Identify some materials that have a crystalline structure.

Answer: Materials typically possessing a crystalline structure are metals and ceramics other than glass. Some plastics have a partially crystalline structure.

Ques.9: Identify some materials that possess a noncrystalline structure.

Answer: Materials typically having a noncrystalline structure include glass (fused silica), rubber, and certain plastics (specifically, thermosetting plastics).

Ques.10: What is the basic difference in the solidification (or melting) process between crystalline and noncrystalline structures?

Answer: Crystalline structures undergo an abrupt volumetric change as they transform from liquid to solid state and vice versa. This is accompanied by an amount of energy called the heat of fusion that must be added to the material during melting or released during solidification. Noncrystalline materials melt and solidify without the abrupt volumetric change and heat of fusion.

Ques.11: State Hooke's Law.

Answer: Hooke's law defines the stress-strain relationship for an elastic material: $\sigma = E\varepsilon$, where E = a constant of proportionality called the modulus of elasticity.

Ques.12: What is the difference between engineering stress and true stress in a tensile test?

Answer: Engineering stress divides the load (force) on the test specimen by the original area; while true stress divides the load by the instantaneous area which decreases as the specimen stretches.

Ques.13: Define tensile strength of a material.

Answer: The tensile strength is the maximum load experienced during the tensile test divided by the original area.

Ques.14: Define yield strength of a material.

Answer: The yield strength is the stress at which the material begins to plastically deform. It is usually measured as the 0.2% off set value - the point at which the stress-strain for the material intersects a line which is offset from the elastic region of the stress-strain curve by 0.2%.

Ques.15: Why cannot a direct conversion be made between the ductility measures of elongation and reduction in area using the assumption of constant volume?

Answer: Because of necking that occurs in the test specimen.

Ques.16: What is work hardening?

Answer: Strain hardening is the increase in strength that occurs in metals when they are strained.

Ques.17: In what case does the strength coefficient have the same value as the yield strength?

Answer: When the material does not strain harden.

Ques.18: How does the change in cross-sectional area of a test specimen in a compression test differ from its counterpart in a tensile test specimen?

Answer: In a compression test, the specimen cross-sectional are increases as the test progresses;while in a tensile test, the cross-sectional area decreases.

Ques.19: Tensile testing is not appropriate for hard brittle materials such as ceramics. What is the test commonly used to determine the strength properties of such materials?

Answer: A three-point bending test is commonly used to test the strength of brittle materials. The test provides a measure called the transverse rupture strength for these materials.

Ques.20: How is the shear modulus of elasticity G related to the tensile modulus of elasticity E, on average?

Answer: G = 0.4 E, on average.

Ques.21: How is shear strength S related to tensile strength TS, on average?

Answer: S = 0.7 TS, on average.

Ques.22: What is hardness and how is it generally tested?

Answer: Hardness is defined as the resistance to indentation of a material. It is tested by pressing a hard object (sphere, diamond point) into the test material and measuring the size (depth, area) of the indentation.

Ques.23: Why are different hardness tests and scales required?

Answer: Different hardness tests and scales are required because different materials possesswidely differing hardnesses. A test whose measuring range is suited to very hard materials is not sensitive for testing very soft materials.

Ques.24: Define the recrystallization temperature for a metal.

Answer: The recrystallization temperature is the temperature at which a metal recrystallizes (forms new grains) rather than work hardens when deformed.

Ques.25: What is viscoelasticity, as a material property?

Answer: Viscoelasticity refers to the property most commonly exhibited by polymers that defines the strain of the material as a function of stress and temperature over time. It is a combination of viscosity and elasticity.

Ques.26: What is the difference in melting characteristics between a pure metal element and an alloy metal?

Answer: A pure metal element melts at one temperature (the melting point), while an alloy begins melting at a certain temperature called the solidus and finally completes the transformation to the molten state at a higher temperature called the liquidus.

Between the solidus and liquidus, the metal is a mixture of solid and liquid.

Ques.27: Describe the melting characteristics of a noncrystalline material such as glass.

Answer: In the heating of a noncrystalline material such as glass, the material begins to soften as temperature increases, finally converting to a liquid at a temperature defined for these materials as the melting point.

Ques.28: What is the thermal conductivity of a material?

Answer: Thermal conductivity is the capacity of a material to transfer heat energy through itself by thermal movement only (no mass transfer).

Ques.29: Define thermal diffusivity.

Answer: Thermal diffusivity is the thermal conductivity divided by the volumetric specific heat.

Ques.30: What are the important variables that affect mass diffusion?

Answer: According to Fick's first law, mass diffusion depends on: diffusion coefficient which rises rapidly with temperature (so temperature could be listed as an important variable), concentration gradient, contact area, and time.

Ques.31: Define the resistivity of a material.

Answer: Resistivity is the material's capacity to resist the flow of an electric current.

Ques.32: Why are metals better conductors of electricity than ceramics and polymers?

Answer: Metals are better conductors because of metallic bonding, which permits electrons to move easily within the metal. Ceramics and polymers have covalent and ionic bonding, in which the electrons are tightly bound to particular molecules.

Ques.33: What is the dielectric strength of a material?

Answer: The dielectric strength is defined as the electrical potential required to break down the insulator per unit thickness.

Ques.34: What is an electrolyte?

Answer: An electrolyte is an ionized solution capable of conducting electric current by movement of the ions.

Ques.35: Identify some of the changes and injuries that can occur at or immediately below the surface of a metal.

Answer: The changes and injuries include: cracks, craters, variations in hardness near the surface, metallurgical changes resulting from heat, residual stresses, intergranular attack, etc.

Ques.36: Name some manufacturing processes that produce very poor surface finishes.

Answer: Processes that produce poor surfaces include: sand casting, hot rolling, sawing, and thermal cutting (e.g., flame cutting).

Ques.37: Name some manufacturing processes that produce very good or excellent surface finishes.

Answer: Processes that produced very good and excellent surfaces include: honing, lapping, polishing, and superfinishing.

Ques.38: What are some of the general properties that distinguish metals from ceramics and polymers?

Answer: Metallic properties include: high strength & stiffness, good electrical & thermal conductivity, and higher density than ceramics or polymers.

Ques.39: What are the two major groups of metals? Define them.

Answer: Ferrous metals, which are based on iron; and nonferrous, which includes all others.

Ques.40: What is the definition of an alloy?

Answer: An alloy is a metal comprised of two or more elements, at least one of which is metallic.

Ques.41: What is a solid solution in the context of alloys?

Answer: A solid solution is an alloy in which one of the metallic elements is dissolved in another to form a single phase.

Ques.42: Distinguish between a substitutional solid solution and an interstitial solid solution.

Answer: A substitutional solid solution is where the atoms of dissolved element replace atoms of the solution element in the lattice structure of the metal. An interstitial solid solution is where the dissolved atoms are small and fit into the vacant spaces (the interstices) in the lattice structure of the solvent metal.

Ques.43: What is an intermediate phase in the context of alloys?

Answer: An intermediate phase is an alloy formed when the solubility limit of the base metal in the mixture is exceeded and a new phase, such as a metallic compound (e.g., Fe_3C) or intermetallic compound (e.g., Mg_2Pb) is formed.

Ques.44: The copper-nickel system is a simple alloy system, as indicated by its phase diagram. Why is it so simple?

Answer: The Cu-Ni alloy system is simple because it is a solid solution alloy throughout its entire composition range.

Ques.45: What is the range of carbon percentages which defines an iron-carbon alloy as a steel?

Answer: The carbon content ranges from 0.02% to 2.0%.

Ques.46: What is the range of carbon percentages which defines an iron-carbon alloy as cast iron?

Answer: The carbon content ranges from 2.11 to about 5%.

Ques.47: Identify some of the common alloying elements other than carbon in low alloy steels.

Answer: The common alloying elements in low alloy steel are Cr, Mn, Mo, Ni, and V .

Ques.48: What are some of the mechanisms by which the alloying elements other than carbon strengthen steel ?

Answer: All of the alloying elements other than C strengthen the steel by solid solution alloying. Cr, Mn, Mo, and Ni increase hardenability during heat treatment. Cr and Mo improve hot hardness. Several of the alloying elements (Cr, Mo, V) form hard carbides with C, which increases wear resistance. Vanadium inhibits grain growth during heat treatment which improves strength and toughness.

Ques.49: What is the mechanism by which carbon strengthens steel in the absence of heat treatment?

Answer: If no heat treatment is done then carbon strengthens by creating a two-phase structure in the steel.

Ques.50: What is the predominant alloying element in all of the stainless steels?

Answer: Chromium.

Ques.51: Why is austenitic stainless steel called by that name?

Answer: It is called austenitic because this alloy exists in its austenitic phase at room temperature. The reason is that nickel has the effect of enlarging the austenitic temperature range to include room temperature.

Ques.52: Identify some of the properties for which aluminum is noted?

Answer: Aluminum is noted for its low density, high electrical & thermal conductivity, formability, good corrosion resistance due to the formation of a tough oxide film on its surface, and ability to be alloyed & strengthened to achieve good strength-to-weight ratios.

Ques.53: What are some of the noteworthy properties of magnesium?

Answer: Magnesium is noted for its very low density (lightest of the structural metals), propensity to oxidize (which can cause problems in processing), and low strength; however, it can be alloyed and strengthened by methods similar to those used for aluminum alloys to achieve respectable strength-to-weight ratios.

Ques.54: What is the most important engineering property of copper which determines most of its applications?

Answer: Its high electrical conductivity.

Ques.55: What elements are traditionally alloyed with copper to form (a) bronze and (b) brass?

Answer: (a) tin, (b) zinc.

Ques.56: What are some of the important applications of nickel?

Answer: The important applications are:

1. as an alloying ingredient in steel, e.g., stainless steel;
2. for plating of steel to resist corrosion; and
3. to form nickel-based alloys noted for high temperature performance and corrosion resistance.

Ques.57: What are the noteworthy properties of titanium?

Answer: Titanium is noted for its high strength-to-weight ratio, corrosion resistance (due to the formation of a thin but tough oxide film), and high temperature strength.

Ques.58: Identify some of the important applications of zinc.

Answer: The important applications are:

1. die castings - zinc is an easy metal to cast;

2. as a coating in galvanized steel;
3. as an alloying element with copper to form brass.

Ques.59: What important alloy is formed from lead and tin?

Answer: Solder.

Ques.60: (a) Name the important refractory metals.

(b) What does the term refractory mean?

Answer: (a) The refractory metals include columbium (Cb), molybdenum (Mo), tantalum (Ta and tungsten (W). Mo and W are the most important.

(b) Refractory means the capability to withstand high temperature service.

Ques.61: (a) Name the four principal noble metals.

(b) Why are they called noble metals?

Answer: (a) The principal noble metals are copper, gold, platinum, and silver.

(b) Nobel metals are so-named because they are chemically inactive.

Ques.62: The superalloys divide into three basic groups, according to the base metal used in the alloy. Name the three groups.

Answer: The three groups are: (1) iron-based alloys, (2) nickel-based alloys, and (3) cobalt-based alloys.

Ques.63: What is so special about the superalloys? What distinguishes them from other alloys?

Answer: The superalloys are generally distinguished by their strength and resistance to corrosion and oxidation at elevated temperatures.

Ques.64: What are the three basic methods by which metals can be strengthened?

Answer: The three basic methods are:

(1) alloying to form solid solutions and two-phase structures which are stronger than the elemental metals;

(2) cold working, in which the strain-hardened metal is stronger and harder than the unstrained metal; and

(3) heat treatment - most of the commercial heat treatments are designed to increase the strength of the metal.

Ques.65: What is a ceramic ?

Answer: A ceramic is an inorganic, nonmetallic compound, usually formed into useful products by a heating process.

Ques.66: What are the four most common elements in the earth's crust?

Answer: Oxygen, silicon, aluminum, and iron.

Ques.67: What is the difference between the traditional ceramics and the new ceramics?

Answer: Traditional ceramics are based primarily on clay products (*e.g.*, pottery, bricks) while new ceramics are more recently developed ceramics which are generally simpler in chemical composition (*e.g.*, oxides, carbides).

Ques.68: What is the feature that distinguishes glass from the traditional and new ceramics?

Answer: Glass is noncrystalline (amorphous), while most other ceramics assume a crystalline structure.

Ques.69: Why are graphite and diamond not classified as ceramics?

Answer: Because they are not compounds; they are alternative forms of the element carbon.

Ques.70: What are the general mechanical properties of ceramic materials?

Answer: Usually high hardness, brittle, no ductility.

Ques.71: What are the general physical properties of ceramic materials?

Answer: Usually electrical and thermal insulators, medium density (typically below the density of metals), high melting temperatures, thermal expansion usually less than metals.

Ques.72: What type of atomic bonding characterizes the ceramics?

Answer: Covalent and ionic bonding.

Ques.73: What is clay, used in making ceramic products?

Answer: Clay most commonly consists of hydrous aluminum silicate, the usually kaolinite ($Al_2(Si_2O_5)(OH)_4$).

Ques.74: What is glazing, as applied to ceramics?

Answer: Glazing involves the application of a surface coating of oxides such as alumina and silica,usually to a porous ceramic product such as earthenware, to make the product more impervious to moisture and more attractive.

Ques.75: What does the term refractory mean?

Answer: Refractories are heat resistant ceramic materials. The term is sometimes also applied to metals that are heat resistant.

Ques.76: What are some of the principal applications of the cemented carbides, such as WC-Co?

Answer: Important applications of WC-Co include: cutting tool inserts, drawing dies, rock drilling bits, dies for powder metallurgy, and other applications where hardness is a critical factor.

Ques.77: What is one of the important applications of titanium nitride ?

Answer: As a thin coating on cutting tools to prolong tool life.

Ques.78: What elements comprise the ceramic material Sialon?

Answer: Silicon, aluminum, oxygen, and nitrogen.

Ques.79: Define glass.

Answer: Glass is an inorganic, nonmetallic material which cools to a rigid solid without crystallization.

Ques.80: What is the primary mineral in glass products?

Answer: Silica, or silicon dioxide (SiO_2).

Ques.81: What are the functions of the ingredients that are added to glass in addition to silica?

Answer: The functions of the additional ingredients include:

1. acting as flux (promoting fusion)during heating;
2. increasing fluidity in the molten glass during processing;
3. retarding devitrification - the tendency to crystallize from the glassy state;
4. reducing thermal expansion in the final product;
5. increasing the chemical resistance against attack by acids, basic substances,orwater;
6. adding color to the glass; and
7. altering the index of refraction for optics applications (*e.g.*, lenses).

Ques.82: What does the term devitrification mean?

Answer: Devitrification is the transformation from the glassy state into a polycrystalline state.

Ques.83: What is graphite?

Answer: Graphite is carbon in the form of hexagonal crystalline layers, in which covalent bonding exists between atoms in the layers, and the (parallel) layers are bonded by Van der Waals forces, thus leading to highly anisotropic properties.

Ques.84: What is a polymer?

Answer: A polymer is a compound that consists of long-chain molecules. The molecules consist of repeating units, called mers, connected end to end.

Ques.85: What are the three basic categories of polymers?

Answer: The categories are: (1) thermoplastics, (2) thermosetting and (3) elastomers.

Ques.86: How do the properties of polymers compare with those of metals?

Answer: In general, polymers have lower strength, hardness, stiffness, density, and temperature resistance compared to metals. In addition, polymers are low in electrical and thermal conductivity.

Ques.87: What are the two methods by which polymerization occurs?

Answer: The two types of polymerization are: (1) addition or chain polymerization and (2) step polymerization, also known as condensation polymerization.

Ques.88: What does the degree of polymerization indicate?

Answer: The degree of polymerization indicates the average number of mers or repeating units in the polymer molecule.

Ques.89: Define the term tacticity as it applies to polymers.

Answer: Tacticity refers to the way the atoms or atom groups replacing H atoms in the molecule are arranged.

Ques.90: What is cross-linking in a polymer and what is its significance?

Answer: Cross-linking is the formation of connections between the long-chain molecules in a polymer. It causes the polymer structure to be permanently altered. If the amount of cross-linking is low, the polymer is transformed into an elastomer; if cross-linking is significant, the polymer is transformed into a thermosetting polymer.

Ques.91: What is a copolymer?

Answer: A copolymer is a polymer made up of two different types of mers, such as ethylene and propylene.

Ques.92: The arrangement of repeating units in a copolymer can vary. What are some of the possible arrangements?

Answer: There are four possible arrangements of the mers along the chain: (1) alternating, (2) random, (3) block, and (4) graft.

Ques.93: What is a terpolymer?

Answer: A terpolymer is a polymer with three different mer types. An example is ABS (acrylonitrile-butadiene-styrene) plastic.

Ques.94: How are a polymer's properties affected when it takes on a crystalline structure?

Answer: Density, stiffness, and melting temperature increase.

Ques.95: Can any polymer ever become 100% crystalline?

Answer: No.

Ques.96: What are some of the factors that influence a polymer's tendency to crystallize?

Answer: Factors are: (1) only linear polymers can form crystals; (2) copolymers do not form crystals; (3) stereoregularity - isotactic polymers always form crystals, atactic polymers never form crystals, and syndiotactic polymers sometimes form crystals; (4) slow cooling from the molten states promotes crystal formation; (5) plasticizers inhibit crystal formation; and (6) stretching the polymer tends to promote crystallization.

Ques.97: Why are fillers added to a polymer?

Answer: Fillers are added to increase strength or simply to reduce the cost of the polymer.

Ques.98: What is a plasticizer?

Answer: A plasticizer is a chemical added to the polymer to make it softer and more flexible. It is often added to improve the polymer's flow characteristics for shaping.

Ques.99: In addition to fillers and plasticizers, what are some other additives used with polymers?

Answer: Other additives include: lubricants - to reduce friction & improve flow; flame retardents; colorants; cross-linking agents, antioxidants, and ultraviolet light absorbers.

Ques.100: Describe the difference in mechanical properties as a function of temperature between a highly crystalline thermoplastic (TP) and an amorphous thermoplastic.

Answer: A highly crystalline TP retains rigidity during heating until just before its Tm is reached. An amorphous TP shows a significant drop in deformation resistance at its Tg as temperature is raised; it becomes increasingly like a liquid as temperature continues to increase.(T_m is melting point, T_g is glass transition temperature)

Ques.101: What is unique about the polymer cellulose?

Answer: Cellulose is a polymer that grows in nature. Wood fiber contains about 50% cellulose and cotton fiber is about 95% cellulose.

Ques.102: What is the basic difference between low density(LDPE) & high density(HDPE polyethylene?

Answer: LDPE has a branched structure and is amorphous. HDPE is linear and highly crystalline These differences account for HDPE higher density, stiffness, and melting point.

Ques.103: How do the properties of thermosetting polymers differ from those of thermoplastics?

Answer: Thermosets are more rigid, brittle, capable of higher service temperatures, and cannot be remelted.

Ques.104: Cross-linking (curing) of thermosetting plastics is accomplished by one of three ways. Name the three ways.

Answer: The three ways are:

(1) temperature-activated systems, in which elevated temperatures accomplish curing;
(2) catalyst-activated systems, in which small amounts of a catalyst cause cross-linking; and
(3) mixing-activated systems, in which two reactive components are mixed and curing occurs by their chemical reaction.

Ques.105: Elastomers & thermosetting polymers are both cross- linked. Why are their properties so different?

Answer: Elastomers are lightly cross-linked, whereas thermosets are highly cross-linked. Light cross-linking allows extensibility; a highly cross-linked structure makes the polymer rigid.

Ques.106: What happens to an elastomer when it is below its glass transition temperature (Tg)?

Answer: An elastomer is hard and brittle below its Tg.

Ques.107: What is the primary polymer ingredient in natural rubber?

Answer: Polyisoprene.

Ques.108: How are thermoplastic elastomers different from conventional rubbers?

Answer: TPEs are different in two basic ways:

(1) they exhibit thermoplastic properties, and
(2) their extensibility derives from physical connections between different phases in the polymer.

Ques.109: What is a composite material?

Answer: A composite material is a materials system consisting of two or more distinct phases whose combination results in properties that differ from those of its constituents.

Ques.110: Identify some of the characteristic properties of composite materials.

Answer: Typical properties include:

(1) high strength-to- weight & stiffness-to-weight ratios;

(2) good fatigue properties & toughness;

(3) anisotropic properties in many cases; and

(4) other properties and features that are difficult or impossible to obtain with metals, ceramics, or polymers alone.

Ques.111: What does the term anisotropic mean?

Answer: Anisotropic means that the properties of a material vary depending on the direction in which they are measured.

Ques.112: How are traditional composites distinguished from synthetic composites?

Answer: Traditional composites have been used for decades or centuries; some of them are obtained from sources in nature, such as wood. Synthetic composites are manufactured.

Ques.113: Name the three basic categories of composite materials.

Answer: Metal matrix composites (MMCs), ceramic matrix composites (CMCs), and polymer matrix composites (PMCs).

Ques.114: What are the common forms of the reinforcing phase in composite materials?

Answer: The forms are:

1. fibers,
2. particles and flakes, and
3. an infiltrated phase in skeletal structures.

Ques.115: What is a whisker?

Answer: A whisker is a thin, hairlike crystal of very high strength.

Ques.116: What are the two forms of sandwich structure among laminar composite structures?

Answer: The two forms are:

(1) foamed-core sandwich, in which the core is polymer foam between two solid skins; and

(2) honeycomb, in which the core is a honeycomb structure sandwiched between two solid skins.

Ques.117: Give some examples of commercial products which are laminar composite structures.

Answer: Examples are: automotive tires, honeycomb sandwich structures, fiber reinforced polymer structures such as boat hulls, plywood, printed circuit boards, snow skis made from fiber reinforced polymers, and windshield glass.

Ques.118: What are the three general factors that determine the properties of a composite material?

Answer: Three factors are :

(1) the component materials;

(2) the geometric shapes of the constituents - the reinforcing phase in particular & the resulting structure of the material; and

(3) the interaction of the phases.

Ques.119: What is the rule of mixtures?

Answer: The rule of mixtures applies to certain properties of composite materials; it states that the property value is a weighted average of the property values of the components, the weighting being by proportions of the components in the composite.

Ques.120: What is a cermet?

Answer: A cermet is a composite material consisting of a ceramic and a metal. It is defined as a composite consisting of ceramic grains imbedded in a metallic matrix.

Ques.121: Cemented carbides are what class of composites?

Answer: Although the cemented carbide industry does not generally think of cemented carbides as cermets, they fit within the definition.

Ques.122: What are some of the weaknesses of ceramics that might be corrected in fiber-reinforced ceramic matrix composites?

Answer: Weaknesses of ceramics include: low tensile strength, poor toughness, and susceptibility to thermal cracking.

Ques.123: What is the most common fiber material in fiber-reinforced plastics?

Answer: E-glass.

Ques.124: What does the term advanced composites mean?

Answer: An advanced composite is a PMC (Polymer Matrix Composites) in which carbon, Kevlar, or boron fibers are used as the reinforcing material.

Ques.125: What is a hybrid composite?

Answer: A hybrid composite is a fiber-reinforced PMC in which two or more fibers materials are combined in the FRP.

Ques.126: Identify some of the important properties of fiber- reinforced plastic (FRP) composite materials.

Answer: Properties include: high strength-to-weight ratio, high modulus-to-weight ratio, low density, good fatigue strength, good corrosion resistance, and low thermal expansion for many FRPs.

Ques.127: Name some of the important applications of FRPs (Fibre Reinforced Plastics).

Answer: FRPs are used in modern aircraft as skin parts, automobile body panels, printed circuit boards, tennis rackets, boat hulls, and a variety of other items.

Ques.128: What is meant by the term interface in the context of composite materials?

Answer: The interface is the boundary between the component phases in a composite material.

Ques.129: Identify some of the important advantages of shape casting processes.

Answer: Advantages include:

1. complex part geometries are possible;
2. some casting operations are net shape processes, meaning that no further manufacturing operations are needed to accomplish the final part shape;
3. very large parts are possible;
4. applicable to any metal that can be melted; and
5. some casting processes are suited to mass production.

Ques.130: What are some of the limitations and disadvantages of casting?

Answer: Disadvantages include:

1. limitations on mechanical strength properties;
2. porosity;
3. poor dimensional accuracy;
4. safety hazards due to handling of hot metals; and
5. environmental problems.

Ques.131: What is the difference between an open mold and a closed mold?

Answer: An open mold is open to the atmosphere at the top; it is an open container in the desired shape which must be flat at the top. A closed mold has a cavity that is entirely enclosed by the mold, with a passageway (called the gating system) leading from the outside to the cavity. Molten metal is poured into this gating system to fill the mold.

Ques.132: Name the two basic mold types that distinguish casting processes.

Answer: The two types are:

(1) expendable molds and
(2) permanent molds.

Ques.133: Which casting process is the most important commercially?

Answer: The most important casting process is sand casting.

Ques.134: What is the difference between a pattern and a core in sand molding?

Answer: The pattern determines the external shape of the casted part, while a core determines its internal geometry.

Ques.135: What is meant by the term superheat?

Answer: Superheat is the temperature difference above the melting point at which the molten metal is poured. The term also refers to the amount of heat that is removed from the molten metal between pouring and solidification.

Ques.136: Why should turbulent flow of molten metal into the mold be avoided?

Answer: Turbulence causes several problems:

(1) accelerates formation of oxides in the solidified metal, and
(2) mold erosion or gradual wearing away of the mold due to impact of molten metal.

Ques.137: What is the continuity law as it applies to the flow of molten metal in casting?

Answer: The continuity law, or continuity equation, indicates that the volumetric flow rate is constant throughout the liquid flow.

Ques.138: What are some of the factors affecting the fluidity of a molten metal during pouring into a mold cavity?

Answer: Factors include:

1. pouring temperature,
2. metal alloy composition,
3. viscosity of liquid metal, and
4. heat transfer to the surroundings.

Ques.139: What does heat of fusion mean in casting?

Answer: Heat of fusion is the amount of heat energy required to transform the metal from solid state to liquid state.

Ques.140: How does solidification of alloys differ from solidification of pure metals?

Answer: Pure metals solidify at a single temperature equal to the melting point. Most alloys (exceptions are eutectic alloys) start to solidify at the liquidus and complete solidification at the solidus, where the liquidus is a higher temperature than the solidus.

Ques.141: What is a eutectic alloy?

Answer: A eutectic alloy is a particular composition in an alloy system for which the solidus and liquidus temperatures are equal. The temperature is called the eutectic temperature. Hence, solidification occurs at a single temperature, rather than over a temperature range.

Ques.142: What is the relationship known as Chvorinov's Rule in casting?

Answer: Chvorinov's Rule is summarized: $TST = C_m(V/A)^2$, where TST = total solidification time, C_m = constant, V = volume of casting, and A = surface area of casting.

Ques.143: Identify the three sources of contraction in a metal casting after pouring.

Answer: The three contractions occur due to:

1. contraction of the molten metal after pouring,
2. solidification shrinkage during transformation of state from liquid to solid, and
3. thermal contraction in the solid state.

Ques.144: What is a chill in casting?

Answer: A chill is a heat sink placed to encourage rapid freezing in certain regions of the casting.

Ques.145: Name the two basic categories of casting processes .

Answer: The two categories are:

(1) expendable mold processes, and

(2) permanent mold processes.

Ques.146: There are various types of patterns used in sand casting. What is the difference between a split pattern and a match- plate pattern?

Answer: A split pattern is a pattern that consists of two pieces; a match-plate pattern consists of the two split patterns attached to opposite sides of a plate.

Ques.147: What is a chaplet?

Answer: Chaplets are metal supports of various designs used to hold the core in place in the sand mold.

Ques.148: What properties determine the quality of a sand mold for sand casting?

Answer: The usual properties are:

1. strength - ability to maintain shape in the face of the flowing metal,
2. permeability - ability of the mold to allow hot air and gases to escape from the cavity,
3. thermal stability - ability to resist cracking and buckling when in contact with the molten metal,
4. collapsibility - ability of the mold to give way during shrinkage of the casting,
5. reusability - can the sand be reused to make other molds?

Ques.149: What is the Antioch process?

Answer: The Antioch process refers to the making of the mold. The mold is 50% sand and 50% plaster heated in an autoclave and then dried. This mold has greater permeability than a plaster mold.

Ques.150: What is the difference between vacuum permanent-mold casting and vacuum molding?

Answer: Vacuum permanent mold casting is a form of low pressure casting in which a vacuum is used to draw molten metal into the cavity.

Vacuum molding is sand casting in which the sand mold is held together by vacuum pressure rather than a chemical binder.

Ques.151: What are the most common metals processed using die casting?

Answer: Common die cast metals include: zinc, tin, lead, aluminum, brass, and magnesium.

Ques.152: Which die casting machines usually have a higher production rate, cold-chamber or hot-chamber, and why?

Answer: Hot-chamber machines are faster because cold chamber die casting machines require molten metal to be ladled into the chamber from an external source.

Ques.153: What is flash in die casting?

Answer: Flash is a thin portion at the exterior of a casting that results from molten metal being squeezed into the spaces between the die halves of the mold at the parting line, or into the clearances around the cores and ejector pins.

Ques.154: What is the difference between true centrifugal casting and semi centrifugal casting?

Answer: In true centrifugal casting, a tubular mold is used and a tubular part is produced. In semicentrifugal casting, the shape is solid; an example is a railway wheel. The mold is rotated so that centrifugal force is used to distribute the molten metal to the exterior of the mold so that the density of the final metal is greater at the outer sections.

Ques.155: What is a cupola?

Answer: A cupola is a vertical cylindrical furnace equipped with a tapping spout near its base. Cupolas are used for melting cast irons.

Ques.156: What are some of the operations required of sand castings after removal from the mold?

Answer: These operations include:

1. trimming, in which the sprues, runners, risers, and flash are removed,
2. core removal,
3. surface cleaning,
4. inspection,
5. repair if needed,
6. heat treatment, and
7. machining.

Ques.157: What are some of the general defects encountered in casting processes?

Answer: General defects include:

1. misruns,
2. cold shuts,
3. cold shots,
4. shrinkage cavity,
5. microporosity, and
6. hot tearing.

Ques.158: Glass is classified as a ceramic material; yet glass is different from the traditional and new ceramics. What is the difference?

Answer: Glass is vitreous - it is in the glassy state, whereas traditional & new ceramics are, by and large, polycrystalline materials.

Ques.159: What are the three basic steps in the glassworking sequence?

Answer: (1) raw materials preparation & melting,

(2) shaping, and

(3) heat treatment.

Finishing operations (*e.g.*, grinding, polishing, etching) are performed on some glass products, if needed.

Ques.160: Melting furnace for glassworking can be divided into four types. Name them.

Answer: The four types are:

1. pot furnaces.
2. day tanks.
3. continuous tank furnaces, and
4. electric furnaces.

Ques.161: Describe the spinning process in glassworking.

Answer: Spinning in glassworking is similar to centrifugal casting in metalworking. A gob of molten glass is dropped in to a conical mold which spins, causing centrifugal force to spread the glass upward onto the mold surface.

Ques.162: What is the main difference between the press-and-blow & the blow-and-blow shaping processes in glassworking?

Answer: In the press-and-blow process, the initial forming step is pressing of the part, while the first step in the blow-and-blow process is blowing.

Ques.163: There are several ways of shaping plate or sheet glass. Name and briefly describe one of them.

Answer: The methods described in this test are:

(1) rolling, in which the hot glass is squeezed between opposing cylindrical rolls; and

(2) the float process, in which the melted glass flows onto a molten tin surface to achieve uniform thickness and smoothness.

Ques.164: Describe the Danner process?

Answer: In the Danner process, molten glass flows around a rotating hollow mandrel through which air is blown while the glass is being drawn. The temperature of the air and its volumetric flow rate as well as the drawing velocity determine the diameter and wall thickness of the tubular cross-section. During hardening, the glass tube is supported by a series of rollers extending beyond the mandrel.

Ques.165: Name the two processes for forming glass .

Answer: The two processes are:

(1) drawing, in which fine glass fibers are pulled through small orifices in a heated plate; and

(2) centrifugal spraying, in which molten glass is forced to flow through small orifices in a rapidly rotating bowl to form glass fibers.

Ques.166: What is the purpose of annealing in glassworking?

Answer: Annealing is performed on glass to remove internal stresses that result from shaping and solidification.

Ques.167: Describe how a piece of glass is heat treated to produce tempered glass.

Answer: The glass is heated to a temperature above the annealing temperature and the surfaces are then quenched by air jets to cool and harden them while the interior of the piece remains plastic; as the interior cools and contracts, it puts the previously hardened surfaces in compression,which strengthens it.

Ques.168: Describe the type of material that is commonly used to make windshields for automobiles.

Answer: Laminated glass, in which two sheets of glass are laminated on either side of a polymer sheet. This has good impact resistance and does not splinter when broken.

Ques.169: What are some of the design recommendations for glass parts?

Answer: The guidelines are:

5. subject ceramic parts to compressive, not tensile loads;
6. ceramics are brittle, so avoid impact loading;
7. use large radii on inside and outside corners;
8. screw threads should be course.

Ques.170: What are some of the reasons why the plastic shaping processes are important?

Answer: Some of the reasons are:

1. many of the processes are net shape processes;
2. ingeneral, less energy employed than in metalworking processes;
3. lower temperatures are required to process plastics than metals or ceramics;
4. great flexibility in geometry; and
5. painting and other finishing processes generally not required.

Ques.171: Identify the main categories of plastics shaping processes, as classified by the resulting product geometry.

Answer: The categories are:

1. extrusion,
2. molding,
3. forming of continuous sheets & films,
4. fibers,
5. foamed products, and
6. discrete formed sheets & films.

Ques.172: Viscosity is an important property of a polymer melt in plastics shaping processes. Upon what parameters does viscosity depend?

Answer: Viscosity of a polymer melt depends on:

1. temperature, and
2. shear rate and,
3. the molecular weight of the polymer .

Ques.173: How does the viscosity of a polymer melt differ from most fluids that are Newtonian.

Answer: A polymer melt exhibits pseudoplasticity, which means that its value decreases with increasing shear rate.

Ques.174: Besides viscosity, what other properties of a polymer melt are important in plastics processing?

Answer: Other properties include: viscoelasticity, a combination of viscous and elastic properties which cause the melt to exhibit the tendency to return to its previous shape, as exhibited by die swell in extrusion.

Ques.175: Define die swell in extrusion.

Answer: Die swell is the tendency of the extrudate to expand in cross-section dimensions immediately on exiting the die orifice. It results from the viscoelastic properties of the polymer melt.

Ques.176: Briefly describe the plastic extrusion process.

Answer: In plastic extrusion, a polymer melt is compressed to flow through a die orifice and thus the continuous length of the plastic assumes a cross-sectional shape that is approximately the same as the profile of the orifice.

Ques.177: The barrel and screw of an extruder are generally divided into three sections; identify the sections.

Answer:
(1) feed section, in which the feed stock is fed from the hopper and heated;
(2) compression section, in which the polymer changes to a viscous fluid; and
(3) metering section, inwhich pressure is developed to pump the plastic through the die orifice.

Ques.178: What are the various forms of extruded shapes and corresponding dies?

Answer:

1. solid profiles, such as rounds & L-shapes;
2. hollow profiles, such as tubes;
3. wire & cable coating;
4. sheet & film; and
5. filaments (continuous fibers).

Ques.179: What is the distinction between plastic sheet and film?

Answer: Sheet has a thickness greater than 0.5 mm , while film is less than 0.5 mm thick.

Ques.180: What is the blown-film process for producing film stock?

Answer: The blown-film process is a widely used process for making thin polyethylene film for packaging. It combines extrusion and blowing to produce a tube of thin film. The process begins with the extrusion of a tube which is immediately drawn upward while still molten and simultaneously expanded in size by air inflated into it through the die mandrel.

Ques.181: Describe the calendering process.

Answer: Calendering is a process for producing sheet and film stock out of rubber or rubbery thermoplastics such as plasticized PVC. In the process, the initial feedstock is passed through a series of rolls to work the material and reduce its thickness to the desired gage.

Ques.182: Polymer fibers and filaments are used in several applications; what is the most important application?

Answer: Textiles.

Ques.183: Technically, what is the difference between a fiber and a filament?

Answer: A fiber is a long, thin strand of material whose length is at least 100 times its diameter; a filament is a fiber of continuous length.

Ques.184: Among the synthetic fiber materials, which are the most important?

Answer: Polyester is the most important commercially, followed by nylon, acrylics, and rayon.

Ques.185: An injection molding machine is divided into two principal components; identify them.

Answer: The components of an injection molding machine are:

(1) the injection unit and

(2) the clamping unit.

Ques.186: What are the two basic types of clamping units?

Answer: The clamping units are:

(1) mechanical toggle clamp and

(2) hydraulic. In addition, there are hydromechanical units which combine hydraulic and mechanical actuations.

Ques.187: Gates in injection molds have several functions; name them.

Answer: The functions of gates in an injection mold are:

1. to increase shear rate to increase viscosity & temperature of the polymer melt,
2. the thinner cross- section of the gate freezes more rapidly to seal off the cavity; and
3. parts can be more easily broken off the runner at the gate.

Ques.188: What are the advantages of a three-plate mold over a two- plate mold in injection molding?

Answer: As the mold opens, the three-plate mold automatically separates the molded part(s) from the runner system.

Ques.189: Discuss some of the defects that can occur in plastic injection molding.

Answer: The defects include: short shots, in which the polymer melt solidifies before filling the cavity; flashing, in which the polymer melt is squeezed into the parting surfaces between the mold halves & around ejection pins; sink marks, in which the surface is drawn into the molding by contraction of internal material; and weld lines where the melt has flowed around a core .

Ques.190: Describe structural foam molding.

Answer: Structural foam molding is an injection molding process in which a gas or gas-producing ingredient is mixed with the polymer melt prior to injection into the mold cavity; this results in the part having a tough outer skin surrounded by a foam core.

Ques.191: What are the significant differences in the equipment and operating procedures between injection molding of thermoplastics and injection molding of thermosets?

Answer: The differences in injection molding of thermosets are:

1. shorter barrel length,
2. lower temperatures in the barrel, these first two reasons to prevent premature curing; and
3. use of a heated mold to cause cross-linking of the TS (Thermosetting).

Ques.192: What is reaction injection molding?

Answer: Reaction injection molding involves the mixing of two highly reactive liquid ingredients and immediately injecting the mixture into a mold cavity where chemical reactions leading to solidification occur. The two ingredients form the components used in catalyst-activated or mixing activated thermoset systems .

Ques.193: What kinds of products are produced by blow molding?

Answer: Blow molding is used to produce hollow, seamless containers, such as bottles.

Ques.194: What is the starting material form in thermoforming?

Answer: Thermoforming starts with a thermoplastic sheet or film.

Ques.195: What is the difference between a positive mold and a negative mold in thermoforming?

Answer: A positive mold has a convex shape; a negative mold has a concave cavity.

Ques.196: Why are the molds generally more costly in mechanical thermoforming than in pressure or vacuum thermoforming?

Answer: In mechanical thermoforming, matching mold halves are required; while in other thermoforming processes, only one mold form is required.

Ques.197: What are the processes by which polymer foams are produced?

Answer: There are several foaming processes: (1) mechanical agitation - mixing a liquid resin with air , then hardening the polymer by means of heat or chemical reaction; (2) mixing a physical blowing agent with the polymer - a gas such as nitrogen (N_2) or pentane (C_5H_{12}) which can be dissolved in the polymer melt under pressure, so that the gas comes out of solution and expands when the pressure is subsequently reduced; and (3) mixing the polymer with chemical compounds, called chemical blowing agents, that decompose at elevated temperatures to liberate gases such as CO_2 or N_2 within the melt.

Ques.198: What are some of the general considerations that product designers must keep in mind when designing components out of plastics?

Answer: Some of the general considerations are :

1. Plastics are not as strong or stiff as metals and should not be used in applications where high stresses will be encountered.
2. Impact resistance of plastics is generally good, better than many ceramics.
3. Service temperatures of plastics are limited relative to engineering metals & ceramics.
4. Thermal expansion is greater for plastics than metals; so dimensional changes due to temperature variations are much more significant than for metals.

5. Many types of plastics degrade from sunlight & certain other forms of radiation. Also, some plastics degrade in oxygen and ozone atmospheres. Finally, plastics are soluble in many common solvents.

Ques.199: How is the rubber industry organized?

Answer: The rubber industry is organized into three parts:

(1) rubber growing plantations produce natural rubber,

(2) the petrochemical industry produces synthetic rubber, and

(3) fabricators take the natural & synthetic rubbers and produce finished rubber goods.

Ques.200: How is raw rubber recovered from the latex that is tapped from a rubber tree?

Answer: The rubber is usually recovered as follows:

1. the latex is collected into tanks and diluted to half natural concentration;
2. formic or acetic or other acid is added to the solution which causes the rubber to coagulate;
3. the coagulum is then squeezed through rolls to drive off water; and
4. the resulting sheets are dried in smokehouses for several days. The resulting raw rubber is called ribbed smoked sheet.

Ques.201: What is the sequence of processing steps required to produce finished rubber goods?

Answer: The typical sequence is:

1. production of the raw rubber,
2. compounding,
3. mixing,
4. shaping, and
5. vulcanization.

Ques.202: What are some of the functions of the additives that are combined with rubber during compounding?

Answer: The additives and functions are: vulcanizing chemicals, reinforcing fillers, extenders to reduce cost, antioxidants, coloring pigments, plasticizers to soften the rubber, and blowing agents to make foam rubber.

Ques.203: Name the four basic categories of processes used to shape rubber.

Answer: The categories are:

1. extrusion,
2. calendering,
3. coating, and
4. molding.

Ques.204: What are some of the operations used to coat rubber onto a fabric to produce reinforced rubber?

Answer: Calendering, skimming, dipping, and spraying .

Ques.205: What does vulcanization do to the rubber?

Answer: Vulcanization causes cross-linking of the rubber molecules; this strengthens and stiffens the rubber while extensibility is retained.

Ques.206: Name the three basic tyre constructions and briefly identify the differences in their construction.

Answer: (a) diagonal ply,

(b) belted bias, and

(c) radial ply. Diagonal ply and belted bias both have their carcass plys running in a diagonal direction relative to the tyre circumference; radial ply has its carcass plies running in a radial direction; belted bias and radial ply tyres use belts, which are additional plies around the outside circumference of the tyre; whereas diagonal ply tyres do not have these belts.

Ques.207: What are the three basic steps in the manufacture of a pneumatic tire?

Answer: The three steps are:

(1) preform the components,

(2) building the carcass and adding the rubber for the sidewall and treads, and

(3) molding and curing.

Ques.208: What is the purpose of the bead coil in a pneumatic tire?

Answer: The bead coil provides a rigid support for the tire when it is mounted onto the wheel rim.

Ques.209: What is a TPE?

Answer: TPE stands for thermoplastic elastomer; it is a thermoplastic polymer that behaves like a rubber.

Ques.210: Many of the design guidelines that are applicable to plastics are also applicable to rubber. However, the extreme flexibility of rubber results in certain differences. What are some examples of these differences?

Answer: Three examples:

(1) no draft is needed on the part for mold removal;

(2) holes should be molded into rubber parts rather than machined, whereas holes can be machined or molded in a plastic part; and

(3) screw threads are not normally used on rubber parts.

Ques.211: What are the principal polymers used in fiber-reinforced polymers?

Answer: Principal polymer matrices in FRPs are unsaturated polyesters and epoxies.

Ques.212: What is the difference between a roving and a yarn?

Answer: A roving consists of untwisted filaments, while a yarn consists of twisted fibers.

Ques.213: In the context of fiber reinforcement, what is a mat?

Answer: A mat is a felt consisting of randomly oriented fibers held loosely together in a binder.

Ques.214: What is sheet molding compound (SMC)?

Answer: SMC consists of TS polymer resin, fillers, and chopped glass fibers, all rolled into a sheetof typical thickness = 6.250 mm .

Ques.215: What is preform molding?

Answer: Preform molding is a compression molding process in which a precut mat is placed into the lower half of a mold together with a charge of thermosetting resin; the materials are then pressed between heated molds to cure the resin and produce a fiber-reinforced molding.

Ques.216: With what kinds of products is tube rolling associated?

Answer: Typical products include: bicycle frames and space trusses.

Ques.217: How are FRPs cut?

Answer: Uncured FRPs are cut by methods which include: knives, scissors, power shears, steel-rule blanking dies, laser beam cutting, and water jet cutting. Cured FRPs are cut by WC & HSS cutting tools, diamond cutting tools, and water jet cutting.

Ques.218: Name some of the reasons for the commercial importance of powder metallurgy (PM) technology.

Answer: PM is important because:

1. parts can be made to net or near net shape,
2. parts can be made with a controlled level of porosity,
3. certain metals difficult to process by other methods can be processed by PM,
4. PM allows the formulation of unusual alloys not easily obtained by traditional alloying methods.

Ques.219: What are some of the disadvantages of PM methods?

Answer: Disadvantages include:

1. high tooling costs,
2. metal powders are expensive,
3. difficulties in storing & handling,
4. certain limitations on part geometry imposed by the uniaxial press methods, and
5. variations in density in a PM component can be troublesome.

Ques.220: In the screening of powders for sizing, what is meant by the term mesh count?

Answer: The mesh count of the screen is the number of openings per linear inch.

Ques.221: What is the difference between open pores and closed pores in a metallic powders?

Answer: Open pores are air spaces between particles, while closed pores are voids internal to a particle.

Ques.222: What is meant by the term aspect ratio for a metallic particle?

Answer: The aspect ratio of a particle is ratio of the maximum dimension to the minimum dimension of the given particle.

Ques.223: How would one measure the angle of repose for a given amount of metallic powder?

Answer: One measure would be to let the powders flow through a small funnel and measure the angle taken by the resulting pile of powders relative to the horizontal.

Ques.224: Define bulk density and true density for metallic powders.

Answer: Bulk density refers to the weight per volume of the powders in the loose state, while true density is the weight per volume of the true volume of metal in the powders (the volume that would result if the powders were melted).

Ques.225: What are the principal methods used to produce metallic powders?

Answer: The methods are:

1. atomization - the conversion of molten metal into droplets which solidify into powders;
2. chemical reduction - reducing metallic oxides by use of reducing agents which combine with the oxygen to free the metals in the form of powders; and
3. electrolysis -use of an electrolytic cell to deposit particles of the metal onto the cathode in the cell.

Ques.226: What are the three basic steps in the conventional powder metallurgy shaping process?

Answer: The steps are:

1. blending and/or mixing,
2. pressing, and
3. intering.

Ques.227: What is the technical difference between mixing and blending in powder metallurgy?

Answer: Mixing refers to the combining of metal powders of different chemistries, while blending means combining particles of the same chemistry but different sizes.

Ques.228: What are some of the ingredients usually added to the metallic powders during blending and/or mixing?

Answer: The additives are: (1) lubricants, (2) binders, and (3) deflocculants.

Ques.229: What is meant by the term green compact?

Answer: The green compact is the pressed but not yet sintered PM part.

Ques.230: What happens to the individual particles during compaction ?

Answer: Starting with the initial powder arrangement, the particles are first repacked into a more efficient arrangement, followed by deformation of the particles as pressure is increased.

Ques.231: Which of the following most closely typifies the sintering temperatures in PM?

(a) 0.5 Tm, **(b) 0.8 Tm,** **(c) Tm.**

Answer: (b)

Ques.232: What are the three steps in the sintering cycle in PM?

Answer: The three steps in the cycle are:

1. preheat, in which lubricants & binders are burned off,
2. sintering, and
3. cool down.

Ques.233: What are some of the reasons why a controlled furnace is desirable in sintering?

Answer: Some of the purposes of a controlled atmosphere furnace are:

1. oxidation protection,
2. provide a reducing atmosphere to remove existing oxides,
3. provide a carburizing atmosphere,and
4. remove lubricants & binders from pressing.

Ques.234: What are the advantages of infiltration in PM?

Answer: Advantages of infiltration are: resulting structure is nonporous structure, improved toughness & strength.

Ques.235: What is the difference between powder injection molding and metal injection molding?

Answer: Metal injection molding is a subset of powder injection molding, in which the powders are metallic. The more general term includes powders of ceramic.

Ques.236: How is isostatic pressing distinguished from conventional pressing and sintering in PM?

Answer: Isostatic pressing applies hydrostatic pressure to all sides of the mold, whereas conventional pressing is uniaxial.

Ques.237: Describe liquid phase sintering.

Answer: Liquid phase sintering occurs when two metals of different melting temperatures are sintered at a temperature between their melting points. Accordingly, one metal melts, thoroughly wetting the solid particles and creating a strong bonding between the metals upon solidification.

Ques.238: What are the two basic classes of metal powders as far as chemistry is concerned?

Answer: The two classes are:

(1) elemental powders - powders of pure metal such as iron or copper, and

(2) prealloyed powders - powders of alloys such as stainless steel or brass.

Ques.239: Why is PM (Powder Metallurgy) technology so well suited to the production of gears and bearings?

Answer: The reasons are:

1. the geometries of these parts lend themselves to PM pressing, and
2. the porosity allows impregnation of the PM parts with lubricants.

Ques.240: Describe the slip casting process in traditional ceramics processing.

Answer: In slip casting , a slurry of clay is poured into a plaster of Paris mold, whereupon water is absorbed from the slurry into the plaster to form a clay layer against the mold wall. The remaining slurry is usually poured out to leave a hollow part.

Ques.241: List and briefly describe some of the plastic forming methods used to shape traditional ceramics products.

Answer: The plastic forming methods include:

1. hand modeling, molding, & throwing;
2. jiggering, which is a mechanized extension of hand throwing used to manufacture bowls & plates;
3. plastic pressing, in which a clay slug is pressed in a mold; and
4. extrusion, in which the clay is compressed through a die opening to make long sections of uniform cross-sectional shape.

Ques.242: What is the process of jiggering?

Answer: Jiggering is a clay forming process that uses a convex mold on a potters wheel. The clay is first pressed into rough shape and then rotated & formed with a jigger tool to final shape. It is suited to the manufacture of flatware.

Ques.243: What is the difference between dry pressing and semi-dry pressing of traditional ceramics parts?

Answer: The difference is in the starting clay. For semi- dry pressing, the clay has a typical water content of 10 to 15%.

For dry pressing, the water content is usually less than 5%. Dry clay has virtually no plasticity, and so this imposes certain limitations on part geometry in dry pressing.

Ques.244: What happens to a ceramic material when it is sintered?

Answer: Sintering of green ceramics (or powdered metals) causes bonding between the ceramic grains, which is accompanied by densification and reduction of porosity.

Ques.245: What is the name given to the furnace used to fire ceramic ware?

Answer: Kiln.

Ques.246: Why is the drying step, so important in the processing of traditional ceramics, usually not required in processing of new ceramics?

Answer: Because water is usually not one of the ingredients in the new ceramics during forming. Drying is only needed when the green piece contains water.

Ques.247: What is the freeze drying process used to make certain new ceramic powders?

Answer: In freeze drying, salts are dissolved in water and sprayed into small droplets which are immediately frozen; the water is then removed from the droplets in a vacuum chamber, and the freeze-dried salt is decomposed by heating to form the ceramic powders.

Ques.248: Describe the doctor-blade process.

Answer: In the doctor-blade process, a ceramic slurry is flowed onto a moving film which flows under a wiper blade, so that the resulting ceramic is in the form of a thin green sheet which is dried and reeled onto a spool for subsequent shaping and sintering.

Ques.249: Liquid phase sintering is used for WC-Co compacts, even though the sintering temperatures are below the melting points of either WC or Co. How is this possible?

Answer: The melting point of cobalt is reduced when WC is dissolved in it. At the sintering temperatures used for WC-Co, WC gradually dissolves in the cobalt, reducing its melting point to the sintering temperature. Thus does liquid phase sintering occur in the WC-Co system.

Ques.250: What are the characteristics that distinguish bulk deformation processes from sheet metal processes?

Answer: In bulk deformation, the shape changes are significant, and the work parts have a low area-to-volume ratio. In sheet metal processes, the area-to-volume ratio is high.

Ques.251: Extrusion is a fundamental shaping process. Describe it.

Answer: Extrusion is a compression process in which the work material is forced to flow through a die orifice, thereby forcing its cross-section to assume the profile of the orifice.

Ques.252: Why is the term pressworking often used for sheet metal processes?

Answer: Because these operations are generally performed on presses.

Ques.253: What is the difference between deep drawing and bar drawing?

Answer: Deep drawing is a sheet metal forming process used to fabricate cup-shaped parts; bar drawing is a bulk deformation process used to reduce the diameter of a cylindrical workpart.

Ques.254: Indicate some of the advantages of cold working relative to warm and hot working.

Answer: Advantages of cold working are:

1. better accuracy,
2. better surface finish,
3. increased strength due to work hardening,
4. possible directional properties due to grain flow, and
5. no heating of work required.

Ques.255: What is isothermal forming?

Answer: An isothermal forming operation is performed in such a way as to eliminate surface cooling and thermal gradients in the workpart. This is accomplished by preheating the forming tools.

Ques.256: Describe the effect of strain rate in metal forming.

Answer: Increasing strain rate tends to increase the resistance to deformation. The tendency is especially prominent in hot forming operations.

Ques.257: Why is friction generally undesirable in metal forming operations?

Answer: Reasons why friction is undesirable in metal forming:

1. inhibits metal flow duringdeformation, causing residual stresses & product defects;
2. increases forces & power required; and
3. wearing of the tools.

Ques.258: What is sticking friction in metalworking?

Answer: Sticking friction is when the work surface adheres to the surface of the tool rather than slides against it; it occurs when the friction stress is greater than the shear flow stress of the metal.

Ques.259: What are the reasons why the bulk deformation processes are important commercially and technologically?

Answer: Reasons why the bulk deformation processes are important include:

1. they are capable of significant shape change when hot working is used,
2. they have a positive effect on part strength when cold working is used, and
3. most of the processes produce little material waste; some are net shape processes.

Ques.260: List some of the products produced on a rolling mill.

Answer: Rolled products include flat sheet & plate stock, round bar & rod stock, rails, structural shapes such as I-beams & channels.

Ques.261: Identify some of the ways in which force in flat rolling can be reduced.

Answer: Ways to reduce force in flat rolling include:

1. use hot rolling,
2. reduce draft in each pass, and
3. use smaller diameter rolls.

Ques.262: What is a two-high rolling mill?

Answer: A two-high rolling mill consists of two opposing rolls between which the work is compressed.

Ques.263: What is a reversing mill in rolling?

Answer: A reversing mill is a two-high rolling mill in which the direction of rotation of the rolls can be reversed to allow the work to pass through from either side.

Ques.264: Besides flat rolling and shape rolling, identify some additional bulk forming processes that use rolls to effect the deformation.

Answer: Some other processes that use rolls are ring rolling, thread rolling, gear rolling, roll piercing, and roll forging.

Ques.265: One way to classify forging operations is by the degree to which the work is constrained in the die. By this classification, name the three basic types.

Answer: The three basic types are:

(1) open die forging,

(2) impression die forging, and

(3) closed die forging.

Ques.266: Why is flash desirable in impression die forging?

Answer: Because its presence constrains the metal in the die to fill the details of the die cavity.

Ques.267: What are the two basic types of forging equipment?

Answer: The two types of forging machines are hammers, which impact the work part, and presses, which apply a gradual pressure to the work.

Ques.268: What is isothermal forging?

Answer: Isothermal forging is a hot forging operation in which the die surfaces are heated to reduce heat transfer from the work into the tooling.

Ques.269: Name some products that are produced by extrusion.

Answer: Products produced by continuous extrusion include: structural shapes (window frames, shower stalls, channels), tubes & pipes, and rods of various cross-section. Products made by discrete extrusion include: toothpaste tubes, aluminum beverage cans, and battery cases.

Ques.270: What does the centerburst defect in extrusion have in common with the roll piercing process?

Answer: They are both examples of how compressive stresses applied to the outside surface of a solid cylindrical cross-section can create high tensile stresses in the interior of the cylinder.

Ques.271: In a wire drawing operation, why must the drawing stress never exceed the yield strength of the work metal?

Answer: Because if the drawing stress exceeded the yield strength, the metal on the exit side of the draw die would stretch rather than force metal to be drawn through the die opening.

Ques.272: Identify the three basic types of sheet metalworking operations.

Answer: The three operations are: (1) cutting, (2) bending, and (3) drawing.

Ques.273: In blanking of a round sheet metal part, indicate how the clearance should be applied to the punch and die diameters.

Answer: For a blanking operation, die size = blank size, and the punch is smaller by twice the clearance.

Ques.274: What is the difference between a cutoff operation and a parting operation?

Answer: A cutoff operation separates parts from a strip by shearing one edge of each part in sequence. A parting operation cuts a slug between adjacent parts in the strip.

Ques.275: Describe V-bending and edge bending.

Answer: In V-bending, a simple punch and die which have the included angle are used to bend the part. In edge bending, the punch force a cantilevered sheet metal section over a die edge to obtain the desired bend angle

Ques.276: What is springback in sheet metal bending?

Answer: Springback is the elastic recovery of the sheet metal after bending; it is usually measured as the difference between the final included angle of the bent part and the angle of the tooling used to make the bend, divided by the angle of the tooling.

Ques.277: What are some of the simple measures used to assess the feasibility of a proposed cup drawing operation?

Answer: Measures of drawing feasibility include: (1) drawing ratio, DR = D/Dp; (2) reduction, r =(D - Dp)/D; and (3) thickness-to-diameter ratio, t/D; where t = stock thickness, D = blank diameter, and Dp = punch diameter.

Ques.278: Distinguish between redrawing and reverse drawing.

Answer: In redrawing, the shape change is significant enough (e.g., drawing ratio greater than 2.0) that it must be carried out in two drawing steps, probably with annealing between the steps.

In reverse drawing, two draws are accomplished on the part, one in one direction, the second in the opposite direction.

Ques.279: What are some of the possible defects in drawn sheet metal parts?

Answer: Drawing defects include: (1) wrinkling, (2) tearing, (3) earing, and (4) surface scratches.

Ques.280: What is stretch forming?

Answer: Stretch forming of sheet metal involves stretching and simultaneous bending of theworkpart to achieve shape change.

Ques.281: What are the two basic categories of structural frames used in stamping presses?

Answer: Two press frame types are:

(1) gap frame, also called C-frame, and

(2) straight-sided frame.

Ques.282: What are the relative advantages and disadvantages of mechanical versus hydraulic presses in sheet metalworking?

Answer: Advantage of mechanical presses: faster cycle rates. Advantages of hydraulic presses: longer ram strokes and uniform force throughout stroke.

Ques.283: What is the Guerin process?

Answer: The Guerin process is a sheet metal forming process that uses a rubber die which flexes to force the sheet metal to take the shape of a form block (punch).

Ques.284: Identify a major technical problem in tube bending?

Answer: A major technical problem in tube bending is collapse of the tube walls during the process.

Ques.285: Distinguish between roll bending and roll forming.

Answer: Roll bending involves the forming of large sheet and plate metal sections into curved forms. Roll forming involves feeding a lone strip or coil through rotating rolls so that the shape of the rolls is imparted to the strip.

Ques.286: What distinguishes machining from other manufacturing processes?

Answer: In machining, material is removed from the workpart so that the remaining material is the desired part geometry.

Ques.287: Identify some of the reasons why machining is commercially and technologically important.

Answer: The reasons include:

1. its applicability to most materials;
2. its capability to produce a variety of geometries to a part;
3. it can achieve closer tolerances than most other processes; and
4. it can create good surface finishes.

Ques.288: Name the three most common machining processes.

Answer: The three common machining processes are:

(1) turning,

(2) drilling, and

(3) milling.

Ques.289: What are the two basic categories of cutting tools in machining? Give an example of a machining operation that uses each of the tooling types.

Answer: The two categories are:

(1) single -point tools, used in operations such as turning and boring; and

(2) multiple-edge cutting tools, used in operations such as milling and drilling.

Ques.290: Identify the parameters of a machining operation that are included within the scope of cutting conditions.

Answer: Cutting conditions include: speed, feed, depth of cut, and whether or not a cutting fluid is used.

Ques.291: Define the difference between roughing and finishing operations in machining.

Answer: A roughing operation is used to remove large amounts of material rapidly and to produce a part geometry close to the desired shape. A finishing operation follows roughing and is used to achieve the final geometry and surface finish.

Ques.292: What is a machine tool?

Answer: A machine tool can be defined as a power-driven machine that positions and moves a tool relative to the work to accomplish machining or other metal shaping process.

Ques.293: What is an orthogonal cutting operation?

Answer: Orthogonal cutting involves the use of a wedge shaped tool in which the cutting edge is perpendicular to the direction of speed motion into the work material.

Ques.294: Name and briefly describe the three types of chips that occur in metal cutting.

Answer: The three types are:

(1) discontinuous, in which the chip is formed into separated segments;

(2) continuous, in which the chip does not segment and is formed from a ductile metal; and

(3) continuous with built-up edge, which is the same as except that friction at the tool-chip interface causes adhesion of the work material to the tool rake face.

Ques.295: Describe the turning process.

Answer: Turning is a machining process in which a single point tool removes material from the surface of a rotating cylindrical workpiece, the tool being fed in a direction parallel to the axis of work rotation.

Ques.296: What is the difference between threading and tapping?

Answer: A threading operation is performed on a turning machine and produces an external thread, while tapping is normally performed on a drilling machine and produces an internal thread.

Ques.297: How does a boring operation differ from a turning operation?

Answer: Boring produces an internal cylindrical shape from an existing hole, while turning produces an external cylindrical shape.

Ques.298: How does a turret lathe differ from an engine lathe?

Answer: A turret lathe has a tool holding turret in place of a tailstock; the tools in the turret can be brought to work to perform multiple cutting operations on the work without the need to change tools as in operating a conventional engine lathe.

Ques.299: What is the difference between peripheral milling and face milling?

Answer: In peripheral milling, cutting is accomplished by the peripheral teeth of the milling cutter and the tool axis is parallel to the work surface; in face milling, cutting is accomplished by the flat face of the cutter whose axis is perpendicular to the work surface.

Ques.300: Describe profile milling.

Answer: Profile milling generally involves the milling of the outside periphery of a flat part.

Ques.301: Describe the difference between up milling and down milling?

Answer: In up milling, the cutter speed direction is opposite the feed direction; in down milling, the direction of cutter rotation is the same as the feed direction.

Ques.302: How does a universal milling machine differ from a conventional knee-and-column machine?

Answer: The universal milling machine has a worktable that can be rotated about a vertical axis to present the part at any specified angle to the cutter spindle.

Ques.303: How do shaping and planing differ?

Answer: In shaping, the work is stationary during the cut, and the speed motion is performed by the cutting tool; while in planing, the cutting tool is stationary, and the workpart is moved past the tool in the speed motion.

Ques.304: What is the difference between internal broaching and external broaching?

Answer: Internal broaching is accomplished on the inside surface (hole) of a workpart; while external broaching is performed on one of the outside surfaces of the part.

Ques.305: Identify the three basic forms of sawing operation?

Answer: The three forms of sawing are:

(1) hacksawing,

(2) bandsawing, and

(3) circular sawing.

Ques.306: Name the three modes of tool failure in machining.

Answer: The three tool failure modes are:

(1) fracture failure,

(2) temperature failure, and

(3) gradual wear.

Ques.307: What are the two principal locations on a cutting tool where tool wear occurs?

Answer: Wear occurs on the top face of the cutting tool as crater wear and on the side or flank of the tool, called flank wear. Portions of flank wear are often identified separately as notch wear, corresponding to the surface of the work; and nose radius wear, corresponding to the tool point.

Ques.308: Identify the mechanisms by which cutting tools wear during machining.

Answer: The important tool wear mechanisms are:

1. abrasion,
2. adhesion,
3. diffusion, and
4. plastic deformation of the cutting edge.

Ques.309: What are some of the tool life criteria used in production machining operations?

Answer: Production tool life criteria include:

1. complete failure of the tool,
2. visual observation of flank or crater wear,
3. fingernail test to feel flank wear,
4. sound of the tool,
5. chip disposal problems,
6. degradation of finish,
7. power increase,
8. workpiece count, and
9. length of cutting time for the tool.

Ques.310: Identify three desirable properties of a cutting tool material.

Answer: Three desirable properties are:

(1) toughness to resist fracture failure,

(2) hot hardness to resist temperature failure, and

(3) wear resistance to prolong the life of the tool during gradual wear.

Ques.311: What are the principal alloying ingredients in high speed steel (HSS) ?

Answer: Principal alloying ingredients in HSS are:

1. either tungsten or a combination of tungsten and molybdenum,
2. chromium,
3. vanadium, and
4. carbon. Some grades of HSS also contain cobalt.

Ques.312: What is the difference in ingredients between steel cutting grades and non-steel cutting grades of cemented carbides?

Answer: The non-steel cutting grades contain only WC and Co. The steel cutting grades contain TiC and/or TaC in addition to WC-Co.

Ques.313: Identify some of the common compounds that form the thin coatings on the surface of coated carbide inserts.

Answer: The common coatings are: TiN, TiC, and Al_2O_3.

Ques.314: Name the seven elements of tool geometry for a single point cutting tool.

Answer: The seven elements of single point tool geometry are:

1. back rake angle,
2. side rake angle,
3. end relief angle,
4. side relief angle,
5. end cutting edge angle,
6. side cutting edge angle, and
7. nose radius.

Ques.315: Why are ceramic cutting tools generally designed with negative rake angles?

Answer: Ceramics possess low shear and tensile strength but good compressive strength. During cutting, this combination of properties is best exploited by giving the tool a negative rake angle to tend to load the tool in compression.

Ques.316: Name the two main categories of cutting fluid according to function.

Answer: The two functional categories of cutting fluids are:

(1) coolants and

(2) lubricants.

Ques.317: Name the three main types of cutting fluid according to chemistry.

Answer: The three types of cutting fluids are:

(1) cutting oils,

(2) emulsified oils, and

(3) chemical & semi- chemical fluids.

Ques.318: Dry machining is being considered by machine shops because of certain problems inherent in the use of cutting fluids.

(a) What are those problems associated with the use of cutting fluids?

(b) What are some of the new problems introduced by machining dry?

Answer: (a) Cutting fluids become contaminated over time with a variety of contaminants, including tramp oil, garbage, small chips, molds, fungi, and bacteria. In addition to causing odors and health hazards, contaminated cutting fluids do not perform their lubricating function as well as when they are fresh and clean.

(b) Problems with dry machining include:

1. overheating the tool,
2. operating at lower cutting speeds and production rates to prolong tool life, and
3. absence of chip removal benefits in grinding & milling.

Ques.319: Define machinability.

Answer: Machinability can be defined as the relative ease with which a material can be machined using an appropriate cutting tool under appropriate cutting conditions.

Ques.320: What are the criteria by which machinability is commonly assessed in a production machining operation?

Answer: The machinability criteria include:

1. tool wear & tool life,
2. forces & power,
3. surface finish, and
4. ease of chip disposal.

Ques.321: Name some of the important mechanical and physical properties that affect the machinability of a work material.

Answer: The properties include: hardness, strength, and thermal diffusivity.

Ques.322: Why do costs tend to increase when better surface finish is required on a machined part?

Answer: Because additional operations such as grinding, lapping, or similar finishing processes must be included in the manufacturing sequence at higher cost.

Ques.323: What are the basic factors that affect surface finish in machining?

Answer: The factors that affect surface finish are:

1. geometric factors such as type of operation, feed, and tool shape (nose radius in particular);
2. work material factors such as built-up edge effects, and tearing of the work surface when machining ductile materials, which factors are affected by cutting speed; and

(3) vibration and machine tool factors such as setup and work part rigidity, and backlash in the feed mechanism.

Ques.324: Why are abrasive processes technologically and commercially important?

Answer: Important reasons include:

1. applications on all types of materials,
2. very fine finishes, and
3. close tolerances.

Ques.325: What are the five principal parameters of a grinding wheel?

Answer: The parameters are:

1. abrasive material,
2. grit size,
3. bonding material,
4. wheel structure, which refers to the relative spacing of grains, and
5. wheel grade, which refers to the bond strength of the wheel in retaining abrasive grains.

Ques.326: What are some of the principal abrasive materials used in grinding wheels?

Answer: The principal abrasive grit materials include:

1. aluminum oxide,
2. silicon carbide,
3. cubic boron nitride, and
4. diamond.

Ques.327: Name some of the principal bonding materials used in grinding wheels.

Answer: The bonding materials in grinding wheels are:

1. vitrified bond - clay and ceramics,
2. silicate,
3. rubber,
4. resinoid,
5. shellac, and
6. metallic.

Ques.328: What is wheel structure?

Answer: Wheel structure indicates the relative spacing of the abrasive grains in the wheel. An open structure is one in which the grains are far apart, and a dense structure indicates that the grains are close together.

Ques.329: Why are specific energy values so much higher in grinding than in traditional metal cutting processes?

Answer: Reasons for higher specific energy in grinding:

1. size effect - smaller chip size means higher specific energy;
2. extremely negative rake angles on the abrasive particles in a grinding wheel; and
3. not all of the grains in the wheel surface are engaged in cutting; some are plowing or deforming while others are simply rubbing and creating friction.

Ques.330: Grinding creates high temperatures. How is temperature harmful in grinding?

Answer: High temperatures in grinding create surface burns and cracks. High temperatures can also soften the surfaces of workparts that have been heat treated to obtain high hardness.

Ques.331: What are the three mechanisms of grinding wheel wear?

Answer: The mechanisms are:

(1) grain fracture, in which a portion of the grain breaks off during cutting;

(2) attritious wear, in which the grains become dull during cutting; and

(3) bond fracture, in which the grains are pulled out of the bonding material.

Ques.332: What abrasive material would one select for grinding a cemented carbide cutting tool?

Answer: Choose a diamond wheel.

Ques.333: What are the functions of a grinding fluid?

Answer: Functions of a grinding fluid are:

1. reduce friction,
2. remove heat,
3. wash away chips, and
4. reduce workpiece temperature.

Ques.334: What is centerless grinding?

Answer: Centerless grinding is a grinding operation in which cylindrical workparts (*e.g.*, rods) a refed between two rotating wheels:

(1) a high speed grinding wheel and

(2) a low speed regulating wheel which is tilted at a slight angle to control the feed-through rate.

Ques.335: Name some of the abrasive operations available to achieve very good surface finishes.

Answer: High finish abrasive processes include: honing, lapping, superfinishing, buffing, and polishing.

Ques.336: There are four categories of nontraditional machining processes, based on principal energy form. Name the four categories.

Answer: The four categories are:

1. mechanical, but not including conventional machining;
2. electrical;
3. thermal; and
4. chemical.

Ques.337: How does the ultrasonic machining process work?

Answer: In ultrasonic machining, abrasives contained in a slurry are driven at high velocity against the work by a tool vibrating at low amplitude and high frequency.

The tool oscillates in a direction perpendicular to the work surface, and is fed slowly into the work, so that the shape of the tool is formed in the part. The abrasives, impinging against the work surface, perform the cutting.

Ques.338: Describe the water jet cutting process.

Answer: Water jet cutting uses a high-pressure, high-velocity stream of water directed at the work surface to cut of the work.

Ques.339: What is the difference between water jet cutting (WJC) , abrasive water jet cutting (AWJC) , and abrasive jet cutting (AJC) ?

Answer: WJC cuts with a narrow, high velocity water stream; AWJC adds abrasive grits to the water stream; and AJC cuts with abrasive particles that have been added to a high velocity air stream.

Ques.340: Name the three main types of electrochemical machining.

Answer: The three types are: electrochemical machining, deburring, and grinding.

Ques.341: Identify the significant disadvantages of electrochemical machining (ECM) .

Answer: Disadvantages of ECM include:

(1) cost of electrical power to operate the process, and

(2) cost of disposal of electrolyte sludge.

Ques.342: How does increasing discharge current affect metal removal rate and surface finish in electric discharge machining?

Answer: As discharge current increases, metal removal rate increases and surface finish is degraded.

Ques.343: Identify two major disadvantages of plasma arc cutting.

Answer: Two disadvantages are:

(1) rough surface on cut edge, and

(2) metallurgical damage to cut surface.

Ques.344: Name the four principal steps in chemical machining.

Answer: The four steps are:

1. cleaning,
2. masking,
3. etching, and
4. demasking.

Ques.345: Why are metals heat treated?

Answer: Metals are heat treated to effect metallurgical changes that beneficially alter properties.

Ques.346: Identify the important reasons why metals are annealed.

Answer: The purposes of annealing include:

1. control properties,
2. reduce brittleness & improve toughness,
3. recrystallize cold-worked metals, and
4. relieve stresses from prior metal working.

Ques.347: What is the most important heat treatment for hardening steels?

Answer: Martensite formation by heating steel into the austenite region and quenching.

Ques.348: What is the mechanism by which carbon strengthens steel during heat treatment?

Answer: If the steel is heat treated, martensite is formed which depends on the presence of carbon to create the nonequilibrium structure of this phase.

Ques.349: What information is conveyed by the TTT curve?

Answer: The TTT curve indicates what phases in the iron- carbon phase diagram will be produced under various conditions of cooling.

Ques.350: What function is served by tempering?

Answer: Tempering involves heating and soaking of martensite for about one hour, followed by slow cooling to reduce brittleness, relieve stresses, and increase toughness & ductility.

Ques.351: Define hardenability?

Answer: Hardenability is the relative capacity of a steel to be hardened by transformation to martensite.

Ques.352: Name some of the elements which have the greatest effect on the hardenability of steel.

Answer: Important hardenability elements are: chromium, manganese, molybdenum, and nickel.

Ques.353: Indicate how the hardenability alloying elements in steel affect the TTT curve.

Answer: The hardenability alloying elements operate by pushing the nose of the TTT curve to the right, there by permitting slower cooling rates for conversion of austenite to martensite.

Ques.354: Define precipitation hardening?

Answer: Precipitation hardening is a heat treatment in which very fine particles (precipitates) are formed so that dislocation movement is blocked and the metal is thus strengthened & hardened.

Ques.355: How does carburizing work?

Answer: Carburizing adds carbon to the surface of low-C steel, thereby transforming the surface into high-C steel for grater hardening potential.

Ques.356: Identify the selective surface hardening methods.

Answer: The selective surface hardening methods include: flame hardening, induction hardening, high-frequency (HF) resistance heating, electron beam (EB) heating, and laser beam (LB) heating.

Ques.357: What are some of the important reasons why manufactured parts must be cleaned?

Answer: The reasons include:

1. to prepare the surface for subsequent industrial processing,
2. to improve hygiene conditions,
3. to remove contaminants which might chemically react with the surface; and
4. to enhance product appearance and performance.

Ques.358: Mechanical surface treatments are often performed for reasons other than or in addition to cleaning. What are the reasons?

Answer: Reasons for mechanical surface treatments include: deburring, improving smoothness, adding luster, and enhancing surface properties.

Ques.359: What are the basic types of contaminants that must be cleaned from metallic surfaces in manufacturing?

Answer: Basic contaminant types are:

1. oil & grease,
2. solid particles, such as metal chips,abrasive grits, shop dirt, & dust,
3. buffing & polishing compounds, and
4. oxide films, rust,& scale.

Ques.360: Identify some of the mechanical cleaning methods.

Answer: Mechanical cleaning methods include: belt grinding, honing, lapping, polishing, buffing, and mass finishing (such as barrel tumbling). These method also accomplish other functions such as deburring and surface finish improvement.

Ques.361: In addition to surface cleaning, what is the main function performed by shot peening?

Answer: Shot peening is primarily used to improve the fatigue strength of metals by introducing cold working the metallic surface.

Ques.362: Name some of the important chemical cleaning methods.

Answer: The chemical cleaning methods can be categorized as follows:

1. alkaline cleaning,
2. emulsion cleaning,
3. solvent cleaning,
4. acid cleaning and pickling, and
5. ultrasonic cleaning.

Ques.363: What is the difference between diffusion and ion implantation?

Answer: Diffusion coating is a diffusion process in which atoms or molecules move across a boundary between two contacting materials.

Ion implantation produces a similar result, but the process involves penetration of high-velocity ions into the surface of a substrate material.

Ques.364: What is calorizing?

Answer: Calorizing is the diffusion of aluminum into carbon steel, alloy steels, and the alloys of nickel & cobalt. The process is also known as aluminizing.

Ques.365: Why are metals coated?

Answer:

1. Corrosion protection,
2. enhance appearance,
3. provide a specified color, and
4. prepare surface for subsequent processing.

Ques.366: Identify the most common types of coating processes.

Answer: The common coating processes are:

1. plating,
2. chemical surface treatments, such as anodizing,
3. vapor deposition processes
4. organic coating painting,
5. porcelain enameling, and
6. thermal & mechanical treatments.

Ques.367: What are the many reasons why a metallic surface is plated?

Answer: The reasons are:

1. corrosion protection of the substrate metal,
2. attractive appearance,
3. wear resistance,
4. increased electrical conductivity,
5. improved solderability, and
6. enhanced lubricity of the surface.

Ques.368: What are the two basic mechanisms of corrosion protection?

Answer: The mechanisms are

(1) barrier protection, in which the coating simply covers the substrate to protect it, and

(2) sacrificial protection, in which the coating metal corrodessacrificially to protect the substrate.

Ques.369: What is the most commonly plated substrate metal?

Answer: Steel.

Ques.370: How does electroless plating differ from electrochemical plating?

Answer: Electroless plating uses only chemical reactions to form the plating; electroplating uses electrolysis.

Ques.371: What is a conversion coating?

Answer: A conversion coating is a thin coating produced by chemical reaction of the metallic surface. The most common conversion coatings are phosphates, chromates, and oxides.

Ques.372: How does anodizing differ from other conversion coatings?

Answer: Anodizing uses electrochemical processing methods to convert the metallic surface. Best example is aluminum anodizing.

Ques.373: What is physical vapor deposition (PVD) ?

Answer: Physical vapor deposition (PVD) refers to a family of processes in which a material is converted to its vapor phase in a vacuum chamber and condensed onto a substrate surface as a very thin film.

Ques.374: What is the difference between physical vapor deposition (PVD) and chemical vapor deposition (CVD)?

Answer: In PVD, the coating vapors are synthesized by heating the coating material and allowing it to condense as a thin film on the surface of the workpart. In CVD a coating is formed on a heated substrate by the chemical reaction or dissociation of vapors and/or gases; the reaction product nucleates and grows on the substrate surface.

Ques.375: What are some of the applications of PVD?

Answer: PVC applications include: decorative coatings on trophies & automotive trim, antireflection coatings on optical lenses, deposition of metal in electronic connections, and cutting tool coatings (e.g., TiN).

Ques.376: Name the three basic types of PVD.

Answer: The three types are:

(1) vacuum evaporation,

(2) sputtering, and

(3) ion plating.

Ques.377: What is a commonly used coating material deposited by PVD onto cutting tools?

Answer: Titanium nitride (TiN). Titanium carbide (TiC) and aluminum oxide (Al_2O_3) might also be mentioned.

Ques.378: What are some of the advantages of chemical vapor deposition?

Answer: Advantages include:

1. capability to deposit refractory materials at temperatures below their melting or sintering temperatures;
2. grain size control;
3. process is performed at atmospheric pressure; and
4. good bonding to substrate surface.

Ques.379: What are the two most common titanium compounds that are coated onto cutting tools by chemical vapor deposition?

Answer: TiC and TiN.

Ques.380: Identify the four major types of ingredients in organic coatings?

Answer: The major ingredients are: (1) binder, which are polymers, (2) dyes or pigments, which provide color, (3) solvents, and (4) additives such as surfactants & plasticizers.

Ques.381: Describe the principal methods by which organic coatings are applied to a surface.

Answer: The main methods include: brushing and rolling, spraying, immersion (dip coating), and flow coating.

Ques.382: The terms drying and curing have different meanings; indicate the distinction.

Answer: Drying means evaporation of solvents in the organic coating liquid. Curing involves a chemical change in the organic resin (polymerization and/or cross-linking) which hardens the coating.

Ques.383: In porcelain enameling, what is frit?

Answer: Frit is glassy porcelain prepared as fine particles (powders) by crushing and milling.

Ques.384: What does the term mechanical galvanizing refer to?

Answer: Mechanical galvanizing refers to the mechanical plating of zinc onto a substrate, with thicknesses up to 0.075 mm.

Ques.385: What are the advantages and disadvantages of welding compared to other types of assembly operations?

Answer: Advantages:

1. provides a permanent joint, so the parts are joined permanently;
2. joint strength is typically as high as strength of base metals;
3. most economical in terms of material usage;
4. versatile in terms of where it can be accomplished.

Disadvantages:

1. usually performed manually, so labor cost is high and the skilled labor to perform it is sometimes scarce;
2. welding is inherently dangerous;
3. difficult to disassemble;
4. quality defects sometimes difficult to detect.

Ques.386: What is meant by the term faying surface?

Answer: The faying surfaces are the contacting surfaces in a welded joint.

Ques.387: Define the term fusion weld.

Answer: A fusion weld is a weld in which the metal surfaces have been melted in order to cause coalescence.

Ques.388: What is the fundamental difference between a fusion weld and a solid state weld?

Answer: In a fusion weld, the metal is melted. In a solid state weld, the metal is not melted.

Ques.389: What is an autogenous weld?

Answer: It is a fusion weld made without the addition of filler metal.

Ques.390: Discuss the reasons why most welding operations are inherently dangerous.

Answer: Most welding operations are carried out at high temperatures that can cause serious burns on skin and flesh. In gas welding, the fuels are a fire hazard. In arc welding and resistance welding, the high electrical energy can cause shocks which are fatal to the worker. In arc welding, the electric arc emits intense ultraviolet radiation which can cause blinding. Other hazards include sparks,smoke, fumes, and weld spatter.

Ques.391: What is the difference between machine welding and automatic welding?

Answer: An automatic welding operation uses a weld cycle controller which regulates the arc movement and workpiece positioning; whereas in machine welding, a human worker must continuously control the arc and the relative movement of the welding head and the workpart.

Ques.392: Name the five joint types.

Answer: Five joint types are: (1) butt, (2) corner, (3) lap, (4) tee, (5) edge.

Ques.393: Define a fillet weld?

Answer: A fillet weld is a weld joint of approximately triangular cross-section used to fill in the edges of corner, lap, and tee joints.

Ques.394: Define a groove weld?

Answer: A groove weld is a weld joint used to fill in the space between the adjoining edges of butt and other weld types except lap.

Ques.395: Why is a surfacing weld different from the other weld types?

Answer: Because it does not join to distinct parts, but instead adds only filler metal to a surface.

Ques.396: What is the difference between a continuous weld and an intermittent weld as the terms apply to a fillet weld of a lap joint?

Answer: A continuous weld would be made along the entire length of the fillet weld, whereas an intermittent weld would only fill the joint along certain portions (usually equally spaced) of the total length.

Ques.397: Why is it desirable to use energy sources for welding that have high heat densities?

Answer: Because the heat is concentrated in a small region for greatest efficiency and minimum metallurgical damage.

Ques.398: Define and distinguish the two terms heat transfer efficiency and melting efficiency in welding.

Answer: Heat transfer efficiency is the ratio of the actual heat received at the work surface divided by the total heat generated by the source. Melting efficiency is the ratio of heat required for melting divided by the heat received at the work surface.

Ques.399: What is epitaxial grain growth, and how is this form of solidification different from that which occurs in casting?

Answer: Epitaxial grain growth occurs when atoms from the molten pool solidify on already existing lattice sites of the adjacent solid base metal.

Ques.400: What is the heat affected zone (HAZ) in a fusion weld?

Answer: The HAZ is a region of base metal surrounding the fusion zone in which melting has not occurred, but temperatures from welding were high enough to cause solid state microstructural changes.

Ques.401: Name the principal groups of processes included in fusion welding.

Answer: Arc welding, resistance welding, oxyfuel welding, and "other." The other category includes EBW, LBW, thermit welding, and others.

Ques.402: Define what an electrical arc is.

Answer: An electrical arc is a discharge across a gap in a circuit. In arc welding, the arc is sustained by a thermally ionized column of gas through which the current can flow.

Ques.403: Electrodes in arc welding are divided into two categories. Name and define the two types.

Answer: Consumable and non consumable electrodes . The consumable type, in addition to being the electrode for the process, also provide filler metal for the welding joint. The non consumable type are made of materials that resist melting, such as tungsten and carbon.

Ques.404: What are the two basic methods of arc shielding?

Answer:
(1) Shielding gas, such as argon & helium; and
(2) flux, which covers the welding operation and protects the molten pool from the atmosphere.

Ques.405: Why is the heat transfer efficiency greater in arc welding processes that utilize consumable electrodes?

Answer: Because molten metal from the electrode is transferred across the arc and contributes to the heating of the molten weld pool.

Ques.406: Describe the shielded metal arc welding (SMAW) process.

Answer: SMAW is an arc welding process that uses a consumable electrode consisting of a filler metal rod coated with chemicals that provide flux and shielding.

Ques.407: Why is the shielded metal arc welding (SMAW) process difficult to automate?

Answer: Because the stick electrodes must be changed frequently, which would be difficult to do automatically. It is much easier to automate the feeding of continuous filler wire, such as in GMAW, FCAW, SAW (Submerged Arc Welding), or GTAW.

Ques.408: Describe submerged arc welding (SAW).

Answer: SAW is an arc welding process that uses a continuous, consumable bare wire electrode, and arc shielding is provided by a cover of granular flux.

Ques.409: Describe electrogas welding (EGW) process and identify its major application.

Answer: EGW is an arc welding process that uses a continuous consumable electrode, either flux-cored wire or bare wire with externally supplied shielding gas, and molding shoes to contain the molten pool.

Ques.410: Why are the temperatures much higher in plasma arc welding than in other AW(Arc Welding) processes?

Answer: Because the arc is restricted in diameter, thus concentrating the energy into a smaller area, resulting in much higher power densities.

Ques.411: Define resistance welding (RW) .

Answer: RW consists of a group of fusion welding processes that utilize a combination of heat and pressure to accomplish coalescence of the two faying surfaces. Most prominent in the group is resistance spot welding.

Ques.412: What are the desirable properties of a metal that would provide good weldability for resistance welding?

Answer: High resistivity, low electrical & thermal conductivity, and low melting point.

Ques.413: Describe the sequence of steps in the cycle of a resistance spot welding operation.

Answer: The steps are:

1. the parts are inserted between electrodes,
2. squeeze the parts between the electrodes,
3. weld, in which the current is switched on for a brief duration (0.1 to0.4 sec),
4. hold, during which the weld nugget solidifies, and
5. the electrodes are opened & the parts removed.

Ques.414: What is resistance projection welding (RPW) ?

Answer: RPW is a resistance welding process in which coalescence occurs at one or morerelatively small points on the parts; the contact points are designed into the geometry of the parts as embossments or projections.

Ques.415: Describe cross-wire welding.

Answer: Cross-wire welding is a form of resistance projection welding used to fabricate welded wire products such as shopping carts and stove grills.

Ques.416: Why is the oxyacetylene welding process favored over the other oxyfuel welding processes?

Answer: Because acetylene and oxygen burn hotter than other oxyfuels.

Ques.417: Define pressure gas welding (PGW) .

Answer: PGW is a fusion welding process in which coalescence is obtained over the entire contact surfaces of the two parts by heating them with an appropriate fuel mixture and then applying pressure to bond the surfaces.

Ques.418: Electron beam welding (EBW) has a significant disadvantage in high-production applications. What is that disadvantage?

Answer: EBW is usually carried out in a vacuum for a high quality weld. The time to draw the vacuum adds significantly to the production cycle time.

Ques.419: Laser beam welding (LBW) and electron beam welding (EBW) are often compared because they both produce very high power densities. LBW has certain advantages over EBW. What are they?

Answer:

1. No vacuum chamber is required and
2. no x-rays are emitted in LBW; and
3. the laser beam can be focused and directed with conventional optical mirrors & lenses.

Ques.420: There are several modern-day variations of forge welding, the original welding process. Name the variations.

Answer:

1. Cold welding,
2. roll welding,
3. and hot pressure welding.

Ques.421: There are two basic types of friction welding. Describe and distinguish the two types.

Answer: The two types are:

(1) continuous-drive friction welding and

(2) inertia friction welding. In continuous-drive friction welding, one part is rotated at a constant speed and forced into contact with the stationary part with a certain force so that friction heat is generated at the interface; when the right temperature is reached, the rotating part is stopped abruptly and the two parts are forced together at forging pressures. In inertia friction welding, the rotating part is connected to a flywheel which is brought up to proper speed; then the flywheel is disengaged from the drive motor and the parts are forced together, so that the kinetic energy of the flywheel is converted to friction heat for the weld.

Ques.422: Distortion (warpage) is a serious problem in fusion welding, particularly arc welding. What are some of the measures that can be taken to reduce the incidence and extent of distortion?

Answer: The following measures can be used to reduce warpage in arc welding:

1. welding fixtures,
2. presetting of the parts in relative orientations to compensate for warpage,
3. heat sinks,
4. tack welding at several points along the joint rather than continuous weld,
5. balance the weld about the neutral axis of the part,
6. selecting proper welding conditions,
7. preheating of base parts,
8. stress relief of the weldment, and
9. proper design of the weldment to minimize warpage.

Ques.423: What are some of the important welding defects?

Answer:

1. cracks,
2. cavities,
3. solid inclusions,
4. incomplete fusion,
5. imperfect shape or contour of weld cross-section.

Ques.424: What are the three basic categories of inspection and testing techniques used for weldments? Name some typical inspections and/or tests in each category.

Answer: The three categories are:

1. visual inspection, which includes dimensional checks and inspection for warpage, cracks, & other visible defects;
2. nondestructive evaluation, which includes dye-penetrant, magnetic particle, ultrasonic, & radiographic tests; and
3. destructive tests, which includes conventional mechanical tests adapted to weld joints, and metallurgical tests.

Ques.425: Identify the factors that affect weldability.

Answer: Factors that affect weldability include:

1. welding process,
2. metal properties (*e.g.*, melting point, thermal conductivity, coefficient of thermal expansion),
3. whether the base metals are similar or dissimilar (dissimilar base metals are generally more difficult to weld),
4. surface condition (surfaces should be clean and free of oxides, moisture, *etc.*), and
5. filler metal and its composition relative to the base metals.

Ques.426: What are some of the design guidelines for weldments that are fabricated by arc welding?

Answer: The guidelines for weldments by arc welding include:

1. Good fit-up of parts to be welded is important to maintain dimensional control and minimize distortion. Machining is sometimes required to achieve satisfactory fit-up.
2. The design of the assembly must provide access room to allow the welding gun to reach the welding area.
3. Whenever possible, design of the assembly should allow flat welding to be performed, as opposed to horizontal, vertical, or overhead arc welding positions.

Ques.427: How do brazing and soldering differ from the fusion welding processes?

Answer: In brazing and soldering, no melting of the base metal(s) occurs.

Ques.428: How do brazing and soldering differ from the solid state welding processes?

Answer: In brazing and soldering filler metal is used, whereas in solid state welding no filler metal is added.

Ques.429: What is the technical difference between brazing and soldering?

Answer: In brazing the filler metal melts at a temperature above 450^0C . In soldering the filler metal melts at a temperature of 450^0C or below.

Ques.430: Under what circumstances would brazing or soldering be preferred over welding?

Answer: Brazing or soldering might be preferred if:

1. the base metals have poor weldability,
2. the components cannot tolerate the higher heat & temperatures of welding,
3. production rates might be faster & less expensive than welding,
4. joint areas might be inaccessible for welding but brazing or soldering is possible, and
5. the high strength of a welded joint is not a requirement.

Ques.431: What are the two joint types most commonly used in brazing?

Answer: Butt and lap joints.

Ques.432: Certain changes in joint configuration are usually made to improve the strength of brazed joints. What are some of these changes?

Answer: In butt joints, the butting surface areas are increased in various ways such as scarfing or stepping the edges. In brazed or soldered lap joints, the overlap area is made as large as possible.

Ques.433: The molten filler metal in brazing is distributed throughout the joint by capillary action. What is capillary action?

Answer: Capillary action is the physical tendency of a liquid to be drawn into a small diameter tube or other narrow openings in spite of the force of gravity. It is caused by the adhesive attraction between the liquid molecules and the solid surfaces that define the narrow openings.

Ques.434: What are the desirable characteristics of a brazing flux?

Answer: A brazing flux should do the following:

1. inhibit the formation of oxides and other unwanted by products during the brazing,
2. have a low melting temperature,
3. have a low viscosity when melted,
4. be a good wetting agent, and
5. protect the joint until solidification occurs.

Ques.435: What is dip brazing?

Answer: The parts to be brazed are dipped into a molten salt or molten metal bath which supplies the heating for the operation.

Ques.436: Define braze welding.

Answer: Braze welding is used for adding braze metal to a more conventional geometry weld joint, such as a V-joint. It differs from the typical brazing operation in that no capillary action occurs. It differs from a conventional welding operation in that no melting of the base metals occurs.

Ques.437: What are some of the disadvantages and limitations of brazing?

Answer: Disadvantages and limitations:

1. the strength of the brazed joint is generally less than that of a welded joint;
2. high service temperatures may weaken a brazed joint;
3. part sizes are limited; and
4. the color of the filler metal is often different than the color of the base metals.

Ques.438: What are the two most common alloying metals used in solders?

Answer: Tin and lead.

Ques.439: Why are rosins as soldering fluxes losing favor ?

Answer: Since natural rosin is insoluble in water, chemical solvents must be used to remove the flux after the joining operation. These chemical solvents are hazardous to the environment and to humans.

Ques.440: What are the functions served by the bit of a soldering iron in hand soldering?

Answer: The functions include:

1. provide heat to the parts,
2. melt the solder,
3. convey solder to the joint, and
4. withdraw excess solder from the joint.

Ques.441: What is wave soldering?

Answer: Wave soldering involves the flow of molten solder onto the underside of a printed circuit board to provide soldered connections between the component leads that project through holes in the boards.

Ques.442: List the advantages often attributed to soldering as an industrial joining process?

Answer: Advantages of soldering:

1. lower heat energy required than brazing or welding,
2. various heating methods available,
3. good electrical & thermal conductivity of the joint,
4. capable of making air-tight & liquid-tight joints, and
5. ease of repair & rework.

Ques.443: What are the disadvantages and drawbacks of soldering?

Answer: (1) low mechanical strength unless reinforced and (2) elevated service temperatures can weaken the joint.

Ques.444: An adhesive must cure in order to bond. What is meant by the term curing?

Answer: Curing is the chemical reaction in which the adhesive transforms from liquid to solid and in the process forms the surface attachment between the two adherends.

Ques.445: What are some of the methods used to cure adhesives?

Answer: The curing methods include: chemical reaction between two components of the adhesive (*e.g.*, epoxies), heating of the adhesive, use of ultraviolet light, and application of pressure.

Ques.446: Name the three basic categories of commercial adhesives.

Answer: The categories are:

(1) natural adhesives (*e.g.*, starch, collagen);

(2) inorganic adhesives (*e.g.*, sodium silicate); and

(3) synthetic adhesives (*e.g.*, thermoplastic and thermosetting polymers such as epoxies and acrylics).

Ques.447: Identify some of the advantages of adhesive bonding compared to alternative joining methods.

Answer: Advantages of adhesive bonding:

1. applicable to a wide variety of materials - similar or dissimilar,
2. fragile parts can be joined,
3. bonding occurs over entire surface area of joint,
4. certain adhesives are flexible after curing, thus permitting them to tolerate strains encountered in service,
5. low curing temperatures,
6. some adhesives are suited to sealing as well as bonding, and
7. simplified joint design.

Ques.448: What are some of the limitations of adhesive bonding?

Answer: Limitations of adhesive bonding:

1. adhesively bonded joints are generally not as strong as other joining techniques,
2. the adhesive must be compatible with the adherend materials,
3. service temperatures are limited,
4. surfaces to be bonded must be very clean,
5. curing times can limit production rates,
6. inspection of the bond is difficult.

Ques.449: What are the distinguishing characteristics of metallic bonding? Discuss "cohesive energy" and " electron affinity".

Answer: The metallic state can be visualized as an array of positive ions, with a common pool ofelectrons to which all the metal atoms have contributed their outer electrons. These electrons have freedom to move anywhere within the crystal and make the metallic bonds non directional. **'Cohesive energy'** : In a chemical bond, it is the energy required to dissociate a solid into isolated atoms or molecules as appropriate. **'Electron affinity'** : In a system of a neutral atom and an extra electron, when the extra electron is attracted from the infinity to the outer orbit of the neutral atom, the work done is known as electron affinity of the atom.

Ques.450: What are the point, line and surface imperfections found in solid materials

Answer: **Point imperfection**: They are imperfect point-like regions in the crystal. One or two atomic diameters is the typical size of a point imperfection. A substitutional impurity and an interstitial impurity are examples of point imperfection.

Line imperfection: Displacement with a curved boundary produces a mixed dislocation line at the boundary.

Surface Imperfection: These are two dimensional and refer to a region of distortions that lies around a surface having a thickness of a few atomic diameters.

Ques.451: What is the purpose of zone refining? In a binary phase diagram (pressure omitted), what is the maximum number of phases that can coexist for at least one degree of freedom?

Answer: Zone Refining : Zone refining process is based on the fact that the first solid to crystallize in a two component system is generally purer than the liquid as impurity by repeating the sequence of operations a few times. Max. no. of Phases: Two.'

Ques.452: (A) What are major differences in the processes & purposes of hardening (by quenching) and tempering?

Answer: Tempering is a heat treatment technique for metals & alloys. In steels, tempering is done to "toughen" the metal by transforming brittle martensite into bainite or a combination of ferrite and cementite. Precipitation hardening alloys, like many grades of aluminum and super alloys, are tempered to precipitate intermetallic particles which strengthen the metal. The brittle martensite becomes strong & ductile after it is tempered. Carbon atoms are trapped in the austenite when it is rapidly cooled, typically by oil or water quenching, forming the martensite. The martensite becomes strong after being tempered because when reheated, the microstructure can rearrange and the carbon atoms can diffuse out of the distorted BCT structure. After the carbon diffuses, the result is nearly pure ferrite.Precise control of time and temperature during the tempering process are critical to achieve a metal with well balanced mechanical properties. Quenching is most commonly used to harden steel by introducing martensite, in which case the steel must be rapidly cooled through its eutectoid point, the temperature at which austenite becomes unstable. In steel alloyed with metals such as nickel and manganese, the eutectoid temperature becomes much lower, but the kinetic barriers to phase transformation remain the same. This allows quenching to start at a lower temperature, making the process much easier. High speed steel also has added tungsten, which serves to raise kinetic barriers and give the illusion that the material has been cooled more rapidly than it really has. Even cooling such alloys slowly in air has most of the desired effects of quenching.

Ques.453: What is Hall effect? Briefly discuss the physical origin and uses of Hall effect.

Answer: If a conducting bar is placed in a magnetic field B⊥ to its axis and if a current flows through the bar in axial direction, then an electric field E is developed that is ⊥ to both I and B. This effect is known as Hall Effect. It is used to get carrier density and the sign of the carriers involved (whether holes or electrons).

Ques.454: Explain the following:

(i) dielectric loss

(ii) dielectric break down

(iii) local electric field

(iv) polarizability.

Answer: *(i)* **Dielectric loss**: These losses occur due to electrons hopping from one lattice site to another in transition metal oxides.

(ii) **Dielectric Breakdown**: Dielectric Breakdown of a dielectric material is due to the excitation of electrons into the conduction band across the energy gap under conditions of excessive voltage, resulting in an avalanche of conducting electrons and consequent physical breakdown.

(iii) **Local electric field**: It is the sum of the electric field from external sources and the field of the dipoles within the specimen. It is an idealized field measured under certain specified conditions.

(iv) **Polarization**: The dipole moment per unit volume of the solid is the sum of all the individual dipole moments within that volume and is called 'Polarization' P of the solid.

Ques.455: Distinguish soft magnetic material from hard magnetic material in respect of hysteresis losses, eddy current losses & domain wall motion with suitable examples .

Answer: Soft magnetic materials should have low hysteresis losses and low eddy current losses. Easy domain wall motion is a key factor in keeping the hysteresis losses to a minimum. Increasing the electrical resistivity of the magnetic medium reduces eddy current losses. Hard magnetic materials must retain a large part of their magnetization on removal of the applied field. Obstacles to domain wall motion should be provided in permanent magnets so that the energy product high residual induction Br times larger coercive force Hc is large.

Ques.456: What are ferrites and ferrox cubes? How are mixed ferrites prepared for industrial uses? Give an account of the applications of ferrites pointing out their advantages over a ferromagnetic material.

Answer: Ferrites are ceramic compounds containing trivalent metal having three valence electrons, iron and oxide of divalent element *i.e.*, metals having two free electrons e.g., MFe_2O_3 where M indicates transition elements such as Co, Ni, Mn, divalent iron, Zn and Cu. They are mainly refractory materials in which all of the valence electrons can be considered as being tied up in an ionic bonding. They are unavailable for construction and unable to generate eddy currents. Eddy currents not only dissipate energy but also dampen mechanical & electrical vibrations. They therefore reduce the usefulness of ferromagnetic materials for high frequency applications. The dielectric losses of ferrites are very low at high frequencies because they have electrical resistivity. The ferrites are therefore used as ferromagnetic materials for high frequency applications such as TV tubes, memory devices, high-speed switches, transformers, microwave applications. Mixing powdered oxides, compacting and sintering at elevated temperatures make them more useful. .

Ques.457: What are the objectives of annealing? Discuss the different annealing processes? Is spheroidising different from annealing?

Answer: A slow cooling rate from the eutectoid temperature yields course pearlite – a mixture of relatively coarse crystals of ferrite & cementite.Such a slow cooling is called annealing. Here,sufficient time is available for the carbon in the austenite to diffuse and redistribute itself to 0.02% in ferrite & 6.67% in cementite. During annealing, effective transformation temperature is low, where the rate of growth is rapid compared to the rate of nucleation. Thus, spheroidising is obviously different from annealing.

Ques.458: Distinguish with suitable examples the following:

***(i)* Rolling and Forging *(ii)* Extrusion and Wire drawing.**

Answer: *(i)* Rolling and Forging Rolling: A pair of cylindrical rollers made of iron or steel rotate in opposite direction with a gap between them which is smaller than the cross section of the piece which is to be rolled. The work piece is inserted into the gap and as it passes between the rolls, it is squeezed, its cross section being progressively reduced. Since the working volume remains constant, the result of passing through the rolls would be lengthening of the work piece and a corresponding reduction and shaping of the cross section. Rolling provides the cheapest & most efficient method of reducing the cross sectional area of a piece of the material in such a way that the final thickness is uniform throughout the long lengths of the product. Forging: Forging is shaping of metal either by impact or steady compression between a hammer or ram and anvil. Related to hammering or pressing of metal, the minute difference between the two is the speed of pressure application. Hammering process makes use of hammer that is energised by gravity, air or stream and the repeated blows of vertically guided ram on metal resting on the anvil, cause the metal to change its shape. Cold forging processes are used when it is necessary to develop strength & hardness in a component, have a bright, clean finish, eliminate forging scale, and eliminate decarburisation. This method is used for making bolts, nails ,nuts etc .

In the case of hot forging the metal to be forged is heated first. In this case the finish is not bright and clean. Hot forging is used for making gear, crankshaft, bolts, rivets, and couplings.

(ii) Extrusion and Wire drawing- Extrusion: It is a metal working process that produces long lengths of uniform cross sectional area from a metal billet by causing the latter to flow under hydrostatic pressure through a restricted die or opening. It is a hot working process; starting with cast billets and producing wrought sections & tubes in one stage. A heated cylindrical billet is placed in the container and forced out through a steel die by a ram or plunger. Mainly used for manufacturing rods, tubes (circular, rectangular and hollow forms).

Drawing: Refers to shaping in a punch press of flat blanks of sheet metal into various shapes. This process is basically forcing the flat sheet of metal into a die cavity with a punch. The force exerted by the punch must be sufficient to draw the metal over the edge of the tie die opening and into the die. The metal flow is similar to a viscous fluid. Deep drawing is a special case of drawing when the length of the object to be drawn is deeper than its width. For example washing machine tubes, aluminium milk churns.

Ques.459: Describe briefly the following fabrication processes:

***(i)* Metallization.**

***(ii)* Photolithography.**

***(iii)* Single crystal growth.**

***(iv)* Casting.**

Answer:

(i) **Metallization**: It is the process of providing electrical connections between different parts of the circuit. Aluminium is commonly used. It has a high electrical conductivity and a low melting point for easy evaporation during vacuum deposition.

(ii) **Photolithography**: Used in the manufacture of I.C.S, it is the process of transferring geometrical shapes on a mask to the surface of a silicon wafer. Photomask is prepared first.

(iii) **Single crystal growth**: The single crystals are grown either by the Czochralski (CZ) method or by the Float Zone (FZ) method. It consists of a furnace with a gradient in temperature. The main parts are crucible, the susceptor, the heating element, power supply and the seed shaft.

(iv) **Casting**: Fe-C alloys with more than 2% carbon are called cast irons. On crossing the liquidus, proeutectic crystallizes first. On passing through the eutectic temperatures, the liquid of eutectic composition decomposes to a mixture of austenite and Cementite. On further cooling austenite decomposes to pearlite.

Ques.460: Why a covalent bond is directional?

Answer: The covalent bond is formed as a result of pairing of two electrons in the atomic orbitals of two atoms. The bond then should lie along the direction of the overlapping of atomic orbital *i.e.* the bond only occurs in the direction of the shared pairs of electrons. Hence the covalent bonds will have strong directional preferences unlike ionic and metallic bonds.

Ques.461: What are the different types of point defects? How are they caused?

Answer: Point defects are :

(i) Vacancies

(ii) Interstitialcies

(iii) Compositional defects

(iv) Electronic defects defects

(i) **Vacancies:** This refers to a missing atom or a vacant atomic site due to absence of a matrix atom. Missing of one cation and one anion ion in an ionic crystal is called Schottky imperfection. Electrical neutrality is maintained in this type of imperfection which is observed in alkali halides.

(ii) **Interstitialcies:** An extra atom(substantially smaller than the parent atoms) enters the interstitial void or space between the regularly positioned atoms. The vacancy and interestitialcy are, therefore inverse phenomena. Displacement of cation ions from a lattice site into the void space is called Frankel imperfection and this results in creation of vacancy. An imperfection do not affect the overall electrical neutrality of the crystal. Compositional defect:

(iii) **Substitutional impurity:** Presence of foreign atom in place of a matric atom is called substitutional impurity. If a foreign atom occupies a vacant position within the crystal lattice, this defect is known as interstitial impurity.

(iv) **Electronic defects**: Errors in charge distribution in solids are termed as electronic defects. There is departure from the normal regularity of charge distribution. This effect is responsible for the operation of p-n junction and transistors.

The point defects are formed by :

1. Thermal fluctuations during preparation of crystals.
2. Quenching(quick cooling) from a higher temperature .
3. Severe deformation like hammering or rolling.
4. External bombardment by atoms or high energy particles .

Ques.462: How do temperature and impurities affect electrical resistivity of metals?

Answer: Temperature effect on resistivity: Any increase in temperature of a conductor increases thermal agitation of the metallic ions as they vibrate about their mean positions. This reduces the mean free path and restricts the flow of electrons, thus reducing the conductivity of the material & increasing its resistivity

Effect of impurities on resistivity:

A small percentage of impurities can result in significant increase in resistivity i.e. decrease in the conductivity.

Ques.463: What are properties of an ideal electrical insulating material? What are the various products and application of mica?

Answer: The following are the properties of an ideal(electrical) insulating material:

1. Low dielectric constant(high dielectric constant if used in capacitors for storing energy.)
2. Low power factor.
3. High dielectric strength.
4. High insulation resistance.

5. Adequate mechanical strength to face service condition.
6. Resistance to heat and temperature.
7. Low moisture absorption or water proof.
8. Good chemical stability and chemical proof.
9. Fire proof.
10. High volume and surface resistivity.
11. Good surface finish and machining.

Products and applications of mica:

(i) **Sheet mica**: Primarily used in electrical equipment and applications in the form of washers, spacers, sleeves, tubes etc, where high voltages must be withstood. Widely used in high reliability capacitors due to its high dielectric strength are also used in microwave window and x-ray tube applications.

(ii) **Pasted mica**: Prepared by bonding loose mica splittings with various resins and glues into the form of hard plates or flexible sheets. Pasted mica is used as flat segment plate in rotary machinery and as moulding plate where complex shapes are required such as commutators, v-rings and channels.

(iii) **Mica tape and wrappers**: used for insulating high voltage coils, motor armatures and other area of rotating machinery.

(iv) **Heater plate**: used in percolators and similar commercial appliances.

(v) **Reconstituted mica paper**: used in high temperature transformers, capacitors and motors.

(vi) **Glass bonded mica and ceramoplatics**: used in electrical & electronic system where the insulation requirements are preferably low dissipation factor at high frequencies, a high insulation resistance & dielectric breakdown strength along with extreme dimensional stability e.g. telemetering communication plates, moulded printed circuitry relay spacers.

Ques.464: What is dielectric strength? Explain the various causes & processes which give rise to different types of dielectric break down.

Answer: **Dielectric strength**: It is the maximum voltage/ field gradient(voltage per unit thickness) which the dielectric can withstand without failure / breakdown.

Different types of breakdown:

(i) **Intrinsic breakdown**: It is due to the excitation of electrons into the conduction bandacross the energy gap under the condition of excessive voltage. The excite delectrons can excite more electrons in turn, resulting in an avalanche of conducting electrons. The impurities in the dielectric can create additional energy levels that lie in the energy gap and can help in the excitation of electrons into the conduction band.

(ii) **Thermal breakdown**: It is due to the attainment of an excessive temperature in the

dielectric. If the heat dissipated is less than the heat generated, there is a progressive increase in the temperature of the dielectric, which may melt eventually.

(iii) **Defect breakdown**: It is due to cracks and pores at the surface. Moisture from the atmosphere can collect on the surface and result in breakdown. Glazing is done on ceramic insulators to make surface non-absorbent.

Ques.465: What are the most important properties of permanent magnetic materials?

Answer: The most important properties of permanent magnetic materials are :

(i) **Remanence(Residual magnetism(Br)**: It is the intensity of the residual magnetism after the magnetism field(H) has been removed.

(ii) **Coercive force(Hc)**: It is the resistance of magnetic material to demagnetization by electromagnetic techniques.

(iii) Energy product value(BH)max: It is a measure of the amount of magnetic energy stored in a magnet after the magnetising field is removed. The requirement of permanent magnetic material is that the product of B and H must be maximum possible and it occurs when the product of CD and DE on the demagnetising curve is at a maximum. Depending upon the requirement of remanence and coercive force, the permanent magnetic material is selected.

Ques.466: What are ferrites? Describe their properties & uses .

Answer: Ferrites are ceramic compounds containing trivalent metal having three valence electrons, iron and oxide of divalent element i.e. metals having two free electrons. e.g. MFe2O3 , where M indicates transition elements such as Co, Ni, Mn, divalent iron, Zn and Cu.

They are mainly refractory materials with following properties and uses:

1. A high permeability.
2. Very low dielectric loss at high frequencies.
3. Very high electrical resistivity(more than 105Wcm).
4. Low eddy current losses.
5. High hysteresis loss.
6. Brittle in nature.
7. Difficult to machine due to brittle nature. The dielectric losses of ferrites are very low at high frequencies because they have electrical resistivity. The ferrites are therefore used as ferromagnetic materials for high frequency applications such as TV tubes, memory devices, high speed switches, transformers, microwave applications.

Ques.467: What is the basis of classification of hot and cold working? What are the advantages and disadvantages of cold working over hot working?

Answer: **Cold and hot working**: Changing the shape of material by extrusion, forging, rolling, drawing involves plastic deformation of metals. When the deformation is carried out at temperature below the recrystallization temperature, it is called **cold working** like cold rolling, drawing, pressing, spinning, impact extrusion *etc.* When the deformation is at temperature higher than the recrystallization temperature the process is known as **hot working** like forging, rolling, extrusion etc .

Advantages of cold working over hot working :

1. No heat is required.
2. Better surface finish is obtained.
3. Superior dimension control.
4. Improved strength and hardness properties.
5. Contamination problems are minimised.
6. Better reproducibility and interchangeability of parts.
7. No loss of metal.

Disadvantages of cold working over hot working :

1. Higher forces are required for deformation.
2. Heavier and more powerful equipment is required.
3. May produce undesirable residual stresses.
4. Not clean and scale free surfaces.
5. Less ductility is available.
6. Strain hardening occurs(requires immediate annealing).
7. More energy is required than hot working.

Ques.468: What are the functions of oxide layer in high quality IC (Integrated Circuits)?

Answer: Silicon has the unique ability to be oxidized into silica, which produces a chemically stable, protective and insulating layer on the surface of water.

The functions of the oxide layer are :

1. To mask against diffusion or ion-implant,
2. To passivate the surface electrically and chemically.
3. To isolate one device from another, and
4. To act as a component in MOS devices.

Ques.469: Explain the process of extrusion. What are its applications?

Answer: In this process a round heated billet of metal placed in a container is forced out through a die in a press. This process is generally used for manufacture of rods, tubes, brass cartridges, rail, automobile and aircraft parts. Depending upon the material and the end use there are four different types of extrusions as follows:

(i) **Direct extrusion**:The billet is placed in a container & forced through the die by a piston .

(ii) **Indirect extrusion**: The extruded part is first forced through the die and then passes through ram stem. The billet is pressed between the container and the die in case of indirect extrusion. In case of direct extrusion the billet is placed between die and

the ram.

(iii) **Tube extrusion**: The billet is placed inside a container between the die and mandel. The mandel is pressed by ram and the metal is extruded through the die. This process is used for making steel tubes. The billets are heated to about 1300 ^{0}C except low carbon steel tubes which are cold worked.

The advantages of this process are :

(a) It is comparatively a cheap process.

(b) Low wall thickness can be obtained by this process.

(iv) **Impact extrusion**: The blank is placed over the die and pressed by the punch. The metal is squirted upward round the punch. The thickness of the part is fixed by the gap between the punch and the die. This is a cold working process and is mainly used for non ferrous metals like zinc, lead, tin and aluminium alloy for the manufacture of collapsible tubes for shaving cream, tooth pastes etc.

Ques.470: What are the objectives of heat treatment processes? Describe the hardening process and explain its various stages.

Answer: The heat treatment is generally adopted for the following purposes :

1. To refine grain structure
2. To improve machinability
3. To improve hardness and strength
4. To relieve internal stresses developed during cold working, welding, casting, forging etc.
5. To improve mechanical properties like tensile strength, ductibility and toughness etc.
6. To increase heat, wear and corrosion resistance of materials.
7. To reduce corrosion rate.
8. To improve electrical and magnetic properties.
9. To soften metals for further(cold) working as in wire drawing or cold rolling.
10. To homogenise the structure; to remove coring or segregation.
11. To spheroidize tiny particles by diffusion.

Stages of heat treatment process:

All heat treatment processes consist of 3-main steps:

(i) Heating of metal/alloy to the predetermined(definite) temperature.

(ii) Holding(or soaking) of the metal at that temperature for a sufficient period to allow necessary changes(e.g. austenitizing) to occur *i.e.* structure becomes uniform throughout the section.

(iii) Cooling at a predetermined rate necessary to obtain desired properties associated with changes in nature, form, size and distribution of grains .

Ques.471: Differentiate between "chemical vapour deposition" and "lithography" in the fabrication of ICs. How does addition of copper help in reducing electromigration in the process of metallization?

Answer: Chemical vapor deposition (CVD) is a chemical process used to produce high-purity, high performance solid materials. The process is often used in the semiconductor industry to produce thin films. In a typical CVD process,the wafer (substrate) is exposed to one or more volatile precursors, which react and/or decompose on the substrate surface to produce the desired deposit. Frequently, volatile by-products are also produced, which are removed by gas flow through the reaction chamber. Now adding of copper increases the no. of electrons and hence the collisions increase among the electrons. The overall effect results in reducing the electromigration in the process of metallization.

Whereas Photolithography (also called optical lithography), which is one of the kinds of lithography is a process used in micro fabrication to selectively remove parts of a thin film (or the bulk of a substrate). It uses light to transfer a geometric pattern from a photo mask to a light-sensitive chemical (photo resist, or simply "resist") on the substrate. A series of chemical treatments then engraves the exposure pattern into the material underneath the photo resist. In a complex integrated circuit (for example, modern CMOS), a wafer will go through the photolithographic cycle up to 50 times.

Ques.472: Explain the mechanism of ferromagnetism. On the basis of this explanation how will you explain hystereris and curie point?

Answer: Ferromagnetism is the basic mechanism by which certain materials (such as iron) form permanent magnets and/or exhibit strong interactions with magnets. Thus, an ordinary piece of iron generally has little or no net magnetic moment. However, if it is placed in a strong enough external magnetic field, the domains will re-orient in parallel with that field, and will remain re-oriented when the field is turned off, thus creating a "permanent" magnet. This magnetization as a function of the external field is described by a hysteresis curve. Although this state of aligned domains is not a minimal-energy configuration, it is extremely stable and has been observed to persist for millions of years in seafloor magnetite aligned by the Earth's magnetic field (whose poles can thereby be seen to flip at long intervals). The net magnetization can be destroyed by heating and then cooling (annealing) the material without an external field. As the temperature increases,thermal motion,or entropy,competes with the ferromagnetic tendency for dipoles to align.

When the temperature rises beyond a certain point, called the Curie temperature, there is a second-order phase transition and the system can no longer maintain a spontaneous magnetization, though it still responds paramagnetically to an external field. Below that temperature, there is a spontaneous symmetry breaking and random domains form (in the absence of an external field). The Curie temperature itself is a critical point, where the magnetic susceptibility is theoretically infinite and, although there is no net magnetization, domain-like spin correlations fluctuate at all length scales.

Ques.473: What are the main sources of electrical resistance in a metal? Discuss the effect of impurity, temperature and alloying on the electrical conductivity of metal.

Answer: The factors that affect the electrical resistance of a metal are impurity, temperature and alloying. The free mean path of an electron is the mean distance it travels between successive collisions. For an ideal crystal with no impurities and imperfections the mean free path at 0 K is infinite. That is, there are no collisions and the electrical conductivity is ideally infinite. Introduction of solute atoms into the crystal results in collisions, decreasing the mean free path and the conductivity. At temperatures above 0 K the atoms vibrate randomly about their mean positions. These vibrations, destroys the initial periodicity of a crystal and interferes with the electron motion. Consequently,the free mean path & conductivity decreases, with increasing temperature. Pure metal has less electrical resistivity than alloys. Alloys have a higher electrical resistivity. As resistivity increases the conductivity decreases.

Ques.474: Explain why nichrome and not copper is used as heating element where as manganin is used as standard resistance.

Answer: For heating elements, the primary requirements are high melting point, high electrical resistance and low thermal expansion. The last two requirements help in reducing thermal fatique due to repeated heating and cooling. The heating elements should be designed in a way as to allow unhindered expansion and contractions for example, in the form of a coil of wire. Nichrome (80% nickel & 20% chromium) has all above, mentioned requirements and can be used up to 1300°C. So, nichrome and not copper is used as heating element. Also the melting point of copper is 1083°C, which is less than nichrome. For resistor applications, the primary requirements are uniform resistivity, stable resistance, small temperature coefficient of resistance and low thermoelectric potential with respect to copper. A small α minimizes the error in measurement due to variations in ambient temperature. α is defined as $\alpha = 1/R = dR/dT$ where R is the resistance of the alloy at temperature T. For Manganin alloy (87% copper and 13% manganese) α is only 20×10^{-6}/K as against 4000×10^{-5}/K for pure copper. Hence manganin is used as standard resistance.

Ques.475: Distinguish between intrinsic and extrinsic semiconductor.

Answer: In intrinsic semiconductors, the conduction is due to the intrinsic processes characteristics of the crystal, without the influence of impurities. A pure crystal of silicon or germanium is an intrinsic semiconductor. The electrons that are excited from the top of the valence band to the bottom of the conduction band by the thermal energy are responsible for conduction. In extrinsic semiconductors, the conduction is due to the presence of extraneous impurities. The process of deliberate addition of controlled quantities of impurities to a pure semiconductor is called doping. The addition of impurities markedly increases the conductivity of a semiconductor. The conductivity of an intrinsic semiconductor depends on the concentration of charge carriers . The mobility of conduction electrons and holes can be defined as the drift velocity acquired by them under unit field gradient.

Ques.476: Explain the term 'depletion layer' across a p-n junction. How does a p-n junction function as a rectifier?

Answer: The rectifying action of p-n diode can be explained on the basis of the electronic structure of the semiconductor. When a pure semiconductor is doped to become n-type, the Fermi level shifts up from the middle of the energy gap towards the donor level. This is so because the position corresponding to probability of occupation moves up due to the relatively high concentration of donor electrons in the conduction band. If the crystal is p-type, the Fermi level shifts down towards the acceptor level. When the same crystal is doped to become n-type on one side and p-type on the other side, the Fermi level has to be constant throughout the crystal in thermal equilibrium. This results in the electron energy levels at the bottom of the conduction band in the n-part to be lower than those in the p-part, by an amount equal to the contact potential . The contact potential at the junction gives rise to energy barrier or depletion layer. At equilibrium, there is no net current flowing across the p-n junction.The concentration of electrons in the conduction band on the p-side is small.These electrons can accelerate down the potential hill across the junction to the n-side resulting in a current Io, which is proportional to their number.The concentration of electrons in the conduction band on the n-side is large in comparison, due to the donor contribution. However, only a small number of these electrons can flow to the p-side across the junction as they face a potential barrier. If an external voltage Vi is now applied to the crystal such that the p-side becomes positive with respect to the n-side the electron energy levels will change . The barrier at the junction is now lowered by an amount eVi resulting in a greatly enhanced current flow in the forward direction that is from the n-side to p-side. This change in barrier does not affect the flow of electrons in the reverse direction, from the p-side to the n-side, as the flow here is still down the potential hill. So the applied voltage causes a large net current flow in the forward direction. If an external voltage Vi is applied in the reverse direction the potential barrier for electrons at the junction is increased by an amount eVi . This would drastically reduce the current flow from the n-side to p-side. It is seen that the forward current increases exponentially and the reverse current remains a constant at a small value. This characteristic explains how a p-n junction can act as a rectifier.

Ques.477: What is piezoelectricity? What are different applications in which piezoelectricity is used.

Answer: Piezoelectricity: Piezoelectricity provides us a means of converting electrical energy to assume new geometric positions and the mechanical dimensions of the substance are altered. This phenomenon is called electrostriction. The reverse effect i.e. production of polarization by the application of mechanical stresses can take place only if the lattice has no center of symmetry, this phenomenon is known as piezoelectricity. Example are – Rochelle salt, Quartz, Barium titante.

Applications: Piezoelectric materials serve as a source of ultrasonic waves. At sea, they are used to measure depth, distance of shore and position of icebergs & submarines. They are also used in microphones, phonograph pickups, and strain gauges.

Ques.478: Distinguish between ferromagnetic and anti-ferromagnetic materials. Give an example of each class of material. Discuss the various uses of ferrites.

Answer: Ferromagnetic materials are those for which k susceptibility is large and positive. These are strongly attracted by the magnetic field. Examples are iron, cobalt, nickel etc. An important feature of ferromagnetic materials is that they can retain their magnetism even after the magnetising field has been removed, i.e. they can become permanent magnets. Ferromagnetic materials are those with spontaneous magnetic alignment. Ferromagnetism is the property of a material to be strongly attracted to a magnetic field and to become a powerful magnet.

Antiferromagnetic materials are the materials in which almost all magnetic dipoles are linked up antiparallel to each other. Its susceptibility increases with increase in temperature, until a critical temperature is reached, beyond which it becomes paramagnetic. Some of the materials are MnF_2, MnO_2, MnS. These are hardly used in electrical application.

The ferrites are used as ferromagnetic materials for high frequency applications such as TV tubes, memory devices, high-speed switches, transformers, microwave applications.

Ques.479: What are the objectives of heat treatment of metals? What precautions are necessary while heat-treating to avoid defects?

Answer: Objectives of heat treatment : Cause relief of internal stresses developed during cold working, welding, casting,forging etc. Harden and strengthen metals. Improve machinability. Change grain size. Soften metals for further working as in wire drawing or cold rolling. Improve ductility and toughness. Increase, heat, wear and corrosion resistance of materials. Improve electrical and magnetic properties. Homogenise the structure to remove coring or segregation. Spheroidize tiny particles, such as those of Fe2C in steel, by diffusion Some of the precautions to be carried out while heat treating are : The metal/alloy has to be heated to a definite temperature. Holding at that sufficient temperature to allow the change to occur. Tempering relieves residual stresses, improves ductility. Improve toughness, and reduces hardness ; increases % elongation.

Ques.480: In what manner hot worked and cold worked products differ? Describe the hot and cold forging. Compare their properties and economics.

Answer: **Hot working** of a metal is carried out above its re-crystallization temperature. In this case the metal is not strain hardened. Hot worked products have a refined grain structure. Surface finish of hot worked metal is not nearly as good as cold working because of oxidation and scaling. Hot worked products are free from blowholes, internal porosity, and cracks. Hot worked products are more ductile.

Cold working of a metal is carried out below its re-crystallization temperature. In this case the metal is strain hardened. Cold worked products have a distorted grain structure. Surface finish of cold worked metal is good. Cold worked products may have cracks. Cold worked products are less ductile.

Forging is shaping of metal either by impact or steady compression between a hammer or ram and anvil. Related to hammering or pressing of metal, the main difference between the two is the speed of pressure application. Hammering process makes use of a hammer that is energised by gravity, air or steam and the repeated blows of vertically guided ram on metal resting on the anvil, causing the metal to change its shape.

Cold forging processes are used when it is necessary to develop strength and hardness in a component, have a bright, clean finish, eliminate forging scale, and eliminate decarburisation. This method is used for making bolts, nails, nuts.

In the case of **hot forging** the metal to be forged is heated first. In this case the finish is not bright and clean. Hot forging is used for making gear, crankshaft, bolts, rivets and couplings.

Ques.481: Write short notes on the following processes:-

***(i)* Oxidation in processing of electronic materials.**

***(ii)* Epitaxial growth (CVD).**

***(iii)* Ion implantation.**

***(iv)* Soldering and brazing.**

Answer: *(i)* Oxidation – Silicon has the unique ability to be oxidized into silica, which produces a chemically stable, protective and insulating layer on the surface of the wafer. The production of high quality IC's requires a high quality oxide layer. The function of the oxide layer is to mask against diffusion or ion-implant, to passiviate the surface electrically & chemically, isolate one device from another and act as a component in MOS devices. Thermal oxidation is the principal technique and is carried out between 900°C and 1300°C in dry oxygen or steam with the following reactions occurring $Si + O_2 \longrightarrow SiO_2$ Si; $+ 2H_2O \longrightarrow SiO_2 + 2H_2$

(ii) Epitaxial Growth – Refers to the growing of a crystal using another crystal as the substrate, such that the atomic arrangement is continuous across the interface without the formation of a grain boundary. Homoepitaxy refers to silicon grown on silicon; the substrate and the grown crystal usually differ in the level and type of doping. Hetroepitaxy refers to a substantially different composition grown on the substrate. The two common processes of Epitaxial Growth are the chemical vapour deposition (CVD) and the molecular beam epitaxy.

(iii) Ion Implantation – For doping at lower temperatures than those used in diffusion doping, ion implantation is used. Here ionised projectile atoms are introduced into the wafer with enough energy to penetrate the surface. Ion implantation is widely used for shallow doping of n+ regions in the n-channel MOSFET and base region of a bipolar transistor.

(iv) Brazing and Soldering – Both soldering and brazing processes are useful for joining two dissimilar metals. Both brazing alloy and solder have lower melting points than the metals to be joined. Filler metal used in soldering has a melting point below 450^0C whereas that used in brazing has melting point above 450^0C. In brazing process, bonding conditions are set up so that a large amount of diffusion will take place in order to strengthen and improve the bond, whereas in soldering diffusion is of secondary importance. Brazing produces joints stronger than those made by soldering and can be used in service at higher temperatures.

Ques.482: Discuss Bohr's structure of hydrogen atom? Define excitation and ionisation potentials.

Answer: The Rutherford Bohr model of the hydrogen atom (Z = 1) or a hydrogen-like ion (Z > 1), where the negatively charged electron confined to an atomic shell encircles a small positively charged atomic nucleus, and an electron jump between orbits is accompanied by an emitted or absorbed amount of electromagnetic energy h. The orbits that the electron may travel in are black circles; their radius increases as n2,where n is the principal quantum number. The transition depicted here produces the first line of the Balmer series, and for hydrogen (Z = 1) results in a photon of wavelength 656 nm . Excitation is an elevation in energy level above an arbitrary baseline energy state.The ionization potential of an atom or molecule is the energy required to remove an electron from the isolated atom or ion.

Ques.483: Name various types of cubic class of crystalline structures. Explain their characteristics. Give the values of parameters which distinguish them from each other.

Answer: The cubic crystal system (or isometric) is a crystal system where the unit cell is in the shape of a cube. This is one of the most common and simplest shapes found in crystals and minerals. There are three main varieties of these crystals, called "simple cubic" (S.C), "body-centered cubic" (BCC), and "face-centered cubic" (FCC). Note that although the unit cell in these crystals is conventionally taken to be a cube, the primitive unit cell often is not. This is related to the fact that in most cubic crystal systems, there is more than one atom per cubic unit cell.

The simple cubic(S.C) system consists of one lattice point on each corner of the cube. Each atom at the lattice points is then shared equally between eight adjacent cubes, and the unit cell therefore contains in total one atom (1/8 * 8). The body centered cubic system has one lattice point in the center of the unit cell in addition to the eight corner points. It has a net total of 2 lattice points per unit cell ((1/8)*8 + 1).

Finally, the face centered cubic has lattice points on the faces of the cube of which each unit cube gets exactly one half contribution, in addition to the corner lattice points, giving a total of 4 atoms per unit cell ((1/8 for each corner) * 8 corners + (1/2 for each face) * 6 faces). Some of the elementary characteristics of the cubic structures are given below: Characteristics S.C, B.C.C , F.C.C.

Co-ordination number : 6, 8, 12

Atomic radius : a/2, 3a/4, 2/4

Atoms per unit cell : 1, 2, 4

Atomic Packing Factor : 0.52, 0.68, 0.74

Ques.484: Describe various properties & uses of mica and transformer oil.

Answer: Properties and uses of Mica

Mica has a high dielectric strength and excellent chemical stability, making it a favoured material for manufacturing capacitors for radio frequency applications. It has also been used as an insulator in high voltage electrical equipment. It is heat resistant and is commonly used to make quarter and half wave plates. Because mica is resistant to heat it is used instead of glass in windows for stoves and kerosene heaters. It is also used to separate electrical conductors in cables that are designed to have a fire-resistance rating in order to provide circuit integrity. The idea is to keep the metal conductors from fusing in order to prevent a short-circuit so that the cables remain operational during a fire, which can be important for applications such as emergency lighting. K+ ions between layers of mica prevent swelling by blocking water molecules. Aventurine is a variety of quartz with mica inclusions used as a gemstone. Pressed Mica sheets are often used in place of glass in greenhouses. Muscovite mica is the most common substrate for sample preparation for the atomic force microscope. Some brands of toothpaste include powdered white mica. This acts as a mild abrasive to aid polishing of the tooth surface, and also adds a cosmetically pleasing glittery shimmer to the paste. The shimmer from mica is also used in makeup, as it gives a translucent "glow" to the skin or helps to mask imperfections. Mica sheets are used to provide structure for heating wire (like Kanthal, Nichrome, etc.) in heating elements and can withstand up to 900 °C. Another use of Mica is in the production of ultra flat thin film surfaces(e.g. gold surfaces) using mica as substrate. Although the deposited film surface is still rough due to deposition kinetics,the back side of the film at mica-film interface provides ultra flatness, when the film is removed from the substrate.

Mica slices are used in electronics to provide electric insulation between a heat generating component and the heat sink used to cool it.

The Transformer oil helps cool the transformer. Because it also provides part of the electrical insulation between internal live parts, transformer oil must remain stable at high temperatures over an extended period. To improve cooling of large power transformers, the oil-filled tank may have external radiators through which the oil circulates by natural convection. Very large or high-power transformers (with capacities of millions of KVAs) may also have cooling fans, oil pumps, and even oil-to-water heat exchangers. Large, high-voltage transformers undergo prolonged drying processes, using electrical self heating, the application of a vacuum, or both to ensure that the transformer is completely free of water vapor before the cooling oil is introduced. This helps prevent corona formation and subsequent electrical breakdown under load.

Ques.485: Discuss the salient properties and uses of common conducting materials. What types of conductors are used in filament and contact materials?

Answer: The various common conducting materials used are:

(a) **Copper Properties :**

It's highly ductile and malleable.

It can be cast, forged, rolled and machined.

Mechanical working hardens it, but annealing restores it to soft state.

It is a very good conductor of heat and electricity.

Uses :

Soft copper is used to make a variety of winding wires and conductors for producing cables. Hard copper is used to make commutator segments and contact wire.

(b) **Aluminium Properties :**

It possesses high ductility.

It can be readily worked by rolling, drawing, spinning, extruding, and forging.

It has relatively high thermal and electrical conductivities.

It has the lowest resistance per unit weight

Uses :

The steel reinforced aluminium conductors find extensive use in long transmission lines.

Aluminium conductors are particularly suitable for operation in very high temperature range. It is used in domestic wiring, flexible wires, rotor bars of squirrel cage induction motor.

(c) **Tungsten Properties :**

It has very poor resistivity.

It has highest melting point amongst the metals.

It is very hard and has high boiling point.

Uses :

It is suitable for applications requiring high operating conditions such as lamps and valve filament lamps. It is an outstanding material for electrical contacts in certain applications. Tungsten contacts are used in battery ignition systems, vibrators, and electric razors.

Materials with high resistivity and high melting points like Carbon, Tantalum, and Tungsten are used for making the filaments. For contact materials, the factors which affect the choice of the material are contact resistance, contact force, voltage and current. So platinum, palladium, Silver, Gold, Tungsten, Molybdenum, Rhodium are used.

Ques.486: Discuss, briefly, various properties and applications of some common dielectric materials.

Answer: A dielectric material is a substance that is a poor conductor of electricity, but an efficient supporter of electrostatic field. In practice, most dielectric materials are solid. Examples include porcelain (ceramic), mica, glass, plastics, and the oxides of various metals. Some liquids and gases can serve as good dielectric materials. Dry air is an excellent dielectric, and is used in variable capacitors and some types of transmission lines. Distilled water is fairly dielectric. A vacuum is an exceptionally efficient dielectric.

An important property of a dielectric is its ability to support an electrostatic field while dissipating minimal energy in the form of heat. The lower the dielectric loss (the proportion of energy lost as heat), the more effective is a dielectric material. Another consideration is the dielectric constant, the extent to which a substance concentrates the electrostatic lines of flux. Substances with a low dielectric constant include a perfect vacuum, dry air, and most pure, dry gases such as helium and nitrogen. Materials with moderate dielectric constants include ceramics, distilled water, paper, mica, glass and polyethylene. Metal oxides, in general, have high dielectric constants.

The prime asset of high-dielectric-constant substances, such as aluminium oxide, is the fact that they make possible the manufacture of high-value capacitors with small physical volume. But these materials are generally not able to withstand electrostatic fields as intense as low dielectric- constant substances such as air. If the voltage across a dielectric material becomes too great -- that is, if the electrostatic field becomes too intense -- the material will suddenly begin to conduct current. This phenomenon is called dielectric breakdown. In components that use gases or liquids as the dielectric medium, this condition reverses itself if the voltage decreases below the critical point. But in components containing solid dielectrics, dielectric breakdown usually results in permanent damage.

Ques.487: How does "holding or soaking" time affect the properties of steel during heat treatment process?What are the objectives of annealing?

Answer: Holding or Soaking time can be defined as the time for which prolonged heating of a metal is done at a selected temperature. Soaking at a lower solution temperature, however, requires longer holding times to achieve adequate solution treatment within the alloy. A higher soaking temperature allows one to shorten the total holding time while still achieving adequate solution treatment within the alloy.

Basically increasing or decreasing the holding or soaking time in any heat treatment process like that of steel affects and modifies the mechanical properties of the material undergoing the heat process. The holding or soaking time varies the hardness of the material and its strength also.

Annealing, is a heat treatment wherein a material is altered, causing changes in its properties such as strength and hardness. It is a process that produces conditions by heating and maintaining a suitable temperature, and then cooling. Annealing is used to induce ductility, relieve internal stresses, refine the structure and improve cold working properties.

Ques.488: What are ferrites? Where are they used? Discuss two important applications of ferrites.

Answer: Ferrites are a class of chemical compounds with the formula AB2O4, where A and B represent various metal cations, usually including iron. These ceramic materials are used in magnetic components in microelectronics. Ferrites are a class of spinels, materials that adopt a crystal motif consisting of cubic close-packed (FCC) oxides (O^{2-}) with A cations occupying one eighth of the octahedral holes and B cations occupying half of the octahedral holes. The magnetic material known as "ZnFe" has the simple formula $ZnFe_2O_4$, with Fe^{3-} occupying the octahedral sites and half of the tetrahedral sites. The remaining tetrahedral sites in this spinel are occupied by Zn^{2+}. Soft ferrites :Ferrites that are used in transformer or electromagnetic cores contain nickel, zinc, or manganese compounds. They have a low coercivity and are called soft ferrites. Because of their comparatively low losses at high frequencies, they are used in the cores of Switched-Mode Power Supply (SMPS) and RF transformers & inductors. A common ferrite, abbreviated "MnZn" is composed of the oxides of manganese and zinc.

Hard ferrites : In contrast, permanent ferrite magnets (or "hard ferrites"), which have a high remanence after magnetization, are composed of iron and barium or strontium oxides. In a magnetically saturated state they conduct magnetic flux well and have a high magnetic permeability. This enables these so-called ceramic magnets to store stronger magnetic fields than iron itself. They are the most commonly used magnets in radios. The maximum magnetic field B is about 0.35 tesla and the magnetic field strength H is about 30 to 160 kilo ampere turns per meter (400 to 2000 oersteds).

Uses of ferrites :

Ferrite cores are used in electronic inductors, transformers, and electromagnets where the high electrical resistance of the ferrite leads to very low eddy current losses. They are commonly seen as a lump in a computer cable, called a ferrite bead, which helps to prevent high frequency electrical nöise (radio frequency interference) from exiting or entering the equipment.

Early computer memories stored data in the residual magnetic fields of hard ferrite cores, which were assembled into arrays of core memory. Ferrite powders are used in the coatings of magnetic reccrding tapes. One such type of material is iron (III) oxide. Ferrite particles are also used as a component of radar-absorbing materials or coatings used in stealth aircraft and in the expensive absorption tiles lining the rooms used for electromagnetic compatibility measurements. Most common radio magnets, including those used in loudspeakers,are ferrite magnets. Ferrite magnets have largely displaced Alnico magnets in these applications.

It is a common magnetic material for electromagnetic instrument pickups, because of price and relatively high output. However, such pickups lack certain sonic qualities found in other pickups, such as those that use Alnico alloys or more sophisticated magnets.

Ques.489: Outline the domain theory of ferromagnetic materials. How are the domains formed and are influenced by temperature changes?

Answer: The long range order which creates magnetic domains in ferromagnetic materials arises from a quantum mechanical interaction at the atomic level. This interaction is remarkable in that it locks the magnetic moments of neighbouring atoms into a rigid parallel order over a large number of atoms in spite of the thermal agitation which tends to randomize any atomic level order. Sizes of domains range from a 0.1 mm to a few mm. When an external magnetic field is applied, the domains already aligned in the direction of this field grow at the expense of their neighbours. If all the spins were aligned in a piece of iron, the field would be about 2.1 Tesla (T) . A magnetic field of about 1 T can be produced in annealed iron with an external field of about 0.0002 T, a multiplication of the external field by a factor of 5000 . For a given ferromagnetic material the long range order abruptly disappears at a certain temperature which is called the Curie temperature for the material. The Curie temperature of iron is about 1043 K.

Ques.490: Name various types of magnetic materials. List their properties to distinguish them from each other?

Answer: The origin of magnetism lies in the orbital and spin motions of electrons and how the electrons interact with one another. The best way to introduce the different types of magnetism is to describe how materials respond to magnetic fields. This may be surprising, but all matter is magnetic. It's just that some materials are much more magnetic than others.The main distinction is that in some materials there is no collective interaction of atomic magnetic moments, whereas in other materials there is a very strong interaction between atomic moments.

The magnetic behavior of materials can be classified into following five major groups:

— Diamagnetism
— Paramagnetism
— Ferromagnetism
— Ferrimagnetism
— Antiferromagnetism

Materials in the first two groups are those that exhibit no collective magnetic interactions and are not magnetically ordered. Materials in the last three groups exhibit long-range magnetic order below a certain critical temperature. Ferromagnetic and ferrimagnetic materials are usually considered as being magnetic (ie., behaving like iron). The remaining three are so weakly magnetic that they are usually thought of as "nonmagnetic". Diamagnetism

Diamagnetism is a fundamental property of all matter, although it is usually very weak. It is due to the non-cooperative behavior of orbiting electrons when exposed to an applied magnetic field. Diamagnetic substances are composed of atoms which have no net magnetic moments (ie,all the orbital shells are filled & there are no unpaired electrons). However, when exposed to a field, a negative magnetization is produced and thus the susceptibility is negative. Paramagnetism

In this class of materials, some of the atoms or ions have a net magnetic moment due to unpaired electrons in partially filled orbitals. One of the most important atoms with unpaired electrons is iron. However, the individual magnetic moments do not interact magnetically, and like diamagnetism, the magnetization is zero when the field is removed. In the presence of a field, there is now a partial alignment of the atomic magnetic moments in the direction of the field, resulting in a net positive magnetization and positive susceptibility. Ferromagnetism

Unlike paramagnetic materials, the atomic moments in these materials exhibit very strong interactions. These interactions are produced by electronic exchange forces and result in a parallel or antiparallel alignment of atomic moments. Ferrimagnetism

In ionic compounds, like oxides, more complex forms of magnetic ordering can occur as a result of the crystal structure.One type of magnetic ordering is ferrimagnetism. Magnetite is a well known ferrimagnetic material. Antiferromagnetism

If the A and B sublattice moments are exactly equal but opposite, the net moment is zero. This type of magnetic ordering is called antiferromagnetism.

Ques.491: What is hysteresis? What does it signify? Sketch the hysteresis loop for a:

***(i)* Transformer core.**
***(ii)* Strong electromagnet.**
***(iii)* Magnetic tape.**

Answer: Hysteresis is well known in ferromagnetic materials. When an external magnetic field is applied to a ferro magnet, the atomic dipoles align themselves with the external field. Even when the external field is removed, part of the alignment will be retained: the material has become magnetized.

The relationship between magnetic field strength (H) and magnetic flux density (B) is not linear in such materials. If the relationship between the two is plotted for increasing levels of field strength, it will follow a curve up to a point where further increases in magnetic field will result in no further change in flux density. This condition is called magnetic saturation. If the magnetic field is now reduced linearly, the plotted relationship will follow a different curve back towards zero field strength at which point it will be offset from the original curve by an amount called the remanent flux density or remanence.

If this relationship is plotted for all strengths of applied magnetic field, the result is a S shaped loop.The 'thickness' of the middle bit of the S describes the amount of hysteresis, related to the coercivity of the material.

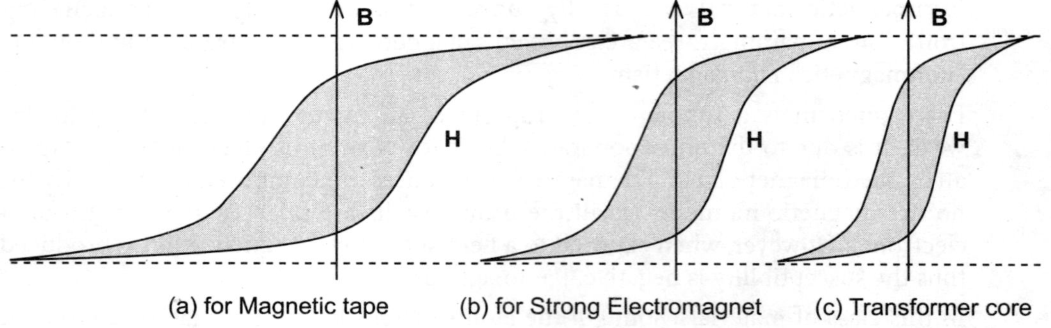

(a) for Magnetic tape (b) for Strong Electromagnet (c) Transformer core

Hysteresis signifies the lagging of magnetization or induction flux density (B) behind the magnetizing force (H) or it is that quality of a magnetic substance due to which energy is dissipated in it on the reversal of its magnetism. The hysteresis loop equals the work which is necessary to reverse the direction of magnetization.

Ques.492: Explain the phenomenon of polarization in dielectric materials. Name various types of polarizations and discuss their frequency dependence.

Answer: Polarization in Dielectric Materials: If a material contains polar molecules, they will generally be in random orientations when no electric field is applied. An applied electric field will polarize the material by orienting the dipole moments of polar molecules. This decreases the effective electric field between the plates and will increase the capacitance of the parallel plate structure. The dielectric must be a good electric insulator so as to minimize any DC leakage current through a capacitor.

The various types of polarization processes are:

a. Electronic polarization

b. Ionic Polarization

c. Orientation Polarization, and

d. Space charge Polarization

Electronic Polarization is very rapid and occurs at all frequencies even in the optical range (~1015 Hz). Ionic Polarization is slower than electronic polarization and if an electric field of frequency in the optical range is now applied, the ions don't respond at all. So at optical frequencies there is no ionic polarization. Orientation Polarization is even slower than the ionic polarization and Orientation polarization occurs, when the frequency of the applied voltage is in the audio range. And the Space charge polarization occurs at machine frequencies only (50-60 Hz).

Ques.493: Outline salient properties of Fermi probability function. Sketch it at two different temperatures. Define Fermi energy. What is its importance?

Answer: At 0 K, the free electrons occupy all the levels up to the Fermi level, leaving all those above it empty. At temperatures above 0 K, due to thermal excitation, there is a finite probability of some of the electrons from below the Fermi level moving to levels above EF.This probability is given by the Fermi-Dirac statistics, which takes into account the quantum restrictions due to the Pauli Exclusion Principle. The probability of occupation P (E) of an energy level E by an electron is given by

$$P(E) = 1 / 1 + \exp[(E - EF)/kT]$$

If EF is independent of temperature, P (E) varies with E. At 0 K, P (E) remains constant at unity with increasing E up to EF, where it follows abruptly to zero. At T1>0 K, some of the electrons just below the Fermi level are thermally excited to higher levels just above EF. So, P (E) is less than one just below the Fermi level and is greater than zero just above the Fermi level. At a still higher temperature, more electrons leave the lower energy levels and occupy higher levels. The cross-over point, where P (E) is 0.5 for different temperatures, occurs at the same value of the energy level which is EF. This is true, provided that EF is independent of temperature, an assumption valid for most ordinary temperatures. Under such conditions, the Fermi level can be defined as that level which has a 50% probability of occupation by an electron at any temperature.

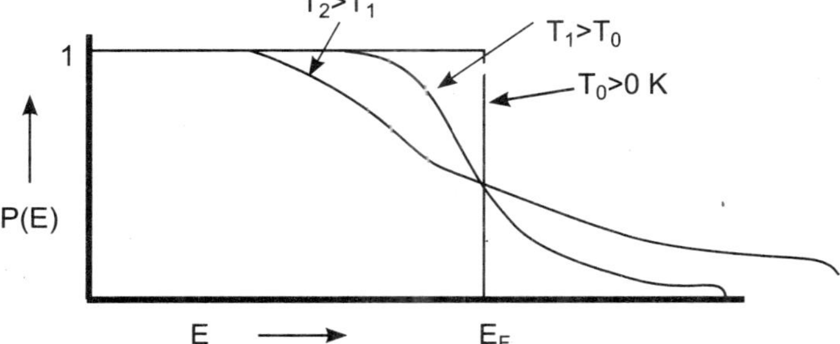

The Fermi energy usually refers to the energy of the highest occupied quantum state in a system of fermions at absolute zero temperature. Thus Fermi Energy refers to the energy of the highest occupied state. What this means is that even if we have extracted all possible energy from a metal by cooling it down to near absolute zero temperature (0 Kelvin), the electrons in the metal are still moving around; the fastest ones would be moving at a velocity that corresponds to a kinetic energy equal to the Fermi energy.

Ques.494: Discuss salient properties of common semiconducting materials. What are their uses? What are semiconducting devices? Where are they used? Give examples.

Answer: The salient properties of semi-conducting materials are:

These are the materials whose valence electrons are bounded somewhat loosely to their atom.

They require energy less than dielectrics and more than good conductors to remove an electron from the parent atom. They have:

1. An empty conduction band .
2. Almost filled valance band .
3. Very narrow energy gap (of about 1eV) separating the two bands. Some of the commonly used semi-conducting materials are:

(a) Germanium Properties ;

It is grey metallic looking lustrous material.

It is brittle and glass like in its mechanical properties.

It crystallizes in the diamond cubic lattice.

Uses :

Germanium is widely used in making semi-conducting devices, rectifiers, photo conductive and photo voltaic cells, memory elements of computers etc.

(b) Silicon Properties :

It's a hard element having bluish-grey metallic lustre .

It has melting point of 1420°C .

It is sensitive to impurities and radiations.

Uses :

Silicon has become an important transistor material. It is used to make devices like Thyristors etc. Also it is used in computer memories. And it can be used in all the applications where germanium is used.

Semiconductor devices exploit the electronic properties of semiconductor materials, principally silicon, germanium, and gallium arsenide. Semiconductor devices have replaced thermionic devices (vacuum tubes) in most applications. They use electronic conduction in the solid state as opposed to the gaseous state or thermionic emission in a high vacuum.Semiconductor devices are manufactured both as discrete devices and as integrated circuits (ICs), which consist of a number of devices manufactured and interconnected on a single semiconductor substrate. Some of the semi conducting devices are diodes, transistors, thyristors, triac etc.

Semiconductor device applications :

All transistor types can be used as the building blocks of logic gates, which are fundamental in the design of digital circuits. In digital circuits like microprocessors, transistors act as on-off switches; in the MOSFET, for instance, the voltage applied to the gate determines whether the switch is on or off.

Transistors used for analog circuits do not act as on-off switches; rather, they respond to a continuous range of inputs with a continuous range of outputs. Common analog circuits include amplifiers and oscillators.

Circuits that interface or translate between digital circuits and analog circuits are known as mixed-signal circuits.

Power semiconductor devices are discrete devices or integrated circuits intended for high current or high voltage applications. Power integrated circuits combine IC technology with power semiconductor technology; these are sometimes referred to as "smart" power devices.

Ques.495: State and explain Hall Effect? What are its applications?

Answer: The Hall Effect refers to the potential difference (Hall voltage) on the opposite sides of an electrical conductor including semiconductors through which an electric current is flowing, created by a magnetic field applied perpendicular to the current.

The ratio of the voltage created to the product of the current and the magnetic field (I*B) divided by the element thickness, is known as the Hall coefficient. It is a characteristic of the material from which the conductor is made, as its value depends on the type, number and properties of the charge carriers that constitute the current.

The Hall Effect comes about due to the nature of the current flow in a conductor.Current consists of the movement of many small charge-carrying "particles" (not necessarily, electrons). Moving charges experience a force, called the Lorentz Force, when a magnetic field is present that is not parallel to their motion.When such a magnetic field is absent, the charges follow an approximately straight, 'line of sight' path. However, when a perpendicular magnetic field is applied, their path is curved so that moving charges accumulate on one face of the material. This leaves equal and opposite charges exposed on the other face, where there is a scarcity of mobile charges. The result is an asymmetric distribution of charge density across the Hall element that is perpendicular to both the 'line of sight' path and the applied magnetic field. The separation of charge establishes an electric field that opposes the migration of further charge, so a steady electrical potential builds up for as long as the current is flowing.

The Hall Effect is very useful as a means to measure both the carrier density and the magnetic field.

Sensors and devices based on Hall Effect are used for:

— Analog multiplication
— Current sensing
— Position and motion sensing
— Automotive ignition and fuel injection
— Wheel rotation sensing
— Electric motor control
— Industrial applications

Ques.496: Discuss the metallurgical factors which affect the quality of a welded joint. Explain soldering and brazing processes.

Answer: The quality of a weld is dependent on the combination of materials used for the base material and the filler material. Not all metals are suitable for welding, and not all filler metals work well with acceptable base materials. Most often, the major metric used for judging the quality of a weld is its strength and the strength of the material around it. Many factors influence this, including the welding method, the amount & concentration of energy input, the base material, the filler material, the flux material, the design of the joint, and the interactions between all these factors.

Soldering is a process in which two or more metal items are joined together by melting and flowing a filler metal into the joint, the filler metal having a relatively low melting point. Soft soldering is characterized by the melting point of the filler metal, which is below 400 °C .The filler metal used in the process is called solder. Soldering is distinguished from brazing by use of a lower melting temperature filler metal; it is distinguished from welding by the base metals not being melted during the joining process. In a soldering process, heat is applied to the parts to be joined, causing the solder to melt and be drawn into the joint by capillary action and to bond to the materials to be joined by wetting action. After the metal cools, the resulting joints are not as strong as the base metal, but have adequate strength, electrical conductivity, and water tightness for many uses.

Brazing is a joining process whereby a filler metal or alloy is heated to melting temperature above 450 °C and is distributed between two or more close fitting parts by capillary action. At its liquid temperature, the molten filler metal and flux interacts with a thin layer of the base metal, cooling to form a strong, sealed joint. By definition the melting temperature of the braze alloy is lower (sometimes substantially) than the melting temperature of the materials being joined. The brazed joint becomes a sandwich of different layers, each metallurgically linked to the adjacent layers.

Ques.497: Explain Casting, Forging and Rolling. Name two components of structures and machines generally made by each of these processes.

Answer: Casting is a manufacturing process by which a liquid material is usually poured into a mold, which contains a hollow cavity of the desired shape, and then allowed to solidify. The solid casting is then ejected or broken out to complete the process. Casting may be used to form hot liquid metals or various materials that cold set after mixing of components (such as epoxies, concrete, plaster and clay). Casting is most often used for making complex shapes that would be otherwise difficult or uneconomical to make by other methods.

Forging is the term for shaping metal by using localized compressive forces. Cold forging is done at room temperature or near room temperature. Hot forging is done at a high temperature, which makes metal easier to shape and less likely to fracture. Warm forging is done at intermediate temperature between room temperature and hot forging temperatures.

Forged parts can range in weight from less than a kilogram to 170 metric tons. Forged parts usually require further processing to achieve a finished part.

Rolling is a metal working process in which metal is deformed by passing it through rollers at a temperature below its recrystallization temperature. Rolling increases the yield strength and hardness of a metal by introducing defects into the metal's crystal structure. These defects prevent further slip and can reduce the grain size of the metal, resulting in Hall-Petch hardening. Rolling is most often used to decrease the thickness of plate and sheet metal.

Ques.498: Write notes on :

***(i)* Eddy current losses.**

***(ii)* Any one alloy of copper & aluminium.**

Answer: *(i)* Eddy Current Losses The core of a generator armature is made from soft iron, which is a conducting material with desirable magnetic characteristics. Any conductor will have currents induced in it when it is rotated in a magnetic field. These currents that are induced in the generator armature core are called EDDY CURRENTS. The power dissipated in the form of heat, as a result of the eddy currents, is considered a loss. Eddy currents, just like any other electrical currents, are affected by the resistance of the material in which the currents flow. And the resistance of any material is inversely proportional to its cross-sectional area.

(ii) Aluminum-Copper Alloys that contain 4 to 5% Cu, with the usual impurities of iron & silicon and sometimes with small amounts of magnesium, are heat-treatable and can reach quite high strength and ductility, especially if prepared from ingot containing less than 0.15% iron. Manganese in small amounts also may be added, mainly to combine with the iron & silicon and reduce their embrittling effect. However, these alloys have poor castability and require very careful gating if sound castings are to be obtained. Such alloys are used mainly in sand casting; when they are cast in metal molds, silicon must be added to increase fluidity and curtail hot shortness, and this addition of silicon substantially reduces ductility.

AI-Cu alloys with somewhat higher copper contents (7 to 8%), formerly the most commonly used aluminum casting alloys, have steadily been replaced by AI-Cu-Si alloys and today are used to a very limited extent. The best attribute of these higher-copper Al-Cu alloys is their insensitivity to impurities, but they have very low strength and only fair castability.

Also in limited use are AI-Cu alloys that contain 9 to 11 % Cu, whose high-temperature strength & wear resistance make them suitable for automotive pistons and cylinder blocks. These alloys usually contain manganese as an impurity because wrought metal scrap is used in preparing them. The manganese has little effect. Very good high-temperature strength is an attribute of alloys containing copper, nickel and magnesium, sometimes with iron in place of part of the nickel.

Ques.499: Explain with suitable examples the ionic, covalent and metallic type of bondings in solids.

Answer: An ionic bond (or electrovalent bond) is a type of chemical bond that can often form between metal and non-metal ions (or polyatomic ions such as ammonium) through electrostatic attraction. Or we can say, it is a bond formed by the attraction between two oppositely charged ions.

For example, common salt is sodium chloride. When sodium (Na) and chlorine (Cl) are combined, the sodium atoms each lose an electron, forming a cation (Na+), and the chlorine atoms each gain an electron to form an anion (Cl-). These ions are then attracted to each other in a 1:1 ratio to form sodium chloride (NaCl).

Na + Cl _ Na+ + Cl- _ NaCl

A covalent bond is a form of chemical bonding that is characterized by the sharing of pairs of electrons between atoms, or between atoms and other covalent bonds. Or we can say that, attraction to repulsion stability that forms between atoms when they share electrons is known as covalent bonding.

Covalent bonding includes many kinds of interactions, including _-bonding, _-bonding, metal-metal bonding, agostic interactions, and three-center two-electron bonds.

For example, in the molecule H_2, the hydrogen atoms share the two electrons via covalent bonding.

Metallic bonding is the electromagnetic interaction between delocalized electrons, called conduction electrons, and the metallic nuclei within metals. In the metallic bond, an atom achieves a more stable configuration by sharing the electrons in its outer shell with many other atoms. Metallic bonds prevail in elements in which valence electrons are not tightly bound with the nucleus, namely metals, thus the name metallic bonding. In this type of bond, each atom in a metal crystal contributes all the electrons in its valence shell to all other atoms in the crystal. This free electron imparts electrical conductivity to metals. For example, metallic bonding is found in metals like zinc.

Ques.500: What are ferrites? Given an account of the applications of ferrites for high frequency transformers and computer memory cores, pointing out their advantages over a ferromagnetic material.

Answer: Ferrites are usually non-conductive ferrimagnetic ceramic compounds derived from iron oxides such as hematite (Fe_2O_3) or magnetite (Fe_3O_4) as well as oxides of other metals. Ferrites are, like most other ceramics, hard and brittle. In terms of the magnetic properties, ferrites are often classified as "soft" and "hard" which refers to their low or high coercivity of their magnetism, respectively.

Ferrite cores are used in electronic inductors, transformers, and electromagnets where the high electrical resistance of the ferrite leads to very low eddy current losses. They are commonly seen as a lump in a computer cable, called a ferrite bead, which helps to prevent high frequency electrical noise (radio frequency interference) from exiting or entering the equipment.

Early computer memories stored data in the residual magnetic fields of hard ferrite cores, which were assembled into arrays of core memory. Ferrite powders are used in the coatings of magnetic recording tapes. One such type of material is iron (III) oxide.

Ferrites have higher resistance than the ferromagnetic material. This makes them suitable for application at high frequencies- even microwave frequencies.

Ques.501: Explain the principle of phase diagram. Discuss some simple binary phase diagrams. What are eutectic systems?

Answer: A phase diagram is a type of graph used to show the equilibrium conditions between the thermodynamically distinct phases. These can be single phase or Binary phase diagram.

Other much more complex types of phase diagrams can be constructed, particularly when more than one pure component are present. In that case concentration becomes an important variable. Phase diagrams with more than two dimensions can be constructed that show the effect of more than two variables on the phase of a substance. Phase diagrams can use other variables in addition to or in place of temperature, pressure and composition, for example the strength of an applied electric or magnetic field and they can also involve substances that take on more than just three states of matter.

The percentage of carbon present and the temperature define the phase of the iron carbon alloy and therefore its physical characteristics & mechanical properties. The percentage of carbon determines the type of the ferrous alloy: iron, steel or cast iron

One type of phase diagram plots temperature against the relative concentrations of two substances in a binary mixture called a binary phase diagram . Such a mixture can be a solid solution, eutectic or peritectic, among others. These two types of mixtures result in very different graphs. An example of a eutectic phase diagram is that of the olivine (forsterite and fayalite) system.A complex phase diagram of great technological importance is that of the iron-carbon system for less than 7% carbon.

The x-axis of such a diagram represents the concentration variable of the mixture. As the mixtures are typically far from being dilute and their density as a function of temperature usually unknown, the preferred concentration measure is mole fraction.

EUTECTIC SYSTEMS

An eutectic mixture is a mixture at such proportions that the melting point is as low as possible, and that furthermore all the constituents crystallize simultaneously at this temperature from molten liquid solution. Such a simultaneous crystallization of a eutectic mixture is known as a eutectic reaction, the temperature at which it takes place is the eutectic temperature, and the composition and temperature at which it takes place is called the eutectic point. In an equilibrium phase diagram for a binary system, the eutectic point is the point at which the liquid phase borders directly on the solid phase.

Ques.502: Discuss different methods of soldering.

Answer: Soldering methods are broadly divided into two types: the partial heating method and the total heating method. In Partial heating method, heat is applied to the package leads and/or PWB in a localized manner.

There are two types of soldering methods:

(a) Soldering iron

(b) Hot air

Soldering with soldering iron : This method consists of using a soldering iron to solder the package leads and printed circuit with wire solder, etc. Determine the soldering temperature of the soldering iron to be used according to the size & shape of the location to be soldered, and the melting point of the solder. If the soldering temperature is too high, the printed circuit may peel off from PWB and various degradation may be caused due to overheating. The actual soldering temperature must be determined according to the thermal capacity of the SMD. It is recommended that the temperature characteristics of the SMD be actually measured to determine the soldering temperature. Whenever possible, use a soldering ironwhose temperature can be adjusted.

Hot air soldering: This method solders the SMD by heating air or N_2 gas with a heater and flowing hot gas from a nozzle onto the joint on the PWB. The temperature is adjusted by adjusting the heat source and/or the flow of gas. Partial heating involves less heat stress on the device and wiring board, but is unsuitable for large volume production. Therefore, this method is mainly used to correct soldering or for devices with a low heat resistance.

In Total heating method, heat is applied to the entire package and/or PWB.

There are five types of such soldering methods:

1. Infrared reflow
2. Convection reflow
3. Infrared convection combined
4. VPS (Vapour Phase Soldering)
5. flow (wave) soldering

Because of excellence in productivity and running cost, these types are widely used. However, this method can place considerable heat stress on the semiconductor device and board.

Ques.503: Give an explanation for the formation of energy bands in a solid.

Answer: The electronic band structure (or simply band structure) of a solid describes ranges of energy that an electron is "forbidden" or "allowed" to have. It is due to the diffraction of the quantum mechanical electron waves in the periodic crystal lattice. The band structure of a material determines several characteristics,in particular its electronic and optical properties.

Any solid has a large number of bands. In theory, it can be said to have infinitely many bands (just as an atom has infinitely many energy levels). However, all but a few lie at energies so high that any electron that reaches those energies escapes from the solid. These bands are usually disregarded.

Bands have different widths, based upon the properties of the atomic orbitals from which they arise. Also, allowed bands may overlap, producing (for practical purposes) a single large band.

Metals contain a band that is partly empty and partly filled regardless of temperature. Therefore they have very high conductivity. The lowermost, almost fully occupied band in an insulator or semiconductor is called the valence band by analogy with the valence electrons of individual atoms. The uppermost, almost unoccupied band is called the conduction band because only when electrons are excited to the conduction band can current flow in these materials. The difference between insulators and semiconductors is only that the forbidden band gap between the valence band and conduction band is larger in an insulator, so that fewer electrons are found there and the electrical conductivity is lower. Because one of the main mechanisms for electrons to be excited to the conduction band is due to thermal energy, the conductivity of semiconductors is strongly dependent on the temperature of the material.

This band gap is one of the most useful aspects of the band structure, as it strongly influences the electrical and optical properties of the material. Electrons can transfer from one band to the other by means of carrier generation & recombination processes. The band gap and defect states created in the band gap by doping can be used to create semiconductor devices such as solar cells, diodes, transistors, laser diodes, and others.

Ques.504: Name two important properties of stainless steel as compared to carbon and alloy steel. What type of stainless steel would you prefer for household utensils?

Answer: Stainless steel is defined as a steel alloy with a minimum of 11.5% chromium content by mass. Stainless steel does not stain, corrode or rust as easily as ordinary steel (it is "stainsless"). It is also called corrosion resistant steel when the alloy type and grade are not detailed, particularly in the aviation industry. There are different grades and surface finishes of stainless steel to suit the environment to which the material will be subjected in its lifetime.Common uses of stainless steel are cutlery and watch straps.

Stainless steel differs from carbon steel by amount of chromium present. Carbon steel rusts when exposed to air and moisture. This iron oxide film is active and accelerates corrosion by forming more iron oxide. Stainless steels have sufficient amount of chromium present so that a passive film of chromium oxide forms which prevents further corrosion.

The stainless steel can be classified as:-

Austenitic: Chromium-nickel-iron alloys with 16-26% chromium, 6-22% nickel (Ni), and low carbon content, with non-magnetic properties (if annealed - working it at low temperatures, then heated and cooled). Nickel increases corrosion resistance. It is hardenable by cold-working (worked at low temperatures) as well as tempering (heated then cooled). Type 304 (S30400) or "18/8" (18% chromium /8% nickel), is the most commonly used grade or composition.

Martensitic: Chromium-iron alloys with 10.5-17% chromium and carefully controlled carbon content, hardenable by quenching (quickly cooled in water or oil) and tempering (heated then cooled). It has magnetic properties. Commonly used in knives. Martensitic grades are strong & hard, but are brittle and difficult to form &weld. Type 420 (S42000) is a typical example.

Ferritic: Chromium-iron alloys with 17-27% chromium and low carbon content, with magnetic properties. Cooking utensils made of this type contain the higher chromium levels. Type 430 is the most commonly used ferritic stainless steel .

All of these stainless steels have following applications:

The austenitic microstructure is most commonly used for knives and cooking utensils. It is very tough, hardened through a process that consists of heating, cooling and heating. It resists scaling and retains strength at high temperatures.

Both ferritics and austenitics are used in kitchenware and household appliances. Austenitics are preferred in the food industry & beverage equipment due to the superior corrosion resistance and ease of cleaning. Type 301, for example, is an austenitic stainless steel, with 17% chromium, 7% nickel, and .05% carbon, and is widely used for institutional food preparation utensils.

Low quality cutlery is generally made out of grades like 409 and 430 (ferritic), while the finest Sheffield cutlery uses specially produced 410 and 420 (martensitic) for the knives, and 304 (austenitic) for the spoons and forks. Grades like the 410/420 can be hardened and tempered so that the knife blades will take a sharp edge, whereas the more ductile 304 stainless is easier to work with and therefore more suitable for objects that have to undergo numerous shaping, buffing and grinding processes.

The best quality stainless steel knife blades have high carbon content, and usually have molybdenum and vanadium in their composition.

Ques.505: What is major difference in the purpose of annealing and normalizing?

Answer: When steel is heated to a point 50 -75^{0}C above its critical range, a condition referred to as "austenite" is produced. If slowly cooled from above its critical temperature, the austenite is broken down and a succession of other conditions are produced, each being normal for a particular range of temperatures. Starting with austenite, these successive conditions are martensite, troostite, sorbite, and finally pearlite.

The most important step in annealing is to raise the temperature of the metal above the critical point, as any hardness that have existed will then be completely removed. Strains that may have been set up through heat treatment will be eliminated when the steel is heated to the critical point, and then restored to its lowest hardness by slow cooling. In annealing, the steel must never be heated more than approximately 28 to 40°C above the critical point. When large articles are annealed, sufficient time must be allowed for the heat to penetrate the metal.

Steel is usually subjected to the annealing process for the following purposes:

— To increase its ductility by reducing hardness and brittleness.

— To refine the crystalline structure and remove residual stresses.

Steel that has been cold worked is usually annealed to increase its ductility. Assuming that the part to be annealed is heated to the proper temperature, the required slow cooling may be accomplished in several ways, depending on the metal and the degree of softness required.

Normalizing, although involving a slightly different heat treatment, may be classified as a form of annealing. This process removes all strains due to machining, forging,bending and welding. Normalizing can only be accomplished with a good furnace, where the temperatures & the atmosphere may be closely regulated and held constant throughout the entire operation. . A reducing atmosphere will normalize the metal with a minimum amount of oxide scale,while an oxidizing atmosphere will leave the metal heavily coated with scale, thus preventing proper development of hardness in any subsequent hardening operation.

The articles are put in the furnace and heated to a point above the critical temperature of the steel. After the parts have been held at this temperature for a sufficient time to allow the heat to penetrate to the centre of the section, they must be removed from the furnace and cooled in still air. Drafts will result in uneven cooling, which will again set up strains in the metal. Prolonged soaking of the metal at high temperatures must be avoided, as this practice will cause the grain structure to enlarge. The length of time required for the soaking temperature will depend upon the mass of metal being treated.

The main difference between annealing and normalizing is that annealed parts are uniform in softness (and machinablilty) throughout the entire part since the entire part is exposed to the controlled furnace cooling. In the case of the normalized part, depending on the part geometry, the cooling is non-uniform resulting in non-uniform material properties across the part. This may not be desirable if further machining is desired, since it makes the machining job somewhat unpredictable. In such a case it is better to do full annealing.

Ques.506: Write notes on any TWO of the following :

(i) Domain theory of hysteresis.

(ii) Role of manganese in cast irons.

(iii) Rolling

Answer: (i) Domain theory of hysteresis : Lot of information can be had about the magnetic properties of a material by studying its hysteresis loop. A hysteresis loop shows the relationship between the induced magnetic flux density (B) and the magnetizing force (H). It is often referred to as the B-H loop.

An example of hysteresis loop is shown below.

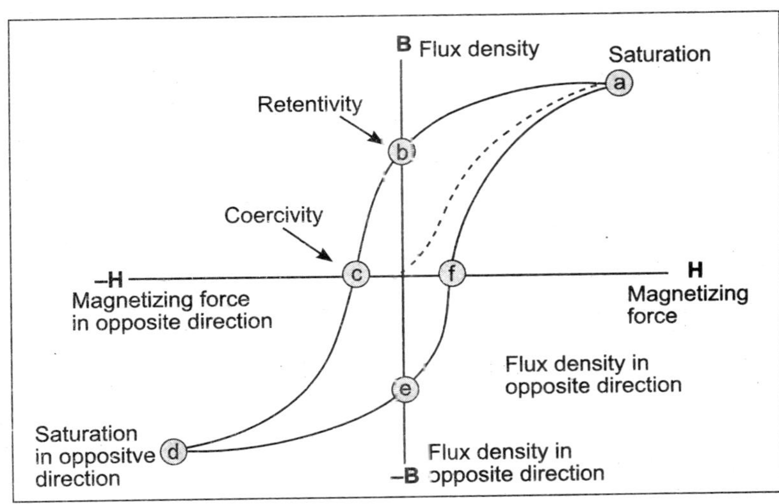

The loop is generated by measuring the magnetic flux of a ferromagnetic material while the magnetizing force is changed. A ferromagnetic material that has never been previously magnetized or has been thoroughly demagnetized will follow the dashed line as H is increased. As the line demonstrates, the greater the amount of current applied (H+), the stronger the magnetic field in the component (B+). At point "a" almost all of the magnetic domains are aligned and an additional increase in the magnetizing force will produce very little increase in magnetic flux. The material has reached the point of magnetic saturation.

When H is reduced to zero, the curve will move from point "a" to point "b." At this point, it can be seen that some magnetic flux remains in the material even though the magnetizing force is zero. This is referred to as the point of retentivity on the graph and indicates the remanence or level of residual magnetism in the material. (Some of the magnetic domains remain aligned but some have lost their alignment.) As the magnetizing force is reversed, the curve moves to point "c", where the flux has been reduced to zero. This is called the point of coercivity on the curve. (The reversed magnetizing force has flipped enough of the domains so that the net flux within the material is zero.) The force required to remove the residual magnetism from the material is called the coercive force or coercivity of the material. As the magnetizing force is increased in the negative direction, the material will again become magnetically saturated but in the opposite direction (point "d"). Reducing H to zero brings the curve to point "e." It will have a level of residual magnetism equal to that achieved in the other direction. Increasing H back in the positive direction will return B to zero. Notice that the curve did not return to the origin of the graph because some force is required to remove the residual magnetism. The curve will take a different path from point "f" back to the saturation point where it will complete the loop.

(ii) **Role of Manganese in Cast Iron :**

Stabilises austenite .

Acts as deoxidizer .

Refines the graphite and pearlite .

Refines the grains.

Increases the basic mechanical properties when added in low percentage.

(iii) **Rolling** :Rolling process is the easiest way of reducing the cross-sectional area of a piece of material to have uniform and long lengths of the product. Most of the steel-building materials are produced by this process. Rolling can be classified as Hot Rolling and Cold Rolling.

Cold rolling is a metal working process in which metal is deformed by passing it through rollers at a temperature below its recrystallization temperature. Cold rolling increases the yield strength and hardness of a metal by introducing defects into the metal's crystal structure. These defects prevent further slip and can reduce the grain size of the metal, resulting in Hall- Petch hardening. Cold rolling is most often used to decrease the thickness of plate and sheet metal.

Hot rolling, also called hot working, is a metallurgical process in which large pieces of metal such as slabs or billets are deformed between roller at high temperature to form thinner cross sections. Hot rolling produces thinner cross sections than cold rolling processes with the same number of stages. Hot rolling will reduce the average grain size of a metal while maintaining an equiaxed microstructure where as cold rolling will produce a hardened microstructure

Ques.507: Differentiate between the following:

(i) Extrinsic and intrinsic properties.

(ii) Phase and component.

Answer: *(i)* A pure semiconductor is called intrinsic semiconductor and it has properties that of an insulator. There are no free electrons available since all the co-valent bonds are complete. Here the no. of electrons and holes are equal thus conductivity is poor. And the resistance decreases with increase in temperature. But when tetravalent or Pentavalent impurity is added to an intrinsic semiconductor it results into an extrinsic semiconductor. On adding impurity it attains the current conducting properties. Here the no. of electrons is not equal to that of holes and so has better conductivity. Their conductivity increases with increase in temperature.

(ii) A Phase is a physically distinct, chemically homogenous and mechanically separable region of a system. The various states of aggregation of matter, namely, the gaseous, the liquid and the solid states, form separate phases. Whereas the Components of a system may be elements, ions or compounds. They refer to the independent chemical species that comprise the system.

Ques.508: Distinguish between ionic and metallic bonds in solids?

Answer: Ionic bond exists due to electrostatic force of attraction between positive and negative ions of different elements whereas Metallic bonds exist due to electrostatic force of attraction between the electron cloud of valence electrons and positive ions of the same or different metal. Ionic bonds have low thermal conductivity whereas metallic bonds have good thermal conductivity.

Ionic bonds are generally stronger than the metallic bonds.

Solids formed with ionic bonding have high hardness whereas solids formed with metallic bonding have high ductility.

Ques.509: Explain the following terms as applied to crystals:

***(i)* Crystal lattice.**

***(ii)* Unit cell.**

***(iii)* Lattice parameters of a unit cell**

***(iv)* Primitive cell**

(v) Packing Factor.

(vi) Atomic radius.

(vii) Co-ordination number.

Answer: ***(i)* Crystal Lattice**

It can be defined as the orderly, regular three-dimensional arrangement of atoms in a crystal. Also it can be considered as an infinite periodic array of points. Example: NaCl crystal lattice structure

(ii) Unit cell

The unit cell is the basic building block of a crystal, repeated infinitely in three dimensions. It is characterised by the three vectors (a, b, c) that form the edges of a parallelepiped and the angles between the vectors (alpha, the angle between b and c; beta, the angle between a and c; gamma, the angle between a and b).

(iii) Lattice parameters of a Unit cell

The size of the unit-cell is described in terms of its unit-cell parameters. The set of six numbers are called the lattice parameters. These are the edge lengths of the unit cell i.e. a, b and c respectively and the angles of the unit-cell (alpha, the angle between b and c; beta, the angle between a and c; gamma, the angle between a and b).

(iv) Primitive cell

A primitive cell is a unit cell built on the basis vectors of a primitive basis of the direct lattice,namely a crystallographic basis of the vector lattice L such that every lattice vector t of L may be obtained as an integral linear combination of the basis vectors, a, b, c. It contains only one lattice point and its volume is equal to the triple scalar product (a, b, c). Non-primitive bases are used conventionally to describe centred lattices. In that case, the unit cell is a multiple cell and it contains more than one lattice point. The multiplicity of the cell is given by the ratio of its volume to the volume of a primitive cell.

(v) Packing factor

Atomic packing factor (APF) or packing fraction is the fraction of volume in a crystal structure that is occupied by atoms. It is dimensionless and always less than unity. For practical purposes, the APF of a crystal structure is determined by assuming that atoms are rigid spheres. For one-component crystals (those that contain only one type of atom), the APF is represented mathematically by :

$$APF = \frac{N_{atoms}\, V_{atoms}}{V_{crystal}}$$

Where Natoms is the number of atoms in the crystal, Vatom is the volume of an atom, and Vcrystal is the volume occupied by the crystal.or one-component structures, the densest arrangement of atoms has an APF of about 0.74. In reality, this number can be higher due to specific intermolecular factors. For multiple-component structures, the APF can exceed 0.74.

(vi) Atomic Radius

Atomic radius is defined as half the distance between nearest neighbours in a crystal of a pure element.

(vii) Co-ordination number

The coordination number of an atom in a molecule or a crystal is the number of its nearest neighbours. This number is determined somewhat differently for molecules and for crystals.

Ques.510: Compare the mechanical properties and uses of mild steel, tool steels and alloy steels.

Answer: Mild steel is the most common form of steel as its price is relatively low while it provides material properties that are acceptable for many applications. Mild steel has low carbon content (up to 0.3%) and is therefore neither extremely brittle nor ductile. It becomes malleable when heated, and so can be forged. It is also often used where large amounts of steel need to be formed, for example as structural steel. Density of this metal is 7,861.093kg/m^3 , the tensile strength is a maximum of 500 MPa and it has a Young›s modulus of 210 GPa.

Tool steel refers to a variety of carbon and alloy steels that are particularly well-suited to be made into tools. Their suitability comes from their distinctive hardness, resistance to abrasion, their ability to hold a cutting edge, and/or their resistance to deformation at elevated temperatures (red-hardness).

With carbon content between 0.7 and 1.4%, tool steels are manufactured under carefully controlled conditions to produce the required quality. The manganese content is often kept low to minimise the possibility of cracking during water quenching. However, proper heat treating of these steels is important for adequate performance, and there are many suppliers who provide tooling blanks intended for oil quenching.

Tool steels are made to a number of grades for different applications. Choice of grade depends on, among other things, whether a keen cutting edge is necessary, as in stamping dies, or whether the tool has to withstand impact loading and service conditions encountered with such hand tools as axes, pickaxes, and quarrying implements. In general, the edge temperature under expected use is an important determinant of both composition and required heat treatment. The higher carbon grades are typically used for such applications as stamping dies, metal cutting tools, etc.

Tool steels are also used for special applications like injection molding, because the resistance to abrasion is an important criterion for a mold that will be used to produce hundreds of thousands of parts.

Low-alloy Steels:

When these steels are designed for welded applications, their carbon content is usually below 0.25 % and often below 0.15 %. Typical alloys include nickel, chromium, molybdenum, manganese, and silicon, which add strength at room temperatures and increase low-temperature notch toughness.

These alloys can, in the right combination, improve corrosion resistance and influence the steel's response to heat treatment. But the alloys added can also negatively influence crack susceptibility, so it's a good idea to use low-hydrogen welding processes with them. Preheating might also prove necessary.

High-alloy Steels: Stainless steel is the most important commercial high-alloy steel. Stainless steels have at least 12 % chromium and many have high nickel contents. The three basic types of stainless are:

— Austenitic
— Ferritic
— Martensitic

Martensitic stainless steels make up the cutlery grades. They have the least amount of chromium, offer high hardenability, and require both pre and post heating while welding to prevent cracking in the heat-affected zone (HAZ).

Ferritic stainless steels have 12 to 27 % chromium with small amounts of austenite forming alloys. Austenitic stainless steels offer excellent weldability, but austenite isn't stable at room temperature. Consequently, specific alloys must be added to stabilize austenite. The most important austenite stabilizer is nickel, and others include carbon, manganese, and nitrogen.

Special properties, including corrosion resistance, oxidation resistance, and strength at high temperatures, can be incorporated into austenitic stainless steels by adding certain alloys like chromium, nickel, molybdenum, nitrogen, titanium, and columbium. While carbon can add strength at high temperatures, it can also reduce corrosion resistance by forming a compound with chromium. It's important to note that austenitic alloys can't be hardened by heat treatment. That means they don't harden in the welding HAZ.

Ques.511: Write short note on Lithography and its applications.

Answer: The word lithography comes from the Greek lithos, meaning stones, and graphia, meaning to write. It means quite literally writing on stones. In the case of semiconductor lithography our stones are silicon wafers and our patterns are written with a light sensitive polymer called a photoresist. To build the complex structures that make up a transistor and the many wires that connect the millions of transistors of a circuit, lithography and etch pattern transfer steps are repeated at least 10 times, but more typically are done 20 to 30 times to make one circuit. Each pattern being printed on the wafer is aligned to the previously formed patterns and slowly the conductors, insulators, and selectively doped regions are built up to fabricate the final device.

Ques.512: Differentiate between welding, soldering, and brazing processes.

Answer: Welding is a fabrication process that joins materials, usually metals, or thermoplastics, by causing coalescence. This is often done by melting the workpieces and adding a filler material to form a pool of molten material (the weld puddle) that cools to become a strong joint, with pressure sometimes used in conjunction with heat, or by itself, to produce the weld. This is in contrast with soldering and brazing, which involve melting a lower-melting-point material between the workpieces to form a bond between them, without melting the workpieces.

Soldering is a process in which two or more metal items are joined together by melting and flowing a filler metal into the joint, the filler metal having a relatively low melting point. Soft soldering is characterized by the melting point of the filler metal, which is below 400 °C. The filler metal used in the process is called solder.

Soldering is distinguished from brazing by use of a lower melting-temperature filler metal; it is distinguished from welding by the base metals not being melted during the joining process. In a soldering process, heat is applied to the parts to be joined, causing the solder to melt and be drawn into the joint by capillary action and to bond to the materials to be joined by wetting action. After the metal cools, the resulting joints are not as strong as the base metal, but have adequate strength, electrical conductivity, and water-tightness for many uses. Brazing is a joining process whereby a filler metal or alloy is heated to melting temperature above 450 °C and distributed between two or more close-fitting parts by capillary action. At its liquid temperature, the molten filler metal and flux interacts with a thin layer of the base metal, cooling to form a strong, sealed joint. The melting temperature of the braze alloy is lower (sometimes substantially) than the melting temperature of the materials being joined. The brazed joint becomes a sandwich of different layers, each metallurgically linked to the adjacent layers.

Ques.513: Compare hot working and cold working processes.

Answer: Plastic deformation which is carried out in a temperature region and over a time interval such that the strain hardening is not relieved is called cold working. In the early stages of plastic deformation, slip is essentially on primary glide planes and the dislocations form coplanar arrays. As deformation proceeds, cross slip takes place. The cold-worked structure forms high dislocation density regions that soon develop into networks. The grain size decreases with strain at low deformation but soon reaches a fixed size.

Cold working will decrease ductility. Hot working refers to the process where metals are deformed above their recrystallization temperature and strain hardening does not occur. Hot working is usually performed at elevated temperatures. Lead, however, is hot-worked at room temperature because of its low melting temperature. At the other extreme, molybdenum is cold-worked when deformed even at red heat because of its high recrystallization temperature.

The resistance of metals to plastic deformation generally falls with temperature. For this reason, larger massive sections are always worked hot by forging, rolling, or extrusion. Metals display distinctly viscous characteristics at sufficiently high temperatures, and their resistance to flow increases at high forming rates. This occurs not only because it is a characteristic of viscous substances, but because the rate of recrystallization may not be fast enough.

Ques.514: Describe properties and applications of Bakelite .

Answer: **Properties of Bakelite :**

Bakelite is a material based on the thermosetting phenol formaldehyde resin. Formed by the reaction under heat and pressure of phenol (a toxic, colourless crystalline solid) and formaldehyde (a simple organic compound), generally with a wood flour filler, it was the first plastic made from synthetic components. It was used for its electrically nonconductive and heat-resistant properties in radio and telephone casings and electrical insulators, and was also used in such diverse products as kitchenware, jewellery, pipe stems, and toys.

Bakelite is manufactured and produced in the form of sheets,rods & tubes for hundreds of industrial applications in the electronics, power generation & aerospace industries, and under a variety of commercial brand names, including Garolite.

Bakelite sheet is a hard, dense material made by applying heat and pressure to layers of paper or glass cloth impregnated with synthetic resin. These layers of laminations are usually of cellulose paper, cotton fabrics, synthetic yarn fabrics, glass fabrics or unwoven fabrics. When heat and pressure are applied to the layers, a chemical reaction (polymerization) transforms the layers into a high-pressure thermosetting industrial laminated plastic. When rubbed, original Bakelite has a telltale odor.

Applications of Bakelite-Although not extensively used as an industrial manufacturing material any more, in the past Bakelite was used in myriad applications, such as saxophone mouthpieces, cameras, solid body electric guitars, rotary-dial telephones, early machine guns, and appliance casings. It was at one point considered for the manufacture of coins, due to a shortage of traditional manufacturing material.

Ques.515: Give properties and uses of following materials :

***(i)* Manganin *(ii)* Nichrome**

Answer: *(i)* Manganin is a trademarked name for an alloy of typically 86% copper, 12% manganese, and 2% nickel. Manganin foil and wire is used in the manufacture of resistors, particularly ammeter shunts, because of its virtually zero temperature coefficient of resistance value and long term stability. Manganin wire is also used as an electrical conductor in cryogenic (low temperature) systems, minimizing heat transfer between points which need electrical connections. Manganin is also used in gauges for studies of high-pressure shock waves (such as those generated from the detonation of explosives) because it has low strain sensitivity but high hydrostatic pressure sensitivity.

(ii) Nichrome is a brand name for a nickel-chromium resistance wire, a non-magnetic alloy of nickel and chromium. A common alloy is 80% nickel and 20% chromium, but there are many others to accommodate various applications. It is silvery-grey in colour, is corrosion resistant, and has a high melting point of about 1400 °C. Due to its relatively high resistivity and resistance to oxidation at high temperatures, it is widely used in heating elements, such as in hair dryers, electric ovens and toasters.

Typically, Nichrome is wound in wire coils to a certain electrical resistance, and current passed through to produce heat. Nichrome is used in the explosives & fireworks industry as a reliable ignition system, such as electric matches and model rocket igniters. The alloy tends to be expensive due to its high nickel content. Other areas of usage include motorcycle silencers, and in certain microbiological lab apparatus.

Ques.516: Briefly describe the process & purpose of the following heat treatment operations :

***(i)* Full annealing.**

***(ii)* Hardening.**

***(iii)* Tempering.**

Answer: *(i)* **Full Annealing**

A full anneal typically results in the most ductile state a metal can assume for metal alloy. To perform a full anneal, a metal is heated to its annealing point (about 50°c above the austenic temperature) and held for sufficient time to allow the material to fully austenitize, to form austenite or austenite-cementite grain structure. The material is then allowed to cool slowly so that the equilibrium microstructure is obtained. In some cases this means the material is allowed to air cool. In other cases the material is allowed to furnace cool. The details of the process depend on the type of metal and the precise alloy involved. In any case the result is a more ductile material that has greater stretch ratio and reduction of area properties but a lower yield strength and a lower tensile strength. This process is also called LP annealing for lamellar pearlite in the steel industry as opposed to a process anneal which does not specify a microstructure and only has the goal of softening the material. Often material that is annealed will be machined and then be followed by further heat treatment to obtain the final desired properties.

(ii) **Work hardening**, strain hardening, or cold work is the strengthening of a material by increasing the material's dislocation density. In metallic crystals, irreversible deformation is usually carried out on a microscopic scale by defects called dislocations, which are created by fluctuations in local stress fields within the material culminating in a lattice rearrangement as the dislocations propagate through the lattice. At normal temperatures the dislocations are not annihilated by annealing. Instead, the dislocations accumulate, interact with one another, and serve as pinning points or obstacles that significantly impede their motion. This leads to an increase in the yield strength of the material and a subsequent decrease in ductility.

(iii) **Tempering** is a heat treatment technique for metals and alloys. In steels, tempering is done to "toughen" the metal by transforming brittle martensite into bainite or a combination of ferrite and cementite. Precipitation hardening alloys, like many grades of aluminum and super alloys, are tempered to precipitate intermetallic particles which strengthen the metal.

Ques.517: Describe the process of extrusion giving its advantages.

Answer: Extrusion is a process used to create objects of a fixed cross-sectional profile. Amaterial is pushed or drawn through a die of the desired cross-section. The two main advantages of this process over other manufacturing processes is its ability to create very complex cross-sections and work materials that are brittle, because the material only encounters compressive and shear stresses. It also forms finished parts with an excellent surface finish. Extrusion may be continuous (theoretically producing indefinitely long material) or semi continuous (producing many pieces). The extrusion process can be done with the material hot or cold. Commonly extruded materials include metals, polymers, ceramics, and foodstuffs.

Extrusion of a round blank through a die :

The process begins by heating the stock material. It is then loaded into the container in the press. A dummy block is placed behind it where the ram then presses on the material to push it out of the die. Afterwards the extrusion is stretched in order to straighten it. If better properties are required,then it may be heat treated or cold worked.

Hot extrusion

Hot extrusion is done at an elevated temperature to keep the material from work hardening and to make it easier to push the material through the die. Most hot extrusions are done on horizontal hydraulic presses that range from 250 to 12,000 tons. Pressures range from 5,000 to 100,000 psi, therefore lubrication is required, which can be oil or graphite for lower temperature extrusions, or glass powder for higher temperature extrusions. The biggest disadvantage of this process is its cost for machinery and its upkeep.

Cold extrusion

Cold extrusion is done at room temperature or near room temperature. The advantages of this over hot extrusion are the lack of oxidation, higher strength due to cold working, closer tolerances, good surface finish, and fast extrusion speeds if the material is subject to hot shortness. Materials that are commonly cold extruded include: lead, tin, aluminum, copper, zirconium, titanium, molybdenum, beryllium, vanadium, niobium, and steel.

Examples of products produced by this process are: collapsible tubes, fire extinguisher cases, shock absorber cylinders, automotive pistons, and gear blanks.

Warm extrusion

Warm extrusion is done above room temperature, but below the recrystallization temperature of the material. It is usually used to achieve the proper balance of required forces, ductility and final extrusion properties.

Ques.518: What are the desirable properties of an insulation material?

Answer: Not all properties are significant for all materials or applications. In some applications, however, omitted properties may assume extreme importance (i.e. when insulations must be compatible with chemically corrosive atmospheres.)

The following properties are referenced according to their significance in meeting design criteria of specific applications.

THERMAL PROPERTIES OF INSULATION

Thermal properties are the primary consideration in choosing insulations.

a. Temperature limits: Upper & lower temperatures within which the material must retain all its properties.

b. Thermal conductance "C": The rate of heat flow for the actual thickness of a material.

c. hermal conductivity "K": The rate of heat flow based on 25 mm thickness.

d. Emissivity "E": Significant when the surface temperature of the insulation must be regulated as with moisture condensation or personnel protection.

e. Thermal resistance "R": The overall resistance of a "system" to the flow of heat.

f. Thermal transmittance "U":The overall conductance of heat flow through a"system".

MECHANICAL & CHEMICAL PROPERTIES OF INSULATION

Properties other than thermal must be considered when choosing materials for specific applications. Among them are:

a. Alkalinity (pH or acidity): Significant when corrosive atmospheres are present. Also insulation must not contribute to corrosion of the system.

b. Appearance: Important in exposed areas and for coding purposes.

c. Breaking load: In some installations the insulation material must "bridge" over a discontinuity in its support.

d. Capillarity: Must be considered when material may be in contact with liquids. *e.*

e. Chemical reaction: Potential fire hazards exist in areas where volatile chemicals are present. Corrosion resistance must also be considered.

f. Chemical resistance: Significant when the atmosphere is salt or chemical laden.

g. Coefficient of expansion & contraction: Enters into the design & spacing of expansion/contraction joints and/or the use of multiple layer insulation applications.

h. Combustibility: One of the measures of a material's contribution to a fire hazard.

i. Compressive strength: Important if the insulation must support a load or withstand mechanical abuse without crushing. If, however, cushioning or filling in space is needed as in expansion/contraction joints, low compressive strength materials are specified.

j. Density: A material's density affects other properties of that material, especially thermal properties.

k. Dimensional stability: Significant when the material is exposed to atmospheric and mechanical abuse such as twisting or vibration from thermally expanding pipe.

l. Fire retardancy: Flame spread and smoke developed ratings should be considered.

m. Hygroscopicity: Tendency of a material to absorb water vapor from the air.

n. Resistance to ultraviolet light: Significant if application is outdoors.

o. Resistance to fungal or bacterial growth: Is necessary in food or cosmetic process areas.

p. Shrinkage: Significant on applications involving cements and mastics.

Ques.519: Sound absorption coefficient: Must be considered when sound attenuation is required, as it is in radio stations, some hospital areas, etc.

Answer: Sound transmission loss value: Significant when constructing a sound barrier. s. Toxicity: Must be considered in food processing plants and potential fire hazard areas.

Ques.520: Explain dielectric loss and loss angle.

Answer: Dielectric loss can be defined as the loss of energy which eventually produces a rise in temperature of a dielectric placed in an alternating electrical field. Dielectric losses result from the heating effect on the dielectric material between the conductors. Power from the source is used in heating the dielectric. The heat produced is dissipated into the surrounding medium.

When there is no potential difference between two conductors, the atoms in the dielectric material between them are normal and the orbits of the electrons are circular. When there is a potential difference between two conductors, the orbits of the electrons change.

The excessive negative charge on one conductor repels electrons on the dielectric toward the positive conductor and thus distorts the orbits of the electrons. A change in the path of electrons requires more energy, introducing a power loss. And the dielectric loss angle is given by the difference between 90° and the dielectric phase angle.

Ques.521: What are Miller indices? What are their significances? Draw a (110) and a(111) plane inside a cubic unit cell.

Answer: Miller indices are a notation system in crystallography for planes and directions incrystal (Bravais) lattices. In particular, a family of lattice planes is determined by three integers h, k, and l, the Miller indices. They are written (h, k, l) and denote planes orthogonal to a direction in the basis of the reciprocal lattice vectors. By convention, negative integers are written with a bar, as in $\bar{3}$ for – 3. The integers are usually written in lowest terms, i.e. their greatest common divisor should be 1.

There are also several related notations. [ ], with square instead of round brackets, denotes a direction in the basis of the direct lattice vectors instead of the reciprocal lattice. The notation { }denotes all planes that are equivalent to by the symmetry of the crystal.The precise meaning of this notation depends upon a choice of lattice vectors for the crystal. Usually, three primitive lattice vectors are used. However, for cubic crystal systems, the cubic lattice vectors are used even when they are not primitive (e.g., as in body-centered and face centered crystals).

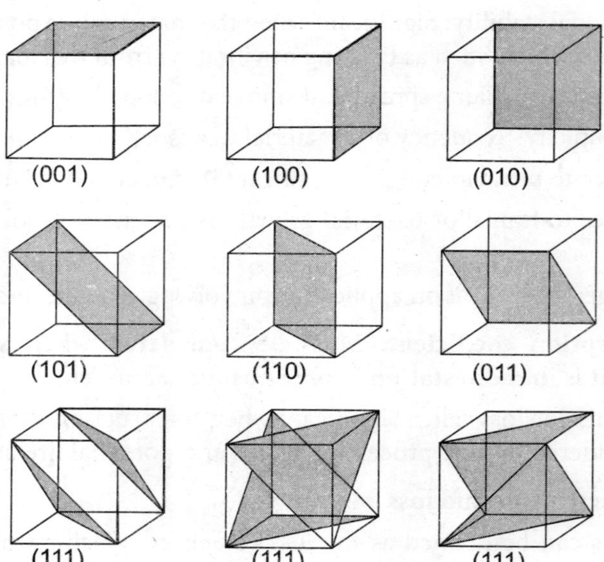

Ques.522: Explain the term 'welding'. Briefly explain the distinguishing features of the following method of welding : *(i)* pressure welding *(ii)* fusion welding; (iii) electric arc welding.

Answer: Welding is the process of joining two metallic pieces by the application of heat or pressure or both. This process utilises the property of the material called fusion. Pressure welding: Also called 'cold welding', pressure is applied to the workpieces through dies or rolls. Because of the plastic deformation involved, it is necessary that at least one (but preferably both) of the mating parts be ductile. Prior to welding, the interface is degreased, wire-brushed, and wiped to remove oxide smudges. Pressure welding can be used to join small workpieces made of soft, ductile metals.

It is known that the force of adhesion following first contact can be augmented by pressing the metals tightly together, increasing the duration of contact, raising the temperature of the workpieces, or any combination of the above. Even for very smooth metals, only the high points of each surface, called asperities, touch the opposing piece. Perhaps as little as a few thousandths of a percent of the total surface is involved.

However, these small areas of taction develop powerful molecular connections; electron microscope investigations of contact points reveal that an actual welding of the two surfaces takes place after which it is impossible to discern the former asperitic interface. If the original surfaces are sufficiently smooth the subatomic attractions between contact points eventually draw the two pieces completely together and eliminate even the macroscopic interface.

Exposure to oxygen or certain other reactive compounds produces surface layers that reduce or completely eliminate the pressure welding effect. This is especially true if, for example, a metal oxide has mechanical properties similar to those of the parent element (or softer), in which case surface deformations do not crack the oxide film. The reason pressure welding does not normally occur between metals on earth is because there is a very fine layer of oxidized metal due to the atmosphere, and even earth-made metals taken to space would retain this layer and not cold-weld.

(ii) **Fusion welding :** Fusion welding involves a heat source and may involve the use of a filler material such as a consumable electrode or a wire fed into the weld pool. These processes also use a protective layer between the atmosphere and the molten metal, either in the form of gas shielding or a flux which melts to give a viscous slag on the weld metal that eventually solidifies and can be removed.

There are several different types of fusion welding processes that can be used e.g. Manual Metal Arc (MMA), Submerged arc welding (SAW), Metal inert gas (MIG), Tungsten inert gas (TIG), Laser, Electron Beam etc.

(iii) **Electric arc welding:** in this type of welding, heat is supplied by the electric arc between electrode and the workpiece, arc caused by the potential difference between electrode and the work piece. The supply source can be AC or DC . And electrode can be of consumable or non consumable type. Eg ; SMAW, MIG, TIG.

Ques.523: What is an energy band? Why does the Fermi level in an intrinsic semiconductor lies in the middle of the energy gap?

Answer: Any solid has a large number of bands. In theory, it can be said to have infinitely many bands (just as an atom has infinitely many energy levels). However, all but a few lie at energies so high that any electron that reaches those energies escapes from the solid. These bands are usually disregarded. Bands have different widths, based upon the properties of the atomic orbitals from which they arise. Also, allowed bands may overlap, producing (for practical purposes) a single large band.

Metals contain a band that is partly empty and partly filled regardless of temperature. Therefore, they have very high conductivity. The lowermost, almost fully occupied band in an insulator or semiconductor is called the valence band by analogy with the valence electrons of individual atoms. The uppermost, almost unoccupied band is called the conduction band because only when electrons are excited to the conduction band can current flow in these materials.

The difference between insulators and semiconductors is only that the forbidden band gap between the valence band and conduction band is larger in an insulator, so that fewer electrons are found there and the electrical conductivity is lower. Because one of the main mechanisms for electrons to be excited to the conduction band is due to thermal energy, the conductivity of semiconductors is strongly dependent on the temperature of the material.

This band gap is one of the most useful aspects of the band structure, as it strongly influences the electrical and optical properties of the material. Electrons can transfer from one band to the other by means of carrier generation and recombination processes. The band gap and defect states created in the band gap by doping can be used to fabricate semiconductor devices such as solar cells, diodes, transistors, laser diodes, and others.

Fermi Level in an Intrinsic Semiconductor

In an Intrinsic Semiconductor, Fermi level (Ef) lies in the middle of energy gap or mid way between the conduction and valence bands.

Let,

nv = No. of electrons in the valence band

nc = No. of electrons in the conduction band

N = No. of electrons in both bands = nv + nc

After considering some assumptions, Let the zero energy reference level is arbitrarily taken at the top of the valence band.

Therefore, no. of electrons in the conduction band, nc = N.P (Eg)

Where P (Eg) = probability of an electro having energy Eg

This shows that in an Intrinsic Semiconductor, Fermi level (Ef) lies in the middle of energy gap or mid way between the conduction and valence bands.

Ques.524: Distinguish between "doping" and "alloying". Which of the two should be resorted to for changing the mechanical properties and why?

Answer: Doping is the process of intentionally introducing impurities into a pure (also referred to as intrinsic) semiconductor in order to change its electrical properties, whereas alloying means making one solid solution or homogeneous mixture of two or more elements, at least one of which is a metal, which itself has metallic properties. It usually has different properties from those of its component elements. Alloying one metal with others often enhances its properties

Ques.525: Explain the process of "forging".

Answer: Forging is the process by which metal is heated and is shaped by plastic deformation by suitably applying compressive force. Usually the compressive force is in the form of hammer blows using a power hammer or a press. Forging refines the grain structure and improves physical properties of the metal. With proper design, the grain flow can be oriented in the direction of principal stresses encountered in actual use. Grain flow is the direction of the pattern that the crystals take during plastic deformation. Physical properties (such as strength, ductility and toughness) are much better in a forging than in the base metal, which has, crystals randomly oriented.

Forgings are consistent from piece to piece, without any of the porosity, voids, inclusions and other defects. Thus, finishing operations such as machining do not expose voids, because there aren't any. Also coating operations such as plating or painting are straightforward due to a good surface, which needs very little preparation.

Ques.526: What are the objectives of annealing? Describe annealing and spheroidising processes.

Answer: Annealing, is a heat treatment wherein a material is altered, causing changes in its properties such as strength and hardness. It is a process that produces conditions by heating and maintaining a suitable temperature, and then cooling. Annealing is used to induce ductility, relieve internal stresses, refine the structure and improve cold working properties.

Annealing occurs by the diffusion of atoms within a solid material, so that the material progresses towards its equilibrium state. Heat is needed to increase the rate of diffusion by providing the energy needed to break bonds. The movement of atoms has the effect of redistributing and destroying the dislocations in metals and (to a lesser extent) in ceramics.This alteration in dislocations allows metals to reform more easily, so increases their ductility.

There are three stages in the annealing process, the first being the recovery phase, which results in softening of the metal through removal of crystal defects (the primary type of which is the linear defect called a dislocation) and the internal stresses which they cause. The second phase is recrystallization, where new grains nucleate and grow to replace those deformed by internal stresses. If annealing is allowed to continue once recrystallization has been completed, grain growth will occur in which the microstructure starts to coarsen and may cause the metal to have less than satisfactory mechanical properties.

Spheroidising is the process that applies more to the hypereutectoid steels (above 0.8% C). The process involves heating the metal to between 600 and 650°C and holding it at the selected temperature for a period of time the cementite changes from a lamella formation to a formation based on an alpha ferrite matrix with particles of spheroidal cementite (Fe3C) are embedded. This resulting steel has improved ductility and toughness compared to the original steel with reduced hardness and strength.

Ques.527: Compare and differentiate the properties of common semi-conducting materials and common dielectric materials.

Answer: **Semi-conducting materials :**

1. These are the materials whose valence electrons are bounded somewhat loosely to their atom.

2. They require energy less than dielectrics and more than good conductors to remove an electron from the parent atom.
3. They have:
   (i) an empty conduction band .
   (ii) almost filled valence band .
   (iii) Very narrow energy gap (of about 1eV) separating the two bands.

Dielectric materials :

1. These are the materials in which valance electrons used are bounded very tightly to their parent atoms.
2. They require very large electric field to remove them from the attraction of the nuclei.
3. They have:
   (i) a full valence band .
   (ii) an empty conduction band .
   (iii) a large energy gap separating the two bands.

Ques.528: Draw sketches illustrating a (100) plane, a (110) plane, and a (111) plane in a cubic unit cell. How many equivalent {100} planes are there in a cubic crystal?

Answer: In totality, 6 equivalent planes to {100} are there in a cubic crystal as shown below.

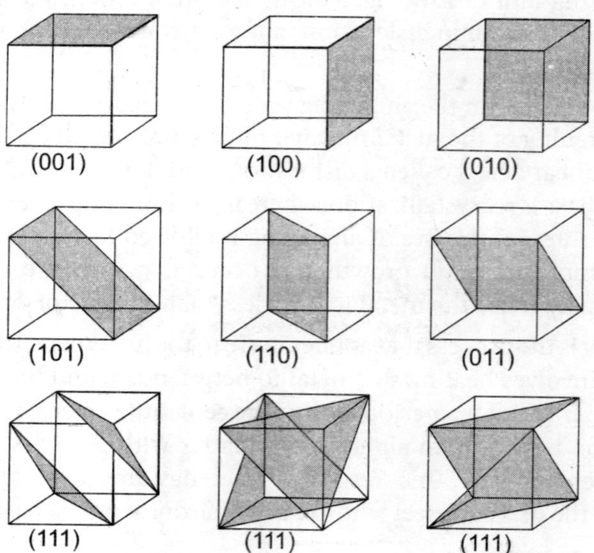

Ques.529: Explain the terms electric polarization and the electric susceptibility in the context of dielectrics.

Answer: Polarization of Dielectric If a material contains polar molecules, they will generally be in random orientations when no electric field is applied. An applied electric field will polarize the material by orienting the dipole moments of polar molecules. This decreases the effective electric field between the plates and will increase the capacitance of the parallel plate structure. The dielectric must be a good electric insulator so as to minimize any DC leakage current through a capacitor.

Electric susceptibility

The electric susceptibility of a dielectric material is a measure of how easily it polarizes in response to an electric field. This, in turn, determines the electric permittivity of the material and thus influences many other phenomena in that medium, from the capacitance of capacitors to the speed of light.

Ques.530: What is meant by crystal imperfections? Classify them in the order of their geometry. Explain the difference between edge and screw dislocations. What is a Burger's vector?

Answer: Crystalline solids have a very regular atomic structure: ie; the local positions of atoms with respect to each other are repeated at the atomic scale. These arrangements are called crystal structures, and their study is called crystallography. However, most crystalline materials are not perfect: the regular pattern of atomic arrangement is interrupted by crystal defects.

The various types of defects are :

(a) **Point defects:** Point defects are defects which are not extended in space in any dimension.There is no strict limit for how small a "point" defect should be, but typically the term isused to mean defects which involve at most a few extra or missing atoms without an ordered structure of the defective positions. Larger defects in an ordered structure are usually considered dislocation loops. For historical reasons, many point defects especially in ionic crystals are called 'centers': for example the vacancy in many ionic solids is called an F-center.

Vacancies are sites which are usually not occupied by an atom and remain unoccupied. If a neighboring atom moves to occupy the vacant site, the vacancy moves in the opposite direction to the site which used to be occupied by the moving atom. The stability of the surrounding crystal structure guarantees that the neighboring atoms will not simply collapse around the vacancy. In some materials, neighboring atoms actually move away from a vacancy, because they can better form bonds with atoms in the other directions. A vacancy (or pair of vacancies in an ionic solid) is sometimes called a **Schottky defect**.

Interstitials are atoms which occupy a site in the crystal structure at which there is usually no atom. They are generally high energy configurations. Small atoms in some crystals can occupy interstices without high energy, such as hydrogen in palladium.

A nearby pair of a vacancy and an interstitial is often called a **Frenkel defect** or Frenkel pair. Impurities occur because materials are never 100% pure. In the case of an impurity, the atom is often incorporated at a regular atomic site in the crystal structure. This is neither a vacant site nor is the atom on an interstitial site and it is called a substitutional defect. The atom is not supposed to be anywhere in the crystal, and is thus an impurity

Anti-site defects occur in an ordered alloy. For example, some alloys have a regular structure in which every other atom is a different species, for illustration assume that type A atoms sit on the cube corners of a cubic lattice, and type B atoms sit in center of the cubes.

If one cube has an A atom at its center, the atom is on a site usually occupied by an atom, but it is not the correct type. This is neither a vacancy nor an interstitial, nor an impurity. Topological defects are regions in a crystal where the normal chemical bonding environment is topologically different from the surroundings. For instance, in a perfect sheet of graphite (graphene) all atoms are in rings containing six atoms. If the sheet contains regions where the number of atoms in a ring is different from six, while the total number of atoms remains the same, a topological defect is formed. An example is the Stone Wales defect in nanotubes, which consists of two adjacent 5-membered and two 7- membered atom rings.

Also amorphous solids may contain defects. These are naturally somewhat hard to define, but sometimes their nature can be quite easily understood. For instance, in ideally bonded amorphous silica all Si atoms have 4 bonds to O atoms and all O atoms have 2 bonds to Si atom. Thus *e.g.* an O atom with only one Si bond can be considered a defect in silica.

Complexes can form between different kinds of point defects. For example, if a vacancy encounters an impurity, the two may bind together if the impurity is too large for the lattice. Interstitials can form 'split interstitial' or 'dumbbell' structures where two atoms effectively share an atomic site, resulting in neither atom actually occupying the site.

(b) Line defects: Line defects can be described by gauge theories. Dislocations are linear defects around which some of the atoms of the crystal lattice are misaligned. There are two basic types of dislocations, the EDGE dislocation and the SCREW dislocation. "MIXED" dislocations combining aspects of both types are also common. Edge dislocations are caused by the termination of a plane of atoms in the middle of a crystal. In such a case, the adjacent planes are not straight, but instead bend around the edge of the terminating plane so that the crystal structure is perfectly ordered on either side. The analogy with a stack of paper is apt: if a half a piece of paper is inserted in a stack of paper, the defect in the stack is only noticeable at the edge of the half sheet.

The screw dislocation is more difficult to visualise, but basically comprises a structure in which a helical path is traced around the linear defect (dislocation line) by the atomic planes of atoms in the crystal lattice.

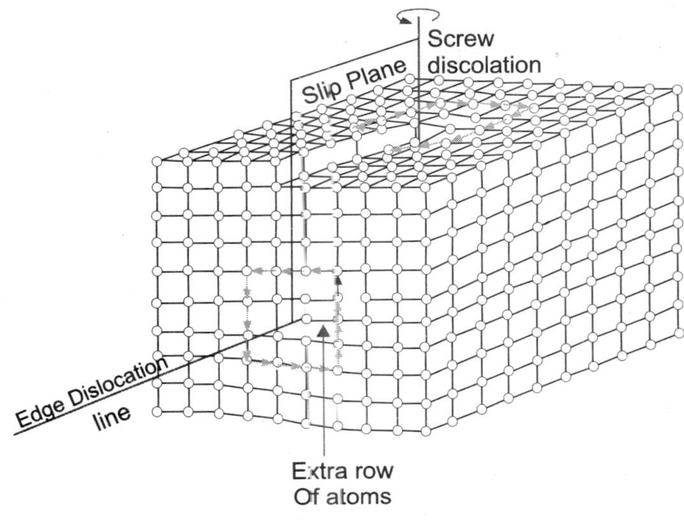

Screw dislocation

The presence of dislocation results in lattice strain (distortion). The direction and magnitude of such distortion is expressed in terms of a Burgers vector (b). For an edge type, b is perpendicular to the dislocation line, whereas in the cases of the screw type it is parallel. In metallic materials, b is aligned with close-packed crytallographic directions and its magnitude is equivalent to one interatomic spacing. Dislocations can move if the atoms from one of the surrounding planes break their bonds and rebond with the atoms at the terminating edge.

It is the presence of dislocations and their ability to readily move (and interact) under the influence of stresses induced by external loads that leads to the characteristic malleability of metallic materials.

Dislocations can be observed using transmission electron microscopy, field ion microscopy and atom probe techniques. Deep level transient spectroscopy has been used for studying the electrical activity of dislocations in semiconductors, mainly silicon. Disclinations are line defects corresponding to "adding" or "subtracting" an angle around a line. Basically, this means that if you track the crystal orientation around the line defect, you get a rotation. Usually they play a role only in liquid crystals.

(c) Planar defects

- Grain boundaries occur where the crystallographic direction of the lattice abruptly changes. This usually occurs when two crystals begin growing separately and then meet.
- Anti phase boundaries occur in ordered alloys: in this case, the crystallographic direction remains the same, each side of the boundary has an opposite phase: For example if the ordering is usually ABABABAB, an anti phase boundary takes the form of ABABBABA.

- Stacking faults occur in a number of crystal structures, but the common example is in close-packed structures. Face-centered cubic (fcc) structures differ from hexagonal close packed (hcp) structures only in stacking order: both structures have close packed atomic planes with six fold symmetry -- the atoms form equilateral triangles. When stacking one of these layers on top of another, the atoms are not directly on top of one another -- the first two layers are identical for hcp and fcc, and labelled AB. If the third layer is placed so that its atoms are directly above those of the first layer, the stacking will be ABA -- this is the hcp structure, and it continues ABABABAB. However there is another location for the third layer, such that its atoms are not above the first layer. Instead, the fourth layer is placed so that its atoms are directly above the first layer. This produces the stacking ABCABCABC, and is actually a cubic arrangement of the atoms. A stacking fault is a one or two layer interruption in the stacking sequence, for example if the sequence ABCABABCAB were found in an fcc structure.

(d) Bulk defects

- Voids are small regions where there are no atoms, and can be thought of as clusters of vacancies.
- Impurities can cluster together to form small regions of a different phase. These are often called precipitates.

BURGER VECTOR

The Burgers vector, often denoted by b, is a vector that represents the magnitude and direction of the lattice distortion of dislocation in a crystal lattice. The direction of the vector depends on the plane of dislocation, which is usually on the closest-packed plane of unit cell.

The magnitude of the Burgers vector for a dislocation is of a magnitude equal to the interatomic spacing of the material, since a single dislocation will offset the crystal lattice by one close-packed crystallographic spacing unit.

In edge dislocations, the Burgers vector and dislocation line are at right angles to one another. In screw dislocations, they are parallel. The Burgers Vector is significant in determining the strength of a material: it affects solute hardening,precipitation hardening and work hardening, all of which affect yield strength.

Ques.531: What is the phase rule and what does it indicate? Using Gibb's phase rule show that eutectoid reaction is non-variant.

Answer: Gibbs' phase rule, is the fundamental rule on which phase diagrams are based. It provides the number of degrees of freedom for a given thermodynamic condition, that is: how many control variables (pressure, temperature, composition, etc) can be altered while maintaining this condition. Mathematically, this number is :

$$P + F = C + 2,$$

Where P is the number of phases in equilibrium coexistence (solid, liquid, gas phases etc) and C is the number of chemical components in the system. The number 2 arises from the two main thermodynamic control parameters, usually temperature & pressure, so that C+2 may be considered as the total number of control parameters necessary to specify a thermodynamic state for a mixture with C components. At Eutectoid temperature, three phases are in equilibrium i.e. F = 2-3+1 = 0. Thus at eutectoid temperature, compositions are fixed & none of them can be varied arbitrarily. Thus with zero degree of freedom, the eutectoid reaction is called as non-variant reaction.

Ques.532: What are the chief characteristics of ferro-electric materials? How do you establish the existence of ferro-electricity in a material?

Answer: Ferroelectric materials have following characteristics:

a. They have a high dielectric constant which is non-linear i.e. it depends to a considerable extent on the intensity of the electrical field.

b. They exhibit hysteresis loops; i.e. the polarization is not linear function of the applied field.

The most important class of magnetic materials is the Ferro magnets: iron, nickel, cobalt and manganese, or their compounds (and a few more exotic ones as well). The magnetization curve looks very different to that of a diamagnetic or paramagnetic material. The magnetization curves for ferromagnetic materials are all strongly dependant upon purity, heat treatment and other factors. A normal magnetization curve because starts from an unmagnetized sample and shows how the flux density increases as the field strength is increased. At an atomic level ferromagnetism is explained by a tendency for neighbouring atomic magnetic moments to become locked in parallel with their neighbours.This is only possible at temperatures below the Curie point,above which thermal disordering causes a sharp drop in permeability and degeneration into paramagnetism. Ferromagnetism is distinguished from paramagnetism by more than just permeability because it also has the important properties of remanence and coercivity.

Remnance occurs within ferromagnetic materials and is the flux density which remains in a magnetic material when any externally applied field is removed . For materials used in permanent magnets we usually need a high value of remnance. For transformers we need low remnance. Remnance is also what makes possible all magnetic recording technologies; including the hard disk drive on which this text was stored until you loaded it into your browser.

Coercivity occurs within ferromagnetic materials and is the field strength which must be applied in order to reduce (or coerce) a remnant flux to zero. Materials with high coercivity (such as those used for permanent magnets) are sometimes called hard. A good example of a hard material is Alnico . Conversely, materials with low coercivity (such as those used for transformers) are called soft.

The materials with a low coercivity tend to have a high permeabilty and vice versa.

Energy loss per cycle is a heat loss caused by the magnetic properties of the armature. When an armature core is in a magnetic field, the magnetic particles of the core tend to line up with the magnetic field. When the armature core is rotating, its magnetic field keeps changing direction. The continuous movement of the magnetic particles, as they try to align themselves with the magnetic field, produces molecular friction. This, in turn, produces heat. This heat is transmitted to the armature windings. The heat causes armature resistances to increase. To compensate for these losses, heat-treated silicon steel laminations are used in most DC generator armatures. After the steel has been formed to the proper shape, the laminations are heated and allowed to cool. This annealing process reduces the loss per cycle to a low value.

Ques.533: Explain clearly the difference between hard and soft magnetic materials. Explain why Fe- Si alloys are used for power transformers whereas Ni-Fe alloys are used for pulse transformers.

Answer: Hard magnetic materials

Permanent magnets are used as passive generators of magnetic field. To produce and maintain the magnetic field, a material should have both high residual induction and high coercivity. The minimum coercivity that a permanent magnet should have is at least of 50 000 A/m.

The quality of a magnet has to be judged in relation to remnence Br, coercive force HcJ & HcB, maximum energy product BHmax and to physical characteristics like temperature dependence, corrosion resistance, resistivity, brittleness, and the cost.

Examples: Alnico alloys, Ferrite, Samarium-Cobalt magnets, NdFeB, Bonded magnets etc.

Applications of permanent magnets

a. Applications based on attraction or repulsion forces

- Compass, magnetometers
- Holding devices
- Magnetic filtration
- Drives and couplings
- Switches
- Magnetic clutches
- Magnetic tools

b. Applications based on Faraday's induction law

- Generators
- Microphones
- Brakes
- Speedometers

c. Application based on Lorentz forces on conductors

- Loudspeakers
- DC motors

- Synchronous motors
- Brushless motors
- Hysteresis motors
- Computer peripherals
- Measuring instruments
- Telephone receivers

d. Application based on Lorentz forces on single charges

- Travelling-wave tube
- Electron microscope
- Klystron
- Magnetron
- Mass spectrometer

Soft magnetic materials

Soft magnetic materials are mainly utilized in those applications in which the soft material has to amplify the flux generated by the electrical current or by a permanent magnet.

The principal characteristics of soft magnetic materials are remanence, coercivity, maximum & initial value of permeability and saturation value of B, H, J.

It is common to subdivide the soft magnetic materials in two categories, depending on the application: DC applications and AC applications.

Soft Magnetic Materials for DC applications

These applications generally require low coercivity Hc and high permeability.The low value of coercivity is necessary to avoid remanence after the use, when the power switches off (for example in an electromagnet), and a high value of permeability is necessary to enhance the field H (being $B = \mu H$).

At the same time, it is also important to know the complete normal magnetization curve that describes the B induction value vs the value of H.

The most common DC applications are:

- Electromagnets and yokes
- Relays
- Magnetic shields
- Electrical measuring devices
- Magnetic amplifiers

Materials used for DC applications are:

- Iron and low-carbon steels (maximum carbon 0.05 %)
- Iron-Nickel alloys (Permalloy)

- Iron-Cobalt alloy (Permendur)

Soft Magnetic Materials for AC applications

These applications generally require low coercivity Hc & low power loss, high permeability μ, high saturation magnetization. The low value of power loss is necessary because in AC applications the material alternatively changes from one polarity to the other, so it is important to have a very thin hysteresis loop, with high saturation and consequently low coercivity (Hc).

Typical AC applications are:

- Transformers
- Motors and generators
- Signal transmitters and receivers

Materials used for AC applications are:

- Iron - Silicon alloys
- Iron - Nickel alloys (Permalloy)
- Soft ferrites

Fe-Si alloys are used for making power transformers because adding silicon helps in reducing the hysteresis loss and these alloys are suitable for operation at power frequencies whereas Fe- Ni alloys are suitable in communication equipment, where high sensitivity and fidelity are required so these are used in making pulse transformers where there is constant core saturation for a particular period and therefore we use Ni-Fe alloys for making them . Ferrites have higher resistance than the ferromagnetic material. This makes them suitable for application at high frequencies- even microwave frequencies.

Ques.534: Write notes on the following:-

***(i)* Properties of filament and contact materials.**

***(ii)* Free electron theory of metals.**

Answer: *(i)* The Filament materials must have low resistance temperature co-efficients, high melting point, high mechanical strength, ductility and no tendency for oxidation. Therefore materials like Carbon, Tantalum and Tungsten are most widely used for making the filaments.

Electrical contacts and contact materials are soft, high-conductivity, and oxidation resistant materials that are used in circuit breakers, relays, switches, and electrical discharge machining (EDM) applications. They often have a second phase in order to provide anti welding and/or arc resistance. Electrical contacts and contact materials are used in many different type of contacts. Battery contacts usually consist of a low-cost base metal that is plated with a layer of high-conductivity metal. Contact probes or spring contacts are brush assemblies with a plunger-style configuration and an internal helical spring. Electrical brushes are used in conjunction with slip rings and other contact surfaces to maintain an electrical connection in mobile or rotary applications. Furnace electrodes are used to heat and melt metals or ceramics in arc furnaces. Welding electrodes are designed for resistance, arc, or plasma welding. Special electrical contacts and contact materials are also commonly available. Electrical contacts and contact materials are available in many different forms.

Examples include: contact tips or buttons, contact rivets, EDM drilling electrodes, electrode blocks or billets, pins or probes, solid rod stock, sockets or receptacles, small stampings, sheets or foils, strips or tapes, tamped connections, tapping electrodes, wires, and wheels.

Contact tips and contact buttons consist of a shaped pad, often with nibs for welding or serrations for brazing. EDM drilling electrodes are solid-rod or hollow-tube electrodes used in drilling and trepanning applications. Electrode blocks and electrode billets are raw stock forms that are used to produce EDM electrodes or arc furnace electrodes. Pins and probes are used for testing printed circuit boards (PCB) or electronic components. Custom, application specific and assembled electrical contacts and contact materials are also available.

For contact materials, the factors which affect the choice of the material are contact resistance, contact force, voltage and current. So platinum, palladium, Silver, Gold, Tungsten, Molybdenum, Rhodium are used.

(ii) **Free Electron theory of metals**

This theory tells that, metals conduct electricity because of the presence of free electrons in it. The outermost shells of metal atoms will be loosely bound with their nucleus. So the electrons in it are free to move anywhere in the solid. These electrons are called free electrons and they are responsible for the conduction of electricity. It can also be considered as a model of a metal in which the free electrons, that is, those giving rise to the conductivity, are regarded as moving in a potential (due to the metal ions in the lattice and to all the remaining free electrons) which is approximated as constant everywhere inside the metal. I is also known as Sommerfeld model.

Sommerfeld theory: The theory was originally proposed in 1900 to describe and correlate the electrical & thermal properties of metals. Later, quantum mechanics became the basis for the theory of most of the general properties of simple metals such as sodium, with one free electron per atom, magnesium with two, and aluminum with three. Transition metals, such as iron, have partially filled electronic d states and are not treated by the free-electron model.

Ques.535: Classify plastic materials into two categories, describe their important properties and give two examples of each.

Answer: Plastic materials can be classified into thermoplastic and thermosetting plastics. Thermoplastic materials:- The properties of these plastic materials do not change considerably if they are melted and then cooled & solidified . They can be repeatedly melted or dissolved in various solvents. They are more elastic, less brittle and do not lose elasticity when subjected to prolonged heating. They are less apt to age thermally. They can be remoulded again and again in any shape after heating. Many of them possess extraordinary high insulating properties and are water repellent. They are polymers of linear structure, i.e. their molecules are elongated and are thread like. This, type of structure is fusible, soluble, highly plastic, capable of forming thin flexible threads and films.

Examples of thermoplastic materials are :

Polytetra Flouroethylene (P.T.F.E. or Teflon), Polyvinyl Chloride (P.V.C.).

Thermosetting Plastic Materials:- They undergo great changes when subjected to high temperatures for quite sometimes. They are said to be baked and no longer can melt or be dissolved. They are less elastic, more brittle and lose their elasticity when subjected to prolonged heating. So they cannot be remoulded in different shapes once they are set and hardened. They are used, when an insulation is to withstand high temperatures without melting or losing its shape & mechanical strength. Thermosetting plastic substances are space-polymers and the molecules branch off in various directions during polymerisation. This structure makes them very rigid, poorly soluble, fusible and incapable of forming elastic threads & films.

Examples of thermosetting materials are :

Phenol formaldehyde (Bakelite), Epoxy resins.

Ques.536: Give four examples of natural insulating materials.

Answer: Examples of the natural insulating materials are cotton, rubber, wood, mica.

Ques.537: State the functions of a fuse. Mention the desirable properties of fuse materials.

Answer: Fuse is a protective device, which consists of a thin wire or strip. This wire or strip is placed with the circuit it has to protect, so that the circuit-current flows through it. When this current is too large, and heat produced due to I^2Rt exceeds the melting point of the fuse, consequently fuse wire or strip melts thus breaking the circuit and interrupting the power supply.

A fuse material should possess the following properties;-

Low resistivity – This means, thin wires can be used, which will give less metal vapour after melting of the wire. Less metal vapour in the arc gives lower conductivity and thus makes quenching of arc easier.

Low conductivity of the metal vapours itself. Low melting point- This means that the temperature of the fuse material for normal currents stays at a low value. Previously lead was used as fuse material because of its low melting point but thick wires of lead are required due to higher resistivity. For rewireable fuses alloys of tin and lead or tinned copper wires are commonly used. In caridge fuses silver and silver alloys are used of lower ratings. Copper alloys are used in fuses of higher ratings.

Ques.538: Differentiate between n and p type semiconductors.

Answer: Features of n – type semiconductor:-

(i) When small amount of pentavalent impurity is added to a pure semiconductor providing a large number of free electrons in it, the extrinsic semiconductor thus

formed is known as n- type semiconductor. The addition of pentavalent impurities such as arsenic and antimony provide a large number of free electrons in the semiconductor crystal. Such impurities, which produce n- type semiconductors, are known as donor impurities because each atom of them donates one free electron to the semiconductor crystal.

(ii) The majority and minority carriers are electrons and holes respectively.

(iii) The impurity level is just below the bottom of conduction band.

Features of p - type semiconductor:-h

(i) When small amount of trivalent impurity is added to a pure semiconductor providing a large number of holes in it, the extrinsic semiconductor thus formed is known as p- type semiconductor. The addition of trivalent impurities such as gallium and indium provide a large number of holes in the semiconductor crystal. Such impurities, which produce p- type semiconductors, are known as acceptor impurities because each atom of them creates one hole, which can accept one electron.

(ii) The majority and minority charge carriers are holes and electrons respectively.

(iii) The impurity level is just above the valence band.

Ques.539: Describe the electrical properties of insulating materials.

Answer: Electrical properties of an insulating material are:

Insulation resistance- It is the property, by the virtue of which, a material resists flow of electrical current. It should be as high as possible.

Insulation resistance is of two types:

(i) Volume resistance;

(ii) Surface resistance.

The resistance offered to the current, which flows through the material is called volume resistance. The resistance offered to the current, which flows over the surface of the insulating material is called surface resistance. Factors that affect the insulation resistance are-temperature variations, exposure to moisture, voltage applied and aging.

Dielectric Strength- It is the minimum voltage which when applied to an insulating material will result in the destruction of its insulating properties. It can also be defined as the maximum potential gradient that the material can withstand without rupture, or without loosing dielectric properties. This value is expressed in volts or kilovolts per unit thickness of the insulating material. This value is greatly affected by the conditions under which the material is operated. Factors affecting the dielectric strength are temperature and humidity.

Dielectric Constant- Every insulating material has got the basic property of storing charge (Q), when a voltage (V) is applied across it. The charge is proportional to the voltage applied i.e. $Q \alpha V$, or $Q = CV$. Where C is called the capacity or capacitance of the material across which the voltage is applied. Every insulating material behaves as a capacitor. Capacitance is different for different insulating material. The property of insulating materials that causes the difference in the value of capacitance, with the physical dimensions remaining the same is called dielectric constant or permittivity (□) and = C/Co, where C is capacity in presence of Dielectric and Co is the capacity in air or vacuum or in the absence of dielectric.

Dielectric loss and Loss angle: When a perfect insulation is subjected to alternating voltage, it is like applying alternate voltage to a perfect capacitor. In a perfect capacitor the charging current would lead the applied voltage by 90 0 exactly. This means that there is no power loss in the insulation. In most insulating materials this is not the case.

There is a definite amount of dissipation of energy when an insulator is subjected to alternating voltage. This dissipation of energy is called dielectric loss. Factors affecting dielectric loss are – frequency of applied voltage, humidity, temperature rise and voltage.

Ques.540: Discuss important properties and uses of :

(i) **Rubber.**

(ii) **Mica and Mica products.**

(iii) **Glass and glass products.**

(iv) **Ceramics**

Answer: *(1)* **Rubber:**

Rubber is polymeric material with high elastic yield strain. The different types of rubber materials are- Natural rubber, Hard rubber and Synthetic rubber.

Natural Rubber – Natural rubber is extracted from milky sap collected from special trees. Water is then evaporated. Additives like sulphur, oxidation inhibitors like aromatic amino compounds, softeners like vegetable oil and fillers like carbon black & zinc oxide are added to it. It is vulcanized by adding sulphur and heating it. Vulcanization improves heat and frost resistance of rubber, making it mechanically stronger. The permittivity and power factor varies depending on the sulphur content and temperature change.

Properties – This rubber is moisture repellent and has good insulating properties. It has good abrasion resistance.

Applications – It is used for the manufacturing of protective clothing such gloves, boots. It is used as an insulation covering for wires and cables.

Hard Rubber – Hard rubber is obtained by addition of more sulphur and by extended vulcanization.

Properties – It has good electrical properties. Water absorption is less. Maximum permissible operating temperature is 60 ^{0}C. It can not be continuously exposed to Sun as it is harmful. It has high tensile strength.

Applications – Hard rubber is used for construction of storage battery housing, panel boards, bushings of various types. It is also used as jacketing material for cables.

Synthetic Rubber – The different types of synthetic rubber are :

(a) **Butadiene rubber** – Its properties are greater resistance to ageing & oxidation, lower tensile & tear strength, lower water absorption, higher heat conductivity.

(b) **Butyl rubber** – Its properties are excellent resistance to vegetable oils and alcoholicsolvents, but it is easily attacked by petroleum oils & greases. It has high resistance to ozone, high thermal & oxidation stability but poor tensile strength. It is used as insulation for wires and cables.

(c) **Chloroprene Rubber (Neoprene Rubber)** – It has better resistance to thermal ageing, oxidation, sunlight and gas diffusion. These rubbers have better thermal conductivity and more flame resistance. They exhibit better adhesion to metals. They possess better resistance to attack by solvents like mineral and vegetable oils but poor resistance to aromatic hydrocarbon liquids. They are inferior in mechanical properties like tear & tensile strength and abrasion resistance. Neoprene rubber is used as insulating material for wires and cables. It is also used as jacketing material for cables.

(d) **Chlorosulphonated Polyethylene (Hypalon)** – It has better electrical properties, high resistance to degradation when exposed to high temperature and oxidation. It can be operated at temperatures as high as 150 0 C. It has poor solvent resistance to hydrocarbons. It is mechanically less tough. It is used as insulating material for wires and cables and also as jacketing material for cables.

(e) **Silicon Rubber** – It has high thermal conductivity. Its tensile strength is low, has good flexibility at low temperatures and resistant to ozone, oxidation & severe atmospheric conditions. It can be used over a wide range of temperatures from –100 to 150 0 C. Silicon rubber is used as insulating material for wires and cables, in the manufacture of moulded parts, as an insulating tape and coating material.

(2) **Mica and Mica Products**

Mica is an inorganic material. It is one of the best insulating materials available. From the electrical point of view, mica is of two types

— Muscovite mica and

— Phologopite mica.

Muscovite mica – Chemical composition is $KH_2Al_3(SiO_4)_3$.

The properties are :

1. Strong, tough and less flexible
2. Colourless, yellow, silver or green in colour
3. Insulating properties are very good
4. Abrasion resistance is high
5. Alkalies do not affect it

Uses

Muscovite mica is used where electrical requirements are severe. Because of high dielectric strength, it is used in capacitors. It is also used in commutators due to high abrasion resistance.

Phologopite mica – Chemical composition is $KH(MgF)_8MgAl(SiO_4)_3$.

The properties are :

1. Amber, yellow, green or grey in colour .
2. Greater structural stability, being tougher and harder than muscovite mica, less rigid .
3. Resistant to alkalies, but less to acids .
4. Greater thermal stability than that of muscovite mica .

Uses

It is used when there is greater need of thermal stability as in domestic appliances like irons, hotplates, toasters.

Mica products

(*a*) **Glass bonded mica** – Ground mica flakes and powdered glass when moulded makes glass bonded mica. This material is impervious to water and chemically stable. This is used in high humidity and high ambient temperatures.

(*b*) **Mica paper and Mica sheet**– Mica is broken into small particles in aqueous solution. Out of this sheets of mica paper are produced which are used as insulation for armature and field coils of rotating machines. Also used as washers, spacers, sleeves, tubes etc.

(*c*) **Manufactured mica** – Mica flakes held together with adhesives is called manufactured mica. It is used in commutators, electrical heating devices, motor slot insulation, transformers, etc.

***(3)* Glass and glass products**

Glass is an inorganic material made by the fusion of different metallic oxides.

The properties of glass are –

1. It is transparent, brittle and hard .
2. Glass is insoluble in water and usual organic solvents .
3. It has low dielectric loss, slow ageing and good mechanical strength .
4. It is susceptible to destruction when sudden & high temperature cycles are applied.

Uses

Glass is used in moulded insulating devices such as electrical bushings, fuse bodies, insulators. Glass is used as a dielectric in capacitors. Radio and television tubes, electrical laminated boards also make use of glass .

Glass Products

(*a*) **Silica glass or fused quartz** – Silica when heated to a temperature of fusion and then cooled is known as silica glass. This material has good electrical properties, low coefficient of expansion and high resistivity.

(*b*) **Borosilicate glass or Pyrex** – This glass requires 28 per cent of boron oxide along with other oxides. They resist the effect of chemicals and moisture better than other glasses.

(*c*) **Fibre glass insulation** – This is capable of withstanding high temperature. For most applications fibre glass is impregnated with materials like synthetic resins or with mineral oil. They possess good electrical & mechanical properties and sufficient flexibility to be moulded into required shapes.

(*d*) **Epoxy glass** – It is made by joining glass fibre layer with a thermosetting compound. It is immune to alkalies and acids and is used in PCB making, terminal holders and instrument cases.

(*iv*) **Ceramics** – Ceramics are materials made by high temperature firing treatment of natural clay and certain inorganic matters.

The properties of ceramics are –

1. Ceramics are hard, strong and dense.
2. Ceramics are not affected by chemical action except by strong acids and alkalies.
3. Stronger in compression than in tension.
4. Excellent dielectric properties.
5. Stable at high temperatures.

Uses

The capacity to withstand high temperature, immunity to moisture, good electrical properties make ceramics valuable for the use in different types of insulators, transformer bushing pins, fuse holders, plugs and sockets.

Main ceramic materials are-

(*a*) **Porcelain** – It is used in low frequency applications due to high dielectric loss factor. It has low electrical resistivity.

(*b*) **Steatite** – These are low dielectric loss porcelains, used for variable capacitor coils, switches, resistor shafts and bushings.It is also used for low voltage and high frequency applications.

(*c*) **Alumina** – It has high mechanical and dielectric strength. It also has high electrical resistivity and low dielectric losses. It is chemically stable and retains its properties over a wide range of temperature and frequencies. It is used for circuit breakers, spark plugs, resistor cores, integrated circuits and power transistors.

(*d*) **Zirconia** – It has poor thermal conductivity and shock resistance. It is available with calcium or yttrium. It is used for high temperature heating elements.

Ques.541: State the advantages and disadvantages of alloying steel with silicon for use as magnetic materials in transformer & electric machines .

Answer: Extensive use, is made of iron-silicon alloy called silicon steel for relatively strong alternating magnetic fields generally used in transformers, electrical rotating machines, reactors, electromagnets and relays. Silicon sharply increases the electrical resistivity of iron thus decreasing the iron losses due to eddy currents. It increases the permeability at low and moderate flux densities but decreases it at high densities. Addition of silicon to iron reduces the hysteresis loss. The magnetostriction effect is also reduced. A, steel with more than 5% silicon may be too hard and brittle to be easily workable. The introduction of silicon in steel was an important development. In the past, iron was used as the core material in the form of sheets, the material gradually deteriorated due to repeated heating and cooling. This difficulty was overcome by using silicon sheet steel as core material. High silicon sheet steel with silicon of about 4% is used in magnetic circuits of power transformers, which operate at flux densities of about 1 Wb/m2. The cores of rotating electrical machines are slotted with the teeth having small cross sectional areas. The flux density in the teeth is much higher, of the order of 1Wb/m^2. To obtain hign flux density in the case of high silicon steel, the magnetising current necessary is very large. Also punching slots in high silicon steel would be difficult because it is harder and more brittle.

Ques.542: Write short notes on PVC (Poly Vinyl Chloride) and its applications .

Answer: PVC is obtained from polymerisation of vinyl chloride in the presence of a catalyst like peroxides at about 50^{0}C. The properties like mechanical strength, porosity, flexibility, moisture absorption and electrical properties can be changed by adding certain materials. PVC has good mechanical and electrical properties. It is hard and brittle. It resists flame, most solvents and sunlight. It is non-hygroscopic. PVC is widely used as insulation and jacketing material for wires, cables etc . PVC films, tapes and sheets are used as insulation for dry batteries and conduit pipes.

Ques.543: What are the desirable properties of an insulation material?

Answer: The desirable properties of an insulation material are:

— Very good dielectric strength as that of mica (upto 80 kV/mm at 25 ^{0}C).

— Volume and surface resistivity equal to that of sulphur.

— Good mechanical strength like that of steel.

— Very high crushing resistance (as that of granite).

— Easy of machining.

— Good fire proofing qualities (as that of silica).

— It should have very high chemical inertness.

— It should have good water proofing qualities similar to that of paraffin wax.

Ques.544: With the help of energy bands explain how conduction takes place in conductors, semiconductors and insulators.

Answer: On the basis of energy , band materials are classified as insulators, conductors, and semiconductors. Insulators: Substance like wood, glass, which do not allow the passage of current through them are known as insulators. The valence band of these substances is full whereas the conduction band is completely empty. The forbidden energy gap between valence band and conduction band is very large (8ev) as shown in the fig (a).

Therefore a large amount of energy, i.e. a very high electric field is required to push the valence electrons to the conduction band. This is the reason, why such materials under ordinary conditions do not conduct at all and are designated as insulators.

Conductors: Substances like copper, aluminium, silver which allow the passage of current through them are conductors. The valence band of these substances overlaps the conduction band as shown in fig (b). Due to this overlapping, a large number of free electrons are available for conduction. This is the reason, why a slight potential difference applied across them causes a heavy flow of current through them.

Semiconductors: Substances like carbon, silicon, germanium whose electrical conductivity lies in between the conductors & insulators are known as semiconductors. The valence band of these substances is almost filled, but the conduction band is almost empty. The forbidden energy gap between valence and conduction band is very small (1ev) as shown in fig (c). Therefore comparatively a smaller electric field is required to push the valence electrons to the conduction band. This is the reason, why such materials under ordinary conditions do not conduct current and behaves as an insulator. Even at room temperature, when some heat energy is imparted to the valence electrons, a few of them cross over to the conduction band imparting minor conductivity to the semiconductors. As the temperature is increased, more valence electrons cross over to the conduction band and the conductivity of the material increases. Thus these materials have negative temperature co-efficient of resistance.

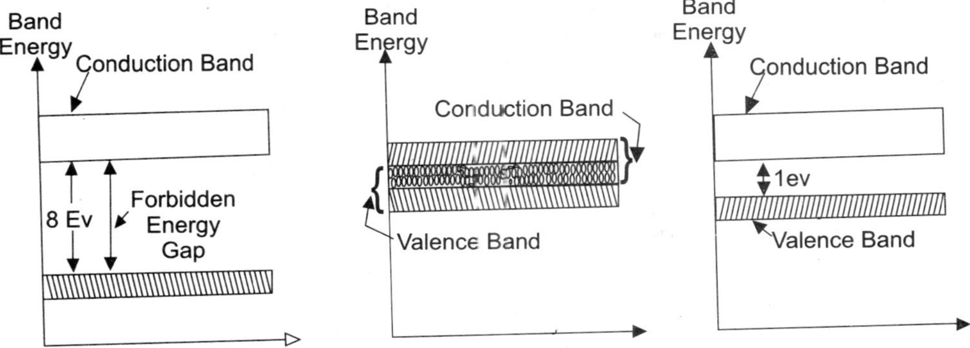

Ques.545: Write a note on superconductors.

Answer: Superconductors - The resistivity of most metals increases with increase in temperature and vice-versa. There are some metals and chemical compounds whose resistivity becomes zero when their temperature is brought near 0 Kelvin (-273 0C). At this stage such metals or compounds are said to have attained superconductivity. Example – mercury becomes superconducting at approximately 4.5 Kelvin (-268.5 0C). Superconductivity was discovered by Heike Kamerlingh Onnes. The transition from normal conductivity to superconductivity takes place almost suddenly; it occurs over a very narrow range of temperature about 0.05 K.

The temperature at which transition takes place from the state of normal conductivity to that of superconductivity is called transition temperature. Superconductors are used for producing very high magnetic fields of about 50 Tesla. Magnetic energy can be stored in large superconductors and drawn as required to counter the voltage fluctuations during peak loading. Superconductors can be used to perform logic & storage functions in computers.

As there is no, I^2R losses in a superconductor, so power can be transmitted through the superconducting cables without any losses. Superconducting property can be destroyed by applying external magnetic field as in fig (b). In fig (b) HC is the critical magnetic field and TC critical temperature.

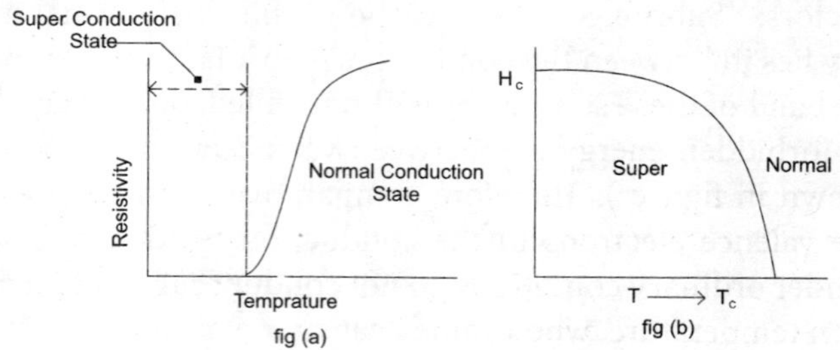

fig (a) fig (b)

Ques.546: Classify the conducting materials. Describe their properties.

Answer: Conducting materials are classified as low resistivity materials and high resistivity materials. Low resistivity materials: The conducting materials having resistivity between 10^{-8} to 10^{-6} ohm-m come under this category and are used in transmission & distribution lines, transformers and motor windings.

Properties of low resistivity materials:

(*a*) **Low temperature coefficient:** For minimum variations in voltage drop and power loss with the change in temperature, these materials should have low temperature coefficient.

(*b*) **Sufficient mechanical strength:** These materials must withstand the mechanical stresses developed during its use for particular applications.

(*c*) **Ductility:** The material to be used for conductors must be ductile so that it can be drawn and moulded into different sizes and shapes.

(*d*) **Solderability:** The conducting materials are required to be joined and the joint must have minimum contact resistance. These materials must have a good solderability.

(*e*) **Resistance to corrosion:** The material should have a high resistance to corrosion so that it should not be corroded when used in different environmental conditions.

High resistivity materials: The conducting materials having resistivity between 10^{-6} to 10^{-3} ohm-m come under this category and are used for making resistance elements for heating devices, precision instruments, rheostats etc.

Properties of high resistivity materials:

(*a*) **Low temperature coefficient:** For minimum variations in voltage drop and power loss with the change in temperature, these materials should have low temperature coefficient.

(*b*) **High melting point:** These materials, which are used as heating elements should have high melting point.

(*c*) **Ductility:** The material to be used for conductors must be ductile so that it can be drawn and moulded into different sizes & shapes.

(*d*) **Oxidation resistance:** The material should have a high oxidation resistance so that it is not get oxidised when used in different environmental conditions.

(e) **High mechanical strength:** These materials must withstand the mechanical stresses developed during its use for particular applications.

Ques.547: Discuss the chemical properties of insulating materials.

Answer: **Chemical resistance:** Presence of gases, water, acids, alkalies and salts affects different insulators differently. Chemically a material is a better insulator if it resists chemical action.

Hygroscopicity: Many insulators come in contact with the atmosphere during manufacture or operation or both. Moisture affects the electrical properties of the insulator.

Effect of contact with other materials: Insulation remains in contact with different types of materials like air, gases, moisture, conducting materials and structural materials. This can adversely affect the insulating properties.

Ageing: Ageing is the long time effect of heat, chemical action, and voltage application. These factors decide the natural life of an insulator.

Ques.548: Describe the properties and applications of Teflon.

Answer: Teflon is obtained by the polymerisation of tetraflloourethylene.

Properties:

— It has good electrical, mechanical and thermal properties.

— It can tolerate very, high temperature, without damage.

— Dielectric constant does not change with time, frequency and temperature.

— Its insulation resistance is very high.

— It is highly resistant to water absorption.

— It melts at 327°C.

— Its maximum usable temperature is 300°C.

Applications: Teflon is used as dielectric material in capacitors. It is used as covering for conductors and cables, which are required to operate at high temperature. It is used as a base material for PCB's.

Ques.549: Describe the properties & application of brass , bronze and asbestos .

Answer: Brass is an alloy of copper and zinc with 60% copper and 40% zinc.

Its properties are-

— Its electrical resistivity is 7.0X 10-8 ohm-m, which is higher than the copper resistivity.

— It is ductile and can be drawn into different shapes.

— It melts at 890°C.

— Its specific gravity is 3.3.

— It has got an excellent corrosion resistance.

— It has got good mechanical properties.

Applications:

Brass is used as a structural and current carrying material in power switches, plugs, sockets, lamp holders, fuse holders, knife switches, sliding contacts for starters & rheostats, wave- guide components, brazing purpose, jewellery and utencils etc.

Bronze: It is an alloy of copper and tin.

Properties :

— This alloy is very hard and brittle.

— Its corrosion resistance is better than brass.

— It is ductile and can be drawn into different shapes.

— In bronzes, which are used as electrical conductor, the content of tin and other metal is usually low as compared to the bronzes, which are used for mechanical applications.

Applications:

It is used for making structural elements for equipments. It is also used for making current carrying springs, sliding contacts, knife switches, bearings, marine work, turbine blading etc.

Asbestos: It is inorganic fibrous material. Two types of asbestos are available.

Chrysotile asbestos: It is hydrated silicate of magnesium.

— Its specific gravity varies between 2-2.8.

— It is highly hygroscopic.

— It has high dielectric losses and dielectric strength.

— The melting temperature is 1525 ^{0}C.

Amphibole asbestos:

It is found in Africa and Alaska.

— Its fibres cannot be woven easily as the fibres are too soft or too hard and brittle.

— It possesses good tensile strength.

— It is highly hygroscopic.

— Its electrical properties are poorer.

Uses of Asbestos:

Asbestos is used in low voltage work as insulation in the form of rope, tape, cloth and board. It is impregnated with liquid or solid resin in all such applications to improve its mechanical and electrical properties. It is used as insulator in wires & cables under high temperature conditions, as conductor insulator &layer insulator in transformer, as arcing barrier in switches and circuit breakers.

Ques.550: Write notes on Applications & properties of aluminium and copper.

Answer: Aluminium:

Properties :

— Pure aluminium is silver white in color.

— It is a ductile metal and can be put to a shape by rolling, drawing and forging.

— It melts at 655 ^{0}C.

— It is resistant to corrosion.

— Its tensile strength is 60MN/m^2.

— It can be alloyed with other elements.

— Annealing can soften it.

— It has a higher thermal conductivity.

Applications:

Aluminium is widely used as conductor for power transmission and distribution. It is used in overhead transmission lines, busbars, ACSR conductors etc.

Copper:

Properties

— It is reddish brown in color.

— It is malleable & ductile and can be cast, forged, rolled, drawn & machined.

— It melts at 1083 ^{0}C.

— It easily alloys with other metals.

— Electrical resistivity of copper is 1.7×10^{-8} Ω-m.

— Tensile strength for copper is 210 MN/m^2 .

— It is highly resistant to corrosion.

— It is a non-magnetic material.

Applications:

Copper is used in conductor wires, coil windings of generators & transformers, cables, bus bars etc.

Ques.551: Differentiate between conductors, semiconductors and insulators .

Answer: Materials, which are commonly used in electrical & electronics engineering, can beclassified as conductors, insulators and semiconductors.

Conductors: These are the material, which allow the current to pass through them. These have very low electrical resistance and are available in a large variety having different properties. Also the number of valence electrons is less than four. The, valence- band and the conduction- band overlap each other. Examples are copper, brass, aluminium, silver, gold, bronze, etc.

Semiconductors: These are the materials, which possess the electrical resistivity in between that of conductors and insulators. They are used for the manufacture of diodes and transistors. Also the number of valence electrons is equal to four. There is a small forbidden energy gap of about 1eV between the conduction and the valence band. Examples: germanium, silicon, selenium, etc.

Insulators: These are the materials, which do not allow the current to pass through them without any appreciable loss. They have very high electrical resistance and are also available in a large variety to cover different applications. Some of the specific insulating materials are used for the purpose of storing of an electrical energy and are called dielectric materials such as mica, ceramic, paper etc. These materials are used as a dielectric in capacitors. Also the number of valence electrons is more than four. The energy gap between valence and conduction band is very large (more than 5-6 eV). *Examples:* Mica, rubber, ceramics, glass, diamond etc.

Ques.552: Discuss the advantages and disadvantages of aluminium as compared to copper as a conductor of electricity.

Answer: Advantages of aluminium as compared to copper as a conductor of electricity: The electrical conductivity of aluminium is next to that of copper.Its electrical resistivity is 2.8×10^{-8} ohm-m, i.e. about 1.6 times higher than copper.

Its density is 2.68 which means that aluminium is much lighter than copper. Its melting point is 655°C. Like copper, it can be easily drawn into thin wires.

Aluminium is soft metal but when alloyed with some other materials like magnesium, silicon or iron, it acquires higher mechanical strength and can be used for overhead transmission lines.

Like copper, aluminium also forms an oxide layer over its surface when exposed to atmosphere and that layer prevents the material from further oxidation and acts as a resistance layer to corrosion.

Disadvantages: As aluminium is a soft material, there is always a possibility of loosecontacts. Due, to the insulating property of aluminium oxide formed on the surface, it isdifficult to solder aluminium wires. However for applications like winding of electrical machines and transformers, it is difficult to substitute aluminium for copper. This is because aluminium wires have lower tensile strength than that of copper.

Ques.553: What is the effect of overheating on the life of an electric insulator?

Answer: An insulator is designed to withstand certain amount of heat. But when an insulator is overheated, dielectric losses will increase. Also overheating will affect the various important properties as electrical properties, mechanical strength, hardness, viscosity, solubility etc.

Ques.554: Discuss the classification of insulating material on the basis of their physical & chemical structure .

Answer: Insulating materials, on the basis of their physical and chemical structure may be classified in various categories as follows: Fibrous materials: They are derived from animal origin or from cellulose, which is the major solid constituent of vegetable plants. The majority of materials are from cellulose. This includes paper, wood, card- board, cotton, jute and silk. Impregnated fibrous material: The fibrous materials are impregnated with suitable impregnated oil, varnish, and epoxy resin to improve its thermal, chemical & hygroscopic properties.

Non-resinous materials: Solid or semisolid insulations which are directly available in nature and are organic based come under this class. These materials are mineral waxes, asphalts, bitumen and chlorinated naphthalene.

Insulating liquids: Apart from working as insulation, they fulfil other important requirements like they offer good heat dissipation media, they are used for extinguishing arcs in certain applications like circuit breakers. They include vegetable oils, fluorinated liquids, mineral insulating oils and synthetic liquids.

Ceramics: They are materials made by high temperature firing treatment of natural clay and certain inorganic matters. They are used as dielectric in capacitors, as insulators etc. Mica and mica products: It is an inorganic mineral and one of the best natural insulating materials available. Mica is used as a dielectric in capacitors, as insulator. Some of the mica products are glass bonded mica, synthetic mica, mica paper, manufactured mica etc .

Asbestos and asbestos products: These are strong and flexible fibres. It finds extensive use in electrical equipment as insulation because of its ability to withstand very high temperatures. Some of the asbestos products are: asbestos roving, asbestos paper, asbestos tapes and asbestos cement.

Glass: Glass is an inorganic material made by the fusion of different metallic oxides. It is normally transparent, brittle and hard. Glass finds use in electrical industry because of its low dielectric loss, slow ageing and good mechanical strength. Glass is used in electrical bushings, fuse bodies, insulators, radio and television tubes.

9. Natural and synthetic rubber: Natural rubber is obtained from the milky sap of rubber trees. It finds limited applications because it is rigid when solid, sticky when warm and gets oxidised, when exposed to atmosphere. Synthetic rubber are of various types such as butadiene rubber, butyl rubber, chloroprene and silicon rubber which are obtained by the polymerisation. Synthetic rubber, are used as insulating material for wires and cables. It is also used as jacketing material for cables.

Insulating resins and their products: Plastic or resins are of two types – one derived from plant & animals the other is synthetic resin obtained from chemical reactions. Natural resins are used as binder material. It is used as thickening agent for manufacture of mineral insulating oils. Synthetic resins are used as insulation, manufacture of switches and instrument mountings, electrical bushings, radio and television cabinets etc.

Laminates, adhesives, enamels and varnishes: Laminates are multiple, thin layers or sheets of insulating materials like that of mica, paper, cloth, glass etc, bonded together.

Adhesives, is a class of material compositions required to carry out bonding between two or more solid surfaces. Adhesives are used in the manufacture of laminated boards, coil winding cylinders, rods, tubes and special shaped insulators.

Enamel is a fusible insulated coating of some organic base material, which is generally applied on conducting surface. Enamel finds extensive use in coating wires used for the windings of low rated motors, transformers, various types of instruments, etc.

Varnish is a liquid, which when applied to a surface dries resulting in hard shining coating which is resistant to air and water. Lacquer is used for protecting wood and metal surface from external weather conditions.

Ques.555: Write a note on *(i)* Corrosion. *(ii)* Bimetals.

Answer: *(i)* **Corrosion :** The process of constant eating (destruction) up of metals (from the surface) by the surrounding is called as corrosion. The metals are corroded when exposed to the atmosphere. The metals are generally converted into their oxides. This oxide covers the surface of the metal, which results in the destruction of the metal. Rusting of iron is the most common example of corrosion in which iron makes iron oxide with reaction with the oxygen of the atmosphere. The iron oxide covers the surface in the form of brownish powder. Therefore the conducting material should be corrosion resistant.

(ii) **Bimetals:** A bimetal is made of two metallic strips of unlike metal alloys with different coefficients of thermal expansion. At a certain temperature the strip will bend and actuate a switch or a lever of a switch. The bimetal can be heatedly directly or indirectly. When heated the element bends so that the metal with the greater coefficient of expansion is on the outside of the arc formed while that with smaller coefficient is on the inside. When cooled the element bends in the other direction. Alloys of iron and nickel with low coefficients of thermal expansion are used as one element of the bimetallic strip. The other element consists of materials having high values of thermal expansion. Examples are iron, nickel, constantan, brass etc. Bimetallic strips are used in electrical apparatus and in devices such as relays & regulators.

Ques.556: Explain *(i)* Eddy currents. *(ii)* Magnetostriction. (iii) Permeability.

Answer: *(i)* **Eddy currents:** Magnetic materials placed in alternating magnetic fields also have eddy currents induced in them. This is because the material is subjected to rate of change of flux linkages and in accordance with Faraday's Law of electromagnetic induction, emfs are induced in the material causing currents, called eddy currents, to flow in the material. These currents cause loss of energy. This results in the heating up of the material. Eddy current loss is proportional to the square of the frequency and the square of the thickness of the material and inversely proportional to the resistivity of the material. In order to reduce this eddy currents loss thin sheets called laminations are used instead of solid core.

(ii) **Magnetostriction:** When ferromagnetic materials are magnetized a small change of dimensions of the material takes place. There is a small extension with corresponding reduction of cross-section of the crystals of which the material is made. When subject to rapidly alternating magnetic fields there is a rapid & continuous extension and contraction of the material. This is called magnetostriction. Magnetostriction is the major cause of hum in transformers and chokes.

(iii) **Permeability:** It is defined as the capability of the material to conduct flux. It is defined as the ratio of magnetic flux 'B' in a medium to the magnetic flux intensity 'H' at the same location in the medium, *i.e.* = B/H, where B is plotted against H, a curve is obtained, called magnetization curve or B-H curve.

Also $\mu = \mu 0 \mu r$, where, $\mu 0$ = absolute permeability of air and μr = relative permeability for air, μr =1.The permeability of any material is not a constant.

The permeability at low value of H is called initial permeability. The common core materials such as low carbon steel, silicon steel have low initial permeability.

Ques.557: Discuss the properties and uses of

(i) Cold rolled grain oriented (CRGO) steel.

(ii) ALNICO.

Answer: *(i)* **Cold rolled grain oriented steel:** The grain orientation of silicon steel is obtained by a special technique called cold rolling. Sheet steel obtained in this process is called Cold Rolled Grain Oriented Steel. CRGO silicon steel is widely used for making transformer cores. The magnetising current required by transformers using CRGO steel is low. If CRGO sheet steel is used as core material for rotating machines, the core will be assembled from a large number of sections but this still will not result in grain orientations completely parallel with the flux path because of the circular nature of cores of rotating machines. Also assembling the core from sections of sheet steel will make the construction difficult and expensive.

(ii) **Alnico:** is an alloy of iron, cobalt, nickel and small amount of aluminium & copper.

Properties of Alnico -

— Saturation flux density is 1.2 Wb/m^2.

— More expensive than Alni, but magnetic properties are better than Alni.

— Available in different grades, each having some different properties.

— Hard & brittle, so it cannot be machined but has to be cast to shape and finished by grinding.

— Magnets made with this alloy are smaller in size and lighter in weight than that made with tungsten steel.

— The hysteresis loop is more rectangular.

— Thermal & mechanical process, modify the crystal size & shape and thereby alter the shape of the hysteresis curve.

Uses: Alnico magnets find applications in loudspeakers, microwave devices, motors, generators, separators, vending machines and communication devices.

Ques.558: Explain the factors which change the resistivity of a conducting material .

Answer: Factors, which change the resistivity of conducting materials are : -

Temperature- The electrical resistance of most metals increases with increase of temperature while those of semiconductors and electrolytes decreases with increase of temperature. Alloying- Alloying is another factor, which affects the resistivity of a material. By the addition of some impurity to a metal, its resistivity can be changed. Alloys have higher resistivity than that of pure metal.

Mechanical stressing-When a material undergoes a mechanical treatment, its resistivity changes due to mechanical distortion of the crystal structure. Age Hardening- Age hardening increases the resistivity of an alloy.

Ques.559: Explain the effect of magnetic field on superconductors.

Answer: A large number of metals become superconducting below a temperature, which is the characteristic of the particular metal. They have zero resistivity and the temperature at which this change takes place is called superconducting transition temperature. Metals, which are good conductors at room temperature like gold, silver, tin, do not exhibit superconducting properties.Whereas metals & compounds which have superconducting properties at certain temperatures are insulators at room temperature.

It is possible to destroy superconductivity by the application of a strong magnetic field.

When, the magnetic field exceeds a certain critical value, the superconducting state disappears, the magnetic field penetrates the material and electrical resistance is restored.

Ques.560: Give the properties and uses of silicon iron alloy and nickel iron alloy.

Answer: **Silicon Iron alloy:** Pure iron has low resistivity, which results in higher eddy current losses. These losses can be minimized by increasing the resistivity of the material, which is achieved by adding 1 to 4 % of silicon to iron. Silicon increases the electrical resistivity of iron. It reduces hysteresis loss. The magnetostriction effect is also reduced. Silicon also improves resistance to corrosion & oxidation and increases hardenability.

Silicon Iron alloy is used in the form of thin sheets called laminations. These laminations are used in transformers, small machines and large turbo generators.

Nickel Iron alloy: A group of iron alloys containing between 40 to 90 % nickel have much higher permeabilities at low flux densities and lower losses than ordinary iron. The important alloys are permalloy and mumetal. Mumetal has lower permeability but higher resistivity.

Addition of small amounts of other elements to nickel iron alloys improves their magnetic properties. Nickel improves strength, toughness and resistance to fatigue. It also lowers the critical cooling rate and hence increases hardenability.

Nickel Iron alloy is widely used in transformer cores and loading coils for telephone circuits, instrument transformers, for magnetic circuits of instruments, for magnetic screens of electronic equipments.

Ques.561: Discuss the applications of carbon and steel.

Answer: Carbon:

Carbon materials used in the field of electrical engineering are manufactured from graphite and other forms of carbon like coal etc. It has very high value of resistivity. It has negative temperature coefficient of resistance. It is pressure sensitive and has low surface friction.

Applications: Carbon is used in applications like brushes for electrical machines & apparatus, electrodes for electric arc furnaces, in automatic voltage regulators for making the pressure sensitive pile resistors, components for telecommunication equipment, battery cell element, arc lamps etc.

Steel:

Steel contains iron with a small percentage of carbon added to it. Iron itself is not very strong but when carbon is added it assumes very good properties. Mild steel contains 0.25% of carbon; medium steel contains 0.45% of carbon and high carbon steel about 0.70% of carbon and above. The resistivity of steel is 8 to 9 times higher than that of copper.

Applications: Galvanised steel is used in overhead telephone wires and earth-wires. Steel alloyed with chromium and aluminium is used for making starter rheostats in which lightness combined robustness and good heat dissipation are important.

Ques.562: Give some applications of constantan, German silver, manganin, and nichrome.

Answer: Constantan: It is an alloy of copper and nickel and is used in precision resistors and thermocouples, motor starters.

German silver: It is an alloy of copper, nickel and zinc. It is used in electrical measuring instruments.

Manganin: Alloy of manganese, copper and nickel. It is used for making precision resistances and shunts in measuring instruments and for making their coils.

Nichrome: It is an alloy of nickel, chromium, iron and manganese. It is used for making heating elements for electric iron and other heating appliances.

Ques.563: Give the classification of magnetic materials and explain each with examples.

Answer: Diamagnetic Materials: These are the materials whose atoms do not carry permanent magnetic dipoles. If an external magnetic field is applied to a diamagnetic material, it induces a magnetization in the direction opposite to the applied field intensity. For these materials the relative permeability is negative. These are hardly used as magnetic material in electrical/ electronic engineering applications. Ex. Aluminium oxide, copper, gold, barium chloride.

Paramagnetic Materials: The atoms of these materials contain permanent magnetic dipoles. Individual dipoles are oriented in random fashion such that resultant magnetic field is zero or negligible. For these materials relative permeability is slightly greater than unity and it is independent of magnetizing force. Ex. Chromium chloride, chromium oxide, manganese sulphate, air.

Ferromagnetic Materials: These are materials in which magnetic dipoles interact in such a manner that they tend to line up in parallel. A ferromagnetic substance consists of a number of regions or domains, which are spontaneously magnetized. The direction of magnetization varies from domain to domain. The resultant magnetization is zero or nearly zero. The relative permeability is very high. The ferromagnetic materials are widely used in industries. Ex. Iron, nickel, cobalt.

Antiferromagnetism: The magnetic dipoles are aligned antiparallel to each other.

Antiferromagnetic materials are not so well known as the common ferromagnetic materials. Only a few, such as manganese, chromium, ferrous, nickel oxides exist. The phenomenon occurs below a certain temperature known as the Neel temperature.

Ferrimagnetic Materials: In these materials unequal magnetic dipoles are lined up antiparallel to each other. Permeabilities are of the order that of ferromagnetic materials. Under ordinary conditions the magnetic characteristics of ferromagnetic materials are similar to those of ferromagnetic materials, as they show spontaneous magnetization below a certain temperature . Example: magnetite, nickel ferrite.

Ques.564: What are thermocouples? Give two examples of the most commonly used thermocouples.

Answer: Thermocouples are used for the measurement of temperature. When two wires of different metals are joined together an emf exists across the junction which is dependent upon the types of metals or alloys used and also directly proportional to the temperature of the junction. Depending on the range of temperature to be measured, proper materials are to be chosen for a thermocouple. Its one junction called the cold junction is held at a known constant temperature,the e.m.f. produced becomes measure of the temperature of the other junction. The emf produced by the thermocouple is very small but it can be measured with reasonable accuracy by a sensitive moving coil mili voltmeter, which can be calibrated in terms of temperature. Some of the materials used for thermocouples are copper/constantan, iron/constantan , platinum/rhodium etc .

Ques.565: Discuss polymers and their applications.

Answer: Polymeric materials or plastics comprise a large group of organic or organo metallic high molecules compound. The common properties of these materials are their ability to soften and even melt, ability to pass into a liquid state, insolubility in water and solubility in one or more organic solvents.

The mechanical properties of these materials vary widely, some can be spun into fibers like nylon and terrylene , other can be moulded and are hard & glass like in mechanical properties. Other group shows rubber like properties.

Plastic are synthetic resins obtained basically in two different ways, by linear polymerization and by polycondensation. The polymers obtained by linear polymerization are known as thermoplastics since they can be repeatedly melted or dissolved in various solvents. The properties of the thermoplastics do not change considerably , if they are melted and then cooled and solidified. Example of thermoplastics is polystyrene.

Second types of polymers are thermosetting and are used as binding agents and varnish base. Thermosetting plastic melt on heating but are converted into a rigid solid mass if maintained at an elevated temperature. Examples of these polymer is phenol formaldehyde. Synthetic resins are widely used in electrical industry as insulating and structural components. Some of the important synthetic resins commonly used are polystyrenes, P.V.C.(polyvinyl chloride.) etc

Ques.566: Discuss applications of dielectrics.

Answer: Application of dielectrics:

For different applications different properties of dielectric materials are required such as: electrical, mechanical, thermal and chemical. Electrically the insulating material should have high resistivity to reduce the leakage current and high dielectric strength to enable it to withstand higher voltage without being punctured. Further, the insulator should have small dielectric loss. Since the insulators are used on the basis of volume and not weight, a low density is preferred.

Liqid and gaseous insulators are used also as coolants.

For example: transformer oil, hydrogen and helium are used both for insulation & cooling purpose. For such materials good thermal conductivity is a desirable property.

Materials with large electronic and ionic Polariability and therefore large permittivity are used for making dielectric capacitors. Titanium oxide which has a permittivity of about 100 is a good example of such a material.

Some of the important dielectric materials and their applications are :

Mica: Mica sheets are used for insulating leaves between commutator segments.

Porcelain: It is used for high voltage transmission line insulators, conductor rail support on railways and high voltage switch parts.

Bakelite: It is used for small moulded parts such as lamp holder, terminal blocks, switch covers etc.

PVC: Cables with PVC insulation are not affected by oil and petrol and are therefore widely used in aircraft and factories.

Cotton and silk: Cotton covered wire is widely used for the winding of small magnet coils, armature winding of small and medium sized machines, small transformer coil choke etc. silk is more expensive than cotton but takes up less space and is therefore used for winding in fractional horsepower machines.

Ques.567: Classify the engineering materials. Explain the different types of plain carbon steels with their properties and applications.

Answer: The engineering materials are those materials which are used for engineering and structural purpose, such as to make bridge, road, machine, building, etc. These can be classified as:

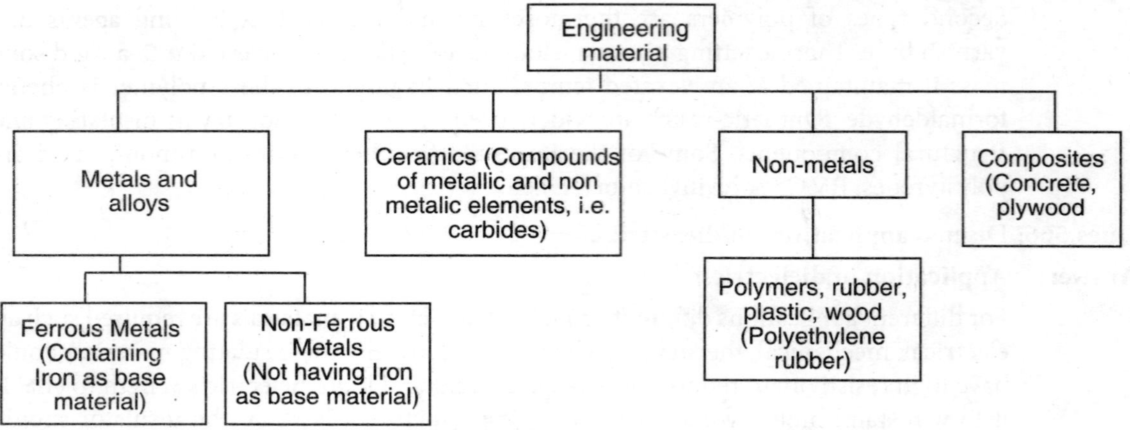

Industrial application of engineering materials:

Sr. No.	Classification	Engineering Material	Application
1	Metals	Steel	Used for making of buildings, automobiles, trains, machine and their components.
		Gold	Used for making of ornaments, computers, coins.
2	Non-metals	Plastics	Used in the making of pipelines, cable insulations, chairs.
		Wood	Doors, windows, bridges, furniture.
3	Ceramic	Aluminum oxide	Used in the grinding wheels, abrasives
		Silicon carbide	Used in the cutting tools.
4	Composites	Fiberglass	Used in the making of bulletproof jackets.
		CFRP	Used for aerospace and sports applications.
5	Advanced Materials	Semiconductors, Bio- Materials, Smart Materials,Shape memory alloy	Used in the making of electronic equipments, Used in the replacement of damaged body parts, Used in the sensors and actuators.

Ques.568: Describe various mechanical properties of materials .

Answer: Various mechanical properties of materials are described below :

Strength: Strength is the measure of the ability of the material to with stand or to resist external forces. Higher the strength of the material, higher will be the resistance offered by the material to the applied external force.

For example a wooden bar will break easily and iron bar will not break easily because the resistance offered by the iron bar is very high.

Depending upon the type of loading the strength may be **tensile, compressive, shear, torsional and bending Strength.**

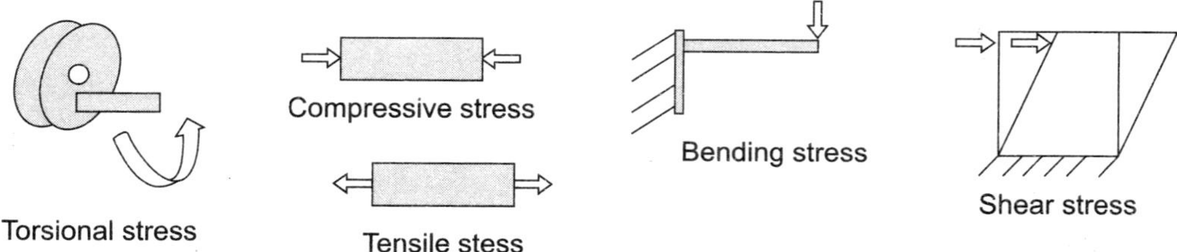

Strength of the material is represented or measured by the stress. The stress is defined as the internal resistance set up by the molecules of the material to resist the deformation on application of the external force.

$$\text{Stress} = \frac{\text{Force or Load Applied}}{\text{Cross Sectional Area}}$$

Elasticity: Elasticity is the property of material because of which the material is able to regain its original shape and size after removal of deforming forces. As in case of a rubber ball, if the ball is pressed between two hands it gets deformed and as soon as we release the force of the hands the ball regains its original shape and size. Steel is more elastic than rubber.

Plasticity: Plasticity is the property of the material because of which the material is able to retain (maintain) the deformed shape even after removal of deforming forces.

Stiffness: Stiffness is also called as the rigidity and is the ability of the material to resist the deflection or Elastic deformation. Stiffness is generally related to the springs, thicker springs are stiffer.

Stiffness = Load Deflection

Malleability: Malleability is the ability of a metal to be hammered into thin

sheets. It is the ability of a solid to bend or be hammered into other shapes without breaking. Examples of malleable metals are gold, iron, aluminum, copper (to a degree) and lead.When a piece of hot iron is hammered it takes the shape of a sheet.

Ductility: *Ductility is the property of the material by virtue of which the material can be drawn into thin wires without rupture or fracture.*It is

the ability of a metal to withstand elongation (deformation in

longitudinal direction) under tension without rupture.

***Example of ductile material**: Gold, silver, aluminum.*

Brittleness: Brittleness is the property of the material because of which the material fails or breaks or fractures or ruptures without any appreciable amount of deformation, before failure.In other words, lack of ductility is brittleness. An elongation of less than 5% is often taken to indicate a brittle material.

***Example of brittle material**: Chalk, glass, cast iron etc.*, it will not stretch and bend before breaking.

Toughness: *Toughness is a measure of the amount of energy a material can absorb before failure or fracture takes place.* The toughness of a material is expressed as energy absorbed

(Nm) per unit volume of material (m^3) or Nm/m^3. *Toughness is related to impact strength i.e. resistance to shock loading and it is desirable in the components subjected to shock and impacts such as gears, hammer, anvil. The area under the stress and strain curve of brittle and ductile material represents the toughness of the brittle (X-Y-Z) and ductile (X-A- B-C-D-E-F-G) materials.*

Example: If a load is suddenly applied to a piece of mild steel and then to a piece of glass or chalk, the mild steel will absorb much more energy before failure occurs. This indicates that mild steel is tougher than glass.

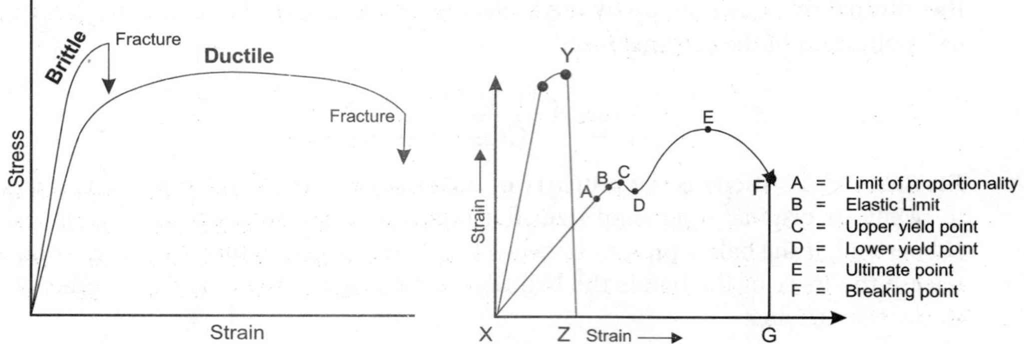

Stress-stain curve for duche material (e.g. mid steel).And Brittle material (e.g. cast iron)

Hardness: Hardness is usually defined as resistance of material to penetration, indentation or scratching action. Hard materials resist wear and scratches.

Example: Diamond is the hardest known material.Stone, tungsten carbide, etc. Hard materials are chosen for the manufacturing of cutting tools.

Resilience: Resilience is the ability of a material to absorb energy when it is deformed elastically, and release that energy upon unloading. The deformation is carried out within the elastic limits (Point B in the stress strain diagram).

Machine ability: The Machine ability represents the ease or suitability with which the material can be cut or removed to make a desired and size. Machine ability is expressed as a comparative index by comparing the metal under consideration with free cutting steels.

Formability: Formability represents the ease or suitability with which the metal can undergo plastic deformation without fracture. Formability improves at higher temperatures for most of the metals.

Weld Ability: Weld ability represents the ease or suitability with which the metal can be weld by welding processes. The welding processes are the joining or fabrication processes.

Stress strain diagram: The stress strain diagram is the curve plotted on the paper when the metal specimen is subjected to tensile stressesupto the fracture point.The stress strain diagram for the ductile (Curve from X-A-B-C-D-E to F)and brittle (X to Y) material is shown in figure. The various points of interest are:

Proportional Limit or limit of proportionality: It is the maximum stress up to which Hook's law is valid i.e. upto this point stress is directly proportional to strain.

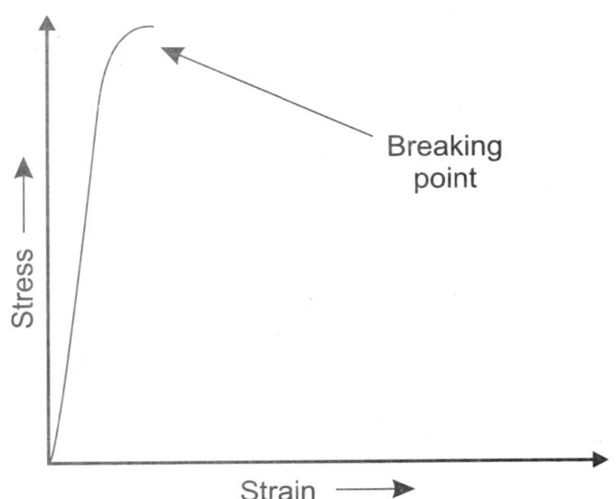

Stress and strain curve for Brittle material

Elastic Limit: The elastic limit is the maximum stress which the material can withstand without causing permanent deformation *i.e.* after the removal of load; material regains its original dimensions.

Yield Point: If the material is stressed beyond elastic limit, plastic stage will reach *i.e.* with the removal of load; the material will not be able to recover its original size and shape.

Beyond elastic limit, strain increases at a faster rate with very little increase in stress. At point C material yields and there is an appreciable strain without any increase in stress. For mild steel, there are two yield points C and D. The point C and D are called upper and lower yield point respectively. The stress corresponding to yield point is known as yield point stress.

Ultimate Tensile Stress (UTS): If further load is applied beyond point D, the stress goes on increasing till the point E is reached. At E, the stress attains the maximum value known as ultimate tensile stress.

Breaking Stress: After the specimen has reached the ultimate stress, a neck is formed, which decreases the cross-sectional area of the specimen. The stress corresponding to point F (**fracture point**) is known as breaking stress.

Difference between (properties of) Ductile and Brittle materials

Ductile material	Brittle materials
Ductile materials can be converted into thin wires.	These cannot be converted into thin wires.
These materials have high toughness.	These materials have low toughness.
These materials show appreciable amount of deformation before failure.	The materials fail without any appreciable amount of deformation
Failure takes place with cup and cone feature	The failure surface is shard and reflects light
Example: Gold, silver, copper	Example: Chalk, glass, Concrete,

Ques.569: Explain fatigue, creep and fracture of materials with neat sketches .

Answer: **Fatigue:** *Fatigue is the behavior of the material when subjected to fluctuating, cyclic or repeated loads, which is different from the one when under steady loads. In this case the* *material fails much below the yield strength of the material.* **The fatigue failures occur due to repeated, cyclic, alternating or reversing loading/ stresses.**

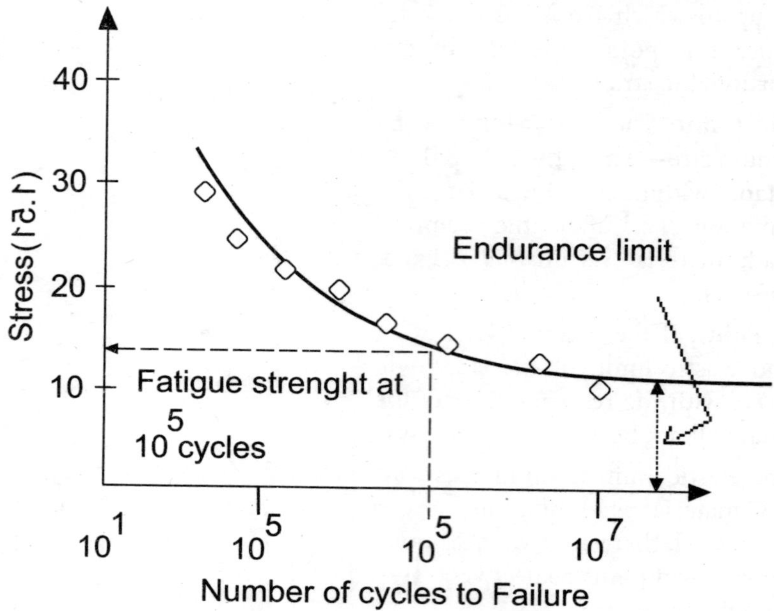

Fatigue curve OR stress and number of cycles before failure (S/N Curve)

The fatigue properties of a material determine its behavior when subjected to thousands or even millions of cyclic load applications in which the maximum stress developed in each cycle is well within the elastic rangeof the material. Under these conditions, failure may occur when the material continues to give service indefinitely. Bridges and, many components of high speed aero and turbine engines have failed because of fatigue.

The stress at which the material fails due to fatigue is known as fatigue strength, corresponding to the number of loading cycles.

Endurance limit or fatigue limit is the limiting value of stress, much below the normal yield stress, at which the material will not fail, regardless of number of load cycles and this may taken to be infinite life.

Creep: *Creep is a time dependent phenomenon and observed after a long period of time. Creep is the slow plastic deformation of metal with time, under constant stresses usually at high temperature.* It can take place and lead to failure at static stresses much smaller than those which will fail the specimen by loading it quickly. Metals generally exhibit creep at high temperatures, whereas plastics, rubbers are very temperature sensitive to creep.

Creep can be classified as primary, secondary and tertiary creep depending upon the rate of deformation. The plot between the *strain developed in the specimen* with respect to *time* is known as creep curve. The creep curve has four distinct regions such as:

Instantaneous Strain: This is produced instantly as soon as the load is applied.

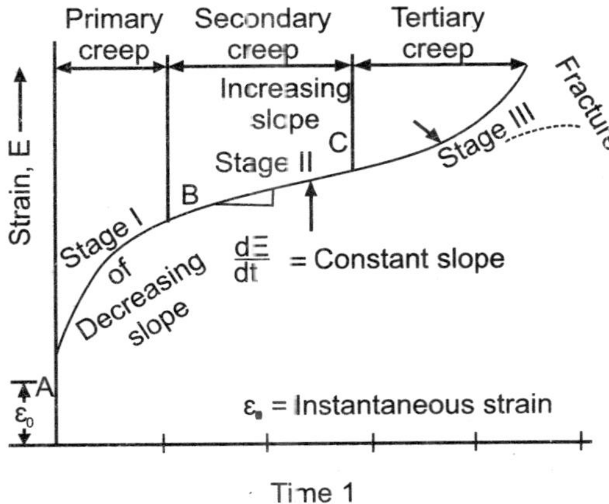

The creep curve showing the four stages in creep curve instantaneous deformation, primary, secondary and tertiary stages of creep

Primary creep: The initial portion of creep curve extending from A to B with a decreasing rate of deformation and this covers a short part of total period.

Secondary creep OR steady state creep: The middle part of the creep curve (B to C) with a long span and nearly constant rate of

deformation. The constant rate is obtained

because of the balance between the strain hardening and softening due to temperature.

Tertiary creep: The last part of the creep curve (C to fracture point) with an increasing rate of deformation and final failure or fracture.

Effect of temperature (or stress) on creep: The temperature and stresses decrease the time period and the failure takes place earlier with increasing rate of deformation.

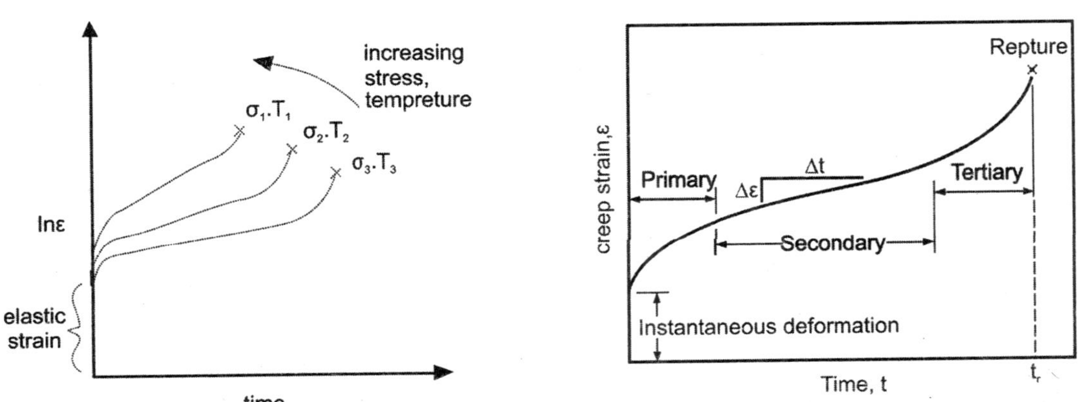

Effect of temperature (or stress) on creep

Fracture: Fracture means the failure, of material under the application of external load/stress/force by, breaking into two or more pieces.

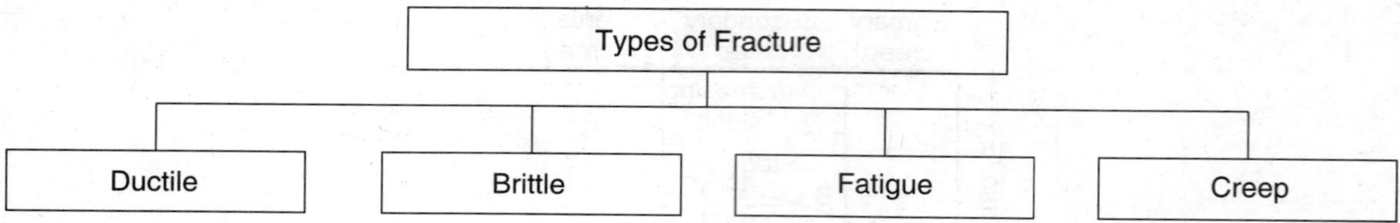

Ductile fracture is the failure of material after considerable amount of plastic deformation and associated with neck formation & cup and cone formation.

Brittle fracture is the failure of material without appreciable plastic deformation. If the two

broken pieces are placed together, then the original shape can be obtained.

Fatigue and creep fractureare fractures which takes place when the part is subjected to fatigue and creep respectively. Fatigue fractures take place in brittle manner whereas the creep fracture take place in ductile manner.

Differentiation/ comparison/ characteristics of ductile and brittle fracture

Ductile fracture	Brittle fracture
Associated with large plastic deformation	No or Minimum plastic deformation
A rough & dirty surface	Sharp and shiny surface
Cup and cone formation	Fracture surface is generally Plane
Two broken pieces will not regain original shape if placed together	Two broken pieces will regain original shape if placed together

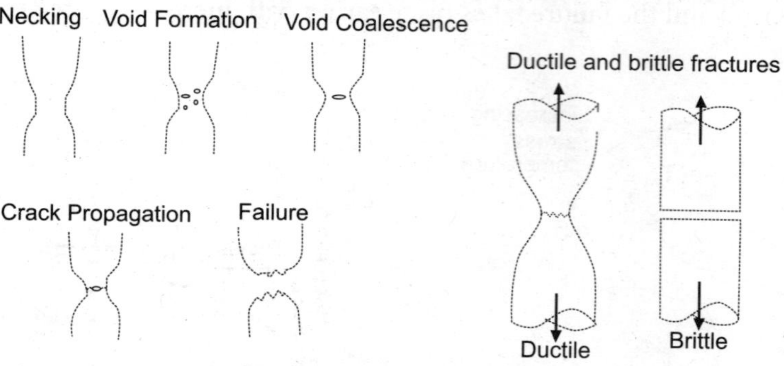

The Ductile and brittle fractures

The ductile and brittle fractures

Ques.570: Classify steels based on percentage of carbon with their applications. How does steel become stainless steel?

Answer: Steel is an alloy of iron, carbon and silicon; it may contain carbon upto 2%. (Practically upto 1.7%). Steel may be categorized into plain carbon and alloy steels.

Plain carbon steels (Carbon steels) are those steels which contain primarily Iron and Carbon. Silicon, Manganese, Sulphur considered as impurities. The increase in carbon content. and Phosphorus are present in small quantity but these are properties of plain carbon steels are greatly influenced by an

Stainless steel (OR Corrosion resistant steels): The addition of chromium in the low carbon steels makes them corrosion resistant, chromium reacts with the oxygen and forms a oxide layer on the metal surface and protects the metal from corrosion. The amount of chromium added may be from 10 to 18%. **Classification:**Depending upon the percentage of carbon the steels may be classified as and their applications and properties are represented by the figure.

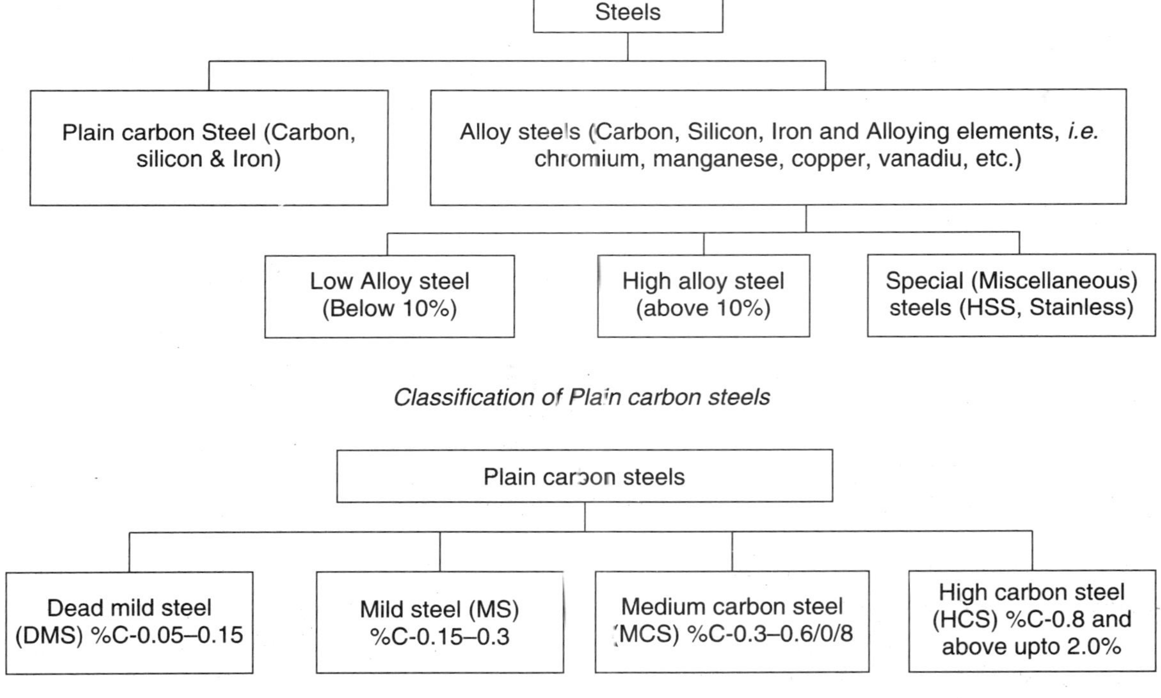

Classification of Plain carbon steels

Type	% of Carbon	Property	Uses/ Application
Dead mild steel (DMS)	**Below -0.15**	**Good ductility and formability**	**Rods and sheets**
Mild steel (MS)	**0.15-0.3**	**Good weld ability**	**Structural sections, making of doors and angles**
Medium Carbon steel (MCS)	**0.3-0.6/0.8**	**Good strength and toughness**	**Axles, springs and gears, Hand tools**
High carbon steel (HCS)	**0.8 and upto 2.0% above**	**High wear resistance & hardness**	**Dies, drills and cutting tools**

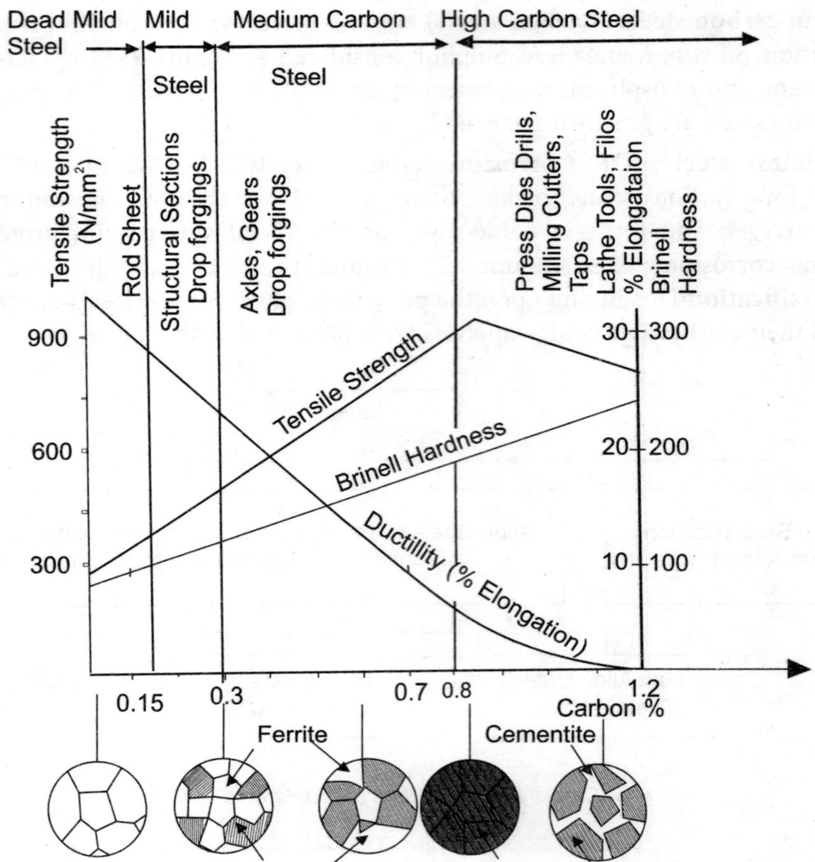

Classification, properties and uses of plain carbon steel as the % of carbon

Ques.571: Write short notes on Cast Iron .

Answer: **Cast Iron** is primarily an alloy of iron and carbon. The carbon percentage varies from 2 to 4.5 per cent. It also contains small amount of silicon, manganese, phosphorus and sulphur. The cast iron is obtained by re-melting pig iron with coke, limestone and steel scrap in a furnace known as **cupola**.

Properties of cast iron:

Cast iron is brittle.

1. Cast iron is weak in tension so it cannot be used for making bolts and machine parts which are liable to tension.
2. It is having low cost.
3. It is having good casting characteristics.
4. High compressive strength.
5. High wear resistance.
6. Good corrosion resistance

Uses of Cast Iron:

1. In the making of castings, hand pump
2. Water pipes
3. Sewerage pipelines
4. Beds of machines such as lathe machine, drill machine, etc.
5. Bodies of motors and engines

Classification: Cast Iron may be classified as grey cast iron, white cast iron, mottled cast iron, nodular cast iron, and malleable cast iron.

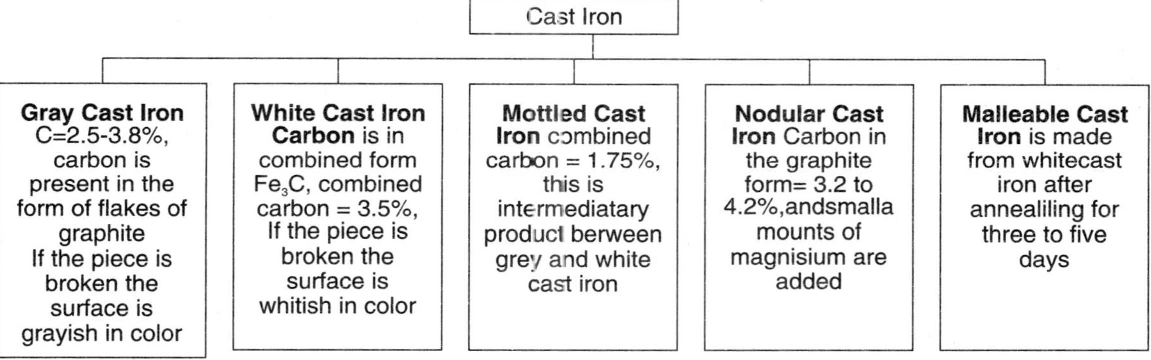

Ques.572: What is alloy steel and why alloying is required? Describe the importance of five alloying elements in steel.

Answer: When certain special properties are required, some alloying elements are added to carbon steels. The alloying element may be nickel, chromium, manganese, vanadium, tungsten etc. The steels thus obtained are called alloy steels.

Purpose of Alloying is to improve the following properties of steel :

1. Corrosion resistance.
2. Wear resistance.
3. Ability to retain shape at high temperature.
4. Machine ability
5. Ability to resist distortion at elevated temperature
6. Cutting ability.
7. Hardness, toughness and tensile strength

Alloying elements and their importance:

Chromium: Improves corrosion resistance

Vanadium: Improves toughness and wear resistance

Molybdenum: Improves hardness and strength

Tungsten: Improves high temperature resistance, toughness, wear resistance

Magnesium: Improves machine ability

Stainless steel (OR Corrosion resistant steels): The addition of chromium in the low carbon steels makes them corrosion resistant, chromium reacts with the oxygen and forms a oxide layer on the metal surface and protects the metal from corrosion. The amount of chromium added may be from 10% to 18%. These are used in the making of utensils and medical instruments.

Tool steel: Tool steel refers to a variety of carbon and alloy steels that are particularly well-suited for making cutting tools. Their suitability comes from their distinctive hardness, resistance to abrasion, their ability to hold a cutting edge, and/or their resistance to deformation at elevated temperatures (red-hardness). Tool steel is generally used in a heat-treated state. The carbon content in the tool steels is between 0.7% and 1.5%.these are used for making of cutting tools.

Ques.573: What is heat treatment and the objective of heat treatment.?

Answer: ***Heat treatment*** *is controlled heating and controlled cooling of a metal or alloy in the solid state with the purpose material. of changing the properties of the*

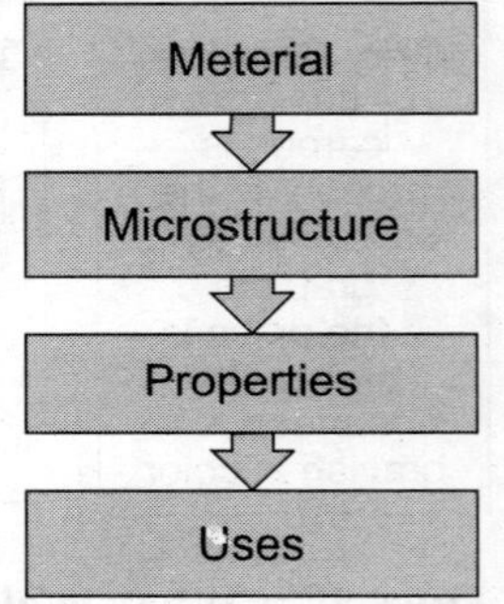

Heat treatment consists of three steps (1) Heating of the metal (2) Soaking (3) Cooling of the metal. or holding

Purpose or objectives of heat treatment

1. To improve machine-ability.
2. To improve mechanical properties, *e.g.,* tensile strength, ductility, hardness, shock resistance, resistance to corrosion etc.
3. To relieve the stresses induced during hot or cold working.
4. To change or refine grain size.
5. To improve magnetic and electrical properties.
6. To improve heat resistance, wear resistance.
7. To improve weld-ability

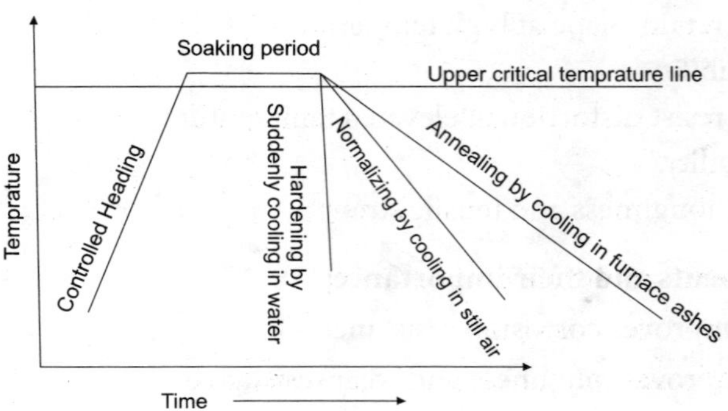

Different types of heat treatment processes on time and temperature plot

Different types of heat treatment processes on time and temperature plot

Ques.574: Describe the classification of heat treatment processes.What are various steps involved in heat treatment ?

Answer:

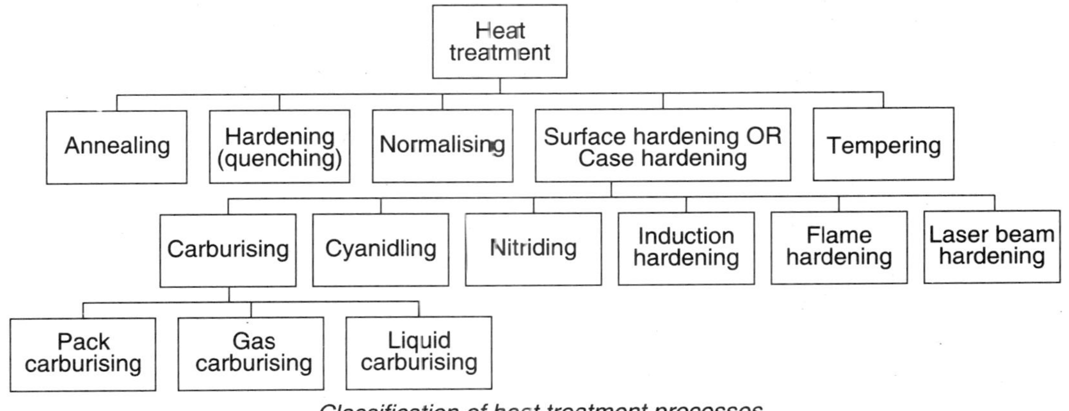

Classification of heat treatment processes

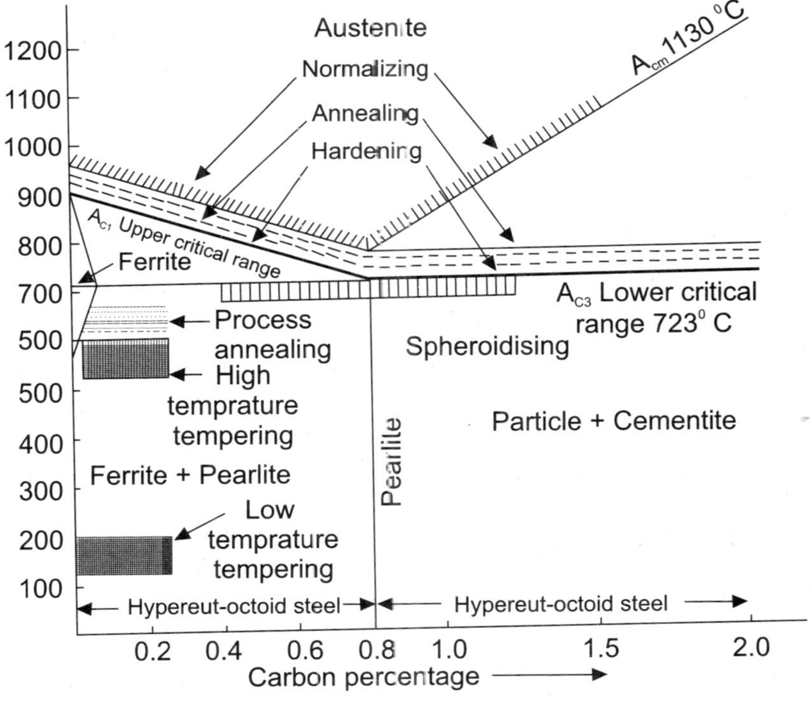

The process of heat treatment consists of the following steps:

Controlled Heating of Metal/Alloy (Steel)

Heating temperature of metal depends upon its grade, grain size, type and shape of a metal or alloy. Generally, the metal is heated upto its upper critical temperature.

Holding or Soaking period

Metal or alloy to be given heat treatment, is held at the specified temperature for a specified period so that there is a uniformity of temperature throughout the mass. The period of heating depends upon the size and shape of the component.

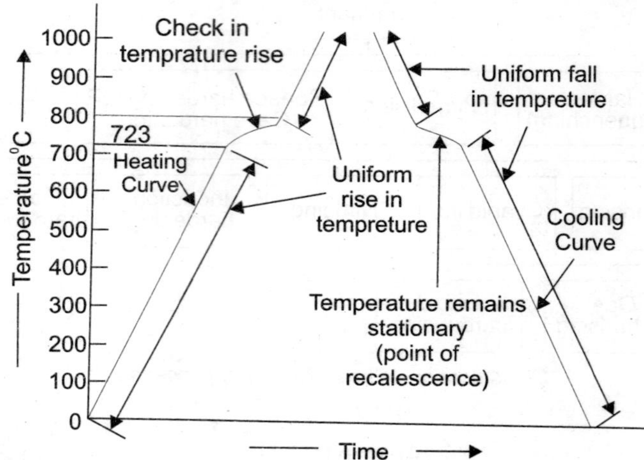

Figure : Heating and cooling curves for steel showing upper and lower critical temperature

3. Controlled Cooling of the Metal (Steel)

Main changes in the properties of metal or alloy take place in the cooling process. Change in properties or transformation depends mostly on the rate at which the cooling takes place. Cooling of metal is also known as quenching. Generally five quenching media adopted for quenching are: caustic soda solution, brine, and air. water, oil etc.

Ques.575: What happens during heat treatment?

Answer: During heat treatment process the internal microstructure of the metal is changed. The heat supplied during heating brings the structural and chemical changes in the metal. In between the upper and lower critical temperatures the carbon is in solid solution of iron. At normal temperature the steel consists of pure iron(ferrite) combined with iron carbide (cementite).

Ques.576: What are different microstructures of steel?

Answer: Different microstructure of steel are :

1. *Ferrite:* It is very soft and magnetic .
2. *Cementite:* It is combined carbon with iron also known as cementite and it is very hard.
3. *Pearlite:* It is the mixture of ferrite and cementite.
4. Austenite: It is solid solution of carbon and iron and it is stable only above upper critical point. If cooled below 723°C it transforms into pearlite and ferrite.
5. *Martensite:* It is very hard and brittle.

6. *Troostite:*
7. *Sorbite:*

Ques.577: What are various purposes of annealing ? Describe various annealing processes.

Answer: The various purposes of annealing are:

1. To soften the metal to improve machine ability.
2. To refine grain size and structure to improve mechanical properties.
3. To relieve internal stresses.
4. To remove trapped gases.
5. To modify electrical, magnetic and physical properties.
6. To increase ductility of metal.
7. To prepare the steel for further treatment.

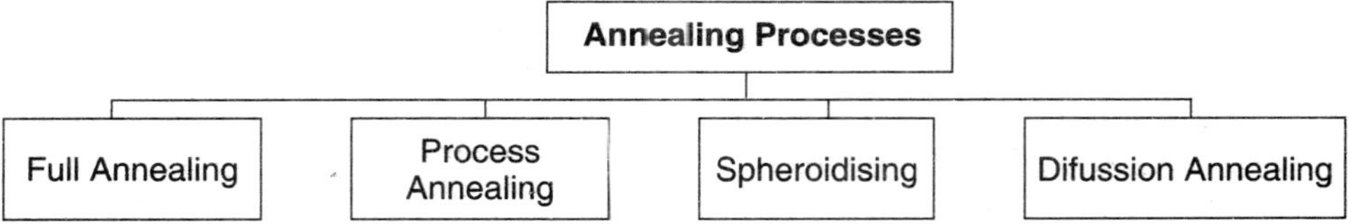

Full Annealing Process

Annealing removes all structural imperfections by complete re- crystallization. The purpose of full annealing is to soften the metal, to refine the grain structure, and to relieve the stresses. Full annealing process consists of the upper critical temperature.

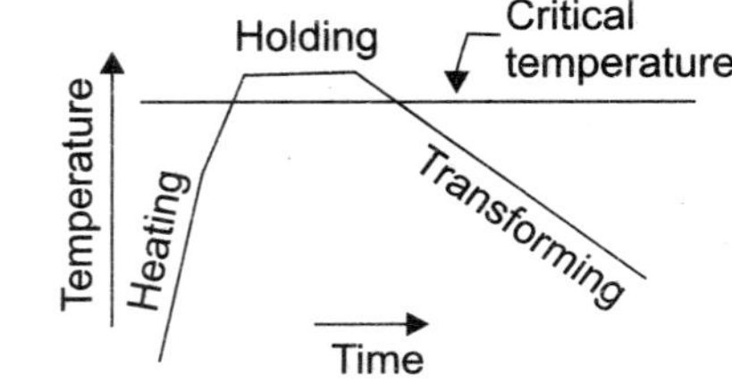

(a) Heating the steel above (about 50°-60°C).

(b) Holding it at this temperature for a sufficient time depending upon the thickness of work

(c) Slowly cooling it in the furnace ashes to room temperature. The metal is held at annealing temperature for a sufficient period of time so that proper structurechanges can take place.

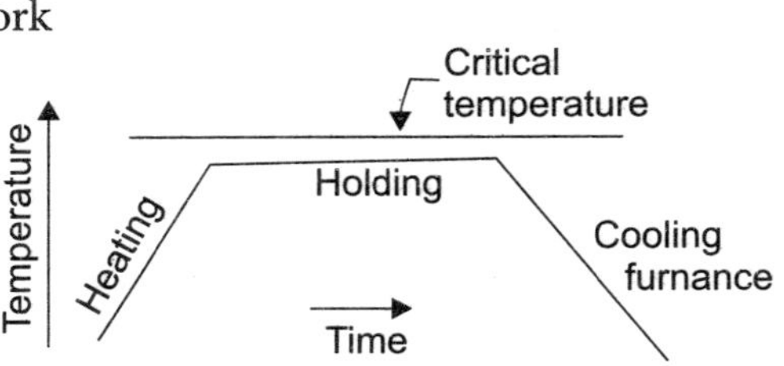

Process Annealing

In this case the metal is heated below the lower critical temperature and the rest of the procedure is as for the full annealing process. This is applicable to low carbon steels only.

Ques.578: Describe the process of steel hardening. Why steels are required to be tempered after it has been hardened ?

Answer: **Quenching**

Quenching is the process of suddenly cooling of the material by dropping in the cooling medium. Sometime this refers to hardening process also.

Hardening

In this process, steel is heated to temperature above the critical point (30°C to 50°C), held at this temperature for a considerable time, depending upon its thickness and then quenched (rapidly cooled) in water, oil or molten salt baths.

The purposes of hardening are:

1. To increase the hardness of the metal so that it can resist wear.
2. To increase the wear resistance so that it can resist wear.
3. To make it suitable for cutting tools

Normalizing

In normalizing process, steel is heated to a temperature above the critical point (40°C to 50°C), held at this temperature for a relatively shorter period (About 15 minutes) and then cooled down to room temperature in the still air.

The purposes or the objectives of normalizing are:

1. Internal stresses caused during previous operations are removed
2. Internal structure is refined
3. Mechanical properties of steels are improved

Tempering

The hardened steel (Martensitic structures) formed by direct quenching of high carbon steel are hard and strong, but **unfortunately are also brittle**. Since they are brittle and possess very little toughness so they are unable to resist impact loads. The hardness and strength is sacrificed to obtain suitable ductility and toughness to make the steels usable for different applications.This is done by the tempering process.

The objectives of tempering process are:

1. To make steel tough to resist shock and fatigue
2. To remove internal stresses
3. To stabilize the structure
4. To decrease the hardness.
5. To remove the ill effects of hardening process and make the steel usable.

Tempering Process: In tempering process the steels are heated to certain temperature in the range of 150 to 650°C and held at that temperature for some time may be 1 to 3 hours and then cooled to room temperature. The cooling may be done in open air. Depending upon the temperature the tempering may be classified as:

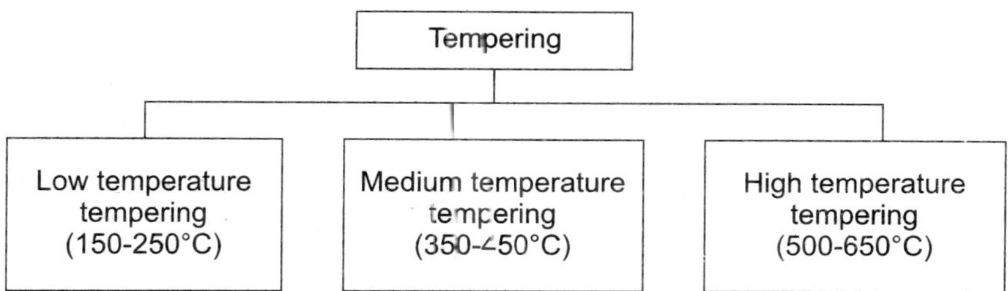

Ques.579: .What is surface or case hardening ? Describe various processes of surface/case hardening .

Answer: **Surface Hardening or Case Hardening**: *The process by which the surface of an object or part can be made hard and leaving the interior as it is, is called as surface or case hardening process. In many engineering applications, it is required that a steel being used should have a hard surface so that it can resist wear and tear. At the same time it should have soft and tough interior so that steel is able to absorb any shocks, impact, etc. This is possible only when the surface is hard while rest or interior of the metal is left as soft and ductile.* This type of treatment is called surface hardening or case hardening and is applied to gears, ball bearings, railway wheels etc.

Various types of surface hardening or case hardening processes are shown in figure:

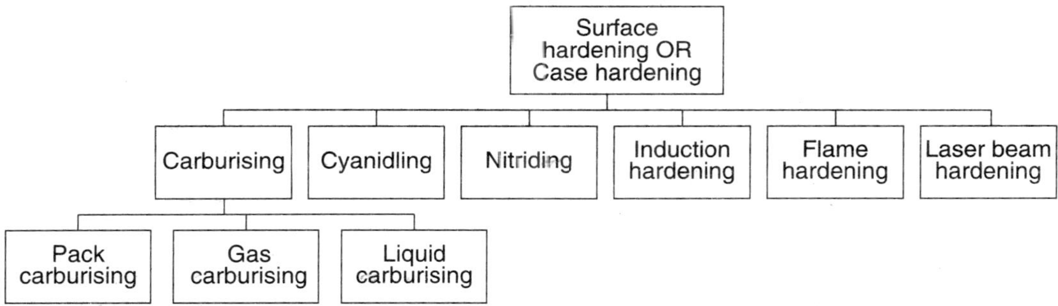

Surface or Case hardening processes

a. **Carburizing:** *The process of inducing of additional carbon to low carbon steels in order to give it a hard surface is known as carburizing.*_The surface is made hard only upto a certaindepth using the any one of below:

(i) **Pack or Solid Carburizing:** *The article to be carburized is placed in a carburizing box and surrounded by solid carbonaceous materials—usually a mixture of charcoal, leather, horn and barium carbonate as catalyst. The boxes are sealed with clay to exclude air and are placed in an oven or furnace where they are heated to a temperature between 900-950°C for several days depending upon the extent of carburizing action desired.*_In this way, carbon from the carburizing compound soaks or diffuses into the surface of the hot- steel. After carburizing, steel is reheated to a temperature just above its critical point followed by quenching in water, brine or oil. This hardens the skin and at the

same time refines the core. This steel is also given a second heat treatment at a lower temperature range *i.e.* 750-770°C, in order to improve the ductility and impact resistance of the core and case.

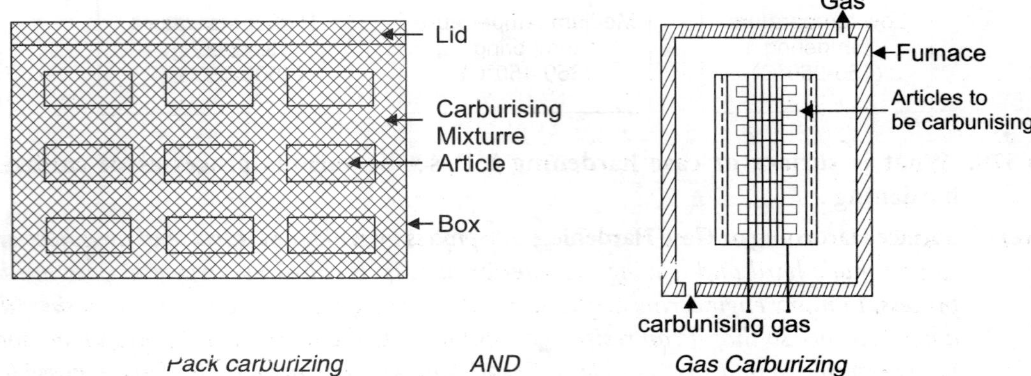

Pack carburizing AND *Gas Carburizing*

(ii) Liquid Carburizing: *The articles to be carburized are heated in a container filled with a molten salt, such as sodium carbonate.* If only selected portions of the components are to be carburized, then the remaining portions are covered by copper plating.

Advantages:

1. Very little deformation of article.
2. Time saving process.
3. Greater depth of penetration possible.
4. Selective carburizing is possible if needed.
5. Ease of carburizing a wider range of products.
6. Uniform heating.
7. Parts leave the bath with a clean and bright finish. There is no scale as there in pack hardening.

(iii) Gas Carburizing: In gas carburizing, the article is heated and surrounded by a hydrocarbon gas (such as methane, ethane, carbon monoxide, etc) in the furnace.

A. Cyaniding: It is used to give a very thin but hard outer case. The articles are heated in the salt bath of Sodium cyanide or potassium cyanide in the temperature ranging between 700°C to 820°C and both carbon and nitrogen are added to the outer surface resulting in a surface harder than that produced by the carburizing process.

B. Nitriding: Nitriding is a case hardening process in which nitrogen is added instead of carbon, in the skin or outer surface of the steel. The article is placed in a container and heated in the range of 450°C to 550°C and ammonia gas is circulated over the articles.

Advantage of nitriding: The articles may be used directly without ant secondary machining operations because the steel is not exposed to elevated temperatures.

C. Induction hardening and Flame hardening: In Induction hardening and gas flame hardening the articles are heated using induction heating and flames and followed by cooling with spray of water.

Ques.580: What is the ore of copper? Write the properties and uses of copper. What are the alloying elements of brass and bronze?

Answer: Copper is extracted from copper ores such as copper glance (Cu_2S), copper pyrites ($CuFeS_2$).

Properties:

1. Copper is a reddish brown metal.
2. Pure copper is one of the good conductors of heat and electricity.
3. It can withstand severe bending and forging without failure.
4. Copper can be welded at red heat.
5. It is highly resistant to corrosion.
6. Melting point = 1084°C, boiling point = 2595°C.
7. Pure copper does not cast well.
8. If copper is heated to red heat and cooled slowly it becomes brittle, but if cooled readily it is soft, malleable and ductile.

Uses:

1. It is largely used in electric cables and wires, electrical machinery, motor winding etc.
2. Used in electroplating
3. Copper tubes are used in mechanical engineering.
4. It is used for household utensils.
5. It is used to form alloys like brass, bronze and gun metal.
6. Used in boilers, condensers, roofing etc.

Alloying elements of brass and bronze:

Brass is an alloy of copper and zinc.

Common types of brass are

Cartridges brass (70%Cu +30% Zn)

Naval brass ((60%Cu +39% Zn+ 1%Tin)

Bronze:

Bronze is basically an alloy of **copper and tin**. Various properties of bronzes are:

1. Bronze has higher strength than brasses.

2. Bronze has better corrosion resistance than brasses.
3. Bronze has antifriction or bearing properties.
4. Bronze is costlier than brass.
5. Bronze is most ductile when it contains about 5% of tin. As the amount of tin increases about 5%, the ductility gradually decreases and practically it disappears with about 20% of tin.
6. Addition of phosphorus increases the tensile strength, elasticity and resistance to fatigue

Phosphor Bronze

When bronze contains phosphorus, it is called phosphor bronze. A common type of phosphor bronze has the following composition: Cu = 89 to 94%+Tin = 5 to 10% + Phosphorous = 0.1 to 0.3%.

Ques.581: Write short notes on Duralumin.

Answer: It is an alloy of aluminum and it contains Al = 94%, Cu = 4%; Mg, Mn, Si, Fe - 0.5% each.

Properties:

1. It can be cast, forged and stamped.
2. It has high tensile strength.
3. It has high electrical conductivity.
4. It hardens spontaneously when exposed to room temperature.
5. The alloy is soft enough for a workable period after it has been quenched.
6. It is light in weight as compared to its strength.

Uses:

1. It is used for sheets, tubes, rivets, nuts, bolts etc.
2. Because of lighter weight, it is used in automobile and aircraft components.
3. It is also employed in surgical and orthopedic work and for non-magnetic and other instrument parts.
4. Utensils, window frames, engine bodies, etc.

Ques.582: Describe the objectives & classification of of metal forming or mechanical working. Also discuss the features and advantages of hot working and cold working .

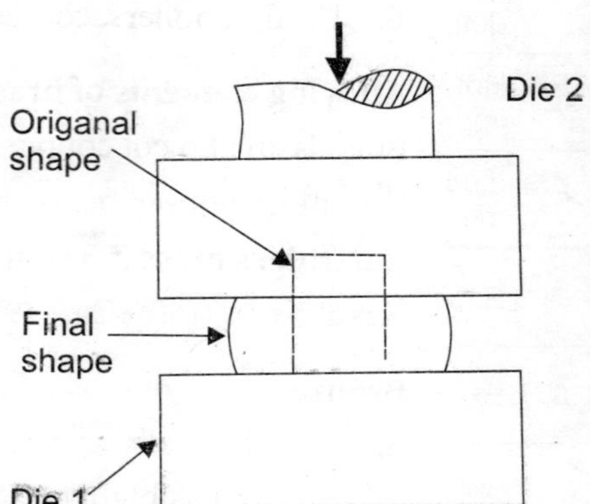

Answer: **Metal forming or mechanical working** is a process of intentional deformation of metals for shaping the metals into required shape or form by plastic deformation with application of external forces. In these processes the metals are given a required shape, size, dimensions and properties without any significant amount of wastage of material. The plastic deformation takes place once the forces are applied beyond the yield point stress. The figure shows the process of metal working, a metal block shown by dashed lines is given required shape by application of external force using Die-1 & Die-2 and the final shape obtained is shown by solid lines.

Objectives of metal forming:

1. To refine grain structure
2. To induce the directional properties into the material
3. To achieve high dimensional accuracy
4. To improve mechanical properties of the metal

Re-crystallization is the process of formation of new grains or crystal structure and the temperature at which the old crystal structure is destroyed and formation of new crystal structure is complete, is known as **Re-crystallization Temperature**.

Classification: The metal forming processes/ operations can be classified, depending upon the temperature, into Hot and Cold working processes. A comparison of hot working and cold working is presented below.

Metal Forming OR Mechanical working		
Hot working	**Cold working**	**Warm Working**
1. It is carried out above the recrystal-lisation temperature but below the melting or burning point. Hence, deformation of metal and recovery take place simultaneously.	1. It is carried out below the recrystaal-lisation temperature. As such, there is no appreciable recovery.	
2. No internal or residual stresses are set up in the metal in hot working.	2. In this process, internal or residual stresses are set-up in the metal.	
3. Due to recrystallisation and recovery, no or very negligible hardening of metal takes place.	3. Since this is done below recrystaal-lisation temperature, the metal gets work hardened.	
4. Due to higher deformation temperatures used, the stress (load) required for effecting deformation is less.	4. The stress required to cause deformation is much higher.	
5. Hot working refines metal grains, resulting in improved mechanical properties.	5 Most of the cold working processes lead to distortion of grains.	
6. If cracks and blow holes are present in the metal, they are finished through hot working.	6. In cold working the existing cracks propagate and new cracks may develop.	
7. Rate of production in Hot higher.	7. Rate of production is lower as compared toHot Working	
8. Example: Wire making, Tube making,	8. Example: Forging, Extriusion, Hor Rolling Blanking, Punching, etc	

Advantages of hot and cold working:

Advantages of hot working	Advantages cold working
1.Large deformation can be accomplished	1. Better dimensional accuracy can be obtained
2. Porosity of metal gets reduced	2. Good surface finish
3. Grain structure gets refined	3. Strength and Hardness increase
4. No residual stresses are induced	4. Easier to handle cold material

The advantage of one process is the disadvantage of the other, and in this manner we can understand that each process has its own advantage and disadvantages.

Ques.583: What are the steps involved in the rolling process for making a round bar of steel ?

Answer: Rolling is the process of compressing the metal by passing it between the two or more revolving rolls or cylinders. As a result the area of cross section gets reduced and length is increased. The process of rolling a steel bar is shown in the figure.

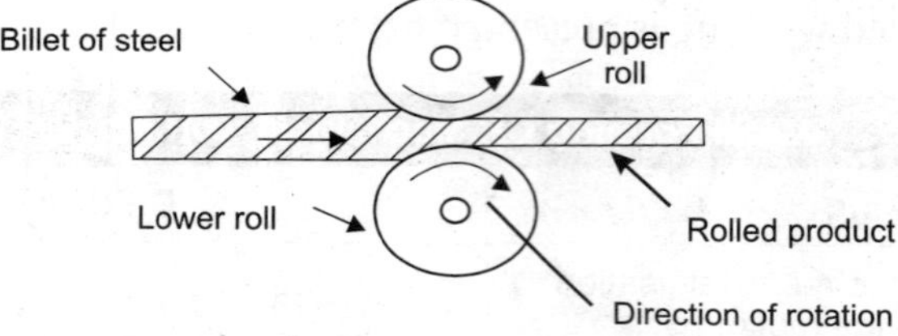

Schematic diagram for rolling process

Steps Involved: Initially a billet with the square cross-sectional area is taken and passed through set of rollers which reduces the area of cross section and also its shape, after a number of passes through the set of rollers with different cross sections and orientations the bar is finally reduced to a round bar as shown in the figure.

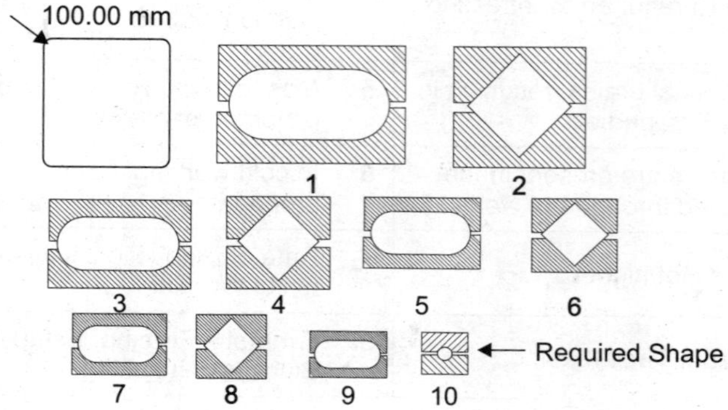

Steps involved in the rolling process for making a round bar of steel

Ques.584: Explain with neat sketch, basic working principle of rolling. Describe its applications in industry.

Answer: Rolling is the process of shaping the metal by compressing the metal, by squeezing or passing it between the two or more revolving rolls or cylinders. As a result the area of cross section gets reduced and length is increased. Rolling is normally a hot working process unless specifically mentioned as cold working and associated with re- crystallization of the grains after rolling.

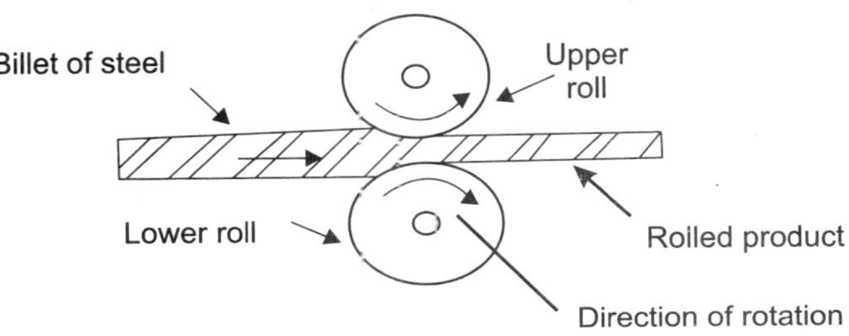

Rolling process: Heated metal is passed between two rolls, rotating in the opposite direction; the gap between the rolls is somewhat less than the thickness of metal at entrance. Thus the rolls will squeeze the ingot material and reduce the area of cross section & also increase the length. Original coarse grain structure is initially elongated and then the re-crystallization takes place and the finer grains are produced.

Principle or Characteristics of rolling process:

1. The frictional forces are pulling and drawing the metal inside the rolls after initial engagement of metal. The amount of area reduction depends upon frictional forces.
2. The metal stock enters the rolls with a lesser speed than the peripheral speed of the rolls.
3. The metal after being squeezed exits the rolls at higher speed.
4. The area of cross section at entry is more than the area at exit; however the quantity entering is equal to the exit quantity. This can be presented by the following equations and figure:

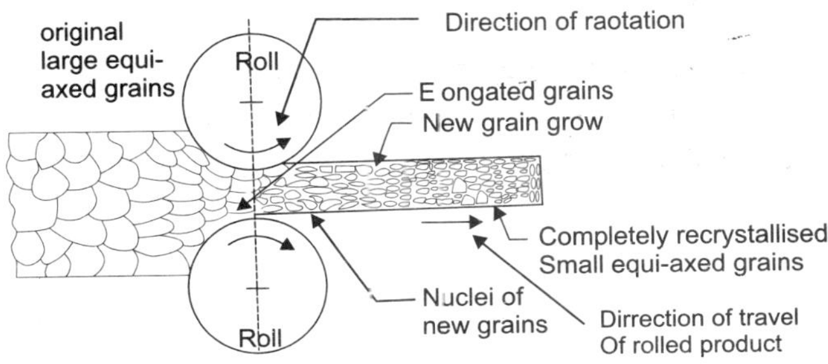

Schematic presentation of rolling process

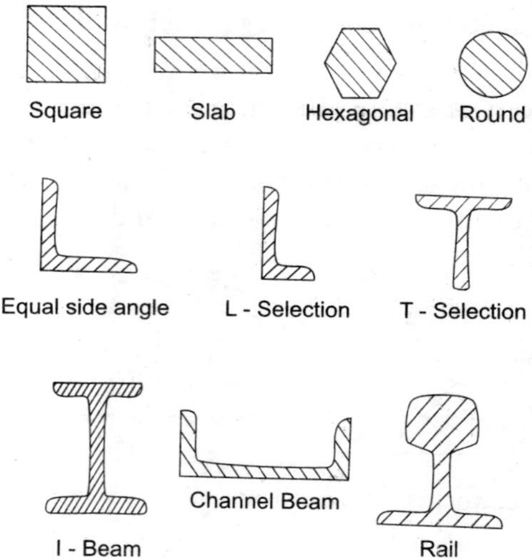

Different cross- sections produced by rolling **(Products of rolling process)**

Applications Or Products: The different shapes which can be rolled are plates, sheets, bars, structural shapes, I, L, T, or channel section such as shown above left side.

Ques.585: Describe various Types of Rolling Mill arrangements:

Answer: There are different rolling mill arrangements as shown in the figure.

(a) **Two high rolling mill:** Consists of two heavy horizontal rolls, supported on the bearings.

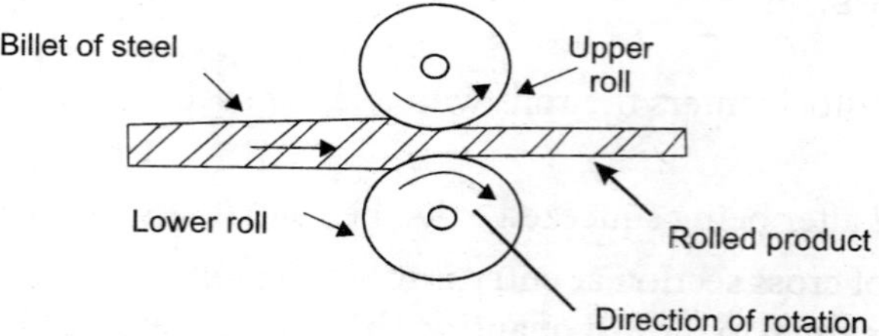

The bearings are placed in the side frames called as stands. The space between the rolls is adjustable.

(b) **Two high reversible rolling mill:** In this case the rolls have the capacity to rotate in both direction depending upon the requirement.

(c) **Three high rolling mill:** This set up is used to give two continuous passes to the metal.

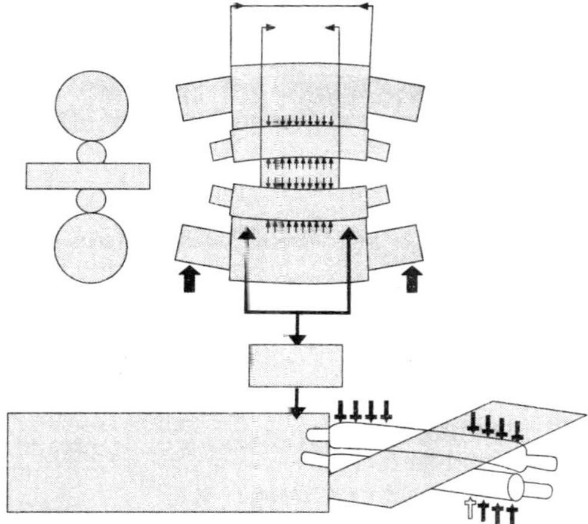

(d) **Four high rolling mills:** In this set up four rollers are used and two are smaller called as working rollers and other two are large and called as Back uprolls.

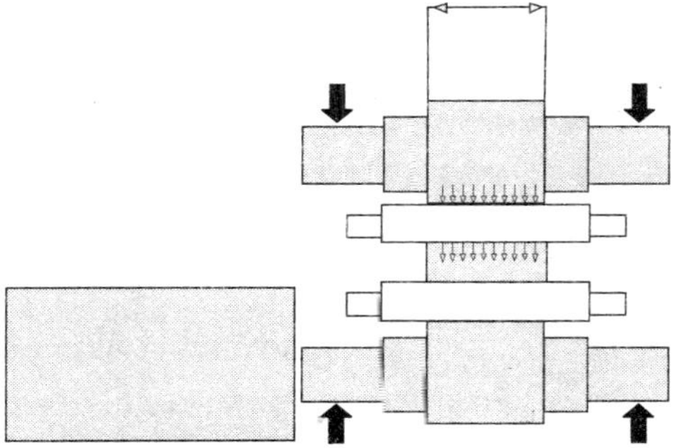

(e) **Cluster rolling mill:** This is improvement over the four high rolling mill and provides better supports.

(f) **Planetary rolling mill:**This provides anumber of free rolls over the periphery of large back up roll and suitable for large reduction in area in one pass.

(g) **Continuous rolling mill:** It consists of many non-reversing two-high rolling mills arranged on after the other, so that the material can be passed through all of them in a sequence. It is suitable for mass production.

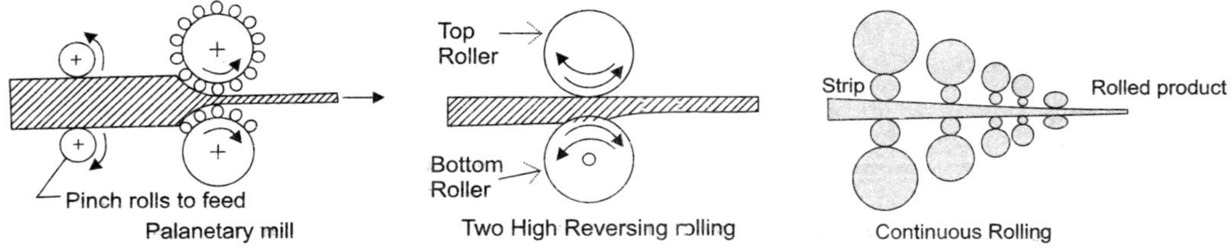

Comparison of different rolling mill arrangements

Two High Rolling mill	Two High Reversible Rolling mill	Three high rolling mill	Four high rolling mill	Cluster rolling mill	Planetary Rolling mill
Two rolls rotate in opposite direction.	Two rolls can rotate in the reverse direction to increase the production.	These increase the rate of production.	Working rollers are supported by two Back up rolls.	This is improvement over the four high rolling mill and provides better supports by additional rollers	A number of free rolls over the periphery of large back up roll
		Direction of rotation of middle roll is opposite to top and bottom roll.	The rollers deflection is reduced by back up rollers.		Suitable for large reduction in area in one pass.
Used for bars and general sections	Used for bars and general sections	For rods and billets	Used for Sheets and plates	Used for cold rolling of Sheets and plates	

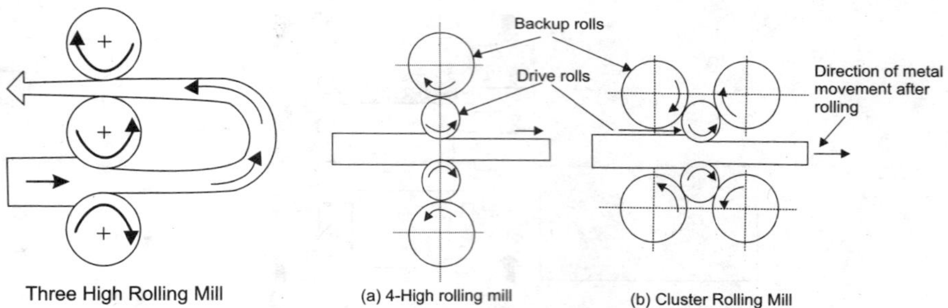

Three High Rolling Mill (a) 4-High rolling mill (b) Cluster Rolling Mill

Ques.586: Draw a neat sketch and explain metal Extrusion process. Give the example of four extruded products. What are different types of extrusion?

Answer: **Basic working principle of extrusion:** Extrusion is the process in which the metal is forced to flow through a shaped opening or die, so that metal will take the shape of the opening. The operation of extrusion is similar to squeezing of the tooth paste tube as a result the tooth paste comes out in the shape of the opening. The heated metal is placed in the cylinder and pressure is applied by ram, while keeping a dummy block in between. The metal is forced to pass through the die/ orifice opening.

Extruded Shapes/Applications: (Examples of extruded products)

- Rods, tubes, splined shafts, structural sections and bars of different cross sections.
- Almost any length can be produced but with constant cross section only.
- tubes such as shaving cream, tooth paste by impact extrusion.

(1) Direct extrusion OR forward extrusion: If the direction of extruded product is same as the direction of force application then the extrusion is called as direct extrusion.

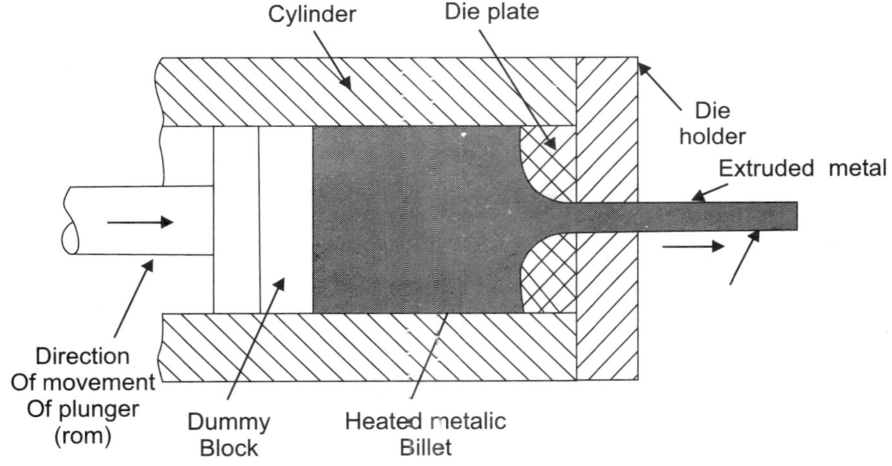

Figure:direct Extrusion

Classification: Extrusion can be classified as:

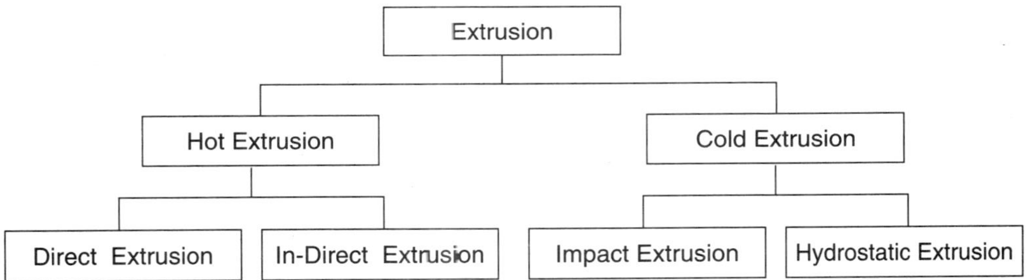

In-Direct extrusion OR backward extrusion: If the direction of extruded product is opposite to the direction of force application, then the extrusion is called as **In-Direct extrusion** as shown in figure:

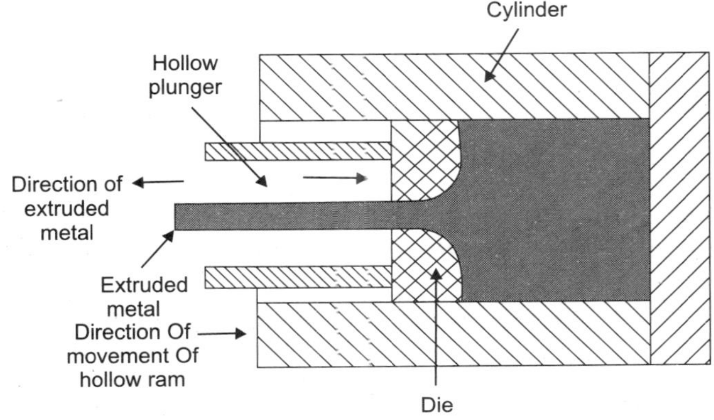

In-Direct extrusion

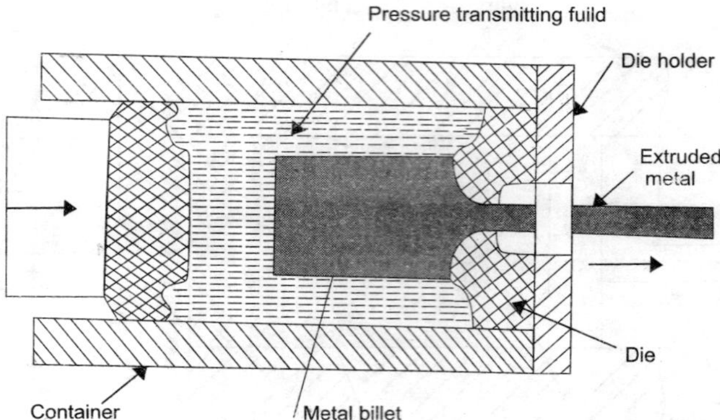

Hydrostatic Extrusion

Hydrostatic extrusion: In this type of extrusion the force is applied indirectly using a liquid medium and force is uniformly applied because of presence of fluid/liquid. The difficult to extrude materials can be extruded by this method and the process is schematically shown in the figure.

Impact extrusion: Impact extrusion is most suitable for producing collapsible tubes of soft materials, aluminum and copper etc. This may be tooth paste tubes and cans etc. The impact extrusion is performed using a punch and die as shown in the figure.The material is placed on the die & punch is struck from top with high pressure & speed, the metal flows up and forms cup shape over the punch which is removed from the punch using the compressed air.

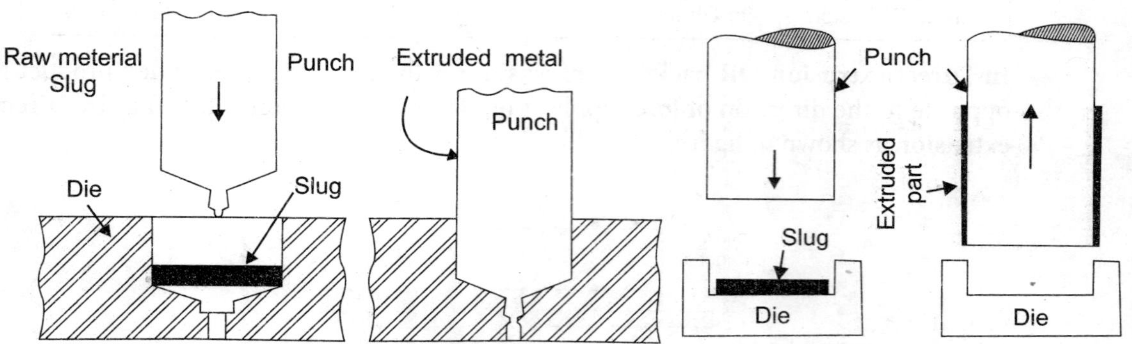

Figures showing the process of impact extrusion for tooth paste tube and cup shape

Comparative merits and demerits of different types of extrusion

Direct extrusion	In-Direct extrusion	Impact extrusion	Hydrostatic extrusion
Problem of friction is dominant	Problem of friction is eliminated	High speeds are involved	No Problem of friction
High force is required	Less force is required	Applicable to soft materials only	Difficult to extrude materials may be worked

Lubricants are required	Ram gets weakened		Additional Lubricant is not required
Brittle material cannot be extruded	Brittle materials can not be extruded	Brittle materials cannot be extruded	Brittle materials may be extruded

Ques.587: What is forging? Classify the different methods of forging.

Answer: Forging is the process of compressing or squeezing the material between two or more dies to get the required shape and size. In this the shaping or intentional deformation is induced by localized compressive forces. In hot forging process the metal is heated above re-crystallization temperature and pressure is applied with help of hammers or presses to get the required shape.

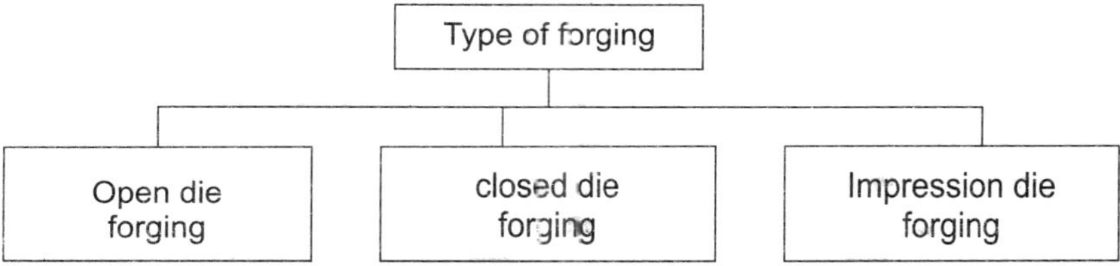

Steps involved in open die forging

a. The metal is heated in the furnace.

b. The process may be carried out manually or mechanical hammers may be used.

c. The heated metal is squeezed or compressed between two flat or open dies, with the help of hammers.

d. The metal may have to be heated and rotated during the process for many times.

e. Many hammer blows may be required to get the final shape.

Steps involved in closed die/ impression die forging

a. The hot material stock is placed in the die block cavity.

b. Pressure is applied using machine/ press to compress the die set.

c. Material flows to fill the die cavity.

d. Flash may be produced in case of impression die forging.

e. Final shape is obtained.

Applications of forging: The following are the applications of forging processes:

a. Bars and Step shafts.

b. Hooks, agricultural equipments.

c. Blocks, discs, spanners, bolts, etc.

d. Brackets, fittings, elbows, etc.

Advantages of forging process:

a. Refinement of metal grain structure takes place

b. The metal improves in directional properties

c. The time consumed is reduced in comparison to other shaping processes

d. Reasonable degree of dimensional accuracy is obtained

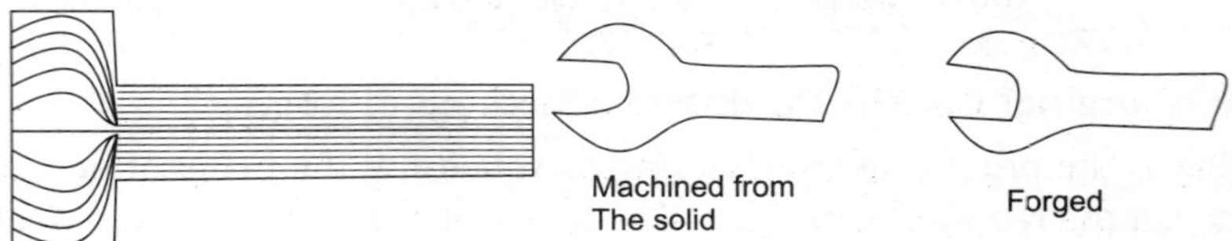

A bolt and spanner made by forging process ; figure shows the grain flow

Different types of forging Operations:

The different types of forging operations are as:

a. Upsetting or jumping: Cross-sectional area is increased with reduction in length

b. Drawing out or drawing down: Cross-sectional area is decreased with increased in length; using fullers or swages or hammer

c. Cutting: Metal is cut in two pieces using the chisels

d. Bending: The bars or flats can be bend to required shape

e. Punching: this is a process of making a hole in the bar or work piece

f. Forge welding: the two pieces of metal are joined together by this process

Comparison of forging processes with machining and casting

Smith forging OR Hammer forging	Drop forging	Press forging	Machining & Casting
Forging is carried out manually in case of smith forging and by mechanical hammers in case of hammer forging	Forging is carried out with help of drop hammer and two halves of dies, one kept on the anvil and other attached to the hammer	Forging is carried out with help of press which may be mechanical or hydraulic, pressure is applied continuously	In case of machining and casting the directional properties are not induced
Skilled labor is required	Highly skilled labor is not required	Highly skilled labor is not required	The parts will be weaker in tension compared to produced by forging
Open Dies are used	Closed or impression dies are used	Closed or impression dies are used	The parts will have low fatigue life
Any shape can be produced	Only die shape can be produced	Only die shape can be produced	The crack initiation and propagation is easy in case of machined and cast parts
Less dimensional accuracy	High dimensional accuracy	Very high dimensional accuracy	Production rates are low in case of machining and casting

Blows are repeated	Blows are repeated	Continuous pressure is applied	
Small size work can be carried out	Medium size work can be carried out	Large work can be carried out	
Operation is not safe	Operation is not safe	Operation is safer	
Operation is slow	Operation is faster, noisy and with vibrations	Operation is faster and quieter without vibrations	

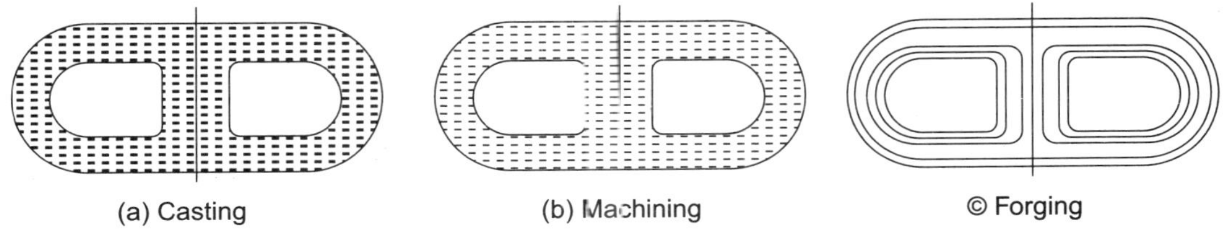

(a) Casting (b) Machining © Forging

A part made by three differnet processes, showing grain flow

Comparison of types of FORGING OPERATIONS open die, closed die &impression die forging

Open die forging	Closed die forging		Impression die forging
The metal being shaped during open die forging is not completely confined during the process, which derives the name open die forging.	Close die forging is a special form of impression-die forging, in this case flash dose not form and the work piece completely fills the die cavities.		Impression die forging is a special form of die forging, in this case the flash is formed and metal is squeezed out from the die cavities.
Many hammer blows are required to get the final shape	**Drop forging** Drop forging has several blows or compressions.	**Press forging** Uses steady pressure instead of impacting pressure. The energy is transmitted into the surface layers of the workpiece. Part gets produced in only a single stroke.	In impression-die forging, the work piece acquires the shape of the die cavities, while being forged between two
Machining may be required	The force comes from the ram [hammer free fall by		forging, some of the material flows outward and forms a flash.
	gravity]. The		
	ram is lifted by a		Once the part is

The dimensional accuracy is poor	single-acting steam(or air) cylinder.	**Advantages:** High strength-to-weight ratio. Surface is smooth. Good dimensional accuracy. Product is tougher than casting. Have little or no waste. No machining required after part is produced by closed die forging	produced then flash is removed by machining. Dimensional accuracy is very good

Forging Methods

- **Open-die drop hammer or smith Forging** The process may be performed manually by the iron smith using open or flat faced dies OR by Drop hammers.
- **Impression die drop hammer forging** The process is performed by the machine and series of blows is applied over the dies after placing the heated material inside the closed dies or impression die.
- **Press Forging** The process is performed by the machine and continuous pressure is applied over the heated material inside the closed impression die. The pressure may be applied using hydraulic or mechanical presses
- **Machine forging or Upset forging** The process is performed by the machine; continuous pressure is applied for upsetting the bar into a shape similar to bolt.
- **Roll forging** The rollers with shaped grooves are used to reduce the thickness of bar stock and give required shape.

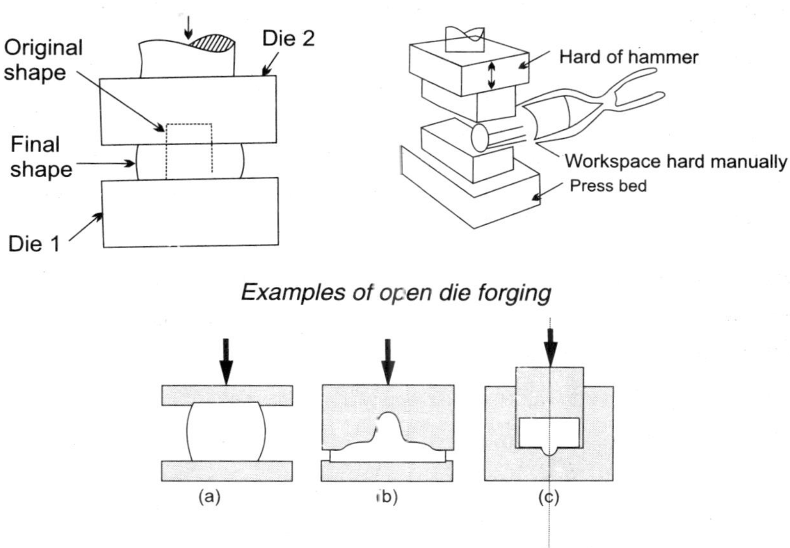

Examples of open die forging

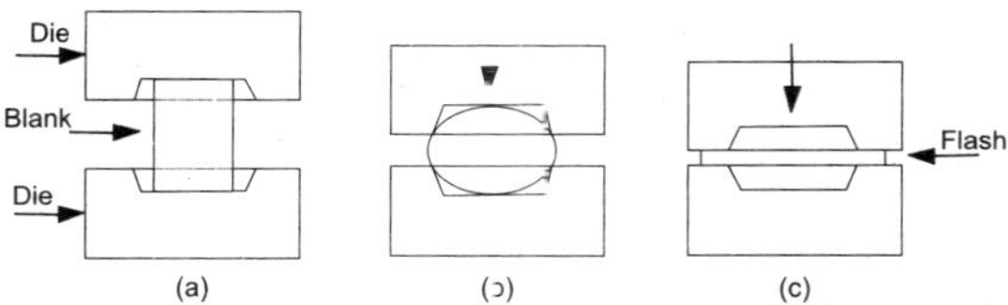

The figure shows the open die, impression die and closed die forging operations

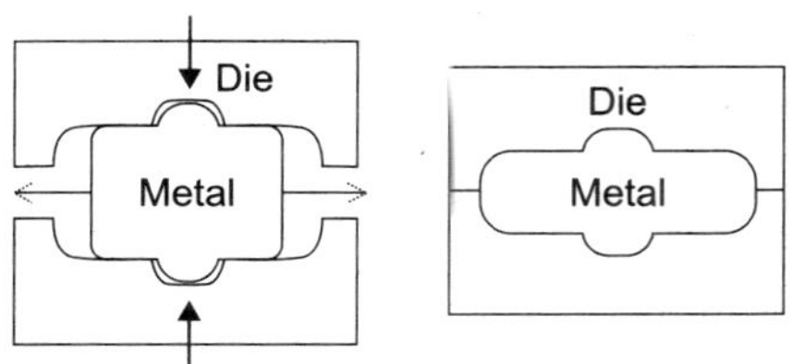

The figure shows the steps in impression die forging operation

The processes of Closed die forging

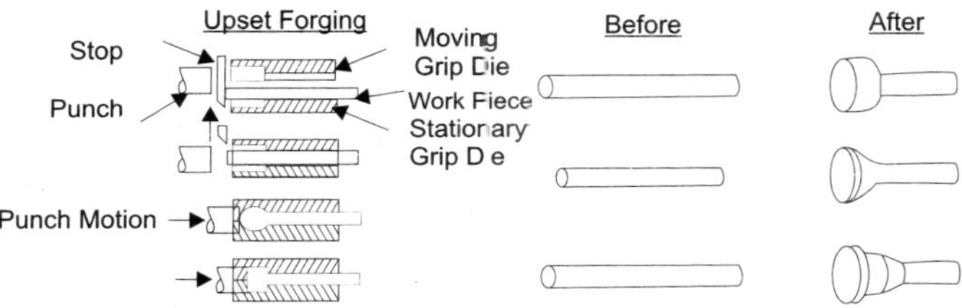

Process of upset forging and its product shapes

Ques.588: Describe with neat sketch the wire and tube drawing/ making process.

Answer: **Wire/ bar drawing** is a metalworking process used to reduce the cross-section of a wire/ bar by pulling (applying tensile force on) the wire through a single, or series of, drawing die(s).Drawing is usually performed at room temperature, thus classified as a cold working process, but it may be performed at elevated temperatures for large/ thick wires to reduce forces. The figure below shows the process of wire drawing and dies used for the purpose.

Process of wire or bar drawing:

a. The wire/ bar are prepared by shrinking the beginning of wire, by hammering, filing, rolling or swaging.

b. Then wire is passed through the die and lubricant is used to reduce friction.

c. The wire is then pulled through the die by applying the pulling force.

d. As the wire is pulled through the die, its volume remains the same, so as the diameter decreases, the length increases.

e. Usually the wire will require more than one draw, through successively smaller dies, to reach the desired size.

f. Annealing may be required in between the number of passes.

Applications: There are many applications for wire drawing, including electrical wiring, cables, tension-loaded structural components, springs, paper clips, spokes for wheels, and stringed musical instruments.

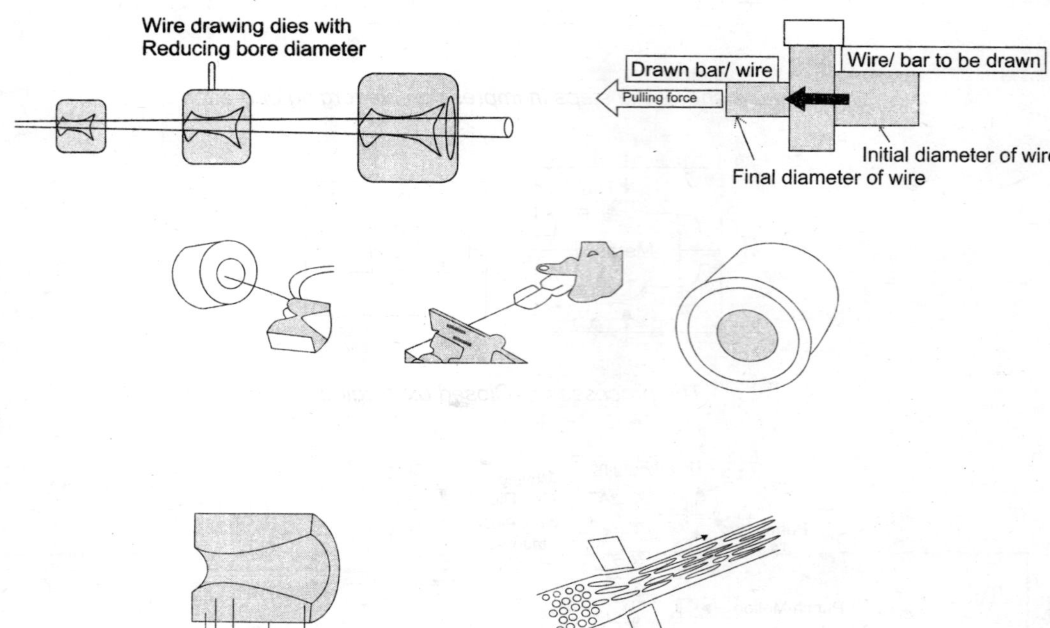

Wire/ bar making process with the views of dies drawing arrangement and grain flow direction

TUBE Making/ Drawing is a metal working process to reduce the diameter of tubes or tube shells by shrinking the large diameter tube into a smaller one, by drawing the tube through a die.

Basic Process of tube making/ drawing:

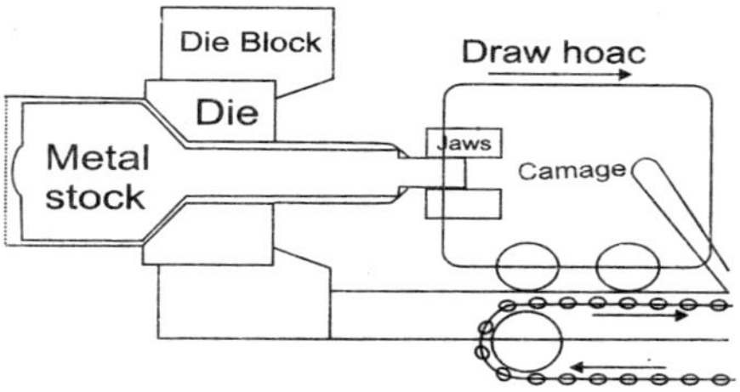

a. The tube with larger diameter is prepared by shrinking the beginning of tube by hammering, filing, rolling or swaging.

b. Then tube is passed through the die and lubricant is used to reduce friction.

c. The floating plug or mandrel or fixed plug is inserted.

d. The tube is then pulled through the die by applying the pulling force.

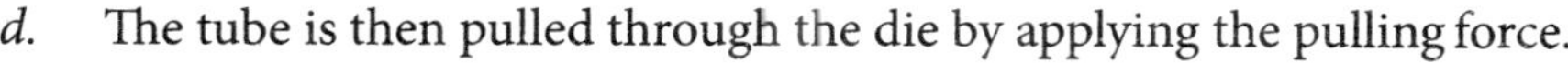

e. Usually the tube will require more than one draw, through successively smaller dies and plugs to reach the desired size.

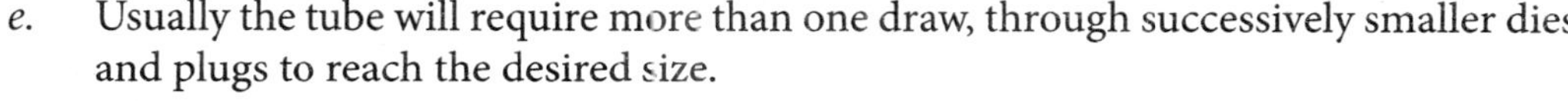

f. Annealing may be required in between the number of passes.

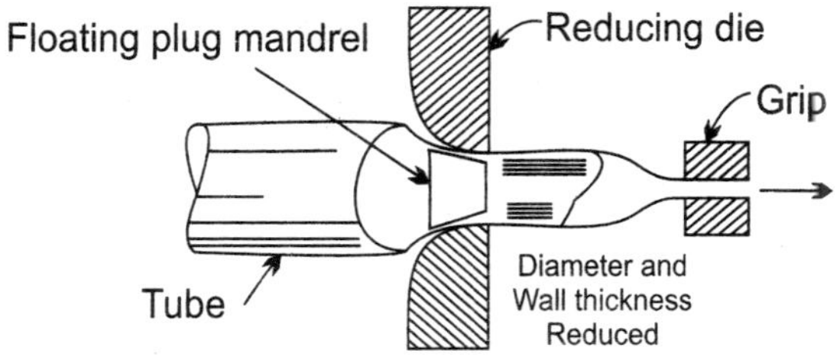

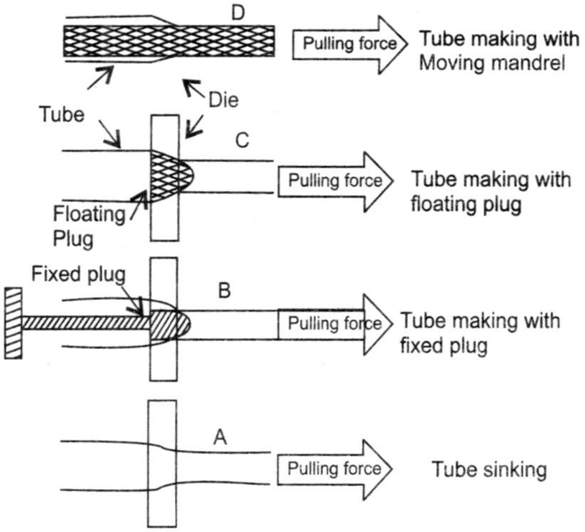

Types of Tube Drawing/Making Process

- Tube Sinking
- Tube drawing with fixed plug
- Tube drawing with floating plug
- Tube drawing with moving mandrel

The comparison of tube making processes

Tube Sinking	Tube drawing with fixed plug	Tube drawing with floating plug	Tube drawing with moving mandrel
Tube sinking, also known as free tube drawing, reduces the diameter of the tube without a mandrel or float inside the tube.	Fixed plug drawing, also known as stationary/ fixed mandrel drawing, uses a mandrel at the end of the die to shape the ID of the tube.	Floating plug drawing, also known as floating mandrel drawing, uses a mandrel that is not anchored fixed.	The tube is drawn with a mandrel. The mandrel consists of along hard rod or wire that extends over the entire length of the tube and is drawn through the die with the tube.
The inner diameter ID is determined by the inner and outer diameter of the stock tube, the outer diameter of the final product, the length of the die landing, the amount of back tension, and the friction between the tube and the die. Simple to perform	This process is slow	The mandrel is held	In this design, the area
	and the area reductions are limited, but it gives	in by the friction forces between the mandrel and the tube.	reduction can be 50 per cent. However, after drawing, the mandrel must be
	the best inner surface	This axial force is	removed from the tube
	finish of any of the	given by friction and	by rolling (reeling),
	processes. This is the oldest tube drawing method.	pressure.	which increases the tube diameter slightly and disturbs the dimensional accuracy.
		The greatest advantage of this is that it can be used on extremely long lengths	

Ques.589: Describe the various processes in sheet metal forming & working and draw a neat sketch of die & punch assembly.

Answer: **Drawing process:** Drawing can refer to two different operations, if the starting metal is in the sheet form, then it refers to forming of cup shaped objects, produced by punch and die with plastic deformation of metal sheet on application of pressure.

If the starting metal is in the bulk form, such as wire, rod, or tube then drawing refers to the process of reducing the cross sectional area by pulling it through a die.

Various items produced by drawing are tubes, rods, wires, cup, metallic glass, and cylindrical objects.

Sheet metal forming processes are those processes in which force is applied to a piece of sheet metal to modify its geometry rather than remove any material. The applied force produces the stresses in the metal beyond its yield strength, causing the material to plastically deform, but not to fail. By doing so, the sheet can be bent or stretched into a variety of complex shapes.

Sheet metal working or forming processes are the processes used to give required shape and size to the metal sheets, using different operations.

Principle: The materials are sheared or formed between a die and punch without formation of chips.

Press: The press is a sheet metal forming machine tools designed to shape or cut the metal by applying mechanical force or pressure.

Details or construction of Press: It is a hand operated press. It has a fly wheel, with operating handle. The screw of the press operates in a nut, fixed in the vertical columns of press body. A ram is attached to the lower end of the screw. When the screw is operated by a handle; the ram moves in a vertical direction in the slides on the press column face. A punch is fixed in the ram. A die is fixed on a bolster plate which is mounted on the table.

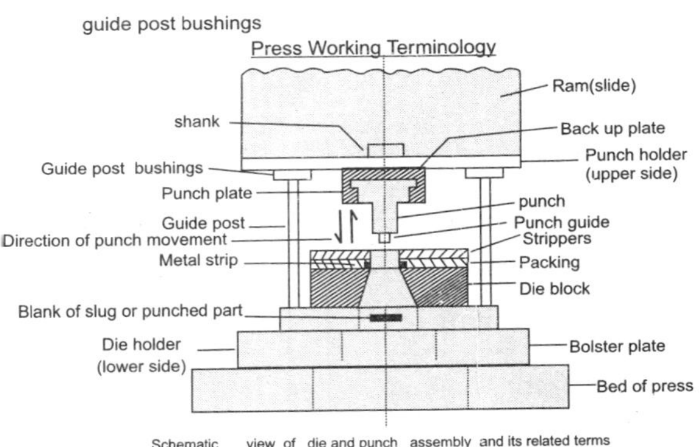

Schematic view of die and punch assembly and its related terms

A simple cutting die used for punching and blanking operation in the press is shown in the figure with different terms associated with press work and press machine:

a. **Bed:** The bed is the lower part of the press frame that serves as a table to which a Bolster plate is mounted.

b. **Bolster Plate:** This is a thick plate secured to the press bed, which is used for locating and supporting the die assembly.

c. **Guide post and bushings:** To hold the punch and keep alignment.

d. **Die Block:** It is a block or a plate which contains a die cavity

e. **Die holder:** This is generally mounted on the bolster plate of the press. The die block is mounted on it, also the guide post are mounted on it.

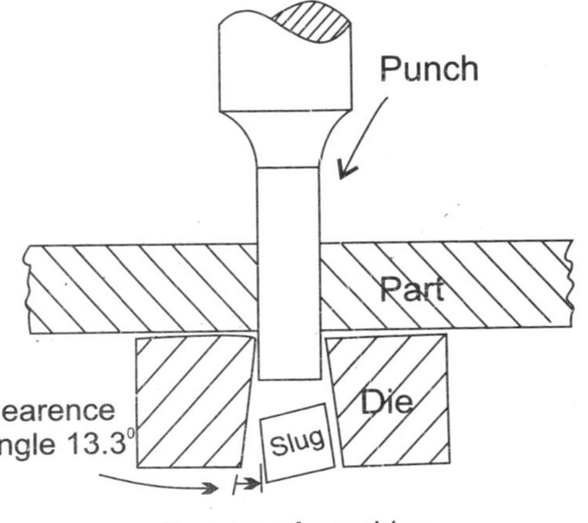

Process of punching

f. **Punch:** This is male component of a die assembly, which is directly or indirectly moved by and fastened to the press ram or slide.

g. **Punch holder & Plate:** The punch plate or punch retainer fits closely over the body of the punch and holds it in proper relative position.

h. **Stripper:** It is a plate which is used to strip the metal strip from Punch during returning.

Working of press: The sheet metal to be worked is placed over the die. Arm is pulled to give a quick rotation by means of handle. This enables the rotation of fly wheel to store kinetic energy, for giving further movement to the screw. These moves ram and punch downwards, which in turn, gives enough thrust on the sheet to do the required operation.

Sheet Metal Operations: The different sheet metal operations based upon operation can be classified as (another classification is based on type of dies a. Simple b. Progressive c. Combination and d. Compound):

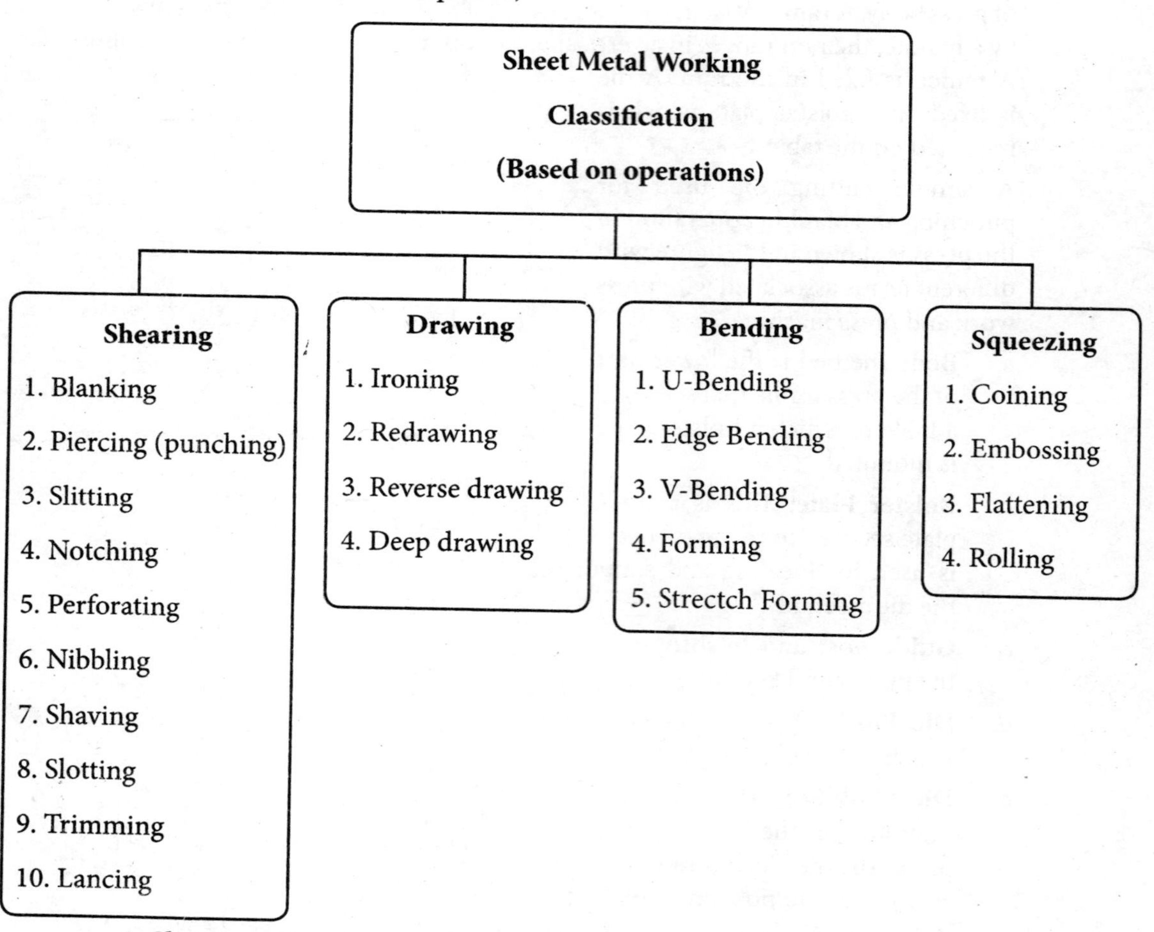

1. Shearing: Shearing is a sheet metal cutting operation without formation of chips or the use of burning or melting.

1.2 Piercing: The piercing is the operation of production of hole in a sheet metal by the punch and the die. The materials punched out to form the hole constitute the waste.

1.3 Blanking and punching (Difference)

Blanking and punching are similar, sheet metal operations that involve cutting the sheet metal, by a die and punch, along a closed outline. If the part that is cut out is the desired product, the operation is called blanking and the product is called blank. If the remaining stock is the desired part, the operation is called punching. Both operations are illustrated by the example of producing a washer.

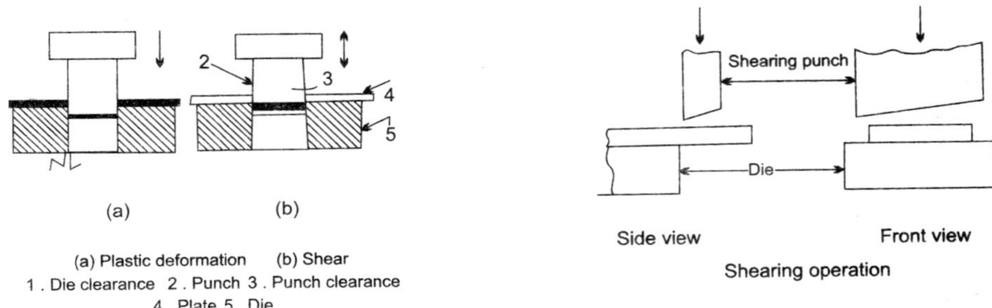

(a) Plastic deformation (b) Shear
1 . Die clearance 2 . Punch 3 . Punch clearance
4 . Plate 5 . Die

Shearing operation

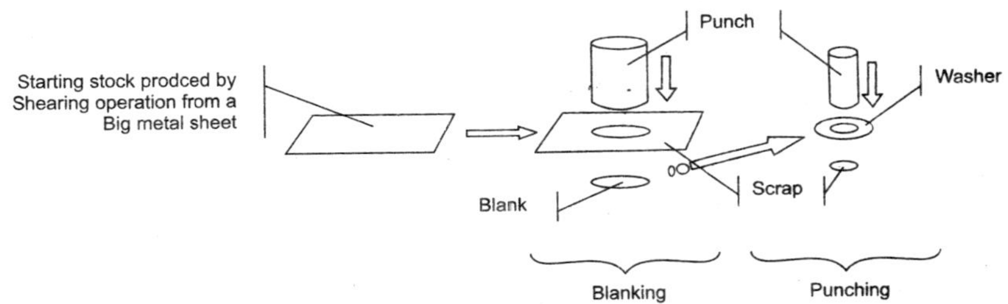

1.4 Cutting off: The cutting off is the operation of severing a piece from a sheet of metal or a bar with a cut along a single line. The cutting off operation illustrated in Fig. (a) (b) (c) can be performed along a straight line or a curve.

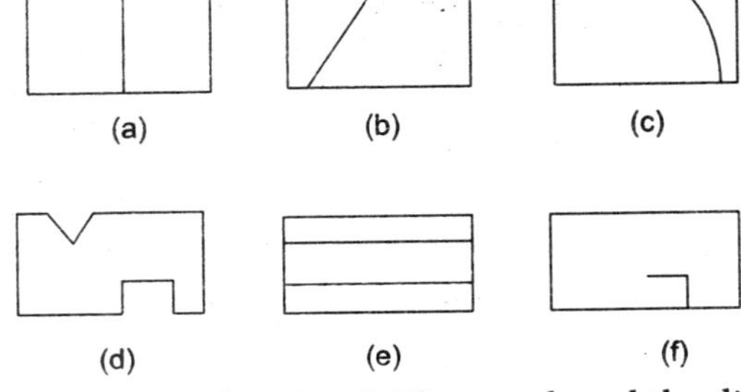

1.5 Notching: The notching is the operation of removal of the undesired shape from the edge of a plate. The operation is illustrated in Fig. d. The punch and the die set up are similar to the piercing or punching operation.

1.6 Slitting: The slitting is the operation of cutting a sheet metal in a straightline along the length. Generally used for making many coils of small width from one wide coil of the metal. The slitting operation is shown in Fig. e above and below.

1.7 Lancing: The lancing is the operation of cutting a sheet metal through partof its length and then bending the cut portion. The operation is illustrated inFig. f and below.

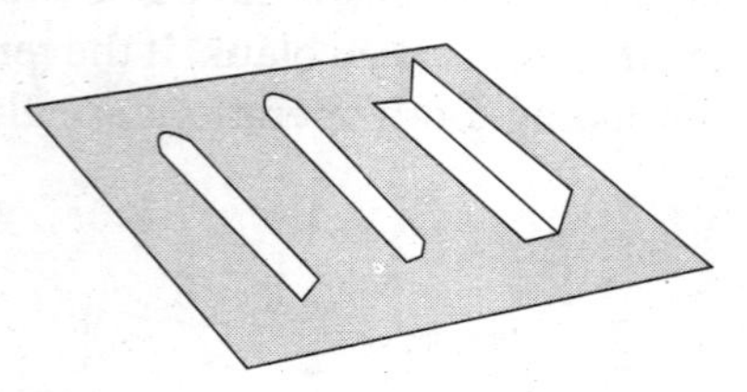

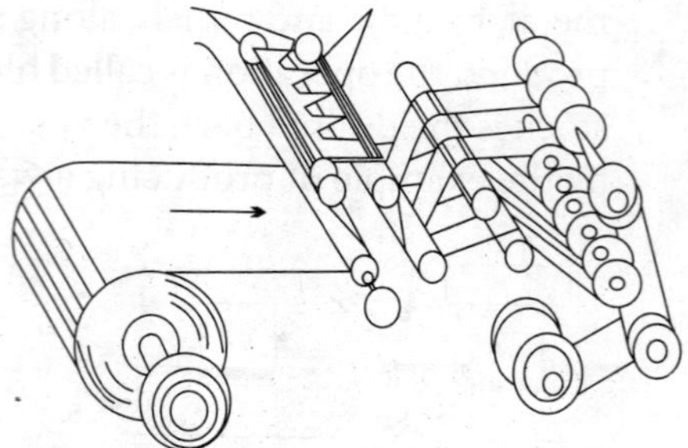

Three steps to show the lancing process

2. Drawing

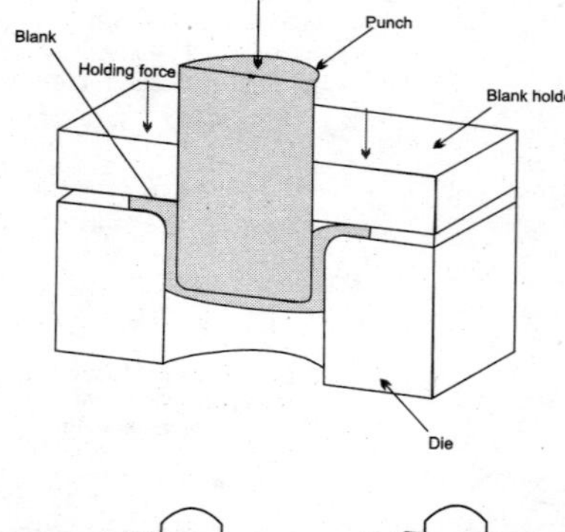

Drawing is a sheet-metal operation to make hollow-shaped parts from a sheet metal such as cup. The drawing is the operation of production of cup shaped parts from flat sheet metal blanks by bending and plastic flow of the metal.

The operation is also known as **cupping.**Different drawing operations are: **Deep drawing** operation is performed to draw very deep parts (When the depth is more than the diameter) from the metal strips through plastic deformation (example of deep drawing is the aluminum cap of the electric bulbs,**Ironing** is the operation of reducing the wall thickness), **redrawing** (drawing is performed many times to reduce the diameter of earlier drawn cup shape), **Reverse drawing** (the drawn cup is subjected to opposite direction). The force in drawing operation is illustrated atedin below in.

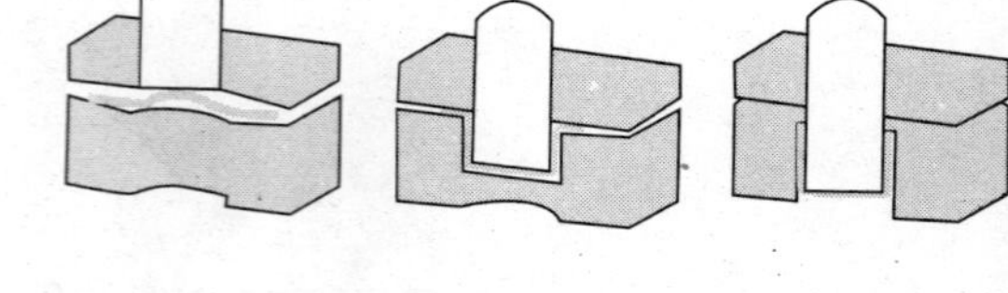

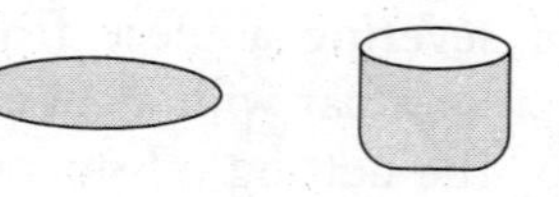

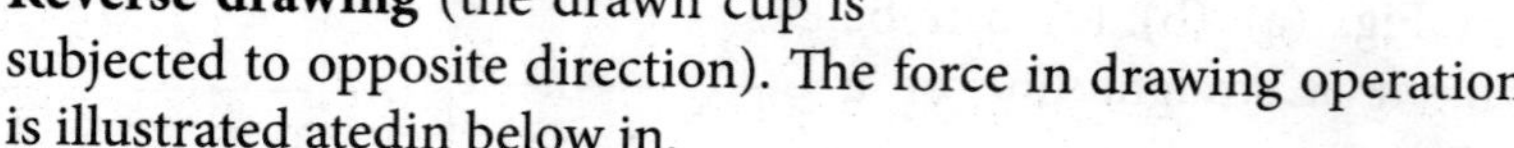

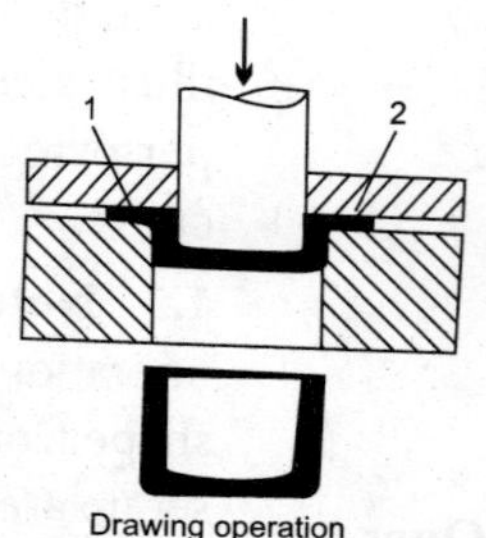

Drawing operation
1. Blank, 2. Pressure pad

3. Bending

Bending is defined as the plastic deformation about a linear axis with little or no change in the surface area. Figure illustrates the bending operation on a metal sheet. The different bending operations may be angle or simple bending, edge bending, U & V bending, forming and stretch forming. Stretch forming is a metal forming process in which a piece of sheet metal is stretched and bent simultaneously over a die in order to form large contoured parts.

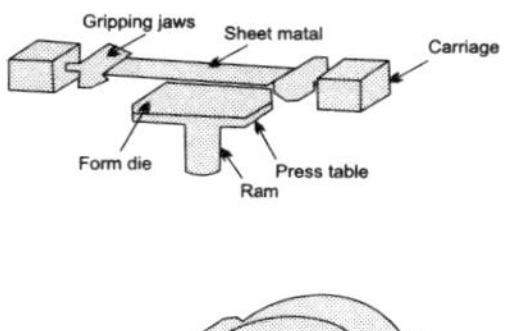

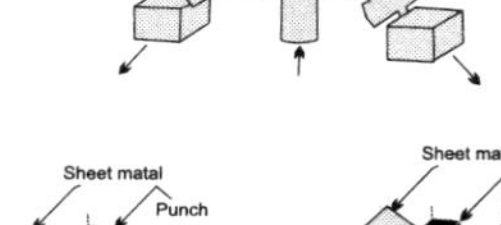

4. Squeezing: In the squeezing operation, large amount of pressure is required to squeeze a metal which is made to flow in a cold state within the cavity of the die and the punch to attain the desired shape. For this reason the squeezing operation is performed in a hydraulic press. The metal may be in the bulk or sheet form. Rolling is also a squeezing operation. The different squeezing operations are described below:

4.1 Coining: The coining is the operation of production of coins, medals or other ornamental parts by squeezing operation.

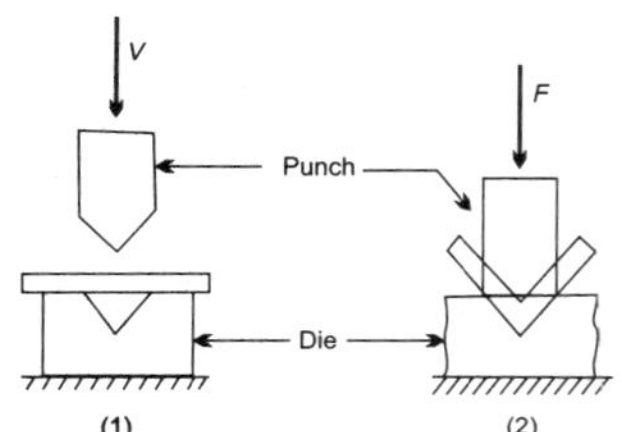

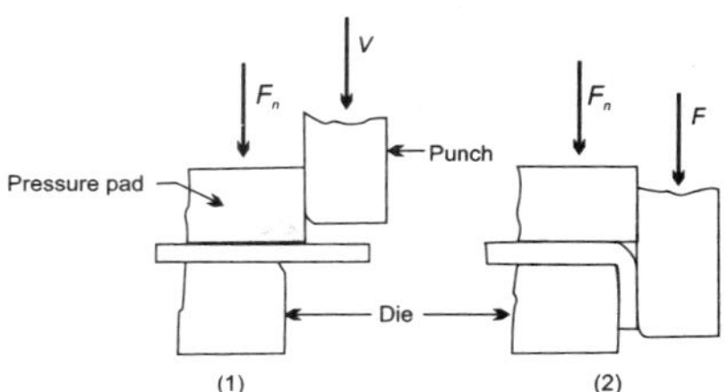

4.2 Embossing: The embossing is the operation of giving impressions of figures, letters or designs on sheet metal parts. The punch, or the die, or both of them may have the engravings which are marked on the sheet metal by squeezing and plastic flow of the metal. Figure below illustrates the coining and embossing operations.

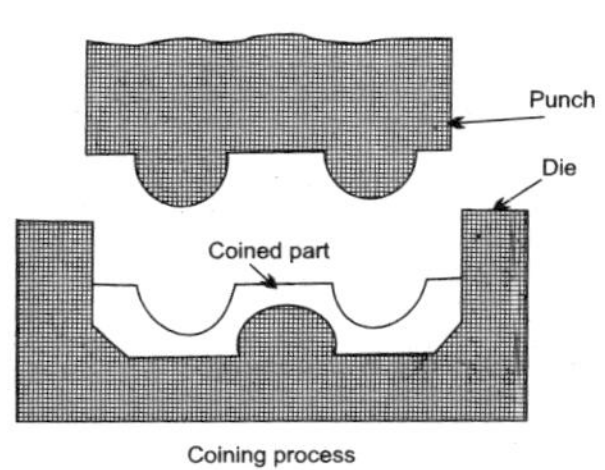

Coining process

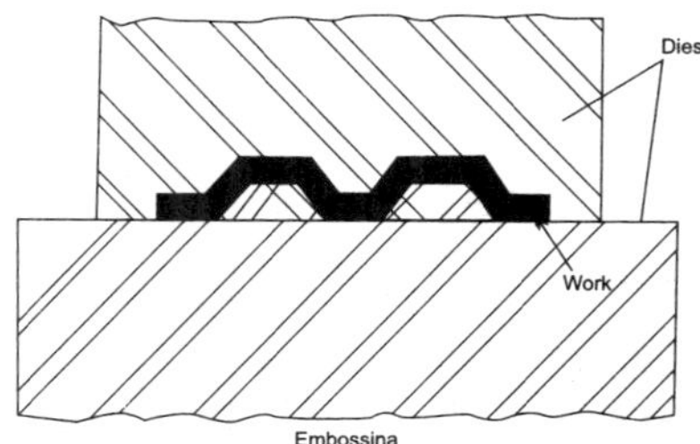

Embossina

Applications of sheet metal working

Coin & medal making, Sheets for Roofs, Air conditioning Ducts, and Vehicles body buildings like 3 wheelers, 4 wheelers, ships, aircrafts, Furniture, Household utensils, pans, and Railway equipment

Ques.590: Describe with neat sketch, the various operations in casting process. Describe the various kinds of patterns in use. What are the allowances provided, when making a pattern? How does the pattern differ from casting required?

Answer: Casting is a process and also it is a product of the process. The casting process is carried out in the foundry and the product of the casting process is called as casting.

Principle: The molten metal is poured into a cavity (called as mould) of required shape & size and allowed to cool & solidify. The solidified product is called as Casting and it is a bit smaller in size from that of the cavity because of contraction of the metal on cooling.

How the pattern is different from casting: The pattern is different from the casting because the pattern is used to make the cavity for the casting, and the metal may shrink or expand on solidification. Also other factors (known as allowances) influence the dimensions of the casting and are listed under the differences in the casting and pattern.

Classification of casting process

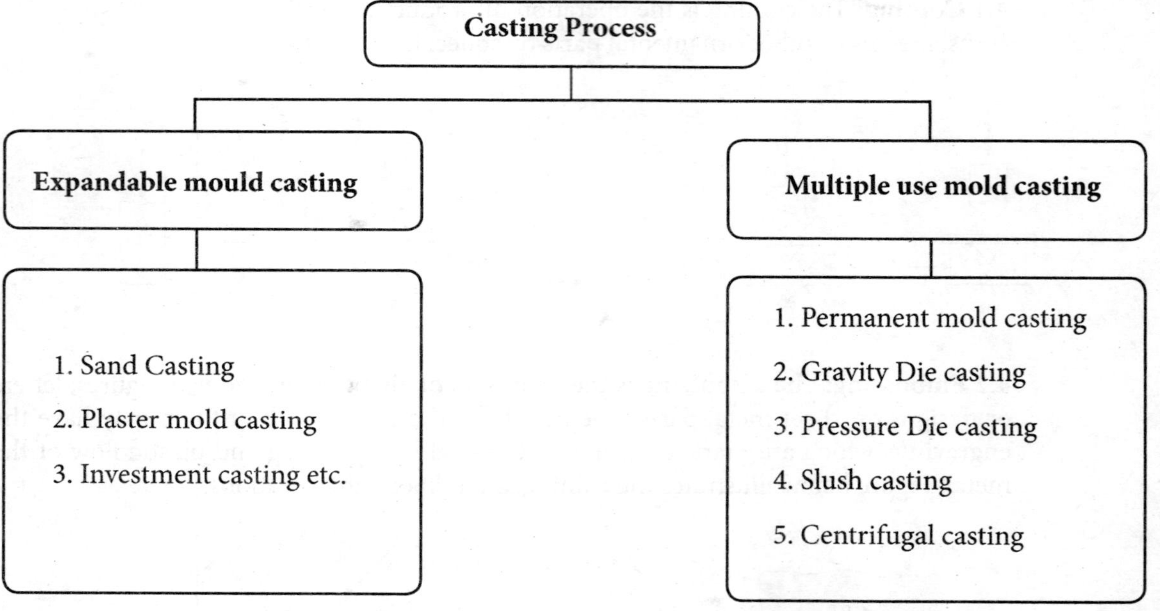

Sand Casting: In this casting process the sand is used as mould making material.

Plaster mold casting: The mould is made from Plaster of paris.

Investment casting: The process involves making of wax patterns and used for high precision and complex products i.e. turbine blades, jeweleries, dentistry.

Permanent mold casting: Reusable mould is used and molten metal flows in the cavity by gravity.

Slush casting: In this process the material is allowed to stay in the mould cavity only till a thin layer of the metal is solidified to form a shell and then the mould is inverted to drain the metal.

Die casting: Reusable mould is used and molten metal is forced in the cavity by applying large pressure in case of pressure die casting and by gravity in the gravity die casting.

Centrifugal casting: The centrifugal force is used to distribute the metal in the cavity and the mould is rotated at large speed approximately at 500 rpm.

Advantages of casting Process

a. Intricate shapes internal or external can be produced in one step.

b. Objects can be cast in single piece without needing machining or f[illegible]ing.

c. Thin sections of the complex shapes are possible.

d. High production rate.

e. Greater surface finish.

f. The labour cost involved is less.

g. A number of non-ferrous alloys can be die cast.

Pattern: The pattern is the repplica OR duplicate of the casting to be produced. A pattern is a model of the desired product (called casting), constructed in such a way that it can be used for forming an impression called mould (cavity) in damp sand. When this mould (cavity) is filled with molten metal, it solidifies and produces a casting (desired product). Patterns are used to mold the sand mixture into the shape of the casting. They may be made of wood, plastic, or metal. The selection of a pattern material depends on the size & shape of the casting, the dimensional accuracy, the quantity of castings required, and the molding process.

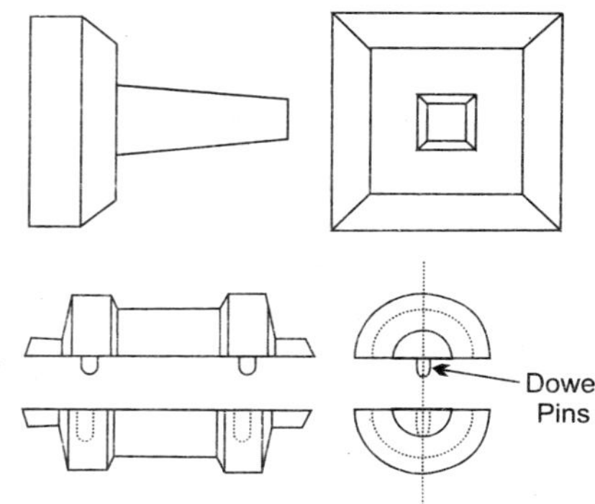

Master Pattern: This is a type of pattern and used in the process of investment casting for making of wax patterns.

Pattern materials: The patterns are generally made from following materials: (i) Wood (ii) Metals (iii) Cast iron (iv) Brass (v) Plaster of Paris (vi) Plastic (vii) Wax

Types of Patterns

(a) Solid or single piece pattern: This pattern is made without joints, partings or any loose pieces in its construction and is called a single piece or solid pattern.

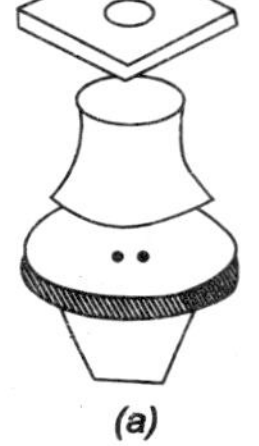
(a)

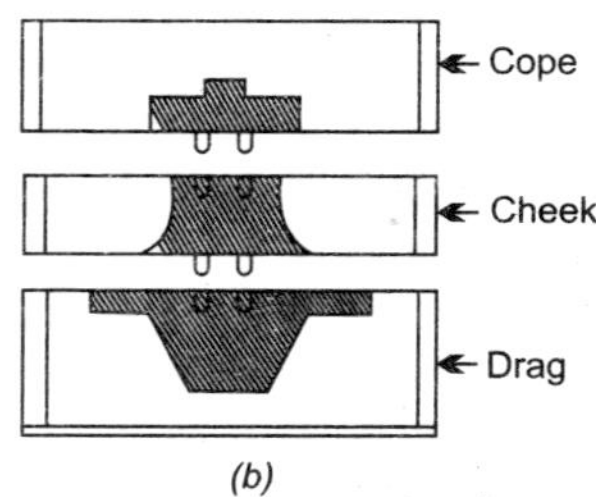

(b)

(b) Split (Two piece) pattern: Sometimes patterns cannot be made in single piece because of the difficulties encountered in molding or difficulty in withdrawal from the mould. Because of these, patterns are usually made in two pieces called split patterns. One part will produce the lower half of the mould and the other part will produce the upper half of the mould. These two pieces are held in their proper relative position by means of dowel-pins fastened in one piece and fitting holes bored in the other. The surface which is formed at the line of separation of the two parts, usually at the centre line of the pattern is called as parting surface or parting line.

(c) Multi piece pattern: Sometimes the patterns are made in more than two parts if the shape is complex. Then these patterns are called multi-piece pattern which may consist of 3, 4 or more number of parts depending upon their design.

(d) Match plate pattern: If the split patterns are mounted with one half on one side of a plate (match plate) and the other half directly opposite to the other side of the plate, the pattern is called a match plate pattern.These patterns are advantageous in Machine molding, Mass production, Greater dimensional accuracy and Minimum machining requirement.

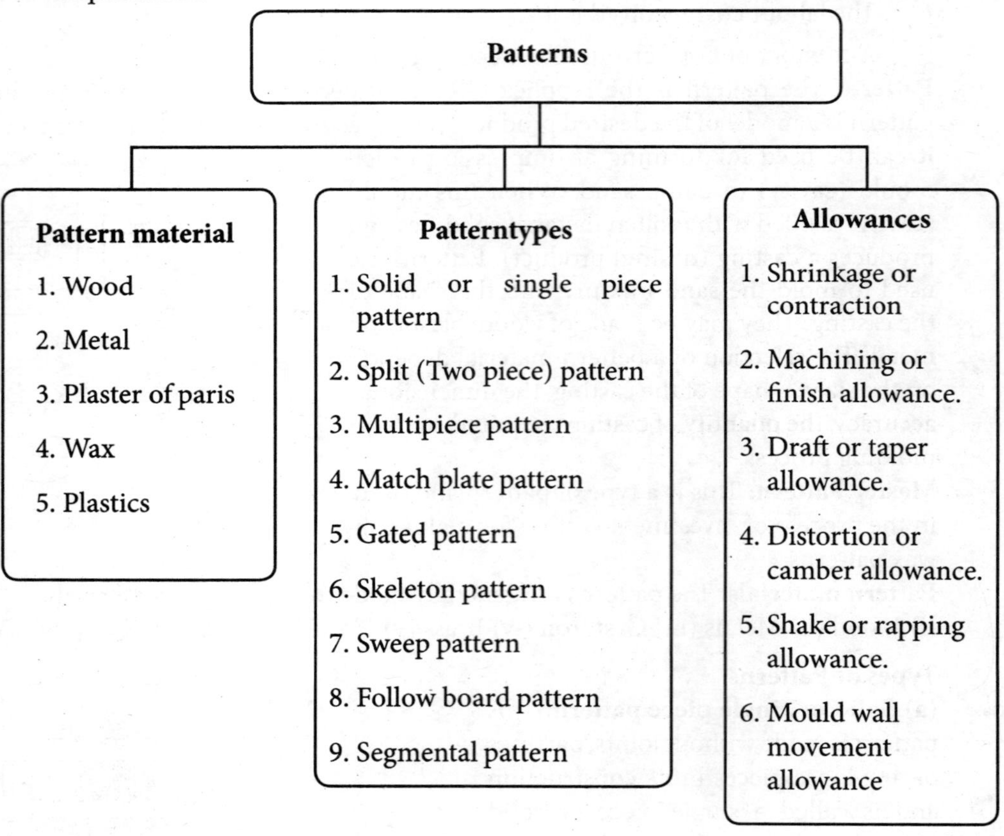

A single pattern or a number of patterns may be mounted on a match plate. The match plate may be of wood, steel, magnesium or aluminum. Gates and runners are also attached to the plate along with the pattern. So when the match plate is lifted off the mould, patterns are also withdrawn. The gates and runners are also completed in one operation because patterns for runner and gates are also attached with the match plate along with the patterns of the desired castings. So lot of manual work is reduced.

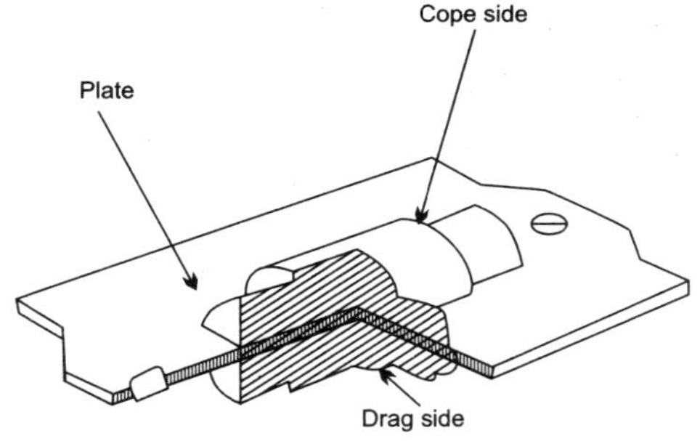

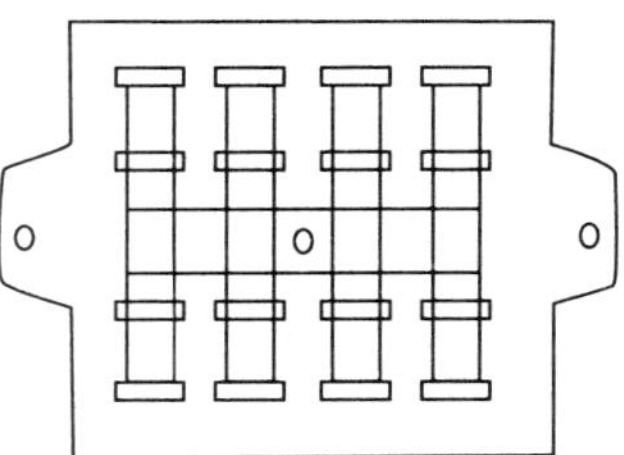

Match plate patterns

(e) Gated pattern: The provision of gating is provided in the gated patterns.In gated patterns, a single sand mould carries a number of cavities. Patterns for these cavities are connected to each other by means of gate formers which provide suitable channels (path) or gates in sand for feeding the molten metal to these cavities. A single runner can be used for feeding all the cavities. This enables a considerable saving in molding time and a uniform feeding of molten metal. For small quantities, these patterns may be made of wood, but for large production metallic patterns are preferred. To produce good casting, it is necessary to ensure that full supply of molten metal flows into every part of the mould. Provision for easy passage of the flowing metal in the mould is called gating and is provided in the gated patterns.

Visual presentation of gated pattern

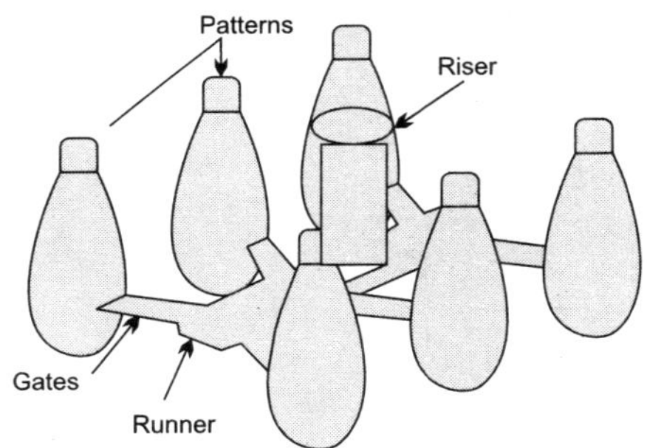

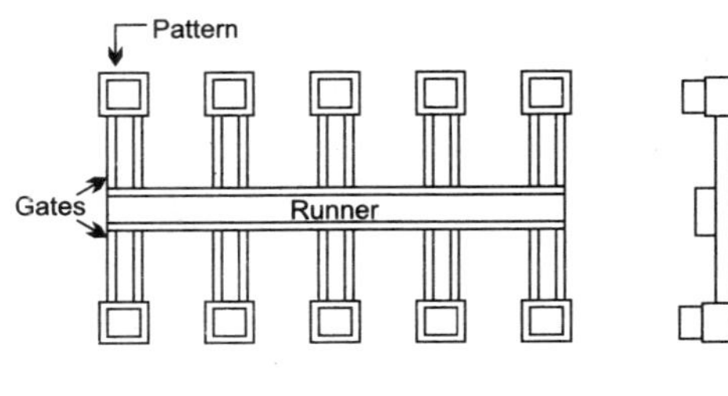

(f) Skeleton pattern: Patterns for very large castings would require a large amount of pattern material. In such cases skeleton patterns are used. Skeleton patterns are hollow patterns and made up of a wooden frame and strips. The frame work is filled & rammed with loam sand and stickle board is used to remove the excess sand. Stickle board is having the same contour as desired so it can remove the extra sand by passing over the skeleton pattern.

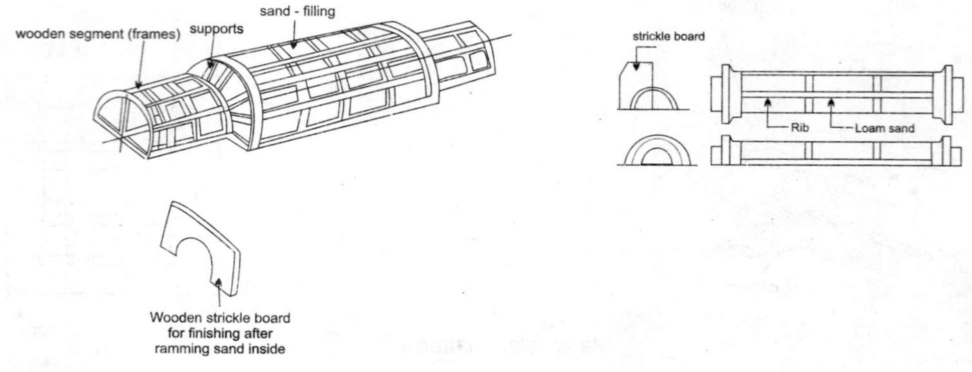

Skeleton pattern

(g) Sweep pattern: Sweep pattern is just a form or shape made on a wooden board which sweeps the shape of the casting into the sand all round the circumference. The sweep pattern rotates about the post. The moulds of large size and symmetrical in shape particularly of circular section can be easily prepared by using a sweep instead of a full pattern. Once the mould is ready, sweep pattern and the post (spindle) can be removed. Sweep pattern avoids the necessity of making a full, large circular and costly three-dimensional pattern. Sweep patterns are used for large patterns of cast iron, ridges etc. The figure is shown below:

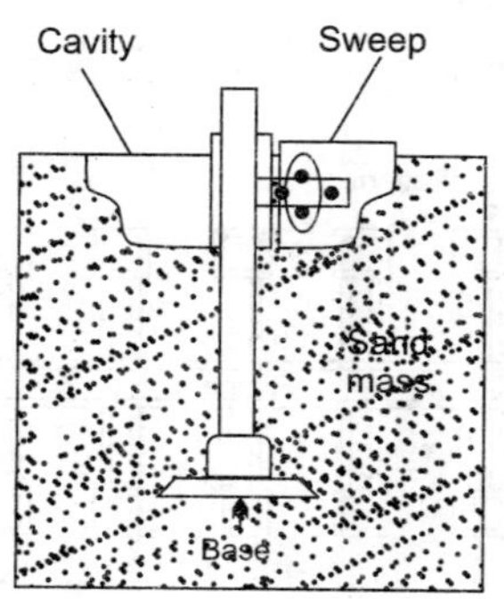

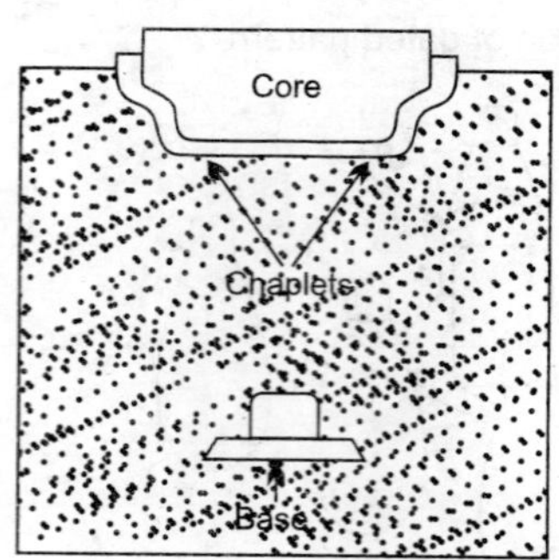

(h) Follow board pattern: A follow board pattern is a wooden board and is used for supporting a pattern which is very thin & fragile and which may give away & collapse under the pressure when the sand above the pattern is being rammed. In addition to supporting a thin sectioned pattern, a follow board forms the natural parting line of the mould or the casting.

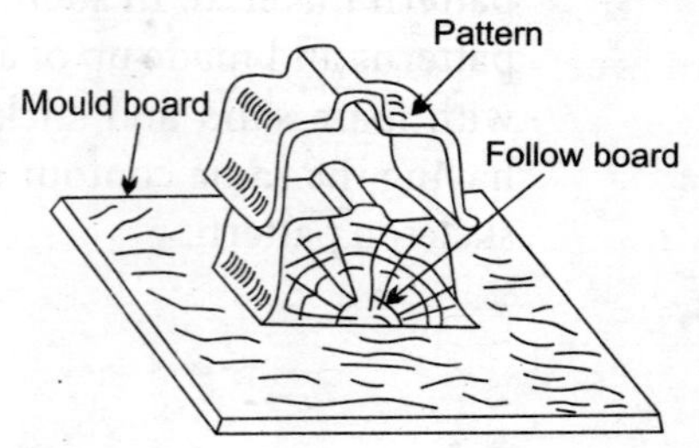

(i) Segmental pattern: Segmental patterns are sections of a pattern so arranged as to form a complete mould. A segmental pattern completes one portion (stage) of the mould and then moved to the next position to make the next part of the mould and so on till the mould is completed. A segmental pattern differs from a sweep pattern in the sense that it does not revolve, continuously about the post to make the mould rather it prepares the mould by parts. For example making of a flywheel using only one segment of the flywheel.

Difference between casting and Pattern: The various differences between casting and pattern are as tabulated below:

Factor	Pattern	Casting
Material	The material of the pattern is not necessarily same as that of the casting. Pattern may be made from wood.	Casting is made of required material, i.e. cast iron, aluminum
Color	The colour of the pattern may not be same as that of the casting.	Depends upon casting metal.
Allowances	Pattern carries an additional allowance to compensate for metal shrinkage.	No allowances provided.
Allowances	It carries additional allowance for machining.	No allowances provided.
Allowances	It carries the necessary draft to enable its easy removal from the sand mass.	No allowances provided.
Allowances	It carries distortions allowance. Due to distortion allowance, the shape of casting is opposite to pattern	No allowances provided.
Core Prints	Pattern may carry additional projections, called core prints to produce seats for cores.	Core prints Not present in casting.
Pieces	Pattern may be in pieces (more than one piece) whereas casting is in one piece.	Casting is always in single piece.
Surface finish	Surface finish may not be same as that of casting.	Surface finish will change on machining.
Sharp corners	Sharp changes are not provided on the patterns. These are provided on the casting with the help of machining.	These sharp corners if required are produced by machining.

Ques.591: Describe various Operations or steps involved in the making of casting or Casting process .

Answer : Casting process comprises of the following steps:

(1) Making a pattern: Pattern is the model/ replica/ duplicate of the desired product (called casting), constructed in such a way that it can be used for forming an impression/ cavity called mould (cavity) in damp sand. Various allowances such as shrinkage, machining, draft, shaking, distortion etc., are provided. Sometimes core prints are also provided to the patterns to make a core seat in the damp sand.

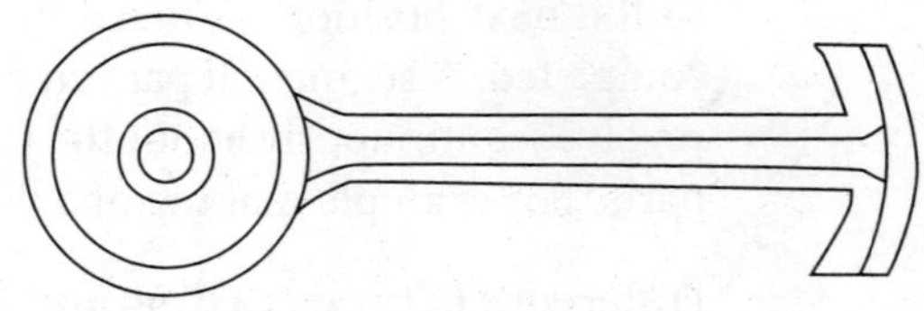

(2) Preparing Molding Sand: Sand is the principal molding material in a foundry shop. The quality of the casting depends upon the selection and mixing of sand, which may be natural or synthetic and is used for mould & core making.

(3) Preparing a mould and core making: Moulds are prepared with the help of pattern to produce a cavity of desired shape. Usually the mould is made of sand and used only once. But sometimes permanent metal moulds are also used. For obtaining hollow portions, cores are prepared separately in core boxes. Moulds and cores are baked to impart strength.

(4) Melting the Metal: The required quantity of the metal with proper composition is melted in a suitable furnace.

(5) Pouring the metal into the mould: When the molten metal attains pouring temperature, it is taken into laddles and poured into the moulds.

(6) Cooling i.e., Solidification: After pouring the molten metal into the mould cavity it is allowed to cool down so that the metal solidifies.

(7) Removing the Solidified casting from the mould: The solidified casting are extracted by breaking the mould and cleaned by removing adhering sand.

(8) Fettling: The unwanted projection in the form of gates, risers etc. are cut off and the entire surface is cleaned and made uniform.

(9) Heat Treatment: The castings may need heat treatment depending on the specific properties required.

(10) Testing and Inspection: Finally the casting is inspected to ensure that it is free from casting defects and is as per desired specifications.

Making a pattern

⇩

Preparing Molding Sand

⇩

Preparing a mould and core

⇩

Melting the Metal

⇩

Pouring the metal into the mould

⇩

Cooling i.e., Solidification

⇩

⇩*Removing the Solidified casting from the mould*

Fettling OR Cleaning

⇩

Heat Treatment

⇩

Testing and Inspection

The flow chart indicating the main steps in the making of a casting

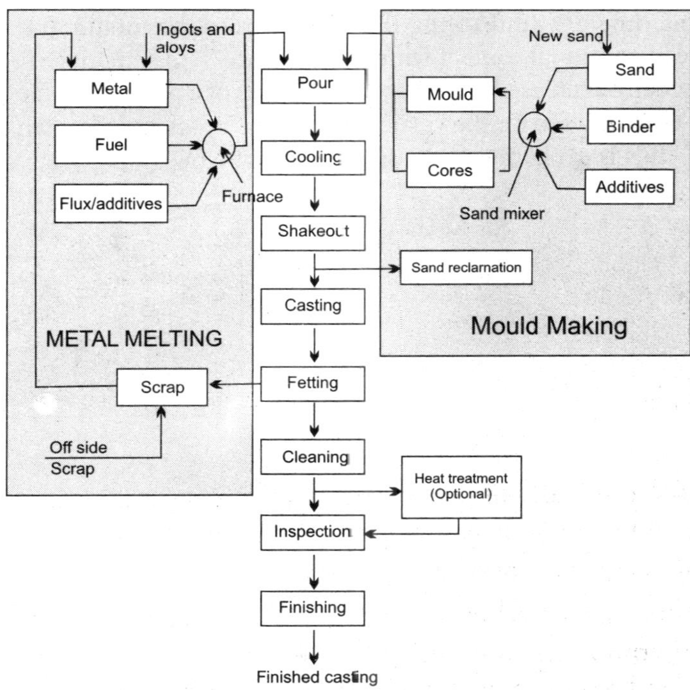

Figure presents all the steps involved in the making of a casting

Ques.592: Explain various allowances provided in pattern making and the reasons for it.

Answer : **Pattern allowances:** Pattern is having different size/ dimension as compared to casting because it carries certain allowances or deviation from the actual dimension of the casting due to metallurgical, mechanical and process variables.The difference between the dimensions of pattern and casting is known as pattern allowance. The allowances are provided so that the produced final castings are of the correct dimensions. The various allowances used when making a pattern are:

(a) Shrinkage or contraction allowance: As metal solidifies and cools, it shrinks & contracts in size. To compensate for this, a pattern is made larger than a finished casting by means of a shrinkage or contraction allowance.Contraction is different for different metals.Wood patterns used to make metallic patterns are given double allowance, one for the shrinkage of the metal of the pattern and the other for that of metal to be cast. The total contraction is volumetric but the shrinkage allowance is added to the linear dimensions.

(b) Machining or finish allowance: For good surface finish, machining of casting is required. For machining, extra metals are needed. This extra metal is called machining or finishing allowance.The amount of this allowance depends upon the type of casting metal, size & shape of casting, method of casting used, method of machining to be employed and the degree of finish required.This allowance is added in addition to shrinkage allowance

(c) Draft or taper allowance: It is provided for easy removal of pattern from the sand mold.At the time of withdrawing the pattern from the mould, the vertical faces of the pattern are in continual contact with the sand, which may damage the mould cav:ty. This danger is greatly decreased if the vertical surfaces of a pattern are tapered inward slightly. The slight taper inward on the vertical surface of a pattern is known as the draft or taper allowance. This is explained in the figures given below.

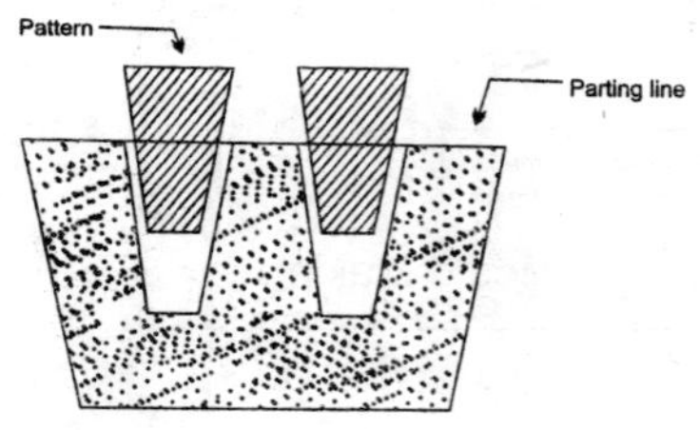

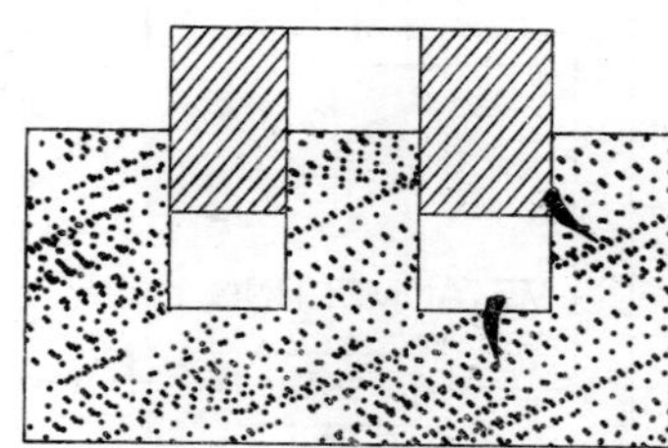

(d) Distortion or camber allowance: If the shape of the casting changes after the production that is called distortion of the casting. A casting will distort or warp, if it is of irregular shape.

(i) All its parts do not shrink uniformly.

(ii) It has long flat casting.

(iii) The arms possess unequal thickness.

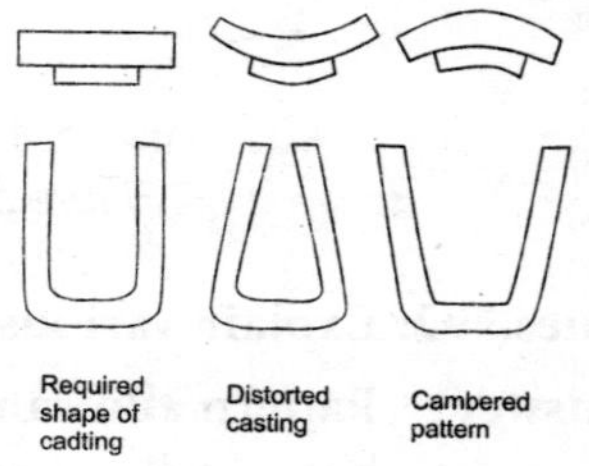

Distortion can be practically eliminated by providing an allowance and constructing the pattern initially distorted. e.g., For U shape casting, legs will diverge so we make the pattern having legs converging so that after casting, the product is having legs parallel.

(e) Shake or rapping allowance: When a pattern is rapped (shaked) in the mould before it is withdrawn, the cavity in the mould is slightly increased. So in order to compensate this, pattern is made slightly smaller than the actual. This allowance is called shaking or rapping allowance. Rapping allowance is negative.

(f) Mould wall movement allowance: Movement of mould wall in sand moulds takes place because of heat and the static pressure exerted on the walls of the mould which comes in contact with the molten metal.Because of this, the size of the mould (cavity) increases. In order to compensate this, the size of the pattern is made smaller so that the casting produced has an accurate size (i.e., desired size).

The pattern allowances are provided to compensate for the various expansion and contraction in dimensions of the metal on solidification and of also of pattern. Also the other parameters of casting process are considered; so that the final casting produced is of correct dimensions.

Functions of a pattern

a. A pattern prepares a mould cavity for the purpose of making a casting.

b. To produce seats for cores in the moulds, so need core prints on the pattern.

c. Runner, gates and riser may form a part of the pattern.

d. Pattern establishes the parting line and parting surfaces in the mould.

e. Patterns properly made and having finished & smooth surface reduce casting defects

Ques.593: Describe with neat sketch, the various operations in mould making with use of a core. What is the function of core in casting?

Answer : Function of cores: Core is used for making cavities (holes) and hollow projections in the casting. For all those castings, where coring is required, provision should be made on the pattern to support the core inside the mould.

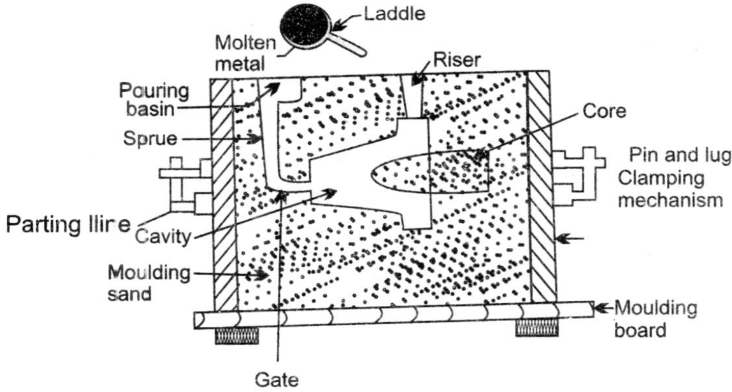

Mould: A mould can be described as a void or hollow or cavity created in a compact sand mass with the help of pattern which (Mould), when filled with molten metal will produce a casting.

Molding: The process of making this cavity or mould in the compact sand is called molding.

Steps involved in making of a mould: The mould ready for the pouring of molten metal is shown on the figure and the steps involved are mentioned in this flow chart.

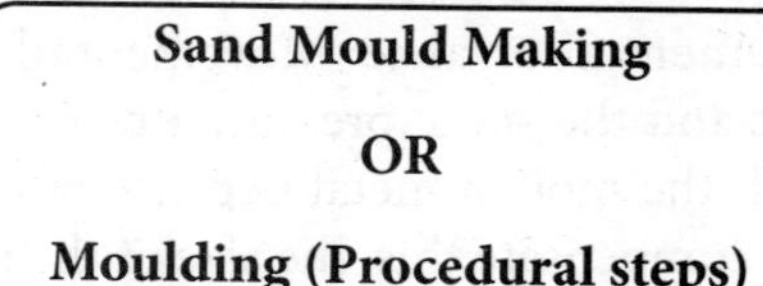

Floor moulding

for large to medium castings

Bench moulding

1. Select the suitable flask and prepare to sand.
2. Place drag part of flask upside down on the board.
3. Put the drag half of pattern
4. Fill the sand and compact
5. Reverse the flask and place the cope half of flask, then place the cope half of the pattern over the drag half
6. Spray the facing and parting sand
7. Place the riser, sprue pins and fill the sand carefully
8. Separate the two halves
9. Remove the riser, sprue pine
10. Cut the gating system . i.e. runner, gate, etc.
11. Place the corn in the location.
12. Replace and clsoe the both flask halves over another carefully aligning the pins
13. The mould is ready for pouring

Note: Redressing and repair may be rquired before the closing of the mould

Pit moulding

for very large casting and done by digging a pit

machine moulding

done by machines

The different terms related to the sand mould and casting are and shown in the figure:

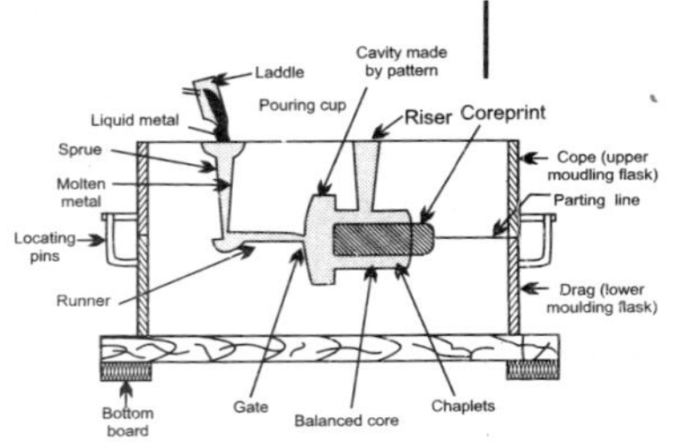

a. **Core**: A sand shaped insert placed in the mold cavity to produce internal features on the part.

b. **Cope:** The upper half of the sand mold, flask or pattern as referred.

c. **Drag:** The lower half of the sand mold, flask or pattern as referred.

d. **Flask:** A box made of wood or metal to contain the sand.

e. **Gates:** Multiple openings in the mold to allow the molten metal to flow into the mold cavity.

f. **Gating System:** A passage from where the molten metal flows into the mold. The gating system is made up of the pouring cup, runner, sprue and gates.

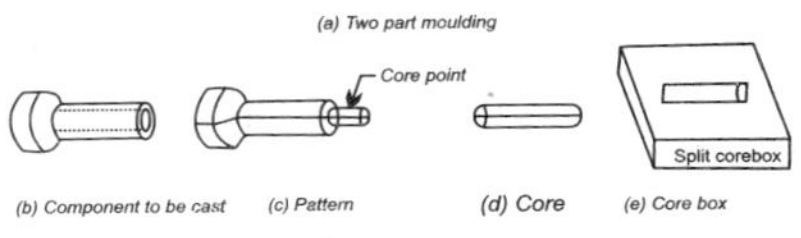

g. **Mould:** A cavity by which molten metal is into the shaped desired product.

h. **Parting Line:** The line where the top and bottom halves of the mold meet.

i. **Pattern:** A representation or replica or duplicate of the final product used to make cavity / imprint of the shape into the sand.

j. **Risers:** Reservoir is called risers inside the mold, which is filled with the molten metal to compensate for shrinkage or "feed" the mold cavity during the solidification process.

k. **Runner:** The horizontal part of the gating system, which supplies molten metal to the gates.

l. **Sprue:** The vertical part of the gating system, which is connected to the pouring cup at the top and feeds the runner with molten metal at the bottom.

m. **Vent:** An opening in the mold to allow the escape of hot gases.

Ques.594: Why gating system is required in the casting? Explain the purpose of different gating system.

Answer : **Gating system** means the system of all the passages/ ways through which the molten metal enters the mould cavity *i.e.* it includes the pouring basin, runner, gate and riser etc. The manner in which the molten metal enters the mould has a great impact on the quality of the casting produced. So the gating system should be carefully designed and produced.

A good Gating System should meet the following requirements:

a. The velocity of molten metal entering into the mould cavity should be as low as possible, so that there is no erosion of mould.

b. Gating system should ensure the complete filling of the mould cavity.

c. Gating system should prevent the molten metal from entrapping/ absorbing air/ steam while flowing through it.

d. Gating system should prevent the formation of oxides.

e. Gating system should prevent the entry of oxides, slag, dross etc.

f. Gating system should assist in directional solidification of the casting.

g. Gating system design should be practicable and economical

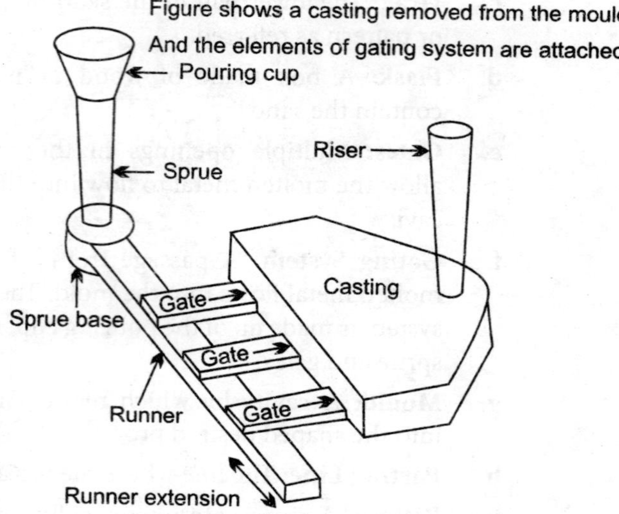

Elements of a Gating System(the elements of gating system are shown in figure)

a. **Pouring Basin**: Molten metal is poured into a pouring basin which acts as a reservoir from which it moves smoothly into the sprue. The pouring basin is also able to stop the slag from entering the mould cavity by means of a skimmer.

b. **Sprue**: It is the channel through which the molten metal is brought into the parting plane where it enters the runners and then gates. Sprue should be conical in shape because:

(i) The molten metal when moving from top of the cope to the parting plane gains in velocity; requires a smaller area of cross section for the same amount of metal to flow at the bottom.

(ii) Liquid tries to attain the minimum area at the bottom so there may be air gap between the liquid jet and sprue wall so air inspiration will be there which will cause problem.

c. **Sprue Base/ Well**: This is a reservoir for metal at the bottom of the sprue to reduce the momentum of the molten metal.

d. **Runner**: It is generally located in the horizontal plane (parting plane) which connects the sprue to gates.

e. **Runner Extension**: This extension is provided to trap the slag in the molten metal.

f. **Gates**: These are the openings through which the molten metal enters the mould cavity.

g. **Riser**: Most of the foundry metals/ alloys shrink during solidification. As a result of this volumetric shrinkage during solidification, voids are likely to form in the castings so these (voids) needs the additional molten metal. Hence a reservoir of molten metal is to be maintained from which the metal can flow readily into the casting when the need arises. The riser work as reservoirs. These may be blind or open type.

Functions of Riser: Metals and their alloys shrink as they cool and solidify. It creates a partial vacuum within the casting.Partial vacuum will lead to a shrinkage void.This shrinkage void will grow and form shrinkage cavity, if extra liquid metal from outside the mould (cavity) is not supplied.

a. The primary function of the riser (attached with the mould) is to feed the metal to the solidifying casting so that shrinkage cavities are not formed. Shrinkage is a very common casting defect.

b. A riser permits the escape of air and mould gases as the mould cavity is being filled with the molten metal.

c. A riser full of molten metal indicates that the mould cavity has already been completely filled up with the same.

d. Riser promotes directional solidification.

Advantage of riser: A casting solidifying under the liquid metal pressure of the riser is comparatively sound.

Directional solidification refers to the process of solidification when the casting is being cooled. It is desired that the cooling starts from the bottom and continues in gradual manner to the top. The use of CHILLS is made to achieve this. Chills are the made of metals and accelerate the cooling rate, so that cooling progresses gradually.

Ques.595: What are various types of casting defects.

Answer : Defect can be defined as imperfection or a fault. Casting defects are the defects developed in the casting during the process of making the casting; they occur because some steps in the manufacturing cycle are not properly controlled.

The factors, which are normally responsible for the occurrence of these defects are:

a. Design of casting

b. Design of pattern equipment

c. Molding and core making equipments

d. Mould and core material

e. Gating and risering

f. Melting and pouring.

g. Melting and core making techniques

h. Metal composition

Various casting defects may be categorized as:

a. **Metallic projections**: joints flash or fins: These are formed where two parts of the gating system meet and intersect.

b. **Cavities:** Blow holes, pinholes: These are formed as holes or cavities because of gas entrapment or excess of moisture.

c. **Discontinuities:** Shrinkage, cracks: these are formed because of improper temperature of molten metal, discontinuity during the pouring and improper directional cooling.

d. **Defective surface:** Flow marks: These are formed near the side wall of casting because of metal erosion during the filling process.

e. **Incomplete casting:** Poured short: The molten metal poured is less in quantity.

f. **Incorrect dimension or shape:** Distortion

g. **Inclusion:** Metallic inclusions

Ques.596: What do you understand by die casting? Explain the cold & hot chamber die casting and write the advantages of die casting process.

Answer : In die casting, the molten metal is forced into permanent mould (die) cavity under pressure or due to gravity (then known as gravity die casting). The pressure varies from 20 to 2000 kgf/cm^2 and is maintained till solidification stage is reached. The pressure is generally obtained by compressed air or hydraulically.

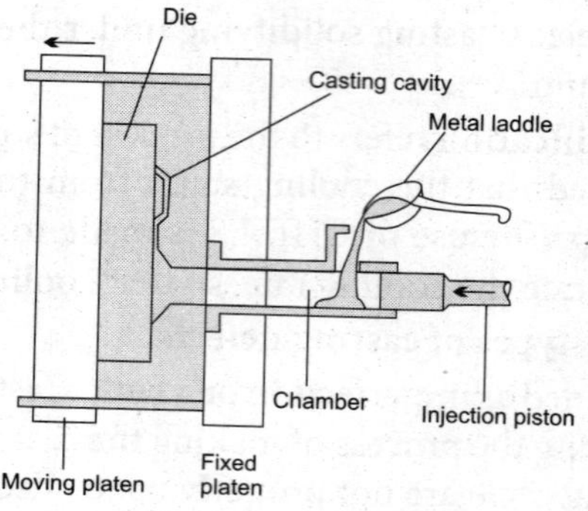

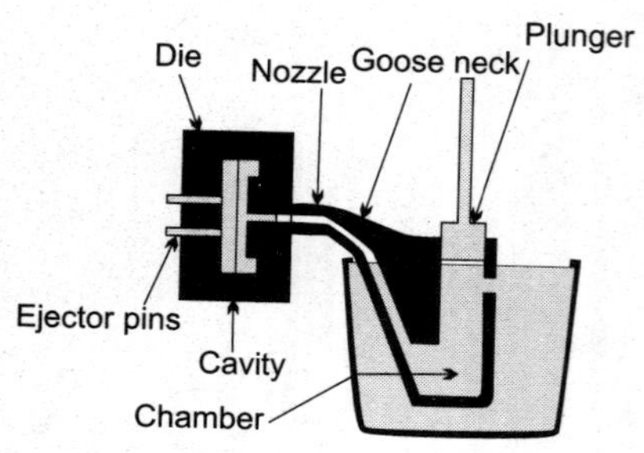

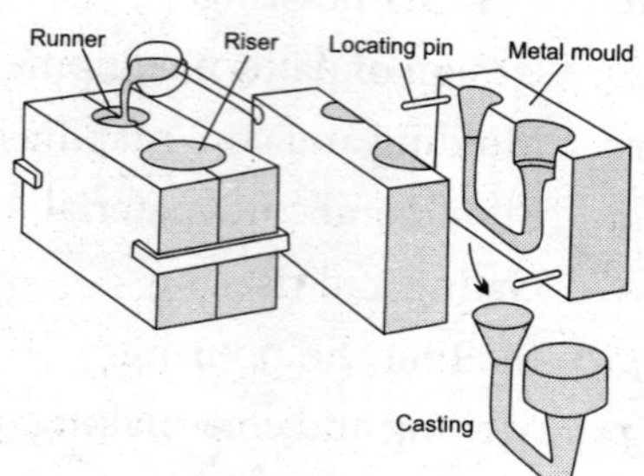

Cold Chamber Die Casting

In cold chamber die casting, the metal is melted separately in a furnace and transferred by means of small hand ladle. After closing the die, the molten metal is forced into the die cavity by a hydraulically operated plunger and pressure is maintained till solidification. These machines can either have vertical plunger or horizontal plunger for forcing molten metal into die. These machines are widely used for casting of aluminum alloys and brasses.

Hot Chamber Die Casting

In hot chamber die casting machines, the melting unit is integral part of the machine itself that is why it is called hot chamber die casting machine.The molten metal needs less pressure to force the liquid metal into die.It has further two types of arrangements:

(i) Goose neck or air injection type (or direct air pressure).

(ii) Submerged plunger type.

Submerged plunger type

In this machine, the goose neck type container always remains immersed in the metal pot. The molten metal from the metal container is forced inside the die with the help of a plunger submerged in the molten metal and operates hydraulically. When the plunger moves up, the molten metal comes up and fills the cylinder and when the plunger moves down, the metal is forced into the die. The movable die platen is synchronized such that when plunger is moving up, the movable die platen moves away and the casting is removed. The figure presents the steps involved the hot chamber diecasting.

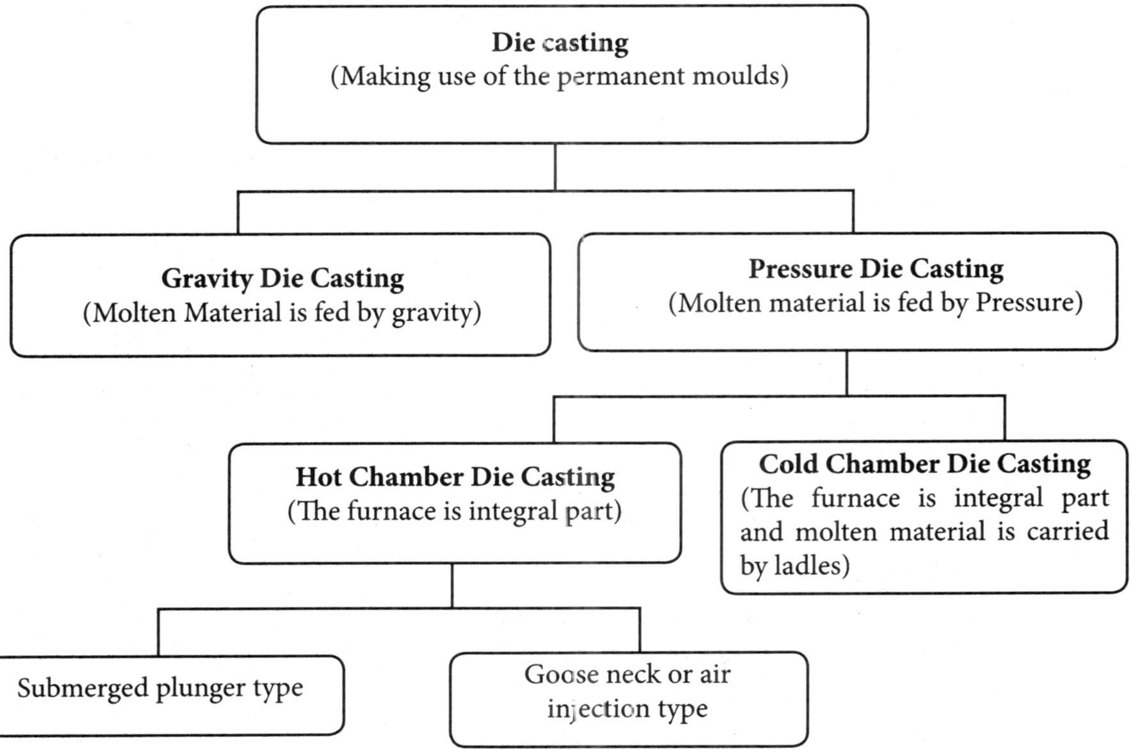

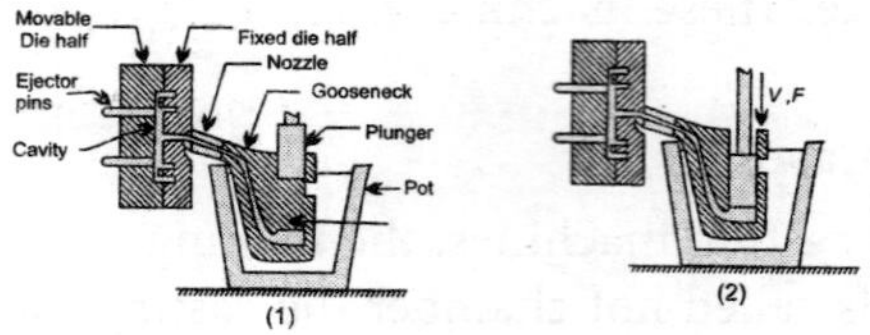

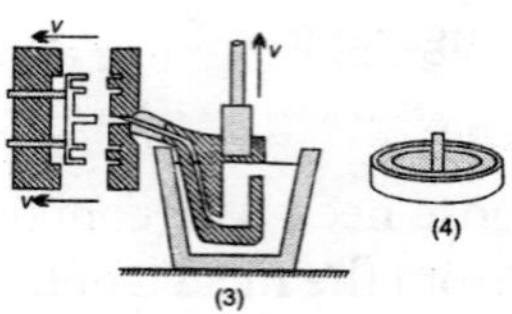

Hot chamber die casting showing the different steps

Ques.597: What are different foundry equipments, explain cupola furnace and its operation. What are the different zones of cupola furnace? Explain with a neat sketch.

Answer : A foundry is a place where castings are produced, and for performing different operations of mould making, like pattern making, machining, lifting, shifting, inspecting, testing etc. a variety of hand tools and equipments are needed. The tools and equipments can be classified as:

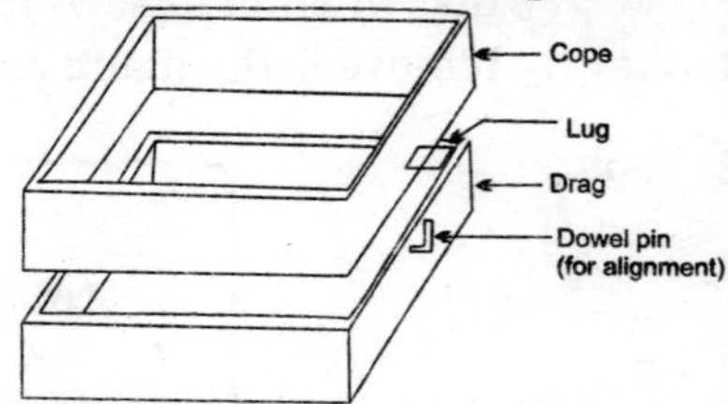

Two piece molding flask or box

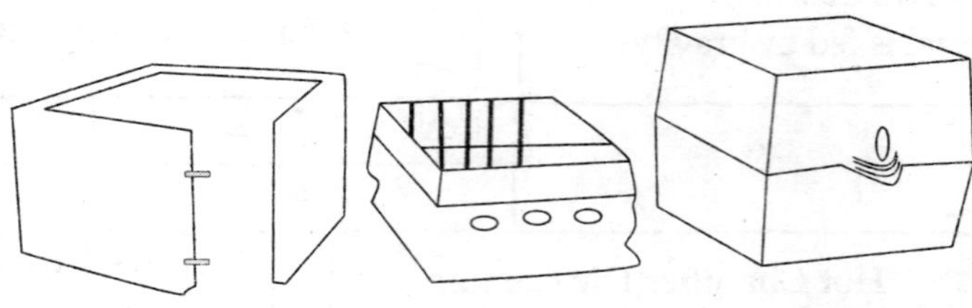

Molding flasks

Cupola Furnace

Cupola is a furnace generally used for production and melting of cast iron and other metals can also be melted. The **advantages** of cupola are as follows:

a. The cost of melting is low.

b. The control of chemical composition is better.

c. Temperature control is easier.

d. Molten metal can be tapped from the cupola at regular intervals.

e. It consumes the easily available and less expensive fuels.

f. Cupola furnace has low initial cost as compared to other furnace of same capacity.

g. Continuous production can be obtained from the cupola furnace once started.

h. Less floor space requirements as compared to other furnaces of same capacity.

i. It has nigh degree of efficiency.

The main disadvantage of Cupola is that it is not possible to produce iron below 2.8% carbon in this furnace. So for producing white cast iron (containing below 2.7% carbon) the duplex process is employed. Moreover, molten iron and coke comes in contact with each other, certain elements such as Si, Mn are lost while others are included in the metal.

Foundry Eqipments

Hand tools

1. Showel
2. Vent wire
3. Trowels
4. lifters
5. Swab
6. Rammers
7. Spruep in
8. Clamps, etc.

Moulding boxes or Flask

The boxes used in sand moulding may be made of wood, cast iron or steel.

It is a container in which sand is packed and rammed.

Moulding boxes consists of one, two or more parts. The upper part is called 'cope'. The lower part is called 'drag' and intermediate parts are called 'cheeks'. These are provided with suitable clamps to help in locating the right position during casting.

Pouring equipments

1. Ladles
2. Crucible

Ladles are used to transport molten metal from the melting furnace to the mould.

Melting equipments

1. Cupola
2. Induction furnace etc.

Moulding machines

1. Jolt machine
2. Squeezing Machine

Construction and description of Cupola Furnace (Shown in the figure below)

Shell and refractory brick lining: Shell is vertical & cylindrical in shape and made of steel. It is lined inside with refractory bricks and clay. Refractory bricks and clay used for cupola lining consist of silica (SiO_2) and Alumina (Al_2O_3). Cupola diameter varies from 1 to 2 meters and the height is 4 to 6 times of the diameter.

Foundation: The shell is mounted either on a brick works foundation or on steel columns. The bottom of the shell consists of a drop bottom door, through which debris consisting of coke, slag etc. can be discharged at the end of a melting.

Tuyere: Air for combustion of fuel is delivered through the tuyeres which are provided at the height of between 0.6 to 1.2 meters above the working bottom or sand bed.

Wind Belt or box: The air is delivered to the tuyere from a wind belt which is a steel plate duct mounted on the outer shell of the cupola.

Blower: A high pressure fan or blower supplies the air to the wind belt through a blast pipe.

Slag Hole: It is located at a level below (about 250 mm) the centers of the tuyeres. It is used to remove or tap the slag.

Charging Door: It is situated above (3 to 4 meters) the tuyere level. Through this hole, metal, coke and flux are fed into the furnace in layers.

Chimney or Stack: The shell is usually continued for 4 meters to 5 meters above the charging hole to form a chimney.

Zones in a Cupola

The entire section or height of the cupola is divided into the following zones:

Crucible Zone or Well: It is between top of the sand bed and bottom of the tuyeres. The molten metal accumulates here

Combustion or Oxidizing Zone: Iti situated normally above (150 to 300 mm) the tuyere zone. Heat is evolved in this zone because of the following oxidation reactions:

$$C + O_2 \rightarrow CO_2 + \text{Heat}$$

$$Si + O_2 \rightarrow SiO^2 + \text{Heat}$$

$$Fe + O_2 \rightarrow FeO + \text{Heat}$$

$$Mn + C_2 \rightarrow 2MnO + \text{Heat}$$

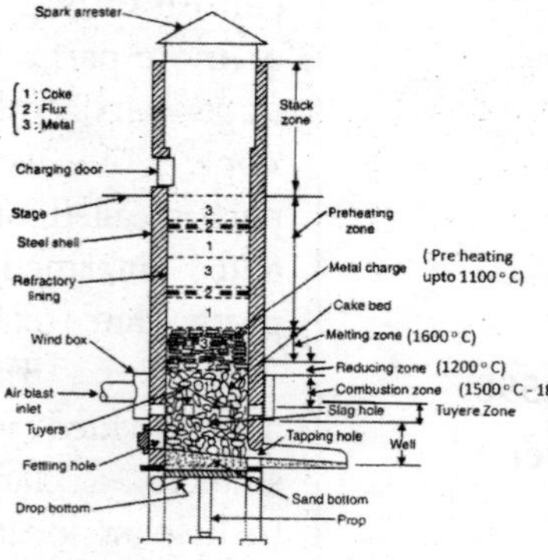

Reducing zone:Reducing zone: This zone [CO_2+C(coke)=2CO -Heat] starts from the end of combustion zone and is upto the top of the coke bed . In this zone (reaction heat is absorbed,thus temperature falls) the temperature drops to about 1200 0C.

Cupola furnace: showing internal cross section details & temperatures with zones

Melting Zone: The zone starts from the top of the coke bed and extends upto a height of 900 mm. The temperature in this zone is very high-about 1600⁰C.

$3Fe + 2CO \rightarrow Fe_3C + CO_2$

Preheating Zone or Charging Zone: It starts from the top of the melting zone and extends upto the charging door. Charging materials are fed in this zone and get preheated.

Stack Zone: It starts from the charging zone and extends upto the top of the cupola. The gases generated within the furnace are carried to the atmosphere by this zone through chimney.

Operation of Cupola Furnace

Preparation of Cupola: Clean out the slag and repair the damaged lining with the mixture of fire clay & silica sand. After this bottom doors are raised and bottom sand is introduced, the surface of the sand bottom is sloped from all directions towards the tap hole. Slag hole is also formed to remove the slag.

Firing the Cupola: A fire of soft wood is ignited on the sand bottom. After proper burning of wood, coke is added to a level slightly above the tuyeres. Air blast at a slower rate is turned on. After having red spots over the fuel bed, extra coke is added to the predeter mined height.

Charging the cupola: Once proper burning of fire starts in the cupola, alternate layers of pig iron (scrap), coke and flux (limestone) are added from the charging door. Flux will be added to prevent the oxidation as well as to remove the impurities. (Flux is 2 to 3% of the metal charge by weight). Metal (pig iron, steel scrap) flux and coke are added in alternate layers until the cupola is full to the charging door.

Soaking of iron: After the furnace is charged fully, metal charge starts heating slowly.

Opening of Air Blast: At the end of the soaking period, the air blast is opened. Tap hole is closed to accumulate the sufficient amount of metal. The rate of charging should be equal to the rate of melting so that the furnace remains full.

Pouring of Molten Iron: When sufficient metal collects in the well, the slag hole is opened, and the slag is removed;After this tap hole is opened to collect the molten metal.

Closing the Cupola: At the end of the operation, the charge feeding is stopped, air supply is cut off and the prop is removed (when the cupola is in operation, the bottom door is supported by a bottom prop so that it may not collapse due to the large weight of the charge, coke etc., it carries). As soon as the prop is removed, door swing down providing a clear space for the coke fire, residue of the molten metal with slag and the sand bed to fall down and, thus, the fire inside ceases gradually.

The ashes can be cooled by water.

Ques.598: What are the desirable properties of molding sand ?.

Answer : Good and proper molding sand must possess the following properties:

a. Porosity or permeability: This is the property of sand because of which the sand allows the gasses to pass through it.Molten metal always contains a certain amount of dissolved gases which are evolved when the metal freezes. Molten metal also coming in contact with the moist sand; generates steam or water vapors. Thus to provide a path for free escape of the gases, the molding sand should be permeable or porous. If these gases and water vapors evolved by the molding sand don't find opportunity to escape completely through the mould, they will form gas holes and pores in the casting called porosity defect. Sands which are coarse or have rounded grains exhibit more permeability. Soft ramming and clay addition in lower amount also improves permeability. Hard ramming and addition of more binder decreases permeability. Addition of more water decreases the permeability.

b. Flow ability (Plasticity): It is the ability of the molding sand to get compacted to a uniform density. Flowability assists molding sand to flow & pack all around the pattern and take up the required shape. Flow ability increases as clay and water content increases. The sand must retain its shape when the pressure is removed.

c. Refractoriness: It is the ability of the molding sand to withstand high temperature of the molten metal without fusion, cracking or buckling.Refractoriness is measured by the sinter point of the sand rather than its melting point.The degree of refractoriness depends upon the quartz content and the shape & grain size of the particles. The higher the quartz content, rougher the grain shape, higher is the refractoriness of the sand.

d. Adhesiveness: It is that property of the sand due to which it adheres or clings to another body or material (i.e., sides of the molding box). It is due to this property that the sand mass can be successfully held in a molding box and it does not fall out of the box when it is tilted (roll over).

e. Cohesiveness: This is the ability of sand particles to stick together. It may be defined as the strength of the molding sand. It is of three types :

(i) Green Strength: The property (strength) of the sand in its green or moist state is known as green strength.

(ii) Dry Strength: The strength of the sand that has been dried or baked is called dry strength.

(iii) Hot Strength: After the moisture has evaporated, the sand may be required to possess strength at some elevated temperature (above 100°C). The strength of the sand at elevated temperature is called hot strength.

f. Collapsibility: It is that property of the sand due to which the sand mould breaks (collapse) automatically (or with very less forces) after the solidification of the casting occurs.If the mould or core does not collapse easily, it may restrict free contraction of the solidifying metal and cause the same (casting) to tear or crack.If the molding sand have more strength, then breaking of molding sand around the casting becomes difficult or in other sense molding sand have a poor collapsibility.Addition of binders increases the adhesiveness and cohesiveness i.e., strength of the molding sand but decreases the collapsibility of the molding sand. So the molder has to add the optimum amount of the binder in the molding sand so that molding sand have required strength as well as required collapsibility.

g. Durability: The molding sand should possess the capacity to withstand repeated cycles of heating and cooling during casting operations. This ability of sand is known as durability.

h. Fineness: Finer sand mould (grain size small) resists metal penetration and produces smooth casting surface.Fineness and permeability are inversely proportional. They must be balanced for good results.

I Bench Life: It is the ability of the molding sand to retain its properties during storage or while standing (i.e., in case of any delay).Sometimes molding sand gets hardened because of the exposure to the atmospheric air i.e., molder is not able to use that sand. So sand should have good bench life i.e., once it is prepared, it can be used for long hours.

j. Co-efficient of expansion: Molding sand should possess low co-efficient of expansion.

k. Chemical stability: Molding sand should be chemically stable/ neutral.

l. Reusability: Molding sand should be reusable.

m. Cheap and **Easy Availability.**

The principle constituents or ingredients of molding sand are:

Silica Sand: The silica sand is found in nature on the bottoms and bank of rivers, lakes and large bodies of water.It is a product of the breaking up of quartz rocks or the decomposition of granite, which is composed of quartz and feldspar.

Binder: The purpose of adding a binder to the molding sand is to impart the sufficient strength and cohesiveness. However, it produces an adverse effect on the permeability of sand mould. Cement, linseed oil, clay, etc. are used as binder.

Additives: Additives are those materials which are added to the molding sand to improve the existing properties of sand.Corn flour, graphite, wood flour, etc.are used as additives.

Water: Naturally available and is present from 2-8%.

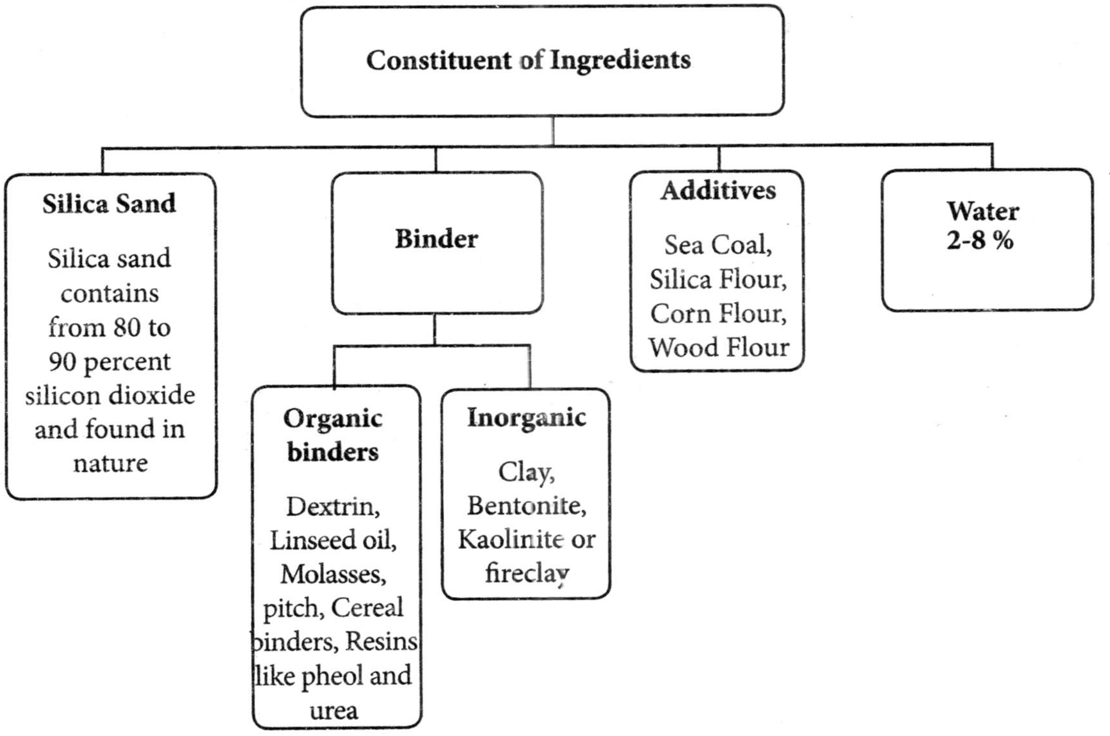

Effect of particle size on surface quality of casting

Fine grains of sand produce good surface finish but give low permeability. These are useful for small and intricate castings. Medium size sand is suitable for normal casting in bench working. Coarse grained sand is used for large castings and it posses good refractoriness but the surface quality is not good.

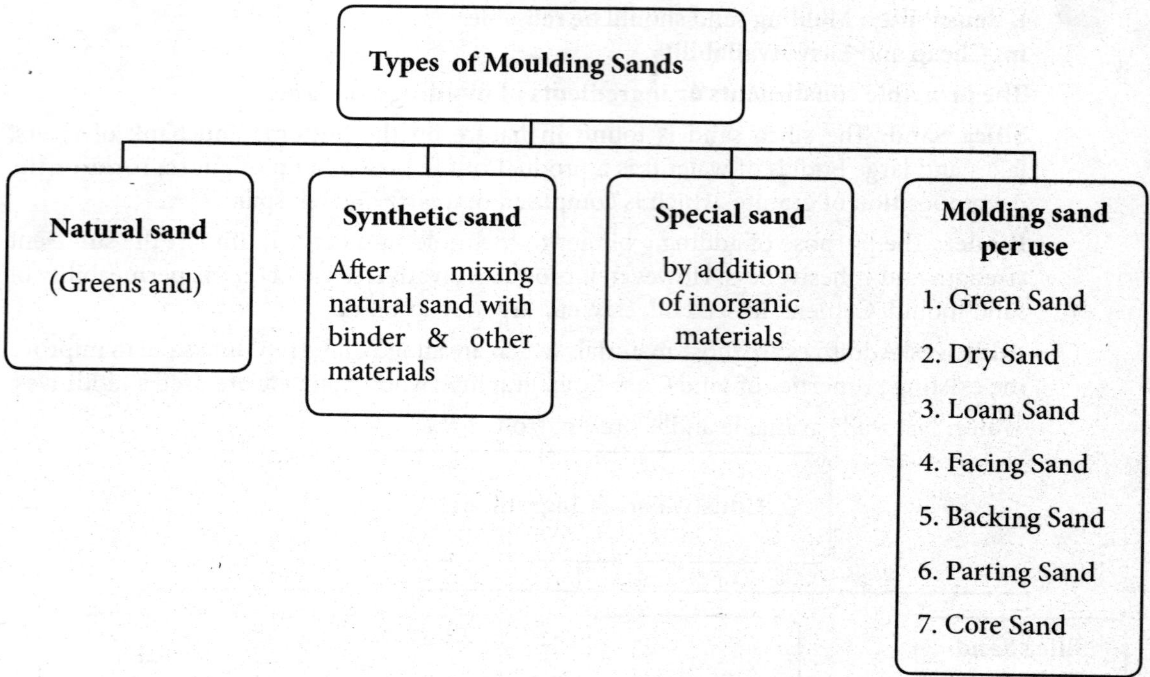

Types of molding sand

1. Natural Sand (Green sand)

These sands are taken from river beds and are dug from pits & are purely natural. They possess an appreciable amount of clay (acts as a binder) and moisture.Natural molding sands are also obtained by crushing and milling soft yellow sand stones. During milling operation, clay aggregates breaks down and are uniformly distributed over the sand grains.Due to their low cost and easy availability, these are used for most of the ferrous & non-ferrous castings.

2. Synthetic Sand

It is artificial sand obtained by mixing relatively clay free sand, binder and other materials as required. It is a better molding sand as its properties can be easily controlled by varying the mixture content.

3. Special Sand

It contains the mixtures of inorganic compounds. Cost of these sands are more but they offer high temperature stability, better cast surfaces etc.Special sands used are zircon, olivine etc.

The molding sands according to their use, are further classified as below:

1. Green Sand: The sand in its natural or moist state is called green sand. It is a mixture of silica sand with 18 to 30 percent clay, having total amount of water 6 to 8 percent.The molten metal is poured in the green sand moulds withoout any prior baking (Heating). It is used for simple, small and medium size castings.

2. Dry Sand: The green sand moulds when baked or dried before pouring the molten metal are called dry sand mould. The sand in this condition is called dry sand. Dry sand has more strength, rigidity and thermal stability as compared to green sand. These moulds are used for large and heavy castings.

3. Loam Sand: Loam sand contains much more clay as compared to ordinary molding sand. The clay content is of the order of 50%.Sweep or skeleton patterns may be used for loam molding.It is used for loam molding of large grey iron castings.

4. Facing Sand: This sand is used directly next to the surface of the pattern and comes in contact with the molten metal when the mould is poured.It is fresh sand i.e., without the addition of used sand.It must possess high strength and refractoriness.The layer of facing sand in a mould usually ranges from 20 to 30 mm.

5. Backing Sand: It is the sand which backs up the facing sand. It is the floor sand which is repeatedly used.Backing sand has black colour due to the addition of coal dust and burning on coming in contact with molten metal.Before use, the backing sand should, however, be cleaned off the foreign matter like fins, nails etc.

6. System Sand: System sand is one which is used in a mechanical sand preparation and handling system (mechanized foundries).In mechanized foundries, no facing sand is used, rather the complete otherwise used sand is cleaned and reactivated by the addition of waters, binders & special additives.Since the whole mould is made of this system sand, the strength, permeability and refractoriness of this sand must be higher than those of backing sand.

7. Parting Sand: This sand is clay free sand and consists of dried silica sand, sea sand or burnt sand.It is used to keep the green sand from sticking to the pattern and also to allow the sand on the parting surface of the cope and drag to separate without clinging.

8. Core Sand: The sand which is used for the preparation of the cores is called core sand. It is also called oil sand.It is the silica sand mixed with linseed oil or any other oil as binder.

Ques.599: Describe the basic working principle and important parts of Lathe machine. What are the different operations performed on lathe machine? Explain each of them.

Answer : **Machining** may be defined as the operation or process of removing the excess metal by means of a cutting tool, in the **machine tool**; to obtain the desired shape, size and surface finish.

A machine, which performs the material removal operation with tools, to produce desired shape and size of the work-piece, is known as **machine tool.**

Purpose of Machine tools:

a. To improve production rates
b. To reduce cost of production
c. To reduce fatigue of workers
d. To achieve better quality
e. To reduce wastage.

Basic Conditions for machining (Machining Principle) :

a. The material of tool should be harder than metal to be machined
b. Tool should be strong and held rigidly
c. Shape of tool is designed properly
d. There must be a relative movement between the tool and work piece

Basic relative movements/ motions of the tool and work-piece must a have relative movement against each other for the machining to take place and these movements/ motions may be Circular (Rotary) or Reciprocating (Straight line).

Lathe machine : The lathe machine is one of the earliest machine tools and one of the most versatile & widely used as machine tool for performing the machining operations.

Principle: The job to be machined is held and rotated in a lathe chuck; a single point cutting tool is advanced which is stationary against the rotating job. Since the cutting tool material is harder than the work-piece, so metal is easily removed from the job.

The principal parts of lathe are:

(1) Bed	(2) Head stock stock
(3) Tail Feed	(4) Carriage
(5) mechanism	(6) Legs

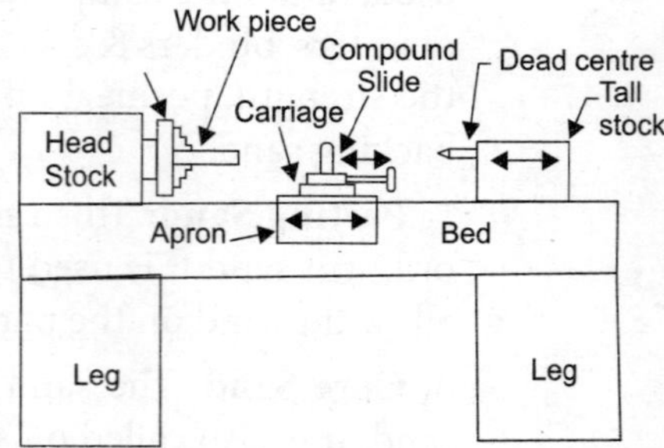

1. Bed

It is the base or foundation of the lathe. It is heavy, rugged and single piece casting made to support the working parts of the lathe. On the top of the bed, there are two sets of guide ways—outer ways and inner ways. Outer ways is for the carriage and the inner ways for the tailstock. The guide ways are of two types wide flat guide ways and inverted V-guide ways. With flat guide ways, chip accumulation is a problem but life is more but for V-guide ways combination of both the flat and inverted V-guide ways are used.

2. Headstock

The headstock is permanently fastened on the inner ways at the left side of the bed. The headstock spindle, a hollow cylindrical shaft supported by bearings, provides a drive from the motor to the work holding device. A live centre & sleeve, a face plate, or a chuck can be fitted to the spindle nose to hold and drive the work. All lathes receive their power with the help of a head stock. The power transmission device may be step cone pulleys or a geared head drive.

3. Tailstock

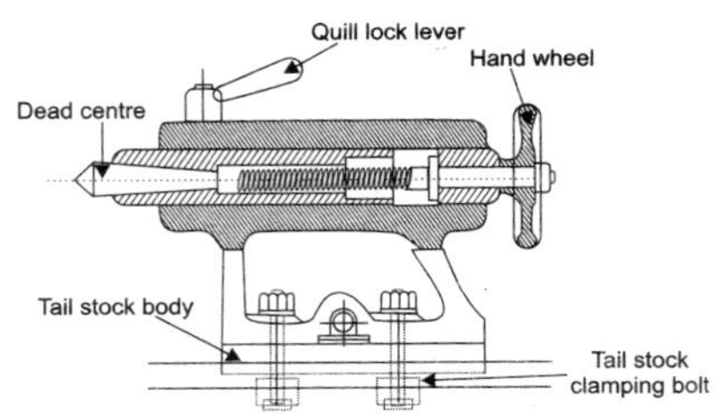

It is situated at the right hand end of the bed and is mounted on the inner guide ways. It can be moved towards or away from the operator. Tailstock can be locked in any position along the bed of the lathe by tightening the clamp lever or nut. The tailstock spindle is a hollow tapered shaft (left side end). It can be used to hold the dead centre or other tools having the same tapers such as drills and reamers. The tailstock hand wheel is used to move the tail stock spindle in or out of the tailstock casting and a spindle binding (clamping) lever or lock handle is used to hold the tailstock spindle in a fixed position.

4. Carriage

The carriage controls and supports the cutting tool. By the help of this, tool moves away or towards the headstock. It has five major parts:

a. **Saddle**: It is an H-shaped casting mounted on the top of the lathe ways so it slides along the ways between the headstock and tailstock. On the top it supports the cross slide.

b. **Cross Slide**: It is mounted on the saddle. The cross slide has a dovetail that fits over the saddle.

Carriage

c. **Dovetail**. It provides the cross movement (towards or away from the operator) to the cutting tool. It supports the compound rest.

d. **Compound Rest**: It is mounted on the top of the cross-slide and is used to support the cutting tool. It can be swiveled to any angle for taper turning operations.

e. **Tool Post**: It is mounted above the compound rest. A T-slot is machined in the compound rest to accommodate the tool post. It clamps the cutting tool or cutting tool holder in a desired position.

f. **Apron**: It is fastened to the saddle and contains the feeding mechanism. The apron hand wheel can be turned by hand to move the carriage along the bed of the lathe. The automatic feed lever is used to engage power feeds to the carriage and the cross slide.

5. Feed Mechanism

The movement of the tool relative to the work is termed as feed. The combination of different parts to ensure the feed is known as feed mechanism. A lathe tool may have three types of feed.

a. **Longitudinal Feed:** When the tool moves parallel to the work i.e. towards or away from the headstock.

b. **Cross Feed:** When the tool moves perpendicular to the work i.e. towards or away from the operator.

c. **Angular Feed:** When the tool moves at an angle to the work. It is obtained by swiveling the compound slide.

Cross and longitudinal feed are both hand and power operated but angular feed is only hand operated.

a. **Lead screw** is used for cutting of the threads in combination with the split nut. Split nut **(Half nut)** ensures that carriage moves without any slippage.

b. Feed Rod is used for powered longitudinal movement of the carriage and cross slide.

(6) Legs

Legs support the entire weight of the machine. These are made of cast iron and may be with the lathe bed. These are firmly secured to the foundation by bolts.

Lathe machine operations

The most common operations which can be performed on a lathe machine are described below:

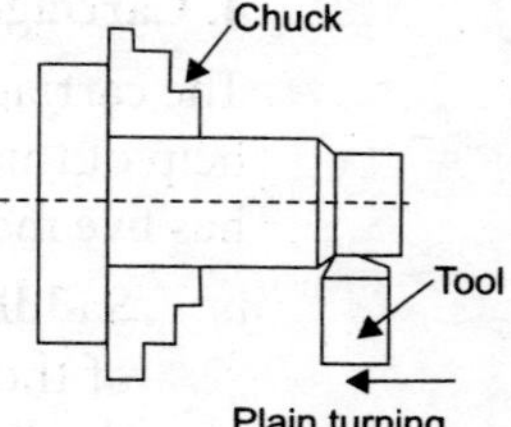

Plain turning

a. **Plain Turning:** It is an operation of removing excess amount material from the surface (cylindrical surface or circumference) of the cylindrical workpiece. This operation is done to reduce the diameter of the workpiece.

b. **Step Turning:** It is an operation of producing various steps of different diameters in the workpiece.

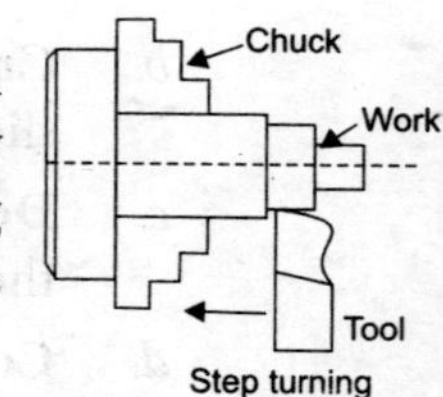

Step turning

c. **Taper Turning:** It is an operation of producing an external conical surface on a work-piece. Tapernig can be performed by the tail stock set over, by sweiveling the compound rest, or using the taper turning attachment. Most commonly used method is by swiveling the compound rest.

$$\tan \alpha = \frac{(D-1)}{2L}$$

Where, $\propto$ = angle of swiveling of compound slide

D= Larger diameter

d= small diameter

L= Length of the bar

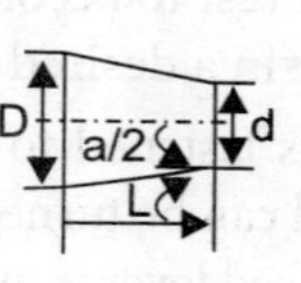

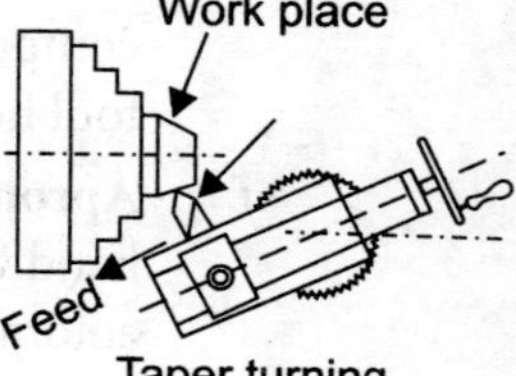

Taper turning

Steps Involved In the Taper Turning of a bar(How taper turning is done?)

(i) Fix the bar in the chuck of head stock

(ii) Fix the single point cutting tool on the tool post

(iii) Adjust the taper angle on the compound rest by loosening the two nuts on the cross slide and tighten them

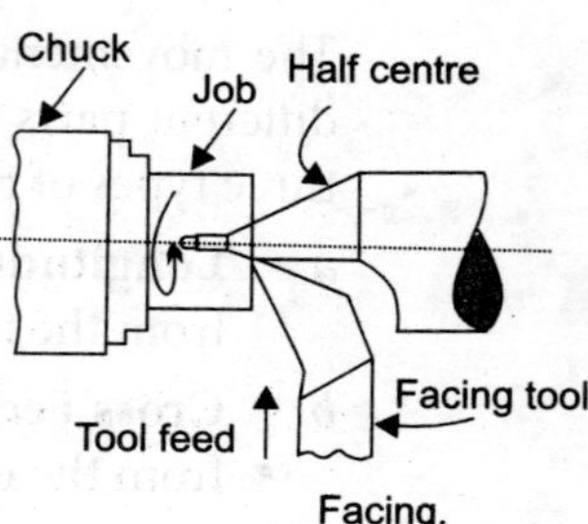

Facing.

(iv) Now adjust the depth of cut and start the lathe machine

(v) Move the wheel on the compound rest to give the feed to the cutting tool

(vi) Repeat the procedure (iv and v)till the required taper is obtained

Job Drill Tail stock end
Fixture Tool movement
Drilling.

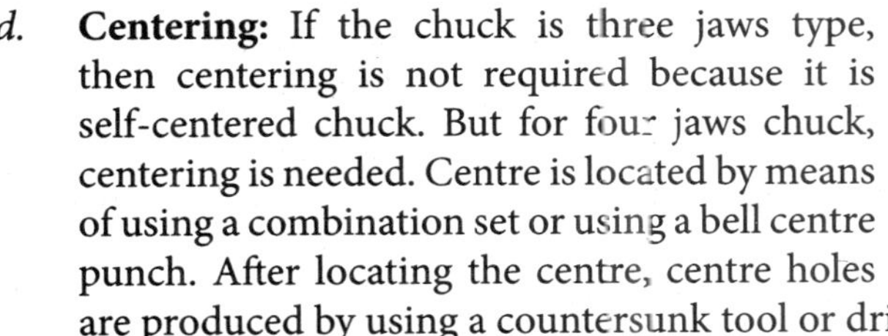

d. **Centering:** If the chuck is three jaws type, then centering is not required because it is self-centered chuck. But for four jaws chuck, centering is needed. Centre is located by means of using a combination set or using a bell centre punch. After locating the centre, centre holes are produced by using a countersunk tool or drill.

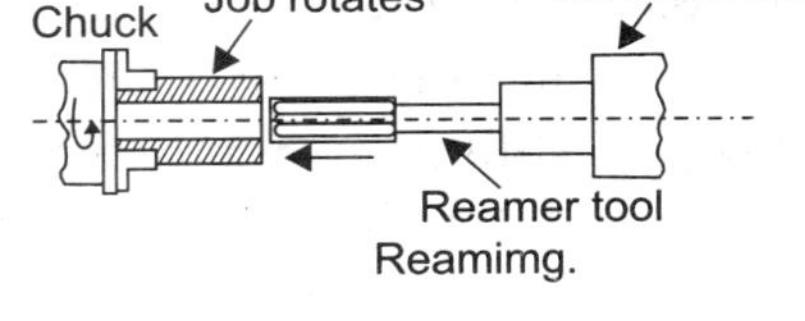

Reamimg.

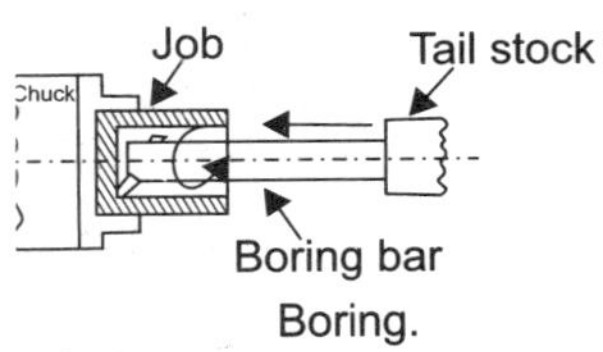

Boring.

e. **Facing:** Facing is the operation of machining the ends of a piece of work to produce the flat surface. The facing tool is fed from the centre of the work towards the outer surface or from the outer to the centre, with the help of a cross slide.

f. **Drilling:** It is an operation of making a round hole in the workpiece with the help of a drill. The work is held in a chuck and the drill is held in the tails. The drill is fed manually, into the rotating workpiece, by rotating the tailstock handwheel.

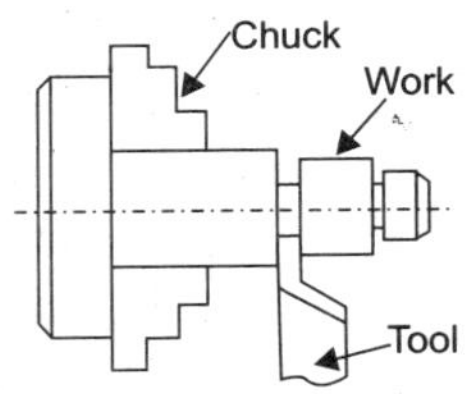

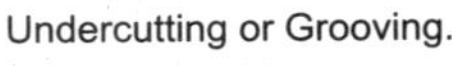
Undercutting or Grooving.

g. **Reaming:** It is an operation of finishing the previously drilled role. In this case, reamer is held in the tailstock and is fed into the hole by rotating the tailstock hand wheel.

h. **Boring:** It is an operation of enlarging a hole made in a workpiece. In this, boring tool is held in the tool post and is fed into the work similarly as as in the plain turning.

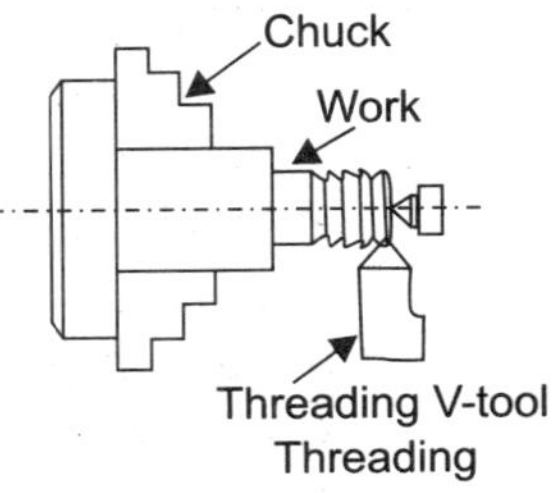

Threading

i. **Undercutting or Grooving:** It is an operation of reducing the diameter of a workpiece over a very narrow surface. In this case, the proper cutting tool is fed into the revolving work upto the desired depth at right angle to the workpiece.

j. **Threading:** Both external and internal threads can be cut on lathe. For this operation some lathes are provided with quick change gear box, which enables the establishment of the required speed ratio very quickly. A chart is provided on the

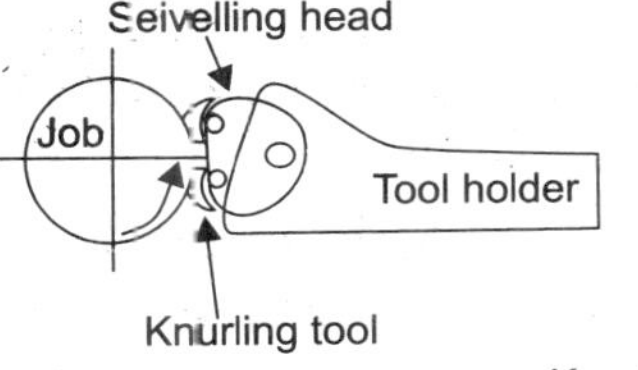

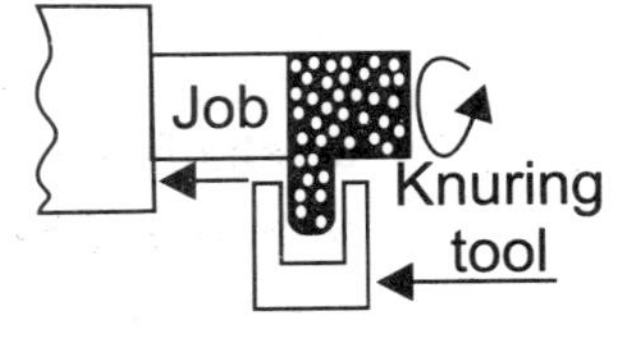

Knurling

gearbox, which carries the complete information of speed and recommended feed. With the help of gear change lever, required speed and feed can be obtained

k. **Knurling:** Knurling produces a regularly shaped, roughened surface on a workpiece. The knurling tool is pressed against the workpiece, which causes the slight outward and lateral displacement of the metal so as to form the knurl, a raised diamond pattern.

Ques.600: With the help of schematic sketch, describe the basic working principle of milling machine. Differentiate between up milling and down milling.

Answer: **Milling** is a process of metal removal/ cutting by means of a multi-teeth (point) rotating tool called milling cutter. The form of each tooth of cutter is same as that of the single point cutting tool, used in the lathe machine. Milling process is of two types:

Milling Procss

Up or convention milling	**Down orclimb milling**

Up Milling: In up milling also known as conventional milling, the cutter rotates against the direction in which the work is being fed.

Down Milling: In down milling also known as climb milling; the cutter rotates in the same direction in which the work is being fed.

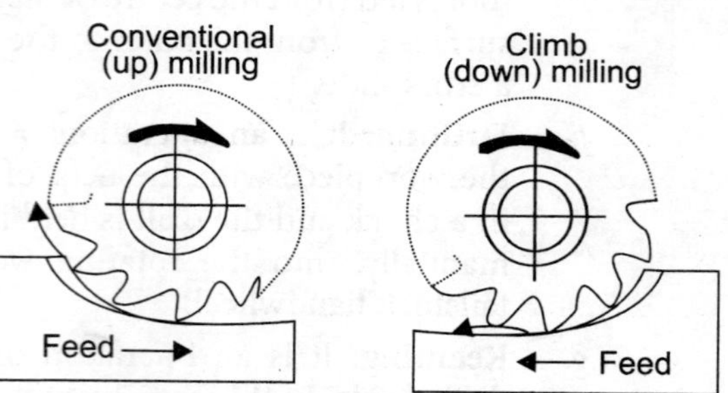

down milling Up milling

Working principle of milling process is that the work is rigidly held by the clamps (Vice) on the table and a revolving milling cutter mounted on the arbor or spindle removes the metal. The cutter revolves at high speed and each tooth of the milling cutter removes the metal in the form of chips. The work may be fed to the cutter, longitudinally, transversely or vertically, the cutter is set to a certain depth of cut by raising the table.

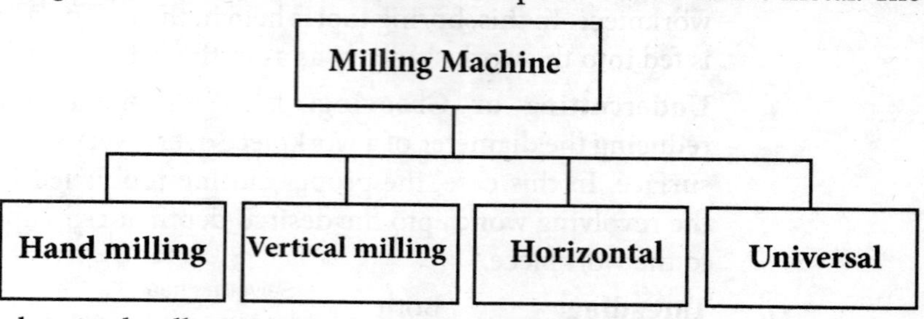

Milling machine is a machine tool in which metal is removed by means of a revolving cutter with many teeth (multipooint cutting tool known as **milling cutter**). Milling machines are of many types and the universal milling machine has a swivel-able housing on which the table is mounted.

Differentiation / Comparison between up milling and down milling

UP MILLING	DOWN MILLING
The cutter rotates against the direction in which the work is being fed.	The cutter rotates in the same direction as that in which the work is being fed.
It is also known as conventional milling.	Job-tool motion is in the same direction.
Job-tool motion is in the opposite direction.	Chip thickness varies from maximum to minimum
Cutting forces vary from zero to maximum.	Cutting forces vary form maximum to zero.
Surface finish is poorer compared to down milling.	Surface finish is better, if it is free from backlash error.
Use of cutting fluid is difficult.	Use of cutting fluid is easy.
There is tendency to lift the job so more clamping forces are needed to fix the job.	Forces are sufficient on the job to press down.

Work holding Devices: These are the devices used for holding and fixing the workpiece on the table of milling machine for performing the milling operations.

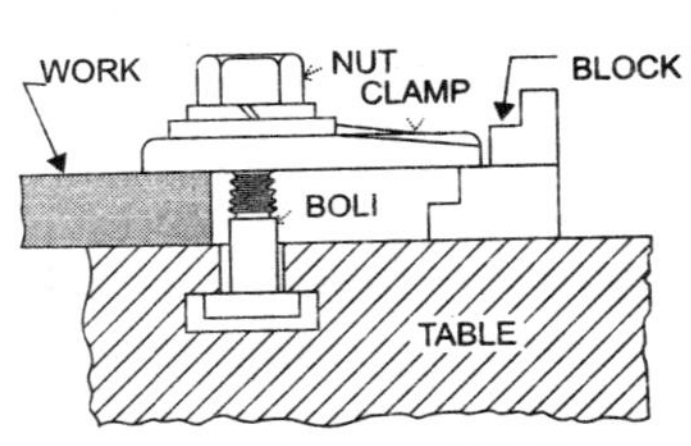

In case of simple shape, size and weight many work-pieces can be directly held conveniently with help of vices but in case of complex shapes the work-pieces are secured to the machine table with help of the **Vice** (used to hold round and square work-pieces)***Clamps, strap clamps (are cheap,easy to use, flexible force can be applied by tightening a nut by hand or wrench), jacks, step blocks (to provide support to outer end of the strap clamps) and clamping bolts, etc.*** All these clamps, straps contain a hole which is aligned and clamping bolt is passed through it to fix the work-piece. Toggle clamps can accommodate small variation in thickness and provide excellent & consistent clamping force. Toggle clamps are fast, adjustable, inexpensive, provide high ratio of leverage and can be automated.

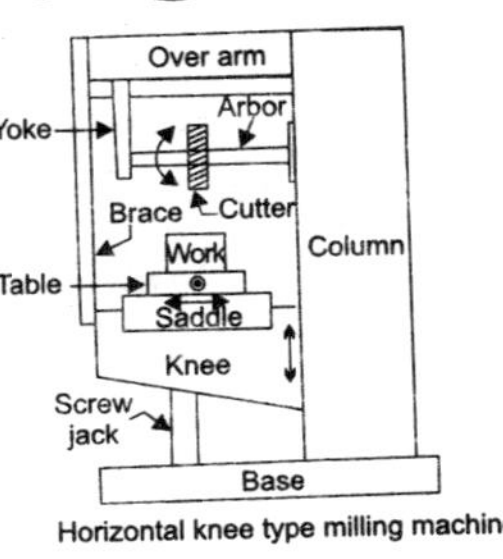

Horizontal knee type milling machine

A milling machine has the following Principal parts:

(a) Base (b) Column (c) Knee (d) Saddle
(e) Table (f) Over Arm (g) Spindle (h) Arbor

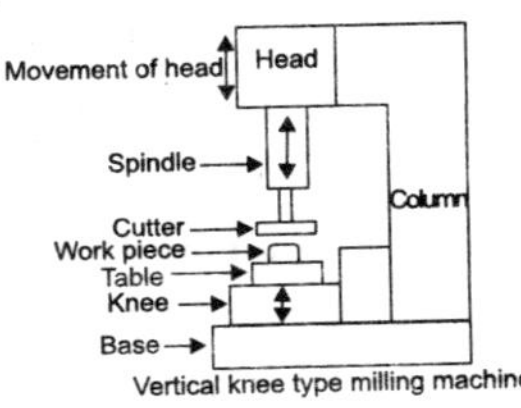

Vertical knee type milling machine

Base: It is the foundation of the machine upon which all other parts are mounted. It is generally made of gray cast iron to absorb shock and vibration. Sometime it also serves as a reservoir for cutting fluid.

Column: It is the main supporting frame mounted vertically on one side of the base. The motor and other driving mechanisms are contained in it. It supports and guides the knee in its vertical travel. It carries the jack for elevating the knee.

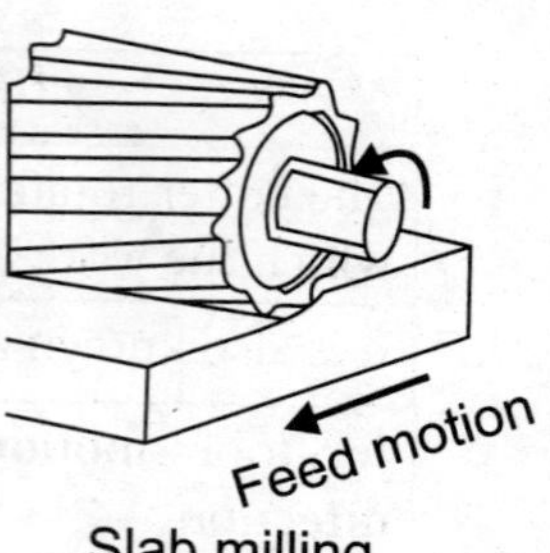

Slab milling

Knee: The knee projects from the column and slides up and down on its face. It supports the saddle and table. It is partially supported by the elevating screw which adjusts its height. It carries the table feed mechanism and controls to feed in longitudinal, cross, vertical and rotation etc. by hand power or machine power.

Saddle: The saddle supports & carries the table and is adjustable transversely on ways on top of the knee. It is provided with graduation for exact movement and can be operated by hand or power.

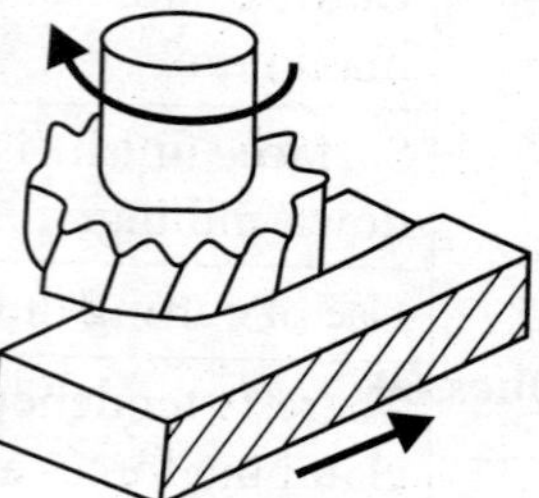
Face milling

Table: The table rests on ways on the saddle and travels longitudinally in a horizontal plane. It supports the workpiece, fixtures etc.

Over-arm: The over arm is mounted on and guided by the top of the column. It is adjusted in & out by hand to the position of maximum support for the arbor and then clamped. (not in the vertical type of milling machines)

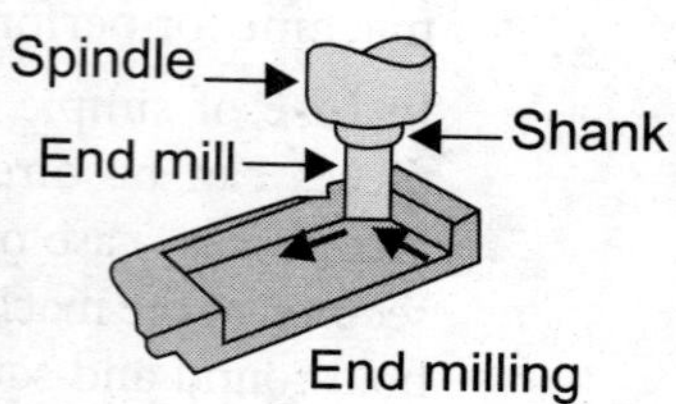

End milling

Spindle: The spindle is mounted on the upper part of the column. It receives power from the motor through belts, gears, clutches etc. and can be rotated at different speeds by the step cone pulley drive or by gearing arrangement and transmits it to arbor or sub-arbor. (In the vertical type of milling machines)

Arbor: The arbor is the extension of the spindle on which all the various cutters are mounted. It is tapered at one end to fit the spindle nose and has two slots to fit the nose keys for locating & driving it.

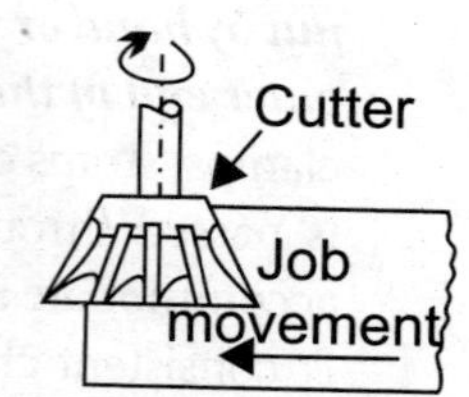

Angular or bevel milling

Operations carried out on the milling machine are:

Plain or Slab Milling: The plain milling operation is the production of flat or horizontal surface parallel to the axis of the cutter.

Face Milling: The face milling operation is also the production of flat surface which is at right angle to the axis of rotation of the face milling cutter.

Angular or Bevel Milling: The angular milling operation is production of flat surface, which is at an angle to the axis of the cutter.

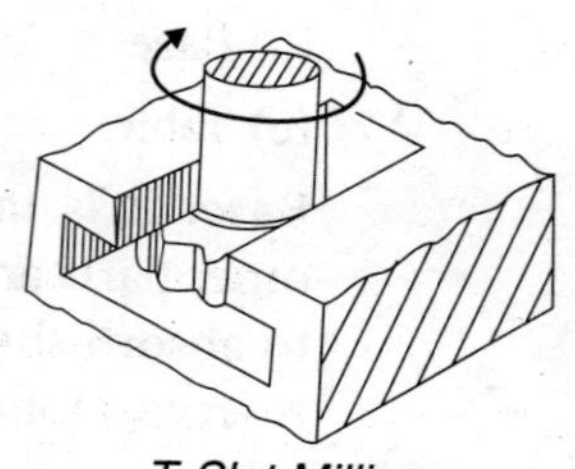
T-Slot Milling

Side Milling: The side milling operation is the production of a vertical flat surface on the side face of a job by using a side milling cutter.

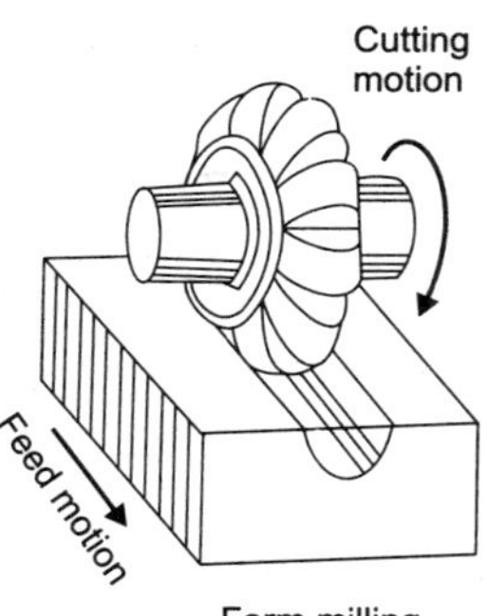

Form milling

End Milling: The End milling operation is the production of both peripheral and face milling operations simultaneously, generates vertical, horizontal or angular surfaces by using an end milling cutter.

Gang Milling: Gang milling is operation of producing many surfaces of on a job simultaneously by feeding the table against a number of required cutters. The surfaces produced may be flat horizontal or vertical surfaces and are produced simultaneously.

Form Milling: The form milling operation is the production of irregular contours by using the cutters having the same profile corresponding to the surface to be generated.

T-Slot milling: It refers to the formation of T-Slots.

Ques.601: With the help of schematic sketch describe the basic working principle of drilling machine. What are the different operations performed on drilling machine?

Answer : **Drilling/ drilling process:** It is an operation of producing a circular hole in a workpiece by forcing a drill also called drill bit, in the workpiece.

Drilling machine is a machine tool designed for drilling round holes in metallic and nonmetallic materials. The cutting tool is a multi-point cutting tool, known as drill or drill bit.

Types of drilling machines: The various types of drilling machines are:

Portable drilling machines are very small, compact and self contained units with an electric motor inside.

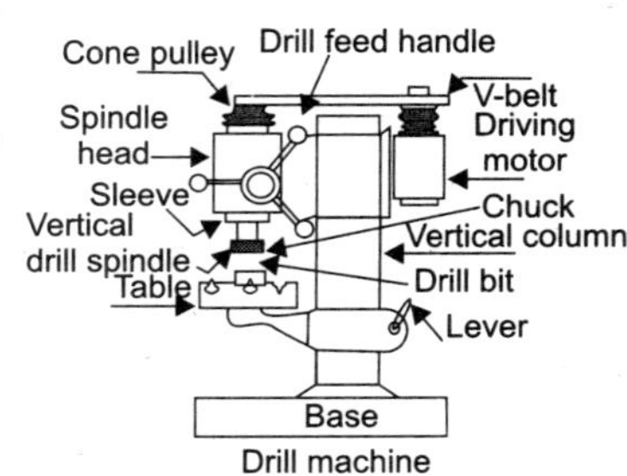

Drill machine

Bench drilling machines are commonly used for light & precision work and are available in training workshops. This type of machine is discussed next.

Radial drilling machines have radial arm with a capacity of tool movement, and used for the job which cannot be manipulated easily.

Principal Parts of the Drilling Machine

1. **Head or spindle head:** Head contains the electric motor, V- pulleys and V- belt which transmit rotary motion to the drill spindle from the motor.
2. **Spindle:** Spindle is made up of alloy steel. It rotates as well as moves up and down in a sleeve.
3. **Drill Chuck:** It is held at the end of the drill spindle and in turn it holds the drill bit.
4. **Adjustable Table:** It is supported on the column of the drilling machine and can be moved vertically and horizontally. It also carries slots for bolts clamping.
5. **Base:** It supports the column, which, in turn, supports the table, head etc. and made of cast iron.

6. Column: It is a vertical round or box section, which rests on the base and supports the head & the table.

Working Principle & Operation of a Drilling Machine:

1. The drill is held in the chuck and workpiece is fixed in the vice on the table.
2. On switching 'ON' the machine, the drill rotates.
3. The linear motion is given to the drill towards the workpiece by drill feed handlewhich is called feed.
4. In order to remove the chips from the hole, drill is taken out from the hole frequently and coolant is used to keep the drill cool. The combination of rotary and linear motion produces the round hole in the workpiece.

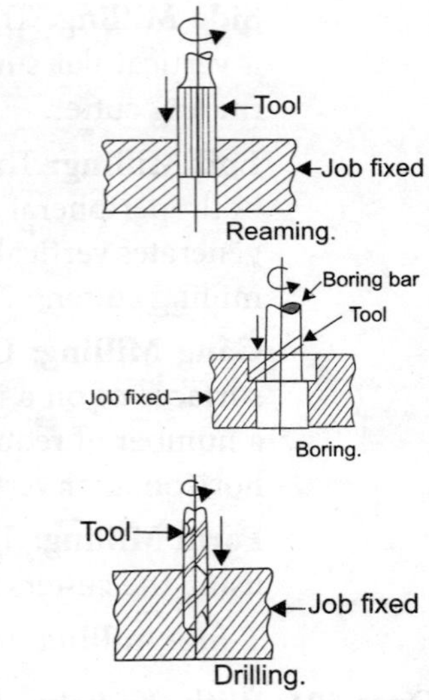

Reaming.
Boring.
Drilling.

Operations performed on Drilling machine:

1. **Drilling:** It is an operation of producing a circular hole in a workpiece by forcing a drill in the workpiece.
2. **Boring:** It is an operation of enlarging a hole that has already been drilled. Single point cutting tool is used in boring.
3. **Reaming:** Reaming is done with reamers. It is done to generate the hole of proper size and finish after drilling.
4. **Tapping:** It is an operation of producing internal threads in a hole by means of a tap.
5. **Counter Boring:** It is an operation of enlarging the entry of a drilled hole to accommodate the bolt head etc. Counter boring tool does it.
6. **Spot Facing:** It is an operation done on the drilled hole to provide smooth seat for bolt head.
7. **Counter Sinking:** It is an operation to bevel the top of a drilled hole for making a conical seat. A counter sunk drill as shown in Fig is used in this operation.

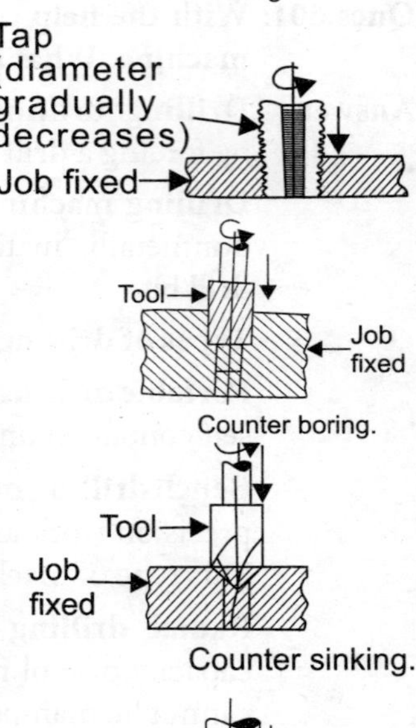

Counter boring.
Counter sinking.
Spot Facing.

Ques.602: With the help of neat sketch explain the basic components of shaper machine and various operations performed on it.

Answer :

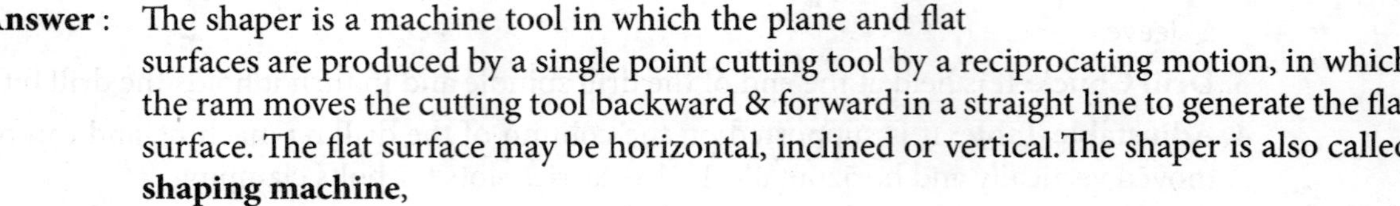

The shaper is a machine tool in which the plane and flat surfaces are produced by a single point cutting tool by a reciprocating motion, in which the ram moves the cutting tool backward & forward in a straight line to generate the flat surface. The flat surface may be horizontal, inclined or vertical.The shaper is also called **shaping machine**,

Working Principle and Operation

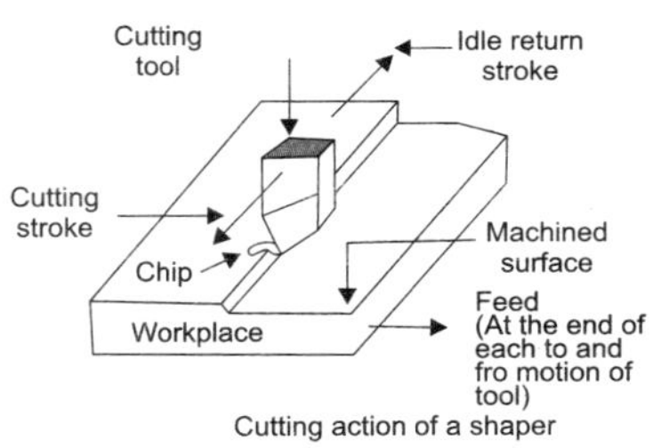

Cutting action of a shaper

In a shaper machine, a single point cutting tool reciprocates over the stationary work piece. The workpiece is rigidly held in a vice or clamped directly on the table. The tool is held in the tool head mounted on the ram of the machine. When the ram moves forward, cutting of material takes place. So, it is called cutting stroke. When the ram moves backward, no cutting of material takes place; so called idle stroke. The time taken during the return stroke is less as compared to forward stroke and this is obtained by quick return mechanism. The depth of cut is adjusted by moving the tool downward towards the work piece.

Principal Parts of a Shaper machine or shaper Base: It is a heavy and robust cast iron body which acts as a support for all other parts of the machine which are mounted over it.

Shaper machine

Column (Body or frame): It is a box type iron body mounted upon the base. It acts as housing for the operating mechanism of the machine, electrical motor, quick return mechanism, cross rail and ram. On the top it is having two guide ways on which the ram reciprocates.

Cross-rail: It is a heavy cast iron construction, attached to the column in front of the machine & on the vertical guide ways. It carries two mechanisms, one for elevating the table and the other for cross travel of the table.

Table: It is made of cast iron and used for holding the work piece. T slots are provided on its top and sides for securing the work on to it. It slides along the cross rail to provide feed to the work.

Ram: It reciprocates on the guide ways provided above the column. It carries the tool head and mechanism for adjusting the stroke length.

Tool Head: It is attached to the front portion of the ram and is used to hold the tool rigidly. It also provides the vertical and angular movement to the tool for cutting.

Vice: It is job holding device and mounted on the table, job can be directly clamped to table also.

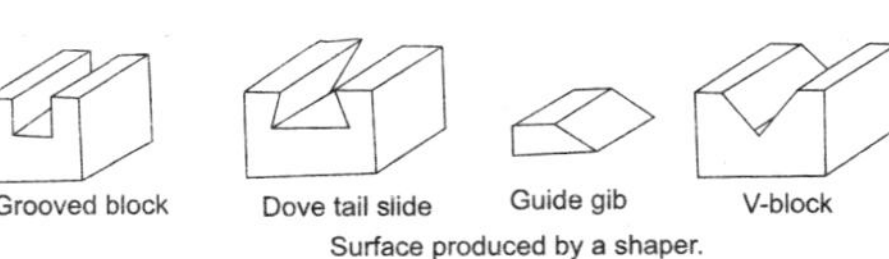

Surface produced by a shaper.

Operations performed or Surfaces Produced by a **Shaper machine**: All type of flat surfaces which may be horizontal, inclined or vertical can be produced by shaper machine. Some of the products are shown in the figure.

Planer or planer machine

The planer (also called **planning machine**) is used to produce plane and flat surfaces by a single point cutting tool when the work piece is **very large** in size. It is similar to shaper but its size is very large; so used for producing flat surfaces on much larger work than a shaper. The major difference between shaper and planer is that in a planer, the tool remains stationary and the work reciprocates whereas in the shaper, the work remains stationary and the tool reciprocates.

A **planer** consisting of following main parts is shown in block diagram below.

Bed: The bed of the planer is the heavy cast iron structure which provides the foundation for the machine and supports the housing and all other moving parts. At its top, V- type guide ways are provided on which the table slides.

Table: It is a box type structure made up of cast iron. It reciprocates along the ways of the bed and supports the work. At its top, it carries longitudinal T-slots and holes to accommodate the clamping bolts and other devices.

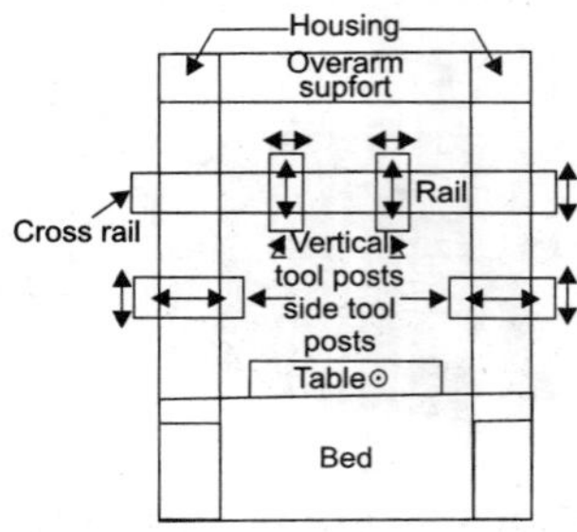

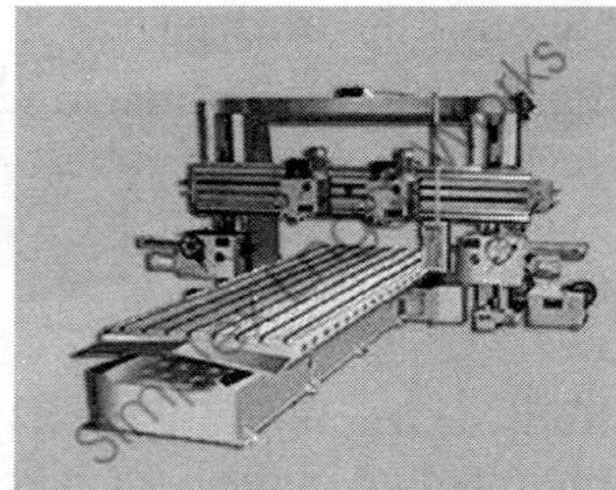

figure of a Planer

Housing or Columns: The housings are rigid castings placed on each side of the bed in case of double housing planer and on one side only in case of open side planer. It carries cross rail elevating screws, vertical feed shaft and cross feed bar to transmit the power to the upper parts of the machine. The front face carries the vertical ways along which the cross rail slides up and down.

Cross Rail: The cross rail is mounted on the vertical guide ways of the two housings. It can be raised or lowered. Accurately finished ways are provided at the front of the cross rail for two vertical tool heads.

Tool Heads: Two tool heads are mounted on the cross rail and the other two on the vertical columns. Each column carries one side tool head. All the four tool heads work independently, such that they can operate separately or simultaneously, as desired.

Operations performed on planer

The common operations performed on planer are as follows:

a. Machining horizontal flat surfaces.

b. Machining vertical flat surfaces.

c. Machining angular surfaces, including dovetails.

d. Machining different types of slots and grooves.

e. Machining curved surfaces.

f. Machining along contours.

Differences between shaper and planer

PLANER	SHAPER
1. It is heavier, more rigid and costlier machine.	1. It is a comparatively lighter and cheaper machine.
2. It requires more floor area.	2. It reqiores less floor area.
3. It is adopted	3. It is used for small works.
4. Tool is fixed and work moves.	4. Work is fixed and the tool moves.
5. More than one cutting tool can be used at a time.	5. Only one cutting tool is used at a time.
6. The tools used on a planer are larger, heavier and stronger.	6. The tool used on a shaper is small in size as compared to planer tool.
7. Heavier feeds area applied.	7. Lighter feeds area applied.
8. It can take deep cut.	8. It cannot take deep cuts.
9. Work setting requires much of skill and time.	9. Work may be clamped easily and quickly.
10. Indexed feed is given to the tool during work during the idle stroke of the work table.	10. Indexed feed is given to the stroke of the ram.

Drive mechanisms

The reciprocating movement of ram and quick return mechanisms of the shaper machine is obtained by any of the following mechanisms:

1. Crank and slotted link mechanism
2. Whitworth quick return mechanism
3. Hydraulic

Crank and slotted link mechanism

The figure shows the crank and slotted link mechanism, the motor drives the crank and crank is connected to sliding block, engaged in the slotted lever. The block can slide in the slotted lever.

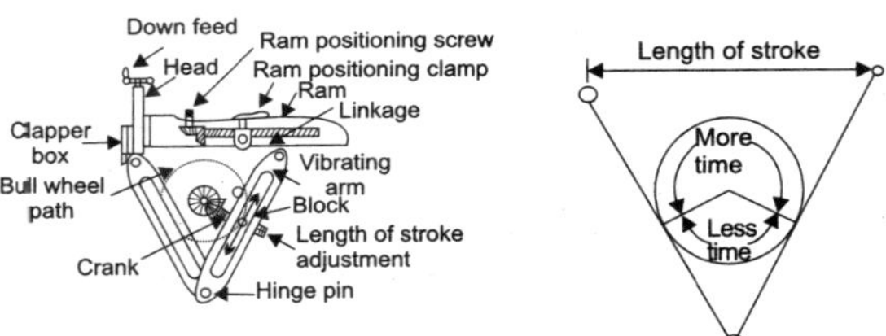

Figure: Crank and slotted link mechanism

Let us consider that the arm is in the right side end and the crank is rotating anti- clock-wise with uniform angular speed. The timer taken by the arm or the slotted lever to reach towards left side will be more (Forward stroke), and the further rotation of the crank will bring the arm back in less time (Return stroke). The rotation angle for forward stroke is more and for return stroke is less.

Ques.603: With the help of schematic sketch, describe the basic working principle of grinding process.

Answer : Grinding is a process of removing material by the abrasive action of a Revolving Wheel (**called as grinding wheel**) from the surface of a workpiece, in order to bring it to the required shape and size. It consists of Sharp Crystals, called Abrasives, held together by a Binding material or Bond. The Wheel may be a Single piece or Solid type or may be composed of Several Segments of Abrasive Blocks joined together. In most cases, it is a **Finishing Operation** and a **very small** amount of material is removed from the surface during the operation.

A Grinding Wheel essentially consists of the following two materials:

1. Bond: This material of the grinding wheel acts as a binder to hold the Abrasive grains together.

2. Abrasives: It is that material of the Grinding Wheel which does the cutting action. These are extremely hard materials consisting of very small particles called grains, which carry a number of sharp Cutting Edges and Corners. Examples are silicon carbide, aluminium oxide, etc.

A **grinding wheel** is an expendable wheel that is composed of an abrasive compound used for various grinding (abrasive cutting) and abrasive machining operations. They are used in grinding machines. The wheels are generally made from a matrix of coarse particles pressed 7 bonded together to form a solid, circular shape; various profiles and cross sections are available depending on the intended usage for the wheel. They may also be made from a solid steel or aluminum disc with particles bonded to the surface.

Characteristics

There are five characteristics of a cutting wheel: **material, grain size, wheel grade, grain pacing, and bond** type. These are indicated by codes on the wheel's label.

Material, for the actual abrasive, is selected according to the hardness of the material being cut. The frequently used materials are: Aluminium Oxide, Silicon Carbide, Diamond, etc.

Grain size (Grit), from 8 (coarsest) to 600 (finest), determines the physical size of the abrasive grains in the wheel. A larger grain will cut freely, allowing fast cutting but poor surface finish. Ultra-fine grain sizes are for precision finish work.

Wheel grade, from A (soft) to Z (hard), determines how tightly the bond holds (strength) the abrasive. Softer wheels are preferred for harder materials and vice-versa.

Grain spacing, or **structure**, from 1 (densest) to 16 (least dense). Density is the ratio of bond and abrasive to air space. A less-dense wheel will cut freely, and has a large effect on surface finish. It is also able to take a deeper or wider cut with less coolant, as the chip clearance on the wheel is greater.

Wheel bond, indicates that the material by which the wheel holds the abrasives, this affects finish, coolant, and minimum/maximum wheel speed.

Vitrified (V), Resinoid (B), Silicate (S), Shellac (E), Rubber (R), Metal (M).

Difference between grade and structure

Grade represents the strength of the bond whereas the structure represents the spacing between the grains or the density of the abrasive particles (i.e dense (1) to least dense (16). Brittle and hard materials need a grinding wheel with dense structure.

Types/ Classification of grinding wheels

Straight wheel

To the right is an image of a **straight wheel**. These are by far the most common style of wheel and can be found on bench or pedestal grinders. They are used on the periphery only and therefore produce a slightly concave surface (hollow ground) on the part. This can be used to advantage on many tools such as chisels. Straight Wheels are generally used for cylindrical, centre-less, and surface grinding operations. Wheels of this form vary greatly in size, the diameter and width of face naturally depending upon the class of work for which is used and the size & power of the grinding machine.

Figure: *Grinding wheel*

Grinding Machines (Grinders)

Roughing or non-precision grinder	Precision grinder
1. Bench grinder	1. Cylindrical grinder
2. Pedestal grinder	2. Surface grinder
3. Portable grinder	3. Internal grinder

Cylinder or wheel ring

Cylinder wheels provide a long, wide surface with no center mounting support (hollow). They can be very large, up to 12" in width. They are used only in vertical or horizontal spindle grinders.

Cylinder or wheel ring is used for producing flat surfaces, the grinding being done with the end face of the wheel.

Tapered wheel

A straight wheel which tapers towards the center of the wheel is called tapered wheel. This arrangement is stronger than straight wheels and can accept higher lateral loads. Tapered face straight wheel is primarily used for grinding thread, gear teeth etc.

Diamond wheel

Diamond wheels are grinding wheels with industrial diamonds bonded to the periphery. They are used for grinding extremely hard materials such as carbide cutting tips, gemstones or concrete. The saw pictured to the right is a slitting saw and is designed for slicing hard materials, typically gemstones.

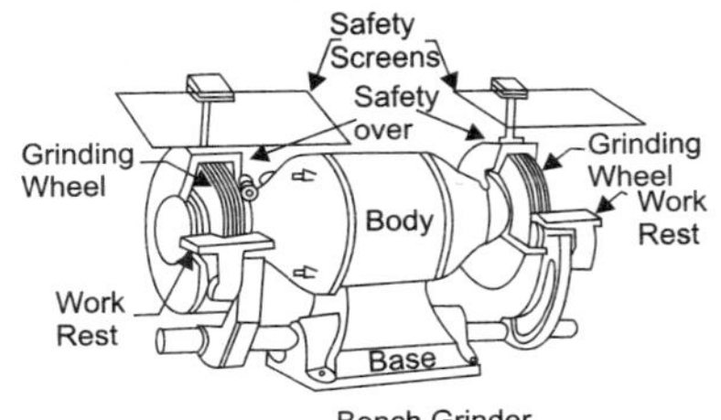

Bench Grinder

Cut off wheels

Cut off wheels, also known as parting wheels, are self-sharpening wheels that are thin in width and often have radial fibers reinforcing them. They are often used in the construction industry for cutting reinforcement bars, protruding bolts or anything that needs quick removal or trimming.

Grinder or grinding machine

These are the machines to perform the grinding operations. These may be of different types as shown.

Bench grinder consists of body having electrically driven motor, base to fix on the table or bench, grinding wheels mounted on the motor shaft, safety cover & screen and work rest. The figure below shows the bench grinder.

The different operations performed by the grinders are follows:

a. **Cylindrical grinding:** It is used for Grinding of Outside cylindrical and tapered surfaces.

b. **Internal grinding:** It means a method of grinding the internal surfaces of cylindrical and the tapered holes.

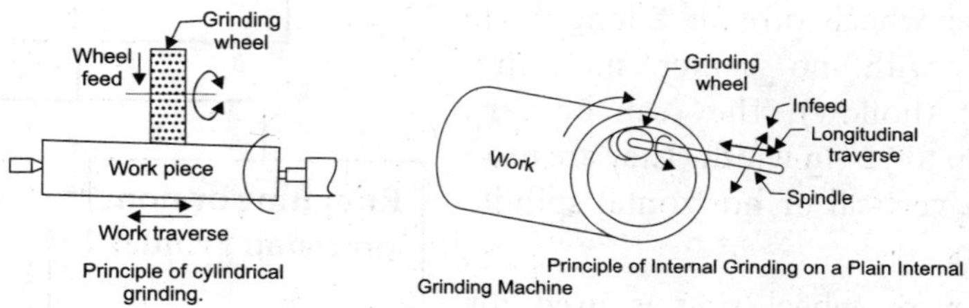

Principle of cylindrical grinding.

Principle of Internal Grinding on a Plain Internal Grinding Machine

c. **Surface grinding:** It is a method of grinding Horizontal flat surfaces. The wheel spindle can be horizontal or vertical.

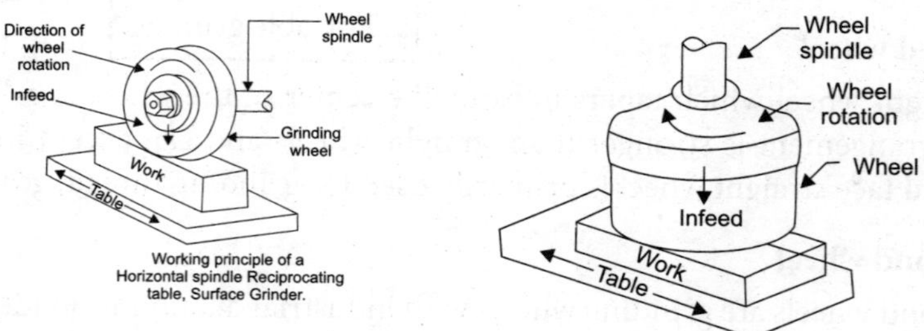

Working principle of a Horizontal spindle Reciprocating table, Surface Grinder.

d. **Face grinding:** It is a method of grinding Vertical flat surface.

e. **Form grinding:** The formed surfaces can be produced by this method of grinding using a formed wheel.

f. **Centered Grinding:** In this method the work piece is held from the sides between the supports and grinding wheel is brought in the contact for machining operation.

Centred Grinding

g. **Centre-less grinding:** In this method, the external cylindrical surfaces are generated by supporting the work on the regulating wheel, grinding wheel and work rest (support)blade.

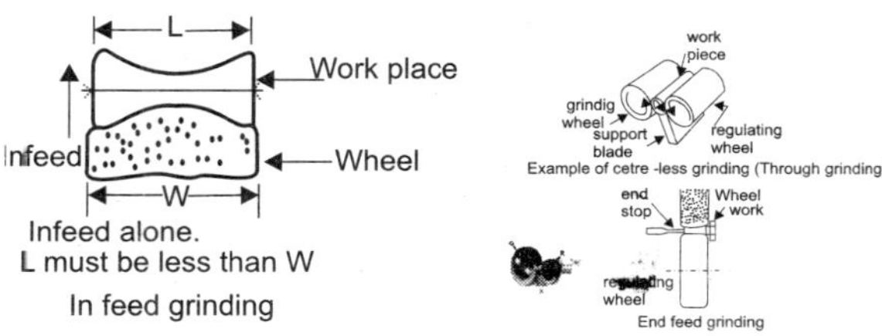

Centre- less grinding is of three types:

(i) Through-feed grinding,

(ii) In-feed grinding

(iii) End-Feed Grinding

h. **Off-hand grinding:** It is roughening operation. In this method,the work is held by hands and grinding is performed on the items on which accuracy is not of prime importance.

Ques.604: Explain fusion as it relates to welding operations. How will you classify the welding processes? Explain with neat sketch the features and uses of neutral, reducing and oxidizing flames in case of oxy- acetylene gas welding.

Answer : The **welding** is a process of joining two similar or dissimilar metals by **fusion** (application of heat), with or without the application of **pressure** and with or without the use of **filler metal**.

Weldability of a metal is the ability (or the ease) of the metals with which two similar or dissimilar metals are joined by fusion with or without the application of pressure and with or without the use of filler metal.

The fusion, of metal takes place by application of heat. The heat may be obtained from electric arc, electric resistance, chemical reaction, friction or radiant energy. The importance of fusion is clear from the observation that without melting and mixing of the two metals pieces, joining will not be successful.

Base Metal: The metal to be joined or cut is termed as the base metal.

Filler metal: It is the metal (in the form of wire or rod) added to supplement the molten metal pool to make a good and strong joint.

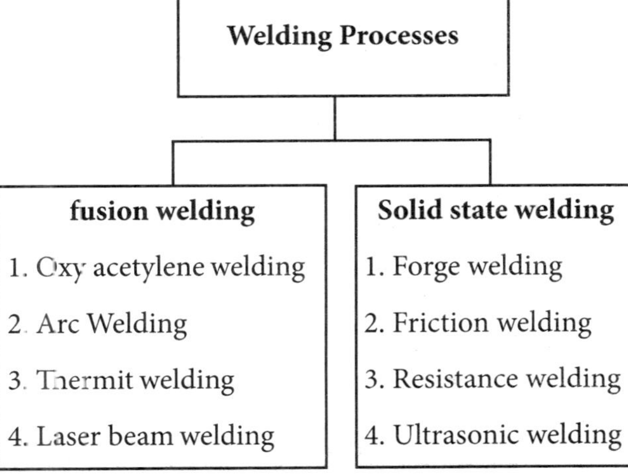

Heat sources

The sources of heat for welding operation may be combustion of gas & coal, electric arc, electric resistance, and chemical reaction. The fuel gas and electricity are most used power sources. The gas welding can be performed for thin section and without much expenditure. In case of electric power source welding is suitable for thick sections and provides clean joints but heavy & costly equipment is required. The welding process using the chemical reaction as power source can be carried out at remote locations where it is not practicable to carry heavy equipments i.e. railway track welding.

Classification of welding process

The welding processes may be classified on the basis of:

(i) Metals joined and filler material, Autogenous Welding: The process of joining similar metals by melting the edges together, without the addition of filler metal, is called Autogenous welding.

Homogeneous Welding: The process of joining similar metals with the help of filler rod of the same metal is called homogeneous welding.

Heterogeneous Welding: The process of joining dissimilar metals using filler rod is called heterogeneous welding.

(ii) **The state of metals at the time of welding**, i.e. if the metals are heated to the melting temperature then the welding is called as **fusion welding**, and in case the metal is heated only upto such a temperature to bring it to plastic state then welding is called as solid state or pressure welding.

Solid state welding: In forge or pressure welding, the work-pieces are heated to plastic state and then, the work-pieces are joined together by applying external pressure on them. The examples are:1. Forge welding, 2. Friction welding,3.Ultrasonic welding, & 4.Spot welding, 5. Seam welding, 6. Projection welding, 7. Resistance butt welding, 8. Flash butt welding9.Percussion welding, etc.

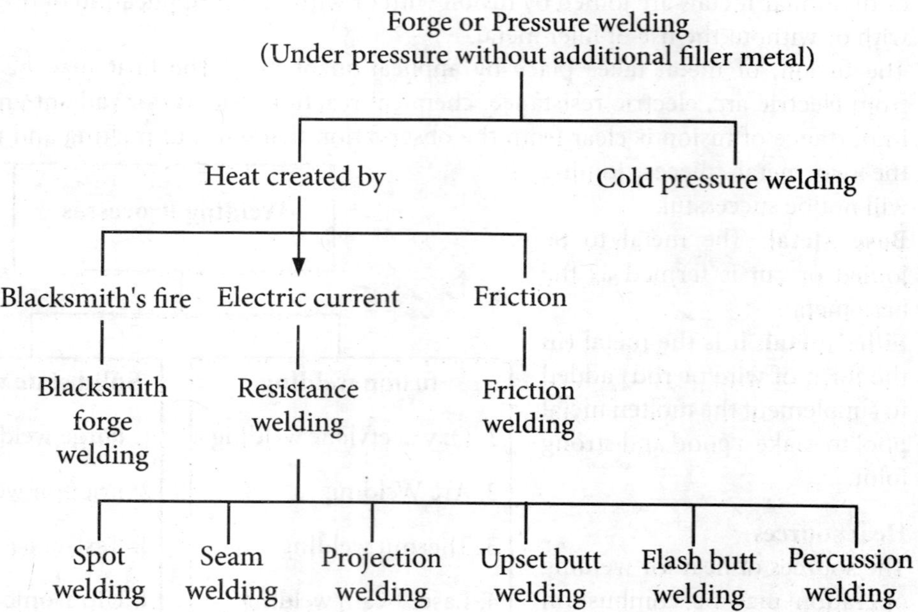

Fusion or Non-pressure Welding: In this welding, the material at the joint is heated to a molten state and then allowed to solidify. The filler material may or may not be added. The examples are:

1. Air-acetylene welding, 2. Oxy acetylene welding, 3. Arc Welding, 4. Carbon arc welding, 5. Shielded metal arc welding, 6.. TIG (Tungsten Inert Gas) welding, and 7.MIG (Metal Inert Gas) 8.Thermit welding 9. Laser beam welding, etc

ADVANTAGES OF WELDING

a. A good weld is as strong as the base metal.

b. A large number of metals/alloys can be joined by welding.

c. Repair by welding is very easy

d. Welding can be easily mechanized.

e. Portable welding equipment is available.

f. General welding equipment is not very costly.

g. Total joining cost is less in case of welding joint.

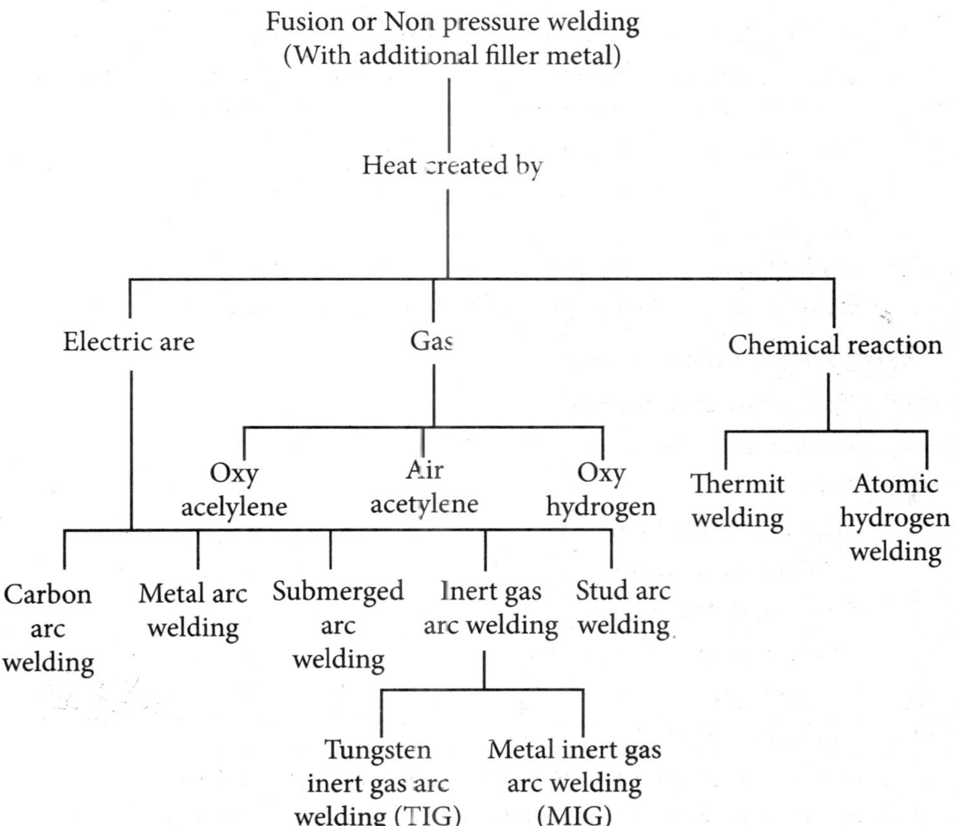

DISADVANTAGES OF WELDING

a. Welding produces the harmful radiation, fumes and spatter.
b. A skilled welder is required.
c. Welding heat produces metallurgical changes.
d. Cost of equipment (initial cost) is high.
e. Edge preparation is required before welding.
f. More safety devices are required.
g. Jigs and fixtures are required to hold the parts to be welded

WELDING AS COMPARED TO RIVETING AND CASTING

a. Welding is more economical and is a much faster process as compared to both casting and riveting.
b. As compared to riveting and casting, fewer persons are involved in a fabrication process.
c. Welding involves less cost of handling.
d. Welding produces less noise as compared to riveting.
e. Welding designs involve less cost and are flexible also (we can change easily). Cost of pattern making, storing, maintenance, repairs is eliminated.
f. Welding structures are comparatively lighter.
g. Welding can be carried out at any point on a structure, but riveting requires enough clearance.
h. Drilling holes to accommodate rivets weakens a riveted structure.
i. Welding can produce a 100% efficient joint, which is difficult to make by riveting.

Applications of welding

a. Automobile construction.
b. Railroad equipment.
c. Ships.
d. Aircraft construction.
e. Building construction.
f. Bridges construction.
g. Pressure vessels.
h. Storage tanks.
i. Piping and pipe lines.
j. Fabrication of jigs, fixtures and machine tools.
k. Repair of broken and damaged parts.
l. Household furniture.
m. Material handling equipments etc.

Types of welded joints

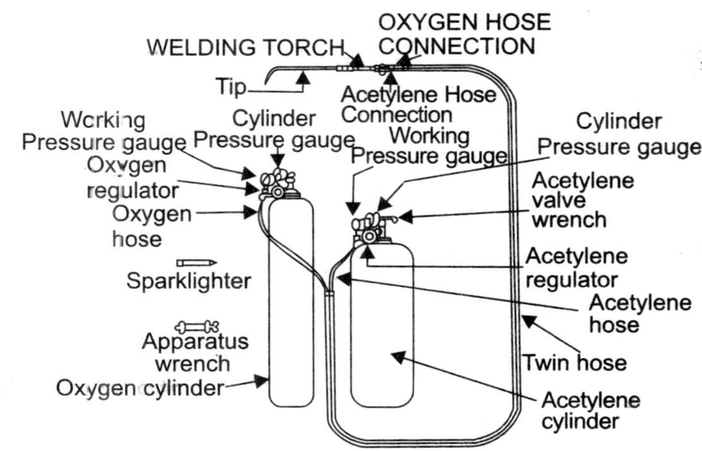

The following are the five basic types of joints commonly used in fusion welding:

1. **Lap Joint:** The lap joint is obtained by overlapping the plates and then welding the edges of the plates.
2. **Butt Joint:** The butt joint is obtained by welding the ends or edges of the two plates.
3. **Corner Joint:** The corner joint is obtained by joining the edges of two plates whose surfaces are at an angle of apprcx. 90 degrees to each other.
4. **Edge Joint:** The edge joint is obtained by joining two parallel plates. It is economical for plates having thickness less than 6 mm.
5. **T-Joint:** The T-joint is obtained by joining two plates whose surfaces are approximately at right angles to each other. These joints are suitable upto 3 mm thickness.

Gas welding

Gas welding also called an oxy-fuel gas welding, and it obtains the heat for welding from the combustion of a fuel gas such as acetylene in combination with oxygen. The process is a fusion welding process wherein joint is completely melted to obtain the fusion.

Principle of operation

The acetylene gas is mixed with oxygen in correct proportions in the welding torch and ignited, the flame is produced which is sufficiently hot to melt and join (upon cooling) the parent metal. Temperature of flame is about 2800-3400°C. A filler rod is generally added to build up the seam for greater strength.

Oxy-acetylene welding may be classified as:

High Pressure Oxy-acetylene Welding: In case of high pressure Oxy-acetylene gas welding, the acetylene gas is supplied from the acetylene cylinder (Painted maroon) in compressed form. The acetylene is dissolved in the porous mass soaked in acetone. The oxygen gas is supplied from the commercial cylinders (Painted black) in both types of gas welding.

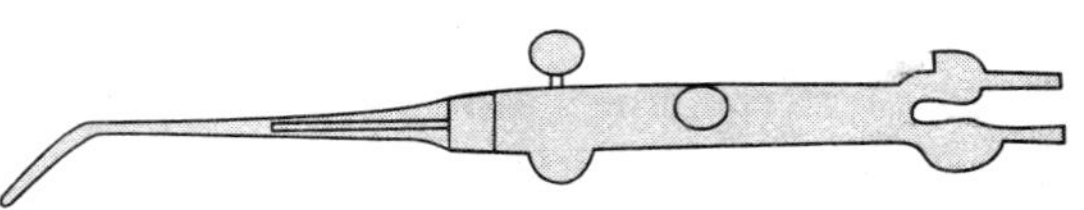

Gas Welding Torch

Low Pressure Oxy-acetylene Welding: In case of low pressure Oxy-acetylene gas welding, the acetylene gas is supplied from the generator at low pressure. In the generator calcium carbide stone is added in the chamber in which water is already present. Calcium carbide stone reacts with the water and produce acetylene gas. This gas can be easily collected from the top of the water and can be used for welding purpose.

Chemical reactions acetylene gas reacts and produces heat. T

Types of oxy-acetylene welding flames :

1. Neutral flame

a. A neutral al flame is produced when oxygen to acetylene ratio is 1.1 to 1.

b. The temperature is of the order of about (3000-3200°C).

c. The flame has nicely defined inner cone (light blue in colour) and is surrounded by outer envelope which is dark blue in colour than the inner cone.

d. It is called neutral because it will **not oxidize or carburize** the metal.

e. It is used for welding of Copper, Cast iron.

2. Oxidizing flame

After the neutral flame, if the supply of oxygen is further increased, the result will be an oxidizing flame.

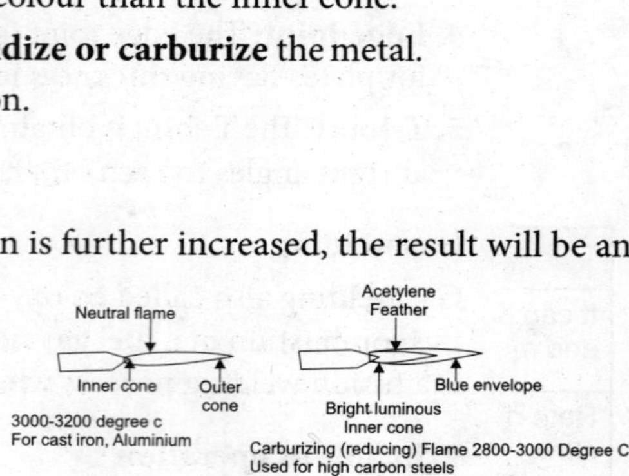

a. Its inner cone is more pointed and outer flame envelope is much shorter.

b. It burns with a loud roar.

c. The temperature is of the order of about 3400 ^{0}C (because of excess of O_2 so complete combustion takes place).

d. This flame is harmful for steels, because it oxidizes the steels.

e. Only in the welding of copper and copper based alloys, oxidizing flame is desirable, because in those cases a thin protective layer of slag forms over the molten metal.

Reducing flame

a. If the volume of oxygen supplied to the neutral flame is reduced, the resulting flame will be a carburizing or reducing flame i.e. rich in acetylene.

b. In this flame, acetylene feather exists between the inner cone and outer envelope.

c. Temperature is of the order of about 2800°C (less because it does not completely consume the available carbon).

d. Metals that tend to absorb carbon should not be welded with reducing flame.

e. Carburizing flame contains more acetylene than a reducing flame.

f. Carburizing flame is used for the welding of lead and for carburizing (surface hardening) purposes.

g. Reducing flame is used, with low alloy steel rod, for welding high carbon steel.

Flux is a chemical cleaning agent, flowing agent, or purifying agent. Fluxes may have more than one function at a time. They are used in both extractive metallurgy and metal joining. During welding process the metal is heated/ melted in air, oxygen from the air combines with the metal to form oxides which results in poor quality, low strength welds and in some cases may even make welding impossible.

Purpose of flux The role of a flux in joining processes (welding, soldering & brazing) is multiple as such:

a. It dissolves of the oxides on the metal surface, which facilitates wetting by molten metal, and

b. Acts as an oxygen barrier by coating the hot surface, preventing its oxidation.

c. In some applications molten flux also serves as a heat transfer medium, facilitating heating of the joint by the soldering tool or molten solder.

d. It improves the fluidity of the metal and capillary action.

e. Examples of flux are sodium carbonate, borax and ammonium chloride.

Advantages of Gas Welding	Disadvantages of Gas Welding
It can be applied to a wide variety of manufacturing and maintenance situations	Flame temperature temperature of the arc is less than the
Rate of heating and cooling of weld deposit and joo is slow.	Refractory metals (e.g. tungsten, molybdenum, tantalum etc.) and reactive metals (titanium and zirconium) cannot be gas welded.
No electric current is required	Gas flame takes a long time to heat up the metal than an arc
Equipment is having less cost	Heat affected zone is wider
Operator is having better control because sources of heat and filler metals are separate	
Cost and maintenance equipment is low of the welding	

Ques.605: Differentiate between welding, soldering and brazing .

Answer : Soldering and brazing provide permanent joint to metal pieces. Soldering and brazing process lie in between fusion welding and solid state welding. These processes have some advantages over welding process. These can join the metal having poor weldability, dissimilar metals, and very less amount of heating is needed.

The major disadvantage is joint made by soldering and brazing has low strength as compared to Joints made by welding.

Brazing : Brazing is the metal joining process with the help of filler metal called as brass (an alloy of copper 60% and Zinc 40%). Filler metal is melted and distributed by capillary action between the mating surfaces of the metallic parts being joined. In this case only filler metal melts. There is no melting of workpiece metal. The filler metal (brazing metal) has the melting point more than 450 ^{0}C. Its melting point should be lesser than the melting point of workpiece metal. The metallurgical bonding between work & filler metal and geometric constrictions imposed on the joint by the workpiece metal make the joint stronger than the filler metal out of which the joint has been formed.

Soldering: Soldering is metal joining process with help of solder (filler metal). (Solder is an alloy of lead 60% and tin 40%). Filler metal has the melting temperature lower than 450 ^{0}C. The surfaces to be soldered must be pre-cleaned so that these are free of oxides, oils, etc. An appropriate flux must be applied to the mating surfaces and then surfaces are heated. Filler metal called solder is added to the joint, which distributes between the closely fitted surfaces. Strength of soldered joint is much lesser than welded joint and lesser than a brazed joint.

Soldering Procedure

Following sequential steps should be carried out as soldering procedure.

1. **Work Preparation:** Workpieces which are to be joined together should be perfectly clean. There should not be any dirt, dust, rust, paint or grease. This is so that the solder or spelter can stick to the joint with proper strength. Cleaning is done with the help of a file or sandpaper of Joint. Workpieces should be clamped to avoid any relative movement between them that may disturb the joint making. At the joint smaller grooves are made on the workpieces to facilitate better flow of molten solder and so good strength of the joint.

2. Fluxing: The flux is applied.

3. Tinning: In this step of soldering procedure, the bit of solder iron is cleaned, application of flux is done over it. It is brought in contact of solder wire so the bit carries sufficient amount of molten solder over it. After that it is used to make tags of solder at various places throughout the joint.

4. Cooling The joint is filled with molten solder and allowed to cool & solidify.

The differences between soldering, brazing and welding are tabulated below.

Welding	Soldering	Brazing
In welding filler metal melting point is same as that of base metal	In soldering the filler metal is solder and melting point is below 450 °C	In brazing the filler metal is brass and melting point is above 450 °C
NO capillary action between the base and filler metal and the joint takes place due to fusion of metals.	Joint takes place due to capillary action between the base and filler metal	Joint takes place due to capillary action between the base and filler metal.

These are the strongest joints used to bear the load. Strength of a welded joint may be more than the strength of base metal.	These are weakest joint out of three. Not meant to bear the load. Use to make electrical contacts generally.	These are stronger than soldering but weaker than welding. These can be used to bear the load upto some extent.
Temperature required is upto 3800 °C of welding zone.	Temperature requirement is upto 450 °C or below.	It may go from 450 °C to 600 °C in brazing.
Workpiece to be joined need to be heated till their melting point.	No need to heat the workpieces till melting point.	Workpieces are heated but below their melting point.
Mechanical properties of base metal may change at the joint due to heating and cooling.	No change in mechanical properties after joining.	May change in mechanical properties of joint but it is almost negligible
High cost is involved and high skill level is required.	Cost involved and skill requirements are very low.	Cost involved and skill required are in between others two.
Heat treatment is generally required to eliminate undesirable effects of welding.	No heat treatment is required.	No heat treatment is required after brazing.
No preheating of workpiece is required before welding as it is carried out at high temperature.	Preheating of workpiece before soldering is good for making good quality joint.	Preheating is desirable to make strong joint as brazing is carried out at relatively low temperature.
Distortion is more	Distortion is least	Distortion is less
Used for joining steel , mild steel, aluminium, etc	Used for electrical applications, radiator tubes, utensils, etc	Used for carbide tools, pipe fittings, dissimilar metals, tubes, utensils, etc

Ques.606: Explain the resistance welding. What are the different types of resistance welding? Draw neat sketches of each process.

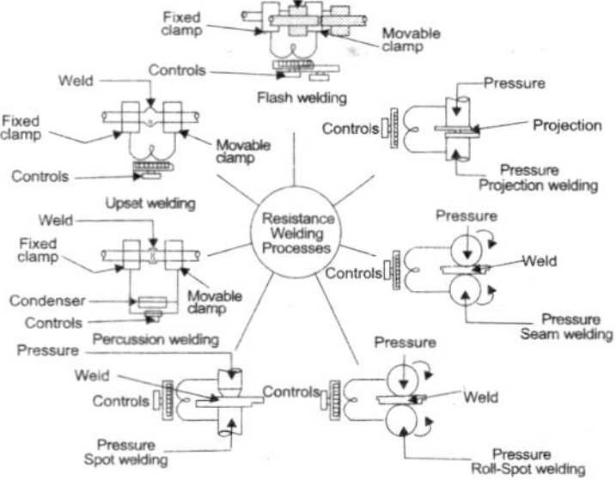

Answer: **Resistance welding** is a group of welding processes in which heat for welding is obtained from the resistance of the work-piece (metal) to the flow of electric current and with the application of pressure. No filler metal or flux is added.

In resistance welding both heat and pressure are used to make the joint or establish the coalescence.

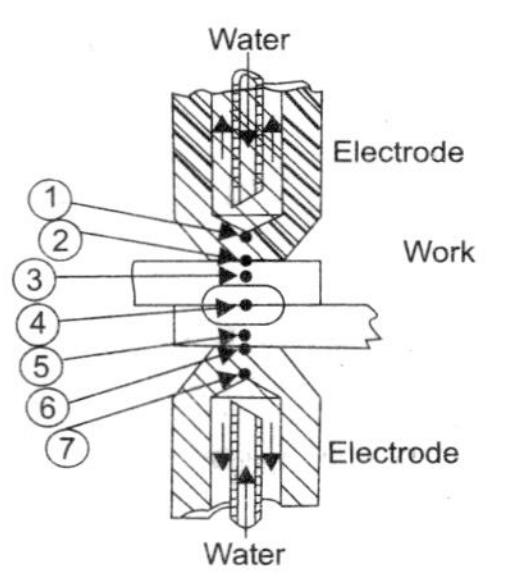

Types of Resistance Welding

a. Spot welding
b. Seam welding
c. Projection Welding
d. Resistance butt welding
e. Flash butt welding
f. Percussion welding.

Principle of Resistance Welding

In resistance welding, a low voltage (typically 0.5 to 10 volt) and very high current (typically 15000 A) is passed through the joint for a very short time (typically 0.25 sec). This high amperage heats the joint.

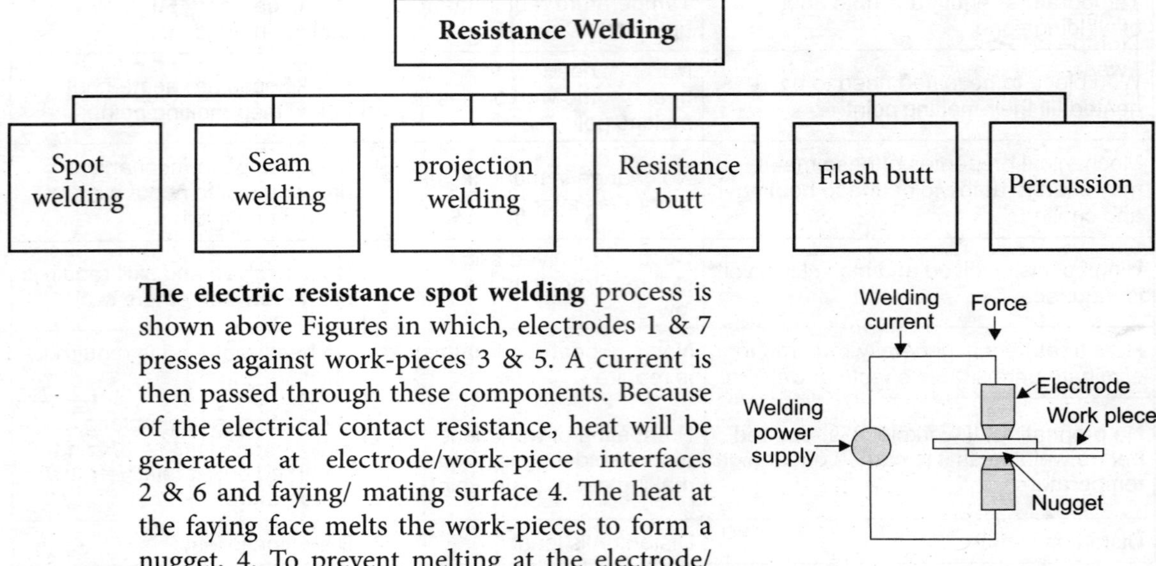

The electric resistance spot welding process is shown above Figures in which, electrodes 1 & 7 presses against work-pieces 3 & 5. A current is then passed through these components. Because of the electrical contact resistance, heat will be generated at electrode/work-piece interfaces 2 & 6 and faying/ mating surface 4. The heat at the faying face melts the work-pieces to form a nugget, 4. To prevent melting at the electrode/ work-piece interface, water is circulated in the cooling chamber of the electrodes.

Electrodes for Resistance Welding

The electrodes in resistance welding have to carry large amount of current, pressure and also help to remove the heat from the weld zone thus preventing overheating & surface fusion of work. The electrodes should have higher electrical conductivity as well as higher hardness. Hence, copper in alloyed form is used for making electrodes. The electrodes may be water cooled internally.

Advantages of Resistance Welding	Disadvantages of Resistance Welding
Very little skill is required to operate the resistance welding machine	The initial cost of equipment is high.
High production rate so; well suited for mass production.	Certain resistance welding operations are limited only to lap joints
Heating of the work-piece is confined to a very small part, which results in less distortion	Skilled persons are needed for maintenance
No filler rod and flux is needed.	Bigger job thicknesses cannot be welded
It is possible to weld dissimilar metals as well as metal plates of different thickness.	

Semi-automatic equipments are available.	
Very little skill is required to operate the resistance welding machine.	
There are no consumables used in this process except for the electrical power and a relatively small electrode wear. As a result, it is a very economical process.	

Applications of Resistance Welding

Joining of sheets, bars, tubes, making of metal furniture. making fuel tanks of cars, tractors, making of containers. welding aircraft & automobile parts etc.

Spot welding

Spot welding is a resistance welding process in which overlapping sheets are joined by local fusion at one or more spots by the heat and pressure is applied by the electrodes one above & other below the work-pieces. The heat is generated because of the resistance to the flow of electric current through work-piece.

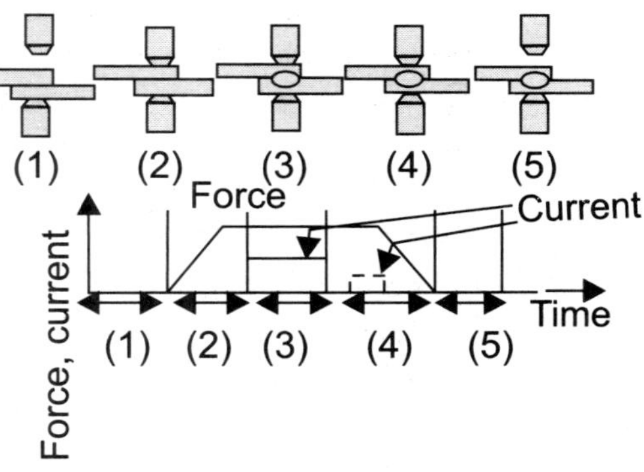

Procedure: (The steps involved in spot welding are represented by the figure)

a. The job should be clean.

b. It should be free from grease, dirt, paint, scale, oxide etc.

c. Clean the electrode tip surface. Very fine emery cloth may be used for routine cleaning.

d. Water is kept running through the electrodes in order to cool the weld and save the electrodes from getting overheated.

e. Proper welding current is set. Proper time has been set on weld timer.

f. Electrodes are brought together against the overlapping work-pieces and pressure is applied so that the surfaces of the two work-pieces come in physical contact with each other.

g. Welding current is switched on for a definite period of time.

h. As the current passes, a small area where the work-pieces are in contact is heated and spot weld takes place. The temperature of this weld zone is approx. 815°C to930°C.

i. After the welding takes place, the welding current is cut off. Extra electrode force is then applied or the original force is prolonged. Hold until the metal cools down and gains strength.

j. After that, electrode pressure is released to remove the spot welded work-pieces.

Advantages of Spot Welding:

Low cost: Less skilled worker can perform this welding.

Higher productivity: Operation may be made automatic or semiautomatic. No edge preparation is needed.

Application of Spot Welding:

Welding of low carbon steels, high speed steels, stainless steels, nickel, nickel alloys, automobile, aircraft industry, household furniture, Containers, etc.

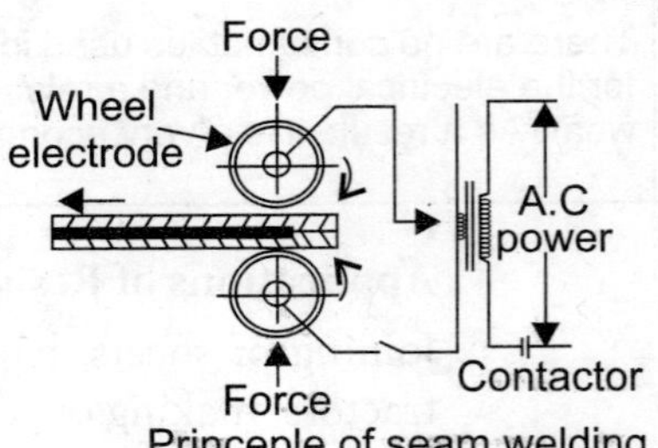

Principle of seam welding

Seam Welding

Seam welding is a resistance welding process in which overlapping sheets are joined by local fusion progressively, along a joint, by two rotating circular electrodes. Fusion takes place because of heat, which is generated, from the resistance to electric current flow through the work parts which are held together under pressure by electrodes.

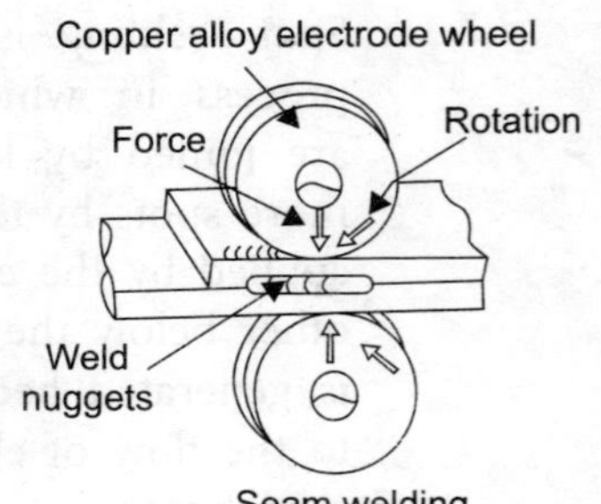

Seam welding

Principle of Operation (Procedure)

a. The work-pieces to be seam welded are cleaned, overlapped suitably and placed between the two circular electrodes which hold the work-pieces together by the pressure on electrode force.

b. Switch on the coolant supply (in some machines, the electrodes are cooled by external spray of water; in others, the electrodes are cooled by refrigerant fluid that flow inside the working electrodes).

c. Switch on the current supply. As the first current impulse is applied, the power driven circular electrodes are set in rotation and the work-pieces steadily move forward.

d. If the current is **put off and on quickly**, a continuous fusion zone made up of overlapping nuggets is obtained. It is known as stitch welding.

e. If individual spot welds are obtained by constant and regularly timed interruption of the welding current, the process is known as **roll (spot) welding.**

Advantages of Seam Welding	Disadvantages of Seam Welding
It can produce gas tight or liquid tight joints.	Cost of equipment is high as compared to spot welding set
Overlap can be less than spot or projections welds	Welding can be done only along a straight or uniformly curved line.

Several parallel seams may be produced	It is difficult to weld thickness greater than 3 mm.

Applications of Seam Welding:

It is used for welding of stainless steels, steels alloys, nickel & its alloys, magnesium alloys etc.

Projection Welding

Projection welding is another advancement of spot welding, where one of the sheets to be joined is provided with a number of projections to help localise the current at a predetermined spot. Thus, the surfaces of the workpieces are in contact with each other only at the projections.

Steps involved in the projection welding

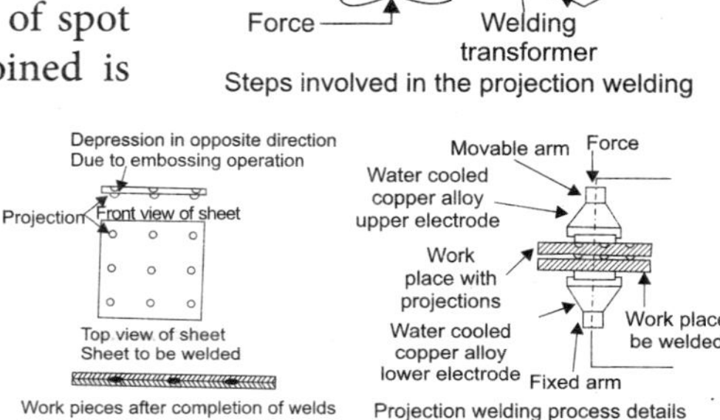

Projection welding process details

Projection Welding Procedure

a. The projections are made in one of the sheet.

b. The sheets are placed between the electrodes.

c. As the current is switched on, it will pass through these projections. Because of heat produced due to resistance to the flow of electric current, these projections are melted.

d. The electrodes are pressed together to complete the weld, by pressing the upper electrode downward. The melted projections form the weld.

e. Welding can be done at several points simultaneously.

f. This completes one cycle of welding. The projections are generally very small of the order of 0.8 mm and are obtained by means of embossing.

Advantages of Projection Welding	Disadvantages of Projection Welding
It is possible to weld more than one spot at a given time	Making of projections is an extra operation
The welds may be placed closer than spot welding	All projections should be of same height
Proper heat balance can be easily obtained	Metals, which cannot support projections, cannot be welded satisfactorily
Projections are to be made in thicker plate or in the plate which is having higher electrical conductivity	
Life of electrode is much longer than the life of electrode in spot welding	
The uniformity and appearance of the weld is better as compared to spot welding	

Application of Projection Welding

a. Small fasteners, nuts etc. can be welded to larger components.

b. It is used for welding of refrigerator condensers, joining of wires etc.

c. Welding of stainless steel parts, titanium alloys, etc.

d. Used for joining and making of utensils.

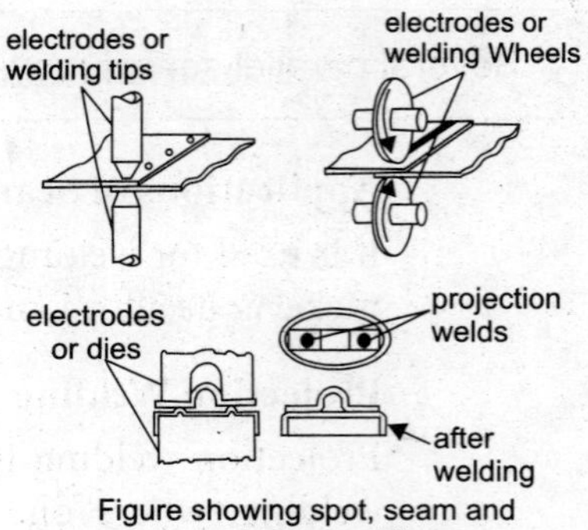

Figure showing spot, seam and projection welding

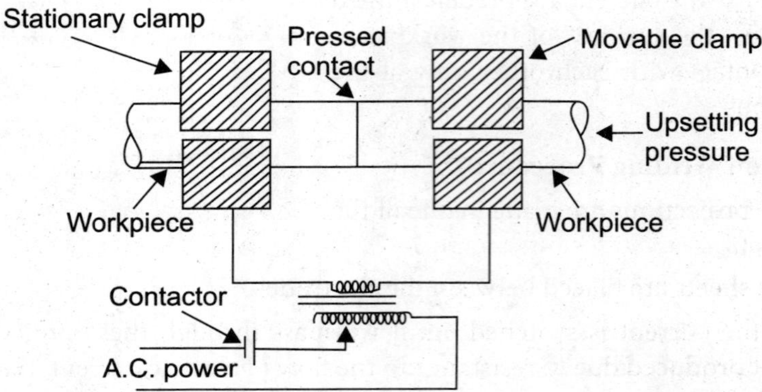

Principle of Resistance Butt welding (UPSET)

Resistance Butt Welding (Upset)

This is used for joining the rods or bars face to face using the electric resistance or heat generation.

Resistance Butt Welding (Upset) Procedure

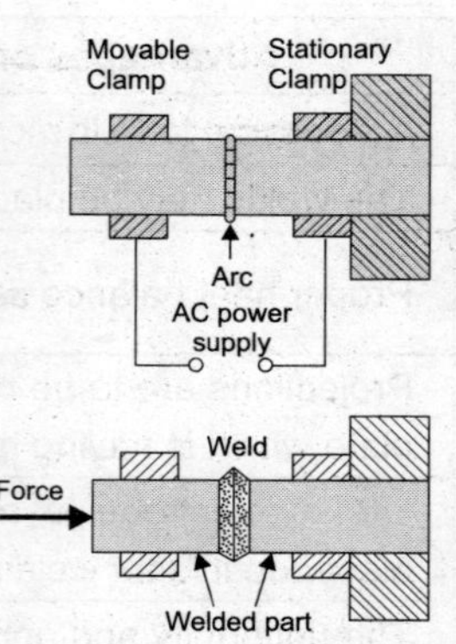

a. One workpiece to be joined is clamped in stationary clamping block and other is clamped in moving clamping block.

b. The two workpieces are brought together (butt joint) by pressure.

c. A heavy current is passed through these workpieces. The heat is generated because of resistance to the flow of electric current.

d. When the welding temperature has been reached, the pressure is increased so that proper joint takes place.

e. After that welding current is cut off.

f. Force is released when the welded joint has reached the desired temperature (normal temperature).

g. After that workpieces are unclamped.

Application of Resistance Butt Welding:

a. Used in wire drawing industries.

b. Used for producing butt joints in tubes, pipes, rods etc.

Flash Butt Welding

Flash butt welding is similar to upset welding except that the heat required for melting is obtained by means of an arc rather than the simple resistance heating.

Procedure

a. One workpiece to be joined is clamped in moving platen while other is mounted on fixed platen.

b. Current is switched 'ON'.

c. The workpiece held in the movable platen is moved towards the workpiece which is held in stationary platen as they come closer, flashing is established.

d. This flashing produces the welding heat. When sufficient heat is produced, more pressure is applied so that sound joint takes place.

e. After that, welding current is cut 'OFF' and workpieces are unclamped.

Advantages of Flash Butt Welding	Disadvantages of Flash Butt Welding
It consumes less welding current than upset butt weld process.	Concentricity and straightness of workpieces during welding is difficult to maintain
Flash welding offers strength factor upto 100%.	The shapes of the workpieces should be similar which is not always possible
Preparation of weld surface is not required.	Chances of fire hazards are there
The process is cheap	Flashes may cause eye trouble
It is a faster process	

Applications of Flash Butt Welding

a. It is used for the welding of bars, rods and tubes.

b. It is also used for the welding of saw blades into continuous loops, taps and reamers to alloy steel shanks.

Percussion Welding

In this resistance welding, welding heat is obtained from an arc produced by a rapid discharge of stored electrical energy. Pressure is applied rapidly during and immediately following the electric discharge. The electrostatic capacitors/condensers are used to store the electrical energy.

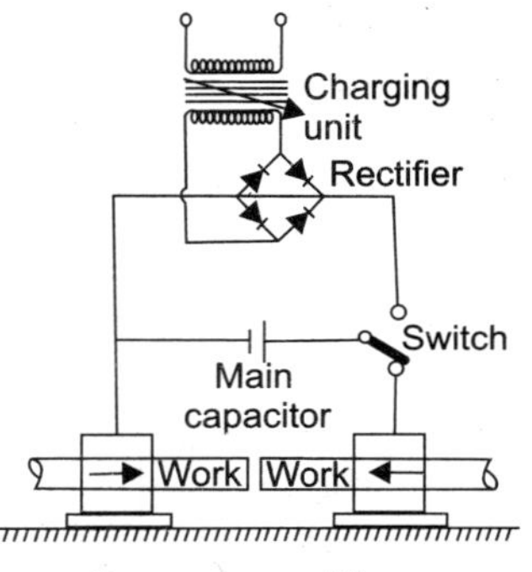

Percussion welding

Procedure

a. In this welding process, workpieces to be joined are clamped in moving platens.

b. Workpieces are brought near to each other.

c. The two workpieces are brought nearer.

d. An arc is produced by sudden discharge of the Condensers/ capacitors and rapidly pressure is applied.

e. At this stage arc is extinguished. Pressure is released when the weld cools.

f. After that workpieces are unclamped.

Advantages of Percussion Welding	Disadvantages of Percussion Welding
Arc temperature is more as compared to flash butt welding	The process is limited to butt welds only
Strong joints are produced	The equipment used for this purpose is quite expensive
In this welding, there is no or very less upsetting	The equipment must be extremely rugged and provided with accurate holding fixtures and sensitive timing devices etc
Can be used for joining different metals	

Applications of Percussion Welding:

a. It is used for welding stellite tips to tools, silver contact tips to copper, copper to aluminium etc.

b. It is used in telephone industries.

c. It is also used for welding fine wire leads to filaments in lamps, bulbs, etc.

Ques.607: Explain the process and principle of electric arc welding with a neat sketch.

Answer : Arc welding is a metal joining process, in this process the heat is obtained from an electric arc. Arc welding is carried out without the application of pressure and with or without the use of filler metal, depending upon the base plate thickness.

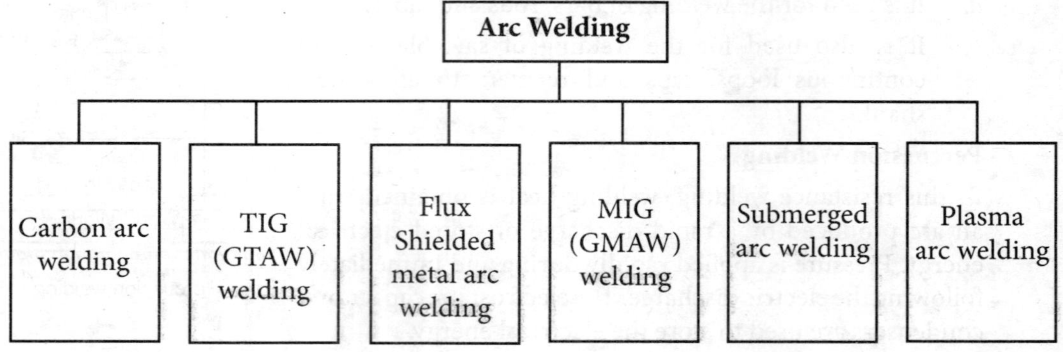

Basics of Arc Welding

The arc is struck between the electrode and the metal. It then heats the metal to a melting point. The electrode is then removed, breaking the arc between the electrode and the metal. This allows the molten metal to "freeze" or solidify.

What is arc and How an arc is formed?

The arc is like a flame of intense heat that is generated as the electrical current passes through a highly resistant air gap.

Procedure: The arc is formed (after switching ON the electric arc welding machine and connecting the electrodes) by striking the end of the welding rod on the work piece and a spark is produced, simultaneously the gap (approximately to 2 mm) is maintained between the welding rod and work piece, this result in the continuous arc and heats the workpiece.

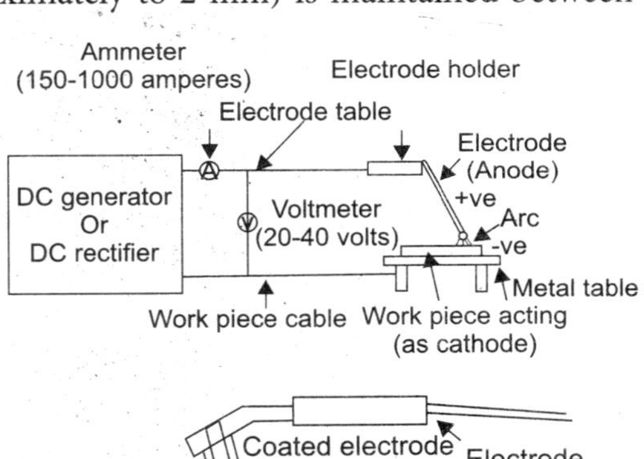

Arc Welding Principle

In arc welding, arc is generated between the electrode and the workpiece connected to electric supply which may be DC or AC. When these two poles are brought together, and separated for a small distance (0.8 to 2 mm) then the current starts to flow through a path of ionized particles, called plasma, and an electric arc is formed. Heat is generated as the ions strike the cathode (Negative). About 60% heat is librated at positive terminal (Anode).

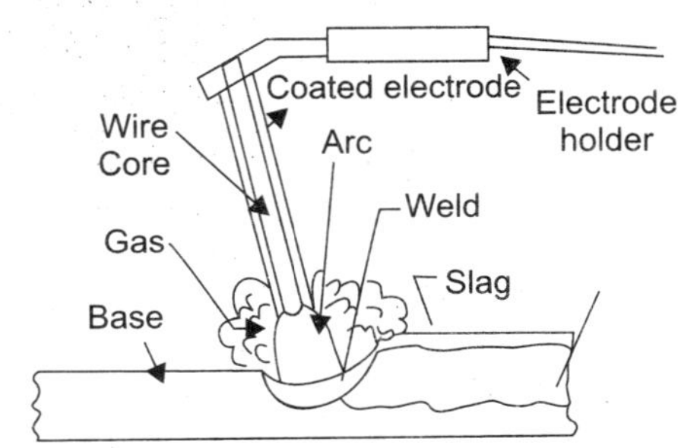

Metal arc welding with coated electrode

Various arc welding processes are:

a. Carbon-arc welding
b. Metal arc welding
c. Shielded metal arc welding (SMAW)
d. TIG (Gas Tungsten Arc Welding) welding
e. MIG (Gas Metal Arc Welding) welding
f. Submerged arc welding
g. Plasma arc welding

Carbon-arc welding

In carbon arc welding the arc is established between the work piece and the carbon electrode the filler material may be used separately.

Metal arc welding

In metal arc welding the arc is established between the work piece and the metal electrode which acts as the filler material also.

Metal arc welding Procedure

a. The work piece is connected to one terminal of the welding machine.(Positive terminal in case of straight polarity)

b. The electrode is attached to the electrode holder.

c. The machine is switched 'ON' and current adjusted.

d. The metal electrode is struck to the workpiece and a gap of 0.8-1.2 mm is maintained to establish the arc.

e. The developed arc melts the metals and forms a pool of molten metal, called puddle. The molten metal forms a joint on cooling and solidification.

f. The joint is complete on solidification.

Shielded metal arc welding (SMAW)

a. Also referred to as "Stick Welding"

b. Commonly used method and can be seen in the local market and small workshops, for everything from pipeline welding, farm repair & complex fabrication processes.

c. Uses a "stick" shaped coated electrode.

d. The procedure is same as that of metal arc welding but the electrodes or the stick is coated with flux which shields and protects the weld pool or heated metal from oxidation.

e. Can weld steel, cast iron, stainless steel, etc.

MIG (Metal Inert gas welding)

The welding is performed by electric arc using a continuously fed metal wire and the work piece. An inert gas is continuously supplied on the weld pool to make the protective atmosphere, such as argon or helium.

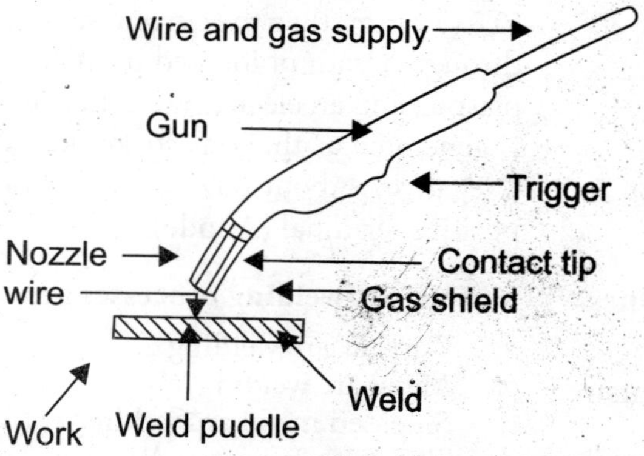

Metal insert gas welding

Applications of arc welding

a. Used for fabrication work and repairs

b. Used in ship building, pipe lines, bridge construction, tanks, boilers, etc

c. Domestic repairs and making of doors/ gates.

d. Manufacturing of domestic equipments such as chair, table, etc.

Advantages and disadvantages of arc welding

Advantages	Limitations or disadvantages
Most efficient way to join metals	Manually applied, therefore high labor cost.
Lowest-cost joining method	Need high energy causing danger
Affords lighter weight through betterutilization of materials	Not convenient for disassembly.
Joins all commercial metals	Defects are hard to detect at joints
Provides design flexibility	

Comparison of A.C. and D.C. power sources in arc welding

Alternating Current (from Transformer)	Direct Current (from Generator)
More efficiency	Less efficiency
Power consumption less	Power consumption more
Cost of equipment is less	Cost of equipment is more
Higher voltage – hence not safe	Low voltage – safer operation
Not suitable for welding non ferrous metals	suitable for both ferrous non ferrous metals
Not preferred for welding thin sections	preferred for welding thin sections
Any terminal can be connected to the work or electrode	Positive terminal connected to the work

Arc welding equipments:

The equipments required for arc welding are:

a. Power Source is a welding machine or generator (D.C.) or Transformer (A.C.)
b. Two cables- one for work and one for electrode
c. Electrode holder
d. Electrode
e. Protective shield
f. Gloves
g. Wire brush
h. Chipping hammer
i. Goggles or Safety Mask/ glass

Ques.608: Describe the process,applications and advantages & disadvantages of powder metallurgy .

Answer. Powder Metallurgy Process

Powder Metallurgy can be described as process or technique of manufacturing commercial articles *powdered metals* by placing these powders in moulds and applying Pressure.

Powder Metallurgy Process

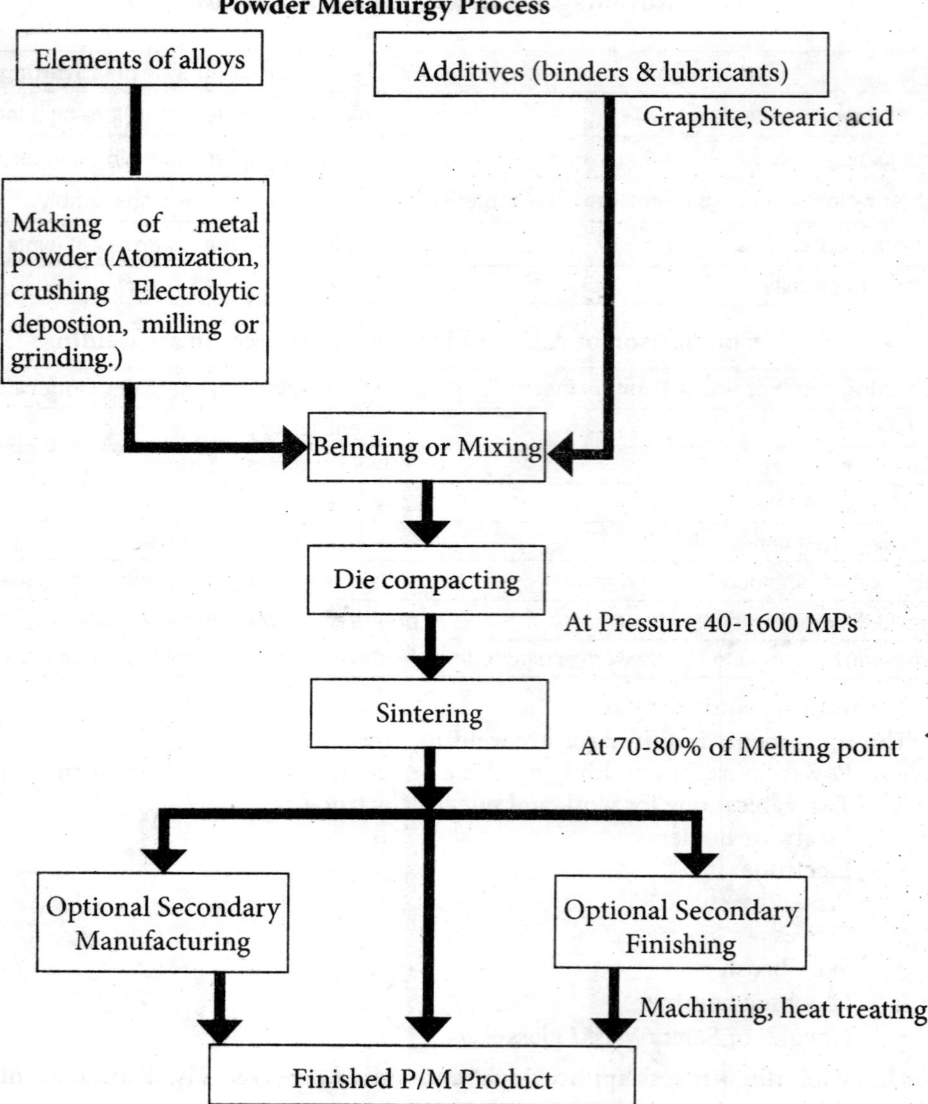

These compressed parts are then heated to bind particles the together and improve their strength and other properties. This process is known as **Sintering.** The temperature during this process is kept below the melting point of the powder.

The Products made through this process are very costly on account of the high cost of metal powders as well as of the dies used. A few typical examples of such products are Tungsten carbide Cutting Tools, Oil less Self Lubricating Bearings, special Electrical Contacts and Turbine-Blades, having high temperature strength. The Components made through this technique may be made from a single Metal powder or alloyed Metal powders.

Powder Metallurgy Process

The entire Powder metallurgy process mainly consists of the following **basic steps**:

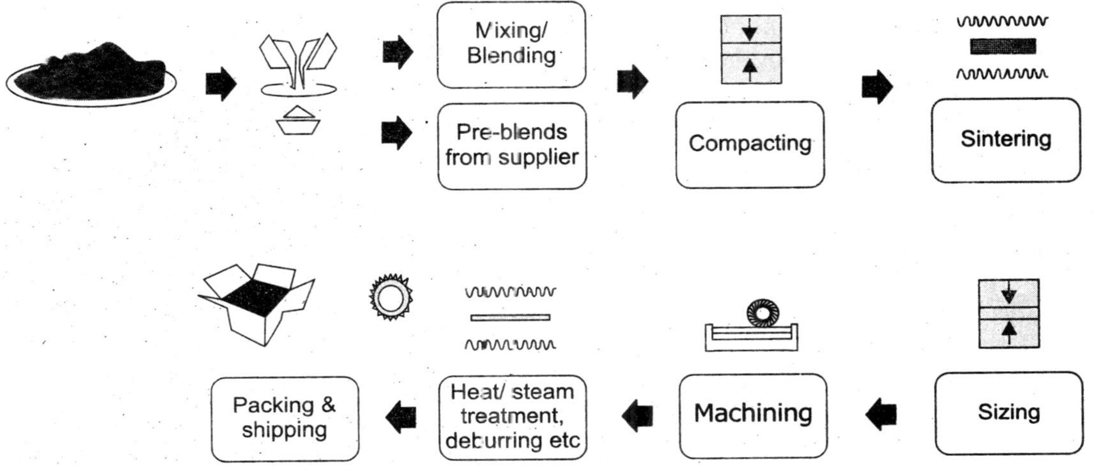

a. Production of metal powders.

b. Mixing or Blending of the metal powders in required proportions.

c. Pressing and compacting the powders into desired shapes and sizes.

d. Sintering the compacted parts in a furnace of controlled atmosphere.

e. Subjecting the sintered parts to Secondary Processing, if required.

f. Heat treat the components if, required.

The complete powder metallurgy process of making components from the powders is shown in the above figures.

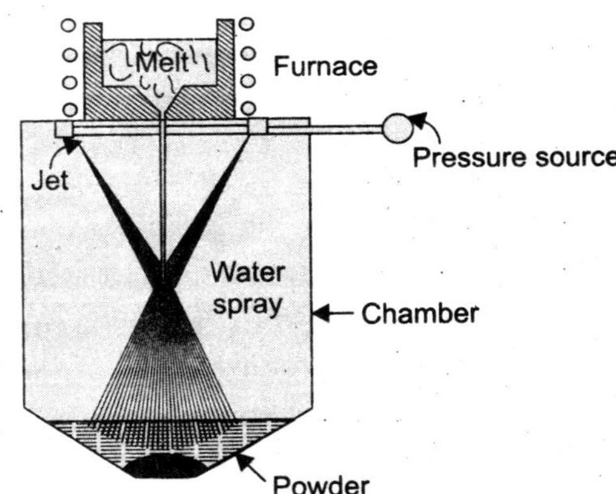

a. Production of Metal Powders

The following Processes are commonly used for producing these powders:

Atomization: It consists of forcing the Molten Metal through a Nozzle into a stream of **Air** or **Water**. As it comes in contact with the stream, the metal solidifies into small particles of various irregular shapes and sizes. Size of Nozzle, Rate of flow of metal and Temperature & Pressure of the air or water stream control the size of these particles. The Powders produced through this method are normally of Spherical or Pear shape.

The maximum amount of Metal powders is produced through Water Atomization Process, shown in Figure. The setup shown in the diagram consists of a Hopper like Container, called **Tundish** with a furnace, at the top. It carries a small Orifice at its bottom. This Container (Tundish) is fitted on the top of a larger container, which carries provision for Water jets inside. Molten metal is poured into **Tundish from a Ladle at the top of it**. As this metal flows down in the form of a stream through the nozzle, Water Streams from both sides impinge on it. This results in instantaneous cooling of the flowing metal in the form of small metal particles, which fall down and are collected at the bottom.

In **Air Atomization**, an air stream replaces the water jets. In either case, the particles will be oxidized. If, however, oxidized particles are not acceptable, an Inert gas stream can be used; the process then being known as **Gas Atomization.**

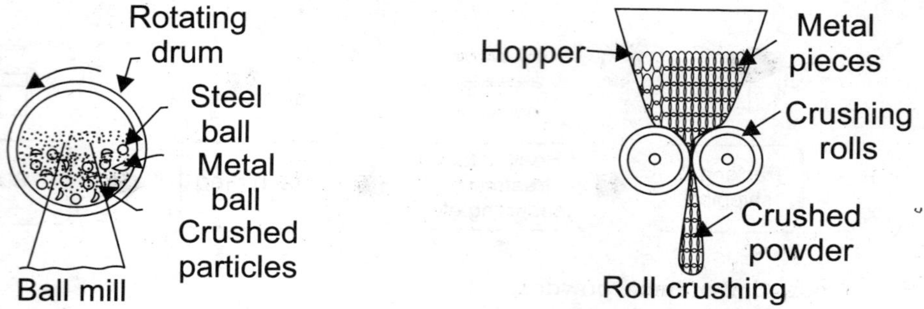

Figure: *Ball mill and roll crushing method of powder making*

Electrolysis or Electrolytic Process: Metal powders may be deposited electrolytically under controlled conditions. The deposited material is Mulled or Ground for fineness, Pulverized for desired grain size & Annealed to drive out hydrogen and remove the effect of work hardening. The Powders produced through this process are the Purest.

Milling and Grinding Mechanical Pulverization : It consists of breaking down metal into small particles by the Crushing and Impact through Ball Mills, Stampers, Crushers, etc.

The process involves pulverising (grind/ crush) the metal mechanically so that it will disintegrate into small particles, as shown in Figure. Brittle materials suit best to this method. It is also used for Malleable and Ductile metals.

Machining: It is mainly used for producing Magnesium and Beryllium particles. The particles produced are, however, coarse, which can be converted into fine powder through Ball milling and Impact grinding.

Properties/ characteristics desirable in the metal powders

(i) Particle shape: particles may be angular, flat or spherical and this will affect the compactness, porosity & strength of part. A mixture of all shape/type of particle is desirable.

(ii) Particle size and distribution: it may be fine and coarse. The mixture of both is desirable for good part production.

(iii) Flowability: the powders must be flowable to occupy the each corner of the mould easily.

(iv) Compressibility: The powders should possess good compressibility which is the ratio of initial volume to the final volume after compaction.

(v) Purity: the powders should not have any impurity as this will lowers the quality and mechanical properties of the parts.

(vi) Sintering ability: It should be sufficiently high for good products.

b. Blending or Mixing

The process consists of a thorough mixing of the constituents either wet or dry. Wet Mixing reduces dust and minimizes the dangers of explosion. Lubricants are added during blending to reduce friction during pressing and enable easy ejection of pressed parts. Common lubricants are Graphite, Stearic acid and Lithium stearate. Blending also facilitates uniform distribution of particle size.

c. Compacting

It is the process of converting loose powder into a Green Compact of accurate shape and size. It is done in steel dies and punches. Due to inter particle friction, the pressure applied from one side is not uniformly distributed throughout the mass. Two punches are, therefore, employed, one from the top and the other from the bottom of the powder, as shown in Figure. The lower punch also works as an Ejector for the compressed part. Due to low flowability of metal powders, the density variation is kept at a minimum by both-side pressing.

d. Sintering

Sintering of compacted parts is the process of heating to high temperatures, but below the melting points of the materials being sintered; it may be carried out in two steps, using pre- sintering to a lower temperature. The actual values of sintering temperatures for most materials range between 70 to 80 per cent of their melting temperatures.

The main objectives of Sintering can be summarized as follows:

a. Achieving high strength

b. Achieving good bonding of powder particles

c. Producing a dense and compact structure

d. Producing parts free of oxides

e. Causing metallurgical diffusion and facilitate alloying of constituent materials.

f. Obtaining desired structure and improved mechanical properties.

e. Secondary Processes

Sizing: It consists of placing the sintered part in a Die, made to correct dimensions, and repressing the same to bring it to the required size. It also takes care of certain distortions which might have occurred during Sintering.

Machining: In most of the cases, the parts produced through powder metallurgy are finished to required dimensions & tolerances and no further machining is required on them. However, sometimes, in order to hold very close tolerances or for producing such features like Small holes, Threads, Undercuts and Grooves, etc., which cannot be produced through Powder Metallurgy process, machining may have to be done.

Impregnation: The process is performed to provide Self-Lubricating Properties to the sintered parts. For this, the parts are placed in a tank, carrying the lubricant, and heated to about 93°C for about 15 to 20 minutes. The lubricant enters the pores of the part through capillary action and is retained there.

Heat Treatment: All sintered metal parts can be heat treated through the usual methods, although the results are not as impressive as with solid metal parts.

Applications of powder metallurgy

Some Typical applications of metal powder parts are:

a. Porous Metal Bearings, made from Brass, Iron, Aluminum, Bronze, etc. which are later impregnated with lubricants.

b. Machine parts are produced from powders of Iron, Brass and Bronze.

c. Filament wire for electric bulbs is made from Tungsten Carbide Powders with Cobalt as binder.

d. Products where metals and non-metals are combined

e. Combination of metals and ceramics

f. Used for making the grinding wheels

g. Used for making electrical contacts

h. Used in medical services for medicine and making of equipments

Advantages and Disadvantages of Powder Metallurgy

Advantages	Disadvantages and Limitations
It facilitates production of many such parts which cannot be produced through other methods, such as Sintered carbides and Self lubricating bearings	The metal powders and the equipment used are very costly
It also facilitates mixing of both metallic and non- metallic powders to give products of special characteristics	There is a limitation to the size of the product as the same will depend on the capacity of the press used and the compression ratio of the powders
The products carry very high dimensional accuracy, thus eliminating the need for further machining in most cases. If at all needed, it is not much	Storing of powders offers great difficulties because of the possibilities of fire and explosion hazards
Layers of different metal powders can be molded together to obtain multi-metallic products	Design restrictions, due to low flowability of metal powders, restrict the production of intricate shapes
The products of powder metallurgy are highly pure	Sintering of low melting point metal powders, like those of lead, zinc and tin etc. offers serious difficulties
The process facilitates saving in material as no material loss occurs during fabrication	

It is possible to ensure uniformity of composition, since exact proportions of constituent metal powders can be used	A completely dense and compact metal structure cannot be produced through this process.
The rate of production is quite high	
It enables production of parts from such alloys which possess poor castability	
The process does not require highly skilled workmen	The process is not found economical for small scale production
Hard to process materials, like diamond and ceramics, can be converted into usable components and tools through this process	Physical properties of parts produced through this method are generally not comparable to cast or wrought parts

Ques.609: Explain the following:

(a) Electroplating

(b) Galvanizing

(c) Ceramics

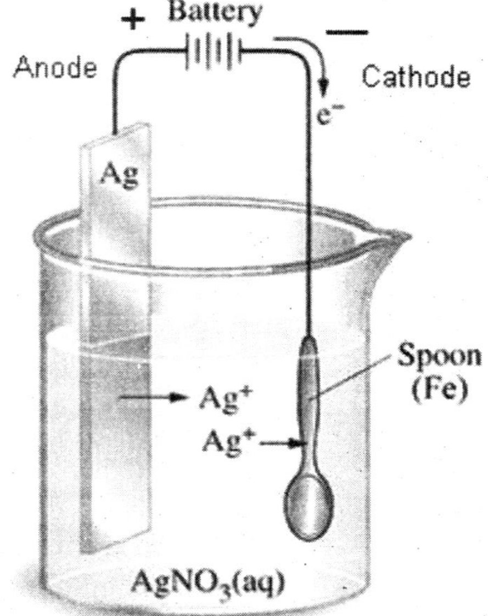

Answer : **(a) Electroplating**: Electroplating is a very popular process of providing metallic coatings by means of Electrolysis. The article to be coated may be metallic or non-metallic (Plastics). If metallic, the Base metal and the coated metal may be same or different. Coatings of several different Metals and Alloys can be provided through this technique. Some common metals & alloys, of which coatings can be provided on metallic and non-metallic surfaces, include Chromium, Tin, Nickel, Copper, Cadmium, Silver, Gold, Platinum, Brass, Bronze, etc.

The **Basic Principle** involved in Electroplating of all metals is the same. The main elements involved in this process are an Electrolyte, a Tank containing this electrolyte, an Anode (+ve terminal), a Cathode (-ve terminal) and a low voltage Direct Current (d.c.) supply. The Electrolyte used is a metallic salt solution i.e.,dissolved salts of the Plating Metal. The Work-pieces to be plated are suspended & immersed in the electrolyte filled in the tank and electrically, they are connected to the Negative Terminal so as to act as Cathode in the circuit. The material to be deposited to provide the Coating is connected to the Positive terminal, so as to act as Anode. When plastics are to be plated, a primary coat of an electrically conductive material is first provided over them and then they are electroplated.

(b) Galvanizing: Galvanizing or galvanization is a process which employs an Electrochemical Action for providing a Coating of a highly corrosion resistant material on the surface of another metal.This process (Galvanizing) is largely used for providing a coating of Zinc on Iron and Steel. It is a fairly low-cost process and is widely used for corrosion resistant coatings on sheet metal, house hold items of daily use made from iron and steel i.e. buckets, tubs & other containers, machine parts, tools, ships, tanks, wires, different threaded fasteners, etc. Galvanized metal sheets, pipes & wires are probably the most popular galvanized items and find application both in industrial work as well as in articles made for domestic use, i.e. hand pump or submersible pump pipes.

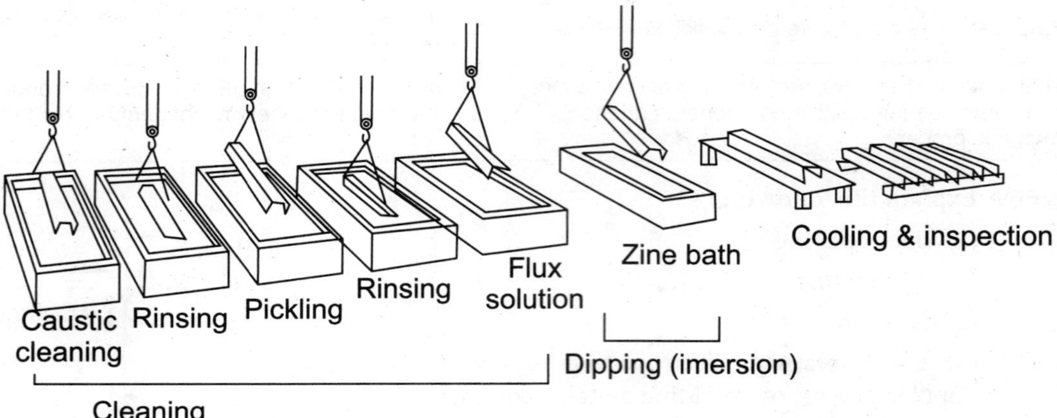

Hot-dipping Galvanizing Procedure

a. The first step in applying this process is cleaning of the parts or items to be galvanized. This may involve degreasing, acid rinsing in cold water, etc. depending upon the condition of the work-surfaces. The basic idea is to obtain an absolutely clean surface. If the item to be galvanized is sheet metal, it is annealed after surface cleaning and then cooled in an oxide free atmosphere.

b. The sheet is then dipped in the molten zinc bath consisting of a minimum of 98% pure molten zinc. The bath temperature is maintained at about 450 °C.

c. If perfect and uniform thickness is an important consideration, the sheets are drawn through a set of rollers, immediately following the coating process. This process drives away the Excess Zinc and allows only the required amount of zinc to remain on the surface, which is just enough to provide the coating of desired thickness.

d. This is followed by drying & inspection.

e. The sheets are ready for shipment.

f. The complete procedural steps are shown with help of the figure above.

(c) Ceramics: Ceramic is a compound which consists of metallic and non- metallic constituents. This word is derived from a Greek word Keramicos, which means Potter Earth or Clay. It is found that they carry more than one type of bond in a single material, i.e., an ionic bond as well as the covalent Bond. Due to these bonds, Ceramics are characterized by their: High Brittleness, High hardness, high Electrical insulation, Chemical inertness and High melting points.

Examples: Common Ceramic Materials include Glasses, Insulating Materials, Enamels, Abrasives, etc. All metal oxides, nitrides, silicates, borides and Carbides are also considered as Ceramics.

Classification of Ceramics: Ceramics may be considered as materials made from naturally occurring clay or earth. Ceramics are non-metallic and inorganic materials which are processed at high temperature. Ceramics may be classified as natural ceramics (Silica, Silicates) and manufactured ceramics (Silicon carbide, Aluminium oxide, etc.)

Properties of Ceramics

a. They are hard.

b. They are brittle.

c. Ceramics have high melting point.

d. Ceramics have low electrical and thermal conductivity.

e. They have good creep resistance.

f. They have high corrosion resistance.

g. Ceramics have good chemical and thermal stability.

h. Ceramics have high compressive strength.

i. Ceramics have excellent resistance to wear.

j. Porous ceramics are resistant to thermal shocks.

k. Ceramics have low density.

l. Ceramics have low tensile strength.

Applications of Ceramics

a. Clay Products :

(1) Earthen-ware : Porous drainage pipes, ceramic filter, wall, tiles and bricks.

(2) Fine China : Table ware.

(3) Stone ware : Glazed pipes, roofing tiles and table ware.

(4) Porcelain : Fine table ware, scientific equipment and spark plug insulators for automobiles

b. Refractory bricks

c. Cutting tool materials

d. Abrasives

e. Electrical switches and capacitors

Ques.610: Write notes on the following:

(a) Plastics and their applications

(b) Composite materials & applications

Answer : **Plastics:** Plastics are the organic materials and derived from the materials which consist of carbon combined with hydrogen oxygen and other non-metallic substances.

Also the Plastics are synthetic organic materials termed as polymers, which are formed from monomers after the process of polymerization.

Properties of Plastics and Advantages

a. They are light weight having low density and at the same time possess good strength & rigidity.

b. They are less brittle when compared to glass and yet can be made equally transparent & smooth.

c. The high heat & electric insulation make them suitable to be applied in the radio & electrical engineering industries as dielectrics and as substitutes for porcelain, ebonite, mica, shellac etc.

d. They possess good toughness.

e. They have good chemical stability, when exposed to the action of solvents, and certain oxidizing agents i.e., they resist corrosion and action of chemicals.

f. Plastics have good damping capacity.

g. Plastics have good surface finish.

h. They possess the property of low moisture absorption.

i. They can be easily molded to desired shapes.

j. Simple processing of plastics is required to obtain machine parts.

k. Plastics possess low melting point and can easily flow into a mould.

l. They possess high strength to weight ratio.

m. Mass production of plastics is easy and so they are extensively used as wrappers and bags.

n. Plastics possess weather-ability, colour-ability, heat resistance and deformability

Disadvantages of Plastics

(1) Comparatively higher costs of materials.

(2) Inability of most of the plastics to withstand moderately high temperature

Types of Plastics

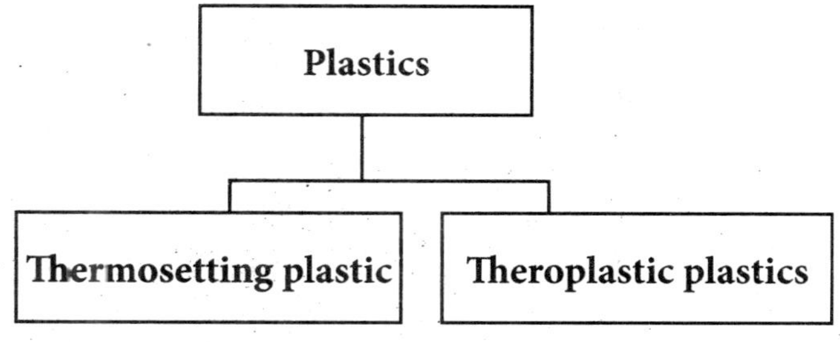

Thermosetting Plastics: These plastics, when subjected (once only) to heat and pressure required for forming, change into hard & rigid substances. Once these Thermosetting Resins are changed into rigid substance they cannot be softened again by application of heat.

Thermoplastic Plastics: These plastics can be re-softened and re-molded by application of heat and pressure. These plastics harden on cooling and can be re-softened under heat. Thus, they retain their fusibility, solubility and capability of being repeatedly shaped. Thermoplastic resins are linear & branched linear polymers and are sensitive to temperature & sunlight.

Injection molding

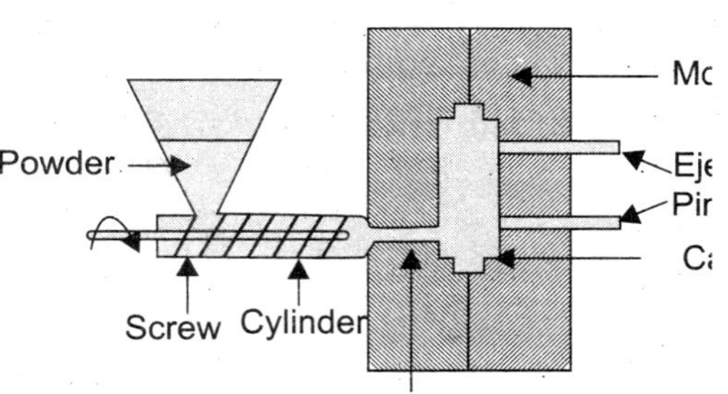

Plastic Products Manufacturing

The following methods are commonly employed for manufacturing of plastics into various usable articles:

a. Extrusion
b. Injection molding
c. Reaction Injection molding
d. Blow molding
e. Casting
f. Slush molding

Injection Molding

It is very commonly used for Thermoplastic Plastics. The process is illustrated in Figure. The Powdered plastic Compound is first heated to drive off Moisture and then fed into the Hopper. The material (Plastic) is fed into the cavity by the screw. This compresses the material and forces it forward into the water cooled mould through the nozzle. During heating in the Chamber, the temperature of the material rises to between 177°C and 274°C. After the Plastic has been cooled and sufficiently hardened in the mould, the mould is opened and the produced part knocked out. It is a faster process and suits best for large-quantity Production.

Reaction injection molding

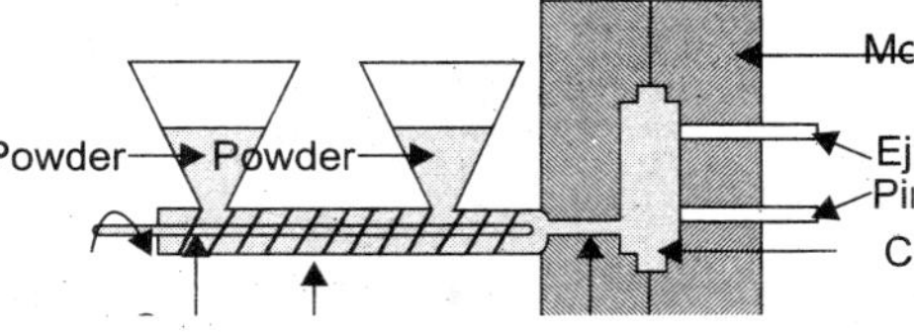

Reaction injection molding

This method is used for thermosetting plastics and the arrangement is similar to injection molding. The powders of two materials are heated & mixed and forced in the cavity. The arrangement uses two hoppers.

Blow moulding

Blow moulding a shape is a common industrial process. The example shown below is of the production of a plastic bottle. The plastics normally used in this process are; polythene, PVC and polypropylene.

The process is similar to injection moulding and extrusion.

a. The plastic is fed in granular form into a 'hopper' that stores it

b. A large thread is turned by a motor which feeds the granules through a heated section.

c. In this heated section the which forces the become a liquid and the liquid is fed into a mould.

d. Air is forced into the mould which forces the plastic to the sides, giving the shape of the bottle.

e. The mould is then cooled and is removed.

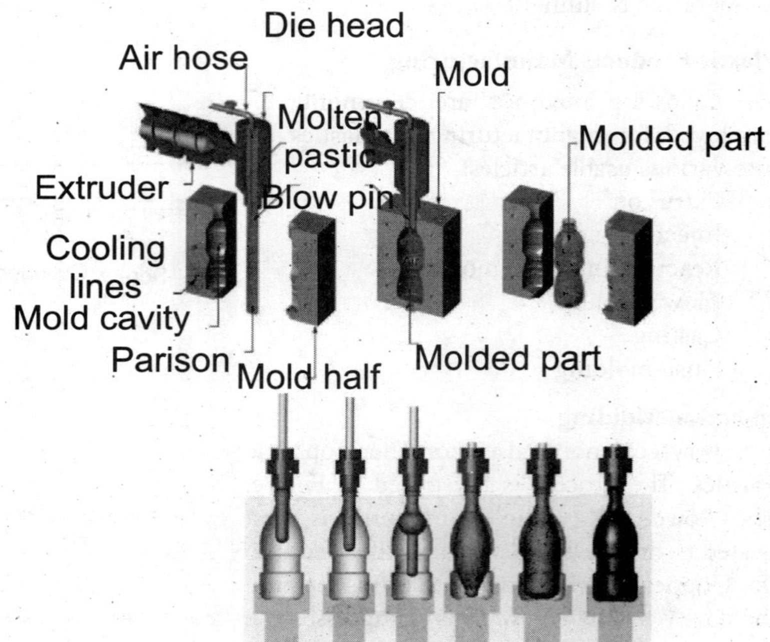

(b) Composite Materials

Composite Material is "Macroscopic combination of two or more distinct materials, having a recognizable interface between them".

Composite materials form a material system, composed of a mixture or a combination of two or more macro constituents that differ in from & chemical composition and are insoluble in each other. These constituents maintain their identity even after mixing & being together and give entirely different properties to the mixture.

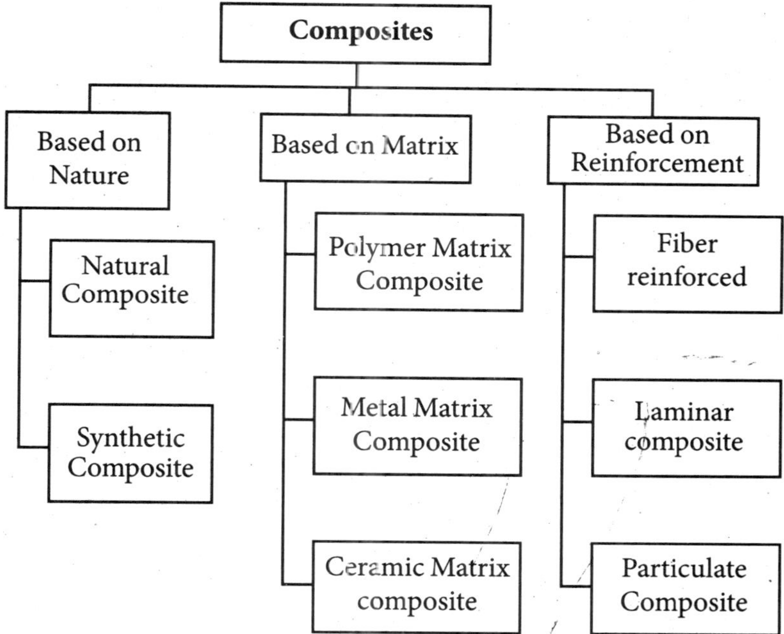

Figure: *Classification of composite materials*

Components (Parts or Constituents)of Composites

There are TWO major components:

(a) Reinforcement, i.e., material that provides strength to the matrix

(b) Matrix, i.e., material that holds the reinforcements in place

Properties and Advantages of Composites

a. Low density and very light weight with high strength, hardness & stiffness

b. Corrosion resistance

c. Weather resistance

d. Low thermal conductivity

e. Low coefficient of thermal expansion

f. High dielectric strength

g. Non-magnetic

h. Radar transparency

i. They can withstand elevated temperatures in the corrosive environment.

j. Exceptional formability.

k. Outstanding durability.

l. Inherent damping.

Disadvantages of Composites

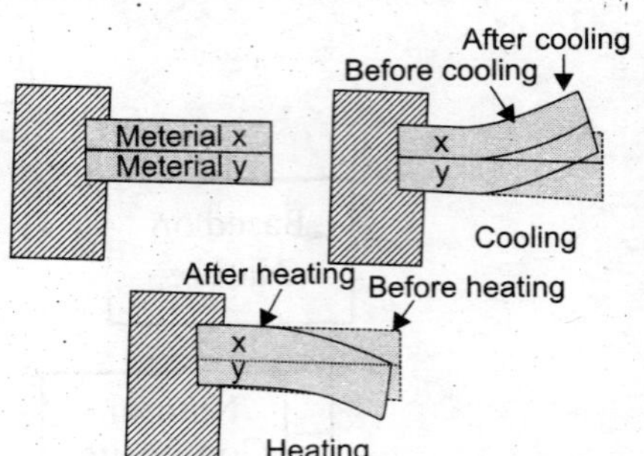

a. High cost of raw materials and fabrication.
b. Composites are more brittle than wrought iron and thus easily damaged.
c. Matrix is weak, therefore, low toughness.
d. Matrix is subject to environment degradation.

Applications of composites

a. Used as a replacement material in various engineering applications such as for cylinder block liners, vehicle drive, shafts, automotive pistons, bicycle frames, etc.
b. Hockey sticks, vaulting poles, golf carts etc.
c. Fuel and chemical storage tanks.
d. Aerospace applications: helicopter blades, engine couplings, ducts etc.
e. Marine: Boats, ships, hulls etc.
f. Electronic and recreational industries.
g. High voltage power transmission lines and heat sinks for electronic components.
h. As bio-compatible materials
i. Used in electrical appliance as thermal switch.

Ques.611: Write a short note on rubber.

Answer : Rubber is a polymer material and is very elastic in nature. It is obtained from tree known as heveabrasilliensis in the form of emulsion called as Latex. Rubber may be classified as:

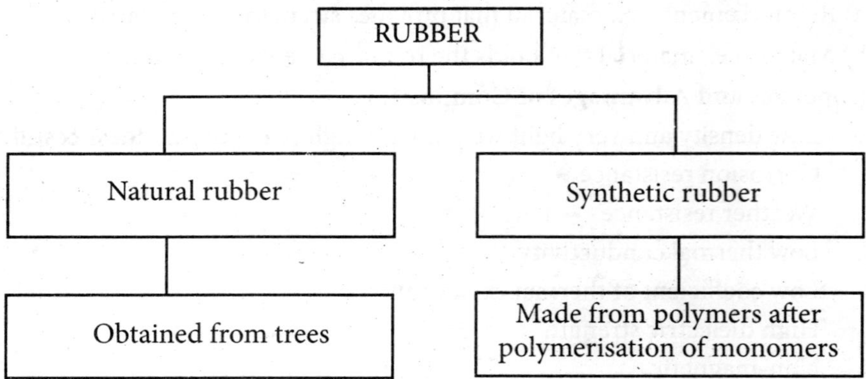

Properties: Rubber has the following properties:

1. Good elasticity
2. Good texture
3. Good resilience
4. Good insulator of electricity

Uses: Rubber finds application/uses in the following:

1. Automobiles
2. Rubber hoses for water and air
3. Rain coats, floor mats, belts, electrical insulators, tyre and tubes, etc

Vulcanization: Vulcanization is the process of heating rubber with sulpher to provide the durability and strength to rubber. The vulcanized rubber is used for making of tyre and tubes.

Ques.612: Classify ferrous materials. Describe the composition, properties & applications of various types of steel.

Answer:

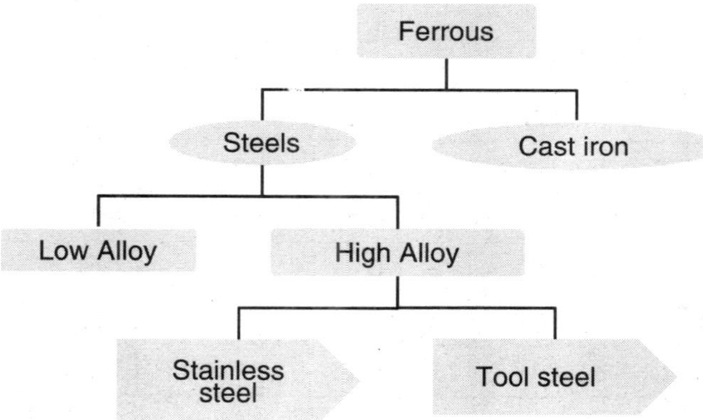

- Steels - alloys of iron-carbon. May contain other alloying elements.
- Several grades are available.
- Low Alloy (<10wt%)
  - Low Carbon (<0.25 wt% C)
  - Medium Carbon (0.25 to 0.60 wt%)
  - High Carbon (0.6 to 1.4 wt%)
- High Alloy
  - Stainless Steel (> 11 wt% Cr)
  - Tool Steel

Low Carbon Steel

- Plain carbon steels- very low content of alloying elements and small amounts of Mn.
- Most abundant grade of steel is low carbon steel - greatest quantity produced; least expensive.

- Not responsive to heat treatment; cold working needed to improve the strength.
- Good Weldability and machinability
- High Strength, Low Alloy (HSLA) steels - alloying elements (like Cu, V, Ni and Mo) up to 10 wt %; have higher strengths and may be heat treated.

Compositions of some low carbon and low alloy steels

Designation		Composition (wt%)		
AISI/SAE or ASTM Number	UNS Number	C	Mn	Others
Plain Low-Carbon Steels				
1010	G10100	0.10	0.45	
1020	G10200	0.20	0.45	
A36	K02600	0.29	1.00	0.20 Cu (min)
A516 Grade 70	K02700	0.31	1.00	0.25 Si
High-Strength, Low-Alloy Steels				
A 440	K12810	0.28	1.35	0.30 Si (max), 0.20 Cu (min)
A633 Grade E	K12002	0.22	1.35	0.30 Si, 0.08 V, 0.02 N. 0.03 Nb
A656 Grade 1	K11804	0.18	1.60	0.60 Si, 0.1 V. 0.20 Al, 0.015 N

Properties and typical application of some low carbon and low alloys steels

AISI/SAE or ASTM Number	Tensile Strength [MPa (ksi)]	Yield Strength [MPa (psi)]	Ductility [%EL in 50 mm (2 in.)]	Typical Applications
Plain Low-Carbon Steels				
1010	325 (47)	180 (26)	28	Automobile panels, nails, and wire
1020	380 (55)	210(30)	25	Pipe: structural and sheet steel
A36	400 (58)	220 (32)	23	Structural (bridges and buildings)
A516 Grade 70	485 (70)	260 (38)	21	Low-temperature pressure vessels
High-Strength, Low -Alloy Steels				
A440	435 (63)	290 (42)	21	Structures that are bolted or riveted
A633 Grade E	520 (75)	380 (55)	23	Structures used at low ambient temperatures

A656 Grade 1	655 (95)	552 (80)	15	Truck frames and railway cars

Medium Carbon Steel

- Carbon content in the range cf C.3 – 0.6%.
- Can be heat treated - austenitizing, quenching and then tempering.
- Most often used in tempered condition – tempered martensite
- Medium carbon steels have low hardenability.
- Addition of Cr, Ni, Mo improves the heat treating capacity
- Heat treated alloys are stronger but have lower ductility
- Typical applications – Railway wheels and tracks, gears, crankshafts.

Composition of some alloyed medium carbon steels

AISI//SAE Designation	VNS Designation	Composition Ranges (wt% of Alloying Elements in Addition to C)			
		Ni	Cr	Mo	Other
lOxx, Plain carbon	GlOxx0				
1 lxx. Free machining	GllxxO				0.08-0.33S
12xx. Free machining	G12xx0				0.10-0.35S.
					0.04-0.12P
13xx	G13xx0				1.60-1.90Mn
40xx	G40xx0			0.20-0.30	
4 lxx	G41xx0		0.80-1.10	0.15-0.25	
43xx	G43xx0	1.65-2.00	0.40-0.90	0.20-0.30	
46xx	G46xx0	0.70-2.00		0.15-0.30	
48xx	G48xx0	3.25-3.75		0.20-0.30	
51xx	GSlxx0		0.70-1.10		
61xx	G61xxC		0.50-1.10		0.10-0.15V
86xx	G86xxC	0.40-0.70	0.40-0.60	0.15-0.25	
92xx	G92xx0				1.80-2.20Si

High Carbon Steel

- High carbon steels – Carbon content 0.6 – 1.4%
- High C content provides high hardness and strength. Hardest and least ductile.
- Used in hardened and tempered condition

- Strong carbide formers like Cr, V, W are added as alloying elements to from carbides of these metals.
- Used as tool and die steels owing to the high hardness & wear resistance property.

Compositions and Application of some Tool steels

AISI Number	UNS Number	Composition (wt.%)						Typical Allpications
		C	Cr	Ni	Mo	W	V	
MI	T11301	0.85	3.75	0.30 max	8.70	1.75	1.20	Drills, saws, lathe and planer tools
A2	T30102	1.0	5.15	0.30 max	1.15	-	0.35	Punches, embossing dies
D2	T30402	1.5	12	0.30 max	0.95	-	1.10 max	Cutlery, drawing dies
01	T31501	0.95	0.50	0.30 max	-	0.50	0.30 max	Shear blades, cutting tools
SI	T41901	0.50	1.40	0.30 max	0.50 max	2.25	0.25	Pipe cutters, concrete drills
W1	T72301	1.10	0.15 max	0.20 max	0.10 max	0.15 max	0.10 max	Balcksmith tools

Ques.613: Describe the Effects of various Alloying Elements on Steel .

Answer:

- **Manganese** – Imparts strength & hardness; decreases ductility & weldability; effects hardenability of steel.
- **Phosphorus** – increases strength & hardness and decreases ductility & notch impact toughness of steel.
- **Sulfur** – decreases ductility & notch impact toughness; Weldability decreases. Found in the form of sulfide inclusions.
- **Silicon** – one of the principal deoxidizers used in steel
- **Making**–In low-carbon steels, silicon is generally detrimental to surface quality.
- **Copper** – detrimental to hot-working steels; beneficial to corrosion resistance (Cu>0.20%)
- **Nickel** - ferrite strengthener; increases the hardenability and impact strength of steels.
- **Molybdenum**– Increases the hardenability; enhances the creep resistance of low-alloy steels

Ques.614: What are various categories of Stainless steel ?Describe composition , properties & applications of various stainless steel .

Answer:

Stainless steels - A group of steels that contain at least 11% Cr. Exhibits extraordinary corrosion resistance due to formation of a very thin layer of Cr_2O_3 on the surface.

Categories of stainless steels:

- Ferritic Stainless Steels – Composed of ferrite (BCC)
- Martensitic Stainless Steels – Can be heat treated.
- Austenitic Stainless Steels – Austenite (γ) phase field is extended to room temperature. Most corrosion resistant.
- Precipitation-Hardening (PH) Stainless Steels – Ultra high strength due to precipitation hardening.
- Duplex Stainless Steels – Ferrite + Austenite

Composition and Properties of some stainless steels are given below.

				Mechanical Properties			
AISI Number	**UNS Number**	**Composition (wt. %)**	**Condition**	**UTS (MPa)**	**YS (MPa)**	**Ductility <%EL)**	**Typical Applications**
409	S40900	0.0S C. 9.0 Cr. 1.0 Mn. 0.5 Ni. 0.75 Ti	Annealed	Ferritic 380	205	20	Automotive exhaust components, tanks for agricultural sprays
446	S44600	0.20 C, 25 Cr. 1.5 Mn	Annealed	515	275	20	Valves (high temperature), glass molds, combustion chambers
			Austenitic				
304	S30400	0.08 C, 19 Cr, 9.0 Ni. 2.0 Mn	Annealed	515	205	40	Chemical and food processing equipments, cryogenic vessels
316L	S31603	0.03 C. 17 Cr. 12 Ni, 2.5 Mo, 2.0 Mn	Annealed	485	170	40	Welding construction
			Martensitic				
410	S41000	0.15 C, 12.5 Cr, 1.0 Mn	Annealed Q& T	485 825	275 620	20 12	Rifle barrels, cutlery, jet engine parts
440A	S44002	0.70 C. 17 Cr, 0.75 Mo. 1.0 Mn	Annealed Q&T	725 1790	415 1650	205	Cutlery, bearings, surgical tools

Precipitation Hardenable							
17-7PH	SI7700 7	0.09 C, 17 Cr, Ni. 1.0A1.1.0 Mn	Precipitation hardened	1450	1310	1-6	Springs, knives, pressure vessels

Ques.615: What are different types of cast iron? What are applications of cast iron ?

Answer: Characteristics of cast iron are as follows:

- Carbon: 2.1- 4.5 wt% and Si (normally 1-3 wt%).
- Lower melting point (about 300 °C lower than pure iron) due to presence of eutectic point at 1153 °C and 4.2 wt% C.
- Low shrinkage and good fluidity & casting ability.

Types of cast iron: grey, white, nodular, malleable and compacted graphite.

Grey Cast Iron:

- Grey cast iron contains graphite in the form of flakes. Named after its grey fractured surface. C:3.0 – 4.0 wt%, Si: 1.0 – 3.0 %
- Microstructure: graphite flakes in a ferrite or pearlite matrix
- Weak & brittle in tension (the graphite flake tips act as stress concentration sites). Stronger in compression,
- Excellent damping capacity, wear resistance.
- Microstructure modification by varying silicon content and cooling rate
- Casting shrinkage is low

Nodular or Ductile Iron:

- Addition of Mg and/or Cerium to grey iron converts the graphite flakes to nodules.
- Normally a pearlite matrix.
- Castings are stronger and much more ductile than grey iron as the stress concentration points existing at the flake tips are eliminated.

White Cast Iron:

- White cast iron – C: 2.5 – 3 wt.%, Si: 0.5 – 1.5%. Most of the carbon is in the form of cementite. Named after its white fracture surface.
- Results from faster cooling. Contains pearlite + cementite, not graphite. Thickness variation may result in nonuniform microstructure from variable cooling
- Very hard and brittle
- Used as intermediate to produce malleable cast iron.

Malleable Cast Iron:

- Malleable cast iron – Carbon: 2.3 – 2.7 wt%, Si: 1.0 – 1.75 %
- Obtained by heat treating white iron for a prolonged period that causes decomposition of cementite into graphite.

- Heat treatment : Two stages – Isothermal holding at 950 °C and then holding at 720 °C.
- graphite forms in the form of rosettes in a ferrite or pearlite matrix.
- Reasonable strength and improved ductility (malleable)

Compact Graphite Iron (CGI):

- CGI graphite occurs as blunt flakes or with a worm-like shape (vermicular). Carbon: 3.1 – 4.0 wt%, Silicon: 1.7 – 3.0 wt %.

Microstructure and properties are between gray and ductile iron.

- Alloying addition may be needed to minimize the sharp edges and formation of spheroidal graphite. Matrix varies with alloy additions or heat treatment.
- As castable as grey iron, but has a higher tensile strength and some ductility.
- Relatively high thermal conductivity, good resistance to thermal shock, lower oxidation at elevated temperatures.

Applications of Cast iron

Cast irons are used in wide variety of application owing to the properties like good fluidity, ease of casting, low shrinkage, excellent machinability,wear resistance and damping capacity.

- Applications –
  - Car parts – cylinder heads, blocks and gearbox cases.
  - Pipes, lids (manhole lids)
  - Foundation for big machines (good damping property)
  - Bridges, buildings
  - Cook wares – Excellent heat retention

Ques.616: Describe the composition, properties and applications of important Nonferrous Metals & their alloys .

Answer : Important non-ferrous metals & their alloys are described below.

Copper: Copper is one of the earliest metals discovered by man. The boilers on early steamboats were made from copper. The copper tubing used in water plumbing in Pyramids was found in serviceable condition after more than 5,000 years. Cu is a ductile metal. Pure Cu is soft and malleable, difficult to machine. Very high electrical conductivity – second only to silver. Copper is refined to high purity for many electrical applications. Excellent thermal conductivity – Copper cookware most highly regarded – fast and uniform heating. Electrical and construction industries are the largest users of Cu.

- U.S. nickel is actually 75% copper.
- The various Euro coins are made of Cu-Ni, Cu-Zn-Ni or Cu-Al-Zn-Sn alloys.

COPPER ALLOYS

- Brasses and Bronzes are most commonly used alloys of Cu. Brass is an alloy with Zn. Bronzes contain tin, aluminum, silicon or beryllium.
- Other copper alloy families include copper-nickels and nickel silvers. More than 400 copper-base alloys are recognized.

Family of Cu Alloys

Alloy	Alloying element
Brass	Zinc (Zn)
Phosphor bronze	Tin (Sn)
Aluminium bronzes	Aluminium (Al)
Silicon bronzes	Silicon (Si)
Copper nickel, nickel silvers	Nickel (Ni)

COPPER ALLOYS - BRASS

- Brass is the most common alloy of Cu – It's an alloy with Zn
- Brass has higher ductility than copper orzinc.
- Easy to cast - Relatively low melting point and high fluidity
- Properties can be tailored by varying Zn content.
- Some of the common brasses are yellow, naval and cartridge.
- Brass is frequently used to make musical instruments (good ductility and acoustic properties).

BRONZE

- Copper alloys containing tin, lead, aluminum, silicon and nickel are classified as bronzes.
- Cu-Sn Bronze is one of the earliest alloy to be discovered as Cu ores invariably contain Sn.
- Stronger than brasses with good corrosion & tensile properties; can be cast, hot worked and cold worked.
- Wide range of applications: ancient Chinese cast artifacts, skateboard ball bearings, surgical and dental instruments.

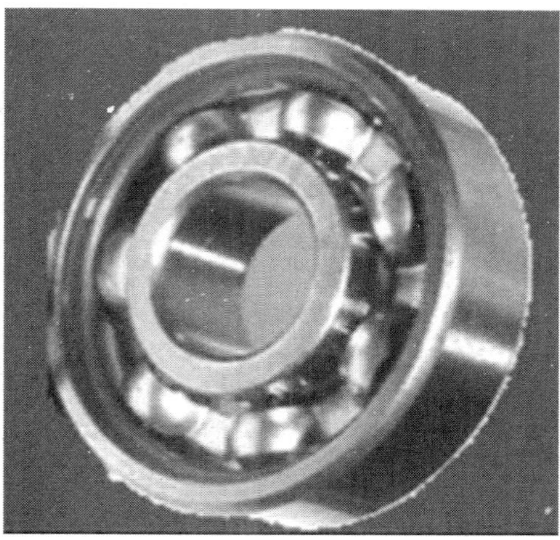

Bronze bearing

BERYLLIUM COPPER

- Cu-Be alloys are heat treatable. Max solubility of Be in Cu is 2.7% at 866 °C. Decreasing solubility at lower temp. imparts precipitation hardening ability.
- Cast alloys - higher Be. Wrought alloys – lower Be and some Co
- Cu-Be is ductile, weldable and machinable. Also resistant to non-oxidizing acids (HCl or H_2CO_3), abrasive wear and galling.
- Thermal conductivity is between steels and aluminum.

Applications

- Used in springs, load cells and other parts subjected to repeated loading. Low-current contacts for batteries and electrical connectors. Cast alloys are used in injection molds. Other applications include jet aircraft landing gear bearings & bushings and percussion instruments.

Compositions, Properties and Application of some Cu Alloys

Name	composition (wt.%)	Condition	YS (MPa)	UTS (MPa)	%El	Applications
Eletrolytic Copper	0.04 O	Annealed	69	220	45	Electrical wires, roofing, nails, rivets
Cartridge brass	30.0 Zn	Cold-	435	525	8	Automotive radiator core, lamp fixture, ammunition.
Phosphor bronze	5.0 Sn, 0.25 P	Annealed	130	325	64	Bellows, clutch disk, diaphragm, fuse clips, springs
Yellow bras (Leaded)	29 Zn, 3.0 Pb, 1 Sn	As cast	83	234	35	Furniture, radiator fittings, battery clamps, light fixtures
Al bronze	11 Al, 4 Fe	As cast	241	586	18	Bearings, bushings, valve seats and guards
Beryllium copper	1.9 Be 0.2 Be	Precipitate hardened	965	1140	10	Electrical, valves, pumps
Cu - Ni	30 Ni	Annealed	125	380	36	Condenser, heat-exchanger, piping, valves
Tin bronze	10 Sn, 2 Zn	As cast	152	310	25	Bearings, bushing,piston rings, gears

ALUMINUM

- Aluminum is a light metal (density=2.7 gm/cc); is easily machinable; has wide variety of surface finishes; good electrical & thermal conductivities; highly reflective to heat & light.
- Versatile metal - can be cast, rolled, stamped, drawn, spun, roll-formed, hammered, extruded and forged into many shapes.
- Aluminum can be riveted, welded, brazed, or resin bonded.
- Corrosion resistant - no protective coating needed, however it is often anodized to improve surface finish, appearance.
- Al and its alloys - high strength-to-weight ratio (high specific strength) owing to low density.
- Such materials are widely used in aerospace & automotive applications where weight savings are needed for better fuel efficiency & performance.
- Al-Li alloys are lightest among all Al alloys and find wide applications in the aerospace industry.

ALUMINUM ALLOYS

- Aluminum alloys are classified into two categories – Cast and Wrought alloys.
- Wrought alloys can be either heat-treatable or non-heat treatable.
- Alloys are designated by a 4 digit number. Wrought – the 1st digit indicates the major alloying element. Cast – The last digit after the decimal indicates product from(casting - 0 or ingot -1).

Wrought

Alloy Series	Principal Alloying Element
1xxx	Minimum 99.00% Aluminum
2xxx	Copper
3xxx	Manganese
4xxx	Silicon
5xxx	Magnesium
6xxx	Magnesium and Silicon
7xxx	Zinc
8xxx	other Elements

As Cast

Alloy Series	Principal Alloying Element
1xx.x	Aluminum, 99.00% or greater
2xx.x	Copper
3xx.x	Silicon with Copper and/or Magnesium
4xx.x	Silicon
5xx.x	Magnesium
6xx.x	Unused Series
7xx.x	Zinc
8xx.x	Tin
9xx.x	Other Elements

Compositions, Properties and Application of some Al Alloys

Composition (wt.%)	Condition	YS (MPa)	UTS (MPa)	%El	Applications
0.12 Cu	Annealed (O)	35	90	45	Food/chemical handling equipment, heat exchangers light reflectors
0.12 Cu,1.2 Mn, 0.1 Zn	Annealed	40	110	30	Utensils, pressure vessels and piping
2.5 Mg, 0.25 Cr	Strain-hardn. (H32)	195	230	14	Bellows, clutch disk, diaphragm, fuse clips, springs
4.4 Cu, 1.5 Mg, 0.6Mn	Heat treated (T4)	325	470	20	Aircraft structure, rivets, truck wheels, screw
1 Mg, 0.6 Si, 0.3 Cu, 0.2 Cr	T4	145	240	22	Trucks, canoes, railroad cars, furniture, pipelines
5.6 Zn, 2.5 Mg, 1.6 Cu, 0.23 Cr	Peak-aged (T6)	505	570	11	Aircraft structures and other highly loaded applications
7 Si, 0.3 Mg	T6	164	228	4	Aircraft pump parts, automotive transmission cases, cylinder blocks
2.0 Li. 1.3 Cu, 0.95 Mg, 0.12 Zr	Heat treated cold-worked (T651	360	465	-	Damage tolerant aircraft structures

TITANIUM

- Pure titanium melts at 1670 °C and has a low density of 4.51 g/cc (40% lighter than steel and 60% heavier than aluminum).
- Titanium has high affinity to oxygen – strong deoxidiser. Can catch fire and cause severe damage

- Ti is stronger than Al - high strength and low weight makes titanium very useful as a structural metal.
- Excellent corrosion resistance due to a presence of a protective thin oxide surface film. Can be used as biomaterial.
- Can be used in elevated temperature components.
- Limitation of pure titanium is its lower strength. Alloying is done to improve strength.
- Oxygen, nitrogen, and hydrogen can cause titanium to become more brittle. Care should be taken during processing.
- Titanium can also be cast using a vacuum furnace.
- Because of its high strength to weight ratio and excellent corrosion resistance, titanium is used in a variety of applications:
- Aircraft – Body structure, Engine parts
- Sporting equipment, chemical processing, desalination, turbine engine parts, valve and pump parts, marine hardware
- Medical implants - prosthetic devices.
- Recently use of Ti in bikes and automotives is increasing

TITANIUM ALLOYS

- Pure Ti exhibits two phases – Hexagonal α-phase at room temperature and BCC β phase above 882 °C.
- Strength of Titanium is improved by alloying. Alloying elements are either α or β stabilizer.
- Elements with electron/atoms ratio < 4 – β stabilizer (Al, O, Ga), = 4 – neutral (Sn, Zr) and > 4 – β stabilizer (V, Mo,Ta,W).
- (α + β) two-phase alloys can be obtained with right proportions of alloying elements.
- α alloys have low density, moderate strength, reasonable ductility and good creep resistance.
- Metastable β alloys are heavier, stronger and less ductile than α alloys. Creep strength reduces with increasing β content
- (α + β) alloys show a good strength-ductility combination

Compositions, Properties and Application of some Ti Alloys

Alloy type	Comp . wt.% (UNS No)	Condition	YS (MPa)	UTS (MPa)	%El	Applications
CP Ti	99.1Ti	Annealed	414	484	45	Airframe skins, marine and chemical processing equipments

α	Ti-5Al-2.5Sn	Annealed	784	826	16	Gas turbine engine casing and rings, chemical processing equipment
Near-α	Ti-8Al-1Mo-1V	Annealed	890	950	15	Forged jet engine components – compressor disc, plate, hubs
α – β	Ti-6Al-4V	Annealed	877	947	14	Prostheticim plants, airframe components
α – β	Ti-6Al-6V - 2Sn	Annealed	985	1050	14	Rocket engine case, airframe structure
β	Ti-10V -2Fe-3Al	Heat treated (aging)	1150	1223	10	High-strength airframe components, parts requiring uniform tensile stresses

NICKEL

- Nickel is a high-density, high-strength metal with good ductility and excellent corrosion resistance & high temperature properties.
- Ni has may unique properties including its excellent catalytic property. Nickel Catalyst for Fuel Cells: Nickel- cobalt is seen as a low-cost substitute for platinum catalysts.
- Two-thirds of all nickel produced goes into stainless steel production. Also used extensively in electroplating various parts in variety of applications.
- Ni-base super alloys are a unique class of materials having exceptionally good high temperature strength,creep and oxidation resistance. Used in many high temperature applications like turbine engines.
- Shape Memory Alloys: Ni base (Ni-Ti) and Ni containing (Cu-Al-Ni) shape memory alloys that can go back to original form, are an important class of engineering materials finding widespread use in many applications.
- Nickel-containing materials are used in buildings & infrastructure, chemical production, communications, energy (batteries: Ni-Cd, Ni-metal hydrides), environmental protection, food preparation, water treatment and transportation.

Applications of Nickel

Turbine engine

Electroplating

Batteries

MAGNESIUM

- Magnesium - Lightest among commonly used metals (ρ 1.7 = g/cm^3). Melting point is 650 °C and it has HCP structure.
- Is very reactive and readily combustible in air. Can be used as igniter or fire starter.
- Thermal conductivity is less than Al while their CTE is almost same.
- Pure Mg has adequate atmospheric resistance and moderate strength.
- Properties of Mg can be improved substantially by alloying.
- Can be alloyed with many elements. Most widely used alloying elements are Al, Zn, Mn and Zr.
- Mg Alloys – Cast, Wrought
- Wrought alloys are available in rod, bar, sheet, plate, forgings and extrusions.

MAGNESIUM ALLOYS

- Mg alloys: Impact and dent resistant, have good damping capacity - effective for high-speed applications.
- Due to its light weight, superior machinability and ease of casting, Mg & its alloys are used in many applications:– Auto parts, sporting goods, power tools, aerospace equipment, fixtures, electronic gadgets, and material handling equipment.
- Automotive applications include gearboxes, valve covers, alloy wheels, clutch housings, and brake pedal brackets.

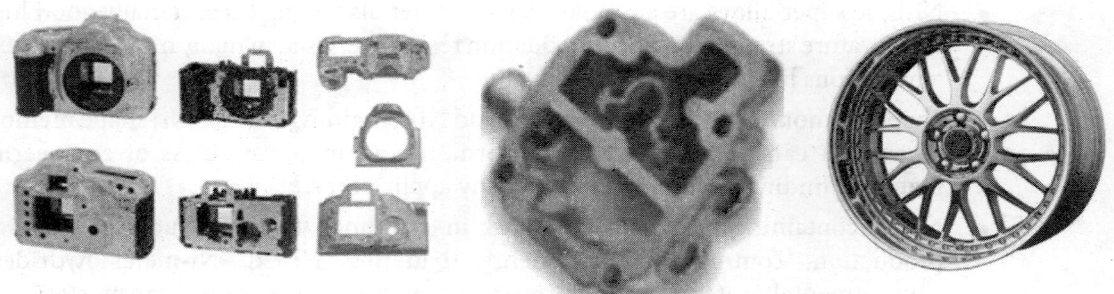

Compositions, Properties and Application of some Mg Alloys

ASTM No	composition (wt.%)	Condition	YS (MPa)	UTS (MPa)	%El	Applications
WroughtAlloys						
AZ31B	3.0 Al, 1.0Zn, 0.2 Mn	Extruded	200	262	15	Structure and tubing, cathodic protection
HK31A	3.0Th, 0.6Zr	Strain hard. Annealed	200	255	9	High temp applications (high strength to 315 °C)

ZK60A	5.5 Zn, 0.45Zr	Aged	285	350	11	Forging of max strength for aircrafts
CastAlloys						
AZ91D	9.0Al, 0.15 Mn, 0.7 Zn	As cast	150	230	3	Die-cast parts for automobile, luggage and electronic devices
AM60A	9.0Al, 0.13Mn,	As cast	130	220	6	Automotive wheels
AS41A	4.3Al, 1.0 Si, 0.35Mn	As cast	140	210	6	Die-cast parts requiring good creep strength

Ques.617: Classify Ceramics Materials.Describe features of various ceramic materials.

Answer:

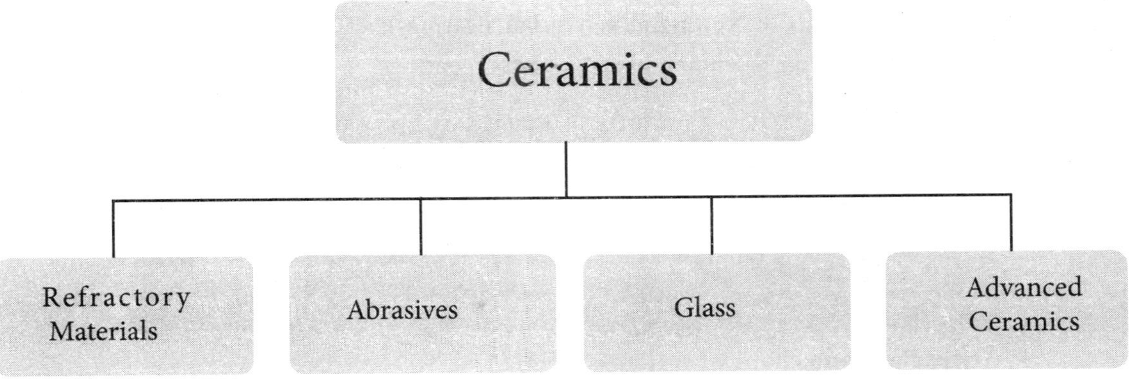

Refractory Materials

- Refractory - retains its strength at high temperatures > 500°C.
- Must be chemically & physically stable at high temperatures. Need to be resistant to thermal shock, should be chemically inert, and have specific ranges of thermal conductivity and thermal expansion.
- Are used in linings for furnaces, kilns, incinerators, crucibles and reactors.
- Aluminium oxide (alumina), silicon oxide (silica), calcium oxide (lime), magnesium oxide (magnesia) and fireclays are used to manufacture refractory materials.
- Zirconia - extremely high temperatures.
- SiC and Carbon – also used in some very severe temperature conditions, but cannot be used in oxygen environment, as they will oxidize and burn.

Composition of some common refractory materials

	Composition (wt%)							
Refractory	**Al_2O_3**	**SiO_2**	**MgO**	**Cr_2O_3**	**Fe_2O_3**	**CaO**	**TiO_2**	**% Porosity**

Fireclay	25-45	50-70	0-1		0-1	0-1	1-2	10-25
High-alumina fireclay	90-50	10-45	0-1		0-1	0-1	1-4	18-25
Silica Periclase	0.2 1.0	96.3 3.0	0.6 90	0.3	3.0	2.2 2.5		25 22
Periclase -	9.0	5.0	73	8.2	2.0	2.2		21

Advanced Ceramics: Automobile Engine parts

Advantages:

Operate at high temperatures – high efficiencies; Low frictional losses; Operate without a cooling system; Lower weights than current engines

Disadvantages:

Ceramic materials are brittle; Difficult to remove internal voids (that weaken structures); Ceramic parts are difficult to form and machine

Potential materials: Si_3N_4 (engine valves, ball bearings), SiC (MESFETS), &ZrO_2(sensors),

Possible engine parts: engine block & piston coatings

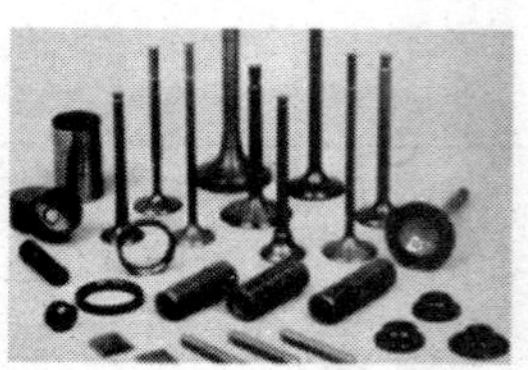

Microelectromechanical systems (MEMS)

- MEMS – These micron-sized structures such as beams, cantilevers, diaphragms, valves, plates and switches that can function as tiny sensors and actuators.
- Fabricated by integrated circuit (IC) manufacturing processes: bulk and surface micromachining.
- Thousands of micromachines can be fabricated on a single silicon wafer with supporting circuits integrated on the chip. Can be mass- produced in the millions at low prices.
- Low-cost, commercial MEMS devices developed for: Corrosion detectors & monitors; Instrumentation for automotive & aerospace; Biological & medical devices; Chemical 7 environmental sensors; Manufacturing & process control devices ;Virtual reality systems

Abrasive Ceramics

- Abrasives are used in cutting and grinding tools.
- Diamonds - natural and synthetic, are used as abrasives, though relatively expensive. Industrial diamonds are hard and thermally conductive. Diamonds unsuitable as gemstone are used as industrial diamond

- Common abrasives – SiC, WC, Al_2O_3 (corundum) and silica sand.
- Either bonded to a grinding wheel or made into a powder and used with a cloth or paper.

Glass

- **Glass** - inorganic, non-crystalline (amorphous)material.
- **Range** - soda-lime silicate glass for soda bottles to the extremely high purity silica glass for optical fibers.
- Widely used for windows, bottles, glasses for drinking, transfer piping and receptacles for highly corrosive liquids, optical glasses, windows for nuclear applications.
- The main constituent of glass is silica(SiO_2). The most common form of silica used in glass is sand.
- Sand fusion temperature to produce glass -1700°C. Adding other chemicals to sand can considerably reduce the fusion temperature.
- Sodium carbonate (Na_2CO_3) or soda ash, (75% SiO_2 + 25% Na_2O) will reduce the fusion temperature to 800°C.
- Other chemicals like Calcia (CaO) and magnesia (MgO) are used for stability. Limestone ($CaCO_3$) and dolomite ($MgCO_3$) are used for this purpose as source of CaO andMgO.

Key Properties of Glass

- Glass-ceramic materials should have:
  - Relatively high mechanical strengths
  - Low coefficients of thermal expansion
  - Relatively high temperature capabilities
  - Good dielectric properties
  - Good biological compatibility
  - Thermal shock resistance

Compositions and Characteristics of some common Glasses

Composition (wt%)							
Glass type	SiO_2	Na_2O	CaO	Al_2O_3	B_2O_3	Other	Remarks
Fused Silica	>99.5						High Mp, low CTE, thermal shock resistant
96% Silica	96				4		Thermal shock and chemically resistant laboratory ware
Borosilicate	81	3.5		2.5	13		Thermal shock and chemically resistant Oven ware

Container	74	16	5	1		4Mgo	Low Mp, formable, and durable
Fiberglass	55		16	15	10	4Mgo	Ease of drawing used in FRP
Optical flint	54	1				37Pbo, $8K_2O$	High density, high refractive index optical lenses
Glass ceramic	43.5	14		30	5.5	$6.5Tio_2,0.5As_2O_3$	Strong, thermal shock resistant Oven ware

Ques.618: What are Polymers ?Describe the characteristics & applications of important Thermoplastics,Thermosetting polymers & elastomers.

Answer:

- **Polymers** – Chain of H-C molecules. Each repeat unit of H-C is a monomer e.g. ethylene (C_2H_4), Polyethylene – (–CH2 – CH2)n
- **Polymers:** Thermosets – Soften when heated and harden on cooling – totally reversible. Thermoplasts – Do not soften on heating
- **Plastics** – moldable into many shape and have sufficient structural rigidity. Are one of the most commonly used classof materials.
- Are used in clothing, housing, automobiles, aircraft, packaging, electronics, signs, recreation items, and medical implants.
- **Natural plastics** – hellac, rubber, asphalt, and cellulose.

Characteristics and Applications of some common Thermoplastics

Material	Characteristics	Applications
Polyethylene	Chemically resistant, tough,low friction coeff., low strength	Flexible bottles, toys, battery parts, ice trays, film wrapping materials
Polyamide (Nylon)	Good strength and toughness, abrasion resistant, liquid absorber, low friction coeff.	Bearings, gears, cams, bushings and jacketing for wires and cables
Fluorocarbon (Teflon)	Chemically inert, excellent electrical properties, relatively weak	Anticorrosive seals, chemical pipes and valves, bearings, anti-adhesive coatings, high temp electronic parts
Polyester (PET)	Tough plastic film, excellent fatigue and tear strength, corrosion resistant	Recording tapes, clothing, automotive tyre cords, beverage containers
Vinyl	Low-cost general purpose material, rigid, can be made flexible	Floor coverings, pipe, electrical al wire insulation, garden hose, phonograph records
Polystyrene	Excellent electrical properties and optical clarity, good thermal and dimensional stability	Wall tile, battery cases, toys, lighting panels, housing appliances

Characteristics and Applications of some common Thermosetting Polymers

Material	Characteristics	Applications
Epoxy (Araldite)	Excellent mechanical properties and corrosion resistance, good electrical prop.. good adhesion and dimensional stability	Electrical moldings, sinks, adhesives, protective coatings, fiber reinforced plastic (FRP), laminates
Phenolic (Bakelite)	Excellent thermal stability (>150°C), inexpensive, can be compounded with many resins	Motor housings, telephones, auto distributors, electrical fixtures
Polyester (Aropol)	Excellent electrical properties, low cost, can formulated for room or high temperature, often fiber reinforced	Helmets, fiberglass boats, auto body components, chair fans

Elastomers

- Elastomer – a polymer with rubber-like elasticity.
- Each of the monomers that link to form the polymer is usually made of carbon, hydrogen, oxygen and/or silicon.
- Cross-linking in the monomers provides the flexibility.
- Glass transition temperature, Tg, is the temperature at which transition from rubbery to rigid state takes place in polymers.
- Elastomers are amorphous polymers existing above their Tg. Hence, considerable segmental motion exists in them.
- Their primary uses are in seals, adhesives and molded flexible parts.

Characteristics and Applications of some commercial Elastomers

Material	Characteristics	Applications
Natural rubber (NR)	Useful temp. range : - 60 – 120°C, good resistance to cutting abrasion, resistant to oil, ozone, elongation 500-700%	Pneumatic tyres and tubes, heels and soles, gaskets
Styrene- butadiene copolymer (SBR))	Temp. range: - 60 – 120°C. Good physical properties, elongation 450 – 500%	Same as natural rubber

Acrylonitrile-butadiene copolymer (NBR)	Temp. range: - 50 – 150°C. Excellent resistance to oils, elongation 400 – 600%	Gasoline, chemicals and oil hose, seals and O-rings, soles and heels
Chloroprene (CR)	Temp. range: - 50 – 105°C. Excellent resistance to high and low temp. excellent electrical prop. elongation 100 – 800%	High and low temp. insulation, seals, diaphragm, tubing for food and medical uses

Liquid Crystal Polymers (LCP)

- LCPs are a group of chemically complex structure having unique properties. Primarily used in LCDs (liquid crystal displays) on watches, flat panel computer monitors, televisions and clocks.
- Advantages - LCDs are thinner & lighter and consume much less power than cathode ray tubes (CRTs).
- The name "liquid crystal" arises out of their characteristics. It takes a fair amount of heat to change a suitable substance from a solid into a liquid crystal, and it only takes a little more heat to turn the liquid crystal into a real liquid.
- LCDs use these liquid crystals because they react predictably to electric current in such a way as to control light passage.
- A particular sort of nematic liquid crystal, called twisted nematics (TN), is naturally twisted. Applying an electric current to these liquid crystals will untwist them to varying degrees, depending on the applied electrical potential.

Advanced Polymers

Ultra high Molecular Weight Polyethylene (UHMWPE)

- Molecular weight ca. 4 x 10^6g/mol
- Outstanding properties:
  - ❑ high impact strength
  - ❑ resistance to wear/abrasion
  - ❑ low coefficient off riction
  - ❑ self-lubricating surface
- Important applications
  - ❑ bullet-proof vests
  - ❑ golf ball covers
  - ❑ hip implants (acetabularcup)

Ques.619: Discuss the concepts of stress & strain in materials.

Answer: To compare specimens of different sizes, the load is calculated per unit area.

Engineering stress: $\sigma = F/Ao$

F is load applied perpendicular to specimen cross- section; A0 is cross-sectional area (perpendicular to the force) before application of the load.

Engineering strain: $\varepsilon = \otimes l / lo \ (\times 100\ \%)$

⊗l is change in length, lo is the original length.

These definitions of stress and strain allow one to compare test results for specimens of different cross- sectional area A_0 and of different length l_0.

Stress and strain are positive for tensile loads, negative for compressive loads.

(shear and torsion)

Shear stress: $\tau = F / Ao$

F is load applied parallel to the upper and lower faces each of which has an area A_0.

Shear strain: $\gamma = t_g X$ ($\times$ 100%)

θ is strain angle

Torsion is variation of pure shear. The shear stress in this case is a function of applied torque T, shear strain is related to the angle of twist, ϕ .

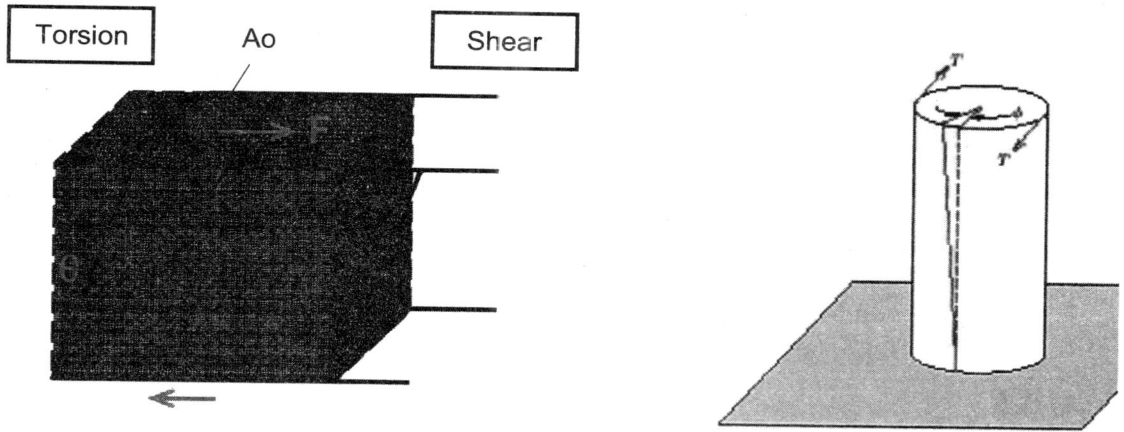

Elastic deformation

Reversible: when the stress is removed, the material returns to the dimensions it had before the loading.

Usually strains are small (except for the case of some plastics, e.g. rubber).

Plastic deformation

Irreversible: when the stress is removed, the material does not return to its original dimensions.

In tensile tests, if the deformation is elastic, the stress- strain relationship is called Hooke's law:

$$\sigma = E\varepsilon$$

E is the Young's modulus of elasticity and E is has the same units as σ , N/m^2 or Pa

In some materials (many polymers, concrete...), elastic deformation is not linear, but it is still reversible.

So far we have assumed that elastic deformation is time independent (i.e. applied stress produces instantaneous elastic strain)

However, in reality elastic deformation takes time (finite rate of atomic/molecular deformation processes) - continues after initial loading, and after load release. This time dependent elastic behavior is known as anelasticity.

The effect is normally small for metals but can be significant for polymers ("visco-elastic behavior").

Materials subjected to tension shrink laterally. Those subjected to compression, bulge. The ratio of lateral and axial strains is called the Poisson's ratio υ . Lateral strain is in opposite sense to longitudinal strain.

υ is dimensionless.

Relationship of shear stress to shear strain:

$\tau = G\ \gamma$, where: $\gamma = \text{tg}\ \theta = \otimes y / zo$

G is Shear Modulus (Units: N/m^2 or Pa)

For isotropic material: $E = 2G(1+\upsilon) \rightarrow G \sim 0.4E$

(Note: single crystals are usually elastically anisotropic: the elastic behavior varies with crystallographic direction,)

Plastic deformation:

- stress and strain are not proportional to each other
- the deformation is not reversible
- deformation occurs by breaking and re-arrangement of atomic bonds (in crystalline materials primarily by motion of dislocations,)

The yield stress is a measure of resistance to plastic deformation

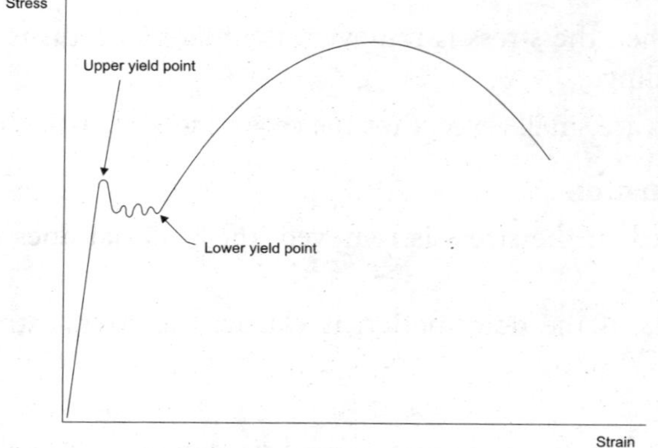

Figure. *For a low-carbon steel, the stress-versus-strain curve includes both an upper and lower yield point.*

In some materials (e.g. low-carbon steel), the stress vs. strain curve includes two yield points (upper and lower). The yield strength is defined in this case as the average stress at the lower yield point.

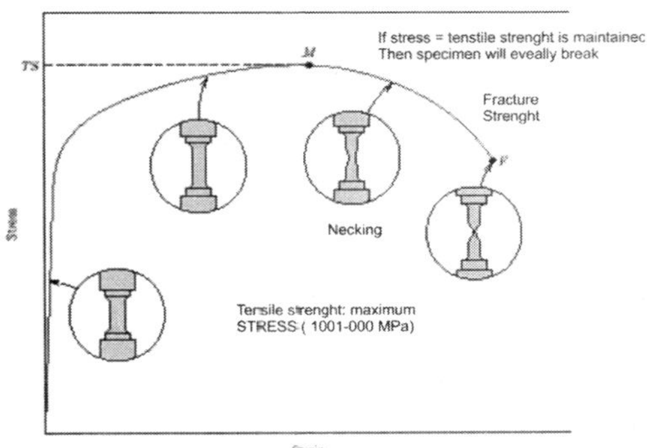

For structural applications, the yield stress is usually a more important property than the tensile strength, since once the yield stress has passed, the structure has deformed beyond acceptable limits.

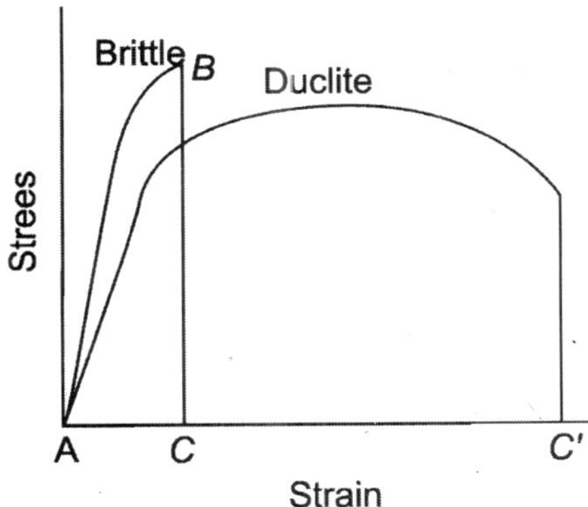

Ductility is a measure of the deformation at fracture defined by percent elongation

Ques.620: Describe Mohs Hardness Scale.

Answer: The Mohs hardness scale for minerals has been used since 1822. It simply consists of 10 minerals arranged in order from 1 to 10. Diamond is rated as the hardest and is indexed as 10; talc as the softest with index number 1. Each mineral in the scale will scratch all those below it as follows:

Diamond	10
Corundum	9
Topaz	8
Quartz	7
Orthoclase (Feldspar)	6

Apatite	5
Fluorite	4
Calcite	3
Gypsum	2
Talc	1

Ques.621: Describe various hardness tests.

Answer: The Brinell Hardness Test

The Brinell hardness test method consists of indenting the test material with a 10 mm diameter hardened steel or carbide ball subjected to a load of 3000 kg. For softer materials the load can be reduced to 1500 kg or 500 kg to avoid excessive indentation. The full load is normally applied for 10 to 15 seconds in the case of iron & steel and for at least 30 seconds in the case of other metals. The diameter of the indentation left in the test material is measured with a low powered microscope. The Brinell harness number is calculated by dividing the load applied by the surface area of the indentation.

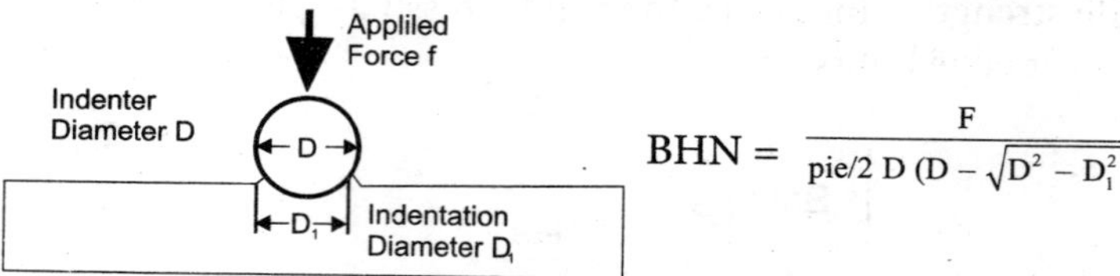

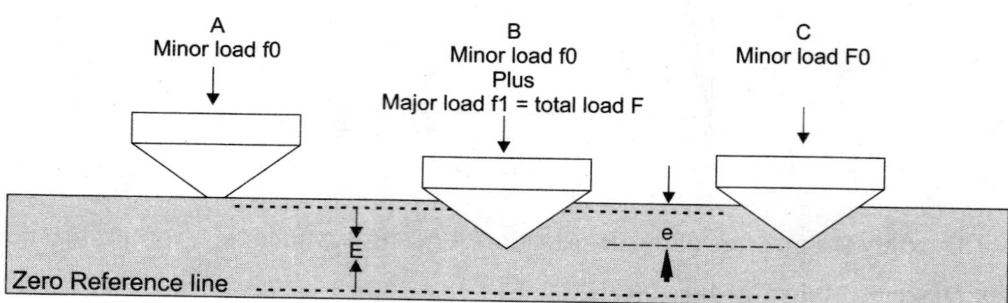

Rockwell Hardness Test

The Rockwell hardness test method consists of indenting the test material with a diamond cone or hardened steel ball indenter. The indenter is forced into the test material under a preliminary minor load F(Fig. 1A) usually 10 kgf. When equilibrium has been reached, an indicating device, which follows the movements of the indenter and so responds to changes in depth of penetration of the indenter is set to a datum position. While the preliminary minor load is still applied an additional major load is applied with resulting increase in penetration. When equilibrium has again been reached, the additional major load is removed but the preliminary minor load is still maintained. Removal of the additional major load allows a partial recovery, so reducing the depth of penetration. The permanent increase in depth of penetration, resulting from the application and removal of the additional major load is used to calculate the Rockwell hardness number. Advantages of the Rockwell hardness method include the direct Rockwell hardness number readout and rapid testing time.

Vickers Hardness Test

The Vickers hardness test method consists of indenting the test material with a diamond indenter, in the form of a right pyramid with a square base and an angle of 136 degrees between opposite faces subjected to a load of 1 to 100 kgf. The full load is normally applied for 10 to 15 seconds. The two diagonals of the indentation left in the surface of the material after removal of the load are measured using a microscope and their average calculated. The area of the sloping surface of the indentation is calculated. The Vickers hardness is the quotient obtained by dividing the kgf load by the square mm area of indentation.

When the mean diagonal of the indentation has been determined the Vickers hardness may be calculated from the formula, but is more convenient to use conversion tables. The Vickers hardness should be reported like 800 HV/10, which means a Vickers hardness of 800, was obtained using a 10 kgf force. Several different loading settings give practically identical hardness numbers on uniform material, which is much better than the arbitrary changing of scale with the other hardness testing methods. The advantages of the Vickers hardness test are that extremely accurate readings can be taken, and just one type of indenter is used for all types of metals and surface treatments. Although thoroughly adaptable and very precise for testing the softest & hardest of materials, under varying loads, the Vickers machine is a floor standing unit that is more expensive than the Brinell or Rockwell machines.

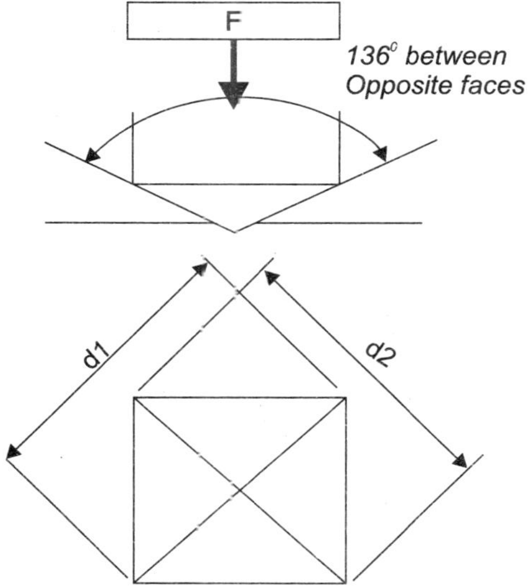

F = Load in kgf

d =Arithmetic mean of the two diagonals, d_1 and d_2 in mm, HV = Vickers hardness

$$HV = \frac{2F\sin\frac{136^\circ}{2}}{d^2} \quad HV = 1.854\,\frac{F}{d^2} \text{ approximately}$$

Ques.622: Describe various Micro-hardness Tests .

Answer: Knoop Hardness Test

The term microhardness test usually refers to static indentations made with loads not exceeding 1 kgf. The indenter is either the Vickers diamond pyramid or the Knoop elongated diamond pyramid. The procedure for testing is very similar to that of the standard Vickers hardness test, except that it is done on a microscopic scale with higher precision instruments. The surface being tested generally requires a metallographic finish; the smaller the load used, the higher the surface finish required. Precision microscopes are used to measure the indentations; these usually have a magnification of around X500 and measure to an accuracy of +0.5 micrometres. Also with the same observer differences of +0.2 micrometres can usually be resolved. It should, however, be added that considerable care and experience are necessary to obtain this accuracy.

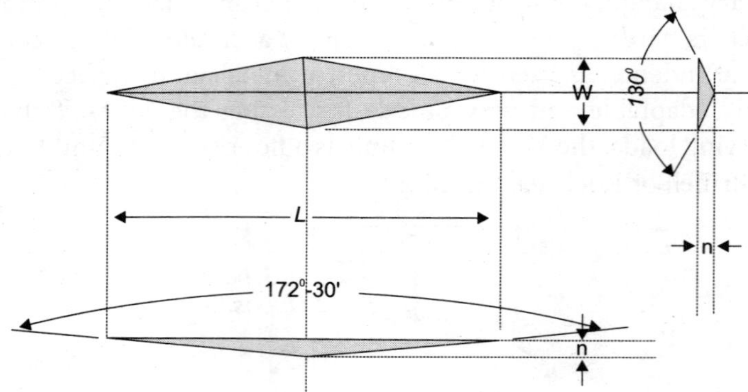

The Knoop hardness number KHN is the ratio of the load applied to the indenter, P (kgf) to the unrecovered projected area A (mm^2)

$$KHN = F/A = P/CL^2$$

Where:

F = applied load in kgf

A = the unrecovered projected area of the indentation in mm2

L = measured length of long diagonal of indentation in mm

C = 0.07028 = Constant of indenter relating projected area of the indentation to the square of the length of the long diagonal.

The Scleroscope Hardness Test

The Scleroscope test consists of dropping a diamond tipped hammer, which falls inside a glass tube under the force of its own weight from a fixed height, onto the test specimen. The height of the rebound travel of the hammer is measured on a graduated scale. The scale of the rebound is arbitrarily chosen and consists on Shore units, divided into 100 parts, which represent the average rebound from pure hardened high-carbon steel. The scale is continued higher than 100 to include metals having greater hardness. In normal use the shore scleroscope test does not mark the material under test. The Shore Scleroscope measures hardness in terms of the elasticity of the material and the hardness number depends on the height to which the hammer rebounds, the harder the material, the higher the rebound. Advantages of this method are portability and non marking of the test surface.

The Durometer

The Durometer is a popular instrument for measuring the indentation hardness of rubber and rubber-like materials. The most popular testers are the Model A used for measuring softer materials and the Model D for harder materials.The operation of the tester is quite simple. The material is subjected to a definite pressure applied by a calibrated spring to an indenter that is either a cone or sphere and an indicating device measures the depth of indentation.

Ques.623: Describe various types of material failures/fracture .Discuss their mechanisms and features.

Answer: Fracture is a form of failure where the material separates in pieces due to stress, at temperatures below the melting point. The fracture is termed ductile or brittle depending on whether the elongation is large or small. Steps in fracture (response to stress):

- Crack formation
- Crack propagation

Ductile Fracture

Stages of ductile fracture:

- Initial necking
- Small cavity formation (microvoids)
- Void growth (ellipsoid) by coalescence into a crack
- Fast crack propagation around neck. Shear strain at45^{o}
- Final shear fracture (cup and cone)

The interior surface is fibrous, irregular, which signify plastic deformation.

Brittle Fracture

There is no appreciable deformation, and crack propagation is very fast. In most brittle materials, crack propagation (by bond breaking) is along specific crystallographic planes (cleavage planes). This type of fracture is transgranular (through grains) producing grainy texture (or faceted texture) when cleavage direction changes from grain to grain. In some materials, fracture is intergranular.

Fracture occurs due to stress concentration at flaws, like surface scratches, voids,

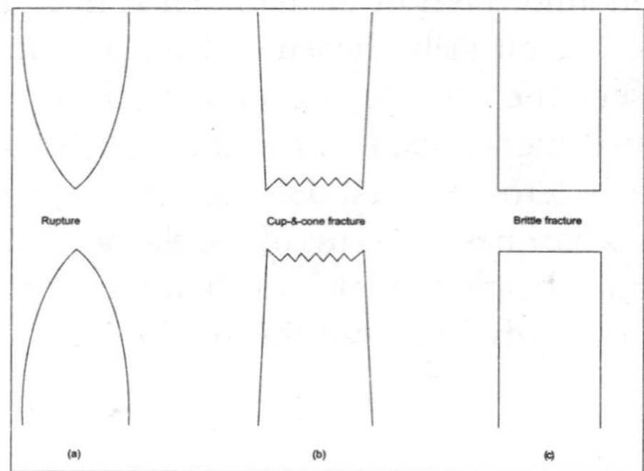

Impact Fracture:

Impact fractures can best be described as a flute or strip of material that was cleanly sheared from a projectile point. The most common type of impact fracture starts at the tip of a point and runs down one blade edge possibly reaching the shoulder of a point. Some points were reworked into a useable point after having been damaged by an impact fracture. Normalized tests, like the Charpoy and Izod tests measure the impact energy required to fracture a notched specimen with a hammer mounted on a pendulum. The energy is measured by the change in potential energy (height) of the pendulum. This energy is called notch toughness.

Ductile brittle transition

Ductile to brittle transition occurs in materials when the temperature is dropped below a transition temperature. Alloying usually increases the ductile-brittle transition temperature, for ceramics, this type of transition occurs at much higher temperatures than for metals.

The notched-bar impact test can be used to determine whether or not a material experiences a ductile-to-brittle transition as the temperature is decreased. In such a transition, at higher temperatures the impact energy is relatively large since the fracture is ductile. As the temperature is lowered, the impact energy drops over a narrow temperature range as the fracture becomes more brittle.

The transition can also be observed from the fracture surfaces, which appear fibrous or dull for totally ductile fracture, and granular & shiny for totally brittle fracture. Over the ductile-to- brittle transition features of both types will exist.

While for pure materials the transition may occur very suddenly at a particular temperature, for many materials the transition occurs over a range of temperatures. This causes difficulties when trying to define a single transition temperature and no specific criterion has been established.

The ductile-brittle transition is exhibited in bcc metals, such as low carbon steel, which become brittle at low temperature or at very high strain rates. Fcc metals, however, generally remain ductile at low temperatures.

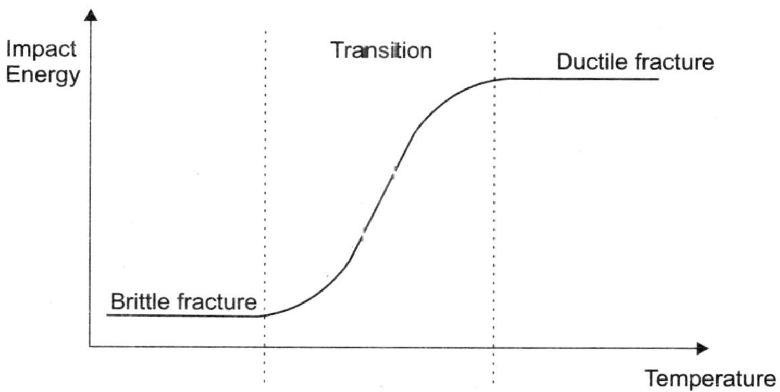

Fatigue failure

Fatigue is the catastrophic failure due to dynamic (fluctuating) stresses. It can happen in bridges, airplanes, machine components, etc. The characteristics are:

- long period of cyclic strain
- The most usual (90%) of metallic failures (happens also in ceramics and polymers)
- Is brittle-like even in ductile metals, with little plastic deformation
- It occurs in stages involving the initiation and propagation of cracks.

Cyclic Stresses

These are characterized by maximum, minimum and mean stress, the stress amplitude, and the stress ratio.

Crack Initiation and Propagation

Stages is fatigue failure:

I. crack initiation at high stress points (stress raisers)
II. propagation (incremental in each cycle)
III. final failure by fracture

Stage I - initiation

- slow
- along crystallographic planes of high shear stress
- flat and featureless fatigue surface

Stage II - propagation

Crack propagates by repetitive plastic blunting and sharpening of the crack tip.

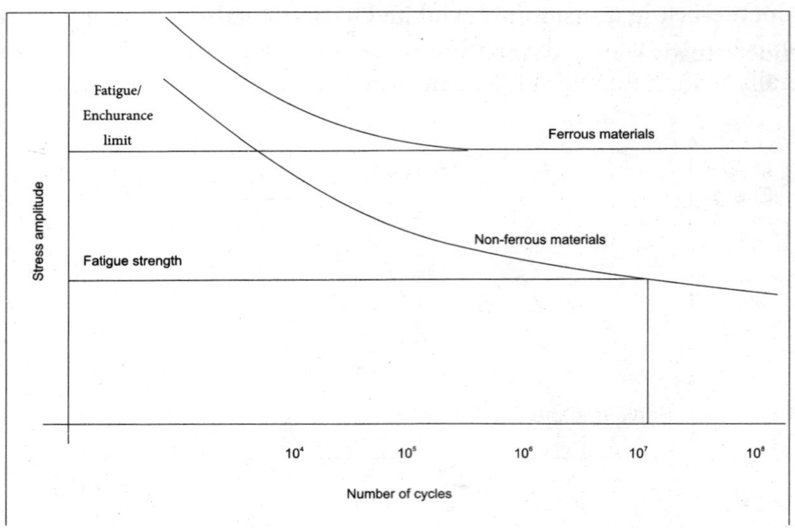

Creep failure

Creep is the time-varying plastic deformation of a material stressed at high temperatures. Examples: turbine blades, steam generators. Keys are the time dependence of the strain and the high temperature.

The Creep Curve

Creep in metals is defined as time dependent plastic deformation at constant stress (or load) and temperature. The form of a typical creep curve of strain versus time is in Figure. The slope of this curve is the creep rate $d\varepsilon/dt$.The curve may show the instantaneous elastic and plastic strain that occurs as the load is applied, followed by the plastic strain which occurs over time. Three stages to the creep curve may be identified:

Primary creep: in which the creep resistance increases with strain leading to a decreasing creep strain rate.

Secondary (Steady State) creep: in which there is a balance between work hardening and recovery processes, leading to a minimum constant creep rate.

Tertiary creep: in which there is an accelerating creep rate due to the accumulating damage, which leads to creep rupture, and which may only be seen at high temperatures & stresses and in constant load machines.

The minimum secondary creep rate is of most interest to design engineers, since failure avoidance is required and in this region some predictability is possible. In the USA two Standards are commonly used: *(i)* The stress to produce a creep rate of 0.0001% per hour (1% in 10,000 hours). *(ii)* The stress to produce a creep rate of 0.00001% per hour (1% in 100,000 hours or approximately 11.5 years). The first requirement would be typical of that for gas turbine blades, while the second for steam turbines. Constant load machines simulate real engineering situations more accurately, but as the specimen extends its cross section area reduces, leading to a rising stress. Machines designed to reduce the load to compensate for the reduced area and maintain constant stress may produce an extended steady state region.

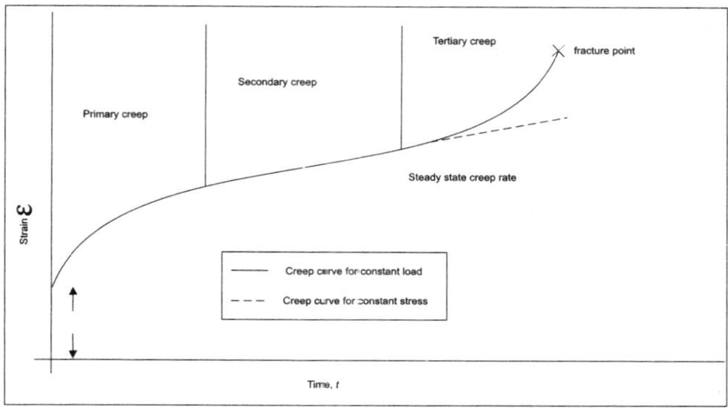

Stress and Temperature Effects

Both temperature and the level of the applied stress influence the creep characteristics. The results of creep rupture tests are most commonly presented as the logarithm of stress versus the logarithm of rupture lifetime. Creep becomes more pronounced at higher temperatures. There is essentially no creep at temperatures below 40% of the melting point. Creep increases at higher applied stresses.

Ques.624: Describe the concept of Dislocations and Strengthening Mechanisms in metals .

Answer : Dislocations can be edge dislocations, screw dislocations and exist in combination of the two. Their motion (slip) occurs by sequential bond breaking and bond reforming . The number of dislocations per unit volume is the dislocation density; in a plane they are measured per unit area.

Characteristics of Dislocations

There is strain around a dislocation which influences how they interact with other dislocations, impurities, etc. There is compression near the extra plane (higher atomic density) and tension following the dislocation line. Dislocations interact among themselves. When they are in the same plane, they repel if they have the same sign and annihilate if they have opposite signs (leaving behind a perfect crystal). In general, when dislocations are close and their strain fields add to a larger value, they repel, because being close increases the potential energy (it takes energy to strain a region of the material). The number of dislocations increases dramatically during plastic deformation. Dislocations spawn from existing dislocations, and from defects, grain boundaries and surface irregularities.

Plastic Deformation

Slip directions vary from crystal to crystal. When plastic deformation occurs in a grain, it will be constrained by its neighbors, which may be less favorably oriented. As a result, polycrystalline metals are stronger than single crystals (the exception is the perfect single crystal, as in whiskers.)

Mechanisms of Strengthening in Metals

Ability to deform plastically depends on ability of dislocations to move. Strengthening consists in hindering dislocation motion. We discuss the methods of grain-size reduction, solid-solution alloying and strain hardening. These are for single phase metals. We discuss others when treating alloys. Ordinarily, strengthening reduces ductility.

Strengthening by Grain Size Reduction

This is based on the fact that it is difficult for a dislocation to pass into another grain, especially if it is very misaligned. Atomic disorder at the boundary causes discontinuity in slip planes. For high-angle grain boundaries, stress at end of slip plane may trigger new dislocations in adjacent grains. Small angle grain boundaries are not effective in blocking dislocations.

The finer the grains, the larger the area of grain boundaries that impedes dislocation motion. Grain-size reduction usually improves toughness as well. Grain size can be controlled by the rate of solidification and by plastic deformation.

Solid-Solution Strengthening

Adding another element that goes into interstitial or substitutional positions in a solution increases strength. The impurity atoms cause lattice strain which can "anchor" dislocations. This occurs when the strain caused by the alloying element compensates that of the dislocation, thus achieving a state of low potential energy. It costs strain energy for the dislocation to move away from this state (which is like a potential well). The scarcity of energy at low temperatures is why slip is hindered. Pure metals are almost always softer than their alloys.

Strain Hardening

Phenomenon where ductile metals become stronger and harder when they are deformed plastically is called strain hardening or work hardening. Increasing temperature lowers the rate of strain hardening. Hence materials are strain hardened at low temperatures, thus also called cold working. During plastic deformation, dislocation density increases. And thus their interaction with each other resulting in increase in yield stress. Strain hardening (work hardening) is the reason for the elastic recovery. The reason for strain hardening is that the dislocation density increases with plastic deformation (cold work) due to multiplication. The average distance between dislocations then decreases and dislocations start blocking the motion of each one

Ques.625: What are various Non-Destructive testing (NDT) methods of materials ? Discuss their features and applicability .

Answer: NDT is the method of detection and measurement of properties or condition of materials, structures, machines without damaging or destroying their operational capabilities. Examples of NDT are: magnetic dust method, penetrating liquid method, ultrasonic test and radiography. All NDTs are used to detect various types of flaws on the surface of material or internal inclusions of impurities and these techniques are also very useful during preventive maintenance & repair. There are few techniques which do not require any special apparatus & are quite simple to handle and only a moderate skill being required. Some of the applications of NDTs are detecting: *(i)* surface cracks *(ii)* material composition *(iii)* internal inclusions *(iv)* internal voids & discontinuities and (v) condition of internal stresses.

Ultrasonic Test

High frequency ultrasonic (sound) waves are applied to the test piece by a Piezoelectric crystal. If the test piece is free from cracks, or flawless, then it reflects ultrasonic waves without distortion. If there are any flaws in the specimen, the time taken by the ultrasonic waves will be less as the reflection of these waves will be from flaw points and not from the bottom of the specimen. Cathode ray oscilloscope (CRO) is used to receive the sound signals, whose time base circuit is connected to it. Knowing the time interval between the transmission of the sound pulse and the reception of the echo signal, we can calculate the depth of the crack. This test is a very fast method of inspection and often used to test aerospace components and automobiles. This test is generally used to detect internal cracks like shrinkage cavities, hot tears, zones of corrosion & non-metallicin clusions.

Liquid-Penetration test

This test is employed for detection of small defects which are very small to detect with the naked eye. This test is used to detect surface cracks or flaws in non-ferrous metals. This test employs a visible colour contrast dye penetrant technique for the detection of open surface flaws in metallic and non-metallic objects. The penetrants are applied by spraying over the surface of material to be inspected. The excess penetrant is then washed or cleaned. Absorbent powder is then applied to absorb the penetrants in the cracks, voids which reveals the flaws. This test reveals flaws such as shrinkage cracks, porosity, fatigue cracks, grinding cracks, forging cracks, seams, heat treatment cracks and leaks etc., on castings, weldings, machined parts, cutting tools, pipes & tubes. If the fluorescent penetrant is used, the developed surface must be examined under ultra violet light to see the presence of defects. This technique is used for non-porous and non- absorbent materials. Care may be taken to clean the surface so that it is free from dust, scale, etc.to have better results. Penetrants are highly toxic & flammable and hence proper precautions should be taken both during use & storage of penetrants.

Ques.626: Describe various steps involved in Microstructural Examination of a material .

Answer : Microstructure is defined as the structure of a prepared surface or thin foil of material as revealed by a microscope above 25× magnification. The microstructure of a material can strongly influence physical properties such as strength, toughness, ductility, hardness, corrosion resistance, high/low temperature behavior, wear resistance, and so on, which in turn govern the application of these materials in industrial practice.

Various steps involved in Microstructural Examination of a material are described below.

(a) Sectioning and cutting

The areas of interest forming the metallography specimens need to be sectioned for ease of handling. Depending on the type of material, the sectioning operation can be done by using abrasive cutter (for metal and metallic composite), diamond wafer cutter (ceramics, electronics and minerals) or thin sectioning with a microtome (plastics). In order not to damage the specimen, proper cutting requires the correct selection of abrasive cutting wheel, proper cutting speed & cutting load and the use of coolant.

(b) Mounting

The mounting operation accomplishes three important functions:

1. To protects the specimen edge and maintain the integrity of materials surface features.
2. Fills voids in porous materials.
3. Improves handling of irregular shaped samples.

Samples for microstructure evaluation are typically encapsulated in a plastic mount for handling during sample preparation. Large sample or samples for macrostructure evaluation can be prepared without mounting.

The metallography specimen mounting is done by encapsulating the specimen into:

1. A compression/hot mounting compound (thermosets – e.g. phenolics, epoxies or thermoplastics – e.g.acrylics)
2. A castable resin/cold mounting (e.g. acrylics resins, epoxy resins and polyester resins)

(c) Grinding

Grinding is required to ensure the surface is flat & parallel and to reduce the damage created during sectioning. Grinding is accomplished by decreasing the abrasive grit size sequentially to obtain the required fine surface finish prior to polishing. It is important to note that the final appearance of the prepared surface is dependent on the machine parameters such as grinding/polishing pressure, relative velocity distribution and the direction of grinding/polishing.

(d) Polishing

For microstructure examination a mirror/reflective finish is needed whereas a finely ground finish is adequate for macrostructure evaluation. Polishing can be divided into two main steps:

Rough polishing

The purpose is to remove the damage produced during grinding. Proper polishing will maintain the specimen flatness & retain all inclusions or secondary phases by eliminating the previous damage and maintaining the specimen integrity.

Fine polishing

The purpose is to remove only surface damage.

(e) Etching

Etchants are specially formulated for the specific material and evaluation objectives. Etching alters the microstructural features based on composition; stress or crystal structure and it will develop the surface topology, which can be visible in the microscope.

Typically, chemical etching involve immersing the polished surface in the prepared chemical solution for a specified time (usually seconds) followed by rinsing the etched specimen under running tap water and drying.

(f) Microscopic Analysis

For microscopic analysis, a reflective surface is required. The analysis can be done by using a metallurgical microscope.

Grain size determination

The grain size is often determined when the properties of a polycrystalline material are under consideration. In this regard, there exist a number of techniques by which size is specified in terms of average grain volume, diameter, or area. Grain size may be estimated by using an intercept method, described as follows. Straight lines of the same length are drawn through several photo micrographs that show the grain structure. The grains intersected by each line segment are counted; the line length is then divided by an average of the number of grains intersected, taken over all the line segments. The average grain diameter is found by dividing this result by the linear magnification of the photo micrographs.

Grain size is measured with a microscope by counting the number of grains within a given area, by determining the number of grains that intersect a given length of random line, or by comparison with standard charts. A very common method of measuring grain size is to compare the grains at a fixed magnification with the American Society for Testing and Materials (ASTM) grain-size charts.

ASTM#	Grains IN^2 @ 100X	Grains/mm^2	Grains/mm^3	Avg. Grain Dia. (mm)
-3	0.06	1	0.7	1.00
-2	0.12	2	2	0.75
-1	0.25	4	5.6	0.50
0	0.5	8	16	0.35
1	1	16	45	0.25
2	2	32	128	0.18
3	4	64	360	0.125
4	8	128	1020	0.091
5	16	256	2990	0.062
6	32	512	8200	0.044
7	64	1024	23000	0.032
8	128	2048	65000	0.022
9	256	4096	185000	0.016
10	512	8200	520000	0.011
11	1024	16400	1200000	0.008
12	2048	32800	1500000	0.00096

Ques.627: Describe the features of The Iron–Carbon PhaseDiagram .

Answer : The Iron–Iron Carbide (Fe–Fe_3C) Phase Diagram

This is one of the most important alloys for structural applications. The diagram Fe—C is simplified at low carbon concentrations by assuming that it is the Fe—Fe_3C diagram. Concentrations are usually given in weight percent. The possible phases are:

- α-ferrite (BCC) Fe-C solution
- γ-austenite (FCC) Fe-C solution
- δ-ferrite (BCC) Fe-C solution
- liquid Fe-C solution
- Fe_3C (iron carbide) or cementite. An intermetallic compound.

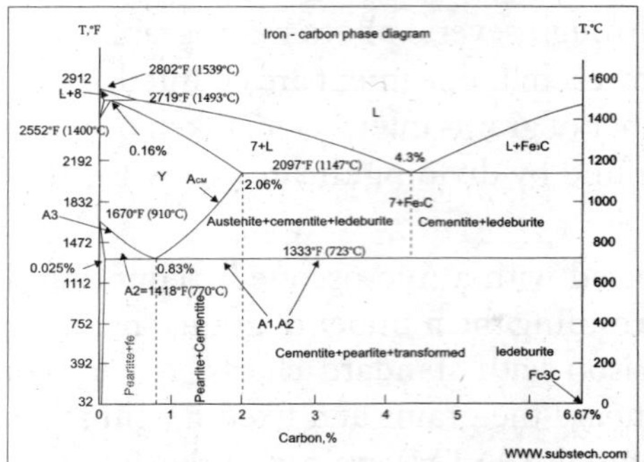

The maximum solubility of C in α- ferrite is 0.022 wt%. δ–ferrite is only stable at high temperatures. It is not important in practice. Austenite has a maximum C concentration of 2.14 wt %. It is not stable below the eutectic temperature (727 °C) unless cooled rapidly. Cementite is in reality metastable, decomposing into α-Fe and C when heated for several years between 650 and 77°C.

- **δ-ferrite** – It is solid solution of carbon in δ-iron. Maximum concentration of carbon in δ- ferrite is 0.09% at 1493°C which is the temperature of the peritectic transformation. The crystal structure of δ-ferrite is BCC (cubic body centered).
- **Austenite** – Austenite is interstitial solid solution of carbon in γ-iron. Austenite has FCC (cubic face centered) crystal structure, permitting high solubility of carbon i.e. up to 2.06% at 1147 °C. Austenite does not exist below 723 °C and maximum carbon concentration at this temperature is 0.83%.
- **α-ferrite** – It is solid solution of carbon in α-iron. α-ferrite has BCC crystal structure and low solubility of carbon up to 0.025% at 723°C. α-ferrite exists at room temperature.
- **Cementite** – Cementite also known as iron carbide, is an intermetallic compound of iron and carbon, having fixed composition Fe_3C. Cementite is a hard & brittle substance, influencing the properties of steels and cast irons.

Critical temperatures

- Upper critical temperature (point) A_3 is the temperature, below which ferrite starts to form as a result of ejection from austenite in the hypo-eutectoid alloys.
- Upper critical temperature (point) ACM is the temperature, below which cementite starts to form as a result of ejection from austenite in the hyper-eutectoid alloys.
- Lower critical temperature (point) A_1 is the temperature of the austenite-to-Pearlite eutectoid transformation. Below this temperature austenite does not exist.
- Magnetic transformation temperature A_2 is the temperature below which α-ferrite is ferromagnetic.

Phase compositions of the iron-carbon alloys at room temperature

- Hypoeutectoid steels (carbon content from 0 to 0.83%) consist of primary (proeutectoid) ferrite (according to the curve A_3) and Pearlite.
- Eutectoid steel (carbon content 0.83%) entirely consists of Pearlite.
- Hypereutectoid steels (carbon content from 0.83 to 2.06%) consist of primary (proeutectoid) cementite (according to the curve A_{CM}) and Pearlite.
- Cast irons (carbon content from 2.06% to 4.3%) consist of cementite ejected from austenite according to the curve A_{CM} , Pearlite and transformed ledeburite (ledeburite in which austenite is transformed to pearlite).

When the liquid of eutectic composition is cooled, at or below eutectic temperature this liquid transforms simultaneously into two solid phases (two terminal solid solutions, represented by α and β). This transformation is known as eutectic reaction and is written symbolically as:

$$\textbf{Liquid (L)} \leftrightarrow \textbf{solid solution-1 } (\alpha) + \textbf{solid solution-2 } (\beta)$$

In the solid state analog of a eutectic reaction, called a eutectoid reaction, one solid phase having eutectoid composition transforms into two different solid phases. Another set of invariant reactions that occur often in binary systems are - peritectic reaction where a solid phase reacts with a liquid phase to produce a new solid phase.

For their role in mechanical properties of the alloy, it is important to note that: Ferrite is soft and ductile Cementite is hard & brittle. Thus, combining these two phases in solution an alloy can be obtained with intermediate properties. (Mechanical properties also depend on the microstructure, that is, how ferrite and cementite are mixed.)

Development of Microstructures in Iron—Carbon Alloys

The eutectoid composition of austenite is 0.8 wt %. When it cools slowly it forms perlite, a lamellar or layered structure of two phases: α-ferrite and cementite (Fe_3C). Hypoeutectoid alloys contain proeutectoid ferrite plus the eutectoid pearlite. Hypereutectoid alloys contain proeutectoid cementite plus pearlite. Since reactions below the eutectoid temperature are in the solid phase, the equilibrium is not achieved by usual cooling from austenite.

The Influence of Other Alloying Elements

Alloying strengthens metals by hindering the motion of dislocations. Thus, the strength of Fe–C alloys increase with C content and also with the addition of other elements.

Ques.628: Discuss the significance of TTT diagram.

Answer: Time-temperature transformation (TTT) diagrams measure the rate of transformation at a constant temperature. In other words a sample is austenitised and then cooled rapidly to a lower temperature and held at that temperature whilst the rate of transformation is measured, for example by dilatometry. Obviously a large number of experiments is required to build up a complete TTT diagram.

- An increase in carbon content shifts the TTT curve to the right (this corresponds to an increase in hardenability as it increases the ease of forming martensite - i.e. the cooling rate required to attain martensite is less severe).
- An increase in carbon content decreases the martensite start temperature.
- An increase in Mo content shifts the TTT curve to the right and also separates the ferrite + pearlite region from the bainite region making the attainment of abainitic structure more controllable.

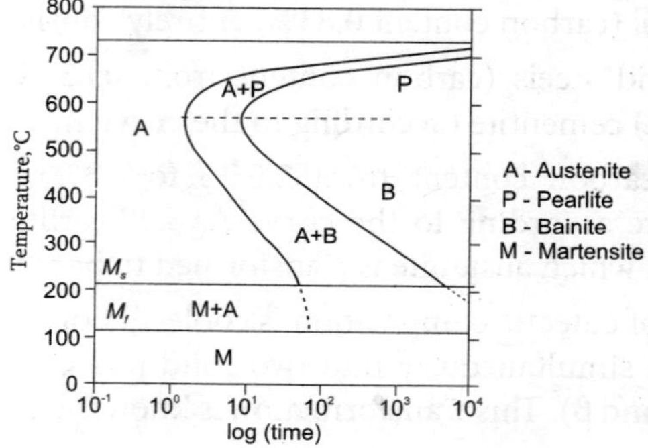

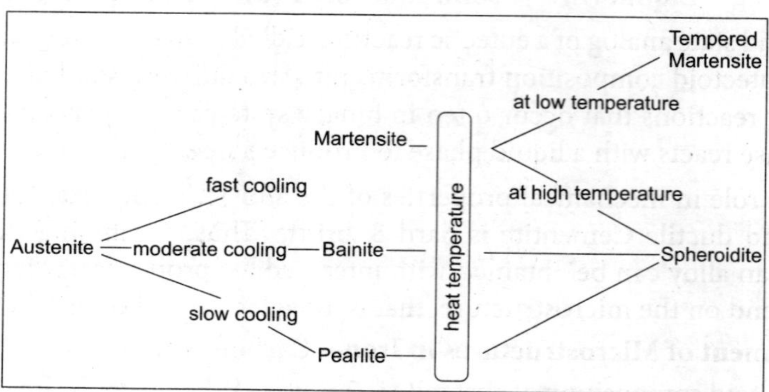

Microstructure and Property Changes in Fe-C Alloys

Isothermal Transformation Diagrams

We use as an example the cooling of an eutectoid alloy (0.8 % C) from the austenite (γ-phase) to pearlite, that contains ferrite (α) plus cementite (Fe_3C or iron carbide). When cooling proceeds below the eutectoid temperature (727 °C) nucleation of pearlite starts. The S-shaped curves (fraction of pearlite vs. log. time,) are displaced to longer times at higher temperatures showing that the transformation is dominated by nucleation (the nucleation period is longer at higher temperatures) and not by diffusion (which occurs faster at higher temperatures).

The family of S-shaped curves at different temperatures can be used to construct the TTT (Time- Temperature-Transformation) diagrams. For these diagrams to apply, one needs to cool the material quickly to a given temperature T_0 before the transformation occurs, and keep it at that temperature over time. The horizontal line that indicates constant temperature T_0 intercepts the TTT curves on the left (beginning of the transformation) and the right (end of the transformation); thus one can read from the diagrams when the transformation occurs. The formation of pearlite indicates that the transformation occurs sooner at low temperatures, which is an indication that it is controlled by the rate of nucleation. At low temperatures, nucleation occurs fast and grain growth is reduced(since it occurs by diffusion,which is hindered at low temperatures). This reduced grain growth leads to fine-grained microstructure (fine pearlite). At higher temperatures, diffusion allows for larger grain growth, thus leading to coarse pearlite.

At lower temperatures nucleation starts to become slower, and a new phase is formed, bainite. Since diffusion is low at low temperatures, this phase has a very fine (microscopic) microstructure.

Spheroidite is a coarse phase that forms at temperatures close to the eutectoid temperature. The relatively high temperatures causes a slow nucleation but enhances the growth of the nuclei leading to large grains.

A very important structure is martensite, which forms when cooling austenite very fast (quenching) to below a maximum temperature that is required for the transformation. It forms nearly instantaneously when the required low temperature is reached; since no thermal activation is needed, this is called an athermal transformation. Martensite is a different phase, a body- centered tetragonal (BCT) structure with interstitial C atoms. Martensite is metastable and decomposes into ferrite and pearlite but this is extremely slow (and not noticeable) at room temperature.

In the examples, we used an eutectoid composition. For hypo- and hypereutectoid alloys, the analysis is the same, but the proeutectoid phase that forms before cooling through the eutectoid temperature is also part of the final microstructure.

Ques.629: Describe theClassification of engineering materials according to various parameters.

Answer : It is the systematic arrangement or division of materials into groups on the basis of some common characteristic ;

1. According to General Properties

2. According to Nature of Materials

3. According to Applications

1. According to General Properties

a. Metals (e.g. iron, aluminium, copper, zinc, lead, etc)

Iron as the base metal, and range from plain carbon (> 98 % Fe).

(i). Ferrous: high alloy steel (< 50 % alloying elements), e.g. cast iron, wrought iron, steel, alloys like high-speed steel, spring steel, etc

(ii). Non-Ferrous: Rest of the all other metals and their alloys, e.g. copper, aluminium, zinc lead, alloys like brass, bronze, duralumin, etc

b. Non-Metals (e.g. leather, rubber, asbestos, plastics, etc)

2. According to Nature of Materials

a. Metals: e.g. Iron & Steel, Alloys &Superalloys, Intermetallic Compounds, etc

b. Ceramics: e.g. Structural Ceramics (high-temperature load bearing), Refractories (corrosion-resistant, insulating), Whitewares (porcelains), Glass, Electrical Ceramics (capacitors, insulators, transducers), Chemically Bonded Ceramics (cement & concrete)

c. Polymers: e.g. Plastics, Liquid Crystals, Adhesives

d. Electronic Materials: e.g. Silicon, Germanium, Photonic materials (solid-state lasers, LEDs)

e. Composites: e.g. Particulate composites (small particles embedded in a different material), Laminate composites (golf club shafts, tennis rackets), Fiber reinforced composites (fiberglass)

f. Biomaterials: e.g. Man-made proteins (artificial bacterium), Biosensors, etc

g. Advanced / Smart Materials: e.g. materials in computers (VCRs, CD Players, etc), fibre optic systems, spacecrafts, aircrafts, rockets, shape-memory alloys, piezoelectric ceramics, magnetostrictive materials, optical fibres, micro electro mechanical (MEMs) devices, electro rheological / magnet or heological fluids, Nanomaterials, etc

3. According to Applications

a. Electrical Materials: e.g. conductors, insulators, dielectrics, etc

b. Electronic Materials: e.g. conductors, semi-conductors, etc

c. Magnetic Materials: e.g. ferromagnetic, paramagnetic & diamagnetic materials, etc

d. Optical Materials: e.g. glass, quartz, etc

e. Bio Materials: e.g. man-made proteins, artificial bacterium

Ques.630: List various Material Properties.

Answer:

- Physical: e.g. appearance, shape, weight, boiling point, melting point, freezing point, density, glass transition temperature,permeability
- Mechanical: e.g. strength (tensile, compressive, shear, torsion, bending), elasticity, plasticity, ductility, malleability, rigidity, toughness, hardness, brittleness, impact, fatigue, creep, strain hardening, Bauschinger effect, strain rate effect, vibration resistance,wear
- Thermal: e.g. thermal conductivity, expansion coefficient, resistivity, thermal shock resistance, thermal diffusivity

- Electrical: e.g. conductivity, resistivity, dielectric strength, thermoelectricity, superconductivity, electrichysteresis
- Magnetic: e.g. ferromagnetism, paramagnetism, diamagnetism, magnetic permeability, coercive force, curie temperature, magnetic hysteresis
- Chemical: e.g. reactivity, corrosion resistance, polymerization, composition, acidity, alkalinity
- Optical: e.g. reflectivity, refractivity, absorptivity, transparency, opaqueness, color, luster

Metallurgical: e.g. grain size, heat treatment done / required, anisotropy, hardenability

Ques.631: Discuss the differences in the properties of metals and non-metals.

Answer: Differences in the properties of metals and non-metals are tabulated below.

Property	Metals	Non-Metals
Structure	Crystalline	Amorphic
State	Generally solids at room temp.	Gaseous & solid at ordinary temp.
Luster	Metallic luster	No metallic luster (except iodine & graphite)
Conductivity	Good conductor heat & electricity	Bad conductors
Malleability	Malleable	Not malleable
Ductility	Ductile	Not ductile
Hardness	Generally hard	Hardness varies
Electrolysis	Form anions	Form anioins
Exitation of valance electron by e.m.f	Easy	Difficult
Density	High	Low

Ques.632: Discuss various Crystal Geometry of materials. Discuss various crystal types and principal metal crystal structure.

Answer:

- Materials may be classified as Crystalline or *Non-Crystalline structures.*
- Crystalline solid can be either Single Crystal Solid (crystal lattice of entire sample is continuous and unbroken to edges of sample with no grain boundary) or Poly Crystal Solid (aggregate of many crystals separated by *well-defined boundaries*)
- Cluster of crystals with identical structure (same crystallographic planes & directions) are known as Grains separated by *Grain Boundaries*
- X-ray diffraction analysis shows that atoms in metal crystal are arranged in a regular, repeated 3-D Pattern known as *Crystalline Structure*

Example:

Ceramic: Inorganic non-metallic crystalline (regular internal Structure) materials

Glass: Inorganic non-metallic non-crystalline / amorphous (completely disordered form) material.

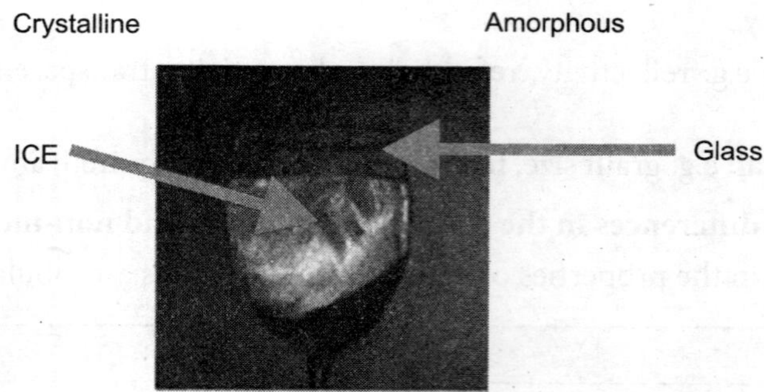

- Arrangement of atoms can be most simply portrayed by Crystal Lattice, in which atoms are visualized as, Hard Balls located at particular locations.
- Space Lattice / Lattice: Periodic arrangement of points in space with respect to three dimensional network of lines.
- Each atom in lattice when replaced by a point is called Lattice Point, which are the intersections of above network of lines.
- Arrangement of such points in 3-D space is called Lattice Array and 3-D space is called Lattice Space.
- Tiny block formed by arrangement of small group of atoms is called Unit Cell. It is chosen to represent the symmetry of crystal structure, and may be defined as:
  - Finite representation of infinite lattice
  - Small repeat entity
  - Basic structural unit
  - Building block of crystal structure
  - Can generate entire crystal by translation

Lattice Parameters

- Six lattice parameters a, b, c, α, β, γ
- Typically in the order of few Angstroms (few tenths of nanometer)
- Example: Cubic structure has following lattice parameters: a = b = c and $\alpha = \beta = \gamma = 90^\circ$

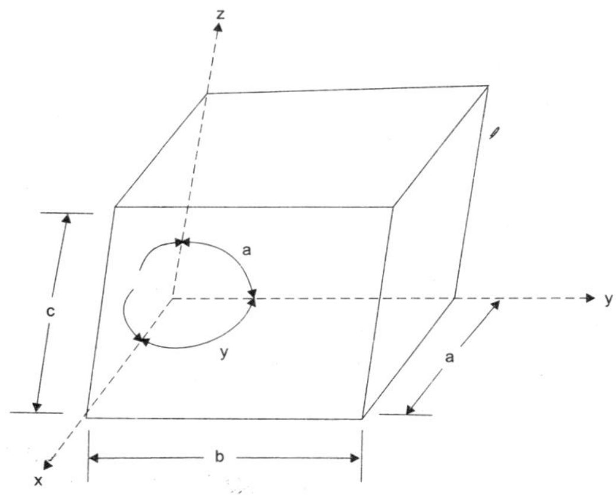

Types of Crystal

Cubic, Monoclinic, Triclinic, Tetragonal, Orthorhombic, Rhombohedral and Hexagonal

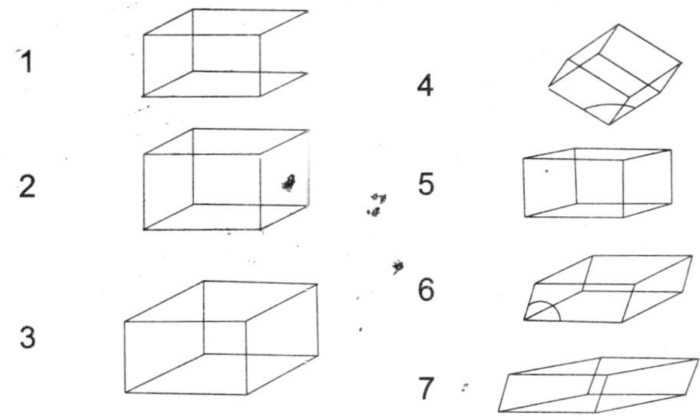

No.	Crystal Type	Lattice Parameters	Examples
1.	Cubic	$a = b = c,\ \alpha = \beta = \gamma = 90^\circ$	Fluorite, Garnet, Pyrite
2.	Tetragonal	$a = b \neq c,\ \alpha = \gamma = 90^\circ = \beta$	Zircon
3.	Orthorhombic	$a \neq b \neq c,\ \alpha = \gamma = \beta = 90^\circ$	Topaz
4.	Rhombohedral	$a = b = c,\ \alpha = \beta = \gamma \neq 90^\circ$	Tourmaline
5.	Hexagonal	$a = b \neq c,\ \alpha = \beta = 90^\circ,\ \gamma = 120^\circ$	Corundum
6.	Monoclinic	$a \neq b \neq c,\ \alpha = \gamma = 90^\circ \neq \beta$	Kunzite
7.	Triclinic	$a \neq b \neq c,\ \alpha \neq \gamma \neq \beta \neq 90^\circ$	Amazonite

Principal Metal Crystal Structures

- Simple Cubic Lattice (SC) Structure
- Body Centered Cubic (BCC) Structure
- Face Centered Cubic (FCC) Structure

Simple Cubic(SC) Lattice

- Most elementary crystal structure with three mutually perpendicular axes arbitrarily placed through one of the corners of ace ll
- Each corners occupied with one atom
- Example: alpha polonium

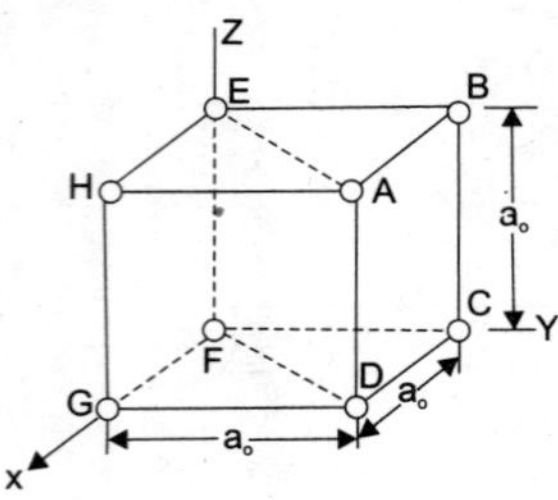

SC Structure

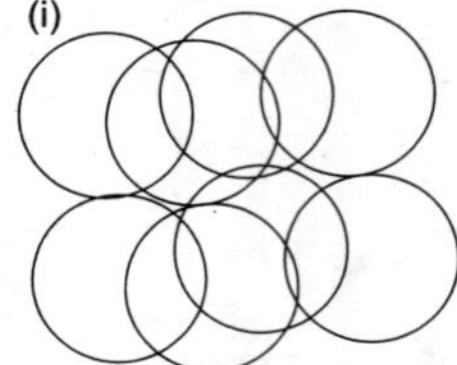

(I). Aggregate of atoms

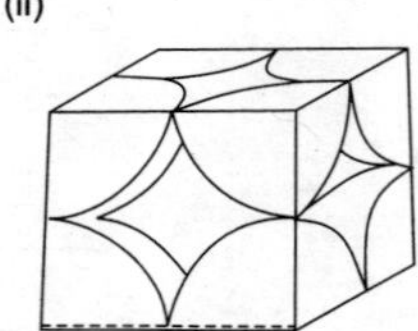

(ii). Hard Sphere UnitCell

Body Centered Cubic (BCC) Structure

- BCC cell has an atom at each corner and another atom at body center of cube
- Each atom at corner is surrounded by eight adjacent atoms
- Example: alpha iron, chromium, molybdenum & tungsten

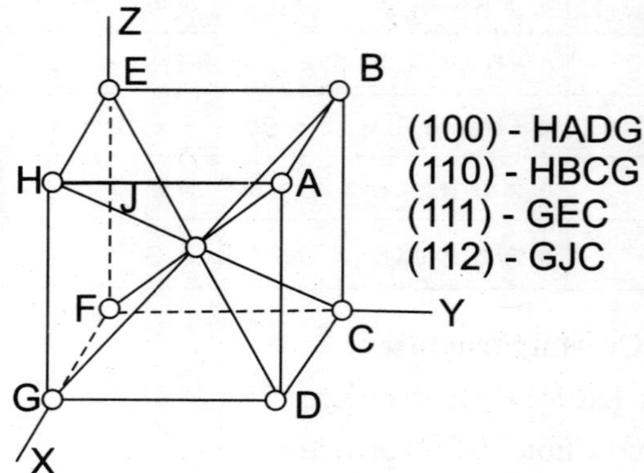

BCC Structure

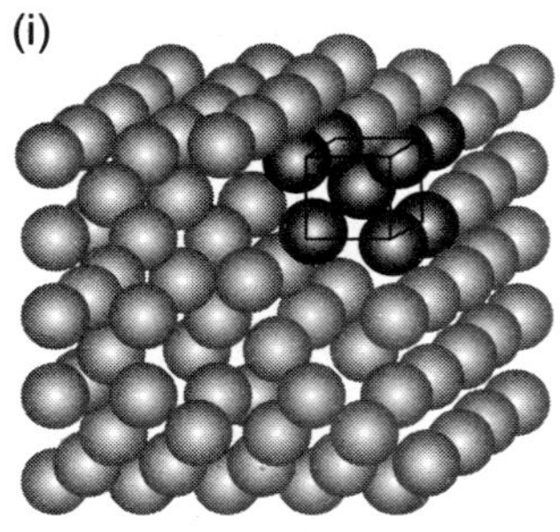

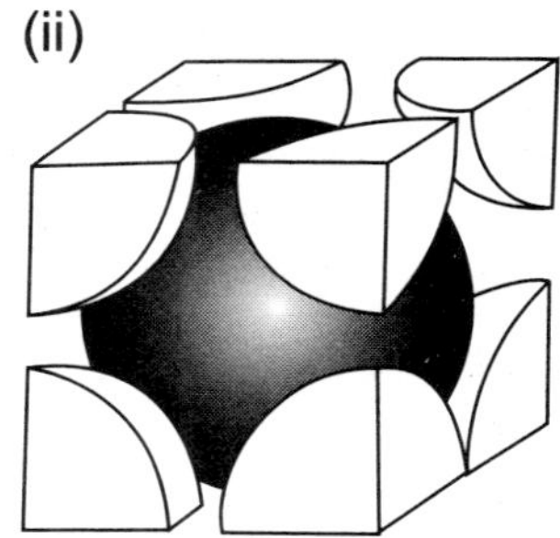

(i) Aggregate of atom

(ii) HardSphereUnit Cell

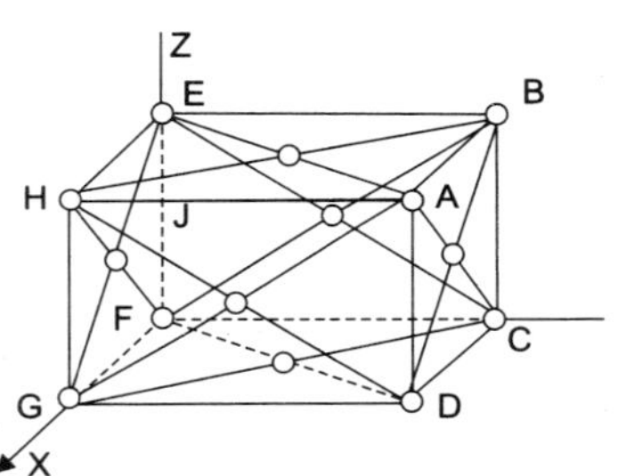

Face Centered Cubic (FCC)Structure

- Atoms at each corner of cube, and in addition there is an atom at the center of each cube's face
- Example: aluminium, copper, gold, lead, silver and nickel

FCC Structure

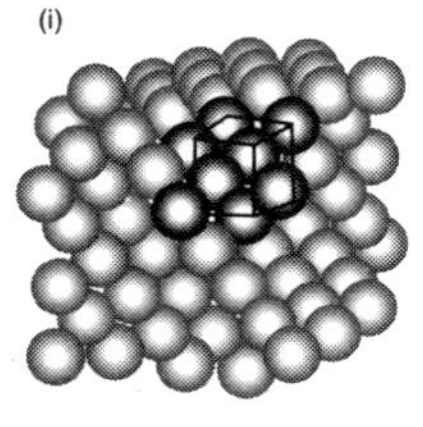

(i). Aggregateofatoma

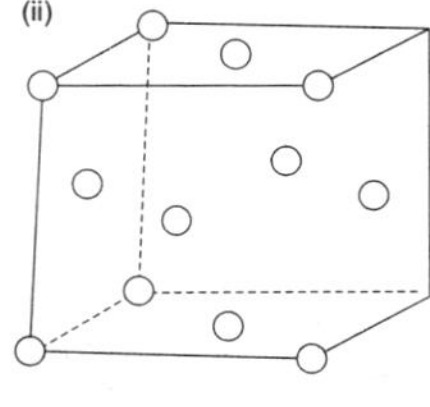

(ii). Hard Sphere UnitCell

Hexagonal Close-packed (HCP) Structure

- Unit cell has an atom at each of twelve corners of hexagonal prism, with one atom at centre of each of two hexagonal faces and three atoms in body of cell
- Example: zinc, lithium, magnesium,beryllium

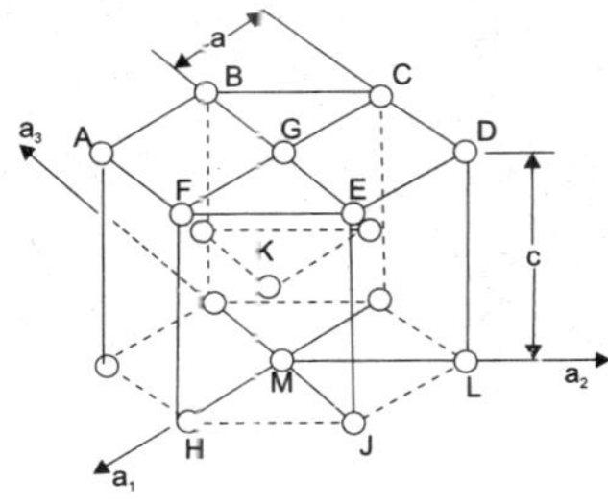

Comparison

Sl. No.	APF (Atomic Packing Factor) (Ratio of volume of atoms in unit cell to unit cell volume)	Co-ordination Number (No. of nearest neighbor atoms)
SC	0.52	6
BCC	0.68	8
FCC	0.72	12

Close-Packed Structures

- How can metal atoms be stacked to fill empty spaces in thelattice?
- Can these 2-D layers be stacked to make 3-D structures?

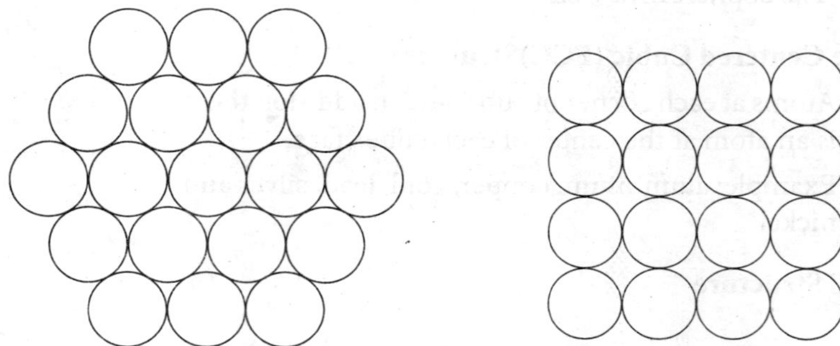

Example: Close-Packed FCC Structure

ABC Stacking Sequence Example: Close-Packed HCP Structure

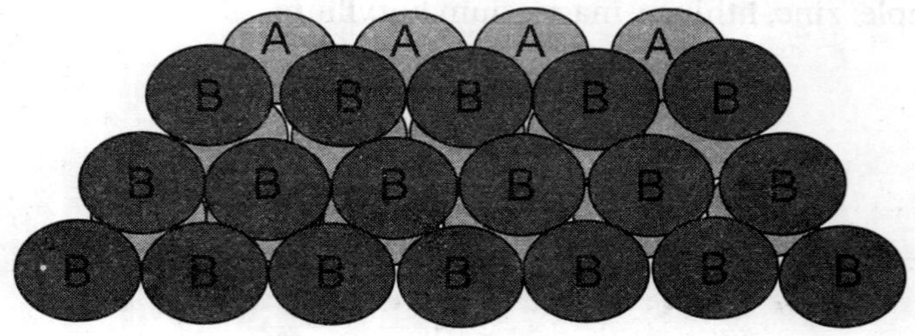

AB Stacking Sequence

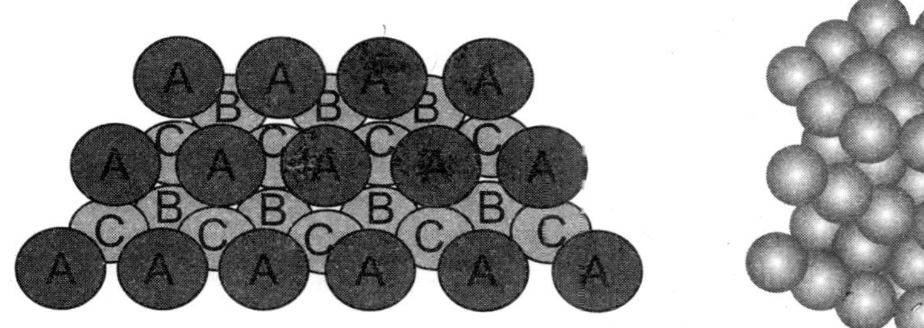

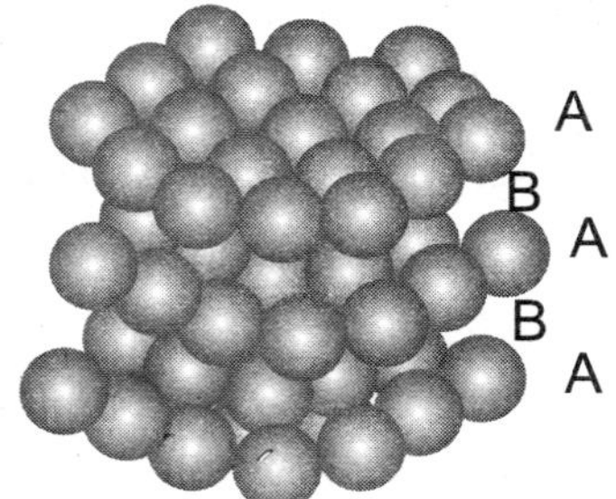

ABC Stacking Sequence

Crystallographic Planes & Directions

- Crystallographic Planes &Directions are specified with respect to the reference axes in terms of 'Miller Indices', which is the system of notation of planes within a crystal of space-lattice.
- Example: SC structure is chosen forunder standing

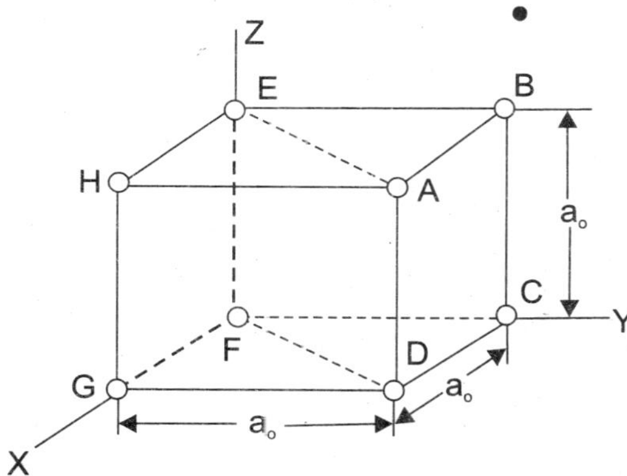

- Crystallographic Plane is specified in terms of length of its intercepts on three mutually perpendicular axes. To simplify it, reciprocals of these intercepts reduced to lowest common denominator (LCM) are used as miller indices,e.g. plane ABCD in figure is parallel to x and z axes and intersects y-axis at one interatomic distance 'a_o', therefore miller indices of the plane are $1/\infty$, $1/a_o$, $1/\infty$ or (hkl) = (010)
- Plane EBCF is designated as (00) plane, where bar indicates that plane intersects the axis in negative direction
- Crystallographic Directions are indicated by integers in brackets like [uvw], where reciprocals are never used, e.g. direction 'FD' is obtained by moving out from origin, a distance 'a_o' along positive x-axis and moving same distance along positive y-axis also, therefore, the crystallographic direction will be given as[110]
- **Example:** Ionic crystals like NaCl, LiF (not for metals)

Note: For simple cubic lattice only, direction is always perpendicular the plane having same indices.

Index system for crystal directions and planes

Crystal directions:

The direction is specified by the three integers [n,n,n]. If the numbers n1n2n3 have a common factor, this factor is removed. For example, [111] is used rather than [222], or [100], rather than [400]. When we speaka bout directions, we mean a whole set of parallel lines, which are equivalent due to transnational symmetry. Opposite orientation is denoted by the negative sign over a number. For example:

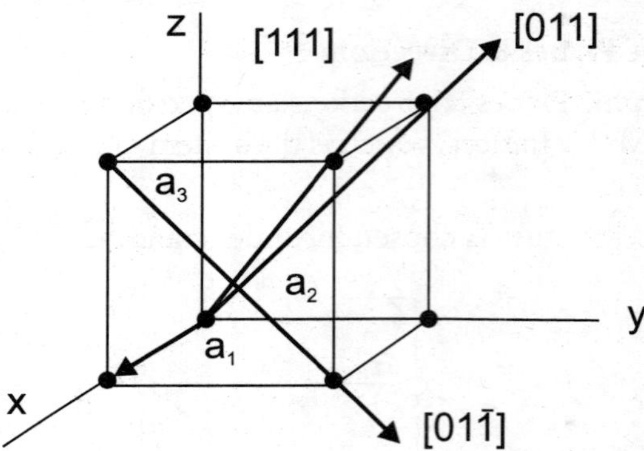

Crystal planes: The orientation of a plane in a lattice is specified by Miller indices. They are defined as follows. We find intercept of the plane with the axes along the primitive translation vectors a_1, a_2 and a_3. Let's these intercepts be x, y, and z, so that x is fractional multiple of a_1,y is a fractional multiple of a_2 and z is a fractional multiple of a_3. Therefore we can measure x, y, and z in units a_1, a_2 and a_3 respectively. We have then a triplet of integers (x y z). Then we invert it (1/x 1/y1/z) and reduce this set to a similar one having the smallest integers by multiplying by a common factor. This set is called Miller indices of the plane (hkl). For example, if the plane intercepts x, y, and z in points 1, 3, and 1, the index of this plane will be (313).

The orientation of a crystal plane is determined by three points in the plane, provided they are not collinear. If each point lay on a different crystal axis, the plane could be specified by giving the coordinates of the points in terms of the lattice constants a, b, c. A notation conventionally used to describe lattice points (sites), directions and planes is known as Miller Indices.

A crystal lattice may be considered as an assembly of equidistant parallel planes passing through the lattice points and are called lattice planes. In order to specify the orientation one employs the so called Miller indices.

Crystal direction is the direction (line) of axes or line from the origin and denoted as [111], [100], [010] etc.

How to find Miller Indices:

To determine the indices for the plane p in Figure :

first we have to find the intercepts with the axes along the basis vector be x, y, z. We form the fractional triplet.Let these intercepts-

- Take reciprocal to this set.
- Then reduce this set to a similar one having the smallest integers multiplying by common factor.

To determine the Miller indices:

(i) Find the intercepts on the axes along the basis vector →→→ cba, in terms of the lattice constants a,b and c. The axes may be those of a primitive or non primitive cell. Let these intercepts be x, y, z. We form the fractional triplet.

(ii) Take the reciprocals of these numbers.

(iii) Reduce the numbers to three smallest integers by multiplying the numbers with the same integral multipliers.

This last set is enclosed in parentheses (h k l), is called the index of the plane or Miller Indices.The Miller indices specify not just one plane but an infinite set of equivalent planes. Note that for cubic crystals the direction [hkl] is perpendicular to a plane (hkl) having the same indices, but this is not generally true for other crystal systems. Examples of the planes in a cubic system:

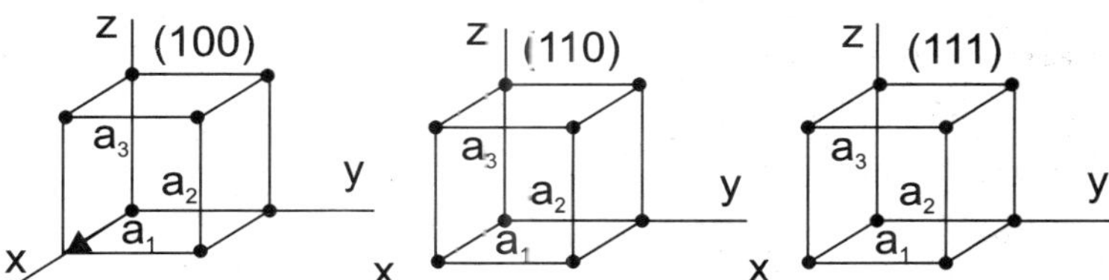

Allotropy / Polymorphism

- Two or more distinct crystal structures for the same material at different temperature & pressure
- Examples: iron (α-iron & γ-iron), carbon (graphite & diamond), tin (α-tin & β-tin)

Body-centered tetragonal $\xrightarrow[\text{cooling}]{13.2^{0}}$ Diamond cubic crystal structure

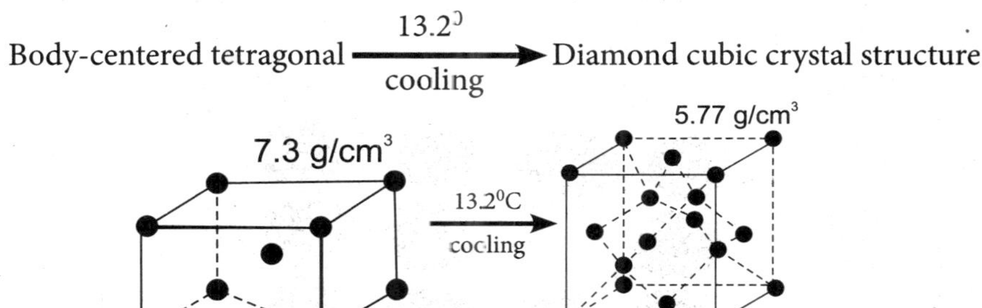

Ques.633: What are different Imperfections/ defects in Crystals ?

Answer : In actual crystals, imperfection or defects are always present, which are important to understand, as they influence the properties of material.

Classifications of Defects

(i). Point Defects (Zero Dimensional Defects)

(ii). (a) Vacancy

(b) Schottky Imperfections (c). Interstitialcy

(d) FrenkelDefect

(e) CompositionalDefect

- SubstitutionalDefect
- InterstitialImpurity

(f) ElectronicDefect

(iii). Line Defects / Dislocations (One Dimensional Defects)

(iv). (a). Edge Dislocations

(b). Screw Dislocations

(v). Surface / Plane Defects (Two Dimensional Defects)

(vi). (a). ExternalDefects

(b). Internal Defects

- Grain Boundary Defect
- Tilt Boundary Defect
- Twin Boundary Defect
- Stacking Fault

(vii). Volume Defects (Three Dimensional Defects)

(i). Point Defects (Zero Dimensional Defects)

- Imperfect point like regions in crystal (size is one or two atomic diameter)
- Completely local in effect, i.e. vacant lattice site
- Always present in crystals
- Created by thermal fluctuations, quenching (high rate of cooling), severe deformation of crystal lattice (hammering or rolling) or external bombardment by atoms / high energy particles

(a). Vacancy

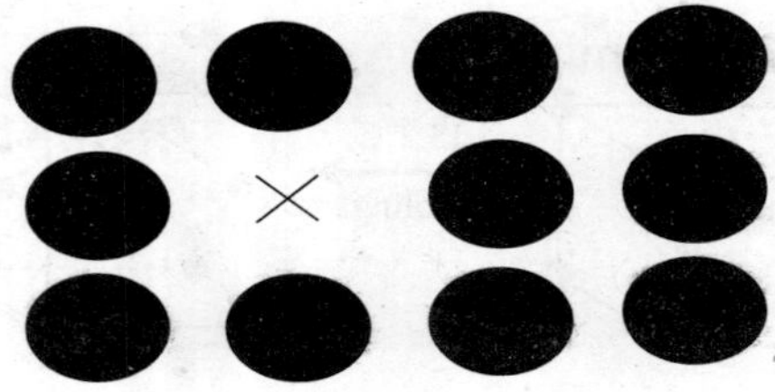

- Simplest type of point defect involving a missing atom within crystal lattice of metal, which results due to imperfect packing during original crystallization process or by thermal vibrations of atoms at very high temperature

(b). Schottky Imperfections

Involve vacancies in pair of ions having opposite charges and are found in compounds, which maintain charge balance, e.g. alkali halides

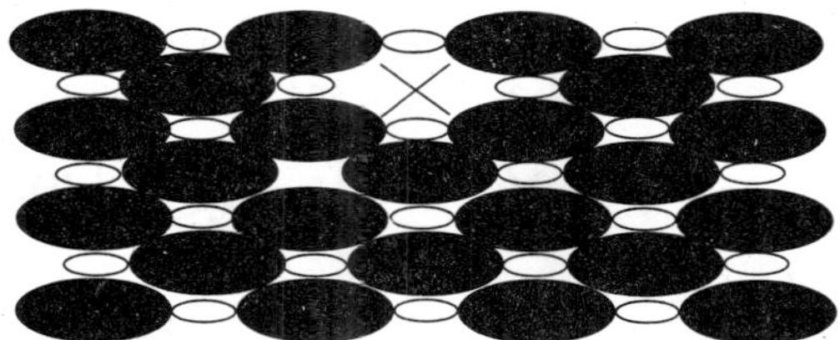

(c). Interstitialcy

Addition of an extra atom in the crystal lattice, when atomic packing density / factor is low and the foreign atom may be an impurity or alloying atom (generally very less in number as additional energy is required to force atom / ion to occupy new position)

(d). Frenkel Defect

Ion dislodged from its crystal lattice into an interstitial site (generally very less in number as additional energy is required to force the atom / ion to occupy the new position).

(e). Compositional Defect

Arises due to presence of impurity atom in crystal lattice of metal during original crystallization process (responsible for functioning of semi-conductor devices)

- Substitutional Defect:Original parent atom from its lattice is replaced
- Interstitial Impurity:Small impurity occupies interstitial spaces in galattice

(f). Electronic Defect

Errors in charge distribution in solids, e.g. diodes and transistors devices, where charges diffuse in opposite regions leading to change of their concentration in that region

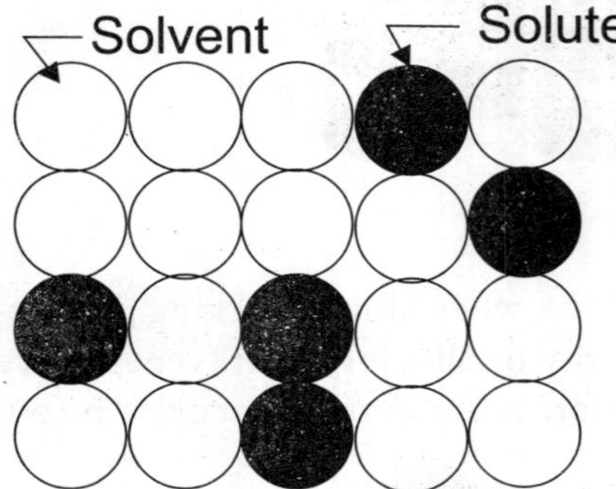

Substitutional Defect

Solvent
Solute

Interstitial Defect

***(ii).* Line Defects (One Dimensional Defects)**

- Linear disturbance of atomic arrangement, which can move very easily on slip plane through crystal
- Occurs during recrystallization process or during slip
- Created along a line, which is also boundary between slipped and unslipped regions of crystals
- Defect is known as 'dislocation' and boundary is known as 'DislocationLine'
- Region near dislocation, where distortion is extremely large is called 'Core of Dislocation' (very high local strain)

(a). Edge Dislocation

- Any extra plane of atoms within a crystal structure is edge dislocation.
- Accompanied by zones of compression & tension and there is a net increase in energy along dislocations.
- Displacement distance for atoms around dislocation is called 'Burger Vector', which is at right angle to edge dislocation
- Berger vector is determined by drawing a rectangle in region by connecting an equal number of atoms on opposite sides (circuit fails to complete), i.e. PP' as shown in figure.

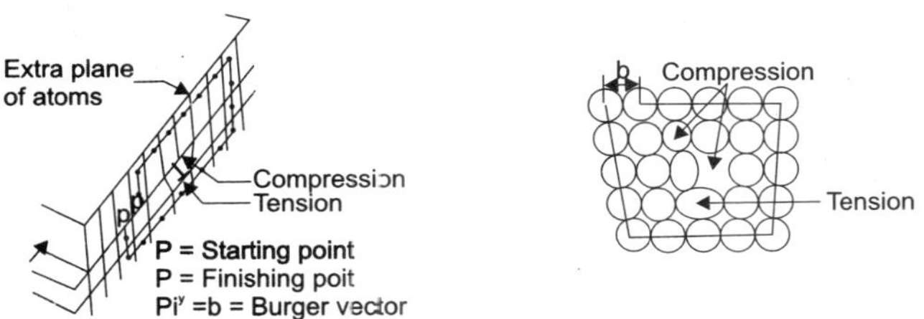

Edge Dislocation

(b). ScrewDislocation

- Originate from partial slipping of a section of crystal plane
- Shear stresses are associated with adjacent atoms and extra energy is involved along the dislocation
- Successive atom planes are transformed into the Surface of Helix of Screw (plane is distorted), which accounts for its name as screwdis location
- Displacement vector is parallel to the line defect.

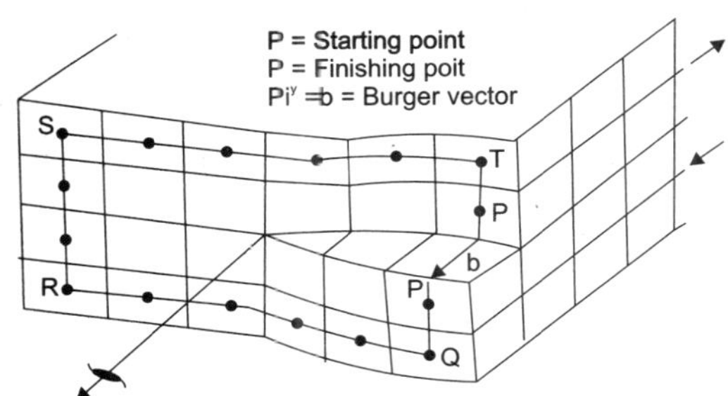

Screw Dislocation

Comparison of edge dislocation and screw dislocation is tabulated below.

Edge Dislocation	Screw Dislocation
Arise due to introduction or elimination of an row of atoms	Arise due to partial slipping of section of crystal where planes of atoms are transformed into suhelix of screw
Tensile, compressive or shear stress field may be present.	Only shear stress field is present
Lattice disturbance extends along an edge inside crystal.	Lattice disturbance extends into two separate at right angles to each other
Burger's vector is always perpendicular to dislocation line.	Burger's vector is parallel to dislocation line
Dislocations can climb and glide.	Dislocation can only glide
Force required is less as compared to that for screw as compared to that dislocation	Force required is more as compared to that for edge dislocation.

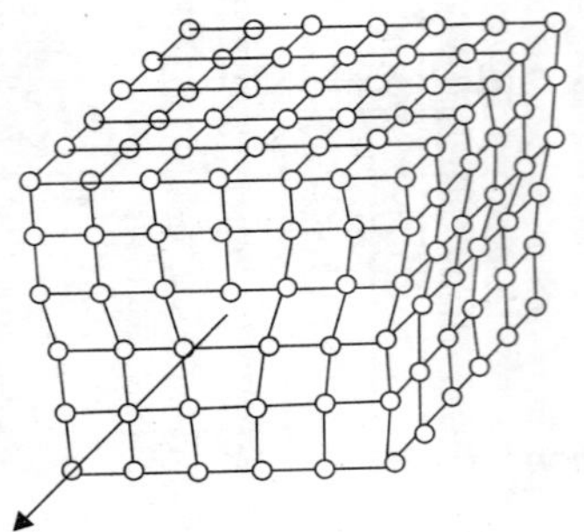

Edge Dislocation

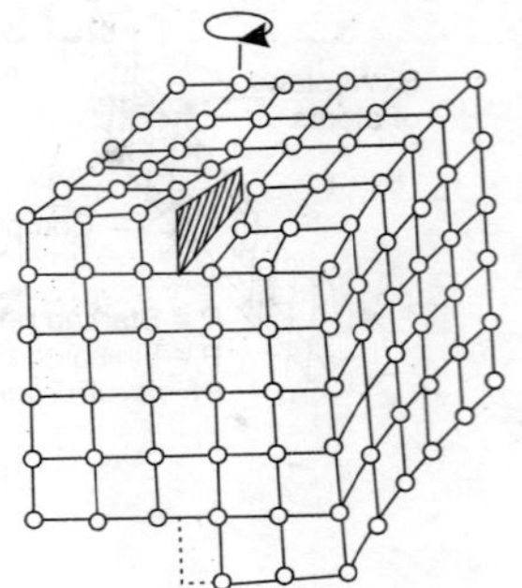

Screw Dislocation

Geometry of Dislocation

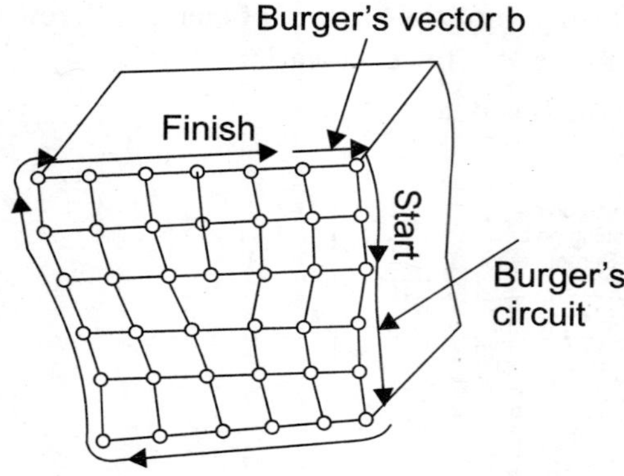

Edge Dislocation

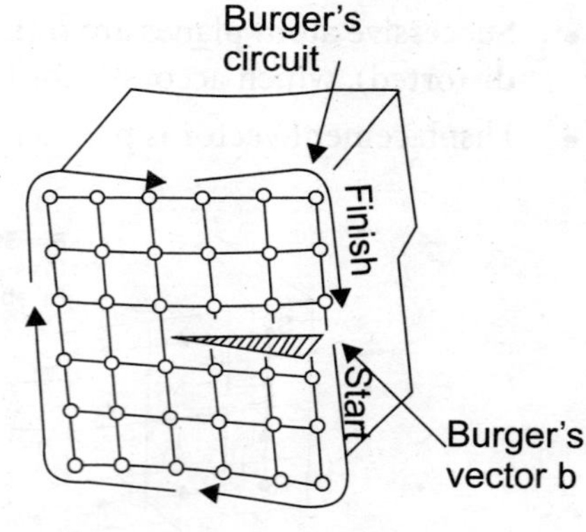

Screw Dislocation

Burger's Vector

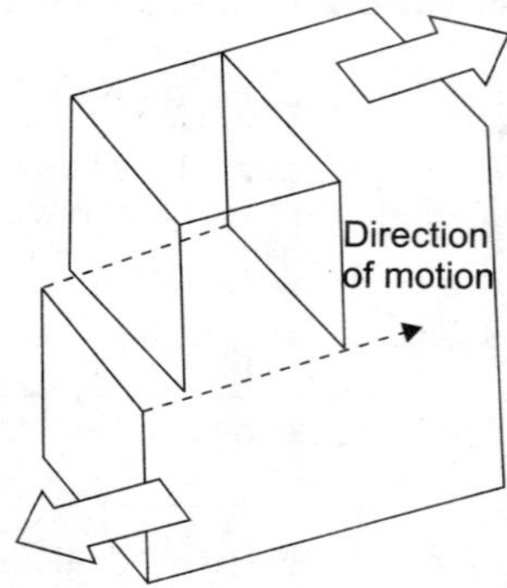

Edge Dislocation

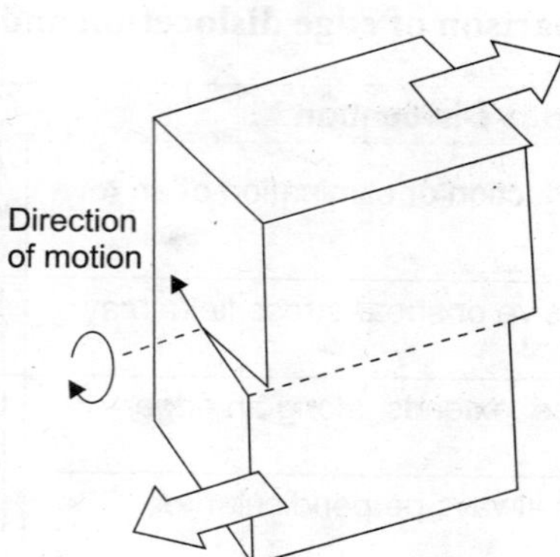

Screw Dislocation

Formation of Step on Crystal Surface

***(iii)*. Surface / Plane Defects (Two Dimensional Defects)**

- Two-dimensional regions in crystal and arise from change in stacking of atomic planes on or across boundary

(a). External Defects

- Defects or imperfections represented by a boundary
- External surface of material itself is an imperfection because atomic bonds do not extend beyond it
- Surface atoms do not have neighboring atoms on one side (as compared to atoms inside the material) and thus, have high energy (in the order of $1 J/m^2$)

(b). Internal Defects

Grain Boundary Defect

- Imperfections, which separate crystals or grains of different orientation in polycrystalline aggregation during nucleation or crystallization
- Individual crystals with different orientation are called as 'Grains', and boundary separating grains are called as 'Interface / Interface Boundary'

Example: copper metal

- Atoms within a grain are arranged with one orientation and pattern, which does not aligns with orientation of atoms in neighboring grains
- Misalignment between adjacent grains may be of various degrees (small or high angle grain boundaries)
- Always a transition zones between two neighboring grains
- Leads to less efficient packing of atoms at boundary and thus, atoms at boundary have higher energy as compared to those at inside of grain boundaries.

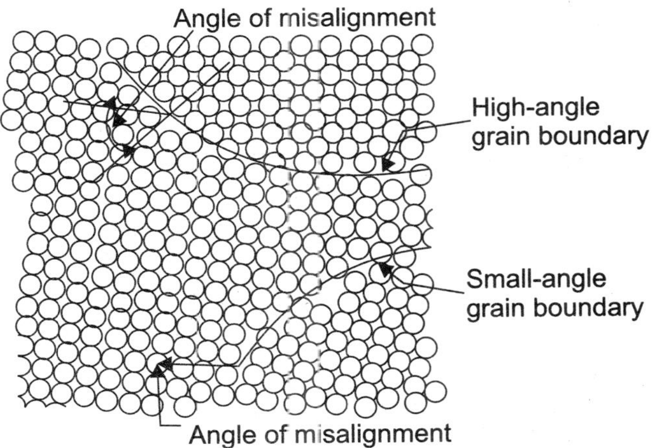

Grain Boundary Defect

Tilt BoundaryDefect

- Series of aligned dislocations, which tend to anchor (fasten) dislocation movements normally contributing to plastic deformation.Associated with little energy
- Orientation difference between two neighboring grains is less than 100, thus, it may be called as small / low angle grain boundary

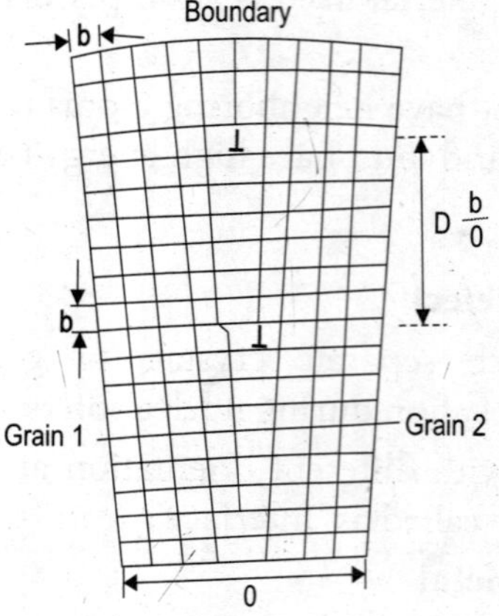

Tilt Boundary Defect

Twin Boundary Defect

- Special type of grain boundary across which, there is specific mirror lattice of symmetry, i.e. atoms on one side of boundary are located in mirror-image positions of atoms on other side.
- Region of material between these boundaries is called as 'Twin', which results due to basic twinning.

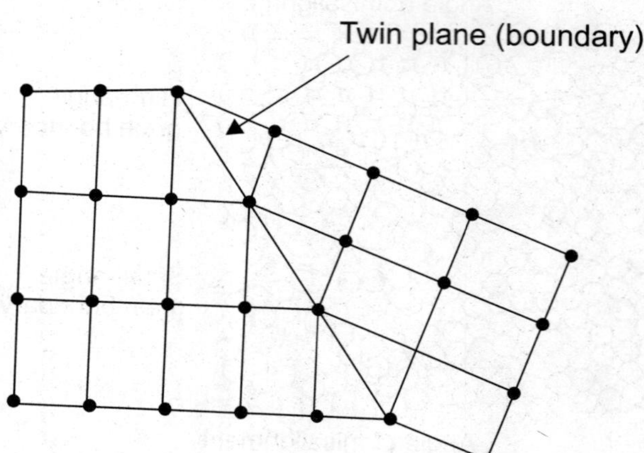

Stacking Fault

- Arise due to stacking of one atomic plane out of sequence on another, while lattice on either side of fault is perfect, i.e. one or more than one atomic plane may be missing from usual conventional style of stacking

(iv). Volume Defects (Three Dimensional Defects)

- Bulk defects, which includes pores, cracks, foreign inclusions or otherphases

Slip by Dislocation Motion

- Slip is plastic deformation process produced by dislocation motion(Note:
- Dislocations Can Move)
- Dislocation motion is analogous to the mode of locomotion employed by caterpillar
- Caterpillar forms a hump near its posterior (later) end by pulling in its pair of legs unit leg distance and then this hump is propelled forward by repeating lifting and shifting of leg pairs
- When a hump moves forward, entire caterpillar moves forward by leg separation distance corresponding to extra half plane of atoms in its dislocation model of plastic deformation

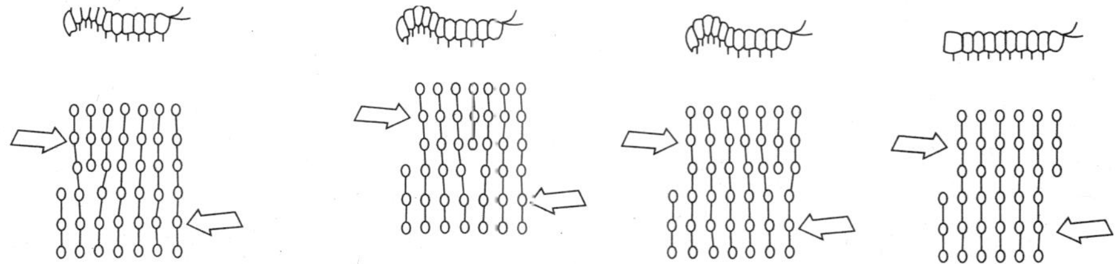

Analogy between Caterpillar & Dislocation

Ques.634: Discuss the concepts of Elastic & Plastic Deformation of metals .

Answer : Classification of Deformation

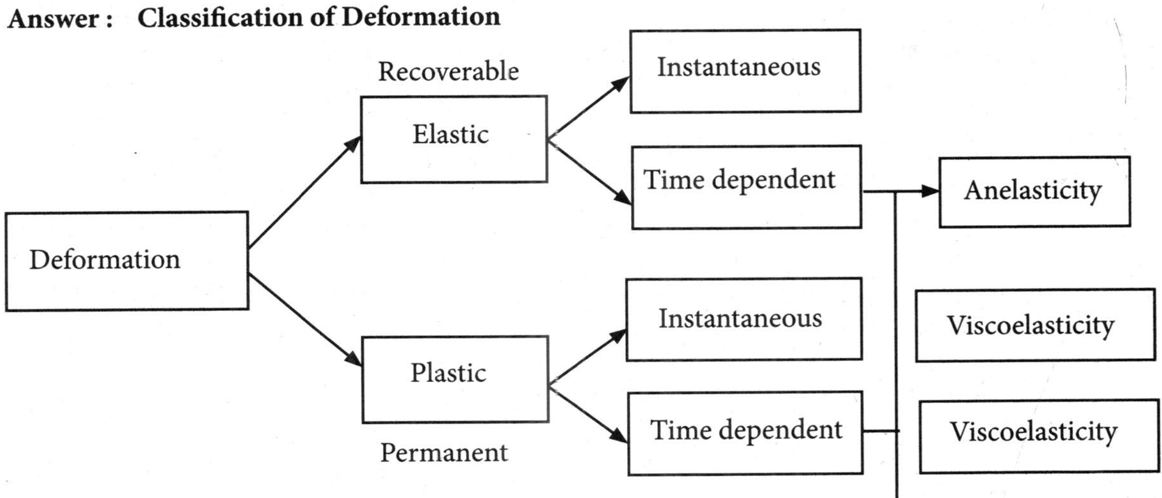

Elastic Deformation

- Disappears, when load is removed & takes place before plastic deformation.
- Strain is proportional to stress(Hooke's Law).
- Ratio of stress to strain is Young's Modulus of Elasticity (E), which is a material characteristic (magnitude depends upon force of attraction between atoms of metal).
- Shear stress produces shear strains and ratio of shear stress to shear strain is Shear Modulus of Rigidity(G).
- Ratio of volumetric stress to volumetric stain is Bulk Modulus of Elasticity(K).
- Perfect elastic material is one which regains back its original shape and size completely, after deforming force is removed.
- During elastic deformation, material stores energy (strain or elastic energy), which is recovered, when load is removed.
- Elastic strain energy is area under stress-strain curve (up to elastic limit).
- Materials subjected to tension shrink laterally & those subjected to compression, bulge.
- Ratio of lateral and axial strains is called the Poisson's ratio (ν), which is a dimensionless quantity (sign shows that lateral strain is in opposite sense to longitudinal strain).
- Theoretical value is 0.25 & maximum value is 0.50 (typical value lies in between 0.24 to 0.30.

Plastic Deformation

- Permanent deformation, which persists even when deforming load is removed (function of stress, temperature & rate of straining)
- Caused by loads exceeding elastic limit & occurs after elasticde formation
- Non-linear relation between stress and strain (nature varies with material & deformation condition)
- Exhibit yield point phenomenon (yield stress)
- Continued plastic deformation leads to fracture (instantaneous fracture for brittle material & necking followed by fracture for ductile material)

Yield Point Phenomenon

- Yielding starts at 1st higher point known as Upper Yield Point and 2nd lower point is known as Lower Yield Point
- Stress required to deform metal after yield point is always higher due to strain hardening, which causes curve to gradually rise upwards uptil ultimate tensile point (corresponding stress known as Ultimate Tensile Strength). After this specimen fails by fracture

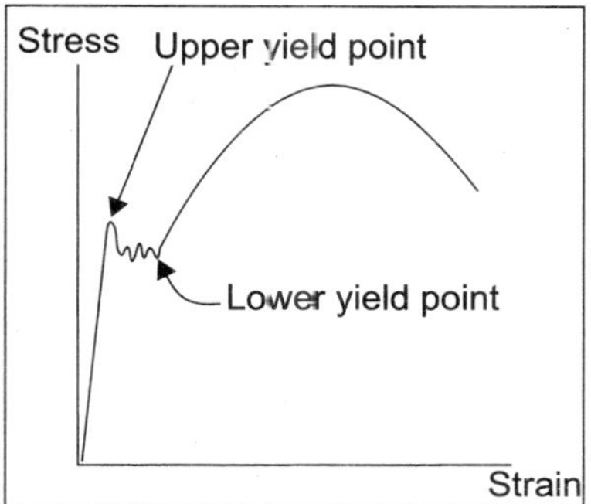

Stress-Strain curve for Mild Steel

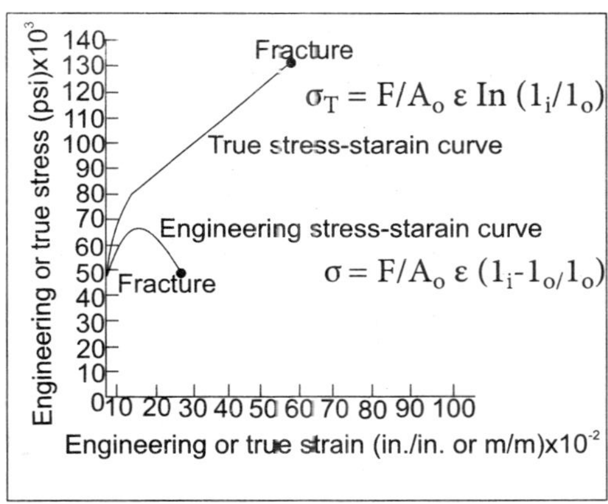

True & Engineering Stress-Strain

Ques.635: What do you understand by strain hardening of materials ?What is Bauschinger effect.

Answer: Strain Hardening

- Increase in yield stress of material when loaded subsequently in same direction again and again below recrystallization temperature.
- When material is loaded in any direction (say tensile direction), it experiences linear elastic deformation till elastic limit. If load is removed, material comes back to its original shape & size and curve will retrace back loading path.
- If material is further loaded (beyond yield stress 'σ 0'), slight plastic deformation occurs leading to small permanent deformation. Now, if load is removed, material will retail this small plastic / permanent deformation and curve will not retrace original path, but will follow path parallel to original loading path.

- If same material is loaded in same (tensile) direction again, it will start yielding at new yield stress ' σ 0, which will be higher than initial yield stress of material' σ 0'.
- This is because material becomes harder during plastic deformation by storing strain energy (some % of total deformation energy)
- Reduces ductility and plasticity of material
- At micro level, dislocation piles up during plastic deformation at slip planes, which interact with each other and create barriers to further motion / movement of dislocations through crystal lattice

BauschingerEffect / Elastic Hysteresis

- Decrease in yield stress of material when loaded initially in one direction (tensile) and then subsequently in opposite direction (compressive) below recrystallization temperature
- When a material is loaded in any direction (say tensile direction), it experiences linear elastic deformation till elastic limit. If it is further loaded (beyond yield stress 'σ 0'), slight plastic deformation will occur leading to small permanent deformation
- Now, even if load is removed, material will retain this small plastic / permanent deformation and curve will not retrace original path, but will follow path parallel to original loading path

Bauschinger Effect

- Further, loaded in opposite direction (compressive), it will start yielding at new yield stress 'σ 0', which will be very less than initial yield stress of material'σ 0'.
- This is because residual tensile stress stored in material during tensile loading need to be removed / compensated before elastic deformation in reversal (compressive) direction starts.
- At micro level, dislocations piles up during tensile loading along slip planes, which can be easily moved in opposite direction (compressive loading) by residual stresses.

Anelasticity:
Time-dependent, recoverable low strain with linear behavior and relatively low damping
Viscoelasticity:
Time-dependent, recoverable higher strains with nonlinear behavior and higher damping

Viscoplasticity: Time-dependent, non-recoverable strain with nonlinear behavior and higher damping.

Ques.636: Write notes onRecovery, Recrystallization & Grain Growth .

Answer: Recovery:

- Recovery takes place at low temperatures of annealing.
- Apparently no change in microstructure.

- During recovery, some of the stored internal strain energy is relieved through dislocation motion due to enhanced atomic diffusion at the elevated temperatures.
- Leads to reduction in the number of dislocations.
- This process also removes the residual stresses formed due to cold working.
- The recovery of physical & mechanical properties varies with the temperature and time.
- Recovery is a relaxation process with the following characteristics:
  - There is no incubation period.
  - Recovery rate is large at the beginning, and then it slows down till it is near zero
  - Recovery has a limit value varying with temperature; the higher the temperature, the greater is the limit value and the shorter is the time needed to reach the limit value.
  - The greater the deformation, the greater is the initial recovery rate, and decrease in grain size helps to accelerate the recovery process.

Recrystallization

- After recovery is complete, the grains are still in a relatively high strain energy state.
- Recrystallization is the formation of a new set of strain-free and uniaxial grains that have low dislocation densities.
- The temperature at which materials are recrystallized is known as the recrystallization temperature, Trecrystallization = (0.3 – 0.5) T_m.
- The driving force to produce the new grain structure is the internal energy difference between strained and unstrained material.
- The new grains form as very small nuclei and grow until they consume the parent material.
- Recrystallization is a heterogeneous process and dependent on the deformation state of steels.
- The kinetics of recrystallization depends on nucleation rate, N, and growth rate, G.

Grain Growth

- After recrystallization, the strain-free grains will continue to grow if the metal specimen is left at elevated temperatures. Grains begin to grow via grain boundary immigration; this phenomenon is called grain growth.
- Grain growth is driven by the tendency to decrease the total boundary surface energy by decreasing the grain boundary area.
- Large grains grow at the expense of smaller grains.

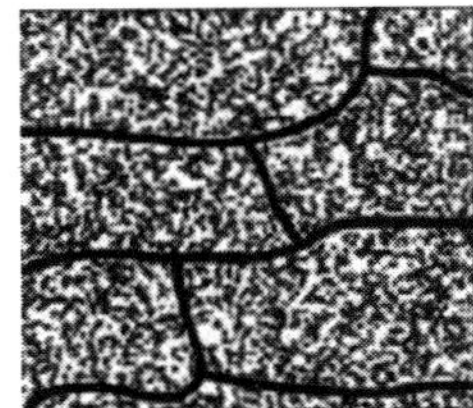 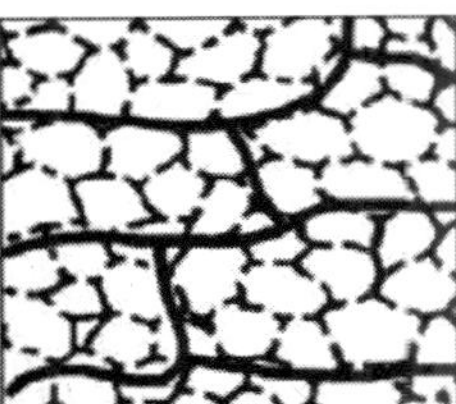 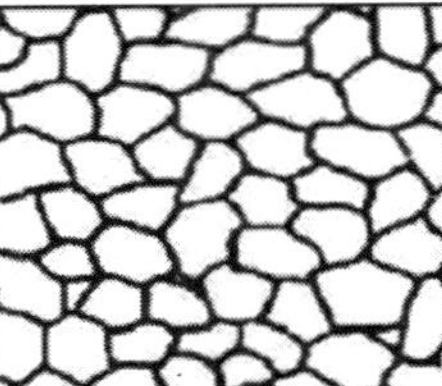 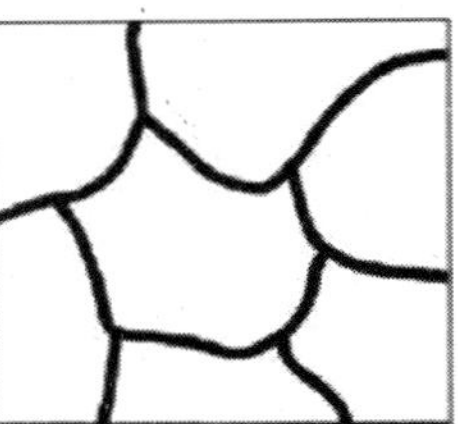

(a) Work hardening, (b) recovery, (c) recrystallization, and (d) grain growth during annealing.

Ques.637: Describe the salient features of various General heat treatment processes .
Answer:

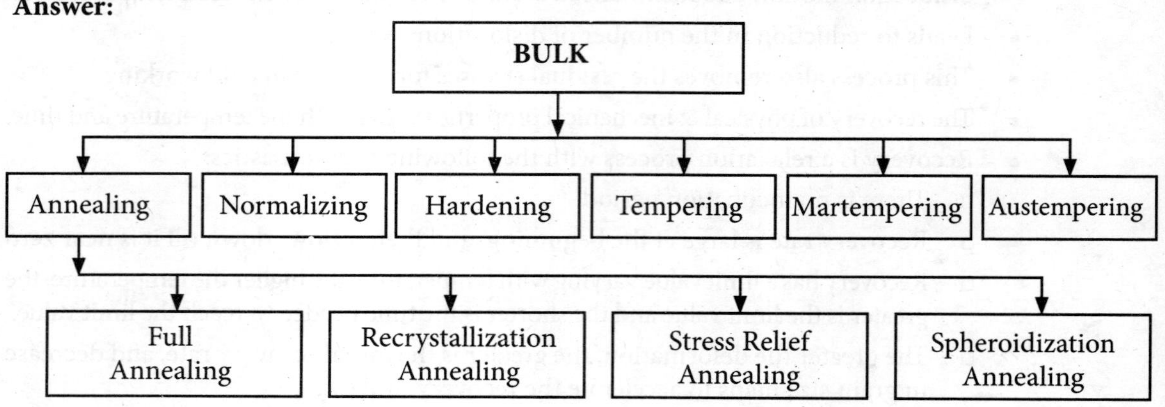

Annealing

Annealing cycles are varied according to the temperature and the method of cooling used. Annealing temperatures may be classified as:

- Subcritical annealing (below the lower critical temperature, A_1)
- Intercritical annealing (above A_1 but below the upper critical temperature, A_3 in hypoeutectoid steels, or A_{cm} in hypereutectoid steels)
- Supercritical annealing to achieve full austenitization of either hypo- or hyper-eutectoid steels.

Some commonly used heat treatment cycle superimposed on Fe-C equilibrium diagram is shown in Figbelow .

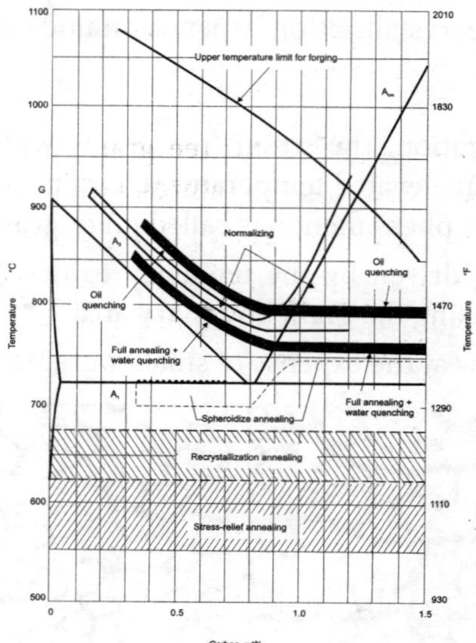

Fig.- Approximate temperature ranges for various heat treatments applied to steels

Full Annealing

- The temperatures for full annealing are typically 50°C above the upper critical temperature (A_3) for hypoeutectic steels and the lower critical temperature (A_1) for hypereutectoid steels.
- It is referred to as full annealing because it achieves full austenitization of hypoeutectoid steels.
- The indicated temperatures for full annealing are a supercritical anneal for hypoeutectoid steels, but are in the range of an intercritical anneal for hypereutectoid steels.
- Full annealing produces a microstructure that is softer and more amenable to other processing such as forming or machining.
- In addition, stainless and high-alloy steels may be austenitized (fully annealed) and quenched to minimize the presence of grain boundary carbides or to improve the ferrite distribution.
- The temperatures for full annealing are in approximately the same range of austenitization temperatures for water quenching but are below austenitization temperatures for either normalizing or hardening by oil quenching.

Recrystallization Annealing

- When cold worked metals are heated to a sufficiently high temperature, the badly deformed cold worked grains are replaced by new strain-free grains.
- This process is referred to as recrystallization, which is distinct from the recovery process during stress relief. At a constant temperature, recovery starts rapidly and then decreases with time. On the other hand, recrystallization, which is a nucleation and growth process, starts slowly and then builds up to a maximum rate before rapidly leveling off. Recrystallization annealing of steel is done at subcritical temperatures.
- Recrystallization annealing of cold worked metal can produce a new grain structure without inducing a phase change. The recrystallization temperature is often defined as the temperature required for the microstructure to undergo 50% recrystallization in 30 min, and for complete recrystallization in approximately one hour.
- Although there is a trade-off between time and temperature, temperature is a much more dominant variable than time. Recrystallization occurs more rapidly at higher temperatures. For most kinetic processes, increasing the temperature by approximately 10°C doubles the reaction rate. Once recrystallization is complete, further heating causes grain growth to occur. The recrystallized grain size is dependent on both the recrystallization time and temperature, particularly the temperature. Higher temperatures tend to promote larger grain sizes.
- The temperature required for recrystallization is not exact—it depends on the composition of the alloy and in particular on the amount of cold work performed. Therefore, recovery may affect the recrystallization temperature, because the tendency toward recrystallization is lowered when appreciable recovery has

occurred; ie, a higher temperature may then be required to cause recrystallization. In addition, grain size affects the nucleation rate and the recrystallization temperature. For equal amounts of cold work, a lower recrystallization temperature and shorter recrystallization time occurs for fine-grained metal than in coarse grained metal.

Stress Relief Annealing

- Annealing at subcritical temperatures (below Ac_1) includes three different temperature regions. The first is the temperature range for stress-relief annealing.
- Annealing involves uniform heating of a structure, or portion thereof, and holding at this temperature for a predetermined period of time, followed by uniform cooling. Stress relief temperatures are sufficiently high to help activate the microstructural process of recovery, such that stored internal strain energy is relieved by rearrangement of dislocations into lower-energy configurations without any change in the shape and orientation of grains.
- Stress-relief heat treating is used to relieve stresses from cold working. Care must be taken to ensure uniform cooling, particularly when a component is composed of variable section sizes. If the rate of cooling is not constant and uniform, new residual stresses can result that are equal to or greater than those that the heat-treating process was intended to relieve.
- Stress-relief heat treating can reduce distortion and high stresses from welding that can affect service performance. The presence of residual stresses can lead to stress-corrosion cracking near welds and in regions of a component that has been cold strained during processing.

Spheroidizing

- Depending on the time-temperature exposure of a steel, the cementite carbides in a steel can form different morphologies, such as the lamellar cementite of pearlite, or a network of cementite along prior austenite grain boundaries in a hypereutectoid steel.
- The shape and distribution of the carbides can be modified by heat treatment, and spheroidization treatments are used to produce carbides with a more spherical morphology.
- Because spherical shapes have lower surface energy than angular shapes, the lamellar shape of cementite in pearlite changes morphology to form spheroids. Portions of the lamellae "pinch off" (dissolve) to form a spheroid from the remaining portions of lamellae. This process can occur over a long period of time, or it can be accelerated by heat treatment. Depending on the steel, spheroidized carbides can lead to improved machinability, toughness, and formability.
- In hypoeutectoid steels, prolonged holding at a temperature just below Ae_1 accelerates the process of spheroidization. The process may take several hours. To improve on kinetics for full spheroidization, some treatments involve heating & cooling alternately between temperatures that are just above Ac_1 and just below Ar_1.

- In low-carbon steels, the typical purpose of spheroidizing is to improve the cold formability of steels. The formability of steel is altered significantly depending on whether carbides are spheroids or present as lamellae in pearlite. Low-carbon steels are seldom spheroidized for machining, because in the spheroidized condition they are excessively soft and "gummy," cutting with long, tough chips.
- In hypereutectoid steels (C > 0.77%) & tool steels, spheroidization is done to improve machinability and improve toughness. Heating of hypereutectoid steels above Ac_1 is done to create dispersed cementite particulates. Heating to dissolve the carbide prevents reformation of a carbide network. If a temperature slightly above Ac_1 is to be used, good loading characteristics & accurate temperature controls are required for proper results; otherwise, it is conceivable that Ac_1 may not be reached and that austenitization may not occur.

Normalizing

- Steel is normalized to refine grain size, make its structure more uniform, make it more responsive to hardening, and to improve machinability.
- When steel is heated to a high temperature, the carbon can readily diffuse, resulting in a reasonably uniform composition from one area to the next. The steel is then more homogeneous and will respond to the heat treatment more uniformly.
- Normalizing is an austenitizing heating cycle followed by cooling in still or slightly agitated air. Typically, the temperatures for normalizing are approximately 55°C above the upper critical line (that is, above Ac_3 for hypoeutectoid steels and above A_{cm} for hypereutectoid steels).
- To be properly classed as a normalizing treatment, the heating portion of the process must produce a homogeneous austenitic phase prior to cooling. Figure below compares the time-temperature cycle of normalizing to that of full annealing.

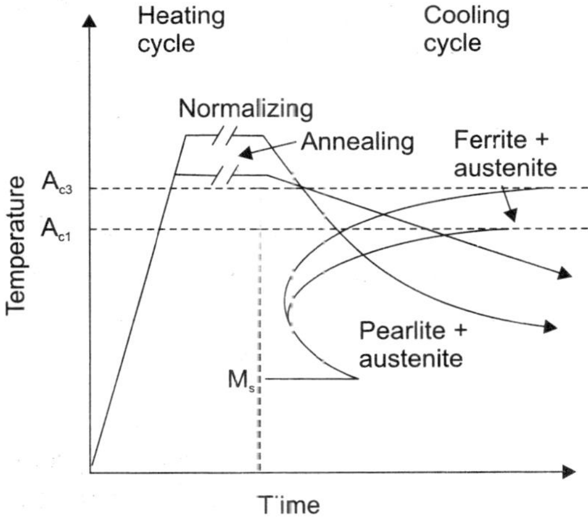

Fig. Comparison of normalizing and full annealing heat treatment cycle

Hardening

- The maximum hardness of any steel is associated with a fully martensitic structure. This microstructure can be produced only if the cooling rate applied is higher than the critical cooling rate for the corresponding steel.
- Hardenability is defined as the "susceptibility to hardening by rapid cooling", or as "the property, in ferrous alloys, that determines the depth and distribution of hardness produced by quenching". Both of these definitions emphasize hardness. As discussed previously, the source of hardening is the formation & presence of martensite, and therefore a third definition of hardenability, "the capacity of a steel to transform partially or completely from austenite to some percentage of martensite at a given depth when cooled under some given conditions," more accurately describes the physical process underlying hardening.

Severity of Quench

- The effectiveness of a given quenching medium is ranked by a parameter referred to as its "severity of quench." This measure of cooling or quenching power is identified by the letter "H" and is determined experimentally by quenching a series of round bars of a given steel.
- Table I lists H values for a number of commonly used quenches. Note that apart from the nature of the quenching medium, the vigorousness of the shake determines the severity of the quench. When a hot solid is put into a liquid medium, gas bubbles form on the surface of the solid (interface with medium). As gas has a poor conductivity the quenching rate is reduced. Providing agitation (shaking the solid in the liquid) helps in bringing the liquid medium in direct contact with the solid; thus improving the heat transfer (and the cooling rate).

Table I. Severity of quench (H) for various quenching media

	Air	Oil	Water	Brine
No ciruclation of fluid or agitation of piece	0.02	0.25-0.30	0.9-1.0	2
Mild cirulation (or agitation)	---	0.30-0.35	1.0-1.1	2-2.2
Moderate circulation	---	0.35-0.40	1.2-1.3	---
Good circulation	---	0.4-0.5	1.4-1.5	---
Strong circulation	0.05	0.5-0.8	1.6-2.0	---
Violent cirulation	---	0.8-1.1	4	5

Tempering

- Martensite in steels can be very strong and in its "as quenched" condition rather brittle. It is then necessary to modify its mechanical properties by heat treatment in the range 150–700°C. This process is called tempering, in which the microstructure approaches equilibrium under the influence of thermal activation.

- The general trend during the tempering of martensite therefore begins with the rejection of excess carbon to precipitate carbides but the substitutional solutes do not diffuse during this process. The end result of tempering is a dispersion of coarse carbides in a ferritic matrix which bears little resemblance to the original martensite.
- It should be borne in mind that in many steels, the martensite reaction does not go to completion on quenching, resulting in varying amounts of retained austenite which might decompose during the tempering process.
- The effect of tempering temperature on the hardness of steel is shown in figure below.

Tempering of Plain Carbon Steels

On reheating as-quenched martensite, the tempering takes place in four distinct but overlapping stages: Stage 1, up to 250°C: precipitation of ε-iron carbide; partial loss of tetragonality in martensite. Stage 2, between 200°C and 300°C: decomposition of retained austenite. Stage 3, between 200°C and 350°C: replacement of ε-iron carbide by cementite; martensite loses tetragonality. Stage 4, above 350°C: cementite coarsens and spheroidizes; recrystallization of ferrite.

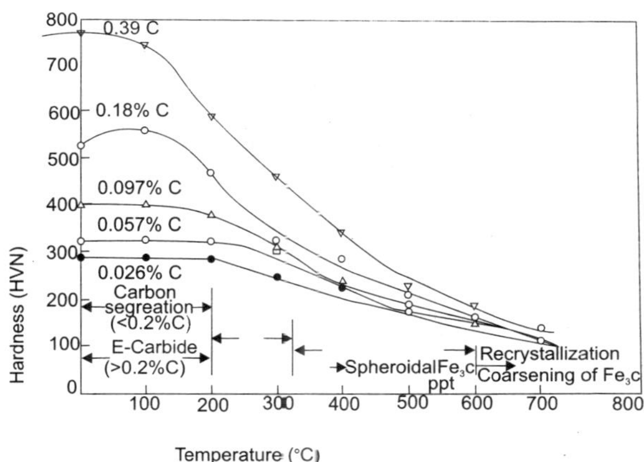

Fig. Hardness of iron–carbon martensites tempered for 1 hr at 100-700 °C.

Martempering

- Martempering is a heat treatment for steel involving austenitisation followed by step quenching, at a rate fast enough to avoid the formation of ferrite, pearlite or bainite to a temperature slightly above the martensite start (Ms) point. A schematic of above the process is shown in figure below.
- The advantage of martempering lies in the reduced thermal gradient between surface and center as the part is quenched to the isothermal temperature and then is air cooled to room temperature.
- Residual stresses developed during martempering are lower than those developed during conventional quenching because the greatest thermal variations occur

while the steel is in the relatively plastic austenitic condition and because final transformation & thermal changes occur throughout the part at approximately the same time. Martempering also reduces or eliminate ssusceptibility to cracking.

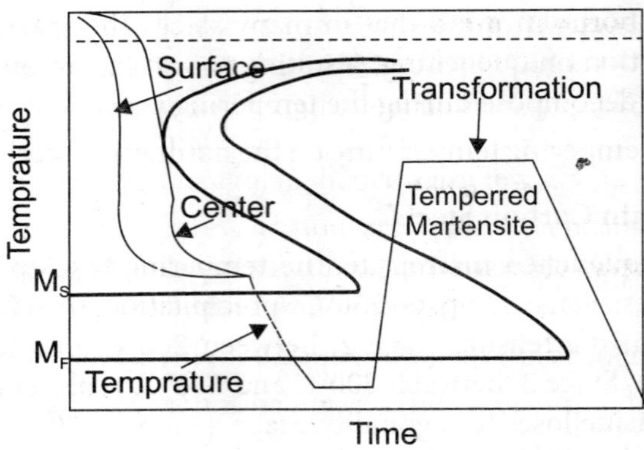

Fig. Schematic representation of the Martempering process

Austempering

- Austempering process is similar to the Martempering process except that the sample after the homogenization of temperature from surface and temperature, instead of quenching, held at that temperature above M_s for sufficient time (in the bainite transformation range) to produce banitic microstructure.
- A schematic of this process is shown in figure below. An advantage of this process over Martempering is that the tempering can be avoided. Other advantages include, Higher ductility, Resistance to shock, Uniform hardness, Tougher & more wear restiance, Higher impact & Fatigue Strengths.

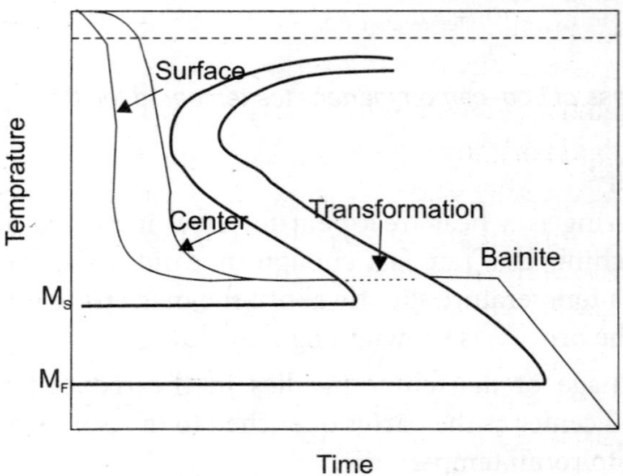

Fig. Schematic representation of the Martempering process.

Surface Hardening

- Numerous industrial applications require a hard wear resistant surface called the case, and a relatively soft, tough inside called the core. Example: Gears
- They are two different categories. They are thermo chemical and thermo mechanical treatments. Thermo chemical treatment is related to change in chemical composition and in Thermo mechanical treatment, there is no change of chemical composition of the steel and are essentially shallow- hardening methods.
- A detailed flow chart is given below related to surface hardening treatments.

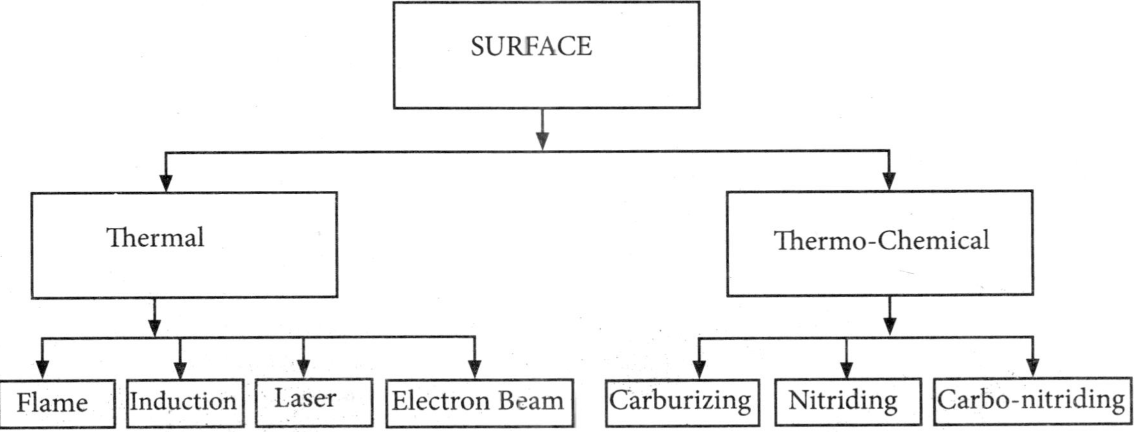

Carburizing

- Carburizing is the most widely used method of surface hardening. Here, the surface layers of a low carbon steel (<0.25) is enriched with carbon up to 0.8-1.0%. The source of carbon may be a solid medium, a liquid or a gas.
- In all cases, the carbon enters the steel at the surface and diffuses into the steel as a function of time at an elevated temperature. Carburizing is done at 920-950°C.
- Where Fe(C) represents carbon dissolved in austenite. the rate of diffusion of carbon in austenite, at a given temperature is dependent upon the diffusion coefficient and the carbon concentration gradient.
- The carburizing equation given previously, $Fe+2CO = Fe(C) + CO_2$ is reversible and may proceed to the left, removing carbon from the surface layer if the steel is heated in an atmosphere containing carbon dioxide (CO_2). This is called decarburization.
- Decarburization may be prevented by using an endothermic gas atmosphere in the furnace to protect the surface of the steel from oxygen, carbon dioxide and water vapor.
- An endothermic gas atmosphere is prepared by reacting relatively rich mixtures of air and hydrocarbon gas (usually natural gas) in an externally heated generator in the presence of a nickel catalyst.

Liquid carburizing

- Liquid carburizing is a method of case hardening steel by placing it in a bath (8% NaCN, 82% $BaCl_2$ and 10% NaCl) of molten cyanide so that carbon will diffuse from the bath into the metal and produce a case comparable to one resulting from pack or gas carburizing.
- Liquid carburizing may be distinguished from cyaniding by the character and composition of the case produced. The cyanide case is higher in nitrogen and lower in carbon; the reverse is true of liquid carburized cases.

$$BaCl_2 + 2NaCN \text{------}\rightarrow Ba(CN)_2 + 2NaCl$$

$$Ba(CN)_2 + Fe \text{------}\rightarrow Fe© + Ba(CN)_2$$

Pack carburizing

- In pack carburizing, the articles to be carburized are packed in a box, embedding them in a powdery mixture of 85% charcoal and 15% of energizers such as $BaCO_3$ The box is sealed with fireclay and loaded into the furnace kept at 930°C. The residual air in the box combines with carbon to produce CO. The energizer decomposes as below :

$$BaCO_3 \text{------}\rightarrow BaO + CO_2$$

$$CO_2 + C \text{------}\rightarrow 2CO$$

- The carbon enters the steel through the following reaction:

$$Fe + 2CO \text{------}\rightarrow Fe© + CO_2$$

- If selective carburization is to be done, copper is electroplated to a thickness of ~0.05 mm in regions where carburization is not desired. Alternatively, a refractory paste of fireclay mixed with asbestos can be applied. Control of temperature & penetration depth is less in pack carburizing as compared to liquid and gas carburizing. Also, direct quench from the carburizing temperature to harden the surface is not possible.

Gas Carburizing

- The steel is heated in contact with carbon monoxide and/or a hydrocarbon which is readily decomposed at the carburizing temperature.
- A mixture consisting of 5-15% methane (or propane) in a neutral carrier gas is used. The methane decomposes according to the following reaction:

$$CH_4 + Fe \text{------}\rightarrow 2H_2 + Fe©$$

- The carbon potential of the gas mixture increases with increasing concentration of methane. Too large a concentration or too high a gas velocity releases carbon faster than it can be absorbed and may result in soot formation on the surface.
- Closer control of temperature and case depth is possible in gas carburizing, as compared to pack carburizing. Also, post quenching can be done directly.

Heat treatment after carburizing

- Since steel is carburized in the austenite region, direct quenching from the carburizing temperature will harden both the case and core, will harden if the cooling rate is greater than the critical cooling rate. Direct quenching of coarse grained steels often leads to brittleness and distortion, so that this treatment should be applied only to fine grained steels. A diagrammatic representation of various hardening treatments for carburized steels shown below.

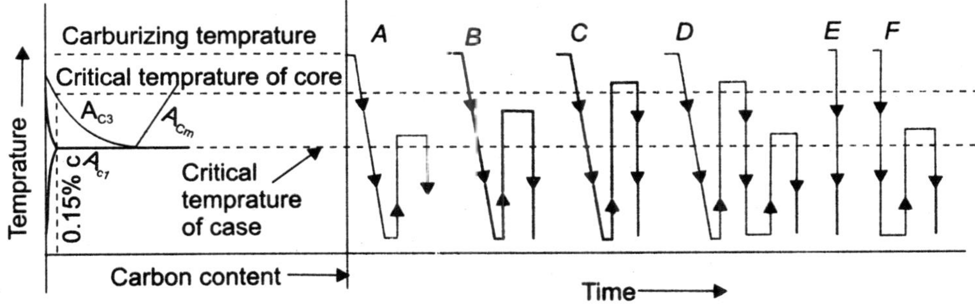

TREATMENT	CASE	CORE
best adapted to fine grained steels	Refined; Excess carbide not dissolved	Unrefined; soft and machinable
B-best adapted to fine grained steels	Slightly coarsened; some solution of excess carbide	Partially refined; stronger and tougher than A
C-best adapted to fine grained steels	Some what coarsened; solution of excess carbide favored; austenite retention promoted in highly alloyed steels.	Refined: maximum core strength and hardness; better combination of strength and ductility than B
D-best treatment for coarse grained steels	Refined solution of excess carbide favored; austenite retention minimized	Refined; soft and machinable; maximum toughness and ressistance to impact .
E-adapted to fine graines teels only	Unrefined with excess carbide dissolved; austenite retained; distortion minimized	Unrefined but hardened
F-adapted to fine grained steels only	Refined; solution of excess carbide favored; austenite retention minimized	Unrefined; fair toughness

Cyaniding and Carbonitriding

- In this, case that contain both carbon and nitrogen are produced in liquid salt baths (cyaniding) or by use of gas atmospheres (carbonitriding). The temperatures used are generally lower than those used in carburizing, being between 750-900°C.
- Exposure is for a shorter time, and thinner cases are produced, up to 0.010 in for cyaniding and up to 0.030 in for carbonitriding.
- Cyaniding is done in a liquid bath of NaCN, with the concentration varying between 30 and 97%. Both carbon and nitrogen enter the steel via the following reactions:

$$2NaCN + O_2 \dashrightarrow 2NaCNO$$

$$3NaCNO \dashrightarrow NaCN + Na_2CO_3 + C + 2N$$

- The temperature used for cyaniding is lower than that for carburizing and in the range of 800-870°C. the time of cyaniding is 0.5-3 hour to produce a case depth of 0.25 mm or less.

- Carbonitriding is a case-hardening process in which a steel is heated in a gaseous atmosphere of such composition that carbon and nitrogen are absorbed simultaneously.
- This process is also known as dry cyaniding or gas cyaniding. It is the gas carburizing process modified by the addition of anhydrous ammonia.
- The decomposition of ammonia provides the nitrogen, which enters the steel along with carbon.
- A typical gas mixture consists of NH_3, 5% CH_4, and 80% of neutral carrier gas. The temperature used is 750-900°C. With increasing temperature, a greater proportion of carbon enters the steel.
- The presence of nitrogen in the austenite accounts for the major differences between carbonitriding and carburizing. Carbon-nitrogen austenite is stable at lower temperatures than plain-carbon austenite and transforms more slowly on cooling. Carbonitriding therefore can be carried out at lower temperatures and permits slower cooling rates than carburizing in the hardening operation.

Nitriding

- In contrast to the processes described before, nitriding is carried out in the ferrite region. Consequently, no phase change occurs after nitriding .
- The part to be nitrided should possess the required core properties prior to nitriding. Pure ammonia decomposes to yield nitrogen which enters the steel.
- The solubility of nitrogen in ferrite is small. Most of the nitrogen that enters the steel forms hard nitrides (e.g. Fe_3N). A typical nitriding steel contains alloying elements of 1%Al, 1.5%Cr and 0.2%Mo. Al, Cr, and Mo form very hard and wear resistant nitrides.
- The temperature of nitriding is 500-590°C; the time for a case depth of 0.02mm is about 2 hour. In addition to wear resistance, it also increases the resistance of a carbon steel to corrosion in moist atmospheres.
- A common problem encountered in nitriding is the formation of γ' nitride (Fe_4N) on the outer layers of the case , known as the "white layer", as it looks white under the microscope. This layer is very brittle and tends to crack. It must be removed by final grinding operation. Its formation can be minimized by maintaining the correct ratio of NH_3/ H_2 in the gas mixture during the heat treatment.

Thermal Energy Treatments

- For large work pieces, such as mill rolls, large gears and complicated cross sections, in such cases, flame hardening is done by means of an oxyacetylene torch. Heating should be done rapidly by the torch and the surface quenched, before appreciable heat transfer to the core occurs .
- Four methods are generally used for Flame Hardening:
  - Stationary (Spot): Torch and work is stationary
  - Progressive: Torch moves over a work piece
  - Spinning: Torch is stationary while work piece rotates
  - Progressive-spinning: Torch moves over a rotating work piece.

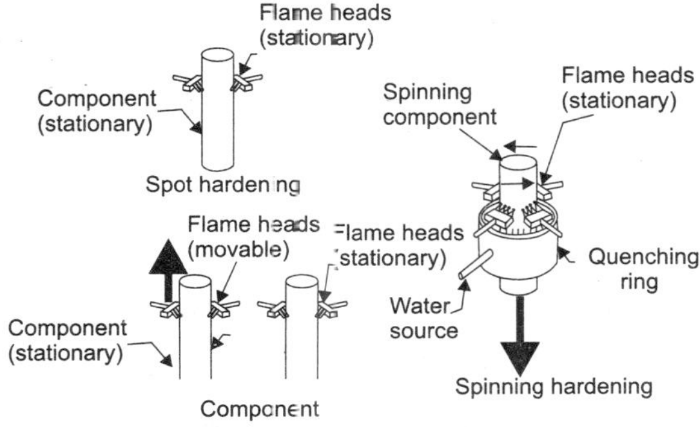

Progressing hardening

Induction Hardening

- Here, an alternating current of high frequency passes through an induction coil enclosing the steel part to be heat treated. The induced emf heats the steel. The depth up to which the heat penetrates and rises the temperature above Ac_3 is inversely proportional to the square root of the AC frequency.
- Correspondingly, the hardened depth decreases with increasing frequency in induction hardening, the heating time is usually a few seconds. Immediately after heating water jets are activated to quench the surface .Martensite is produced at the surface, making it hard and wear resistant. The microstructure of the core remains unaltered. Induction hardening is suitable for mass production of articles of uniform cross section.

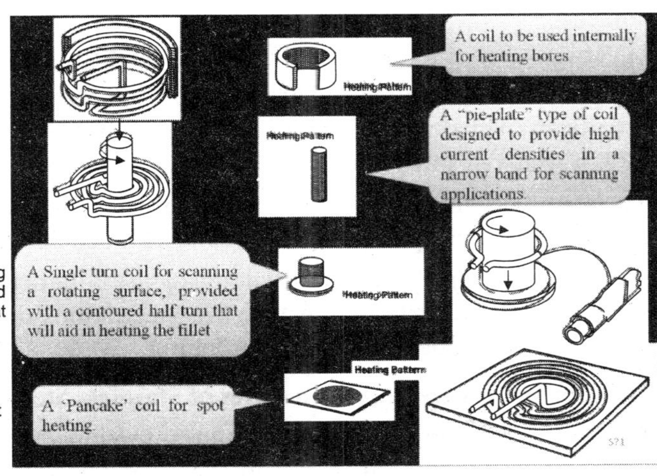

A Coil to be used internally for heatintg bores

A "pie-plate" type of coil designed to provide high current densities in a narrow band for scanning applications.

A Single turn coil for scanning a roasting surface, provided with a contoured half turn that will aid in heating the fillet

A 'Pancake' coil for spot heating.

Laser Hardening

- Laser hardening treatment is widely used to harden localized areas of steel and cast iron machine components. This process is sometimes referred to as laser transformation hardening to differentiate it from laser surface melting phenomena.

- There is no chemistry change produced by laser transformation hardening, and the process, like induction & flame hardening, provides an effective technique to harden ferrous materials selectively.
- As laser beams are of high intensity, a lens is used to reduce the intensity by producing a defocused spot of size ranging from 0.5 to 25 mm. Proper control of energy input is necessary to avoid melting.
- Laser transformation hardening produces thin surface zones that are heated and cooled very rapidly, resulting in very fine Martensitic microstructures, even in steels with relatively low hardenability. High hardness and good wear resistance with less distortion result from this process.
- Laser hardening has the advantage of precise control over the area to be hardened, an ability to harden reentrant surfaces, very high speed of hardening and no separate quenching step (the quench is effected by the mass of the unheated material).
- The disadvantage is that the hardening is shallower than in induction and flame hardening.

Electron Beam (EB) Hardening

- Electron Beam (EB) hardening is like laser treatment, is used to harden the surfaces of steels. The EB heat treating process uses a concentrated beam of high-velocity electrons as an energy source to heat selected surface areas of ferrous parts. Electrons are accelerated and are formed into a directed beam by an EB gun.
- After exiting the gun, the beam passes through a focus coil, which precisely controls beam density levels (spot size) at the work piece surface and then passes through a deflection coil.
- To produce an electron beam, a high vacuum of 10-5 torr is needed in the region where the electrons are emitted and accelerated. This vacuum environment protects the emitter from oxidizing and avoids scattering of the electrons while they are still traveling at a relatively low velocity.
- Like laser beam hardening, the EB process eliminates the need for quenchants but requires a sufficient work piece mass to permit self quenching.
- A mass of up to eight times that of the volume to be EB hardened is required around and beneath the heated surfaces. Electron beam hardening does not require energy absorbing coatings, as does laser beam hardening.

Ques.638: Write a short note on heat treatment and uses of Stainless Steel.

Answer : Stainless Steels are a large group of special alloys developed primarily to withstand corrosion. These steels contain chromium in excess of 11% by weight which imparts "stainless" characteristics to iron alloys.

Martensitic Stainless Steel

- These are primarily straight chromium steels with 11.5 to 18% Cr, 0.15% C, 1.25% Mn, and 1%Si.

- Common examples are 403, 410, 416, 420, 440A, 501 & 502 (AISI grades).
- Used for turbine blades and corrosion resistant applications.

Heat Treatment

- Process Annealing → 650 - 760 °C, increase ductility and machinability
- Austenitizing→ 925 -1065 °C followed by oil quenching or air cooling, increase corrosion resistance and strength

Austenitic Stainless Steel

- These are chrome-nickel (3xx type) or chrome -nickel- manganese(2xx type) alloys.
- Total content of Ni and Cr is at least 23%.
- Difficult to machine but can be improved by addition of selenium or sulfur.
- Best high temperature strength and scaling resistance. Hence shows best corrosion resistance
- Used in chemical industry and for household and sanitary fittings.

Heat Treatment

- Cold working causes work hardening but hot working can easily be done.
- Annealing at high temperatures → Recrystallization and carbide solution
- Solution treatment → Dissolution of chromium carbides

Tool Steel

- Tool steel refers to a variety of carbon and alloy steels that are particularly well-suited to be made into tools.
- Characteristics include high hardness, resistance to abrasion (excellent wear), an ability to hold a cutting edge, resistance to deformation at elevated temperatures (red-hardness).
- Tool steel are generally used in a heat-treated state.

Ferritic Stainless Steel

- These are primarily straight chromium steels with 14 to 27% Cr. Carbon is restricted both to maintain high toughness & ductility and to prevent austenite formation (As carbon expands the gamma loop)
- These steels can be cold formed easily and hence are used for deep drawn parts such as vessels for food & chemical industries and for architectural & automotive trim.

Heat Treatment

- Can be cold or hot worked.
- Annealing → 760 - 966 °C, Recrystallization of cold-worked structures → achieves maximum softness, ductility and corrosion resistance.

Uses

- Shock resisting tool steels → Intended for applications requiring toughness & resistance to shock-loading such as hammers, chisels, punches, driver bits and others.
- Water hardening tool steels → Shallow hardened and relatively low resistance to softening. They are suitable for woodworking tools, hand-metal cutting tools such as taps & reamers and cutlery.
- Steels for Room Temperature Use (Classified according to their quenching media)
- Water hardened grades (W) → Plain carbon steels with 0.6-1.0 %C. These have a low hardenability, ie., martensite only to a depth of 0.5 in. V can be added (forms V_4C_3) to improve the hardness and wear resistance of these steels.
- Shock resistant grades (S) → Contain small amounts of Cr or Mo and are quenched in oil. They have lower C contents (0.5%) to improve impact strength.
- Oil hardened grades (O) → Small percentages of Cr and W with 0.9 %C. The have medium hardness and are used to short run cold forming dies.
- Air hardening grades (A) → Greater amounts of Cr & Mo and 1 %C. Used for complicated shapes and thread rolling. Mo and W are relatively expensive so they are only added in small amounts to give much improved hardenability.
- High carbon, high Cr grades (D grade) → 12 %Cr and 1.5-2.25 %C are extremely wear resistant and used for long run dies & for gauges. Chromium is a relatively low cost addition for increasing hardenability with the excess Cr, $Cr_{23}C_6$ is also formed, which improves wear resistance.

Steels for High Temperature Use

- Chromium hot working steels (H grades) → 5-7 % Cr, 0.4-1.0% V, 1.5-7.0% W, 1.5% Mo, 0.35% C. Medium hot working for Mg and Al extrusion die-casting dies.
- Tungsten hot working steels (H) → 9.5-12 % W, 3.5-12.0 % Cr, 0.35 % C. Hot working, extrusion and forging dies for brass, nickel & steel.
- Tungsten high speed steel (T) → 12-18 % W, 4.0 % Cr, 1-5 % V, 0.7-1.5 % C. Original high speed (HS) cutting steel with excellent HT wear resistance.
- Molybdenum HS steel (M) → 3.5-8.0 % Mo, 1.5-6.0 % W, 4.0 % Cr, 1-5 % V, 5 % Co, 0.8-1.5 % C. Used for 85% of US cutting steels before the advent of ceramic cutting tools.

High Speed Steel (HSS)

- HSS is a subset of tool steel, commonly used in tool bits and cutting tools. It is often used in power saw blades and drill bits.
- They are characterized by high carbon contents, sometimes up to 1.5%, and major additions of strong carbide forming elements such as chromium, molybdenum, tungsten & vanadium. Up to 12% Co is also included in some of the more complex grades.

- It can withstand higher temperatures without losing its temper (hardness) which allows it to cut faster than high carbon steel, hence the name.
- Other characteristics include high hardness, resistance to abrasion (excellent wear), an ability to hold a cutting edge, resistance to deformation at elevated temperatures (red-hardness).
- Modulus Of elasticity: 221GPa, Density : 8767 kg/m3 (↑ durability & hardness), Thermal Conductivity: 21W/m/K
- HSS are mainly of two types: Tungsten based ones (T grades), Molybdenum based ones (M grades).

Glossary

A

Abrasion

The displacement and/or detachment of metallic particles from a surface as a consequence of being exposed to flowing solids, fluids or gases.

Abrasion Resistance

Degree of resistance of a material to abrasion or wear.

Acid Embrittlement

Embrittlement during pickling due to absorption of hydrogen.

Age Hardening

Hardening by aging, usually after rapid cooling or cold working.

Aging

A change in properties of metals & alloys which occurs slowly at room temperature and will proceed rapidly at higher temperatures. The change in properties is often, but not always, due to a phase change (precipitation), but never involves a change in chemical composition of the metal or alloy.

Air Quenching

Accelerated cooling of alloy in an air stream from temperatures above the Ac3 temperature.

Alloy

A substance having metallic properties and composed of two or more chemical elements of which at least one is metal. Usually possesses qualities different from those of the components.

Alloy Steel

Steel containing significant quantities of alloying elements other than carbon and the commonly accepted amounts of manganese, silicon, sulfur, and phosphorus.

Alpha - ferrite

Body-centered cubic type of pure iron stable below 910°C .

Alpha Martensite

A form or stage of martensite representing the least developed and most distorted stage in the transformation of austenite to martensite at ordinary temperatures.

Annealing

Heating to and holding at a suitable temperature, followed by cooling at a suitable rate to lower the hardness or alter other mechanical or physical properties.

Anti carburizing Compounds

Compounds applied to metallic surfaces to prevent surface carburization.

Argon Oxygen Decarburization (AOD)

A secondary refining process in which argon, oxygen and nitrogen are injected into a molten bath of steel. The AOD process improves metal cleanliness and thus gives superior mechanical properties.

Artificial Aging

An aging treatment above room temperature.

As Cast

Referring to metal which has not received finishing (beyond gate removal or sand blasting) or treatment of any kind including heat treatment after casting. Similarly, as drawn, as forged, and as rolled.

ASTM

American Society for Testing and Materials.

Austenite

The face-centered-cubic phase of iron and steel, also referred to as gamma iron. In steel, a solid solution in which gamma iron is the solvent.

Austenitic Steel

Any steel containing sufficient alloy to produce a stable austenitic (gamma iron) crystalline structure at ambient temperatures.

Batch

Amount or quantity of core or mold sand or other material prepared at one time.

Bead

(a) Half-round cavity in a mold, or half-round projection or molding on a casting, (b) a single deposit of weld metal produced by fusion.

BHN

Brinell Hardness Number.

Bimetal

Casting, usually centrifugal, made of two different metals, fused together.

Blacking Carbon

Carbonaceous materials such as plumbago, graphite or powdered coke usually mixed with a binder and frequently carried in suspension in water or other liquid; used as thin facing applied to surfaces of molds or cores to improved casting finish.

Blasting (Blast Cleaning)

A process for cleaning or finishing metal objects by use of an air blast or centrifugal wheel that throws abrasive particles against the surface of the work pieces. Small, irregular particles of steel or iron are used as the abrasive in grit blasting, and steel or iron balls in shot blasting.

Blister

A shallow blow with a thin film of the metal over it appearing on the surface of a casting.

Blow Holes

(a) Holes in the head plate or blow plate of a core-blowing machine through which sand is blown from the reservoir into the core box. (b) Irregular shaped cavities with smooth walls produced in a casting when gas is entrapped during mold filling. The gas sources may be air, binder decomposition products or gases dissolved in the molten steel.

Boil

Agitation of a bath of metal caused by the liberation of a gas beneath its surface. May be deliberately induced by the addition of oxidizing material to a bath containing excess carbon. In the later case it is called a carbon boil and CO or CO_2 are liberated.

Bond

(a) Bonding substance or bonding agents - any material other than water, which, when added to foundry sands, imparts bond strength, (a) the overlapping of brick so as to give both longitudinal and transverse strength.

Bond Strength

Property of a foundry sand to offer resistance to deformation.

Boring

A machining method using single point tools on internal surfaces of revolution.

Bottom Running or Pouring

Filling of the mold cavity from the bottom by means of gates from the runner.

Bright Annealing

A process carried out usually in a controlled furnace atmosphere, so surface does not oxidize, remaining bright.

Brinell Hardness

The value of hardness of a metal on an arbitrary scale representing kg/mm^2, determined by measuring the diameter of the impression made by a ball of given diameter applied under a known load. Values are expressed in Brinell Hardness Numbers, BHN.

Brittle Fracture

Fracture with little or no plastic deformation. Smoothing machined holes or outside surfaces of castings by drawing, pushing on, or more broaches (special cutting tools) through the roughed out hole.

Bulk Density

The ratio of the weight of a material to its over-all volume (including any inherent porosity).

Burned-On-Sand

A misnomer usually indicating metal penetration into sand resulting in a mixture of sand and metal adhering to the surface of a casting.

Burnishing

Developing a smooth finish on a metal by tumbling or rubbing with a polished hand tool.

C

Carbide

A compound of carbon with one or more metallic elements.

Carbon

Element occurring as diamond and as graphite. Carbon reduces many metals from their oxides when heated with the latter, and small amounts of it greatly affect the properties of iron. Though classed as a nonmetallic, metallurgically, like boron, it is treated as a metal.

Carbonitriding

A process in which a ferrous alloy is case hardened by first being heated in a gaseous atmosphere of such composition that the alloy absorbs carbon & nitrogen simultaneously, and then being cooled at a rate that will produce desired properties.

Carburizing

A form of case hardening that produces a carbon gradient inward from the surface, enabling the surface layer to be hardened by either quenching directly from the carbonizing temperature or by cooling to room temperature, then reaustenitizing and quenching.

Case Hardening

A process of hardening a ferrous alloy so that the surface layer or case is made substantially harder than the interior or core. Typically case hardening process are carburizing, carbonitriding, and nitriding.

Cementite

A compound of iron and carbon commonly known as iron carbide and having the approximate chemical structure, Fe_3C. Cementite is characterized by an orthorhombic crystal structure.

Centrifugal Casting

Casting made in molds which are rotating so as to produce a centrifugal force in the molten metal.

Charpy Impact Test

A pendulum-type single-blow impact test in which the specimen, usually notched, is supported at both ends as a simple beam and broken by a falling pendulum. The energy absorbed in fracture, is impact strength or notch toughness.

Chill (External)

Metal, graphite or carbon blocks that are incorporated into the mold or core to locally increase the rate of heat removal during solidification and reduce shrinkage defects.

Chill (Internal)

A metallic device / insert in molds or cores at the surface of a casting or within the mold to increase the rate of heat removal, include directional solidification and reduce shrinkage defects. The internal chill may then become a part of the casting.

CNC

Computer Numerical Controlled Machine Tools.

Coefficient of Expansion

Unit increase in size resulting from a unit increase in temperature; measured in millimeter per millimeter per degree Celsius (mm / mm/ °C).

Coining

(a) A process of straightening and sizing casting by die pressing, (b) a process for shaping metal.

Cold Cracking

Cracks in cold or nearly cold metal due to excessive internal stress caused by contraction. Often brought about when the mold is too hard or casting is of unsuitable design.

Cold Lap

Wrinkled markings on the surface of an ingot or casting from incipient freezing of the surface.

Cold Shot

Small globule of metal embedded in but not entirely fused with the casting.

Cold Shut

Casting defect caused by imperfect fusing or discontinuity of molten metal coming together from opposite directions in a mold, or due to folding of the surface. It may have the appearance of a crack or seam with smooth, rounded edges.

Cold Work

Plastic deformation of a metal at room temperature. Substantial increases in strength and hardness may occur

Collapsibility

The requirement that a sand mixture break down under the pressure & temperatures developed during casting, in order to avoid hot tears or facilitate the separation of the sand and the casting.

Color Etching

A micro-etch resulting from the formation of a thin film of a definite compound of the metal.

Columnar Structure

A coarse structure of parallel columns of grains, which is caused by highly directional solidification.

Compression Test

Imposing a dead load on a small cylindrical test piece to determine compressive strength, expressed in kg / cm^2.

Compressive Strength

The maximum stress in compression that can be withstood without plastic deformation or failure.

Conductivity (Thermal & electrical)

(Thermal).The quantity of heat that flows through a material measured in heat units per unit time per unit of cross-sectioned area per unit of length, (Electrical).the quantity of electricity that is transferred through a material of know cross-section and length.

Constituent

A micrographically distinguishable part of an alloy or mixture.

Contamination

(a) presence of small percentages of deleterious elements in an alloy adversely affecting the alloy's mechanical properties and/or casting soundness. (b) Radioactive deposition of radioactive material in any place where it is not desired, and particularly in any place where its presence may be harmful. The harm may be in vitiating the validity of an experiment or a procedure, or in actually being a source of danger to personnel

Contraction

The volume change occurring in metals (except antimony and bismuth) and alloys on solidification and cooling to room temperature.

Contraction Cracks

Cracks formed by restriction of the metal while contracting in the mold; may occur just after solidification (called a hot tear) or a short time after the casting has been removed from the mold.

Controlled Atmosphere

Any gas or mixture of gases that prevents or retards oxidation and decarburization.

Converter

A furnace in which a gas, usually air, is blown through the molten bath or crude metal for the purpose of oxidizing impurities.

Conveyor, Vibratory

A materials-handling device used usually with shakeout operations, to help clean sand from the castings as they are moved from one place to another in the foundry and as a feeding device to regulate materials flow.

Cooling Curve

A curve showing the relationship between time and temperature during the solidification and cooling of a metal sample. Since most phase changes involve evolution or absorption of heat, there may be abrupt changes in the slope of the curve.

Cooling, Controlled

A process of cooling from an elevated temperature in a predetermined manner used to produce a desired microstructure to avoid hardening, cracking or internal damage.

Cope

Upper or topmost section of a flask, mold or pattern.

Core

A performed sand aggregate inserted in a mold to shape the interior or that part of a casting which cannot be shaped by the pattern.

Core Binder

Any material used to hold the grains of core sand together.

Core Blow

A gas pocket in a casting adjacent to a core cavity caused by entrapping gases from the core.

Core Box, Combination

Core box and core dryers from the same pattern. One half is used as a half core box and a core drier.

Core Sand

Sand for making cores to which a binding material has been added to obtain good cohesion and permeability after drying. Usually low in clays.

Core Shift

A variation from specified dimensions of a cored section due to a change in position of the core or misalignment of cores in assembling.

Core Vents

(a) holes made in the core for escape of gas. (b) A metal screen or slotted piece used to form the vent passage in the core box employed in a core-blowing machine. (c) A wax product, round or oval in form, used to form the vent passage in a core.

Coring (Metallurgical)

Variable composition due to the solidification characteristics of an alloy. Typically these compositional differences occur on a micro scale, the distances between compositional extremes being controlled by the solidification structure of the alloy.

Corrosion

(a) Gradual chemical or electrochemical attack on a metal by atmosphere, moisture or other agents, (b) chemical attack of furnace linings by gases, slags, ashes or other fluxes occurring in various melting practices.

Corrosion Wear

Wear in which chemical or electrochemical reaction with the environment is significant.

Cover

A protective blanket laid on a melt to exclude oxidizing atmosphere and in the case of magnesium to prevent its igniting. Neutral covers simply protect metal from atmosphere; reacting covers contain an agent such as a deoxidizer.

Cover Core

A core set in place during the ramming of a mold to cover and complete a cavity partly formed by the withdrawal of a loose part of the pattern. Also used to form part or all of the cope surface of the mold cavity. A core placed over another core to create a flat parting line.

Crack, Hot Tear

A rupture occurring in a casting at or just below the solidifying temperature by a pulling apart of the soft metal, caused by thermal contraction stresses.

Cracking Strip

A fin of metal molded on the surface of a casting to prevent cracking.

Creep

The flow or plastic deformation of metals held for long periods of time at stresses lower than the normal yield strength. The effect is particularly important if the temperature of stressing is in the vicinity of the recrystallization temperature of the metal.

Creep Limit

The maximum stress that will result in creep at a rate lower than an assigned rate.

Critical Cooling Rate

The minimum rate of continuous cooling just enough to prevent undesired transformations.

Crucible

A ceramic pot or receptacle made of materials such as graphite or silicon carbide, with relatively high thermal conductivity, bonded with clay or carbon, and used in melting metals; sometimes applied to pots made of cast iron, steel, or wrought steel.

Crucible Furnace

A furnace fired with coke, oil, gas, or electricity in which metals are melted in a refractory crucible.

Crystal

A physically homogeneous solid in which the atoms, ions, or molecules are arranged in a three-dimensional repetitive pattern.

Crystal Lattice

The way atoms are arranged in a crystal. Spacewise, there are only 14 different lattices.

Crystalline Fracture

Fracture of a brittle metal, showing definite crystal faces in the fractured surface.

Curing Time

That period of time needed before a sand mass reaches maximum hardness.

D

Datum Points

In layout and machining operations the reference points on a datum plane from which dimensions are measured.

Decarburization

Loss of carbon from the surface of a ferrous alloy as a result of heating in a medium, usually oxygen, that reacts with carbon.

Deep Etching

Macroetching; etching for examination at a low (less than 10X) magnification, in a reagent that attacks the metal to a much greater extent than normal for microscopic examination. Gross features may be developed; i.e., abnormal grain size, segregation, cracks, or grain flow.

Defect

A discontinuity in the product whose severity is judged unacceptable in accordance with the applicable product specification.

Degasser

A material employed for removing gases from molten metals and alloys.

Degassing

Usually a chemical reaction resulting from a compound added to molten metal to remove gases from the metal. Often inert gases are used in this operation.

Dendrite

A crystal of branched appearance, formed during solidification of alloys, the branching habit being controlled by specific crystallographic directions.

Deoxidation

Removal of excess oxygen from molten metal, usually accomplished by adding materials with a high affinity for oxygen, the oxides of which are either gaseous or readily form slags.

Dephosphorization

Elimination of phosphorus from molten steel.

Desulfurization

Removal of sulfur from the molten metal by addition of suitable compounds.

Dewaxing

The process of melting out the expendable wax pattern from an investment mold by the application of heat, usually at temperatures less than 120 °C .

Die

A metal block used in forming materials by casting, molding, stamping, threading, or extruding.

Die Assembly

The parts of a die stamp or press that hold the die and locate it for the punches.

Die Casting

A rapid, water-cooled permanent mold casting process limited to nonferrous metals. There are three types: the plunger-type operated hydraulically, mechanically, or by compressed air with or without a gooseneck; the direct-air injection which forces metal from a goose-neck into the die, and the Cold-Chamber Machine. All force the metal into the die with a pressure greater than that of gravity flow.

Differential Heat Treatment

A heating process by which the temperature is varied within the object so that, after cooling, various parts may have different properties as desired.

Dip Coat

In solid and shell mold investment casting, a fine ceramic coating applied as a slurry to the pattern to produce maximum surface smoothness, followed by a cheaper conventional investment.

Direct Casting

Teeming from the ladle into the casting mold without the use of a tundish.

Direct-Arc Furnace

An electric arc furnace in which the metal being melted is one of the poles.

Dispersed Shrinkage

Small shrinkage cavities dispersed through the casting, which are not necessarily cause for rejection.

Disruptive Strength

Maximum strength of a metal when subjected to three principal tensile stresses at right angles to one another and of equal magnitude.

Dissolved Carbon

Carbon in solution in steel in either the liquid or solid state.

Double Annealing

As applied to hypo eutectoid steel, a process of heating to above the upper critical point (AC_3) and holding at that temperature until complete solution of the carbide has been achieved then cooling rapidly and reheating immediately to above A_3 and slowly cooling.

Double Tempering

A retempering operation sometimes necessary for steel containing retained austenite which breaks down during cooling from the first tempering to form a new, and hence, untempered martensite.

Drag

Lower or bottom section of a mold or pattern.

Draw

A term used for (a) to temper, (b) to remove pattern from mold, (c) an external contraction defect on surface of mold.

Drier

A material, as alcohol ammonium nitrate, sodium perborate and manganese oleate, added to a core or mold mixture to remove or reduce the water content.

Dry Sand Casting

The process in which the sand molds are dried at above 100 °C before using.

Dry Sand Mold

A mold from which the moisture has been removed by heating.

Dry Strength, or Dry Bond Strength

The maximum compressive, shear, tensile, or transverse strength of a sand mixture which has been dried at 105 to 110°C and cooled to room temperature.

Dual Metal Centrifugal Casting

Centrifugal castings produced by pouring a different metal into the rotating mold after the first metal poured.

E

Eddy Current Testing

The detection of discontinuities by observation of the interaction between electromagnetic fields and metals.

Elastic Limit

Maximum stress that a material will withstand without permanent deformation.

Elasticity

The property of recovering original shape and dimensions upon removal of a deforming force.

Electrode

Compressed graphite or carbon cylinder or rod used to conduct electric current in electric arc furnaces, arc lamps, carbon arc welding, etc.

Elongation

Amount of permanent extension in the vicinity of the fractures in the tensile test; usually expressed as percentage of original gage length.

Embrittlement

Loss of ductility of a metal due to chemical or physical change.

End-Quench Hardenability Test

A standardized method for comparing the hardenability of different steels.

Engineering Strain

The average linear strain, obtained by dividing the elongation of the length of the specimen by the original gage length.

Engineering Stress

The load divided by the original area.

Eutectic

(a) An isothermal reversible reaction in which a liquid solution decomposes, on cooling, into two or more intimately mixed solids. The number of solids formed are the same number of components in the system. (b) An alloy having the chemical composition indicated by the eutectic point on an equilibrium diagram.

Eutectoid

(a) An isothermal reversible reaction in which a solid solution on cooling is converted into two or more intimately mixed solids. The number of solids formed are the same number of components in the system. (b) An alloy having the same chemical composition indicated by the eutectoid point on an equilibrium diagram.

F

Fabrication

The joining, usually by welding, of two or more parts to produce a finished assembly. The components of the assembly may be a combination of cast and wrought materials.

Facing Sand

Specially prepared molding sand mixture used in the mold adjacent to the pattern to produce a smooth casting surface.

Fatigue

The loss of load-bearing ability of a material under repeated load application, as opposed to a single load.

Fatigue Crack or Failure

A fracture starting from a nucleus where there is an abnormal concentration of cyclic stress. The fracture surface is smooth and frequently shows concentric markings with a nucleus as a center.

Fatigue Limit (Endurance Limit)

Maximum stress that a material will endure without failure for an infinite number of load cycles.

Fatigue Strength

Maximum stress that a material will endure without failure for a specified number of load cycles.

Feeding

The process of supplying molten metal to compensate for volume shrinkage while the casting is solidifying.

Ferrite

A solid solution of one or more elements in the body-center-cubic phase of iron or steel.

Ferritic Steels

Steels in which ferrite is the predominant phase. These steels are magnetic.

Fillet

A concave corner piece used on foundry patterns, a radius joint replacing sharp inside corners.

Finish Allowance

The amount of stock left on the surface of a casting for machining.

Finish Welding

Production welding carried out in order to ensure the agreed quality of the casting.

Flash

A thin section of metal formed at the mold, core, or die joint or parting in a casting due to the cope & drag not matching completely, or where core & core print do not match.

Flask

A metal frame used for making or holding a sand mold. The upper part is the cope and the bottom half is the drag.

Fluidize

To impart fluid like properties to powders or sands e.g. fluidized beds.

Foundry Returns

Metal in the form of sprues, gates, runners, risers and scrapped castings, with known chemical composition that are returned to the furnace for remelting. Sometimes referred to as "revert".

Frictional Wear

The displacement and/or detachment of metallic particles from a surface as a consequence of being in contact with another moving component.

Gate

The end of a runner in a mold where molten metal enters the mold cavity.

Gating System

The complete assembly of sprues, runners and gates in a mold through which steel flows before entering the casting cavity.

Granular Fracture (Crystalline Fracture)

A type of irregular surface produced when metal is broken.

Green Sand

A naturally bonded sand, or a compounded molding sand mixture which has been tempered with water for use while still in the damp or wet condition.

Green Sand Core

A sand core used in the unbaked condition, also a core made from green sand and used as rammed.

Hardenability

In a ferrous alloy, the property that determines the depth and distribution of hardness induced by quenching.

Hardness

Resistance of a material to indentation as measured by such methods as Brinell, Rockwell, and Vickers. The term hardness also refers to stiffness of a material, or its resistance to scratching, abrasion, or cutting.

Heat

The total amount of metal produced which can be represented by one analysis sample and one set of mechanical tests.

Heat Treatment

A combination of heating & cooling operations applied to a metal or alloy to produce desired properties and microstructures.

High-Alloy Steel

Ferrous alloy with more than 12 weight % of non carbon additions.

Hollow Drill Test (Trepanning)

Removing a cylindrical sample from a metal section or structure to determine soundness of the section.

Homogenizing

A process of heat treatment at high temperature intended to eliminate or decrease chemical segregation by diffusion.

Horizontal Axis Casting Machine

A centrifugal casting machine in which the axis of rotation of the mold is horizontal.

Hot Tear

A crack or fracture formed prior to completion of metal solidification as a result of hindered contraction. A hot tear is frequently open to the surface of the casting and is commonly associated with design limitations.

Hydrogen Embrittlement

A condition of low ductility resulting from the absorption of hydrogen. A time dependent fracture process which results in a loss of ductility.

I

Impact Strength

The resistance to impact loads; usually expressed as the foot pounds of energy absorbed in breaking a standard specimen.

Impact Value

Total energy needed to break a standard specimen by a single blow under standard conditions; e.g., Charpy Impact Test.

Impregnation

The treatment of castings with a sealing medium to stop pressure leaks, such as soaking under pressure with or without prior evacuation, and either with hot or cold application. Mediums used include silicate of soda, drying oils with or without styrene, plastics, and proprietary compounds.

Impurity

An element unintentionally allowed in a metal or alloy. Some impurities have little effect on properties; others will grossly damage the alloy.

Inclusions

Nonmetallic materials in a metal matrix. Sources include reoxidation, refractories, slag, and deoxidization products.

Indentation Hardness

The resistance of a material to indentation. This is the usual type of hardness test, in which a pointed or rounded indenter is pressed into a surface under a substantially static load.

Indirect-Arc Furnace

An AC (Alternating Current) electric-arc furnace in which the metal is not one of the poles.

Induction Furnace

An AC melting furnace which utilizes the heat of electrical induction.

Induction Hardening

A surface hardening process involving the localized use of pulsating magnetic currents to achieve heating above the austenite transformation temperature, Ac_3, followed by quenching.

Induction Heating

Process of heating by electrical resistance and hysteresis losses induced by subjecting a metal to the varying magnetic field surrounding a coil carrying an alternating current.

Inert Gas

A gas that will not support combustion or sustain any chemical reaction; e.g., argon or helium.

Ingot

A mass of metal cast to a convenient size and shape for remelting or hot working.

Insert

A part usually formed from metal, which is placed in a mold and may become an integral part of the casting.

Insulating Pads and Sleeves

As opposed to chills, insulating material, such as gypsum, diatomaceous earth, etc., used to lower the rate of solidification. As sleeves on open risers, they are used to keep the metal liquid, thus increasing the feed efficiency.

Intergranular Corrosion

Corrosion in a metal taking place preferentially along the grain boundaries.

Internal Shrinkage

A void or network of voids within a casting caused by inadequate feeding of that section during solidification.

Internal Stresses (or Thermal Stresses)

Generally stresses which occur during the cooling of a part.

Interrupted Quench

Removing the casting from a quenching bath before it has reached the temperature of the bath.

Investment Casting

Casting produced in a mold obtained by investing an expendable pattern with a refractory to produce a shell. The expendable pattern may consist of wax, plastic, or other material and is removed prior to filling the mold with liquid metal.

Iron

(a) A metallic element, mp 1535°C (b) irons not falling into the steel categories, as Gray Iron, Ductile Iron, Malleable Iron, White Iron, and Wrought Iron.

Iron, Malleable

A mixture of iron and carbon, including smaller amounts of silicon, manganese, phosphorus, and sulfur, which after being cast (white iron, carbon in combined form as carbides) is converted structurally by heat treatment into a matrix of ferrite containing nodules of temper carbon (graphite).

Isothermal

Pertaining to changes or other phenomena occurring at a constant temperature.

Isothermal Annealing

A process in which a ferrous alloy is heated to produce a structure partly or wholly austenitic, and is then cooled to and held at a temperature that causes transformation of the Austenite to a relatively soft ferrite-carbide aggregate.

Isothermal Transformation

(a) The process of transforming Austenite in a ferrous ally to Ferrite or a ferrite-carbide aggregate at any constant temperature within the transformation range, (b) transformation of one phase in an alloy system to another phase at any constant temperature.

Jobbing Foundry

A foundry engaged in the manufacture of numerous types of castings.

Joint Welding

Production welding used to weld cast components together to obtain an integral unit.

K Factor

Tensile strength in pounds per square inch divided by the Brinell Hardness number.

Kayser Hardness Test

A method for determining the true hardness of metals at high temperatures.

Keyhole Specimen

A type of notched impact test specimen which has a hole-and-slot notch shaped like a keyhole.

Knockout Pins (Ejector Pins)

Small diameter pins affixed to a pattern back-up plate for removing cured mold in the shell-molding process.

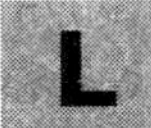

Ladle

Metal receptacle frequently lined with refractories used for transporting and pouring molten metal. Types include hand bull, crane, bottom-pour, holding, teapot, shank, lip-pour.

Ladle, Bottom-Pour

Ladle from which metal flows through a nozzle in the bottom.

Ladle, Lip-Pour

Ladle in which the metal is poured over a lip.

Ladle, Teapot

A ladle in which, by means of an external spout, metal is removed from the bottom rather than the top of the ladle.

Lateral Expansion

A measured property used in Charpy Impact Testing. Refers to the increase width of the specimen after fracture.

Lining

Inside refractory layer of firebrick, clay, sand, or other material in a furnace or ladle.

Liquid Penetrant Testing

A non destructive testing method suitable for evaluating the surface integrity of non-magnetic and ferro-magnetic parts.

Magnetic Particle Inspection

A nondestructive method of inspecting the surface integrity of ferromagnetic materials.

Martempering (Interrupted Quenching)

A hardening treatment of a steel involving a slow cool through the martensitic transformation range to reduce stresses associated with the quenching of austenite. An important aspect of martempering is that no transformation product other than martensite should form.

Martensite

A generic term used for microstructures formed by diffusionless phase transformations. A constituent found in hardened steel; has a needle like microstructure.

Martensitic Stainless Steels

A corrosion-resistant ferrous alloy with a predominant martensitic phase.

Mass Effect

The effect that the mass of a component has on the properties of the material from which the part is made. In castings such effects may arise due to the effect of mass on the solidification and on the rate of temperature change heat treatment.

Mechanical Properties

Properties of a material that reveal its strength and elastic behavior.

Metallographic Structure

The nature, distribution, and amounts of the metallographic constituents in a metal.

Metalloid

(a) An element intermediate between metals and nonmetals possessing both metallic and nonmetallic properties, as arsenic, (b) sometimes applied to elements commonly bonded in small amounts in steel, as carbon, manganese, boron, silicon, sulfur, and phosphorus.

Metallurgical Bond

The bond between two metals whose interface is free of voids, oxide films, or discontinuities.

Metallurgy

The science and technology of metals, a broad field that includes, but is not limited to, the study of internal structures & properties of metals and the effects on them of various processing methods.

Metals Comparator

An instrument for testing or identifying metallic and nonmetallic parts. Parts are placed in an electromagnetic field and a standard parts in a matched electromagnetic field. Distortions of the magnetic fields are compared on an oscilloscope.

Meyer Hardness Test

A test to determine tendency of a metal to harden when deformed plastically. A series of indentations are made in the metal using a fixed-diameter ball and progressively increasing loads.

Microetching

Etching of metal samples for examination under the microscope.

Microformer

A type of extensometer for measuring elongation of test piece in a tensile test.

Micrography

Examination by means of a microscope.

Microhardness

The hardness of microconstituents of a material.

Microradiography

The process of passing x-rays through a thin section of an alloy in contact with a photographic emulsion, and then magnifying the radiograph 50 to 100 times to observe the distribution of alloying constituents & voids.

Microscopic

Minute object or structures which are invisible, or not clearly distinguished, without the use of a microscope.

Microsection

A metal specimen whose surface has been polished and etched to reveal the microstructure.

Microshrinkage

Very finely divided porosity resulting from inter dendritic shrinkage, resolved only by use of the microscope; may be visible on radiographic films as mottling. Etching shows they occur at intersections of convergent dendritic directions.

Microspectroscopy

A method of identifying metallic constituents using spectrographic arc.

Microstructure

The structure of polished and etched metal & alloy specimens, as revealed by the microscope at magnifications over 10 diameters.

Mild Steel

Plain carbon steel of about 0.25% carbon or less.

Mill Scale

Iron oxide scale formed on steel during hot working processes, cooled in air.

Milling

Removing metal with a milling cutter.

Modulus of Elasticity

In tension , it is the ratio of stress to the corresponding strain, within the limit of elasticity (Yield Point) of a material.

Modulus of Resilience

The amount of strain energy per unit volume required to stress a material from zero to the yield stress limit. The modulus of resilience is proportional to the area under the elastic portion of the stress-strain diagram. Units are Pascal .

Modulus of Rigidity

In a torsion test, the ratio of the unit shear stress to the displacement caused by it per unit length in the elastic range.

Modulus of Rupture

Used in both bending and torsion testing. In bending, the modulus of rupture is the bending moment at fracture, divided by the section modulus. In torsion, modulus of rupture is the torque at fracture divided by the polar section modulus.

Modulus of Toughness

Amount of work per unit volume of a material, required to carry that material to failure under static loading. Equal to the area under the entire stress-strain curve. Units are Pascal .

Moh's Scale

A scratch hardness test for determining comparative harness, using ten standard minerals from talc to diamond.

Mold Cavity

The space in a mold which is filled with liquid metal to form the casting upon solidification. The channels through which liquid metal enters the mold cavity (sprue, runner, gates) and reservoirs for liquid metal (risers), are not considered part of the mold cavity proper.

Mold Coating (Mold Facing, Dressing)

(a) Coating to prevent surface defects on permanent mold castings and die castings, (b) coating on sand molds to prevent metal penetration and improve metal finish.

Mold Cover Half (Cover Die)

(a) The top half of the mold, the cope, (b) in die casting, the front half of the die, which remains stationary as the die is opened.

Mold Shift

A casting discontinuity resulting from misalignment of the cope and drag halves.

Moldability

Ability of sand to flow into a flask and around a pattern; measured in the amount of sand falling through an inclined screen or slot.

Molding Material

A material suitable for making molds into which molten metal can be cast.

Molding Sands

Sands containing over 5% natural clay (usually between 8 and 20%).

Molding, Floor

Making sand molds from loose or production patterns of such size that they cannot be satisfactorily handled on a bench or molding machine; the equipment being located on the floor during the entire operation of making the mold.

Molding, Pit

Molding method in which the drag is made in a pit or hole in the floor.

Mushy Stage

The state between solid and liquid in alloys, which freeze over a wide range of temperatures.

N

Necking

Reducing the cross sectional area of the metal in an area by stretching.

Needles

Elongated acicular crystals tapering at each end to a fine point as martensite.

Negative Quenching (Negative Hardening)

Accelerated cooling in water or oil from a temperature below the critical range.

Nimonic

Class of nickel-base cast alloy resistant to stress and to oxidation at high temperatures.

Nital

A solution of nitric acid in alcohol use as an etching agent in ferrous metallography.

Nitriding

A surface hardening process involving heating in an atmosphere of ammonia, or in contact with a nitrogen-bearing material, so as to promote the absorption of nitrogen.

Nitrogen Flush

Bubbling nitrogen gas through a metal melt under vacuum (as with valve bronze), to improve tensile properties and pressure tightness.

Nobake Binder

A synthetic liquid resin sand binder that hardens completely at room temperature, generally not requiring baking, used in Cold-Setting process.

Nodular Iron

Iron of a normally gray cast iron type that has been suitably treated with a nodularizing agent so that all, or the major portion of its graphitic carbon, has a nodular or spherulitic form as cast. Often referred to as Ductile Iron.

Nondestructive Testing (Inspection)

Testing or inspection that does not destroy the object being tested or inspected.

Nonferrous

Refers to alloy in which the predominate metal or solvent is not iron.

Normal Segregation

Concentration of alloying constituents that have low melting points in those portions of a casting that solidify last.

Normalizing

Heating a ferrous alloy to a suitable temperature above the transformation temperature Ac_3, followed by cooling at a suitable rate, usually in still air to a temperature substantially below the transformation range.

Nucleation

(a) (homogeneous) the initiation of solid crystals from the liquid stage, or initiation of solid crystals from the liquid stage, or a new phase within a solid, without outside interference - rarely occurs, (b) (heterogeneous) foreign particles altering the liquid-solid interface energy during phase changes.

O

Open Grain Structure

A defect wherein a casting, when machined or fractured, appears to be coarse grained and porous; usually due to a shrink area.

Orange Peel

A pebble-grained surface that develops in the mechanical forming of sheet metals with coarse grains.

Overaging

Aging a precipitation-hardening alloy under conditions of time and temperature greater than those required to obtain maximum strength or hardness

Overheated

A term applied when, after exposure to an excessively high temperature, a metal develops an undesirable coarse grain structure, but is not necessarily damaged permanently. Unlike burned structure, the structure produced by overheating can be corrected by suitable heat treatment, by mechanical work, or by a combination of the two.

Overstressing

Permanently deforming a metal by subjecting it to stresses that exceed the elastic limit.

Oxidation

Any reaction of an element with oxygen. In a narrow sense, oxidation means the taking on of oxygen by an element or compound, and on the basis of the electron theory, it is a process in which an element loses electrons.

Oxidizing Atmosphere

An atmosphere resulting from the combustion of fuels in an atmosphere where excess oxygen is present, and with no unburned fuel lost in the products of combustion.

P

Packing

Sand, gravel, mill scale or similar materials used to support castings packed in annealing pots to prevent possible warpage under high temperatures.

Padding

The process of adding metal to a cross section of a casting wall, usually extending from a riser, to ensure adequate feed to a localized area where a shrink would occur if the added metal were not present.

Parted Pattern

A pattern made in two or more parts.

Parting Line

A line on a pattern or casting corresponding to the separation between the cope and drag portions of a sand mold.

Passivity

The property of some metals to become abnormally inactive towards certain reagents.

Pattern

A form of wood, plastic, metal, or other material around which molding material is placed to make a mold.

Pattern Draft

The taper on vertical elements in a pattern which allows easy separation of pattern from compacted sand mixture.

Pattern Layout

Full-sized drawing of a pattern showing its arrangement and structure features.

Pattern maker

A craftsman engaged in production of foundry patterns from wood, plastic, or metals, such as aluminum, brass, etc.

Perlite

A highly siliceous volcanic rock which can be expended by heating into a porous mass of particles. Perlite can be used as an insulation in foundry sand mixtures. Not to be confused with Pearlite.

Permanent Mold

A metal mold of two or more parts; not an ingot mold. It is used repeatedly for the production of many castings of the same form.

Phase Diagram

A graphic representation of the equilibrium temperature and composition limits of phase fields reactions in an alloy system. In a binary system, temperature is usually the ordinate and composition the abscissa. Ternary and more complex systems require several two-dimensional diagrams to show the temperature - composition variables completely. In alloy systems, pressure is usually considered constant, although it may be treated as an additional variable.

Photomicrograph

A photograph of the grain structure of a metal as observed when optically magnified more than 10 diameters. The term micrograph may be used.

Physical Metallurgy

The science concerned with the physical and mechanical characteristics of metals & alloys.

Pitting

A form of wear characterized by the presence of surface cavities, the formation of which is attributed to processes such as fatigue, local adhesion, cavitation or corrosion.

Plane Strain

A stress condition in linear elastic fracture mechanics in which there is zero strain in a direction normal to both the axis of applied tensile stress, and the direction of crack growth. Under plane strain conditions, the plane of fracture instability is normal to the axis of the principal tensile stress.

Plastic Deformation

Permanent distortion of a material under the action of applied pressure.

Post heating

A process used immediately after welding whereby heat is applied to the weld zone, either for tempering, or for providing a controlled rate of cooling, in order to avoid a hard or brittle structure.

Pouring

Transfer of molten metal from furnace to ladle, ladle to ladle, or ladle into molds.

Pouring Cup

The flared section of the top of the downsprue. It can be shaped by hand in the cope, or be a shaped part of the pattern used to form the downsprue; also may be baked core cup placed on the top of the cope, over the downsprue.

Precipitation Hardening

A process of hardening an alloy in which a constituent precipitates from a supersaturated solid solution.

Precipitation Heat Treatment

Any of the various aging treatments conducted at elevated temperatures to improve certain mechanical properties through precipitation from solid solution.

Preheating

A general term for heating material, as a die in die casting, as a preliminary to operation, to reduce thermal shock and prevent adherence of molten metal.

Primary Choke (Choke)

That part of the gating system which most restricts or regulates the flow of metal into the mold cavity.

Proeutectoid

The constituent that separates out of a solid solution before the formation of eutectoid.

Profile Tolerances

A system of locating and tolerancing developed to control the orientation of rough parts in machine fixtures. From locating points on the casting a "perfect profile" is established for all surfaces and features. A tolerance envelope surrounding that profile defines the limitations of an acceptable part.

Quenching

Rapid cooling or hardening; normally achieved by immersion of the object to be hardened in water, oil, or solutions of salt or organic compounds in water.

Radioactive Isotopes

Varieties of an element possessing the same chemical characteristics, but emitting detectable radiations, by means of which they can be identified and traced.

Radioactive Material

Any compound or element which may emit any or all of the following; alpha and beta particles, electrons, photons neutrons and gamma, and all other emissions which produce ionization directly or indirectly.

Ramming

Packing sand in a mold by raising and dropping the sand pattern flask on a table. Jolt squeezers, jarring machines, and jolt rammers are machines using this principle.

Recrystallization

A process whereby the distorted grain structure of cold-worked metals is replaced by a new, strain-free grain structure, during annealing above a specific minimum temperature.

Recrystallization Temperature

The lowest temperature at which the distorted grain structure of a cold-worked metal is replaced by a new, strain-free grain structure, during prolonged annealing. Time, purity of the metal, and prior deformation are important factors.

Reduction

The removal of oxygen or addition of hydrogen.

Refractory

(a) Heat-resistant material, usually non-metallic, used for furnace linings etc., (b) the quality of resisting heat.

Repair Welding

Any welding carried out after delivery to the end user, e.g., after the casting has been in service.

Revert

Recycled sprues, gates, risers, defective castings and machine chips.

Rigging

Gates, risers, loose pieces, etc., needed on the pattern to produce a sound casting.

Riser

Reservoir of molten metal from which casting feeds as it shrinks during solidification.

Riser Height

The distance from the top of the riser when liquid rises to the top of the riser neck. Riser height when solid is usually several inches less than when liquid because of contraction and loss of feed metal to the casting.

Riser, Blind

A riser that does not break through the top of the cope, and is entirely surrounded by sand; opened to the atmosphere by means of a firecracker core.

Riser, Open

Conventional form of riser usually located at the heaviest section of the casting, and extending through the entire height of the cope.

Riser-Gating

Practice of running metal for the casting through the riser to help directional solidification.

Runner

A channel through which molten metal or slag is passed from one receptacle to another; in a mold, the portion of the gate assembly that connects the downgate, or sprue, with the casting ingate or riser. The term also applies to similar portions of master patterns, pattern dies, patterns, investment molds, and finished castings.

Runner Extension

In a mold, that part of a runner which extends beyond the farthest ingate as a blind end. It acts as a dirt trap since the first rush of metal along the runner will pick up any loose particles of sand or dirt, and carry them into the extension (not into the mold cavity).

Runner Riser

A conventional runner, usually in the horizontal plane, which permits flow of molten metal to the ingate, and is large enough to act as a reservoir to feed the casting.

Runout

A casting defect caused by incomplete filling of the mold due to molten metal draining or leaking out of some part of the mold cavity during pouring; escape of molten metal from a furnace, mold, or melting crucible.

S

Sag

A decrease in metal section in casting due to sagging of the cope or core.

Sand

In metal casting, a loose, granular material high in SiO_2, resulting from the disintegration of rock. The name sand refers to the size of grain, and not to mineral composition. Diameter of the individual grains can vary from approximately 6, to 270 mesh. Most foundry sands are made up principally of the mineral quartz (silica). Reason for this is that sand is plentiful, refractory, and cheap; miscellaneous sands include zircon, olivine, chromite, $CaCO_3$, black sand (lava grains), titanium minerals, and others.

Sand Casting

Metal castings produced in sand molds.

Sand Control

Procedure whereby various properties of foundry sand, such as fineness, permeability, green strength, moisture content, etc., are adjusted to obtain castings free from blows, scabs, veins, and similar defects.

Sand Porosity

Volume of the pore spaces or folds in a sand. (Not synonymous with permeability).

Sand Reclamation

Processing of used foundry sand grains by thermal, attraction or hydraulic methods, so that it may be used in place of new sand without substantially changing current foundry sand practice.

Scaling (Scale)

Surface oxidation, partially adherent layers of corrosion products, left on metals by heating or casting in air or in other oxidizing atmospheres.

Scanning Electron Microscope

An instrument used for obtaining microstructure images, using an electron beam. The micrographs obtained give depth perception of the metal being observed.

Scrap (Metal)

Metal to be remelted; includes scrapped machinery fabricated items such as rail or structural steel and rejected castings.

Seam

A surface defect on a casting related to but of lesser degree than a Cold Shut; a ridge on the surface of a casting caused by a crack in the mold face.

Segregation

A concentration of alloying elements at specific regions, usually as a result of the primary crystallization of one phase with the subsequent concentration of other elements in the remaining liquid.

Shakeout

(a) The operation of removing castings from the mold (b) a mechanical unit for separating the molding materials from the solidified metal casting.

Shear

A type of deformation in which parallel planes in the metal crystals slide so as to retain their parallel relation.

Shear Modulus

In a torsion test, the ratio of the unit shear stress to the displacement caused by its per unit length in the elastic range. Units are Pascal .

Shear Strain

Elastic displacement produced by pure shear loading.

Shear Strength

Maximum shear stress a material is capable of withstanding without failure.

Shear Stress

Load per unit area parallel to the plane of contact.

Shell Molding

A process for forming a mold from resin-bonded sand mixtures brought in contact with pre-heated metal patterns, resulting in a firm shell with a cavity corresponding to the outline of the pattern.

Shift

A casting defect caused by mismatch of cope and drag, or of cores and mold.

Shot blasting (Shot Peening)

Casting cleaning process employing a metal abrasive (grit or shot) propelled by centrifugal or air force.

Shrink

The difference in volume between liquid metal and solid metal, or the void (shrink hole) left in a casting because of it.

Shrinkage

(a) Liquid, contraction in volume as metal cools to solidification, (b) solidification, contraction in volume when the metal passes from the liquid to the sold at the freezing point (may expend over a range), (c) solid, the contraction on cooling from freezing point to normal temperature, (d) the decrease in dimension in clays occurring when drying at 100°C and even more so on firing, (e) reduction in dimension of refractory material during heating.

Shrinkage Cracks

Cracks that form in metal as result of the pulling apart of grains by contraction before complete solidification.

Silica Sand

Sand with a minimum silica content of 95% used for forming casting molds.

Skimming

Removing or hold back dirt or slag from the surface of the molten metal before or during pouring.

Skin

A thin surface layer different chemically or structurally from the main mass of a metal object.

Slag Trap

An enlargement, dam, or extrusion in the gating or runners system in a mold for the purpose of preventing molten slag particles from entering the mold cavity.

Slurry

A term loosely applied to any clay-like dispersion. It may be used to wash ladles or other refractory linings to impart a smooth surface; as a bonding addition to molding sand, as a thin loam over specially made molds, or as a mixture to fine joints or cracks of a core, etc.

Smelting

A metallurgical thermal process in which a metal is separated in fused form from nonmetallic materials, or other undesired metals with which it is associated.

Softening

A process used to soften metals through annealing or tempering.

Solidification, Shrinkage

The decrease in size accompanying the freezing of a molten metal.

Spalling

Buckling or flaking off of the surface material.

Specific Heat

Equivalent to thermal capacity, or the quantity of heat required to produce a unit change in the temperature of a unit mass.

Spout

A trough through which the metal flows from the furnace to the ladle.

Sprue Hole

The opening through which the metal is poured into the cope to run into the casting cavity.

Stainless Steel

A wide range of steels containing chromium, or chromium and nickel, exhibiting high resistance to corrosion.

Steel

An alloy of iron and carbon that may contain other elements, and in which the carbon content does not exceed about 1.7%; it must be malleable at some temperature while in the as-cast state.

Stopper Rod

A device in a bottom-pour ladle for controlling the flow of metal through the nozzle into the casting. The stopper rod consists of a steel rod, protecting sleeves, and a graphite stopper head. It may also be a single piece manufactured from graphite.

Streamline Flow

Steady flow of liquid without turbulence. Generally, not experienced in metal casting.

Stress, Relieving

A heat treatment to reduce residual stresses, followed by sufficiently slow cooling to minimize development of new residual stresses.

Stress, Residual

Those stresses set up in a metal as a result of non uniform plastic deformation, or the unequal cooling of a casting.

Stress-Corrosion Cracking

Spontaneous failure of metals by cracking under combined conditions of corrosion and stress, either residual or applied.

Superalloy

An alloy developed for very high temperature use, where relatively high stresses are encountered, and where oxidation resistance is needed.

Supercooling

Lowering the temperature of a molten metal below its liquidus during cooling.

Superheat

Any increment of temperature above the melting point of a metal; sometimes construed to be any increment of temperature above normal casting temperatures introduced for the purpose of refining, alloying, or improving fluidity.

Supersaturated

Metastable solution in which the dissolved material exceeds the amount the solvent can hold in normal equilibrium at the temperature and under the other conditions that prevail.

Surfacing

Depositing a filer metal on a metal surface by any method to obtain certain desired properties or dimensions.

Swell

A casting defect consisting of an increase in metal section due to the displacement of sand by metal pressure.

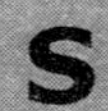

Temper

(a) Reheating hardened, normalized, or mechanically worked steel to a temperature below the critical range to soften it and improve impact strength. (b) The moisture content of a sand at which any certain physical test value is obtained, e.g., temper with respect to green compressive strength, permeability, retained compressive strength, etc. (c) To mix material with enough liquid to develop desired molding properties.

Temper Brittleness

Brittleness that results when certain steels are held within or cooled slowly through a certain range of temperature below the transformation range. The brittleness is revealed by notched-bar impact tests at room temperature or lower temperatures.

Temper Stressing

Quenching in water from the tempering temperature to improve fatigue strength.

Temperature, Holding

(a) Temperature above the critical phase transformation range at which castings are held as a part of the heat treatment cycle, (b) The temperature maintained when metal is held in a furnace, usually prior to pouring.

Temperature, Pouring

The temperature of the metal as it is poured into the mold.

Tempered Martensite

Martensite that has been heated to produce to BCC iron and a fine dispersion of iron carbide.

Tensile Strength

The maximum stress in uniaxial tension testing, which a material will withstand prior to fracture. The ultimate tensile strength is calculated from the maximum load applied during the test, divided by the original cross-sectional area.

Thermal Conductivity

The property of matter by which heat energy is transmitted through particles in contact. For engineering purposes, the amount of heat conducted through refractories is usually given in Kcal/hr/m^2 for a temperature difference of one degree centigrade, and for a thickness of one metre, Kcal/hr $\cdot$m$\cdot{}^0$C/m .

Thermal Contraction

The decrease in a linear dimension and volume of a material accompanying a change of temperature.

Thermal Expansion

The increase in a linear dimension and volume of a material accompanying a change of temperature.

Thermal Fatigue

Failure resulting from rapid cycles of alternate heating and cooling.

Thermal Shock

Stress developed by rapid and uneven heating of a material.

Thermography

The technique of obtaining a photographic record of heat distribution in a solid or fluid.

Tolerance

The permissible deviation of a dimension from the nominal or desired value. Minimum clearance between mating parts.

Tool Steel

Any high-carbon or alloy steel used to make a cutting tool for machining metals and for metal-casting dies.

Toughness

The ability of the metal to absorb energy and to deform plastically during fracture. Toughness values obtained in testing depend upon the test temperature, the rate of loading, the size of the test specimen, as well as the presence of a notch and its acuity.

Transfer Ladle

A ladle that may be supported on a monorail or carried in a shank and used to transfer metal from the melting furnace to the holding furnace, or from furnace to pouring ladles.

Transformation (Temperature) Range

The critical temperature at which a change in phase occurs. To distinguish between the critical points in heating and cooling those in heating are referred to as the Ac points (c for Chauffage or heating) and those in cooling, Ar. (r for Refroidissement).

Turntable

The base on which a centrifugal casting mold rests.

Ultrasonic Testing

A nondestructive method of testing metal for flaws based on the fact that ultrasonic waves are reflected and refracted at the boundaries of a solid medium.

Upgrading

In castings, the removal and repair of discontinuities to raise the quality level of the casting beyond that which can be economically achieved by good foundry practice.

Upper Yield Point (also Yield Point)

Denoted in yield point phenomenon as a distinct break from the elastic region accompanied by a drop in load, yet prior to plastic deformation in the stress-strain curve in a low-carbon steel.

Vacuum Casting

A casting in which metal is melted and poured under very low atmospheric pressure; a form of permanent mold casting where the mold is inserted into liquid metal, vacuum is applied, and metal drawn up into the cavity.

Vacuum Degassing

The use of a vacuum technique to remove dissolved gases from molten alloys.

Vacuum Refining

Melting in a vacuum, usually by electrical induction, to remove gaseous contaminants from the metal.

Venting

Perforation with a vent wire of the sand over and around a mold cavity to assist in the escape of the gases.

Virgin Metal (Primary Metal)

Metal extracted directly from the ore; not previously used.

Void

A shrinkage cavity produced in casting during solidification.

Warpage

Deformation other than contraction that develops in a casting between solidification and room temperature; also, distortion occurring during annealing, stress relieving, and high-temperature service.

Water Test

To subject a casting to water pressure in such a manner that any porous areas will show leakage.

Wax Pattern

(a) A precise duplicate, allowing for shrinkage, of the casting and required gates, usually formed by pouring or injecting molten wax into a die or mold, (b) wax molded around the parts to be welded by a termite welding process.

Wear

The undesired deterioration of a component by the removal of material from its surface.

Weld Bead

The built-up portion of a fusion weld, formed either from the filler metal or the melting of the parent metal.

Welding

A process used to join metals by the application of heat. Fusion welding, which includes gas, arc, and resistance welding, requires that the parent metals be melted.

Welding Electrode

A metal or alloy in rod or wire forms used in electric arc welding to maintain the arc, and at the same time supply molten metal or alloy at the point where the weld is to be accomplished.

Welding Shielded-Arc

Electric-arc welding in which the molten weld metal is protected from the atmosphere. An inert gaseous atmosphere or flux coated electrode may be employed.

Welding Stress

That stress resulting from localized heating and cooling of metal during welding.

Welding, Arc

Welding accomplished by using an electric arc that can be formed between a metal or carbon electrode and the metal being welded; between two separate electrodes, as in atomic hydrogen welding or between the two separate pieces being welded, as in flash welding.

Welding, Autogenous

Method of uniting two pieces of metal by melting their edges together without solder or any added welding metal, as by the thermite process that employs a medium of finely divided aluminum powder and oxide, or iron by which a temperature of some 2982.2°C is obtained.

X-Ray

Form of radiant energy with wavelength shorter than that of visible light, and with the ability to penetrate materials that absorb or reflect ordinary light. X-rays are usually produced by bombarding a metallic target with electrons in a high vacuum. In nuclear reactions it is customary to refer to photons originating in the nucleus as gamma rays and to those originating in the extra nuclear part of the atom as x-rays.

Yield

Comparison of casting weight to total weight of metal poured into mold.

Yield Ratio

The ratio of yield strength to ultimate tensile strength.

Yield Strength

The stress at which a material exhibits a specified limiting permanent strain.

Young's Modulus

In tension it is the ration of stress to the corresponding strain.

Section D

MULTIPLE CHOICE QUESTION (MCQ) & ANSWER

GATE (Graduate Aptitude Test in Engineering)

FROM 2018 to 2021

Section

D

21

GATE-2018

Q.1 – Q.5 carry one mark each.

1. "When she fell down the ________, she received many ________ but little help."

   The words that best fill the blanks in the above sentence are

   (a) stairs, stares (b) stairs, stairs

   (c) stares, stairs (d) stares, stares

2. "In spite of being warned repeatedly, he failed to correct his __________ behaviour."

   The word that best fills the blank in the above sentence is

   (a) rational (b) reasonable

   (c) errant (d) good

3. For $0 \le x \le 2\pi$, sin x and cos x are both decreasing functions in the interval __________.

   (a) $\left(0, \frac{\pi}{2}\right)$ (b) $\left(\frac{\pi}{2}, \pi\right)$

   (c) $\left(\pi, \frac{3\pi}{2}\right)$ (d) $\left(\frac{3\pi}{2}, 2\pi\right)$

4. The area of an equilateral triangle is $\sqrt{3}$. What is the perimeter of the triangle?

   (a) 2 (b) 4

   (c) 6 (d) 8

5. Arrange the following three-dimensional objects in the descending order of their volumes:

   (i) A cuboid with dimensions 10 cm, 8 cm and 6 cm

   (ii) A cube of side 8 cm

   (iii) A cylinder with base radius 7 cm and height 7 cm

   (iv) A sphere of radius 7 cm

   (a) (i), (ii), (iii), (iv)

   (b) (ii), (i), (iv), (iii)

   (c) (iii), (ii), (i), (iv)

   (d) (iv), (iii), (ii), (i)

Q.6 – Q.10 carry two marks each.

6. An automobile travels from city A to city B and returns to city A by the same route. The speed of the vehicle during the onward and return journeys were constant at 60 km/h and 90 km/h, respectively. What is the average speed in km/h for the entire journey?

   (a) 72 (b) 73

   (c) 74 (d) 75

7. A set of 4 parallel lines intersect with another set of 5 parallel lines. How many parallelograms are formed?

   (a) 20 (b) 48

   (c) 60 (d) 72

8. To pass a test, a candidate needs to answer at least 2 out of 3 questions correctly. A total of 6,30,000 candidates appeared for the test. Question A was correctly answered by 3,30,000 candidates. Question B was answered correctly by 2,50,000 candidates. Question C was answered

Answers

1. (a)	2. (c)	3. (b)	4. (c)	5. (d)	6. (a)
7. (c)	8. (d)				

correctly by 2,60,000 candidates. Both questions A and B were answered correctly by 1,00,000 candidates. Both questions B and C were answered correctly by 90,000 candidates. Both questions A and C were answered correctly by 80,000 candidates. If the number of students answering all questions correctly is the same as the number answering none, how many candidates failed to clear the test?
(a) 30,000 (b) 2,70,000
(c) 3,90,000 (d) 4,20,000

9. If $x^2 + x - 1 = 0$ what is the value of $x^4 + \frac{1}{x^4}$?
(a) 1 (b) 5
(c) 7 (d) 9

10. In a detailed study of annual crow births in India, it was found that there was relatively no growth during the period 2002 to 2004 and a sudden spike from 2004 to 2005. In another unrelated study, it was found that the revenue from cracker sales in India which remained fairly flat from 2002 to 2004, saw a sudden spike in 2005 before declining again in 2006. The solid line in the graph below refers to annual sale of crackers and the dashed line refers to the annual crow births in India. Choose the most appropriate inference from the above data.

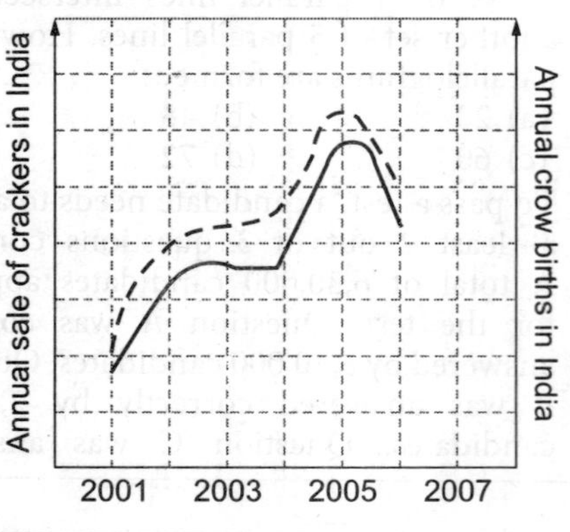

(a) There is a strong correlation between crow birth and cracker sales.
(b) Cracker usage increases crow birth rate.
(c) If cracker sale declines, crow birth will decline.
(d) Increased birth rate of crows will cause an increase in the sale of crackers.

Q.11 – Q.35 carry one mark each.

11. For a laminar flow of a liquid metal over a flat plate, the thicknesses of the velocity and thermal boundary layers are δ_v and δ_t respectively. Kinematic viscosity (viscosity/density) of liquid metal is significantly lower than its thermal diffusivity [thermal conductivity / (density × specific heat)]. Based on this information, pick the correct option.
(**Note:** The temperature of the liquid metal is different from that of the plate).
(a) $\delta_v < \delta_t$ (b) $\delta_v > \delta_t$
(c) $\delta_v = \delta_t$
(d) Information insufficient

12. What is the most-abundant anion in a $2CaO.SiO_2$ melt?
(a) $(SiO_4)^{4-}$ (b) $(Si_2O_7)^{6-}$
(c) $(Si_3O_{10})^{8-}$ (d) $(Si_4O_{14})^{10-}$

13. Analysis of a flow phenomenon in a system requires the following variables:
(i) Pressure [$M L^{-1} T^{-2}$]
(ii) Velocity of the fluid [$L T^{-1}$]
(iii) Size of the system [L]
(iv) Density of the fluid [$M L^{-3}$]
(v) Viscosity of the fluid [$M L^{-1} T^{-1}$]
According to Buckingham Pi theorem (dimensional analysis) what is the number of independent DIMENSIONLESS variables needed to describe this system?
(a) 2 (b) 3
(c) 4 (d) 5

14. In froth flotation, the primary purpose of adding collectors is to:
(a) make the surface of the mineral hydrophobic.
(b) make the surface of the mineral hydrophilic.

Answers

9. (c)	10. (a)	11. (a)	12. (a)	13. (a)	14. (a)

(c) stabilize the froth.
(d) adjust the pH.

15. Arrange the following in the correct sequence of operations in an integrated steel plant:
(i) Basic Oxygen Furnace (BOF)
(ii) Blast Furnace (BF)
(iii) Ruhrstahl Heraeus Degassing Process (RH)
(iv) Ladle Furnace Process (LF)
(a) BOF → BF → LF → RH
(b) BF → LF → BOF → RH
(c) RH → LF → BF → BOF
(d) BF → BOF → LF → RH

16. During decarburization in a steel bath at 1550°C, the compositions of dissolved C (wt.%C) and dissolved O (wt.%O) follow the relation:

$$(\text{wt.\%C})(\text{wt.\%O}) = K$$

When the partial pressure of CO(p_{CO}) is 1 atm, K = 0.002. If p_{CO} = 0.1 atm, what is the value of K, at the same temperature? Note: Assume Henry's law is applicable.
(a) 0.06 (b) 0.002
(c) 0.02 (d) 0.0002

17. During upset forging, and considering friction, which of the following profiles represents the axial compressive stress?

Note: In these profiles, the absolute value of the axial compressive stress is plotted on the y-axis.

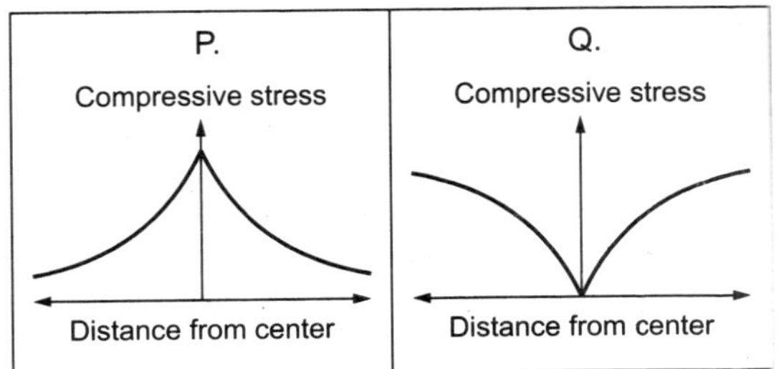

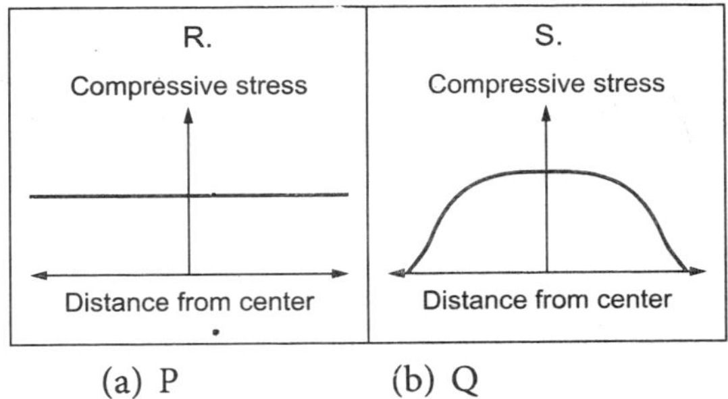

(a) P (b) Q
(c) R (d) S

18. Consider the following engineering components:
P. Gas turbine blades
Q. Tungsten-based heavy alloy penetrators
R. Self-lubricating bearings
S. Engine block of an automobile

Which of the following two components are produced by powder metallurgy?
(a) P & Q (b) Q & R
(c) Q & S (d) P & S

19. Which of the following materials are protected by passivation (i.e., formation of a thin adherent film on the surface) from corrosion?
P. Aluminium alloys
Q. Mild steel R. Stainless steel
S. Silver
(a) P & Q (b) Q & R
(c) Q & S (d) P & R

20. The c/a ratio of Zn (hcp) is 1.856. Slip at room temperature occurs most easily on which of the following slip systems in Zn:

Note: In hcp metals, the ideal c/a ratio is 1.633.
(a) $\{1\bar{1}00\}\langle 11\bar{2}0\rangle$ (b) $\{1\bar{1}00\}\langle 0002\rangle$
(c) $\{0001\}\langle 11\bar{2}0\rangle$ (d) $\{10\bar{1}\bar{1}\}\langle 11\bar{2}3\rangle$

21. At equilibrium, the maximum number of phases in a three-component system at CONSTANT PRESSURE is:

Answers

15. (d)	16. (d)	17. (a)	18. (b)	19. (d)	20. (c)
21. (d)					

(a) 1 (b) 2
(c) 3 (d) 4

22. For a single phase system with K components, the exact differential of Gibbs free energy is given as :

$$dG = -SdT + VdP + \sum_{i=1}^{K} \mu_i dn_i$$

Which one of the following can be derived from the above expression as a Maxwell's relation?

(a) $\left.\frac{\partial S}{\partial P}\right|_{T, n_i} = -\left.\frac{\partial V}{\partial T}\right|_{P, n_i}$

(b) $\left.\frac{\partial S}{\partial V}\right|_{T, n_i} = \left.\frac{\partial V}{\partial T}\right|_{P, n_i}$

(c) $\left.\frac{\partial S}{\partial T}\right|_{T, n_i} = -\left.\frac{\partial V}{\partial P}\right|_{T, n_i}$

(d) $\left.\frac{\partial S}{\partial P}\right|_{T, n_i} = \left.\frac{\partial V}{\partial P}\right|_{T, n_i}$

23. For intrinsic Si at room temperature, the electron concentration = the hole concentration = 1×10^{16} m^{-3}, and the electron and hole mobilities are 0.1 and 0.2 $m^2.V^{-1}.s^{-1}$ respectively. What is the room temperature electrical conductivity?

The charge of an electron is 1.6×10^{-19}C.

(a) 0 $(\Omega.m)^{-1}$ (b) 1.6×10^{-4} $(\Omega.m)^{-1}$
(c) 3.2×10^{-4} $(\Omega.m)^{-1}$
(d) $4.8 \times 10^{-4}(\Omega.m)^{-1}$

24. To minimize refractory loss during BOF steel-making, slag is supersaturated with which of the following:

(a) Al_2O_3 (b) Fe_2O_3
(c) SiO_2 (d) MgO

25. Which of the following welding processes is NOT suitable for joining thin metal sheets?

(a) Gas Metal Arc Welding
(b) Electro Slag Welding
(c) Electron Beam Welding
(d) Laser Beam Welding

26. Which of the following categories of materials is suitable as a filler for brazing of steel sheets?

(a) Epoxy resins
(b) Near-eutectic alloys
(c) Refractory alloys
(d) Large freezing range alloys

27. Arrange the following processes in INCREASING magnitude of surface roughness of the casting.

P. Investment casting
Q. Permanent mould casting
R. Pressure die casting
S. Sand casting

(a) R < S < Q < P (b) P < S < Q < R
(c) R < P < Q < S (d) Q < P < R < S

28. In a Jominy end-quench test of the eutectoid plain-carbon steel, which of the following represents the sequence of microstructures observed from the quenched end of the specimen?

(a) Fine Pearlite, Coarse Pearlite, Martensite and Pearlite, Martensite
(b) Martensite, Martensite and Pearlite, Fine Pearlite, Coarse Pearlite
(c) Coarse Pearlite, Pearlite and Martensite, Fine Pearlite, Martensite
(d) Martensite, Martensite and Pearlite, Coarse Pearlite, Fine Pearlite

29. A long oil pipeline made of steel is suspected to have developed a scale on the inner surface due to corrosion. Which of the following non-destructive techniques is the most suitable for detecting and quantifying such a defect?

(a) Dye penetrant test

Answers

22. (a)	23. (d)	24. (d)	25. (b)	26. (b)	27. (c)
28. (b)	29. (c)				

(b) Magnetic particle inspection
(c) Ultrasonic inspection
(d) Acoustic emission

30. A steel is plastically worked in the temperature range below the nose of the TTT curve and above Ms, followed by quenching to produce fine martensite. What is this process called?
(a) Martempering (b) Ausforming
(c) Inter-critical forming
(d) Normalizing

31. What is the requirement for equilibrium at a triple junction as shown in the schematic, with γ_{12}, γ_{23} and γ_{13} as grain boundary tensions and θ_1, θ_2 and θ_3 as dihedral angles?

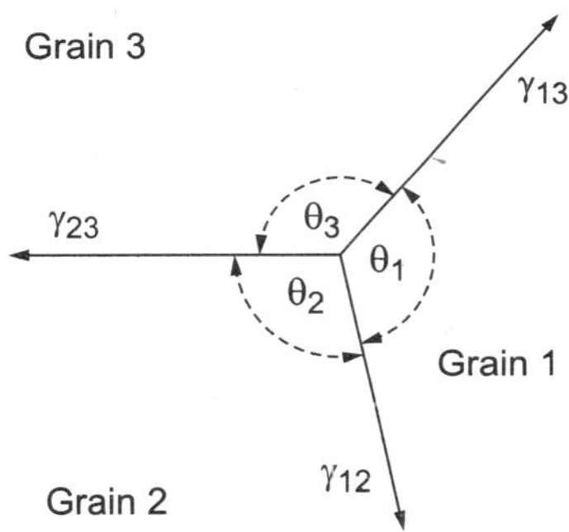

(a) $\frac{\gamma_{23}}{\cos(\theta_1)} = \frac{\gamma_{13}}{\cos(\theta_2)} = \frac{\gamma_{12}}{\cos(\theta_3)}$

(b) $\frac{\gamma_{23}}{\sin(\theta_1)} = \frac{\gamma_{13}}{\sin(\theta_2)} = \frac{\gamma_{12}}{\sin(\theta_3)}$

(c) $\frac{\gamma_{23}}{\sin(\theta_2)\cdot\sin(\theta_3)} = \frac{\gamma_{13}}{\sin(\theta_1)\cdot\sin(\theta_3)} = \frac{\gamma_{12}}{\sin(\theta_1)\cdot\sin(\theta_2)}$

(d) $\gamma_{23}\sin(\theta_1) = \gamma_{13}\sin(\theta_2) = \gamma_{12}\sin(\theta_3)$

32. In the A-rich end of the A-B binary eutectic phase diagram (shown below), the solidus and liquidus are straight lines.

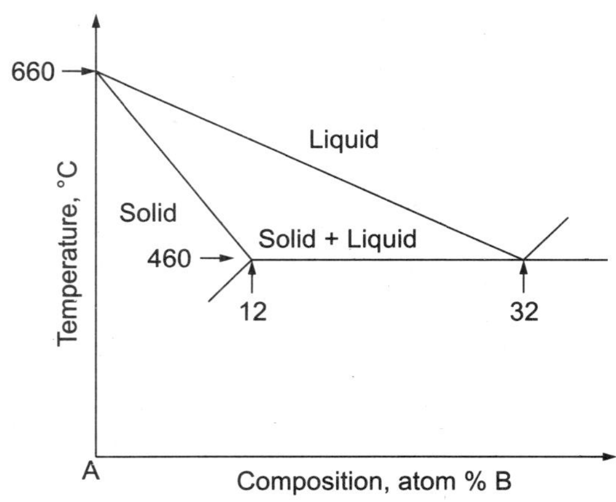

The freezing range of the alloy with 16% B is _______ (in °C *to one decimal place*)

33. A vector field is given as follows:

$$\vec{f} = (-x^2y + xy^2)\hat{i} + (x^2y)\hat{j}$$

The divergence of this field evaluated at (x, y) = (2, 2) is _______.

34. A plate of thickness h = 120 mm is cold rolled in a mill with a roll diameter of 200 mm. If the coefficient of friction μ is 0.1, the maximum possible reduction Δh in a single pass is _______ (in mm *to one decimal place*)

35. A copper-aluminium diffusion couple develops a certain concentration profile after an isothermal treatment at 600°C for 10 hours. The time required to achieve the same concentration profile at 500°C is _______ (in hours *to 1 decimal place*)

Given: The interdiffusion coefficient for copper in aluminium at 500°C and 600°C are $4 \times 10^{-14}\ m^2s^{-1}$ and $8 \times 10^{-13}\ m^2s^{-1}$.

Q.36 – Q.65 carry two marks each.

36. The molar free energy (J mol^{-1}) of a liquid solution of a binary A-B alloy as a function of temperature (T) and composition (x, the mole fraction of B) is given by:

Answers

30. (b) 31. (b) 32. (95 to 105) 33. (–0.1 to 0.1) 34. (0.9 to 1.1)
35. (180 to 220) 36. (a)

$$G^L(T, x) = (1 - x)G_A^{0,L} + xG_B^{0,L} + RT[x \ln x + (1 - x)\ln(1 - x)] + 4000x(1 - x).$$

where $G_A^{0,L}$ and $G_B^{0,L}$ are the molar free energies of pure liquid A and pure liquid B. What is the excess molar free energy, $G^{XS,L}$, for an alloy with x = 0.5 at T = 1000K?

Given: Gas constant R = 8.314 $JK^{-1} mol^{-1}$

(a) 1000 J mol^{-1} (b) – 2000 J mol^{-1}
(c) 4763 J mol^{-1} (d) – 5763 J mol^{-1}

37. Determine the correctness (or otherwise) of the following Assertion [A] and the Reason [R]

Assertion [A]: For a material exhibiting Coble creep, a reduction in grain size results in a significant increase in creep rate

Reason [R]: Grain boundaries act as a barrier to motion of dislocations

(a) Both [A] and [R] are true and [R] is the correct reason for [A]
(b) Both [A] and [R] are true, but [R] is not the correct reason for [A]
(c) Both [A] and [R] are false
(d) [A] is true but [R] is false

38. Determine the correctness (or otherwise) of the following Assertion [A] and the Reason [R]

Assertion [A]: Refractory BCC metals like W and Mo are less ductile than FCC metals like Ni and Pt at room temperature

Reason [R]: BCC metals have fewer independent slip systems than FCC metals

(a) Both [A] and [R] are true and [R] is the correct reason for [A]
(b) Both [A] and [R] are true, but [R] is not the correct reason for [A]
(c) Both [A] and [R] are false
(d) [A] is true but [R] is false

39. A glass fibre of 5 micron diameter is subjected to a tensile stress of 20 MPa. The surface energy and elastic modulus of this material are 0.3 Jm^{-2} and 70 GPa, respectively. Pick the correct answer based on the information provided above:

Note: The glass fibre contains a population of flaws of different lengths.

(a) The fibre will undergo brittle fracture
(b) The fibre will undergo plastic deformation, but not fracture
(c) The fibre will undergo elastic deformation, but not fracture
(d) The fibre will undergo buckling

40.

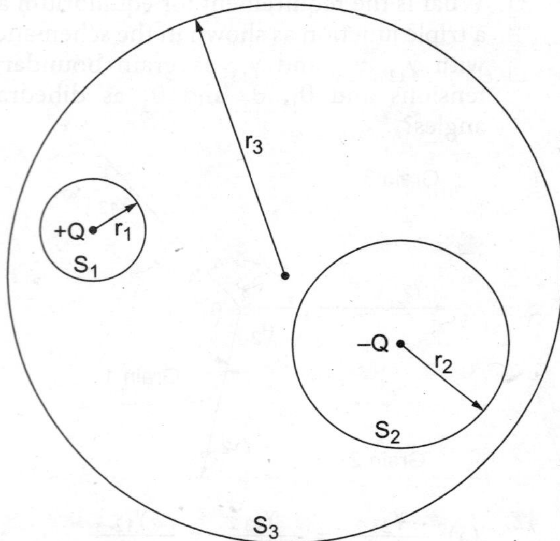

Two equal and opposite point charges +Q and –Q are located as shown in the figure above. A surface integral, F_i is defined on surface S_i of a sphere of radius r_i as follows:

$$F_i = \oint_{S_i} (\vec{E} \cdot \hat{n}) dS,,$$

where $\vec{E}$ is the electric field, and $\hat{n}$ is the unit normal to the surface of integration. If $r_1 : r_2 : r_3$ are in the ratio 1 : 2 : 5, use the Gauss divergence theorem to determine the ratio $F_1 : F_2 : F_3$.

(a) 1 : 2 : 5 (b) 5 : 2 : 1
(c) 1 : –1 : 0 (d) 1 : 1 : 0

Answers

37. (b) 38. (d) 39. (c) 40. (c)

41. Match the four tensile stress-strain curves (P, Q, R, S) with the materials listed in the box:

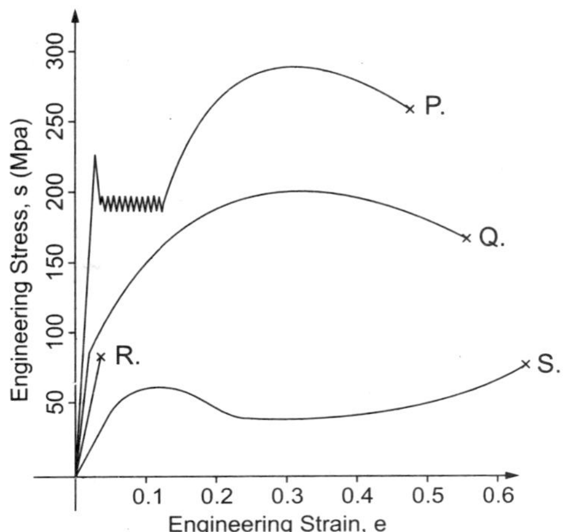

1. Polyester (PET)
2. High purity Copper
3. Mild Steel
4. Soda-lime Glass

(a) P-1, Q-2, R-3, S-4
(b) P-3, Q-1, R-4, S-2
(c) P-3, Q-2, R-4, S-1
(d) P-2, Q-3, R-4, S-1

42. Match the metal in Group 1 with the appropriate extractive process in Group 2

	Group 1		Group 2
P.	Fe	1.	Metallothermic Reduction
Q.	Ni	2.	Carbothermic reduction
R.	Al	3.	Matte Smelting
S.	Cr	4.	Fused Salt Electrolysis

(a) P-1, Q-2, R-4, S-3
(b) P-2, Q-4, R-1, S-3
(c) P-2, Q-3, R-4, S-1
(d) P-2, Q-1, R-4, S-3

43. Fe is produced in a reactor using pure Fe_2O_3, C and O_2. For every mole of Fe produced, 2.38 moles of C is used. The exit gas from the reactor contains CO and CO_2 in the molar ratio of 1 : 1. How many moles of O_2 is consumed for every mole of Fe produced?

(a) 1.035 (b) 2.072
(c) 0.513 (d) 4.147

44. What is the voltage required to electrolytically refine impure copper of activity $a_{Cu} = 0.9$ (Raoultian standard state) to pure copper at 300 K

Given: Gas constant R = 8.314 J $mol^{-1}K^{-1}$, and Faraday's constant F = 96500 C mol^{-1}.

(a) 1.36 mV (b) 2.72 mV
(c) 0 mV (d) 5.44 mV

45. Consider the following Ordinary Differential Equation:

$$\frac{d}{dx}\left[c\frac{dc}{dx}\right] = 0$$

In a domain $0 \le x \le t$, with boundary conditions c(0) = 0.5 and c(t) = 1.0, pick the appropriate choice for c(x)from the following options:

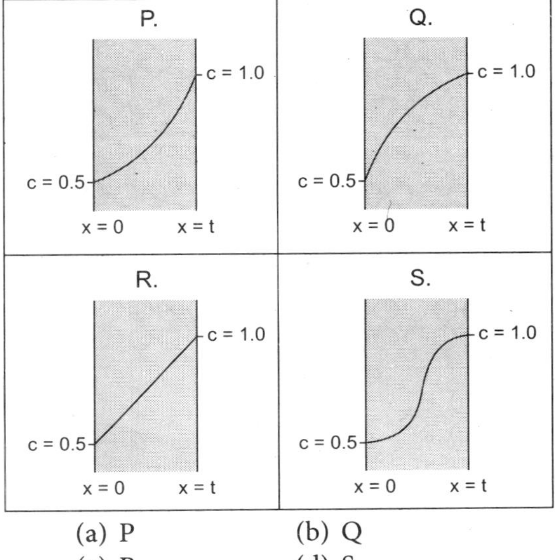

(a) P (b) Q
(c) R (d) S

Answers

41. (c) 42. (c) 43. (a) 44. (a) 45. (b)

46. Match the manufacturing processes in Group 1, with the types of cracks in Group 2:

Group 1	Group 2
P. Arc welding	1. Edge crack
Q. Extrusion	2. Chevron crack
R. Drawing	3. Surface crack
S. Rolling	4. Liquation crack

(a) P-1, Q-2, R-3, S-4
(b) P-1, Q-3, R-4, S-2
(c) P-4, Q-3, R-2, S-1
(d) P-4, Q-2, R-1, S-3

47. Consider a dilute substitutional solid solution of X in a metal A. The powder diffraction pattern of this alloy reveals that all the peaks have shifted to the left when compared to those for pure A (with no splitting of peaks). If such a solute interacts and segregates to an edge dislocation, which of the following positions around the dislocation will it preferentially occupy?

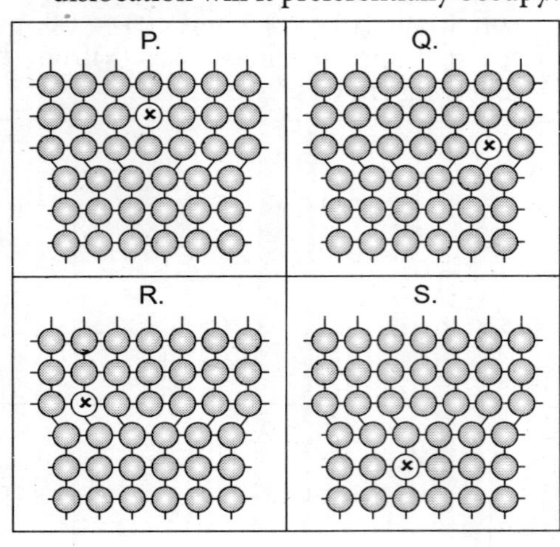

(a) P (b) Q
(c) R (d) S

48. A classroom of 20 students can be categorized on the basis of blood-types: 5 students each with "A", "B", "AB", and "O" blood-types. If four students are selected at random from this class, what is the probability that each student has a different blood-type?
(a) 0.2500 (b) 0.1289
(c) 0.0625 (d) 0.0156

49. If the solid-liquid interfacial energy increases by 10%, the energy barrier for homogeneous nucleation of a spherical solid from the liquid, will change by:
(a) 21% (b) 33%
(c) –10% (d) 46%

50. Consider the following stress state imposed on a material:

$$\sigma = \begin{bmatrix} 90 & 50 & 0 \\ 50 & -20 & 0 \\ 0 & 0 & 140 \end{bmatrix} \text{MPa.}$$

If the material responds elastically with a volumetric strain $\Delta = 3.5 \times 10^{-4}$, what is its bulk modulus?
(a) 150 GPa (b) 350 GPa
(c) 200 GPa (d) 400 GPa

51. A continuous and aligned carbon-fiber composite consists of 25 vol. % of fibers in an epoxy matrix. The Young's modulus of fiber and matrix, respectively are $E_f = 250$ GPa and $E_m = 2.5$ GPa.
If the composite is subjected to longitudinal loading (iso-strain condition and assuming elastic response), the fraction of load borne by the reinforcement is _______ (*to two decimal places*)

52. Pure iron transforms from body centered cubic (BCC) to face centered cubic (FCC) crystal structure at 912°C. If the lattice parameter of the BCC phase is 0.293 nm and that of the FCC phase is 0.363 nm, the associated volume change is _______ (in % *to one decimal place*)

Answers

46. (c) 47. (d) 48. (b) 49. (b) 50. (c)
51. (0.96 to 0.98) 52. (–5.5 to –4.5)

53. As shown in schematic below, an alloy is cast as a rectangular slab between two thick mould walls that differ in their thermal conductivities. Shrinkage defects are found at a distance L1 from Mould-1 with thermal conductivity k_1 and distance L2 from Mould-2 with thermal conductivity k_2. If the ratio L1 : L2 = 3 : 2, and assuming 1-D heat transfer, the ratio $(k_1/k_2)$is _______ (*to two decimal places*)

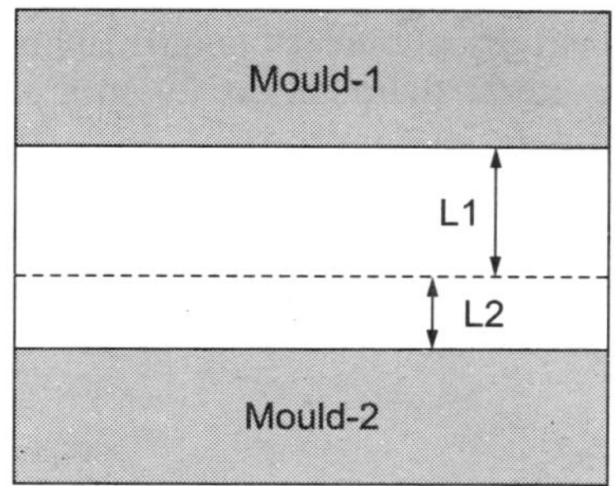

54. The temperature profile (T in Kelvin) of an arc weld across its width is given as $T = 2000 \exp(-0.3x^2)$where x (in mm) is the distance from the weld centre. The melting point of the base material is 1500 K. The width of the fusion zone is _______ (in mm to *two decimal places*).

55. The terminal velocity (v) of a spherical inclusion of diameter D = 50 micrometers rising in liquid steel is _______ (in mm s^{-1} to *two decimal places*)

Assume Stokes law; i.e., drag force $F_d = 3\pi\mu Dv$, where μ is the viscosity of steel.

Given: Density of liquid steel = 7900 kgm^{-3}; Viscosity of liquid steel = 0.0079 Pa s;

Density of the inclusion = 2500 kg m^{-3}; Acceleration due to gravity = 9.8 m s^{-2}

56. A single crystal of aluminium is subjected to 10 MPa tensile stress along the [321] crystallographic direction. The resolved shear stress on the $(11\bar{1})[101]$slip system is _______ (in MPa to *two decimal places*)

57. If the net magnetic moment of an Fe atom in BCC structure is 2 μ_B, then the saturation magnetization of Fe is _______ (kA m^{-1} to *one decimal place*)

Given: $\mu_B = 9.273 \times 10^{-24}$ A m^2; Lattice parameter of BCC iron is 0.287 nm

Note: kA is kiloAmperes

58. If 2 moles of Au and 3 moles of Ag are mixed to form a single-phase ideal solid solution, the total entropy of mixing is _______ (on J K^{-1} to *one decimal place*)

Given: Gas constant R = 8.314 J K^{-1} mol^{-1}

59. A spherical liquid metal droplet of diameter 1 mm is solidified in a stream of gas at 300 K. Assuming that the metal droplet remains at its melting point of 900 K and neglecting radiative losses, the time to complete the solidification is _______ (in seconds to *one decimal place*).

Given: The enthalpy of fusion for the metal is 4000 kJ kg^{-1}; The gas-droplet convective heat transfer coefficient is 200 W $m^{-2}K^{-1}$; Density of liquid metal is 2700 kg m^{-3}.

60. At a temperature of 710 K, the vapour pressure of pure liquid Zn is given by:

$$p_{Zn}(X_{Zn} = 1.0) = 3.6 \times 10^{-4} \text{ atm.}$$

The Raoultian activity coefficient (γ_{Zn}) of Zn in Zn-Cd alloy liquid at 710 K is approximated by:

$$\ln(\gamma_{Zn}) = 0.875(1 - X_{Zn})^2$$

The ratio $\left[\dfrac{p_{Zn}(X_{Zn} = 0.7)}{p_{Zn}(X_{Zn} = 1.0)}\right]$ for a liquid alloy with $X_{Zn} = 0.7$ is _______ (to two *decimal places*).

61. For the reaction: $4Ag(s,pure) + O_2(g) \rightarrow 2Ag_2O(s,pure)$, the standard enthalpy change, $\Delta H^0 = -61080$ J, and the standard entropy change, $\Delta S^0 = -132.22$ JK^{-1}, in the temperature range from 298 K to 500 K. The temperature above which Ag_2O

Answers

53. (2.2 to 2.3) 54. (1.9 to 2.0) 55. (0.9 to 1.0) 56. (4.5 to 4.8) 57. (1500 to 1600)
58. (24.9 to 29.0) 59. (14.9 to 15.1) 60. (0.74 to 0.78) 61. (427 to 432)

decomposes in an atmosphere containing oxygen at a partial pressure p_{O_2} = 0.3 atm is _______ (in K to *one decimal place*).

Given: Gas constant R = 8.314 J K^{-1} mol^{-1}

62. In a powder diffraction experiment on BCC iron, the first peak occurs at 2θ = 68.7°. The wavelength of X-rays is _______ (in nm to *three decimal places*).

    Given: The lattice parameter of iron = 0.287 nm

63. A 1 mol piece of copper at 400 K is brought in contact with another 1 mol piece of copper at 300 K, and allowed to reach thermal equilibrium. The entropy change for this process is _______ (in J K^{-1} to *three decimal places*).

    Given: Specific heat capacity of copper (between 250 K and 500 K) is 22.6 J K^{-1} mol^{-1}.

    Assume that the system containing the two pieces of copper remains isolated during this process.

64. Using the trapezoidal rule with two equal intervals (n = 2, Δx = 1), the definite integral $\int_2^4 \frac{1}{\ln(x)} dx$ = _______ (to *two decimal places*).

65. The ideal plastic work involved in extruding a cylindrical billet of length 100 mm, from an initial diameter of 20 mm to a final diameter of 16 mm is _______ (in J to *one decimal place*).

    The flow stress in compression is 40 MPa, and remains constant throughout the process.

Answers

62. (0.225 to 0.235) 63. (0.450 to 0.480) 64. (1.95 to 2.05)
65. (555.5 to 565.5)

22

GATE-2019

Q.1 – Q.5 carry one mark each.

1. The fishermen, ________ the flood victims owed their lives, were rewarded by the government.
   (a) whom (b) to which
   (c) to whom (d) that

2. Some students were not involved in the strike.
   If the above statement is true, which of the following conclusions is/are logically necessary?
   1. Some who were involved in the strike were students.
   2. No student was involved in the strike.
   3. At least one student was involved in the strike.
   4. Some who were not involved in the strike were students.

   (a) 1 and 2 (b) 3
   (c) 4 (d) 2 and 3

3. The radius as well as the height of a circular cone increases by 10%. The percentage increase in its volume is ________.
   (a) 17.1 (b) 21.0
   (c) 33.1 (d) 72.8

4. Five numbers 10, 7, 5, 4 and 2 are to be arranged in a sequence from left to right following the directions given below
   1. No two odd or even numbers are next to each other.
   2. The second number from the left is exactly half of the left-most number.
   3. The middle number is exactly twice the right-most number.

   Which is the second number from the right?
   (a) 2 (b) 4
   (c) 7 (d) 10

5. Until Iran came along, India had never been ________ in kabaddi.
   (a) defeated (b) defeating
   (c) defeat (d) defeatist

Q.6 – Q.10 carry two marks each.

6. Since the last one year, after a 125 basis point reduction in repo rate by the Reserve Bank of India, banking institutions have been making a demand to reduce interest rates on small saving schemes. Finally, the government announced yesterday a reduction in interest rates on small saving schemes to bring them on par with fixed deposit interest rates.
   Which one of the following statements can be inferred from the given passage?
   (a) Whenever the Reserve Bank of India reduces the repo rate, the interest rates on small saving schemes are also reduced
   (b) Interest rates on small saving schemes are always maintained on par with fixed deposit interest rates

Answers

1. (c) 2. (c) 3. (c) 4. (c) 5. (a) 6. (c)

(c) The government sometimes takes into consideration the demands of banking institutions before reducing the interest rates on small saving schemes
(d) A reduction in interest rates on small saving schemes follow only after a reduction in repo rate by the Reserve Bank of India

7. In a country of 1400 million population, 70% own mobile phones. Among the mobile phone owners, only 294 million access the Internet. Among these Internet users, only half buy goods from e-commerce portals. What is the percentage of these buyers in the country?
(a) 10.50 (b) 14.70
(c) 15.00 (d) 50.00

8. The nomenclature of Hindustani music has changed over the centuries. Since the medieval period *dhrupad* styles were identified as *baanis*. Terms like *gayaki* and *baaj* were used to refer to vocal and instrumental styles, respectively. With the institutionalization of music education the term *gharana* became acceptable. *Gharana* originally referred to hereditary musicians from a particular lineage, including disciples and grand disciples.

Which one of the following pairings is NOT correct?
(a) dhrupad, baani
(b) *gayaki*, vocal
(c) baaj, institution
(d) *gharana*, lineage

9. Two trains started at 7AM from the same point. The first train travelled north at a speed of 80km/h and the second train travelled south at a speed of 100 km/h. The time at which they were 540 km apart is ______ AM.
(a) 9 (b) 10
(c) 11 (d) 11.30

10. "I read somewhere that in ancient times the prestige of a kingdom depended upon the number of taxes that it was able to levy on its people. It was very much like the prestige of a head-hunter in his own community."
Based on the paragraph above, the prestige of a head-hunter depended upon ______.
(a) the prestige of the kingdom
(b) the prestige of the heads
(c) the number of taxes he could levy
(d) the number of heads he could gather

Q.11 – Q.35 carry one mark each.

11. One of the eigenvalues for the following matrix is ________.

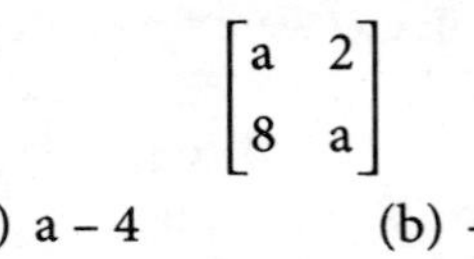

$$\begin{bmatrix} a & 2 \\ 8 & a \end{bmatrix}$$

(a) a – 4 (b) – a – 4
(c) 4 (d) – 4

12. The curl of vector fields shown below is **not** zero for ________.

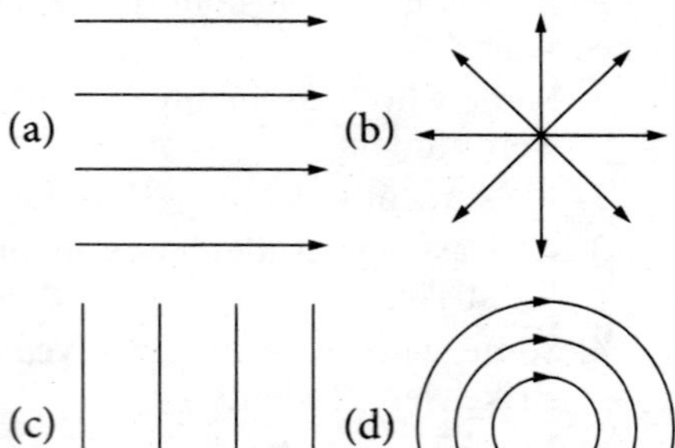

13. The smallest period of function $f(x) = \sin\left(\frac{nx}{k}\right)$ is ________.
(a) 2π (b) $\frac{k}{n}$
(c) $\frac{2\pi k}{n}$ (d) $\frac{2\pi n}{k}$

Answers

7. (a)	8. (c)	9. (b)	10. (d)	11. (a)	12. (d)
13. (c)					

14. The directional derivative of $\phi = x^2 + y$ along the unit vector $\hat{u} = \frac{1}{5}(3\hat{i} + 4\hat{j})$ at (1, 1) is ________.
(a) 3 (b) 2
(c) 1 (d) 0

15. Liquid steel is kept in a graphite crucible for sufficiently long time such that equilibrium is established. Activity of carbon (with respect to graphite as the standard state) in the liquid is ________.
(a) 0.0 (b) 0.5
(c) 1.0 (d) 2.0

16. The variation of standard free energies for two oxides AO and BO with temperature are shown below.

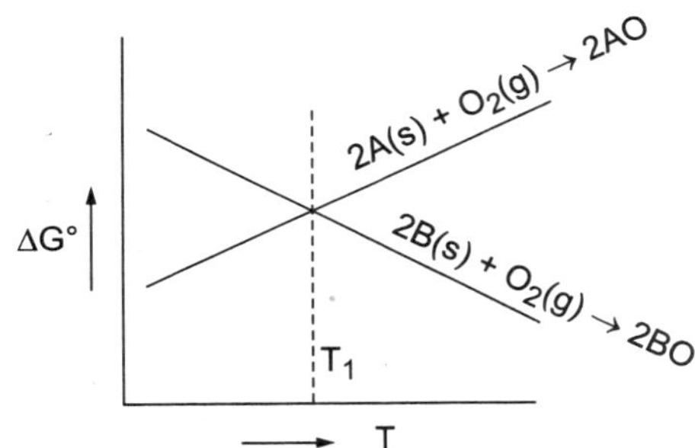

The correct statement with reference to the above figure is ________.
(a) AO is a gas (b) BO is a gas
(c) A will reduce BO above temperature T_1
(d) B will reduce AO below temperature T_1

17. Terminal rise velocity of a spherical shaped solid in a liquid obeys the following functional relationship:
$U = f(d, W, \mu, \rho)$
Where, U is the terminal rise velocity, d is the diameter of the solid, W is the apparent weight of the solid, μ is the viscosity of liquid and ρ is the density of liquid.
According to Buckingham II theorem, the number of independent dimensionless variables needed to describe the phenomenon is ________.
(a) 1 (b) 2
(c) 3 (d) 5

18. Consider electrodeposition of copper on a copper electrode from an aqueous solution containing 0.5×10^{-3} mol.cm^{-3} $CuSO_4$.
Given: Faraday constant, F = 96500 Coulomb per gram equivalent.
Assume transport of reactant is rate limiting, and mass transfer coefficient is 10^{-4} cm.s^{-1}.
The limiting current density (in mA.cm^{-2}) is ________.
(a) 4.83 (b) 9.65
(c) 19.30 (d) 38.60

19. The correct sequence of steelmaking operations is ________.
[BOF: Basic Oxygen Furnace; LF: Ladle Furnace; VD: Vacuum Degassing; CC: Continuous Casting].
(a) BOF → LF → VD → CC
(b) CC → BOF → LF → VD
(c) VD → CC → BOF → LF
(d) LF → BOF → VD → CC

20. The common ore of titanium is ________.
(a) Bauxite (b) Chalcopyrite
(c) Cassiterite (d) Ilmenite

21. Coarse suspended particles are to be separated from a flowing gas in a dust catcher. All factors remaining same, maximum gas-solid separation is expected from the dust catcher ________ shown below.

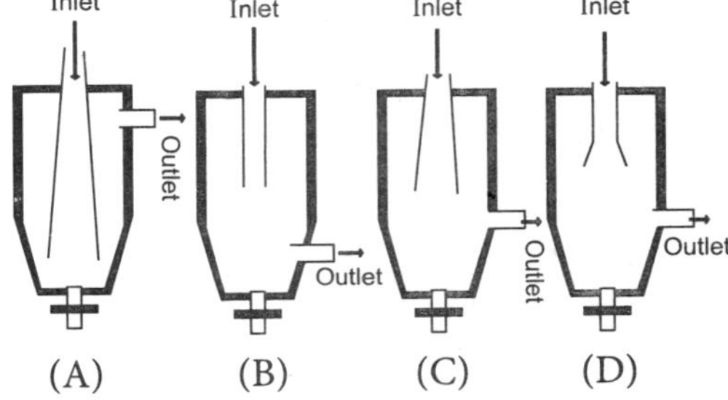

Answers

14. (b)	15. (c)	16. (b)	17. (b)	18. (b)	19. (a)
20. (d)	21. (a)				

22. The Boudouard (or, carbon gasification) reaction is ________.

(a) $C(s) + \frac{1}{2}O_2(g) \rightarrow CO(g)$
(b) $C(s) + O_2(g) \rightarrow CO_2(g)$
(c) $C(s) + CO_2(g) \rightarrow 2CO(g)$
(d) $CO(g) + \frac{1}{2}O_2(g) \rightarrow CO_2(g)$

23. The fastest diffusing element in iron at 1100°C is ________.

(a) Ni (b) Co
(c) Cr (d) C

24. Hydrogen bonds are stronger than ________.

(a) Ionic bonds (b) Metallic bonds
(c) van der Waals bonds
(d) Covalent bonds

25. The carbide that is primarily responsible for intergranular corrosion in austenitic stainless steel is ________.

(a) $Cr_{23}C_6$ (b) Fe_3C
(c) SiC (d) Mn_3C

26. During low strain rate (≤ 0.1 per second) deformation of a metal at room temperature, the one that deforms by twinning mode is ________.

(a) Fe (b) Mg
(c) Al (d) Ni

27. In a tensile creep test of a metal, Nabarro-Herring mechanism is favored over Coble mechanism for ________.

(a) larger grain size and lower temperature
(b) smaller grain size and higher temperature
(c) larger grain size and higher temperature
(d) smaller grain size and lower temperature

28. Beach marks are commonly observed on the fractured surfaces of metals after a ________.

(a) Creep test (b) Fatigue test
(c) Impact test (d) Compression test

29. The length of internal cracks in two samples of the same glass is measured to be $c_1 = 0.5$ mm and $c_2 = 2$ mm. The ratio $\left(\frac{\sigma_1}{\sigma_2}\right)$ of the fracture strength of the two samples is ________.

(a) 0.5 (b) 1.0
(c) 2.0 (d) 4.0

30. Two blocks of the same metal are to be welded. The configuration that will undergo the least distortion after welding is ________.

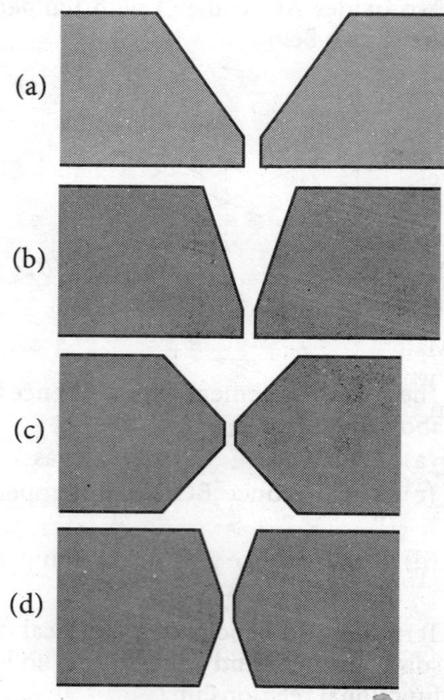

31. The most suitable non-destructive testing method for detecting small internal flaws in a dense bulk material is ________.

(a) Dye penetrant method
(b) Ultrasonic inspection
(c) Eddy current testing
(d) Magnetic particle inspection

Answers

22. (c)	23. (d)	24. (c)	25. (a)	26. (b)	27. (c)
28. (b)	29. (c)	30. (d)	31. (b)		

32. Alligatoring is a defect commonly observed in ________.
   (a) Extrusion
   (b) Deep drawing
   (c) Sheet metal forming
   (d) Rolling

33. The standard deviation (rounded off to one decimal place) of the following set of five numbers is ________.
   6, 8, 8, 9, 9

34. A FCC crystal with a lattice parameter of 0.3615 nm is used to measure the wavelength of monochromatic X-rays. The Bragg angle (θ) for the reflection from (111) planes is 21.68°.
   The wavelength of X-rays (in nm, rounded off to three decimal places) is ________.

35. A plate of width 100 cm and thickness 5 cm is rolled to a thickness of 3 cm. If the entry velocity is 10 $cm.s^{-1}$, the exit velocity of the plate (in $cm.s^{-1}$, rounded off to one decimal place) is ________.
   Assume no change in the width of the plate.

Q.36 – Q.65 carry two marks each.

36. Match the reactors/refining sites in Column I with the corresponding refining processes in Column II.

Column I Reactors/ refining sites	Column II Refining processes
P. Blast furnace runner	1. De-carbur-ization
Q. AOD	2. External De-sulfur-ization
R. Torpedo car	3. De-phosphor-ization
S. BOF	4. External De-silicon-ization

   (a) P-4, Q-1, R-2, S-3
   (b) P-4, Q-2, R-3, S-1
   (c) P-2, Q-1, R-4, S-3
   (d) P-1, Q-3, R-2, S-4

37. Match the injection metallurgy techniques in Column I with the corresponding objectives in Column II.

Column I Injection metallurgy techniques	Column II Objectives
P. Aluminium wire feeding	1. Inclusion modification
Q. Calcium treatment	2. Mixing of liquid steel
R. Argon rinsing	3. De-sulphur-ization
S. Lime powder injection	4. Deoxidation

   (a) P-2, Q-1, R-3, S-4
   (b) P-4, Q-3, R-2, S-1
   (c) P-3, Q-4, R-1, S-2
   (d) P-4, Q-1, R-2, S-3

38. The table (see options below) providing correct information about crystal structure, co-ordination number and packing fraction is ________.
   [**Note:** FCC: Face centered cubic; BCC: Body centered cubic; DC: Diamond cubic. CN: Coordination number; PF: Packing fraction]

(a)

Crystal structure	CN	PF
FCC	12	0.74
BCC	8	0.68
DC	4	0.34

(b)

Crystal structure	CN	PF
FCC	8	0.74
BCC	4	0.68
DC	6	0.34

(c)

Crystal structure	CN	PF
FCC	8	0.52
BCC	12	0.68
DC	12	0.74

Answers

32. (d) 33. (1.1 to 1.6) 34. (0.153 to 0.155) 35. (16.0 to 17.4) 36. (a)
37. (d) 38. (a)

(d)

Crystal structure	CN	PF
FCC	12	0.74
BCC	8	0.68
DC	4	0.74

39. Match the phase transformation in Column I with the corresponding reaction in Column II.
[**Note:** α, β, γ are solid phases; L, L_1, L_2 are liquid phases.]

Column I Phase transformation	Column II Reaction
P. Peritectic	1. $\gamma \rightarrow \alpha + \beta$
Q. Monotectic	2. $L_1 + L_2 \rightarrow \alpha$
R. Eutectoid	3. $L_1 \rightarrow L_2 + \alpha$
S. Syntectic	4. $L + \alpha \rightarrow \beta$

(a) P-4, Q-3, R-1, S-2
(b) P-3, Q-4, R-2, S-1
(c) P-1, Q-3, R-4, S-2
(d) P-4, Q-2, R-3, S-1

40. In a typical scanning electron microscope (SEM) image, information about topography and atomic contrast are obtained from ________.
(a) secondary electron and auger electron, respectively.
(b) primary electron and secondary electron, respectively.
(c) secondary electron and back-scatter electron, respectively.
(d) back-scatter electron and auger electron, respectively.

41. Match the ceramics in Column I with corresponding application in Column II.

Column I Ceramics	Column II Application
P. Mullite	1. Cutting tools
Q. Spinel ferrites	2. Refractories
R. Tungsten carbide	3. Piezoelectric
S. Barium titanate	4. Soft magnet

(a) P-2, Q-3, R-1, S-4
(b) P-4, Q-1, R-2, S-3
(c) P-3, Q-4, R-1, S-2
(d) P-2, Q-4, R-1, S-3

42. An aluminium single crystal is loaded in tension along $[1\bar{1}0]$ axis. Among the following slip systems, the one that will be activated first is ________.
(a) $(1\bar{1}\bar{1})[0\bar{1}1]$ (b) $(\bar{1}\bar{1}1)[011]$
(c) $(\bar{1}\bar{1}1)[1\bar{1}0]$ (d) $(\bar{1}\bar{1}1)[101]$

43. The correct Mohr's circle construction for the stress state given below is ________.

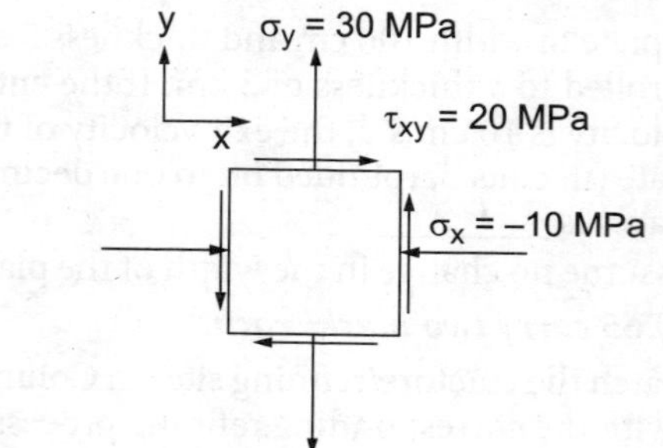

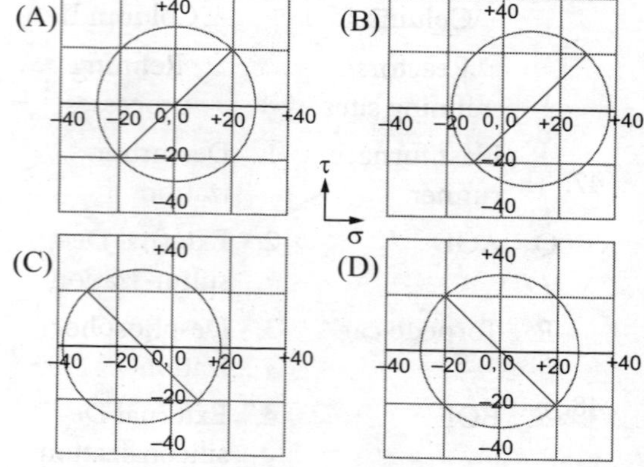

44. Match the automobile components in Column I with the corresponding manufacturing processes in Column II.

Answers

39. (a) 40. (c) 41. (d) 42. (a) 43. (b) 44. (b)

	Column I		Column II
	Automobile components		Manufacturing process
P.	Engine block	1.	Forging
Q.	Brake pad	2.	Sheet metal forming
R.	Connecting rod	3.	Casting
S.	Door panel	4.	Powder metallurgy

(a) P-1, Q-2, R-3, S-4
(b) P-3, Q-4, R-1, S-2
(c) P-3, Q-2, R-4, S-1
(d) P-4, Q-1, R-3, S-2

45. The equilibrium constant for the following reaction at 300 K is ________.
$C(graphite) + 2H_2(g) \rightarrow CH_4(g)$
Given: At 300 K, $\Delta H° = -74,900$ J.mol^{-1}; $\Delta S° = -80$ J.mol^{-1}.K^{-1}; R = 8.314 J.mol^{-1}K^{-1}.
(a) 5.6×10^6 (b) 3.6×10^7
(c) 4.0×10^8 (d) 7.3×10^8

46. A 50 cm long rod is placed against a vertical wall such that the bottom of the rod is 30 cm away from the wall. If the bottom of the rod is pulled horizontally away from the wall at 4 cm.s^{-1}, the top of the rod starts sliding down the wall with an instantaneous velocity (in cm.s^{-1}, rounded off to two decimal places) of magnitude ________.

47. The probability of solving a problem by Student A is (1/3), and the probability of solving the same problem by Student B is (2/5). The probability (rounded off to two decimal places) that at least one of the students solves the problem is ________.

48. Numerical value of work done (rounded off to the nearest integer) by a position-dependent force $\vec{F} = x\hat{i} + 5xy\hat{j}$ (where $\hat{i}$ and $\hat{j}$ are unit vectors) along the path, $y = \frac{x^2}{2}$, from (0, 0) to (2, 2) in the xy plane is ________.

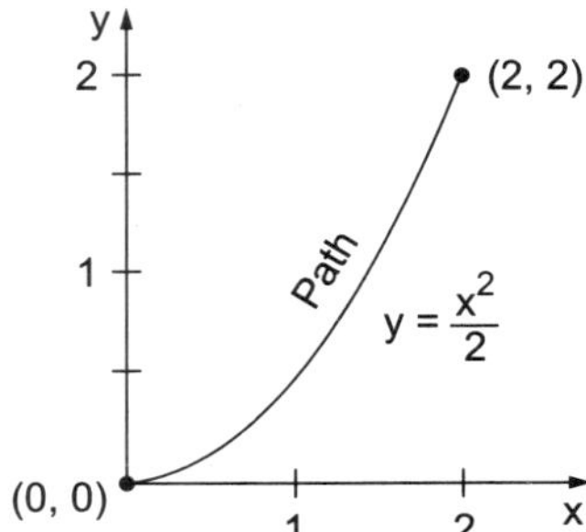

49. The estimated value of the cube root of 37 (rounded off to two decimal places) obtained from the Newton-Raphson method after two iterations (x_2) is ________.
[Start with an initial guess value of $x_0 = 1$].

50. The partial pressure of zinc (in torr, rounded off to two decimal places) in equilibrium with liquid lead containing 0.03 mole % zinc at 900 K is ________.
Given: Vapour pressure of pure zinc (p°_{Zn}) at 900 K is 0.027 atm. Henry's law coefficient (γ°_{Zn}) for zinc at infinite dilute solution in lead on Raoultian scale is 8.55.
1 torr = 1.316×10^{-3} atm.

51. Steady state radial heat conduction through a hollow, infinitely long zirconia cylinder is governed by the following ordinary differential equation:
$$\frac{1}{r}\frac{d}{dr}\left(rk\frac{dT}{dr}\right) = 0$$
Where, T is the temperature and r is the radial distance. The inner surface of the hollow cylinder is maintained at 1473 K and the outer surface at 973 K. The rate of heat loss per unit length through the outer surface of the hollow cylinder (in W.m^{-1}, rounded off to the nearest integer) is ________.

Answers

45. (d) 46. (2.90 to 3.10) 47. (0.59 to 0.61) 48. (17 to 19) 49. (8.50 to 9.00)
50. (0.04 to 0.06) 51. (18670 to 18690)

Given: Inner radius of cylinder = 0.05 m; outer radius of cylinder = 0.07 m and thermal conductivity of zirconia (k) = 2 $W.m^{-1}.K^{-1}$.

52. A 50 mm (diameter) sphere of solid nickel is oxidized in a gas mixture containing 60% argon and 40% oxygen by volume. The rate of oxidation of nickel is controlled by the rate of transport of oxygen through the concentration boundary layer. The rate of oxidation (in moles of nickel per minute (mol/min), rounded off to two decimal places) is ________.

Given: Total pressure = 1 atm.; Temperature = 1173 K; Concentration of oxygen at the surface of the solid = 0; Boundary layer mass transfer coefficient = 0.03 $m.s^{-1}$; Universal gas constant, R = 8.205 $\times 10^{-5}$ $m^3.atm.K^{-1}.mol^{-1}$. Assume ideal gas behavior.

53. Equilibrium concentration of dissolved nitrogen (in wt.%, rounded off to three decimal places) in pure liquid iron, exposed to atmospheric air at 1873 K, is ________.

Given: Sieverts' law constant as a function of temperature (T; in Kelvin) is $\log_{10} K_{[N]} = \left[-\frac{518}{T} - 1.063\right]$, where $K_{[N]}$ has the dimension of $atm^{-(1/2)}$.

Assume $[h_N]$ = [wt.%N], where $[h_N]$ is the activity of nitrogen with respect to 1 wt.% Henrian standard state.

54. Pressure drop in the granular zone of a blast furnace is 300 mm of water per meter of the bed height. The bed permeability is 0.8 $m^4.N^{-1}.s^{-1}$. The volumetric flow rate of gas per unit area through the bed [in $(m^3.s^{-1}).m^{-2}$, rounded off to the nearest integer] is ________.

Assume Darcy's law is applicable and g = 9.8 $m.s^{-2}$. Density of water = 1000 $kg.m^{-3}$.

55. A blast furnace charged with hematite containing 90 wt.% Fe_2O_3 produces liquid iron with 3.6 wt.% carbon. Coke containing 90 wt.% carbon is charged at a rate of 500 kg per Metric ton of liquid iron produced. The blast furnace top gas contains (by volume), 22% CO, 18% CO_2, and balance nitrogen. The volume of the blast furnace top gas [in m^3 (NTP), rounded off to the nearest integer] is ________.

Given: Assume ideal gas behavior, and molar volume of the gas at NTP is 22.4 liter.

56. 100 Metric tons of copper concentrate containing 21 wt.% Cu is to be processed in 6 months with 25 working days per month and 8 working hours per day. The concentrate is leached by sulphuric acid and electrolyzed in 10 cells arranged in series. The minimum current rating (Ampere per month per cell, rounded off to the nearest integer) is ________.

Given: Faraday constant = 96500 Coulomb per gram equivalent.

Atomic weight of copper is 63.

57. Cold working of iron leads to increase in dislocation density from 10^{10} to 10^{15} m^{-2}. The associated stored energy (in $MJ.m^{-3}$, rounded off to one decimal place) is ________.

Given: Shear modulus of iron = 82 GPa,

Burger's vector, $\vec{b} = \frac{a_0}{2}[111]$,

a_0 = 0.2856 nm.

58. The critical radius (in nm, rounded off to one decimal place) of nickel nucleus during solidification at 1673 K is ________.

Given: Enthalpy of fusion of nickel = 2.65 $\times 10^9$ $J.m^{-3}$;

Liquid-solid interfacial energy = 0.5 $J.m^{-2}$, and

Equilibrium melting temperature of nickel = 1728 K.

59. A material, made up of alternating layers of metals A and B, is loaded as shown below. If the volume % of B is 25%, the elastic

Answers

52. (0.11 to 0.13) 53. (0.039 to 0.042) 54. (2347 to 2357) 55. (1927 to 1937)
56. (1485 to 1495) 57. (4.0 to 5.14) 58. (11.0 to 12.5) 59. (155.0 to 165.0)

modulus (in GPa, rounded off to one decimal place) of the material is ________.

Given: Elastic moduli of A and B are 200 GPa and 100 GPa, respectively.

60. The S-N curve for a steel is shown below. If the stress ratio, $(\sigma_{min}/\sigma_{max}) = -0.8$, the maximum stress (in MPa, rounded off to the nearest integer) that the steel can withstand for infinite fatigue life is ________.

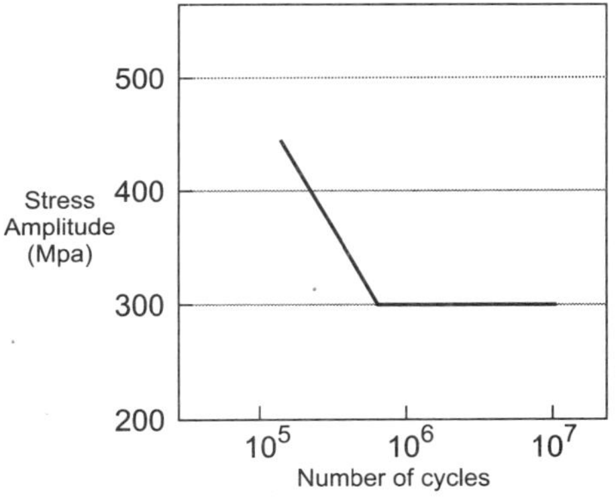

61. True stress - true strain behavior of a metal is given by the flow curve equation $\sigma = 1750\ \varepsilon^{0.37}$, where σ is in MPa. The true stress at necking (in MPa, rounded off to the nearest integer) is ________.

62. In sand-mold casting of a metal, it takes 180 seconds for complete solidification of a 27 cm^3 cube-shaped casting. All other parameters remaining constant, the total solidification time (in seconds, rounded off to one decimal place) for a cylinder-shaped [radius = 1 cm and height = 6 cm] casting of the same metal, is ________.

63. If the solid-solid interfacial energy (γ_{SS}) is 0.87 $J.m^{-2}$ and solid-liquid interfacial energy (γ_{SL}) is 0.5 $J.m^{-2}$, the dihedral angle (ϕ, in degree, rounded off to one decimal place) during sintering is ________.

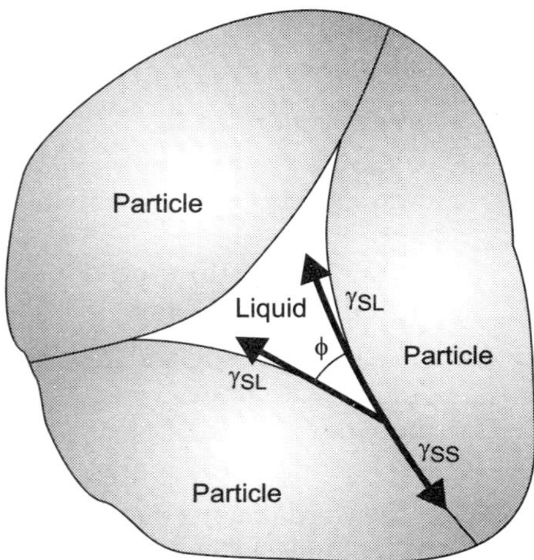

64. The maximum possible reduction (in mm, rounded off to one decimal place) of a 100 mm thick slab during rolling is ________.

Given: The coefficient of friction between roll and the slab is 0.2, and the roll diameter is 200 mm.

65. An arc welding is performed at 400 Amperes, 20 V at a traverse speed of 5 $mm.s^{-1}$. If the heat transfer efficiency is 0.6, the energy input per unit length (in $J.mm^{-1}$, rounded off to the nearest integer) during the process is ________.

Answers

60. (330 to 335) 61. (1160 to 1260) 62. (130. to 134.0) 63. (58.0 to 60.2) 64. (3.6 to 4.2) 65. (955 to 965)

Section

D

23

GATE-2020

Q.1 - Q.5 carry one mark each.

1. He is known for his unscrupulous ways. He always sheds _______ tears to deceive people.
   (a) fox's (b) crocodile's
   (c) crocodile (d) fox
2. Jofra Archer, the England fast bowler, is _______ than accurate
   (a) more fast (b) faster
   (c) less fast (d) more faster
3. Select the word that fits the analogy:
   Build : Building : : Grow :
   (a) Grown (b) Grew
   (c) Growth (d) Growed
4. I do not think you know the case well enough to have opinions. Having said that, I agree with your other point.
   What does the phrase "having said that" mean in the given text?
   (a) as opposed to what I have said
   (b) despite what I have said
   (c) in addition to what I have said
   (d) contrary to what I have said
5. Define [x] as the greatest integer less than or equal to x, for each $x \in (-\infty, \infty)$. If $y = [x]$, then area under y for $x \in [1, 4]$ is _______.
   (a) 1 (b) 3
   (c) 4 (d) 6

Q.6 - Q.10 carry two marks each.

6. Crowd funding deals with mobilisation of funds for a project from a large number of people, who would be willing to invest smaller amounts through web-based platforms in the project.
   Based on the above paragraph, which of the following is correct about crowd funding?
   (a) Funds raised through unwilling contributions on web-based platforms.
   (b) Funds raised through large contributions on web-based platforms.
   (c) Funds raised through coerced contributions on web-based platforms.
   (d) Funds raised through voluntary contributions on web-based platforms.
7. P, Q, R and S are to be uniquely coded using α and β. If P is coded as αα and Q as αβ, then R and S, respectively, can be coded as
   (a) βα and αβ (b) ββ and αα
   (c) αβ and ββ (d) βα and ββ
8. The sum of the first n terms in the sequence 8, 88, 888, 8888, ... is
   (a) $\frac{81}{80}(10^n - 1) + \frac{9}{8}n$
   (b) $\frac{81}{80}(10^n - 1) - \frac{9}{8}n$
   (c) $\frac{80}{81}(10^n - 1) + \frac{8}{9}n$

Answers

1. (c)	2. (a)	3. (c)	4. (b)	5. (d)	6. (d)
7. (d)	8. (d)				

(d) $\frac{80}{81}(10^n - 1) - \frac{8}{9}n$

9. Select the graph that schematically represents BOTH $y = x^m$ and $y = x^{1/m}$ properly in the interval $0 \le x \le 1$, for integer values of m, where $m > 1$.

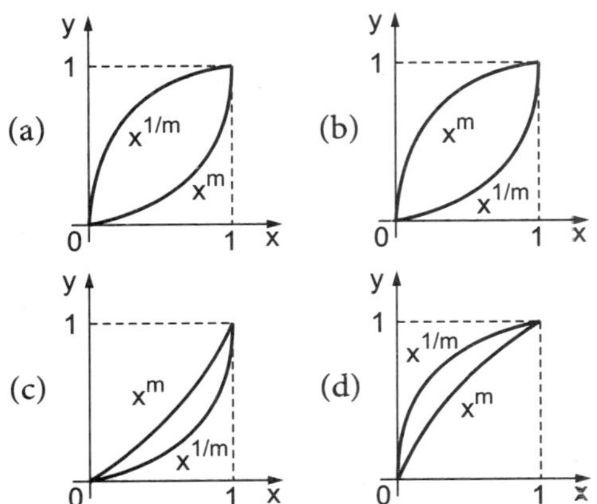

10. The bar graph shows the data of the students who appeared and passed in an examination for four schools P, Q, R and S. The average of success rates (in percentage) of these four schools is ________.

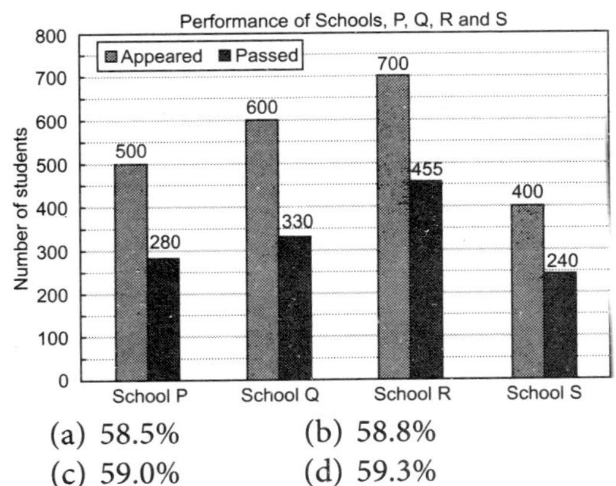

(a) 58.5% (b) 58.8%
(c) 59.0% (d) 59.3%

Q.11 - Q.35 carry one mark each.

11. The general solution to the following homogeneous ODE,

$$\frac{d^2y}{dt^2} + 4\frac{dy}{dt} + 3y = 0,$$

is, $y(t) = c_1 e^{\lambda_1 t} + c_2 e^{\lambda_2 t}$.
The values of λ_1 and λ_2 are:
(a) −1 and −3 (b) −3 and −3
(c) 1 and −3 (d) 1 and 3

12. The number of independent elastic constants of an isotropic material is:
(a) 1 (b) 2
(c) 3 (d) 4

13. A slip system consists of a slip plane and a slip direction. Which one of the following is NOT a valid slip system in a FCC copper crystal?
(a) $(111)[1\bar{1}0]$ (b) $(\bar{1}11)[01\bar{1}]$
(c) $(1\bar{1}1)[10\bar{1}]$ (d) $(11\bar{1})[101]$

14. A dielectric material is:
(a) Electrical conductor
(b) Metallic magnet
(c) Two coupled electrical conductors
(d) Electrical insulator

15. Which one of the following processes is an example of an electrolytic cell?
(a) Corrosion of a metal rod in ambient atmosphere
(b) Charging of a rechargeable battery
(c) Discharging of a rechargeable battery
(d) Sacrificial cathodic protection system

16. Which one of the following statements regarding selective leaching of a binary alloy is TRUE?
(a) The lower atomic weight element is leached.
(b) The element having a higher diffusivity is leached.
(c) The more electronegative element is leached.
(d) The element with lower density is leached.

Answers

9. (a)	10. (c)	11. (a)	12. (b)	13. (b)	14. (d)
15. (b)	16. (c)				

17. In green sand casting, which one of the following is NOT a part of the gating system?
    (a) Runner (b) Sprue
    (c) Riser (d) Pouring

18. For a material to exhibit superplasticity, one of the requirements is:
    (a) Coarse-grained microstructure
    (b) High strain-rate sensitivity
    (c) Low strain-hardening exponent
    (d) High modulus of elasticity

19. The dye penetrant test for detecting flaws is based on:
    (a) Magnetism
    (b) Sound propagation
    (c) X-ray absorption
    (d) Capillary action

20. When 1 mole of C_3H_8 at 300 K is burnt with stoichiometric amount of oxygen at 300 K to form CO_2 and H_2O, the adiabatic flame temperature is 5975 K. If C_3H_8 is burnt under the same conditions but with excess oxygen, the adiabatic flame temperature will be
    (a) equal to 5975 K irrespective of the amount of excess oxygen.
    (b) higher than 5975 K irrespective of the amount of excess oxygen.
    (c) lower than 5975 K irrespective of the amount of excess oxygen.
    (d) higher or lower than 5975 K depending on the amount of excess oxygen.

21. Two solid spheres X and Y of identical diameter are made of different materials having thermal diffusivities 100×10^{-6} $m^2 s^{-1}$ and 25×10^{-6} m^2s^{-1} respectively. Both spheres are heated in a furnace maintained at 1000 K. If the center of the sphere X reaches 800 K in 1 hour, time required for the center of sphere Y to reach 800 K is
    (a) 1 hour (b) 2 hours
    (c) 4 hours (d) 16 hours

22. Select the correct spectra (shown on a log-log scale in the figures) for emission from a gray surface and a black body, both maintained at 1000 K.

(a)

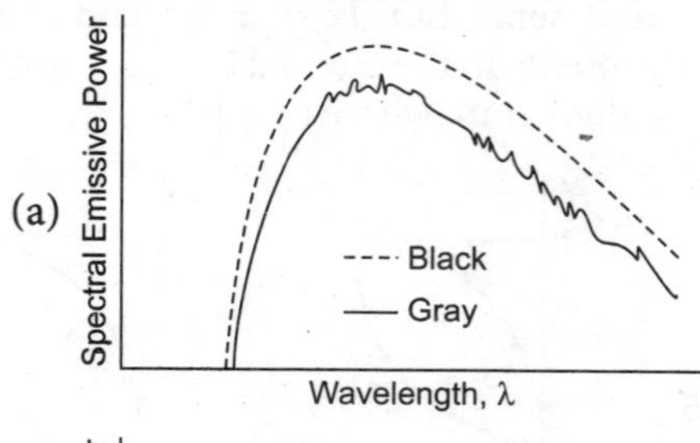

(b)

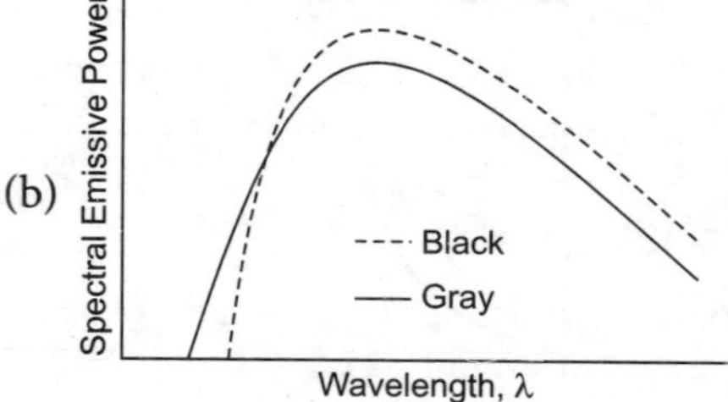

(c)

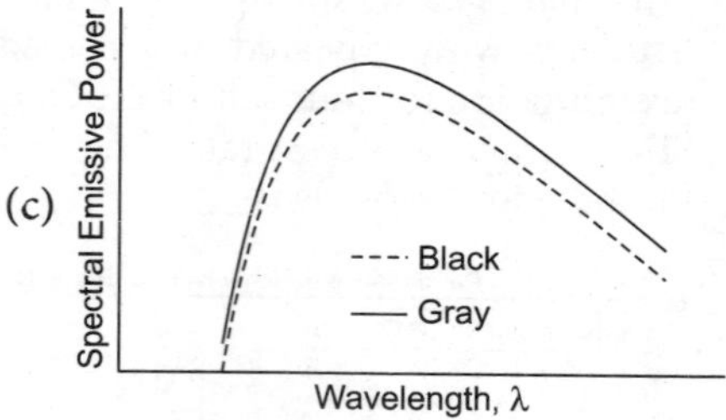

(d)

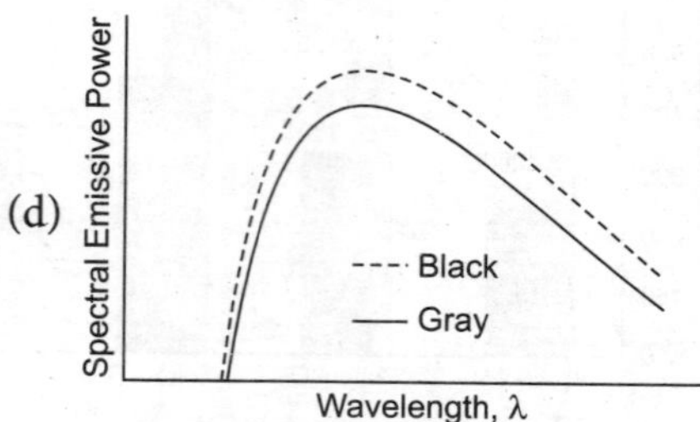

23. Given the three vectors X = – i – j + k, Y = – i + 2j + k and Z = i + k, which one of the following statements is TRUE?
    (a) X, Y and Z are mutually perpendicular.
    (b) X, Y and Z arc coplanar.

Answers

17. (a) 18. (b) 19. (d) 20. (c) 21. (c) 22. (d)
23. (a)

(c) X makes an angle of 30° with the normal to the plane containing Y and Z.

(d) Z makes an angle of 60° with the normal to the plane containing X and Y.

24. Angle between two neighboring tetrahedral bonds in Si having a diamond cubic structure is:

(a) 102.5° (b) 109.5°
(c) 120° (d) 135.5°

25. The sequence of precipitation during aging of Al – 4 wt.% Cu alloy is:

(a) GP zone $\rightarrow \theta'' \rightarrow \theta' \rightarrow \theta$
(b) GP zone $\rightarrow \theta \rightarrow \theta' \rightarrow \theta''$
(c) GP zone $\rightarrow \theta' \rightarrow \theta'' \rightarrow \theta$
(d) $\theta'' \rightarrow \theta' \rightarrow$ GP zone $\rightarrow \theta$

26. The indenter used in Rockwell hardness measurements on C scale is

(a) diamond cone.
(b) 10 mm steel ball.
(c) diamond pyramid.
(d) 1/16-in steel ball

27. For the function $y = a^x$, the derivative $\frac{dy}{dx}$ at x = 1 is:

(a) 1 (b) a
(c) a^2 (d) a ln a

28. Cupola is a furnace used to produce

(a) cast irons. (b) plain carbon steels.
(c) copper alloys. (d) aluminium alloys.

29. The functions $y = e^x$ and $y = e^{-x}$ intersect at the point:

(a) (1, 3) (b) (–2, 2)
(c) (0, 1) (d) (–1, –1)

30. A heavily cold-worked metal will

(a) yield a coarser recrystallized grain size.
(b) possess a lower driving force for recrytallization.
(c) have a higher energy barrier for nucleation of recrystallized grains.
(d) recrystallize at lower temperatures.

31. For the function f(x) given in the figure, the value of $\int_0^1 (1-f(x))dx$ is _______ (round off to one decimal place).

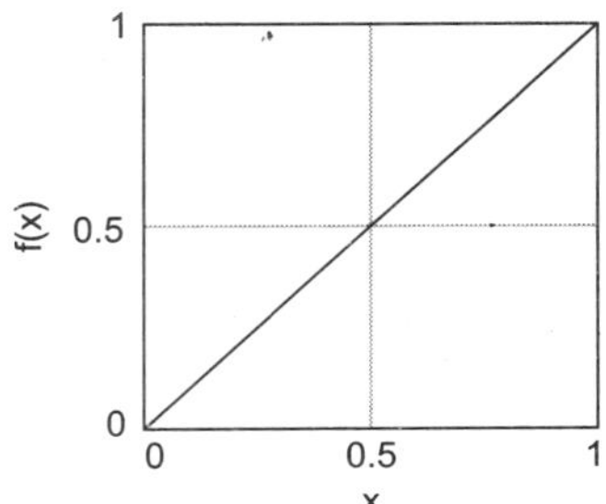

32. A component subjected to tensile stress in a mechanical device is monitored periodically for cracks by NDT. The NDT technique can only detect cracks (both surface and internal) which are larger than 1 mm. Keeping a 10% margin of safety, the maximum allowed tensile stress on the component will be _______ MPa (round off to the nearest integer).
Given, fracture toughness K_{IC} = 30 MPa $m^{1/2}$ and assume crack geometry factor of unity.

33. An iron plate with a total exposed surface area of 50 cm^2 undergoes atmospheric corrosion. If 200 g of weight is lost over a period of 10 years, then the corrosion rate is _______ $kg.m^{-2}.year^{-1}$ (round off to the nearest integer).

34. In cold-rolling, for the sheet to be drawn into rolls, the angle of contact (or angle of bite) should be less than or equal to _______ degree (round off to one decimal place).
Given, the coefficient of friction between sheet and roll is 0.1.

Answers

24. (b)	25. (a)	26. (a)	27. (d)	28. (a)	29. (c)
30. (d)	31. (0.5 to 0.5)	32. (480 to 494)	33. (4 to 4)	34. (5.6 to 5.8)	

35. The number of atoms per unit area in (100) plane of Pb is ______ nm^{-2} (round off to the nearest integer).
Given, crystal structure and atomic radius of Pb are FCC and 0.175 nm respectively.

Q.36 - Q.65 carry two marks each.

36. In the edge dislocation configuration given in the figure, dislocations X and Y are fixed and separated by a distance 2h on the same slip plane. Dislocation Z is free to glide on a parallel slip plane. The two slip planes are separated by a distance h. Which one of the following statements is TRUE regarding the stability of dislocation Z at positions 1, 2 and 3?
Assume all dislocations have identical Burgers vector.

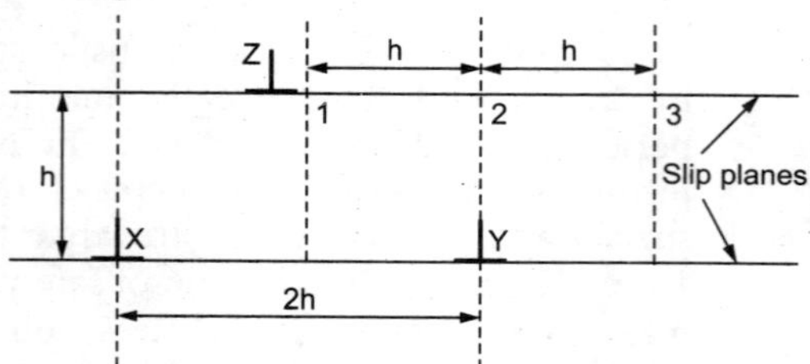

(a) Position 1: unstable equilibrium; Position 2: unstable; Position 3: unstable
(b) Position 1: stable equilibrium; Position 2: unstable; Position 3: unstable
(c) Position 1: unstable equilibrium; Position 2: stable; Position 3: unstable
(d) Position 1: stable equilibrium; Position 2: unstable; Position 3: stable

37. Which one of the following dislocation reactions is NOT feasible in a FCC crystal?
(a) $\frac{1}{2}[0\bar{1}1] \rightarrow \frac{1}{6}[1\bar{2}1] + \frac{1}{6}[\bar{1}\bar{1}2]$
(b) $\frac{1}{2}[1\bar{1}0] + \frac{1}{2}[1\bar{1}0] \rightarrow [1\bar{1}0]$
(c) $\frac{1}{6}[11\bar{2}] + \frac{1}{3}[111] \rightarrow \frac{1}{2}[110]$
(d) $\frac{1}{2}[\bar{1}01] \rightarrow \frac{1}{6}[\bar{2}11] + \frac{1}{6}[\bar{1}\bar{1}2]$

38. A galvanic cell is formed by connecting $Zn(E^0_{Zn^{2+}/Zn} = -0.76\ V)$ and $Fe\ (E^0_{Fe^{2+}/Fe} = -0.44\ V)$ wires immersed in their respective ion solutions. The cell discharges spontaneously with a voltage of 0.5 V. The ratio of the concentration of $[Fe^{2+}]$ to $[Zn^{2+}]$ ions in the cell is of the order of:
Given, $R = 8.314\ J.mol^{-1}.K^{-1}$, $F = 96500 C.mol^{-1}$, $T = 298\ K$
(a) 10^{-6} (b) 10^{-5}
(c) 10^{6} (d) 10^{7}

39. The divergence of the vector field $(x^3 + y^3)i + 3xy^2j + 3zy^2k$ is:
(a) $3y^2 + 6xy + 6x^2$
(b) $3x^2 + 6y^2 + 9xy + 6yz$
(c) $12xyz$ (d) $3(x + y)^2$

40. Match the products in Column I with the manufacturing processes in Column II.

Column I	Column II
P. Blades of a gas turbine	1. Sand casting
Q. Seamless tubing	2. Extrusion
R. Automotive cylinder blocks	3. Powder metallurgy and wire drawing
S. Tungsten filament	4. Investment casting

(a) P-1, Q-2, R-3, S-4
(b) P-2, Q-3, R-1, S-4
(c) P-4, Q-1, R-2, S-3
(d) P-4, Q-2, R-1, S-3

41. $f(x) = x \ln(x) + (1 - x)\ln(1 - x) + 3x(1 - x)$ has ________ at $x = 0.5$.
(a) a local minimum
(b) a local maximum

Answers

35. (7 to 9) 36. (a) 37. (b) 38. (c) 39. (d) 40. (d)
41. (b)

(c) a point of inflection
(d) a non-zero slope

42. Match the processes in Column I with the most appropriate mechanisms in Column II.

	Column I		Column II
P.	Blast furnace iron making process	1.	Metallothermic reduction
Q.	Hall-Heroult's process	2.	Oxidation
R.	Basic oxygen furnace steel making process	3.	Carbothermic reduction
S.	Kroll's process	4.	Fused salt electrolysis

(a) P-1, Q-4, R-2, S-3
(b) P-3, Q-1, R-2, S-4
(c) P-3, Q-4, R-2, S-1
(d) P-1, Q-2, R-3, S-4

43. Match the processes in Column I with the corresponding products in Column II.

	Column I		Column II
P.	COREX	1.	Sponge iron
Q.	MIDREX	2.	Copper matte
R.	Flash smelting reactor	3.	Hot metal or pig iron
S.	Submerged arc furnace	4.	Ferrochrome

(a) P-1, Q-3, R-2, S-4
(b) P-3, Q-4, R-2, S-1
(c) P-3, Q-1, R-2, S-4
(d) P-3, Q-1, R-4, S-2

44. X-ray diffraction pattern from an elemental metal with a FCC crystal structure shows the first peak at a Bragg angle $\theta = 24.65°$. The lattice parameter of this metal is _______ nm.
Given, wavelength of the X-ray used is 0.1543 nm.
(a) 0.185 (b) 0.262
(c) 0.320 (d) 0.370

45. Match the materials in Column I with their common applications in Column II.

	Column I		Column II
P.	Gray iron	1.	Cladding for uranium fuel in nuclear reactor
Q.	Ductile iron	2.	Base structure of heavy machines
R.	Zirconium alloy	3.	Valves and pump bodies
S.	Beryllium-Copper alloy	4.	Jet aircraft landing gear bearings

(a) P-1, Q-3, R-2, S-4
(b) P-4, Q-2, R-1, S-3
(c) P-2, Q-1, R-4, S-3
(d) P-2, Q-3, R-1, S-4

46. The Mg-Sn phase diagram exhibits two eutectics on either side of the high melting intermetallic line compound, Mg_2Sn, as given below.
At 561°C: L(36.9 wt% Sn) $\rightarrow$ α(14.48 wt.% Sn) + Mg_2Sn
At 203°C: L(97.87 wt.% Sn) $\rightarrow$ β-Sn(almost 100 wt.% Sn) + Mg_2Sn
After the eutectic reaction has gone to completion and equilibrium has been attained at a temperature just below 561°C, the amount of eutectic constituent present in the alloy, Mg-50 wt.% Sn, is approximately _______ (in wt.%).
Given, atomic weight of Sn is 118.7 and Mg is 24.3.
(a) 25 (b) 38
(c) 62 (d) 75

Answers

42. (c) 43. (c) 44. (c) 45. (d) 46. (c)

47. Determine the correctness or otherwise of the following Assertion [a] and the Reason [r]

Assertion [a]: Low-alloy steels used for medium-temperature creep resistance often have additions of strong carbide-Conning elements.

Reason [r]: During creep deformation, the particles with higher misfit with the matrix, lose coherency.

(a) Both [a] and [r] are true and [r] is the correct reason for [a].
(b) Both [a] and [r] are true but [r] is not the correct reason for [a].
(c) Both [a] and [r] are false.
(d) [a] is true but [r] is false.

48. Determine the correctness or otherwise of the following Assertion [a] and the Reason [r]

Assertion [a]: The rate of homogenization in a dilute substitutional solid solution of B in A is controlled by the diffusivity of B.

Reason [r]: Atomic migration cannot occur along dislocations and grain boundaries.

(a) Both [a] and [r] are true and [r] is the correct reason for [a]
(b) Both [a] and [r] are true but [r] is not the correct reason for [a]
(c) Both [a] and [r] are false
(d) [a] is true but [r] is false

49. Match the elements in Column I with their electronic behaviour given in Column II.

Column I	Column II
P. Copper	1. Ferromagnetic
Q. Iron	2. Superconducting
R. Mercury	3. Semiconducting
S. Silicon	4. Diamagnetic

(a) P-1, Q-2, R-3, S-4
(b) P-3, Q-4, R-1, S-2
(c) P-4, Q-1, R-2, S-3
(d) P-4, Q-3, R-1, S-2

50. Radius of the largest interstitial atom that can be accommodated in an octahedral void in BCC iron without distorting the lattice is _______ nm (round off to three decimal places).

Assume hard sphere model and radius of Fe atom as 0.124 nm.

51. The production process of cylindrical pipes results in a statistical scatter in their diameter which is modelled by a normal distribution with a mean value of 10 mm. If the area under the normal curve between 9 mm and 10 mm is 0.35, then the probability of producing pipes of diameter greater than 11 mm is _______ (round off to two decimal places).

52. The solution (using trapezoidal rule) of the integral

$$\int_0^1 e^{-x^2}\, dx$$

by dividing the range 0 to 1 into two intervals is _______ (round off to two decimal places).

53. Iron is corroding in fresh water which was dissolved oxygen concentration of 15 mM. The anodic current density at an overpotential of 120 mV is _______ $A.cm^{-2}$ (round off to three decimal places).

Given:

1. Anodic Tafel slope is 0.06 V.
2. Diffusion coefficient of oxygen is 2.42 $\times 10^{-5}$ $cm^2.s^{-1}$.
3. Diffusion layer thickness is 0.06 cm.

54. A metal oxidizes at 1200 K with a parabolic rate constant of 3×10^{-6} $g^2.cm^{-4}.s^{-1}$. Time taken for the oxide film to grow to a thickness of 2 μm is _______ s (round off to two decimal places).

Given, density of oxide is 6.5 $g.cm^{-3}$.

55. Two plates of composition, Fe–10 wt.% Ni and Fe–20 wt.% Cr–5 wt.% Ni are fusion-welded using a filler rod of composition

Answers

47. (b) 48. (d) 49. (c) 50. (0.018 to 0.020) 51. (0.14 to 0.16 or 14 to 16)
52. (0.71 to 0.75) 53. (0.190 to 0.238) 54. (0.54 to 0.58) 55. (15 to 15)

20 wt.% Ni–80 wt.% Cr. Contribution to dilution of the weld pool is 20% from each plate. The Ni content in the weld pool is _______ wt.% (round off to the nearest integer).

56. Figure shows schematic of a venturimeter. The cross sectional area is 100 mm^2 at A and is 50 mm^2 at B. If air is flowing through the venturimeter at a flow rate of 10^{-3} $m^3.s^{-1}$, the height H in the air-over-water manometer is _______ mm (round off to the nearest integer).

Assume:

1. Incompressible flow with no friction losses.
2. Density of air is 1 kg m^{-3}.
3. Density of water is 1000 kg m^{-3}.
4. Acceleration due to gravity is 9.8 m s^{-2}.

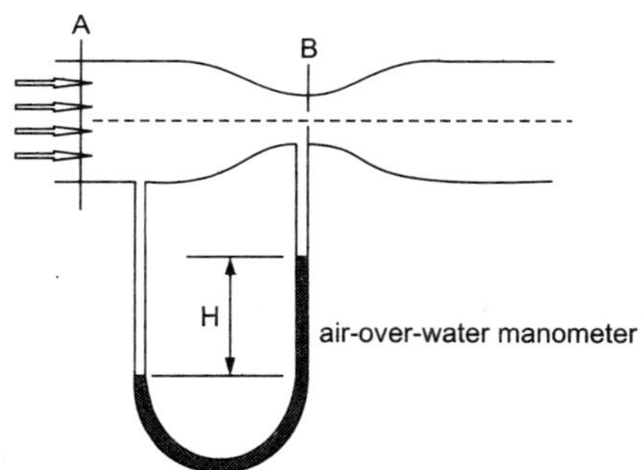

57. For effective communition in a ball mill, it is desired that the balls travelling along the mill wall leave the wall at point C and travel freely in air along the path CDA, as shown in the figure. If ∠BOC is 120°, the rotational speed of the mill is _______ rpm (rounded off to one decimal place) by performing suitable force balance at point C.

Assume:

1. There is no slip between the ball and mill wall.
2. O is the rotational axis of the mill and OB is parallel to the vector g.
3. Inner diameter of ball mill is 3.26 m.
4. Acceleration due to gravity g is 9.8 ms^{-2}.

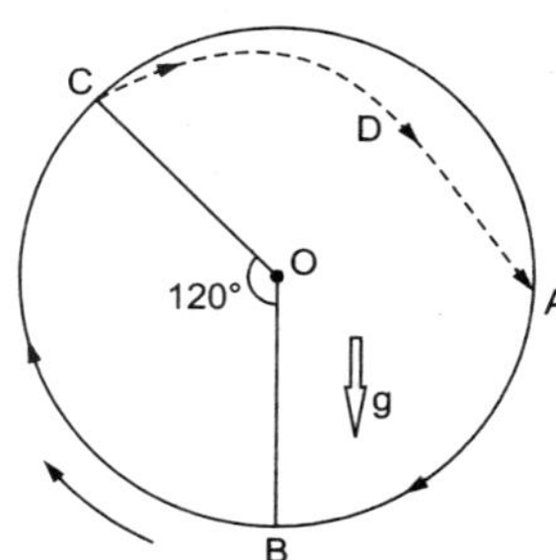

58. If liquid copper is cooled to 1353 K, magnitude of the driving force for liquid to transform to solid is _______ $J.mol^{-1}$ (round off to one decimal place).

Given, melting temperature and enthalpy of melting of copper are 1356 K and 13 $kJ.mol^{-1}$ respectively.

59. 1000 kg of liquid steel containing 0.03 wt.% S needs to be desulphurized using a slag to bring the sulphur content down to 0.015 wt.%. The quantity of slag needed is _______ kg (round off to the nearest integer).

Assume:

1. Thermodynamic equilibrium
2. No sulphur in the slag prior to desulphurization treatment

Given the equilibrium sulphur partition ratio between slag and steel,

$$\frac{[\text{wt. \% S}]_{\text{in slag}}}{[\text{wt. \% S}]_{\text{in steel}}} \text{ is 50.}$$

60. Zone refining of Si results in residual P content of 0.1 parts per billion by weight. The electrical conductivity of this zone refined Si is _______ $\Omega^{-1}.m^{-1}$ (round off to two decimal places).

Given:

1. Avogadro number is 6.02×10^{23}.
2. Density of Si is 2.33 $g.cm^{-3}$.
3. Atomic weight of P is 30.97.
4. Charge of electron is 1.6×10^{-19} A.s
5. Mobility of electron is 0.2 $m^2.V^{-1}.s^{-1}$

Answers

56. (14 to 16) 57. (15.6 to 17.6) 58. (28.6 to 29.0) 59. (19 to 21) 60. (0.14 to 0.16)

61. The steady state creep rate of a material increases by a factor of 20 when the temperature is increased from 890 K to 980 K. The creep rate at a temperature of _______ K (round off to the nearest integer) will be 5 times the creep rate at 890 K.

62. Crack growth is being continuously measured in a test specimen subjected to constant amplitude cyclic stress with a mean stress of zero. The crack growth rate is related to the stress intensity range, ΔK as

$$\frac{da}{dN} \propto (\Delta K)^3,$$

where, a is the crack length and N is the number of cycles. When the crack length increases by a factor of two, the crack growth rate will increase by a factor of_______ (round off to one decimal place).

63. In a top gated mold, liquid metal enters the mold cavity as a freely falling stream under gravity from a height of 0.5 m. Ignore the fluid friction duo to viscosity and the drag due to changes in the direction of flow. If the volume of the mold cavity is 10 m^3, then the time required to fill the mold is _______ s (round off to nearest integer).

Given:

1. Acceleration due to gravity is 9.8 $m.s^{-2}$,
2. Cross-sectional area of gate is 0.2 m^2.

64. A Basic Oxygen Furnace operator, at the end of oxygen blow, measures the dissolved oxygen content in the steel as 0.03 wt.% and the steel temperature as 1800 K. The carbon content [C] in the steel is _______ wt.% (round off to two decimal places).

Assume:

1. Equilibrium between dissolved carbon [C], dissolved oxygen [O], and CO (gas) at 1 atmosphere,
2. Henry's law is valid for both [C] and [O]

Given:

$$[C]_{1\text{ wt.\% Henrian Std. State}} + [O]_{1\text{ wt.\% Henrian Std. State}} \longrightarrow (CO)_{1\text{ atm. Std. State}}$$

$$\Delta G^\circ = -19840 - 40.65\,T\ J$$
$$R = 8.314\ J.mol^{-1}.K^{-1}$$

65. M and N are 3 × 3 matrices. If the det(M) is –9 and the det(N) is –14, then the det(NM) is _______ (round off to the nearest integer).

Answers

61. (933 to 939) 62. (2.6 to 3.0) 63. (14 to 18) 64. (0.06 to 0.08) 65. (126 to 126)

Section

D

24

GATE-2021

Q.1 - Q.5 carry one mark each.

1. Five persons P, Q, R, S and T are to be seated in a row, all facing the same direction, but not necessarily in the same order. P and T cannot be seated at either end of the row. P should not be seated adjacent to S. R is to be seated at the second position from the left end of the row. The number of distinct seating arrangements possible is:
   (a) 2 (b) 3
   (c) 4 (d) 5
2. Consider the following sentences:
   (i) The number of candidates who appear for the GATE examination is staggering.
   (ii) A number of candidates from my class are appearing for the GATE examination.
   (iii) The number of candidates who appear for the GATE examination are staggering.
   (iv) A number of candidates from my class is appearing for the GATE examination.

   Which of the above sentences are grammatically CORRECT?
   (a) (i) and (ii) (b) (i) and (iii)
   (c) (ii) and (iii) (d) (ii) and (iv)
3. A digital watch X beeps every 30 seconds while watch Y beeps every 32 seconds. They beeped together at 10 AM.

   The immediate next time that they will beep together is ________
   (a) 10.08 AM (b) 10.42 AM
   (c) 11.00 AM (d) 10.00 PM
4. If $\oplus \div \odot = 2; \oplus \div \Delta = 3; \odot + \Delta = 5; \Delta \times \otimes = 10$,
   Then, the value of $(\otimes - \oplus)^2$, is:
   (a) 0 (b) 1
   (c) 4 (d) 16
5. The front door of Mr. X's house faces East. Mr. X leaves the house, walking 50 m straight from the back door that is situated directly opposite to the front door. He then turns to his right, walks for another 50 m and stops. The direction of the point Mr. X is now located at with respect to the starting point is ________
   (a) South-East (b) North-East
   (c) West (d) North-West

Q.6 - Q.10 carry two marks each.

6. Given below are two statements 1 and 2, and two conclusions I and II.
   Statement 1: All entrepreneurs are wealthy.
   Statement 2: All wealthy are risk seekers.
   Conclusion I: All risk seekers are wealthy.
   Conclusion II: Only some entrepreneurs are risk seekers.
   Based on the above statements and conclusions, which one of the following options is CORRECT?

Answers

1. (b)	2. (a)	3. (a)	4. (b)	5. (d)	6. (c)

(a) Only conclusion I is correct
(b) Only conclusion II is correct
(c) Neither conclusion I nor II is correct
(d) Both conclusions I and II are correct

7. A box contains 15 blue balls and 45 black balls. If 2 balls are selected randomly, without replacement, the probability of an outcome in which the first selected is a blue ball and the second selected is a black ball, is ________

(a) $\frac{3}{16}$ (b) $\frac{45}{236}$
(c) $\frac{1}{4}$ (d) $\frac{3}{4}$

8. The ratio of the area of the inscribed circle to the area of the circumscribed circle of an equilateral triangle is ________

(a) $\frac{1}{8}$ (b) $\frac{1}{6}$
(c) $\frac{1}{4}$ (d) $\frac{1}{2}$

9. Consider a square sheet of side 1 unit. The sheet is first folded along the main diagonal. This is followed by a fold along its line of symmetry. The resulting folded shape is again folded along its line of symmetry. The area of each face of the final folded shape, in square units, equal to ________

(a) $\frac{1}{4}$ (b) $\frac{1}{8}$
(c) $\frac{1}{16}$ (d) $\frac{1}{32}$

10. The world is going through the worst pandemic in the past hundred years. The air travel industry is facing a crisis, as the resulting quarantine requirement for travelers led to weak demand.
In relation to the first sentence above, what does the second sentence do?
(a) Restates an idea from the first sentence.
(b) Second sentence entirely contradicts the first sentence.
(c) The two statements are unrelated.
(d) States an effect of the first sentence.

Q.11 - Q.35 carry one mark each.

11. For the matrix given below, the eigenvalues are:

$$\begin{pmatrix} 1 & 0 & -1 \\ 0 & 1 & 0 \\ -1 & 0 & 1 \end{pmatrix}$$

(a) 0, 2, 2 (b) 1, 1, 2
(c) 0, 1, 2 (d) 0, 1, 3

12. Which one of the following is a homogeneous function of degree three?
(a) $x^3 + 2x^2y^2$ (b) $y^2x + 2yx^2$
(c) $y^3 + 2x^2$ (d) $xy^2 + 3xy$

13. The divergence of a vector field $\vec{V}$ (x, y, z), where its three components (V_x, V_y, V_z) are functions of x, y, z, is:

(a) $\frac{\partial V_x}{\partial x} + \frac{\partial V_y}{\partial y} + \frac{\partial V_z}{\partial z}$

(b) $\left(\frac{\partial V_z}{\partial y} - \frac{\partial V_y}{\partial z}\right)i + \left(\frac{\partial V_x}{\partial z} - \frac{\partial V_z}{\partial x}\right)j + \left(\frac{\partial V_y}{\partial x} - \frac{\partial V_x}{\partial y}\right)k$

(c) $\frac{\partial V_x}{\partial x}i + \frac{\partial V_y}{\partial y}j + \frac{\partial V_z}{\partial z}k$

(d) $\frac{\partial^2 V_x}{\partial x^2} + \frac{\partial^2 V_y}{\partial y^2} + \frac{\partial^2 V_z}{\partial z^2}$

Answers

7. (b) 8. (c) 9. (b) 10. (d) 11. (c) 12. (b)
13. (a)

14. Which one of the following is 'center split' defect in rolling operation?

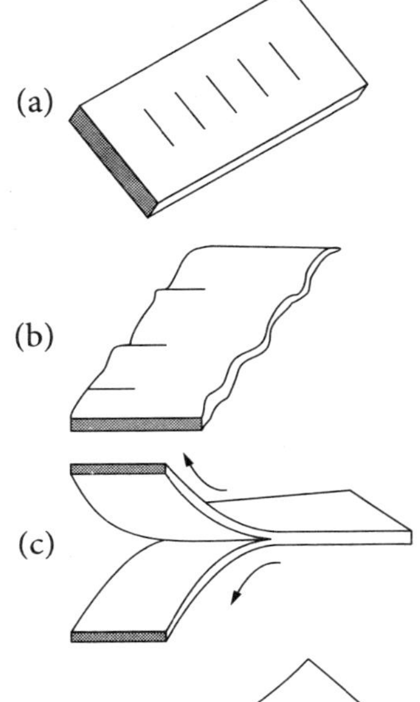

15. Single crystal turbine blades of nickel-based superalloys for aero-engines are manufactured using:
    (a) Investment casting
    (b) Die casting
    (c) Squeeze casting
    (d) Directional solidification

16. Elements A and B have the same crystal structure. For a dilute solution of B in A, which one of the following is true?
    (Given: ΔH_{mix} – Mixing enthalpy, a_B – Activity of B and X_B – Mole fraction of B)
    (a) If $\Delta H_{mix} = 0$, then $a_B < X_B$
    (b) If $\Delta H_{mix} = 0$, then $a_B > X_B$
    (c) If $\Delta H_{mix} > 0$, then $a_B < X_B$
    (d) If $\Delta H_{mix} < 0$, then $a_B < X_B$

17. For uniaxial tensile stress-strain behaviour of polycrystalline aluminium, which one of the following statements is FALSE?
    (a) True stress is always higher than the engineering stress.
    (b) At the ultimate tensile stress point on the true stress - strain curve, $\frac{d\sigma}{d\varepsilon} = 0$
    (c) Resilience is the area under the elastic region of the engineering stress - strain curve.
    (d) Maximum true stress does not correspond to the maximum load.

18. Which one of the following is FALSE for creep deformation?
    (a) The minimum creep rate is obtained in the primary stage (stage I).
    (b) Creep resistance decreases with decrease in grain size.
    (c) Coble creep occurs via grain boundary diffusion.
    (d) Nabarro-Herring creep occurs via lattice diffusion.

19. Which one of the following elements alloyed with iron is a ferrite stabilizer?
    (a) Nickel (b) Manganese
    (c) Carbon (d) Silicon

20. Which one of the following is the correct decreasing sequence of Quenching Power for quenchants used in heat treatment of steels?
    (a) Oil > Water > Brine > Air
    (b) Brine > Oil > Water > Air
    (c) Brine > Water > Oil > Air
    (d) Water > Brine > Oil > Air

21. For a zeroth order chemical reaction, which one of the following is FALSE?
    (a) Concentration versus time plot is a straight line.
    (b) Increase in concentration of reacting species increases the rate of reaction.

Answers

14. (d)	15. (d)	16. (d)	17. (b)	18. (a)	19. (d)
20. (c)	21. (b)				

(c) Half-life depends on the initial concentration and zero-order rate constant.

(d) Rate of reaction depends on temperature.

22. Which one of the following elements oxidizes first in basic oxygen steel making process?
(a) Silicon (b) Carbon
(c) Manganese (d) Phosphorus

23. Which one of the following is a hydrometallurgical operation?
(a) Roasting (b) Leaching
(c) Zone refining (d) Smelting

24. The value of $\lim_{x \to \infty} \frac{\sin^2 5x}{\sin^2 x}$ is: _______ (round off to nearest integer).

25. The grain size (X) of annealed specimens follows a symmetric distribution with a mean (μ) of 5 μm and a standard deviation (σ) of 0.5 μm. The percentage of specimens with grain size in the range 5 to 6 μm is expected to be: _______ (round off to nearest integer).

Given: For the symmetric distribution: Probability $P(X \leq \mu + 2\sigma) = 0.98$.

26. If $E^\circ_{Ni^{2+}/Ni} = -0.25$ V, the value of $\mu^\circ_{Ni^{2+}}$ (in J mol-1) at 298 K is: _______ (round off to nearest integer).

Given: F = 96500 C mol^{-1}

27. Melting point of Cu is 1358 K and its enthalpy of melting is 13400 J mol^{-1}. The value of free energy change (in J mol^{-1}) for liquid to solid transformation at 1058 K is: _______ (round off to nearest integer).

Assume: $C_P^{liquid} = C_P^{solid}$

28. A body is subjected to a state of stress given by the following stress tensor:

$$\begin{pmatrix} 50 & 0 & 0 \\ 0 & 200 & 0 \\ 0 & 0 & 100 \end{pmatrix} \text{MPa.}$$

If yielding is predicted by the Tresca Criterion, the uniaxial tensile yield stress (in MPa) of the body should be less than or equal to: _______ (round off to nearest integer).

29. Consider homogeneous nucleation of a spherical solid in liquid. For a given undercooling, if surface energy of a nucleus increases by 20 %, the corresponding increase (in percent) in the critical radius of the nucleus is: _______ (round off to nearest integer).

30. If saturation magnetization of iron at room temperature is 1700 kA m^{-1}, the magnetic moment (in A m^2) per iron atom in the crystal is: _______ $\times 10^{-23}$ (round off to 1 decimal place).

(Given: Lattice parameter of iron at room temperature = 0.287 nm)

31. In the X-ray diffraction pattern of a FCC crystal, the first reflection occurs at a Bragg angle (θ) of 30°. The Bragg angle (in degree) for the second reflection will be: _______ (round off to 1 decimal place).

32. A 0.6 wt.% C steel sample is slowly cooled from 900 °C to room temperature. The fraction of proeutectoid ferrite in the microstructure is: _______ (round off to 2 decimal places).

Given: Eutectoid composition: 0.8 wt.% C
Maximum solubility of carbon in α-Fe: 0.025 wt.% C

33. If the degree of polymerization of polyethylene is 30000, the average molecular weight (in g mol^{-1}) is: _______ (round off to nearest integer).

(Given: Atomic weights of carbon and hydrogen are 12 and 1, respectively)

Answers

22. (a) 23. (b) 24. (25 to 25) 25. (48 to 48) 26. (– 48251 to – 48240)
27. (– 2962 to – 2958) 28. (150 to 150) 29. (20 to 20) 30. (1.7 to 2.3)
31. (34.8 to 36.1) 32. (0.22 to 0.30) 33. (840000 to 840000)

34. Water flows over a plate of finite length. At $x = x_1$ from the leading edge, the velocity of the flow is $V_x = 0.5y - 0.5y^3$. The thickness, δ (in meter) of the boundary layer at $x = x_1$ is: _______ (round off to 2 decimal places). Given: V_∞ is the free stream velocity.

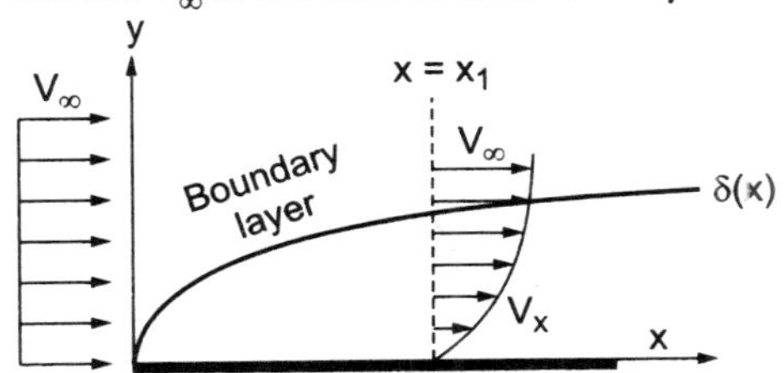

35. The vacancy concentration in a crystal doubles upon increasing the temperature from 27 °C to 127 °C. The enthalpy (in kJ mol^{-1}) of vacancy formation is: _______ (round off to 2 decimal places). Given: $R = 8.314 \text{ J mol}^{-1} \text{ K}^{-1}$

Q.36 - Q.65 carry two marks each.

36. The minimum value of y for the equation $y = x^2 - 2x + 4$ is _______.

(a) 3 (b) 1
(c) 4 (d) 6

37. Match the forming process (in Column I) with its name (in Column II):

Column I

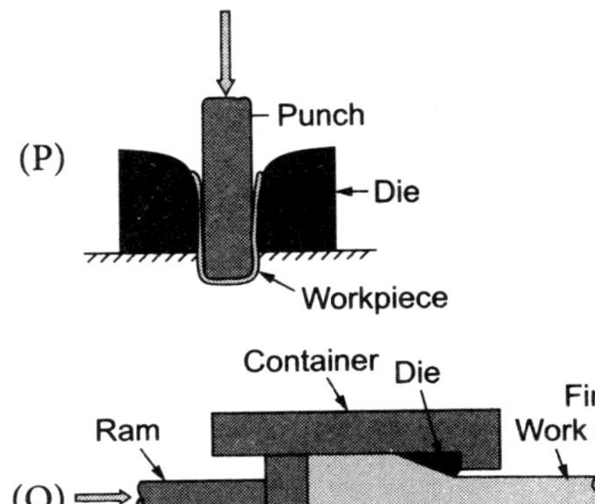

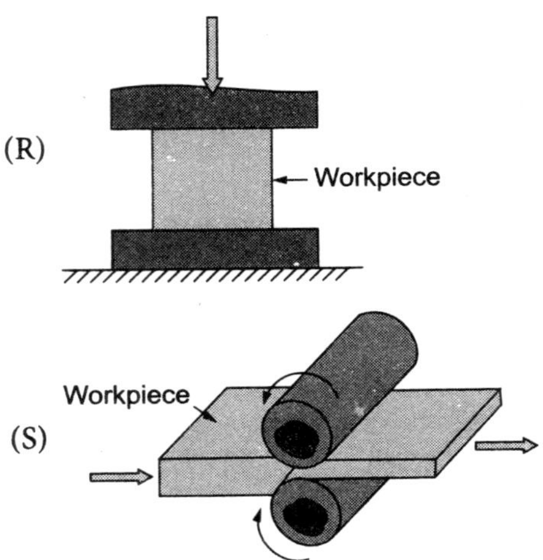

Column II
1. Extrusion
2. Rolling
3. Deep drawing
4. Open die forging

(a) P–1, Q–2, R–3, S–4
(b) P–3, Q–1, R–4, S–2
(c) P–3, Q–4, R–1, S–2
(d) P–1, Q–4, R–3, S–2

38. Match the nondestructive technique (in Column I) with its underlying phenomenon (in Column II):

Column I	Column II
(P) Dye penetrant test	1. X-ray absorption
(Q) Radiography	2. Capillary action
(R) Eddy current test	3. Elastic waves reflection
(S) Ultrasonic inspection	4. Electromagnetic induction

(a) P–4, Q–3, R–2, S–1
(b) P–2, Q–1, R–3, S–4
(c) P–2, Q–1, R–4, S–3
(d) P–3, Q–2, R–1, S–4

Answers

34. (0.53 to 0.59) 35. (6.85 to 7.00) 36. (a) 37. (b) 38. (c)

39. Number of degrees of freedom for the following reacting system is:

$M (s) + CO_2 (g) = MO (s) + CO (g)$

(a) 0 (b) 1
(c) 2 (d) 3

40. The condition for getting the binary phase diagram of A-B (shown below) is:

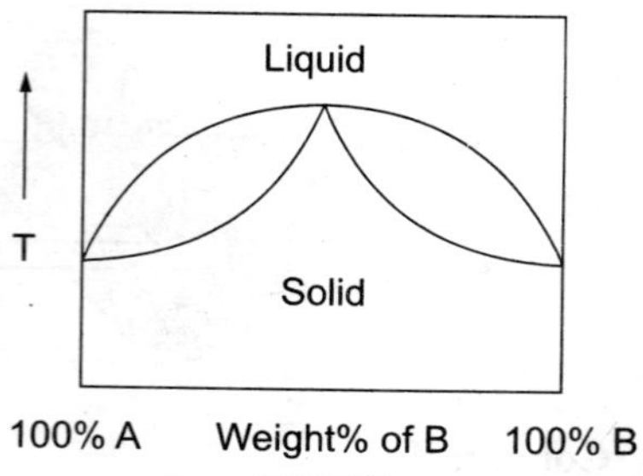

Given: ΔH^{solid}_{mix} – Enthalpy of mixing of solid
ΔH^{liquid}_{mix} – Enthalpy of mixing of liquid

(a) $\Delta H^{solid}_{mix} = 0$ and $\Delta H^{liquid}_{mix} = 0$
(b) $\Delta H^{solid}_{mix} \ll 0$ and $\Delta H^{liquid}_{mix} = 0$
(c) $\Delta H^{solid}_{mix} > 0$ and $\Delta H^{liquid}_{mix} = 0$
(d) $\Delta H^{solid}_{mix} = 0$ and $\Delta H^{liquid}_{mix} \ll 0$

41. In the absence of any external stress, which one of the following statements related to the interaction of point defect and a dislocation is FALSE:

(a) An oversized solute atom would preferentially migrate below the slip plane of an edge dislocation.
(b) A spherically symmetric point defect can interact with both the hydrostatic and shear stress fields of a dislocation.
(c) A point defect can locally modify the elastic modulus and thereby can change the interaction energy.
(d) Vacancies are attracted towards the compressive region of dislocation.

42. A single crystal aluminium sample is subjected to uniaxial tension along [112] direction. If the applied tensile stress is 100 MPa and the critical resolved shear stress (CRSS) is 25 MPa, which one of the following slip systems will be activated?

(a) $[\bar{1}01](111)$ (b) $[\bar{1}10](111)$
(c) $[101](11\bar{1})$ (d) $[011](11\bar{1})$

43. One-dimensional steady-state temperature distribution in two adjacent refractory blocks (with thermal conductivities, k_1 and k_2) of unit cross- sectional area are shown below. The temperature T_1 and thermal contact resistance of the interface, respectively, are:

$k_1 = 2.0\ W\ m^{-1}\ K^{-1}$ $k_2 = 1.0\ W\ m^{-1}\ K^{-1}$

1000 K
Interface
800 K
600 K
1 m
1 m
T_1 (K)
Temperature (K) →
Distance (m) →

(a) 200 K, 0.5 K W^{-1}
(b) 400 K, 1.0 K W^{-1}
(c) 200 K, 0.25 K W^{-1}
(d) 500 K, 0.5 K W^{-1}

44. For a fully developed 1-D flow of a Newtonian fluid through a horizontal pipe of radius R (see figure), the axial velocity (v_z) is given by:

$$v_z = \left[\frac{\Delta P}{L}\right]\left(\frac{R^2 - r^2}{4\mu}\right)$$

where, ΔP is the pressure difference (P_1 – P_2), μ is the viscosity, r is the radial distance from the axis and L is the length of the tube. The shear stress exerted by the fluid on the tube wall is:

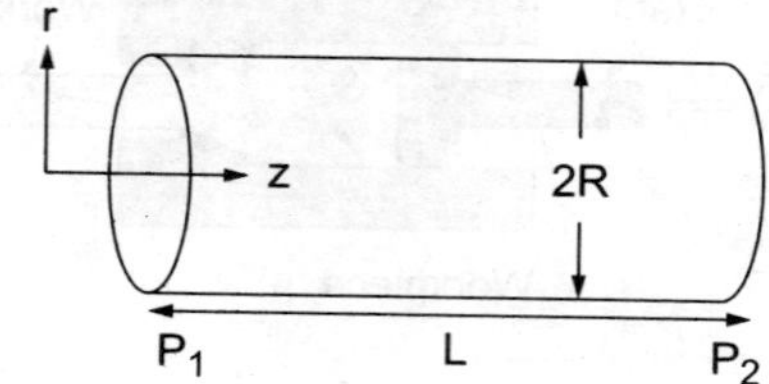

Answers

39. (c) 40. (b) 41. (b) 42. (a) 43. (a) 44. (a)

(a) $\frac{\Delta PR}{2L}$ (b) $\frac{\Delta PR}{L}$

(c) $\frac{3\Delta PR}{2L}$ (d) $\frac{2\Delta PR}{2L}$

45. Match the terms (in Column I) with the unit process (in Column II)

	Column I		Column II
(P)	Submerged Entry Nozzle	1.	Ladle Furnace
(Q)	Electric Heating	2.	Continuous Casting
(R)	Raceway Zone	3.	LD Converter
(S)	Oxygen Lancing	4.	Blast Furnace

(a) P–2, Q–1, R–4, S–3
(b) P–4, Q–1, R–2, S–3
(c) P–4, Q–3, R–1, S–2
(d) P–2, Q–3, R–4, S–1

46. A blast furnace uses hematite ore with 80% Fe_2O_3 and 20% gangue materials. It uses 600 kg coke per ton of hot metal. The coke contains 85% C and 15% ash. The composition of hot metal is 95.5% Fe and 4.5% C.

The weight of iron ore used and slag produced per ton of hot metal respectively, are:

Given: Atomic weight: O = 16, C = 12, N = 14, Fe = 56

All the compositions are in wt.%.

1 ton = 1000 kg

Assume that the gangue materials of the ore and ash content of coke form slag while Fe_2O_3 in the ore is consumed in making hot metal.

(a) 1705 kg, 431 kg (b) 2131 kg, 546 kg
(c) 1705 kg, 331 kg (d) 1500 kg, 431 kg

47. Consider the function f(x) = x – cos x. Using Newton-Raphson method, the estimated root of f(x) after the first iteration is: _______ (round off to 3 decimal places).

Assume: Initial guess of the root = 0.5 radians.

48. The work done by a force $\vec{F} = 2xi + 3yj$ along a straight line from point (0, 0) to (1, 2) is: _______ (round off to nearest integer).

49. A coin is tossed three times. Given that there are more heads than tails, the probability of getting exactly one tail is: _______ (round off to 2 decimal places).

50. A continuous fillet weld is made using a 3000 W welding machine. At a travel speed of 6 mm s^{-1}, the cross-sectional area (in mm^2) of the weld is: _______ (round off to nearest integer).

Given: The unit energy required to melt the metal is 6 J mm^{-3}.

Heat transfer factor = 0.6

Melting factor = 0.5

51. Liquid iron is cast into a spherical sand mold (6 cm radius) and a cubical sand mold (12 cm edge length). If solidification time is 60 minutes in the spherical casting, the time (in minutes) required to solidify in the cubical casting is: _______ (round off to nearest integer).

52. True strain for 60% height reduction of a sample subjected to hot forging is: _______ (round off to 2 decimal places).

53. For the equilibrium reaction: 2Cu (s) + SO_2 (g) = Cu_2S (s) + O_2 (g), the value of $\ln\left(\frac{P_{O_2}}{P_{SO_2}}\right)$ at 973 K is: _______ (round off to 2 decimal places).

Given: 2Cu (s) + $0.5S_2$ (g) = Cu_2S (s)
$\Delta G°$ at 973 K = – 100 kJ

SO_2 (g) = $0.5S_2$ (g) + O_2 (g)
$\Delta G°$ at 973 K = 292 kJ

R = 8.314 J mol^{-1} K^{-1}

Assume: Cu and Cu_2S are pure solids.

54. One mole of an ideal gas at 10 atm. and 300 K undergoes reversible adiabatic expansion

Answers

45. (a)	46. (a)	47. (0.745 to 0.770)	48. (7 to 7)
49. (0.75 to 0.75)	50. (25 to 25)	51. (58 to 62)	52. (– 0.94 to – 0.90)
53. (– 24.00 to – 23.50)		54. (2240 to 2256)	

to a pressure of one atm. The work done (in Joule) by the gas is: _______ (round off to nearest integer).
Given: R = 8.314 J mol^{-1} K^{-1}; 1 atm. = 101325 Pa; Cp = 2.5R

55. The figure shows the entropy versus temperature (S-T) plot of a reversible cycle of an engine. If T_1 = 200 K and T_2 = 600 K, the efficiency of the engine (in percent) is: (round off to 2 decimal places).

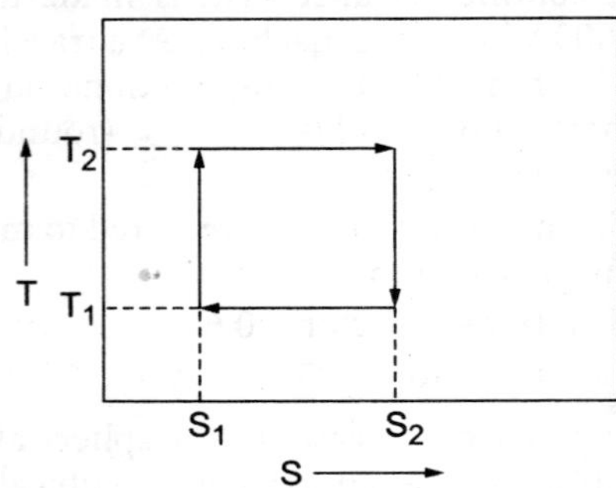

56. Two dislocation lines parallel to z-axis lying in the x-z plane are shown in the figure. The glide force (in Newton) exerted by the edge dislocation on the screw dislocation is: _______ (round off to nearest integer).

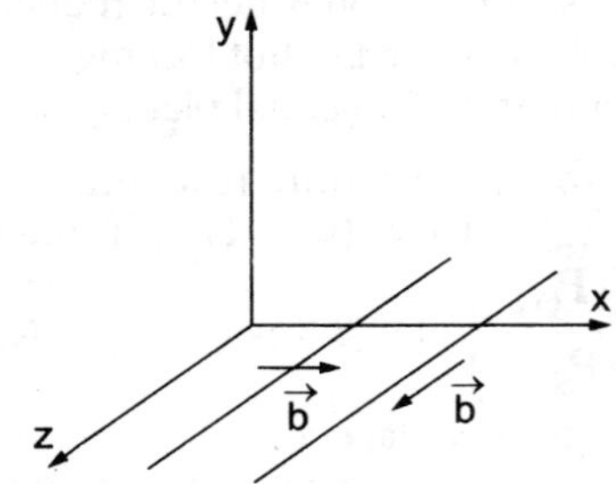

For the edge, the shear stress component is given by:

$$\tau_{xy} = \frac{Gb}{2\pi(1-\nu)} \frac{x(x^2-y^2)}{(x^2+y^2)^2}$$

Given: Shear modulus, G = 28 GPa
Poisson's ratio, ν = 0.3
Burgers vector, b = 0.29 nm
Distance between the two dislocations = 0.5 nm

57. In a material, a shear stress of 100 MPa is required to bow a dislocation line between precipitates with a spacing of 0.2 μm. If the spacing between the precipitates is increased to 0.5 μm, the shear stress (in MPa) to bow the dislocation would be: _______ (round off to nearest integer).

58. A metal plate is in a state of plane strain (ε_{zz} = 0) with $\sigma_{xx} = \sigma_{yy} \neq 0$ and $\tau_{xy} = \tau_{xz} = \tau_{yz}$ = 0. If the Poisson's ratio is 0.3, the ratio, σ_{zz}/σ_{xx} is _______ (round off to 1 decimal place).

59. An infinite metal plate has a central through-thickness crack of length $\frac{80}{\pi}$ mm. The maximum applied stress (in MPa) that the plate can sustain in mode I is: _______ (round off to nearest integer).
Assume: Linear elastic fracture mechanics is valid
Given: Fracture toughness, K_{IC} = 20 MPa m$^{1/2}$

60. A hypothetical binary eutectic phase diagram of A – B is shown below. An alloy with 5 wt.% B solidifies with no convection. Assuming steady state, the critical temperature gradient (in K mm^{-1}) required to maintain planar solidification front is: _______ (round off to nearest integer).

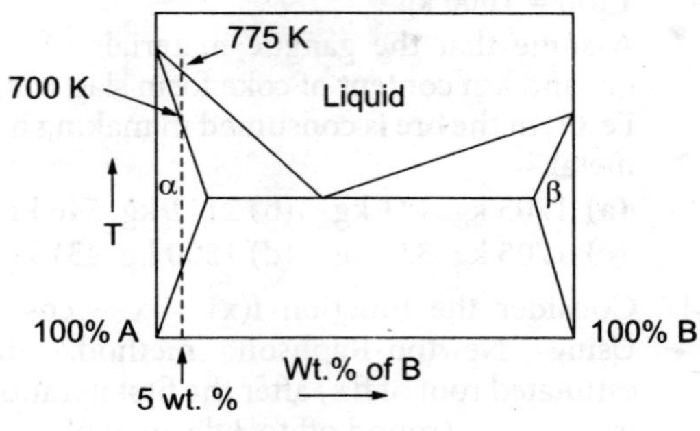

Answers

55. (63.00 to 71.00) 56. (0 to 0) 57. 40 to 40 58. (0.6 to 0.6)
59. (98 to 102) 60. (298 to 302)

Given: Diffusivity of B in liquid = 10^{-9} m^2 s^{-1}

Velocity of solidification front = 4 μm s^{-1}

61 A thick steel plate containing 0.1 wt.% C is carburized at 950 °C. The plate's surface carbon concentration is maintained at 1.1 wt.% C. After 9 hours, the depth (in mm) below the surface at which the carbon concentration is 0.6 wt.% C will be: _______ (round off to 2 decimal places).

Given: Diffusivity of carbon in γ-Fe at 950 °C = 1.6×10^{-11} m^2 s^{-1}

Error function table:

z	0.35	0.40	0.45	0.50	0.55	0.60
erf(z)	0.3794	0.4755	0.4755	0.5205	0.5633	0.6039

62. At 25 °C, iron corrodes in a deaerated acid of pH 3 with a corrosion current density of 4 μA cm^{-2}. The corrosion potential (V) is: _______ (round off to 2 decimal places).

Given: β_c = 0.1 V per decade of current density

Exchange current density of hydrogen on iron surface = 10^{-9} A cm^{-2}

R = 8.314 J mol^{-1} K^{-1}, F = 96500 C mol^{-1}

All potentials are with reference to standard hydrogen electrode.

63. The radius of an interstitial atom which just fits (without distorting the structure) inside an octahedral void of a bcc-iron crystal (in nm) is: _______ (round off to 3 decimal places).

Assume the radius of Fe atom to be 0.124 nm.

64. Nickel undergoes isothermal oxidation at 800 K for a duration of 400 s resulting in a weight gain of 2 mg cm^{-2}. The weight gain (mg cm^{-2}) after a duration of 1600 s is: _______ (round off to nearest integer).

Assume: Weight gain is proportional to square root of time.

65. A solid sphere (0.5 m radius) is enclosed within a larger hollow sphere (1 m radius), as shown in figure. The radiation exchange takes place between the outer surface (surface 1) of the small sphere and the inner surface (surface 2) of the bigger sphere. The value of the view factor, F_{22} is: _______ (round off to 2 decimal places).

Given: View factor (F_{ij}) is the fraction of the radiation leaving surface i that is intercepted by surface j.

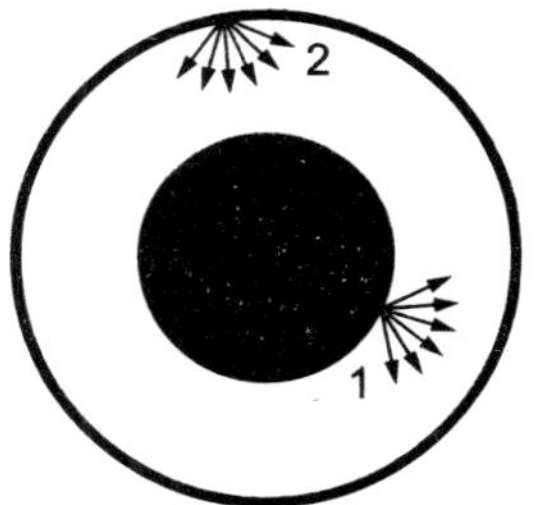

Answers

61. (0.65 to 0.75) 62. (– 0.60 to – 0.50) 63. (0.017 to 0.023) 64. (4 to 4)
65. (0.74 to 0.76)

Khanna's Multichoice Questions & Answers in

METALLURGICAL ENGINEERING

Om Prakash Gupta

FEEDBACK FORM

1. Which topics from your examination point of view are inadequately or not covered in this book?

___

___

2. Is there any misprints/ mistakes/ factual inaccuracy in this book? If so, please specify.

___

___

3. What is your assessment of this book as regards the presentation of the subject-matter, expression, precision and price in relation to the other recommended books available on this subject?

___

___

4. Which competing books you regard as better than this? Please mention their author and publishers.

___

___

5. Any further suggestions/comments you would like to make for the improvement of the book?

___

___

PS: If needed, please attach a separate sheet.

Name : ___

College : ___

Address : ___

Email : ___

Mobile : ___

Please mail the feedback form at:

KHANNA BOOK PUBLISHING CO. (P) LTD.

4C/4344, Ansari Road, Daryaganj, New Delhi-110002

You can also email this feedback form at contact@khannabooks.com or Whatsapp at +91 – 99109 09320

BEST FEEDBACK WILL BE REWARDED SUITABLY